D1223488

Encyclopedia of Human Computer Interaction

Claude Ghaoui
Liverpool John Moores University, UK

IDEA GROUP REFERENCE
Hershey · London · Melbourne · Singapore

Acquisitions Editor:	Michelle Potter
Development Editor:	Kristin Roth
Senior Managing Editor:	Amanda Appicello
Managing Editor:	Jennifer Neidig
Copy Editors:	Shanelle Ramelb and Sue VanderHook
Typesetters:	Diane Huskinson
Support Staff:	Sharon Berger, Amanda Kirlin, and Sara Reed
Cover Design:	Lisa Tosheff
Printed at:	Yurchak Printing Inc.

Published in the United States of America by
Idea Group Reference (an imprint of Idea Group Inc.)
701 E. Chocolate Avenue, Suite 200
Hershey PA 17033
Tel: 717-533-8845
Fax: 717-533-8661
E-mail: cust@idea-group.com
Web site: http://www.idea-group-ref.com

and in the United Kingdom by
Idea Group Reference (an imprint of Idea Group Inc.)
3 Henrietta Street
Covent Garden
London WC2E 8LU
Tel: 44 20 7240 0856
Fax: 44 20 7379 0609
Web site: http://www.eurospanonline.com

Library of Congress Cataloging-in-Publication Data

Encyclopedia of human computer interaction / Claude Ghaoui, Editor.
p. cm.
Summary: "This encyclopedia presents numerous experiences and insights, of professional from around the world, on human computer interaction issues and perspectives"--Provided by publisher.
Includes bibliographical references and index.
ISBN 1-59140-562-9 (hardcover) -- ISBN 1-59140-798-2 (ebook)
1. Human-computer interaction--Encyclopedias. I. Ghaoui, Claude.
QA76.9.H85E 52 2006
004'.019--dc22

2005031640

British Cataloguing in Publication Data
A Cataloguing in Publication record for this book is available from the British Library.

All work contributed to this encyclopedia set is new, previously-unpublished material. The views expressed in this encyclopedia set are those of the authors, but not necessarily of the publisher.

Editorial Advisory Board

List of Contributors

Abhishek / *Indian Institute of Technology, Kharagpur, India* 295
Albers, Michael J. / *The University of Memphis, USA* 338
Al-Hanbali, Nedal / *Al-Balqa' Applied University, Jordan* 242
Alkhalifa, Eshaa M. / *University of Bahrain, Bahrain* 79
Alpert, Sherman R. / *IBM T.J. Watson Research Center, USA* 100
Andrade, Javier / *University of A Coruña, Spain* 280, 381
Ares, Juan / *University of A Coruña, Spain* 280, 381
Atladóttir, Gyda / *University of Iceland, Iceland* 220
Ball, Linden J. / *Lancaster University, UK* 211
Barboni, Eric / *Université Paul Sabatier, France* 154
Bardone, Emanuele / *University of Pavia, Italy* 1, 408
Bastide, Rémi / *Université Paul Sabatier, France* 603
Basu, Anupam / *Indian Institute of Technology, Kharapur, India* 295
Becker, Shirley Ann / *Florida Institute of Technology, USA* 457
Beekhuyzen, Jenine / *Griffith University, Australia* 680
Benyon, David / *Napier University, UK* 344
Berlanga, A. J. / *University of Salamanca, Spain* 504
Bevan, Nigel / *Serco Usability Services, UK* 362
Blecker, Thorsten / *Hamburg University of Technology, Germany* 722
Bocchiola, Michele / *LUISS University, Italy* 408
Boechler, Patricia M. / *University of Alberta, Canada* 574, 648
Boucouvalas, Anthony C. / *Bournemouth University, UK* 227, 324, 523
Bourguet, Marie-Luce / *Queen Mary University of London, UK* 451
Campos, José Creissac / *Universidade do Minho, Braga, Portugal* 45
Carroll, John M. / *The Pennsylvania State University, USA* 567
Chalmers, Patricia A. / *Air Force Research Laboratory, USA* 179
Chen, Chaomei / *Drexel University, USA* 24, 273
Christou, Georgios / *Cyprus College, Cyprus* 668

Contents

Preface

OVERVIEW

Human computer interaction (HCI) evolved as a recognized discipline that attracts innovation and creativity. For the last 25 years, it inspired new solutions, especially for the benefit of the user as a human being, making the user the focal point that technology should serve rather than the other way around. The advent of the Internet, combined with the rapidly falling prices and increasing capability of personal computers, among other things, made the 1990s a period of very rapid change in technology. This has major implications on HCI research and advances, where peoples' demands and expectations as users of technology increased.

There is currently no agreement upon definition of the range of topics which form the area of human-computer interaction. Based on the definition given by the ACM Special Interest Group on Computer-Human Interaction Curriculum Development, which is also repeated in most HCI literature, the following is considered as an acceptable definition:

Human-computer interaction is a discipline concerned with the design, evaluation and implementation of interactive computing systems for human use in a social context, and with the study of major phenomena surrounding them.

A significant number of major corporations and academic institutions now study HCI. Many computer users today would argue that computer makers are still not paying enough attention to making their products "usable". HCI is undoubtedly a multi-disciplinary subject, which draws on disciplines such as psychology, cognitive science, ergonomics, sociology, engineering, business, graphic design, technical writing, and, most importantly, computer science and system design/software engineering.

As a discipline, HCI is relatively young. Throughout the history of civilization, technological innovations were motivated by fundamental human aspirations and by problems arising from human-computer interactions. Design, usability and interaction are recognised as the core issues in HCI.

Today, profound changes are taking place that touch all aspects of our society: changes in work, home, business, communication, science, technology, and engineering. These changes, as they involve humans, cannot but influence the future of HCI since they relate to how people interact with technology in an increasingly dynamic and complex world. This makes it even more essential for HCI to play a vital role in shaping the future.

Therefore, preparing an encyclopedia of HCI that can contribute to the further development of science and its applications, requires not only providing basic information on this subject, but also tackling problems that involve HCI issues in a wider sense, for example, by addressing HCI in and for various applications, that is, e-learning, health informatics, and many others.

CHALLENGES, CONTENT AND ORGANISATION

The following are some challenges in the HCI field, which were taken into consideration when compiling this encyclopedia:

- HCI is continually evolving with the fast change in technology and its cost. We, therefore, covered basic concepts/issues and also new advances in the field.
- The need to strike a balance between covering theory, methods/models, applications, experiences, and research. The balance was sought to provide a rich scientific and technical resource from different perspectives.
- The most important purpose of an encyclopedia in a particular discipline is to be a basic reference work for readers who need information on subjects in which they are not experts. The implication of "basic" is that an encyclopedia, while it should attempt to be comprehensive in *breadth* of coverage, cannot be comprehensive in the *depth* with which it treats most topics. What constitutes breadth of coverage is always a difficult question, and it is especially so for HCI, a relatively new discipline that has evolved over the past three decades and is still changing rapidly.
- An encyclopedia should, however, direct the reader to information at a deeper level, as this encyclopedia does through bibliographic references, indexed keywords, and so forth.
- This encyclopedia differs from other similar related references in that it covers core HCI topics/issues (that we see in most standard HCI books) as well as the use of HCI in various applications and recent advances and research. Thus the choice of specific topics for this encyclopedia has required our judgment of what is important. While there may be disagreement about the inclusion or exclusion of certain topics, we hope and believe that this selection is useful to a wide spectrum of readers. There are numerous articles that integrate the subject matter and put it into perspective. Overall, the encyclopedia is a general reference to HCI, its applications, and directions.

In order to meet these challenges, we invited professionals and researchers from many relevant fields and expertise to contribute. The resulting articles that appear in this volume were selected through a double-blind review process followed by rounds of revision prior to acceptance. Treatment of certain topics is not exclusive according to a given school or approach, and you will find a number of topics tackled from different perspectives with differing approaches. A field as dynamic as HCI will benefit from discussions, different opinions, and, wherever possible, a consensus.

An encyclopedia traditionally presents definitive articles that describe well-established and accepted concepts or events. While we have avoided the speculative extreme, this volume includes a number of entries that may be closer to the "experimental" end of the spectrum than the "well-established" end. The need to do so is driven by the dynamics of the discipline and the desire, not only to include the established, but also to provide a resource for those who are pursuing the experimental. Each author has provided a list of key terms and definitions deemed essential to the topic of his or her article. Rather than aggregate and filter these terms to produce a single "encyclopedic" definition, we have preferred instead to let the authors stand by their definition and allow each reader to interpret and understand each article according to the specific terminology used by its author(s).

Physically, the articles are printed in alphabetical order by their titles. This decision was made based on the overall requirements of Idea Group Reference's complete series of reference encyclopedias. The articles are varied, covering the following main themes: 1) Foundation (e.g., human, computer, interaction, paradigms); 2) Design Process (e.g., design basics, design rules and guidelines, HCI in software development, implementation, evaluation, accessible design, user support); 3) Theories (e.g., cognitive models, social context and organisation, collaboration and group work, communication); 4) Analysis (e.g., task analysis, dialogue/interaction specification, modelling); and 5) HCI in various applications (e.g., e-learning, health informatics, multimedia, Web technology, ubiquitous computing, mobile computing).

This encyclopedia serves to inform practitioners, educators, students, researchers, and all who have an interest in the HCI field. Also, it is a useful resource for those not directly involved with HCI, but who want to understand some aspects of HCI in the domain they work in, for the benefit of "users". It may be used as a general reference, research reference, and also to support courses in education (undergraduate or postgraduate).

CONCLUSION

Human computer interaction will continue to strongly influence technology and its use in our every day life. In order to help develop more "usable" technology that is "human/user-centred", we need to understand what HCI can offer on these fronts: theoretical, procedural, social, managerial, and technical.

The process of editing this encyclopedia and the interaction with international scholars have been most enjoyable. This book is truly an international endeavour. It includes 109 entries and contributions by internationally-talented authors from around the world, who brought invaluable insights, experiences, and expertise, with varied and most interesting cultural perspectives in HCI and its related disciplines.

It is my sincere hope that this volume serves not only as a reference to HCI professionals and researchers, but also as a resource for those working in various fields, where HCI can make significant contributions and improvements.

Claude Ghaoui
Liverpool John Moores University, UK

Acknowledgments

I would like to express my gratitude and deep appreciation to the members of the Editorial Advisory Board of this publication for their guidance and assistance throughout this project. They helped tremendously in the overwhelming reviewing process of articles received originally and after revised. They provided their expert, rigorous, blind review and assessment of the manuscripts assigned to them, which helped in identifying and selecting quality contributions. One hundred other volunteering reviewers have contributed to this process to whom I am grateful, with special thanks to Dr. Jacek Gwizdka, Toronto, and Dr. Barry Ip at the University of Wales Swansea. Much appreciation and thanks go to all of the authors for their excellent contributions. Finally, thanks to Mehdi Khosrow-Pour and his team at Idea Group Inc., who were very helpful and patient throughout the completion of this publication.

About the Editor

Claude Ghaoui, PhD, is a senior lecturer in computer systems (since 1995) at the School of Computing and Mathematical Sciences, Liverpool JMU, UK. Her research interests and expertise are mainly in human computer interaction, multimedia/Internet technology, and their applications in education. She is a UK correspondent for EUROMICRO (since 1998) and served on programme committees for several international HCI/multimedia conferences. Since 2000, she has been an advisor for eUniservity (UK-based), which promotes and provides online learning. She is the editor of two books, "E-Educational Applications: Human Factors and Innovative Approaches" (2004), published by IRM Press, and "Usability Evaluation of Online Learning Programs" (2003), published by Information Science Publishing. She is the editor-in-chief of the *International Journal of Interactive Technology and Smart Education.*

Abduction and Web Interface Design

Lorenzo Magnani
University of Pavia, Italy

Emanuele Bardone
University of Pavia, Italy

INTRODUCTION

According to Raskin (2000), the way we interact with a product, what we do, and how it responds are what define an interface. This is a good starting definition in one important respect: an interface is not something given or an entirely predefined property, but it is the dynamic interaction that actually takes place when a product meets the users. More precisely, an interface is that interaction that mediates the relation between the user and a tool explaining which approach is necessary to exploit its functions. Hence, an interface can be considered a mediating structure.

A useful exemplification of a mediating structure is provided by the so-called *stigmergy*. Looking at the animal-animal interactions, Raskin (2000) noted that termites were able to put up their collective nest, even if they did not seem to collaborate or communicate with each other. The explanation provided by Grassé (Susi et al., 2001) is that termites do interact with each other, even if their interactions are mediated through the environment. According to the stigmergy theory, each termite acts upon the work environment, changing it in a certain way. The environment physically encodes and stores the change made upon it so that every change becomes a clue that affects a certain reaction from it. Analogously, we might claim that an interface mediates the relation between the user and a tool affording him or her to use it a certain way[1]. Understanding the kind of mediation involved can be fruitfully investigated from an epistemological point of view. More precisely, we claim that the process of mediating can be understood better when it is considered to be an inferential one.

BACKGROUND

Several researchers (Kirsh, 2004; Hollan et al., 2000) recently have pointed out that designing interface deals with displaying as many clues as possible from which the user can infer correctly and quickly what to do next. However, although the inferential nature of such interactions is acknowledged, as yet, no model has been designed that takes it into account. For instance, Shneiderman (2002) has suggested that the value of an interface should be measured in terms of its consistency, predictability, and controllability. To some extent, these are all epistemological values. In which sense could an interaction be predictable or consistent? How can understanding the inferential nature of human-computer interaction shed light on the activity of designing good interfaces? Here, the epistemological task required is twofold: first, investigating what kind of inference is involved in such an interaction; and second, explaining how the analysis of the nature of computer interaction as inferential can provide useful hints about how to design and evaluate inferences.

Regarding both of these issues, in both cases we shall refer to the concept of abduction as a keystone of an epistemological model.

THE ROLE OF ABDUCTION IN DESIGNING INTERFACES

More than one hundred years ago, Charles Sanders Peirce (1923) pointed out that human performances are inferential and mediated by signs. Here, signs can be icons or indexes but also conceptions, images, and feelings. Analogously to the case of stigmergy, we have signs or clues that can be icons but also symbols and written words from which certain conclusions are inferred.

According to Peirce (1972), all those performances that involve sign activities are abductions. More precisely, abduction is that explanatory process of inferring certain facts and/or hypotheses that explain or discover some phenomenon or observation (Magnani, 2001). Abductions that solve the problem at hand are considered inferences to the best explanation. Consider, for example, the method of inquiring employed by detectives (Eco & Sebeok, 1991). In this case, we do not have direct experience of what we are taking about. Say, we did not see the murderer killing the victim, but we infer that given certain signs or clues, a given fact must have happened. Analogously, we argue that the mediation activity brought about by an interface is the same as that employed by detectives. Designers that want to make their interface more comprehensible must uncover evidence and clues from which the user is prompted to infer correctly the way a detective does; this kind of inference could be called *inference to the best interaction*.

We can conclude that how good an interface is depends on how easily we can draw the correct inference. A detective easily can discover the murderer, if the murderer has left evidence (clues) from which the detective can infer that that person and only that person could be guilty. Moreover, that an inference could be performed easily and successfully also depends upon how quickly one can do that. Sometimes, finding the murderer is very difficult. It may require a great effort. Therefore, we argue that how quick the process is depends on whether it is performed without an excessive amount of processing. If clues are clear and well displayed, the inference is drawn promptly. As Krug (2000) put it, it does not have to make us think.

In order to clarify this point even more, let us introduce the important distinction between theoretical and manipulative abduction (Magnani, 2001). The distinction provides an interesting account to explain how inferences that exploit the environment visually and spatially, for instance, provide a quicker and more efficient response. Sentential and manipulative abductions mainly differ regarding whether the exploitation of environment is or is not crucial to carrying out reasoning. Sentential abduction mostly refers to a verbal dimension of abductive inference, where signs and clues are expressed in sentences or in explicit statements. This kind of abduction has been applied extensively in logic programming (Flach & Kakas, 2000) and in artificial intelligence, in general (Thagard, 1988).

In contrast, manipulative abduction occurs when the process of inferring mostly leans on and is driven by the environment. Here, signs are diagrams, kinesthetic schemas, decorated texts, images, spatial representations, and even feelings. In all those examples, the environment embodies clues that trigger an abductive process, helping to unearth information that otherwise would have remained invisible. Here, the exploitation of the environment comes about quickly, because it is performed almost tacitly and implicitly. According to that, many cases have demonstrated that problem-solving activities that use visual and spatial representation, for instance, are quicker and more efficient than sentential ones. We can conclude that, in devising interfaces, designers have to deal mostly with the latter type of abduction. Interfaces that lean on the environment are tacit and implicit and, for this reason, much quicker than sentential ones.

Investigating the activity of designing interfaces from the abductive epistemological perspective described previously helps designers in another important respect: how to mimic the physical world within a digital one to enhance understanding.

As we have seen previously, the environment enables us to trigger inferential processes. But it can do that if and only if it can embody and encode those signs from which one can infer what to do next. For example, if you are working in your office and would appreciate a visit from one of your colleagues, you can just keep the door open. Otherwise, you can keep it closed. In both cases, the environment encodes the clue (the door kept open or closed), from which your colleagues can infer whether you do or don't want to be disturbed. Here are the questions we immediately come up: How can we encode those signs in a digital world? How can we enrich it so as to render it capable of embodying and encoding clues?

The question of how to enrich the digital world mainly concerns how to mimic some important features of the physical world in the digital one. Often, common people refer to an interface as easy-to-use, because it is more intuitive. Therefore, we don't need to learn how the product actually works. We just analogically infer the actions we have to perform from ordinary ones. More generally, metaphors are important in interface design, because they relate digital objects to the objects in the physical world, with which the user is more familiar.[2]

In the history of computer interface, many attempts have been made to replace some physical features in the digital one. For instance, replacing command-driven modes with windows was one of the most important insights in the history of technology and human-computer interaction (Johnson, 1997). It enabled users to think spatially, say, in terms of "where is what I am looking for?" and not in terms of "Wat sequence of letters do I type to call up this document?"

Enriching the digital world deals to some extent with faking, transforming those features embedded in the physical world into illusions. For example, consider the rule of projection first invented by Filippo Brunelleschi and then developed by such painters as Leon Battista Alberti and Leonardo da Vinci. In Peircean terms, what these great painters did was to scatter those signs to create the illusion of three-dimensional representations. It was a trick that exploited the inferential nature of visual construction (Hoffman, 1998).[3]

Now, the question is, how could we exploit inferential visual dimensions to enhance the interaction in the digital world? In the window metaphor, we do not have rooms, edges, or folders, such as in the physical world. They are illusions, and they are all produced by an inferential (abductive) activity of human perception analogously to what happens in smashing three to two dimensions. Here, we aim at showing how visual as well as spatial, temporal, and even emotional abductive dimensions can be implemented fruitfully in an interface. Roughly speaking, we argue that enriching the digital world precisely means scattering clues and signs that in some extent fakes spatial, visual, emotional, and other dimensions, even if that just happens within a flat environment.

We argued that the nature of signs can be verbal and symbolic as well as visual, spatial, temporal, and emotional. In correspondence with these last cases, one can recognize three abductive dimensions: visual, spatial, and emotional abduction.[4] We will discuss each of them in detail, providing examples taken from Web designs.

Abductive Inference

Visual dimension is certainly one of the most ubiquitous features in Web interaction. Users mainly interact with Web pages visually (Kirsh, 2004; Shaik et al., 2004). Here, signs and clues are colors, text size, dotted lines, text format (e.g., bold, underline, italics); they convey visual representations and can assign weight and importance to some specific part. Consider, for example, the navigation menu in Figure 1.

Here, colors, capital letters, and text size provide visual clues capable of enhancing the processing of information. The attention immediately is drawn to the menu header that represents its content (conference and research); capital letters and colors serve this function.

Then, the dotted list of the same color of the menu header informs the user about the number of the items. Hence, the fact that items are not marked visibly as menu headers gives a useful overview (Figure 2). Once the user has chosen what to see (conference or research), he or she can proceed to check each item according to his or her preference (Figure 3).

In this example, the user is guided to draw the correct inference; it enables the user to understand what he or she could consult.

Figure 1.

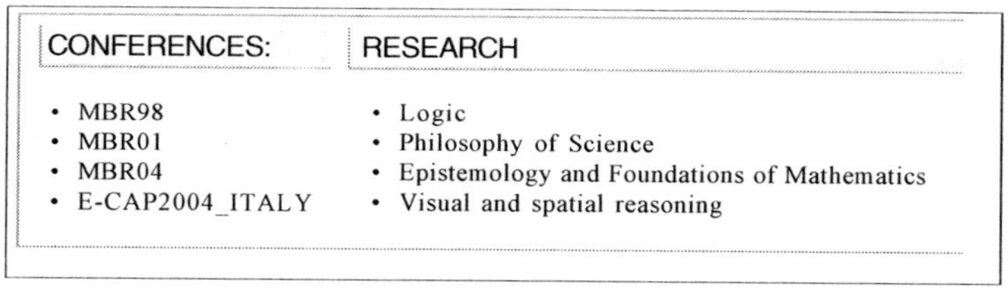

CONFERENCES:	RESEARCH
• MBR98	• Logic
• MBR01	• Philosophy of Science
• MBR04	• Epistemology and Foundations of Mathematics
• E-CAP2004_ITALY	• Visual and spatial reasoning

Figure 2.

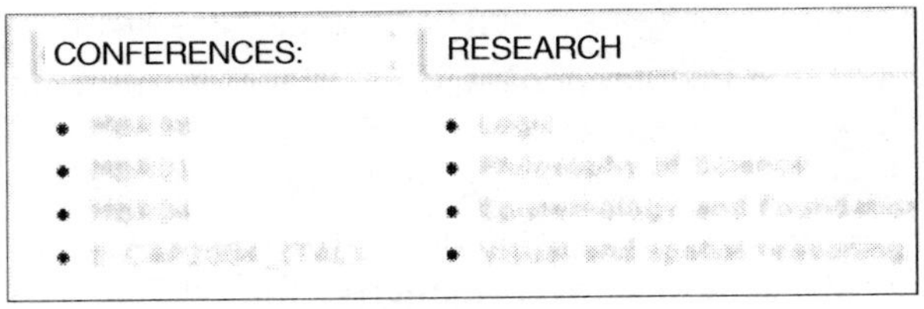

CONFERENCES: RESEARCH

Figure 3.

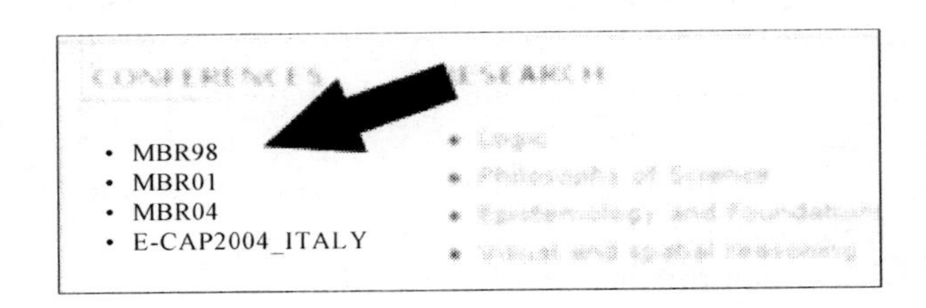

Figure 4.

Conferences: MBR98, MBR01, MBR04, Ecap2004; Research: Logic, Philosopy of Science, Epistemoogy and Foundations of Mathematics

Figure 5.

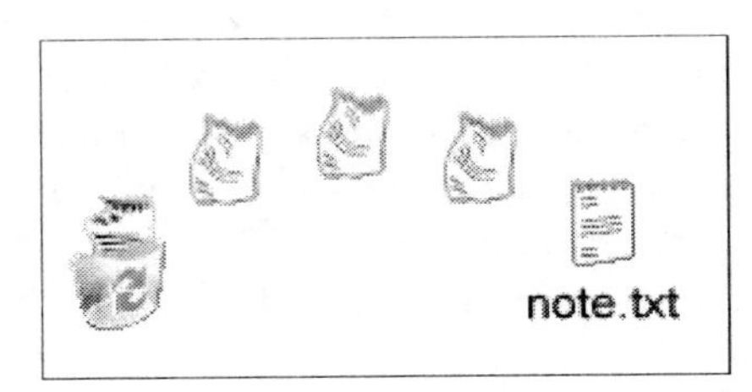

Figure 6.

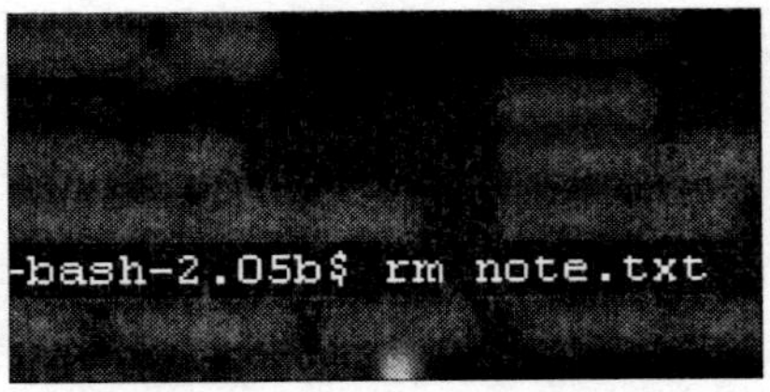

In contrast, consider, for example, the same content represented in Figure 4.

In this case, even if the content is identical, the user does not have any visual clue to understand what he or she is going to consult. The user should read all the items to infer and, hence, to understand, that he or she could know something about past and future conferences and about the research topics. If one stopped the user's reading after the third item (MBR04), the user could not infer that this page deals also with philosophy of science, with epistemology, and so forth. The user doesn't have enough clues to infer that. In contrast, in the first example, the user is informed immediately that this Web site contains information about conferences and research.

Spatial Abductive Inference

As already mentioned, the windows metaphor is certainly one of the most important insights in the history of interface technology. This is due to the fact that, as Johnson maintains, it enables the user to think in terms of "where is what I am looking for?" and not in terms of "what sequence of letters do I type to call up this document?" as in a command line system (Johnson, 1997). The computer becomes a space where one can move through just double-clicking on folders or icons, or dragging them. The difference is described well in Figure 5 and Figure 6.

In Figure 5, the file named *note.txt* is deleted by dragging it to the bin (i.e., the task of deleting is accomplished by a movement analogous to that used in the physical setting). Whereas, in Figure 6, the task is carried out by typing a command line composed by the command itself (*rm*, which stands for *remove*) and the file to be deleted (*note.txt*).

In designing Web pages, spatial dimension can be mimicked in other ways. One of the most well known examples is represented by the so-called *tab*. Tabs usually are employed in the real world to keep track of something important, to divide whatever they stick out of into a section, or to make it easy to open (Krug, 2000). In a Web site, tabs turn out to be very

Figure 7.

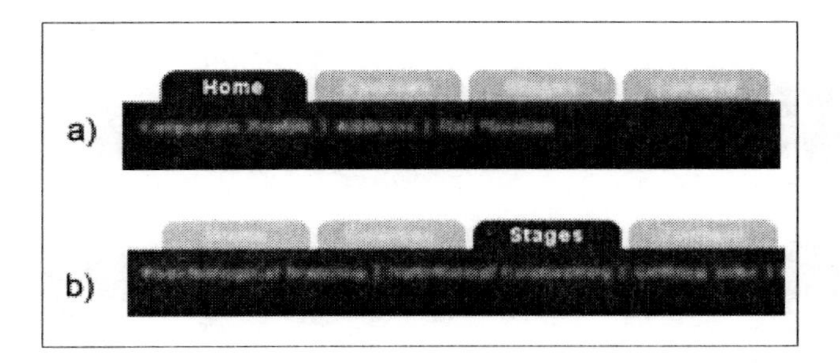

important navigation clues. Browsing a Web site, users often find themselves lost. This happens especially when the Web pages they are consulting do not provide spatial clues from which the user can easily infer where he or she is. For instance, several Web sites change their layout almost on every page; even if provided with a navigation menu, they are not helpful at all. In contrast, tabs enhance spatial inference in one important respect.

Consider the navigation bar represented in Figure 7.

In this example, when the user is visiting a certain page (e.g., a homepage) (Figure 7), the correspondent tab in the navigation bar becomes the same color of the body page. As Krug (2000) noted, this creates the illusion that the active tab actually moves to the front. Therefore, the user immediately can infer where he or she is by exploiting spatial relations in terms of background-foreground.

Emotional Abductive Inference

Recently, several researchers have argued that emotion could be very important to improve usability (Lavie & Tractinsky, 2004; Norman, 2004; Schaik & Ling, 2004). The main issue in the debate is how could emotionally evocative pages help the user to enhance understanding?

Here, abduction once again may provide a useful framework to tackle this kind of question. As Peirce (1923) put it, emotion is the same thing as a hypothetic inference. For instance, when we look at a painting, the organization of the elements in colors, symmetries, and content are all clues that trigger a certain reaction.

Consider, for example, the way a computer program responds to the user when a forbidden operation is trying to be performed. An alert message suddenly appears, often coupled with an unpleasant sound (Figure 8).

In this case, the response of the system provides clues (sounds and vivid colors such as red or yellow) from which we can attribute a certain state to the computer (being upset) and, hence, quickly react to it. Moreover, engaging an emotional response renders that reaction instantaneous; before reading the message the user already knows that the operation requested cannot proceed. Thus, a more careful path can be devised. Exploiting emotional reactions also can be fruitful in another respect. It conveys a larger amount of information. For instance, university Web sites usually place a picture of some students engaged in social activity on their homepage. This does not provide direct information about the courses. However, this triggers a positive reaction in connection with the university whose site is being visited. Any way icons are drawn aims at emotionally affecting the user. Even if they do not strictly resemble the physical feature, they can prompt a reaction. Consider the icon in Figure 9.

The winded envelope suggests rapidity, quickness, and all the attributes that recall efficiency and trustworthiness.

FUTURE TRENDS

In the last section, we have tried to provide a sketch about the role of abduction in human-computer interaction. Several questions still remain. We have illustrated some examples related to visual, spatial, and emotional abduction. However, other abductive

Figure 8.

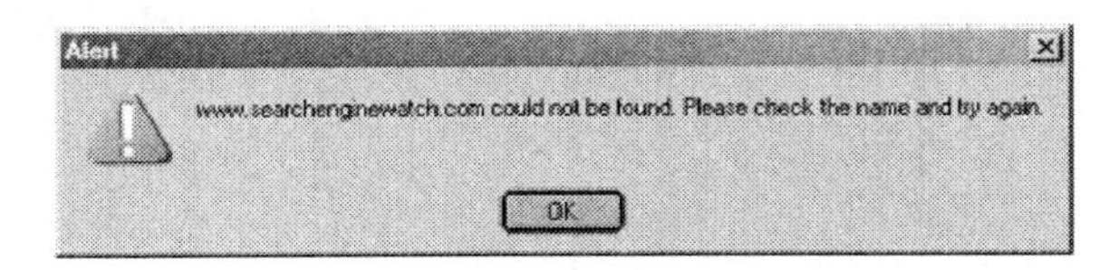

Figure 9.

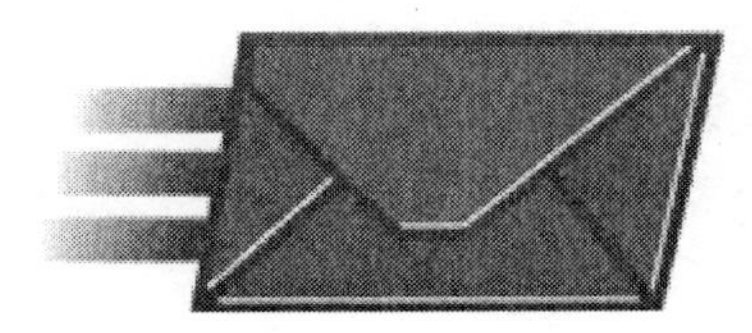

aspects can be addressed. For instance, it would be useful to investigate the temporal dimension. How is it possible to keep good track of the history of a user's actions?

Another question is related to the role of metaphors and analogies in designing interfaces. We have pointed out that metaphors are important in interface design, because they relate the digital objects to the objects in the physical world. However, the physical objects that we may find in the real world also are cultural ones; that is, they belong to a specific cultural setting. Something that is familiar in a given context may turn out to be obscure or even misleading in another one. Here, the task required is to investigate how cultural aspects may be taken into account to avoid misunderstandings.

CONCLUSION

In this article, we have claimed that interfaces play a key role in understanding the human-computer interaction. Referring to it as a mediating structure, we also have shown how human-computer interactions can be understood better using an inferential model. The interface provides clues from which the user can infer correctly how to cope with a product. Hence, we have referred to that inferential process as genuinely abductive. In the last section, we suggested possible future trends relying on some examples from Web interfaces design.

REFERENCES

Collins, D. (1995). *Designing object-oriented interfaces*. Benji/Lamming Addison-Wesley.

Eco, U., & Sebeok, T. (Eds.). (1991). *The sign of three: Dupin, Holmes, Pierce*. Indianapolis: Indiana University Press.

Flach, P., & Kakas, A. (2000). *Abductive and inductive reasoning: Essays on their relation and integration*. Dordrecht: Kluwer Academic Publishers.

Gibson, J. J. (1979). *The echological approach to visual perception*. Hillsdale, NJ: Erlbaum.

Hoffman, D. D. (1998). *Visual intelligence*. New York: W. W. Norton.

Hollan, J., Hutchins, E., & Kirsh, D. (2000). *Distributed cognition: Toward a new foundation for human-computer interaction research.* Retrieved from http://hci.ucsd.edu/lab/publications.htm

Hutchins, E. (1995). *Cognition in the wild*. Cambridge, MA: MIT Press.

Johnson, S. (1997). *Culture interface*. New York: Perseus Books Group.

Kirsh, D. (2004). *Metacognition, distributed cognition and visual design*. Retrieved from http://consci.ucsd.edu/~kirsh/publications.html

Krug, S. (2000). *Don't make me think*. New York: New Riders Publishing.

Lavie, T., & Tractinsky, N. (2004). Assessing dimensions of perceived visual aesthetics of Web sites. *International Journal Human-Computer Studies*, *60*, 269-298.

Magnani, L. (2001). *Abduction, reason, and science: Processes of discovery and explanation*. New York: Kluwer Academic.

Norman, D. (2004). *Emotional design: Why we love (or hate) everyday things*. New York: Basic Books.

Peirce, C. S. (1923). *Chance, love and logic. Philosophical essays.* New York: Harcourt.

Peirce, C. S. (1972). *Charles S. Peirce: The essential writings*. New York: Harper & Row.

Raskin, J. (2000). *The humane interface*. New York: Addison-Wesley.

Shneiderman, B. (2002). *Leonardo's laptop. Human needs and the new computing technologies*. Cambridge, MA: MIT Press.

Susi, T., & Ziemke, T. (2001). Social cognition, artefacts, and stigmergy. *Journal of Cognitive Systems Research*, *2*, 273-290.

Thagard, P. (1988). *Computational philosophy of science*. Cambridge, MA: MIT Press.

Thagard, P., & Shelley, C.P. (1997). Abductive reasoning: Logic, visual thinking, and coherence. In M. L. Dalla et al. (Eds.), *Logic and scientific method* (pp. 413-427). Dordrecht: Kluwer Academic Publishers.

van Schaik, P., & Ling, J. (2004). The effects of screen ratio on information retrieval in Web pages. *Computers in Human Behavior*. Display, 24, 187-195.

KEY TERMS

Abduction: The explanatory process of inferring certain facts and/or hypotheses that explain or discover some phenomenon or observation.

Affordance: Can be viewed as a property of an object that supports certain kinds of actions rather than others.

Enriching Digital World: The process of embedding and encoding those clues or signs within an interface from which the user is enabled to exploit the functionalities of a certain product.

Inference to the Best Interaction: Any interface provides some clue from which the user is enabled to perform the correct action in order to accomplish tasks with a product. The process of inferring the correct action can be called inference to the best interaction.

Interface: The way a user interacts with a product, what he or she does, and how it responds.

Mediating Structure: What coordinates the interaction between a user and a tool providing additional computational resources that simplify the task.

Stigmergy: The process that mediates the interaction between animals and animals through the environment and provides those clues from which any agent is able to infer what to do next.

ENDNOTES

1. The concept of affordance is akin to that of stigmergy. We might say that the change stored within the environment affords a certain reaction rather than others. For more information about the concept of affordance, see Gibson (1979).
2. There are several problems related to metaphors in interface design. For further information on this topic, see Collins (1995).
3. About the inferential role of perception, see Thagard and Shelley (1997).
4. For further details about visual, spatial, and temporal abduction, see Magnani (2001). About emotion as an abductive inference, see Peirce (1972).

Adaptable and Adaptive Web-Based Educational Systems

Elena Gaudioso
Artificial Intelligence Department, UNED, Spain

Miguel Montero
Centro de Profesores y Recursos de Albacete, Spain

INTRODUCTION

Nowadays, the use of computers and Internet in education is on the increase. Web-based educational systems (WES) are now widely used to both provide support to distance learning and to complement the traditional teaching process in the classroom.

To be very useful in the classroom, one of the characteristics expected in a WES is the ability to be aware of students' behaviors so that it can take into account the level of knowledge and preferences of the students in order to make reasonable recommendations (Hong, Kinshuk, He, Patel, & Jesshope, 2001).

The main goal of adaptation in educational systems is to guide the students through the course material in order to improve the effectiveness of the learning process.

Usually, when speaking of adaptive Web-based educational systems, we refer also to adaptable systems. Nevertheless, these terms are not really synonyms. Adaptable systems are abundant (Kobsa, 2004). In these systems, any adaptation is predefined and can be modified by the users before the execution of the system. In contrast, adaptive systems are still quite rare. In adaptive systems, any adaptation is dynamic which changes while the user is interacting with the system, depending on users' behaviors.

Nowadays, adaptable and adaptive systems recently gained strong popularity on the Web under the notion of *personalized systems*. A system can be adaptable and adaptive at the same time.

In educational context, adaptable systems include also those systems that allow the teacher to modify certain parameters and change the response that the system gives to the students.

In this situation, we claim that, in educational context, it is important to provide both types of personalization. On one hand, it is necessary to let teachers control the adaptation to students. On the other hand, due to a great diversity of interactions that take place in a WES, it is necessary to help teachers in the assessment of the students' actions by providing certain dynamic adaptation automatically performed by the system. In this article, we will present how we can obtain adaptable and adaptive systems. Next, we will briefly present how we combine both types of personalization in PDINAMET, a WES for Physics. Finally we will describe some future trends and conclusions.

BACKGROUND

To provide personalization, systems store the information needed in the so-called *models for adaptation*. These models contain information about users' characteristics and preferences (the so-called *user model*). Educational systems also need information about the domain that is being taught (the so-called *domain model*) and the pedagogical strategies that will be followed when guiding students (the so-called *pedagogical model*). The first systems in incorporating these models were the Intelligent Tutoring systems (Wenger, 1987).

These models usually make use of an attribute-value representation. The value of each attribute can be obtained directly from the users by means of initial questionnaires (for example, to acquire personal data). Other attributes can be directly obtained from the data that the system logs from the users' interaction (for example, number of course pages visited) (Gaudioso & Boticario, 2002). Nevertheless, certain attributes can neither be directly obtained from the user nor from the data logged, and they must be automatically inferred by the system.

Various methods can be used to infer these attributes (Kobsa, 2004). These methods include simple rules that predict user's characteristics or assign users to predetermined user groups with known characteristics when certain user actions are being observed (the so-called *profiles* or *stereotypes*).

The main disadvantage of the rule-based approach is that the rules (and the profiles) have to be pre-defined, and so all the process is very static. To make the process more dynamic, there exist other methods to obtain the value of those attributes (Kobsa, 2004). Probabilistic reasoning methods take uncertainty and evidences from users' characteristics and interaction data into account (Gertner, Conati, & VanLehn, 1998). Plan recognition methods aim at linking individual actions of users to presumable underlying plans and goals (Darío, Méndez, Jiménez, & Guzmán, 2004). Machine learning methods try to detect regularities in users' actions (and to use the learned patterns as a basis for predicting future actions) (Soller & Lesgold, 2003).

These systems can be considered adaptive since the adaptation is dynamic and it is not controlled by users. For example, once a rule is defined, it usually cannot be modified. Another review (Brusilovsky & Peylo, 2003) differentiates between adaptive and intelligent WES; authors consider adaptive systems as those that attempt to be different for different students and groups of students by taking into account information accumulated in the individual's or group's student models. On the other hand, they consider intelligent systems as those that apply techniques from the field of artificial intelligence to provide broader and better support for the users of WES. In many cases, Web-based adaptive educational systems fall into the two categories. According to our classification, both adaptive and intelligent systems can be considered as adaptive. We think that our classification is more appropriate from a user's point of view. A user does not usually first care about how the adaptation is being done but if she or he can modify this adaptation. A complete description of the personalization process (distinguishing between adaptive and adaptable systems) can be found at Kobsa, Koenemann, and Pohl (2001) and Oppermann, Rashev, and Kinshuk (1997).

As mentioned earlier, in the educational domain, it is necessary to let teachers control the personalization process. Thus it seems necessary to combine capabilities of both adaptive and adaptable systems.

A HYBRID APPROACH TO WEB-BASED EDUCATIONAL SYSTEMS

In this section, we present PDINAMET, a system that provides both types of personalization.

PDINAMET is a Web-based adaptive and adaptable educational system directed to the teaching of dynamics within the area of the physics. In PDINAMET, we maintain three types of models: student model, domain model, and pedagogical model.

Besides personal data, the student model contains information about the students' understanding of the domain. The domain model includes information about the contents that should be taught. Finally, pedagogical model includes information about instructional strategies.

In PDINAMET, we consider an *adaptation task* as any support provided by the system to learners and teachers taking into account learners' personal characteristics and knowledge level. In PDINAMET, two types of adaptation tasks are considered: *static* (that makes PDINAMET adaptable) and *dynamic* (that makes PDINAMET adaptive). Static adaptation tasks are those based on predefined rules. These include assigning students to pre-defined profiles (that can be modified by teachers) in which PDINAMET based certain recommendations (e.g., recommend a suitable exercise). To come up from the lack of coverage of pre-defined rules for every situation that could arise during the course, some dynamic tasks are performed in PDINAMET. These tasks include: diagnosis of student models, intelligent analysis of students' interactions, and recommend instructional strategies (Montero & Gaudioso, 2005).

FUTURE TRENDS

We have seen that teachers should have access to the models in order to inspect or modify them. From this point of view, an open question is how and when we should present the models to a teacher in a friendly and comprehensible manner and how we can allow interactive refinement to enrich these models.

From the point of view of technology, educational domain poses some challenges in the development of adaptive systems. In educational domain, there exists a great amount of prior knowledge provided by teachers, which must be incorporated in the adaptation mechanisms. As an example, when using machine learning, it is necessary to efficiently combine knowledge and data. A long-standing fundamental problem with machine learning algorithms is that they do not provide easy ways of incorporating prior knowledge to guide and constrain learning.

CONCLUSION

In this article, we have described the differences between adaptable and adaptive systems. Adaptable systems allow the users to modify the personalization mechanism of the system. For example, Web portals permit users to specify the information they want to see and the form in which it should be displayed by their Web browsers. Nevertheless, this process is very static. On the other hand, adaptive systems automatically change the response given to the users taking into account users' characteristics, preferences and behavior. In WES, there is a necessity of combining both capabilities to let teachers control the guidance given to the students and to help them in the assessment of the students' actions. Nevertheless, there exist some open issues mainly regarding the way in which we can present the teachers with this functionality and how we can dynamically introduce the feedback from the teachers in the models for adaptation. Probably, if we make progress in this direction, we will be able to provide teachers with more useful Web-based adaptive systems.

REFERENCES

Brusilovsky, P., & Peylo, C. (2003). Adaptive and intelligent Web-based educational systems. *International Journal of Artificial Intelligence in Education, 13*, 156-169.

Darío, N., Méndez, D., Jiménez, C., & Guzmán, J. A. (2004, August 22-23). Artificial intelligence planning for automatic generation of customized virtual courses. In the *Proceedings of the ECAI'04 Workshop on Planning and Scheduling: Bridging Theory to Practice*, Valencia, Spain (pp. 52-62).

Gaudioso, E., & Boticario, J.G. (2002). User data management and usage model acquisition in an adaptive educational collaborative environment. In the *Proceedings of the International Conference on Adaptive Hypermedia and Adaptive Web-based Systems,* Lecture Notes in Computer Science (LNCS), Málaga, Spain (pp. 143-152).

Gertner, A. S., Conati, C., & VanLehn, K. (1998). Procedural help in Andes: Generating hints using a Bayesian network student model. In the *Proceedings of the 15th National Conference on Artificial Intelligence,* Madison, Wisconsin (pp. 106-111).

Hong, H., Kinshuk, He, X., Patel, A., & Jesshope, C. (2001, May 1-5). Adaptivity in Web-based educational system. In the *Poster Proceedings of the Tenth International World Wide Web Conference*, Hong Kong, China (pp. 100-101).

Kobsa, A. (2004). Adaptive interfaces. *Encyclopaedia of Human-Computer Interaction* (pp. 9-16). Great Barrington, MA: Berkshire Publishing.

Kobsa, A., Koenemann, J., & Pohl, W. (2001). Personalized hypermedia presentation techniques for improving online customer relationships. *The Knowledge Engineering Review, 16*(2), 111-155.

Montero, M., & Gaudioso, E. (2005). P-Dinamet: A Web-based adaptive learning system to assist learners and teachers. To appear in L. Jain & C. Ghaoui (Eds.), *Innovations in knowledge-based virtual education*. Germany: Springer Verlag.

Oppermann, R., Rashev, R., & Kinshuk (1997). Adaptability and adaptivity in learning systems. In A. Behrook (Ed.), *Knowledge transfer* (vol. 2, pp. 173-179). Pace, London.

Soller, A., & Lesgold, A. (2003, July). A computational approach to analysing online knowledge sharing interaction. In the *Proceedings of the 11th International Conference on Artificial Intelligence in Education*, Sydney, Australia (pp. 253-260). Amsterdam: IOS Press.

Wenger, E. (1987). *Artificial intelligence and tutoring systems*. Los Altos, CA: Morgan Kaufmann Publishers.

KEY TERMS

Adaptable Systems: Systems, which offer personalization that is pre-defined before the execution of the system and that may be modified by users.

Adaptive Systems: Systems, which offer personalization that is dynamically built and automatically performed based on what these systems learn about the users.

Domain Model: A model that contains information about the course taught in a WES. A usual representation is a concept network specifying concepts and their relationships.

Pedagogical Model: A model that contains information about the pedagogical strategies which will be followed when making recommendations.

Personalization: Ability of systems to adapt and provide different responses to different users, based on knowledge about the users.

Student Model: A user model in educational systems that also includes information about the student, for example, the student's level of knowledge.

User Model: A model that contains information about users' characteristics and preferences.

Web-Based Educational System: An educational system that supports teaching through the Web.

Agent-Based System for Discovering and Building Collaborative Communities

Olga Nabuco
Centro de Pesquisas Renato Archer—CenPRA, Brazil

Mauro F. Koyama
Centro de Pesquisas Renato Archer—CenPRA, Brazil

Edeneziano D. Pereira
Centro de Pesquisas Renato Archer—CenPRA, Brazil

Khalil Drira
Laboratoire d'Analyse et d'Architecture des Systèmes du CNRS (LAAS-CNRS), France

INTRODUCTION

Currently, organizations are under a regime of rapid economic, social, and technological change. Such a regime has been impelling organizations to increase focus on innovation, learning, and forms of enterprise cooperation. To assure innovation success and make it measurable, it is indispensable for members of teams to systematically exchange information and knowledge.

McLure and Faraj (2000) see an evolution in the way knowledge exchange is viewed from "knowledge as object" to "knowledge embedded in people," and finally as "knowledge embedded in the community."

The collaborative community is a group of people, not necessarily co-located, that share interests and act together to contribute positively toward the fulfillment of their common goals. The community's members develop a common vocabulary and language by interacting continuously. They also create the reciprocal trust and mutual understanding needed to establish a culture in which collaborative practices pre-dominate. Such practices can grasp and apply the tacit knowledge dispersed in the organization, embodied in the people's minds. Tacit knowledge is a concept proposed by Polanyi (1966) meaning a kind of knowledge that cannot be easily transcripted into a code. It can be profitably applied on process and/or product development and production. Therefore, community members can powerfully contribute to the innovation process and create value for the organization. In doing so, they become a fundamental work force to the organization.

BACKGROUND

A collaborative community emerges on searching for something new. It can rise spontaneously or in response to a firm request. In both cases, each volunteer can evaluate whether it is interesting to become a member of the group or not.

Whenever a firm needs to make a decision whether it is feasible to develop a new product, usually it asks its senior engineers (experts) technical opinions about the undertaking. The best solution depends on information such as: fitness of the current production processes considering the new product features, design requirements, characteristics of the materials needed, time constraints, and so forth. In short, it requires assessment and technical opinions from many firms' experts.

Depending on the product's complexity, priority, constraints, and so forth, experts start exchanging opinions with those to whom they truly know to be competent in the subject. As the forthcoming news about the new product spreads, a potential collaborative community can emerge and make the firm's technical experience come afloat. This occurs as the experts evaluate how much the firm's production processes fit the new product requirements, how many similar products it already has designed, what are the product's parts that could be assembled on a partnership schema, and so on. Such opinions strongly support the decision making process related to the new product feasibility.

Every community can develop both individual and collective competence by creating, expanding, and exchanging knowledge. Mutual collaboration among members of a community, as well as with other groups, pre-supposes the ability to exchange information and to make the knowledge available to others. Therefore, collaborative communities should be pervasive, integrated, and supported by the enterprise rules that limit and/or direct people's actions in organizations.

Recommender systems (Table 1) search and retrieve information according to users' needs; they can be specialized on users' profiles or on the users' instantaneous interests (e.g., when users browse the Web). "Recommender systems use the opinions of members of a community to help individuals in that community identify the information or products most likely to be interesting to them or relevant to their needs" (Konstan, 2004, p. 1). By discovering people's interests and comparing whether such interests are the same or similar, recommender systems aid individuals to form collaborative communities.

Usually, this kind of system is based on artificial intelligence technologies such as machine learning algorithms, ontologies, and multi-agent systems. Such technologies can be used separately or combined in different ways to find and deliver information to the people who require it. Ontologies are well-suited for knowledge sharing as they offer a formal base for describing terminology in a knowledge domain (Gruber, 1995; McGuinness, 2001).

Recommender systems can be classified as collaborative filtering with and without content analysis and knowledge sharing. Collaborative filtering with content analysis is based on information from trusted people, which the system recognizes and also recommends. Examples of this kind of system are: GroupLens (Konstan et al., 1997), ReferralWeb (Kautz et al., 1997) and Yenta (Foner, 1999). Collaborative filtering without content analysis examines the meta-information and classifies it according to the user's current context. It can recommend multimedia information that otherwise could be too complex to be analyzed. PHOAKS (Terveen et al., 1997) is an example of a system that associates

Table 1. Recommender systems comparison

	Collaborative filtering		Knowledge sharing	Multi-agent	Uses ontology	Recommends
Recommender systems	With content analysis	Without content analysis				
Yenta (Foner, 1999)	x				x	Scientific papers
GroupLens (Konstan, Miller, Hellocker, Gordon, & Riedl, 1997)	x					Usenet news
Referral Web (Kautz, Selman, & Shah, 1997)	x				x	Scientists
Phoaks (Terveen, Hill, Amento, McDonald, & Creter, 1997)		x		x		Scientists
OntoShare (Davies, Dukes, & Stonkus, 2002)			x	x		Opinions
QuickStep (Middleton, De Roure, & Shadbolt, 2001)			x	x	x	Web pages
OntoCoPI (Alani, O'Hara, & Shadbolt, 2002)			x	x		Communities of practice
Sheik (Nabuco, Rosário, Silva, & Drira, 2004)			x	x	x	Similar profiles

scientific profiles by chaining bibliographic references with no comprehension of the article's contents.

BUILDING COLLABORATIVE COMMUNITIES

This work focuses on the building (or establishment) phase of a collaborative community. It is supposed that the initial set of candidates to participate in such a community are already motivated to that purpose; sociological or psychological elements involved in their decision are not in the scope of this work.

The question faced is: How to fulfill a demand for dealing with a particular event that arises and drives the candidates to search for partners or peers? One potential collaborative community could emerge, triggered by that event and being composed of a subset derived from the initial set of candidates. Inside such subset, the candidates must recognize themselves as peers, as persons sharing particular interests that characterize the community.

The selected approach was to develop a recommender system that can compare candidate's profiles and recommend those candidates that share similar interests as being potential members of a community. From each candidate, the system must gather its profile, characterized as a combination of static information on location (Name, Project, Organization, Address) and dynamic information on skills, described as a set of keywords. These profiles are then processed, aiming at two objectives: to allow the automatic organization of the informal vocabulary used by the candidates into formal knowledge domain categories and to establish a correlation between candidates by selecting those with overlapping areas of interest (formal knowledge domains) enabling the recommending process.

If a subset of candidates exists, that one shares one or more knowledge domains. Such a subset is a potential community, and those candidates are informed of the existence of such a community, receiving information on others candidates' locations and on the overlapping knowledge domains. Now candidates can proceed in contacting their potential peers at will, using their own criteria of personal networking.

One important remark on this process of community building is that candidates are not forced to participate in the process, and, if they want to, the sole requirement is to share their profile with other candidates. Privacy is a very important issue on the process of collaboration. Candidates are assured that their location information will be forwarded only to the future potential communities, and that their complete profile will not be publicly available.

An experimental recommender system, named SHEIK (Sharing Engineering Information and Knowledge), was designed to act as a community-building system. SHEIK is based on agents and ontologies' technologies and implemented using Java language. It advises candidates on the potential collaborative community they could belong to. Ontologies are used to relate knowledge domains and a candidate's profile while agents capture each candidate's interests and provide him/her with useful information.

SHEIK ARCHITECTURE

SHEIK was designed in two layers: an interface layer responsible for gathering candidates' profiles and a processing layer that consolidates the data upon each candidate and enables the process of peer finding.

The interface layer is composed of a set of agents, one for each candidate, each agent residing in the associated candidate's computer. The agents use knowledge acquisition techniques to search the candidate's workspace (in a way, configured by the candidate) and communicate results to the processing layer in a timely basis (usually daily). So the interface agent automatically updates candidate's general information to the system, unburdening the candidate from repetitive tasks he/she is unfamiliar to. Also, as there is one interface agent for each candidate, all of the interface agents operate autonomously and in parallel increasing system power.

The processing layer is composed of agents, named Erudite agents, forming a federation. Each Erudite agent is associated to a particular knowledge domain and has the ability to process candidate's profiles, using a knowledge base, constructed over a model that associates one ontology (related to its knowledge domain) and a data model for the candidate's profiles.

SHEIK BEHAVIOR

For the sake of clarity, let us consider an example of forming an engineering team to develop a new product in a consortium of enterprises. This situation is becoming frequent and permitting to illustrate some interesting issues. First of all, the candidates to compose the collaborative community probably do not recognize their potential peers. Even if they could have an initial idea of the candidates set, based on their social interaction inside their own enterprise, they normally could not have the same feeling outside it, as they could not know in advance other external engineering teams.

Secondly, the use of existing filtering systems, based on Internet technologies, is usually not enough to find candidates for partnership as, because of privacy, enterprises protect their data about both engineering processes and personnel qualifications. Using trial and error procedures for searching the Web is difficult because the available information is unstructured and comes in huge quantities.

From a candidate's perspective, the functioning of SHEIK can be described as follows:

After the necessity of creating a new engineering team for developing a product arises, managers start a SHEIK system, activating a Erudite agent containing knowledge in the main area required to the development process. Each candidate, inside each enterprise, is informed of SHEIK's existence and advised to register into the system. Candidates do their registration; during this brief process, they agree to have an interface agent residing in their machines, to provide location data, and also to determinate where the agent is allowed to search for information (usually their computer's workspace). The interface agent is started and automatically analyzes the candidate's workspace, extracting a series of keywords that can approximately describe the candidate's interests. The agent repeats this operation periodically (for instance daily), so it can capture changes in the candidate's interests, and forwards the information to the Erudite agent. Many knowledge acquisition techniques can be used; for the first prototype, the KEA tool (Witten, Paynter, Frank, Gutwin, & Nevill-Manning, 1999) from Waikato University, New Zealand, was selected.

There is one interface agent for each candidate; they work automatically and asynchronously gathering information on their candidates and pushing it to the Erudite agent.

The Erudite agent, upon receiving the information from a particular interface agent, processes that information and classifies the candidate (associated to the agent) into one or more areas from its main knowledge domain. As already mentioned, many interface agents are flowing information in parallel to the Erudite agent, that is continuously updating its knowledge base. After classifying the candidate, the Erudite agent is able to perform searches in its knowledge base and discover for each candidate who the other candidates are that have related interests. The result of such a search is a potential collaborative community in the retrieved sub area, organized as a list of candidates and their compatible knowledge sub domains. The candidates pertaining to that potential community (list) are informed of their potential peers via an e-mail message. Candidates could analyze the message and decide if they want to contact the recommended candidates, and, after some interaction, they could decide to formalize the new community. The Erudite agent uses the Protégé system (Gennari et al., 2003), developed at the Stanford Medical Informatics (SMI), Stanford University, USA. Protégé is an ontology editor and a knowledge-base editor capable of answering queries.

An experiment demonstrated the potential of SHEIK as a tool in a scientific environment. It is envisaged to broaden its use and application in other domains. SHEIK is being experimented to support collaboration among scientific researchers, in the field of manufacturing, involving academic institutions from Brazil (Nabuco et al., 2004).

FUTURE TRENDS

Since the advent of Internet browsers, the amount of electronic information available has been growing in an astonishing fashion, but such information sometimes cannot be brought to use in its bare form because of its lack of structure. Basically searching engines use syntactic rules to obtain their results; they do not consider the information within its context. Efforts are in course to enhance the information through some sort of pre-classification inside knowledge domains, done via "clues" hidden in the Internet page's codes. One problem that arises is that there are too many knowledge domains and international standardized knowledge classifications are almost inexistent. This is a challenge too big to be solved for only one standardization organism. It must be treated by a diversity of organizations bringing to light another problem: how to describe knowledge in a way that permits us to interchange and complement its meaning among such diversity. Research has been done in this area but to date no standard for knowledge representation has found widespread acceptance.

The problem dealt with in this work is building a collaborative community—such a community being characterized by having interests pertaining to certain knowledge domains. The problem of formally characterizing knowledge domains, aiming to ease the interaction and interoperation between knowledge systems, is an open issue, and such a problem is considered as a future path to be followed, having the use of best practices and de-facto standards as guidelines.

In particular as the Erudite agents are organized as a federation, each agent being responsible for one knowledge domain, it is of interest to study the problem of knowledge domain determination, a difficult task because many recognized knowledge domains overlap with each other (are not self-excluding).

The interface agents were designed to accept different knowledge acquisition techniques, with minor code modification; experiencing with some of those techniques could increase SHEIK's capacity.

CONCLUSION

An approach for aiding the establishment of a collaborative community was proposed, based on the automatic determination of a candidate's profile that can be resolved to knowledge domains (and sub domains). The use of ontologies was advocated as a way of formally characterizing the knowledge domains, easing, in a certain way, the job of knowledge treatment and exchange. As a pre-requisite system, friendliness was enforced, meaning that the system to support such an approach must act automatically, aiding the users (candidates to belong to a new community) in the process of finding peers to fulfill their purposes of collaboration. As a natural outcome of this requisite, the prototype system was designed as being a multi-agent system, being scalable and capable of autonomous operation.

A prototype system was constructed, using pre-existing knowledge system and knowledge acquisition tool, compatible with de-facto standards. In operation, the system is capable of dynamically following users' interests as they change with time, so it can also aid the self-emergence of new communities regardless of managers' requests.

Initial results were encouraging, and a greater experiment is of course in the manufacturing knowledge domain.

REFERENCES

Alani, H., O'Hara, K., & Shadbolt, N. (2002, August 25-30). ONTOCOPI: Methods and tools for identifying communities of practice. *Proceedings of the IFIP 17th World Computer Congress—TC12 Stream on Intelligent Information Processing, IIP-2002*, Montreal, Canada (pp. 225-236). Deventer, The Netherlands: Kluwer, B.V.

Davies, J., Dukes, A., & Stonkus, A. (2002, May 7). OntoShare: Using ontologies for knowledge sharing. *Proceedings of the International Workshop on the Semantic Web, WWW2002*, Hawaii. Sun SITE Central Europe (CEUR).

Foner, L. N. (1999, April). *Political artifacts and personal privacy: The Yenta multi-agent distributed matchmaking system.* Boston: Media Arts and Sciences. PhD Thesis. Massachusetts Institute of Technology, Cambridge (p. 129).

Gennari, J. H., Musen, M. A., Fergerson, R. W., Grosso, W. E., Crubézy, M., Eriksson, H., Noy, N. F., & Tu, S. W. (2003). The evolution of Protégé: An environment for knowledge-based systems development. *International Journal of Human-Computer Studies*, *58*(1), 89-123.

Gruber, T. R. (1995). Towards principles for the design of ontologies used for knowledge sharing, *International Journal of Human-Computer Studies*, *43*(5-6), 907-928.

Kautz, H., Selman, B., & Shah, M. (1997, Summer). The hidden Web. *The AI Magazine, 18*(2), 27-36.

Konstan, J. A. (2004). Introduction to recommender systems: Algorithms and evaluation. *ACM Transactions on Information Systems, 22*(1), 1-4.

Konstan, J. A., Miller, B. N., Hellocker, J. L., Gordon, L. P., & Riedl, J. (1997). Applying collaborative filtering to Usenet News. *Communications of ACM, 40*(3), 77-87.

McGuinness, D. L. (2001). Ontologies and online commerce. *IEEE Intelligent Systems, 16*(1), 8-14.

McLure, M., & Faraj, S. W. (2000). "It is what one does": Why people participate and help others in electronic communities of practice. *Journal of Strategic Information Systems, 9,* 155-173.

Middleton, S., De Roure, D., & Shadbolt, N. (2001, October 22-23). Capturing knowledge of user preferences: Ontologies in recommender systems. *Proceeding of the 1st International Conference on Knowledge Capture, KCAP2001*, Victoria, British Columbia, Canada (pp. 100-107). New York: ACM Press.

Nabuco, O., Rosário, J. M., Silva, J. R., & Drira, K. (2004, April 5-7). Scientific collaboration and knowledge sharing in the virtual manufacturing network. *11th IFAC Symposium on Information Control Problem in Manufacturing, INCOM 2004*, Salvador, Bahia, Brazil (pp. 254-255). New York: ACM Press.

Polanyi, M. (1966). *The tacit dimension.* Garden City, NY: Doubleday & Co. Reprinted Peter Smith, Gloucester, Massachusetts.

Terveen, L., Hill, W., Amento, B., McDonald, D., & Creter, J. (1997). J. PHOAKS: A system for sharing recommendations. *Communications of the ACM, 3*(40), 59-62.

Witten, I. H., Paynter, G. W., Frank, E., Gutwin, C., & Nevill-Manning, C. G. (1999, August 11-14). KEA: Practical automatic keyphrase extraction. *Proceedings of the Fourth ACM Conference on Digital Libraries, DL'99*, Berkeley, CA (pp. 254-265). New York: ACM Press.

KEY TERMS

Agent: A software agent is a programmable artifact capable of intelligent autonomous action toward an objective.

Collaborative Community: A group of people sharing common interests and acting together toward common goals.

Erudite Agent: Acts as a broker locating compatible candidates according to specific similarities found in their profiles. Erudite has two roles: one is upon the requesting of interface agents, it queries its knowledge base searching for other candidates that seem to have the same interests; the other, is to keep its knowledge base updated with the candidate's specific information provided by the resident interface agents. In doing so, it implements the processing part of SHEIK system's architecture.

Filtering: A technique that selects specific things according to criteria of similarity with particular patterns.

Knowledge Base: A knowledge repository, organized according to formal descriptive rules, permitting to perform operations over the represented knowledge.

Machine Learning: A computer embedded capability of data analysis with the purpose of acquiring selected characteristics (attributes, patterns, behavior) of an object or system.

Multi-Agent System: A set of software agents that interact with each other. Inside this system, each agent can act individually in cooperative or competitive forms.

Ontology: In the computer science community, this term is employed in the sense of making explicit the concepts pertaining to a knowledge domain.

Recommender System: A computer program that aids people to find information by giving recommendations on the searched subject. It is up to the user to select useful data among the recommended ones.

Tacit Knowledge: Means the knowledge that is embodied in people's mind, not easily visible and difficult to be transmitted in words by skilled people. The concept was brought to the light by Polanyi (1966).

Agent-Supported Interface for Online Tutoring

Leen-Kiat Soh
University of Nebraska, USA

INTRODUCTION

Traditionally, learning material is delivered in a textual format and on paper. For example, a learning module on a topic may include a description (or a tutorial) of the topic, a few examples illustrating the topic, and one or more exercise problems to gauge how well the students have achieved the expected understanding of the topic. The delivery mechanism of the learning material has traditionally been via textbooks and/or instructions provided by a teacher. A teacher, for example, may provide a few pages of notes about a topic, explain the topic for a few minutes, discuss a couple of examples, and then give some exercise problems as homework. During the delivery, students ask questions and the teacher attempts to answer the questions accordingly. Thus, the delivery is interactive: the teacher learns how well the students have mastered the topic, and the students clarify their understanding of the topic. In a traditional classroom of a relatively small size, this scenario is feasible. However, when e-learning approaches are involved, or in the case of a large class size, the traditional delivery mechanism is often not feasible.

In this article, we describe an interface that is "active" (instead of passive) that delivers learning material based on the usage history of the learning material (such as degree of difficulty, the average score, and the number of times viewed), the student's static background profile (such as GPA, majors, interests, and courses taken), and the student's dynamic activity profile (based on their interactions with the agent). This interface is supported by an intelligent agent (Wooldridge & Jennings, 1995). An agent in this article refers to a software module that is able to sense its environment, receive stimuli from the environment, make autonomous decisions, and actuate the decisions, which in turn change the environment. An intelligent agent in this article refers to an agent that is capable of flexible behaviour: responding to events timely, exhibiting goal-directed behaviour, and performing machine learning. The agent uses the profiles to decide, through case-based reasoning (CBR) (Kolodner, 1993), which learning modules (examples and problems) to present to the students. Our CBR treats the input situation as a problem, and the solution is basically the specification of an appropriate example or problem. Our agent also uses the usage history of each learning material to adjust the appropriateness of the examples and problems in a particular situation. We call our agent Intelligent Learning Material Delivery Agent (ILMDA). We have built an end-to-end ILMDA infrastructure, with an *active* GUI front-end—that monitors and tracks every interaction step of the user with the interface, an agent powered by CBR and capable of learning, and a multi-database backend.

In the following, we first discuss some related work in the area of intelligent tutoring systems. Then, we present our ILMDA project, its goals and framework. Subsequently, we describe the CBR methodology and design. Finally, we point out some future trends before concluding.

BACKGROUND

Research strongly supports the user of technology as a catalyst for improving the learning environment (Sivin-Kachala & Bialo, 1998). Educational technology has been shown to stimulate more interactive teaching, effective grouping of students, and cooperative learning. A few studies, which estimated the cost effectiveness, reported time saving of about 30%. At first, professors can be expected to struggle with the change brought about by technology. However, they will adopt, adapt, and eventually learn to use technology effortlessly and creatively (Kadiyala & Crynes, 1998). As summarized in Graesser, VanLehn, Rosé, Jordan, and Harter (2001), intelligent tutoring systems (ITSs) are clearly one of the successful enterprises in artificial intelligence (AI). There is a long list of ITSs that have been tested on humans and have proven to facilitate learning. These ITSs use a variety of computational modules that are familiar to those of us in AI: production systems, Bayesian networks, schema templates, theorem proving, and explanatory reasoning. Graesser et al. (2001) also pointed out the weaknesses of the current state of tutoring systems: First, it is possible for students to guess and find an answer and such shallow learning will not be detected by the system. Second, ITSs do not involve students in conversations so students might not learn the domain's language. Third, to understand the students' thinking, the GUI of the ITSs tends to encourage students to focus on the details instead of the overall picture of a solution.

There have been successful ITSs such as PACT (Koedinger, Anderson, Hadley, & Mark, 1997), ANDES (Gertner & VanLehn, 2000), AutoTutor (Graesser et al., 2001), and SAM (Cassell et al., 2000), but without machine learning capabilities. These systems do not generally adapt to new circumstances, do not self-evaluate and self-configure their own strategies, and do not monitor the usage history of the learning material being delivered or presented to the students. In our research, we aim to build intelligent tutoring agents that are able to learn how to deliver appropriate different learning material to different types of students and to monitor and evaluate how the learning material are received by the students. To model students, our agent has to monitor and track student activity through its interface.

APPLICATION FRAMEWORK

In the ILMDA project, we aim to design an agent-supported interface for online tutoring. Each topic to be delivered to the students consists of three components: (1) a tutorial, (2) a set of related examples, and (3) a set of exercise problems to assess the student's understanding of the topic. Based on how a student progresses through the topic and based on his or her background profile, our agent chooses the appropriate examples and exercise problems for the student. In this manner, our agent customizes the specific learning material to be provided to the student. Our design has a modular design of the course content and delivery mechanism, utilizes true agent intelligence where an agent is able to learn how to deliver its learning material better, and self-evaluates its own learning material.

The underlying assumptions behind the design of our agent are the following. First, a student's behaviour in viewing an online tutorial, and how he or she interacts with the tutorial, the examples, and the exercises, is a good indicator of how well the student understands the topic in question, and this behaviour is observable and quantifiable. Second, different students exhibit different behaviours for different topics such that it is possible to show a student's understanding of a topic, say, T1, with an example E1, and at the same time, to show the same student's lack of understanding of the same topic T1 with another E2, and this differentiation is known and can be implemented. These two assumptions require our agent to have an *active interface*—an interface that monitors and tracks its interaction with the user.

Further, we want to develop an integrated, flexible, easy-to-use database of courseware and ILMDA system, including operational items such as student profiles, ILMDA success rates, and so forth, and educational items such as learner model, domain expertise, and course content. This will allow teachers and educators to monitor and track student progress, the quality of the learning material, and the appropriateness of the material for different student groups. With the ability to self-monitor and evaluate, our agent can identify how best to deliver a topic to a particular student type with distinctive behaviours. We see this as valuable knowledge to instructional designers and educational researchers as ILMDA not only is a testbed for testing hypotheses, but it is also an active decision maker that can expose knowledge or patterns that are previously unknown to researchers.

MODEL

Our ILMDA system is based on a three-tier model, as shown in Figure 1. It consists of a graphical user interface (GUI) front-end application, a database backend, and the ILMDA reasoning in between. A student user accesses the learning material through the GUI. The agent captures the student's interac-

Figure 1. Overall methodology of the ILMDA system

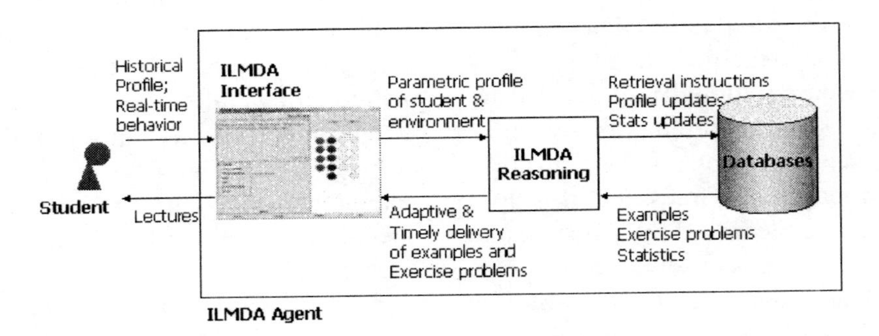

tions with the GUI and provides the ILMDA reasoning module with a parametric profile of the student and environment. The ILMDA reasoning module performs case-based reasoning to obtain a search query (a set of search keys) to retrieve and adapt the most appropriate example or problem from the database. The agent then delivers the example or problem in real-time back to the user through the interface.

Overall Flow of Operations

When a student starts the ILMDA application, he or she is first asked to login. This associates the user with his or her profile information. The information is stored in two separate tables. All of the generally static information, such as name, major, interests, and so forth, is stored in one table, while the user's dynamic information (i.e., how much time, on average, they spend in each section; how many times they click the mouse in each section, etc.) is stored in another table. After a student is logged in, he or she selects a topic and then views the tutorial on that topic. Following the tutorial, the agent looks at the student's static profile, as well as the dynamic actions the student took in the tutorial, and searches the database for a similar case. The agent then adapts the output of that similar case depending on how the cases differ, and uses the adapted output to search for a suitable example to give to the student. After the student is done looking at the examples, the same process is used to select an appropriate problem. Again, the agent takes into account how the student behaved during the example, as well as his or her background profile. After the students complete an example or a problem, they may elect to be given another. If they do so, the agent notes that the last example or problem it gave the student was not a good choice for that student, and tries a different solution. Figure 2 shows the interaction steps between our ILMDA agent and a student.

Figure 2. GUI and interactions between ILMDA and students

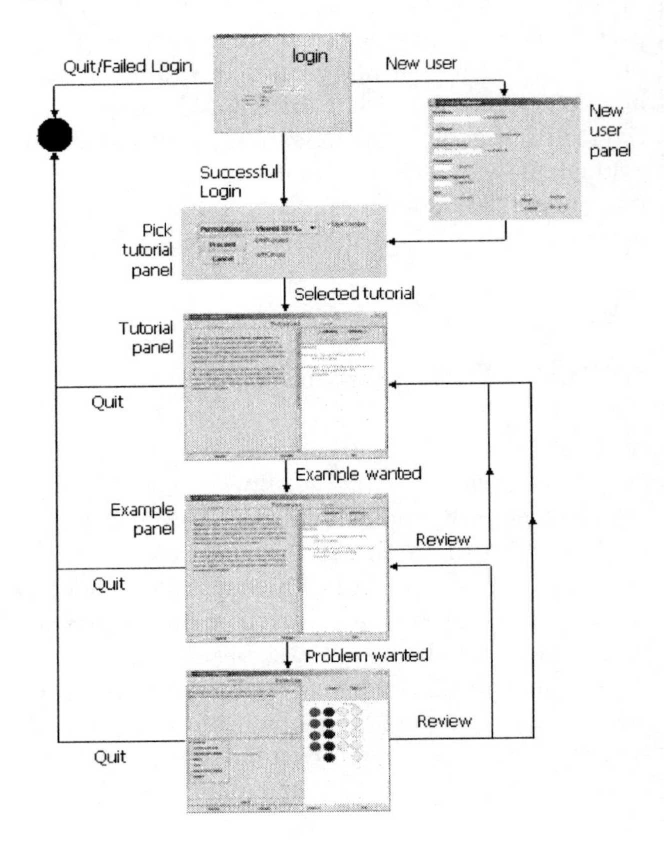

Learner Model

A learner model is one that tells us the metacognitive level of a student by looking at the student's behaviour as he or she interacts with the learning material. We achieve this by profiling a learner/student along two dimensions: student's background and activity. The background of a student stays relatively static and consists of the student's last name, first name, major, GPA, goals, affiliations, aptitudes, and competencies. The dynamic student's profile captures the student's real-time behaviour and patterns. It consists of the student's online interactions with the GUI module of the ILMDA agent including the number of attempts on the same learning material, number of different modules taken so far, average number of mouse clicks during the tutorial, average number of mouse clicks viewing the examples, average length of time spent during the tutorial, number of quits after tutorial, number of successes, and so on.

In our research, we incorporate the learner model into the case-based reasoning (CBR) module as part of the problem description of a case: Given the parametric behaviour, the CBR module will pick the best matching case and retrieve the set of solution parameters that will determine which examples or exercise problems to pick.

Case-Based Reasoning

Each agent has a case-based reasoning (CBR) module. In our framework, the agent consults the CBR module to obtain the specifications of the appropriate types of examples or problems to administer to the users. The learner model discussed earlier and the profile of the learning material constitute the problem description of a case. The solution is simply a set of search keys (such as the number of times the material has been viewed, the difficulty level, the length of the course content in terms of the number of characters, the average number of clicks the interface has recorded for this material, etc.) guiding the agent in its retrieval of either a problem or an exercise.

Note that CBR is a reasoning process that derives a solution to the current problem based on adapting a known solution to a previously encountered, similar problem to the current one. Applied to our application, the information associated with a student and a particular topic is the situation statement of a case, and the solution of the case is basically the characteristics of the appropriate examples or problems to be delivered to the student. When adapting, CBR changes the characteristics of the appropriate examples or problems based on the difference between the current situation and the situation found in the most similar case. CBR allows the agent to derive a decent solution from what the agent has stored in its casebase instead of coming up with a solution from scratch.

Case-Based Learning (CBL)

To improve its reasoning process, our agent learns the differences between good cases (cases with a good solution for its problem space) and bad cases (cases with a bad solution for its problem space). It also meta-learns adaptation heuristics, the significance of input features of the cases, and the weights of a content graph for symbolic feature values.

Our agent uses various weights when selecting a similar case, a similar example, or a similar problem. By adjusting these weights, we can improve our results, and hence, learn from our experiences. In order to improve our agent's independence, we want to have the agent adjust the weights without human intervention. To do this, the agent uses simple methods to adjust the weights called learning modules. Adjusting the weights in this manner gives us a layered learning system (Stone, 2000) because the changes that one module makes propagate through to the other modules. For instance, the changes we make to the similarity heuristics will alter which cases the other modules perceive as similar.

Active Graphical User Interface

The ILMDA front-end GUI application is written in Java, using the Sun Java Swing library. It is active as it monitors and tracks every interaction step between a student and ILMDA. It stores these activities in its database, based on which the ILMDA agent maintains a dynamic profile of the student and reasons to provide the appropriate learning material.

For our GUI, a student progresses through the tutorials, examples, and problems in the following manner. First, the student logs onto the system. If he or she is a new student, then a new profile is created. The student then selects a topic to study. The agent administers the tutorial associated with the topic to the student accordingly. The student studies the tutorial, occasionally clicking and scrolling, and browsing the embedded hyperlinks. Then when the student is ready to view an example, he or she clicks to proceed. Sensing this click, the agent immediately captures all the mouse activity and time spent in the tutorial and updates the student's activity profile. The student goes through a similar process and may choose to quit the system, indicating a failure of our example, or going back to the tutorial page for further clarification. If the student decides that he or she is ready for the problem, the agent will retrieve the most appropriate one based on the student's updated dynamic profile.

FUTURE TRENDS

We expect the field of intelligent tutoring systems to make great contribution in the near future as the intelligent agent technologies are incorporated to customize learning material for different students, re-configure their own reasoning process, and evaluate the quality of the learning material that they deliver to the students. Interfaces that are more flexible in visualizing and presenting learning material will also be available. For example, a tutorial would be presented in multiple ways to suit different students (based on students' background and the topic of the tutorial). Therefore, an agent-supported interface that is capable of learning would maximize the impact of such tutorials on the targeted students' groups. Interfaces, with intelligent agents, will be able to adapt to students timely and pro-actively, which makes this type of online tutoring highly personable.

CONCLUSION

We have described an intelligent agent that delivers learning material adaptively to different students. We have built the ILMDA infrastructure, with a GUI front-end, an agent powered by case-based reasoning (CBR), and a multi-database backend. We have also built a comprehensive simulator for our experiments. Preliminary experiments demonstrate the correctness of the end-to-end behaviour of the ILMDA agent, and show the feasibility of ILMDA and its learning capability. Ongoing and future work includes incorporating complex learner and instructional models into the agent and conducting further experiments on each learning mechanism, and investigating how ILMDA adapts to a student's behaviour, and how ILMDA adapts to different types of learning material.

REFERENCES

Cassell, J., Annany, M., Basur, N., Bickmore, T., Chong, P., Mellis, D., Ryokai, K., Smith, J., Vilhjálmsson, H., & Yan, H. (2000, April 1-6). Shared reality: Physical collaboration with a virtual peer. *Proceedings of the ACM SIGCHI Conference on Human Factors in Computer Systems*, The Hague, The Netherlands (pp. 259-260).

Gertner, A. S., & VanLehn, K. (2000, June 19-23). ANDES: A coached problem-solving environment for physics. *Proceedings of the Fifth International Conference on Intelligent Tutoring System (ITS'2000)*, Montreal, Canada (pp. 131-142).

Graesser, A. C., VanLehn, K., Rosé, C. P., Jordan, P. W., & Harter, D. (2001). Intelligent tutoring systems with conversational dialogue. *AI Magazine*, *22*(4), 39-51.

Kadiyala, M., & Crynes, B. L. (1998, November 4-7). Where's the proof? A review of literature on effectiveness of information technology in education. *Proceedings of the 1998 Frontiers in Education Conference*, Tempe, Arizona (pp. 33-37).

Koedinger, K. R., Anderson, J. R., Hadley, W. H., & Mark, M. A. (1997). Intelligent tutoring goes to school in the big city. *Journal of Artificial Intelligence in Education, 8*(1), 30-43.

Kolodner, J. (1993). *Case-based reasoning.* San Francisco: Morgan Kaufmann.

Sivin-Kachala, J., & Bialo, E. (1998). *Report on the effectiveness of technology in schools, 1990-1997*. Software Publishers Association.

Stone, P. (2000). *Layered learning in multiagent systems*. MIT Press.

Wooldridge, M., & Jennings, N. R. (1995). Intelligent agents: Theory and practice. *The Knowledge Engineering Review*, *10*(2), 115-152.

KEY TERMS

Active Interface: An interface that monitors and tracks its interaction with the user.

Agent: A module that is able to sense its environment, receive stimuli from the environment, make autonomous decisions, and actuate the decisions, which in turn change the environment.

Casebase: A collection of cases with each case containing a problem description and its corresponding solution approach.

Case-Based Learning (CBL): Stemming from case-based reasoning, the process of determining and storing cases of new problem-solution scenarios in a casebase.

Case-Based Reasoning (CBR): A reasoning process that derives a solution to the current problem based on adapting a known solution to a previously encountered, similar problem to the current one.

Intelligent Agent: An agent that is capable of flexible behaviour: responding to events timely, exhibiting goal-directed behaviour and social behaviour, and conducting machine learning to improve its own performance over time.

Intelligent Tutoring System (ITS): A software system that is capable of interacting with a student, providing guidance in the student's learning of a subject matter.

Analyzing and Visualizing the Dynamics of Scientific Frontiers and Knowledge Diffusion

Chaomei Chen
Drexel University, USA

Natasha Lobo
Drexel University, USA

INTRODUCTION

Estimated numbers of scientific journals in print each year are approximately close to 70,000-80,000 (Rowland, McKnight, & Meadows, 1995). Institute of Scientific Information (ISI) each year adds over 1.3 million new articles and more than 30 million new citations to its science citation databases of 8,500 research journals. The widely available electronic repositories of scientific publications, such as digital libraries, preprint archives, and Web-based citation indexing services, have considerably improved the way articles are being accessed. However, it has become increasingly difficult to see the big picture of science.

Scientific frontiers and longer-term developments of scientific disciplines have been traditionally studied from sociological and philosophical perspectives (Kuhn, 1962; Stewart, 1990). The *scientometrics* community has developed quantitative approaches to the study of science. In this article, we introduce the history and the state of the art associated with the ambitious quest for detecting and tracking the advances of scientific frontiers through quantitative and computational approaches. We first introduce the background of the subject and major developments. We then highlight the key challenges and illustrate the underlying principles with an example.

BACKGROUND

In this section, we briefly review the traditional methods for studying scientific revolutions, and the introduction of quantitative approaches proposed to overcome cumbersome techniques for visualizing these revolutions.

The concept of *scientific revolutions* was defined by Thomas Kuhn in his *Structure of Scientific Revolutions* (Kuhn, 1962). According to Kuhn, science can be characterized by normal science, crisis, and revolutionary phases. A scientific revolution is often characterized by the so-called *paradigm shift*.

Many sociologists and philosophers of science have studied revolutions under this framework, including the continental drift and plate tectonics revolution in geology (Stewart, 1990) and a number of revolutions studied by Kuhn himself. Scientists in many individual disciplines are very interested in understanding revolutions that took place at their doorsteps, for example, the first-hand accounts of periodical mass extinctions (Raup, 1999), and superstring revolutions in physics (Schwarz, 1996).

Traditional methods of studying scientific revolutions, especially sociological and philosophical studies, are time consuming and laborious; they tend to overly rely on investigators' intimate understanding of a scientific discipline to interpret the findings and evidence. The lack of large-scale, comparable, timely, and highly repeatable procedures and tools have severely hindered the widespread adaptation and dissemination such research. Scientists, sociologists, historians, and philosophers need to have readily accessible tools to facilitate increasingly complex and time-consuming tasks of analyzing and monitoring the latest development in their fields.

Quantitative approaches have been proposed for decades, notably in scientometrics, to study science itself by using scientific methods, hence the name science of science (Price, 1965). Many expect that quantitative approaches to the study of science may enable analysts to study the dynamics of science. Information science and computer science have become the major driving forces behind the movement. Commonly used sources of input of such studies include a wide variety of scientific publications in books, periodicals, and conference proceedings. Subject-specific repositories include the ACM Digital Library for computer science, PubMed Central for life sciences, the increasing number of open-access preprint archives such as www.arxiv.org, and the World Wide Web.

THE DYNAMICS OF RESEARCH FRONTS

Derek Price (1965) found that the more recent papers tend to be cited about six times more often than earlier papers. He suggested that scientific literature contains two distinct parts: a *classic* part and a *transient* part, and that the two parts have different *citation half-lives*. Citation half-lives mimic the concept of half-life of atoms, which is the amount of time it takes for half of the atoms in a sample to decay. Simply speaking, classic papers tend to be longer lasting than transient ones in terms of how long their values hold. The extent to which a field is largely classic or largely transient varies widely from field to field; mathematics, for example, is strongly predominated by the classic part, whereas life sciences tend to be highly transient.

The notion of *research fronts* is also introduced by Price as the collection of highly-cited papers that represent the frontiers of science at a particular point of time. He examined citation patterns of scientific papers and identified the significance of the role of a quantitative method for delineating the topography of current scientific literature in understanding the nature of such moving frontiers.

It was Eugene Garfield, the founder of the Institute for Scientific Information (ISI) and the father of *Science Citation Index* (SCI), who introduced the idea of using cited references as an indexing mechanism to improve the understanding of scientific literature. Citation index has provided researchers new ways to grasp the development of science and to cast a glimpse of the big picture of science. A *citation* is an instance of a published article $\boldsymbol{a}$ made a reference to a published item $\boldsymbol{b}$ in the literature, be a journal paper, a conference paper, a book, a technical report, or a dissertation. A citation is directional, $\boldsymbol{a} \rightarrow \boldsymbol{b}$. A *co-citation* is a higher-order instance involving three articles, $\boldsymbol{a}$, $\boldsymbol{b}_i$, and $\boldsymbol{b}_j$ if we found both $\boldsymbol{a} \rightarrow \boldsymbol{b}_i$ and $\boldsymbol{a} \rightarrow \boldsymbol{b}_j$. Articles $\boldsymbol{b}_i$ and $\boldsymbol{b}_j$ are co-cited. A citation network is a directed graph, whereas a co-citation network is an undirected graph.

Researchers have utilized co-citation relationships as a clustering mechanism. As a result, a cluster of articles grouped together by their co-citation connections can be used to represent more evasive phenomena such as specialties, research themes, and research fronts. Much of today's research in co-citation analysis is inspired by Small and Griffith's (1974) work in the 1970s, in which they identified specialties based on co-citation networks. A detailed description of the subject can be found in Chen (2003). A noteworthy service is the ISI Essential Science Indicators (ESI) Special Topics, launched in 2001. It provides citation analyses and commentaries of selected scientific areas that have experienced recent advances or are of special current interest. A new topic is added monthly. Other important methods include *co-word analysis* (Callon, Law, & Rip, 1986). A fine example of combining co-citation and co-word analysis is given by Braam, Moed, and van Raan (1991).

KNOWLEDGE DIFFUSION

Knowledge diffusion is the adaptation of knowledge in a broad range of scientific and engineering research and development. Tracing knowledge diffusion between science and technology is a challenging issue due to the complexity of identifying emerging patterns in a diverse range of possible processes (Chen & Hicks, 2004; Oppenheimer, 2000).

Just as citation indexing to modeling and visualizing scientific frontiers, understanding patent citations is important to the study of knowledge diffusion (Jaffe & Trajtenberg, 2002). There are a number of extensively studied knowledge diffusion, or knowledge spillover, cases, namely liquid crystal display (LCD), and nanotechnology. Knowledge diffusion between basic research and technological innovation is also intrinsically related to the scientific revolution. Carpenter, Cooper, and Narin (1980) found that nearly 90% of journal references made by patent applicants and examiners refer to basic or applied scientific journals, as opposed to engineering and technological literature. Research in universities, government laboratories, and various non-profit research institutions has been playing a pivot role in technological inventions and innovations (Narin, Hamilton, & Olivastro, 1997). On the other hand, Meyer (2001) studied patent-to-paper citations between nano-science and nanotechnology and concluded that they are as different as two different disciplines.

Earlier research highlighted a tendency of geographical localization in *knowledge spillovers*, for example, knowledge diffusion (Jaffe & Trajtenberg, 2002). Agrawal, Cockburn, and McHale (2003) found that the influence of social ties between collaborative inventors may be even stronger when it comes to account for knowledge diffusion: Inventors' patents are continuously cited by their colleagues in their former institutions.

The Role of Social Network Analysis in Knowledge Diffusion

Social network analysis is playing an increasingly important role in understanding knowledge diffusion pathways. Classic social network studies such as the work of Granovetter (1973) on weak ties and structural holes (Burt, 1992) provide the initial inspirations. Singh (2004) studied the role of social ties one step further by taking into account not only direct social ties but also indirect ones in social networks of inventors' teams based on data extracted from U.S. Patent Office patents. Two teams with a common inventor are connected in the social network. Knowledge flows between teams are analyzed in terms of patent citations. Socially proximate teams have a better chance to see knowledge flows between them. The chance shrinks as social distance increases. More importantly, the study reveals that social links offer a good explanation why knowledge spillovers appear to be geographically localized. It is the social link that really matters; geographic proximity happens to foster social links.

The Key Player Problem

The *key player problem* in social network analysis is also relevant. The problem is whether a maximum or a minimum spread is desired in a social network (Borgatti & Everett, 1992). If we want to spread, or diffuse, something as quickly or thoroughly as possible through a network, where do we begin? In contrast, if the goal is to minimize the spread, which nodes in the network should we isolate? Borgatti found that because off-the-shelf centrality measures make assumptions about the way things flow in a network, when they are applied to the "wrong" flows, they get the "wrong" answers. Furthermore, few existing measures are appropriate for the most interested types of network flows, such as the flows of gossip, information, and infections. From the modelling perspective, an interaction between flow and centrality can identify who gets things early and who gets a lot of traffic (between-ness). The example explained later in this article utilizes the between-ness centrality metric to identify pivotal points between thematic clusters.

Analysis Using Network Visualization

Freeman (2000) identifies the fundamental role of network visualization in helping researchers to understand various properties of a social network and to communicate such insights to others. He pointed to interesting trends such as increasingly higher-dimensional visualizations, changing from factor analysis to scaling techniques such as principle component analysis and correspondence analysis, more widely used layout algorithms such as spring embedder, more and more interactive images with color and animation. He envisaged that the greatest need for further social network analysis is integrative tools that enable us to access network datasets, compute, and visualize their structural properties quickly and easily—all within a single program! An increasing number of social network analysis software becomes available, including Pajek, InFlow, and many others.

The information visualization community has also produced a stream of computer systems that could be potentially applicable to track knowledge diffusion. Examples of visualizing evolving information structure include disk trees and time tubes (Chi, Pitkow, Mackinlay, Pirolli, Gossweiler, & Card, 1998), which display the changes of a Web site over time. Chen and Carr (1999) visualized the evolution of the field of hypertext using author co-citation networks over a period of nine years. Animated visualization techniques have been used to track competing paradigms in scientific disciplines over a period of 64 years (Chen, Cribbin, Macredie, & Morar, 2002).

APPLICATIONS

The CiteSpace application is discussed here.

CiteSpace is an integrated environment designed to facilitate the modeling and visualization of structural and temporal patterns in scientific literature (Chen, 2004b). CiteSpace supports an increasing number of input data types, including bibliographic data extracted from the scientific literature, grant awards data, digital libraries, and real-time data streaming on the Web. The conceptual model of Citespace is the panoramic expansion of a series of snapshots over time. The goal of CiteSpace is to represent the most salient structural and temporal properties of a subject domain across a user-specified time interval T. CiteSpace allows the user to slice the time interval T into a number of consecutive sub-intervals T_i, called time slices. A hybrid network N_i is derived in each T_i. The time series of networks N_i provides the vehicle for subsequent analysis and visualization. Network analysis techniques such as network scaling can be applied to the time series. CiteSpace helps the user to focus on pivotal points and critical pathways as the underlying phenomenon goes through profound changes. CiteSpace is still evolving as it embraces richer collections of data types and supports a wider range of data analysis, knowledge discovery, and decision support tasks.

An Example of Trend Analysis in CiteSpace

The example is motivated by the question: What are the leading research topics in the scientific literature of terrorism research? In this case, we expect CiteSpace will reveal themes related to the aftermath of the terrorist attacks on September 11, 2001, and some of the earlier ones.

The initial dataset was drawn from the Web of Science using the query "terrorism." The dataset was processed by CiteSpace. Figure 1 is a screenshot of a resultant visualization, which is a chain of subnetworks merged across individual time slices. The merged network consists of two types of nodes: new terms found in citing articles (labeled in red) and

Figure 1. Emerging trends and clusters in the terrorism research literature (Source: Chen, 2004a)

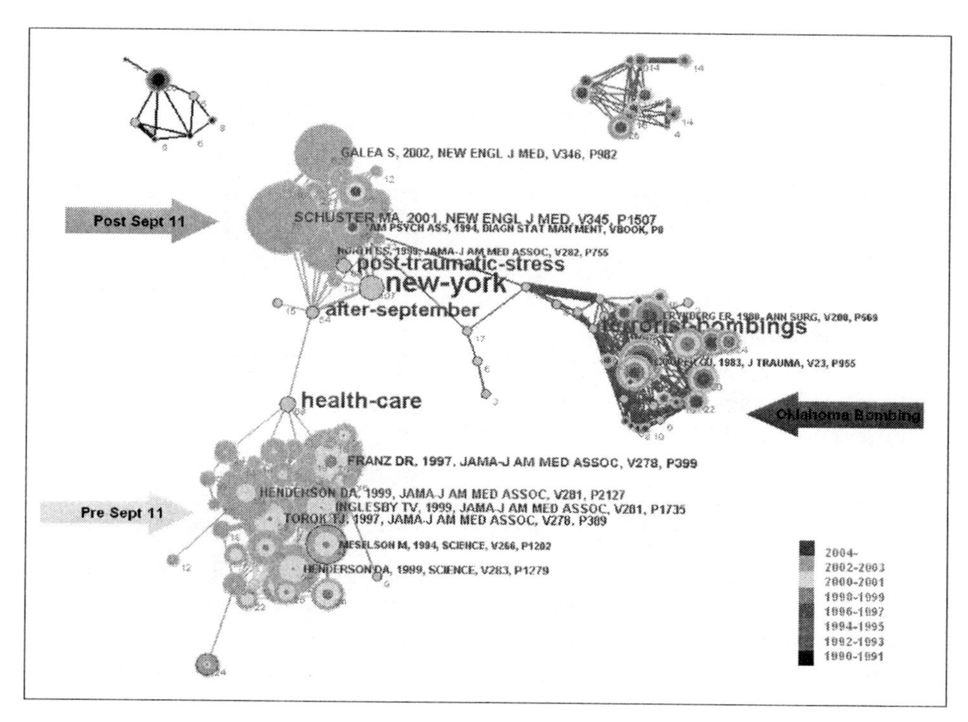

Figure 2. Additional functions supported by CiteSpace

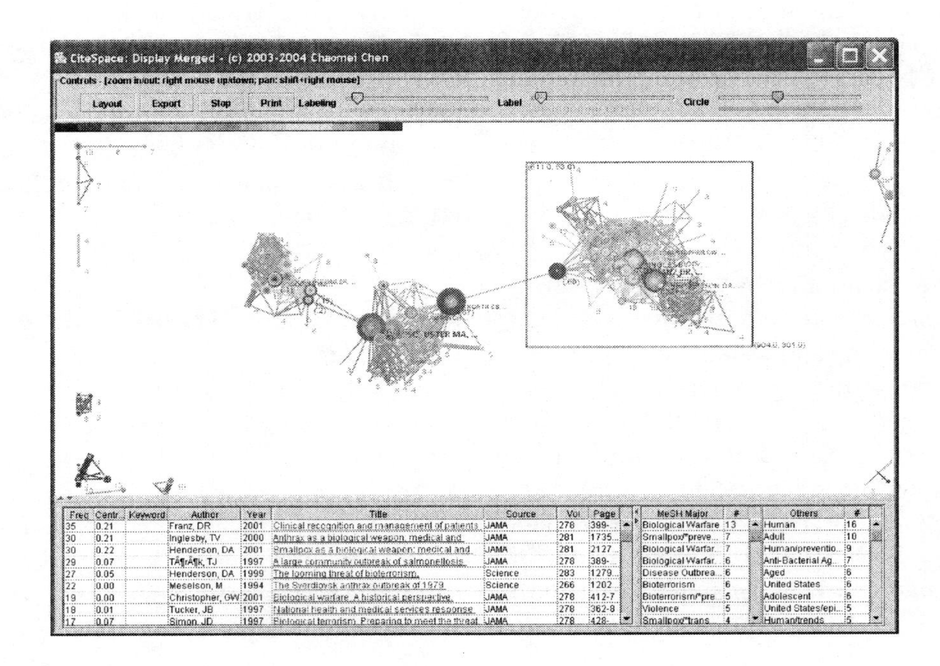

articles being cited in the dataset (labeled in blue). The network also contains three types of links: term-term occurrence, article-article co-citation, and term-to-article citation. The color of a link corresponds to the year in which the first instance was found. There are three predominating clusters in Figure 1, labeled as follows: (A) post-September 11th, (B) pre-Sept 11, and (C) Oklahoma bombing. Automatically extracted high-fly terms identified the emerging trends associated with each cluster. Cluster A is associated with *post-traumatic stress disorders* (PTSD); Cluster B is associated with heath care and bioterrorism; and Cluster C is associated with terrorist bombings and body injuries.

In Figure 2, the merged network shown in CiteSpace. Purple-circled nodes are high in between-ness centrality. The red rectangle indicates the current marquee selection area. Articles in the selected group are displayed in the tables along with matching medical subject heading (MeSH) indexing terms retrieved from PubMed. The nature of the selected group is characterized by the top-ranked MeSH major terms—biological warfare (assigned to 13 articles in the group). The most frequently cited article in this group is in the top row of the table by Franz et al. (2001). Its centrality measure of 0.21 is shown in the second column.

The red rectangle in Figure 2 marks an area selected by the user. The operation is called marquee selection. Each node falls into the area is selected. In this case, CiteSpace launches instant search in PubMed, the largest medical literature resource on the Web. If there is a match, the MeSH terms assigned by human indexers to the matched article will be retrieved. Such terms serve as gold standard for identifying the topic of the article. Frequently used terms across all articles in the marquee selection are ranked and listed in the tables located in the lower right area in the screen display. For example, the most frequently used MeSH term for this cluster is biological warfare, which was assigned to 13 articles in the cluster.

FUTURE TRENDS AND CONCLUSION

Major challenges include the need to detect emergent trends and abrupt changes in complex and transient systems accurately and efficiently, and the need to represent such changes and patterns effectively so that one can understand intuitively the underlying dynamics of scientific frontiers and the movement of research fronts. Detecting trends and abrupt changes computationally poses a challenge, especially at macroscopic levels if the subject is a discipline, a profession, or a field of study as a whole. Although many algorithms and systems have been developed to address phenomena at smaller scales, efforts aiming at macroscopic trends and profound changes have been rare. The emerging knowledge domain visualization (KDviz) studies the evolution of a scientific knowledge domain (Chen, 2003). The need to assemble data from a diverse range of sources also poses a challenge. IBM's DiscoveryLink is a promising example of a middleware approach to address this challenge in the context of life sciences.

Effective means of visually representing the structural and dynamical properties of complex and transient systems are increasingly attracting researchers' attention, but it is still in an early stage. Scalability, visual metaphor, interactivity, perceptual and cognitive task decomposition, interpretability, and how to represent critical changes over time are among the most pressing issues to be addressed.

There is a wide-spread need for tools that can help a broad range of users, including scientists, science policy decision makers, sociologists, and philosophers, to discover new insights from the increasingly complex information. As advocated in Norman (1998), designing task-centered and human-centered tools is a vital step to lead us to a new paradigm of interacting with complex, time-variant information environments. Human-computer interaction holds the key to some of the most critical paths towards integrating the new ways of analysis, discovery, and decision-making.

REFERENCES

Agrawal, A., Cockburn, I., & McHale, J. (2003). *Gone but not forgotten: Labor flows, knowledge spillovers, and enduring social capital.* NBER Working Paper No. 9950.

Borgatti, S. P., & Everett, M. G. (1992). The notion of position in social network analysis. In P. V. Marsden (Ed.), *Sociological methodology* (pp. 1-35). London: Basil Blackwell.

Braam, R. R., Moed, H. F., & van Raan, A. F. J. (1991). Mapping of science by combined co-citation and word analysis, I: Structural aspects and II: Dynamical aspects. *Journal of the American Society for Information Science, 42*, 233-266.

Burt, R. S. (1992). *Structural holes: The social structure of competition.* Cambridge, MA: Harvard University Press.

Callon, M., Law, J., & Rip, A. (Eds.). (1986). *Mapping the dynamics of science and technology: Sociology of science in the real world.* London: Macmillan Press.

Carpenter, M. P., Cooper, M., & Narin, F. (1980). Linkage between basic research literature and patents. *Research Management, 23*(2), 30-35.

Chen, C. (2003). *Mapping scientific frontiers: The quest for knowledge visualization.* London: Springer.

Chen, C. (2004a). *Information visualization: Beyond the horizon* (2nd ed.). Springer.

Chen, C. (2004b). Searching for intellectual turning points: Progressive knowledge domain visualization. In the *Proceedings of the National Academy of Sciences of the United States of America (PNAS), 101*(Suppl. 1) (pp. 5303-5310).

Chen, C., & Carr, L. (1999, October 24-29). Visualizing the evolution of a subject domain: A case study. Paper presented at the *Proceedings of the IEEE Visualization '99 Conference*, San Francisco (pp. 449-452). Los Alamos, CA: IEEE Computer Society.

Chen, C., Cribbin, T., Macredie, R., & Morar, S. (2002). Visualizing and tracking the growth of competing paradigms: Two case studies. *Journal of the American Society for Information Science and Technology, 53*(8), 678-689.

Chen, C., & Hicks, D. (2004). Tracing knowledge diffusion. *Scientometrics, 59*(2), 199-211.

Chi, E., Pitkow, J., Mackinlay, J., Pirolli, P., Gossweiler, R., & Card, S. (1998, April 18-23). Visualizing the evolution of Web ecologies. Paper presented at the *Proceedings of CHI'98*, Los Angeles (pp. 400-407). New York: ACM Press.

Franz, D. R., Jahrling, P. B., McClain, D. J., Hoover, D. L., Byrne, W. R., Pavlin, J. A., Christopher, G.

W., Cieslak, T. J., Friedlander, A. M., & Eitzen, E. M., Jr. (2001) Clinical recognition and management of patients exposed to biological warfare agents. *Clinical Lab Med., 21*(3), 435-73.

Freeman, L. C. (2000). Visualizing social networks. *Journal of Social Structure.* Retrieved October 20, 2005 from http://www.cmu.edu/joss/content/articles/volume1/Freeman.html

Granovetter, M. (1973). Strength of weak ties. *American Journal of Sociology, 8*, 1360-1380.

Jaffe, A., & Trajtenberg, M. (2002). *Patents, citations & innovations*. Cambridge, MA: The MIT Press.

Kuhn, T. S. (1962). *The structure of scientific revolutions*. Chicago: University of Chicago Press.

Meyer, M. (2001). Patent citation analysis in a novel field of technology. *Scientometrics, 51*(1), 163-183.

Narin, F., Hamilton, K. S., & Olivastro, D. (1997). The increasing linkage between U.S. technology and public science. *Research Policy, 26*(3), 317-330.

Norman, D. A. (1998). *The invisible computer: Why good products can fail, the personal computer is so complex, and information appliances are the solution*. Cambridge, MA: MIT Press.

Oppenheimer, C. (2000). Do patent citations count? In B. Cronin & H. B. Atkins (Eds.), *The Web of knowledge: A Festschrift in honor of Eugene Garfield* (pp. 405-432). ASIS.

Price, D. D. (1965). Networks of scientific papers. *Science, 149*, 510-515.

Raup, D. M. (1999). *The nemesis affair: A story of the death of dinosaurs and the ways of science* (Revised and Expanded ed.). New York: W. W. Norton & Company.

Rowland, F., McKnight, C., & Meadows, J. (Eds.). (1995). *Project Elvyn: An experiment in electronic journal delivery. Facts, figures, and findings*. Sevenoaks, Kent: Bowker-Saur.

Schwarz, J. H. (1996). *The second superstring revolution*. Retrieved October 1, 2002, from http://arxiv.org/PS_cache/hep-th/pdf/9607/9607067.pdf

Singh, J. (2004, January 9). *Social networks as determinants of knowledge diffusion patterns*. Retrieved March 24, 2004, from http://www.people.hbs.edu/jsingh/academic/jasjit_singh_networks.pdf

Small, H. G., & Griffith, B. C. (1974). The structure of scientific literatures I: Identifying and graphing specialties. *Science Studies, 4*, 17-40.

Stewart, J. A. (1990). *Drifting continents and colliding paradigms: Perspectives on the geoscience revolution*. Bloomington: Indiana University Press.

KEY TERMS

Citation Indexing: The indexing mechanism invented by Eugene Garfield, in which cited work, rather than subject terms, is used as the part of the indexing vocabulary.

Knowledge Diffusion: The adaptation of knowledge in a broad range of scientific and engineering research and development.

Knowledge Domain Visualization (KDviz): An emerging field that focuses on using data analysis, modeling, and visualization techniques to facilitate the study of a knowledge domain, which includes research fronts, intellectual basis, and other aspects of a knowledge domain. KDviz emphasizes a holistic approach to treat a knowledge domain as a cohesive whole in historical, logical, and social contexts.

Paradigm Shift: The mechanism of scientific revolutions proposed by Kuhn. The cause of a scientific revolution is rooted to the change of a paradigm, or a view of the world.

Research Fronts: A transient collection of highly-cited scientific publications by the latest publications. Clusters of highly co-cited articles are regarded as a representation of a research front.

Scientific Revolutions: Rapid and fundamental changes in science as defined by Thomas Kuhn in his Structure of Scientific Revolutions (Kuhn, 1962).

Art as Methodology

Sarah Kettley
Napier University, UK

INTRODUCTION

HCI has grown up with the desktop; as the specialized tools used for serious scientific endeavor gave way first of all to common workplace and then to domestic use, so the market for the interface has changed, and the experience of the user has become of more interest. It has been said that the interface, to the user, is the computer—it constitutes the experience—and as the interface has become richer with increasing processing power to run it, this experiential aspect has taken center stage (Crampton-Smith & Tabor, 1992). Interaction design has focused largely on the interface as screen with point-and-click control and with layered interactive environments. More recently, it has become concerned with other modes of interaction; notably, voice-activated controls and aural feedback, and as it emerges from research laboratories, haptic interaction. Research on physicalizing computing in new ways, on the melding of bits and atoms, has produced exciting concepts for distributed computing but simultaneously has raised important questions regarding our experience of them. Work in tangible and ubiquitous computing is leading to the possibility of fuller sensory engagement both with and through computers, and as the predominance of visual interaction gives way to a more plenary bodily experience, pragmatism alone no longer seems a sufficient operative philosophy in much the same way that visual perception does not account solely for bodily experience.

Interaction design and HCI in their interdisciplinarity have embraced many different design approaches. The question of what design is has become as important as the products being produced, and computing has not been backward in learning from other design disciplines such as architecture, product design, graphics, and urban planning (Winograd, 1992). However, despite thinkers writing that interaction design is "more like art than science" (Crampton-Smith & Tabor, 1992, p. 37), it is still design with a specific, useful end. It is obvious, for example, how user-centered design in its many methods is aimed at producing better information systems. In knowing more about the context of use, the tasks the tool will be put to, and the traits of the users, it hopes to better predict patterns and trajectories of use. The holy grail in the design of tools is that the tool disappears in use. Transparency is all; Donald Norman (1999) writes that "technology is our friend when it is inconspicuous, working smoothly and invisibly in the background ... to provide comfort and benefit" (p. 115).

It is tempting to point to the recent trend for emotional design as a step in the right direction in rethinking technology's roles. But emotional design does not reassess design itself; in both its aims and methods, emotional design remains closely tied to the pragmatic goals of design as a whole. Both are concerned with precognition—good tools should be instantly recognizable, be introduced through an existing conceptual framework, and exhibit effective affordances that point to its functionality; while emotional design seeks to speak to the subconscious to make us feel without knowing (Colin, 2001). These types of design activity thus continue to operate within the larger pragmatic system, which casts technology as a tool without questioning the larger system itself. More interesting is the emerging trajectory of HCI, which attempts to take account of both the precognitive and interpretive to "construct a broader, more encompassing concept of 'usability'" (Carroll, 2004, pp. 38-40).

This article presents art as a critical methodology well placed to question technology in society, further broadening and challenging the HCI of usability.

BACKGROUND

Artists work to develop personal visual languages. They strive toward unified systems of connotative signifiers to create an artistic whole. They draw and redraw, make and remake, engaging directly with sources of visual and sensory research and with materials, immanently defining their own affective responses, and through a body of work present their world for open reading (Eco, 1989; Eldridge, 2003; Greenhalgh, 2002). Artists of all kinds commonly keep notebooks or sketchbooks of ideas for development along with explorations for possible expression of those ideas. They habitually collect and analyze source material and work through strands of thought using sketches and models, simultaneously defining the aspect of experience they are interested in representing and finding ways of manifesting that representation.

What is Represented?

Debate about what is represented in art tends to highlight issues surrounding Cartesian duality. Commonly, processes of depiction and description might seem, through their use of semiotic systems, to be centered around the object out there; the desktop metaphor in HCI is a good example. This apparent combination of objectivity with the manifold subjectivity involved in reading art poses philosophical problems, not least of which is the nature of that which is represented in the work.

Merleau-Ponty defines the phenomenological world as "not the bringing to explicit expression of a pre-existing being, but the laying down of being," and that art is not the "reflection of a pre-existing truth" but rather "the act of bringing truth into being" (Merleau-Ponty, 2002, pp. xxii-xxiii). Thus, when we talk about a representation, it should be clear that it is not symbolic only of an existing real phenomenon, whether object or emotion, but exists instead as a new gestalt in its own right.

Bearing this in mind, we may yet say that the artist simultaneously expresses an emotion and makes comment upon it through the means of the materiality of the work. Both these elements are necessary for art to exist—indeed, the very word indicates a manipulation. Without either the emotional source (i.e., the artist's reaction to the subject matter) or the attendant comment (i.e., the nature of its materiality), there would appear to be "no art, only empty decorativeness" (Eldridge, 2003, pp. 25-26). This is where design can be differentiated as pragmatic in relation to art: although it may be a practice "situated within communities, ... an exploration ... already in progress prior to any design situation" (Coyne, 1995, p.11), design lacks the aboutness of art, which is why the position for HCI as laid out here is critical as opposed to pragmatic.

What is Read?

Meaning making is an agentive process not only for the artist but also for the audience; a viewer in passive reception of spectacle does not build meaning or understanding in relation to his or her own lifeworld; the viewer is merely entertained. The created artwork is experienced in the first instance as a gestalt; in a successful work, cognitive trains of thought are triggered, opening up "authentic routes of feeling" in the viewer (Eldridge, 2003, p.71). The difficulties in talking about art have been explicated by Susanne Langer as hinging on its concurrent status as expression for its maker and as impression for its audience (Langer, 1953). However, this is the nature of any language, which is manipulated not just to communicate explicit information but as a social activity geared toward consensual understanding (Winograd & Flores, 1986). The "working through undertaken by the artist" is "subsequently followed and recapitulated by the audience" (Eldridge, 2003, p.70). Just as phenomenology sees language as a socially grounded activity (e.g., in the speech acts theory) (Winograd & Flores, 1986), so art as a language is also primarily a process of activity among people. The artwork is a locus for discourse, engaged with ordinary life and, indeed, truth (Farrell Krell, 1977; Hilton, 2003; Ziarek, 2002), as is phenomenology, expressing and inviting participation in the social activity of meaning making (Eldridge, 2003; Greenhalgh, 2002; McCarthy & Wright, 2003). The temptation to see artists' disengagement from society as irrelevant to more user-centered practices is, therefore, misconceived. Empowerment of the user or audience occurs within both processes, only at different points, and with ramifications for the nature of the resulting artifacts. It is argued here that involving the user directly in the design process correspondingly lessens the need for the user to actively engage with the final artifact and, conversely, that removing the user from the process in turn demands the user's full emotional and cognitive apprehension in interaction with the product. The user/audience thus is invited into discourse with the artist and other viewers rather than with themselves, as the work provides a new ground or environment for social being.

Simultaneity: How Things Are Read

The way in which artists, their audiences, and users apprehend artifacts is of central importance to the way in which HCI may attempt to understand interaction. This section briefly introduces a new direction in phenomenologically informed methodologies as a promising way forward.

Phenomenology says we cannot step outside the world, yet this is supposedly what artists do as a matter of course. People cannot remove themselves from the worlds they find themselves in, because those worlds and individuals create each other simultaneously through action (Merleau-Ponty, 2002). But this occurs at a macro level within which, of course, humans continue to engage in cognitive processes. The artist indeed does attempt to examine his or her own sensual and emotional reactions to phenomena in the world through agentive attention to them and uses intellectual means to produce sensual work. In this respect, the artist popularly might be said to stand back from his or her own immediate reactions, but the artist does not and cannot be said to step outside of his or her world as a phenomenological unity (Winograd & Flores, 1986), in which case, the artist more precisely could be said to be particularly adept at disengagement within his or her world and practiced at shifting his or her domain of concern within an experience:

The emotion is, as Wordsworth puts it, "recollected in tranquillity". There is a sense of working through the subject matter and how it is appropriate to feel about it ... Feeling here is mediated by thought and by artistic activity. The poet must, Wordsworth observes, think 'long and deeply' about the subject, "for our continued influxes of feeling are modified and directed by our thoughts. (Eldridge, 2003, p.70)

This process has come to be understood as one of deeper engagement within the feeling process, within the forming of consciousness itself. Here is Augusto Boal, Brazilian theater director, on the technicalities of acting, echoing contemporary theories of consciousness:

The rationalisation of emotion does not take place solely after the emotion has disappeared, it is immanent in the emotion, it also takes place in the course of an emotion. There is a simultaneity of feeling and thinking. (Boal, 1992, p. 47)

This finds corroboration in Dennett's theory of consciousness, which likens the brain to a massive parallel processor from which narrative streams and sequences emerge, subjected to continual editing (Dennett, 1991). In light of this, we also might describe the artist not so much as distanced from his or her observed world but as practiced at probing his or her own changing consciousness of phenomena and reflecting on the value of the resulting narratives.

We have seen, then, that art is a process not only of representation but also of philosophical questioning and narrative, and that as a provider of grounds for discourse and meaning making, it plays a crucial role in the healthy progression of society. The question, of course, is how this might be of practical use to HCI.

FUTURE TRENDS

It goes without saying that we are not all professional artists, and that of those who are, not all are able to create successful works like those we have just attempted to describe. If the description of the art process and its products is a struggle, putting the theory into action can only be more so. As it already does in the domain of tools, HCI, nevertheless, should seek to do two things: to understand the creation process of this type of computational artifact and to understand the perceptions of the people who interact with them. As a basis of any methodology in understanding the arts, Susanne Langer (1953) pointed to the qualitative and even phenomenological, emphasizing the need for us to "know the arts, so to speak, from the inside ... it is in fact impossible to talk about art without adapting to ... the language of the artist" (p. ix). The trend for phenomenologically informed methodologies in HCI is also growing. Recent developments, especially in the growing relations between phenomenology and cognitive science, are now setting a precedent for a "first person treatment of HCI ... a phenomenologically-informed account of the study of people using ... technologies" (Turner, 2004; Ihde, 2002).

Understanding art may not be as daunting as it first appears. HCI already knows how to conduct phenomenologically informed inquiries, is used to methods revolving around the analysis of conversation and language, and has a history of cross-disciplinary teams. Product designers frequently are involved in teams looking at tools; artists and art theorists now should be able to provide valuable insight and be placed in teams engaged in the development of non-tools. This section now looks at the work of Anthony Dunne followed by that of the author as practical examples of putting this broad approach into practice.

Anthony Dunne

Through conceptual genotypes, Anthony Dunne (1999) has sought to develop an aesthetics of use based upon a product's behavior rather than just its form, and to extend preconceptions of our "subjective relationship with the world, our active relationship with other people" (p. 5). *Thief of Affections* was conceptualized as being based on an alternative user persona, an "otaku," or obsessive individual. Seeking intimacy, this user would be able to "technologically grope the victim's heart" (Dunne, 1999, p. 97). Dunne's (1999) creative process followed two lines—an investigation of the *how*, and the *what like*. In keeping with his criteria for value fictions as opposed to science fictions (Dunne, 1999), the technology had to be believable although not implemented. The final concept has the thief stealing weak radio signals from unsuspecting victims' pacemakers, using the narrative prop developed with a sensitivity to the connotative aspects of its materials. The design concept is presented in a series of photographs rather than as a conventional prototype, emphasising its "psycho-social narrative possibilities" (Dunne, 1999, p. 100). In this and in a later work with Fiona Raby can be seen a concern with bracketing or framing concept designs in such a way as to encourage the audience both to contextualize them within their own lives and to question those technologized lives. Dunne's (1999) problems with audience perceptions of his work when shown in galleries led to later works like the Placebo project to be placed with participants within their own homes over a period of time. The self-selecting participants, or "adopters" of the products, filled in application forms detailing experiences and attitudes to electronic products, and at the end of their allotted time, they were interviewed (although it is not made clear if the gallery audience also was interviewed). Ideas and details from these interviews then were translated into a series of photographs expressing the central findings of the project (Dunne & Raby, 2002, 2002a).

Interactive Jewelery

This project differs in its process from Dunne's (1999) in its emphasis on source materials and works through sketches and models (Figure 1) toward a finished working product: in this case, interactive jewelery using ProSpeckz prototype technology (see SpeckNet.org.uk). The aims of the project are to investigate how contemporary craft as an art discipline may benefit HCI reciprocally. The demonstration pendant (Figure 2) is fabricated from acrylic, formica, mild steel, and gold leaf, and interacts with other pieces in the same series to map social interaction, specifically modes of greeting, through dynamic LED displays. The design is user-centered only in the sense that fashion is user-centered; that is, there is a target audience (a friendship group of

Figure 1. Source and sketches

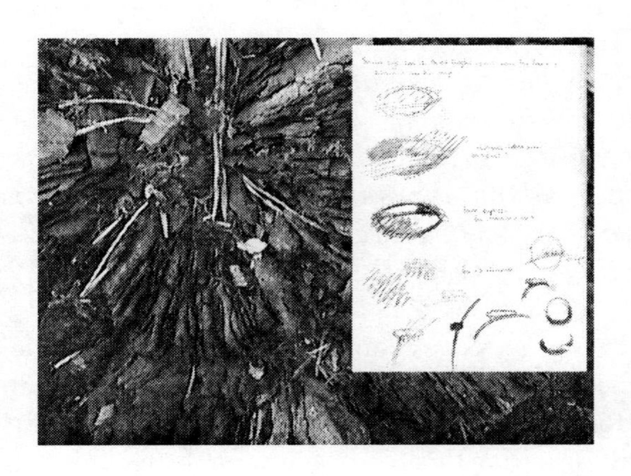

Figure 2. Interactive pendant

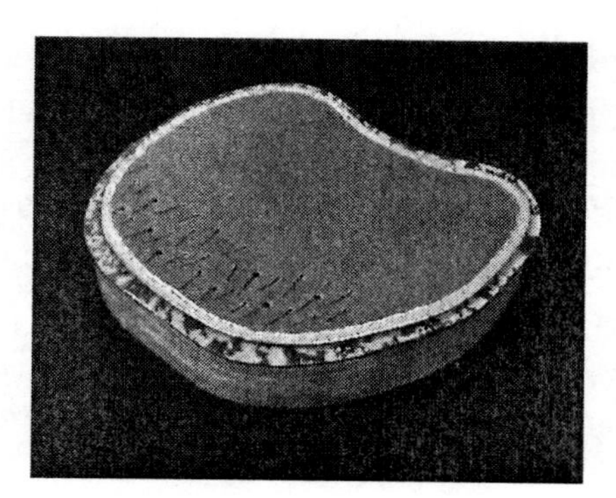

six women), and common elements of lifestyle and style preference are observed, including a notable consumption of art. The designs thus start with source material of interest to the group (e.g., gardens, landscapes) and bear in mind colors that this age group more commonly wears (e.g., pale blue, mauves, greens, greys), but beyond this, there is no iterative design process as HCI would currently understand it. Instead, the pieces of jewelery are presented as finished and tested through observation of the out-of-the-box experience and of longitudinal usage as a group. In the manner of an art object, they are more or less provocative and seductive, and are succeeded quickly by further work.

Other research presents critical design examples as the main deliverable, with the central aim of the research as the provocation of dialogue within the HCI community. These often promote computational elements as material in their own right (Hallnäs et al., 2001; Hallnäs & Redström, 2002; Hallnäs et al., 2002a; Heiner et al., 1999; Orth, 2001; Post et al., 2000) or reflect on the application of a methodology based on an inversion of the form-follows-function leitmotif (Hallnäs et al., 2002b).

While this shift in attention to the experiential is relatively new, it is apparent that conflicts arise in the balancing of roles of designer and user. Most obvious is that the methods of a user-centered approach simply cannot be transposed to meet the needs of the artist-centered methodology. The re-examination of the end goals in design requires no less a thorough reworking of the means to their realization. This is illustrated amply by Vitaly Komar and Alex Melamid's scientific approach to art. Asking people questions like *What's your favorite color?* and *Do you prefer landscapes to portraits?* they produced profoundly disturbing exhibitions of perfectly user-centered art (Wypijewski, 1999; Norman, 2004).

CONCLUSION

Methodologies in HCI have always been borrowed from other disciplines and combined with still others in new ways. The challenge now, caused by a corresponding paradigm shift occurring in philosophy and cognitive science, is to examine the end goals of usability, transparency, and usefulness in HCI and to understand that the processes of reaching our goals have as much import to our ways of living and experience as the goals themselves. In recognition of the artwork as both expressive and impressive, and in view of HCI's twofold interest in development processes and trajectories of use, the following are suggested for consideration toward a complementary methodology for human-computer interaction:

- That the creative process be far less user-centered than we have been used to, placing trust in the role of the artist.
- That this creative process be studied through phenomenologically informed methods.
- That computational elements be approached as material.
- That the work is seen as open for subjective reading, while retaining the voice of its author or the mark of its maker.
- That a body of theory and criticism be built to support the meaning of this type of work, much in the way discourse is created in the art world.
- That trajectories of consumption or use be described through phenomenologically informed methods.

Tools help to do things, but art helps us see why we do them. As a basis for a complementary methodology, art offers an understanding of ourselves in action in a way that instrumentality does not.

REFERENCES

Boal, A. (1992). *Games for actors and non-actors* (trans. A. Jackson). London: Routledge.

Carroll, J. (2004). Beyond fun. *Interactions, 11*(5).

Colin, K. (2001). *Fabrications*. Iver Heath: ElekTex.

Coyne, R. (1995). *Designing information technology in the postmodern age*. Cambridge, MA: MIT Press.

Crampton-Smith, G., & Tabor, P. (1992). The role of the artist designer. In T. Winograd (Ed.), *Bringing design to software*. Boston: Addison-Wesley.

Dennett, D. C. (1991). *Consciousness explained.* London: Penguin Books.

Dunne, A. (1999). *Hertzian tales*. London: RCA CRD.

Dunne, A., & Raby, F. (2002). *Design noir*. August/Birkhauser: Basel.

Dunne, A., & Raby, F. (2002a). The placebo project. *Proceedings of the DIS2002.*

Eco, U. (1989). *The Open Work* (trans. Cancogni). Cambridge, MA: Harvard University Press

Eldridge, R. (2003). *An introduction to the philosophy of art*. Cambridge: Cambridge University Press.

Farrell Krell, D. (Ed.). (1977). *Martin Heidegger: Basic writings*. New York: Harper & Row.

Greenhalgh, P. (2002). Craft in a changing world. In P. Greenhalgh (Ed.), *The persistence of craft*. London: A&C Black.

Hallnäs, L., Jaksetic, P., Ljungstrand, P., Redström, J., & Skog, T. (2001). Expressions: Towards a design practice of slow technology. *Proceedings of Interact 2001, IFIP TC.13 Conference on Human-Computer Interaction*, Tokyo.

Hallnäs, L., Melin, L., & Redström, J. (2002a). Textile displays: Using textiles to investigate computational technology as a design material. *Proceedings of the NordCHI 2002.*

Hallnäs, L., Melin, L., & Redström, J. (2002b). A design research program for textiles and computational technology. *Nordic Textile Journal, 1*, 56-63.

Hallnäs, L., & Redström, J. (2002). Abstract information appliances; Methodological exercises in conceptual design of computational things. *Proceedings of the DIS2002.*

Heiner, J.M., Hudson, S.E., & Tanaka, K. (1999). The information percolator: Ambient information display in a decorative object. *CHI Letters, 1*(1), 141-148.

Hilton, R. (2003). Remarks about painting. In C. Harrison, & P. Wood (Eds.), *Art in theory* (pp. 771-773). Oxford: Balckwell Publishing.

Ihde, D. (2002). *Bodies in technology*. Minneapolis: University of Minnesota Press.

Jordan, P. (2000). *Designing pleasurable products*. London: Taylor & Francis.

Langer, S. (1953). *Feeling and form*. New York: Charles Scribner's Sons.

McCarthy, J., & Wright, P. (2003) The enchantments of technology. In M.A. Blythe, K. Overbeeke, A. F. Monk, & P.C. Wright (Eds.), *Funology, from usability to enjoyment* (pp. 81-91). London: Kluwer Academic Publishers.

Merleau-Ponty, M. (2002). *Phenomenology of perception*. London: Routledge Classics.

Norman, D. (1999). *The invisible computer*. Cambridge, MA: MIT Press.

Norman, D. (2004). *Emotional design: Why we love (or hate) everyday things*. New York: Basic Books.

Orth, M.A. (2001). *Sculpted computational objects with smart and active computing materials.* Doctoral thesis. Retrieved from http://web.media.mit.edu/~morth/thesis/thesis.html

Post, E. R., Orth, M. A., Russo, P. R., & Gershenfeld, N. (2000). E-broidery: Design and fabrication of textile-based computing. *IBM Systems Journal, 39*(3, 4), 840-860.

Turner, P., & Turner, S. (2006). First person perspectives on PDA use. *Interacting with Computers*. Forthcoming.

Winograd, T. (Ed.). (1992). *Bringing design to software*. Boston: Addison-Wesley.

Winograd, T., & Flores, F. (1986). *Understanding computers and cognition*. Norwood, NJ: Ablex Corporation.

Wypijewski, J. (Ed.). (1999). *Painting by numbers; Komar and Melamid's scientific guide to art*. Berkeley: California University Press.

Ziarek, K. (2002). Art, power, and politics: Heidegger on machenschaft and poiesis. *Contretemps 3*. Retrieved from http://www.usyd.edu.au/contretemps/dir/about.html

KEY TERMS

Aesthetics of Use: The behavioral aspects of an artifact, system, or device that precipitate an aesthetic experience based on temporal as well as spatial elements.

Artwork: An actively created physical instance of a narrative. Created by a working through of sensory perception and presented as an invitation to social discourse.

Author: Much of the philosophy extended in analyses of technology stems from literary theory. *Author* has connotations synonymous with *artist* and *designer* and may be useful in future discussions regarding terminology across more or less user-centered processes.

Genotype: Anthony Dunne's alternative to the prototype—a non-working yet complete product specifically aimed at provoking fictive, social, and aesthetic considerations in an audience.

Materiality: The way in which the manner of material realization of an idea on the part of the artist or designer implicates subsequent experience of it.

Phenomenology: A strand of philosophy that accounts for human action without mental representation. Martin Heidegger (1889–1976), currently demanding rereading in the HCI community, is one of the most important thinkers in this field.

Trajectories of Use: No designer, artist, or developer can predict the outcomes of the many and consequential readings or uses to which a public will put his work. The work is a starting point of social interaction, not an end in itself.

Transparency: Also known as *disappearance*, transparency is largely considered the hallmark of good interaction design, wherein the user is able to complete tasks without cognitive interference caused by the interface. The user is said to act through the computer rather than to interact with it.

User: The normative term used in the development of information devices, tools, and systems. Has been criticized for its impersonal, non-performative implications. Alternative suggestions include *actor*, *audience*, *reader*, and *observer*, depending on context. No satisfactory term has been agreed upon that might take into account all of these contexts.

Attention Aware Systems

Claudia Roda
American University of Paris, France

Julie Thomas
American University of Paris, France

INTRODUCTION

Much information science research has focused on the design of systems enabling users to access, communicate, and use information quickly and efficiently. However the users' ability to exploit this information is seriously limited by finite human cognitive resources. In cognitive psychology, the role of attentional processes in allocating cognitive resources has been demonstrated to be crucial. Attention is often defined as the set of processes guiding the selection of the environmental stimuli to be attended. Access to information therefore is not only regulated by its availability but also by the users' choice to attend the information—this choice being governed by attentional processes. Recently several researchers and practitioners in Human Computer Interaction (HCI) have concentrated on the design of systems capable of adapting to, and supporting, human attentional processes. These systems, that often rely on very different technologies and theories, and that are designed for a range of applications, are called *attention-aware* systems (AAS). In the literature, these systems have also been referred to as *Attentive User Interfaces* (Vertegaal, 2003). However, we prefer using the former name as it stresses the fact that issues related to attention are relevant to the design of the system as a whole rather than limited to the interface. The recent interest in this field is testified by the publication of special issues in academic journals (e.g., *Communication of the ACM, 46*(3), 2003; *International Journal of Human-Computer Studies, 58*(5), 2003) and by the organisation of specialised fora of discussion (e.g., the workshop on "Designing for Attention"; Roda & Thomas, 2004).

In this article, we discuss the rationale for AASs and their role within current HCI research, we briefly review current research in AASs, and we highlight some open questions for their design.

BACKGROUND: RATIONALE FOR AND ROLE OF ATTENTION-AWARE SYSTEMS

In this section, we analyze the rationale for AASs and we discuss their role in HCI research.

Why Attention-Aware Systems?

Information overload is one of the most often mentioned problems of working, studying, playing, and generally living in a networked society. One of the consequences of information overload is the fast shift of attention from one subject to another or one activity to another. In certain situations, the ability to quickly access several information sources, to switch activities, or to change context is advantageous. In other situations, it would be more fruitful to create and maintain a focus while offering the possibility to switch attention to other contents or activities only as a background low-noise open choice. System awareness about the cost/benefits of attentional shifts with respect to the users' goals is essential in environments where (1) attentional switches are very often solicited, or (2) where the users' lack of experience with the environment makes it harder for them to select the appropriate attentional focus, or (3) where an inappropriate selection of attentional focus may cause serious damage to the system, its users, or third parties. Systems relying highly on multi-user interaction, such as virtual communities and certain systems supporting cooperative work, are examples of environments where attentional switches are often solicited. Online educational systems are examples of environments where the lack of knowledge and experience of users with the subject at hand makes it harder for them to select the appropriate attentional focus and may easily cause a loss of focus. Life critical systems are examples of environments where an inappropriate selection of attentional focus may cause serious damage. The need for *AAS*s is quite widespread especially if one considers that assessing, supporting, and maintaining users' attention may be desirable in other environments such as entertainment and e-commerce.

Attention-Aware Systems in HCI Research

A large portion of research on human attention in digital environments is based on the findings of cognitive psychology. For example, Raskin (2000) analyses how single locus of attention and habit formation have important consequences on human ability to interact with computers. He proposes that habit creation is a mechanism that can be used to shift the focus of users from the interface to the specific target task.

This study follows the classic "direct manipulation" school (Shneiderman, 1992, 1997) which aims at supporting the attentional choices of the user by making the device "transparent" so that the user can focus on the task rather than on the interface. The wide range of systems designed with this aim is often referred to as *transparent systems*, a term also employed in ubiquitous computing (Abowd, 1999; Weiser, 1991).

Another area of research focuses instead on designing interfaces and systems capable of guiding the users in the choice of attentional focus. The system is seen as proactive, visible, and capable of supporting the users in their choices. These types of systems are often designed as artificial agents (Bradshaw, 1997; Huhns & Singh, 1997) acting as proactive helpers for the user (Maes, 1994; Negroponte, 1997), and they are frequently referred to as *proactive/adaptive systems*.

The two approaches are often regarded as divergent: (1) responding to different needs and (2) requiring different design choices. However this is not necessarily the case, as it should become apparent from the following discussion of these two alleged differences on users' needs and design choices. Concerning the ability to respond to user needs, consider for example, one of the metaphors most often used for proactive systems: Negroponte's English butler (Negroponte, 1995). "The best metaphor I can conceive of for a human-computer interface is that of a well-trained English butler. The 'agent' answers the phone, recognizes the callers, disturbs you when appropriate, and may even tell a white lie on your behalf. The same agent is well trained in timing, versed in finding the opportune moments, and respectful of idiosyncrasies. People who know the butler enjoy considerable advantage over a total stranger. That is just fine" (p. 150). Isn't this proactive/adaptive system an exquisite example of a transparent system? The English butler certainly knows to disappear when it is the case, but he is there when required and is capable of proactive behavior such as selecting the calls you may want to receive or even telling a joke if appropriate! Concerning the design choices, a few considerations should be made. First of all, any system needs to be proactive in certain situations (e.g., reporting errors) and transparent in others. Secondly, certain applications, in particular those where the user has a good knowledge of the most effective attentional focus, require mostly transparent interfaces, while certain others, where the user is more in need of guidance, require more proactive interfaces. Also the users' needs, the system's functionality, and the use that is made of the system, may change with time. Therefore, it may be desirable for a system, that is initially very proactive, to slowly become transparent, or vice-versa. Finally, applications exist where the user is expected to focus on the system/interface itself, that is, digital art. As a consequence, just as proactive adaptive behaviors may not always be desirable, transparency itself may, under certain conditions, not be desirable.

This brings us to another reason for studies related to AASs. In the last two decades, there has been a shift on the use and market of Information and Communication Technologies (ICT) from strictly task oriented (work related) to more of a pervasive personal and social use of these technologies. Performing a task or achieving a goal may not be the main target of the user who instead may turn to ICT artifacts for their symbolic or affective value, entertainment, or pleasure in general; see, for example, Lowgren's arguments for Interactive Design versus classic HCI in Löwgren (2002). Capturing and maintaining user attention may then actually be the ultimate goal of the system.

The real challenge of modern interface design is therefore at the meta-level. We should not aim at designing transparent or proactive systems. Rather we should aim at designing systems capable of reasoning about users' attention, and consequently decide how best to disappear or to gain and guide users' attention. Focusing on attentional mechanisms also provides a framework that reconciles the direct manipulation user interfaces approach and the interface agents approach as clearly presented and exemplified by Horvitz (1999).

HUMAN ATTENTION AND SYSTEMS CAPABLE OF SUPPORTING IT

This section briefly reviews the work done so far in AASs; for a more extensive review, see Roda and Thomas (2006). It should be noted that attention has not often been prioritised as a specific subject of research in HCI (with some notable exceptions including the *Attentional User Interface* project at Microsoft research [Horvitz, Kadie, Paek, & Hovel, 2003]). As a consequence, much of the work relevant to the development of AASs appears in the context of other research frames. This is especially the case as attention processes are related to, and necessary for, the successful accomplishment of many diverse activities.

Human attention has been widely researched in cognitive psychology and, more recently, in neuropsychology. Although there is no common agreement on a definition of "attention", attention is generally understood as the set of processes allowing humans to cope with the, otherwise overwhelming, stimuli in the environment. Attention therefore refers to the set of processes by which we select information (Driver, 2001; Uttal, 2000). These processes are mainly of two types: endogenous (i.e., guided by volition) and exogenous (i.e., guided by reaction to external stimuli). Given this view of attention as a selection of external stimuli, it is obvious that attention is somehow related to human sensory mechanisms. *Visual attention*, for example, has been widely studied in cognitive psychology, and it is particularly relevant to HCI since the current predominant modality for computer-to-human communication is screen display. Using the results of psychological studies in visual attention, some authors have proposed visual techniques for notification displays that aim at easy detection while minimising distraction (Bartram, Ware, & Calvert, 2003). Attention on modalities other than visual, as well as attention across modalities, have not been investigated to the same extent as visual attention. However, Bearne and his colleagues (Bearne, Jones, & Sapsford-Francis, 1994) propose guidelines for the design of multimedia systems grounded in attentional mechanisms.

Systems capable of supporting and guiding user attention must, in general, be able to: (1) assess the current user focus, and (2) make predictions on the cost/benefits of attention shifts (interruptions). We conclude this section with a review of the work done so far in these two directions.

Several sensory-based mechanisms for the detection of users' attention have been employed, including *gaze tracking* (Hyrskykari, Majaranta, Aaltonen, & Räihä, 2000; Vertegaal, 1999; Zhai, 2003), gesture tracking (Hinckley, Pierce, Sinclair, & Horvitz, 2000), head pose and acoustic tracking (Stiefelhagen, 2002). Horvitz and his colleagues (2003) propose that sensory-based mechanisms could be integrated with other cues about the current users' focus. These cues could be extracted from users' scheduled activities (e.g., using online calendars), users' interaction with software and devices, and information about the users and their patterns of activity and attention. In any case, even when employing mechanisms capable of taking into account all these cues, a certain level of uncertainty about users' focus, activities, goals, and best future actions will always remain and will have to be dealt with within the system (Horvitz et al., 2003).

The problem of evaluating the cost/benefit of *interruptions* has been researched mostly in the context of *notification systems* (Brush, Bargeron, Gupta, & Grudin, 2001; Carroll, Neale, Isenhour, Rosson, & McCrickard, 2003; Czerwinski, Cutrell, & Horvitz, 2000; Hudson et al., 2003; McCrickard, Chewar, Somervell, & Ndiwalana, 2003b; McCrickard, Czerwinski, & Bartram, 2003c). This research aims at defining the factors determining the likely utility of a given information, for a given user, in a given context, and the costs associated with presenting the information in a certain manner, to the user, in that context. McCrickard and Chewar (2003) integrate much of the research in this direction and propose an attention-utility trade-off model.

FUTURE TRENDS

AASs will be crucial for the development of applications in a wide variety of domains including education, life critical systems (e.g., air traffic control), support to monitor and diagnosis, knowledge management, simulation of human-like characters, games, and e-commerce. In order to unleash the whole potential of these systems however, there are many fundamental aspects of attention, of the mechanisms that humans use to manage it, and of their application in digital environments that require further exploration. As it will result obvious from the description below, this exploration would greatly benefit from a more interdisciplinary approach to the design of AASs. First, although a very significant amount of research on human attention has been undertaken in psychology, several HCI researchers agree that the reported theories are often too far removed from the specific issues relevant to human computer interaction to be easily applied to this field of research (McCrickard et al., 2003c) and that more focused research in this direction is needed (Horvitz et al., 2003).

A second important issue in the design of AASs is the definition of parameters against which one could measure their efficiency. In their work on notification systems, McCrickard and his colleagues (McCrickard, Catrambone, Chewar, & Stasko, 2003a) advance a proposal in this direction; however, further discussion is needed in order to achieve an agreement on parameters that are generally accepted.

Third, although the visual modality has been extensively researched in cognitive psychology and HCI, this work is mostly focused on still images. How would the principles apply to moving images?

Fourth, much work remains to be done on modalities other than visual. In particular, research on attention in speech (from phonetics to semantics and rhetoric) (Argyle & Cook, 1976; Clark, 1996; Grosz & Sidner, 1990) could be fruitfully applied to HCI research in AASs. Distribution of attention over several modalities is a field that also deserves further research.

Fifth, most of the work on the evaluation of the cost/benefits of interruptions has been done taking the point of view of the user being interrupted; such analysis, however, should also take into account the cost/benefit to the interrupter, and the joint cost/benefit (Hudson, Christensen, Kellogg, & Erickson, 2002; O'Conaill & Frohlich, 1995).

Sixth, certain aspects of human attention related to social and aesthetic processes have been largely disregarded in current research. How could these processes be taken into consideration? Furthermore, most of the target applications in AASs assume that the user is in a "work"/task-oriented situation. How would AAS design apply to different situations (play, entertainment)?

CONCLUSION

AASs are systems capable of reasoning about user attention. In a task-oriented environment, such systems address the problem of information overload by striving to select and present information in a manner that optimizes the cost/benefit associated with users' shifts of attentional focus between contexts and tasks. In this article, we have reviewed the work done so far in this direction. We have also indicated some issues related to the future development of AASs. Among these, the most significant ones are the need to further investigate the application of AASs in environments that are not task-oriented, and the need to take into account collaborative situations when evaluating the cost/benefit of attentional shifts.

REFERENCES

Abowd, G. D. (1999, May 16-22). Software engineering issues for ubiquitous computing. In the *Proceedings 21st International Conference on Software Engineering (ICSE'99)*, Los Angeles (pp. 75-84). Los Alamitos, CA: IEEE Computer Society.

Argyle, M., & Cook, M. (1976). *Gaze and mutual gaze*. Cambridge, MA: Cambridge University Press.

Bartram, L., Ware, C., & Calvert, T. (2003). Moticons: Detection, distraction and task. *International Journal of Human-Computer Studies, 58*(5), 515-545.

Bearne, M., Jones, S., & Sapsford-Francis, J. (1994, October 15-20). Towards usability guidelines for multimedia systems. In the *Proceedings of the Second ACM International Conference on Multimedia*, San Francisco (pp. 105-110). ACM Press.

Bradshaw, J. (1997). *Software agents*. AAAI Press/ The MIT Press.

Brush, B., Bargeron, D., Gupta, A., & Grudin, J. (2001). *Notification for shared annotation of digital documents.* Technical Report MSR-TR-2001-87. Microsoft Research Microsoft Corporation.

Carroll, J. M., Neale, D. C., Isenhour, P. L., Rosson, M. B., & McCrickard, D. S. (2003). Notification and awareness: Synchronizing task-oriented collaborative activity. *International Journal of Human-Computer Studies, 58*(5), 605-632.

Clark, H. H. (1996). *Using language*. Cambridge, MA: Cambridge University Press.

Czerwinski, M., Cutrell, E., & Horvitz, E. (2000, September 5-8). Instant messaging: Effects of relevance and time. In the *Proceedings HCI 2000—14th British HCI group Annual Conference* (pp. 71-76). University of Sunderland.

Driver, J. (2001). A selective review of selective attention research from the past century. British *Journal of Psychology*, 92, 53-78.

Grosz, B. J., & Sidner, C. L. (1990). Plans for discourse. In P. R. Cohen, J. Morgan, & M. Pollack (Eds.), *Intentions in communication* (pp. 417-444). Cambridge, MA: MIT Press.

Hinckley, K., Pierce, J., Sinclair, M., & Horvitz, E. (2000, November). Sensing techniques for mobile interaction. In the *Proceedings ACM UIST 2000 Symposium on User Interface Software and Technology*, San Diego, CA (pp. 91-100). New York: ACM Press.

Horvitz, E. (1999, May). Principles of mixed-initiative user interfaces. In the *Proceedings ACM SIGCHI 1999*, Pittsburgh, Pennsylvania (pp. 159-166). NY: ACM Press.

Horvitz, E., Kadie, C., Paek, T., & Hovel, D. (2003). Models of attention in computing and communication: From principles to applications. *Communications of the ACM, 46*(3), 52-59. New York: ACM Press.

Hudson, J. M., Christensen, J., Kellogg, W. A., & Erickson, T. (2002). "I'd be overwhelmed, but it's just one more thing to do": Availability and interruption in research management. In the *Proceedings SIGCHI Conference on Human Factors in Computing Systems*, Minneapolis, Minnesota (pp. 97-104). New York: ACM Press.

Hudson, S. E., Fogarty, J., Atkeson, C. G., Avrahami, D., Forlizzi, J., Kiesler, S., Lee, J. C., & Yang, J. (2003, April 5-10). Predicting human interruptibility with sensors: A Wizard of Oz feasibility study. In the *Proceedings CHI 2003*, Ft. Lauderdale, Florida (pp. 257-264). New York: ACM Press.

Huhns, M. N., & Singh, M. P. (1997). *Readings in agents*. Morgan Kaufmann.

Hyrskykari, A., Majaranta, P., Aaltonen, A., & Räihä, K.-J. (2000, November). Design issues of iDict: A gaze-assisted translation aid. In the *Proceedings ETRA 2000, Eye Tracking Research and Applications Symposium*, Palm Beach Gardens, FL (pp. 9-14). New York: ACM Press.

Löwgren, J. (2002). *Just how far beyond HCI is interaction design? Boxes and Arrows.* Retrieved on November 20, 2004, from http://www.boxesandarrows.com/archives/just_how_far_beyond_hci_is_interaction_design.php

Maes, P. (1994). Agents that reduce work and information overload. *Communications of the ACM, 37*(7), 30-40.

McCrickard, D. S., Catrambone, R., Chewar, C. M., & Stasko, J. T. (2003a). Establishing tradeoffs that leverage attention for utility: Empirically evaluating information display in notification systems. *International Journal of Human-Computer Studies, 58*(5), 547-582.

McCrickard, D. S., & Chewar, C. M. (2003). Attuning notification design to user goals and attention costs. *Communications of the ACM, 46*(3), 67-72.

McCrickard, D. S., Chewar, C. M., Somervell, J. P., & Ndiwalana, A. (2003b). A model for notification systems evaluation—Assessing user goals for multitasking activity. *ACM Transactions on Computer-Human Interaction (TOCHI), 10*(4), 312-338.

McCrickard, D. S., Czerwinski, M., & Bartram, L. (2003c). Introduction: Design and evaluation of notification user interfaces. *International Journal of Human-Computer Studies, 58*(5), 509-514.

Negroponte, N. (1995). *Being digital.* New York: Vintage Books, Random House Inc.

Negroponte, N. (1997). Agents: From direct manipulation to delegation. In J. Bradshaw (Ed.), *Software agents* (pp. 57-66). AAAI Press/The MIT Press.

O'Conaill, B., & Frohlich, D. (1995). Timespace in the workplace: Dealing with interruptions. In the *Proceedings CHI '95 Conference Companion*, Denver, Colorado (pp. 262-263). ACM Press.

Raskin, J. (2000). *The humane interface: New directions for designing interactive systems.* Addison-Wesley.

Roda, C., & Thomas, J. (2004). Designing for attention. In the *Proceedings HCI2004 Designing for Life—18th British HCI group Annual Conference, Vol. 2*, Leeds, UK (pp. 249-250). Bristol, UK: Research Press International.

Roda, C., & Thomas, J. (2006). Attention aware systems: Theory, application, and research agenda. *Computers in Human Behavior*, Forthcoming.

Shneiderman, B. (1992). *Designing the user interface: Strategies for effective human-computer interaction.* NY: ACM Press.

Shneiderman, B. (1997). Direct manipulation versus agents: Path to predictable, controllable, and comprehensible interfaces. In J. Bradshaw (Ed.), *Software agents* (pp. 97-106). NY: AAAI Press/MIT Press.

Stiefelhagen, R. (2002, October 14-16). Tracking focus of attention in meetings. In the *Proceedings 4th IEEE International Conference on Multimodal Interfaces*, Pittsburgh, PA (pp. 273-280).

Uttal, W. R. (2000). Summary: Let's pay attention to attention. *The Journal of General Psychology, 127*(1), 100-112.

Vertegaal, R. (1999, May 18-20). The GAZE groupware system: Mediating joint attention in multiparty communication and collaboration. In the *Proceedings SIGCHI Conference on Human Factors in Computing Systems*, Pittsburgh, Pennsylvania, May 18-20 (pp. 294-301). New York: ACM Press.

Vertegaal, R. (2003). Attentive user interfaces. *Communications of the ACM, 46*(3), 30-33.

Weiser, M. (1991). The computer of the 21st century. *Scientific American, 265*(3), 66-75.

Zhai, S. (2003). What's in the eyes for attentive input. *Communications of the ACM, 46*(3), 34-39.

KEY TERMS

Direct Manipulation User Interfaces: Interfaces that aim at making objects and actions in the systems visible by [graphical] representation. They were originally proposed as an alternative to command line interfaces. The system's objects and actions are often represented by metaphorical icons on screen (e.g., dragging a file to the recycle bin for deleting a file). Designers of direct manipulation user interface strive to provide incremental reversible operations and visible effects.

Endogenous Attentional Processes: Refers to the set of processes of voluntary (conscious) control of attention. These processes are also referred to as top-down or goal-driven. An example of endogenous attentional mechanism is the attention you are paying at this page as you are reading. Endogenous attention is voluntary; it requires explicit effort, and it is normally meant to last.

Exogenous Attentional Processes: Refers to the set of processes by which attention is captured by some external event. These processes are also referred to as bottom-up or stimulus-driven. An example of this mechanism would be the attention shift from your reading due to a sudden noise. Exogenous attention is triggered automatically, and it normally lasts a short time before it is either shifted or becomes controlled by endogenous processes.

Gaze Tracking: The set of mechanisms allowing to record and analyse human eye-gaze. Gaze tracking is normally motivated by the assumption that the locus of eye-gaze may, to some extent, correspond to the locus of attention, or it can help capturing user interests. Several techniques exist for eye tracking varying in their level of intrusion (from requiring the user to wear special lenses to just having camera-like devices installed on the computer), their accuracy, and ease to use. Normally devices need to be calibrated before use (some systems allow to memorise calibrations for specific users).

Gesture Tracking: The set of mechanisms allowing to record and analyse human motion. Gesture may be tracked either in 2D or 3D. Gesture tracking ranges from the recording and analysis of postures (e.g., head, body) to that of more detailed elements such as hand-fine movement or facial expression. The aims of gesture tracking in HCI span from recognising the user's current activity (or lack of), to recognising emotional states. Gesture tracking is often used in combination with gaze tracking.

Locus of Attention: Among all sensory input, the locus of attention is the input to which one allocates mental resources. Input that falls outside the locus of attention may go absolutely unnoticed. An example of locus of attention is a specific section of a computer screen.

Visual Attention: The process by which we select the visual information most relevant to our current behaviour. In general, of all the visual stimuli we receive, we only attend to a few, this determines what we "see." Visual attention controls the selection of appropriate visual stimuli both by pruning irrelevant ones and by guiding the seeking of relevant ones. Research in visual attention aims at understanding the mechanisms by which human sensory and cognitive systems regulate what we see.

Automated Deduction and Usability Reasoning

José Creissac Campos
Universidade do Minho, Braga, Portugal

Michael D. Harrison
University of Newcastle upon Tyne, UK

INTRODUCTION

Building systems that are correct by design has always been a major challenge of software development. Typical software development approaches (and in particular interactive systems development approaches) are based around the notion of prototyping and testing. However, except for simple systems, testing cannot guarantee absence of errors, and, in the case of interactive systems, testing with real users can become extremely resource intensive and time-consuming. Additionally, when a system reaches a prototype stage that is amenable to testing, many design decisions have already been made and committed to. In fact, in an industrial setting, user testing can become useless if it is done when time or money is no longer available to substantially change the design.

To address these issues, a number of discount techniques for usability evaluation of early designs were proposed. Two examples are heuristic evaluation, and cognitive walkthroughs. Although their effectiveness has been subject of debate, reports show that they are being used in practice. These are largely informal approaches that do not scale well as the complexity of the systems (or the complexity of the interaction between system and users) increases. In recent years, researchers have started investigating the applicability of automated reasoning techniques and tools to the analysis of interactive systems models. The hope being that these tools will enable more thorough analysis of the designs.

The challenge faced is how to fold human factors' issues into a formal setting as that created by the use of such tools. This article reviews some of the work in this area and presents some directions for future work.

BACKGROUND

As stated earlier, discount usability analysis methods have been proposed as a means to achieve some degree of confidence in the design of a system from as early as possible in development. Nielsen and Molich (1990) proposed a usability inspection method based on the assumption that there are a number of general characteristics that all usable systems should exhibit. The method (heuristic evaluation) involves systematic inspection of the design by means of guidelines for good practice. Applying heuristic evaluation involves setting up a team of evaluators to analyze the design of the user interface. Once all evaluators have performed their analysis, results are aggregated thus providing a more comprehensive analysis of the design. To guide analysis, a set of design heuristics is used based on general purpose design guidelines. Over the years, different sets of heuristics have been proposed for different types of systems. The set proposed by Nielsen (1993) comprises nine heuristics: *simple and natural dialog*; *speak the user's language*; *minimize user memory load*; *be consistent*; *provide feedback*; *provide clearly-marked exits*; *provide short cuts*; *good error messages*; and *prevent errors*.

Usability inspection provides little indication of how the analyst should check whether the system satisfies a guideline. Cognitive walkthrough (Lewis, Polson, Wharton, & Rieman, 1990) is one technique that provides better guidance to the analyst. Its aim is to analyze how well the interface will guide the user in performing tasks. User tasks must first be identified, and a model of the interface must be built that covers all possible courses of action the user might take. Analysis of how a user would execute the task is performed by asking three questions at each stage of the interaction: *Will the correct action be made sufficiently evident to users?*;

Will users connect the correct action's description with what they are trying to achieve?; and *Will users interpret the system's response to the chosen action correctly?* Problems are identified whenever there is a "no" answer to one of these questions.

Informal analytic approaches such as those described pose problems for engineers of complex interactive systems. For complex devices, heuristics such as "prevent errors" can become too difficult to apply and validate. Cognitive walkthroughs provide more structure but will become extremely resource intensive as systems increase in complexity and the set of possible user actions grows.

To address these issues, researchers have started looking into the application of automated reasoning techniques to models of interactive systems. These techniques are generally more limited in their application. This happens both because of the cost of producing detailed initial models and because each tool performs a specific type of reasoning only. Nevertheless, they have the potential advantage that they can provide a precise description that can be used as a basis for systematic mechanical analysis in a way that would not otherwise be possible.

Automated theorem proving is a deductive approach to the verification of systems. Available theorem provers range from fully interactive tools to provers that, given a proof, check if the proof is correct with no further interaction from the user. While some systems provide only a basic set of methods for manipulating the logic, giving the user full control over the proof strategy, others include complex tactics and strategies, meaning the user might not know exactly what has been done in each step. Due to this *mechanical* nature, we can trust a proof done in a theorem prover to be correct, as opposed to the recognized error prone manual process. While this is an advantage, it also means that doing a proof in a theorem prover can be more difficult, as *every little bit* must be proved.

Model checking was proposed as an alternative to the use of theorem provers in concurrent program verification (Clarke, Emerson, & Sistla, 1986). The basic premise of model checking was that a finite state machine specification of a system can be subject to exhaustive analysis of its entire state space to determine what properties hold of the system's behavior. By using an algorithm to perform exhaustive state space analysis, the analysis becomes fully automated. A main drawback of model checking has to do with the size of the finite state machine needed to specify a given system: useful specifications may generate state spaces so large that it becomes impractical to analyze the entire state space. The use of symbolic model checking somewhat diminishes this problem. Avoiding the explicit representation of states and exploiting state space structural regularity enable the analysis of state spaces that might be as big as 10^{20} states (Burch, Clarke, & McMillan, 1990). The technique has been very successful in the analysis of hardware and communication protocols designs. In recent years, its applicability to software in general has also become a subject of interest.

AUTOMATED REASONING FOR USABILITY EVALUATION

Ensuring the quality (usability) of interactive systems' designs is a particularly difficult task. This is mainly due to the need to consider the human side of the interaction process. As the complexity of the interaction between users and devices increases, so does the need to guarantee the quality of such interaction. This has led researchers to investigate the applicability of automated reasoning tools to interactive systems development.

In 1995, Abowd, Wang, and Monk (1995) showed how models of interactive systems could be translated into SMV (Symbolic Model Verifier) models for verification. SMV (McMillan, 1993) is a symbolic model checker, at the time being developed at Carnegie Mellon University, USA (CMU). They specified the user interface in a propositional production systems style using the action simulator tool (Curry & Monk, 1995). The specification was then analyzed in SMV using computational tree logic (CTL) formulae. The authors proposed a number of templates for the verification of usability related properties. The questions that are proposed are of the type: "*Can a rule somehow be enabled?*"; "*Is it true that the dialogue is deadlock free?*"; or "*Can the user find a way to accomplish a task from initialization?*".

The modeling approach was quite naive and enabled the expression of models at a very high level of abstraction only. Roughly at the same time, Paternò (1995), in his D.Phil thesis, proposed an approach based on the LOTOS specification language (Language of Temporal Ordering Specifications). Device models were derived from the task analysis and translated into the Lite tool (LOTOS Integrated Tool Environment). The models that could be verified with this approach were far more elaborate than with Abowd et al.'s (1995), but the translation process posed a number of technical difficulties. The language used to express the Lite models (Basic LOTOS) was less expressive than the language used for the modeling of the system (LOTOS interactors). Nevertheless, a number of property templates were proposed for checking the specification. These were divided into interactor, system integrity, and user interface properties. Regarding user interface properties, templates fell into three broad classes: reachability, visibility, and task related. Reachability was defined as: "... given a user action, it is possible to reach an effect which is described by a specific action." (Paternò 1995, p. 103). All properties are expressed in terms of actions. This was due to there being no notion of a system state in the models and logic used.

The main difference between both approaches comes exactly from the specification notations and logics used. Abowd et al. (1995) adopted a simple and easy to use approach. The approach might be too simple, however. In fact, for the verification to be useful, it must be done at an appropriate level of detail, whereas action simulator was designed for very high level abstract specifications. Paternò (1995) avoids this problem by using a more powerful specification notation. This, however, created problems when specifications needed to be translated into the model checker's input language (which was less expressive).

In the following years, a number of different approaches was proposed, using not only model checking but also theorem proving. Most of this work was reported on the DSV-IS series of workshops. d'Ausbourg, Durrieu, and Roché (1996) used the data flow language Lustre. Models were derived from UIL descriptions and expressed in Lustre. Verification is achieved by augmenting the interface with Lustre nodes modeling the intended properties and using the tool Lesar to traverse the automaton generated from this new system. The use of the same language to model both the system and its properties seems to solve some of the problem of translation in Paternò's approach, but the language was limited in terms of the data types available.

Bumbulis, Alencar, Cowan, and Lucena (1996) showed how they were using HOL (a Higher Order Logic theorem prover) in the verification of user interface specifications. They specified user interfaces as sets of connected interface components. These specifications could then be implemented in some toolkit as well as modeled in the higher order logic of the HOL system for formal verification. An immediately obvious advantage of this approach is that the formalism used to perform the analysis, Higher Order Logic, was, at the same level of expressiveness of the formalism, used to write the specification. So, again, the translation problems of Paternò's approach could be avoided. The logic used, however, could not easily capture temporal properties. What was specified was not so much the interaction between the users and the interface, but the interface architecture and how the different components communicate with each other. Although the approach used a powerful verification environment, it had two main drawbacks. The specification style and the logic used did not allow reasoning about some of the important aspects of interaction, and the verification process was quite complex.

Dwyer, Carr, and Hines (1997) explored the application of abstraction to reverse engineer toolkit-based user interface code. The generated models were then analyzed in SMV. This is a different type of approach since it does not rely on developers building models for verification. Instead, models are derived from the code.

Doherty, Massink, and Faconti (2001) applied HyTech, a tool for reachability analysis in hybrid automata, to the analysis of the flight deck instrumentation concerning the hydraulics subsystem of an aircraft. The use of hybrid automata enabled the analysis of continuous aspects of the system.

One of the characteristics of model checking is that all possible interactions between user and device will be considered during the verification step. While this enables a more thorough analysis of the design, in many situations only specific user behaviors will be of interest. To address this, Doherty et al. (2001) propose that a model of the user be explicitly built. However, the user model used was very simplistic: it corresponded simply to all the actions that can be performed by the user.

Rushby (2002) also used a model of the user in his work. In this case, the user model was built into a previously-developed model of the system, and it defined the specific sequences of actions the user is expected to carry out. The analysis was performed in Murø (the Murø verification system), a state exploration tool developed at Stanford University, USA (Dill, 1996), and the author used it to reason about automation surprise in the context of an aircraft cockpit. Also in the context of the analysis of mode confusion in digital flight decks, there has been work carried out at NASA Langley Research Center (Lüttgen & Carreño, 1999). The models used related to the inner working of the system's mode logic, while the goal of the other approaches mentioned herein is to build the models of the user interface. While this latter view might be criticized from the point of view that not all application logic is presented at the interface, it allows better exploration of the interaction between the user and the system, and not simply of how the system reacts to commands.

Campos and Harrison (Campos, 1999; Campos & Harrison, 2001) used SMV, but models were expressed in Modal Action Logic and structured around the notion of interactor (Duke & Harrison, 1993). This enabled richer models where both state information and actions were present. Originally, only device models were built for verification. Whenever specific user behaviors needed to be discarded from the analysis, this was done in the properties to be verified instead of in the model. More recently, Campos has shown how it is possible to encode task information in the model so that only behaviors that comply with the defined tasks for the system are considered (Campos, 2003).

A different style of approach is proposed in Blandford and Good (1998) and Thimbleby (2004). In this case, the modeling process is centered around a cognitive architecture (Programmable User Models) that is supposed to simulate a user. This architecture is programmed with knowledge about the device, and it is then run together with a device model. Observation of the joint behavior of both models is performed in order to identify possible errors.

The approach is not based on model checking nor theorem prover, rather on simulation. Hence, it cannot provide the thoroughness of analysis of the other approaches. The main drawback, however, is the cost of programming the user model. According to the authors, it seldom is cost effective to develop a full model. Instead they argue that the formalization process alone gives enough insight into the design without necessarily having to build a running model.

Thimbleby (2004) uses matrices to model user interfaces, and matrix algebra to reason about usability related properties of the models. Instead of using model checking to reason about the properties of finite state machines representing the user interfaces, matrices are used to represent the transition relation of those finite state machines. The author reverse engineers the user interface of three handheld devices, and shows how they can be analyzed using matrix algebra. The author argues that the approach is simpler and requires less expertise than working with model checking or theorem proving tools.

Finally, in all of the previously-mentioned approaches, the main emphasis is in the applicability of the model checking technology to the task of analyzing properties of interactive system's models. Little or no effort is devoted to making the approaches usable for a wider audience. Loer and Harrison have moved in that direction with IFADIS (Loer, 2003), a tool for the analysis of user interface models. IFADIS uses OFAN Statecharts as the modeling notation, and Statemate as the tool to edit the models. Models are than translated for verification in SMV. The tools provides an environment for modeling, definition of properties, and analysis of the results of the verification process.

FUTURE TRENDS

With the exception of the IFADIS tool, work on the application of automated reasoning tools to the verification of interactive systems has, so far, attempted mainly to prove the technical viability of the approaches. A large number of design notations and tools for model checking have been proposed. Developing an understanding of usability and verification technology to go with it is not enough to guarantee a useful and usable approach to verification.

Interactive systems have specificities that make using typical model checking tools difficult. The richness of the interaction between user and system places specific demands on the types of models that are needed. The assumptions that must be made about the users' capabilities affects how the models should be built and the analysis of the verification results. Tools are needed that support the designer/analyst in modeling, expressing properties, and reasoning about the results of the verification from an interactive systems perspective.

A possibility for this is to build layers on top of existing verification tools, so that the concepts involved in the verification of usability-related properties are made more easily expressed. The use of graphical notations might be a useful possibility. The area opens several lines of work. One is research on interfaces for the verification tools. The STeP prover, for example, uses diagrams to represent proof strategies. Another is the need to support the formulation of properties. Where properties are expressed as goals, maybe graphical representation of start and target interface states could be used.

Another area that needs further research is that of including models of users, work, and context of use in the verification process. Some work has already been done, but the models used are typically very simple. Increasing the complexity of the models, however, means larger state spaces which in turn make verification more difficult.

Finally, some work has already been done on reverse engineering of user interface code. This area deserves further attention. It can help in the analysis of existing systems, it can help verify implementations against properties already proved of the models, or it can help cut down on the cost of building the models during development.

CONCLUSION

This article has reviewed the application of automated reasoning tools to the usability analysis of interactive systems. Reasoning about usability is a difficult task whatever the approach used. The application of automated reasoning to this field still has a long way to go.

Early approaches were based around very simple models of the interactive system. As more complex models started being considered, recognition grew of the need to include considerations about the user or context of usage in the verification process. Some authors did this directly in the models; others encoded that information in the properties to be proved. More recently, the need has been recognized for better tool support, and some steps have been given in that direction.

Applying automated reasoning tools will always mean incurring in the costs of developing adequate models and having adequate expertise. Tool support should help decrease these costs. Increased recognition of the relevance of good usable designs, especially when considering safety-critical and mass market systems, should help make the remaining cost more acceptable.

REFERENCES

Abowd, G. D., Wang, H.-M., & Monk, A. F. (1995, August 23-25). A formal technique for automated dialogue development. In *Proceedings of the First Symposium of Designing Interactive Systems — DIS'95*, Ann Arbor, MI, August 23-25 (pp. 219-226). New York: ACM Press.

Blandford, A., & Good, J. (1998). *Introduction to programmable user modelling in HCI.* British HCI Group. HCI'98 AM3 Tutorial Notes.

Bumbulis, P., Alencar, P. S. C., Cowan, D. D., & Lucena, C. J. P. (1996, June 5-7). Validating properties of component-based graphical user interfaces. In Bodart, F., & Vanderdonckt, J. (Eds.), *Proceedings of the Eurographics workshop on Design, specification and verification of interactive systems '96*, Namur, Belgium, June 5-7 (pp. 347-365). Berlin: Springer-Verlag.

Burch, J. R., Clarke, E. M., & McMillan, K. L. (1990, June 4-7). Symbolic model checking: 1020 states and beyond. In *Proceedings of the Fifth Annual IEEE Symposium on Logic In Computer Science*, Philadelphia, PA, USA (pp. 428-439). Los Alamitos: IEEE Computer Society Press.

Campos, J. C. (1999). *Automated deduction and usability reasoning.* DPhil thesis, Department of Computer Science, University of York. Available as Technical Report YCST 2000/9.

Campos, J. C. (2003). Using task knowledge to guide interactor specifications analaysis. In J. A. Jorge, N. J. Nunes, & J. F. Cunha (Eds.), *Interactive systems: Design, specification and verification*. 10th International Workshop, DSV-IS 2003, volume 2844 of Lecture Notes in Computer Science (pp. 171-186). Heidelberg: Springer-Verlag.

Campos, J. C., & Harrison, M. D. (2001). Model checking interactor specifications. *Automated Software Engineering, 8*(3-4), 275-310.

Clarke, E. M., Emerson, E. A., & Sistla, A. P. (1986). Automatic verification of finite-state concurrent systems using temporal logic specifications. *ACM Transactions on Programming Languages and Systems, 8*(2), 244-263.

Curry, M. B., & Monk, A. F. (1995). Dialogue modelling of graphical user interfaces with a production system. *Behaviour & Information Technology, 14*(1), 41-55.

d'Ausbourg, B., Durrieu, G., & Roché, P. (1996, June 5-7). Deriving a formal model of an interactive system from its UIL description in order to verify and to test its behaviour. In F. Bodart & J. Vanderdonckt (Eds.), *Proceedings of the Eurographics workshop on Design, Specification and Verification of Interactive Systems '96*, Namur, Belgium (pp. 105-122). Berlin: Springer-Verlag.

Dill, D. L. (1996) The Murø verification system. In R. Alur, & T. Henzinger (Eds.), *Computer-Aided Verification, CAV'96, Lecture Notes in Computer Science* (Vol. 1102, pp. 390-393). Heidelberg: Springer-Verlag.

Duke, D. J., & Harrison, M. D. (1993) Abstract interaction objects. *Computer Graphics Forum, 12*(3), 25-36.

Doherty, G., Massink, M., & Faconti, G. (2001). Using hybrid automata to support human factors analysis in a critical system. *Formal Methods in System Design, 19*(2), 143-164.

Dwyer, M. B., Carr, V., & Hines, L. (1997). Model checking graphical user interfaces using abstractions. In M. Jazayeri & H. Schauer (Eds.), *Software Engineering — ESEC/FSE '97, Lecture Notes in Computer Science* (Vol. 1301, pp. 244-261). Heidelberg: Springer-Verlag.

Lewis, C., Polson, P., Wharton, C., & Rieman, J. (1990, April 1-5). Testing a walkthrough methodology for theory-based design of walk-up-and-use interfaces. In *CHI '90 Proceedings*, Seattle, Washington (pp. 235-242). New York: ACM Press.

Loer, K. (2003). *Model-based automated analysis for dependable interactive systems*. PhD thesis. Department of Computer Science, University of York.

Lüttgen, G., & Carreño, V. (1999). *Analyzing mode confusion via model checking*. Technical Report NASA/CR-1999-209332, National Aeronautics and Space Administration, Langley Research Center, Hampton, Virginia (pp. 23681-2199).

McMillan, K. L. (1993). *Symbolic model checking*. Norwell: Kluwer Academic Publishers.

Nielsen, J. (1993). *Usability engineering*. Boston: Academic Press.

Nielsen, J., & Molich, R. (1990, April 1-5). Heuristic evaluation of user interfaces. In *CHI '90 Proceedings*, Seattle, Washington (pp. 249-256). New York: ACM Press.

Paternò, F. D. (1995). *A method for formal specification and verification of interactive systems*. PhD thesis. Department of Computer Science, University of York. Available as Technical Report YCST 96/03.

Rushby, J. (2002). Using model checking to help discover mode confusions and other automation surprises. *Reliability Engineering and System Safety, 75*(2), 167-177.

Thimbleby, H. (2004). User interface design with matrix algebra. *ACM Transactions on Computer-Human Interaction, 11*(2), 181-236.

KEY TERMS

Automated Theorem Prover: A software tool that (semi-)automatically performs mathematical proofs. Available theorem provers range from fully interactive tools to provers that, given a proof, check if the proof is correct with no further interaction from the user.

Cognitive Walkthrough: A model-based technique for evaluation of interactive systems designs. It is particularly suited for "walk up and use" interfaces such as electronic kiosks or ATMs. Its aim is to analyze how well the interface will guide first time or infrequent users in performing tasks. Analysis is performed by asking three questions at each stage of the interaction: *Will the correct action be made sufficiently evident to users?*; *Will users connect the correct action's description with what they are trying to achieve?*; and *Will users interpret the system's response to the chosen action correctly?*

DSV-IS (Design, Specification and Verification of Interactive Systems): An annual international workshop on user interfaces and software engineering. The first DSV-IS workshop was held in 1994 in Carrara (Italy). The focus of this workshop series ranges from the pure theoretical aspects to the techniques and tools for the design, development, and validation of interactive systems.

Heuristic Evaluation: A technique for early evaluation of interactive systems designs. Heuristic evaluation involves systematic inspection of the design by means of broad guidelines for good practice. Typically, 3 to 5 experts should perform the analysis independently, and afterwards combine and rank the results. A well-known set of heuristics is the one proposed by Nielsen: *visibility of system status; match between the system and the real world; user control and freedom; consistency and standards; error prevention; recognition rather than recall; flexibility and efficiency of use; aesthetic and minimalist design; help users recognize, diagnose, and recover from errors; help and documentation.*

IFADIS (Integrated Framework for the Analysis of Dependable Interactive Systems): A tool for the analysis of user interface models developed at the University of York (UK).

Lesar: A Lustre model checker developed a Verimag (France).

Lite (LOTOS Integrated Tool Environment): An integrated tool environment for working with LOTOS specifications. It provides specification, verification/validation, and implementation support. The tools in **LITE** have been developed by participants in the LOTOSPHERE project (funded by the Commission of the European Community ESPRIT II programme).

LOTOS (Language of Temporal Ordering Specifications): A formal specification language for specifying concurrent and distributed systems. LOTOS' syntax and semantics is defined by ISO standard 8807:1989. LOTOS has been used, for example, to specify the Open Systems Interconnection (**OSI**) architecture (ISO 7498).

Lustre: A synchronous data-flow language for programming reactive systems. Lustre is the kernel language of the SCADE industrial environment developed for critical real-time software design by CS Vérilog (France).

Modal User Interface: A user interface is said to be modal (or to have modes) when the same user action will be interpreted by the system differently depending on the system's state and/or the output of the system means different things depending on system state.

Mode Error: A mode error happens when the user misinterprets the mode the system is in. In this situation, actions by the user will be interpreted by the system in a way which will not be what the user is expecting, and/or the user will interpret the information provided by the system erroneously. Mode error typically leads to the user being confounded by the behavior of the system.

Model Checker: A tool that automatically checks a temporal logic formula against a state machine. In the case of symbolic model checking, the tool does not handle the states in the state machine directly. Instead, it handles terms that define sets of states. In this way, it is possible to work with much larger state machines since it is not necessary to explicitly build it.

OFAN: A Statecharts' based task modelling framework developed at the Georgia Institute of Technology (USA).

SMV (Symbolic Model Verifier): A symbolic model checker originally developed at Carnegie Mellon University (USA). Currently, two versions exist: Cadence SMV, being developed by Cadence Berkeley Laboratories (USA) as a research platform for new algorithms and methodologies to incorporate into their commercial products, and NuSMV, a re-implementation and extension of the original tool being developed as a joint project between the ITC- IRST (Italy), Carnegie Mellon University (USA), the University of Genova (Italy), and the University of Trento (Italy).

The STeP Prover (Stanford Temporal Prover): A tool to support the formal verification of reactive, real-time, and hybrid systems. SteP combines model checking with deductive methods to allow for the verification of a broader class of systems.

Task Model: A description of how the system is supposed to be used to achieve pre-defined goals. Task models are usually defined in terms of the actions that must be carried out to achieve a goal.

UIL (User Interface Language): A language for specifying user interfaces in Motif, the industry standard graphical user interfaces toolkit for UNIX systems (as defined by the IEEE 1295 specification).

Usability: The ISO 9241-11 standard defines usability as "the extent to which a product can be used by specified users to achieve specified goals with effectiveness, efficiency, and satisfaction in a specified context of use."

User Model: A model that captures information about users. User models range from simple collections of information about users to cognitive architectures that attempt to simulate user behavior.

A

Automatic Evaluation of Interfaces on the Internet

Thomas Mandl
University of Hildesheim, Germany

INTRODUCTION

Empirical methods in human-computer interaction (HCI) are very expensive, and the large number of information systems on the Internet requires great efforts for their evaluation. Automatic methods try to evaluate the quality of Web pages without human intervention in order to reduce the cost for evaluation. However, automatic evaluation of an interface cannot replace usability testing and other elaborated methods.

Many definitions for the quality of information products are discussed in the literature. The user interface and the content are inseparable on the Web, and as a consequence, their evaluation cannot always be separated easily. Thus, content and interface are usually considered as two aspects of quality and are assessed together. A helpful quality definition in this context is provided by Huang, Lee, and Wang (1999). It is shown in Table 1.

The general definition of quality above contains several aspects that deal with human-computer interaction. For example, the importance of accessibility is stressed. The user and context are important in human-computer interaction, and the information-quality definition also considers suitability for the context as a major dimension.

The automatic assessment of the quality of Internet pages has been an emerging field of research in the last few years. Several approaches have been proposed under various names. Simple approaches try to assess the quality of interfaces via the technological soundness of an implementation, or they measure the popularity of a Web page by link analysis. Another direction of research is also based on only one feature and considers the quality of free text. More advanced approaches combine evidence for assessing the quality of an interface on the Web. Table 2 shows the main approaches and the discipline from which they originated.

Table 1. Categories of information quality (IQ) (Huang et al., 1999)

IQ Category	IQ Dimensions
Intrinsic IQ	Accuracy, objectivity, believability, reputation
Contextual IQ	Relevancy, value-added, timeliness, completeness, amount of information
Representational IQ	Interpretability, ease of understanding, concise representation, consistent representation
Accessibility IQ	Access, security

Table 2. Disciplines and their approaches to automatic quality evaluation

Approach	Discipline
HTML Syntax checking	Web software engineering
Link analysis	Web information retrieval
Indicators for content quality	Library and information science
Interface evaluation	HCI
Text quality	Human language technology

These approaches are discussed in the main sections. The indicators for content quality have not resulted in many implementations and are presented together with the interface evaluation in the subsection "Page and Navigation Structure."

BACKGROUND

In the past, mainly two directions of research have contributed to establish the automatic evaluation of Internet resources: bibliometrics and software testing.

Link analysis applies well-known measures from bibliometrics to the Web. The number of references to a scientific paper has been used as an indicator for its quality. For the Web, the number of links to a Web page is used as the main indicator for the quality of that page (Choo, Detlor, & Turnbull, 2000). Meanwhile, the availability of many papers online and some technical advancement have made bibliometric systems for scientific literature available on the Internet (Lawrence, Giles, & Bollacker, 1999). The availability of such measures will eventually lead to an even greater importance and impact of quantitative evaluation.

Software testing has become an important challenge since software gets more and more complex. In software engineering, automatic software testing has attracted considerable research. The success of the Internet has led to the creation of testing tools for standard Internet languages.

MAIN ISSUES IN THE AUTOMATIC EVALUATION OF INTERFACES ON THE INTERNET

HTML Syntax Checking

Syntax-checking programs have been developed for programming languages and markup languages. Syntax checkers for HTML and other Web standards analyze the quality of Web pages from the perspective of software engineering. However, some systems also consider aspects of human-computer interaction.

One of the first tools for the evaluation of HTML pages was Weblint (Bowers, 1996). It is a typical system for syntax checking and operates on the following levels.

- Syntax (Are all open tags closed? Are language elements used syntactically correct?)
- HTML use (Is the sequence of the headings consistent?)
- Structure of a site (Are there links that lead one hierarchy level up?)
- Portability (Can all expressions be interpreted correctly by all browsers?)
- Stylistic problems (Is alternative text provided for graphics? Do words like *here* appear in link text?)

The examples also illustrate how syntax-checking programs are related to human-computer interaction. Some rules cover only the syntactical correctness. Others address the user experience for a page. For example, missing alternative text for images poses no syntax problem, but it may annoy users of slow-loading pages. In their generality, these simple rules do not apply for each context. For instance, a link upward may not be useful for nonhierarchical sites.

A more comprehensive system than Weblint is available from the National Institute of Standards and Technology (NIST, http://zing.nscl.nist.gov/WebTools/). Its system WebSAT is part of the Suite Web Metrics. WebSAT is based on guidelines from the IEEE and checks whether tags for visually impaired users are present. It also tests whether forms are used correctly and whether the relation between links and text promises good readability.

An overview of systems for code checking is provided by Brajnik (2000). Such systems can obviously lead only to a limited definition of quality. However, they will certainly be part of more complex systems in the future.

Link Analysis

One of the main challenges for search engines is the heterogeneous quality of Internet pages concerning both content quality and interface design (Henzinger, Motwani, & Silverstein, 2002). In Web information retrieval, the common approach to automatically measure the quality of a page has been link analysis. However, link analysis is a heuristic method. The number of links pointing to a page is considered as the main quality indicator (Brin & Page, 1998). From the perspective of human-computer interaction, it is assumed that authors of Web pages prefer to link to pages that are easy to use for them.

A large variety of algorithms for link analysis has been developed (Henzinger, 2000). The most well-known ones are probably the PageRank algorithm and its variants (Haveliwala, 2002; Jeh & Widom, 2003). The basic assumption of PageRank and similar approaches is that the number of in- or back-links of a Web page can be used as a measure for the authority and consequently for the quality of a page including its usability. PageRank assigns an authority value to each Web page that is primarily a function of its back-links. Additionally, it assumes that links from pages with high quality should be weighed higher and should result in a higher quality for the receiving page. To account for the different values each page has to distribute, the algorithm is carried out iteratively until the result converges. PageRank may also be interpreted as an iterative matrix operation (Meghabghab, 2002).

As mentioned above, link analysis has its historical roots in the bibliometric analysis of research literature. Meanwhile, link analysis is also applied to measure the quality of research institutes. For example, one study investigates the relationship between the scientific excellence of universities and the number of in-links of the corresponding university pages (Thelwall & Harries, 2003).

The Kleinberg algorithm (1998) is a predecessor of PageRank and works similarly. It assigns two types of values to the pages. Apart from the authority value, it introduces a so-called hub value. The hub value represents the authority as an information intermediate. The Kleinberg algorithm assumes that there are two types of pages: content pages and link pages. The hub value or information-provider quality is high when the page refers to many pages with high authority. Accordingly, the topical authority is increased when a page receives links from highly rated hubs (Kleinberg, 1998). Unlike PageRank, which is intended to work for all pages encountered by a Web spider of a search engine, the Kleinberg algorithm was originally designed to work on the expanded result set of a search engine.

Link analysis has been widely applied; however, it has several serious shortcomings. The assignment of links is a social process leading to remarkable stable patterns. The number of in-links for a Web page follows a power law distribution (Adamic & Huberman, 2001). For such a distribution, the median value is much lower than the average. That means, many pages have few in-links while few pages have an extremely high number of in-links. This finding indicates that Web-page authors choose the Web sites they link to without a thorough quality evaluation. Rather, they act according to economic principles and invest as little time as possible for their selection. As a consequence, social actors in networks rely on the preferences of other actors (Barabási, 2002). Pages with a high in-link degree are more likely to receive further in-links than other pages (Pennock, Flake, Lawrence, Glover, & Giles, 2002). Another reason for setting links is thematic similarity (Chakrabarti, Joshi, Punera, & Pennock, 2002). Definitely, quality assessment is not the only reason for setting a link.

A study of university sites questions the assumption that quality is associated with high in-link counts. It was shown that the links from university sites do not even lead to pages with scientific material in most cases. They rather refer the user to link collections and subject-specific resources (Thelwall & Harries, 2003).

Large-scale evaluation of Web information retrieval has been carried out within TREC (Text Retrieval Conference, http://trec.nist.gov; Hawking & Craswell, 2001). TREC provides a test bed for information-retrieval experiments. The annual event is organized by NIST, which maintains a large collection of documents. Research groups apply their retrieval engines to this corpus and optimize them with results from previous years. They submit their results to NIST, where the relevance of the retrieved documents is intellectually assessed. A few years ago, a Web track was introduced in which a large snapshot of the Web forms the document collection. In this context, link-based measures have been compared with standard retrieval rankings. The results show that the consideration of link structure does not lead to better retrieval performance. This result was observed in the Web track at TREC 2001. Only for the home-page-finding task have link-based authority measures like the PageRank algorithm led to an improvement (Hawking & Craswell, 2001).

Link analysis is the approach that has attracted the most research within automatic quality evaluation. However, it has its shortcomings, and good usability is only one reason to set a link. As a consequence, link-analysis measures should not be used as the only indicator for the quality of the user interface.

Quality of Text

If systems for assessing the quality of text succeed, they will play an important role for human-computer interaction. The readability of text is an important issue for the usability of a page. Good readability leads to fast and more satisfying interaction. The prototypes developed so far are focused on the application of teacher assistance for essay grading. However, the use of such systems for Web resources will soon be debated. The automatic evaluation of the quality of text certainly poses many ethical questions and will raise a lot of debate once it is implemented on a larger scale.

Two approaches are used in prototypes for the automatic evaluation of texts. The first approach is to measure the coherence of a text and use it as a yardstick. The second typical approach is to calculate the similarity of the texts to sample texts that have been evaluated and graded by humans. The new texts are then graded according to their similarity to the already graded texts.

The Intelligent Essay Assessor is based on latent semantic indexing (LSI; Foltz, Laham, & Landauer, 1999). LSI is a reduction technique. In text analysis, usually each word is used for the semantic description of the text, resulting in a large number of descriptive elements. LSI analyzes the dependencies between these dimensions and creates a reduced set of artificial semantic dimensions. The Intelligent Essay Assessor considers a set of essays graded by humans. For each essay, it calculates the similarity of the essay to the graded essays using text-classification methods. The grade of the most similar cluster is then assigned to the essay. For 1,200 essays, the system reached a correlation of 0.7 to the grades assigned by humans. The correlation between two humans was not higher than that.

A similar level of performance was reached by a system by Larkey (1998). This prototype applies the Bayesian *K*-nearest neighbor classifier and does not carry out any reduction of the word dimensions (Larkey).

The readability of text depends to a certain extent on the coherence within the text and between its parts and sentences. In another experiment, LSI was used to measure the similarity between following sentences in a text. The average similarity between all sentence pairs determines the coherence of the text. This value was compared to the typical similarity in a large corpus of sentences. The results obtained are coherent with psychological experiments estimating the readability (Larkey, 1998).

In Web pages, some text elements are more important than others for navigation. The text on interaction elements plays a crucial role for the usability. It is often short and cannot rely as much on context as phrases within a longer text. These text elements need special attention. In one experiment, similar measures as presented above were used to determine the coherence between link text and the content of the pages to which the link points (Chi et al., 2003). These measures are likely to be applied to Web resources more extensively within the next few years.

PAGE AND NAVIGATION STRUCTURE

Advanced quality models that take several aspects into account are still at an experimental stage. However, they are the most relevant for human-computer interaction. These systems go beyond syntax checking and analyze the design of pages by extracting features from the HTML code of pages. Prototypes consider the use of interaction elements and design, for example, by looking for a balanced layout.

Whereas the first approaches evaluated the features of Web pages intellectually (Bucy, Lang, Potter, & Grabe, 1999), features are now extracted more and more automatically. Most of these prototypes are based on intellectual quality assessment of pages and use these decisions as a yardstick or training set for their algorithms.

An experiment carried out by Amento, Terveen, and Hill (2000) suggests that the human perception of the quality of Web pages can be predicted equally well by four formal features. These four features include link-analysis measures like the PageRank value and the total number of in-links. However, simple features like the number of pages on a site and the number of graphics on a page also correlated highly with the human judgments (Amento et al).

The system WebTango extracts more than 150 atomic and simple features from a Web page and tries to reveal statistical correlations to a set of sites rated as excellent (Ivory & Hearst, 2002). The extracted features are based on the design, the structure, and the HTML code of a page. WebTango includes the ratings of the Weblint system discussed above. The definition of the features is based on hypotheses on the effect of certain design elements on usability. As a consequence, the approach is restricted by the validity of these assumptions. The human ratings could be reproduced to a great extent by the statistical approach.

The information scent analysis in the Bloodhound project uses link analysis to compare the anchor text and its surroundings with the information on the linked page (Chi et al., 2003). The system simulates log files by automatically navigating through a site and determines the quality measure of the site as the average similarity between link text and the text on the following link. The integration with log files shows the interaction focus of the approach. Link analysis has also been combined with log analysis. For example, the approach called usage-aware page rank assigns a bias for pages often accessed (Oztekin, Ertöz, & Kumar, 2003). Usage-aware page rank and Bloodhound are limited to one site.

When comparing the systems, it can be observed that no consensus has been reached yet about which features of Web pages are important for the quality decisions of humans. Much more empirical research is necessary in order to identify the most relevant factors.

FUTURE TRENDS

The literature discussed in the main section shows a clear trend from systems with objective definitions of quality toward systems that try to capture the subjective perspective of individual users. The most recent approaches rely mainly on statistical approaches to extract quality definitions from already assessed resources and apply them to new pages. Future systems will also rely on more and more criteria in order to provide better quality decisions.

Web log files representing the actual information behavior may play a stronger role in these future systems. The assessment of the quality of texts will probably gain more importance for Internet resources.

Quality will also be assessed differently for different domains, for different types of pages, and for various usability aspects. As individualization is an important issue in Web information systems, efforts to integrate personal viewpoints on quality into quality-assessment systems will be undertaken.

Automatic evaluation will never replace other evaluation methods like user tests. However, they will be applied in situations where a large number of Internet interfaces need to be evaluated.

CONCLUSION

The automatic evaluation of Internet resources is a novel research field that has not reached maturity yet. So far, each of the different approaches is deeply rooted within its own discipline and relies on a small number of criteria in order to measure a limited aspect of quality. In the future, systems will need to integrate several of the current approaches in order to achieve quality metrics that are helpful for the user. These systems will contribute substantially to the field of human-computer interaction.

Automatic evaluation allows users and developers to evaluate many interfaces according to their usability. Systems for automatic evaluation will help to identify well-designed interfaces as well as good interface elements. The evaluation of a large number of Internet interfaces will reveal trends in interface design and identify design elements often and successfully used.

REFERENCES

Adamic, L., & Huberman, B. (2001). The Web's hidden order. *Communications of the ACM, 44*(9), 55-59.

Amento, B., Terveen, L., & Hill, W. (2000). Does "authority" mean quality? Predicting expert quality ratings of Web documents. *Proceedings of the Annual International ACM Conference on Research and Development in Information Retrieval* (pp. 296-303).

Barabási, A. -L. (2002). *Linked: The new science of networks.* Cambridge, MA: Perseus.

Bowers, N. (1996). Weblint: Quality assurance for the World-Wide Web. *Proceedings of the Fifth International World Wide Web Conference (WWW5).* Retrieved from http://www5conf.inria.fr/fich_html/papers/P34/Overview.html

Brajnik, G. (2000). Towards valid quality models for Websites. *Proceedings of the Sixth Conference on Human Factors and the Web (HFWEB),* Austin, Texas. Retrieved from http://www.tri.sbc.com/hfweb/brajnik/hfweb-brajnik.html

Brin, S., & Page, L. (1998). The anatomy of a large-scale hypertextual Web search engine. *Computer Networks and ISDN Systems, 30*(1-7), 107-117.

Bucy, E., Lang, A., Potter, R., & Grabe, M. (1999). Formal features of cyberspace: Relationships between Web page complexity and site traffic. *Journal of the American Society for Information Science, 50*(13), 1246-1256.

Chakrabarti, S., Joshi, M., Punera, K., & Pennock, D. (2002). The structure of broad topics on the Web. *Proceedings of the Eleventh International World Wide Web Conference (WWW 2002).* Retrieved from http://www2002.org/CDROM/refereed/338/

Chi, E., Rosien, A., Supattanasiri, G., Wiliams, A., Royer, C., Chow, C., et al. (2003). The Bloodhound project: Usability issues using the InfoScentTM simulator. *Proceedings of the ACM Conference on Human Factors in Computing Systems (CHI '03)* (pp. 505-512).

Choo, C. W., Detlor, B., & Turnbull, D. (2000). *Web work: Information seeking and knowledge work on the World Wide Web.* Dordrecht et al.: Kluwer.

Foltz, P., Klintsch, W., & Landauer, T. (1998). The measurement of textual coherence with latent semantic analysis. *Discourse Processes, 25*(2-3), 285-307.

Foltz, P., Laham, D., & Landauer, T. (1999). The intelligent essay assessor: Applications to educational technology. *Interactive Multimedia: Electronic Journal of Computer-Enhanced Learning, 1*(2). Retrieved from http://imej.wfu.edu/articles/1999/2/04/printver.asp

Haveliwala, T. (2002). Topic-sensitive PageRank. *Proceedings of the Eleventh International World Wide Web Conference 2002 (WWW 2002).* Retrieved from http://www2002.org/CDROM/refereed/127/

Hawking, D., & Craswell, N. (2001). Overview of the TREC-2001 Web track. *The Tenth Text Retrieval Conference (TREC 10).* Retrieved from http://trec.nist.gov/pubs/trec10/t10_proceedings.html

Henzinger, M. (2000). Link analysis in Web information retrieval. *IEEE Data Engineering Bulletin, 23*(3), 3-8.

Henzinger, M., Motwani, R., & Silverstein, C. (2002). Challenges in Web search engines. *SIGIR Forum, 36*(2), 11-22.

Huang, K., Lee, Y., & Wang, R. (1999). *Quality information and knowledge.* Upper Saddle River, NJ: Prentice Hall.

Huberman, B., Pirolli, P., Pitkow, J., & Lukose, R. (1998). Strong regularities in World Wide Web surfing. *Science, 280*, 95-97.

Ivory, M., & Hearst, M. (2002). Statistical profiles of highly-rated sites. *Proceedings of the ACM CHI Conference on Human Factors in Computing Systems* (pp. 367-374).

Jeh, G., & Widom, J. (2003). Scaling personalized Web search. *Proceedings of the Twelfth International World Wide Web Conference (WWW 2003)* (pp. 271-279). Retrieved from http://www2003.org/cdrom/papers/refereed/p185/html/p185-jeh.html

Kleinberg, J. (1998). Authoritative sources in a hyperlinked environment. *Proceedings of the Ninth ACM-SIAM Symposium on Discrete Algorithms* (pp. 668-677). Retrieved from http://citeseer.ist.psu.edu/kleinberg99authoritative.html

Larkey, L. (1998). Automatic essay grading using text categorization techniques. *Proceedings of the 21st Annual International ACM SIGIR Conference on Research and Development in Information Retrieval* (pp. 90-95).

Lawrence, S., Giles, C. L., & Bollacker, K. (1999). Digital libraries and autonomous citation indexing. *IEEE Computer, 32*(6), 67-71.

Meghabghab, G. (2002). Discovering authorities and hubs in different topological Web graph structures. *Information Processing and Management, 38*, 111-140.

Oztekin, B., Ertöz, L., & Kumar, V. (2003). Usage aware PageRank. *Proceedings of the Twelfth International World Wide Web Conference (WWW 2003).* Retrieved from http://www2003.org/cdrom/papers/poster/p219/p219-oztekin.html

Pennock, D., Flake, G., Lawrence, S., Glover, E., & Giles, L. (2002). Winners don't take all: Characterizing the competition for links on the Web. *Proceedings of the National Academy of Sciences, 99*(8), 5207-5211.

Thelwall, M., & Harries, G. (2003). The connection between the research of university and counts of links to its Web pages: An investigation based upon a classification of the relationship of pages to the research of the host university. *Journal of the American Society for Information Science and Technology, 54*(7), 594-602.

KEY TERMS

Authority: Link analysis considers Web pages of high quality to be authorities for their topic. That means these pages contain the best, most convincing, most comprehensive and objective information for that topic.

Bibliometrics: Bibliometrics studies the relationship amongst scientific publications. The most important application is the calculation of impact factors for publications. During this process, a high number of references is considered to be an indicator for high scientific quality. Other analyses include the structure and the development of scientific communities.

Hub: Hub is a term for Web pages in link analysis. In contrast to an authority page, a hub page does not contain high-quality content itself, but links to the authorities. A hub represents an excellent information provider and may be a clearinghouse or a link collection. The high quality of these pages is shown by the information sources they contain.

Information Retrieval: Information retrieval is concerned with the representation of knowledge and subsequent search for relevant information within these knowledge sources. Information retrieval provides the technology behind search engines.

Latent Semantic Indexing: Latent semantic indexing is a dimension-reduction technique used in information retrieval. During the analysis of natural-language text, each word is usually used for the semantic representation. As a result, a large number of words describe a text. Latent semantic indexing combines many features and finds a smaller set of dimensions for the representation that describes approximately the same content.

Link Analysis: The links between pages on the Web are a large knowledge source that is exploited by link-analysis algorithms for many ends. Many algorithms similar to PageRank determine a quality or authority score based on the number of incoming links of a page. Furthermore, link analysis is applied to identify thematically similar pages, Web communities, and other social structures.

PageRank: The PageRank algorithm assigns a quality value to each known Web page that is integrated into the ranking of search-engine results. This quality value is based on the number of links that point to a page. In an iterative algorithm, the links from high-quality pages are weighted higher than links from other pages. PageRank was originally developed for the Google search engine.

Automatic Facial Expression Analysis

Huachun Tan
Tsinghua University, Beijing, China

Yujin Zhang
Tsinghua University, Beijing, China

INTRODUCTION

Facial expression analysis is an active area in human-computer interaction. Many techniques of facial expression analysis have been proposed that try to make the interaction tighter and more efficient.

The essence of facial expression analysis is to recognize facial actions or to perceive human emotion through the changes of the face surface. Generally, there are three main steps in analyzing facial expression. First, the face should be detected in the image or the first frame of image sequences. Second, the representation of facial expression should be determined, and the data related to facial expression should be extracted from the image or the following image sequences. Finally, a mechanism of classification should be devised to classify the facial expression data.

In this article, the techniques for automatic facial expression analysis will be discussed. The attempt is to classify various methods to some categories instead of giving an exhausted review.

The rest of this article is organized as follows. Background is presented briefly firstly. The techniques used in the three steps, which are detecting the face, representing the facial expression, and classifying the facial expression, are described respectively. Then some facial expression databases are discussed. The challenges and future trends to facial expression analysis are also presented. Finally, the conclusion is made.

BACKGROUND

During the past decade, the development of image analysis, object tracking, pattern recognition, computer vision, and computer hardware has brought facial expression into human-computer interaction as a new modality, and makes the interaction tighter and more efficient. Many systems for automatic facial expression have been developed since the pioneering work of Mase and Pentland (1991). Some surveys of automatic facial expression analysis (Fasel & Luettin, 2003; Pantic & Rothkrantz, 2000a) have also appeared.

Various applications using automatic facial expression analysis can be envisaged in the near future, fostering further interest in doing research in different areas (Fasel & Luettin, 2003). However, there are still many challenges to develop an ideal automatic facial expression analysis system.

DETECTING FACE

Before dealing with the information of facial expression, the face should be located in images or sequences. Given an arbitrary image, the goal of face detection is to determine whether or not there are faces in the image, and if present, return the location and extent of each face. Two good surveys of face detection have been published recently (Hjelmas, 2001; Yang, Kriegman, & Ahuja, 2002).

In most of the systems for facial expression analysis, it is assumed that only one face is contained in the image and the face is near the front view. Then, the main aim of this step is to locate the face and facial features.

In face detection, the input can be either a static image or an image sequence. Because the methods are totally different, we discuss them separately in the following paragraphs.

The techniques of face detection from static images can be classified into four categories (Yang et al., 2002), although some methods clearly overlap the category boundaries. These four types of techniques are listed as follows:

- **Knowledge-Based Methods:** These rule-based methods encode human knowledge about what constitutes a typical face. Usually, the rules capture the relationships between facial features. These methods are designed mainly for face localization.
- **Feature-Invariant Approaches:** These algorithms aim to find structural features that exist even when the pose, viewpoint, or lighting conditions vary, and then use these features to locate faces. Usually, the facial features, such as the edge of the eye and mouth, texture, skin color, and the integration of these features, are used to locate faces. These methods are designed mainly for face localization.
- **Template-Matching Methods:** Several standard patterns of faces are stored to describe the face as a whole or the facial features separately. The correlations between an input image and the stored patterns are computed for detection. Usually, predefined face templates and deformable templates are used. These methods have been used for both face localization and detection.
- **Appearance-Based Methods:** In contrast to template matching, models (or templates) are learned from a set of training images that should capture the representative variability of facial appearance. These learned models are then used for detection. Many learning models are studied, such as eigenface, the distribution-based method, neural networks, support vector machines, the Naïve Bayes classifier, the hidden Markov model, the information-theoretical approach, and so forth. These methods are designed mainly for face detection.

The face can also be detected by the motion information from image sequences. The approaches based on image sequences attempt to find the invariant features through face or head motion. They can be classified into two categories:

- **Accumulated Frame Difference:** In this type of approach, moving silhouettes (candidates) that include facial features, a face, or body parts are extracted by thresholding the accumulated frame difference. Then, some rules are set to measure the candidates. These approaches are straightforward and easy to realize. However, they are not robust enough to detect noise and insignificant motion.
- **Moving Image Contour:** In the approach, the motion is measured through the estimation of moving contours, such as optical flow. Compared to frame difference, results from moving contours are always more reliable, especially when the motion is insignificant (Hjelmas, 2001).

REPRESENTING FACIAL EXPRESSION

After determining the location of the face, the information of the facial expression can be extracted. In this step, the fundamental issue is how to represent the information of facial expression from a static image or an image sequence.

Benefiting from the development of image analysis, object and face tracking, and face recognition, many approaches have been proposed to represent the information of facial expression. These methods can be classified into different classes according to different criteria. Five kinds of approaches are discussed as follows.

According to the type of input, the approaches for representing facial expression can be classified into two categories:

- **Static-Image-Based Approaches:** The system analyzes the facial expression in static images. Typically, a neutral expression is needed to find the changes caused by facial expressions (Buciu, Kotropoulos, & Pitas, 2003; Chen & Huang, 2003; Gao, Leung, Hui, & Tanada, 2003; Pantic & Rothkrantz, 2000b, 2004).
- **Image-Sequence-Based Approaches:** The system attempts to extract the motion or changes of the face or facial features, and uses the spatial trajectories or spatiotemporal information to represent the facial expression information (Cohn, Sebe, Garg, Chen, & Huang, 2003; Essa & Pentland, 1997; Tian, Kanade, &

Cohn, 2001; Zhang, 2003). Because facial expression is a spatial-temporal phenomenon, it is reasonable to believe that the approaches that deal with the image sequences could obtain more reliable results.

According to the space where the information is represented, the representations of facial expression information can be categorized into three types:

- **Spatial Space-Based Approaches:** The deformation or the differences between the neutral face or facial features and the current face or facial features is used to represent the facial expression information (Buciu et al., 2003; Chen & Huang, 2003; Gao et al., 2003; Pantic & Rothkrantz, 2004). The methods used in face recognition are usually adopted in this type of approaches.
- **Spatial Trajectory-Based Approaches:** The spatial trajectory from the neutral face or facial features to the current face or facial features is used to represent the facial expression information (Tian et al., 2001). The fundamental issue of this kind of method is the motion of the face or facial features.
- **Spatiotemporal Trajectory-Based Approaches:** The temporal information is also used besides the spatial trajectory (Cohen et al., 2003; Essa & Pentland, 1997). These methods are usually represented inside the spatiotemporal models, such as hidden Markov models.

According to the regions of a face where the face is processed, the representations can be categorized into local approaches, holist approaches, or hybrid approaches:

- **Local Approaches:** The face is processed by focusing on local facial features or local areas that are prone to change with facial expression (Pantic & Rothkrantz, 2000b, 2004; Tian et al., 2001). In general, intransient facial features such as the eyes, eyebrows, mouth, and tissue texture, and the transient facial features such as wrinkles and bulges are mainly involved in facial expression displays.
- **Holist Approaches:** The face is processed as a whole to find the changes caused by facial expressions (Chen & Huang, 2003; Gao et al., 2003).
- **Hybrid Approaches:** Both the local facial features and the whole face are considered to represent the facial expression (Buciu et al., 2003; Cohen et al., 2003; Essa & Pentland, 1997; Lyons, Budynek, & Akamatsu, 1999). For example, a grid model of the whole face is used to represent the whole face, and the properties of local facial features are also used in the approach (Lyons et al.).

According to information of the face, the approaches can be classified into image-based, 2-D-model-based, and 3-D-model-based approaches.

- **Image-Based Approaches:** The intensity information is used directly to represent the deformation or the motion of the face or facial features (Chen & Huang, 2003; Donato, Bartlett, Hager, Ekman, & Sejnowski, 1999).
- **2-D-Model-Based Approaches:** The face is described with the aid of a 2-D face model, including the facial features or the whole face region, without attempting to recover the volumetric geometry of the scene (Buciu et al., 2003; Cohen et al., 2003; Gao et al., 2003; Pantic & Rothkrantz, 2004; Tian et al., 2001).
- **3-D-Model-Based Approaches:** The face is described by a 3-D face model. These techniques have the advantage that they can be extremely accurate, but have the disadvantage that they are often slow, fragile, and usually must be trained by hand (Essa & Pentland, 1997).

According to whether the representation is modeled on the face surface or not, methods can be classified into appearance-based and muscle-based approaches.

- **Appearance-Based Approaches:** The facial expressions are all represented by the appearance of the face or facial features, and the information extracted from the appearance is used to analyze the facial expression (Buciu et al., 2003; Chen & Huang, 2003; Cohen et al., 2003; Tian et al., 2001).

- **Muscle-Based Approaches:** The approaches focus on the effects of facial muscle activities and attempt to interfere with muscle activities from visual information (Essa & Pentland, 1997; Mase & Pentland, 1991). This may be achieved by using 3-D muscle models that allow mapping of the extracted optical flow into muscle actions. The muscle-based approaches are able to easily synthesize facial expressions. However, the relationships between the motion of the appearance and motion of the muscle are not so easily dealt with.

After determining the representation of the facial expression, the information of the facial expression can be extracted according to the representation approach. In general, the methods of extracting facial expression information are determined by the type of facial expression representation.

Each approach has its advantages and disadvantages. In a facial expression analysis system, the facial expression information is always represented using the combination of some types of facial expression representation.

Because facial expression is a spatial-temporal phenomenon, it is more reasonable to believe that the approaches that deal with image sequences and use the spatiotemporal trajectory-based approaches could obtain more reliable results. The facial expression is the effect of the whole face, and any small change of facial appearance means the change of facial expression. Though affected by accuracy and the time consumed, the approaches using hybrid approaches or 3-D models to represent the face may be more promising. The muscle-based approaches seem to be sound; however, the relationship of the appearance and the motion of muscle is not so clear, and is affected by the development of the image processing. Most of the approaches that deal with the changes of appearance directly are more reasonable.

CLASSIFYING FACIAL EXPRESSION

After the information of facial expression is obtained, the next step of facial expression analysis is to classify the facial expression conveyed by the face. In this step, a set of categories should be defined first. Then a mechanism can be devised for classification.

Facial Expression Categories

There are many ways for defining the categories of facial expressions. In the area of facial expression analysis research, two ways for categorizing are usually used. One is the Facial Actions Coding System (FACS; Ekman & Friesen, 1978), and the other is prototypic emotional expressions (Ekman, 1982). Usually, the process of classifying the facial expression into action units (AUs) in FACS is called facial expression recognition, and the process of classifying the facial expression into six basic prototypic emotions is called facial expression interpretation (Fasel & Luettin, 2003). Some systems can classify the facial expression to either AUs (Pantic & Rothkrantz, 2004; Tian et al., 2001) or one of six basic emotions (Buciu et al., 2003; Chen & Huang, 2003; Cohen et al., 2003; Gao et al., 2003). Some systems perform both (Essa & Pentland, 1997; Pantic & Rothkrantz, 2000b).

Mechanism for Classification

Like other pattern-recognition approaches, the aim of facial expression analysis is to use the information or patterns extracted from an input image and classify the input image to a predefined pattern. Generally, rule-based methods and statistic-based methods are applied to classify facial expressions.

- **Rule-Based Methods:** In this type of method, rules or facial expression dictionaries are determined first by human knowledge, then the examined facial expression is classified by the rules or dictionaries (Pantic & Rothkrantz, 2000b, 2004).
- **Statistic-Based Methods:** Statistic-based methods are the most successful approaches in pattern recognition. Most of the approaches in the literature classify facial expressions using statistic-based methods (Buciu et al., 2003; Chen & Huang, 2003; Cohen et al., 2003; Essa & Pentland, 1997; Tian et al., 2001). Since the early 1980s, statistic pattern recognition has

experienced rapid growth, especially in the increasing interaction and collaboration among different disciplines, including neural networks, machine learning, statistics, mathematics, computer science, and biology. Each examined facial expression is represented as a point in an N-dimensional feature space according to the representation of the facial expression information. Then, given a set of training patterns from each class, the objective is to establish decision boundaries in the feature space that separate patterns belonging to different classes. Of course, according to the learning methods, the statistic-based methods could be classified into many subcategories. For more details, refer to the papers of Fasel and Luettin (2003), Jain, Duin, and Mao (2000), and Pantic and Rothkrantz (2000a).

FACIAL EXPRESSION DATABASE

Many facial expression analysis systems have been developed. Without a uniform facial expression database, these systems could not be compared. Some considerations for a facial expression database are the following (Kanada, Cohn, & Tian, 2000).

1. Level of description
2. Transitions among expressions
3. Deliberate vs. spontaneous expression
4. Reliability of expression data
5. Individual difference among subjects
6. Head orientation and scene complexity
7. Image acquisition and resolution
8. Relation to nonfacial behavior

Several databases have been developed to evaluate facial expression analysis systems. However, most of them are not open. A few databases that can be obtained freely or by purchase are as follows:

- **CMU-Pittsburgh AU-Coded Face Expression Image Database (Kanada et al., 2000):** The database is the most comprehensive test bed to date for comparative studies of facial expression analysis. It includes 2,105 digitized image sequences from 182 adult subjects of varying ethnicity and performs multiple tokens of most primary FACS action units.
- **CMU AMP Face Expression Database:** There are 13 subjects in this database, each with 75 images, and all of the face images are collected in the same lighting condition. It only allows human expression changes.
- **The Japanese Female Facial Expression (JAFFE) Database:** The database contains 213 images of seven facial expressions (six basic facial expressions and one neutral) posed by 10 Japanese female models. Each image has been rated on six emotion adjectives by 60 Japanese subjects. The database was planned and assembled by Miyuki Kamachi, Michael Lyons, and Jiro Gyoba (Lyons et al., 1999).
- **Japanese and Caucasian Facial Expression of Emotion (JACFEE) and Neutral Faces:** The database consists of two sets of photographs of facial expression. JACFEE shows 56 different people, half male and half female, and half Caucasian and half Japanese. The photos are in color and illustrate each of seven different emotions. JACNEUF shows the same 56 subjects with neutral faces.

FUTURE TRENDS AND CHALLENGES

In spite of the applications, we think the future trends would deal with the problems mentioned above and try to establish an ideal facial expression analysis system. Of course, the development of facial expression analysis would benefit from other areas, including computer vision, pattern recognition, psychological studies, and so forth.

The challenge to facial expression analysis can be outlined by an ideal facial expression system, which is proposed to direct the development of facial expression analysis systems, as shown in Table 1.

Many researchers have attempted to solve the challenges mentioned above. However, in currently existing systems of facial expression analysis, few of them did.

In many systems, strong assumptions are made in each step to make the problem of facial expression analysis more tractable. Some common assumptions

Table 1. Challenges to facial expression analysis

For all steps	Automatic processing
	Real-time processing
Detecting face	Deal with subjects of any age, ethnicity and outlook
	Deals with variation in imaging conditions, such as lighting and camera characteristics
	Deal with variation in pose or head motion
	Deal with various facial expressions
	Deals with partially occluded faces
Representing facial expression	Deal with both static images and image sequences
	Deal with various size and orientation of the face in input image
	Deal with various pose of face and head motion
	Deal with occluded face/facial features
Classifying facial expression	Distinguishes all possible expressions and their combinations
	Distinguishes unlimited interpretation categories
	Quantifies facial expressions or facial actions
	Deal with inaccurate facial expression data
	Deal with unilateral facial changes
	Features adaptive learning facility

are the use of a frontal facial view, constant illumination, a fixed light source, no facial hair or glasses, and the same ethnicity. Only the method proposed by Essa and Pentland (1997) in the literature deals with subjects of any age and outlook. In representing facial expression, it is always assumed that the observed subject is immovable.

None of the methods could distinguish all 44 facial actions and their combinations. The method developed by Pantic and Rothkrantz (2004) deals with the most classes, that is, only 32 facial actions occurring alone or in combination, achieving an 86% recognition rate.

Though some methods claim that they can deal with the six basic emotion categories and neutral expression (Cohen et al., 2003), it is not certain that all facial expressions displayed on a face can be classified under the six basic emotion categories in psychology. This makes the problem of facial expression analysis even harder.

CONCLUSION

Facial expression is an important modality in human-computer interaction. The essence of facial expression analysis is to recognize the facial actions of facial expression or perceive the human emotion through the changes of the face surface. In this article, three steps in automatic facial expression analysis, which are detecting the face, representing the facial expression, and classifying the facial expression, have been discussed. Because of the lack of a uniform facial expression database, the facial expression analysis systems are hard to evaluate. Though some developments have been achieved, many challenges still exist.

ACKNOWLEDGMENT

This work has been supported by Grant RFDP-20020003011.

REFERENCES

Buciu, I., Kotropoulos, C., & Pitas, I. (2003). ICA and Gabor representation for facial expression recognition. *IEEE International Conference on Image Processing, 2*, 855-858.

Chen, X., & Huang, T. (2003). Facial expression recognition: A clustering-based approach. *Pattern Recognition Letters, 24*, 1295-1302.

Cohen, I., Sebe, N., Garg, A., Chen, L., & Huang, T. (2003). Facial expression recognition from video sequences: Temporal and static modeling. *Computer Vision and Image Understanding, 91*(1-2), 160-187.

Donato, G., Bartlett, S., Hager, C., Ekman, P., & Sejnowski, J. (1999). Classifying facial actions. *IEEE Transactions on Pattern Analysis and Machine Intelligence, 21*(10), 974-989.

Ekman, P. (1982). *Emotion in the human face.* Cambridge University Press, UK.

Ekman, P., & Friesen, W. V. (1978). *Facial Action Coding System (FACS): Manual.* Palo Alto, CA: Consulting Psychologists Press.

Essa, I., & Pentland, A. (1997). Coding, analysis, interpretation and recognition of facial expressions. *IEEE Transactions on Pattern Analysis and Machine Intelligence, 19*(7), 757-763.

Fasel, B., & Luettin, J. (2003). Automatic facial expression analysis: A survey. *Pattern Recognition, 36*, 259-275.

Gao, Y., Leung, M., Hui, S., & Tanada, M. (2003). Facial expression recognition from line-based caricatures. *IEEE Transactions on Systems, Man, and Cybernetics Part A: Systems and Humans, 33*(3), 407-412.

Hjelmas, E. (2001). Face detection: A survey. *Computer Vision and Image Understanding, 83*, 236-274.

Jain, A. K., Duin, R. P. W., & Mao, J. (2000). Statistical pattern recognition: A review. *IEEE Transactions on Pattern Analysis and Machine Intelligence, 22*(1), 4-37.

Kanada, T., Cohn, J. F., & Tian, Y. (2000). Comprehensive database for facial expression analysis. *Proceedings of Fourth IEEE International Conference on Automatic Face and Gesture Recognition* (pp. 46-53).

Lyons, M. J., Budynek, J., & Akamatsu, S. (1999). Automatic classification of single facial images. *IEEE Transactions on Pattern Analysis and Machine Intelligence, 21*(12), 1357-1362.

Mase, K., & Pentland, A. (1991). Recognition of facial expression from optical flow. *IEICE Transactions, E74*(10), 3474-3483.

Pantic, M., & Rothkrantz, L. (2000a). Automatic analysis of facial expressions: The state of the art. *IEEE Transactions on Pattern Analysis and Machine Intelligence, 22*(12), 1424-1445.

Pantic, M., & Rothkrantz, L. (2000b). Expert system for automatic analysis of facial expression. *Image and Vision Computing, 18*(11), 881-905.

Pantic, M., & Rothkrantz, L. (2004). Facial action recognition for facial expression analysis from static face images. *IEEE Transactions on Systems, Man, and Cybernetics, 34*(3), 1449-1461.

Tian, Y., Kanade, T., & Cohn, J. (2001). Recognizing action units for facial expression analysis. *IEEE Transactions on Pattern Analysis and Machine Intelligence, 23*(2), 97-115.

Yang, M. H., Kriegman, D. J., & Ahuja, N. (2002). Detecting faces in images: A survey. *IEEE Transactions on Pattern Analysis and Machine Intelligence, 24*(1), 34-58.

Zhang, Y., & Ji, Q. (2003). Facial expression understanding in image sequences using dynamic and active visual information fusion. *Proceedings of the IEEE International Conference on Computer Vision* (vol. 2, pp. 1297-1304).

KEY TERMS

Face Detection: Given an arbitrary image, the goal of face detection is to determine whether or not there are any faces in the image, and if present, return the image location and extent of each face.

Face Localization: Given a facial image, the goal of face localization is to determine the position of a single face. This is a simplified detection problem with the assumption that an input image contains only one face.

Face Model Features: The features used to represent (model) the face or facial features, such as the width, height, and angle in a template of the eye, or all nodes and triangles in a 3-D face mesh model.

Facial Action Coding System (FACS): It is the most widely used and versatile method for measuring and describing facial behaviors, which was developed originally by Ekman and Friesen (1978) in the 1970s by determining how the contraction of each facial muscle (singly and in combination with other muscles) changes the appearance of the face.

Facial Expression Recognition: Classifying the facial expression to one facial action unit defined in FACS or a combination of action units, which is also called FACS encoding.

Facial Expression Representation: Classifying the facial expression to one basic emotional category or a combination of categories. Often, six basic emotions defined by Ekman (1982) are used, which are happiness, sadness, surprise, fear, anger, and disgust.

Facial Features: The prominent features of the face, which include intransient facial features, such as eyebrows, eyes, nose, mouth, chin, and so forth, and transient facial features, such as the regions surrounding the mouth and the eyes.

Intransient Facial Features: The features that are always on the face, but may be deformed due to facial expressions, such as eyes, eyebrows, mouth, permanent furrows, and so forth.

Transient Facial Features: Different kinds of wrinkles and bulges that occur with facial expressions, especially the forefront and the regions surrounding the mouth and the eyes.

A Case Study on the Development of Broadband Technology in Canada

Rocci J. Luppicini
University of Ottawa, Canada

INTRODUCTION

The Development of Broadband

Broadband commonly refers to Internet connection speeds greater than narrowband connection speed of 56kbs. Digital subscriber lines (DSL) and cable modems were the most popular forms of broadband in public use over the last 10 years. In 2004, over 80% of U.S. homes were equipped with cable modems, and up to 66% of U.S. households were able to receive DSL transmissions. It is expected that the impact of broadband technologies will continue to play an important role in the U.S. and the rest of the world. It is predicted that the number of broadband-enabled homes will exceed 90 million worldwide by 2007 (Jones, 2003). Canada and Korea currently are the two countries leading the way in broadband saturation. The following discussion focuses on the Canadian case of broadband development.

Canadian Broadband

A bandwidth revolution is underway in Canada driven by an explosion in computing power and access to the world's fastest research network. (Lawes, 2003, p. 19)

As is the case almost everywhere, the development of broadband in Canada began with narrowband Internet. Canada's main broadband initiative, CANARIE (Canadian Network for the Advancement of Research, Industry and Education), can be traced to regional-federal cooperative network principles established by NetNorth (forerunner to Ca*net) in the late 1980s and growing public and private sector interest in developing high-speed networks during the early 1990s (Shade, 1994). By 1993, CANARIE emerged as a not-for-profit federally incorporated organization consisting of public and private sector members. Its goal was to create a networking infrastructure that would enable Canada to take a leading role in the knowledge-based economy. The initial three-phase plan to be carried out within an eight-year period was expected to cost more than $1 billion with more than $200 million coming from the federal government. The objectives of the first phase were to promote network-based R&D, particularly in areas of product development, with expected gains in economic trade advancement. The objectives of the second phase were to extend the capabilities of CA*net to showcase new technology applications that advance educational communities, R&D, and public services. The objective in the third phase were to develop a high-speed test network for developing products and services for competing internationally in a knowledge-based economy. CANARIE's overarching aim in the first three phases was to leverage Canada's information technology and telecommunication capacities in order to advance the Canadian information economy and society. By the end of CANARIE's three phases, high-speed optical computing networking technology connected public and private institutions (i.e., universities, research institutes, businesses, government agencies and laboratories, museums, hospitals, and libraries, both nationally and internationally) (Industry Canada, 2003). CANARIE's contribution to sustaining the Ca*net 4 broadband network (now in its fourth generation) made it possible for networks to share applications, computing power, and other digital resources nationwide and internationally.

CANARIE also provided funding for a number of organizations carrying out innovative initiatives requiring broadband technology, including Absolu Technologies Inc., Shana Corporation, HyperCore Technology Inc., Cifra Médical Inc., Broadband Networks Inc., Callisto Media Systems Inc., The Esys

Corporation, PacketWare Inc., NBTel InterActive, Nautical Data International Inc., and Miranda Technologies Inc. CANARIE has funded more than 200 projects involving 500 Canadian companies and providing an average of 30% of total project costs (CANARIE, 2003).

BACKGROUND

Recent broadband-based research and development initiatives in areas of interinstitutional networking and learning object and learning object repository development are particularly relevant to the field of human-computer interaction (HCI). A growing number of broadband-based research and development projects is appearing worldwide, such as, ICONEX (UK), JORUM (UK), JISC Information Environment (UK), AESharenet (AU), COLIS project (Australia), TALON Learning Objects System (US), and Multimedia Educational Resource for Learning and Online Teaching (International).

Over the last decade, a number of important interinstitutional networking and learning object repository initiatives were spearheaded in Canada. Through the advancement of grid computing, satellite communications, and wireless networks, computers in research labs around the world and in the field could be connected to a computer network, allowing users to share applications, computer power, data, and other resources. Canada's broadband network provided a technology infrastructure for a wide range of large-scale research and development initiatives, such as virtual astrophysics communities (Canadian Virtual Observatory), microelectronic online testing (National Microelectronics and Photonics Testing Collaboratory), remote satellite forest monitoring (SAFORAH), and brain map database sharing (RISQ), SchoolNet, and the Canadian Network of Learning Object Repositories. SchoolNet was a federal government institutional networking project developed in 1994 to increase connectivity to public schools and to promote social equity by allowing all Canadian schools and public libraries to be interconnected, regardless of geographical distance. Through this project, Canada became the first country in the world to connect all of its public schools to the Information Highway (School Net, 1999). Another major initiative was the Canadian Network of Learning Object Repositories (EduSource Canada) created in 2002 to develop interoperable learning object repositories across Canada. EduSource Canada sponsored a number of learning object repository projects, including Broadband-Enabled Lifelong Learning Environment (BELLE), Campus Alberta Repository of Educational Objects (CAREO), and Portal for Online Objects in Learning (POOL).

NON-TECHNICAL ASPECTS OF BROADBAND TECHNOLOGY

Key Non-Technical Problems of Broadband Technology

A selected review of federal government databases on major broadband initiatives in Canada over the last decade reveals a number of problems highlighted in government documents and news reports on government broadband efforts. Particularly salient are problems revolving around public knowledge, education, and systemic organization.

Problem of Public Knowledge

Although more than $1 billion was invested in CANARIE's projects, very few people know of its existence. With little public knowledge of its existence, CANARIE is in danger of being eliminated through economic cutbacks. Efforts to gain media attention have not been effective. Despite the presence of numerous CANARIE-sponsored gophers, Web sites, and press releases, there is a paucity of public information available in popular media.

Problem of Education

The success of projects like SchoolNet was measured in terms of how many computers there were in schools and libraries and how many were connected to the Internet. One major criticism was that efforts from project leaders to promote public interest overemphasized the physical aspects of computers and connectivity and underemphasized how individuals employ technology for educational ends. This partly explains resistance from local network users to participate in many learning object repository initiatives currently in progress. There is little point in amassing thousands of learning objects for users if they do not see how the objects enhance learning.

Problem of Organization

The organization of broadband and broadband-based initiatives can create problems, particularly where public support is vital. There is strong potential for public interest in the development in Canadian learning object repositories and repository networks accessible by all Canadians. The problem is that EduSource (established in 2002) and some of the learning repository initiates emerged almost a decade after broadband initiatives were funded. From an engineering perspective, it made sense to invest infrastructure development first and content second. From a public interest perspective, however, earlier development of learning repository networks could have raised public interest sooner by providing open access. For instance, learning object repositories could have been created on narrowband networks while broadband network development was still underway.

Problem of Funding

Another problem is related to project advancement is funding and funding continuity. For instance, the future of CANARIE and CANARIE-sponsored initiatives was uncertain in 2003 due to delays in federal funding commitment to renew the project (Anonymous, 2003). The trickledown effect of funding uncertainty for umbrella projects like CANARIE affected researchers and developers across the county. The suspension of funding slowed and, in some cases, eliminated projects that previously were funded.

FUTURE TRENDS

Lessons Learned

A focus on non-technical aspects of broadband technology and broadband-based developments revealed problems revolving around public knowledge, education, organization, and funding. Based on the aforementioned problems, valuable lessons can be derived to advance future development within Canada and in similar contexts. First, the existing problem of public knowledge suggests that increased public understanding of broadband technology is needed. One option is to gain greater private-sector participation in promotional activities connected to CANARIE (i.e., supporting charities, scholarships, donations to public institutions, etc.), since companies are interested in gaining popular support. Greater private-sector interest in CANARIE could increase media attention. On the other hand, there also must be careful consideration of the value structure governing stakeholder participation in broadband-based initiatives in the public sphere. Parrish (2004) also raises similar concerns in discussing the value of metadata development recommendations from organizational entities like the ADRIADNE Foundation and the Advanced Distributed Learning (ADL) Initiative, which have invested financial interests. Next, the problem of education and the underemphasis on how diverse individuals in different contexts employ technology for educational ends suggests that that more efforts must be invested into building pedagogical elements into broadband technology development that satisfy the needs of diverse populations in different contexts. This is particularly important in the case of learning object and learning object repositories, where learning is a major concern in project development. Whiley (2003) and Friesen (2003) make similar points in recent criticisms of learning object repository trends. Also, existing problems of organization suggest that the sequence of developmental efforts is also an important consideration. Assuming that public support is vital, the organization of broadband and broadband-related development efforts should take public interest into consideration when planning developmental phases. In the case of Canada, if more effort were placed on learning object repository development in the early 1990s, then stronger public support of current broadband technology could have occurred. Moreover, existing financial problems suggest that more stable funding investment for large-scale projects like CANARIE is crucial.

The aforementioned problems concerning broadband development are not limited to the Canadian context. For instance, financial difficulties in the

U.S. arising from limited public interest in broadband services (i.e., Excite@Home) destabilized broadband development by the end of the 1990s. As was the case in Canada, limited public knowledge and interest were key elements. Despite the efforts of a small group of researchers, developers, and companies to explore the potential of broadband, the general population was unaware of the benefits and the challenges that broadband introduced into their lives.

Defining Features of Non-Technical Dimension of Broadband

Based on the discussion of broadband technology, broadband-based developments, and non-technical aspects of broadband, this section identifies key non-technical criteria that must be satisfied in order to develop a broadband application that will be useful in a national context. Toward this end, Table 1 describes four key non-technical dimensions of broadband development: public knowledge, education, organization, and funding continuity.

The dimensions listed in Table 1 are not intended to be an exhaustive list. Rather, they are to be used as a foundation for exploring other non-technical questions related to broadband technology and developmental strategies for broadband application planning. Broadband and broadband-based applications have the potential to change the way people work, learn, entertain themselves, and access fundamental government services (Ostry, 1994). However, the general public is still unaware of what broadband can do for them. If people are to capitalize on these new technologies in order to improve their standard of living and their quality of life, then greater attention to non-technical aspects of broadband technology in the field of human-computer interaction (HCI) is required.

CONCLUSION

The purpose of this article was to extend understanding of non-technical aspects of broadband technology by focusing on Canadian broadband technology and broadband-based application development. It was based on the apparent lack of attention to non-technical human aspects of broadband development. The article explored how problems of public awareness, education, organization, and funding influenced broadband development in Canada. Based on lessons learned, the article posited four important, non-technical dimensions of broadband development intended to be used as a foundation for exploring other non-technical questions related to broadband technology and developmental strategies for broadband application planning.

Table 1. Four important, non-technical dimensions of broadband development

Non-Technical Dimension	Defining Features
Public Knowledge	• Public knowledge of broadband technology and its application • Public knowledge of broadband applicability to diverse cultures and contexts • Public knowledge of public access opportunities to broadband-based services
Education	• Integration of diverse perspectives to guide broadband development and its conceptualization • Inclusion of pedagogical and instructional design considerations in new developments • Emphasis on broadband applications in areas of learning and skills development
Organization	• Involvement from all stakeholder groups in project planning and development • Indication of multi-level broadband planning and development
Funding	• Indication of adequate social and economic benefits of broadband • Indication of adequate funding and funding continuity to support new developments

REFERENCES

Anonymous. (2003). CANARIE seeking transitional funding to bridge uncertain funding environment. *Research Money, 17*(9), 4.

CANARIE. (2003). *Shaping the future: Success stories from the CANARIE files.* Retrieved February 15, 2004, from http://canarie.gc.ca

Friesen, N. (2003). *Three objections to learning objects.* Retrieved February 1, 2004, from http://phenom.educ.ualberta.ca/nfriesen/

Industry Canada. (2003). *The new national dream: Networking the nation for broadband access: Report of the national broadband task force.* Retrieved February 1, 2004, from http://broadband.gc.ca

Lawes, D. (2003). Extreme bandwidth: CA*net 4 puts Canadian science in the driver's seat. *Research Horizons, 2*(2), 19-22.

Ostry, B. (1994). *The electronic connection: An essential key to Canadians' survival.* Ottawa: Industry Canada.

Parrish, P. (2004). The trouble with learning objects. *Educational Technology Research and Development, 52*(1), 1042-1062.

SchoolNet. (2003). *What is SchoolNet? Industry Canada.* Retrieved February 1, 2004, from http://www.schoolnet.ca/home/e/whatis.asp

Shade, L. (1994). Computer networking in Canada: From CA*net to CANARIE. *Canadian Journal of Communication, 19,* 53-69.

Wiley, D. A. (Ed.). (2002). *The instructional use of learning objects.* Bloomington, IN: Agency for Instructional Technology and Association for Educational Communications and Technology.

Wiley, D. A. (2003). *Learning objects: Difficulties and opportunities.* Retrieved July 21, 2003, from http://wiley.ed.usu.edu/docs/lo_do.pdf

KEY TERMS

Broadband: Refers to Internet connection speeds greater than narrowband connection speed of 56kbs.

Grid Computing: Linking computers from different locations to a computer network, which allows users to share applications, computer power, data, and other resources.

Human-Computer Interaction: The study of relationships between people and computer technologies and the application of multiple knowledge bases to improve the benefits of computer technology for society.

Learning Objects: Learning objects are reusable digital assets that can be employed to advance teaching and learning.

Learning Object Repositories: Digital resources within a structure accessible through a computer network connection using interoperable functions.

Satellite Communications: The amplification and transmission of signals between ground stations and satellites to permit communication between any two points in the world.

Telecommunications: The exchange of information between computers via telephone lines. This typically requires a computer, a modem, and communications software.

Wireless Networks: Computer networking that permits users to transmit data through a wireless modem connecting the remote computer to Internet access through radio frequency.

C

Cognitive Graphical Walkthrough Interface Evaluation

Athanasis Karoulis
Aristotle University of Thessaloniki, Greece

Stavros Demetriadis
Aristotle University of Thessaloniki, Greece

Andreas Pombortsis
Aristotle University of Thessaloniki, Greece

INTRODUCTION

Interface evaluation of a software system is a procedure intended to identify and propose solutions for usability problems caused by the specific software design. The term evaluation generally refers to the process of "gathering data about the usability of a design or product by a specified group of users for a particular activity within a specified environment or work context" (Preece et al., 1994, p. 602). As already stated, the main goal of an interface evaluation is to discover usability problems. A usability problem may be defined as anything that interferes with a user's ability to efficiently and effectively complete tasks (Karat et al., 1992).

The most applied interface evaluation methodologies are the expert-based and the empirical (user-based) evaluations. Expert evaluation is a relatively cheap and efficient formative evaluation method applied even on system prototypes or design specifications up to the almost-ready-to-ship product. The main idea is to present the tasks supported by the interface to an interdisciplinary group of experts, who will take the part of would-be users and try to identify possible deficiencies in the interface design.

According to Reeves (1993), expert-based evaluations are perhaps the most applied evaluation strategy. They provide a crucial advantage that makes them more affordable compared to the empirical ones; in general, it is easier and cheaper to find experts rather than users who are eager to perform the evaluation. The main idea is that experts from different cognitive domains (at least one from the domain of HCI and one from the cognitive domain under evaluation) are asked to judge the interface, everyone from his or her own point of view. It is important that they all are experienced, so they can see the interface through the eyes of the user and reveal problems and deficiencies of the interface. One strong advantage of the methods is that they can be applied very early in the design cycle, even on paper mock-ups. The expert's expertise allows the expert to understand the functionality of the system under construction, even if the expert lacks the whole picture of the product. A first look at the basic characteristics would be sufficient for an expert. On the other hand, user-based evaluations can be applied only after the product has reached a certain level of completion.

BACKGROUND

This article focuses on the expert-based evaluation methodology in general and on the walkthrough methodologies in particular. The Cognitive Graphical Jogthrough method, described in detail in Demetriades et al. (1999) and Karoulis et al. (2000), belongs to the expert-based evaluation methodologies. Its origin is in Polson et al.'s (1992) work, where the initial Cognitive Walkthrough was presented (Polson et al., 1992; Wharton et al., 1994) and in the improved version of the Cognitive Jogthrough (Aedo et al., 1996; Catenazzi et al., 1997; Rowley & Rhoades, 1992). The main idea in Cognitive Walkthroughs is to present the interface-supported tasks to a group of four to six experts who will play the role of would-be users and try to identify any possible deficiencies in the interface design. In order to assess the interface, a set of tasks has to be defined that characterizes the method as task-based. Every task consists of a number of actions that complete the task. The methods utilize an appropriately structured questionnaire to record the evaluators' ratings. They also are characterized as cognitive to denote that the focus is on the cognitive dimension of the user-interface interaction, and special care should be given to understand the tasks in terms of user-defined goals, not just as actions on the interface (click, drag, etc.).

The evaluation procedure takes place as follows:

- A presenter describes the user's goal that has to be achieved by using the task. Then the presenter presents the first action of the first task.
- The evaluators try to (1) pinpoint possible problems and deficiencies during the use of the interface and (2) estimate the percentage of users who will possibly encounter problems.
- When the first action is finished, the presenter presents the second one and so forth, until the whole task has been evaluated. Then, the presenter introduces the second task, following the same steps. This iteration continues until all the tasks are evaluated.
- The evaluators have to answer the following questions in the questionnaire:
  1. How many users will think this action is available?
  2. How many users will think this action is appropriate?
  3. How many users will know how to perform the action? (At this point, the presenter performs the action)
  4. Is the system response obvious? Yes/No
  5. How many users will think that the system reaction brings them closer to their goal?

These questions are based on the CE+ theory of exploratory learning by Polson et al. (1992) (Rieman et al., 1995). Samples of the evaluators' questionnaire with the modified phrasing of the questions derived from the studies considered here can be obtained from http://aiges.csd.auth.gra/academica.

THE COGNITIVE GRAPHICAL WALK- AND JOG-THROUGH METHODS (CGW/CGJ)

The basic idea in modifying the walk- and jog-through methods was that they both focus on novice or casual users who encounter the interface for the first time. However, this limits the range of the application of the method. Therefore, the time factor was introduced by recording the user's experience while working in the interface. This was operationalized through the embodiment of diagrams in the questionnaires to enable the evaluators to record their estimations. The processing of the diagrams produces curves, one for each evaluator; so, these diagrams graphically represent the intuition and the learning curve of the interface. The learning curve in its turn is considered to be the main means of assessing the novice-becoming-expert pace, which is the locus of this modification.

Two main types of diagrams are suggested in Figure 1.

The differentiation of the diagrams refers mainly to their usability during the sessions, as perceived by the evaluators. The main concern of the applications was to pinpoint the easiest diagram form to use.

THE FOUR APPLICATIONS

Application I: The Network Simulator

The modified method of the Graphical Jogthrough was first applied for the evaluation of an educational simulation environment, the Network Simulator. Any simulation is a software medium that utilizes the interactive capabilities of the computer and delivers a properly structured environment to the learner, where user-system interaction becomes the means for knowledge acquisition (Demetriades et al., 1999). Consequently, the main characteristics of a simulation interface that can and must be evaluated are intuitiveness (using proper and easily understandable metaphors), transparency (not interfering with the learning procedure) (Roth & Chair, 1997), as well as easy mapping with the real world (Schank & Cleary, 1996).

Figure 1.

Almost all users					
Most users					
About half the users					
Some users					
Almost no one					
	1st attempt	Few (2-3) attempts	Some (4-6) attempts	More (7-8) attempts	Many (>8) attempts

A. The digital form of the diagram (utilizing boxes) and countable assessment (horizontal axis in attempts)

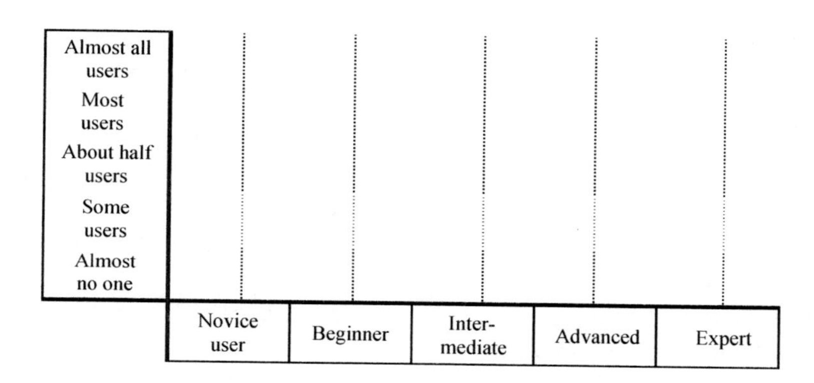

B. The analog type of the diagram (utilizing lines) and internal assessment (horizontal axis depicts experience)

Application II: Perivallon

The second application of the modified method concerned a software piece called Perivallon. This educational software addressed to junior high school students is aimed at supporting the multimedia tuition of biology and is described analytically in Karavelaki et al. (2000).

In this particular evaluation, there is additional evidence concerning the reliability of the method, since the results of this session have been compared and combined with the results of an empirical (i.e., user-based) evaluation session that followed. The details of this approach as well as its results can be found in Karoulis and Pombortsis (2000). In brief, the results proved the method to be unexpectedly reliable. The goal-task-action approach was followed in both sessions. This approach led to an accordance of over 60% of the expert-based results when compared to the empirical ones.

Application III: Orestis

The next application of the method was done in order to evaluate an educational CD called Orestis, which was produced in our laboratory within the framework of an EU-funded program. The software to be evaluated lies in the category of CBL (Computer-Based Learning) software.

This session was organized using a minimalist concept; namely, with the participation of only four evaluators and with the fewest possible resources in order to assess the efficiency of the method, in the case where only the bare essentials are used.

Application IV: Ergani CD

This particular evaluation took place during the first software usability seminar at the Department of Informatics of the University of Athens. The software under evaluation is called Business Plan: How to Start and Manage Your Own Business. It is vocational software that was constructed by the company Polymedia. The main characteristic of this session was that all the evaluators were informatics experts, and all except one HCI were experts.

OVERALL RESULTS AND DISCUSSION

What was actually performed in the present work is an empirical observational evaluation of the modified CGJ method. The evaluators actually acted as users who perceived the method as a tool for achieving their goal, which was the evaluation of the software under consideration. So, through the utilization of this tool in four applications, qualitative results were collected. Therefore, the results of this study are, in fact, the results of an empirical (user-based) evaluation of the modified methods of CGW/CGJ.

Throughout all the applications, the use of the diagrams was characterized as successful. After a short introduction in which the presenter described the method and the use of the diagrams, the evaluators were able to easily record their assessments during the session by ticking the appropriate boxes. There was no indication that assessing the augmentation of the user's experience during the use of the interface would be a difficult task; on the contrary, the diagrams were considered to be the appropriate tool for assessing this variable, and they quickly became transparent in their use.

The minimized conduct of one of the sessions brought controversial results. The unexpected occurrence of a dispute concerning the notion of the expert evaluator made it necessary to have a camera clear; however, the whole minimalist approach of this session speeded up the procedure. A tentative conclusion is that this minimalist design provides the best cost/performance combination; however, it is inadequate in case a dispute occurs.

A research question in this study was also whether these modified versions are applicable to interfaces of a broader scope. The research has shown that the modified methods performed efficiently in both multimedia and educational interfaces, and valuable results with fewer resources (time, money, effort, etc.) were obtained.

However, one always should bear in mind that these methods belong to the category of expert-based methodologies, and consequently, they show the respective advantages and disadvantages. The advantages of these methods are that they are inexpensive and efficient in relation to the resources that are needed, since only a few experts are able to pinpoint significant problems. Moreover, they can be conducted at almost every stage of the design process, from the system specification stage to the construction phase of the product.

On the other hand, there are certain drawbacks. The evaluators must be chosen carefully so that they are not biased. Moreover, the nature of the method is such that it focuses on the specific tasks and actions, often missing the point in the overall facet of the system.

FUTURE TRENDS

There are some suggestions for further investigation regarding the general reliability of the method. A fundamental issue that should be examined is the validation of the suggestion regarding the augmentation of the reliability of the methods by means of its combination with empirical sessions. Therefore, there is a need a for more combinatory evaluations evidence, as suggested in Karoulis and Pombortsis (2000).

Another issue for further investigation is the ideal composition of the evaluating team. Which disciplines are considered to be of most importance? Which are the desired qualities and skills of the evaluators? However, often the appropriate evaluators are not available. In this case, what are the compromises one can make, and how far can the results be threatened in the case of having evaluators with limited qualities and skills? Finally, are teachers of informatics, because of their additional cognitive background, an appropriate group for selecting the evaluators? Of course, these are all major questions not yet addressed, the answers to which might significantly enhance the potential of the methods.

CONCLUSION

About the Cognitive Graphical Jogthrough

While the main aim of the original cognitive walkthrough was only to locate problems and malfunctions during the use of the interface, the incorporation of the diagrams adds the possibility of assessing the intuition and the overall usability of the interface. So, in the modified CGW/CGJ methods presented in this article, the aforementioned drawbacks either are diminished or eliminated. Therefore, it can be claimed that the methods provide both augmented reliability compared to all previously reported studies and augmented usability during their application.

The Final Proposal

The final form that the method can take now can be proposed. The following can be suggested:

- The Cognitive Graphical Walkthrough with the optional use of a video camera. The taped material needs to be processed only in case of an emergency, such as an important dispute.
- The conductors of the session are reduced to only one person, the presenter, since it has been proved (during the last two applications) that one person is sufficient.
- The evaluation questionnaire must be printed double-sided, so that the evaluators are not discouraged by its size.
- The analog type of diagram should be used primarily, since it seems to be more favored by the evaluators.
- A verbal modification of the questions is necessary, as follows:
    1. How many users will think that this action is **available**, namely, that the system can do what the user wants and simultaneously affords the mode for it to be accomplished?
    2. How many users will consider this action, and not some other, to be **appropriate** for the intended goal?
    3. How many users will know how to **perform** this action?
    4. Is the system response **obvious**?
    5. How many users will consider that the system response brings them **closer to their goal**?

These modified questions are still in accordance with the CE+ theory, the theory of exploratory learning by Polson et al. (1992) (Rieman et al., 1995), on which all mentioned methods are based.

The graphical gradations in the text (bold type and smaller-sized characters) are considered to be important, since they also contribute in their own way to clarifying the meaning of the questions.

Irrespective of the final form of the evaluation, the importance of the appropriate evaluating team must be emphasized once more. Cognitive science experts are indispensable, since they can lead the session along the right path, using their comments and also focusing on the cognitive dimension of the evaluation. The (anticipated) lack of double experts (i.e., those with expertise in the cognitive domain as well as in HCI) is a known weakness of the methods.

REFERENCES

Aedo, I., Catenazzi, N., & Diaz, P. (1996). The evaluation of a hypermedia learning environment: The CESAR experience. *Journal of Educational Multimedia & Hypermedia*, *5*(1), 49-72.

Catenazzi, N., Aedo, I., Diaz, P., & Sommaruga, L. (1997). The evaluation of electronic book guidelines from two practical experiences. *Journal of Educational Multimedia & Hypermedia*, *6*(1), 91-114.

Demetriades, S., Karoulis, A., & Pombortsis, A. (1999). "Graphical" Jogthrough: Expert based methodology for user interface evaluation, applied in the case of an educational simulation interface. *Computers and Education, 32*, 285-299.

Karat, C., Campbell, R., & Fiegel, T. (1992, May 3-7). Comparison of empirical testing and walkthrough methods in user interface evaluation. *Proceedings of the ACM CHI '92*, Monterey, CA (pp. 397-404).

Karavelaki, M., Papapanagiotou, G., & Demetriadis, S. (2000). Perivallon: A multidisciplinary multimedia educational enviroment. Information & Communication Technologies in Education, *Proceedings of the 2nd Panhellenic Conference With International Participation,* Patras, Greece.

Karoulis, A., Demetriades, S., & Pombortsis, A. (2000). The cognitive graphical jogthrough—An evaluation method with assessment capabilities. *Proceedings of the Applied Informatics 2000 Conference*, Innsbruck, Austria.

Karoulis, A., & Pombortsis, A. (2000). Evaluating the usability of multimedia educational software for use in the classroom using a "combinatory evaluation" approach. *Proceedings of the Eden 4th Open Classroom Conference,* Barcelona, Spain.

Polson, P.G., Lewis, C., Rieman, J., & Warton, C. (1992). Cognitive walkthroughs: A method for theory-based evaluation of user interfaces. *International Journal of Man-Machine Studies, 36*, 741-773.

Preece, J., et al. (1994). *Human-computer interaction.* New York: Addison-Wesley.

Reeves, T.C. (1993). Evaluating technology-based learning. In G.M. Piskurich (Ed.), *The ASTD handbook of instructional technology.* New York: McGraw-Hill.

Rieman, J., Frazke, M., & Redmiles, D. (1995, May 7-11). Usability evaluation with the cognitive walkthrough. *Proceedings of the CHI-95 Conference*, Denver, Colorado (pp. 387-388).

Roth, W., & Chair, L. (1997). Phenomenology, cognition and the design of interactive learning environments. *Proceedings of the ED-MEDIA & ED-TELECOM 97,* Charlottesville, Virginia.

Rowley, D., & Rhoades, D. (1992). The cognitive jogthrough: A fast-paced user interface evaluation procedure. *Proceedings of the ACM CHI '92,* Monterey, California.

Schank, R., & Cleary, C. (1996). *Engines for education.* Hillsdale: Lawrence Erlbaum.

Wharton, C., Rieman, J., Lewis, C., & Polson, P. (1994). The cognitive walkthrough: A practitioner's guide. In J. Nielsen, & R.L. Mack (Eds.), *Usability inspection methods* (pp. 105-139). New York: John Wiley & Sons.

KEY TERMS

Cognitive Graphical Walkthrough: A modification of the initial «Cognitive Walkthrough» interface evaluation method, which materializes diagrams to enable the evaluators to assess the time variable as well as accelerating the evaluation procedure.

Cognitive Jogthrough: A modification of the initial «Cognitive Walkthrough» in order to speed up the procedure. A video camera now records the evaluation session.

Cognitive Walkthrough: An expert-based interface evaluation method. Experts perform a walkthrough of the interface according to pre-specified tasks, trying to pinpoint shortcomings and deficiencies in it. Their remarks are recorded by a recorder and are elaborated by the design team.

Empirical Interface Evaluation: The empirical evaluation of an interface implies that users are involved. Known methods, among others, are «observational evaluation», «survey evaluation», and «thinking aloud protocol».

Expert-Based Interface Evaluation: Evaluation methodology that employs experts from different cognitive domains to assess an interface. Known methods included in this methodology are (among others) «heuristic evaluation», «cognitive walkthrough», and «formal inspection».

Interface Evaluation: Interface evaluation of a software system is a procedure intended to identify and propose solutions for usability problems caused by the specific software design.

Usability Evaluation: A procedure to assess the usability of an interface. The usability of an interface usually is expressed according to the following five parameters: easy to learn, easy to remember, efficiency of use, few errors, and subjective satisfaction.

Cognitively Informed Multimedia Interface Design

Eshaa M. Alkhalifa
University of Bahrain, Bahrain

INTRODUCTION

The rich contributions made in the field of human computer interaction (HCI) have played a pivotal role in shifting the attention of the industry to the interaction between users and computers (Myers, 1998). However, technologies that include hypertext, multimedia, and manipulation of graphical objects were designed and presented to the users without referring to critical findings made in the field of cognitive psychology. These findings allow designers of multimedia educational systems to present knowledge in a fashion that would optimize learning.

BACKGROUND

The long history of human computer interaction (HCI) has witnessed many successes represented in insightful research that finds its way to users' desktops. The field influences the means through which users interact with computers—from the introduction of the mouse (English et al., 1967) and applications for text editing (Meyrowitz & Van Dam, 1982) to comparatively recent areas of research involving multimedia systems (Yahaya & Sharifuddin, 2000).

Learning is an activity that requires different degrees of cognitive processing. HCI research recognized the existence of diversity in learning styles (Holt & Solomon, 1996) and devoted much time and effort toward this goal. However, Ayre and Nafalski (2000) report that the term *learning styles* is not always interpreted the same way and were able to offer two major interpretations. The first group believes that learning styles emerge from personality differences, life experiences, and student learning goals, while the second group believes that it refers to the way students shape their learning method to accommodate teacher expectations, as when they follow rote learning when teachers expect it.

The first interpretation includes, in part, a form of individual differences but does not explicitly link them to individual cognitive differences, which, in turn, caused researchers more ambiguities as to interpreting the different types of learning styles. In fact, these differences in interpretations caused Stahl (1999) to publish a critique, where he cites five review papers that unite in concluding the lack of sufficient evident to support the claim that accommodating learning styles helps to improve children's learning when acquiring the skill to read. He criticized Carbo's reading style inventory and Dunn and Dunn's learning inventory because of their reliance on self-report to identify different learning styles of students, which, in turn, results in very low replication reliability.

These criticisms are positive in that they indicate a requirement to base definitions on formal replicable theory. A candidate for this is cognitive learning theory (CLT), which represents the part of cognitive science that focuses on the study of how people learn the information presented to them and how they internally represent the concepts mentally in addition to the cognitive load that is endured during the learning process of the concepts.

Some of the attempts that were made to take advantage of the knowledge gained in the field include Jonassen (1991), van Jooligan (1999), and Ghaoui and Janvier (2004).

Jonassen (1991) advocates the constructivist approach to learning, where students are given several tools to help them perform their computation or externally represent text they are expected to remember. This allows them to focus on the learning task at hand. Jonassen (1991) adopts the assumption originally proposed by Lajoie and Derry (1993) and

Derry (1990) that computers fill the role of cognitive extensions by performing tasks to support basic thinking requirements, such as calculating or holding text in memory, and thus allowed computers to be labeled cognitive tools. Jonassen's (1991) central claim is that these tools are offered to students to lower the cognitive load imposed during the learning process, which, in turn, allows them to learn by experimentation and discovery.

Van Jooligan (1999) takes this concept a step further by proposing an environment that allows students to hypothesize and to pursue the consequences of their hypotheses. He did this through utilizing several windows in the same educational system. The system was composed of two main modules: the first supports the hypothesis formation step by providing menus to guide the process; the second provides a formatted presentation of experiments already tested and their results in a structured manner. They also added intelligent support to the system by providing feedback to students to guide their hypothesis formation approach.

Ghaoui and Janvier (2004) presented a two-part system. The first part identified the various personality types, while the second either had an interactive or non-interactive interface. They report an increase in memory retention from 63.57% to 71.09% that occurred for the students using the interactive interface. They also provided a description of the learning style preferences for the students tested, which exhibited particular trends, but these were not analyzed in detail.

Montgomery (1995) published preliminary results of a study aimed at identifying how multimedia, in particular, can be used to address the needs of various learning styles. Results indicate that active learners appreciate the use of movies and interaction, while sensors benefit from the demonstrations.

Although a glimmer of interest in CLT exists, there is a distinct lack of a clear and organized framework to help guide educational interface designers.

ALIGNMENT MAP FOR MULTIMEDIA INSTRUCTIONAL INTERFACE

The problems that arose with learning styles reveal a need for a more fine-grained isolation of various cognitive areas that may influence learning. Consequently, an alignment map, as shown in Table 1, may offer some guidelines as to what aspects of the multimedia interface design would benefit from what branch of the theory in order to gain a clearer channel of communication between the designer and the student.

CASE STUDY: DATA STRUCTURES MULTIMEDIA TUTORING SYSTEM

The alignment map presents itself as an excellent basis against which basic design issues of multimedia systems may be considered with the goal of making the best possible decisions.

The multimedia tutoring system considered here (Albalooshi & Alkhalifa, 2002) teaches data structures and was designed by considering the various design issues as dictated by the alignment map that was specifically designed for the project and is shown in Table 1. An analysis of the key points follows:

1. **Amount of Media Offered:** The system presents information through textual and animated presentation only. This is done to avoid cognitive overload caused by redundancy (Jonassen, 1991) that would cause students to find the material more difficult to comprehend.
2. **How the Screen is Partitioned:** The screen grants two-thirds of the width to the animation window that is to the left of the screen, while the verbal description is to the right. Although the language used for the textual description is in English, all students are Arabs, so they are accustomed to finding the text on the right side of the screen, because in Arabic, one starts to write from the right hand side. This design, therefore, targeted this particular pool of students to ensure that both parts of the screen are awarded sufficient attention. It presents an interface that requires divided attention to two screens that complement each other, a factor that, according to Hampson (1989), minimizes interference between the two modes of presentation.
3. **Parallel Delivery of Information:** Redundancy is desired when it exists in two different

Table 1. Alignment map from multi-media design questions to various cognitive research areas that may be of relevance

Multimedia Design Issues	Cognitive Areas That May Be of Relevance
1. Amount of media offered	1. Cognitive load 2. Limited attention span 3. Interference between different mental representations
2. How the screen is partitioned	1. Perception and recognition 2. Attention
3. Parallel delivery of information	1. Redundancy could cause interference 2. Limited working memory (cognitive load issues) 3. Limited attention span 4. Learner difference
4. Use of colors	1. Affects attention focus 2. Perception of edges to promote recall
5. Use of animation	1. Cognitive load reduction 2. Accommodates visualizer/verbalizer learners
6. Use of interactivity	1. Cognitive load reduction 2. Raises the level of learning objectives
7. Aural media	1. Speech perception issues like accent and clarity 2. Interference with other media
8. Verbal presentation of material	1. Clarity of communication 2. Accommodates verbal/serialist learners

media, because one re-enforces the other. It is not desired when it exists within the media, as when there is textual redundancy and things are explained more than once. Consequently, the textual description describes what is presented in the animation part, especially since only text and animation media exist in this case, which means that cognitive load issues are not of immediate concern (Jonassen, 1991).

4. **Use of Colors:** Colors were used to highlight the edges of the shapes and not on a wide scale to ensure that attention is drawn to those. By doing so, focus is expected to be directed toward the object's axes, as suggested by Marr and Nishihara (1978), in order to encourage memory recall of the shapes at a later point in time.
5. **Use of Animation:** The animated data structures are under the user's control with respect to starting, stopping, or speed of movement. This allows the user to select whether to focus on the animation, text, or both in parallel without causing cognitive overload.
6. **Use of Interactivity:** The level of interactivity is limited to the basic controls of the animation.
7. **Aural Media:** This type of media is not offered by the system.
8. **Verbal Presentation of Material:** The verbal presentation of the materials is concise and fully explains relevant concepts to a sufficient level of detail, if considered in isolation of the animation.

EVALUATION OF THE SYSTEM

The tool first was evaluated for its educational impact on students. It was tested on three groups: one exposed to the lecture alone; the second to a regular classroom lecture in addition to the system; and the third only to the system. Students were distributed among the three groups such that each group had 15 students with a mean grade similar to the other two groups in order to ensure that any learning that occurs is a result of the influence of what they are exposed to. This also made it possible for 30 students to attend the same lecture session composed of the students of groups one and two, while 30 students attended the same lab session composed of the students of groups two and three in order to avoid any confounding factors.

Results showed a highly significant improvement in test results of the second group when their post-classroom levels were compared to their levels following use of the multimedia system, with an overall improvement rate of 40% recorded with F=9.19, with $p < 0.005$ from results of an ANOVA test after ensuring all test requirements had been satisfied. The first and third groups showed no significant differences between them.

Results shown indicate that learning did occur to the group that attended the lecture and then used the system, which implies that animation does fortify learning by reducing the cognitive load. This is especially clear when one takes the overall mean grade of all groups, which is around 10.5, and checks how many in each group are above that mean. Only six in group one were above it, while 11 were above it in group two and 10 in group three. Since group three was exposed to the system-only option and achieved a number very close to group two, which had the lecture and the system option, then clearly, multimedia did positively affect their learning rate.

Since one of the goals of the system is to accommodate learner differences, a test was run on group two students in order to identify the visualizers from the verbalizers. The paper-folding test designed by French et al. (1963) was used to distinguish between the two groups. The test requires each subject to visualize the array of holes that results from a simple process. A paper is folded a certain number of folds, a hole is made through the folds, and then the paper is unfolded. Students are asked to select the image of the unfolded paper that shows the resulting arrangement, and results are evaluated along a median split as high vs. low visualization abilities.

These results then were compared with respect to the percentage of improvement, as shown in Table 2. Notice that the question numbers in the pre-test are mapped to different question numbers in the post-test in order to minimize the possibility of students being able to recall them; a two-part question also was broken up for the same reason.

Results indicate that, although the group indeed was composed of students with different learning preferences, they all achieved comparable overall improvements in learning. Notice, though, the difference in percentage improvement in Question 4. The question is: List and explain the data variables that are associated with the stack and needed to operate on it. This particular question is clearly closer to heart to the verbalizer group than to the visualizer group. Therefore, it should not be surprising that the verbalizer group finds it much easier to learn how to describe the data variables than it is for students who like to see the stack in operation. Another point to consider is that the visualizer group made a bigger improvement in the Q1+Q8 group in response to the question: Using an example, explain the stack concept and its possible use. Clearly, this question is better suited to a visualizer than to a verbalizer.

FUTURE TRENDS

CLT already has presented us with ample evidence of its ability to support the design of more informed and, therefore, more effective educational systems. This article offers a map that can guide the design process of a multimedia educational system by highlighting the areas of CLT that may influence design. The aim, therefore, is to attract attention to the vast pool of knowledge that exists in CLT that could benefit multimedia interface design.

Table 2. The percentage improvement of each group from the pretest to the posttest across the different question types

	Q1 PLUS Q8 MAPPED TO Q1	Q3 MAPPED TO Q2	Q4 MAPPED TO Q3	Q6 MAPPED TO Q6
VISUALIZER GROUP	**27.8%**	**18.6%**	**9.72%**	**9.76%**
T-TEST RESULTS	**.004**	**.003**	**.09**	**.01**
VERBALIZER GROUP	**20.7%**	**22.8%**	**21.4%**	**15.7%**
T-TEST RESULTS	**.004**	**.005**	**.003**	**.009**

CONCLUSION

This article offers a precise definition of what is implied by a computer-based cognitive tool (CT) as opposed to others that were restricted to a brief definition of the concept. Here, the main features of multimedia were mapped onto cognitive areas that may have influence on learning, and the results of an educational system that conforms to these design requirements were exhibited.

These results are informative to cognitive scientists, because they show that the practical version must deliver what the theoretical version promises. At the same time, results are informative to educational multimedia designers by exhibiting that there is a replicated theoretical groundwork that awaits their contributions to bring them to the world of reality.

The main conclusion is that this is a perspective that allows designers to regard their task from the perspective of the cognitive systems they wish to learn so that it shifts the focus from a purely teacher-centered approach to a learner-centered approach without following the route to constructivist learning approaches.

REFERENCES

AlBalooshi, F., & Alkhalifa, E. M. (2002). Multimodality as a cognitive tool. *Journal of International Forum of Educational Technology and Society, IEEE, 5*(4), 49-55.

Ayre, M., & Nafalski, A. (2000). Recognising diverse learning styles in teaching and assessment of electronic engineering. *Proceedings of the 30th ASEE/IEEE Frontiers in Education Conference*, Kansas City, Missouri.

Derry, S. J. (1990). Flexible cognitive tools for problem solving instruction. *Proceedings of the Annual Meeting of the American Educational Research Association*, Boston.

English, W. K., Engelbart, D. C., & Berman, M. L. (1967). Display selection techniques for text manipulation. *IEEE Transactions on Human Factors in Electronics*, *8*(1), 5-15.

French, J. W., Ekstrom, R. B., & Price, L. A. (1963). *Kit of reference tests for cognitive factors*. Princeton, NJ: Educational Testing Services.

Ghaoui, C., & Janvier, W. A. (2004). Interactive e-learning. *Journal of Distance Education Technologies*, *2*(3), 23-35.

Hampson, P. J. (1989). Aspects of attention and cognitive science. *The Irish Journal of Psychology*, *10*, 261-275.

Jonassen, D. H. (1991). Objectivism vs. constructivism: Do we need a new philosophical paradigm shift? *Educational Technology: Research and Development*, *39*(3), 5-13.

LaJoie, S. P., & Derry, S. J. (Eds.). (1993). *Computers as cognitive tools*. Hillsdale, NJ: Lawrence Erlbaum Associates.

Marr, D., & Nishihara, K. (1978). *Representation and recognition of the spatial organization of three-dimensional shapes*. London: Philosophical Transactions of the Royal Society.

Meyrowitz, N., & Van Dam, A. (1982). Interactive editing systems: Part 1 and 2. *ACM Computing Surveys*, *14*(3), 321-352.

Montgomery, S. M. (1995). Addressing diverse learning styles through the use of multimedia. *Proceedings of the 25th ASEE/IEEE Frontiers in Education Conference*, Atlanta, Georgia.

Myers, B. A. (1998). A brief history of human computer interaction technology. *ACM Interactions*, *5*(2), 44-54.

Stahl, S. (1999). Different strokes for different folks? A critique of learning styles. *American Educator*, *23*(3), 27-31.

Yahaya, N., & Sharifuddin, R. S. (2000). Concept-building through hierarchical cognition in a Web-based interactive multimedia learning system: Fundamentals of electric circuits. *Proceedings of the Ausweb2k Sixth Australian World Wide Web Conference*, Cairns, Australia.

KEY TERMS

Alignment Map: A representation on a surface to clearly show the arrangement or positioning of relative items on a straight line or a group of parallel lines.

Attention: An internal cognitive process by which one actively selects which part of the environmental information that surrounds them and focuses on that part or maintains interest while ignoring distractions.

Cognitive Learning Theory: The branch of cognitive science that is concerned with cognition and includes parts of cognitive psychology, linguistics, computer science, cognitive neuroscience, and philosophy of mind.

Cognitive Load: The degree of cognitive processes required to accomplish a specific task.

Learner Differences: The differences that exist in the manner in which an individual acquires information.

Multimedia System: Any system that presents information through different media that may include text, sound, video computer graphics, and animation.

C

Communication + Dynamic Interface = Better User Experience

Simon Polovina
Sheffield Hallam University, UK

Will Pearson
Sheffield Hallam University, UK

INTRODUCTION

Traditionally, programming code that is used to construct software user interfaces has been intertwined with the code used to construct the logic of that application's processing operations (e.g., the business logic involved in transferring funds in a banking application). This tight coupling of user-interface code with processing code has meant that there is a static link between the result of logic operations (e.g., a number produced as the result of an addition operation) and the physical form chosen to present the result of the operation to the user (e.g., how the resulting number is displayed on the screen). This static linkage is, however, not found in instances of natural human-to-human communication.

Humans naturally separate the content and meaning that is to be communicated from how it is to be physically expressed. This creates the ability to choose dynamically the most appropriate encoding system for expressing the content and meaning in the form most suitable for a given situation. This concept of interchangeable physical output can be recreated in software through the use of contemporary design techniques and implementation styles, resulting in interfaces that improve accessibility and usability for the user.

BACKGROUND

This section accordingly reviews certain theories of communication from different disciplines and how they relate to separating the meaning being communicated from the physical form used to convey the meaning.

Claude Shannon (1948), a prominent researcher in the field of communication theory during the 20th century, put forward the idea that meaning is not transmitted in its raw form, but encoded prior to transmission. Although Shannon was primarily working in the field of communication systems and networks such as those used in telephony, his theory has been adopted by those working in the field of human communications. Shannon proposed a five-stage model describing a communication system. Beginning with the first stage of this model, the sender of the communication creates some content and its intended meaning. In the second stage, this content is then encoded into a physical form by the sender and, in the third stage, transmitted to the receiver. Once the communication has been received by the receiver from the sender, it is then at its fourth stage, whereby it is decoded by the receiver. At the fifth and final stage, the content and meaning communicated by the sender become available to the receiver.

An example of how Shannon's (1948) model can be applied to human communication is speech-based communication between two parties. First, the sender of the communication develops some thoughts he or she wishes to transmit to the intended receiver of the communication. Following on from the thought-generation process, the thoughts are then encoded into sound by the vocal cords, and further encoded into a particular language and ontology (i.e., a set of mappings between words and meaning) according to the sender's background. This sound is subsequently transmitted through the air, reaching the receiver's ears where it is decoded by the receiver's auditory system and brain, resulting in the thoughts of the sender finally being available to the receiver.

This split between meaning, its encoding, and the physical transmission of the meaning is recognised in psychology. Psychology considers that there are three stages to receiving data: (a) the receiving of sensory stimuli by a person, (b) the perception of these stimuli into groups and patterns, and (c) the cognitive processing of the groups and patterns to associate cognitively the meaning with the data (Bruno, 2002). Thus, for example, a receiver may see a shape with four sides (the data) and associate the name *square* (the meaning) with it. There is accordingly a split between the input a person receives and the meaning he or she cognitively associates with that input.

Consider, for example, the words on this page as an example of the psychological process through which meaning is transmitted. The first stage of the process is where the reader receives sensory stimuli in the form of black and white dots transmitted to the eyes using light waves of varying wavelength. Upon the stimuli reaching the reader, the brain will perceptually group the different dots contained within the received stimuli into shapes and, ultimately, the reader will cognitively associate the names of letters with these shapes and extract the meaning conveyed by the words.

Semiotics, which is the study of signs and their meanings (French, Polovina, Vile, & Park, 2003; Liu, Clarke, Anderson, Stamper, & Abou-Zeid, 2002), also indicates a split between meaning and its physical presentation. Within semiotics, the way something is presented, known as a sign, is considered to be separate from the meaning it conveys. Accordingly, in semiotics there are three main categories of signs: icons, indexes, and symbols. This delineation is, however, not mutually exclusive as a particular sign may contain elements of all categories. Vile and Polovina (2000) define an icon as representative of the physical object it is meant to represent; a symbol as being a set of stimuli, that by agreed convention, have a specific meaning; and indexes as having a direct link to a cause, for example, the change of a mouse pointer from an arrow shape to an hourglass to reflect the busy state of a system.

This classification of the physical representation according to its relationship with the content and meaning it conveys provides further opportunities to distinguish content and meaning from its physical presentation, and to classify the different elements of presentation. For example, a shop selling shoes may have a sign outside with a picture of a shoe on it. The image of the shoe is the sign, or the physical presence of the meaning, which in this case is an icon, while the fact that it is a shoe shop is the intended meaning. Equally, this could be represented using the words *shoe shop* as the physical sign, in this case a symbol of the English language, while the meaning is again that of a shoe shop.

This split of content and meaning from its physical presentation, which occurs naturally in human communication, allows for the same content and meaning to be encoded in a variety of different forms and encoding methods. For example, the meaning of "no dogs allowed" can be encoded in a variety of visual images. For instance, there might be (a) an image of a dog with a cross through it, (b) the words "no dogs allowed," (c) an auditory sequence of sounds forming the words "no dogs allowed," or (d) the use of tactile alphabets such as Braille, which is used to encode printed writing into a form for the blind. However the content and meaning is conveyed, it remains the same regardless of how it is physically presented.

SOFTWARE ARCHITECTURES FOR CONTENT SEPARATION

For the true separation of presentation from content to occur therefore in software, the content (namely the data or information itself as well as the application's operations, i.e., its business logic as indicated earlier) is stored in a neutral format. This neutrality is achieved when the content is untainted by presentation considerations. This allows any given content to be translated and displayed in any desired presentation format (e.g., through an HTML [hypertext markup language] Web browser such as Microsoft's Internet Explorer, as an Adobe Acrobat PDF [Portable Document Format], as an e-book, on a mobile phone, on a personal digital assistant [PDA], or indeed on any other device not mentioned or yet to be invented). The theories of detaching content and meaning from its physical presentation thus give a framework to separate content from presentation. Once that conceptual separation can be made, or at least continually realisable ways toward it are achieved, then this approach can actually be deployed in the design and implementation of computer systems.

There are a number of methods offered by contemporary software languages and architectures to achieve this detachment between the content and meaning, and how the content can thus be displayed. In the sphere of Web development, the extensible markup language (XML) is one such example. XML provides a useful vehicle for separating presentation from content (Quin, 2004a). Essentially, unlike HTML in which the tags are hard coded (e.g., *Head*, *Body*, *H1*, *P*, and so forth), XML allows designers or developers to define their own tags particular to their domain (e.g., *Name*, *Address*, *Account-number*, *Transactions*, *Debits*, *Credits*, and so forth in, say, a banking scenario). How this content is presented has, of course, to be defined by the designer or developer; he or she can no longer rely on the browser to format it by simply recognising the hard-coded HTML tags. The extensible stylesheet language (XSL) is the vehicle to achieve this (Quin, 2004b). Equally, the scaleable vector graphics (SVG) format, based on XML, is another World Wide Web format capable of separating content and meaning from presentation. SVG specifies drawing objects, their dimensions, colour, and so forth, but leaves the determination of presentation modality to the client viewer application (Ferraiolo, Jun, & Jackson, 2003).

Within enterprise systems, this separation can be achieved through the use of object-orientated and *n*-tier design methodologies. Object orientation works through its embodiment of the four goals of software engineering (Booch, 1990; Meyer, 1988; Polovina & Strang, 2004). These four goals of software engineering, namely (a) abstraction, (b) cohesion, (c) loose coupling, and (d) modularity, determine the principled design of each object that makes up the system. They seek to ensure that the object only performs the functions specific to its role, for example, to display a piece of information or to perform a calculation. Accordingly, these goals seek to ensure that presentation objects only present the information, while logic objects only perform calculations and other business-logic operations. These content objects thus do not concern themselves with how the information is presented to the user; instead these content objects communicate their information via presentation objects to perform this function.

In addition to embodying the four goals of software engineering, object orientation builds on these by providing three further principles: (a) encapsulation, (b) inheritance, and (c) polymorphism (Booch, 1990; Meyer, 1988; Polovina & Strang, 2004). Inheritance allows an object to inherit the characteristics and behaviours of another object. Utilising this feature, it is possible to extend the functionality of an object to include new functionality, which may be new buttons or other interface elements within a user interface. Polymorphism is used to select an object based on its ability to meet a given set of criteria when multiple objects perform similar functions. For example, there may be two objects responsible for displaying the same interface element; both display the same content and meaning, but using different languages. In this scenario, the concept of polymorphism can be used to select the one appropriate for the language native to the user. Thus, object-orientated design can be used to naturally compliment the process of separating content and meaning from its method of presentation.

A common practice within the field of software engineering is to base software designs on common, predefined architectures, referred to as patterns. One pattern, which lends itself well to the separation of content and meaning from its method of presentation, is the *n*-tier architecture. The *n*-tier architecture separates the objects used to create the design for a piece of software into layers (Fowler, 2003). The objects contained within each layer perform a specific group of functions, such as data storage. In the three-tier architecture, for example, one layer is responsible for handling the software's input and output with the user, another handles its business-logic processes, and the final layer handles the persistent storage of information between sessions of the software being executed. Through the use of an *n*-tier architecture and the separation of the different areas of an application's design that it creates, it is possible to separate the content from its mode of presentation within software design.

Software engineering's ability to separate content and meaning from its physical presentation can be aided by some contemporary implementation methods. These methods are based on component architectures that aim to create reusable segments of code that can be executed. This enhances object orientation, which seeks to create reusable segments of software at the source-code level. While there is not much difference in the design, having reusable segments of executable code translates to faster time to change segments, further enhancing the plug-and-play nature of software. Microsoft's Component Object Model (COM) is a client-side Windows-based component architecture (Microsoft Corporation, 1998). This architecture enables programs to be built as individual components that are linked together using a client application to form a complete software program. This approach to software implementation provides the ability to construct similar pieces of software using the same components, where the functionality is common between the pieces of software. For example, if the storage and logic elements of a piece of software were to remain the same but the user interface were to be changed due to the differing needs of user groups, the same components forming the storage and logic sections could be used for all versions of the software. Furthermore, this could occur while different components were created to provide the different user interfaces required. This method would reduce the time taken to build and deploy the software amongst a group of diverse users.

Another implementation technique, built around distributed components located on different physical machines, are Web services (MacDonald, 2004). Instead of the components used to build the software being located on the same machine, different components can be placed on different machines. This results in users being able to share and access the same physical instance of objects. This enhances COM, which although it gives access to the same components, forces each user to use different instances of them. One advantage of Web services is that they allow the existence of different user interfaces while letting users access the same physical objects used for the logic and storage processes. This type of deployment will ensure that all users are accessing the same data through the same logic processes, but allows the flexibility for each user or user group to use an interface that is the most optimal for their needs, be they task- or device-dependant needs.

THE HUMAN-COMPUTER INTERACTION BENEFITS

The human race rarely uses fixed associations between content or meaning and its physical representation. Instead, people encode the meaning into a form appropriate for the situation and purpose of the communication. Communication can be encoded using different ontologies such as different languages and terminology. Communication is thus able to take different physical channels (e.g., sound through the air, or writing on paper), all of which attempt to ensure that the content or meaning is communicated between the parties in the most accurate and efficient manner available for the specific characteristics of the situation. Currently, this is not the case with computer interfaces; contemporary interfaces instead tend to adopt a "one size fits all" approach for the majority of the interface.

In taking this one-size-fits-all approach, content and meaning may not be transmitted to the user in the most accurate form, if it is communicated at all. The characteristics of the situation and participants are not taken into account. This makes the interface harder to use than might be, if it can be used at all. Some users, such as those with a sensory disability or those with a different native language, may not be able to access the information as it has been encoded using an inaccessible physical form (e.g., visual stimuli are inaccessible for the blind). Or it has been encoded using a foreign language, which the user does not understand. This immediately prevents the user from accessing the content and meaning conveyed by that form of presentation.

Equally, terminology can be prohibitive to the ease of use of a user interface. The set of terms that we know the meaning for (i.e., ontology) is based on factors such as the cultural, educational, and social background of the user as well as the geographic area the user inhabits. This leads to different groups of people being familiar with different terminology from those in other groups, although there is some degree of overlap in the ontologies used by the different groups. The user is forced to learn the terminology built into the interface before they can extract the meaning that it conveys. This imposes a learning curve on the user, unless they are already familiar with the particular set of terms used. Hence, by using a one-size-fits-all user interface, some users will find it difficult or impossible to use.

By utilising the facilities offered by contemporary software-engineering practices, it is possible to avoid this one-size-fits-all approach and its inherent disadvantages in terms of human-computer interaction. By allowing the encoding scheme used to present software interfaces to change with different users, interfaces will begin to mimic the processes used to encode content and meaning that are found in natural human-to-human communication. This change will result in interfaces that are accessible by those who could not previously access them, and will also result in greater ease of use for those who previously had to learn the terminology used within the interface, hence improving interface usability.

FUTURE TRENDS

One emerging trend is the use of explicit user modeling to modify the behaviour and presentation of systems based on a user's historic use of that system (Fischer, 2001). Explicit user modeling involves tracking the preferences and activities of a user over time, and building a model representing that behaviour and associated preferences. This, coupled with the concept of presenting the content and meaning in the form most suitable for the user, holds the ability to tailor the content to a specific individual's needs. By monitoring how a user receives different types of information over time, a historic pattern can be developed that can subsequently be used to present the content and meaning based on an individual's actual requirements, not on a generalized set of requirements from a specific group of users.

CONCLUSION

Currently, by entwining the association between content and meaning and the physical form used to represent it, software user interfaces do not mimic natural human-to-human communication. Within natural communication, the content and meaning that is to be conveyed is detached from its physical form, and it is only encoded into a physical form at the time of transmission. This timing of the point at which the content and meaning are encoded is important. It gives the flexibility to encode the content and meaning in a form that is suitable for the characteristics of the situation (e.g., the channels available, the languages used by the parties, and the terminology that they know). This ensures that humans communicate with each other in what they consider to be the most appropriate and accurate manner, leading to encoding schemes from which the parties can access the content and meaning in an easy method.

This is not currently the case for software user interfaces, which use a too tightly coupled association between the content and meaning and the physical form used to encode it. By utilising contemporary Web-based or object-orientated component architectures, this problem of fixed encoding schemes can be overcome. Therefore, software user interfaces can more closely mimic natural language encoding and gain all the benefits that it brings.

REFERENCES

Booch, G. (1990). *Object oriented design with applications*. Redwood City, CA: Benjamin-Cummings Publishing Co., Inc.

Bruno, F. (2002). *Psychology: A self-teaching guide*. Hoboken, NJ: John Wiley & Sons.

Ferraiolo, J., Jun, F., & Jackson, D. (Eds.). (2003). *SVG 1.1 recommendation*. The World-Wide Web Consortium (W3C). Retrieved October 4, 2005 from http://www.w3.org/TR/SVG

Fischer, G. (2001). User modeling in human-computer interaction. *User Modeling and User-Adapted Interaction, 11*(1-2), 65-86.

Fowler, M. (2003). *Patterns of enterprise application architecture*. Boston: Addison Wesley Press.

French, T., Polovina, S., Vile, A., & Park, J. (2003). Shared meanings requirements elicitation (SMRE): Towards intelligent, semiotic, evolving architectures for stakeholder e-mediated communication. *Using Conceptual Structures: Contributions to ICCS 2003*, 57-68.

Liu, K., Clarke, R., Anderson, P., Stamper, R., & Abou-Zeid, E. (2002). Organizational semiotics: Evolving a science of information systems. *Proceedings of IFIP WG8.1 Working Conference*.

MacDonald, M. (2004). *Microsoft .NET distributed applications: Integrating XML Web services and .NET remoting*. Redmond, WA: Microsoft Press.

Meyer, B. (1988). *Object-oriented software construction*. Upper Saddle River, NJ: Prentice-Hall.

Microsoft Corporation. (1998). *Microsoft component services: A technical overview*. Retrieved August 29, 2004, from http://www.microsoft.com/com/wpaper/compsvcs.asp

Polovina, S., & Strang, D. (2004). Facilitating useful object-orientation for visual business organizations through semiotics. In K. Liu (Ed.), *Virtual, distributed and flexible organisations: Studies in organisational semiotics* (pp. 183-184). Dordrecht, The Netherlands: Kluwer Academic Publishers.

Quin, L. (2004a). *Extensible markup language (XML)*. Retrieved March 26, 2004, from http://www.w3.org/XML/

Quin, L. (2004b). *The extensible stylesheet language family (XSL)*. Retrieved April 29, 2004, from http://www.w3.org/Style/XSL/

Shannon, C. (1948). A mathematical theory of communication. *The Bell System Technical Journal, 27*, 379-423, 623-656.

Vile, A., & Polovina, S. (2000). *Making accessible Websites with semiotics*. Retrieved August 15, 2004, from http://www.polovina.me.uk/publications/Semiotics4AWDshortpaper.PDF

KEY TERMS

Accessibility: The measure of whether a person can perform an interaction, access information, or do anything else. It does not measure how well he or she can do it, though.

Content: The information, such as thoughts, ideas, and so forth, that someone wishes to communicate. Examples of content could be the ideas and concepts conveyed through this article, the fact that you must stop when a traffic light is red, and so on. Importantly, content is what is to be communicated but not how it is to be communicated.

Encoding: Encoding is the process by which the content and meaning that is to be communicated is transformed into a physical form suitable for communication. It involves transforming thoughts and ideas into words, images, actions, and so forth, and then further transforming the words or images into their physical form.

Object Orientation: A view of the world based on the notion that it is made up of objects classified by a hierarchical superclass-subclass structure under the most generic superclass (or root) known as an object. For example, a car is a (subclass of) vehicle, a vehicle is a moving object, and a moving object is an object. Hence, a car is an object as the relationship is transitive and, accordingly, a subclass must at least have the attributes and functionality of its superclass(es). Thus, if we provide a generic user-presentation object with a standard interface, then any of its subclasses will conform to that standard interface. This enables the plug and play of any desired subclass according to the user's encoding and decoding needs.

Physical Form: The actual physical means by which thoughts, meaning, concepts, and so forth are conveyed. This, therefore, can take the form of any physical format, such as the writing or displaying of words, the drawing or displaying of images, spoken utterances or other forms of sounds, the carrying out of actions (e.g., bodily gestures), and so forth.

Software Architecture: Rather like the architecture of a building, software architecture describes the principled, structural design of computer software. Contemporary software architectures are multitier (or *n*-tier) in nature. Essentially, these stem from a two-tier architecture in which user-presentation components are separated from the information-content components, hence the two overall tiers. Communication occurs through a standard interface between the tiers. This enables the easy swapping in and out of presentation components, thus enabling information to be encoded into the most appropriate physical form for a given user at any given time.

Usability: A measure of how well someone can use something. Usability, in comparison to accessibility, looks at factors such as ease of use, efficiency, effectiveness, and accuracy. It concentrates on factors of an interaction other than whether someone can perform something, access information, and so forth, which are all handled by accessibility.

Computer Access for Motor-Impaired Users

Shari Trewin
IBM T.J. Watson Research Center, USA

Simeon Keates
IBM T.J. Watson Research Center, USA

INTRODUCTION

Computers can be a source of tremendous benefit for those with motor impairments. Enabling computer access empowers individuals, offering improved quality of life. This is achieved through greater freedom to participate in computer-based activities for education and leisure, as well as increased job potential and satisfaction.

Physical impairments can impose barriers to access to information technologies. The most prevalent conditions include rheumatic diseases, stroke, Parkinson's disease, multiple sclerosis, cerebral palsy, traumatic brain injury, and spinal injuries or disorders. Cumulative trauma disorders represent a further significant category of injury that may be specifically related to computer use. See Kroemer (2001) for an extensive bibliography of literature in this area.

Symptoms relevant to computer operation include joint stiffness, paralysis in one or more limbs, numbness, weakness, bradykinesia (slowness of movement), rigidity, impaired balance and coordination, tremor, pain, and fatigue. These symptoms can be stable or highly variable, both within and between individuals. In a study commissioned by Microsoft, Forrester Research, Inc. (2003) found that one in four working-age adults has some dexterity difficulty or impairment. Jacko and Vitense (2001) and Sears and Young (2003) provide detailed analyses of impairments and their effects on computer access.

There are literally thousands of alternative devices and software programs designed to help people with disabilities to access and use computers (Alliance for Technology Access, 2000; Glennen & DeCoste, 1997; Lazzaro, 1995). This article describes access mechanisms typically used by individuals with motor impairments, discusses some of the trade-offs involved in choosing an input mechanism, and includes emerging approaches that may lead to additional alternatives in the future.

BACKGROUND

There is a plethora of computer input devices available, each offering potential benefits and weaknesses for motor-impaired users.

Keyboards

The appeal of the keyboard is considerable. It can be used with very little training, yet experts can achieve input speeds far in excess of handwriting speeds with minimal conscious effort. Their potential for use by people with disabilities was one of the factors that spurred early typewriter development (Cooper, 1983).

As keyboards developed, researchers investigated a number of design features, including key size and shape, keyboard height, size, and slope, and the force required to activate keys. Greenstein and Arnaut (1987) and Potosnak (1988) provide summaries of these studies.

Today, many different variations on the basic keyboard theme are available (Lazzaro, 1996), including the following.

- Ergonomic keyboards shaped to reduce the chances of injury and to increase comfort, productivity, and accuracy. For example, the Microsoft® Natural Keyboard has a convex surface and splits the keys into two sections, one for each hand, in order to reduce wrist flexion for touch typists. The Kinesis® Ergonomic Keyboard also separates the layout into

right- and left-handed portions, but has a concave surface for each hand designed to minimise the digit strength required to reach the keys and to help the hands maintain a flat, neutral position.
- Oversized keyboards with large keys that are easier to isolate.
- Undersized keyboards that require a smaller range of movement.
- One-handed keyboards shaped for left- or right-handed operation. These may have a full set of keys, or a reduced set with keys that are pressed in combinations in the same way a woodwind instrument is played.
- Membrane keyboards that replace traditional keys with flat, touch-sensitive areas.

For some individuals, typing accuracy can be improved by using a key guard. Key guards are simply attachments that fit over the standard keyboard with holes above each of the keys. They provide a solid surface for resting hands and fingers on, making them less tiring to use than a standard keyboard for which the hands are held suspended above. They also reduce the likelihood of accidental, erroneous key presses. Some users find that key guards improve both the speed and accuracy of their typing. Others find that key guards slow down their typing (McCormack, 1990), and they can make it difficult to see the letters on the keys (Cook & Hussey, 1995).

The Mouse

A mouse is a device that the user physically moves across a flat surface in order to produce cursor movement on the screen. Selection operations are made by clicking or double clicking a button on the mouse, and drag operations are performed by holding down the appropriate button while moving the mouse. Because the buttons are integrated with the device being moved, some people with motor impairments experience difficulties such as unwanted clicks, slipping while clicking, or dropping the mouse button while dragging (Trewin & Pain, 1999). Tremors, spasms, or lack of coordination can cause difficulties with mouse positioning.

Trackball

Trackballs offer equivalent functionality to a mouse, but are more straightforward to control. This device consists of a ball mounted in a base. The cursor is moved by rolling the ball in its casing, and the speed of movement is a function of the speed with which the ball is rolled. Buttons for performing click and double-click operations are positioned on the base, which makes it easier to click without simultaneously moving the cursor position. For dragging, some trackballs require a button to be held down while rolling the ball, while others have a specific button that initiates and terminates a drag operation without needing to be held down during positioning.

Thumb movement is usually all that is required to move the cursor to the extremities of the screen, as compared to the large range of skills necessary to perform the equivalent cursor movement with a mouse.

Joystick

The joystick is a pointing device that consists of a lever mounted on a base. The lever may be grasped with the whole hand and have integrated buttons, or may be operated with the fingers, with buttons mounted on the base. The cursor is moved by moving the lever in the desired direction. When the lever is released, it returns to its original, central position. Of most relevance are models in which the cursor moves at a fixed or steadily accelerating rate in the direction indicated by lever movement and retains its final position when the lever is released. The buttons are often located on the base of such models, and a drag button is generally included since it is difficult to hold down a button while moving the lever with a single hand.

Isometric Devices

Isometric devices measure force input rather than displacement. An example is the TrackPoint device supplied with IBM laptops: a small red button located in the center of the keyboard. These devices do not require any limb movement to generate the input, only muscle contractions. As it has been postulated that some spasticity in particular is brought on by limb movement, these devices offer a means of avoiding that.

Studies performed using isometric joysticks (Rao, Rami, Rahman, & Benvunuto, 1997), and an adapted Spaceball Avenger (Stapleford & Maloney, 1997), have shown that the ability to position the cursor is improved by using isometric devices.

Touch Pad

The touch pad is a flat, touch-sensitive surface representing all or part of the screen. The cursor is moved by sliding a finger across the surface in the desired direction. It requires only a small range of motion. Buttons for object selection are located near the touch surface, and a drag button may or may not be available.

Switch Input

The most physically straightforward input device to operate is a single switch. Switches can come in many different formats (Lazzaro, 1995) and be activated by hand, foot, head, or any distinct, controlled movement. There are also mouth switches, operated by tongue position or by sudden inhalation or exhalation (sip-puff switches). If a user is capable of generating several of these motions independently, then it is possible to increase the number of switches to accommodate this and increase the information transfer bandwidth.

Given their cheapness and the relatively low level of movement required to operate them, switches have become extremely popular as the preferred method of input for the more severely impaired users. However, they do have drawbacks.

It is necessary to use switches in conjunction with some kind of software adaptation to generate the full range of input of a keyboard-mouse combination. The most frequently used method for this is scanning input. This involves taking a regular array of on-screen buttons, be they symbols, letters, or keys, and highlighting regions of the screen in turn. The highlighting dwells over that region of the screen for a predetermined duration, then moves to another part of the screen, dwells there, and so on until the user selects a particular region. This region is then highlighted in subregions and this continues until a particular button is selected. Therefore, this process can involve several periods of waiting for the appropriate sections of the screen to be highlighted, during which the user is producing no useful information.

Brewster, Raty, and Kortekangas (1996) report that for some users, each menu item must be highlighted for as much as five seconds. There has been much research on efficient scanning mechanisms, virtual keyboard layouts, and other ways of accelerating scanning input rates (e.g., Brewster et al., 1996; Simpson & Koester, 1999).

A switch can also be used to provide input in Morse code. This can be faster than scanning, but requires more accurate control of switch timing and the ability to remember the codes.

Head-Motion Transducers

Head-pointing systems operate by detecting the user's head position and/or orientation using ultrasonic, optical, or magnetic signals, and using that information to control the cursor. Nisbet and Poon (1998) describe a number of existing systems and note them to be easy to use, providing both speed and accuracy. The majority of these systems are ultrasound based, such as the Logitech 6D mouse and the HeadMaster system (Prentke-Romich). An alternative system is the HeadMouse (Origin Instruments), which involves the user wearing a small reflective patch on either the forehead or the bridge of a pair of spectacles.

As with most of the mouse-replacement systems, no software-interface modifications are necessary to access most existing applications. However, some kind of switch device is needed to make selections. This is often a mouth-mounted sip-puff switch as most users of these systems do not have sufficiently good arm movement to operate a hand switch.

Learning to use head movements to control the cursor can take a little while as there is a lack of tactile feedback from the input device, but once used to it, users can control the cursor quite successfully.

Eye Gaze Input

A review of these and other input devices in the context of wheelchairs and environmental controls is presented by Shaw, Loomis, and Crisman (1995). Edwards (1995) reports that the most successful eye-gaze systems operate by detecting infrared light bounced off the user's retina, but there are still many unsolved problems with this technology, such as coping with head movements. Today's eye-gaze systems may be either head mounted or remote, can be accurate to within 1 cm, and can be used to control off-the-shelf applications (e.g., ERICA, http://www.eyeresponse.com/ericasystem.html).

Speech Recognition Systems

Speech input is widely touted as the eventual successor to the keyboard, being a natural form of human communication. Speech is a potentially very good input medium for motor-impaired users, although speech difficulties accompanying motor impairments may impede the recognition process. Nisbet and Poon (1998) also note that some users have reported voice strain from using this input technique.

Besides the technical difficulties of the actual recognition process, environmental considerations also have to be addressed. Users with motor impairments may be self-conscious and wish to avoid drawing attention to themselves. An input system that involves speaking aloud fails to facilitate this. However, there have been cases in which speech recognition systems have been found to be a good solution.

Speech recognition systems can be programmed to offer verbal cursor control and so can replace both the keyboard and mouse in the interaction process. However, true hands-free operation is not provided in most of today's products.

SOFTWARE SUPPORTING PHYSICAL ACCESS

Software programs can be used to alter the behaviour of input devices or the input requirements of applications. Software modifications can tackle input errors by changing an input device's response to specific inputs. They can reduce fatigue by reducing the volume of input required, or they can provide alternatives to difficult movements. They can also minimise the input required of the user, thus reducing effort and opportunities for error. Some examples of useful facilities that can be implemented in software are the following.

- The keyboard and mouse configuration, or the way the keyboard or mouse reacts to a given input, can be changed. For example, the delay before a key starts to repeat can be altered, or the cursor can be made to move more slowly relative to the mouse. Another option sets the computer to ignore repeated key presses within a set time less than a particular threshold value. This filters the input for tremor cases in which the user depresses a key more than once for a single character input. Another powerful option is Sticky Keys. This registers the pressing of keys such as Shift and Control and holds them active until another key is pressed. This removes the need for the user to operate several keys simultaneously to activate keyboard shortcuts, hence simplifying the degree of coordination demanded for the input. Simple alterations like these can be very effective (Brown, 1992; Nisbet & Poon, 1998; Trewin & Pain, 1998).
- For those who do not use a keyboard but have some form of pointing device, an on-screen keyboard emulator can be used to provide text input. On-screen keyboards are built into many modern operating systems. Several commercial versions also exist, such as WiViK.
- For users who find input slow or laborious, macros can be used to perform common sequences of operations with a single command. For example, a user who always enters the same application and opens the same file after logging on to a system could define a macro to open the file automatically. Many word-processing packages also include macro facilities to allow commonly used text to be reproduced quickly. For example, a user could create a macro representing his or her address as it appears at the top of a letter.
- When knowledge about the user's task is available, more advanced typing support can be

provided through word prediction. Many word-prediction systems have been developed, and a number are reviewed in Millar and Nisbet (1993). As a user begins to type a word, the prediction system offers suggestions for what the word might be. If the desired word is suggested, the user can choose it with a single command. In practice, word-prediction systems have been observed to reduce the number of keystrokes required by up to 60% (Newell, Arnott, Cairns, Ricketts, & Gregor, 1995). Newell et al. report that using the PAL word-prediction system, some users were able to double their input speed. However, studies with disabled users have also shown that a reduction in keystrokes does not necessarily produce an increase in input rate (for more detailed summaries, see Horstmann, Koester, & Levine, 1994; Horstmann & Levine, 1991). Word prediction is most useful for very slow typists, particularly switch users. Those who type at a rate of greater than around 15 words a minute may find that the time spent searching the lists of suggestions for the right word is greater than the time saved (Millar & Nisbet). Nevertheless, faster users may still find word prediction helpful in reducing fatigue, reducing errors, or improving spelling (Millar & Nisbet).

Input acceleration and configuration are complementary approaches. The former improves accuracy and/or comfort while the latter reduces the volume of input required, thus increasing the input rate. For users with slow input rates, or those who tire easily, both techniques can be useful.

CHOOSING AN APPROPRIATE INPUT MECHANISM

Finding the best input mechanism for a given individual requires the analysis and adjustment of many interrelated variables, including the choice of device, its position and orientation, the need for physical control enhancers such as mouth sticks, and the available configuration options of the device itself (Cook & Hussey, 1995; Lee & Thomas, 1990). This is best achieved through professional assessment. In an assessment session, an individual may try several devices, each in a number of different positions, with different control enhancers.

Users often prefer to use standard equipment whenever possible (Edwards, 1995). Alternative input devices can often be slower to use and may not provide access to the full functionality of applications the user wishes to use (Anderson & Smith, 1996; Mankoff, Dey, Batra, & Moore, 2002; Shaw et al., 1995). Also, skills for standard equipment learned at home can be transferred to machines at work, college, or other public places. Finally, their use does not identify the user as different or disabled.

Many less severely impaired users can use standard input devices with minor modifications. For some people, positioning is very important. Adjustable tables allow keyboard and screen height to be adjusted, and this in itself can have a dramatic effect on input accuracy. The keyboard tilt can also be adjusted. For those who find it tiring to hold their arms above the keyboard or mouse, arm supports can be fitted to tables. Wrist rests can also provide a steadying surface for keyboard and mouse use. Some users wear finger or hand splints while others use a prodder or head stick to activate keys.

The potential effectiveness of physical modifications to input devices is illustrated by Treviranus, Shein, Hamann, Thomas, Milner, and Parnes (1990), who describe three case studies of users for whom modification of standard pointing devices was required. They define the physical demands made by direct manipulation interfaces, and the difficulties these caused for three users with disabilities. In all cases, the final solution involved a combination of pointing devices or minor modifications to a standard device.

FUTURE TRENDS

Clearly, the field of input device technology is an evolving one, with new technologies emerging all the time. For example, some of the most exciting developments in computer input in recent years have been in the field of brain-computer interfaces. For reviews of recent research progress, see Moore (2003) and Wolpaw, Birbaumer, McFarland, Pfurtscheller, and Vaughan (2002). A brain-computer interface is a system in which electrical brain activity is measured and interpreted by a computer in order to provide computer-based control without reliance on muscle movement. Contrary to popular opinion, such interfaces do not read a person's thoughts. Instead, the person learns to control an aspect of his or her brain signals that can be detected and measured.

Such interfaces represent what may be the only possible source of communication for people with severe physical impairments such as locked-in syndrome. Brain-computer interfaces have been shown to enable severely impaired individuals to operate environmental-control systems and virtual keyboards, browse the Web, and even make physical movements (Moore, 2003; Perelmouter & Birbaumer, 2000; Wolpaw, Birbaumer, et al., 2002). Clinical trials are under way for at least one commercial system, the BrainGate by Cyberkinetics Inc. (http://www.cyberkineticsinc.com).

A related computer-control system already on the market, Cyberlink Brainfingers, is a hybrid brain- and body-signal transducer consisting of a headband that measures brain and muscle activity in the forehead. Information is available from Brain Actuated Technologies Inc. (http://www.brainfingers.com).

Computer input devices will also have to evolve to match changes in software user interfaces. For example, the next generation of Microsoft's ubiquitous Windows operating system will apparently move from two-dimensional (2-D) interfaces to three-dimensional (3-D) ones. While the dominant output technologies, that is, monitors and LCD panels, remain two-dimensional, it is likely that 2-D input devices such as the mouse will continue to be used. However, when three-dimensional output technologies become more common, there will be a need to migrate to 3-D input devices. If the 3-D outputs are genuinely immersive, this may benefit motor-impaired users as the targets could be enlarged, allowing for larger gross movements for selecting them. However, if the outputs remain comparatively small, then the difficulties of locating, selecting, and activating targets in two dimensions on the screen are going to be further compounded by the addition of a third dimension. Consequently, the jury is still out on whether the move to 3-D will be beneficial or not for motor-impaired users.

CONCLUSION

Many computer access options are available to people with motor impairments. For those individuals who prefer to use standard computer input devices, accuracy and comfort can be improved through modifications to device positioning, the use of control enhancers such as wrist rests, appropriate device configuration, and software to accelerate input rates. Where standard devices are not appropriate, the above enhancements can be used in conjunction with alternative devices such as trackballs and head or eye gaze systems. When choosing an input setup, professional assessment is highly beneficial.

Speech input is useful for some individuals but has significant drawbacks. Brain-computer interfaces show great promise, offering hope to individuals with severe physical impairments.

REFERENCES

Alliance for Technology Access. (2000). *Computer and Web resources for people with disabilities: A guide to exploring today's assistive technologies* (3rd ed.). CA: Hunter House Inc.

Anderson, T., & Smith, C. (1996). "Composability": Widening participation in music making for people with disabilities via music software and controller solutions. *Proceedings of the Second Annual ACM Conference on Assistive Technologies* (pp. 110-116).

Baecker, R., Grudin, J., Buxton, W., & Greenberg, S. (1995). *Readings in human computer interaction: Towards the year 2000* (2nd ed.). CA: Morgan Kaufmann Publishers Inc.

Brewster, S., Raty, V., & Kortekangas, A. (1996). Enhancing scanning input with non-speech sounds. *Proceedings of the Second Annual ACM Conference on Assistive Technologies* (pp. 10-14).

Brown, C. (1992). Assistive technology computers and people with disabilities. *Communications of the ACM, 35*(5), 36-45.

Cook, S., & Hussey, A. (1995). *Assistive technologies: Principles and practice.* Mosby-Year Book, Inc.

Cooper, W. (1983). *Cognitive aspects of skilled typewriting.* New York: Springer-Verlag.

Edwards, A. D. N. (1995). *Extra-ordinary human-computer interaction: Interfaces for users with disabilities.* Cambridge University Press.

Forrester Research, Inc. (2003). *The wide range of abilities and its impact on computer technology.* Cambridge, MA: Author.

Glennen, S., & DeCoste, D. (1997). *Handbook of augmentative and alternative communication.* San Diego, CA: Singular Publishing Group.

Greenstein, J., & Arnaut, L. (1987). Human factors aspects of manual computer input devices. In G. Salvendy (Ed.), *Handbook of human factors* (chap. 11.4, pp. 1450-1489). New York: John Wiley & Sons.

Horstmann, H. M., Koester, H., & Levine, S. (1994). Learning and performance of able-bodied individuals using scanning systems with and without word prediction. *Assistive Technology, 6*(1), 42-53.

Horstmann, H. M., & Levine, S. P. (1991). The effectiveness of word prediction. *Proceedings of RESNA '91* (pp. 100-102).

Jacko, J., & Sears, A. (Eds.). (2003). *The human computer interaction handbook: Fundamentals, evolving technologies and emerging applications.* Lawrence Erlbaum Associates.

Jacko, J., & Vitense, H. (2001). A review and reappraisal of information technologies within a conceptual framework for individuals with disabilities. *Universal Access in the Information Society, 1*, 56-76.

Kroemer, K. (2001). Keyboards and keying: An annotated bibliography of the literature from 1878 to 1999. *Universal Access in the Information Society, 1*, 99-160.

Lazzaro, J. (1996). *Adapting PCs for disabilities.* Addison Wesley.

Mankoff, J., Dey, A., Batra, U., & Moore, M. (2002). Web accessibility for low bandwidth input. *Proceedings of ASSETS 2002: The Fifth International ACM Conference on Assistive Technologies* (pp. 17-24).

McCormack, D. (1990). The effects of keyguard use and pelvic positioning on typing speed and accuracy in a boy with cerebral palsy. *American Journal of Occupational Therapy, 44*(4), 312-315.

Millar, S., & Nisbet, P. (1993). *Accelerated writing for people with disabilities.* University of Edinburgh, CALL Centre and Scottish Office Education Department.

Moore, M. (2003). Frontiers of human-computer interaction: Direct-brain interfaces. *Frontiers of Engineering: Reports on Leading-Edge Engineering from the 2002 NAE Symposium on Frontiers of Engineering*, 47-52.

Newell, A., Arnott, J., Cairns, A., Ricketts, I., & Gregor, P. (1995). Intelligent systems for speech and language impaired people: A portfolio of research. In A. D. N. Edwards (Ed.), *Extra-ordinary human-computer interaction: Interfaces for users with disabilities* (chap. 5, pp. 83-101). Cambridge University Press.

Nisbett, P., & Poon, P. (1998). *Special access technology.* University of Edinburgh, CALL Centre and Scottish Office Education Department.

Perelmouter, J., & Birbaumer, N. (2000). A binary spelling interface with random errors. *IEEE Transactions on Rehabilitation Engineering, 8*(2), 227-232.

Potosnak, K. (1988). Keys and keyboards. In M. Helander (Ed.), *Handbook of HCI* (pp. 475-494). North-Holland, Elsevier Science Publishers B.V.

Rao, R., Rami, S., Rahman, T., & Benvunuto, P. (1997). Evaluation of an isometric joystick as an interface device for children with CP. *RESNA '97*, 327-329.

Sears, A., & Young, M. (2003). Physical disabilities and computing technology: An analysis of impairments. In J. Jacko & A. Sears (Eds.), *The human-computer interaction handbook: Fundamentals, evolving technologies and emerging applications* (pp. 482-503). NJ: Lawrence Erlbaum.

Shaw, R., Loomis, A., & Crisman, E. (1995). Input and integration: Enabling technologies for disabled users. In A. D. N. Edwards (Ed.), *Extra-ordinary human-computer interaction: Interfaces for users with disabilities* (chap. 13, pp. 263-278). Cambridge University Press.

Simpson, R., & Koester, H. H. (1999). Adaptive one-switch row-column scanning. *IEEE Transactions on Rehabilitation Engineering, 7*(4), 464-473.

Stapleford, T., & Mahoney, R. (1997). Improvement in computer cursor positioning performance for people with cerebral palsy. *Proceedings of RESNA'97* (pp. 321-323).

Treviranus, J., Shein, F., Hamann, G., Thomas, D., Milner, M., & Parnes, P. (1990). Modifications of direct manipulation pointing devices. In *Resna '90: Proceedings of the 13th Annual Conference* (pp. 151-152).

Trewin, S., & Pain, H. (1998). A study of two keyboard aids to accessibility. In H. Johnson, L. Nigay, & C. Roast (Eds.), *People and Computers XIII: Proceedings of HCI 98*, 83-97.

Trewin, S., & Pain, H. (1999). Keyboard and mouse errors due to motor disabilities. *International Journal of Human-Computer Studies, 50*, 109-144.

Vaughan, T. M., Heetderks, W. J., Trejo, L. J., Rymer, W. Z., Weinrich, M., Moore, M. M., et al. (2003). Brain-computer interface technology: A review of the second international meeting. *IEEE Transactions on Neural Systems & Rehabilitation Engineering, 11*, 94-109.

Wolpaw, J. R., Birbaumer, N., McFarland, D., Pfurtscheller, G., & Vaughan, T. (2002). Brain-computer interfaces for communication and control. *Clinical Neurophysiology, 113*, 767-791.

Wolpaw, J. R., McFarland, D. J., Vaughan, T. M., & Schalk, G. (2003). The Wadsworth Center Brain-Computer Interface (BCI) Research and Development Program. *IEEE Transactions on Neural Systems & Rehabilitation Engineering, 11*, 204-207.

KEY TERMS

Accessibility: A characteristic of information technology that allows it to be used by people with different abilities. In more general terms, accessibility refers to the ability of people with disabilities to access public and private spaces.

Assessment: A process of assisting an individual with a disability in the selection of appropriate assistive technology devices and/or configurations of standard information technology devices.

Input Acceleration: Techniques for expanding user input, allowing a large volume of input to be provided with few user actions.

Motor Impairment: A problem in body motor function or structure such as significant deviation or loss.

Transducer: An electronic device that converts energy from one form to another.

Computer-Based Concept Mapping

Sherman R. Alpert
IBM T.J. Watson Research Center, USA

INTRODUCTION

A *concept map* (also known as a *knowledge map*) is a visual representation of knowledge of a domain. A concept map consists of nodes representing concepts, objects, events, or actions connected by directional links defining the semantic relationships between and among nodes. Graphically, a node is represented by a geometric object, such as a rectangle or oval, containing a textual name; relationship links between nodes appear as textually labeled lines with an arrowhead at one or both ends indicating the directionality of the represented relation. Together, nodes and links define propositions or assertions about a topic, domain, or thing. For example, an arrow labeled *has* beginning at a node labeled *bird* and ending at a *wings* node represents the proposition "A bird has wings" and might be a portion of a concept map concerning birds, as portrayed in Figure 1.

BACKGROUND: CONCEPT MAPS AS KNOWLEDGE REPRESENTATION

Representing knowledge in this fashion is similar to semantic network knowledge representation from the experimental psychology and AI (artificial intelligence) communities (Quillian, 1968). Some have argued that concept maps accurately reflect the content of their authors' knowledge of a domain (Jonassen, 1992) as well as the structure of that knowledge in the authors' cognitive system (Anderson-Inman & Ditson, 1999). Indeed, in addition to the structured relationships among knowledge elements (nodes and links) that appear in a single map, some concept mapping tools allow for multiple layer maps. The structure of such maps is isomorphic to the cognitive mechanisms of abstraction, wherein a single node at one level of a map may represent a chunk of knowledge that can be further elaborated by any number of knowledge elements at a more

Figure 1. A concept map in the Webster concept mapping tool. Nodes in this concept map portray a variety of representational possibilities: A node may contain a textual description of a concept, object, event, or action, or may be an image or a link to a Web site, audio, video, spreadsheet, or any other application-specific document.

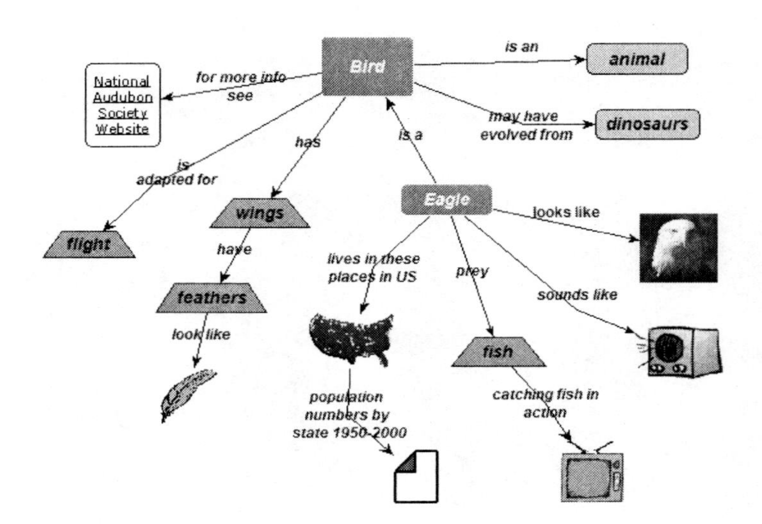

detailed level of the overall map (Alpert, 2003). Concept maps, thus, can be viewed as knowledge visualization tools.

CONCEPT MAPS AS COGNITIVE TOOL

Concept maps have been used in educational settings since the early 1970s as both pedagogical and evaluation tools in virtually every subject area: reading and story comprehension, science, engineering, math word problems, social studies, and decision making (see, e.g., Bromley, 1996; Chase & Jensen, 1999; Fisher, Faletti, Patterson, Thornton, Lipson, & Spring, 1990; Novak, 1998). Concept maps permit students to demonstrate their knowledge of a domain; act as organizational and visualization tools to aid study and comprehension of a domain, a story, or an expository text; and support the generation and organization of thoughts and ideas in preparation for prose composition. They are also used as instructional materials whereby teacher-prepared maps present new materials to learners, showing the concepts and relationships among concepts of a domain new to the students. Concept maps constructed by students help those students to learn and exercise the metacognitive practice of reflecting on what they know to explain or demonstrate their knowledge to others. Such activities may lead to self-clarification and elaboration of their knowledge. There is considerable anecdotal and experimental evidence that the use of graphical knowledge-visualization tools such as concept maps helps improve student comprehension and enhance learning. For example, Fisher et al. (1990) have reported that providing concept maps constructed by domain experts to present new information to learners and illustrating how an expert organizes concepts of the domain results in demonstrable pedagogical benefits. Dunston (1992) and Moore and Readance (1984) have shown that concept maps are pedagogically effective when students create their own maps to reflect on and demonstrate their own knowledge.

In educational environments, the use of concept maps has evolved from paper-and-pencil to computer-based tools. A number of computer-based concept-mapping tools have been reported by researchers (e.g., Alpert & Grueneberg, 2000; Fisher et al., 1990; Gaines & Shaw, 1995b; Kommers, Jonassen, & Mayes, 1992), and there exist shareware programs as well as commercial products for this activity (e.g., Inspiration,[1] Axon,[2] Decision Explorer,[3] SemNet,[4] SMART Ideas,[5] and the IHMC CmapTools[6]). With such tools, users using a mouse and keyboard can create, position, organize, modify, evolve, and store and retrieve the nodes and links that comprise concept maps. Concept-mapping software offers the same sorts of benefits that word processors provide over composing written works on paper. That is, such software facilitates revision of existing work, including additions, deletions, modifications, or reorganizations. In fact, students often revisit their existing maps to revise them as their knowledge of a subject evolves (Anderson-Inman & Zeitz, 1993).

Computer-based concept mapping tools have also been used outside educational settings. In business settings, for example, concept-map tools have been used for organizing notes taken during meetings and lectures, and for the preparation of presentations and written works. There have also been efforts to use concept maps as organizing vehicles for both designers and end users of Web sites and other hypermedia environments (e.g., Gaines & Shaw, 1995a; Zeiliger, Reggers, & Peeters, 1996). In this context, concept maps have provided visualizations of the structure of the pages, documents, or resources of a site and the hyperlink relationships among them, as well as a mechanism for directly navigating to specific pages.

FUTURE TRENDS

More recently, concept-mapping tools have been enhanced to enable the representation of information or knowledge that is neither textual nor proposition based. In many tools, for example, a node may be an image rather than a geometric shape with an embedded textual description. In the Inspiration product, nodes in a concept map may also reference media files, such as video and audio, and application-specific files, such as spreadsheet or presentation documents. The Webster knowledge mapping tool (Alpert & Grueneberg, 2000) offers a Web-enabled version of these facilities, in which nodes in a concept map may reference any media that can be presented or played in a Web browser (including Web-based tools and applications, such as Flash™interactive programs). As new sense-based resources, such as tactile, haptic, and aroma-based media, become available on inexpensive computers, concept maps should also be capable of incorporating nodes representing such information for end-user navigation and playback with the goal of having concept maps comprehensively represent knowledge of a domain (Alpert, 2005).

Concept-map tools have now integrated other Web facilities as well, such as nodes that act as hyperlinks to Web sites. This innovation allows concept maps to incorporate the vast amount of knowledge and information available on the World Wide Web. For learning, users need content organized in some fashion, focused on a particular topic or domain. Rather than a generic search engine to, hopefully, find relevant content and its resulting flat view of information, a concept map provides a centralized place to access knowledge and information. Such a tool visually organizes relevant content in lucidly structured ways while providing semantic links between knowledge and information elements. Concept maps can thereby help students by imposing order on the perhaps overwhelming amounts and complexity of information germane to a domain. This can be especially useful when that information is distributed across numerous locations on the Web. Concept maps can thereby serve as personal knowledge management tools for students and other knowledge workers.

Concept maps are also emerging as visualization tools for the nascent area of *topic maps*. Topic maps are used to organize, for end-user navigation, resources available on the Web that are germane to a particular domain of interest and/or multiple topically related domains. A topic map is defined as a model "for representing the structure of information resources used to define topics, and the associations (relationships) between topics" (Pepper & Moore, 2001). Thus, the conceptual connection to concept maps is obvious. Topic maps are consistent with the ideas expressed above regarding knowledge management: The value of a topic map is that it organizes for the user, in a single location, resources that reside in multiple disparate locations on the Web. However, topic maps are defined using a textual language and typically presented to end users textually as well. The XML (extensible markup language) Topic Maps (XTM) specification is an ISO/IEC standard that defines an XML-based language for defining and sharing Web-based topic maps (International Organization for Standardization & International Electrotechnical Commission, 2002). But the specification does not specify or suggest tools for the visualization of a map's topics, Web-based resources, or their interrelationships. In practice to date, topic maps are often displayed for users using a textual (plus hyperlink) format. However, several developers are beginning to apply the concept map visualization model to provide users with a graphical mechanism for both understanding topic maps and navigating to specific pages contained therein (e.g., HyperGraph[7] and Mondeca[8]).

CONCLUSION

Concept maps are a form of knowledge representation and knowledge visualization portraying knowledge and information about a domain of interest in a visual and organized fashion. Concept maps have evolved from paper-and-pencil tools, to computer-based text-only applications, to computer-based tools that permit the incorporation of any sort of knowledge or information source available in any computational form. As new forms of sensory output become available in digital form, concept maps should provide facilities for the inclusion of such media in order to fully represent and share knowledge of any domain. Concept maps have been used as cognitive tools, especially in educational settings, for organizing thoughts and ideas, for knowledge elicitation, and for learning.

REFERENCES

Alpert, S. R. (2003). Abstraction in concept map and coupled outline knowledge representations. *Journal of Interactive Learning Research, 14*(1), 31-49.

Alpert, S. R. (2005). Comprehensive mapping of knowledge and information resources: The case of Webster. In S. Tergan & T. Keller (Eds.), *Knowl-*

edge visualization and information visualization: Searching for synergies (pp. 208-225). Heidelberg: Springer-Verlag.

Alpert, S. R., & Grueneberg, K. (2000). Concept mapping with multimedia on the Web. *Journal of Educational Multimedia and Hypermedia, 9*(4), 313-330.

Anderson-Inman, L., & Ditson, L. (1999). Computer-based cognitive mapping: A tool for negotiating meaning. *Learning and Leading with Technology, 26,* 6-13.

Anderson-Inman, L., & Zeitz, L. (1993). *Computer-based concept mapping: Active studying for active learners. The Computing Teacher, 21*(1), 6-19.

Brachman, R. J., & Levesque, H. J. (Eds.). (1985). *Readings in knowledge representation.* Los Altos, CA: Morgan Kauffman.

Bromley, K. D. (1996). *Webbing with literature: Creating story maps with children's books* (2nd ed.). Needham Heights, MA: Allyn & Bacon.

Chase, M., & Jensen, R. (Eds.). (1999). *Meeting standards with Inspiration®: Core curriculum lesson plans.* Beaverton, OR: Inspiration Software.

Dunston, P. J. (1992). A critique of graphic organizer research. *Reading Research and Instruction, 31*(2), 57-65.

Fisher, K. M., Faletti, J., Patterson, H., Thornton, R., Lipson, J., & Spring, C. (1990). Computer-based concept mapping: SemNet software: A tool for describing knowledge networks. *Journal of College Science Teaching, 19*(6), 347-352.

Gaines, B. R., & Shaw, M. L. G. (1995a). Concept maps as hypermedia components. *International Journal of Human-Computer Studies, 43*(3), 323-361. Retrieved from http://ksi.cpsc.ucalgary.ca/articles/ConceptMaps

Gaines, B. R., & Shaw, M. L. G. (1995b). WebMap: Concept mapping on the Web. *World Wide Web Journal, 1*(1), 171-183. Retrieved from http://ksi.cpsc.ucalgary.ca/articles/WWW/WWW4WM/

International Organization for Standardization & International Electrotechnical Commission (ISO/IEC). (2002). *ISO/IEC 13250: Topic maps* (2nd ed.). Retrieved from http://www.y12.doe.gov/sgml/sc34/document/0322_files/iso13250-2nd-ed-v2.pdf

Jonassen, D. H. (1992). Semantic networking as cognitive tools. In P. A. M. Kommers, D. H. Jonassen, & J. T. Mayes (Eds.), *Cognitive tools for learning* (pp. 19-21). Berlin, Germany: Springer-Verlag.

Kommers, P. A. M., Jonassen, D. H., & Mayes, J. T. (Eds.). (1992). *Cognitive tools for learning.* Berlin, Germany: Springer-Verlag.

Moore, D. W., & Readance, J. E. (1984). A quantitative and qualitative review of graphic organizer research. *Journal of Educational Research, 78,* 11-17.

Novak, J. D. (1998). *Learning, creating, and using knowledge: Concept maps as facilitative tools in schools and corporations.* Mahwah, NJ: Lawrence Erlbaum Associates.

Pepper, S., & Moore, G. (Eds.). (2001). *XML topic maps (XTM) 1.0.* Retrieved from http://www.topicmaps.org/xtm/index.html

Quillian, M. R. (1968). Semantic memory. In M. Minsky (Ed.), *Semantic information processing* (pp. 227-270). Cambridge, MA: MIT Press.

Zeiliger, R., Reggers, T., & Peeters, R. (1996). Concept-map based navigation in educational hypermedia: A case study. *Proceedings of ED-MEDIA 96.* Retrieved from http://www.gate.cnrs.fr/~zeiliger/artem96.doc

KEY TERMS

Concept Map: A visual representation of knowledge of a domain consisting of nodes representing concepts, objects, events, or actions interconnected by directional links that define the semantic relationships between and among nodes.

Knowledge Management: Organizing knowledge, information, and information resources and providing access to such knowledge and information in such a manner as to facilitate the sharing, use, learnability, and application of such knowledge and resources.

Knowledge Map: See Concept Map.

Knowledge Representation: According to Brachman and Levesque (1985, p. xiii), it is "writing down, in some language or communicative medium, descriptions or pictures that correspond...to the world or a state of the world."

Knowledge Visualization: A visual (or other sense-based) representation of knowledge; a portrayal via graphical or other sensory means of knowledge, say, of a particular domain, making that knowledge explicit, accessible, viewable, scrutable, and shareable.

Proposition: A statement, assertion, or declaration formulated in such a manner that it can be judged to be true or false.

Semantic Network: A knowledge-representation formalism from the cognitive-science community (understood by cognitive psychologists to represent actual cognitive structures and mechanisms, and used in artificial-intelligence applications) consisting primarily of textually labeled nodes representing objects, concepts, events, actions, and so forth, and textually labeled links between nodes representing the semantic relationships between those nodes.

ENDNOTES

1. Flash™is a trademark of Macromedia, Inc. Inspiration® is a registered trademark of Inspiration Software, Inc. http://www.inspiration.com/
2. Axon is a product of Axon Research. http://web.singnet.com.sg/~axon2000/index.htm
3. Decision Explorer® is a registered trademark of Banxia® Software. http://www.banxia.com/demain.html
4. SemNet® is a registered trademark of the SemNet Research Group. http://www.biologylessons.sdsu.edu/about/aboutsemnet.html
5. SMART Ideas™ is a trademark of SMART Technologies, Inc. http://www.smarttech.com/Products/smartideas/index.asp
6. Institute for Human & Machine Cognition's CmapTools, http://cmap.ihmc.us/
7. http://hypergraph.sourceforge.net/example_tm.html
8. http://www.mondeca.com/english/documents.htm

C

Computer-Supported Collaborative Learning

Vladan Devedžić
University of Belgrade, Serbia and Montenegro

INTRODUCTION

In computer-supported collaborative learning (CSCL), information and communication technologies are used to promote connections between one learner and other learners, between learners and tutors, and between a learning community and its learning resources. CSCL is a coordinated, synchronous activity of a group of learners resulting from their continued attempt to construct and maintain a shared conception of a problem (Roschelle & Teasley, 1995).

CSCL systems offer software replicas of many of the classic classroom resources and activities (Soller, 2001). For example, such systems may provide electronic shared workspaces, on-line presentations, lecture notes, reference material, quizzes, student evaluation scores, and facilities for chat or online discussions. This closely reflects a typical collaborative learning situation in the classroom, where the learners participating to learning groups encourage each other to ask questions, explain and justify their opinions, articulate their reasoning, and elaborate and reflect upon their knowledge, thereby motivating and improving learning.

These observations stipulate both the *social context* and the *social processes* as an integral part of collaborative learning activities. In other words, CSCL is a natural process of *social interaction* and *communication* among the learners in a group while they are learning by solving common problems.

BACKGROUND

Theory

Collaborative learning is studied in many learning theories, such as Vygotsky's socio-cultural theory—zone of proximal development (Vygotsky, 1978), in constructivism, self-regulated learning, situated cognition, cognitive apprenticeship, cognitive flexibility theory, observational learning, distributed cognition, and many more (see Andriessen, Baker, & Suthers, 2003; Dillenbourg, Baker, Blaye, & O'Malley, 1996; Roschelle & Teasley, 1995; TIP, 2004, for a more comprehensive insight). A number of researchers have shown that effective collaboration with peers is a successful and powerful learning method—see, for example Brown and Palincsar (1989), Doise, Mugny, and Perret-Clermont (1975), Dillenbourg et al. (1996), and Soller (2001). However, there is an important prerequisite for collaborative learning to result in improved learning efficiency and bring other learning benefits—the group of learners must be active and well-functioning. Just forming a group and placing the students in it does not guarantee success. The individual learners' behaviour and active participation is important, and so are their roles in the group, their motivation, their interaction, and coordination. Soller (2001) makes an important observation that "while some peer groups seem to interact naturally, others struggle to maintain a balance of participation, leadership, understanding, and encouragement."

One should differentiate between cooperative and collaborative learning. In cooperative learning, the learning task is split in advance into sub-tasks that the partners solve independently. The learning is more directive and closely controlled by the teacher. On the other hand, collaborative learning is based on the idea of building a consensus through cooperation among the group members; it is more student-centered than cooperative learning.

The Goals of CSCL

The goals of CSCL are three-fold:

- **Personal:** By participating in collaborative learning, the learner attains elimination of misconceptions, more in-depth understanding of

the learning domain, and development of self-regulation skills (i.e., metacognitive skills that let the learner observe and diagnose his/her self-thinking process and self-ability to regulate or control self-activity);
- **Support Interaction:** Maintaining interaction with the other learners, in order to attain the personal goal associated with the interaction; this leads to learning by self-expression (learning by expressing self-thinking process, such as self-explanation and presentation), and learning by participation (learning by participating as an apprentice in a group of more advanced learners);
- **Social:** The goals of the learning group as a whole are setting up the situation for peer tutoring (the situation to teach each other), as well as setting up the situation for sharing cognitive or metacognitive functions with other learners (enabling the learners to express their thinking/cognitive process to other learners, to get advise from other learners, discuss the problem and the solution with the peers, and the like).

Web-Based Education

Web-based education has become a very important branch of educational technology. For learners, it provides access to information and knowledge sources that are practically unlimited, enabling a number of opportunities for personalized learning, tele-learning, distance-learning, and collaboration, with clear advantages of classroom independence and platform independence. On the other hand, teachers and authors of educational material can use numerous possibilities for Web-based course offering and teleteaching, availability of authoring tools for developing Web-based courseware, and cheap and efficient storage and distribution of course materials, hyperlinks to suggested readings, digital libraries, and other sources of references relevant for the course.

Adaptivity and Intelligence

Typically, an adaptive educational system on the Web collects some data about the learner working with the system and creates the learner model (Brusilovsky, 1999). Further levels of adaptivity are achieved by using the learner model to adapt the presentation of the course material, navigation through it, its sequencing, and its annotation, to the learner. Furthermore, collaborative Web-based educational systems use models of different learners to form a matching group of learners for different kinds of collaboration. This kind of adaptivity is called *adaptive collaboration support*. Alternatively, such systems can use intelligent class monitoring to compare different learner models in order to find significant mismatches, for example, to identify the learners who have learning records essentially different from those of their peers. These learners need special attention from the teacher and from the system, because their records may indicate that they are progressing too slow, or too fast, or have read much more or much less material than the others, or need additional explanations.

Intelligence in a Web-based educational system nowadays usually means that the system is capable of demonstrating some form of knowledge-based reasoning in curriculum sequencing, in analysis of the learner's solutions, and in providing interactive problem-solving support (possibly example-based) to the learner. Most of these intelligent capabilities exist in traditional intelligent tutoring systems as well, and were simply adopted in intelligent Web-based educational applications and adapted to the Web technology.

CSCL Model

CSCL technology is not a panacea. Learners who use it need guidance and support online, just as students learning in the classroom need support from their instructor. Hence, developers of CSCL tools must ensure that collaborative learning environments support active online participation by remote teachers, as well as a variety of means for the learners to deploy their social interaction skills to collaborate effectively.

In order for each CSCL system to be effective, it must be based on a certain model, such as the one suggested by Soller (2001) that integrates the following four important issues:

- indicators of effective collaborative learning;
- strategies for promoting effective peer interaction;
- technology (tools) to support the strategies and;
- a set of criteria for evaluating the system.

CSCL system should recognize and target group interaction problem areas. It should take actions to help the learners collaborate more effectively with their peers, improving individual and group learning.

Figure 1 shows the typical context of CSCL on the Web. A group of learners typically uses a CSCL system simultaneously. The system runs on one or more *educational servers*. The learners' activities are focused on solving a problem in the CSCL system domain collaboratively. A human teacher can participate in the session too, either by merely monitoring the learners' interactions and progress in solving problems, or by taking a more active role (e.g., providing hints to the learners, suggesting modes of collaboration, discussing the evolving solution, and so on). Intelligent *pedagogical agents* provide the necessary infrastructure for knowledge and information flow between the clients and the servers. They interact with the learners and the teachers and collaborate with other similar agents in the context of interactive learning environments (Johnson, Rickel, & Lester, 2000). Pedagogical agents very much help in locating, browsing, selecting, arranging, integrating, and otherwise using educational material from different educational servers.

MAIN ISSUES IN CSCL

The Types of Interaction

Since the issue of interaction is central to CSCL, it is useful to introduce the types of interaction the learner typically meets when using such systems (Curtis & Lawson, 2001):

- interaction with resources (such as related presentations and digital libraries);
- interaction with teachers (teachers can participate in CSCL sessions);
- interaction with peers (see the above description of the goals of CSCL) and;
- interaction with interface (this is the most diverse type of interaction, ranging from limited text-only interactions, to the use of specific software tools for dialogue support, based on dialogue interaction models, to interaction with pedagogical agents [see Figure 1]).

The Kinds of CSCL

Starting from the theory of collaborative learning and applying it along with AI techniques to CSCL systems on the Web, the research community has made advances in several directions related to collaboration in learning supported by Web technologies:

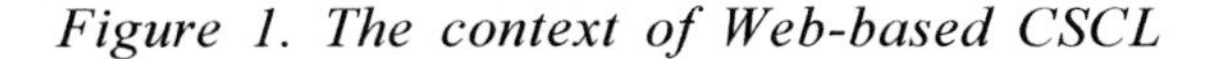

Figure 1. The context of Web-based CSCL

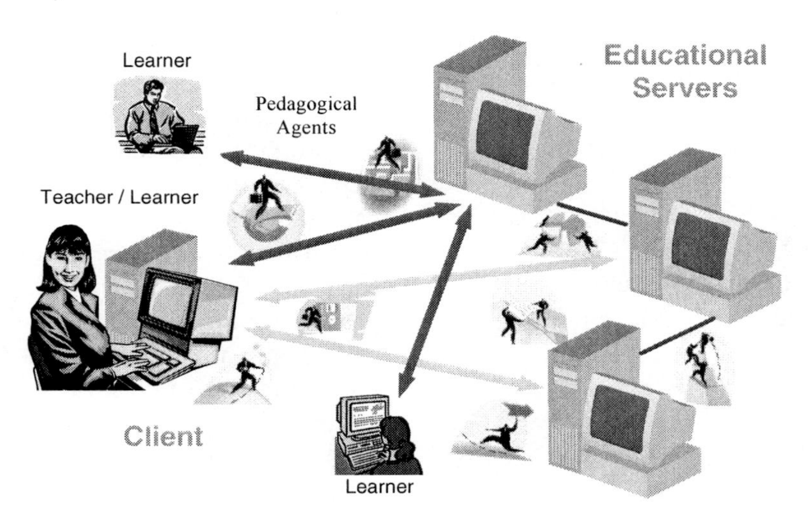

- **Classical CSCL:** This comprises setting up CSCL in Web classrooms, as well as infrastructure for CSCL in distance learning;
- **Learning Companions:** These are artificial learners, for example, programs that help human learners learn collaboratively if they want so, even when no other peer learners are around;
- **Learning Communities:** Remote learners can communicate intensively not only by solving a problem in a group, but also by sharing common themes, experiences, opinions, and knowledge on the long run;
- **Web Services:** This general and extremely popular recent technology is nowadays used in learning situations as well;
- **Hybrid Modes:** Some, or even all of the above capabilities can be supported (at least to an extent) in an intelligent Web-based CSCL system.

Design Issues

It is quite understood that the learning process is more effective if the user interface is designed to be intuitive, easy-to-use, and supportive in terms of the learners' cognitive processes. With CSCL systems, additional flexibility is required. The learners have to work collaboratively in a shared workspace environment, but also use private workspaces for their own work. Moreover, since work/learning happens in small groups, the interface should ideally support the group working in one environment, or in synchronous shared environments. It also must support sharing of results, for example, exchanging settings and data between the groups and group members, as well as demonstrating the group's outcomes or conclusions. A suitable way to do it is by using a public workspace.

This division of learning/work into shared and private workspaces leads to the idea of workspaces that can contain a number of transparent layers (Pinkwart, 2003). The layers can have "solid" objects (synchronizeable visual representations), that is, handwriting strokes or images. Also, the layers can be private or shared, for example, a private handwriting layer used for personal annotations.

Group Formation

If for any reason a learner wants to participate in collaborative learning on the Web, the learning efficiency depends on joining an appropriate learning group. Hence the question, "How to form a group?" is important.

Opportunistic group formation (OGF) is a framework that enables pedagogical agents to initiate, carry out, and manage the process of creating a learning group when necessary and conducting the learner's participation to the group. Agents in OGF support individual learning, propose shifting to collaborative learning, and negotiate to form a group of learners with appropriate role assignment, based on the learners' information from individual learning.

In OGF, collaborative learning group is formed dynamically. A learner is supposed to use an intelligent, agent-enabled Web-based learning environment. When an agent detects a situation for the learner to shift from individual to collaborative learning mode (a "trigger," such as an impasse or a need for review), it negotiates with other agents to form a group. Each group member is assigned a reasonable learning goal and a social role. These are consistent with the goal for the whole group.

APPLICATIONS

Two practical examples of CSCL systems described in this section illustrate the issues discussed earlier.

COLER

COLER is an intelligent CSCL system for learning the principles of entity-relationship (ER) modelling in the domain of databases (Constantino-González, Suthers, & Escamilla de los Santos, 2003). The learners using the system through the Web solve a specific ER modelling problem collaboratively. They see the problem's description in one window and build the solution in another one, which represents a shared workspace. Each learner also has his/her own private workspace in which he/she builds his/her own solution and can compare it to the evolving group solution in the shared workspace. A learner can invoke a personal coach (an intelligent pedagogical agent) to help him/her solve the problem and contribute to the group solution. In addition to such guidance, there is also a dedicated HELP button to retrieve the basic principles of ER modelling if necessary. At any time during the problem solving, a learner can see in a specially-designated panel on the screen which team-mates are already connected and can ask for floor control. When granted control, the learner contributes to the group solution in the shared workspace by, for example, inserting a new modelling element. He/she can also express feelings about the other team-mates' contributions through a designated opinion panel, and can also engage in discussion with them and with the coach through a chat communication window.

Cool Modes

The idea of using transparent layers in the design of user interface is best exemplified in the intelligent Web-based CSCL called Cool Modes (COllaborative Open Learning, MOdelling and DEsigning System) (Pinkwart, 2003). The system supports the Model Facilitated Learning (MFL) paradigm in different engineering domains, using modelling tools, construction kits, and system dynamics simulations. The focus of the learning process is on the transformation of a concrete problem into an adequate model. The shared workspace, Figure 2, is public and looks the same to all the learners in the group. However, the handwritten annotations are placed in private layers and can be seen only by individual learners. Cool Modes also provides "computational objects to think with" in a collaborative, distributed framework. The objects have a specified domain-related functionality and semantics, enriched with rules and interpretation patterns. Technically, Cool Modes is integrated with visual modelling languages and has a set of domain-specific palettes of such objects (see the palette on the right-hand side of Figure 2). The palettes are defined externally to encapsulate domain-dependent semantics and are simply plugged-in the system when needed. Currently, the system has palette support for modelling stochastic processes, system dynamics, Petri nets, and other engineering tools, as well as for learning Java.

FUTURE TRENDS

Open Distributed Learning Environments

There is an important trend in software architectures for CSCL—*open distributed learning environments* (Muehlenbrock & Hoppe, 2001). The idea here is that learning environments and support systems are not conceived as self-containing, but as embedded in realistic social and organizational environments suitable for group learning, such as Web classrooms. Different Web classrooms can be interconnected among themselves letting the learners communicate with the peers and teachers not physically present in the same classroom, but logged onto

Figure 2. A screenshot from Cool Modes (after Pinkwart, 2003)

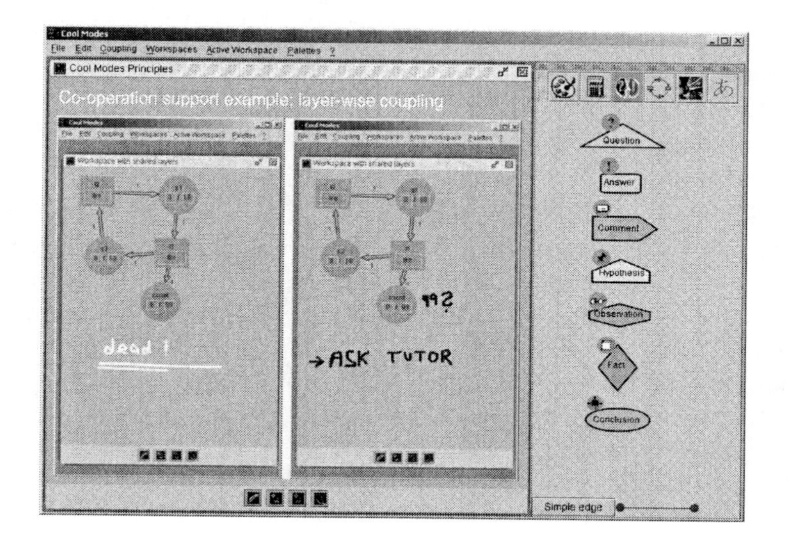

the same network/application. Moreover, the CSCL application can link the learners with other relevant learning resources or remote peers and tutors through the Internet.

CSCL and the Semantic Web

Semantic Web (SemanticWeb.org, 2004) is the new-generation Web that makes possible to express information in a precise, machine-interpretable form, ready for software agents to process, share, and reuse it, as well as to understand what the terms describing the data mean. It enables Web-based applications to interoperate both on the syntactic and semantic level. The key components of the Semantic Web technology are ontologies of standardized terminology that represent domain theories; each ontology is a set of knowledge terms, including the vocabulary, the semantic interconnections, and some simple rules of inference and logic for some particular topic.

At the moment, educational ontologies are still scarce—developing ontologies of high usability is anything but easy, and the Semantic Web is around for just a couple of years. Still, CSCL community has already ventured in developing CSCL ontologies. In their pioneering but extremely important work, Supnithi, Inaba, Ikeda, Toyoda, and Mizoguchi (1999) have made a considerable effort towards developing the *collaborative learning ontology* (CLO). Although still not widely used, CLO clarifies the concepts of a collaborative learning group, and the relations among the concepts. It answers general questions like:

- What kinds of groups exist in collaborative learning?
- Who is suitable for attaining the group?
- What roles should be assigned to the members?
- What is the learning goal of the whole group?

CONCLUSION

This article surveyed important issues in CSCL in the context of intelligent Web-based learning environments. Current intelligent Web-based CSCL systems integrate a number of Internet and artificial intelligence technologies. This is not to say that learning theories and instructional design issues should be given lower priority than technological support. On the contrary, new technology offers more suitable ways for implementing and evaluating instructional expertise in CSCL systems.

REFERENCES

Andriessen, J., Baker, M., & Suthers, D. (Eds.). (2003). *Arguing to learn: Confronting cognitions in computer-supported collaborative learning environments*. Kluwer book series on Computer Supported Collaborative Learning, Pierre Dillenbourg (Series Editor). Dordrecht: Kluwer.

Brown, A., & Palincsar, A. (1989). Guided, cooperative learning and individual knowledge acquisition. In L. Resnick (Ed.), *Knowledge, learning and instruction* (pp. 307-336). Mahwah, NJ: Lawrence Erlbaum Associates.

Brusilovsky, P. (1999). Adaptive and intelligent technologies for Web-based education. *Künstliche Intelligenz,* 4, 19-25.

Constantino-Gonzalez, M. A., Suthers, D. D., & Escamilla de los Santos, J. G. (2003). Coaching Web-based collaborative learning based on problem solution differences and participation. *International Journal of Artificial Intelligence in Education,* 13, 263-299.

Curtis, D. D., & Lawson, M. J. (2001). Exploring collaborative online learning. *Journal of Asynchronous Learning Networks, 5*(1), 21-34.

Dillenbourg, P. (1999). What do you mean by "collaborative learning?" In P. Dillenbourg (Ed.), *Collaborative learning: Cognitive and computational approaches* (pp. 1-19). Amsterdam, The Netherlands: Elsevier Science.

Dillenbourg, P., Baker, M. J., Blaye, A., & O'Malley, C. (1996). The evolution of research on collaborative learning. In P. Reimann, & H. Spada (Eds.), *Learning in humans and machines: Towards an interdisciplinary learning science* (pp. 189-211). Oxford: Pergamon.

Dillenbourg, P., & Schneider, D. (1995). Collaborative learning and the internet. [Online]. Retrieved

November 2004, from http://tecfa.unige.ch/tecfa/research/CMC/colla/iccai95_1.html

Doise, W., Mugny, G., & Perret-Clermont A. (1975). Social interaction and the development of cognitive operations. *European Journal of Social Psychology, 5*(3), 367-383.

Johnson, W. L., Rickel, J., & Lester, J. C. (2000). Animated pedagogical agents: Face-to-face interaction in interactive learning environments. *International Journal of Artificial Intelligence in Education, 11*, 47-78.

Muehlenbrock, M., & Hoppe, U. (2001, May). A collaboration monitor for shared workspaces. In the *Proceedings of the International Conference on Artificial Intelligence in Education AIED-2001*, San Antonio, Texas (pp. 154-165). Amsterdam, The Netherlands: IOS Press.

Pinkwart, N. (2003). A plug-in architecture for graph based collaborative modelling systems. In the *Proceedings of the 11th Conference on Artificial Intelligence in Education*, Sydney, Australia (pp. 535-536). Amsterdam, The Netherlands: IOS Press.

Roschelle, J., & Teasley, S. (1995). The construction of shared knowledge in collaborative problem solving. In C. O'Malley (Ed.), *Computer-supported collaborative learning* (pp. 69-97). Berlin: Springer-Verlag.

SemanticWeb.org (2004). *The semantic Web community portal.* [Online]. Retrieved December 2004, from http://www.semanticweb.org

Soller, A. L. (2001). Supporting social interaction in an intelligent collaborative learning system. *International Journal of Artificial Intelligence in Education, 12*, 54-77.

Supnithi, T., Inaba, A., Ikeda, M., Toyoda, J., & Mizoguchi, R. (1999, July). Learning goal ontology supported by learning theories for opportunistic group formation. In the *Proceedings of the International Conference on Artificial Intelligence in Education AIED-1999*, Le Mans, France (pp. 263-272). Amsterdam, The Netherlands: IOS Press.

TIP (2004). *Explorations in learning & instruction: The theory into practice database.* [Online]. Retrieved December 2004, from http://tip.psychology.org/theories.html

Vygotsky, L. (1978). *Mind in society: The development of higher psychological processes.* Cambridge, MA: Harvard University Press.

KEY TERMS

Adaptive Collaboration Support in CSCL: Using models of different learners to form a matching group of learners for different kinds of collaboration.

Computer-Supported Collaborative Learning (CSCL): The process related to situations in which two or more subjects build synchronously and interactively a joint solution to some problem (Dillenbourg, 1999; Dillenbourg et al., 1996; Dillenbourg & Schneider, 1995).

Group Formation: The process of creating a suitable group of learners to increase the learning efficiency for both the individual peers and the group as a whole.

Pedagogical Agents: Autonomous software entities that support human learning by interacting with learners and teachers and by collaborating with other similar agents, in the context of interactive learning environments (Johnson et al., 2000).

Private Workspace: Part of the CSCL system, usually represented as a designated window in the system's GUI, where a member of the learning group builds his/her own solution of the problem the group solves collaboratively, and where he/she can also take notes, consider alternative solutions, and prepare contributions to the group solution.

Shared Workspace: Part of the CSCL system, usually represented as a designated window in the system's GUI, where the members of the learning group build the joint solution to the problem they solve collaboratively.

Conceptual Models and Usability

Ritchie Macefield
Staffordshire University, UK

INTRODUCTION

The study of conceptual models is both a complex and an important field within the HCI domain. Many of its key principles resulted from research and thinking carried out in the 1980s, arguably in the wake of Norman (1983). Since then, the importance of conceptual models in affecting the usability of an Information and Communication Technology (ICT) system has become well-established (e.g., they feature prominently in the widely cited design guidelines for interfaces defined by Norman [1988], which are summarized in Figure 1 by Lienard [2000]).

Today, most HCI professionals are able to attribute significant meaning to the term *conceptual model* and to recognize its importance in aiding usability. However, two problems seem to prevail. First, some HCI researchers and practitioners lack a precise understanding of conceptual models (and related ideas), and how they affect usability. Second, much of the research in this field is (necessarily) abstract in nature. In other words, the study of conceptual models is itself highly conceptual, with the result that practitioners may find some of the theory difficult to apply.

This article is designed to help both researchers and practitioners to better understand the nature of conceptual models and their role in affecting usability. This includes explaining and critiquing both contemporary and (possible) future approaches to leveraging conceptual models in the pursuit of improved usability.

BACKGROUND

Key to understanding the role of conceptual models in promoting usability are clear definitions of these terms, related ideas, and their appropriate contextualization within the HCI domain.

Definitions of Usability

Probably the first widely cited definition of *usability*, as it applies to ICT systems, was established by Shackel (1991) and is shown in Figure 2.

The definition provided by Shackel (1991) is reasonably comprehensive, which is one reason it remains useful today. However, a more concise

Figure 1. Design guidelines for interfaces defined by Norman (1988), summarized by Lienard (2000)

A. Good **visibility** means you can: • tell the state of the system by looking at it • tell what the alternatives for actions are • identify controls to make the system perform the available actions	+
B. Good **conceptual models** provide: • consistent presentation of the system's state • consistent controls, possible actions, and results	**System image =**
C. Good **mappings** mean you can determine the: • relationship between actions and results • relationship between controls and actions • system state from what is visible	**User's model of system**
D. Good **feedback** involves: • full and continuous presentation of the results of actions • timely (i.e., rapid) response times	**A "good" user model makes the user feel:** • **in control of the system** • **confident of getting the required result(s)**

Figure 2. Definition of usability by Shackel (1991)

Effectiveness Improvement in task performance in terms of speed and/or error rate by a given percentage of the population within a given range of the user's tasks (related to the user's environment) **Learnability** Within some specified time from commissioning and start of user training based upon some specified amount of user support and within some specified relearning time each time for intermittent users **Flexibility** With flexibility allowing adaptation to some specified percentage variation in task and/or environments beyond those first specified **Attitude** Within acceptable levels of human cost in terms of tiredness, discomfort, frustration and personal effort, so that satisfaction causes continued and enhanced usage of the system

definition was established in ISO 9241-11:1998 and is summarized by Maguire (1998):

- **Effectiveness:** How well the user achieves the goals he or she sets out to achieve using the system.
- **Efficiency:** The resources consumed in order to achieve his or her goals.
- **Satisfaction:** How the user feels about his use of the system.

These definitions are widely cited. However, the ISO 9241-11:1998 arguably has superseded that of Shackel (1991) and is, therefore, used throughout the remainder of this article.

Definitions of a Conceptual Model

The word *model* implies an abstraction of the subject matter, or artefact, being modeled. This is true whether that artefact is an ICT system, a motorcycle, or a house. Inevitably, a model lacks the full detail present within an actual artefact, so, in producing the model, some properties of the artefact will be ignored or simplified. The particular abstraction will depend on the (intended) use of the model (e.g., a technical drawing of a motorcycle used in its manufacture abstracts different properties from that of an artist's sketch used in a sales brochure). Similarly, a usability engineer may model only those properties of an ICT system concerned with its interface, while a technical architect might model in terms useful to coding the system. In both cases, the subject matter is common, yet the abstractions and resulting models are very different.

The word *conceptual* stems from the word *concept*. This also implies some form of abstraction and, hence, a model. In psychology-oriented fields, this term may be used synonymously with the word *idea* and, therefore, has connotations relating to cognition, perception, innovation, and, most importantly, models stored in the *mind*. Alternatively, in (product) design-oriented fields, a *conceptual model* is more likely to be interpreted as an abstraction concerned only with the key or fundamental properties of an artefact (i.e., a model considerably lacking detail). Further, such models typically are expressed in concrete terms (e.g., a designer's sketch, clay model, or engineer's prototype).

The HCI domain incorporates principles related to both psychology and (product) design (an ICT system is a product). Similarly, both of the two (overlapping) definitions of a conceptual model presented have relevance here.

Mental Models

More in keeping with a (product) design-oriented view of conceptual models, we might define and express the conceptual model of an ICT system in concrete terms, using devices such as storyboards and Entity Relationship Diagrams (ERDs). However, these conceptual models only can be utilized once inside our minds (i.e., once converted into a *mental model*). Indeed, most cognitive scientists agree that our entire perception of the world, including ourselves, is constructed from models within our minds. Further, we only can interact with the world through these mental models. This is an insight generally credited to Craik (1943), although its origins can be traced back to Plato.

Mental models inevitably are incomplete, constantly evolving, and contain errors (Khella, 2002). They usefully can be considered as an *ecosystem*, a term used by Ratey (2002) to describe the brain, which, of course, stores and processes our mental models. This means that particular models can come and go, increase or decrease in accuracy, and constantly mutate and adapt, both as a result of internal processing and in response to external stimuli.

A person may maintain simultaneously two or more compatible mental models of the same subject matter, as with the example of the technical drawing and the artist's sketch used for a motorcycle (Khella, 2002). Similarly, it is possible for a person to maintain simultaneously two or more contradictory mental models, a condition known as *cognitive dissonance* (Atherton, 2002).

In the early 1980s, the idea was established that a person may acquire and maintain two basic types of mental models for an ICT system: a functional model and a structural model. *Functional models*, also referred to as task-action mapping models, are concerned with how users should interact with the system in order to perform the desired tasks and achieve their goals. ICT professionals such as usability engineers typically are concerned with this type of model. *Structural models* are concerned more with the internal workings and architecture of the system and on what principles it operates (i.e., how a system achieves its functionality). This type of model is generally the concern of ICT professionals such as systems architects and coders. Of course, this fits well with the idea that an individual may maintain simultaneously two or more compatible mental models—an informed user of an ICT system may have both a good functional and structural mental model of the system.

MENTAL MODELS AND USABILITY

The arguments for a user possessing a good mental model of the ICT system they are using can be expressed using the elements of usability defined in ISO 9241-11:1998:

- **Efficiency:** Users with a good mental model will be more efficient in their use of an ICT system, because they already will understand the (optimal) way of achieving tasks; they will not have to spend time learning these mechanisms.
- **Effectiveness:** Users will be more effective due to their understanding of the system's capabilities and the principles by which these capabilities may be accessed.
- **Satisfaction:** As a result of increased efficiency and effectiveness and because users can predict more successfully the behavior of the system, they also are likely to be more satisfied when using the system.

Conversely, users with a largely incomplete, distorted, or inaccurate mental model may experience one or more of the following usability problems, which again, are categorized using the elements of usability defined in ISO 9241-11:1998:

- **Efficiency:** The user may execute functions in a (highly) suboptimal way (e.g., not utilizing available shortcuts).
- **Effectiveness:** The user's understanding of the system will be limited detrimentally in scope, so (potentially) useful functionality might be hidden from him or her.
- **Satisfaction:** The user's mental model may work (to a certain degree) until task complexity increases or until completely new tasks are required. At this stage, catastrophic model failure may occur (e.g., a user fails to predict the consequences of a particular input to the system and gains an outcome very different from that expected). Such failure can be quite disastrous, leaving users confused and demotivated. Indeed, this is one pathology of so-called *computer rage*, a term popular at the time of writing.

These factors explain why the highest level of usability occurs when an individual operates a system that they have designed themselves. Here, the designer and user's model of the system can be identical (since they belong to the same person). This also explains why these user-designed systems very often fail when rolled out to other users (Eason, 1988). These systems often are successful in the designer's hands, but this success is attributed to the design of the system itself rather than simply being a consequence of the fact that the designer inevitably has an excellent mental model of (how to use) the system.

Given this, it seems obvious that a user's mental model of an ICT system benefits from being as comprehensive and accurate as possible. This is why Norman (1988) included this element in his design guidelines for interfaces (summarized in Figure 1 by Lienard [2000]). The question then is how do we develop and exploit these models to promote usability?

The Intuitive Interface Approach to Developing Users' Mental Models

Today, many ICT systems are designed with the anticipation (or hope) that users will be able to (learn how to) operate them within a matter of minutes or even seconds. Further, these users often are anonymous to the system designers. This is particularly true in the case of pervasive ICT systems such as those found on the World Wide Web (WWW).

Many contemporary practitioners propose that quick and easy access to pervasive ICT systems can be achieved by designing an *intuitive interface*. This term is widely interpreted to mean that a user can operate a system by some innate or even nearly supernatural ability (Raskin, 1994). However, this notion is ill founded in the HCI domain (Norman, 1999; Raskin, 1994) and lacks supporting empirical evidence (Raskin, 1994). Rather, so-called intuitive interfaces simply rely on the fact that users already possess mental models that are sufficiently relevant to a system such that they can (at least) begin to use the system. In other words, the term *intuitive* simply implies *familiar* (Raskin, 1994). This familiarity is often exploited through the use of metaphors, whereby a mental model that was developed for use in the physical world (e.g., Windows) is leveraged by the system designers to aid its use. Important in this approach is the idea (or hope) that, from a position of (some) familiarity, users then are able to develop their understanding of the system by self-exploration and self-learning, or *self-modeling*.

Norman (1981) hypothesized that if users are left to self-model in this way, they always will develop a mental model that explains (their perception of) the ICT system, and research carried out by Bayman and Mayer (1983) supports this hypothesis. Norman (1981) also argued that, in these situations, the mental models developed by the users are likely to be incorrect (e.g., a user simply may miss the fact that a metaphor is being used or how the metaphor translates into the ICT system (a scenario that is likely to result in the usability problems cited earlier). Another problem with this approach is that with pervasive ICT systems where users are often anonymous, it can be extremely difficult to predict accurately what mental models already are possessed by the target user group.

The Conceptual Model Approach to Developing Users' Mental Models

An alternative approach to exploiting mental models is to explicitly provide users with a conceptual model of the ICT system that accurately reflects its true properties. This conceptual model approach was advocated in hypothetical terms by Norman (1981), Carroll and Olson (1988), and Preece (1994). In practice, this approach generally is realized by presenting the users with suitable schemas or metaphors relevant to the system. In relation to this, Norman (1983) offered some revised definitions of modeling terminology—he distinguished between the tangible conceptual model that is provided to the user as an explanation of the system, which might use, for example, a story board or ERD and the resulting mental model that is formed in the user's mind. This distinction is useful and will be used throughout the remainder of this article.

With the conceptual model approach, it seems unlikely that a user's mental model will overlap completely with the conceptual model presented. This is because the formation of mental models is a highly complex process involving human beings with all of their individual abilities, preferences, and idiosyncrasies. Further, determining the degree of overlap is somewhat problematic, since it is notoriously difficult to elicit and understand the mental model a user has of an ICT system and how it is being exploited during interaction. Indeed, many studies have attempted to do this and have failed (Preece, 1994; Sasse, 1992). Given this, it is difficult to prove a casual link between the conceptual model approach and increased usability. However, a large number of studies has demonstrated that, when users are explicitly presented with accurate conceptual models, usability can be improved significantly. These studies include Foss, et. al. (1982), Bayman and Mayer (1983), Halasz and Moran (1983), Kieras and Bovair (1984), Borgman (1986), and Frese and Albrecht (1988). As an example, Borgman (1986) showed how users could better operate a library database system after being provided with a conceptual model of the system that utilized a card index metaphor, as compared with users in a control group who were taught in terms of the operational procedures required to achieve specific goals and tasks. Further, research from Halasz (1984) and Borgman (1986) demonstrated that the usability benefits of the conceptual model approach increase with task complexity.

While these studies demonstrated well the usability benefits of the conceptual model approach, they were limited in that the conceptual models were explained through some form of face-to-face teaching. This presents two interrelated problems within the context of modern-day pervasive ICT systems. First, this type of education is expensive. Second, the user population may be diverse and largely unknown to the system vendors. In summary, this sort of face-to-face approach is often unviable within this context.

FUTURE TRENDS

Progression in this field might be sought in two important ways. First, to address the limitations of the conceptual model approach cited previously, it would be useful to establish viable means of presenting users with conceptual models when the ICT system is pervasive. Second, the opportunity exists to develop better conceptual models with which to explain ICT systems.

Online Conceptual Models

A means of providing conceptual models where the ICT system is pervasive is through the use of online digital presentations. These might constitute a type of mini-lecture about the conceptual model for the system, perhaps utilizing (abstractions of) the very schemas used to design the system (e.g., storyboards and ERDs or suitable metaphors). This is a similar idea to online help, except that the user support is much more conceptual and self-explanatory in nature.

Many organizations produce (online) digital presentations to complement their products and services. However, such presentations typically are exploited to sell the system rather than to explain its use. There are digital presentations that are educationally biased. These include a vast amount of computer-based training (CBT) resources available both online (WWW) and in compact disk (CD) format. However, these tend to focus on how particular tasks are performed rather than developing a deep understanding of the concepts that underpin the system. Similarly, there are some good examples of how digital presentations have been used to communicate concepts (e.g., EDS). However, these presentations generally are not directed at using a specific ICT system or for use by the typical user.

Site maps in WWW-based systems are (arguably) an example of online devices designed to convey conceptual understanding. However, while their use is now widespread and these devices are sometimes useful, they are limited in the scope of what they convey and their ability to explain themselves. This is in contrast to an online presentation specifically designed to explain a suitable conceptual model of a system.

Better Conceptual Models

Works related to both of the approaches discussed in the previous section (intuitive interface and conceptual model) share some similarity in that they both focus on the user's knowledge of the system's interface and, therefore, the development of functional mental models. This might be expected in an HCI-related field. However, over reliance on (just) functional mental models inevitably limits the user's understanding of the system, and it can be argued that this consequentially limits usability.

Some researchers (e.g., Preece, 1994) have hypothesised that users might benefit greatly from (also) acquiring *structural* mental models of the systems they use. Such models may better help the users to anticipate the behavior of the system, particularly in new contexts of use and when the system is being used nearly at its performance limits. Indeed, the benefit of having a good structural mental model helps to explain why the user-designed ICT systems discussed earlier are so usable—the designer of a system inevitably has a good structural mental model. The problem with this approach is that simply inferring structural models through self-modeling is extremely difficult and likely to fail (Miyake, 1986; Preece, 1994). However, it may be possible to develop useful structural mental models in users by providing them with appropriate conceptual models.

Further, if users are provided with both structural and functional conceptual models with which to form their mental model, *triangulation* can take place—a user is able to think critically about whether the functional and structural models are complementary or are the source of cognitive dissonance; in which case users may seek greater clarity in their understanding. Indeed, such use of triangulation is an established principle of understanding any subject matter (Weinstein, 1995).

CONCLUSION

The merit in users having a good mental model of an ICT system would seem to be universally recognized and has been inferred by many research studies. Some professionals in this field might argue that progression has slowed since the 1980s and that the arguments and conclusions presented here might make a useful agenda for future research and consultancy. Specifically, we might proceed by encouraging users to develop more structural mental models of ICT systems. Similarly, presenting conceptual models using online digital presentations may be of key importance in improving the usability of pervasive ICT systems.

REFERENCES

Atherton, J. S. (2002). *Learning and teaching: Cognitive dissonance*. Retrieved February 19, 2003, from www.dmu.ac.uk/~jamesa/learning/dissonance.htm

Bayman, P., & Mayer, R. E. (1983). A diagnosis of beginning programmers' misconceptions of BASIC programming statements. *Communications of the Association for Computing Machinery (ACM), 26,* 677-679.

Borgman, C. L. (1986). The user mental model of an information retrieval system: An experiment on a prototype online catalogue. *International Journal of Man-Machine Studies, 24,* 47-64.

Carroll, J. M., & Olson J.R. (1988). Mental models in human computer interaction. In M. Helendar (Ed.), *Handbook of human computer interaction* (pp. 45-65). Amsterdam: Elsevier.

Craik, K. (1943). *The nature of explanation*. Cambridge: Cambridge University Press.

Eason, K. D. (1988). *Information technology and organisational change*. London: Taylor and Francis.

EDS. (n.d.). *Web services*. Retrieved February 14, 2004, from www.eds.com/thought/thought_leadership_web.shtml

Foss, D. J., Rosson, M. B., & Smith, P. L. (1982). Reducing manual labor: An experimental analysis of learning aids for a text editor. *Proceedings of the Human Factors in Computing Systems Conference*, Gaithersburg, Maryland.

Frese, M., & Albrecht, K. (1988). The effects of an active development of the mental model in the training process: Experimental results in word processing. *Behaviours and Information Technology, 7,* 295-304.

Halasz, F. G. (1984). *Mental models and problem solving using a calculator*. Doctoral thesis. Stanford, CA: Stanford University.

Halasz, F. G., & Moran, T. P. (1983). Mental models and problem solving using a calculator. *Proceedings of the Conference of the Association for Computing Machinery (ACM), Special Interest Group on Computer and Human Interaction (SIGCHI) and the Human Factors Society*, Boston.

Khella, K. (2002). *Knowledge and mental models in HCI*. Retrieved February 19, 2003, from www.cs.umd.edu/class/fall2002/cmsc838s/tichi/knowledge.html

Kieras, D. E., & Bovair, S. (1984). The role of a mental model in learning to operate a device. *Cognitive Science, 8,* 255-273.

Lienard, B. (2000). *Evaluation scheme*. Retrieved August 27, 2004, from www.ioe.ac.uk/brian/maict/evalGuide.htm

Maguire, M. (1998). *Dictionary of terms*. Retrieved April 27, 2002, from www.lboro.ac.uk/research/husat/eusc/g_dictionary.html

Miyake, N. (1986). Constructive interaction and the iterative process of understanding. *Cognitive Science, 10,* 151-177.

Norman, D. (1981). The trouble with UNIX. *Datamation, 27*, 139-150.

Norman, D. A. (1983). Some observations on mental models. In D. A. Gentner, & A. L. Stevens (Eds.), *Mental models* (pp. 15-34). Hillsdale, NJ: Erlbaum.

Norman, D. A. (1988). *The psychology of everyday things*. New York: Basic Books.

Norman, D. A. (1999). *Thinking beyond Web usability*. Retrieved April 27, 2002, from www.webword.com/interviews/norman.html

Preece, J. (1994). Humans and technology: Humans. In J. Preece et al. (Eds.), *Human computer interaction*. Wokingham, UK: Addison Wesley.

Raskin, J. (1994). Intuitive equals familiar. *Communications of the ACM, 37*(9), 17.

Ratey, J. J. (2002). *A user's guide to the brain: Perception, attention and the four theaters of the brain*. New York: Pantheon.

Sasse, M. A. (1992). Users' mental models of computer systems. In Y. Rogers, A. Rutherford, & P. Bibby (Eds.), *Models in the mind: Theory, perspective, and application* (pp. 226-227). London: Academic Press.

Shackel, B. (1991). Usability—Context, framework, definition, design and evaluation. In B. Shackel, & S. Richardson (Eds.), *Human factors for informatics usability* (p. 27). Cambridge, UK: Cambridge University Press.

Weinstein, K. (1995). *Action learning: A journey in discovery and development*. London: Harper Collins.

KEY TERMS

Cognitive Dissonance: The situation where a person simultaneously holds two contradictory models of the same subject matter (e.g., an ICT system).

Conceptual Model: A model concerned with key, or fundamental, properties of the system. Typically concerned with the rationale and scope of a system, what the system is designed to do, the basic principles on which the system operates. Also, the basic principles utilized in operating the system. Alternatively, the model specifically offered to a user, which explains these ideas (Norman, 1983).

Functional Model: A model concerned primarily with how to interact with a system, how it is operated.

Mental Model: A model stored and processed in a person's mind.

Model: A simplified abstraction that shows properties of some subject matter relevant to a particular purpose, context, or perspective.

Self-Modeling: The process whereby an unaided user develops his or her own mental model of a system to explain its behavior, achieved through exploration or trial and error learning.

Structural Model: A model concerned primarily with the internal workings and architecture of the system and on what principles it operates and how a system achieves its functionality.

Task-Action Mapping Model: Synonym for functional model.

Triangulation: When a subject is viewed from more than one perspective during the learning or perceptual process.

Usability: Defined in ISO 9241-11:1998 as having three elements, as summarized by Maguire (1998): effectiveness—how well users achieve the goals they set out to achieve using the system; efficiency—the resources consumed in order to achieve their goals; and satisfaction—how users feel about their use of the system.

A Cooperative Framework for Information Browsing in Mobile Environment

Zhigang Hua
Chinese Academy of Sciences, China

Xing Xie
Microsoft Research Asia, China

Hanqing Lu
Chinese Academy of Sciences, China

Wei-Ying Ma
Microsoft Research Asia, China

INTRODUCTION

Through pervasive computing, users can access information and applications anytime, anywhere, using any device. But as mobile devices such as Personal Digital Assistant (PDA), SmartPhone, and consumer appliance continue to flourish, it becomes a significant challenge to provide more tailored and adaptable services for this diverse group. To make it easier for people to use mobile devices effectively, there exist many hurdles to be crossed. Among them is small display size, which is always a challenge.

Usually, applications and documents are mainly designed with desktop computers in mind. When browsing through mobile devices with small display areas, users' experiences will be greatly degraded (e.g., users have to continually scroll through a document to browse). However, as users acquire or gain access to an increasingly diverse range of portable devices (Coles, Deliot, & Melamed, 2003), the changes of the display area should not be limited to a single device any more, but extended to the display areas on all available devices.

As can be readily seen from practice, the simplest multi-device scenario is when a user begins an interaction on a first access device, then ceases to use the first device and completes the interaction using another access device. This simple scenario illustrates a general concern about a multi-device browsing framework: the second device should be able to work cooperatively to help users finish browsing tasks.

In this article, we propose a cooperative framework to facilitate information browsing among devices in *mobile environment*. We set out to overcome the display constraint in a single device by utilizing the cooperation of multiple displays. Such a novel scheme is characterized as: (1) establishing a communication mechanism to maintain *cooperative browsing* across devices; and (2) designing a *distributed user interface* across devices to cooperatively present information and overcome the small display area limited by a single device.

BACKGROUND

To allow easy browsing of information on small devices, there is a need to develop efficient methods to support users. The problems that occur in information browsing on the small-form-factor devices include two aspects: (1) how to facilitate information browsing on small display areas; and (2) how to help user's access similar information on various devices.

For the first case, many methods have been proposed for adapting various media on small display areas. In Liu, Xie, Ma, and Zhang (2003), the author proposed to decompose an image into a set of spatial-temporal information elements and generate an automatic image browsing path to display every image element serially for a brief period of time. In Chen, Ma, and Zhang (2003), a novel approach is devised to adapt large Web pages for tailored display on mobile device, where a page is organized into a two-level hierarchy with a thumbnail representation at the top level for providing a global view and index to a set of sub-pages at the bottom level for detail information. However, these methods have not considered utilizing multiple display areas in various devices to help information browsing.

For the second case, there exist a number of studies to search relevant information for various media. The traditional image retrieval techniques are mainly based on content analysis, such as those content-based image retrieval (CBIR) systems. In Dumais, Cutrell, Cadiz, Jancke, Sarin, and Robbins (2003), a desktop search tool called Stuff I've Seen (SIS) was developed to search desktop information including email, Web page, and documents (e.g., PDF, PS, MSDOC, etc.). However, these approaches have not yet taken into account the phase of information distribution in various devices. What's more, user interface needs further consideration such as to facilitate user's access to the information that distributes in various devices.

In this article, we propose a cooperative framework to facilitate user's information browsing in mobile environment. The details are to be discussed in the following sections.

OUR FRAMEWORK

Uniting Multiple Displays Together

Traditionally, the design of user interface for applications or documents mainly focus on desktop computers, which are commonly too large to display on small display areas of mobile devices. As a result, readability is greatly reduced, and users' interactions are heavily augmented such as continual scrolling and zooming.

However, as users acquire or gain access to an increasingly diverse range of the portable devices, the thing changes; the display area will not be limited to a single device any more, but extended to display areas on all available devices. According to existing studies, the user interface of future applications will exploit multiple coordinated modalities in contrast to today's uncoordinated interfaces (Coles et al., 2003). The exact combination of modalities will seamlessly and continually adapt to the user's context and preferences. This will enable greater mobility, a richer user experience of the Web application, and a more flexible user interface. In this article, we focus on overcoming display constraints rather than other *small form factors* (Ma, Bedner, Chang, Kuchinsky, & Zhang, 2000) on mobile devices.

The Ambient Intelligence technologies provide a vision for creating electronic environments sensitive and responsive to people. Brad (2001) proposed to unite desktop PCs and PDAs together, in which a PDA acts as a remote controller or an assistant input device for the desktop PC. They focused on the shift usage of mobile devices mainly like PDAs as extended controllers or peripheries according to their mobility and portability. However, it cannot work for many cases such as people on the move without access to desktop computers.

Though multiple displays are available for users, there still exist many tangles to make multiple devices work cooperatively to improve the user's experience of information browsing in mobile devices. In our framework, we design a distributed interface that crosses devices to cooperatively present information to mobile users. We believe our work will benefit users' browsing and accessing of the available information on mobile devices with small display areas.

Communication Protocol

The rapid growth of wireless connection technologies, such as 802.11b or Bluetooth, has enabled mobile devices to stay connected online easily. We propose a communication protocol to maintain the cooperative browsing with multiple devices. When a user manipulates information in one device, our task is to let other devices work cooperatively. To better illustrate the communication, we introduce two notations as follows: (1) *Master device* is defined as the device that is currently operated on or manipulated by a user; and (2) *Slave device* refers to the device that displays cooperatively according to user's interactions with a master device.

We define a whole set of devices available for users as a cooperative group. A rule is regulated that there is only one master device in a cooperative group at a time, and other devices in the group act as slave devices. We call a course of cooperative browsing with multiple devices as a cooperative session. In such a session, we formulate the communication protocol as follows:

- A user selects a device to manipulate or access information. The device is automatically set as the master device in the group. A cooperative request is then multicast to the slave devices.
- The other devices receive the cooperative request and begin to act as slave devices.
- When the users manipulate the information on the master device, the features are automatically extracted according to the analysis of interactions, and are then transferred to slave devices.
- According to the received features, the corresponding cooperative display updates are automatically applied on the slave devices.
- When a user quits the manipulation of information in the master device, a cooperative termination request is multicast to the slave devices to end the current cooperative session.

Two-Level Browsing Scheme

We set out to construct distributed user interfaces by uniting the multiple display areas on various devices to overcome the display constraint in a single device. In our framework, we propose a two-level cooperative browsing scheme, namely within-document and between-document. If a document itself needs to be cooperatively browsed across devices, we define this case as within-document browsing. Otherwise, if a relevant document needs to be cooperatively browsed across devices, we consider this case as between-document browsing.

1: Within-Document Cooperative Browsing

There exist many studies to improve the browsing experiences on small screens. Some studies proposed to render a thumbnail representation (Su, Sakane, Tsukamoto, & Nishio, 2002) on mobile devices. Though users can browse an overview through such a display style, they still have to use panning/zooming operations for a further view. However, users' experiences have not been improved yet since these operations are difficult to be finished in a thumbnail representation.

We propose a within-document cooperative strategy to solve this problem, where we develop a so-called two-level representation for a large document: (1) presenting an index view on the top level with each index pointing to detailed content portion of a document in the master device; and (2) a click in each index leads to automatic display updates of the corresponding detailed content in the slave devices.

We believe such an approach can help users browse documents on small devices. For example, users can easily access the interesting content portions without scrolling operations but a click on the index view.

2: Between-Document Cooperative Browsing

As shown in previous studies (Hua, Xie, Lu, & Ma, 2004, 2005; Nadamoto & Tanaka, 2003), users tend to view similar documents (e.g., image and Web page) concurrently for a comparative view of their contents. User's experience will be especially degraded in such scenarios due to two reasons. Firstly, it's difficult for users to seek out relevant information on a mobile device, and the task becomes more tedious with the increase of the number of devices for finding. Secondly, it's not feasible to present multiple documents simultaneously on a small display area, and it's also tedious for users to switch through documents for a comparative view.

In our system, we propose a between-document cooperative mechanism to address this problem. Our approach comprises of two steps: (1) relevant documents are automatically searched based on the information a user is currently focusing on the master device; and (2) the searched documents are presented on the slave devices. Therefore, this method can facilitate users' comparative view without manual efforts. Thus, users can easily achieve a comparative view with a simple glimpse through devices.

APPLICATION OF OUR COOPERATIVE FRAMEWORK

To assess the effectiveness of our framework, we apply it to several types of documents that are ubiquitous in mobile devices, including images, text documents (such as e-mail, PDF file, MS documents like DOC or PPT files) and Web pages. In the following sections, we illustrate each in detail.

Cooperative Browsing of Documents

1: Within-Document Cooperative Browsing

Document readability is greatly degraded due to the small display areas on current mobile devices; users have to continually scroll through the content to browse each portion in detail. In this case, we believe our between-document solution is capable of solving this problem: (1) partitioning a document into a series of content sections according to paragraph or passage information; (2) extracting a summary description from each portion using a title or sub-title (summary instead if no titles); and (3) generating an index view for the document where each index points to the relevant detailed content portion in a large document.

Figure 1 shows an example of our solution, where an MSWord document is represented through its outline index, and a click leads to the display of detailed content in its slave devices. The design for slides is really useful in practice. For instance, for a speaker who often moves around to keep close contact with his/her audiences, it's necessary to develop a mechanism to facilitate the speaker's interaction with the slides when he or she moves around. We present an indexed view on small devices like SmartPhone, which can be taken by users, and the interaction with this phone generates the display updates on the screen.

2: Between-Document Cooperative Browsing

In our multi-device solution, we search relevant documents automatically and then deliver them to the slave devices to help browsing. We automatically identify the passages that are currently displayed on the center screen in the master device as the indicative text to find out relevant information. As has been pointed out by many existing studies, it is sometimes better to apply retrieval algorithms to portions of a document text than to all of the text (Stanfill & Waltz, 1992). The solution adopted by our system was to create new passages of every appropriate fixed length words (e.g., 200 words). The system searches a similar document from each device by using the passage-level feature vectors with keywords. The similarities of passages are computed using the Cosine distance between the keyword feature vectors (e.g., TFIDF vector model). In this way, our system searches for similar passages in the available document set, and the document with the greatest number of similar paragraphs becomes the similar page.

Figure 2 shows an example of this case, where (b) is the search results by our approach according to the content information that is extracted from (a).

Figure 1. An example for within-document cooperative browsing

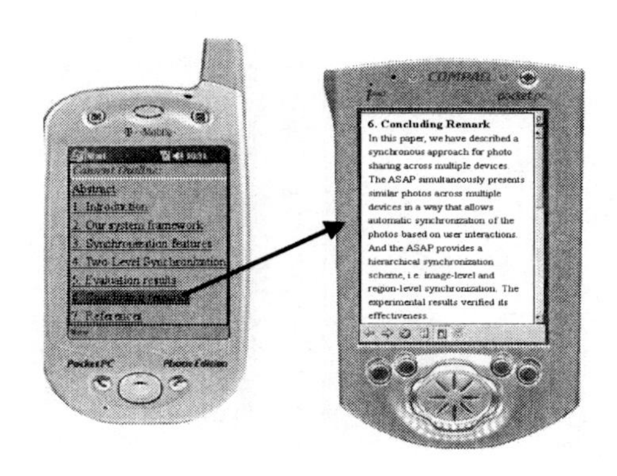

Figure 2. An example for between-document cooperative browsing

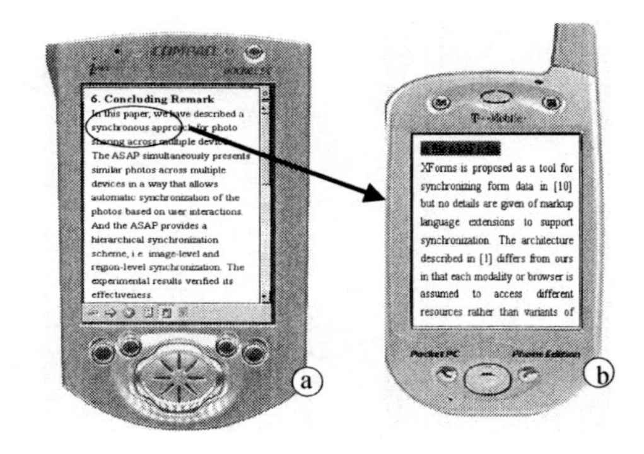

Furthermore, our system automatically scrolls the relevant document to the similar passages that hold the maximal similarity.

Cooperative Browsing of Web Pages

With the pervasive wireless connection in mobile devices, users can easily access the Web. However, Web pages are mostly designed for desktop computers, and the small display areas in mobile devices are consequently too small to display them. Here, we apply our framework to employ a cooperative way to generate a tailored view of large Web pages on mobile devices.

1: Within-Page Cooperative Browsing

Different from documents (e.g., MSWord), Web pages include more structured contents. There exist a lot of studies on page segmentation to partition Web pages into a set of tailored blocks. Here, we adopt the methods by Chen, Xie, Fan, Ma, Zhang, and Zhou (2003), in which each page is represented with an indexed thumbnail with multiple segments and each of them points to a detailed content unit. Figure 3 shows an example of this case. In our system, we deliver the detailed content blocks to various devices. Additionally, each detailed content block is displayed on a most suitable display area.

2: Between-Page Cooperative Browsing

Besides improving page readability in small display areas of mobile devices, users also need to browse relevant pages that contain similar information. In common scenarios, users need to manually search these relevant pages through a search engine or check-through related sites. Here, we develop an automatic approach to present relevant Web pages through a cross-device representation.

Our method to find out similar pages comprises three steps: (1) extracting all links from the page, which are assumed to be potential candidates that contain relevant information; (2) automatically downloading content information for each extracted links, and representing each document as a term-frequency feature vector; and (3) comparing the similarity of extracted pages and current page based on the Cosine distance through the feature vector. Thus, the page with the maximal similarity is selected as the relevant one, and its URL is sent to other devices for an automatic display update. An example is shown in Figure 4.

Cooperative Browsing of Images

Lpictures are not fitful for the display on mobile devices with small display areas. Here we apply our framework to facilitate users' image browsing through a cooperative cross-device representation.

1: Within-Image Cooperative Browsing

In addition to all previous automatic image browsing approaches in a single small device (Chen, Ma, & Zhang, 2003; Liu, Xie, Ma, & Zhang, 2003), we provide in our approach a so-called smart navigation mode (Hua, Xie, Lu, & Ma, 2004). In our approach,

Figure 3. An example for within-page cooperative browsing

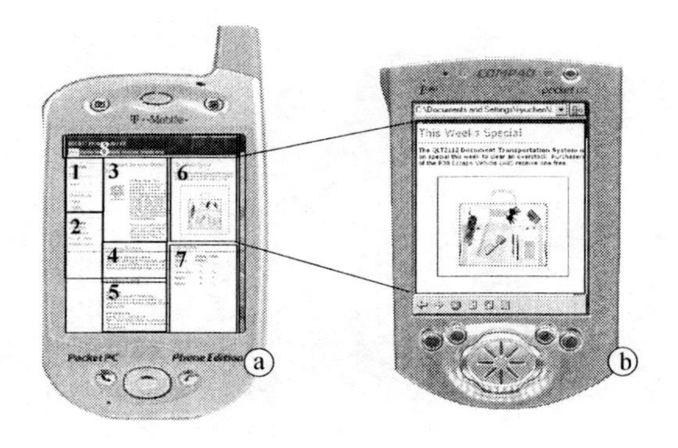

Figure 4. An example for between-page cooperative browsing

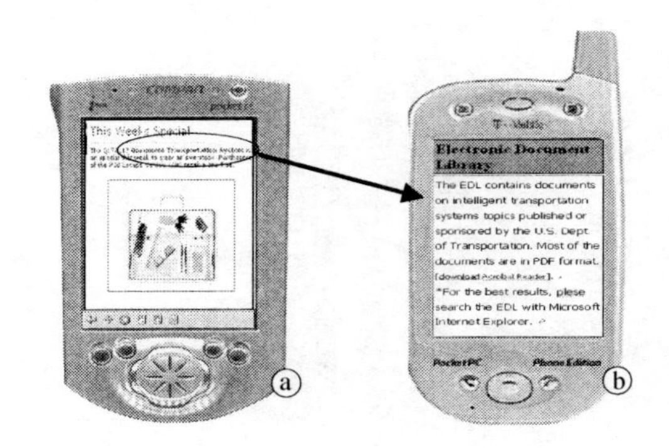

Figure 5. An example for within-image cooperative browsing

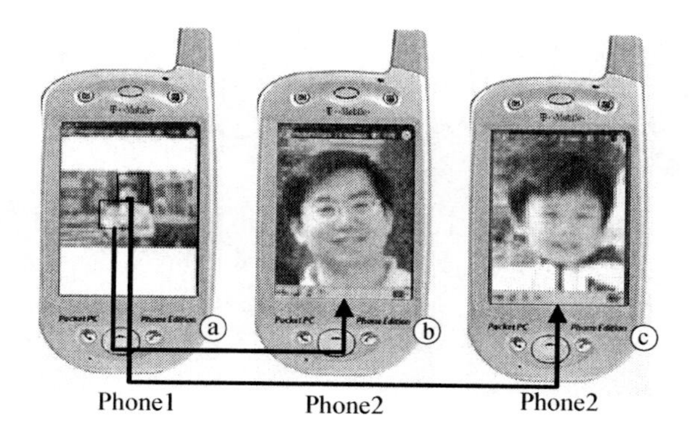

Figure 6. The synchronization between PDA1 and PDA2

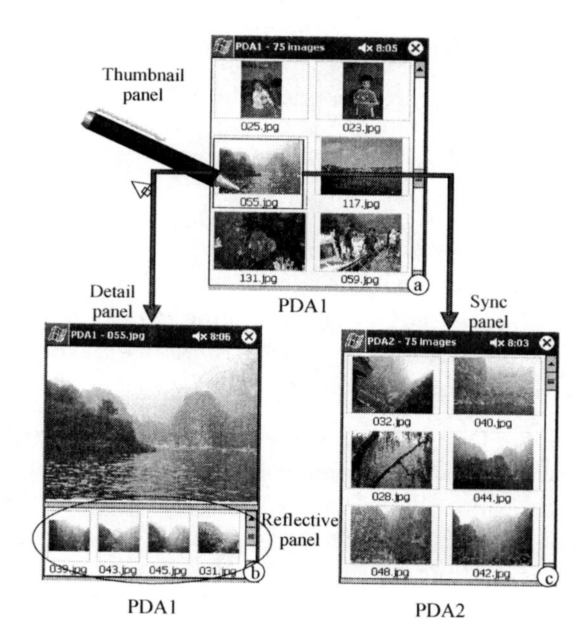

an image is decomposed into a set of attention objects according to Ma and Zhang (2003) and Chen, Ma, and Zhang (2003), and each is assumed to contain attentive information in an image. Switching objects in a master device will result in a detailed rendering of the corresponding attention object on the slave device (e.g., Figure 5).

2: Between-Image Cooperative Browsing

In our previous work (Hua, Xie, Lu, & Ma, 2004), we proposed a synchronized approach called ASAP to facilitate photo viewing across multiple devices, which can simultaneously present similar photos on various devices. A user can interact with either of the available devices, and the user's interaction can automatically generate the synchronized updates in other devices. In Figure 6, there are two PDAs denoted PDA1 and PDA2, and each stores a number of pictures. When a user clicks a photo in PDA1 (Figure 6a), there are two steps to be done simultaneously: (1) PDA1 searches out similar images and displays them (b); (2) PDA1 sends image feature to PDA2, which then search out the similar photos (c).

FUTURE TRENDS

Now, we are planning to improve our work in three aspects. First, we will develop more accurate algorithms to search out relevant information of various media types including image, text and Web page. Second, we plan to devise more advanced distributed interfaces to facilitate users' information browsing tasks across various devices. Third, we plan to apply our work to other applications such as more intricate formats of documents and execution application GUIs. We also plan to conduct a usability evaluation among a wide number of users to collect their feedbacks, which will help us find the points they appreciate and the points that need further improvements.

CONCLUSION

In this article, we developed a cooperative framework that utilizes multiple displays to facilitate information browsing in mobile environment. A two-level browsing scheme is employed in our approach, namely within- and between- document browsing. We apply our framework to a wide variety of applications including documents, Web pages and images.

REFERENCES

Brad, A. (2001). Using handhelds and PCs together. *Communications of the ACM, 44*(11), 34-41.

Chen, L. Q., Xie, X., Fan, X., Ma, W. Y., Zhang, H. J., & Zhou, H. Q. (2003). A visual attention model for adapting images on small displays. *ACM Multimedia Systems Journal, 9*(4), 353-364.

Chen, Y., Ma, W.Y., & Zhang, H. J. (2003, May). Detecting Web page structure for adaptive viewing on small form factor devices. In the *Proceedings of the 12th International World Wide Web Conference*, Budapest, Hungary (pp. 225-233).

Coles, A., Deliot, E., & Melamed, T. (2003, May). A framework for coordinated multi-modal browsing with multiple clients. In the *Proceedings of the 13th International Conference on World Wide Web*, Budapest, Hungary (pp. 718-726).

Dumais, S., Cutrell, E., Cadiz, J., Jancke, G., Sarin, R., & Robbins, D. C. (2003, July). Stuff I've seen: A system for personal information retrieval and re-use. In the *Proceedings of the 26th Annual International ACM SIGIR Conference on Research and Development in Information Retrieval*, Toronto, Canada (pp. 72-79).

Hua, Z., Xie, X., Lu, H. Q., & Ma, W. Y. (2004, November). Automatic synchronized browsing of images across multiple devices. In the *Proceedings of the 2004 Pacific-Rim Conference on Multimedia*, Tokyo, Japan. Lecture Notes in Computer Science (pp. 704-711). Springer.

Hua, Z., Xie, X., Lu, H. Q., & Ma, W. Y. (2005, January). ASAP: A synchronous approach for photo sharing across devices. In the *11th International Multi-Media Modeling Conference*, Melbourne, Australia.

Liu, H., Xie, X., Ma, W. Y., & Zhang, H. J. (2003, November). Automatic browsing of large pictures on mobile devices. In the *Proceedings of ACM Multimedia 2003 Conference*, Berkeley, California (pp. 148-155).

Ma, W. Y., Bedner, I., Chang G., Kuchinsky, A., & Zhang H. J. (2000, January). A framework for adaptive content delivery in heterogeneous network environments. *Multimedia Computing and Networking 2000*, San Jose.

Ma, Y. F., & Zhang, H. J. (2003, November). Contrast-based image attention analysis by using fuzzy growing. In the *Proceedings of ACM Multimedia 2003 Conference*, Berkeley, California (pp. 374-381).

Nadamoto, A., & Tanaka, K. (2003, May). A comparative Web browser (CWB) for browsing and comparing Web pages. In the *Proceedings of the 12th International World Wide Web Conference*, Budapest, Hungary (pp. 727-735).

Stanfill, C., & Waltz D. L. (1992). Statistical methods, artificial intelligence, and information retrieval. In P. S. Jacobs (Eds.), *Text-based intelligent systems* (pp. 215-225). Lawrence Erlbaum.

Su, N. M., Sakane, Y., Tsukamoto, M., & Nishio, S. (2002, September). Rajicon: Remote PC GUI operations via constricted mobile interfaces. *The 8th Annual International Conference on Mobile Computing and Networking*, Atlanta (pp. 251-262).

KEY TERMS

Ambient Intelligence: Represents a vision of the future where people will be surrounded by electronic environments that are sensitive and responsive to people.

ASAP System: The abbreviation of a synchronous approach for photo sharing across devices to facilitate photo viewing across multiple devices, which can simultaneously present similar photos across multiple devices at the same time for comparative viewing or searching.

Attention Object: An information carrier that delivers the author's intention and catches part of the user's attention as a whole. An attention object often represents a semantic object, such as a human face, a flower, a mobile car, a text sentence, and so forth.

Desktop Search: The functionality to index and retrieve personal information that is stored in desktop computers, including files, e-mails, Web pages and so on.

Small Form Factors: Mobile devices are designed for portability and mobility, so the physical size is limited actually. This phase is called small form factors.

TFIDF Vector Model: TF is the raw frequency of a given term inside a document, which provides one measure of how well that term describes the document contents. DF is the number of documents in which a term appears. The motivation for using an inverse document frequency is that terms that appear in many documents are not very useful for distinguishing a relevant document from a non-relevant one.

Visual Attention: Attention is a neurobiological conception. It implies the concentration of mental powers upon an object by close or careful observing or listening, which is the ability or power to concentrate mentally.

CSCW Experience for Distributed System Engineering

Thierry Villemur
LAAS-CNRS, France

Khalil Drira
Laboratoire d'Analyse et d'Architecture des Systèmes du CNRS (LAAS-CNRS), France

INTRODUCTION

Cooperation with computers, or Computer Supported Cooperative Work (CSCW), started in the 1990s with the growth of computers connected to faster networks.

Cooperation and Coordination

CSCW is a multidisciplinary domain that includes skills and projects from human sciences (sociology, human group theories, and psychology), cognitive sciences (distributed artificial intelligence), and computer science (human/computer interfaces; distributed systems; networking; and, recently, multimedia).

The main goal of the CSCW domain is to support group work through the use of networked computers (Ellis et al., 1991; Kraemer et al., 1988). CSCW can be considered a specialization of the Human-Computer Interaction (HCI) domain in the sense that it studies interactions of groups of people through distributed groups of computers.

Two main classes can be defined within the CSCW systems. Asynchronous Cooperations do not require the co-presence of all the group members at the same time. People are interacting through asynchronous media like e-mail messages on top of extended and improved message systems. At the opposite, Synchronous Cooperations create stronger group awareness, because systems supporting them require the co-presence of all the group members at the same time. Exchanges among group members are interactive, and nowadays, most of them are made with live media (audio- and videoconferences).

Groupware (Karsenty, 1994) is the software and technological part of CSCW. The use of multimedia technologies leads to the design of new advanced groupware tools and platforms (Williams et al., 1994), such as shared spaces (VNC, 2004), electronic boards (Ellis et al., 1991), distributed pointers (Williams et al., 1994), and so forth. The major challenge is the building of integrated systems that can support the current interactions among group members in a distributed way.

Coordination deals with enabling and controlling cooperation among a group of human or software distributed agents performing a common work. The main categories of coordination services that can be distinguished are dynamic architecture and components management; shared workspace access and management; multi-site synchronization; and concurrency, roles, and group activity management.

Related Projects

Many researches and developments for distance learning are made within the framework of the more general CSCW domain.

Some projects, such as Multipoint Multimedia Conference System (MMCS) (Liu et al., 1996) and Ground Wide Tele-Tutoring System (GWTTs) (GWTTS project, 1996), present the use of video communications with videoconference systems, communication boards, and shared spaces, built on top of multipoint communication services. The Distance Education and tutoring in heterogeneous teleMatics envirOnmentS (DEMOS) project (Demos project, 1997) uses common public shared spaces to share and to control remotely any Microsoft Windows application. The MultiTeam project (MultiTeam project, 1996) is a Norwegian project to link distributed classrooms over an ATM network through a giant electronic whiteboard the size of an ordinary blackboard. Microsoft NetMeeting (Netmeeting, 2001), together with Intel Proshare (now Polycom company) (Picture Tel, 2004), are very popular and common synchronous CSCW toolkits based on the H.320 and H.323 standards and both composed of a shared electronic board, a videoconference, and an application sharing space. These tools are used most of the time on top of the classical Internet that limits their efficiency due to its not guaranteed quality of service and its irregular rate. Most of their exchanges are based on short events made with TCP/IP protocol in peer-to-peer relationships. The ClassPoint (ClassPoint, 2004) environment has been created by the First Virtual Communications society (formerly White Pine society). It is composed of three tools: a videoconference based on See You See Me for direct video contacts among the distributed group members, the dialogs, and views of the students being under the control of the teacher. An electronic whiteboard reproduces the interactions made by classroom blackboards. A Web browser has been customized by a synchronous browsing function led by the teacher and viewed by the whole-distributed class. This synchronous browsing can be relaxed by the teacher to allow freer student navigation.

BACKGROUND

A Structuring Model for Synchronous Cooperative Systems

Different requirements are identified for the design of networked synchronous CSCW systems. Such systems may be designed to improve the efficiency of the group working process by high-quality multimedia material. The networked system must support both small- and large-scale deployments, allowing reduced or universal access (Demos project, 1997). Defining the requirements of a networked synchronous CSCW system needs multidisciplinary expertise and collaboration. For this purpose, we distinguish three distinct viewpoints: functional, architectural, and technological.

Moreover, several objectives may be targeted by the networked solution retained for the synchronous CSCW system, including adaptability, upgradability, multi-user collaboration, and interaction support. In practice, the design and development of a networked solution first involve general skills such as software architecture, knowledge organization, and other work resources management; and the second development of domain-specific multimedia cooperation tools. We identify three generic interaction levels that are likely to be significant for the different viewpoints: the cooperation level, the coordination level, and the communication level. Their content is summarized in Table1.

For software architecture design, level-based layering allows different technologies to be used for implementing, integrating, and distributing the software components. This separation increases the upgradability of the systems. Layering allows the implemented system to likely guarantee the end-user quality of service while taking advantage of the different access facilities. For adaptability, multi-user collaboration, and interaction support, level-based decomposition allows functional separation between individual behaviors and group interaction rules definition.

Table 1. The three levels and three viewpoints of the structuring model

Interaction Levels/ Viewpoints	Cooperation	Coordination	Communication
Functional view	User-to-user interaction paradigms	User-level group coordination functions (sharing, and awareness)	User-to-user information exchange conventions
Architectural view	Cooperation tools	software-level group coordination services (for tools and components)	Group communication protocols (multipeer protocols)
Technological view	Individual tool implementation technology (interfacing, and processing)	Components integration technology	Component distribution technology (request transport protocols)

From a functional viewpoint, the cooperation level describes the interaction paradigms that underlay the member interactions during the group working process. From an architectural viewpoint, it defines the tool set covering the above paradigms. From a technological viewpoint, the cooperation level describes the different technologies that are used to implement the individual tools, including interfacing and processing.

According to the functional viewpoint, the coordination level lists the functions that are used to manage the user collaboration, providing document sharing and group awareness. From an architectural viewpoint, it describes the underlying user services, as tools group coordination, membership management, and component activation/deactivation. From a technological viewpoint, the coordination level defines the integration technology used to make components interact, including interaction among peers (different components of the same tool), tools, and users.

From a functional viewpoint, the communication level describes the conventions that manage the information exchange between users. From an architectural viewpoint, it enumerates multicast protocols, allowing user-to-user or tool-to-tool group communication. From a technological viewpoint, the communication level defines the protocols used to handle groups and to exchange requests between components, including tools and services.

In the sequel of this article, this multi-view/level model for cooperative applications has been applied for the analysis, comparison, and development of a group-based networked synchronous platform. The application domain of this platform relates to Distributed System Engineering.

DISTRIBUTED SYSTEM ENGINEERING (DSE)

The Distributed System Engineering (DSE) European Project—contract IST-1999-10302—started in January 2000 and ended in January 2002 (DSE project, 2002). It was conducted by an international consortium, whose following members have strong involvement in the space business: Alenia Spazio (ALS), coordinator of the project; EADS Launch Vehicles (ELV); and IndustrieAnlagen BetriebsGesellshaft (IABG). The other members of the DSE project belong to technology providers and research centers: Silogic, Societa Italiana Avionica (SIA), University of Paris VI (LIP6), LAAS-CNRS, and D3 Group.

Research Objectives

The Engineering Life Cycle for complex systems design and development, where partners are dispersed in different locations, requires the setup of adequate and controlled processes involving many different disciplines (Drira et al., 2001; Martelli et al., 2000).

The design integration and the final system physical/functional integration and qualification imply a high degree of cross-interaction among the partners. The in-place technical information systems supporting the life cycle activities are specialized with respect to the needs of each actor in the process chain and are highly heterogeneous among them.

To globally innovate in-place processes, involved specialists will be able to work as a unique team in a Virtual Enterprise model. To this aim, it is necessary to make interoperable the different Technical Information Systems and to define Cooperative Engineering Processes that take into account distributed roles, shared activities, and distributed process controls. DSE is an innovative study in this frame. It addressed this process with the goal of identifying that proper solutions (in terms of design, implementation, and deployment) have been carried out.

DSE Software Platform

The software platform that has been realized to support Cooperative Engineering scenarios is composed of several distributed subsystems, including Commercial Off-The-Shelf (COTS) components and specifically developed components.

Session Management

The Responsibility and Session Management subsystem is the central component (Figure 1) developed at LAAS-CNRS (Molina-Espinosa, 2003). It is in charge of session preparation and scheduling.

Figure 1. Responsibility and session management GUI

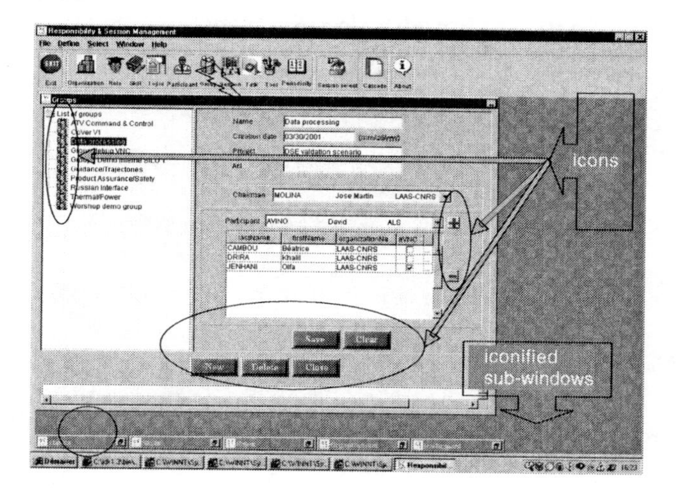

It supports the definition and programming of user profiles, roles, and sessions characteristics, using a remote responsibilities and schedule repository. Both synchronous and asynchronous notification mediums are provided by this environment for troubleshooting, tuning, and negotiation during the preparation and enactment of the collaboration sessions. Its Graphical User Interface (GUI) is implemented with Java Swing.

DSE Architecture

The DSE global collaboration space integrates (i) multipoint communication facilities (Multicast Control Unit [MCU]—the MeeetingPoint server); (ii) several Groupware tools (videoconferences: CU-See-Mee or Netmeeting—document sharing tools: E-VNC); and (iii) domain-specific tools (Computer Aided Design [CAD] tools or System Engineering tools).

The DSE architecture is composed of several clients and servers. The link between clients and servers is made through high-level communication facilities based on CORBA, JAVA-RMI, and XML messages.

The Main DSE server contains an Enterprise Data Repository (EDR) to store documents handled by the session members and an HTTP server as unique front-end.

The EDR is implemented using the following technologies:

- ORACLE Portal 3.0 for EDR features
- ORACLE database 8i release 2

Communications are implemented with:

- DSE WEB Portal (HTML page foreseen to be generated and managed by ORACLE Portal 3.0)
- Servlet Engine: Jserv (Apache's JSP/Servlet Engine)

The DSE front-end provides a unique entry point to all DSE users. It allows transparent access to all the DSE subsystems described in the sequel. The chosen HTTP server is:

- Apache server included with ORACLE Portal 3.0

The DSE Awareness Subsystem is a notification service. It is composed of a notification server based on commercial CORBA 2.3-compliant ORB Service.

The DSE Responsibility and Session (RMS and SMS) management sub-subsystem provides communication facilities for the managing and activating sessions. It is composed of an HTTP server for using servlets and a JAVA middleware that provides RMI communications.

A typical configuration of a DSE client contains:

(i) Groupware tools
H.323 videoconference client: MS NetMeeting or CuSeeMe
Application sharing client: Extended-VNC package (E-VNC) (Client, Proxy with chairman GUI, Server).

(ii) Generic communication tools
Web Browser (Internet Explorer, Netscape, or Mozilla)
Mail client (NT Messenger, Outlook Express, or Eudora)

(iii) Management interfaces
Session and Responsibility Management GUI

(iv) Domain-specific interfaces
A graphical client component for distributed simulation based on High Level Architecture (HLA)

Remark: Some light DSE clients are only communicating with Web-based protocols and are not using CORBA services.

A Preliminary Design Review (PDR) server is added for the domain-specific application. The PDR validation scenario will be presented in a next section.

DSE Deployment Architecture

Figure 2 shows the architecture of the DSE platform with clients and servers distributed among the involved partners. The network connections used by the partners during the Integration and Validation phases are listed below:

- **ALS (Turin):** ISDN 2xBRI with IABG, Public Internet
- **ELV (Paris):** ATM 2Mbps to/from LIP6, Public Internet
- **IABG (Munich):** ISDN 2xBRI with ALS, Public Internet
- **SILOGIC (Toulouse):** Public Internet
- **D3 (Berlin):** Public Internet
- **SIA (Turin):** Public Internet, High Speed local connection with ALS
- **LAAS (Toulouse):** ATM to/from LIP6, Public Internet
- **LIP6 (Paris):** ATM to/from LAAS and ELV, Public Internet

Machines used for validation were running Windows NT 4.0. However, the light DSE clients also can run on PCs with Windows 2000 or on Solaris 2.x SUN workstations.

Validation Scenario

The context of collaborative design and analysis scenarios is based on real design material from the flight segment development program of the Automated Transfer Vehicle (ATV). The ATV missions are to:

Figure 2. The global DSE servers deployment architecture

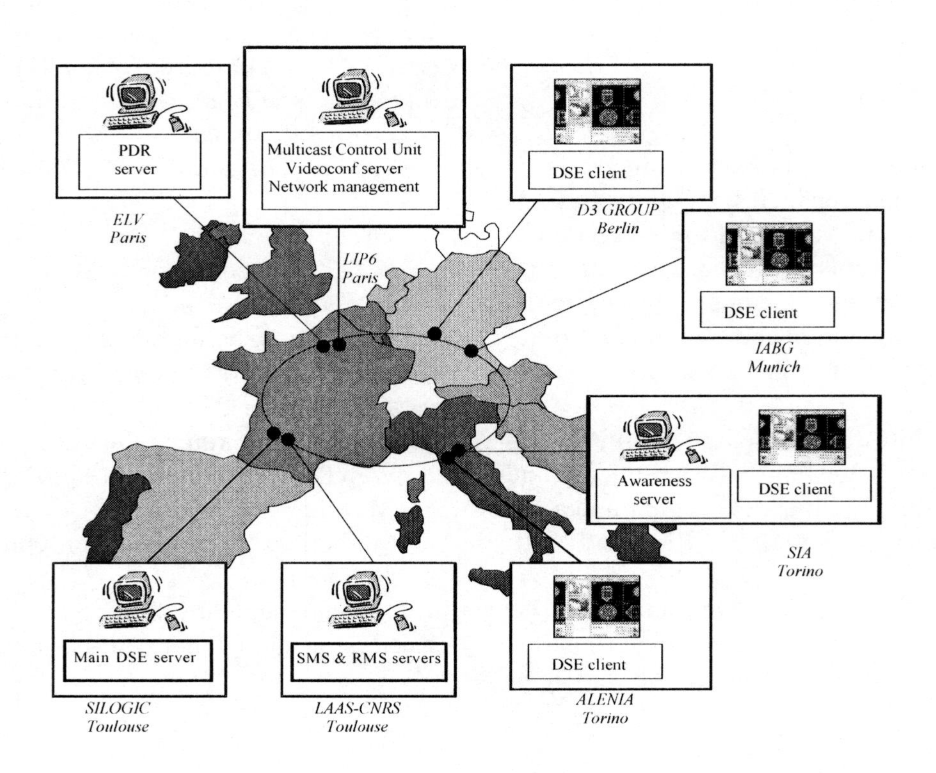

- Deliver freight to the International Space Station (ISS)
- Reboost ISS
- Collect ISS garbage and destroy them

Preliminary Design Review

The precise DSE validation scenario is called Preliminary Design Review (PDR). The PDR process is a very codified design phase for which a set of contractors has to submit a large set of documentation to the customer review team. The PDR Review Objectives can be detailed in:

- Obtaining an independent view over program achievements
- Identifying difficulties and major risks in order to reduce them
- Assessing the progress report
- Putting documents and product in a validated baseline
- Supporting the management in deciding to continue or not

The PDR actors form a large group composed of review committees, project team interfaces, and review boards. The review committees contain 50 to 100 experts split into five to 10 thematic groups. They are controlled by a review chairman and a secretary. They examine documents and make Review Item Discrepancy (RID), which is a kind of codified comments. The project team interfaces are composed of eight to 12 project team members. They answer to committee comments and board recommendations. The review board is composed of the customers with the support of prime contractors. They decide to implement or reject the review committee's recommendations. Then, they decide to begin the next program phase or not.

Distributed PDR Scenario

The main applicative goal of the DSE project is to realize the PDR process in a distributed way. The three phases of the PDR review (prepare the review, execute the review, and process the review results) were supported through the set of distributed networked tools and components that has been developed and integrated. Using this set of components, the session manager was able to organize the review. This included definitions of group members, tasks, roles, and progress level of the review. The second offered service was the management of the data issued during the PDR activity, the Review Items Discrepancies (RID). Using the EDR database from the main DSE server, the RID documents were created remotely, accessed, shared, and modified during the review process.

MODEL-BASED OVERVIEW OF THE DSE ENVIRONMENT

Table 2 represents the design model instance describing the DSE environment. The functional view directly refers to the DSE PDR scenario. The architectural view represents all the tools and components used to support the PDR scenarios. The

Table 2. An overview of the DSE model

Interaction Levels/ Viewpoints	Cooperation	Coordination	Communication
Functional view	- Actions on Review Item Discrepancies (RIDs)	- PDR role attribution (Review Committee, Project Team, Review Board)	- PDR role based: - Comment/Answer - Commit/Reject
Architectural view	- Videoconference - Application Sharing Tool - Text Chat	- Session management services (definition and enactment) - Multi-user coordination	- Audio/Video multicast protocol - IP multicast - Reliable multicast
Technological view	- H.263 video cards - JAVA AWT/SWING	- JAVA applets - JAVA/CORBA objects - XML Canvas	- JAVA RMI - CORBA IIOP protocol - WWW/HTTP protocol

technological view focuses on infrastructure, languages, and protocols used for the development and the integration of the global environment.

FUTURE TRENDS

Future trends will be to apply the proposed design framework in different application contexts where CSCW technology also should be useful. The next application domain considered is distance learning.

We are now involved in a European IST project called Lab@Future, which started in May 2002 and will last for three years. It aims to define a generic and universal platform for mixed and augmented reality. This platform must be able to support interactions through mobile devices and wireless networks.

It will be applied in the framework of new educational theories as activity theory and constructivism to give remote access to schools to advanced laboratory experiments.

These theories, through canvas as real-time problem solving, collaborative, exploratory, and interdisciplinary learning, consider the learning activity as a strongly interactive building process through various knowledge exchanges.

The mixed and augmented reality platform will be deployed in eight European countries. It will be used for laboratory teaching experiments within fluid mechanics, geometry, art and humanities, and human science.

CONCLUSION

In this article, we presented an overview of recent research results in the design of cooperative systems for supporting distributed system engineering scenarios. A layered multi-viewpoint structuring approach is adopted as a design framework. It covers three viewpoints: functional, architectural, and technological views involved in CSCW environments. It also identifies three interaction levels: cooperation, coordination, and communication.

REFERENCES

ClassPoint. (2004). ClassPoint environment. Retrieved from http://www2.milwaukee.k12.wi.us/homehosp/Cpinfo.htm

Demos Project. (1997). Retrieved from http://www.cordis.lu/libraries/en/liblearn.html

Drira, K., Martelli, A., & Villemur, T. (Eds). (2001). Cooperative environments for distributed system engineering [Lecture]. In Computer Science 2236.

DSE Project. (2002). Retrieved from http://cec.to.alespazio.it/DSE

Ellis, C., Gibbs, S., & Rein, G. (1991). Groupware: Some issues and experiences. *Communications of the ACM, 34*(1), 38-58.

GWTTS Project. (1996). Retrieved from http://www.cs.virginia.edu/~gwtts

Karsenty, A. (1994). Le collecticiel: De l'interaction homme-machine à la communication homme-machine-homme. *Technique et Science Informatiques, 13*(1), 105-127.

Kraemer, K.L., & King, J.L. (1988). Computer-based systems for cooperative work and group decision making. *ACM Computing Surveys, 20*(2), 115-146.

Liu, C., Xie, Y., & Saadawi, T.N. (1996). Multipoint multimedia teleconference system with adaptative synchronisation. *IEEE Journal on Selected Areas in Communications, 14*(7), 1422-1435.

Martelli, A., & Cortese, D. (2000). Distributed system engineering—DSE concurrent and distributed engineering environment for design, simulations and verifications in space project. *Proceedings of the 6th International Workshop on Simulation for European Space Programmes*, Noordwijk, The Netherlands.

Molina-Espinosa, J.M. (2003). *Modèle et services pour la coordination des sessions coopératives multi-applications: Application à l'ingénierie système distribuée* [doctoral thesis]. Institut National Polytechnique of Toulouse.

MultiTeam Project. (1996). Retrieved from http://147.8.151.101/ElectronicClassroom/munin.html

Netmeeting. (2001). Microsoft netmeeting 2001. Retrieved from http://www.microsoft.com/windows/netmeeting

Picture Tel. (2004). Retrieved from http://www.picturetel.com

VNC. (2004). *Virtual network computing application.* Retrieved from http://www.realvnc.com/

Williams, N., & Blair, G. S. (1994). Distributed multimedia applications: A review. *Computer Communications, 17*(2), 119-132.

KEY TERMS

Application-Sharing Space: A groupware tool that produces multiple distributed remote views of a particular space. Any single-user application put under the control of the particular space can be viewed remotely and controlled by the group members that have access to this space. Therefore, the application-sharing space transforms any single-user application put under its control into a multi-user shared application.

Asynchronous Cooperation: Members are not present in the same time within the cooperation group (no co-presence). They communicate with asynchronous media (e.g., e-mail messages) on top of extended and improved message systems.

Computer Supported Cooperative Work (CSCW): According to Ellis, et al. (1991), CSCW platforms are "computer-based systems that support groups of people engaged in a common task (or goal) and that provide an interface to a shared environment" (p. 40). Another similar definition (Kraemer et al., 1988) is computer-based technology that facilitates two or more users working on a common task.

Cooperation: A group of people working on a common global task.

Coordination: Enabling and controlling the cooperation among members of a group of human or software-distributed agents. It can be considered as software glue for groupware tools, including architectural and behavioral issues. Coordination includes several synchronization and management services.

Electronic Board: A classic groupware tool that supports the functionalities of a traditional whiteboard (sharing sketches, pointing, annotating) through a set of distributed computers.

Groupware: The software and technological part of CSCW. It contains application studies and platforms adapted to groups and supporting group working.

Synchronous Cooperation: Members are present in the same time within the cooperation group (co-presence). The communications among them are interactive and made with live media, such as videoconferences or application sharing spaces.

System Engineering: The branch of engineering concerned with the development of large and complex systems. It includes the definition and setup of adequate and controlled processes for the design and development of these complex systems.

Cultural Diversity and Aspects of Human Machine Systems in Mainland China

Kerstin Röse
University of Kaiserslautern, Germany

INTRODUCTION

Addressing intercultural considerations is increasingly important in the product development processes of globally active companies.

Culture has a strong relevance for design, which means that culture influences the daily usage of products via design. During humans' "growing up" time and socialization, the daily usage and the interaction with different products are very important. This forms the user behavior because it supports the forming of users' basic interaction styles. Education and interaction styles vary in different cultures. Hence, it should be interesting to look at the relation of users' behavior and culture.

BACKGROUND

For many years, researchers of social sciences have analyzed cross-cultural differences of interpersonal communication styles and behavior (Hofstede, 1991; Trompenaars, 1993). During the last ten years, usability engineers also have focused on intercultural differences of icon/color coding, navigation, and other human-machine interface components (Hoft, 1996; Marcus, 1996; Prabhu & Harel, 1999). The time for product development and the time between redesign efforts both are becoming shorter than ever before. Consequently, to prepare effectively interactive products for the global market, one must engineer their intercultural attributes of such a market. One basic step for intercultural engineering is the analysis of user requirements in different cultures.

Within the framework of this project described in the following, a requirement analysis of user needs in mainland China was conducted as first step of a human machine system localization. But, why do you have to localize your product? Why is it important to know the culture of a target user group?

Bourges-Waldegg (2000) says:

> *...Design changes culture and at the same time is shaped by it. In the same way, globalization is a social phenomenon both influencing and influenced by design, and therefore by culture..., both globalization and technology have an effect on culture, and play a role in shaping them.*

This article describes the analysis of culture-specific information from users in Mainland China and the application of different methods for different design issues, especially in an intercultural context. Selected results of this analysis will also be presented. The analysis and their results are part of the project INTOPS-2: Design for the Chinese market, funded by several German companies.

The project was carried out by the Center for Human Machine Interaction (the University of Kaiserslautern, Germany). The aim of the project Intops-2 was to find out the influence of culture on the design of human machine systems, and to analyze local specifics for the area of machine tools and the requirement analysis of the Chinese user from that area.

USER REQUIREMENT ANALYSIS IN MAINLAND CHINA

Study Outline

The requirement analysis in China was carried out at the end of 2000. During two months, 32 Chinese organizations in Shanghai, Beijing, and Chongqing were visited, of which 26 were Chinese industrial enterprises (including Chinese machine tool producers and some machine users). The other six organizations included some governmental organizations for import administration and some research institutes for machine user-interface design in China. The analysis was conducted by a native speaking Chinese researcher from the Center for Human-Machine-Interaction in Kaiserslautern, Germany.

Study Methods

The following three investigation methods were applied in the INTOPS-2 project: test, questionnaire, and interview. These methods have been followed by another similar previous project, namely INTOPS-1 (see also Zühlke, Romberg, & Röse, 1998). The tests are based on the analysis of the following: Choong and Salvendy (1998); Shih and Goonetilleke (1998); Dong and Salvendy (1999); Piamonte, Abeysekera, and Ohlsson (1999); and Röse (2002a). However, to find out more details of the Chinese user requirements, a few new tests and a more detailed questionnaire and interview checklist have been developed for the INTOPS-2 project. An overview of all the implemented tests is presented in Table 1.

Table 1. Overview of implemented tests

No	Test	Aim	Material	Subject	Analysis
1	Preference to color composition for machine tools	Eliciting preferred color composition and difference to German one	10 cards with differently colored machine tools.	No special requirement	Average preference degree for each composition
2	Recalling performance for graphical information vs. textual information	Testing information processing ability for different information presentation methods	3 pieces of paper with different ways of info. presentations: only Text only picture text & picture	No special requirement	Average recall rate for each method. Characters for better recalled info.
3	Understanding of color coding	Testing the understanding of standard color coding and difference to German one	7 standard colors of IEC 73. 3 groups of concepts in daily life and at work (5 in each one)	Matching for concepts at work only for machine operators	The color association rate for each concept
4	Symbol understanding	Testing the understanding of standard ISO symbols and eliciting the preferred symbol characteristics for information coding	Icons from ISO and Windows. 2 kinds of materials: 18 icons, each with 3 possible meanings; 14 meanings, each with 3 possible icons	Machine operators	Average recognition rate for each icon Character for better matched icon
5	Familiarity with Windows interface	Testing the familiarity with the Windows interface	Integrated with Test 4	Machine operators	Recognition rate for Windows icons
6	Concept of grouping (Type of Card Sorting)	Eliciting the grouping rule and the difference to German one	74 cards with different CNC machine functions	Only with experienced CNC machine operators	Preferred structure for grouping
7	Preference for screen layout	Eliciting the familiar screen layout characters and difference to German one	Over 20 different cards in form and size representing the screen elements	CNC machine operators	Preferred layout for different screen elements
8	Understanding of English terms	Testing the English understanding ability	One table with 54 English technical terms	Machine operators	Average understanding rate Character for better understanding

Figure 1. Part of the questionnaire, originally used in Chinese

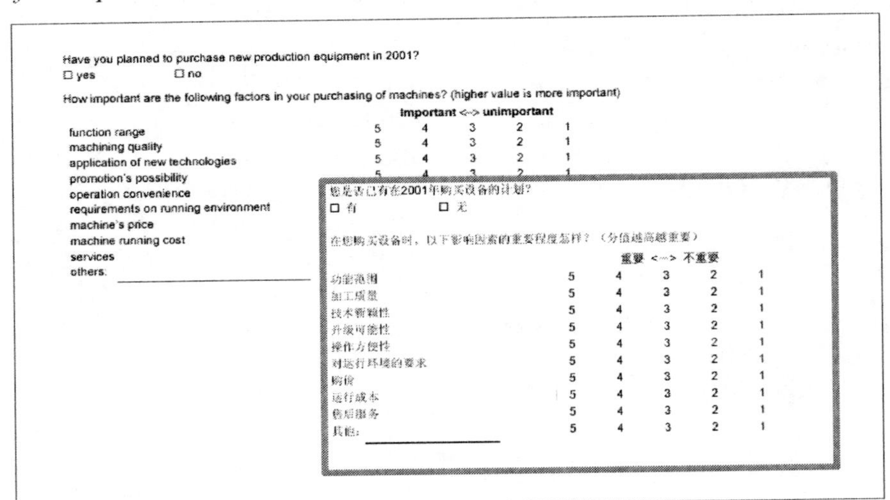

Have you planned to purchase new production equipment in 2001?

☐ yes ☐ no

How important are the following factors in your purchasing of machines? (higher value is more important)

	Important <--> unimportant				
function range	5	4	3	2	1
machining quality	5	4	3	2	1
application of new technologies	5	4	3	2	1
promotion's possibility	5	4	3	2	1
operation convenience					
requirements on running environment					
machine's price					
machine running cost					
services					
others: ____________					

您是否已有在2001年购买设备的计划?

☐ 有 ☐ 无

在您购买设备时，以下影响因素的重要程度怎样？（分值越高越重要）

	重要 <--> 不重要				
功能范围	5	4	3	2	1
加工质量	5	4	3	2	1
技术新颖性	5	4	3	2	1
升级可能性	5	4	3	2	1
操作方便性	5	4	3	2	1
对运行环境的要求	5	4	3	2	1
购价	5	4	3	2	1
运行成本	5	4	3	2	1
售后服务	5	4	3	2	1
其他: ____________	5	4	3	2	1

The main categories of questionnaire and interview checklist are the following:

- **Questionnaire:** Basic information about visited companies, information about machine purchasing, requirements on machine service, requirements on machine user-interface and requirements on machine technical documentation (e.g., Figure 1)
- **Interview:** Basic information about visited companies, information about machine purchasing, application situation of imported machines in practice, requirements on machine user interface, requirements on technical documentation, requirements on service, information about work organization, and training.

Study Conditions

The tests were conducted in the break room. The subjects were all Chinese machine operators. A total of 42 users have participated in the tests, most of them in more than one test. The total test span was controlled within 30 minutes, which has proven in practice as almost the maximum acceptable time for one subject being willing to participate. Due to this constraint, none of the subjects had participated in more than four tests.

The interviews were mainly conducted in the office of the interviewees. All interviews were conducted in Chinese. Since the investigator was a native-speaking Chinese, there was no problem in language communication.

In total, 35 main interviewees were involved. Fifty-eight percent of the interviewees were machine tool users in China. They came from different application areas, like automobile, motorcycle, and machine tool industrial sector. The interviewees were either responsible for machine purchasing decisions (the chief manager, chief engineer, and factory/workshop director) or responsible for actual machine applications in practice (the chief engineer, factory/workshop director, equipment engineer, and technician).

Most of the questionnaires were filled out directly after the interviews by the same interviewees or by other people in charge. A total of 19 questionnaires were obtained in the investigation. About 63% of the questionnaires were filled out by the state owned firms in China. Forty-seven percent of the questionnaires were filled out by machine tool users in China.

Brief Results

As an example for test results, the results of the English understanding test revealed that Chinese machine operators have generally unsatisfactory English understanding for safe and effective machine operation (see test in Table 1 and Rose, Liu, & Zühlke, 2001). Table 2 shows the results for the understanding rate of the English terms.

Table 2.

Terms	OK	OFF	Stop	Home	Help	kg	ON	Start	mm	End	Enter	Del
%	95,7	82,6	82,6	73,9	73,9	73,9	69,6	69,6	69,6	65,2	56,5	52,2
Terms	Exit	Reset	Shift	Menu	Edit	rpm	inch	Disk	Undo	Esc	Icon	PIN
%	43,5	39,1	39,1	39,1	30,4	26,1	26,1	26,1	21,7	21,7	13,0	8,7

Therefore, text coding in English should be restricted in a very limited scope (see also del Galdo, 1990). More detailed requirements at this point are summarized in the following:

1. The most important information should at any time be encoded using Chinese text (and with combination to other information coding methods such as color, shape, and position, etc.).
2. Only when it is really necessary, English texts which express some general machine operations such as On, Off, Start, Stop, and so forth, and are closely related to some daily use such as Ok, Yes, Help, Home, and so forth, could be applied for information coding. But the application of most English abbreviation in text labeling should be avoided.
3. In many cases, other information coding methods should be applied together with Chinese text to ensure that the machine can be well maintained by foreign-service engineers.

Chinese machine operators have often very little experience with the operation of a Microsoft Windows user interface. Therefore, the application of the Windows user interface for machine operation in China will meet some problems at the beginning. Furthermore, the more freely structured dialogue provided by Ms Windows user interface, in comparison to guided dialogue, also could make Chinese machine operators unsure about their operations. The results suggested that the application of the Ms Windows user interface in China is not encouraged for machine operation at present time.

The most obviously different requirements in menu structure come from the different working organization in China. A Chinese operator's task range is narrower than that for a German machine operator. In fact, the interviews have pointed out that the Chinese users require a self-defined user interface to fit their individual production tasks. In conclusion, the menu structure should correspond well to the operation tasks of the Chinese machine operators and the cultural-based organizational and structural diversity. The new menu structure should be characterized by separated (and hierarchical) access rights of different machine operators to different machine functions, with the menu structure for each operator group thus simplified. For machine operators, the operation functions should be much simpler. This could also make the operation of specific machine functions more quickly reachable.

Because the actual working organization for each customer is different, it is impossible for machine producers to provide an individual menu structure for each customer. Generally, there should be one consistent menu structure provided by machine producers and the flexibility of users to configure the structure specifically for a particular working organization. Based on this consideration, the modularization of the menu structure to leave room for further adaptation is a good design strategy. Then, the menu structure for a specific customer could be configured according to individual needs.

Chinese customers have a very low assessment of their own machine operators. The operators themselves also have very low self-evaluations. Both groups often made a remark that operators have quite low qualifications and could not understand complicated machine operations. This status suggests that machine operators could have potential fear to actively interact with the machine and would prefer to follow the definite operation instructions provided by the system. Consequently, the dialogue system must provide more error tolerance for the operation and should have a very clear guide for workers to enable them to follow the operation process.

The questionnaire's results show that the main problem of the Chinese machine operators is based on a bad understanding of different machine functions and different operation processes. Accordingly, the user interface must provide a navigation concept that presents general machine functions and operation processes in an overview to enable operators to establish easily a clearer overall image of machine operation.

FUTURE TRENDS

Finally, this article should mention cultural diversity in user requirements and their relevance in time of globalization. Mainland China is only an example for a new and increasing market with an interesting and diverse user culture.

Product localization is not possible without requirement analysis in the target user culture. Only this can guarantee to meet the user expectations on product use in the future. This method is, however in internationalization or localization context, a needed base to engineer user-oriented human machine systems (see Röse, 2002).

For researchers and developers working on advancing user-oriented design, one must realize that, in time of globalization, the culture-orientation is an essential component for successful usability and user friendliness. Culture is an influence factor on user interface design, and it is also an element of user-experiences. Engineers of products for the global market have to address this issue (see Röse, 2001).

CONCLUSION

This article described the INTOPS-2 project and summarized general advice of lessons learned for such intercultural study.

First, it needed around nine months to prepare the study, which included analyzing cultural specifics, studying regulations and analyzing system documentations, construction of test, and questionnaire materials, building up an industrial-based and founded working group to discuss problems detailing the focus of the study, contacting firms, arranging visits, conducting pre-tests, and many other tasks. Careful preparation is very important for completing an evaluation of an intercultural project. For such field analysis, everything must be made easy

Second, all the test material was translated to Chinese. This is the only acceptable method of localization research and a guarantee for the collection of trustworthy relevant data.

Third, the investigation in Mainland China was carried out by a native-speaking Chinese member of the team. His background knowledge was product design and usability, and he was deeply involved in the preparation of the study.

Without these frame conditions, it is not possible to get real cultural impact data from a user target culture. Culture-oriented design will continue to be a challenge. The sooner software engineers start to integrate cultural diversity of users into international product development, the sooner products become more fit for a global market.

ACKNOWLEDGMENT

Many thanks to the sponsors of this project, the companies: MAN Roland Druckmaschinen AG, IWKA AG, Bühler AG Amriswil-CH and Rittal Rudolf Loh GmbH & Co. KG. A special thanks to the PhD student Long Liu. He collected the data in Mainland China and worked with many engagements in this project.

REFERENCES

Bourges-Waldegg, P. (2000, July 13-15). Globalization: A threat to cultural diversity? In D. Day, E. M. del Galdo, & G. V. Prabhu (Eds.), *Designing for Global Markets 2, Second International Workshop on Internationalisation of Products and Systems, IWIPS 2000*, Baltimore (pp. 115-124). Backhouse Press.

Choong, Y., & Salvendy, G. (1998). Designs of icons for use by Chinese in mainland China. *Interacting with Computers: The Interdisciplinary Journal of Human-Computer-Interaction, 9*(February), 417-430.

del Galdo, E. M. (1990). Internationalization and translation: Some guidelines for the design of human-computer interfaces. In J. Nielsen (Ed.), *Designing user interfaces for international use* (pp. 1-10). Amsterdam: Elsevier Science Publishers.

Dong, J. M., & Salvendy, G. (1999). Designing menus for the Chinese population: Horizontal or vertical? *Behavior & Information Technology, 18*(6), 467-471.

Hofstede, G. (1991). *Cultures and organizations: Software of the mind: Intercultural cooperation and its importance for survival.* New York: McGraw-Hill.

Hoft, N. (1996). Developing a cultural model. In E. M. del Galdo, & J. Nielsen (Eds.), *International user interfaces* (pp. 41-73). New York: John Wiley & Sons.

Marcus, A. (1996). Icon and symbol design issues for graphical user interfaces. In E. M. del Galdo, & J. Nielsen (Eds.), *International user interfaces* (pp. 257-270). New York: John Wiley & Sons.

Piamonte, D. P. T., Abeysekera, J. D. A., & Ohlsson, K. (1999, August 22-26). Testing videophone graphical symbols in Southeast Asia. In H.-J. Bullinger, & J. Ziegler (Eds.), *Human-Computer Interaction: Ergonomics and User Interfaces, Vol. 1, Proceedings 8th International Conference on Human-Computer Interaction, HCI International '99*, Munich, Germany (pp. 793-797).

Prabhu, G., & Harel, D. (1999, August 22-26). GUI design preference validation for Japan and China—A case for KANSEI engineering? In H. -J. Bullinger & J. Ziegler (Eds.), *Human-Computer Interaction: Ergonomics and User Interfaces, Vol. 1, Proceedings 8th International Conference on Human-Computer Interaction, HCI International '99*, Munich, Germany (pp. 521-525).

Röse, K. (2001, September 18-20). *Requirements for the future internationalization and localization of technical products.* In G. Johannsen (Ed.), *8th IFAC/IFIPS/IFORS/IEA Symposium on Analysis, Design, and Evaluation of Human-Machine Systems*, Kassel, Germany (pp. 127-132).

Röse, K. (2002). *Methodik zur Gestaltung interkultureller Mensch-Maschine-Systeme in der Produktionstechnik (Method for the design of intercultural human machine systems in the area of production automation).* Dissertationsschrift zur Erlangung des akademischen Grades, Doktor-Ingenieur' (Dr.-Ing.) im Fachbereich Maschinenbau und Verfahrenstechnik der Universität Kaiserslautern, Fortschritt-Bericht pak, Nr. 5, Universität Kaiserslautern.

Röse, K. (2002a, May 22-25). Models of culture and their applicability for designing user interfaces. In H. Luczak, A. Cakir, & A. Cakir (Eds.), *WWDU 2002, Work With Display Units, World Wide Work. Proceedings of the 6th International Scientific Conference on Work with Display Units*, Berchtesgaden (pp. 319-321).

Röse, K., Liu, L., & Zühlke, D. (2001, October 8-9). The needs of Chinese users on German machines: Localization as one sales aspect. In R. Bernhardt, & H. -H. Erbe (Eds.), *6th IFAC Symposium on Cost Oriented Automation (Low Cost Automation 2001)*, Berlin, Germany (pp. 167-172). Preprints.

Shih, H. M., & Goonetilleke, R. S. (1998). Effectiveness of menu orientation in Chinese. *Human Factors, 40*(4), 569-576.

Trompenaars, F. (1993). *Riding the waves of culture: Understanding cultural diversity in business.* London: The Economist Press.

Zühlke, D., Romberg, M., & Röse, K. (1998, September 16-18). Global demands of non-European markets for the design of user-interfaces. In the *Proceedings of the 7th IFAC/IFIP/IFORS/IEA Symposium, Analysis, Design and Evaluation of Man-Machine Systems*, Kyoto, Japan. Kyoto: IFAC, Japan: Hokuto Print.

KEY TERMS

CNC Machine: Tool machine with a computer numeric control; a standard in the mechanical engineering field.

Culture-Oriented Design: Specific kind of user-oriented design. Also focused on the user as a central element of development, but taking into account the cultural diversity of different target user groups.

INTOPS-1: Requirements of the non-European market on machine design. This project was founded by the German ministry of education and research (1996-1998).

INTOPS-2: Requirements of the user in Mainland China for Human Machine Systems in the area of production automation. This project was founded by several companies from Germany and Switzerland (2000-2001).

Product Localization: Optimization of a product for a specific target culture; could also be the development of a product only and alone for this specific target culture (not so often).

Target User Group: Refers to a focus user group, which a product or a development process specifically targets or aims at. The qualities of such a group are a relevant baseline for the developer (used as orientation and target for the development process and the features of the new designed product). Their requirements are relevant for the orientation of the developer.

User-Oriented Design: Development approach with a focus on user requirements and user needs as a basis for a system or product development.

The Culture(s) of Cyberspace

Leah P. Macfadyen
The University of British Columbia, Canada

INTRODUCTION

Computer-mediated communication between humans is becoming ubiquitous. Computers are increasingly connected via high-speed local and wide-area networks, and via wireless technologies. High bandwidth interaction is increasing communication speed, offering the possibility for transmission of images, voice, sound, video and formatted data as well as text. Computer technologies are creating the possibility of entirely new interfaces of human-machine interaction, and entirely new virtual "spaces" for human-human interaction. As a collectivity, these new spaces of communication are known as cyberspace.

Human-human interaction is the foundation of culture. Vygotsky and Luria's (1994) model of cultural development highlights the need to consider the culture(s) of cyberspace ("cyberculture(s)") in any examination of computer-mediated human communications, because it invokes both the communicative and behavioural practices that humans employ as they interact with their environment.

BACKGROUND

Vygotsky and Luria (1994) propose that human beings use multiple psychological structures to mediate between themselves and their surroundings. Structures classified as *signs* include linguistic and non-linguistic mechanisms of communication; structures classified as *tools* encompass a wide range of other behavioural patterns and procedures that an individual learns and adopts in order to function effectively within a culture or society. Together, signs and tools allow individuals to process and interpret information, construct meaning and interact with the objects, people and situations they regularly encounter. When these elaborate mediating structures, finely honed to navigate a specific environment, encounter a different one, they can malfunction or break down completely.

In the context of the Internet, human beings do not simply interact with digital interfaces. Rather, they bring with them into cyberspace a range of communicative and behavioural cultural practices that impact their ability to interact with technology interfaces, with the culture of the virtual spaces they enter, and with other humans they encounter there. Their individual and group cultural practices may or may not "match" the practices of the virtual culture(s) of cyberspace. Some investigators have gone as far as to suggest that the sociocultural aspects of computer-mediated human interaction are even more significant than technical considerations of the interface in the successful construction and sharing of meaning. This article surveys current theories of the nature and construction of cyberculture(s), and offers some brief thoughts on the future importance of cyberculture studies to the field of HCI.

KEY DEBATES IN CYBERCULTURE STUDIES

Perhaps the most striking feature of the body of current literature on cyberculture is the polarization of debate on almost every issue. A few authors examine these emerging paradoxes directly. Fisher and Wright (2001) and Poster (2001) explicitly compare and contrast the co-existing utopian and dystopian predictions in discourse surrounding the Internet. Lévy (2001a), Poster (2000), and Jordan (1999) go as far as to suggest that the very nature of the Internet itself *is* paradoxical, being universalizing but non-totalizing, liberating *and* dominating, empowering *and* fragmenting, constant only in its changeability. Most writers thus far have tended, however, to theorize *for* one side or the other within polarized debates, as will become evident next.

Utopia or Dystopia?

While not explicitly espousing technological instrumentalism (an assumption that technology is "culture neutral"), a number of writers offer utopian visions for the so-called Information Superhighway. Such theorists predict that the emancipatory potential of Internet communications will help to bring about new forms of democracy and new synergies of collective intelligence within the Global Village of cyberspace (Ess, 1998; Lévy, 2001a, 2001b; Morse, 1997).

Their detractors argue that these writers ignore the reality that culture and cultural values are inextricably linked to both the medium and to language (Anderson, 1995; Benson & Standing, 2000; Bijker & Law, 1992; Chase, Macfadyen, Reeder, & Roche, 2002; Gibbs & Krause, 2000; Pargman, 1998; Wilson, Qayyum, & Boshier, 1998) and that cyberculture "originates in a well-known social and cultural matrix" (Escobar, 1994, p. 214). These theorists more commonly offer dystopian and technologically deterministic visions of cyberspace, where money-oriented entrepreneurial culture dominates (Castells, 2001), which reflects and extends existing hierarchies of social and economic inequality (Castells, 2001; Escobar, 1994; Jordan, 1999, Keniston & Hall, 1998; Kolko, Nakamura, & Rodman, 2000; Luke, 1997; Wilson et al., 1998), and which promotes and privileges American/Western cultural values and the valorization of technological skills (Anderson, 1995; Castells, 2001; Howe, 1998; Keniston & Hall, 1998; Luke, 1997; Wilson et al., 1998).

These and other thematically polarized arguments about cyberculture (such as "Internet as locus of corporate control" versus "Internet as new social space" (Lévy, 2001a) or "Internet as cultural context" versus "Internet as a cultural artifact" (Mactaggart, 2001) are evident in the philosophical arguments underlying work listed in other sections of article.

Modern or Postmodern?

A second major division in theoretical discussions of the nature and culture of the cyberspace is the question of whether the Internet (and its associated technologies) is a modern or postmodern phenomenon. Numerous writers frame the development of Internet technologies, and the new communicative space made possible by them, as simply the contemporary technical manifestation of "modern ideals, firmly situated in the revolutionary and republican ideals of liberty, equality and fraternity" (Lévy, 2001a, p. 230). Emphasizing the coherence of current technologies with ongoing cultural evolution(s), Escobar (1994) discusses the Western cultural foundations of technological development, and Gunkel and Gunkel (1997) theorize that the logic of cyberspace is simply an expansion of colonial European expansionism. Castells (2001) sees cyberculture as emerging from an existing culture of scientific and technological excellence "enlisted on a mission of world domination" (p. 60). Orvell (1998) pointedly argues that "debates about postmodernity have evinced a kind of amnesia about the past" (p. 13) and claims that cyberspace and virtual reality technologies are continuous with the Romantic imagination as it developed in the 1830s and 1840s. Disembodiment, he argues, is not a new product of the modern age, but was the "triumph of the Romantic imagination" (p. 16).

More recently, other writers have begun to envision the cultural sphere of cyberspace as radically new, postmodern, and signifying a drastic break with cultural patterns of community, identity and communication. For example, Webb (1998) suggests that the frontier metaphors of cyberspace symbolize a postmodern shift from human/territorialized to non-human/deterritorialized computer-mediated environments. Poster (2000) claims that Internet technologies have actually brought into being a "second order of culture, one apart from the synchronous exchange of symbols and sounds between people in territorial space" (p. 13). He predicts that the cultural consequences of this innovation must be "devastation for the modern" (p. 13), and (2001) reformulates for this context the propositions of postmodern theorists such as Foucault, Heidegger, Deleuze, Baudrillard and Derrida who challenge modernist notions of progress, definable "authentic" selfhood and the existence of absolute foundations for or structures of, knowledge (for a short review of postmodern thought see Schutz, 2000). Poster argues effectively that postmodern perspectives on life and culture that go beyond old notions of fixed social structures may be most relevant for the fluid, dynamic, and even contradictory cultured environment of cyberspace.

Cybercultural Values

Are the values of cyberculture simply the imported values of existing non-virtual cultures? Or does cyberculture represent a newly evolving cultural milieu? Various authors speculate about the nature and origins of cybercultural values. Anderson (1995) argues that cyberculture values are "speed, reach, openness, quick response" (p. 13). Castells (2001) believes that "hacker culture" is foundational to cyberculture, and carries with it meritocratic values, an early notion of cybercommunity, and a high valuing of individual freedom. Jordan (1999) contends that it is *power* which structures culture, politics and economics, and theorizes the existence of "technopower" as the power of the Internet élite that shapes the normative order of cyberculture. Later (Jordan, 2001), and similarly to Castells, he elaborates on the Anglo-American language and culture bias of cyberspace, which he argues is founded on competition and informational forms of libertarian and anarchist ideologies. Knupfer (1997) and Morse (1997) explore the gendered (masculine) nature of cyberculture, while Kolko et al. (2000) theorize that (the predominantly Anglo-American) participants in new virtual communities bring with them the (predominantly American) values of their home cultures. As a result, and as Star (1995) had already pointed out, "there is no guarantee that interaction over the net will not simply replicate the inequities of gender, race and class we know in other forms of communication" (p. 8). Essays in Shields' (1996) edited collection examine features of cybercultural values and practice such as attitudes to censorship, social interaction, politics of domination and gendered practices of networking.

Others, however, speculate that cyberspace is the site of creation of an entirely new culture. Lévy (2001a) argues, for example, that cyberculture expresses the rise of "a *new* universal, different from the cultural forms that preceded it, because it is constructed from the indeterminateness of global meaning" (p. 100), and Healy (1997) characterizes cyberspace as a middle landscape between civilization and wilderness, where new cultural directions and choices can be selected.

Subcultures of/in Cyberspace

Are subcultures identifiable within the cultural milieu of cyberspace? A number of theorists discuss the online cultures of specific subgroups: Castells (2001) discusses the "hacker culture" in detail, Leonardi (2002) investigates the online culture and website designs of U.S. Hispanics, and Gibbs and Krause (2000) explore the metaphors used by different Internet subcultures (hackers, cyberpunks).

Rather than simply itemizing and describing cyberspace subcultures, however, a growing number of studies are exploring the marginalization in or lack of access to cyberspace of some cultural groups. Howe (1998) argues, for example, that radical differences in cultural values make cyberspace inhospitable for Native Americans. Keniston and Hall (1998) offer statistics on the English language and Western dominance of the Internet; they discuss the reality that 95% of the population of the world's largest democracy—India—are excluded from computer use because they lack English language fluency. Anderson (1995) suggests that the "liberal humanist traditions of Islamic and Arab high culture" (p. 15) are absent from the world of cyberspace, not because they do not translate to new media, but because they are literally drowned out by the cultural values attached to the dominant language and culture of cyberspace as it is currently configured. Similarly, Ferris (1996), Morse (1997) and Knupfer (1997) suggest that the gendered culture of cyberspace has tended to exclude women from this virtual world. Interestingly, Dahan (2003) reports on the limited online public sphere available to Palestinian Israelis relative to the Jewish majority population in Israel. Rather than being the result of Anglo-American cultural domination of the Internet, this author convincingly argues that this imbalance demonstrates that "existing political and social disenfranchisement" (¶ 1) in Israeli society is simply mirrored in cyberspace. More generally, Davis (2000) reports on the potential for disenfranchisement from the "global technoculture" depending on their different manifestations of "a diversity of human affinities and values" (p. 105). Stald and Tufte (2002) meanwhile present a series of reports on cyberspace activities of a number of discrete and identifiable minority communities: rural black African males in a South

African university, South Asian families in London, women in India, Iranian immigrants in London, and young immigrant Danes.

In Search of Utopia: Cultural Impact and Technology Design

If we accept cyberculture as a value system that embodies free speech, individual control, and a breaking down of the barrier of distance, does this imply that existing (or perceived) inequalities can be corrected in pursuit of this utopia? Keniston and Hall (1998) attempt to detail issues that must be faced in such efforts: attention to nationalist reactions to English-speaking élites, development of standardized forms for vernacular languages, and the "real (and imagined)" (p. 331) challenges faced by North American software firms when dealing with South Asian languages. Benson and Standing (2000) consider the role that policy plays in the preservation of cultural values, and present a new evaluation framework for assessing the impact of technology and communication infrastructure on culture. Wilson et al. (1998) propose that a framework based on Chomsky's 1989 analysis of mass media can help determine the extent to which American corporations and institutions dominate (and thus determine the culture of) cyberspace.

Of particular relevance to the field of HCI may be some preliminary efforts to tailor online environments to particular cultural groups. For example, Turk (2000) summarizes recent attempts to establish relationships between the culture of users and their preferences for particular user interfaces and WWW site designs, while Leonardi (2002) reports on a study of the manifestation of "Hispanic cultural qualities" (p. 297) (cultural qualities perceived within the US-American context to derive from "Spanish" cultures) in Web site design, and makes design recommendations for this community. Heaton (1998) draws on Bijker and Law's (1992) notion of "technological frame" to explain how Japanese designers invoke elements of Japanese culture in justifying technical decisions. The theoretical and methodological challenges of this approach to technology design are explored in greater detail in "Internet-Mediated Communication at the Cultural Interface" (Macfadyen, 2006), contained in this encyclopedia.

FUTURE TRENDS

An ongoing tension is apparent in the existing cyberculture literature between theories that assume "importation of pre-existing cultures" and theories that anticipate new "cultural construction" in the emerging communicative spaces of cyberspace. In the former, theorists have tended to identify and characterize or categorize groups of communicators in cyberspace using rather deterministic or essentialist (usually ethnic) cultural definitions, and have then theorized about the ways in which such groups import and impose, or lose, their cultural practices in the cyberspace milieu. Sociopolitical analyses have then built on such classifications by positioning cyberspace communicators as constrained or privileged by the dominant cyberculture. While problematic, application of such static definitions of culture has allowed some preliminary forays into development of so-called culturally appropriate "versions" of human-computer interfaces and online environments.

A continuing challenge, however, is perceived to be the lack of an adequate theory of culture that would allow analysis of the real complexities of virtual cultures and virtual communities (Ess, 1998), and that could guide better technology and interface design. Recently, however, a few theorists have begun to question the utility of static and "classificatory" models or definitions of culture. Abdelnour-Nocera (2002) instead argues that examination of "cultural construction from inside the Net" (¶ 1) is critical. Benson and Standing (2000) offer an entirely new systems theory of culture that emphasizes culture as an indivisible system rather than as a set of categories. Importantly, challenging postmodern theorists such as Poster (2001) argue that the Internet demands a social and cultural theory all its own. Common to these theories is an underscoring of the dynamic nature of culture, and of the role of individuals as active agents in the *construction* of culture—online or off-line.

CONCLUSION

Whenever humans interact with each other over time, new cultures come into being. In cyberspace, networked computer technologies facilitate and shape (or impede and block) the processes of cultural construction. Although debates over the nature of cyberculture continue to rage, one point is increasingly clear: in the field of human-computer interaction, it is no longer sufficient to focus solely on the interface between individual humans and machines. Any effort to examine networked human communications must take into consideration human interaction with and within the cultures of cyberspace that computer technologies bring into being.

REFERENCES

Abdelnour-Nocera, J. (2002). Ethnography and hermeneutics in cybercultural research. Accessing IRC virtual communities. *Journal of Computer-Mediated Communication*, *7*(2). Retrieved from http://www.jcmc.indiana.edu/jcmc/vol7/issue2/nocera.html

Anderson, J. (1995). "Cybarites", knowledge workers and new creoles on the superhighway. *Anthropology Today*, *11*(4), 13-15.

Benson, S., & Standing, C. (2000). A consideration of culture in national IT and e-commerce plans. In F. Sudweeks & C. Ess (Eds.). *Proceedings, Cultural Attitudes Towards Technology and Communication, 2000,* Perth, Australia (pp. 293-303). Australia: Murdoch University.

Bijker, W. E., & Law, J. (Eds.). (1992). *Shaping technology/building society: Studies in sociotechnical change*. Cambridge, MA: MIT Press.

Castells, M. (2001). *The Internet galaxy. Reflections on the Internet, business and society.* Oxford: Oxford University Press.

Chase, M., Macfadyen, L. P., Reeder, K., & Roche, J. (2002). Intercultural challenges in networked learning: Hard technologies meet soft skills. *First Monday*, *7*(8). Retrieved from http://firstmonday.org/issues/issue7_8/chase/index.html

Chomsky, N. (1989). *Necessary illusions*. Toronto, Canada: Anansi Press.

Dahan, M. (2003). Between a rock and a hard place: The changing public sphere of Palestinian Israelis. *Journal of Computer-Mediated Communication*, *8*(2). Retrieved from http://www.ascusc.org/jcmc/vol8/issue2/dahan.html

Davis, D. M. (2000). Disenfranchisement from the global technoculture. In F. Sudweeks, & C. Ess (Eds.), *Proceedings, Cultural Attitudes Towards Technology and Communication, 2000,* Perth, Australia (pp. 105-124). Australia: Murdoch University.

Escobar, A. (1994). Welcome to cyberia: Notes on the anthropology of cyberculture. *Current Anthropology*, *35*(3), 211-231.

Ess, C. (1998, August 1-3). First looks: CATaC '98. In C. Ess, & F. Sudweeks (Eds.), *Proceedings, Cultural Attitudes Towards Communication and Technology, 1998*, London (pp. 1-17). Australia: University of Sydney. Retrieved from http://www.it.murdoch.edu.au/catac/catac98/01_ess.pdf

Ferris, P.S. (1996). Women on-line: Cultural and relational aspects of women's communication in online discussion groups. *Interpersonal Computing and Technology*, *4*(3-4), 29-40. Retrieved from http://jan.ucc.nau.edu/~ipct-j/1996/n3/ferris.txt

Fisher, D. R., & Wright, L. M. (2001). On utopias and dystopias: Toward an understanding of the discourse surrounding the Internet. *Journal of Computer-Mediated Communication*, *6*(2). Retrieved from http://www.ascusc.org/jcmc/vol6/issue2/fisher.html

Gibbs, D., & Krause, K.-L. (2000). Metaphor and meaning: Values in a virtual world. In D. Gibbs, &. K.-L. Krause (Eds.), *Cyberlines. Languages and cultures of the Internet* (pp. 31-42). Albert Park, Australia: James Nicholas Publishers.

Gunkel, D. J., & Gunkel, A. H. (1997). Virtual geographies: The new worlds of cyberspace. *Critical Studies in Mass Communication*, *14*(2), 123-37.

Healy, D. (1997). Cyberspace and place: The Internet as middle landscape on the electronic frontier. In D. Porter (Ed.), *Internet culture* (pp. 55-72). New York: Routledge.

Heaton, L. (1998). Preserving communication context: Virtual workspace and interpersonal space in Japanese CSCW. In C. Ess, & F. Sudweeks (Eds.), *Proceedings, Cultural Attitudes Towards Technology and Communication 1998,* London (pp. 207-230). Sydney, Australia: University of Sydney. Retrieved from http://www.it.murdoch.edu.au/catac/catac98/pdf/18_heaton.pdf

Howe, C. (1998). Cyberspace is no place for tribalism. *WICAZO-SA Review, 13*(2), 19-28.

Jordan, T. (1999). *Cyberculture: The culture and politics of cyberspace and the Internet.* London: Routledge.

Jordan, T. (2001). Language and libertarianism: The politics of cyberculture and the culture of cyberpolitics. *The Sociological Review, 49*(1), 1-17.

Keniston, K., & Hall, P. (1998). Panel: Global culture, local culture and vernacular computing. In C. Ess, & F. Sudweeks (Eds.), *Proceedings, Cultural Attitudes Towards Communication and Technology,* London (pp. 329-331). Australia: University of Sydney. Retrieved from http://www.it.murdoch.edu.au/catac/catac98/pdf/28_hall.pdf

Knupfer, N. N. (1997, February 14-18). New technologies and gender equity: New bottles with old wine. In the *Proceedings of Selected Research and Development Presentations at the 1997 National Convention of the Association for Educational Communications and Technology,* 19th, Albuquerque, NM.

Kolko, B. E., Nakamura, L., & Rodman, G. B. (Eds.) (2000). *Race in cyberspace.* New York: Routledge.

Leonardi, P. M. (2002, July 12-15). Cultural variability in Web interface design: Communicating US Hispanic cultural values on the Internet. In F. Sudweeks, & C. Ess (Eds.), *Proceedings, Cultural Attitudes Towards Technology and Communication, 2002,* Université de Montréal, Canada (pp. 297-315). Australia: Murdoch University.

Lévy, P. (2001a). *Cyberculture.* Minneapolis: University of Minnesota Press.

Lévy, P. (2001b). *The impact of technology in cyberculture.* Minneapolis: University of Minnesota Press.

Luke, W. T. (1997). The politics of digital inequality: Access, capability and distribution in cyberspace. *New Political Science,* 41-42. Retrieved from http://www.urbsoc.org/cyberpol/luke.shtml

Macfadyen, L. P. (2006). Internet-mediated communication at the cultural interface. In C. Ghaoui (Ed.), *Encyclopedia of Human Computer Interaction.* Hershey, PA: Idea Group Reference.

Mactaggart, J. (2001, June). Book review of *Virtual ethnography* (C. Hines, Ed.). Retrieved December 2002, from the Resource Centre for Cyberculture Studies, http://www.com.washington.edu/rccs/bookinfo.asp?ReviewID=125&BookID=109

Morse, M. (1997). Virtually female: Body and code. In J. Terry, & M. Calvert, (Eds.), *Processed lives: Gender and technology in everyday life* (pp. 23-36). London: Routledge.

Orvell, M. (1998). Virtual culture and the logic of American technology. *Revue française d'études americaines,* 76, 12-27.

Pargman, D. (1998). Reflections on cultural bias and adaptation. In C. Ess, & F. Sudweeks (Eds.), *Proceedings, Cultural Attitudes Towards Communication and Technology* London (pp. 81-99). Australia: University of Sydney. Retrieved from http://www.it.murdoch.edu.au/catac/catac98/pdf/06_pargman.pdf

Poster, M. (2000). *The digital culture and its intellectuals: From television, to tape, to the Internet.* Berlin: Freie Universität.

Poster, M. (2001). *What's the matter with the Internet?* Minneapolis: University of Minnesota Press.

Schutz, A. (2000). Teaching freedom? Postmodern perspectives. *Review of Educational Research, 70*(1), 215-251.

Shields, R. (Ed.). (1996). *Cultures of Internet: Virtual spaces, real histories, and living bodies.* London: Sage Publications.

Stald, G., & Tufte, T. (Eds.). (2002). *Global encounters: Media and cultural transformation.* Luton, UK: University of Luton Press.

Star, S. L. (Ed.). (1995). *The cultures of computing.* Oxford: Blackwell.

Turk, A. (2000). A WorldWide Web of cultures, or a WorldWide Web culture? In F. Sudweeks, & C. Ess (Eds.), *Proceedings, Cultural Attitudes Towards Technology and Communication,* Perth (pp. 243-256). Australia: Murdoch University.

Vygotsky, L. S., & Luria, A. R. (1994). Tool and symbol in child development. In R. van der Veer, & J. Valsiner (Eds.), *The Vygotsky reader* (Trans. T. Prout & S. Valsiner) (pp. 97-174). Cambridge. MA: Blackwell Publishers.

Webb, S. (1998). Visions of excess: Cyberspace, digital technologies and new cultural politics. *Information, Communication & Society, 1*(1), 46-69.

Wilson, M., Qayyum, A., & Boshier, R. (1998). WorldWide America? Think globally, click locally. *Distance Education, 19*(1), 109-123.

KEY TERMS

Culture: Multiple definitions exist, including essentialist models that focus on shared patterns of learned values, beliefs, and behaviours, and social constructivist views that emphasize culture as a shared system of problem-solving or of making collective meaning. The key to the understanding of online cultures—where communication is as yet dominated by text—may be definitions of culture that emphasize the intimate and reciprocal relationship between culture and language.

Cyberculture: As a social space in which human beings interact and communicate, cyberspace can be assumed to possess an evolving culture or set of cultures ("cybercultures") that may encompass beliefs, practices, attitudes, modes of thought, behaviours and values.

Cyberspace: While the "Internet" refers more explicitly to the technological infrastructure of networked computers that make worldwide digital communications possible, "cyberspace" is understood as the virtual "places" in which human beings can communicate with each other, and that are made possible by Internet technologies. Lévy (2001a) characterizes cyberspace as "not only the material infrastructure of digital communications but...the oceanic universe of information it holds, as well as the human beings who navigate and nourish that infrastructure."

Dystopia: The converse of Utopia, a Dystopia is any society considered to be undesirable. It is often used to refer to a fictional (often near-future) society where current social trends are taken to terrible and socially-destructive extremes.

Modern: In the social sciences, "modern" refers to the political, cultural, and economic forms (and their philosophical and social underpinnings) that characterize contemporary Western and, arguably, industrialized society. In particular, modernist cultural theories have sought to develop rational and universal theories that can describe and explain human societies.

Postmodern: Theoretical approaches characterized as postmodern, conversely, have abandoned the belief that rational and universal social theories are desirable or exist. Postmodern theories also challenge foundational modernist assumptions such as "the idea of progress," or "freedom."

Technological Determinism: The belief that technology develops according to its own "internal" laws and must therefore be regarded as an autonomous system controlling, permeating, and conditioning all areas of society.

Technological Instrumentalism: The view that technologies are merely useful and "culture-neutral" instruments, and that they carry no cultural values or assumptions in their design or implementation.

Utopia: A real or imagined society, place, or state that is considered to be perfect or ideal.

Design Frameworks

John Knight
University of Central England, UK

Marie Jefsioutine
University of Central England, UK

INTRODUCTION

Design frameworks are a phenomena appearing in the field of new media (e.g., Brook & Oliver, 2003; Fiore, 2003; Dix, Rodden, Davies, Trevor, Friday, & Palfreyman, 2000; Taylor, Sumner, & Law, 1997). They appear to be a response to the multi-disciplinary nature of the field and have a number of things in common. They are usually developed in response to a perceived lack of common understanding or shared reference. Frameworks often advocate a set of principles, a particular ethos, or expound a philosophical position, within which a collection of methods, approaches, tools, or patterns are framed. They aim to support design analysis, decision-making and guide activity, and provide a common vocabulary for multi-disciplinary teams. In contrast to some design methods and models, they tend to be broad and encompass a wider area of application. Rather than prescribe a single "correct" way of doing something, they provide a guiding structure that can be used flexibly to support a range of activity. This article describes one design framework, the experience design framework (Jefsioutine & Knight, 2004) to illustrate the concept.

BACKGROUND

The experience design framework (EDF) illustrates a number of the features of design frameworks identified previously. It was developed in response to the low take-up of user-centred design observed by the authors and identified in the literature (e.g., Landauer, 1996; Nielsen, 1994). For example, Säde (2000, p. 21) points out that some of the large-scale user-centred design (UCD) methods "do not suit the varied and fast paced consulting projects of a design firm." Nielsen suggests that one of the key reasons why usability engineering is not used in practice is the perceived cost. He argues that a "discount usability engineering" approach can be highly effective and describes a set of "simpler usability methods" (Nielsen, 1994, pp. 246-247). Eason and Harker (1988) found that, as well as perceived cost and duration, user-centred methods were not used because designers felt that useful information was either not available when needed or was not relevant and that methods did not fit in with their design philosophy.

The authors thus set about identifying a set of user-centred methods that would be cost effective, flexible enough to apply to any design life cycle and, most importantly, would be useful and relevant to the needs of the designer. Through a combination of literature reviews and application to practice, the authors identified different aspects of designing a user experience and the way in which these aspects can be drawn together to focus design research and practice. The EDF is thus based on the principles of user-centred design and represents a way of using a range of methods to achieve a set of qualities that work at all dimensions of experience.

USER-CENTRED DESIGN PRINCIPLES (UCD)

Human-centred design processes for interactive systems identifies the following characteristics of a user-centred design process: "The active involvement of users and a clear understanding of user and task requirements; An appropriate allocation of function between users and technology; The iteration of design solutions; Multidisciplinary design" (International Organization for Standardization, ISO/IEC

13407, 1999). Additionally, Gould and Lewis (1985) emphasise the importance of early and continual user testing and integrating all aspects of usability.

These principles of UCD set out a clear approach around which to plan a design life cycle, but they focus very much on design for usability. The EDF proposes that the same principles be applied to other qualities of design.

Qualities, Dimensions and Effectors of an Experience

It was felt that one of the reasons UCD methods were seen as irrelevant and limited was that the traditional focus on usability does not capture other aspects of the user-experience. The EDF identifies a broader set of qualities that address the less tangible aspects of an experience, such as pleasure and engagement. It then identifies the different dimensions of experiencing, visceral, behavioural, reflective, and social (from Jordan, 2000; Norman, 2003) that need to be addressed to design a holistic user experience. It identifies a number of aspects that have an effect on an experience, such as who, why, what, where, when, and how, that help to guide research, design, and evaluation.

METHODS AND TOOLS

Product design, HCI, and human factors research are awash with methods and tools that can be used to support user-centred design. Generally, tools have focused on technological aspects of design, either in terms of making coding easier or automating aspects of design. Where tools have related to usability, this has often focused on evaluation. A less developed area is in tools that support the understanding of the user at early stages of design and supporting the entire user-centred design process (some rare examples are HISER, 1994; NIST's WebCAT, 1998).

Jordan (2000) describes a collection of empirical and non-empirical methods suitable for the "new human factors approach" to designing pleasurable products. Rather than prescribing a process or a set of key methods or tools, the EDF suggests that a range of tools and techniques can be employed provided they cover four basic purposes of observing/exploring, participation/empathy, communicating/modelling, and testing/evaluation. Furthermore, by applying these methods in the context of the EDF, a better understanding of the user experience as a whole can be achieved.

Observation and Exploration

These methods are about finding out and can be drawn from demography, ethnography, market research, psychology, and HCI (e.g., task analysis, field observation, interviews, questionnaires, focus groups, affinity diagramming, laddering, and experience diaries). The EDF indicates the kind of information that should be sought, such as the range of user characteristics including personality, motivations, social affiliations, physical or mental disabilities, and so forth.

Communicating and Modelling

These methods serve to communicate the research data, design requirements, and ideas to a multidisciplinary team who may not have a common vocabulary (e.g., user profiles and personas, use cases or task scenarios, scenario-based design, mood boards, written briefs and specifications, storyboarding, and prototypes). Again, the EDF helps to focus the information that is communicated on issues pertinent to the whole user experience.

Participation and Empathy

These methods represent an approach aimed at gaining a deeper understanding and empathy for users, socio-political and quality of life issues (e.g., immersive methods such ethnographic participant-observation and the "eat your own dog food" approach). Other methods such as participatory design advocate designing *with* users rather than *for* them (see Schuler & Namioka, 1993).

Testing and Evaluating

Gould and Lewis (1985) recommend iterative design based on empirical testing (e.g., usability testing through controlled observation and measurement). The EDF broadens the test and evaluative criteria from the traditional focus on cognitive and behavioural measures, like the time taken to complete a task or

Figure 1. The experience design framework

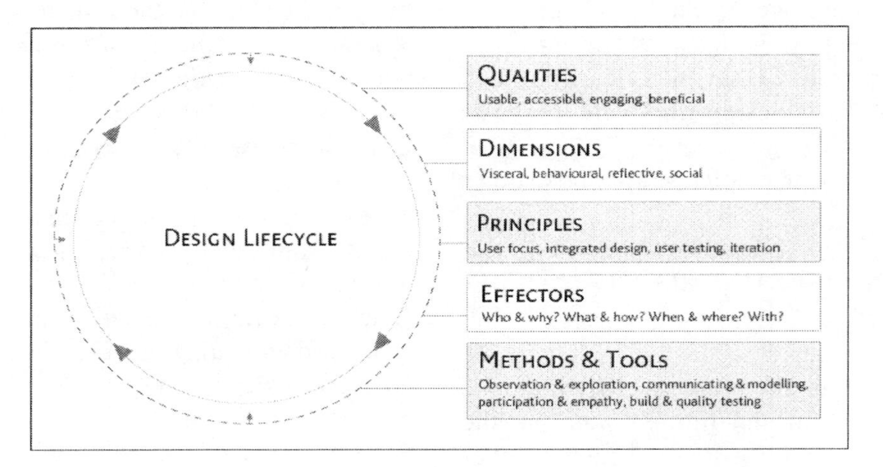

the number of errors or deviations from a critical path, to include methods such as transcript analysis, attitude measurement, and emotional response.

User-Lab has used the EDF to adapt and focus methods for requirements research, brief development, ideation, and testing, and has developed a range of services and training based on this research.

FUTURE TRENDS

The growing number of design frameworks results from an increasingly multi-disciplinary field where any single approach is not sufficient. They are, perhaps, indicative of a rejection of prescriptive approaches to design. Competitive work environments demand flexibility, and practitioners are attracted to tools and methods that can be adapted to their working practices. A design framework provides a general approach for framing these methods and for disseminating multidimensional research in a way that is neither daunting nor demanding extensive study. It is likely that, as the boundaries between disciplines blur, the number of design frameworks will continue to grow.

CONCLUSION

The EDF is an example of a design framework. It was developed to provide a flexible set of methods and tools to guide and support the design process within a principled context, without being prescriptive or restricting, and as such can be used within any design life cycle, and at any stage. It provides a common vocabulary to a multi-disciplinary team and also serves to direct research and the development of new methods and tools. Although designed primarily for digital media product design, it is broad enough to be applied to other areas of product design. It is illustrative of what seems to be a growing trend in multi-disciplinary approaches to design and in bridging the gaps between research and practice.

REFERENCES

Brook, C., & Oliver, R. (2003). Online learning communities: Investigating a design framework. *Australian Journal of Educational Technology, 19*(2), 139-160.

Dix, A., Rodden, T., Davies, N., Trevor, J., Friday, A., & Palfreyman, K. (2000). Exploiting space and location as a design framework for interactive mobile systems. *ACM Transactions on Human Computer Interaction, 17*(3), 285-321.

Eason, K., & Harker, S. (1988). Institutionalising human factors in the development of teleinformatics systems. In R. Speth (Ed.), *Research into Net-*

works and Distributed Applications, European Teleinformatics Conference—EUTECO '88, Vienna, Austria (pp. 15-25). Amsterdam: North-Holland.

Fiore, S. (2003). *Supporting design for aesthetic experience.* Retrieved November 14, 2003, from http://www-users.cs.york.ac.uk/~pcw/KM_subs/Sal_Fiore.pdf

Gould, J. D., & Lewis, C. (1985). Designing for usability: Key principles and what designers think. *Communications of the ACM, 28*(3), 300-311

HISER. (1994). *Usability toolkit.* Retrieved April 26, 2002, from http://www.hiser.com.au

International Organization for Standardization, ISO/IEC 13407. (1999). Human-centred design processes for interactive systems. ISO.

Jefsioutine, M., & Knight, J. (2004). Methods for experience design: The experience design framework. In *Futureground: Proceedings of the Design Research Society Conference, Monash* (in print).

Jordan, P. W. (2000). *Designing pleasurable products.* London: Taylor and Francis.

Landauer, T. K. (1996). *The trouble with computers: Usefulness, usability and productivity.* Cambridge, MA: MIT Press.

Nielsen, J. (1994). Guerrilla HCI: Using discount usability engineering to penetrate the intimidation barrier. In R. Bias, & D. Mayhew (Eds.), *Cost-justifying usability* (pp. 245-270). London: Academic Press.

NIST WebCAT (1998). Category analysis tool. National Institute of Standards and Technology. Retrieved on April 26, 2002, from http://zing.ncsl.nist.gov/WebTools/WebCAT/overview.html

Norman, D. (2003). *Emotional design.* New York: Basic Books.

Säde, S. (2000). Towards user-centred design: A method development project in a product design consultancy. *The Design Journal, 4*(3), 20-32.

Schuler, D., & Namioka, A. (Eds.) (1993). *Participatory design: Principles and practices.* Hillsdale, NJ: Lawrence Erlbaum Associates.

Taylor, J., Sumner, T., & Law, A. (1997). Talking about multimedia: A layered design framework. *Journal of Educational Media, 23*(2 & 3), 215-241. Retrieved January 9, 2004, from http://kn.open.ac.uk/public/document.cfm?documentid=2832

KEY TERMS

Design Framework: An open-ended design methodology that combines research and design activity.

Design Methods: Methods, tools, and techniques employed during research, design, and development.

Design Research: Exploratory activities employed to understand the product, process of design, distribution and consumption, and stakeholders' values and influence.

Discount Usability Methods: A focused set of design and evaluation tools and methods aimed at improving usability with the minimum resources.

Human-Centred Design: An alternative name for user-centred design (UCD) used in ISO process standards.

Multidisciplinary Design: A collaborative approach to design that shares research and design activities among a range of disciplines.

Product Design: An overall term that covers the study and execution of design pertaining to physical products.

Design Rationale for Increasing Profitability of Interactive Systems Development

Xavier Lacaze
Université Paul Sabatier, France

Philippe Palanque
Université Paul Sabatier, France

Eric Barboni
Université Paul Sabatier, France

David Navarre
Université Paul Sabatier, France

INTRODUCTION

User-centred development (Norman & Draper, 1986; Vredenburg, Isensee, & Righi, 2001) processes advocate the use of participatory design activities, end-user evaluations, and brainstorming in the early phases of development. Such approaches work in opposition of some software-engineering techniques that promote iterative development processes such as in agile processes (Beck, 1999) in order to produce software as quickly and as cheaply as possible.

One way of justifying the profitability of development processes promoted in the field of human-computer interaction (HCI) is to not only take into account development costs, but also to take into account costs of use, that is, costs related to employment, training, and usage errors. Gain, in terms of performance (for instance, by providing default values in the various fields of a computer form) or in reducing the impact of errors (by providing undo facilities, for instance), can only be evaluated if the actual use of the system is integrated in the computation of the development costs.

These considerations are represented in Figure 1. The upper bar of Figure 1 shows that development costs (grey part and black part) are higher than the development costs of RAD (rapid application development), represented in the lower bar (grey part). The black part of the upper bar shows the additional costs directly attributed to user-centred design. User-centred development processes compensate additional costs by offering additional payoffs when the system is actually deployed and used.

Figure 1. Comparing the cost of development processes

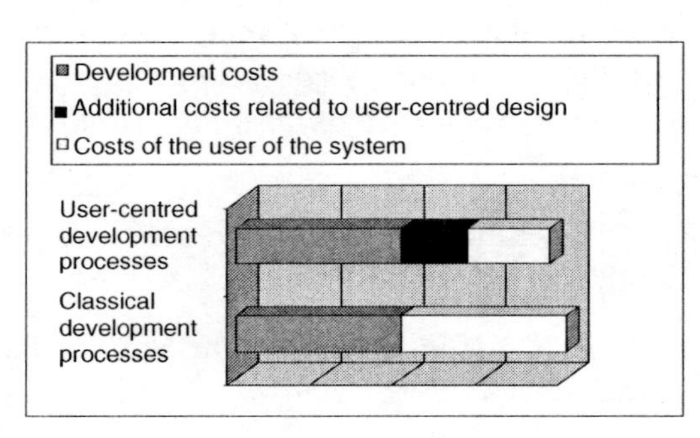

The precise evaluation of costs and payoffs for usability engineering can be found in Mayhew and Bias (1994).

Design-rationale approaches (Buckingham Shum, 1996) face the same problems of profitability as user-centred development processes. As payoffs are not immediately identifiable, developers and designers of software products are still reluctant to either try it or use it in a systematic way.

Design rationale follows three main goals.

1. Provide means (notations, tools, techniques, etc.) for the systematic exploration of design alternatives throughout the development process

2. Provide means to support argumentation when design choices are to be made
3. Provide means to keep track of these design choices in order to be able to justify when choices have been made

Such approaches increase the production of rational designs, that is, where trust in designers' capabilities can be traced back. One of the main arguments for following a rationale-based-design development process is that such processes increase the overall quality of systems. However, when it comes to putting design rationale into practice, that is, within development teams and real projects, more concrete arguments around costs and benefits have to be provided.

Figure 2 reuses the same argumentation process as the one used in Figure 1 for justifying the profitability of user-centred approaches. While user-centred approaches find their profitability when costs related to the actual use of the system are taken into account, design rationale finds its profitability when costs are taken into account amongst several projects. Figure 2 is made up of three bars, each representing a different project. The grey parts of the bars represent the development cost for the project. The black parts represent the additional costs for using a development process following a design-rationale approach. As shown, the lengths of the black parts of the bars remain the same, representing the fact that costs related to design-rationale activities remain the same across projects. According to the projects we have been working on, it is clearly not true for the first project in a given domain.

Figure 2: Profitability related to design rationale

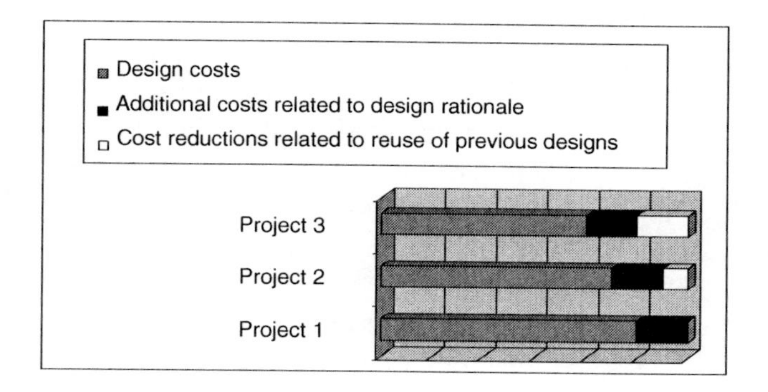

Indeed, the basic elements of design rationale have to be gathered first, such as the pertinent criteria and factors according to the domain and the characteristics of the project. The other interesting aspect of Figure 2 is the fact that the cost of the development of the project decreases according to the number of projects as reuse from design rationale increases accordingly. The white parts of the bars represent the increasing savings due to the reuse of information by using the design-rationale approach of previous projects. This amount is likely to follow a logarithmic curve, that is, to reach a certain level where the cost decrease will reduce. However, our experience of design-rationale approaches is not wide enough to give more precise information about this.

Development processes in the field of safety-critical systems (such as RTCA/DO-178B, 1992) explicitly require the use of methods and techniques for systematically exploring design options and for increasing the traceability of design decisions. DO-178B is a document describing a design process. However, even though such development processes are widely used in the aeronautical domain, the design-rationale part remains superficially addressed.

We believe that this underexploitation of such a critical aspect of the design process lies in two main points.

- There is no integration of current practice in user-centred design processes and design rationale. For instance, no design-rationale notation or tool relates to task modeling, scenarios, dialogue models, usability heuristics, and so forth that are at the core of the discipline.
- There is no adequate tool to support a demanding activity such as design rationale that is heavily based on information storage and retrieval as well as on reuse. In software engineering, similar activities are supported by case tools that are recognised as critical elements for the effective use of notations.

The next section presents a set of design-rationale notations and a tool, based on the QOC (questions, options, criteria) notation (MacLean, Young, Bellotti, & Moran, 1991) that is dedicated to the rationale design of interactive systems.

BACKGROUND

In this section, we briefly describe the design-rationale notation, and then we detail the QOC notation.

IBIS (issue-based information system; Kunz & Rittel, 1970) was designed to capture relevant information with low-cost and fast information retrieval. IBIS has some scalability problems due to the nonhierarchical organisation of the diagram. DRL (decision representation language) was conceived by Lee (1991). The goal was to provide a notation for tracing the process that would lead the designers to choose an alternative. DRL, based on a strict vocabulary, captures more information than necessary for design rationale. Diagrams quickly become incomprehensible.

QOC is a semiformal notation (see Figure 3) introduced by MacLean et al. (1996) that has two main advantages: It is easy to understand but still useful in terms of structuring. QOC was designed for reuse.

The easy-to-understand characteristic is critical for design rationale as the models built using the notation must be understandable by the various actors involved in the development process (i.e., designers, software engineers, human-factors experts, and so forth). A QOC diagram is structured in three columns, one for each element (questions, options, criteria), and features links between columns' elements. For each question that may occur during the development process, the actor may relate one or more relevant options (i.e., candidate design solutions), and to these options, criteria are related (by means of a line) in order to represent the fact that a given option has an impact (if beneficial, the line is thick; if not, the line is dashed). In QOC, an option may lead to another question (as, for instance, Question 2 in Figure 3), thus explicitly showing links between diagrams. In addition, arguments can be attached to options in order to describe further detail: either the content or the underlying rationale for representing the option.

Figure 3. Schematic view of a QOC diagram

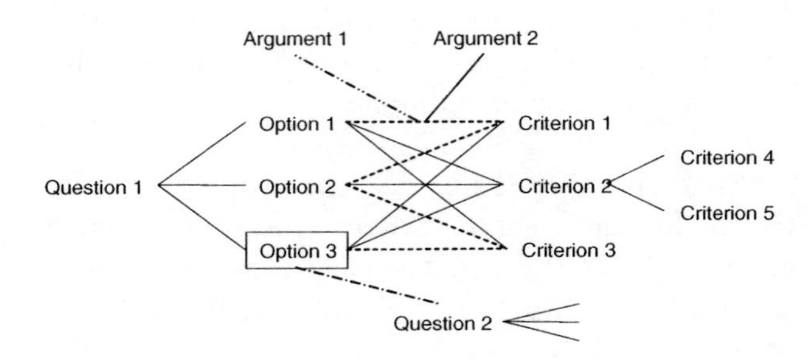

HCI-RELATED EXTENSIONS

In this section, we detail the extensions that we propose. These extensions integrate HCI principles into the notation.

Adding Task Models to QOC Diagrams

This extension (Lacaze, Palanque, & Navarre, 2002) aims at integrating task models in QOC diagrams in order to be able (through scenarios extracted from the task models) to assess the respective performance of the various options under consideration (Lacaze, Palanque, Navarre, & Bastide, 2002).

This extension is critical for an efficient, rationalised development process as it takes into account task analysis and modeling as well as scenarios that are important and expensive activities in user-centred design.

Adding Factors to QOC Diagrams

This extension has been introduced by Farenc and Palanque (1999). In original QOC, there is no way to store and thus to argue with respect to user requirements. However, it is clear that in user-centred development, users take an important role in the decisions made. In these extensions, user requirements are expressed as a set of factors. The factors correspond to high-level requirements such as learnability, safety, and so forth, and the satisfaction of those factors can be checked against their corresponding criteria. The early identification of factors has been based on McCall, Richards, and Walters' (1977) classification that is widely used in software engineering. The elements of the classification are the following.

- **Quality factors:** requirements expressed by the clients and/or users

- **Quality criteria:** Characteristics of the product (technical point of view)
- **Metrics:** The allowing of the actual valuation of a criterion

Figure 4 shows some metrics; however, factors would be directly related to criteria.

TOOL SUPPORT

We have developed a case tool called DREAM (Design Rationale Environment for Argumentation and Modelling) supporting the previous extensions as well as several others that have not been presented here due to space restrictions. A snapshot of this tool is presented in Figure 5. The tool can be accessed at http://liihs.irit.fr/dream/. DREAM proposes several visualisations for the same diagram. In the bottom right-hand corner of Figure 5, the diagram is displayed as a bifocal tree (Cava, Luzzardi, & Freitas, 2002).

FUTURE TRENDS

Design rationale clearly has to be integrated in the design process and merged with UML (Booch, Rumbaugh, & Jacobson, 1999). This will improve the quality of interactive systems (Newman & Marshall, 1991). Interactive systems will be designed and built rationally, and they will not depend solely on the designers' beliefs.

Figure 4. Scenario and task-model extensions to QOC

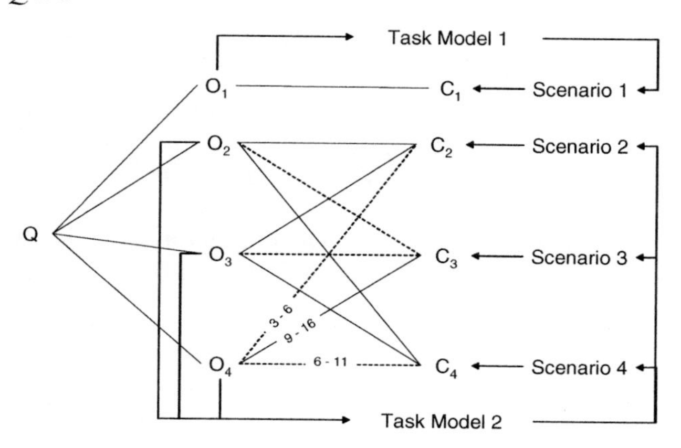

Figure 5. Snapshot of DREAM tool

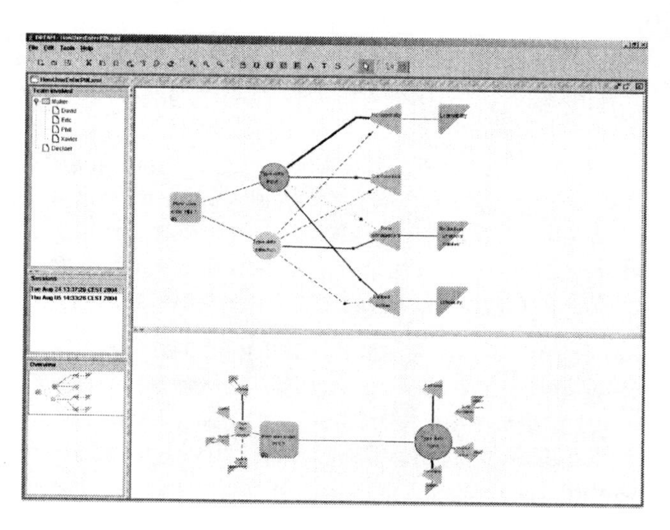

CONCLUSION

The work presented here offers a plea for more extensive use of design-rationale approaches for the design of interactive systems. Notations closely related to current practice in the field of HCI and adequate tools are the ways to achieve this goal.

REFERENCES

Beck, K. (1999). *Extreme programming explained: Embrace change.* Boston: Addison-Wesley Longman Publishing Co.

Booch, G., Rumbaugh, J., & Jacobson, I. (1999). *The unified modeling language user guide.* Boston: Addison Wesley Longman Publishing Co.

Buckingham Shum, S. (1996). Analysing the usability of a design rationale notation. In T. P. Moran & J. Carroll (Eds.), *Design rationale: Concepts, techniques, and use* (pp. 185-215). Mahwah, NJ: Lawrence Erlbaum Associates.

Cava, R. A., Luzzardi, P. R. G., & Freitas, C. M. D. S. (2002). The bifocal tree: A technique for the visualization of hierarchical information structures. *IHC 2002: 5th Workshop on Human Factors in Computer Systems* (pp. 303-313).

Farenc, C., & Palanque, P. (1999). Exploiting design rationale notations for a rationalised design of interactive applications. *IHM'99: 11th French-Speaking Conference on Human Computer-Interaction*, (pp. 22-26).

Freitas, C., Cava, R., Winckler, M., & Palanque, P. (2003). Synergistic use of visualisation technique and Web navigation model for information space exploration. *Proceedings of Human-Computer Interaction International* (pp. 1091-1095).

Gruber, T. R., & Russell, D. M. (1990). *Design knowledge and design rationale: A framework for representation, capture, and use* (Tech. Rep. No. KSL 90-45). Stanford University, Knowledge Systems Laboratory, California.

Kunz, W., & Rittel, H. (1970). *Issues as elements of information systems.* Berkeley: University of California.

Lacaze, X., Palanque, P., & Navarre, D. (2002). Evaluation de performance et modèles de tâches comme support à la conception rationnelle des systèmes interactifs. *IHM'02: 14th French-Speaking Conference on Human Computer-Interaction*, (pp. 17-24).

Lacaze, X., Palanque, P., Navarre, D., & Bastide, R. (2002). Performance evaluation as a tool for quantitative assessment of complexity of interactive systems. In *Lecture notes in computer science: Vol. 2545. DSV-IS'02 9th Workshop on Design Specification and Verification of Interactive Systems* (pp. 208-222). Springer.

Lee, J. (1991). Extending the Potts and Bruns model for recording design rationale. *Proceedings of the 13th International Conference on Software Engineering* (pp. 114-125).

MacLean, A., Young, R. M., Bellotti, V., & Moran, T. (1991a). Questions, options and criteria: Elements of design space analysis. *Journal on Human Computer Interaction, 6*(3-4), 201-250.

MacLean, A., Young, R. M., Bellotti, V. M. E., & Moran, T. P. (1996). Questions, options, and criteria: Elements of design space analysis. In T. P. Moran & J. M. Carroll (Eds.) *Design rationale: Concepts, techniques, and use.* Mahwah, NJ: Laurence Erlbaum Associates.

Martin, J. (1991). *Rapid application development.* UK: Macmillan Publishing Co.

Mayhew, D., & Bias, R. (1994). *Cost-justifying usability.* Pages Academic Press.

McCall, J., Richards, P., & Walters, G. (1977). *Factors in software quality* (Vol. 3, RADC-TR-77-369). Rome Air Development Center (RADC).

Newman, S. E., & Marshall, C. C. (1991). *Pushing Toulmin too far: Learning from an argument representation scheme* (Tech. Rep. No. SSL-92-45). Xerox PARC.

Norman, D. A., & Draper, S. W. (1986). *User-centred system design: New perspectives on human computer interaction.* Hillsdale, NJ: Lawrence Erlbaum Associates.

Paternò, F. (2001). Task models in interactive software systems. In S. K. Chang (Ed.), *Handbook of software engineering and knowledge engineering* (pp. 355-370). London: World Scientific Publishing Co.

RTCA/DO-178B. (1992). *Software considerations in airborne systems and equipment certification.* RTCA Inc.

Vredenburg, K., Isensee, S., & Righi, C. (2001). *User-centered design: An integrated approach.* Prentice Hall.

KEY TERMS

Bifocal Tree: "The Bifocal Tree (Cava et al., 2002) is a visual representation for displaying tree structures based on a node-edge diagram. The technique displays a single tree structure as a focus+context visualisation. It provides detailed and contextual views of a selected sub-tree" (Freitas, Cava, Winckler, & Palanque, 2003, p. 1093).

Criterion: The expressed characteristics of an interactive system. The criterion must be valuable, and it denies or supports options.

Design Rationale: "A design rationale is an explanation of how and why an artefact, or some portion of it, is designed the way it is. A design rationale is a description of the reasoning justifying the resulting design—how structures achieve functions, why particular structures are chosen over alternatives, what behaviour is expected under what operating conditions. In short, a design rationale explains the 'why' of a design by describing what the artefact is, what it is supposed to do, and how it got to be designed that way" (Gruber & Russell, 1990, p. 3).

DREAM (Design Rationale Environment for Argumentation and Modelling): DREAM is a tool dedicated to design-rationale capture by the way of an extended QOC notation.

Factor: The expressed requirement of the customer.

QOC: QOC is a semiformal notation dedicated to design rationale. Problems are spelled out in terms of questions, options to solve the questions, and criteria that valuate each option. See MacLean et al. (1991).

RAD: A software-development process that allows usable systems to be built in as little as 90 to 120 days, often with some compromises.

Task Model: "Task models describe how activities can be performed to reach the users' goals when interacting with the application considered" (Paternò, 2001, p. 359).

The Development of the Personal Digital Assistant (PDA) Interface

Bernard Mark Garrett
University of British Columbia, Canada

INTRODUCTION

The original idea of a portable computer has been credited to Alan Kay of the Xerox Palo Alto Research Center, who suggested this idea in the 1970s (Kay, 1972a, 1972b; Kay & Goldberg, 1977). He envisioned a notebook-sized, portable computer named the Dynabook that could be used for all of the user's information needs and used wireless network capabilities for connectivity.

BACKGROUND

The first actual portable laptop computers appeared in 1979 (e.g., the Grid Compass Computer designed in 1979 by William Moggridge for Grid Systems Corporation [Stanford University, 2003]). The Grid Compass was one-fifth the weight of any model equivalent in performance and was used by NASA on the space shuttle program in the early 1980s. Portable computers continued to develop in the 1980s onwards, and most weighed about 5kg without any peripherals.

In 1984, Apple Computer introduced its Apple IIc model, a true notebook-sized computer weighing about 5kg without a monitor (Snell, 2004). The Apple IIc had an optional LCD panel monitor that made it genuinely portable and was, therefore, highly successful.

In 1986, IBM introduced its IBM Convertible PC with 256KB of memory, which was also a commercial success (Cringely, 1998). For many, this is considered the first true laptop (mainly due to its clamshell design) that soon was copied by other manufacturers such as Toshiba, who also was successful with IBM laptop clones (Abetti, 1997). These devices retained the A4 size footprint and full QWERTY keyboards and weighed between 3 and 4 kg. Following these innovations, Tablet PCs with a flat A4 footprint and a pen-based interface began to emerge in the 1990s.

There were several devices in the 1970s that explored the Tablet, but in 1989, the Grid Systems GRiDPad was released, which was the world's first IBM PC Compatible Tablet PC that featured handwriting recognition as well as a pen-based point-and-select system. In 1992, Microsoft released Microsoft Windows for Pen Computing, which had an Application Programming Interface (API) that developers could use to create pen-enabled applications. Focusing specifically on devices that use the pen as the primary input device, this interface has been most successfully adopted in the new breed of small highly portable personal digital assistants.

In 1984 David Potter and his partners at PSION launched the PSION Organiser that retailed for just under £100 (Troni & Lowber, 2001). It was a battery-powered, 14cm × 9cm block-shaped unit with an alphabetic keyboard and small LCD screen, with 2K of RAM, 4KB of applications in ROM, and a free 8KB data card (which had to be reformatted using ultraviolet light for reuse). Compared to the much larger notebook computers of the time, it was a revolutionary device, but because of its more limited screen size and memory, it fulfilled a different niche in the market and began to be used for personal information management and stock inventory purposes (with a plug-in barcode reader).

In the late 1980s and 1990s, PSION continued to develop commercially successful small computing devices incorporating a larger LCD screen and a new fully multi-tasking graphical user interface (even before Microsoft had Windows up and running). These small devices were truly handheld. The dimensions of the PSION 3c (launched in 1991) were 165mm×85mm×22 mm, with a 480×160-pixel LCD screen; the device weighed less than 400g. A small keyboard and innovative touch pad provided control of the cursor, and graphical icons could be selected to start applications/functions and select items from menus. The small keyboard proved difficult to use, however, and the following 5c model in 1997 used an innovative foldout miniature QWERTY keyboard. These genuinely handheld devices with their interface innovations and ability to synchronize data with a host personal computer made the PSION models particularly successful and firmly established the personal digital assistant (PDA) as a portable computing tool for professionals.

ALTERNATIVE INTERFACES AND THE INTEGRATION OF MULTIMEDIA

The limitations of keyboard-based data entry for handheld devices had been recognized, and following PSION's lead, Apple Computers introduced the Newton Message Pad in 1993. This device was the first to incorporate a touch-sensitive screen with a pen-based graphical interface and handwriting-recognition software. Although moderately successful, the device's handwriting recognition proved slow and unreliable, and in 1998, Apple discontinued its PDA development (Linzmayer, 1999). However, the PDA market now was becoming based firmly upon devices using pen-based handwriting recognition for text entry, and in mid-2001, PSION, with dwindling sales and difficulties with business partnerships, ceased trading. US Robotics launched the Palm Pilot in 1996, using its simple Graffiti handwriting recognition system, and Compaq released the iPAQ in 1997, incorporating the new Microsoft Windows CE/Pocket PC operating system with the first PDA color screen (Wallich, 2002).

Microsoft's relatively late entry into this market reflected the considerable research and development it undertook in developing a user-friendly pocket PC handwriting recognition interface. This remains a highly competitive field, and in November 2002, PalmSource (the new company owning the Palm Operating System) replaced the Graffiti system with Computer Intelligence Corporation's JOT as the standard and only handwriting software on all new Palm Powered devices. Computer Intelligence Corporation (CIC) was founded in conjunction with Stanford Research Institute, based on research conducted by SRI on proprietary pattern recognition technologies (CIC, 1999). The original Graffitti system relied on the user learning a series of special characters, which, though simple, was irksome to many users. The CIC JOT and Microsoft Pocket PC systems have been developed to avoid the use of special symbols or characters and to allow the user to input more naturally by using standard upper and lower case printed letters. Both systems also recognize most of the original Palm Graffiti-based special characters.

The arrival of the Short Messaging Service (SMS), otherwise known as text messaging, for cellular phones in the late 1990s led several PDA manufacturers to adopt an alternative Thumb Board interface for their PDAs. SMS allows an individual to send short text and numeric messages (up to 160 characters) to and from digital cell phones and public SMS messaging gateways on the Internet. With the widespread adoption of SMS by the younger generation, thumb-based text entry (using only one thumb to input data on cell phone keypads) became popular (Karuturi, 2003). Abbreviations such as "C U L8er" for "see you later" and emoticons or smileys to reduce the terseness of the medium and give shorthand emotional indicators developed. The rapid commercial success of this input interface inspired the implementation of Thumb Board keyboards on some PDAs (i.e., the Palm Treo 600) for text interface. Clip-on Thumb Board input accessories also have been developed for a range of PDAs.

Current developments in PDA-based interfaces are exploring the use of multimedia, voice recognition, and wireless connectivity. The expansion of memory capabilities and processor speeds for PDAs has enabled audio recording, digital music storage/playback, and now digital image and video recording/playback to be integrated into these devices. This and the integration of wireless network and cellular phone technologies have expanded their utility considerably.

Audio input has become very attractive to the mobile computer user. Audio is attractive for mobile applications, because it can be used when the user's hands and eyes are occupied. Also, as speech does not require a display, it can be used in conditions of low screen visibility, and it may consume less power than text-based input in the PDA. The latest PDA interface innovations include voice command and dictation recognition (voice to text), voice dialing, image-based dialing (for cell phone use, where the user states a name or selects an image to initiate a call), audio memo recording, and multimedia messaging (MMS). Several devices (e.g., the new Carrier Technologies i-mate) also incorporate a digital camera.

Wireless connectivity has enabled Internet connectivity, enabling users to access e-mail, text/graphical messaging services (SMS and MMS), and the Web remotely (Kopp, 1998). These developments gradually are expanding the PDA's functionality into a true multi-purpose tool.

FUTURE TRENDS

Coding PDA applications to recognize handwriting and speech and to incorporate multimedia requires additional code beyond traditionally coded interfaces. PDA application design and development need to support this functionality for the future.

One of the key limitations of PDA interfaces remains the output display screen size and resolution. This arguably remains a barrier to their uptake as the definitive mobile computing device. As input technologies improve and as voice and handwriting recognition come of age, attention to the display capabilities of these devices will need to be addressed before their full potential can be realized. The display size and resolution already is being pushed to the limits by the latest PDA applications such as global positioning system (GPS) integration with moving map software (deHerra, 2003; Louderback, 2004).

Data and device security are key areas for highly portable networked PDAs, and the first viruses for PDAs have started to emerge (BitDefender, 2004). As multimedia interfaces develop, the specific security issues that they entail (i.e. individual voice recognition, prevention of data corruption of new file formats) also will need to be addressed.

CONCLUSION

Since the early models, manufacturers have continued to introduce smaller and improved portable computers, culminating in the latest generation of powerful handheld PDAs offering fast (400 MHz and faster) processors with considerable memory (64MB of ROM and 1GB of RAM or more). This area of technological development remains highly competitive, and by necessity, the user interface for these devices has developed to fulfill the portable design brief, including the use of pen- and voice-based data input, collapsible LCD displays, wireless network connectivity, and now cell phone integration. Modern PDAs are much more sophisticated, lightweight devices and are arguably much closer to Kay's original vision of mobile computing than the current laptop or tablet computers and possibly have the potential to replace this format with future interface developments. Indeed, if the interface issues are addressed successfully, then it is probable that these devices will outsell PCs in the future and become the major computing platform for personal use.

REFERENCES

Abetti, P.A. (1997). Underground innovation in Japan: The development of Toshiba's word processor and laptop computer. *Creativity and Innovation Management, 6*(3), 127-139.

BitDefender. (2004). *Proof-of-concept virus hits the last virus-resistant Microsoft OS.* Retrieved July 17, 2004, from http://www.bitdefender.com/bd/site/presscenter.php?menu_id=24&n_id=102

Computer Intelligence Corporation (CIC). (1999). *Economic assessment office report: Computer recognition of natural handwriting.* Retrieved August 8, 2004, from http://statusreports-atp.nist.gov/reports/90-01-0210.htm

Cringely, R.X. (1993). *Accidental empires: How the boys of Silicon Valley make their millions, battle foreign competition and still can't get a date.* New York: Harper Business.

deHerra, C. (2003). *What is in the future of Windows mobile pocket PCs.* Retrieved August 12, 2004, from http://www.cewindows.net/commentary/future-wm.htm

GSM World. (2002). *GSM technology: What is SMS?* Retrieved June 24, 2004, from http://www.gsmworld.com/technology/sms/intro.shtml

Karuturi, S. (2003). *SMS history*. Retrieved August 8, 2004, from http://www.funsms.net/sms_history.htm

Kay, A. (1972a). A personal computer for children of all ages. *Proceedings of the ACM National Conference*, Pittsburgh, Pennsylvania.

Kay, A. (1972b). A dynamic medium for creative thought. *Proceedings of the National Council of Teachers of English Conference*, Seattle, Washington.

Kay, A., & Goldberg, A. (1977, March). Personal dynamic media. *IEEE Computer,* 31-41.

Kopp, A. (1998). Get mobile with today's handheld wireless devices. *Power, 142*(4), 41-48.

Linzmayer, O.W. (1999). *Apple confidential: The real story of Apple Computer, Inc.* San Francisco: No Starch Press.

Louderback, J. (2004). *PalmSource CTO on the future of the PDA*. Retrieved August 24, 2004, from http://www.eweek.com/article2/0,4149,1526483,00.asp

Snell, J. (2004). 20 years of the MAC. *Macworld, 21*(2), 62-72.

Stanford University. (2003). *Human computer interaction: Designing technology*. Retrieved August 10, 2004, from http://hci.stanford.edu/cs547/abstracts/03-04/031003-moggridge.html

Troni, P., & Lowber, P. (2001). *Very portable devices (tablet and clamshell PDAs, smart phones and mini-notebooks: An overview.* Retrieved August 10, 2004, from http://cnscenter.future.co.kr/resource/rsc-center/gartner/portabledevices.pdf

Wallich, P. (2002). The ghosts of computers past. *IEEE Spectrum, 39*(11), 38-43.

KEY TERMS

Audio Memo: A recorded audio message of speech. Speech is digitally recorded via a built-in or attached microphone and stored as a digital audio file on the storage media of the PDA.

Emoticon: Text (ASCI) characters used to indicate an emotional state in electronic correspondence. Emoticons or smileys, as they are also called, represent emotional shorthand. For example :-) represents a smile or happiness.

Laptop: A portable personal computer small enough to use on your lap with a QWERTY keyboard and display screen. It usually has an A4-sized footprint in a clamshell configuration and may incorporate a variety of peripheral devices (e.g., trackball, CD-ROM, wireless network card, etc.).

Media Player: A device or software application designed to play a variety of digital communications media such as compressed audio files (e.g., MPEG MP3 files), digital video files, and other digital media formats.

Multimedia: Communications media that combine multiple formats such as text, graphics, sound, and video (e.g., a video incorporating sound and subtitles or with text attached that is concurrently displayed).

Multimedia Messaging Service (MMS): A cellular phone service allowing the transmission of multiple media in a single message. As such, it can be seen as an evolution of SMS with MMS supporting the transmission of text, pictures, audio, and video.

Palmtop: A portable personal computer that can be operated comfortably while held in one hand. These devices usually support a QWERTY keyboard for data input with a small display screen in an A5-sized footprint.

PDA: Personal Digital Assistant. A small handheld computing device with data input and display facilities and a range of software applications. Small keyboards and pen-based input systems are commonly used for user input.

Pen Computing: A computer that uses an electronic pen (or stylus) rather than a keyboard for data input. Pen-based computers often support handwriting or voice recognition so that users can write on the screen or vocalize commands/dictate instead of typing with a keyboard. Many pen computers are handheld devices. It also is known as pen-based computing.

Personal Information Manager (PIM): A software application (i.e., Microsoft Outlook) that provides multiple ways to log and organize personal and business information such as contacts, events, tasks, appointments, and notes on a digital device.

Smartphone: A term used for the combination of a mobile cellular telephone and PDA in one small portable device. These devices usually use a small thumb keyboard or an electronic pen (or stylus) and a touch-sensitive screen for data input.

SMS: Short Message Service. A text message service that enables users to send short messages (160 characters) to other users and has the ability to send a message to multiple recipients. This is known as *texting*. It is a popular service among young people. There were 400 billion SMS messages sent worldwide in 2002 (GSM World, 2002).

Synchronization: The harmonization of data on two (or more) digital devices so that both (all) contain the same data. Data commonly are synchronized on the basis of the date they were last altered, with synchronization software facilitating the process and preventing duplication or loss of data.

Tablet PC: A newer type of format of personal computers. It provides all the power of a laptop PC but without a keyboard for text entry. Tablet PCs use pen-based input and handwriting and voice recognition technologies as the main forms of data entry, and they commonly have an A4-size footprint.

Wireless Connectivity: The communication of digital devices between one another using data transmission by radio waves. A variety of standards for wireless data transmission now exist, established by the Institute of Electrical and Electronics Engineers (IEEE) and including Wi-fi (802.11) and Bluetooth (802.15).

Development Methodologies and Users

Shawren Singh
University of South Africa, South Africa

Paula Kotzé
University of South Africa, South Africa

INTRODUCTION

There are various development methodologies that are used in developing ISs, some more conventional than others. On the *conventional* side, there are two major approaches to systems development methodologies that are used to develop IS applications: the traditional systems development methodology and the *object-oriented (OO)* development approach. The *proponents of HCI and interaction design* propose life cycle models with a stronger user focus than that employed in the conventional approaches. Before the researcher looks at these approaches, he or she needs to ponder about the method of comparing and assessing the various methodologies. There are always inherent problems in comparing various development methodologies (The Object Agency, 1993).

It is, in many instances, difficult to repeat the results of a methodology comparison with any accuracy. Since few (if any) of the comparisons cite page references indicating where a particular methodology comparison item (e.g., a term, concept, or example) can be found in the methodology under review, it is difficult, if not impossible, to verify the accuracy of these methodology comparisons. The researchers did not compare the methodologies step-by-step, but rather in terms of whether and when they address the human element. Researchers have to acknowledge that methodologies are always in a state of flux. In theory, one thing happens, and in practice the methodologies are modified to suit individual business needs.

BACKGROUND

Development Methodologies

This section gives an overview of the three primary groups of development methodologies and the major phases/processes involved. The aim of all these methodologies is to design effective and efficient ISs. But how effective are they when the wider environment is considered? A more contemporary approach is that the information system is open to the world and all stakeholders can interact with it (see Figure 1).

Figure 1. Contemporary approach to business

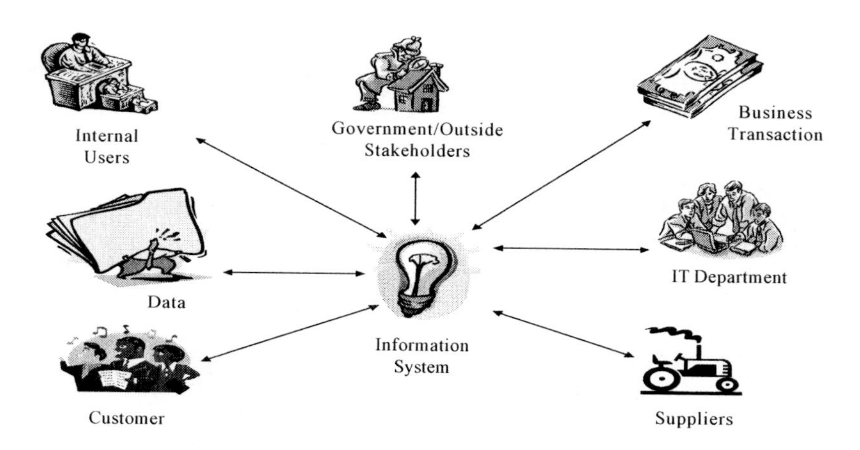

Traditional Systems Development Approaches

Under the traditional development approaches, there are various methodologies. All of these approaches have the following phases in common: *Planning* (why build the system?): Identifying business value, analysing feasibility, developing a work plan, staffing the project, and controlling and directing the project; *Analysis* (who, what, when, where will the system be?): Analysis, information gathering, process modelling and data modelling; *Design* (how will the system work?): Physical design, architecture design, interface design, database and file design and program design; *Implementation* (system delivery): Construction and installation of system. We will look at the Dennis and Wixom Approach (2000).

OO Methodologies

Although diverse in approach, most object-oriented development methodologies follow a defined system development life cycle, and the various phases are intrinsically equivalent for all the approaches, typically proceeding as follows (Schach, 2002): requirements phase; OO analysis phase (determining what the product is to do) and extracting the objects; OO (detailed) design phase; OO programming phase (implementing in appropriate OO programming language); integration phase; maintenance phase; and finally retirement. OO stages are not really very different from the traditional system development approaches mentioned previously.

The OO development approach in general lends itself to the development of more effective user interfaces because of the iterative design process, although this process does not seem to be effectively managed and guidelines for doing so are often absent. The authors analyzed three OO methodologies: The Rumbaugh, Blaha, Premerlani, Eddy, and Lorensen (1991), Coad and Yourdan (1991), and IBM (1999) approaches and their relationship to the aspects illustrated in Figure 1.

HCI-Focused Life Cycle Methodologies

The HCI proponents aim to focus more on the human and end-user aspects. There are four types of users for most computer systems: These are naïve, novice, skilled, and expert users. With the widespread introduction of information and communication technology into our everyday lives, most computer users today have limited computer experience, but are expected to use such systems.

Usability is a measurable characteristic of a product user interface that is present to a greater or lesser degree. One broad dimension of usability is how easy for novice and casual users to learn the user interface (Mayhew, 1999). Another usability dimension is how easy for frequent and proficient users to use the user interface (efficiency, flexibility, powerfulness, etc.) after they have mastered the initial learning of the interface (Mayhew, 1999).

Williges, Williges, and Elkerton (1987) have produced an alternative model of systems development to rectify the problems in the traditional software development models. In their model, interface design drives the whole process. Preece, Rogers, and Sharp (2002) suggest a simple life cycle model, called the Interaction Design Model, consisting of identifying needs/establishing requirements; evaluating; building an interactive version; and (re)designing. Other life cycle models that focus on HCI aspects include the Star Model of Hartson and Hix (1989), the Usability Engineering Life Cycle of Mayhew (1999), Organizational Requirements Definition for Information Technology (ORDIT) method, Effective Technical and Human Implementation of Computer-based Systems (ETHICS), visual prototyping and Hackos and Redish's model (1998). These methods also introduce various strategies for the development of effective user interfaces.

ASSESSING THE METHODOLOGIES

One of the problems with the traditional model for software development and the OO approaches is that they do not, in general, clearly identify a role for HCI in systems development. User interface concerns are "mixed in" with wider development activities. This may result in one of two problems: either HCI is ignored, or it is relegated to the later stages of design as an afterthought. In either case, the consequences can be disastrous. If HCI is ignored, then there is a good chance that problems will occur in the testing and maintenance stages. If HCI is

Table 1. Methodology matrix

Approach	UI Component	Internal Users	Customers	Suppliers	IT Department	Government
The Dennis and Wixom (2000) Approach						
Planning	no	yes	not involved	not involved	actively part of	not involved
Analysis	no	yes	not involved	not involved	actively part of	not involved
Design	yes	yes	not involved	not involved	actively part of	not involved
Implementation	yes	yes	not involved	not involved	actively part of	not involved
The Rumbaugh et al. (1991) model						
Analysis phase	attempts	attempts	not part of	not part of	actively part of	not part of
System design	no	not involved	not part of	not part of	actively part of	not part of
Object design	no	not involved	not part of	not part of	actively part of	not part of
The Coad and Yourdan (1991) Model						
Analysis	no	attempts	not part of	not part of	actively part of	not part of
Design	yes	attempts	not part of	not part of	actively part of	not part of
The IBM (1999) model						
OO design phase	yes	attempts	not part of	not part of	actively part of	not part of
The design the business model phase	no	not part of	not part of	not part of	not part of	not part of
Williges et al. (1987)						
Initial Design	yes	yes	attempts	attempts	actively part of	attempts
Formative Evaluation	yes	yes	attempts	attempts	actively part of	attempts
Summative Evaluation	yes	yes	attempts	attempts	actively part of	attempts
Hackos and Redish (1998) Approach						
Systems development	no	attempts	attempts	attempts	actively part of	attempts
Interface design	yes	yes	attempts	attempts	actively part of	attempts
Design and implementation	yes	yes	attempts	no	actively part of	no
Testing phase	yes	yes	attempts	no	actively part of	no

relegated to the later stages in the development cycle then it may prove very expensive to "massage" application's functionality into a form that can be readily accessed by the end-user and other stakeholders.

If we examine the methodology matrix (Table 1) in more detail in respect of all the components reflected in Figure 1, we find the following with regard to the structured development and OO development methodologies:

a. In the Dennis and Wixom (2000) approach, interface design is only considered in the later stages of development. The components of Figure 1 only partially map onto this approach, with no reference to the customers, suppliers, the IT department specifically, or the governmental issues.
b. In the Rumbaugh et al. (1991) approach, there is no special consideration given to the design of the user interface or any of the other components reflected in Figure 1.

c. In the Coad & Yourdon (1991) model, the human interaction component includes the actual displays and inputs needed for effective human-computer interaction. This model is a partial fit onto Figure 1. While the internal users of the system are well catered for, the other stakeholders are not actively involved in the process at all.
d. The IBM (1999) model considers the users in the development of the system, but Figure 1 is still only a partial fit onto this model. The internal users are considered in the development of the system, but the external users and other stakeholders are sidelined.

It is clear from this that there are several missing components in all these software developments life cycles (SDLCs). The Coad and Yourdan (1991) approach explicitly takes into account the HCI aspect and tends to ignore the other aspects of Figure 1. The same applies to the IBM (1999) approach, but the process is much shorter. The Rumbaugh et al. (1991) approach is very detailed but still ignores the issue of direct mapping to the final user interface application. Although in this approach, use case scenarios are actively employed and users get involved in systems design, it does not map directly onto the system's user interface design.

The root cause of this poor communication is that all the conventional development methodologies (including the traditional development approach and the OO approach) do not devote adequate attention to the human aspect of systems development. Many researchers have proposed ways of improving the systems' interface, but most of this has not been integrated into the development techniques. The researchers' findings are a confirmation of the work of Monarchi and Puhr (1992).

When we consider Table 1 with regard to the HCI-focused development models, we find that: (a) Williges et al. (1987) try to introduce the usability issues at a much earlier stage of the development process, but this model is not widely used; (b) The Hackos and Redish (1998) model seems to be the most comprehensive one we assessed. The shortcoming of this model is, however, that it still ignores the outside stakeholders, unless the corporate objectives phase states categorically that the organization should give special consideration to the external users, such as customers, suppliers, and government. Hackos and Redish (1998) are silent on this issue, however, and do not elaborate on what they mean by "corporate objectives." If the corporate objectives do include the outside stakeholders, this is the only model that we investigated that does this. In fact, if this is the case, the model maps onto Figure 1. The usability engineering is done in parallel with the systems development, and integrated throughout.

FUTURE TRENDS

There are major gaps of communication between the HCI and SE fields: the methods and vocabulary being used by each community are often foreign to the other community. As a result, product quality is not as high as it could be, and (avoidable) re-work is often necessary (Vanderdonckt & Harning, 2003). The development of mutually accepted standards and frameworks could narrow the communication gap between the HCI and SE fields.

CONCLUSION

The shortcomings of all the methodologies are therefore related to the complexity of the wider environment introduced by the issues highlighted in Figure 1, and how these aspects should inform the system's development process. None of the development methodologies addressed the human component or the issue of other stakeholders sufficiently. Both the traditional SDLC and OO approaches fall short on the issue of human aspects and stakeholder involvement. Although we expected the OO approaches to fare better on these issues, the results given earlier clearly illustrate that these methodologies still have a long way to go in fully integrating environmental issues. Which one fared the best? Although the Williges et al. (1987) and Hackos and Redish (1998) approaches focusing on the user go a long way towards achieving this, several shortcomings can still be identified. There has to be a balanced approach to systems development and HCI development components in the overall systems development process.

REFERENCES

Coad, P., & Yourdon, E. (1991). *Object-oriented analysis*. Englewood Cliffs, NJ: Yourdon Press.

Dennis, A., & Wixom, H. B. (2000). *System analysis and design*. New York: John Wiley & Sons.

Hackos, T. J., & Redish, C. J. (1998). *User and task analysis for interface design*. New York: John Wiley & Sons.

Hartson, H. R., & Hix, D. (1989). Human-computer interface development: Concepts and systems for its management. *ACM Computing Surveys, 21,* 5-92.

IBM. (1999). *Object-oriented design: A preliminary approach-document GG24-3647-00*. Raleigh, NC: IBM International Technical Support Centers.

Mayhew, D. J. (1999). *The usability engineering lifecycle: A practitioner's handbook for user interface design*. San Francisco: Morgan Kaufmann.

Monarchi, D. E., & Puhr, G. I. (1992). A research typology for object-oriented analysis and design. *Communication of the ACM, 35*(9), 35-47.

The Object Agency. (1993). *A comparison of object-oriented development methodologies*. The Object Agency. Retrieved August 2, 2003, from http://www.toa.com/pub/mcr.pdf

Preece, J., Rogers, Y., & Sharp, H. (2002). *Interaction design: Beyond human-computer interaction*. New York: John Wiley & Sons.

Rumbaugh, J., Blaha, M., Premerlani, W., Eddy, F., & Lorensen, W. (1991). *Object-oriented modeling and design*. Englewood Cliffs, NJ: Prentice-Hall.

Schach, S. R. (2002). *Object-oriented and classical software engineering* (5th ed.). Boston: McGraw Hill.

Vanderdonckt, J., & Harning, M. B. (2003, September 1-2). *Closing the gaps: Software engineering and human-computer interaction.* Interact 2003 Workshop, Zurich, Switzerland. Retrieved in July 2004, from http://www.interact2003.org/workshops/ws9-description.html.

Williges, R. C., Williges, B. H., & Elkerton, J. (1987). Software interface design. In G. Salvendy (Ed.), *Handbook of human factors* (pp. 1414-1449). New York: John Wiley & Sons.

KEY TERMS

Effective Technical & Human Implementation of Computer-Based Systems (Ethics): A problem-solving methodology that has been developed to assist the introduction of organizational systems incorporating new technology. It has as its principal objective the successful integration of company objectives with the needs of employees and customers.

Object-Oriented Design (OOD): A design method in which a system is modelled as a collection of cooperating objects and individual objects are treated as instances of a class within a class hierarchy. Four stages can be discerned: identify the classes and objects, identify their semantics, identify their relationships, and specify class and object interfaces and implementation. Object-oriented design is one of the stages of object-oriented programming.

Ordit: Based on the notions of role, responsibility, and conversations, making it possible to specify, analyze, and validate organizational and information systems supporting organizational change. The Ordit architecture can be used to express, explore, and reason about both the problem and the solution aspects in both the social and technical domains. From the simple building blocks and modelling language, a set of more complex and structured models and prefabrications can be constructed and reasoned about. Alternative models are constructed, allowing the exploration of possible futures.

System Development Life Cycle (SDLC): The process of understanding how an information system can support business needs, designing the system, building it, and delivering it to users.

Usability: ISO 9241-11 standard definition for usability identifies three different aspects: (1) a specified set of users, (2) specified goals (tasks) which have to be measurable in terms of effectiveness, efficiency, and satisfaction, and (3) the context in which the activity is carried out.

Usability Engineering: Provides structured methods for optimizing user interface design during product development.

A Dynamic Personal Portfolio Using Web Technologies

Michael Verhaart
Eastern Institute of Technology, New Zealand

Kinshuk
Massey University, New Zealand

INTRODUCTION

With the ubiquitous availability of the Internet, the possibility of creating a centralized repository of an individual's knowledge has become possible. Although, at present, there are many efforts to develop collaborative systems such as wikis (Leuf & Cunningham, 2002), Web logs or blogs (Winer, 2002) and sharable content management systems (Wikipedia, 2004), an area that is overlooked is the development of a system that would manage personal knowledge and information. For example, in an educational setting, it has been found that most lecturers customize content to suit their particular delivery styles. This article outlines a framework that uses Web technologies allowing the storage and management of personal information, the sharing of the content with other personal systems, and allows for annotations to be captured *within context* from people who visit the personal knowledge portfolio.

BACKGROUND

Continuing with the case of a lecturer, a vast amount of knowledge will be accumulated. This needs to be organised in a way so that it can be delivered in a variety of contexts. For example, a piece of knowledge about image resizing could be useful in the following domains: Web page design, databases, multimedia, and digital photography. But, this knowledge is continually changing as printed media are read or other people contribute with their comments or observations. Also, knowledge does not exist is one format, for example, an image can be used to illustrate a concept, a video can be used to show directions, and so forth.

With the ability to manage a wide variety of digital formats, Web technologies have become an obvious way to organise an individual's knowledge. In the early 1990s, the Web was primarily made up of many static Web pages, where content and layout were hard coded into the actual page, so managing the ever-changing aspect of content was a time consuming task. In the late 1990s, database technologies and scripting languages such as ASP (active server pages) and PHP (A recursive acronym for personal home page: hypertext pre-processor) emerged, and with these opportunities to develop new ways to capture, manage, and display individual and shared knowledge.

But, what is knowledge? Generally, it is attached to an individual, and can be loosely defined as "what we know" or "what you have between the ears" (Goppold, 2003). Experience shows that as an individual's content is personalized, as mentioned earlier, lecturers tend to customize content to suit their particular delivery style. So combining the notions that knowledge is something dynamic and is attached to an individual, that it may be enhanced and modified by other individuals, and that Web technologies can assist in its management, the "Virtual Me framework" has been developed.

THE Virtual Me FRAMEWORK

Figure 1 illustrates the concept of *Virtual Me* framework. The framework essentially is made up of three parts, the *sniplet model* which includes the *multimedia object model*, and an *annotation* capability.

Figure 1. Rich picture of Me framework

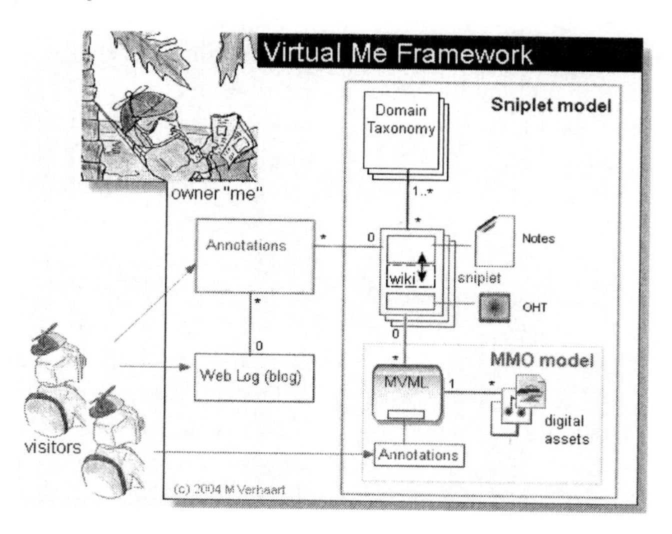

Sniplet Model

In order to manage knowledge, the smallest useful and usable piece of content needs to be defined. Several prototypes indicated that a workable size for the content is that the one which can be represented by a single overhead projection. In order to refer to this, the term *sniplet* was coined (Verhaart, 2002). A sniplet needs to maintain context of the content with respect to everything else in the environment. Hence, they are initially classified in a backbone taxonomy (Guarino & Welty, 2002). The proposed *Virtual Me* framework allows for alternative taxonomies to be created where the content can be used in other domains.

In an electronic sense, an overhead can consist of many media elements, or *digital assets*, such as images, sounds, animations, and videos. But a digital asset also has other issues that need to be considered. For example, an image displayed on a computer screen (at 75 dpi) produces poor quality results when produced in hardcopy (600 dpi and above). If accessibility issues are included, then the ability to represent a digital asset in multiple forms is required. For example, an image needs to be described in text, or alternatively in a sound file, to assist screen readers or for those visitors who have sight impairments. Finally, if the digital asset is to maintain its original context and ownership, some meta-data needs to be attached to it. There are many meta-data standards available, for example, Dublin Core (DCMI, 2004) describes the object, and vCard (1996) describes the creator. *EXtensible Markup Language* (XML) is a portable way for data to be described on the Internet, by providing a structure where data is easily categorized. For example, an XML file could contain <author>Verhaart</author>. vCard is commonly distributed using its own format, but in a paper for the World Wide Web Consortium W3C (Iannella, 2001) described vCard in the Web formats XML and the *Resource Definition framework* (RDF) (Miller, Swick, & Brickley, 2004). Another important feature is that the digital asset has some permanency, that is, in the future it can be located. On the Internet, the Uniform Resource Identifier (of which a Uniform Resource Locator—URL—is a subset) is one way to give a resource an address. The Resource Definition Framework (RDF) takes this a stage further and also structures the resource using eXtensible Markup Language (XML) (W3C, 2004b). This is one of the cornerstones of the semantic Web where objects of the Web maintain some meaning.

To cope with the available standards, a digital asset is described in the *Virtual Me* framework using a multimedia object (MMO) (Verhaart, Jamieson, & Kinshuk, 2004). An MMO essentially is a manifest of files that address the issues described previously. It is made up of the actual files that form the digital asset (multiple images, maybe a sound file) plus a special file that manages the meta-data for the digital asset. In order to describe the meta-data and associated files, a description language (Media Vocabulary Markup Language – MVML) has been developed (Verhaart & Kinshuk, 2004), and is based on the standards mentioned previously: XML, RDF, Dublin Core and vCard.

Annotations

The ability to capture the many types of knowledge possessed by an individual is an integral part of creating a *Virtual Me*. *Implicit* knowledge is the simplest, since it can be easily written down, while *tacit* knowledge (Polanyi, 1958) is often impossible to be communicated in words and symbols (Davidson & Voss, 2002) and is much more difficult to capture. Tacit knowledge is often recalled when a context is presented, and in the physical world can be preceded with "Oh.. I remember…". What about missing knowledge, that is, the knowledge "we don't know we don't know"? This is where visitors can assist. In the real world, this is made up of research or personal contacts filling in the blanks. The *Virtual Me* needs the ability for visitors to add their own annotations at various levels. Annotations can be classified at three levels: creator only (intended to operate like post-it type notes, although they should be viewable by site owner), creator and site owner (a direct communication between them), and public (viewable to any visitor).

Another technology gaining popularity to capture missing knowledge is the wiki. Cortese (2003) indicated that a wiki is Web collaboration software used by informal online groups, and is taking hold in the business realm. First coined in 1995 by Ward Cunningham, wiki means quick in Hawaiian (Jupitermedia Corporation, 2004). It is proposed that in the *Virtual Me*, a wiki type field is available to users for each sniplet (Figure 1). Here, a visitor would make a copy of the sniplet and be able to edit it. The site owner could then update the actual sniplet from the wiki to allow inclusion of missing knowledge.

FUTURE TRENDS

Development of learning resources is a costly and time consuming process. In order to facilitate the sharing and management of content, there is considerable research in the construction and standardization of learning object repositories (McGreal & Roberts, 2001). IEEE Learning Technology Standards Committee (IEEE LTSC) (1999) and Sharable Content Object Reference Model (SCORM) (Advanced Distributed Learning, 2003) compliancy is considered to be the current standard although many countries are developing standards based on these, such as United Kingdom Learning Object Metadata (UK LOM) Core (Cetis, 2004) and the Canadian equivalent, Cancore (Friesen, Fisher, & Roberts, 2004). Content packagers such as RELOAD (2004) are evolving to allow manifests of files to be easily bundled together with learning object meta-data. A survey by Verhaart (2004) indicated that problems such as ownership have caused learning object repositories to become meta-data sites linked to personal content. This is a persuasive argument for the *Virtual Me* framework.

The second major trend is that of the semantic Web. This is the vision of the modern Internet's creator Tim Berners Lee, and the W3C (2004) is investing a significant part of its resources to making this vision a reality. The semantic Web is fundamentally about adding context and meaning to the Web content. Use of the structures suggested in this article will enable media and multimedia objects to retain and maintain their original context even when used in many different contexts.

CONCLUSION

This article presents a discussion where the management of information is returned back to the individual, even though there is a prevailing trend to centralize information. Observations in an educational setting indicate that many content deliverers: tutors and lecturers, prefer to customize material to reflect their personal styles. Further, personal ownership is a powerful motivator in the management and evolution of information.

The *Virtual Me* framework attempts to address this issue, allowing for the creation of a personal learning portfolio where visitors can contribute to the building of the content repository.

REFERENCES

Advanced Distributed Learning. (2003). *SCORM overview*. Retrieved on December 11, 2004, from http://www.adlnet.org/index.cfm?fuseaction=scormabt

Cetis. (2004). *The UK LOM Core home page*. Retrieved on December 11, 2004, from http://www.cetis.ac.uk/profiles/uklomcore

Cortese, A. (2003, May 3). Business is toying with a Web tool. *New York Times* (p. 3). Retrieved December 11, 2004, from http://www.nytimes.com/

Davidson, C., & Voss, P. (2002). *Knowledge management. An introduction to creating competitive advantage from intellectual capital*. Auckland, NZ: Tandem Press.

DCMI. (2004). *Dublin Core Metadata Initiative*. Retrieved December 11, 2004, from http://dublincore.org/

Friesen, N., Fisher, S., & Roberts, A. (2004). *CanCore Guidelines 2.0*. Retrieved on December 11, 2004, from http://www.cancore.ca/en/guidelines.html

Goppold, P. (2003) *Semiotics, biological and cultural aspects, philosophy*. Retrieved on May 21, 2003, from http://www.uni-ulm.de/uni/intgruppen/memosys/infra04.htm

Guarino, N., & Welty, C. (2002). Evaluating ontological decisions with OntoClean. *Communications of the ACM*, *45*(2), 61-65.

Iannella, R. (2001). *Representing vCard Objects in RDF/XML*. Retrieved September 24, 2003, from http://www.w3.org/TR/2001/NOTE-vcard-rdf-20010222/

IEEE Learning Technology Standards Committee (LTSC). (1999). *Learning object metadata*. Retrieved July 10, 2003, from http://ltsc.ieee.org/doc/wg12/LOM3.6.html.

Jupitermedia Corporation. (2004). *Wiki*. Retrieved December 11, 2004, from http://isp.webopedia.com/TERM/W/wiki.html

Leuf, B., & Cunningham, W. (2002). *Welcome visitors*. Retrieved December 11, 2004, from http://wiki.org/

McGreal, R., & Roberts, T. (2001). *A primer on metadata standards: From Dublin Core to IEEE LOM*. Retrieved November 14, 2001, from http://vu.cs.sfu.ca/vu/tlnce/cgi-bin/VG/VF_dspmsg.cgi?ci=130&mi=1

Miller, E., Swick, R., & Brickley, D. (2004). *Resource Description Framework (RDF)*. Retrieved December 11, 2004, from http://www.w3.org/RDF/

Reload. (2004). *Reusable elearning object authoring and delivery*. Retrieved December 11, 2004, from http://www.reload.ac.uk

Polanyi, M. (1958). *Personal knowledge: Towards a post-critical philosophy*. London: Routledge & Kegan Paul.

Vcard. (1996). *The Electronic Business Card Version 2.1*. Retrieved June 16, 2002, from http://www.imc.org/pdi/pdiproddev.html

Verhaart, M. (2002, December 3-6). Knowledge capture at source. Developing collaborative shared resources. Doctoral Student Consortium Papers. *Proceedings International Conference on Computers in Education*, Auckland, New Zealand (pp. 1484-1485). Los Alamitos, CA: IEEE Computer Society.

Verhaart, M. (2004, July 6-9). Learning object repositories: How useful are they? In S. Mann, & T.Clear (Eds.), *Proceedings of the 17th Annual Conference of the National Advisory Committee on Computing Qualifications Conference*, Christchurch, New Zealand (pp. 465-469).

Verhaart, M., Jamieson, J., & Kinshuk (2004). Collecting, organizing and managing non-contextualised data, by using MVML to develop a human-computer interface. In M. Masoodian, S. Jones, & B. Rogers (Eds.), *Lecture Notes in Computer Science, 3101*, (pp. 511-520). Hamilton, NZ: NACCQ.

Verhaart, M., & Kinshuk (2004, August 30-September 1). Adding semantics and context to media resources to enable efficient construction of learning objects. In Kinshuk, C.-K. Looi, E. Sutinen, D.

Sampson, I. Aedo, L. Uden, & E. Kähkönen (Eds.), *Proceedings of the 4th IEEE International Conference on Advanced learning Technologies 2004,* Joensuuu, Finland (pp. 651-653). Los Alamitos, CA: IEEE Computer Society.

W3C. (2004) *Semantic Web*. Retrieved on December 11, 2004, from http://www.w3.org/2001/sw/

W3C. (2004b) *Extensible Markup Language (XML)*. Retrieved on December 11, 2004, from http://www.w3.org/2001/XML/

Wikipedia. (2004) *Content management system*. Retrieved December 11, 2004, from http://en.wikipedia.org/wiki/Web_content_management

Winer, D. (2002) *The history of Weblogs*. Retrieved December 11, 2004, from http://newhome.weblogs.com/historyOfWeblogs

KEY TERMS

Digital Asset: An electronic media element, that may be unstructured such as an image, audio, or video, or structured such as a document or presentation, usually with associated meta-data.

Dublin Core: A set of 15 meta-data fields such as title and author, commonly used by library systems to manage digital assets. (All fields are optional.)

Learning Object: An artifact or group of artifacts with learning objectives that can be used to increase our knowledge.

Media Vocabulary Markup Language (MVML): A XML-based language that describes a media element.

Meta-Data: Commonly *data about data*. For example, a digital asset has meta-data which would include the derived data (size, width) and annotated data (creator, description, context).

Multimedia Object (MMO): A self-describing manifest of files used to encapsulate an electronic media element. Consists of media files conforming to a defined naming standard and an associated MVML file.

Resource Definition Framework (RDF): Part of the Semantic Web, and is a way to uniquely identify a resource whether electronic or not.

Sniplet: A piece of knowledge or information that could be represented by one overhead transparency.

vCARD: A meta-data format that enables a person to be described. This is used extensively in commercial e-mail systems and can be thought of as an electronic business card.

Virtual Me: A framework that uses Internet technologies to structure a personal portfolio and allows external users to add annotations. A sniplet is its basic unit, and digital assets are structured as multimedia objects (MMOs).

Web Log (Blog): An online diary, typically authored by an individual, where unstructured comments are made and annotations can be attached.

Wiki: A publicly modifiable bulletin board, where anyone can change the content. Some provide features so that changes can be un-done. From "wiki" meaning "quick" in Hawaiian, and coined by Ward Cunningham in 1995.

Education, the Internet, and the World Wide Web

John F. Clayton
Waikato Institute of Technology, New Zealand

INTRODUCTION

What is the Internet?

The development of the Internet has a relatively brief and well-documented history (Cerf, 2001; Griffiths, 2001; Leiner et al., 2000; Tyson, 2002). The initial concept was first mooted in the early 1960s. American computer specialists visualized the creation of a globally interconnected set of computers through which everyone quickly could access data and programs from any node, or place, in the world. In the early 1970s, a research project initiated by the United States Department of Defense investigated techniques and technologies to interlink packet networks of various kinds. This was called the Internetting project, and the system of connected networks that emerged from the project was known as the Internet. The initial networks created were purpose-built (i.e., they were intended for and largely restricted to closed specialist communities of research scholars). However, other scholars, other government departments, and the commercial sector realized the system of protocols developed during this research (Transmission Control Protocol [TCP] and Internet Protocol [IP], collectively known as the TCP/IP Protocol Suite) had the potential to revolutionize data and program sharing in all parts of the community. A flurry of activity, beginning with the National Science Foundation (NSF) network NSFNET in 1986, over the last two decades of the 20th century created the Internet as we know it today. In essence, the Internet is a collection of computers joined together with cables and connectors following standard communication protocols.

What is the World Wide Web?

For many involved in education, there appears to be an interchangeability of the terms *Internet* and *World Wide Web* (WWW). For example, teachers often will instruct students to "surf the Web," to use the "dub.dub.dub," or alternatively, to find information "on the net" with the assumption that there is little, if any, difference among them. However, there are significant differences. As mentioned in the previous section, the Internet is a collection of computers networked together using cables, connectors, and protocols. The connection established could be regarded as physical. Without prior knowledge or detailed instructions, the operators of the connected computers are unaware of the value, nature, or appropriateness of the material stored at the node with which they have connected. The concepts underlying the WWW can be seen to address this problem. As with the Internet, the WWW has a brief but well-documented history (Boutell, 2002; Cailliau, 1995; Griffiths, 2001). Tim Benners-Lee is recognized as the driving force behind the development of the protocols, simplifying the process locating the addresses of networked computers and retrieving specific documents for viewing. It is best to imagine the WWW as a virtual space of electronic information storage. Information contained within the network of sites making up the Internet can be searched for and retrieved by a special protocol known as a Hypertext Transfer Protocol (HTTP). While the WWW has no single, recognizable, central, or physical location, the specific information requested could be located and displayed on users' connected devices quickly by using HTTP. The development and refinement of HTTP were followed by the design of a system allowing the links (the HTTP code) to be hidden behind plain text, activated by a click with the mouse, and thus, we have the creation and use of Hypertext Markup Language (HTML). In short, HTTP and HTML made the Internet useful to people who were interested solely in the information and data contained on the nodes of the network and were uninterested in computers, connectors, and cables.

BACKGROUND

Educational Involvement

The use and development of the Internet in the 1970s was almost entirely science-led and restricted to a small number of United States government departments and research institutions accessing online documentation. The broader academic community was not introduced to the communicative power of networking until the start of the 1980s with the creation of BITNET, (Because It's Time Network) and EARN (European Academic and Research Network) (Griffiths, 2001). BITNET and EARN were electronic communication networks among higher education institutes and was based on the power of electronic mail (e-mail). The development of these early networks was boosted by policy decisions of national governments; for example, the British JANET (Joint Academic Network) and the United States NSFNET (National Science Foundation Network) programs that explicitly encouraged the use of the Internet throughout the higher educational system, regardless of discipline (Leiner et al., 2000). By 1987, the number of computer hosts connected to networks had climbed to 28,000, and by 1990, 300,000 computers were attached (Griffiths, 2001). However, the development of the World Wide Web and Hypertext Markup Language, combined with parallel development of browser software applications such as Netscape and Internet Explorer, led to the eventual decline of these e-mail-based communication networks (CREN, 2002). Educational institutions at all levels joined the knowledge age.

FUTURE TRENDS

The advances in and decreasing costs of computer software and hardware in the 1980s resulted in increased use of and confidence in computer technologies by teachers and learners. By the mid-1990s, a number of educational institutions were fully exploiting the power of the Internet and the World Wide Web. Search engines to locate and retrieve information had been developed, and a mini-publication boom of Web sites occurred (Griffiths, 2001). In the early stages, educational institutions established simple Websites providing potential students with information on staff roles and responsibilities; physical resources and layout of the institution; past, present, and upcoming events; and a range of policy documents. As confidence grew, institutions began to use a range of Web-based applications such as e-mail, file storage, and exams, to make available separate course units or entire and programs to a global market (Bonk et al., 1999). Currently, educational institutions from elementary levels to universities are using the WWW and the Internet to supplement classroom instruction, to give learners the ability to connect to information (instructional and other resources), and to deliver learning experiences (Clayton, 2002; Haynes, 2002; Rata Skudder et al., 2003). In short, the Internet and the WWW altered some approaches to education and changed the way some teachers communicated with students (McGovern & Norton, 2001; Newhouse, 2001). There was and continues to be an explosion of instructional ideas, resources, and courses on the WWW during the past decades as well as new funding opportunities for creating courses with WWW components (Bonk, 2001; Bonk et al., 1999; van der Veen et al., 2000). While some educators regard online education with suspicion and are critical that online learning is based on imitating what happens in the classroom (Bork, 2001), advocates of online, Web-assisted, or Internet learning would argue that combining face-to-face teaching with online resources and communication provides a richer learning context and enables differences in learning styles and preferences to be better accommodated (Aldred & Reid, 2003; Bates, 2000; Dalziel, 2003; Mann, 2000). In the not-too-distant future, the use of compact, handheld, Internet-connected computers will launch the fourth wave of the evolution of educational use of the Internet and the WWW (Savill-Smith & Kent, 2003). It is envisaged that young people with literacy and numeracy problems will be motivated to use the compact power of these evolving technologies in learning (Mitchell & Doherty, 2003). These students will be truly mobile, choosing when, how, and what they will learn.

CONCLUSION

The initial computer-programming-led concept of the Internet first mooted in the early 1960s has expanded to influence all aspects of modern society. The development of the Hypertext Transfer Protocol to identify specific locations and the subsequent development of Hypertext Markup Language to display content have enabled meaningful connections to be made from all corners of the globe. As procedures and protocols were established, search facilities were developed to speed up the discovery of resources. At this stage, educationalists and educational institutions began to use the power of the Internet to enhance educational activities. Although in essence, all we basically are doing is tapping into a bank of computers that act as storage devices, the potential for transformation of educational activity is limitless. Increasingly, students will independently search for resources and seek external expert advice, and student-centered learning will have arrived.

REFERENCES

Aldred, L., & Reid, B. (2003). Adopting an innovative multiple media approach to learning for equity groups: Electronically-mediated learning for off-campus students. *Proceedings of the 20th Annual Conference of the Australasian Society for Computers in Learning in Tertiary Education.*

Bates, T. (2000). *Distance education in dual mode higher education institutions: Challenges and changes.* Retrieved March 15, 2003, from http://bates. cstudies.ubc.ca/

Bonk, C. (2001). *Online teaching in an online world.* Retrieved from http://PublicationShare.com

Bonk, C., Cummings, J., Hara, N., Fischler, R., & Lee, S. (1999). *A ten level Web integration continuum for higher education: New resources, partners, courses, and markets.* Retrieved May 1, 2002, from http://php.indiana.edu/~cjbonk/paper/edmdia99.html

Bork, A. (2001). What is needed for effective learning on the Internet. *Education Technology & Society, 4*(3). Retrieved September, 12, 2005, from http://ifets.ieee.org/periodical/vol_3_2001/bork.pdf

Boutell, T. (2002). *Introduction to the World Wide Web.* Retrieved June 14, 2002, from http://www.w3.org/Overview.html

Cailliau, R. (1995). *A little history of the World Wide Web.* Retrieved June 14, 2002, from http://www.w3.org/

Cerf, V.G. (2001). *A brief history of the Internet and related networks.* Retrieved June 4, 2002, from http://www.isoc.org/

Clayton, J. (2002). Using Web-based assessment to engage learners. *Proceedings of the DEANZ: Evolving e-Learning Conference*, Wellington, New Zealand.

CREN. (2002). *CREN history and future.* Retrieved June 11, 2002, from http://www.cren.net/

Dalziel, J. (2003). Implementing learning design: The learning activity management system (LAMS). *Proceedings of the 20th Annual Conference of the Australasian Society for Computers in Learning in Tertiary Education.*

Griffiths, R. T. (2001). *History of the Internet, Internet for historians (and just about everyone else).* Retrieved June 7, 2002, from http://www.let.leidenuniv.nl/history/ivh/frame_theorie.html

Haynes, D. (2002). The social dimensions of online learning: Perceptions, theories and practical responses. *Proceedings of the Distance Education Association of New Zealand*, Wellington, New Zealand.

Leiner, B., et al. (2000). *A brief history of the Internet.* Retrieved May 20, 2002, from http://www.isoc.org/internet/history/brief.shtml# Origins

Mann, B. (2000). Internet provision of enrichment opportunities to school and home. *Australian Educational Computing, 15*(1), 17-21.

McGovern, G., & Norton, R. (2001). *Content critical: Gaining competitive advantage through high-quality Web content.* London: Financial Times Prentice Hall.

Mitchell, A., & Doherty, M. (2003). *m-Learning support for disadvantaged young adults: A midstage review.* Retrieved August 18, 2004, from http://www.m-learning.org/index.shtml

Newhouse, P. (2001). Wireless portable technology unlocks the potential for computers to support learning in primary schools. *Australian Educational Computing, 16*(2), 6-13.

Rata Skudder, N., Angeth, D., & Clayton, J. (2003). All aboard the online express: issues and implications for Pasefica e-learners. *Proceedings of the 20th Annual Conference of the Australasian Society for Computers in Learning in Tertiary Education.*

Savill-Smith, C., & Kent, P. (2003). *The use of palmtop computers for learning: A review of the literature.* Retrieved August 17, 2004, from http://www.m-learning.org/index.shtml

Tyson, J. (2002). *How the Internet works.* Retrieved May 20, 2002, from http://www.howstuffworks.com/

van der Veen, J., de Boer, W., & van de Ven, M. (2000). W3LS: Evaluation framework for World Wide Web learning. *Journal of International Forum of Educational Technology & Society, 3*(4), 132-138.

KEY TERMS

HTML: Hypertext Markup Language (HTML) was originally developed for the use of plain text to hide HTTP links.

HTTP: Hypertext Transfer Protocol (HTTP) is a protocol allowing the searching and retrieval of information from the Internet.

Internet: An internet (note the small i) is any set of networks interconnected with routers forwarding data. The Internet (with a capital I) is the largest internet in the world.

Intranet: A computer network that provides services within an organization.

Node: These are the points where devices (computers, servers, or other digital devices) are connected to the Internet and more often called a host.

Protocol: A set of formal rules defining how to transmit data.

TCP/IP Protocol Suite: The system of protocols developed to network computers and to share information. There are two protocols: the Transmission Control Protocol (TCP) and the Internet Protocol (IP).

World Wide Web: A virtual space of electronic information and data storage.

The Effect of Usability Guidelines on Web Site User Emotions

Patricia A. Chalmers
Air Force Research Laboratory, USA

INTRODUCTION

Two decades ago, the U.S. Air Force asked human factors experts to compile a set of guidelines for command and control software because of software usability problems. Many other government agencies and businesses followed. Now hundreds of guidelines exist. Despite all the guidelines, however, most Web sites still do not use them. One of the biggest resulting usability problems is that users cannot find the information they need. In 2001, Sanjay Koyani and James Mathews (2001), researchers for medical Web information, found, "Recent statistics show that over 60% of Web users can't find the information they're looking for, even though they're viewing a site where the information exists". In 2003, Jakob Nielsen (2003), an internationally known usability expert, reported, "On average across many test tasks, users fail 35% of the time when using Web sites." Now in 2005, Muneo Kitajima, senior researcher with the National Institute of Advanced Industrial Science and Technology, speaks of the difficulties still present in locating desired information, necessitating tremendous amounts of time attempting to access data (Kitajima, Kariya, Takagi, & Zhang, to appear).

This comes at great costs to academia, government, and business, due to erroneous data, lost sales, and decreased credibility of the site in the opinion of users. Since emotions play a great role in lost sales and lost credibility, the goal of this study was to explore the question, "Does the use of usability guidelines affect Web site user emotions?" The experimenter tasked participants to find information on one of two sites. The information existed on both sites; however, one site scored low on usability, and one scored high. After finding nine pieces of information, participants reported their frequency of excitement, satisfaction, fatigue, boredom, confusion, disorientation, anxiety, and frustration. Results favored the site scoring high on usability.

BACKGROUND

In 2003, Sanjay Koyani, Robert W. Bailey, and Janice R. Nall (2003) conducted a large analysis of the research behind all available usability guidelines. They identified research to validate existing guidelines, identify new guidelines, test the guidelines, and review literature supporting and refuting the guidelines. They chose reviewers representing a variety of fields including cognitive psychology, computer science, documentation, usability, and user experience. "The reviewers were all published researchers with doctoral degrees, experienced peer reviewers, and knowledgeable of experimental design" (Koyani et al., 2003, p. xxi). They determined the strength of evidence for each guideline, based on the amount of evidence, type of evidence, quality of evidence, amount of conflicting evidence, and amount of expert opinion agreement with the research (Koyani et al., 2003, pp. xxi-xxii). They then scored each guideline with points for evidence as follows: 5 = strong *research support*, 4 = moderate *research support*, 3 = weak *research support*, 2 = strong *expert opinion*, and 1 = weak *expert opinion*.

The author organizes this article in the following groups throughout to discuss usability topics:

- **Visibility of Location:** Pearrow (2000, p.167) states, "Users want to know where they are in a Web site, especially when the main site contains many microsites." One way to help users know their location is to provide a site map (Danielson, 2002). A site map is an outline of all information on a site. Koyani et al. (2003, p.62) found moderate research support that site maps enhance use if topics reflect the user's conceptual structure. Other aids such as headers and navigation paths may also be useful.

- **Consistency:** WordNet, maintained by the Cognitive Science Laboratory of Princeton University (2005), defines consistency as "a harmonious uniformity or agreement among things or parts". The purpose of consistency is "to allow users to predict system actions based on previous experience with other system actions" (NASA/Goddard Space Flight Center, 1996, p. 1). Ways to make a Web site consistent include placing navigation elements in the same location (Koyani et al., 2003, p. 59), and placing labels, text, and pictures in the same location (Koyani et al., 2003, p. 97). Using the same or similar colors, fonts, and backgrounds for similar information will also provide consistency, as will following business, industry, and government standards.
- **Error Prevention and Error Recovery:** The least costly way to prevent errors is to provide a well-designed Web site at the outset. Even though users tend to accommodate for inconsistencies, Koyani et al. (2003, p. 97) found strong research support showing a relationship between decreased errors and visual consistency. In addition, Asim Ant Ozok and Gavriel Salvendy (2003) found well-written sites decrease comprehension errors. Additional ways to prevent errors may be to provide Undo and Redo commands as well as Back and Forward commands, provide a Frequently-Asked Questions section, and provide help menus and search menus. However, even the best designed site will not prevent all errors. When errors do occur, sites need to provide users with ways to recover from them. Ben Shneiderman (1998, p. 76) advises, "Let the user see the source of error and give specific positive instructions to correct the error."
- **Inverted Pyramid Style:** For the purposes of this paper, the inverted pyramid style refers to putting the most important information at the top of the page. Koyani et al. (2003, p. 47) found moderate research support for the usefulness of putting the most important information on the top of the page and the least used information at the bottom of the page.
- **Speaking the User's Language:** Speaking the user's language refers to speaking the language of the intended audience. Koyani et al. (2003, p. 145) found a strong expert opinion to support avoiding acronyms. It follows that if site owners must use acronyms, they should provide an acronym finder and/or glossary. Other ways to speak the user's language are to avoid jargon and to provide a search engine that recognizes naturalistic language, or language consistent with real-world conventions.
- **Easy Scanning:** Koyani et al. (2003, p. 157) found with moderate evidence that "80% of users scan any new page and only 16% read word-by-word." Therefore, it may be useful to make information easy to scan, especially when presenting large amounts of information. Ways to accomplish this may be to use a bold or italicized font when users need to understand differences in text content. Avoid underling, however, as users may confuse an underlined word or phase with a link. In addition, highlighting may make information easy to visually scan. However, Koyani et al. (2003, p. 77) advise, "Use highlighting sparingly...(use it for)...just a few items on a page that is otherwise relatively uniform in appearance."
- **Proper Printing:** For this study, providing for "proper printing" means providing for the *printed* page to look the same as the *presented* page on the computer screen. For example, a printed copy of a Web page should not show the right side of the page trimmed off, if it does not appear as such to the user. Not all users know how to query the Help function to assist them with this problem. Although Koyani et al. (2003, p. 21) did not find support for providing proper printing in experimental research, they did find a strong expert opinion supporting proper printing.
- **Short Download Time:** Download time refers to the time "to copy data (usually an entire file) from a main source to a peripheral device" (Webopedia, 2005). Koyani et al. (2003, p. 16) found moderate support for the usefulness of minimizing Web page download time. The best

way to decrease download time is to limit the number of bytes per page (Koyani, 2003, p. 17). Stakeholders should determine what download time is desired for their Web site, and designers should aim for that time or shorter.

- **Providing Assistance:** Providing assistance refers to helping users use a Web site and may include: providing shortcuts for frequent users, legends describing icon buttons, a presentation that is flexible to user needs, a search engine with advanced search capabilities, providing for recognition rather than recall, and phone numbers and addresses for technical assistance as well as regular contact.

FOCUS

The redesign team consisted of a graphic designer, a systems administrator, a usability expert, three human factor visualization experts, a computer scientist, a security officer, and a human factors administrative assistant. They met an average of once every three weeks for 12 months. Before and during the redesign process, the team determined and discussed their desired audience and the goals for their site. The team redesigned the site using an iterative process of making changes, then reviewing those changes with the team and potential users, and repeating this process until establishing a consensus.

Evaluation of the Original Web Site

Before redesign, the team asked employees who were representative of the Web site's desired audience what they liked and did not like in sites similar to the division site. To conduct a formative, or initial, evaluation the usability expert devised a "Usability Evaluation Checklist" (Appendix A) based on employee comments, input from the human factors representatives on the redesign team, and common usability guidelines from usability literature. Usability literature included the NASA/Goddard User Interface Guidelines (NASA/Goddard Space Flight Center, 1996), Designing the User Interface (Shneiderman, 1998), Usability Inspection Methods (Nielsen & Mack, 2004), the Web site Usability Handbook (Pearrow, 2000), current journals, and usability sites such as Jakob Nielsen's Alertbox.

Redesign

The team redesigned the original site to correct the problems found in the formative evaluation. They made the following changes.

- **Location Status:** When using the original Web site, some employees reported they felt "lost" and said they wanted to know where they were on the site. To address this, the team added tabs describing the major contents of the site, as well as a site map. Some users also stated they wanted to skip back and forth to topic areas without having to repeatedly key the Back button. To address this, the team provided a frame that appeared at the top of all pages. This frame included tabs for the seven main areas of the site, including a tab for the site map.
- **Consistency:** The original Web site did provide consistent terminology and color. Headers, however, were inconsistent in appearance and placement. Therefore, the team provided similar headers for all pages. The team also provided consistent font type, font size, background color, and layout for all pages.
- **Error Prevention and Error Recovery:** The team added a Help feature and a Frequently-Asked Questions section for user support. The team changed some background and foreground colors to provide adequate contrast for trouble-free reading. In addition, they replaced cluttered interfaces with more "minimalist" interfaces.
- **Inverted Pyramid Style:** The original site scored excellent in this category; therefore, the team made no changes.
- **Speaking the User's Language:** The team identified all acronyms, spelled them out on their initial appearance, and provided an acronym finder. In addition, they added a search engine with naturalistic language capability.
- **Easy Scanning:** The team eliminated the right sidebar to eliminate horizontal scrolling and increase scanning. They also used bolding and italicizing to aid scanning, but used it sparingly to preserve attention.
- **Proper Printing:** Originally, the team eliminated the right sidebar to provide easy scan-

ning capabilities. However, this change had a dual effect because it also eliminated printing problems with trimming off the right sides of pages.

- **Short Download Times:** The team found download time was under three seconds. They avoided elaborate features in the redesign, and again checked download time on interim and summative evaluations. On completion, download times were still less than three seconds on employment computers, and less than seven seconds on a sample of home computers.
- **Assistance:** The team provided providing drop down menus and a search engine with advanced search capability.

Method

The experimenter randomly assigned fifteen participants from the research laboratory's subject pool to either the original Web site or the redesigned site. The experimenter tasked participants to find nine pieces of information on their assigned site. The experimenter gave all participants the same tasks. All participants completed their tasks separately from other participants. Each group's site was the same, except for the redesign changes described earlier. Following the tasks, users completed a questionnaire regarding the frequency of the following emotions during the experiment: fatigue, boredom, confusion, disorientation, anxiety, frustration, satisfaction, and excitement during their task assignment.

Results

An independent statistician ran t-tests on the data. Statistics showed the redesigned site was statistically significant for *higher* frequencies of satisfaction and excitement than the original site. In addition, the redesigned site was statistically significant for *lower* frequencies of frustration, fatigue, boredom, and confusion (see Figure 1). The t-tests showed no significant differences in disorientation or anxiety.

FUTURE TRENDS

There is a debate over the number of participants needed to find usability problems. Jakob Nielsen (1993) and Robert A. Virzi (1992) proposed five participants, while Laura Faulkner (2003) proposed 20. The author recommends future studies include larger groups of 20 to 30 to increase statistical power.

In this study, it is possible that the positive results occurred because of *all* of the changes. However, it is also possible that the results occurred because of only *some* of the changes or *only one* of the changes. It is also possible that another variable other than usability was the cause. Other variables include variables in the design process, such as the consultation of numerous users (cognitive task analyses) during the redesign process. The author recommends further research to determine which variables, or combination of variables, produce the best results.

Figure 1. Comparison of original site to revised site

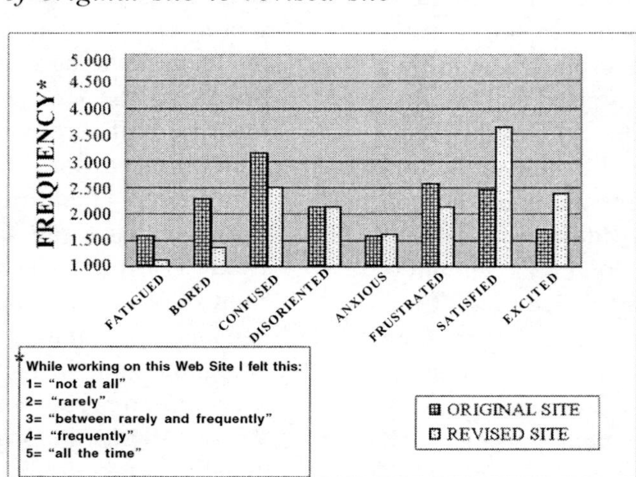

The author also recommends future research to address individual personality differences and situational differences. Examples of questions include: "Do certain guidelines produce better results for situational differences such as chaotic or stressful situations, than they do for calm situations?" In addition, "Do certain guidelines produce better results for different personality types?" For example, some users may enjoy the challenge of finding "hard-to-find" information, while others may be annoyed or frustrated by the same situation.

The author also cautions that self-reported emotions may not be reflective of true emotions. Some participants may have reasons to present themselves as being positive or at ease with a site when they are not. For example, many users feel that ease of use is a sign of intelligence so they may want to appear at ease. They may also want to appear positive or happy to please the tester. One emerging trend in software research that may address problems with self-reporting is biometric research. Biometric data includes heart activity, sweat gland activity, blood pressure, and brain activity. Although these data are more objective than self-reported emotion, users may have different emotions than they normally do simply because of the apparatus. Therefore, the author recommends both subjective and objective studies to capitalize on the benefits of both types of software metrics.

CONCLUSION

Coverage of the topic of *usability* is about 20 years old. Large scale studies of the complex relationships between *emotions and usability*, however, are only recently emerging. This study attempts to understand these relationships. In answer to the question, "Does the Use of Usability Guidelines Affect Web Site User Emotions?" this study answers a tentative "Yes." However, caveats do apply. First, this was a small study with only seven to eight participants in each group. Second, when experimenters measure emotions by self-report, the reports may not be accurate. Third, situational and individual differences need further research in order to generalize to additional types of situations and different personality types. Studying user emotions when users are trying to find information is challenging. However, our users deserve a fulfilling experience. The benefits of helping all Web users find the information they need and want should be well worth the challenge in increased effectiveness, efficiency, and satisfaction.

DISCLAIMER

The views expressed in this article are those of the author and do not reflect the official policy or position of the United States Air Force, Department of Defense, or the U.S. Government.

REFERENCES

Danielson, D. (2002). Web navigation and the behavioral effects of constantly visible site maps. *Interacting with Computers, 14*, 601-618.

Faulkner, L. (2003). Beyond the five-user assumption: Benefits of increased sample sizes in usability testing. *Behavior Research Methods, Instruments, and Computers, 35*(3), 379-383.

Kitajima, M., Kariya, N., Takagi, H., & Zhang, Y. (to appear). *Evaluation of Web site usability using Markov chains and latent semantic analysis.* Institute of Electronics Information and Communication Engineers Transactions on Communications.

Koyani, S., Bailey, R., & Nall, J. (2003). *Research-based Web design and usability guidelines.* National Cancer Institute, National Institutes of Health.

Koyani, S., & Mathews, J. (2001). Cutting to the chase: Delivering usable and useful online health information from Web sites through usability engineering. *Health Promotion Practice, 2*(2), 130-132.

NASA/Goddard Space Flight Center (1996). *User interface guidelines.* Greenbelt, Maryland.

Nielsen, J. (1993). *Usability engineering.* Boston: A. P. Professional.

Nielsen, J. (2003, November 24). Two sigma: Usability and six sigma quality assurance. *Jakob Nielsen's Alertbox.* Retrieved on February 4, 2005, from http://www.useit.com/alertbox/20031124.html

Nielsen, J., & Mack, R. L. (1994). *Usability inspection methods*. New York: John Wiley & Sons.

Ozok, A. A., & Salvendy, G. (2003). The effect of language inconsistency on performance and satisfaction in using the Web: Results of three experiments. *Behavior and Information Technology, 22*(3), 155-163.

Pearrow, M. (2000). *Web site usability handbook*. Rockland, MA: Charles River Media, Inc.

Princeton University Cognitive Science Laboratory. (2005). *WordNet, a lexical database for the English language*. Retrieved on February 4, 2005, from http://wordnet.princeton.edu/

Shneiderman, B. (1998). *Designing the user interface*. Reading, MA: Addison-Wesley Longman.

Virzi, R. A. (1992). Refining the test phase of usability evaluation: How many subjects is enough? *Human Factors, 34*, 457-468.

Webopedia (2005, February). Retrieved on February 4, 2005, from http://webopedia.internet.com/TERM/d/download.html

KEY TERMS

Consistency: Consistency in Web sites refers to keeping similar Web pages similar in their look and feel. Examples of ways to achieve consistency include using the same or similar colors, font, and layout throughout the site.

Easy Scanning: Sanjay Koyani, Robert W. Bailey, and Janice R. Nall (2003, p. 157) found with moderate evidence that "80% of users scan any new page and only 16% read word-by-word." Therefore, it may be useful to make information easy to scan when presenting large amounts of information. Ways to accomplish this may be to use a bold or italicized font so users can quickly pick up differences in content. Avoid underling, however, as the user may confuse an underlined word or phrase with a link. In addition, highlighting may make information easy to visually scan. However, Koyani et al. (2003, p. 77) advise, "Use highlighting sparingly (using) just a few items on a page that is otherwise relatively uniform in appearance."

Minimalist Design: Refers to providing simple and easy to read screen designs. When Web designs are not minimalist, they may cause cognitive overload, or the presence of too much information for users to process. Keeping pages uncluttered and chunking information into categories are examples of ways to provide a minimalist design.

Providing for Recognition rather than Recall: Recognition is easier than recall, as evidenced in most "multiple-choice" questions compared to "fill-in-the-blank" questions. For example, when users return to a Web site, they may not recall where certain information occurred, although they may recognize it when they see it. Examples of ways to provide for recognition rather than recall include providing drop-down menus, providing for book marking, and providing a search engine. When providing a search engine, most experts recommend explaining its use as well as providing for advanced searching. This accommodates the needs of novice users as well as advanced users.

Short Download Time: Download time refers to the time "to copy data (usually an entire file) from a main source to a peripheral device" (Webopedia, 2005). Sanjay Koyani, Robert W. Bailey, and Janice R. Nall (2003, p. 16) found moderate support for the usefulness of minimizing Web page download time. The best way to decrease download time is to limit the number of bytes per page (Koyani, 2003, p. 17).

Speaking the User's Language: Refers to speaking the language of the intended Web audience. It means avoiding jargon, acronyms, or system terms that some of the intended audience may not understand. If you must use jargon, acronyms or system terms, provide a glossary, and/or an acronym finder. Another way to speak the user's language is to ensure your search engine recognizes naturalistic language.

Visibility of Location: In the field of Web usability, visibility of location refers to letting users know where they are in a Web site as well as the status of their inputs and navigation. Examples of ways to increase visibility of location include providing a site map, headers, and navigation paths.

APPENDIX A

E

USABILITY EVALUATION CHECKLIST

Patricia A. Chalmers, Ph.D.
U.S. Air Force

Evaluator: Score every item as a 1, 2, 3, 4, or 5 with points as follows:
1 = Never occurs
2 = Occurs rarely
3 = Occurs sometimes
4 = Occurs much of the time
5 = Occurs all the time

Visibility of Location
The interface:
  enables users to know where they are within the system
  provides a link to the Web site's "home page"
  provides the user with a shortcut to go back to topics

Consistency
The interface uses:
  consistent terminology
  consistent color, for same or similar topic headings
  consistent background for same or similar pages
  consistent layout for pages
  consistent colors for accessed links
  consistent colors for unaccessed links

Error Prevention and Error Recovery
The interface provides:
  a Help Menu
  "Frequently-Asked Questions"
  The interface is uncluttered

The "Inverted Pyramid" Style
The interface locates:
  the most important information at the top
  the least important information at the bottom

Speaking the User's Language
The interface uses:
  language consistent with real-world conventions
  natural language (rather than jargon, acronyms, or system terms)
  If jargon is used, there is a glossary
  If acronyms are used, the interface has an "acronym finder"

Easy Scanning
Key words and phrases are easily visible (for example, by highlighting)
The interface organizes information into manageable chunks
The user can see the screen without scrolling horizontally with 640 × 480 resolution on a 14" monitor

Proper Printing
The user can print selected pages (rather than everything)
Printed contents are free of right side"trimmings"

Short Download Times
The Web site downloads within seven seconds

Provide Assistance
The interface provides:
shortcuts for frequent users
legends describing icon buttons
presentation that is flexible to user needs
a search engine
advanced search capabilities
recognition rather than recall
e-mail addresses for contact
phone numbers for contact
postal addresses for traditional mail contact
e-mail addresses for technical assistance
phone numbers for technical assistance

Elastic Interfaces for Visual Data Browsing

Wolfgang Hürst
Albert-Ludwigs-Universität Freiburg, Germany

INTRODUCTION

In this article, we discuss the concept of elastic interfaces, which was originally introduced by Masui, Kashiwagi, and Borden (1995) a decade ago for the manipulation of discrete, time-independent data. It gained recent attraction again by our own work in which we adapted and extended it in order to use it in a couple of other applications, most importantly in the context of continuous, time-dependent documents (Hürst & Götz, 2004; Hürst, Götz, & Lauer, 2004). The basic idea of an elastic interface is illustrated in Figure 1. Normally, objects are moved by dragging them directly to the target position (direct positioning). With elastic interfaces, the object follows the cursor or mouse pointer on its way to the target position with a speed s that is a function of the distance d between the cursor and the object. They are called elastic because the behavior can be explained by the rubber-band metaphor, in which the connection between the cursor and the object is seen as a rubber band: The more the band is stretched, the stronger the force between the object and the cursor gets, which makes the object move faster. Once the object and cursor come closer to each other, the pressure on the rubber band decreases, thus slowing down the object's movement.

In the next section we describe when and why elastic interfaces are commonly used and review related approaches. Afterward, we illustrate different scenarios and applications in which elastic interfaces have been used successfully for visual data browsing, that is, for skimming and navigating through visual data. First, we review the work done by Masui (1998) and Masui et al. (1995) in the context of discrete, time-independent data. Then we describe our own work, which applies the concept of elastic interfaces to continuous, time-dependent media streams. In addition, we discuss specific aspects considering the integration of such an elastic behavior into common GUIs (graphical user interfaces) and introduce a new interface design that is especially useful in context with multimedia-document skimming.

BACKGROUND

Direct positioning is usually the approach of choice when an object has to be placed at a specific target position. However, elastic interfaces have advantages in situations in which the main goal is not to move the object itself, but in which its movements are mapped to the motion of another object. The

Figure 1. Illustration of the concept of elastic interfaces

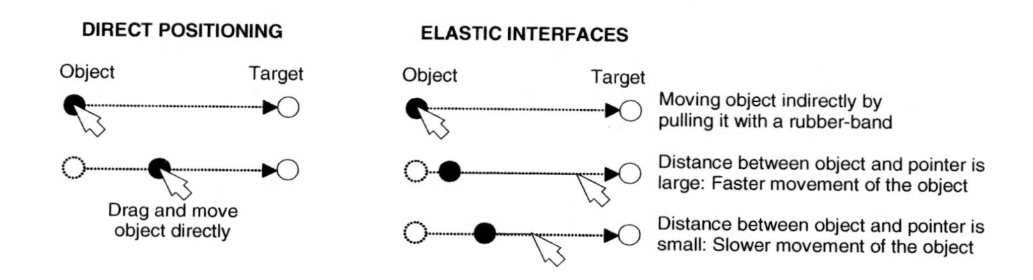

Figure 2. Illustration of the scaling problem of scroll bars and slider

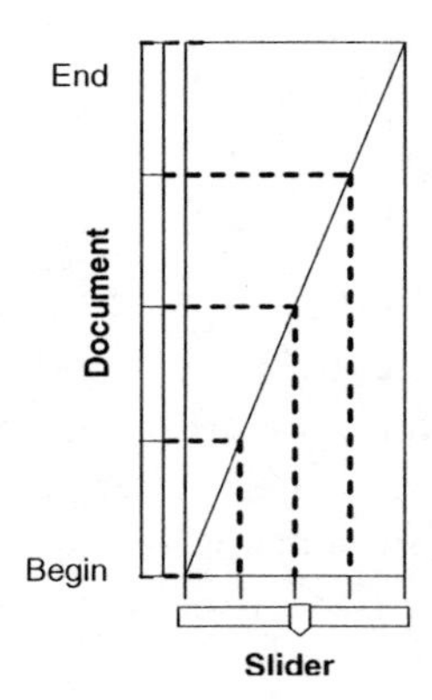

THE SCALING PROBLEM OF SCROLL BARS AND SLIDERS:

If a document is very long, it is impossible to map any position within the document onto the scale of the corresponding slider or scroll bar.

As a consequence, parts of the document's content can not be accessed directly with the slider and the resulting jumps during scrolling lead to a jerky visualization which is usually disliked by the users and considered as disturbing and irritating.

most typical examples for such a case are scroll bar and slider interfaces, for which the dragging of the scroll bar or slider thumb to a target position is mapped to the corresponding movements within an associated document. One common problem with scroll bars and sliders is that a random document length has to be matched to their scale, which is limited by window size and screen resolution (the scaling problem; compare Figure 2). Hence, if the document is very long, specific parts of the file are not accessible directly because being able to access any random position of the document would require movements of the slider thumb on a subpixel level. This is impossible with direct manipulation since a pixel is the smallest unit to display (and thus to manipulate) on the screen. In addition, the movement of the document's content during scrolling becomes rather jerky, which is usually considered irritating and disturbing by users. This is where elastic interfaces come into play: Since the scrolling speed is indirectly manipulated based on the mapping of the distances between cursor and thumb to a corresponding speed, navigation becomes independent of the scroll bar's or slider's scale and thus independent of the actual length of the document. If the function for the distance-to-speed mapping is chosen appropriately, subpixel movements of the thumb and thus slow scrolling on a finer scale can be simulated.

Other solutions to solve the scaling problem have been proposed in the past, mainly as extensions or replacements of slider interfaces. The basic functionality of a slider is to select a single value or entry by moving the slider thumb along the slider bar, which usually represents an interval of values. A typical selection task is, for example, the modification of the three values of an RGB color by three different sliders, one for each component. If visual feedback is given in real-time, sliders can also be used for navigation either in a continuous, time-dependent media file, such as a video clip, or to modify the currently visible part of a static, time-independent document whose borders expand beyond the size of its window (similar to the usage of a scroll bar). In both cases, again, the user drags the thumb along the bar in order to select a single value. In the first case, this value is a specific point in time (or the corresponding frame of the video), and in the second case, it is a specific position in the document (and the task is to position the corresponding content within the visible area of the screen).

Most approaches that try to avoid the scaling problem have been proposed either for selection tasks or for scrolling interfaces that enable navigation in static, time-independent data. The most well known is probably the Alphaslider introduced by Ahlberg and Shneiderman (1994). Here, the thumb of a slider or a scroll bar is split into three different areas, each of which allows for navigation at a different granularity level. Ayatsuka, Rekimoto, and Matsuoka (1998) proposed the Popup Vernier in which the user is able to switch between different scrolling resolutions by using additional buttons or keys. Instead of relying on different granularities for the whole slider scale, the TimeSlider interface introduced by Koike, Sugirua, and Koseki (1995) works with a nonlinear resolution of the scale. This way, users can navigate through a document at a finer level around a particular point of interest.

So far, only few projects have dealt with the scaling problem in the context of navigation through continuous, time-dependent data such as video files. One of the few examples is the Multi-Scale Timeline Slider introduced by Richter, Brotherton, Abowd, and Truong (1999) for browsing lecture recordings. Here, users can interactively add new scales to a slider interface and freely modify their resolution in order to be able to navigate through the corresponding document at different granularity levels. Other approaches have been proposed in the context of video editing (see Casares et al., 2002, for example). However, video editing is usually done on a static representation of the single frames of a video rather than on the continuous signal, which changes over time. Thus, approaches that work well for video editing cannot necessarily be applied for video browsing (and vice versa). In Hürst, Götz, and Jarvers (2004), we describe two slider variants, namely the ZoomSlider and the NLslider, which allow for interactive video browsing at different granularity levels during replay. Both feature some advantages compared to other approaches, but unfortunately they share some of their disadvantages as well.

Most of the approaches described above rely on a modification of the scale of the slider in order to enable slower scrolling or feature selection at a finer granularity. However, this has one significant disadvantage: Adjusting the slider's scale to a finer resolution makes the slider bar expand beyond the borders of the corresponding window, thus resulting in situations that might be critical to handle when the thumb reaches these borders. Approaches such as the Alphaslider circumvent this problem by not modifying the slider's scale explicitly, but instead by doing an internal adaptation in which the movements of the input devices are mapped differently (i.e., with a finer resolution) to the movements of the corresponding pointer on the screen. However, this only works with a device that controls the pointer on the screen remotely, such as a mouse, but it can be critical in terms of usability if a device is used that directly interacts with the object on the screen, such as a pen on a touch screen. In the following two sections, we describe how elastic interfaces can be used successfully to avoid both of the problems just mentioned as well as the scaling problem.

ELASTIC INTERFACES APPLIED TO STATIC, TIME-INDEPENDENT DATA

In their original work about elastic interfaces, Masui et al. (1995) introduced the so-called FineSlider to solve the scaling problem in relation to the task of value selection. The FineSlider is illustrated in Figure 3 together with a possible distance-to-speed mapping that defines how the distance between the current position of the slider's thumb and the mouse pointer is mapped to the actual scrolling speed. Its feasibility and usefulness were proven in a user study in which the test persons had to select a single value from a list of alphabetically sorted text entries of which only one entry was visible at a time (Masui

Figure 3. FineSlider interface and corresponding distance-to-speed mapping

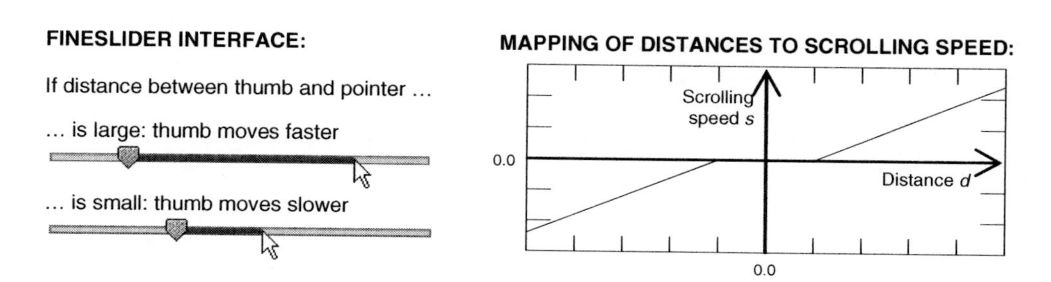

et al.). This task is similar to a random-selection task in which a single value is selected from a somehow sorted list of values. However, the authors describe usages of the FineSlider concept in relation to some other applications as well, including the integration of the elastic scrolling functionality into a regular scroll bar in order to solve the scaling problem when navigating through static documents such as text or images. An actual implementation of this is presented in Masui (1998), in which the LensBar interface for the visual browsing of static data was introduced. Among other functionalities, the LensBar offers elastic navigation by moving a scroll-bar-like thumb and thus enables skimming through a document at a random granularity. In addition to these navigation tasks, Masui et al. describe the usage of elastic interfaces for the modification of the position of single points in a 2-D (two-dimensional) graphics editor. In such a situation, direct positioning is normally used because the task is not to remotely modify an object, but to reposition the object itself (compare Figure 1). However, elastic interfaces can still be useful in such a situation, for example, if accurate positioning on the pixel level is hard to do, which is sometimes the case with pen-based input devices and touch screens.

All these tasks for the selection of values, the navigation in documents, or the repositioning of objects have been discussed and realized in relation to time-independent, static data. However, the concept of elastic interfaces can also be applied to continuous visual data streams, as we describe in the next section.

ELASTIC INTERFACES APPLIED TO TIME-DEPENDENT DATA

If a media player and the data format of the respective file support real-time access to any random position within the document, a slider can be used to visually browse the file's content in the same way as a scroll bar is used to navigate and skim discrete data, for example, a text file. This technique, which is sometimes referred to as random visible scrolling (Hürst & Müller, 1999), has proven to be a very easy, convenient, and intuitive way for the visual browsing of continuous data streams. However, the scaling problem, which appears in relation to discrete documents (compare Figure 2), occurs here as well. In fact, it is sometimes considered even more critical because the resulting jerky visual feedback can be particularly disturbing in the case of a continuous medium such as the visual stream of a video. In Hürst, Götz, et al. (2004), we applied the concept of elastic interfaces described above to video browsing. The conceptual transfer of elastic interfaces from static data to continuous data streams is straightforward. Instead of selecting single values or positions within a static document, a value along a timeline is manipulated. In a video, this value relates to a single frame, which is displayed instantly as a result of any modification of the slider thumb caused by the user. However, because of the basic differences between these two media types (i.e., discrete values or static documents vs. time-dependent, continuous data streams), it is not a matter of course that the concept works as well for continuous media as it does for discrete data in terms of usability. For example, when using a scroll bar to navigate through some textual information, it is not only a single entity such as a line or even just a word that is usually visible to the user, but there is also some context, that is, all the lines of the text that fit into the actual window size. With a continuous visual data stream, for example, a video, just one frame and therefore no context is shown at a time. While this is also true for value selection using a slider, those values are usually ordered in some way, making navigation and browsing much easier. Ramos and Balakrishnan (2003) introduced the PVslider for video browsing, which features some sort of elastic behavior as well. However, the authors assume a slightly different interpretation of the term elastic by measuring the distance (i.e., the tension of a virtual rubber band) between the mouse pointer and a fixed reference point instead of a moving object (the slider thumb), which makes their approach more similar to an elastic fast-forward and rewind functionality than to the idea of an elastic interface as it was introduced originally by Masui et al. (1995).

In an informal study with the interface presented on the left side of Figure 5, we showed the usefulness and feasibility of the original elastic-slider approach in relation to video browsing (Hürst, Götz, et al., 2004). With this implementation, users are able to visually skim through the video very fast (by moving the mouse pointer quickly away from the

Figure 4. Illustration of the elastic panning approach

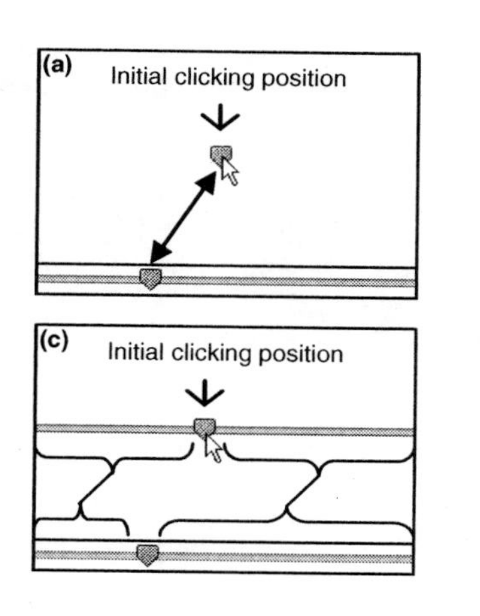

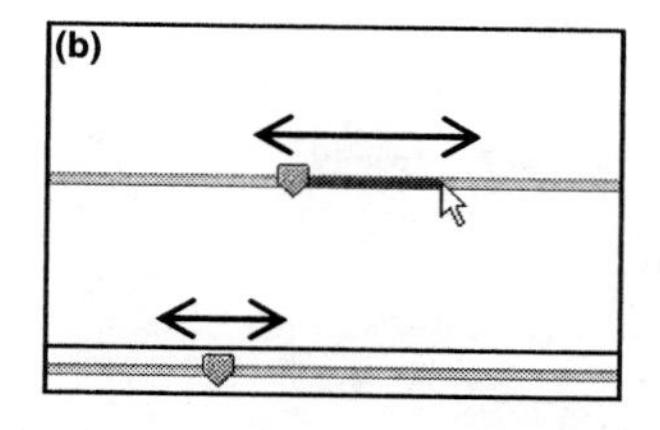

(a) The initial clicking position is associated with the current position of the original thumb
(b) Moving the pointer to the left or right initiates backward or forward browsing, respectively
(c) Mapping actual to virtual slider scale results in a non-linear resolution of the virtual scale

thumb) or very slow (by moving the mouse pointer closer to the thumb), even on the level of single frames. However, we also identified some problems with this approach. Some users complained that they have to focus on two things when browsing the video: the actual content as well as the slider widget. While this is also true with a regular slider, it is more critical in the case of an elastic interface because the scrolling speed depends directly on the pointer's and the thumb's movements. As a consequence, accidental changes of the scrolling direction happen quite frequently when a user tries to reduce the scrolling speed. In such a case, both the thumb and the pointer are moving toward each other. Since a user is more likely to look at the content of the document in such a situation rather than at the slider thumb, it can happen quite easily that the pointer accidentally moves behind the thumb, thus resulting in an unwanted change of the scrolling direction.

These observations were the main reason for us to introduce a modification of the concept of elastic interfaces called elastic panning. With elastic panning, browsing gets activated by clicking anywhere on the window that shows the content of the document. As a consequence, an icon appears on the screen that represents a slider thumb and is associated with the current position within the file (compare Figure 4a). Moving the slider to the right or left enables forward or backward browsing, respectively, along a virtual scale that extends to both sides of the virtual slider thumb (compare Figure 4b). The resulting movements of the thumb, and thus the scrolling speed, are similar to the elastic slider illustrated in Figure 3: Scrolling is slow if the thumb and pointer are close to each other, and the speed increases with the distance between those two objects. Distance is only measured horizontally along the virtual slider scale. Vertical movements of the mouse pointer do not influence scrolling behavior, but only make the visualization of the virtual thumb and scroll bar follow the pointer's movements (i.e., the widgets on the screen are "glued" to the pointer). The beginning and the end of the virtual scale are mapped to the borders of the player window. This, together with the association of the initial clicking position with the actual position in the document at that time, can result in a mismatch of the scales to both sides of the slider thumb (compare Figure 4c). However, since the original motivation of elastic interfaces was to make scrolling independent of the actual length of a document (and thus of the slider's scale), this mismatch does not influence the overall scrolling behavior. It is therefore uncritical as we also confirmed in the evaluation presented in Hürst,

Figure 5. Snapshots of a video player featuring an elastic slider (left) and elastic panning (right), respectively

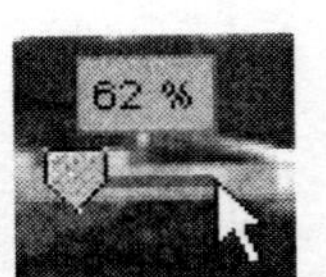

Enlargement of the virtual slider thumb for elastic panning

Götz, et al. (2004). However, it was observed that some users were irritated by the change of the scale, which is why we revised the interface design by showing only the part of the virtual slider bar facing the actual scrolling direction and by not representing the actual values on the scale but only relative positions (compare Figure 5, which shows a snapshot of the actual implementation of the revised interface design). The improvement resulting from this revision was confirmed by the test persons when we confronted them with the revised design. In addition, we ran a couple of experiments, for example, with different distance-to-speed mappings (compare Figure 3, right side), in order to optimize the parameters of the implementation.

In Hürst, Götz, et al. (2004), we present a qualitative usability study in which 10 participants had to perform a search task with elastic panning as well as with a standard slider interface. This comparative study showed the feasibility of the proposed approach as well as its usefulness for video browsing. It circumvents the problem of accidental direction changes that frequently appears with the interface illustrated in the first snapshot in Figure 5. Generally, elastic panning was considered by the participants as intuitive as well as easy to learn and operate. In addition to it solving the scaling problem, most users appreciated that the movements were less jerky with elastic panning than with the original slider. Some noted that they particularly liked the ability to quickly navigate the file and to be able to easily slow down once they get closer to a target position. Another feature that was highly appreciated is the ability to click directly on the window instead of having to use a separate slider widget for browsing. Users liked that they did not have to continuously change between looking at the video and the slider widget, but could focus on the player window all the time. This seems to be a big advantage on small devices as well, for which display space is limited and hence videos are often played in full-screen mode.

In addition to video browsing, we also applied elastic panning successfully to the task of the visual browsing of recorded lectures, that is, visual streams that contain recorded slides used in lectures as well as handwritten annotations made on the slides (Hürst & Götz, 2004). Here, we are dealing with a mixture of static data, that is, the prepared slides that contain text, graphics, and still images, and a continuous signal, that is, the annotations that are made and recorded during the lecture that are replayed by the player software in the same temporal order as they appeared during the live event. Especially if there are a lot of handwritten annotations, for example, if a mathematical proof was written down by the lecturer, the scaling problem can become quite critical. Again, users—in this case students who use recorded lectures for recapitulation and exam preparation—highly appreciated that with this interface they are able to quickly skim through the visual signal of the recording as well as slow down easily to analyze the content in more detail. Being able to scroll and navigate at the smallest possible level allows the users, for example, to access every position in which some event happened in the visual stream, such as a small annotation made in a graphical illustration, and to start replay at the corresponding position in the lecture recording easily.

FUTURE TRENDS

The fact that traditional sliders and scroll bars do not scale to large document sizes is well known and has been studied extensively in the past. Different solutions to solve or circumvent this problem have been proposed throughout the '90s. However, more recent developments in input and output devices as well as an enhanced spectrum of data that we have to deal with today make it necessary to refocus on this issue. For example, portable devices (such as very small laptops) and handheld devices (such as PDAs, portable digital assistants) have a restricted screen resolution, thus amplifying the scaling problem. With pen-based input becoming more popular due to devices such as the tablet PC (personal computer) and PDAs, not all solutions that were proposed in the past are applicable anymore. Last but not least, new media types, especially continuous, time-dependent data, call for different interface designs.

The work on elastic interfaces that we summarized in this article is a first step in supporting these issues because it showed that this interaction approach can be used not only for static, time-independent data, but for continuous, time-dependent data streams as well. In addition, it offers great potential for usage in relation with pen-based input devices and it is also applicable to very small screen sizes such as PDAs, although a final proof of this claim is left for future work. However, first experiments with elastic panning and pen-based interaction on a tablet PC have been very promising. In addition to visual data, which is the focus of this article, today we are more and more faced with digital audio documents as well. In Hürst and Lauer (in press), we describe how the concept of elastic interfaces can also be applied to flexible and easy speech skimming. Detailed descriptions about elastic audio skimming as well as first usability feedback can be found in Hürst, Lauer, and Götz (2004a, 2004b). Probably the most exciting but also the most difficult challenge for future work in this area is to bring these two approaches together in one interface, thus enabling real multimodal navigation in both acoustic as well as visual data at the same time.

CONCLUSION

This article described how elastic interfaces can be used for visual data browsing and feature selection. First, we reviewed the original work done by Masui (1998) and Masui et al. (1995) on elastic interfaces in relation to static, time-independent data. Our own work includes the successful transition of this concept from the static, time-independent domain to continuous, time-dependent media streams, such as video as well as mixed-media streams. In addition, we introduced a modification in the interface design that proved to lead to a clear improvement in usability, especially in relation to continuous visual media streams. Current and future work includes evaluations of the discussed concepts with different input devices, in particular pen-based input, as well as in combination with acoustic-data browsing.

REFERENCES

Ahlberg, C., & Shneiderman, B. (1994). The Alphaslider: A compact and rapid selector. *Proceedings of the SIGCHI Conference on Human Factors in Computing Systems, ACM CHI 1994* (pp. 365-371).

Ayatsuka, Y., Rekimoto, J., & Matsuoka, S. (1998). Popup Vernier: A tool for sub-pixel-pitch dragging with smooth mode transition. *Proceedings of the 11th Annual ACM Symposium on User Interface Software and Technology, UIST 1998* (pp. 39-48).

Casares, J., Myers, B. A., Long, C., Bhatnagar, R., Stevens, S. M., Dabbish, L., et al. (2002). Simplifying video editing using metadata. *Proceedings of Designing Interactive Systems, DIS 2002* (pp. 157-166).

Hürst, W., & Götz, G. (2004). Interface issues for interactive navigation and browsing of recorded lectures and presentations. *Proceedings of ED-MEDIA 2004* (pp. 4464-4469).

Hürst, W., Götz, G., & Jarvers, P. (2004). Advanced user interfaces for dynamic video browsing. *Proceedings of ACM Multimedia 2004* (pp. 742-743).

Hürst, W., Götz, G., & Lauer, T. (2004). New methods for visual information seeking through video browsing. *Proceedings of the 8th International Conference on Information Visualisation, IV 2004* (pp. 450-455).

Hürst, W., & Lauer, T. (in press). Interactive speech skimming via time-stretched audio replay. In C. Ghaoui (Ed.), *Encyclopedia of human computer interaction.* Hershey, PA: Idea Group Reference.

Hürst, W., Lauer, T., & Götz, G. (2004a). An elastic audio slider for interactive speech skimming. *Proceedings of the 3rd Nordic Conference on Human-Computer Interaction, NordChi 2004* (pp. 277-280).

Hürst, W., Lauer, T., & Götz, G. (2004b). Interactive manipulation of replay speed while listening to speech recordings. *Proceedings of ACM Multimedia 2004* (pp. 488-491).

Hürst, W., & Müller, R. (1999). A synchronization model for recorded presentations and its relevance for information retrieval. *Proceedings of ACM Multimedia 1999* (pp. 333-342).

Koike, Y., Sugirua, A., & Koseki, Y. (1995). TimeSlider: An interface to specify time point. *Proceedings of the 10th Annual ACM Symposium on User Interface Software and Technology, UIST 1997* (pp. 43-44).

Masui, T. (1998). LensBar: Visualization for browsing and filtering large lists of data. *Proceedings of 1998 IEEE Symposium on Information Visualization* (pp. 113-120).

Masui, T., Kashiwagi, K., & Borden, G. R., IV. (1995). Elastic graphical interfaces for precise data manipulation. *Conference Companion to the SIGCHI Conference on Human Factors in Computing Systems, ACM CHI 1995* (pp. 143-144).

Ramos, G., & Balakrishnan, R. (2003). Fluid interaction techniques for the control and annotation of digital video. *Proceedings of the 16th Annual ACM Symposium on User Interface Software and Technology, UIST 2003* (pp. 105-114).

Richter, H., Brotherton, J., Abowd, G. D., & Truong, K. (1999). *A multi-scale timeline slider for stream visualization and control* (Tech. Rep. No. GIT-GVU-99-30). Georgia Institute of Technology, USA.

KEY TERMS

Elastic Interfaces: Interfaces or widgets that manipulate an object, for example, a slider thumb, not by direct interaction but instead by pulling it along a straight line that connects the object with the current position of the cursor. Movements of the object are a function of the length of this connection, thus following the rubber-band metaphor.

Elastic Panning: An approach for navigation in visual data, which has proven to be feasible not only for the visual browsing of static, time-independent data, but for continuous, time-dependent media streams as well. Similar to the FineSlider, it builds on the concept of elastic interfaces and therefore solves the scaling problem, which generally appears if a long document has to be mapped on a slider scale that is limited by window size and screen resolution.

FineSlider: A special widget introduced by Masui et al. (1995) for navigation in static, time-independent data. It is based on the concept of elastic interfaces and therefore solves the scaling problem, which generally appears if a long document has to be mapped on a slider scale that is limited by window size and screen resolution.

Random Visible Scrolling: If a media player and a data format support real-time random access to any position within a respective continuous, time-dependent document, such as a video recording, a common slider interface can be used to visually browse the file's content in a similar way as a scroll bar is used to navigate and browse static information, such as text files. This technique is sometimes referred to as random visible scrolling.

Rubber-Band Metaphor: A metaphor that is often used to describe the behavior of two objects that are connected by a straight line, the rubber band, in which one object is used to pull the other one toward a target position. The moving speed of the pulled object depends on the length of the line between the two objects, that is, the tension on the rubber band: Longer distances result in faster movements, and shorter distances in slower movements.

Scaling Problem: A term that can be used to describe the problem of scroll bars and sliders not scaling to large document sizes. If a document is very long, the smallest unit to move the scroll bar or slider thumb on the screen, that is, one pixel, already represents a large jump in the file, thus resulting in jerky visual feedback that is often considered irritating and disturbing, and in the worst case leads to a significant loss of information.

Visual Data Browsing: A term generally used to summarize all kinds of interactions involved in visually skimming, browsing, and navigating visual data in order to quickly consume or identify the corresponding content or to localize specific information. Visual data in this context can be a static document, such as a text file, graphics, or an image, as well as a continuous data stream, such as the visual stream of a video recording.

Engagability

John Knight
University of Central England, UK

INTRODUCTION

Recent trends in HCI have sought to widen the range of use qualities beyond accessibility and usability. The impetus for this is fourfold. First, some argue that consumer behaviour has become more sophisticated and that people expect products to give them a number of life-style benefits. The benefits that products can give people include functional benefits (the product does something) and suprafunctional benefits (the product expresses something). Engagability is thus important in understanding people's preferences and relationships with products. Second, technological advances offer the possibility of designing experiences that are like those in the real world. Engagability is therefore important in providing an evaluative and exploratory approach to understanding "real" and "virtual" experiences. Third, the experiences that people value (e.g., sports) require voluntary engagement. Thus, engagability is important in designing experiences that require discretionary use. Lastly, the product life cycle suggests the need to look beyond design to engagement. Products change from their initial production, through distribution to consumption. Each phase of this life cycle contains decision-making activities (e.g., purchasing, design, etc.). Engagability is an important research focus in explaining stakeholders' values in making these decisions. As such, engagability research seeks to understand the nature of experience in the real and virtual worlds. The activities that people become engaged with are often complex and social and thus challenge the traditional HCI focus on the single task directed user. Important application areas for this inquiry are learning, health and sport, and games.

BACKGROUND

Engagability research has primarily come from outside of HCI. It includes research into motivation, education, and understanding human experience. For example, the feeling of being engaged in experience has been investigated by Csikszentmihalyi (1991, p. 71), who describes the qualities of optimal experience and flow:

A sense that one's skills are adequate to cope with the challenges at hand, in a goal-directed, rule-bound action system that provides clear rules as to how well one is performing. Concentration is so intense that there is no attention left over to think about anything irrelevant, or to worry about problems. Self-consciousness disappears, and the sense of timing becomes distorted.

Norman's work with Andrew Ortony and William Revelle (Norman, 2003) proposes that people are engaged in compelling experiences at three levels of brain mechanism comprising:

The automatic, prewired layer called the visceral level; the part that contains the brain processes that control everyday behaviour, known as the behavioural level and the contemplative part of the brain, or the reflective level. (Norman 2003, p. 6)

Furthermore, *These three components interweave both emotions and cognition.* (Norman 2003, p. 6)

Jordan focuses on hedonic use qualities and states that, "Games are an example of a product type that are designed primarily to promote emotional enjoyment through providing people with a cognitive and physical challenge." He goes on to say that, "well-designed games can engage players in what they are doing. Instead of having the feeling that they are sitting in front of the television controlling animated sprites via a control pad, they may feel that they are playing soccer at Wembley Stadium or trying to escape from a monster in some fantasy world" (Jordan, 2000, p. 45).

Dunne's (1999) "aesthetics of use" and Laurel's concept of engagement (from Aristotle) describe a similar phenomenon: "Engagement ... is similar in many ways to the theatrical notion of the "willing suspension of disbelief," a concept introduced by the early nineteenth century critic and poet Samuel Taylor Coleridge" (Laurel, 1991, p. 113).

Engagement in relation to learning is proposed by Quinn (1997). He suggests that engagement comes from two factors—"interactivity" and "embeddedness." Jones, Valdez, Nowakowski, and Rasmussen (1994) describe engaged learning tasks as "challenging, authentic, and multidisciplinary. Such tasks are typically complex and involve sustained amounts of time... and are authentic." Jones, Valdez, Nowakowski, and Rasmussen (1995) go on to suggest six criteria for evaluating educational technology in the context of engaged learning:

1. Access
2. Operability
3. Organisation
4. Engagability
5. Ease of use
6. Functionality

FUTURE TRENDS

Engagability was first applied to HCI design by Knight and Jefsioutine (2003). The meaning and impact of engagability was explored at the 1st International Design and Engagability Conference in 2004 (Knight & Jefsioutine, 2004). Papers related to design practise and the qualities of engaging experiences. Papers presented examples of engagement in the context of:

1. Community
2. Creativity
3. Design
4. Education
5. Emotion
6. Health
7. Physiology
8. Real and virtual experience
9. Identity
10. Well-being

CONCLUSION

Many researchers argue for design to go beyond usability and there is a consensus to move to hedonic use qualities. The widening of HCI research and design into the realms of emotion is to be welcomed and engaging products and services offer the promise of richer interactions. However, engagement also requires an ethical as well as aesthetic approach to design. Including human values in design means not only better products but also transformative qualities as well.

REFERENCES

Csikszentmihalyi, M. (1991). *Flow: The psychology of optimal experience*. New York: Harper Collins.

Dunne, A. (1999). Hertzian tales: Electronic products, aesthetic experience and critical design (pp. 73-75). London: Royal College of Art.

Jones, B., Valdez, G., Nowakowski, J., & Rasmussen, C. (1994). *Designing learning and technology for educational reform.* Oak Brook, IL: North Central Regional Educational Laboratory. Retrieved June 19, 2004, from http://www.ncrel.org/sdrs/engaged.htm

Jones, B., Valdez, G., Nowakowski, J., & Rasmussen, C. (1995). *Plugging in: Choosing and using educational technology*. Oak Brook, IL: North Central Regional Educational Laboratory. Retrieved on June 19, 2004, from http://www.ncrel.org/sdrs/edtalk/body.pdf

Jordan, P. W. (2000). *Designing pleasurable products.* London: Taylor and Francis.

Knight, J., & Jefsioutine, M. (2003, July 21). *The experience design framework: From pleasure to engagability*. Arts and Humanities and HCI Workshop, King's Manor, York, UK, July 21. Retrieved from http://www-users.cs.york.ac.uk/~pcw/KM_subs/Knight_Jefsioutine.pdf

Knight, J., & Jefsioutine, M. (Eds.) (2004, July 6). *Proceedings of The 1st International Design and*

Engagability and Conference, University of Central England, Birmingham.

Laurel, B. (1991). *Computers as theatre*. Reading, MA: Addison-Wesley Publishing Company.

Norman, D. (2003). *Emotional design: Why we love [or hate] everyday things*. New York: Basic Books.

Quinn, C. N. (1997). Engaging learning. *Instructional Technology Forum* (ITFORUM@UGA.CC.UGA.EDU), Invited Presenter. Retrieved on June 19, 2004, from http://itech1.coe.uga.edu/itforum/paper18/paper18.html

KEY TERMS

Engagability: A product or service use quality that provides beneficial engagement.

Functional Use Qualities: The quality of a product to deliver a beneficial value to the user.

Hedonic Use Qualities: The quality of a product to deliver pleasurable value to the user.

Product Life Cycle: The evolution of a product from conception onward.

Suprafunctional Use Qualities: Qualities experienced in interacting with a product or service that do not have an immediate instrumental value. Suprafunctional user qualities include aesthetics and semantics that influence the user experience but are not the primary goal of use.

Use Quality: The value of the experience of interacting with a product or service.

Ethics and HCI

John Knight
University of Central England, UK

INTRODUCTION

The goal of HCI research and design has been to deliver universal usability. Universal usability is making interfaces to technology that everyone can access and use. However, this goal has been challenged in recent times. Critics of usability (e.g., Eliot, 2002) have argued that usability "dumbs down" the user-experience to the lowest common denominator. The critics propose that focusing on ease of use can ignore the sophistication of expert users and consumers. At the same time, researchers have begun to investigate suprafunctional qualities of design including pleasure (Jordan, 2000), emotion (Norman, 2003), and fun. While recent discussions in HCI have bought these questions to the surfaces, they relate to deeper philosophical issues about the moral implications of design. Molotch (2003, p. 7), states that:

Decisions about what precisely to make and acquire, and when, where, and how to do it involve moral judgements about what a man is, what a woman is, how a man ought to treat his aged parents...how he himself should grow old, gracefully or disgracefully, and so on.

One response to this moral dilemma is to promote well-being rather than hedonism as an ethical design goal.

BACKGROUND

The Western ethical tradition goes back to ancient Greece. Ethics develop the concept of good and bad within five related concepts:

1. Autonomy
2. Benefiance
3. Justice
4. Non-malefiance
5. Fidelity

At an everyday level, ethics (the philosophy of morality) informs people about the understanding of the world. The motivation for ethical behaviour goes beyond the gratification of being a good person. Social cohesion is based on a shared understanding of good and bad. Bond (1996, p. 229) suggests that ethics tries to: "Reconcile the unavoidable separateness of persons with their inherently social nature and circumstances."

DESIGN

Design is the intentional creation of utilitarian objects and embodies the values of the maker. Harvey Molotch (2003, p. 11) argues that products affect people:

At the most profound level, artefacts do not just give off social signification but make meaning of any sort possible...objects work to hold meaning more or less, less still, solid and accessible to others as well as one's self.

The moral responsibility of design has led some (e.g., William Morris) towards an ethical design approach. Ethical design attempts to promote good through the creation of products that are made and consumed within a socially accepted moral framework. Victor Papanek (1985, p. 102) has focused on the ecological impact of products and has demanded a "high social and moral responsibility from the designer." Whiteley (1999, p. 221) describes this evolution of ethical design as: "[Stretching] back to the mid-nineteenth century and forward to the present. However, just what it is that constitutes the ethical dimension has changed significantly over 150 years, and the focus has shifted from such concerns as the virtue of the maker, through the integrity and aesthetics of the object, to the role of the designer—and consumer—in a just society."

Unlike Morris's arts and craft approach, engineering and science based design is often perceived as value free. Dunne (1999) quotes Bernard Waites to counter this apparent impartiality: "All problems...are seen as 'technical' problems capable of rational solution through the accumulation of objective knowledge, in the form of neutral or value-free observations and correlations, and the application of that knowledge in procedures arrived at by trial and error, the value of which is to be judged by how well they fulfil their appointed ends. These ends are ultimately linked with the maximisation of society's productivity and the most economic use of its resources, so that technology....becomes 'instrumental rationality' incarnate...."

HCI

HCI applies scientific research to the design of user-interfaces. While many (e.g., Fogg, 2003) have promoted ethics in HCI, Cairns and Thimbleby (2003, p. 3) go further to indicate the similarities between the two: "HCI is a normative science that aims to improve usability. The three conventional normative sciences are aesthetics... ethics...and logic. Broadly, HCI's approaches can be separated into these categories: logic corresponds to formal methods in HCI and computer science issues; modern approaches, such as persuasive interfaces and emotional impact, are aesthetics; and the core body of HCI corresponds with ethics...HCI is about making the user experience *good*."

In promoting "good" HCI, ethics has concentrated on professional issues and the impact of functionality, ownership, security, democracy, accessibility, communication, and control. Friedman (2003) summarises this work as pertaining to:

1. Accountability
2. Autonomy
3. Calmness
4. Environmental sustainability
5. Freedom from bias
6. Human welfare
7. Identity
8. Informed consent
9. Ownership and property
10. Privacy
11. Trust
12. Universal usability

Guidelines are often used to communicate HCI ethics. Fogg (2003, pp. 233-234) provides guidelines for evaluating the ethical impact of persuasive computing. This requires researchers to:

1. List all stakeholders.
2. List what each stakeholder has to gain.
3. List what each stakeholder has to lose.
4. Evaluate which stakeholder has the most to gain.
5. Evaluate which stakeholder has the most to lose.
6. Determine ethics by examining gains and losses in terms of values.
7. Acknowledge the values and assumptions you bring to your analysis.

Standards are a more mandatory form of ethical guidelines and prescribe processes, quality, and features. Compliance can be informal or through "de jure" agreements (e.g., International Organization for Standardization, 2000). Cairns and Thimbleby (2003, p. 15) offer a less stringent set of HCI ethical principles ethics comprising:

1. A rule for solving problems
2. A rule for burden of proof
3. A rule for common good
4. A rule of urgency
5. An ecological rule
6. A rule of reversibility

Citing Perry (1999) as evidence, Cairns and Thimbleby (2003, p. 15) imply that ethical rules are a poor substitute for knowledge:

Students...generally start from an absolutist position: 'There is one right way to do HCI.' This initial position matures through uncertainty, relativism, and then through stages of personal ownership and reflection. At the highest levels...a student makes a personal commitment to the particular ethical framework they have chosen to undertake their work. However, it is a dynamic activity to develop any such framework in the complex and conflicting real world.

Developing ethical knowledge requires understanding and empathy with users. This becomes harder as the design goal shifts from usable user interface to engaging designed experience. Patrick Jordan's (2000) "new human factors" describes an evolution of consumers where usability has moved from being a "satisfier" to a "dissatisfier." This argument implies that it would be unethical to ignore users' deeper wants and needs, which he maintains are based on pleasure and emotional benefit. An important advance from traditional HCI is Jordan's emphasis on the product life cycle and the changing relationships that people have with their possessions. The emotions involved in buying a mobile telephone are different from those elicited though ownership and long-term use. The key difference with Jordan's approach (to traditional HCI) is the importance placed on the benefit of the users' interaction.

As well as emotional interaction, another recent trend in HCI argues that interfaces are not just used but are experienced. Shedroff's (2001, p. 4) model of experience design offers users an attraction, an engagement, and a conclusion. Experience design is about the whole experience of an activity. Activities occur in time and involve a number of agents, artefacts, and situations. Experiences are predicated by motivation and the reward of the conclusion. Combining Jordan's emphasis on the emotional benefit of interaction and Shedroff's model of experience, an alternative set of design qualities are suggested:

1. Attraction
2. Engagement
3. Benefit

EMOTIONAL ATTRACTION

How are users attracted to an experience? In emotional design, Donald Norman (2003, p. 87) states that "attractiveness is a visceral level phenomenon—the response is entirely to the surface look of an object." While showing how emotions are integral to cognition and decision-making, the design model he proposes (Visceral, Behavioural, and Reflective) diminishes the social and rational basis of emotion. Alternatively, Jordan (2000) suggests that attraction is based on Lionel Tiger's (1992) four pleasures:

1. Socio-pleasure
2. Pyscho-pleasure
3. Physio-pleasure
4. Ideo-pleasure

The focus on pleasure raises a number of ethical dilemmas. Epicurus suggests that pleasure and displeasure need to be measured against their impact on well-being. Singer (1994, p. 188) cites Epicurus' treatise on "The Pursuit of Pleasure":

Every pleasure then because of its natural kinship to us is good, yet not every pleasure is to be chosen: even as every pain also is an evil, yet not all are always of a nature to be avoided ...For the good on certain occasions we treat as bad, and conversely the bad as good.

EMOTIONAL ENGAGEMENT

How are users emotionally engaged in an experience? Laurel (1991, p. 113) applies the concept of engagement from Aristotle. Quinn (1997) uses engagement as a paradigm for learning applications. While synthetic experiences that seize the human intellect, emotions and senses have the potential for good; they could also be harmful. Computer games can be compulsive but they do not necessarily benefit the user? Learning can be boring but it is often the precursor to the reward of greater knowledge and experience. Indeed, Dejean (2002, pp. 147-150) suggests that apparently unpleasant experiences, such as difficulty, challenge, and fatigue, can be rewarding in certain circumstances.

EMOTIONAL BENEFIT

What profit can users derive from an experience? Many researchers argue that users' wants and needs are becoming more sophisticated. Users want more from products than value for money and usability. Products are not just used; they become possessions and weave their way into the fabric of the peoples' lives. Bonapace (2002, pp. 187-217) presents a hierarchical model of users' expectations with safety and well-being, at ground level, functionality, and then usability, leading up to an apex of pleasure. Dunne (1999) challenges the goal of usability, although his critique replaces usability with aesthetics, turning products into art. The foreword (1999, p. 7) to "Hertzian Tales: Electronic Products, Aesthetic Experiences and Critical Design" proposes that:

The most difficult challenge for designers of electronic objects now lies not in technical and semiotic functionality, where optimal levels of performance are already attainable, but in the realms of metaphysics, poetry and aesthetics, where little research has been carried out.

Aesthetic use-values may be of limited profit for people. Owning a beautiful painting can provide pleasure for the individual but may be of little benefit to them or society. In contrast, the goal of ethics, individual and social well-being, are universally beneficial. Bond (1996, p. 209) promotes self-interest as an intrinsically human goal: "Morality, we have said…is a value for all humanity, because it *profits* us, because it is a contribution, a necessary contribution, to our thriving, flourishing, happiness, or well-being or eudaimonia."

FUTURE TRENDS

There is a discernible trend in HCI away from the more "functional" aspects of interaction to the "suprafunctional". As well as emotion and pleasure, there is a growing concern for traditional design issues of aesthetics. Ethics has a role to play in advancing inquiry into these areas but more importantly provides an alternative perspective on the benefit and quality of the user experience.

In focusing on well-being, Bond distinguishes between short-term and long-term hedonism. Fun is balanced against long-term well-being that often involves challenges including learning about the world and the self. This is echoed in the work of Ellis' (1994) whose rational emotive therapy (RET) focuses on self-understanding as a prerequisite to well-being. Sullivan (2000, pp. 2-4) suggests that well-being is based on three criteria: Antonovsky's Sense Of Coherence (Antonovsky, 1993), Self-Esteem, and Emotional Stability (Hills & Argyle, 2001). Alternatively, Hayakawa (1968, pp. 51-69) suggests personal qualities of fully-functioning humans:

1. Nonconformity and individuality
2. Self-awareness
3. Acceptance of ambiguity and uncertainty
4. Tolerance
5. Acceptance of human animality
6. Commitment and intrinsic enjoyment
7. Creativity and originality
8. Social interest and ethical trust
9. Enlightened self-interest
10. Self-direction
11. Flexibility and scientific outlook
12. Unconditional self-acceptance
13. Risk-taking and experimenting.
14. Long-range hedonism.
15. Work and practice

An alternative to designing for hedonic use qualities is to try to build products that promote well-being. This can be shown as a hierarchy of user needs (Figure 1) that includes ethical and aspirational dimensions linked to design goals. Picard (1999) is one of the few designers and researchers to offer an insight into how products might facilitate well-being. Her "Affective Mirror" enables users to see how others see them and so enables them to change their behaviour and perception.

Figure 1. An ethical framework for HCI design and research

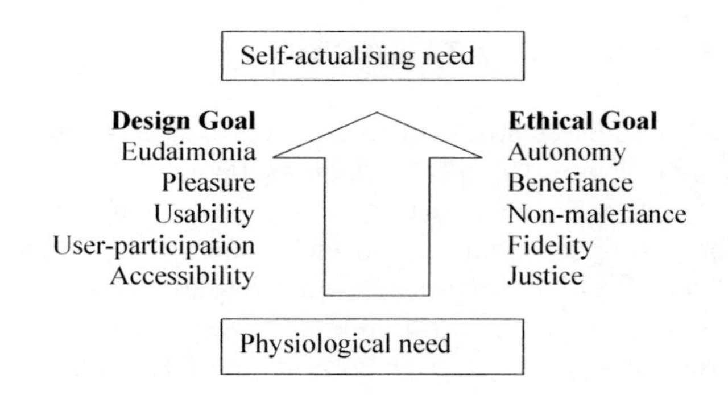

CONCLUSION

The goal of HCI research and design has been to deliver universal usability. Critics of usability have argued that it "dumbs down" the user-experience to the lowest common denominator. In contrast, researchers have recommended emotional design goals. These include pleasure (Jordan, 2000), emotion (Norman, 2003), and fun. While these design goals match a wider range of human capabilities, they also raise ethical issues. In contrast, the author suggests that the ultimate goal of HCI should be to promote the benefit of well-being through a value-centred design approach.

REFERENCES

Antonovsky, A. (1993). The structure and properties of the Sense of Coherence Scale. *Social Science and Medicine, 36,* 725-733.

Bonapace, L. (2002). Linking product properties to pleasure: The sensorial quality assessment method—SEQUAM. In W. S. Green & P. W. Jordan (Eds.), *Pleasure with products: Beyond usability* (pp. 187-217). London: Taylor and Francis.

Bond, E. J. (1996). *Ethics and human well being: An introduction to moral philosophy*. Cambridge, MA: Blackwell.

Cairns, P., & Thimbleby, H. (2003). *The diversity and ethics of HCI.* Retrieved on April 5, 2004, from http://www.uclic.ucl.ac.uk/harold/ethics/tochiethics.pdf

DeJean, P. H. (2002). Difficulties and pleasure? In W. S. Green & P. W. Jordan (Eds.), *Pleasure with products: Beyond usability* (pp. 147-150). London: Taylor and Francis.

Dunne, A. (1999). *Hertzian tales: Electronic products, aesthetic experience and critical design.* London: Royal College of Art.

Eliot, B. (2002). Hiding behind the user. In *Sp!ked-IT*. Retrieved on January 31, 2002, from http://www.spiked-online.com/articles/00000002D3DE.htm.

Ellis, A. (1994). Rational emotive behavior therapy approaches to obsessive-compulsive disorder. *Journal of Rational-Emotive & Cognitive-Behavior Therapy, 12*(2), 121-141.

Epicurus. (1994). The pursuit of pleasure. In P. Singer (Ed.), *Ethics* (p. 188). Oxford, UK: Oxford University Press.

Fogg, B. (2003). *Persuasive technology, using computer to change what we think and do.* San Fransisco: Morgan Kaufmann.

Friedman, B. & Kahn, P. (2003). Human values, ethics, and design. In J. Jacko & A. Sears (Eds.), *The HCI handbook: Fundamentals, evolving technologies and emerging applications* (pp. 1177-1201). Mahwah, NJ: Lawrence Erlbaum Associates.

Hayakawa, S. I. (1968). The fully functioning personality. In S. I. Hayakawa (Ed.), *Symbol, status and personality* (pp. 51-69). New York: Harcourt Brace Jovanovich.

Hills, P., & Argyle, M. (2001). Emotional stability as a major dimension of happiness. *Personality and Individual Differences, 31*, 1357-1364.

International Organization for Standardization (2000). *ISO/IEC FDIS 9126-1: Software Engineering — Product quality—Part 1: Quality model. ISO.* Retrieved on October 18, 2005, from http://www.usability.serco.com/trump/resources/standards.htm#9126-1

Jordan, P. W. (2000). *Designing pleasurable products*. London: Taylor and Francis.

Laurel, B. (1991). *Computers as theatre*. Boston: Addison-Wesley Publishing Company.

Molotch, H. (2003). *Where stuff comes from: How toasters, toilets, cars, computers and many other things come to be as they are.* London: Routledge.

Norman, D. (2003). *Emotional design: Why we love [or hate] everyday things*. New York: Basic Books.

Papanek, V. (1985). *Design for the real world.* London: Thames and Hudson.

Perry, W. G. (1999). *Forms of ethical and intellectual development in the college years*. San Francisco: Josey Bass.

Picard, R. (1997). *Affective computing*. Cambridge, MA: MIT Press.

Quinn, C. N. (1997). *Engaging learning, instructional technology forum*. Retrieved on April 5, 2004 from http://itech1.coe.uga.edu/itforum/paper18/paper18.html

Shedroff, N. (2001). *Experience design 1*. Indianapolis, IN: New Riders Publishing.

Sullivan, S. (2000). The relations between emotional stability, self-esteem, and sense of coherence. In D. Brown (Ed.), *Journal of Psychology and Behavioral Sciences*. Department of Fairleigh Dickinson University at Madison, New Jersey. Retrieved August 18, 2004, from http://view.fdu.edu/default.aspx?id=784

Tiger, L. (1992). *The pursuit of pleasure*. NJ: Transaction Publishers.

Whiteley, N. (1999).Utility, design principles and the ethical tradition. In J. Attfield (Ed.), *Utility reassessed, the role of ethics in the practice of design* (p 221). Manchester, UK: Manchester University Press.

KEY TERMS

Design: The intentional creation of objects.

Ethical Design: Attempts to promote good through the creation of products that are made and consumed within a socially-accepted moral framework.

Ethics: The philosophy of morality.

Experience Design: The intentional creation of a time-based activity that includes physical objects, agents, and situations.

Universal Usability: Concerns research and design activities that enable interfaces to be accessed and used by all users.

Value-Centred Design: An approach to design that involves explicating stakeholder (including designers and developers) values as well as needs. Design then aims to communicate and deliver products and services that meet stakeholders values and needs.

Expectations and Their Forgotten Role in HCI

Jan Noyes
University of Bristol, UK

INTRODUCTION

Technology with its continuing developments pervades the 21st century world. Consequently, HCI is becoming an everyday activity for an increasing number of people from across the population. Interactions may involve personal computers (PCs), household and domestic appliances, public access technologies, personal digital assistants (PDAs), as well as more complex technologies found in the workplace. Given the increasing use of technology by the general public, HCI assumes an ever-growing importance. User interactions need to be taken into account by designers and engineers; if they fail to do this, the opportunities presented by the new technologies will remain unfulfilled and unrealized. Furthermore, it is likely that those interactions that take place will be marred by frustration and irritation, as users fail to achieve the smooth transactions with the technology that they expect and desire. One aspect of HCI that appears to have been recently overlooked is that of expectations. When confronted with a new device or when using a familiar one, we have expectations about how it will or does work. These expectations are part of the interaction process and are important in the sense that they will influence our immediate and later use of the technology/device. It is suggested that in recent times we have neglected expectations and failed to consider them to any great extent in the design process.

BACKGROUND

Fifty years ago, expectations were recognized as having a role to play in human-machine interactions and the design of products. One of the first studies was concerned with the design of telephones (Lutz & Chapanis, 1955). At this time, Chapanis was working at Bell Laboratories, when a project was initiated to develop a telephone operated via a push-button keyset as opposed to a rotary dial. At that time, there was only one existing push-button model—the toll operators' keyset. The numerical part of this keyset comprised two vertical columns of five keys each. The first column included 4, 3, 2, 1, 0, while the second column held the numerals, 5, 6, 7, 8, 9.

However, when Chapanis studied the toll operators at work, he found that lots of miskeying was occurring. Although it was illegal to listen to calls, it was possible to do service observing where the number requested by the caller could be checked against the number dialed. It was found that 13% of long distance calls were being incorrectly dialed. In essence, the toll operators expected the numbers to be elsewhere. This led Chapanis to devise a study on the expected locations of numbers on keysets. The investigation had three aims: namely, to find out where people expected to find numbers and then letters on each of six configurations of 10 keys, and where they expected to find letters, given certain preferred number arrangements. In a questionnaire study, 300 participants filled in numbers/letters on diagrams of keysets according to the arrangements that they felt were the most natural. Analysis of this data allowed Lutz and Chapanis (1955) to deduce where people expected to find the various alphanumeric characters on a keyset.

In the 1950s, the discipline of HCI as we know it today did not exist; it was not until the launch of the first PCs that people began to recognize HCI as a distinct entity. Consequently, by the mid-1980s, there was a lot of information available on HCI for the designers of computer systems and products. One of the most comprehensive sources was the 944 guidelines for designing user interface software compiled by Smith and Mosier (1986). This 287-page report is still available online at: http://www.hcibib.org/sam/. Given the detail and breadth of this report, it is somewhat surprising that the topic of expectations is mentioned rarely; it is only referred to on five occasions, which are listed as follows:

Section on Flowcharts

1. "Flowchart coding within to established conventions and user *expectations*."
2. "... survey prospective users to determine just what their *expectations* may be."
3. Section on Compatibility with User Expectations
4. "... control entry are compatible with user *expectations*"
5. "User *expectations* can be discovered by interview, questionnaire, and/or prototype testing."
6. Section on System Load
7. "But load status information may help in any case by establishing realistic user *expectations* for system performance."

The guidelines suggest that the system should conform to established conventions and user expectations, and the way in which we find out about these expectations is through surveys, interviews, questionnaires, and prototype testing. However, the following comment implies that consistency in design is all that is needed:

Where no strong user expectations exist with respect to a particular design feature, then designers can help establish valid user expectations by careful consistency in interface design. (Section 3.0/16)

Given the relative wealth of information available on HCI in the 1980s compared to earlier years, it is somewhat surprising that expectations have not received more emphasis and interest. Lutz and Chapanis (1955) were certainly aware of their importance in their study, but despite the plethora of new technological developments since then, this aspect of HCI seems generally to have been forgotten.

DEFINING EXPECTATIONS

In terms of HCI, expectations relate to how we expect a product/system to respond and react when we use it. At one level, our expectations are an automatic response (e.g., when we perceive visual stimuli). For example, when we look at perceptual illusions, our past experience and knowledge of the properties of straight lines and circles in our environment determines how we perceive the objects. This may help to explain how we perceive straight lines as bending, circles as moving, and so forth.

At another level, expectations and beliefs are powerful forces in shaping our attitudes and behavior (e.g., schema and scripts). These are socially developed attributes that determine how we behave and what is appropriate/inappropriate behavior in a particular context. In both of these examples, the environment in which we have been nurtured and the corresponding culture will have a major influence. As an example, straight lines are very much a feature of the human-made world and do not exist in nature, so perceptual differences would be expected between those individuals living in a city and those living in the jungle.

The common feature in both of these examples is that expectations are based on past experience and knowledge; they also are quite powerful determinants of how we behave, both on an automatic level (i.e., the perceptual processing of information) and at a behavioral level. These are important considerations in HCI, and because of this, they need to be taken into account in the design process.

FUTURE TRENDS

Population Stereotypes

One example of the cultural influence on design is the population stereotype. These stereotypes are everyday artefacts that have strong associations (e.g., the way in which color is used). The color red suggests danger and is often used to act as a warning about a situation. For example, traffic lights when red warn vehicles of danger and the need to stop at the signal. Likewise, warnings on the civil flight deck are colored red, while cautions are amber. Often, these stereotypes are modeled in nature; for example, berries that are red warn animals not to eat them, as their color implies they are poisonous.

Population stereotypes are determined culturally and will differ between countries. They also change over time. For example, the current stereotype in Europe and North America is for baby boys' clothes to be associated with the color blue and baby girls' clothes to be pink. However, this has not always been the case. In America, up until the 1940s, this stereotype was the opposite, as shown by the following two quotations:

[U]se pink for the boy and blue for the girl, if you are a follower of convention. (*Sunday Sentinal*, March 1914)

[T]he generally accepted rule is pink for the boy and blue for the girl. The reason is that pink being a more decided and stronger color is more suitable for the boy, while blue, which is more delicate and dainty, is pertier for the girl. (*Ladies Home Journal*, June 1918)

Most people would probably agree that blue for a boy and pink for a girl is a strong population stereotype in the UK/North America, and today, it would be quite unusual for a young male to be dressed in pink. Yet, this stereotype has developed only recently.

Consistency

A further consideration is that of consistency (i.e., a match between what people expect and what they perceive when actually using the product/system). Smith and Mosier (1986) alluded to this in their guidelines, although they referred to consistency in interface design, which could be interpreted as being consistent in terms of controls, functions, movements, and so forth when using the software (e.g., the same command is always used for exit). In design terms, consistency also can be interpreted as compatibility (i.e., compatibility between what we expect and what we get). One of the classic studies demonstrating the importance of consistency/compatibility was carried out by Chapanis and Lindenbaum (1959). They looked at four arrangements of controls and burners as found, for example, on electric/gas stoves. They found that the layout, which had the greatest spatial compatibility between the controls and the burners, was superior in terms of speed and accuracy. This study has been repeated many times (Hsu & Peng, 1993; Osborne & Ellingstad, 1987; Payne, 1995; Ray & Ray, 1979; Shinar & Acton, 1978).

Past Experience

The role of past experience and knowledge in developing our expectations has already been mentioned; these are viewed as important determinants in our attitudes toward technology, as demonstrated by the following study. Noyes and Starr (2003) carried out a questionnaire survey of British Airways flight deck crew in order to determine their expectations and perceptions of automated warning systems. The flight deck crew was divided into those who had experience flying with automated warning systems (n=607) and those who did not have this experience (n=571). It was found that automation was more favored by the group who had experience flying using warning systems with a greater degree of automation. Those without experience of automation had more negative expectations than those who had experienced flying with more automated systems. This suggests that we are more negative toward something that we have not tried. These findings have implications for HCI, training, and the use of new technology; in particular, where there already exists a similar situation with which the individual is familiar. Pen- and speech-based technologies provide examples of this; most of the adult population has expectations about using these emerging technologies based on their experiences and knowledge of the everyday activities of writing and talking.

Noyes, Frankish, and Morgan (1995) surveyed people's expectations of using a pen-based system before they had ever used one. They found that individuals based their expectations on using pen and paper; for example, when asked to make a subjective assessment of the sources of errors, most people thought these would arise when inputting the lower-case characters. This is understandable, since if we want to write clearly, we tend to do this by writing in capitals rather than lower-case. However, the pen recognition systems are not working on the same algorithms as people, so there is no reason why errors are more likely to occur with lower-case letters. In fact, this was found to be the case. When participants were asked about their perceptions after having used the pen-based system, they indicated the upper-case letters as being the primary source of errors.

None of the participants in this work had previously used a pen-based system; perhaps lack of familiarity with using the technology was a significant factor in the mismatch between their expectations and perceptions. If this was the case, eliciting people's expectations about an activity with which they are familiar should not result in a mismatch. This aspect recently has been investigated by the author in a study looking at people's expectations/ preconceptions of their performance on activities with which they are familiar. When asked about their anticipated levels of performance when using paper and computers, participants expected to do better on the computer-based task. After carrying out the equivalent task on paper and computer, it was found that actual scores were higher on paper; thus, people's task performance did not match their expectations.

Intuitiveness

An underlying aspect relating to expectations in HCI is naturalness (i.e., we expect our interactions with computers and technology to be intuitive). In terms of compatibility, designs that are consistent with our expectations (e.g., burners that spatially map onto their controls) could be perceived as being more natural/intuitive. Certainly, in the Chapanis and Lindenbaum (1959) study, participants performed best when using the most compatible design. It is reasonable, therefore, to anticipate that intuitive designs are more beneficial than those that are not intuitive or counter-intuitive, and that HCI should strive for naturalness.

However, there are two primary difficulties associated with intuitive design. With emerging technologies based on a primary human activity, intuitiveness may not always be desirable. Take, for example, the pen-based work discussed previously. People's expectations did not match their perceptions, which was to the detriment of the pen-based system (i.e., individuals had higher expectations). This was primarily because people, when confronted with a technology that emulates a task with which they are familiar, base their expectations on this experience—in this case, writing. A similar situation occurs with automatic speech recognition (ASR) interfaces. We base our expectations about using ASR on what we know about human-human communications, and when the recognition system does not recognize an utterance, we speak more loudly, more slowly, and perhaps in a more exaggerated way. This creates problems when we use ASR, because the machine algorithms are not working in the same way as we do when not understood by a fellow human. A further irony is that the more natural our interaction with the technology is, the more likely we are to assume we are not talking to a machine and to become less constrained in our dialogue (Noyes, 2001). If perfect or near-perfect recognition was achievable, this would not be a problem. Given that this is unlikely to be attained in the near future, a less natural interaction, where the speaker constrains his or her speech, will achieve better recognition performance.

In addition to intuitiveness not always being desirable, humans are adaptive creatures, and they adapt to non-intuitive and counter-intuitive designs. Take, for example, the standard keyboard with the QWERTY layout. This was designed in the 1860s as part of the Victorian typewriter and has been shown repeatedly to be a poor design (Noyes, 1998); however, it now dominates computer keyboards around the world. There have been many keyboards developed that fit the shape of the hands and fingertips, but these more natural keyboards have been unsuccessful in challenging the supremacy of QWERTY. Likewise, most hob designs still have incompatible control-burner linkages, but users adapt to them. We also learn to use counter-intuitive controls (e.g., car accelerator and brake pedal controls) that have similar movements but very dissimilar results. This ability to adapt to designs that are not intuitive further weakens the case for naturalness; however, this approach would have considerable human costs, and it would be unreasonable to support the notion that poor design is admissible on these grounds.

CONCLUSION

Expectations are an important aspect of human behavior and performance and, in this sense, need to be considered in HCI. A mismatch between expectations and perceptions and/or ignoring people's expectations could lead them to feel frustrated and irritated,and to make unnecessary mistakes. Finding out about people's expectations is achieved readily, as demonstrated by the studies mentioned here, and will bring benefits not only to the design process but also to training individuals to use new technology.

Expectations are linked intrinsically with naturalness and intuitiveness expressed through consistency and compatibility in design. However, intuitiveness in design is not necessarily always desirable. This is especially the case when technology is being used to carry out familiar everyday tasks (e.g., writing and speaking). The situation is further compounded by the ability of the human to adapt, in particular, to counter-intuitive designs.

REFERENCES

Chapanis, A., & Lindenbaum, L. E. (1959). A reaction time study of four control-display linkages. *Human Factors*, *1*, 1-7.

Hsu, S. -H., & Peng, Y. (1993). Control/display relationship of the four-burner stove: A reexamination. *Human Factors, 35*, 745-749.

Lutz, M. C., & Chapanis, A. (1955). Expected locations of digits and letters on ten-button keysets. *Journal of Applied Psychology*, *39*, 314-317.

Noyes, J. M. (1998). QWERTY—The immortal keyboard. *Computing & Control Engineering Journal, 9*, 117-122.

Noyes, J. M. (2001). Talking and writing—How natural in human-machine interaction? *International Journal of Human-Computer Studies*, *55*, 503-519.

Noyes, J. M., Frankish, C. R., & Morgan, P. S. (1995). Pen-based computing: Some human factors issues. In P.J. Thomas (Ed.), *Personal information systems: Business applications* (pp. 65-81). Cheltenham, UK: Stanley Thornes.

Noyes, J. M., & Starr, A. F. (2003). Aircraft automation—Expectations versus perceptions of flight deck crew. In C. Stephanidis (Ed.), *Adjunct proceedings of HCI International* (pp. 199-200). Heraklion, Greece: Crete University Press.

Osborne, D. W., & Ellingstad, V. S. (1987). Using sensor lines to show control-display linkages on a four-burner stove. *Proceedings of 31st Annual Meeting of the Human Factors Society*, Santa Monica, California.

Payne, S. J. (1995). Naïve judgments of stimulus-response compatibility. *Human Factors, 37,* 495-506.

Ray, R. D., & Ray, W. D. (1979). An analysis of domestic cooker control design. *Ergonomics, 22,* 1243-1248.

Shinar, D., & Acton, M. B. (1978). Control-display relationships on the four-burner range: Population stereotypes versus standards. *Human Factors, 20*, 13-17.

Smith, S. L., & Mosier, J. N. (1986). *Design guidelines for user-system interface software* [Report MTR-9420]. Bedford, MA: Mitre Corporation.

KEY TERMS

Compatibility: Designs that match our expectations in terms of characteristics, function, and operation.

Consistency: Similar to compatibility and sometimes used interchangeably; designs that match our expectations in terms of characteristics, function, and operation, and are applied in a constant manner within the design itself.

Expectations: In HCI terms, how we expect and anticipate a product/system to respond and react when we use it.

Intuitiveness: Knowing or understanding immediately how a product/system will work without reasoning or being taught. Intuitiveness is linked closely to naturalness (i.e., designs that are intuitive also will be perceived as being natural).

Perceptions: In HCI terms, how we perceive a product/system to have responded and reacted when we have used it. Ideally, there should be a good match between expectations and perceptions.

Perceptual Illusions: Misperceptions of the human visual system so that what we apprehend by sensation does not correspond with the way things really are.

Population Stereotypes: These comprise the well-ingrained knowledge that we have about the world, based on our habits and experiences of living in a particular cultural environment.

Schema and Scripts: Mental structures that organize our knowledge of the world around specific themes or subjects and guide our behavior so that we act appropriately and according to expectation in particular situations.

Eye Tracking in HCI and Usability Research

Alex Poole
Lancaster University, UK

Linden J. Ball
Lancaster University, UK

INTRODUCTION

Eye tracking is a technique whereby an individual's eye movements are measured so that the researcher knows both where a person is looking at any given time and the sequence in which the person's eyes are shifting from one location to another. Tracking people's eye movements can help HCI researchers to understand visual and display-based information processing and the factors that may impact the usability of system interfaces. In this way, eye-movement recordings can provide an objective source of interface-evaluation data that can inform the design of improved interfaces. Eye movements also can be captured and used as control signals to enable people to interact with interfaces directly without the need for mouse or keyboard input, which can be a major advantage for certain populations of users, such as disabled individuals. We begin this article with an overview of eye-tracking technology and progress toward a detailed discussion of the use of eye tracking in HCI and usability research. A key element of this discussion is to provide a practical guide to inform researchers of the various eye-movement measures that can be taken and the way in which these metrics can address questions about system usability. We conclude by considering the future prospects for eye-tracking research in HCI and usability testing.

BACKGROUND

The History of Eye Tracking

Many different methods have been used to track eye movements since the use of eye-tracking technology first was pioneered in reading research more than 100 years ago (Rayner & Pollatsek, 1989). Electro-oculographic techniques, for example, relied on electrodes mounted on the skin around the eye that could measure differences in electrical potential in order to detect eye movements. Other historical methods required the wearing of large contact lenses that covered the cornea (the clear membrane covering the front of the eye) and sclera (the white of the eye that is seen from the outside) with a metal coil embedded around the edge of the lens; eye movements then were measured by fluctuations in an electromagnetic field when the metal coil moved with the eyes (Duchowski, 2003). These methods proved quite invasive, and most modern eye-tracking systems now use video images of the eye to determine where a person is looking (i.e., their so-called *point-of-regard*). Many distinguishing features of the eye can be used to infer point-of-regard, such as corneal reflections (known as Purkinje images), the iris-sclera boundary, and the apparent pupil shape (Duchowski, 2003).

How Does an Eye Tracker Work?

Most commercial eye trackers that are available today measure point-of-regard by the corneal-reflection/pupil-center method (Goldberg & Wichansky, 2003). These kinds of trackers usually consist of a standard desktop computer with an infrared camera mounted beneath (or next to) a display monitor, with image processing software to locate and identify the features of the eye used for tracking. In operation, infrared light from an LED embedded in the infrared camera first is directed into the eye to create strong reflections in target eye features to make them easier to track (infrared light is used to avoid dazzling the user with visible light). The light enters the retina, and a large proportion of it is reflected back, making the pupil appear as a bright, well defined disc (known as the bright-pupil effect). The corneal reflection (or

Figure 1. Corneal reflection and bright pupil as seen in the infrared camera image

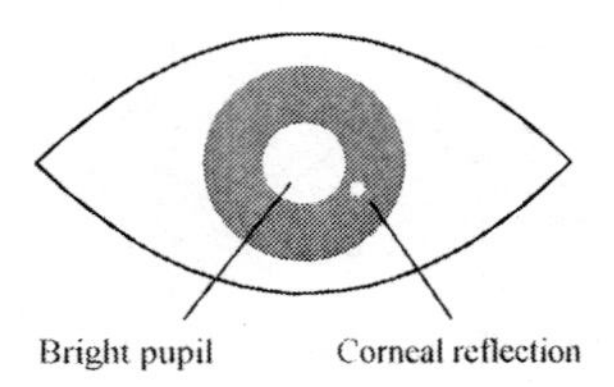

first Purkinje image) is also generated by the infrared light, appearing as a small but sharp glint (see Figure 1).

Once the image processing software has identified the center of the pupil and the location of the corneal reflection, the vector between them is measured, and, with further trigonometric calculations, point-of-regard can be found. Although it is possible to determine approximate point-of-regard by the corneal reflection alone (as shown in Figure 2), by tracking both features, eye movements critically can be disassociated from head movements (Duchowski, 2003, Jacob & Karn, 2003).

Video-based eye trackers need to be fine-tuned to the particularities of each person's eye movements by a calibration process. This calibration works by displaying a dot on the screen, and if the eye fixes for longer than a certain threshold time and within a certain area, the system records that pupil-center/corneal-reflection relationship as corresponding to a specific *x,y* coordinate on the screen. This is repeated over a nine- to 13-point grid pattern to gain an accurate calibration over the whole screen (Goldberg & Wichansky, 2003).

Why Study Eye Movements in HCI Research?

What a person is looking at is assumed to indicate the thought "on top of the stack" of cognitive processes (Just & Carpenter, 1976). This eye-mind hypothesis means that eye movement recordings can provide a dynamic trace of where a person's attention is being directed in relation to a visual display. Measuring other aspects of eye movements, such as fixations (i.e., moments when the eyes are relatively stationary, taking in or encoding information), also can reveal the amount of processing being applied to objects at the point-of-regard. In practice, the process of inferring useful information from eye-movement recordings involves the HCI researcher defining areas of interest over certain parts of a display or interface under evaluation and analyzing the eye movements that fall within such areas. In this way, the visibility, meaningfulness, and placement of specific interface elements can be evaluated objectively, and the resulting findings can be used to improve the design of the interface (Goldberg & Kotval, 1999). For example, in a task scenario where participants are asked to search for an icon, a longer-than-expected gaze on the icon before eventual selection would indicate that it lacks meaningfulness and probably needs to be redesigned. A detailed description of eye-tracking metrics and their interpretation is provided in the following sections.

EYE TRACKING AS A RESEARCH AND USABILITY-EVALUATION TOOL

Previous Eye-Tracking Research

Mainstream psychological research has benefited from studying eye movements, as they can provide insight into problem solving, reasoning, mental imagery, and search strategies (Ball et al., 2003; Just & Carpenter, 1976; Yoon & Narayanan, 2004; Zelinsky & Sheinberg, 1995). Because eye movements provide a window into so many aspects of cognition,

Figure 2. Corneal reflection position changing according to point-of-regard (Redline & Lankford, 2001)

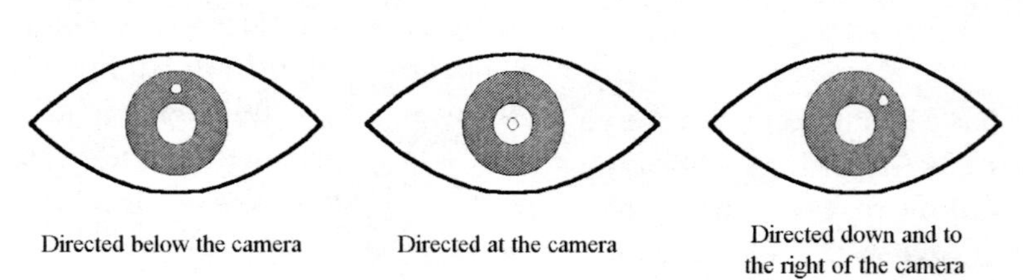

there also are rich opportunities for the application of eye-movement analysis as a usability research tool in HCI and related disciplines such as human factors and cognitive ergonomics. Although eye-movement analysis is still very much in its infancy in HCI and usability research, issues that increasingly are being studied include the nature and efficacy of information search strategies with menu-based interfaces (Altonen et al., 1998; Byrne et al., 1999; Hendrickson, 1989) and the features of Web sites that correlate with effective usability (Cowen et al., 2002; Goldberg et al., 2002; Poole et al., 2004). Additionally, eye trackers have been used more broadly in applied human factors research to measure situation awareness in air-traffic-control training (Hauland, 2003) in order to evaluate the design of cockpit controls to reduce pilot error (Hanson, 2004) and to investigate and improve doctors' performances in medical procedures (Law et al., 2004; Mello-Thoms et al., 2002). The commercial sector also is showing increased interest in the use of eye-tracking technology in areas such as market research, for example, to determine what advert designs attract the greatest attention (Lohse, 1997) and to determine if Internet users look at banner advertising on Web sites (Albert, 2002).

Table 1. Fixation-derived metrics and how they can be interpreted in the context of interface design and usability evaluation (references are given to examples of studies that have used each metric)

Eye-Movement Metric	What it Measures	Reference
Number of fixations overall	More overall fixations indicate less efficient search (perhaps due to sub-optimal layout of the interface).	Goldberg and Kotval (1999)
Fixations per area of interest	More fixations on a particular area indicate that it is more noticeable, or more important, to the viewer than other areas.	Poole et al. (2004)
Fixations per area of interest and adjusted for text length	If areas of interest are comprised of text only, then the mean number of fixations per area of interest can be divided by the mean number of words in the text. This is a useful way to separate out a higher fixation count, simply because there are more words to read, from a higher fixation count because an item is actually more difficult to recognize.	Poole et al. (2004)
Fixation duration	A longer fixation duration indicates difficulty in extracting information, or it means that the object is more engaging in some way.	Just and Carpenter (1976)
Gaze (also referred to as dwell, fixation cluster, and fixation cycle)	Gaze is usually the sum of all fixation durations within a prescribed area. It is best used to compare attention distributed between targets. It also can be used as a measure of anticipation in situation awareness, if longer gazes fall on an area of interest before a possible event occurring.	Mello-Thoms et al. (2004); Hauland (2003)
Fixation spatial density	Fixations concentrated in a small area indicate focused and efficient searching. Evenly spread fixations reflect widespread and inefficient search.	Cowen et al. (2002)
Repeat fixations (also called post-target fixations)	Higher numbers of fixations off target after the target has been fixated indicate that it lacks meaningfulness or visibility.	Goldberg and Kotval (1999)
Time to first fixation on target	Faster times to first fixation on an object or area mean that it has better attention-getting properties.	Byrne et al. (1999)
Percentage of participants fixating on an area of interest	If a low proportion of participants is fixating on an area that is important to the task, it may need to be highlighted or moved.	Albert (2002)
On target (all target fixations)	Fixations on target divided by total number of fixations. A lower ratio indicates lower search efficiency.	Goldberg and Kotval (1999)

Eye-Movement Metrics

The main measurements used in eye-tracking research are fixations (described previously) and *saccades*, which are quick eye movements occurring between fixations. There is also a multitude of derived metrics that stem from these basic measures, including *gaze* and *scanpath* measurements. Pupil size and blink rate also are studied.

Fixations

Fixations can be interpreted quite differently, depending on the context. In an encoding task (e.g., browsing a Web page), higher fixation frequency on a particular area can be indicative of greater interest in the target (e.g., a photograph in a news report), or it can be a sign that the target is complex in some way and more difficult to encode (Jacob & Karn, 2003; Just & Carpenter, 1976). However, these interpretations may be reversed in a search task—a higher number of single fixations, or clusters of fixations, are often an index of greater uncertainty in recognizing a target item (Jacob & Karn, 2003). The duration of a fixation also is linked to the processing time applied to the object being fixated (Just & Carpenter, 1976). It is widely accepted that external representations associated with long fixations are not as meaningful to the user as those associated with short fixations (Goldberg & Kotval, 1999). Fixation-derived metrics are described in Table 1.

Saccades

No encoding takes place during saccades, so they cannot tell us anything about the complexity or salience of an object in the interface. However, regressive saccades (i.e., backtracking eye movements) can act as a measure of processing difficulty during encoding (Rayner & Pollatsek, 1989). Although most regressive saccades (or regressions) are very small, only skipping back two or three letters in reading tasks; much larger phrase-length regressions can represent confusion in higher-level processing of the text (Rayner & Pollatsek, 1989). Regressions equally could be used as a measure of recognition value, in that there should be an inverse relationship between the number of regressions and the salience of the phrase. Saccade-derived metrics are described in Table 2.

Scanpaths

Describes a complete saccade-fixate-saccade sequence. In a search task, typically an optimal scan path is viewed as being a straight line to a desired target with a relatively short fixation duration at the actual target (Goldberg & Kotval, 1999). Scanpaths can be analyzed quantitatively with the derived measures described in Table 3.

Table 2. Saccade-derived metrics and how they can be interpreted in the context of interface design and usability evaluation (references are given to examples of studies that have used each metric)

Eye-Movement Metric	What it Measures	Reference
Number of saccades	More saccades indicate more searching.	Goldberg and Kotval (1999)
Saccade amplitude	Larger saccades indicate more meaningful cues, as attention is drawn from a distance.	Goldberg et al. (2002)
Regressive saccades (i.e., regressions)	Regressions indicate the presence of less meaningful cues.	Sibert et al. (2000)
Saccades revealing marked directional shifts	Any saccade larger than 90 degrees from the saccade that preceded it shows a rapid change in direction. This could mean that the user's goals have changed or the interface layout does not match the user's expectations.	Cowen et al. (2002)

E

Table 3. Scanpath-derived metrics and how they can be interpreted in the context of interface design and usability evaluation (references are given to examples of studies that used each metric)

Eye-Movement Metric	What it Measures	Reference
Scanpath duration	A longer-lasting scanpath indicates less efficient scanning.	Goldberg and Kotval (1999)
Scanpath length	A longer scanpath indicates less efficient searching (perhaps due to a suboptimal layout).	Goldberg et al. (2002)
Spatial density	Smaller spatial density indicates more direct search.	Goldberg and Kotval (1999)
Transition matrix	The transition matrix reveals search order in terms of transitions from one area to another. Scanpaths with an identical spatial density and convex hull area can have completely different transition values—one is efficient and direct, while the other goes back and forth between areas, indicating uncertainty.	Goldberg and Kotval (1999); Hendricson (1989)
Scanpath regularity	Once cyclic scanning behavior is defined, and then deviation from a normal scanpath can indicate search problems due to lack of user training or bad interface layout.	Goldberg and Kotval (1999)
Spatial coverage calculated with convex hull area	Scanpath length plus convex hull area define scanning in a localized or larger area.	Goldberg and Kotval (1999)
Scanpath direction	This can determine a participant's search strategy with menus, lists, and other interface elements (e.g., top-down vs. bottom-up scanpaths). *Sweep* denotes a scanpath progressing in the same direction.	Altonen et al. (1998)
Saccade/ fixation ratio	This compares time spent searching (saccades) to time spent processing (fixating). A higher ratio indicates more processing or less searching.	Goldberg and Kotval (1999)

Blink Rate and Pupil Size

Blink rate and pupil size can be used as an index of cognitive workload. A lower blink rate is assumed to indicate a higher workload, and a higher blink rate may indicate fatigue (Brookings, Wilson, & Swain, 1996; Bruneau, Sasse & McCarthy, 2002). Larger pupils also may indicate more cognitive effort (Pomplun & Sunkara, 2003). However, pupil size and blink rate can be determined by many other factors (e.g., ambient light levels), so they are open to contamination (Goldberg & Wichansky, 2003). For these reasons, pupil size and blink rate are used less often in eye tracking research.

Technical Issues in Eye-Tracking Research

Experimenters looking to conduct their own eye-tracking research should bear in mind the limits of the technology and how these limits impact the data that they will want to collect. For example, they should ensure that if they are interested in analyzing fixations that the equipment is optimized to detect fixations (Karn et al., 2000). The minimum time for a fixation is also highly significant. Interpretations of cognitive processing can vary dramatically according to the time set to detect a fixation in the eye-tracking system. Researchers are advised to set the lower threshold to at least 100ms (Inhoff & Radach, 1998).

Researchers have to work with limits of accuracy and resolution. A sampling rate of 60hz is good enough for usability studies but inadequate for reading research, which requires sampling rates of around 500hz or more (Rayner & Pollatsek, 1989). It is also imperative to define areas of interest that are large enough to capture all relevant eye movements. Even the best eye trackers available are only accurate to within one degree of actual point-of-regard (Byrne et al., 1999). Attention also can be directed up to one degree away from measured point-of-regard without moving the eyes (Jacob & Karn, 2003).

Eye trackers are quite sensitive instruments and can have difficulty tracking participants who have eyewear that interrupts the normal path of a reflection, such as hard contact lenses, bifocal and trifocal glasses, and glasses with super-condensed lenses. There also may be problems tracking a person with very large pupils or a lazy eye such that the person's eyelid obscures part of the pupil and makes it difficult to identify. Once a person is calibrated successfully, the calibration procedure then should be repeated at regular intervals during a test session in order to maintain an accurate point-of-regard measurement.

There are large differences in eye movements between participants on identical tasks, so it is prudent to use a within-participants design in order to make valid performance comparisons (Goldberg & Wichansky, 2003). Participants also should have well-defined tasks to carry out (Just & Carpenter, 1976) so that their eye movements can be attributed properly to actual cognitive processing. Visual distractions (e.g., colorful or moving objects around the screen or in the testing environment) also should be eliminated, as these inevitably will contaminate the eye-movement data (Goldberg & Wichansky, 2003). Finally, eye tracking generates huge amounts of data, so it is essential to perform filtering and analysis automatically, not only to save time but also to minimize the chances of introducing errors through manual data processing.

EYE TRACKING AS AN INPUT DEVICE

Eye movements can be measured and used to enable an individual actually to interact with an interface. Users could position a cursor by simply looking at where they want it to go or click an icon by gazing at it for a certain amount of time or by blinking. The first obvious application of this capability is for disabled users who cannot make use of their hands to control a mouse or keyboard (Jacob & Karn, 2003). However, intention often can be hard to interpret; many eye movements are involuntary, leading to a certain Midas Touch (see Jacob & Karn, 2003), in that you cannot look at anything without immediately activating some part of the interface. One solution to this problem is to use eye movements in combination with other input devices to make intentions clear. Speech commands can add extra context to users' intentions when eye movements may be vague, and vice versa (Kaur et al., 2003).

Virtual reality environments also can be controlled by the use of eye movements. The large three-dimensional spaces in which users operate often contain faraway objects that have to be manipulated. Eye movements seem to be the ideal tool in such a context, as moving the eyes to span long distances requires little effort compared with other control methods (Jacob & Karn, 2003). Eye movement interaction also can be used in a subtler way (e.g., to trigger context-sensitive help as soon as a user becomes confused by performing too many regressions, for example, or while reading text [Sibert et al., 2000]). Other researchers (Ramloll et al., 2004) have used gaze-based interaction to help autistic children learn social skills by rewarding them when they maintain eye contact while communicating.

Some techniques alter a display, depending on the point-of-regard. Some large-display systems, such as flight simulators (Levoy & Whitaker, 1990; Tong & Fisher, 1984), channel image-processing resources to display higher-quality or higher-resolution images only within the range of highest visual acuity (i.e., the fovea) and decrease image processing in the visual range where detail cannot be resolved (the parafovea). Other systems (Triesch, Sullivan, Hayhoe & Ballard, 2002) take advantage of the visual suppression during saccades to update graphical displays without the user noticing. Yet another rather novel use is tracking the point-of-regard during videoconferencing and warping the image of the eyes so that they maintain eye contact with other participants in the meeting (Jerald & Daily, 2002).

FUTURE TRENDS IN EYE TRACKING

Future developments in eye tracking should center on standardizing what eye-movement metrics are used, how they are referred to, and how they should be interpreted in the context of interface design (Cowen et al., 2002). For example, no standard exists yet for the minimum duration of a fixation (Inhoff & Radach, 1998), yet small differences in duration thresholds can make it hard to compare studies on an even footing (Goldberg & Wichansky, 2003). Eye-tracking technology also needs to be improved to increase the validity and reliability of the recorded data. The robustness and accuracy of data capture need to be increased so that point-of-regard measurement stays accurate without the need for frequent recalibration. Data-collection, data-filtering, and data-analysis software should be streamlined, so that they can work together without user intervention. The intrusiveness of equipment should be decreased to make users feel more comfortable, perhaps through the development of smaller and lighter head-mounted trackers. Finally, eye-tracking systems need to become cheaper in order to make them a viable usability tool for smaller commercial agencies and research labs (Jacob & Karn, 2003). Once eye tracking achieves these improvements in technology, methodology, and cost, it can take its place as part of a standard HCI toolkit.

CONCLUSION

Our contention is that eye-movement tracking represents an important, objective technique that can afford useful advantages for the in-depth analysis of interface usability. Eye-tracking studies in HCI are beginning to burgeon, and the technique seems set to become an established addition to the current battery of usability-testing methods employed by commercial and academic HCI researchers. This continued growth in the use of the method in HCI studies looks likely to continue as the technology becomes increasingly more affordable, less invasive, and easier to use. The future seems rich for eye tracking and HCI.

REFERENCES

Albert, W. (2002). Do Web users actually look at ads? A case study of banner ads and eye-tracking technology. *Proceedings of the Eleventh Annual Conference of the Usability Professionals' Association.*

Altonen, A., Hyrskykari, A., & Räihä, K. (1998). 101 spots, or how do users read menus? *Proceedings of the CHI'98.*

Ball, L. J., Lucas, E. J., Miles, J. N. V., & Gale, A. G. (2003). Inspection times and the selection task: What do eye-movements reveal about relevance effects? *Quarterly Journal of Experimental Psychology, 56A*, 1053-1077.

Brookings, J. B., Wilson, G. F., & Swain, C. R. (1996). Psychophysiological responses to changes in workload during simulated air traffic control. *Biological Psychology, 42,* 361-377.

Bruneau, D., Sasse, M. A., & McCarthy, J. D. (2002). The eyes never lie: The use of eye tracking data in HCI research. *Proceedings of the CHI'02 Workshop on Physiological Computing.*

Byrne, M. D., Anderson, J. R., Douglas, S., & Matessa, M. (1999). Eye tracking the visual search of clickdown menus. *Proceedings of the CHI'99.*

Cowen, L., Ball, L. J., & Delin, J. (2002). An eye-movement analysis of Web-page usability. In X. Faulkner, J. Finlay, & F. Détienne (Eds.), *People and computers XVI—Memorable yet invisible* (pp. 317-335). London: Springer-Verlag.

Duchowski, A. T. (2003). *Eye tracking methodology: Theory and practice.* London: Springer-Verlag.

Goldberg, H. J., & Kotval, X. P. (1999). Computer interface evaluation using eye movements: Methods and constructs. *International Journal of Industrial Ergonomics, 24,* 631-645.

Goldberg, H. J., & Wichansky, A. M. (2003). Eye tracking in usability evaluation: A practitioner's guide. In J. Hyönä, R. Radach, & H. Deubel (Eds.), *The mind's eye: Cognitive and applied aspects of eye movement research* (pp. 493-516). Amsterdam: Elsevier.

Goldberg, J. H., Stimson, M. J., Lewenstein, M., Scott, N., & Wichansky, A. M. (2002). Eye tracking in Web search tasks: Design implications. *Proceedings of the Eye Tracking Research and Applications Symposium 2002.*

Hanson, E. (2004). Focus of attention and pilot error. *Proceedings of the Eye Tracking Research and Applications Symposium 2000.*

Hauland, G. (2003). Measuring team situation awareness by means of eye movement data. *Proceedings of the HCI International 2003.*

Hendrickson, J. J. (1989). Performance, preference, and visual scan patterns on a menu-based system: Implications for interface design. *Proceedings of the CHI'89.*

Inhoff, A. W., & Radach, R. (1998). Definition and computation of oculomotor measures in the study of cognitive processes. In G. Underwood (Ed.), *Eye guidance in reading, driving and scene perception* (pp. 29-53). New York: Elsevier.

Jacob, R. J. K., & Karn, K. S. (2003). Eye tracking in human-computer interaction and usability research: Ready to deliver the promises. In J. Hyönä, R. Radach, & H. Deubel (Eds.), *The mind's eye: Cognitive and applied aspects of eye movement research* (pp. 573-605). Amsterdam: Elsevier.

Jerald, J., & Daily, M. (2002). Eye gaze correction for videoconferencing. *Proceedings of the Eye Tracking Research and Applications Symposium 2002.*

Just, M. A., & Carpenter, P. A. (1976). Eye fixations and cognitive processes. *Cognitive Psychology, 8,* 441-480.

Karn, K., et al. (2000). Saccade pickers vs. fixation pickers: The effect of eye tracking instrumentation on research. *Proceedings of the Eye Tracking Research and Applications Symposium 2000.*

Kaur, M., et al. (2003). Where is "it"? Event synchronization in gaze-speech input systems. *Proceedings of the Fifth International Conference on Multimodal Interfaces.*

Law, B., Atkins, M. S., Kirkpatrick, A. E., & Lomax, A. J. (2004). Eye gaze patterns differentiate novice and experts in a virtual laparoscopic surgery training environment. *Proceedings of the Eye Tracking Research and Applications Symposium.*

Levoy, M., & Whitaker, R. (1990). Gaze-directed volume rendering. *Proceedings of the 1990 Symposium on Interactive 3D Graphics.*

Lohse, G. L. (1997). Consumer eye movement patterns on Yellow Pages advertising. *Journal of Advertising, 26,* 61-73.

Mello-Thoms, C., Nodine, C. F., & Kundel, H. L. (2004). What attracts the eye to the location of missed and reported breast cancers? *Proceedings of the Eye Tracking Research and Applications Symposium 2002.*

Pomplun, M., & Sunkara, S. (2003). Pupil dilation as an indicator of cognitive workload in human-computer interaction. *Proceedings of the HCI International 2003.*

Poole, A., Ball, L. J., & Phillips, P. (2004). In search of salience: A response time and eye movement analysis of bookmark recognition. In S. Fincher, P. Markopolous, D. Moore, & R. Ruddle (Eds.), *People and computers XVIII-design for life: Proceedings of HCI 2004.* London: Springer-Verlag.

Ramloll, R., Trepagnier, C., Sebrechts, M., & Finkelmeyer, A. (2004). A gaze contingent environment for fostering social attention in autistic children. *Proceedings of the Eye Tracking Research and Applications Symposium 2004.*

Rayner, K., & Pollatsek, A. (1989). *The psychology of reading.* Englewood Cliffs, NJ: Prentice Hall.

Redline, C. D., & Lankford, C. P. (2001). Eye-movement analysis: A new tool for evaluating the design of visually administered instruments. *Proceedings of the Section on Survey Research Methods of the American Statistical Association.*

Sibert, J. L., Gokturk, M., & Lavine, R. A. (2000). The reading assistant: Eye gaze triggered auditory prompting for reading remediation. *Proceedings of the Thirteenth Annual ACM Symposium on User Interface Software and Technology.*

Tong, H. M., & Fisher, R. A. (1984). *Progress report on an eye-slaved area-of-interest visual display. Report No. AFHRL-TR-84-36.* Brooks

Air Force Base, TX: Air Force Human Resources Laboratory.

Triesch, J., Sullivan, B. T., Hayhoe, M. M., & Ballard, D. H. (2002). Saccade contingent updating in virtual reality. *Proceedings of the Eye Tracking Research and Applications Symposium 2002*.

Yoon, D., & Narayanan, N. H. (2004). Mental imagery in problem solving: An eye tracking study. *Proceedings of the Eye Tracking Research and Applications Symposium 2004*.

Zelinsky, G., & Sheinberg, D. (1995). Why some search tasks take longer than others: Using eye movements to redefine reaction times. In J. M. Findlay, R. Walker, & R. W. Kentridge (Eds.), *Eye movement research: Mechanisms, processes and applications* (pp. 325-336). North-Holland: Elsevier.

KEY TERMS

Area of Interest: An area of interest is an analysis method used in eye tracking. Researchers define areas of interest over certain parts of a display or interface under evaluation and analyze only the eye movements that fall within such areas.

Eye Tracker: Device used to determine point-of-regard and to measure eye movements such as fixations, saccades, and regressions. Works by tracking the position of various distinguishing features of the eye, such as reflections of infrared light off the cornea, the boundary between the iris and sclera, or apparent pupil shape.

Eye Tracking: A technique whereby an individual's eye movements are measured so that the researcher knows where a person is looking at any given time and how the a person's eyes are moving from one location to another.

Eye-Mind Hypothesis: The principle at the origin of most eye-tracking research. Assumes that what a person is looking at indicates what the person currently is thinking about or attending to. Recording eye movements, therefore, can provide a dynamic trace of where a person's attention is being directed in relation to a visual display such as a system interface.

Fixation: The moment when the eyes are relatively stationary, taking in or encoding information. Fixations last for 218 milliseconds on average, with a typical range of 66 to 416 milliseconds.

Gaze: An eye-tracking metric, usually the sum of all fixation durations within a prescribed area. Also called dwell, fixation cluster, or fixation cycle.

Point-of-Regard: Point in space where a person is looking. Usually used in eye-tracking research to reveal where visual attention is directed.

Regression: A regressive saccade. A saccade that moves back in the direction of text that has already been read.

Saccade: An eye movement occurring between fixations, typically lasting for 20 to 35 milliseconds. The purpose of most saccades is to move the eyes to the next viewing position. Visual processing is automatically suppressed during saccades to avoid blurring of the visual image.

Scanpath: An eye-tracking metric, usually a complete sequence of fixations and interconnecting saccades.

E

From User Inquiries to Specification

Ebba Thóra Hvannberg
University of Iceland, Iceland

Sigrún Gunnarsdóttir
Siminn, Iceland

Gyda Atladóttir
University of Iceland, Iceland

INTRODUCTION

The aim of this article is to describe a method that helps analysts to translate qualitative data gathered in the field, collected for the purpose of requirements specification, to a model usable for software engineers.

Requirements specification constitutes three different parts: functional requirements, quality requirements, and nonfunctional requirements. The first one specifies how the software system should function, who are the actors, and what are the input and output of the functions. The second one specifies what quality requirements the software should meet while operating in context of its environment such as reliability, usability, efficiency, portability, and maintainability. Finally, the third part specifies other requirements including context of use and development constraints. Examples of context of use are where and when the system is used, and examples of development constraints are human resources, cost and time constraints, technological platforms, and development methods. The role of the requirements specification is to give software engineers a basis for software design, and, later in the software development life cycle, to validate the software system. Requirements specification can also serve the purpose of validating users' or customers' view of the requirements of the system.

On one hand, there has been a growing trend towards analyzing needs of the user and abilities through participatory design (Kuhn & Muller, 1993), activity theory (Bertelsen & Bødker, 2003), contextual design (Beyer & Holtzblatt, 1998), user-centered design (Gulliksen, Göransson, & Lif, 2001), and co-design and observation as in ethnography (Hughes, O'Brien, Rodden, Rouncefield, & Sommerville, 1995). Common to these methods is that qualitative data (Taylor & Bogdan, 1998) is collected, to understand the future environment of the new system, by analyzing the work or the tasks, their frequency and criticality, the cognitive abilities of the user and the users' collaborators. The scope of the information collected varies depending on the problem, and sometimes it is necessary to gather data about the regulatory, social, and organizational contexts of the problem (Jackson, 1995) to be solved. The temporal and spatial contexts describe when the work should be carried out and where.

On the other hand, software engineers have specified requirements in several different modelling languages that range from semiformal to formal. Examples of semiformal languages are UML (Larman, 2002), SADT, and IDEF (Ross, 1985). Examples of the latter are Z (Potter, Sinclair, & Till, 1996), VDM or ASM. Those are modelling languages for software development, but some languages or methods focus on task or work modelling such as Concurrent Task Trees (Paterno, 2003) and Cognitive Work Analysis (Vicente, 1999). Others emphasize more the specification method than a modelling language. Examples of the former are Scenario-Based Design (SBD) (Rosson & Carroll, 2002) and Contextual Design (Beyer & Holtzblatt, 1998). In practice, many software developers use informal methods to express requirements in text, e.g., as narrations or stories. Agile Development Methods (Abrahamsson, Warsta, Siponen, & Ronkainen, 2003) emphasize this approach and thereby are consistent with their aim of de-emphasizing methods, processes, and languages in favor of getting things to work. A popular approach to requirements elicitation is developing prototypes.

There has been less emphasis on bridging the gap between the above two efforts, for example, to deliver a method that gives practical guidelines on how to produce specifications from qualitative data (Hertzum, 2003). One reason for this gap can be that people from different disciplines work on the two aspects, for example, domain analysts or experts in HCI, and software engineers who read and use requirements specification as a basis for design. Another reason for this gap may be in the difference in the methods that the two sides have employed, that is, soft methods (i.e., informal) for elicitation and analysis and hard methods (i.e., formal) for specification.

In this article, we suggest a method to translate qualitative data to requirements specification that we have applied in the development of a Smart Space for Learning. The process borrows ideas from or uses scenarios, interviews, feature-based development, soft systems methodology, claims analysis, phenomenology, and UML.

FOLLOWING A PROCESS

The proposed process comprises five distinct steps or subprocesses, each with defined input and output (Table 1). The input to a step is the data or information that is available before the step is carried out, and the output is the work product or a deliverable of the step. We have also defined who, for example what roles, will carry out the steps. We will describe individual steps of the processes in more detail in subsequent sections.

Table 1. Steps of an analysis process

Who	Input	Step	Output
Domain analyst	• Feature ideas from designers • Marketing information • CATWOE Root definition	Elicitation design	• Framework for research study in terms of questions and goals • Preliminary life cycle of artefacts in domain • Scenarios describing work in the domain • Scenarios describing work in the domain using new features • Features of new system • User selection • Time and place of data gathering
Domain analyst	• Access to user and context • Output from previous step	Data gathering	• Answers to questions in terms of textual, audio or video information • Modified scenarios, sequences of work • Artefacts • Results of claims analysis
Domain analyst Requirement analyst	Output from previous step	Data analysis	• Matrices • Implications as facts, phenomena, relationships • Conflicts and convergence of phenomena
Requirement analyst	Output from previous step	Model specification	• Entity, Actor, Stimulus, Events, Behavior and Communication model
Domain analysts Requirement analyst	• Output from previous step • Access to user and customer	Validation	Revised model

Elicitation Design

The specific goals of field studies are different from one project to another. Examples of goals include analysing hindrances, facilitators of work, qualities, and context. The general goal is to build a software system that solves a problem and/or fulfills the actor's goals.

To prepare for visiting the work domain, we design a framework that should guide us to elicit information on what type of work is carried out, who does it, in what context, and the quality of work such as frequency, performance, and criticality. Analysts should gather as much background information beforehand as possible, for example, from marketing studies, manuals, current systems, and so forth. To understand the context of the system, a CATWOE analysis (Checkland & Scholes, 1999) is conducted.

Before the field studies, designers may brainstorm some possible features of the system. When creating a system for a new technological environment, designers tend to have some ideas on how it can further advance work. The work product of this step should be an Elicitation Design in a form of a handbook that describes overall goals of the contextual inquiries. In qualitative studies, researchers should not predefine hypotheses before the inquiry, but they can nonetheless be informed. Instead of visiting the work domain empty-handed, we suggest that analysts be prepared with descriptions of hypothetical scenarios or narrations of work with embedded questions on more details or procedures in this particular instance. Fictional characters can be used to make the scenarios more real.

Work is determined by some life cycle of artefacts. The questions can be formed in the framework of such a life cycle. In many domains, a life cycle of artefacts is already a best practice, and the goal can be to investigate how close the particular instance is to the life cycle. To reach this goal, several initial questions can be designed, bearing in mind that more can be added during the field study. Another way to reach this goal is through observation of work with intermittent questions.

A part of the preparation is the selection of clients and planning of the site visits. Analysts should recognize that users could be a limited resource.

Data Gathering

On site, the researcher uses the Elicitation Design framework to inquire about the work domain. The first part of the interview gathers information through questions about the current work domain with the help of hypothetical scenarios. In the second part of data gathering, claims analysis (Rosson & Carroll, 2002) is performed on suggested new features of a system. In claims analysis, people are asked to give positive and negative consequences to users of particular features. An example can be: Apples are sweet and delicious (positive) but they are of many types that make it difficult to choose from.

It is useful to notice best practice in interviewing, and ask the interviewee for concrete examples and to show evidence of work or artefacts. The interviewer should act as an apprentice, show interest in the work and be objective. Data is gathered by writing down answers and recording audio. It is almost necessary to have a team of two persons where one is the writer and the other is the interviewer.

Data Analysis

As soon as the site visit is over, the analyst should review the data gathered and write down additional facts or observations that he/she can remember, and encode the audiotapes to text. After this preparation, the actual analysis takes place, but the analysis can start already at the data-gathering phase. With large amounts of data, as is usual in qualitative studies, there is a need to structure it to see categorizations of phenomena or facts. Data is initially coded to derive classes of instances. There are several ways of doing this, but we propose to use one table for each goal or question, where the lines are individual instances of facts observed from a domain client and the columns are answers. The data is coded by creating a column that can be an answer to a question. The columns can subsequently be categorized, thus building a hierarchy of codes.

The output of the analysis is to derive understanding of phenomena, their characteristics, relationships, and behavior. We can formulate this knowledge in propositions and theories. One way of building theory, attributed to Glaser and Strauss, is called the grounded theory approach (Taylor & Bogdan, 1998).

Model Specification

From the analysis, we derive the model specification. It is described with different element types that are listed in Table 2. A further step can be taken to refine this to a more formal model, using for example, UML or Z.

Validation

The last step of the process is the validation. The analyst walks through the model and reviews it in cooperation with stakeholders.

VALIDATION IN A CASE STUDY

The method has been successfully applied in a case study of eliciting and specifying requirements for a software system that supports corporations with training management. This system is called Smart Space for Learning (SS4L).

This section describes the needs assessment resulting from companies' interviews, but the entire analysis is outside the scope of this article and will be described elsewhere. The aim of SS4L is to support companies in offering effective learning to their employees, which serves the needs of organizations, intelligent support to employees through personalization and learning profiles and open interfaces between heterogeneous learning systems and learning content.

Seven processes of the learning life cycle were introduced for the interviews: Training needs, Training goals, Input controlling, Process controlling, Output controlling, Transfer controlling, and Outcome controlling. These processes were built on the evaluation model of Kirkpatrick (1996) with its four-level hierarchy. He identifies four levels of training evaluation: reaction (do they like it?), learning (do they learn the knowledge?), transfer (does the knowledge transfer?), and influence (do they use it and does it make a difference to the business?). The study was conducted in 18 companies in five different countries.

The aim of this qualitative requirements study was threefold. First, we aimed at assessing the current and future needs of corporate training man-

Table 2. Different types of elements in the model

Elements	Description	Example
Entity	An entity is a representation of something abstract or concrete in the real world. An entity has a set of distinct attributes that contain values.	Knowledge
Actor	Actor is someone that initiates an action by signalling an event.	Human resource manager
Stimulus	Stimulus is what gets the Actor to act. This may be an event or a condition.	Company needs to satisfy customer demand
Input	Input is the information that is needed to complete the Behavior. The Input can be in the form of output from other Behaviours.	Knowledge Grade of Employees
Output	Output is the result of the Behavior.	Knowledge Grade of Department
Communication	Communication is an abstraction of the activity that enables transfer of information between two Behaviors. It can contain input, output data or events.	Communication to Analyze-Knowledge(Department)
Behaviour	Behavior is a sequence of actions. It can also be termed a process.	Analyze-Knowledge (Company)

agement. A second objective was to collect ideas from companies on what type of IT-support they foresee to be useful. Finally, a third objective was to find out positive and negative consequences to users of the features developed in SS4L.

The main results of the requirements studies are:

1. **Training management life cycle as a quality model for learning.** We have seen through the interviews that measuring the success of a course or a seminar implies an overall quality model for learning. First, one has to define the goals, plan the strategy for implementation, and then it can be assessed whether the results meet the goals. The interviews showed that some companies are still lacking this overall quality model. Only some steps of the overall process are performed, and therefore the training management leads to unsatisfactory results.
2. **Functions, that will be useful to have in an IT system for training management, were analyzed.**
   - Assessment of current and needed knowledge.
   - Targeted communication of strategies.
   - Peer evaluation or consulting.
   - Problem driven and on-demand learning.
   - Assessment of transfer during work by experienced colleagues.
   - Budget controlling.
3. **Stakeholders demand certain qualities of a training management system.** During the interview and the claims analysis, users also demanded other qualities. *Confidence* and *quality of data* has to be ensured when implementing a system that includes features that rely on contextual information such as personal or corporate profile and recommendation of peers. In general, all companies that participated in this study revealed a certain concern about training management and the need for supportive systems that allow for a greater *transparency*.

The training management life cycle proved to be a useful framework for the study. The scenarios were useful to set a scope for the interviews and they made the discussion easier. Some practice is required to get interviewees to give evidence through artefacts and explain answers with examples from the domain. As in any qualitative study, the data analysis is quite tedious, but coding the interviews into tables proved useful and a good basis for determining the components of the model as described in Table 2. A good tool for tracking the data, that is, from answers to tables to models will give researchers confidence that no data is lost and that it is correct. Using a combination of interviews and claims analysis for proposed features creates interesting results and may reveal inconsistencies. For example, users express the desire to select training courses based on the evaluation of peers but are reluctant themselves to enter ratings of training courses or other personal information because of threat to privacy. The final steps in the study are to translate the activities to use cases (Larman, 2002) and to prioritize them and determine their feasibility.

FUTURE TRENDS

There seems to be a growing interest in qualitative methods that require analysts to gather data in context. The reason may be that software systems are more than ever embedded into the environment and used in every day life, requiring access for everybody. Developers will expect to make decisions on design based on empirical studies. This may call for combined qualitative and quantitative methods. The cause for the latter is that it may be easier to collect extensive usage information in a larger population and a triangulation of methods will be used to retrieve reliable results.

Although interdisciplinary research has been and continues to be difficult, there is a growing trend to do research at the boundaries in order to discover innovative ideas and to work cross-discipline. The motivation is that there is a growing awareness to examine the boundaries between software systems, humans and their cognitive abilities, artificial physical systems, and biological and other natural systems.

CONCLUSION

This article has described a method for bridging the gap between the needs of the user and requirements specification. In a case study, we have elicited requirements by interviewing companies in various countries, with the aid of tools such as scenarios linked with questions related to a process life cycle. A qualitative analysis has given us insight into current practices and future needs of corporations, thereby suggesting innovative features. Interviews followed by claims analysis of suggested features that indicated positive and negative consequences have helped us analyze what features users welcomed and which have been rejected or need to be improved. Future work will entail further validation of the method for modelling of the functional requirements for developers.

REFERENCES

Abowd, G. D., & Mynatt, E. D. (2001). Charting past, present, and future research in ubiquitous computing. In J. M. Carroll (Ed.), *Human-computer interaction in the new millennium* (pp. 513-535). New York: Addison-Wesley.

Abrahamsson, P., Warsta, J., Siponen, M., & Ronkainen, J. (2003, May 3-10). New directions on agile methods: A comparative analysis. In the *Proceedings of the 25th International Conference on Software Engineering,* Portland, Oregon (pp. 244-254). USA: IEEE.

Bertelsen, O. W., & Bødker, S. (2003). Activity theory. In J. M. Carroll (Ed.), *HCI models, theories and frameworks* (pp. 291-342). Elsevier Science.

Beyer, H., & Holtzblatt, K. (1998). *Contextual design.* Morgan Kaufman

Checkland, P., & Scholes, J. (1999). *Soft systems methodology in action.* UK: Wiley.

Gulliksen, J., Göransson, B., & Lif, M. (2001). A user-centered approach to object-oriented user interface design. In M. van Harmelen (Ed.), *Object modeling and user interface design* (pp. 283-312). Addison Wesley.

Hertzum, M. (2003). Making use of scenarios: A field study of conceptual design. *International Journal of Human-Computer Studies, 58,* 215-239.

Hughes, J., O'Brien, J., Rodden, T., Rouncefield, M., & Sommerville, I. (1995, March 27-29). Presenting ethnography in requirements process. In *Requirements Engineering, Proceedings of the Second IEEE International Symposium,* University of York, UK (pp. 27-34). IEEE.

International Organization for Standardization (2001). ISO/IEC 9126-1. *Software engineering—Product quality—Part 1: Quality model.*

Jackson, M. (1995). *Software requirements & specifications.* Addison-Wesley.

Kirkpatrick, D. (1996, January). Revisiting Kirkpatrick's four-level model. *Training and Development, 50*(1), 54-57.

Kotonya, G., & Sommerville, I. (1998). *Requirements engineering.* Wiley.

Kuhn, S., & Muller, M. (Guest Eds. of Special Issue on Participatory Design) (1993). Participatory design. *Communication of the ACM, 36*(4), 25-28.

Larman, C. (2002). *Applying UML and patterns: An introduction to object-oriented analysis and design and the unified process* (2nd ed.). Prentice-Hall.

Lieberman, H., & Selker, T. (2000). Out of context: Computer systems that adapt to, and learn from, context. *IBM Systems Journal, 39*(3/4), 617-632.

Paterno, F. (2003). ConcurTaskTrees: An engineered notation for task models (Chapter 24). In D. Diaper, & N. Stanton (Eds.), *The handbook of task analysis for human-computer interaction* (pp. 483-503). Mahwah: Lawrence Erlbaum Associates.

Potter, B., Sinclair, J., & Till, D. (1996). *An introduction to formal specification and Z* (2nd ed.). Prentice Hall.

Ross, D. T. (1985). Applications and extensions of SADT. *IEEE Computer, 18*(4), 25-34.

Rosson, M. B., & Carroll, J. (2002). *Usability engineering: Scenario-based development of human computer interaction.* Morgan Kaufmann.

Taylor, S. J., & Bogdan, R. (1998). *Introduction to qualitative research methods* (3rd ed.). Wiley.

Vicente, K. J. (1999). *Cognitive work analysis.* Lawrence Erlbaum Associates.

KEY TERMS

Actor: Someone that initiates an action by signalling an event. An actor is outside a system and can be either another system or a human being.

CATWOE: Clients are the stakeholders of the systems, **A**ctors are the users, **T**ransformation describes the expected transformations the system will make on the domain, **W**orldview is a certain aspect of the domain we have chosen to focus on, **O**wners can stop the development of the system, and **E**nvironment describes the system context.

Context: Everything but the explicit input and output of an application. Context is the state of the user, state of the physical environment, state of the computational environment, and history of user-computer-environment interaction. (Lieberman & Selker, 2000) Put another way, one can say that context of the system includes, Who is the actor, When (e.g., in time) an actor operates it, Where the actor operates it, Why or under what conditions the actor activates the system or What stimulates the use (Abowd & Mynatt, 2002).

Contextual Inquiry: Context, partnership, interpretation, and focus are four principles that guide contextual inquiry. The first and most basic requirement of contextual inquiry is to go to the customer's workplace and observe the work. The second is that the analysts and the customer together in a partnership understand this work. The third is to interpret work by deriving facts and make a hypothesis that can have an implication for design. The fourth principle is that the interviewer defines a point of view while studying work (Beyer & Holtzblatt, 1998).

Prototype: Built to test some aspects of a system before its final design and implementation. During requirements elicitation, a prototype of the user interface is developed, that is, to give stakeholders ideas about its functionality or interaction. Prototypes are either high fidelity, that is, built to be very similar to the product or low fidelity with very primitive tools, even only pencil and paper. Prototypes can be thrown away, where they are discarded, or incremental, where they are developed into an operational software system.

Qualitative Methodology: "The phrase *qualitative methodology* refers in the broadest sense to research that produces descriptive data – people's own written or spoken words and observable behaviour" (Taylor & Bogdan, 1998, p. 7).

Requirements Elicitation: "Requirements elicitation is the usual name given to activities involved in discovering the requirements of the system" (Kotonya & Sommerville, 1998, p. 53).

Requirements Specification: Provides an overview of the software context and capabilities. Formally, the requirements should include:

- Functional Requirements
- Data Requirements
- Quality Requirements
- Constraints

Software Quality: A quality model (ISO/IEC 9126-1, 2001) categorises software quality attributes into the following six characteristics that are again subdivided into sub-characteristics. The characteristics are specified for certain conditions of the software product:

- **Functionality:** The software product provides functions which meet needs.
- **Reliability:** The software product maintains performance.
- **Usability:** The software product should be understood, learned, used, and attractive to user.
- **Efficiency:** The software product provides appropriate performance relative to the amount of resources used.
- **Maintainability:** The software product should be modifiable. Modifications include corrections, improvements, or adaptations.
- **Portability:** The software product can be transferred from one environment to another.

System Model: A system model is a description of a system. Initially, the model describes what problem the system should solve and then it can be gradually refined to describe how the system solves the problem. Finally, when operational, the system can be viewed as a model of some domain behaviour and characteristics.

Fuzzy Logic Usage in Emotion Communication of Human Machine Interaction

Zhe Xu
Bournemouth University, UK

David John
Bournemouth University, UK

Anthony C. Boucouvalas
Bournemouth University, UK

INTRODUCTION

As the popularity of the Internet has expanded, an increasing number of people spend time online. More than ever, individuals spend time online reading news, searching for new technologies, and chatting with others. Although the Internet was designed as a tool for computational calculations, it has now become a social environment with computer-mediated communication (CMC).

Picard and Healey (1997) demonstrated the potential and importance of emotion in human-computer interaction, and Bates (1992) illustrated the roles that emotion plays in user interactions with synthetic agents.

Is emotion communication important for human-computer interaction?

Scott and Nass (2002) demonstrated that humans extrapolate their interpersonal interaction patterns onto computers. Humans talk to computers, are angry with them, and even make friends with them. In our previous research, we demonstrated that social norms applied in our daily life are still valid for human-computer interaction. Furthermore, we proved that providing emotion visualisation in the human-computer interface could significantly influence the perceived performances and feelings of humans. For example, in an online quiz environment, human participants answered questions and then a software agent judged the answers and presented either a positive (happy) or negative (sad) expression. Even if two participants performed identically and achieved the same number of correct answers, the perceived performance for the one in the positive-expression environment is significantly higher than the one in the negative-expression environment (Xu, 2005).

Although human emotional processes are much more complex than in the above example and it is difficult to build a complete computational model, various models and applications have been developed and applied in human-agent interaction environments such as the OZ project (Bates, 1992), the Cathexis model (Velasquez, 1997), and Elliot's (1992) affective reasoner.

We are interested in investigating the influences of emotions not only for human-agent communication, but also for online human-human communications. The first question is, can we detect a human's emotional state automatically and intelligently?

Previous works have concluded that emotions can be detected in various ways—in speech, in facial expressions, and in text—for example, investigations that focus on the synthesis of facial expressions and acoustic expression including Kaiser and Wehrle (2000), Wehrle, Kaiser, Schmidt, and Scherer (2000), and Zentner and Scherer (1998). As text is still dominating online communications, we believe that emotion detection in textual messages is particularly important.

BACKGROUND

Approaches for extracting emotion information from textual messages can be classified into the categories of keyword tagging, lexical affinity, statistical methods, or real-world models (Liu, Lieberman, & Selker, 2003).

We have developed a textual emotion-extraction engine that can analyze text sentences typed by users. The emotion extraction engine has been presented by Xu and Boucouvalas (2002).

The emotion-extraction engine can analyze sentences, detect emotional content, and display appropriate expressions. The intensity and duration of the expressions are also calculated and displayed in real time automatically. The first version of our engine searched for the first person, *I*, and the current tense, therefore the ability of the engine was very limited. In our latest version, the engine applies not only grammatical knowledge, but also takes real-word information and cyberspace knowledge into account. It intends to satisfy the demands of complicated sentence analysis.

The user's mood is defined as the feelings perceived from a user's series are input in the emotion-extraction engine. The current emotion of a user is based totally on the information assessed within a single sentence.

A user's mood may not be consistent with the current emotion of the user. For example, a user may present a sad feeling in one sentence, but previously the user was talking about happy and interesting things. The sad feeling presented may not be a significant emotion and overall the user's mood may be still happy.

To calculate the mood of a user, previous emotions and current emotions need to be analyzed together. We assume that emotions are additive and cumulative. One way of calculating the mood is to average the historic emotions and then find out what category the averaged emotion is in. This approach is described by Xu (2005). Here, an alternative fuzzy-logic approach is presented.

Fuzzy Logic

Fuzzy logic was developed to deal with concepts that do not have well-defined, sharp boundaries (Bezdek, 1989), which theoretically is ideal for emotion as no well-defined boundaries are defined for emotion categories (e.g., happiness, sadness, surprise, fear, disgust, and anger).

The transition from one physiological state to another is a gradual one. These states cannot be treated as classical sets, which either wholly include a given affect or exclude it. Even within the physiological response variables, one set merges into another and cannot be clearly distinguished from another. For instance, consider two affective states: a relaxed state and an anxious state. If classical sets are used, a person is either relaxed or anxious at a given instance, but not both. The transition from one set to another is rather abrupt and such transitions do not occur in real life.

EMOTION EXTRACTION ENGINE

The emotion extraction engine is a generic prototype based on keyword tagging and real-world knowledge. Figure 1 depicts an overview of the architecture of the emotion-extraction engine.

The sentence analysis component includes three components: input analysis, the tagging system, and the parser. The input-analysis function splits textual messages into arrays of words and carries out initial analysis to remove possible errors in the input. The tagging system converts the array of words into an array of tags. The parser uses rewrite rules and AI (artificial intelligence) knowledge to carry out information extraction. The engine classifies emotions into the following categories: happiness, sadness, surprise, fear, disgust, and anger. For further details, please refer to Xu and Boucouvalas (2002) and Xu (2005). This article only discusses the fuzzy-logic components, which can be seen as an extension to the parser. With fuzzy logic methods, the emotion-extraction engine can be used to analyze complex situations.

Conflicting Emotion Detection

The inputs of the conflicting-emotion detection component are the emotion parameters that are passed from the sentence analysis component. As mixed emotions are a common phenomenon in daily life, it is not unusual for a user to type in a sentence, such as, "I am happy that I got a promotion, but it is sad that my salary is cut," that contains mixed emotions in an online chatting environment.

Figure 1. The emotion extraction engine overview

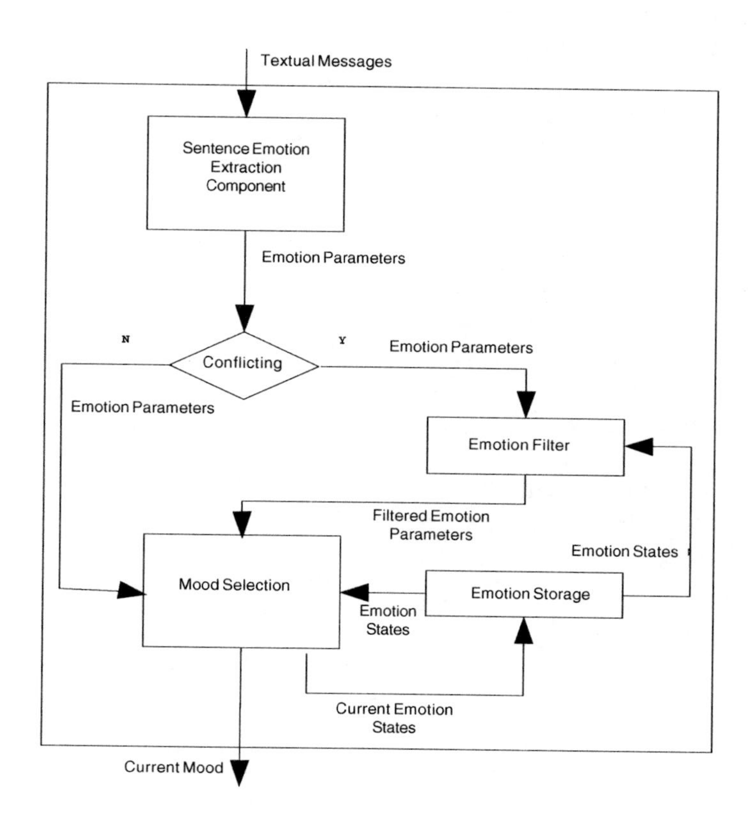

When a sentence contains conflicting emotions, judging which emotion represents the overall emotional feeling is not only based on the current sentence, but also on the mood. For example, in the emotion extraction engine, the mood *happy* indicates that the previous messages an individual typed contain overwhelmingly more happy feelings than others. When the user types a sentence containing both happy and sad emotions, the perceived current mood of the user may still be happy instead of happy and sad. The reason is that the individual was in a predominately happy mood, and the presented sad emotion may not be significant enough to change the mood from happy to sad.

The positive emotion category and negative emotion category are introduced to handle conflicting emotions. Positive emotions include happiness and surprise, while negative emotions are sadness, fear, anger, and disgust. A sentence is treated as a conflicting-emotion sentence only if the sentence contains both positive and negative emotions.

When a sentence with conflicting emotions is found, the emotion-filter component will be called; otherwise, the emotion parameters are passed to the mood-selection component for further operation.

Mood Selection Component

The inputs of the mood-selection component are the emotion parameters from the sentence emotion extraction component and the previous emotions stored in the emotion-storage component.

The aim of the mood selection component is to determine the current mood. To achieve this, the first step of the mood selection component is to convert the emotion parameters into the current emotions by filtering the tense information. For example, the emotion parameter [happiness][middle intensity][present tense] is converted to the current emotion [happiness][middle intensity]. The current emotion is sent to the storage component as well.

The previous emotions of the user are stored in the storage component. The format is [emotion category][intensity]. To covert the emotion data into the format acceptable for a fuzzy system, the following fuzzy-data calculations are carried out.

Fuzzy Data Calculation

An array E is assigned to contain the accumulative intensity values of the six emotion categories. The array elements 0 to 5 in turn represent the accumulative intensity of the emotions happiness, surprise, anger, disgust, sadness, and fear.

The value of each element in array E is calculated by adding the five previous intensities of a specific emotion category with the current intensity of that emotion. Equation 1 applied to calculate the accumulative intensity is shown as follows.

$$E[x] = \sum_{i=-n}^{0} \alpha_i I_i(x), \text{ where } x = 0,1,2,3,4,5 \qquad (1)$$

The values of array E depend on the relative intensity over the last n time periods; n is chosen to be 5 as it is assumed that in a chatting environment users only remember the most recent dialogs. $I_i(x)$ is the intensity of emotion category x at discrete time i, and the value of $I_i(x)$ varies from 0 to 3, which represents the lowest intensity to the highest intensity. When i is 0, $I_i(x)$ contains the intensity values of the current emotions.

Instead of adding up the unweighted previous intensities, the intensities are weighted according to the time. Velasquez (1997) declared that emotions do not disappear once their cause has disappeared, but rather they decay through time. In the FLAME project, El-Nasr, Yen, and Ioerger (2000) follow Velasquez's view and choose to decay positive emotions at a faster rate. However, there is not enough empirical evidence from El-Nasr et al.'s implementation or this implementation to establish the actual rate of the decay. In the emotion extraction engine, the positive and negative emotions are assumed to decay at the same rate and the influence period is chosen to be five sentences. Figure 2 illustrates the assumption.

In Figure 2, Value represents the value of the perceived influence of emotion and t represents time. EMO represents the emotion that occurred at the discrete time point (e.g., the time when a chat message is input). In this figure, the value of EMO at different time points is the same, which means that a user typed six emotional sentences of the same intensity. Zero represents the current time.

Figure 2. The decay of emotion over time

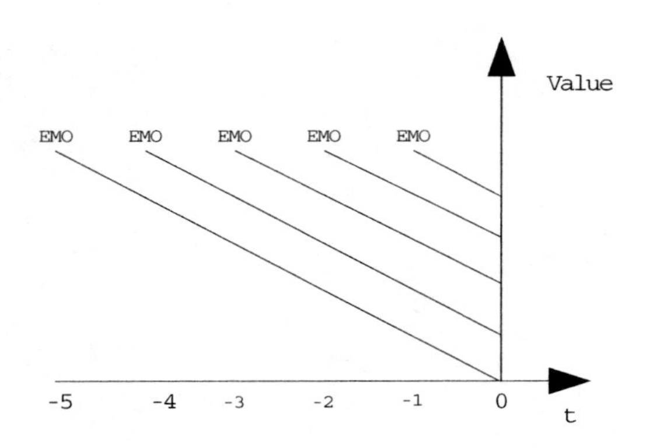

However, at time point 0, the perceived influence of EMO that occurred at time -5 is 0, which means the influence of the emotion input at time -5 has disappeared. The EMO that occurred at time -1 is the least decayed and has the strongest influence at the current time (time 0).

In the fuzzy emotion-analysis component, the emotion decay is represented by the weight parameter, and it is calculated using Equation 2.

$$\alpha_i = \alpha_{i+1} - 0.1 \text{ where } i \subset -5,-4,-3,-2$$
$$\alpha_{-1} = 0.5$$
$$\alpha_0 = 1 \qquad (2)$$

Fuzzy Membership Functions

The following fuzzy membership functions and fuzzy logic rules are applied to assist the mood calculation. Two fuzzy membership functions *high* and *low* (Equations 3 and 4) are defined to guide the analysis of fuzzy functions.

$$high(x) = \frac{E[x]}{\sum_{i=0}^{5} E[i]} \qquad (3)$$

$$low(x) = 1 - \frac{E[x]}{\sum_{i=0}^{5} E[i]} \quad (4)$$

Fuzzy Rules

Fuzzy rules are created based on the high and low membership functions. The rule base includes the following.

Rule 1:
IF the emotion with largest intensity is "high"
AND the other emotions are "low"
THEN the current mood is that emotion

Rule 2:
IF the positive emotions are "high
AND the negative emotions are "low"
THEN the current mood is the highest positive emotion and the intensity is decreased by 0.1

Rule 3:
IF the negative emotions are "high"
AND positive emotions are "low"
THEN the current mood is the highest negative emotion and the intensity is decreased by 0.1

MOOD SELECTION

When a dialog starts, there are no cues as to what mood a user is in. However, it is reasonable to assume that the mood of a new user is neutral. As the emotion extraction engine acquires more data (e.g., the user starts chatting), the above fuzzy rules can be applied and the mood of the user can be calculated. The centre of gravity (COG) point is an important measurement factor in determining the dominant rule. In this implementation, the COG point is calculated as the average of the rules' outputs (Equation 5).

$$\text{COG} = \frac{\sum_{1}^{3} rule(x)}{3} \quad (5)$$

STORAGE COMPONENT

The inputs of the storage component are the current emotions. The storage component is implemented as a first in, first remove (FIFR) stack with a length of five. The structure is shown in Figure 3.

EMOTION FILTER

The emotion filter is designed to analyze the detected conflicting emotions. The inputs of the filter include the current conflicting emotions and the emotions stored in the storage component. Similar fuzzy logic membership functions and rules are applied to analyze the conflicting emotions. The only difference is that here the current emotion data are excluded. The detailed fuzzy membership functions and rules are not discussed here. Readers can find the details in the mood-selection-component section.

FUTURE TRENDS

Fuzzy logic is a popular research field in autocontrol, AI, and human factors. As adaptivity becomes an extremely important interface design criteria, fuzzy logic shows its own advantage in creating adaptive

Figure 3. The structure of the storage component

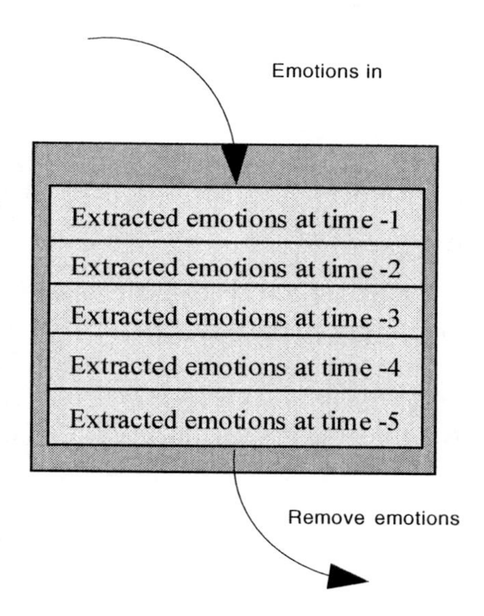

systems: There are no clear category boundaries, and it is easily understood by humans. Fuzzy logic can be integrated with other techniques (e.g., inductive logic, neural networks, etc.) to create fuzzy-neural or fuzzy-inductive systems, which can be used to analyze complex human-computer interactions. For this article, possible future studies may include a comparison of the emotions detected by the engine with emotions by human observers. Also, the applicability of the fuzzy-logic functions with different events should be tested by comparing the performance of the emotion-extraction engine to the different contexts that the emotion-extraction engine is applied.

CONCLUSION

This article presents an overview of the fuzzy logic components applied in our emotion-extraction engine. Fuzzy-logic rules are followed to determine the correct mood when conflicting emotions are detected in a single sentence. The calculation of current mood involves an assessment of the intensity of current emotions and the intensity of previously detected emotions. This article presents an example of fuzzy-logic usage in human-computer interaction. Similar approaches can be tailored to fit situations like emotional-system interaction, telecare systems, and cognition-adaptive systems.

REFERENCES

Bates, J. (1992). The role of emotion in believable agents. *Communications of the ACM, 37*(7), 122-125.

Bezdek, J. C. (1989). The coming age of fuzzy logic. *Proceedings of IFSA.*

Brave, S., & Nass, C. (2002). The human-computer interaction handbook: Fundamentals, evolving technologies and emerging applications archive. In *Human factors and ergonomics archive* (pp. 81-96). Mahwah, NJ: Lawrence Erlbaum Associates.

Elliot, C. (1992). *The affective reasoner: A process model of emotions in a multi-agent system.* Unpublished doctoral dissertation, Northwestern University, Institute for the Learning Sciences, Evanston, IL.

El-Nasr, M. S., Yen, J., & Ioerger, T. (2000). FLAME: A fuzzy logic adaptive model of emotions. *Autonomous Agents and Multi-Agent Systems, 3*(3), 219-257.

Kaiser, S., & Wehrle, T. (2000). Ausdruckspsychologische Methoden [Psychological methods in studying expression]. In J. H. Otto, H. A. Euler, & H. Mandl (Eds.), *Handbuch Emotionspsychologie* (pp. 419-428). Weinheim, Germany: Beltz, Psychologie Verlags Union.

Liu, H., Lieberman, H., & Selker, T. (2003). A model of textual affect sensing using real-world knowledge. *Proceedings of the Seventh International Conference on Intelligent User Interfaces* (pp. 125-132).

Picard, R. W., & Healey, J. (1997). Affective wearables. *Proceedings of the First International Symposium on Wearable Computers* (pp. 90-97).

Velasquez, J. (1997). Modeling emotions and other motivations in synthetic agents. *Proceedings of the AAAI Conference 1997* (pp. 10-15).

Wehrle, T., Kaiser, S., Schmidt, S., & Scherer, K. R. (2000). Studying dynamic models of facial expression of emotion using synthetic animated faces. *Journal of Personality and Social Psychology, 78*(1), 105-119.

Xu, Z., & Boucouvalas, A. C. (2002). Text-to-emotion engine for real time Internet communication. *International Symposium on Communication Systems, Networks and DSPs*, 164-168.

Xu, Z. (2005). Real-time expressive Internet communications. PhD thesis, Bournemouth University, UK.

Zentner, M. R., & Scherer, K. R. (1998). Emotionaler Ausdruck in Musik und Sprache [Emotional expression in music and speech]. *Deutsches Jahrbuch für Musikpsychologie, 13*, 8-25.

KEY TERMS

CMC (Computer-Mediated Communication): The human use of computers as the medium to communicate to other humans.

Current Emotion: The current emotion refers to the emotion contained in the most recent sentence.

Current Mood: The current mood refers to the weighted average emotion in the five most recent sentences.

Emotion Decay: The gradual decline of the influence of emotion over time.

Emotion-Extraction Engine: A software system that can extract emotions embedded in textual messages.

Emotion Filter: The emotion filter detects and removes conflicting emotional feelings.

Fuzzy Logic: Fuzzy logic is applied to fuzzy sets where membership in a fuzzy set is a probability, not necessarily 0 or 1.

GIS Applications to City Planning Engineering

Balqies Sadoun
Al-Balqa' Applied University, Jordan

INTRODUCTION

The rapid progress in information technology (IT) has moved computing and the Internet to the mainstream. Today's personal laptop computer has computational power and performance equal to 10 times that of the mainframe computer. Information technology has become essential to numerous fields, including city and regional planning engineering. Moreover, IT and computing are no longer exclusive to computer scientists/engineers. There are many new disciplines that have been initiated recently based on the cross fertilization of IT and traditional fields. Examples include geographical information systems (GIS), computer simulation, e-commerce, and e-business. The arrival of affordable and powerful computer systems over the past few decades has facilitated the growth of pioneering software applications for the storage, analysis, and display of geographic data and information. The majority of these belong to GIS (Batty et al., 1994; Burrough et al., 1980; Choi & Usery, 2004; Clapp et al., 1997; GIS@Purdue, 2003; Golay et al., 2000; Goodchild et al., 1999; IFFD, 1998; Jankowski, 1995; Joerin et al., 2001; Kohsaka, 2001; Korte, 2001; McDonnell & Kemp, 1995; Mohan, 2001; Ralston, 2004; Sadoun, 2003; Saleh & Sadoun, 2004).

GIS is used for a wide variety of tasks, including planning store locations, managing land use, planning and designing good transportation systems, and aiding law enforcement agencies. GIS systems are basically ubiquitous computerized mapping programs that help corporations, private groups, and governments to make decisions in an economical manner. A GIS program works by connecting information/data stored in a computer database system to points on a map. Information is displayed in layers, with each succeeding layer laid over the preceding ones. The resulting maps and diagrams can reveal trends or patterns that might be missed if the same information was presented in a traditional spreadsheet or plot.

A GIS is a computer system capable of capturing, managing, integrating, manipulating, analyzing, and displaying geographically referenced information. GIS deals with spatial information that uses location within a coordinate system as its reference base (see Figure 1). It integrates common database operations such as query and statistical analysis with the unique visualization and geographic analysis benefits offered by maps. These abilities distinguish GIS from other information systems and make it valuable to a wide range of public and private enterprises for explaining events, predicting out-

Figure 1. A coordinate system (GIS@Purdue 2003)

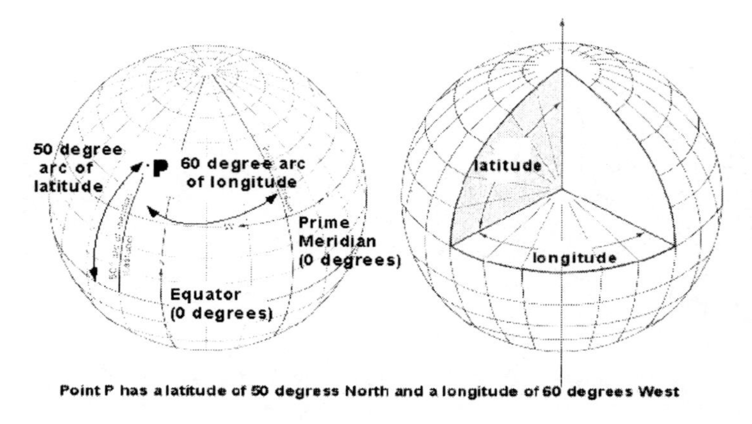

comes, and planning strategies (Batty et al., 1994; Burrough et al, 1980; Choi & Usery, 2004; Clapp et al., 1997; GIS@Purdue, 2003; Golay et al., 2000; Goodchild et al., 1999; IFFD, 1998; Jankowski, 1995; Joerin et al., 2001; Kohsaka, 2001; Korte, 2001; McDonnell & Kemp, 1995; Mohan, 2001; Ralston, 2004; Sadoun, 2003; Saleh & Sadoun, 2004).

BACKGROUND

A working GIS integrates five key components: hardware, software, data, people, and methods. GIS stores information about the world as a collection of thematic layers that can be linked together by geography. GIS data usually is stored in more than one layer in order to overcome technical problems caused by handling very large amounts of information at once (Figure 2). This simple but extremely powerful and versatile concept has proved invaluable for solving many real-world problems, such as tracking delivery vehicles, recording details of planning applications, and modeling global atmospheric circulation. GIS technology, as a human-computer interaction (HCI) tool, can provide an efficient platform that is easy to customize and rich enough to support a vector-raster integration environment beyond the traditional visualization.

Figure 2. Illustration of GIS data layers (GIS@Purdue, 2003)

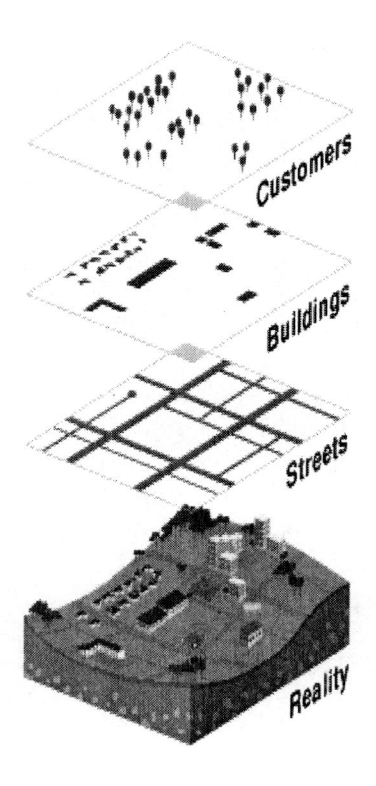

A GIS has four main functional subsystems: (1) data input, (2) data storage and retrieval, (3) data manipulation and analysis, and (4) data output and display subsystem. A data input subsystem allows the user to capture, collect, and transform spatial and thematic data into digital form. The data inputs usually are derived from a combination of hard copy maps, aerial photographs, remotely sensed images, reports, survey documents, and so forth. Maps can be digitized to collect the coordinates of the map features. Electronic scanning devices also can be used to convert map lines and points to digital information (see Figure 3).

The data storage and retrieval subsystem organizes the data, spatial and attribute, in a form that permits them to be retrieved quickly by the user for analysis and permits rapid and accurate updates to be made to the database. This component usually involves use of a database management system (DBMS) for maintaining attribute data. Spatial data usually is encoded and maintained in a proprietary file format.

The data manipulation and analysis subsystem allows the user to define and execute spatial and

Figure 3. A digitizing board with an input device to capture data from a source map (GIS@Purdue, 2003)

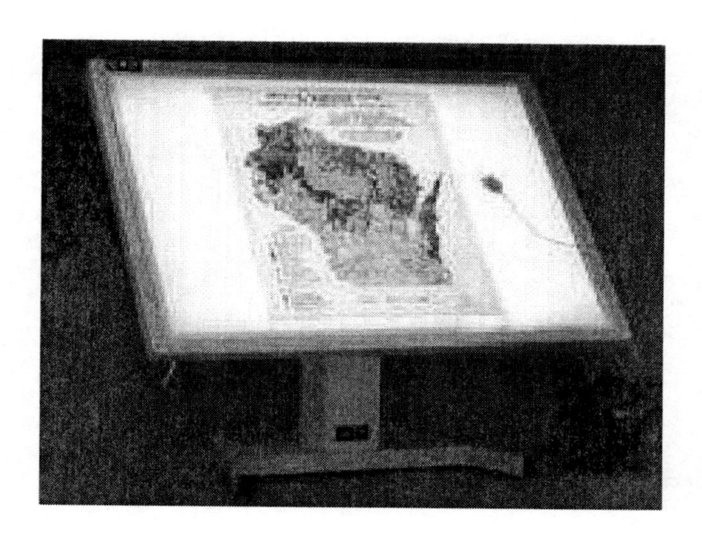

attribute procedures to generate derived information. This subsystem commonly is thought of as the heart of a GIS and usually distinguishes it from other database information systems and computer-aided drafting (CAD) systems. The data output subsystem allows the user to generate graphic displays—normally maps and tabular reports representing derived information products.

The basic data types in a GIS reflect traditional data found on a map. Accordingly, GIS technology utilizes two basic data types: (1) spatial data that describes the absolute and relative location of geographic features and (2) attribute data, which describe characteristics of the spatial features. These characteristics can be quantitative and/or qualitative in nature. Attribute data often is referred to as tabular data.

GIS works with two fundamentally different types of geographic models: vector and raster (see Figure 4). In the vector model, information about points, lines, and polygons is encoded and stored as a collection of *x y* coordinates. The location of a point feature can be described by a single *x y* coordinate. Linear features, such as roads and rivers, can be stored as a collection of point coordinates. Polygonal features, such as sales territories, can be stored as a closed loop of coordinates.

The vector model is extremely useful for describing discrete features but less useful for describing continuously varying features such as soil type or accessibility costs to hospitals. The raster model has evolved to model such continuous features. A raster image comprises a collection of grid cells such as a scanned map/picture. Both vector and raster models for storing geographic data have unique advantages and disadvantages. Modern GISs are able to handle both models.

Figure 4. GIS data model (IFFD, 1998)

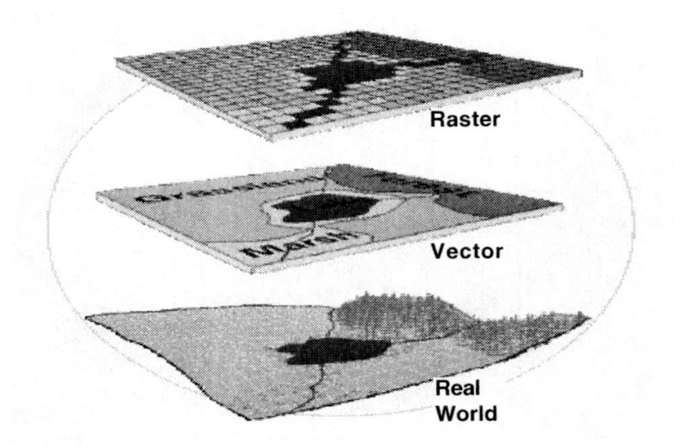

GIS SOFTWARE

A number of GIS software packages exist commercially, providing users with a wide range of applications. Some of them are available online for free. Before selecting a GIS package, the user should find out whether the selected GIS can meet his or her requirements in four major areas: input, manipulation, analysis, and presentation. The major GIS vendors are ESRI, Intergraph, Landmark Graphics, and MapInfo. A brief description of these packages is given next.

ESRI Packages

The Environmental Systems Research Institute (ESRI) provides database design application, database automation, software installation, and support (Korte, 2001; McDonnell et al., 1995; Ralston, 2004). ARC/INFO allows users to create and manage large multi-user spatial databases, perform sophisticated spatial analysis, integrate multiple data types, and produce high-quality maps for publication. ARC/INFO is a vector-based GIS for storing, analyzing, managing, and displaying topologically structured geographic data.

ArcView is considered the world's most popular desktop GIS and mapping software. ArcView provides data visualization, query, analysis, and integration capabilities along with the ability to create and edit geographic data. ArcView makes it easy to create maps and add users' own data. By using ArcView software's powerful visualization tools, it is possible to access records from existing databases and display them on maps. The ArcView network analyst extension enables users to solve a variety of problems using geographic networks such as streets, ATM machines, hospitals, schools, highways, pipelines, and electric lines. Other packages by ESRI include Database Integrator, ArcStorm, ArcTools, and ArcPress.

Intergraph Packages

Intergraph specializes in computer graphics systems for CAD and produces many packages. Intergraph's products serve four types of GIS users: managers, developers, viewers, and browsers. Intergraph bundles most popular components of Modular GIS Environments (MGEs) into a single product called GIS Office. The components are (a) MGE Basic Nucleus for data query and review, (b) MGE Base Mapper for collection of data, (c) MGE Basic Administrator for setup and maintenance of database, and (d) MGE Spatial Analyst for creation, query, displaying of topologically structured geographic data, and spatial analysis (Korte, 2001; McDonnell & Kemp, 1995; Ralston, 2004).

GeoMedia is Intergraph's first GIS product to implement its Jupiter technology. Jupiter functions without a CAD core and uses its object, graphics, and integration capabilities provided by object linking and embedding and component object model standards of the Windows operating system to integrate technical applications with office automation software (Korte, 2001; McDonnell & Kemp, 1995; Ralston, 2004).

MapInfo Packages

MapInfo offers a suite of desktop mapping products that are different from other leading GIS products like MGE and ARC/INFO in that they do not store a fully topologically structured vector data model of map features. MapInfo Professional provides data visualization; step-by-step thematic mapping; and three linked views of data maps, graphs, and tables. It displays a variety of vector and raster data formats. It enables users to digitize maps to create vector images. Moreover, it enables users to perform editing tasks such as selecting multiple nodes for deletion, copying, and clearing map objects and overlaying nodes. MapInfo Professional enables multiple data views with a zoom range of 55 feet to 100,000 miles. It supports 18 map projections, performing map projection display on the fly (Korte, 2001; McDonnell & Kemp, 1995; Ralston, 2004).

MapInfo ProServer enables MapInfo Professional to run on a server. This server side capability enables it to answer the queries over network enabling users to access desktop mapping solutions throughout the enterprise. On the client site, MapInfo Proserver and MapInfo Professional can be handled using an Internet Web browser.

Landmark Graphics Packages

Landmark Information system has been designed mainly for petroleum companies in order to explore and manage gas and oil reservoirs. Its products include ARGUS and Geo-dataWorks (Korte, 2001; McDonnell & Kemp, 1995; Ralston, 2004). ARGUS is a generic petroleum common user interface that consists of a suite of data access tools for management levels and technical disciplines in an organization. Its logical data are independent and object-oriented. It combines Executive Information Systems (EIS) with GIS features to query corporate data and display results virtually. Geo-data Works provides graphical project management. It enables graphical query and selection. It also allows the user to manage multiple projects. Furthermore, it provides user-friendly management and query building capabilities.

GIS APPLICATIONS

GIS applications are increasing at an amazing rate. For instance, GIS is being used to assist businesses in identifying their potential markets and maintaining a spatial database of their customers more than ever before. Among the most popular applications of GIS are city and regional planning engineering. For example, water supply companies use GIS technology as a spatial database of pipes and manholes. Local governments also use GIS to manage and update property boundaries, emergency operations, and environmental resources. GIS also may be used to map out the provision of services, such as health care and primary education, taking into account population distribution and access to facilities.

Firefighters can use GIS systems to track potential damage along the path of forest fires. The Marin County Fire Department in Northern California deploys helicopters equipped with global positioning system (GPS) receivers to fly over an area of land that is ablaze. The receiver collects latitude and longitude information about the perimeter of the fire. When the helicopter lands, that information is downloaded into a PC, which then connects to a database containing information on land ownership, endangered species, and access roads within the area of the fire. Those maps are printed out on mobile plotters at the scene and distributed to firefighters.

Preservation groups use GIS software to assess possible danger caused by environmental changes. In such a case, GIS is used as a tool for integrating data across borders and helping to bring people together to solve problems in an effective manner. For example, the U.S. National Oceanic and Atmospheric Administration and other organizations use GIS systems to create maps of the Tijuana River watershed, a flood-prone area that spans the border of San Diego (USA) and Tijuana (Mexico). The maps include soil and vegetation class structure from each city, which allows city and urban planners to see across the borders when predicting flood danger.

GIS can be used efficiently as human-machine interactive (HMI) tools to realize land-parcel restructuring within a well-defined zone, whose parcels' structure is found to be inadequate for agricultural or building purposes. Golay, et al. (2000) have proposed a new a prototype of an interactive engineering design platform for land and space-related engineering tasks, based on the real-time aggregation of vector and raster GIS data. This concept allows engineers to get real-time aggregate values of a continuously defined spatial variable, such as land value, within structures they are reshaping.

GIS tools can be used to manage land efficiently. Before the advent of GIS, land management services in public administration used to make decisions without analyzing related data properly. These days, such organizations are using GIS to properly manage land based on the data provided by GIS analysis. One example includes the decision support model called MEDUSAT, which proposes a structured application of GIS and multicriteria analysis to support land management. MEDUSAT presents an original combination of these tools (Joerin et al., 2001). In this context, GIS is used to manage information that describes the territory and offer spatial analysis schemes. The multicriteria analysis schemes then are used to combine such information and to select the most adequate solution, considering the decision maker's preference (Jankowski, 1995; Joerin et al., 2001).

GIS techniques can be used along with remote sensing schemes for urban planning, implementation, and monitoring of urban projects. Such a combination has the capability to provide the needed physical input and intelligence for the preparation of basemaps and the formulation of planning proposals. They also can act as a monitoring tool during the implementation phase. Large-scale city and urban projects need decades to complete. Satellite images can be used to maintain a real record of terrain during this period. Clearly, GIS and remote sensing are powerful tools for monitoring and managing land by providing a fourth dimension to the city—time (Joerin et al., 2001; Kohsaka, 2001).

GIS systems can be used to aid in urban economic calculations. A basic goal of urban economics is to analyze spatial relationships. This typically takes the shape of costs to ship customers, employees, merchandise, or services between various locations. Real estate planners often seek to quantify the relations between supply and demand for a particular land type in a given geographical region. Spatial economic theory shows that transportation costs are the most essential factors that determine ease of access. GIS can estimate transportation costs in a better way by computing the distances along roads, weighting interstate roads less than local roads and adding delay factors for construction spots, tunnels, and so forth (Jankowski, 1995; Joerin et al., 2001).

GIS can handle complex network problems, such as road network analysis. A GIS can work out travel times and shortest path between two cities (sites), utilizing one of the shortest path algorithms. This facility can be built into more complicated models that might require estimates of travel time, accessibility, or impedance along a route system. An example is how a road network can be used to calculate the risks of accidents.

GIS can be used to model the flow of water through a river in order to plan a flood warning system. Real-time data would be transmitted by flood warning monitors/sensors, such as rain gauges and river height alarms, which could be received and passed to a GIS system to assess the hazard. If the amount and intensity of rain exceeds a certain limit, determined by the GIS flood model for the area, a flood protection plan could be put into operation with computer-generated maps demarcating the vulnerable areas at any point in time (Golay et al., 2000;

Jankowski, 1995; Joerin et al., 2001; Kohsaka, 2001; Mohan, 2001; Sadoun, 2003; Saleh & Sadoun, 2004).

FUTURE TRENDS

GIS systems are effective HCI tools that have received wide acceptance by organizations and institutions. Many towns and cities worldwide have started or are in the process of using it in planning and engineering services and zones. Many vendors have developed GIS software packages of various features and options. Using GIS software is becoming more and more friendly with the addition of new features in the software such as the Graphical User Interfaces (GUIs), animation, validation, and verification features. GUI helps make the interface between the developer/user and the computer easier and more effective. Animation helps to verify models and plans, as a picture is worth a thousand words. The latter also makes the GIS package more salable. These days, there is a trend to use GIS for almost all applications that range from planning transportation systems in cities and towns to modeling of the flow of water through rivers in order to have a proper flood warning system, where real-time data can be sent by flood warning sensors, such as rain gauges and river height alarms, which could be received and passed to a GIS system to assess the hazard.

During the past few years, law enforcement agencies around the world have started using GIS to display, analyze, and battle crime. Computer-generated maps are replacing the traditional push-pin maps that used to cover the walls of law enforcement agencies. Such a tool has given police officers the power to classify and rearrange reams of data in an attempt to find patterns. In the United States, officers in many precincts are using this technology to track down crimes. Officers gather reports on offenses like car thefts or residential robberies onto weekly hot sheets, which then they enter into a computer that has a GIS software package. The GIS mapping program, in turn, relates each incident to some map information by giving it latitude and longitude coordinates within a map of the suspected area.

Some police departments in developed countries employ GIS computer mapping to persuade residents to get involved in community development. In North Carolina, police use ArcView to follow the relationships between illegal activities and community troubles like untidy lots, deserted trash, and shattered windows.

In India, GIS technology has been applied to analyze spatial information for the environmental change implications, such as the work done for Delhi Ridge (Golay et al., 2000; Jankowski, 1995; Joerin et al., 2001; Mohan, 2001; Sadoun, 2003; Saleh & Sadoun, 2004). Integrated GIS approaches can provide effective solutions for many of the emerging environmental change problems at local, regional, national, and global levels, and may become the preferred environment for ecological modeling. GIS tools have been used to monitor vegetation cover over periods of time to evaluate environmental conditions. The Delhi population has grown by about 25% per year since 1901. Therefore, it is vital to predict the rate of increase of CO_2 emission and ozone depletion in such a heavily populated city. GIS technology can help a lot in predicting possible scenarios and environmental plans and solutions.

CONCLUSION

GIS technology is an important human computer interactive (HCI) tool that has become vital to modern city and urban planning/engineering. Using GIS can produce cost-effective plans/designs that can be modified, tuned, or upgraded, as needed. GIS, as an HCI-based tool, is becoming an interdisciplinary information technology and information science field that has numerous applications to city and urban planning/engineering, ranging from land management/zoning to transportation planning/engineering. Engineering and computer science departments worldwide have started to offer undergraduate and graduate programs/tracks in this important, evolving discipline. These days, GIS is a must for modern city planning and engineering, since it can (a) streamline customer services in an interactive manner; (b) reduce land acquisition costs using accurate quantitative analysis; (c) analyze data and information in a speedy manner, which is important for quick and better decisions-making; (d) build consensus among decision-making teams and populations; (e) optimize urban services provided by local governments; (f) provide visual digital maps and illustrations in a much more flexible manner than traditional manual automated cartography approaches; and (g) reduce pollution and cost of running transportation means by finding the shortest paths to desired destinations.

REFERENCES

Batty, M., et al. (1994). Modeling inside GIS. *International Journal of GIS, 8*, 291-307.

Burrough, P., Evans, A., & McDonned, R. (1980). *Principles of GIS.* Oxford, UK: Oxford University Press.

Choi, J., & Usery, E. (2004). System integration of GIS and a rule-based expert system for urban mapping. *Photogrammetric Engineering & Remote Sensing Journal, 70*(2), 217-225.

Clapp, J. M., Rodriguez, M., & Thrall, G. (1997). How GIS can put urban economic analysis on the map? *Journal of Housing Economics, 6*, 368-386.

GIS@Purdue. (2003). Retrieved from http://pasture.ecn.purdue.edu/~caagis/tgis/overview.html

Golay, F., Gnerre, D., & Riedo, M. (2000). Towards flexible GIS user interfaces for creative engineering. *Proceedings of the International Workshop on Emerging Technologies for Geo-Based Applications*, Ascona, Switzerland.

Goodchild, M., Egenhofer, M., Feheas, R., & Kottman, C. (Eds.). (1999). *Interoperating geographic information systems*. Berlin, Germany: Springer-Verlag.

IFFD. (1998). Draft training manual for ArcView GIS, IFFD.

Jankowski, P. (1995). Integrating geographical systems and multiple criteria decision-making methods. *International Journal of Geographical Information Systems, 9*, 251-273.

Joerin, F., Theriault, M., & Musy, A. (2001). Using GIS and multicriteria analysis for land-use suitability and assessment. *International Journal of Geographical and Information Science, 15*(2), 153-174.

Kohsaka, H. (2001). Applications of GIS to urban planning and management: Problems facing Japanese local governments. *GeoJorunal, 52*, 271-280.

Korte, G.B. (2001). *The GIS book,* (5th ed.). Albany, NY: Onword Press.

McDonnell, R., & Kemp, K. (1995). *International GIS dictionary*. Cambridge, MA: Cambridge University Press.

Mohan, M. (2001). Geographic information system application for monitoring environmental change and restoration of ecological sustainable development over Delhi, ridge. *Proceedings of the International Conference on Spatial Information for Sustainable Development*, Nairobi, Kenya.

Ralston, B. (2004). *GIS and public data.* Albany, NY: Onword Press.

Sadoun, B. (2003). A simulation study for automated path planning for transportation systems. *Proceedings of the 2003 Summer Computer Simulation Conference, SCSC2003*, Montreal, Canada.

Saleh, B., & Sadoun, B. (2005). Design and implementation of a GIS system for planning. *International Journal on Digital Libraries.* Berlin, Germany: Springer-Verlag.

Shannon, R. (1976, December 6-8). Simulation modeling and methodology. *Proceedings of the 76th Bicentennial Conference on Winter Simualtion,* Gaithersburg, Maryland (pp. 9-15).

KEY TERMS

Cartography: The art, science, and engineering of mapmaking.

City and Regional Planning/Engineering: The field that deals with the methods, designs, issues, and models used to have successful plans and designs for cities, towns, and regions.

Coordinate System: A reference system used to gauge horizontal and vertical distances on a planimetric map. It usually is defined by a map projection, a spheroid of reference, a datum, one or more standard parallels, a central meridian, and possible shifts in the x- and y-directions to locate x y positions of point, line, and area features. For example, in ARC/INFO GIS system, a system with units and characteristics defined by a map projection. A common coordinate system is used to spatially register geographic data for the same area.

Data: A collection of attributes (numeric, alphanumeric, figures, pictures) about entities (things, events, activities). Spatial data represent tangible features (entities). Moreover, spatial data are usually an attribute (descriptor) of the spatial feature.

Database Management Systems (DBMS): Systems that store, organize, retrieve, and manipulate databases.

Digital Map: A data set stored in a computer in digital form. It is not static, and the flexibility of digital maps is vastly greater than paper maps. Inherent in this concept is the point that data on which the map is based is available to examine or question. Digital maps can be manipulated easily in GIS package environments.

GIS: A computer system that permits the user to examine and handle numerous layers of spatial data. The system is intended to solve problems and investigate relationships. The data symbolizes real-world entities, including spatial and quantitative attributes of these entities.

GPS: Global Positioning System. GPS is a satellite-based navigation system that is formed from a constellation of 24 satellites and their ground stations. GPS uses these satellites as reference points to calculate positions accurate to a matter of meters. Actually, with advanced forms of GPS, you can make measurements to better than a centimeter! These days, GPS is finding its way into cars, boats, planes, construction equipment, movie-making gear, farm machinery, and even laptop computers.

Information Systems: Information systems are the means to transform data into information. They are used in planning and managing resources.

Model: An abstraction of reality that is structured as a set of rules and procedures to derive new information that can be analyzed to aid in problem solving and planning. Analytical tools in a geographic information system (GIS) are used for building spatial models. Models can include a combination of logical expressions, mathematical procedures, and criteria that are applied for the purpose of simulating a process, predicting an outcome, or characterizing a phenomenon. Shannon defined a model as "the process of designing a computerized model of a system (or a process) and conducting experiments with this model for the purpose either of understanding the behavior of the system or of evaluating various strategies for the operation of the system" (pp. 9-15).

Raster Analysis: Raster analysis implements its spatial relationships mainly on the location of the cell. Raster operations performed on multiple input raster data sets usually output cell values that are the result of calculations on a cell-by-cell foundation. The value of the output for one cell is usually independent on the value or location of other input or output cells.

Spatial Data: Represents tangible or located features, such as a river, a 1,000 by 1,000 meter lot in a grid, a campus, a lake, a river, or a road.

Validation: Refers to ensuring that the assumptions used in developing the model are reasonable in that, if correctly implemented, the model would produce results close to that observed in real systems. Model validation consists of validating assumptions, input parameters and distributions, and output values and conclusions.

Vectro Analysis: In vector analysis, all operations are possible, because features in one theme are located by their position in explicit relation to existing features in other themes. The complexity of the vector data model makes for quite complex and hardware-intensive operations.

Verification: Verification is the process of finding out whether the model implements the assumptions considered. A verified computer program, in fact, can represent an invalid model.

A GIS-Based Interactive Database System for Planning Purposes

Nedal Al-Hanbali
Al-Balqa' Applied University, Jordan

Balqies Sadoun
Al-Balqa' Applied University, Jordan

INTRODUCTION

Decision making in planning should consider state-of-the-art techniques in order to minimize the risk and time involved. Proper planning in developing countries is crucial for their economical recovery and prosperity. Proper database systems, such as the ones based on GIS, are a must for developing countries so that they can catch up and build effective and interactive systems in order to modernize their infrastructures and to help improve the standard of living of their citizens. The huge and fast advancement in computing and information technology make it easy for the developing countries to build their database infrastructures. GIS-technology is one of the best and fastest tools to build such systems, manage resources, encourage businesses, and help to make efficient and cost-effective decisions.

For the purpose of a better informed decision making in planning the improvement of the Bank of Jordan in the city of Amman, Jordan, we had to build a database system and a digital map for the city of Amman, the Bank of Jordan, its branches in Amman, and all other banks and their branches in Amman. We used the popular Geomedia software to allow an interactive time-saving data management; to offer the ability to perform different analysis, including statistical ones; and to provide graphical geospatial results on maps. By using Geomedia software, we built many layers needed for the planning processes and mainly for the region of Amman due to the lack of available digital data in the area. Some layers concern the project and relate to the bank, such as the geographic distribution of the Bank of Jordan branches and its ATMs; and others for the comparison, such as the geographic distribution of all other banks, their branches, and ATMs in Amman. This is to allow the decision makers to compare with all competitive banks in Amman. Besides the geographic location of all existing banks, important attribute data are provided for the Bank of Jordan in specific and all the other banks in general (Batty et al., 1994a, 1994b; Burrough et al., 1980; Doucette et al., 2000; Elmasri & Navathe, 2004; Goodchild, 2003; Longley et al., 1999a, 1999b).

BACKGROUND

The Bank of Jordan started planning for new ATM sites in Amman using the traditional method and, at the same time, the GIS pilot project to support building a quick goespatial information infrastructure that can assess in the decision-making process according to provided criteria, which can be integrated into the GIS analysis process. The real challenge here is to build a digital database to introduce a complete digital map for Amman to help in the analysis process.

Many layers for different purposes are created, including the country boundaries, governorates boundaries, city districts and subdistricts, main and submain streets, blocks and city blocks, government organizations, commercial areas and trading centers with cinemas and theaters, commercial companies, insurance companies, restaurants, hotels, hospitals, gas stations, Jordan Bank branches layer, and the branches of all other banks with their ATMs in the city of Amman.

The design of these layers is based on a specific GIS data model suited for this application. It is based on integrating SPOT image of Amman with many scanned paper maps that provide the needed information. Moreover, integration of Geographical Positioning System (GPS) data into our GIS system is implemented to create many layers required for the analysis.

Once the geospatial database for the city and the banks is ready, the rest of the work is easy and flexible, and the planners can integrate their functions and conditions in no time and will be able to provide better decision making. Moreover, part of the data could be made public and accessible through the Web to help not only in locating the sites of ATMs but also in doing the banking interactions, which is a sort of human computer interaction mechanism as it is done in the developed countries (Batty et al., 1994a; Burrough et al., 1980; Goodchild, 2003; Longley et al., 1999a, 1999b).

METHODOLOGY AND MODELING

Using scanning, digitizing, and registration techniques as well as collected and available attributes of data, many layers were created for the database. Figure 1 illustrates the general procedure for creating the needed layers.

Many basic geospatial data layers were built (by feature) for the project, as follows:

1. Line features such as (street layers) highways, subways, roadways, and railways.
2. Polygon features such as urban, circles, farms, gardens, Jordan Governorates, Amman districts, Amman subdistricts, and so forth.
3. Point features such as banks and/or ATMs, restaurants, hotels, large stores, hospitals, sport clubs, cinemas (movie theaters), cultural and social centers, gas stations, and police stations.

Figure 1. Overview of the GIS project procedure

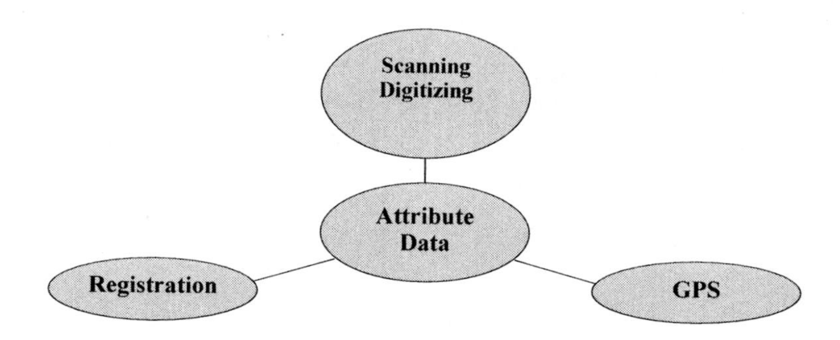

Figure 2 illustrates a descriptive diagram of the GIS data model creation, measuring, development, and implementation stages.

IMPLEMENTATION

SPOT image is used as the registration reference frame for all scanned hardcopy maps, as indicated in the schematic GIS data model in Figure 2. The reference system is the Jordanian Transverse Mercator (JTM). Figure 3 illustrates the reference points in the registration process using the Geomedia software.

Digitization is followed to create the polygon and line layers. Figure 4 shows a digital scanned map image while digitizing, and the drawn redline is the digitized line on the map.

Figure 5 (parts a, b, and c) shows examples of the resulting layers (maps) using line features such as highways, subways, and roads digitized layers, respectively.

Figure 6 (parts a, b, c, and d) shows examples of the resulting layers (maps) using polygon features such as district, subdistrict, urban, and governorate layers, consecutively.

Figure 7 shows some of the created layers for banks, ATMs, hospitals, and gas station locations. Finally, imposing all of the previous layers, a final resulting map was made available to help in the decision-making process. Any kind of information could be provided from any layer to help in the planning, improvement, or finding of the location of an ATM, a new bank branch, and so forth.

Figure 2. The GIS model chart

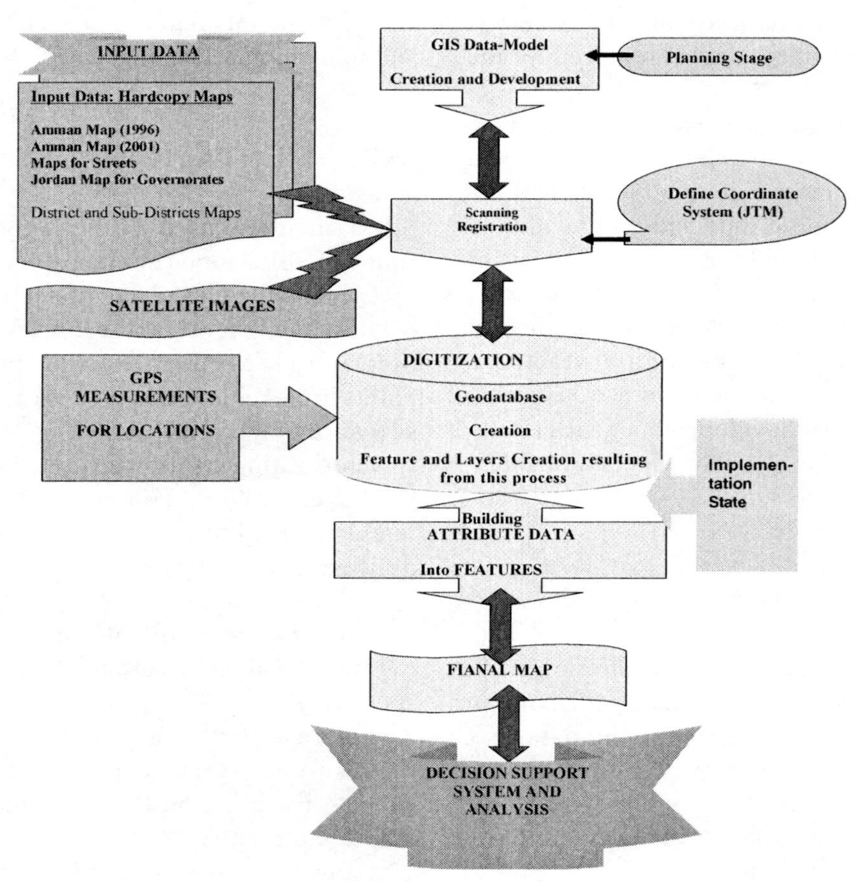

Figure 3. Image registration

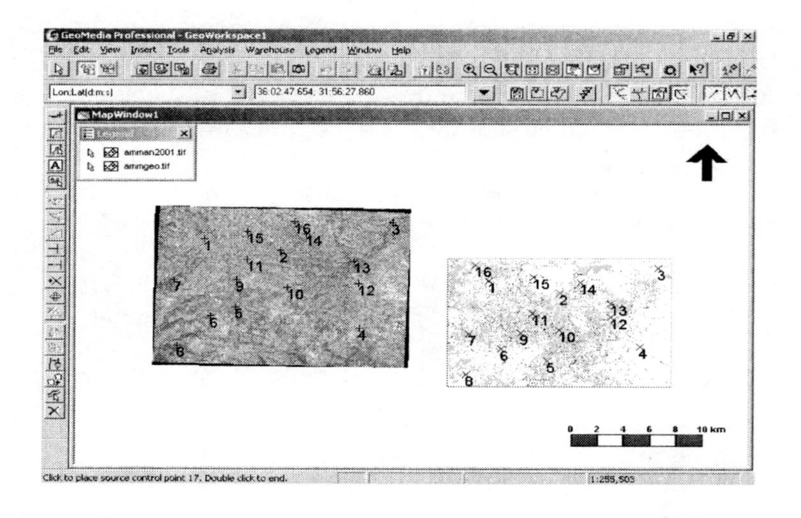

Figure 4. Digitizing process; line is the digitized line on the tourist map

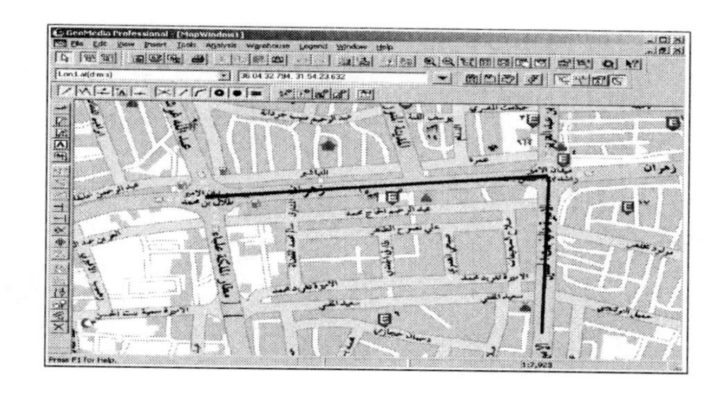

Figure 5(a). Highways layer

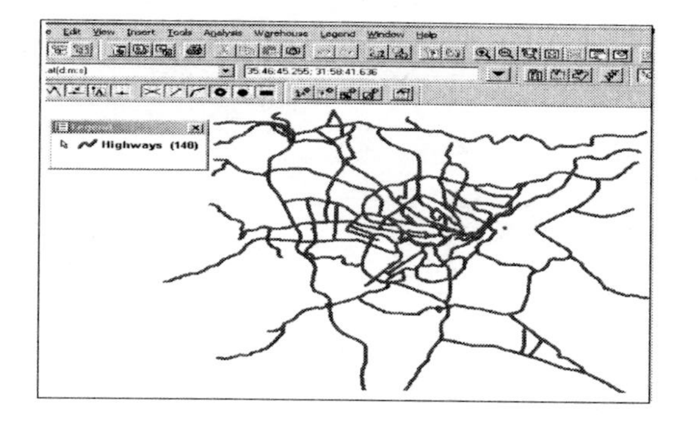

Figure 5(b). Subways layer

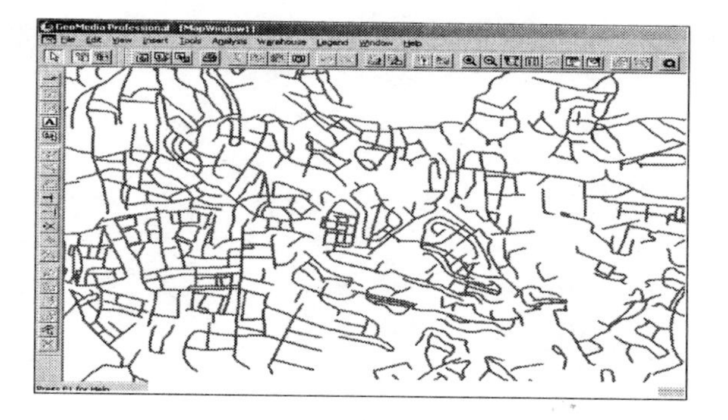

Figure 5(c). Roadways layer

Figure 6(a). District layer

Figure 6(b). Subdistrict layer

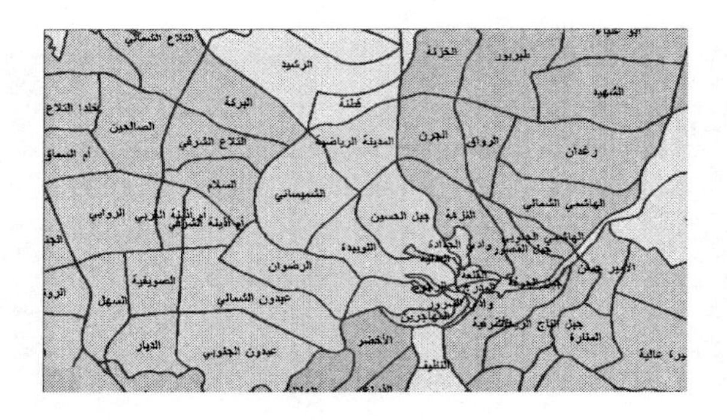

Figure 6(c). Urban layer

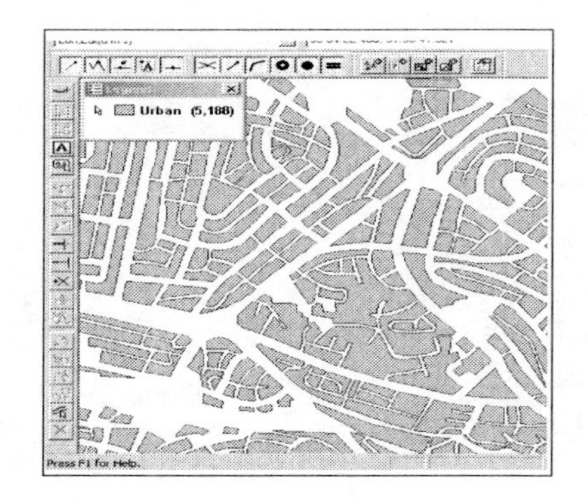

Figure 6(d). Governorates layer

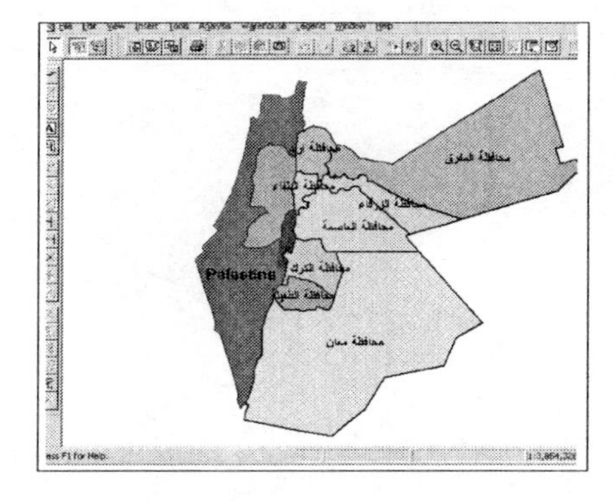

Figure 7(a). Bank of Jordan locations layer

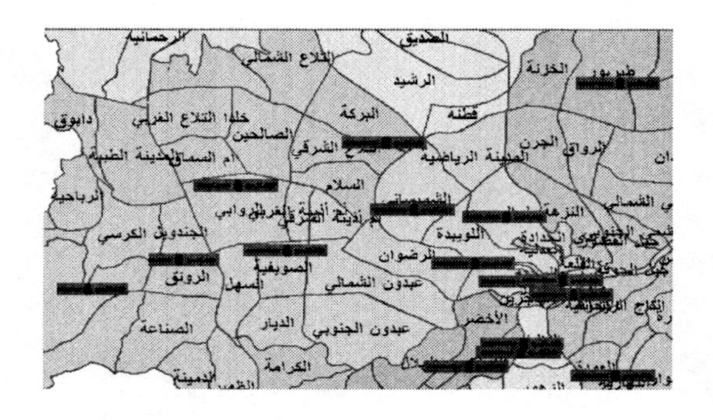

Figure 7(b). ATM locations for all banks layer

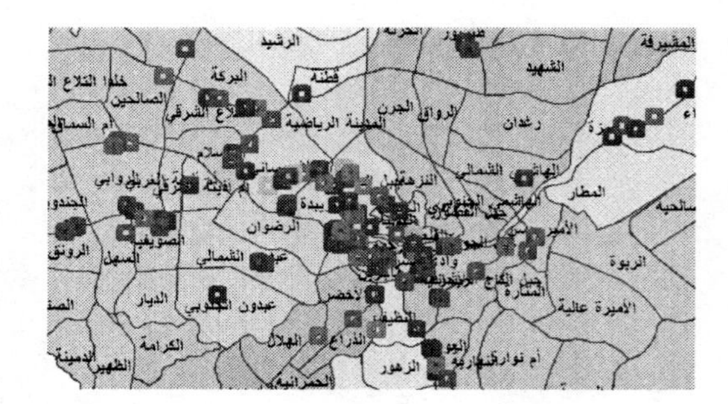

Figure 7(c). All banks locations layer

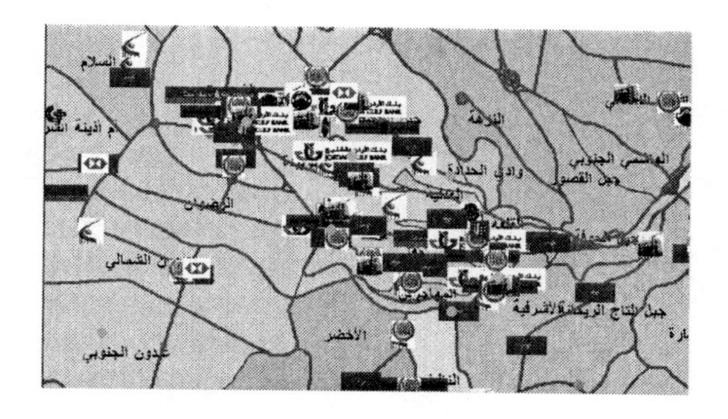

Figure 7(e). Gas station locations layer

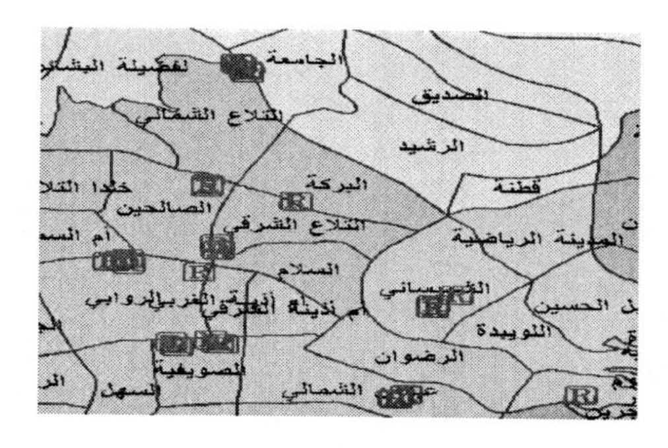

Figure 7(d). Hospital locations layer

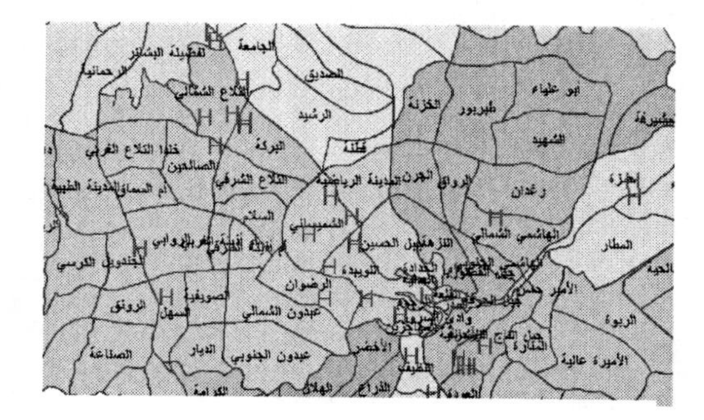

RESULTS AND ANALYSIS

By preparing all required geospatial data layers needed for the project, we can start the analysis, depending on the digital database we acquired. In building our GIS, we used the Geomedia software package to manage our database and to allow the needed analysis. This package is a good example of efficient human interactive tools. All kinds of analysis could be conducted using the software ability, and the results will be geographically posted on maps related to the geographic site. Many analysis techniques could be used, such as thematic maps techniques (a map of all banks and business centers in Amman), classification techniques (classify by color at each subdistrict), querying techniques about the shortest path or the largest population or the smallest area (the distance between two branches of the bank), and buffering (circle buffer around the existing banks to show the served area). Another important analysis that is possible using GIS is the statistical analysis.

A case study as an application on the decision-making process and planning with GIS is to locate sites for ATMs in Amman, as we mentioned earlier. In order to integrate our location constraints in the decision-making process using the GIS data that we created, we first had to define our choice of criterion and then use it for our locations spotting. We used the bank selection criterion for an ATM location to satisfy the following:

1. There should be an ATM at each branch of the Bank of Jordan.
2. There should be an ATM at each subdistrict with a population of about 10,000 people or more (nearby trading centers, gas stations, supermarkets, and so forth).
3. There should be an ATM at busy shopping areas full of stores, restaurants, hotels, or malls.
4. There should be an ATM at popular gas stations that are located on highways or busy streets.
5. Finally, there should be ATMs at a 10-km distance from each other in populated areas (the maximum area coverage for an ATM is within a 5-km radius).

To implement the needed criteria in GIS domain, new layers have to be extracted in order to help in choosing the new ATM locations as follows:

- A query about each subdistrict that has a population of about 10,000 or more has been conducted. For example, Al-Abdaly district in Amman has been chosen. It consists of four subdistricts with a total population of about

Figure 8(a). Abdali area with Bank of Jordan locations

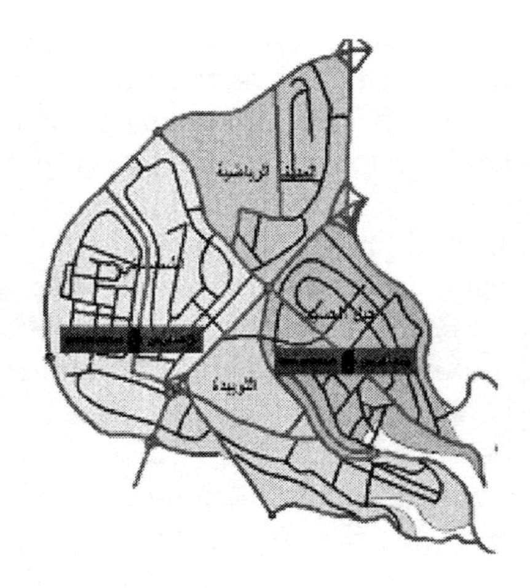

Figure 8 (b). All Point Features of Interest Like Trade Stores, Restaurants, Hotels, Malls, Gas Station, etc.

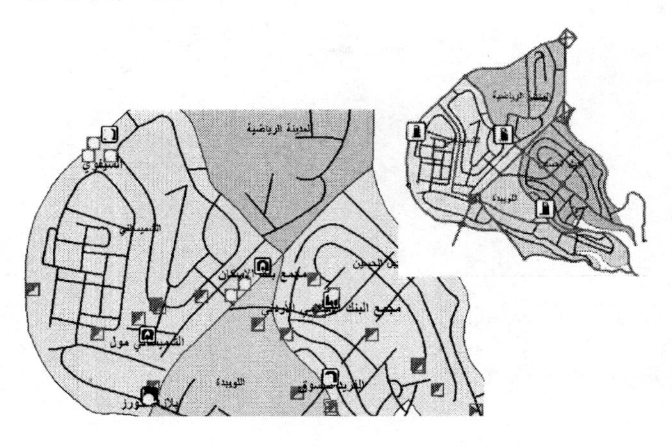

92,080 inhabitants; namely, Al-Shemesany, Al-Madeena, Al-Reyadeyah, Al-Lweebdeh, and Jabal Al-Hussein (see Figure 8(a)).

- Make another query to have a map for all busy commercial areas full of stores, restaurants, hotels, and important gas stations that are located on a highway or busy crowded streets.· Make a query.· Have a map that shows the distribution of all banks in the study area (Figure 8(c)).
- Create a buffer zone layer around the existing banks, branches, and ATMs in order to have a map to show clearly the served areas (inside the circles). Our constraints bank service coverage is within a circular area of a 5-km radius (see Figure 8(d)).

Finally, by combining all of these conditions (map overlay), we can find the best ATMs at new locations to satisfy the bank constraints. Figure 9 shows the suggested ATM locations in Amman.

The results shown in Figure 9 for the new locations match those that resulted from the study of the bank-planning department. This comparison is made to demonstrate to the directors of the bank and the public the capability, effectiveness, and accuracy of the GIS technology as a human computer interactive tool for engineering and planning purposes. The

Figure 8(c). Bank distribution layer in Abdaly area

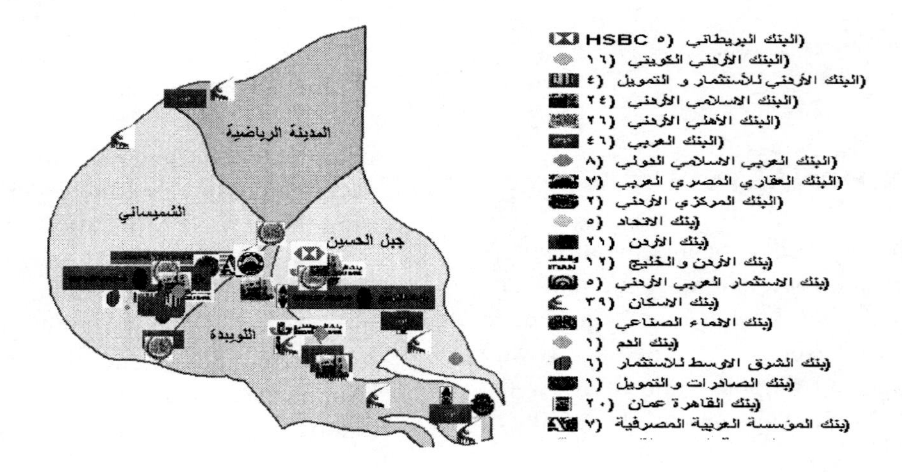

Figure 8(d). Circular buffer zone layer to show the areas covered by banking service

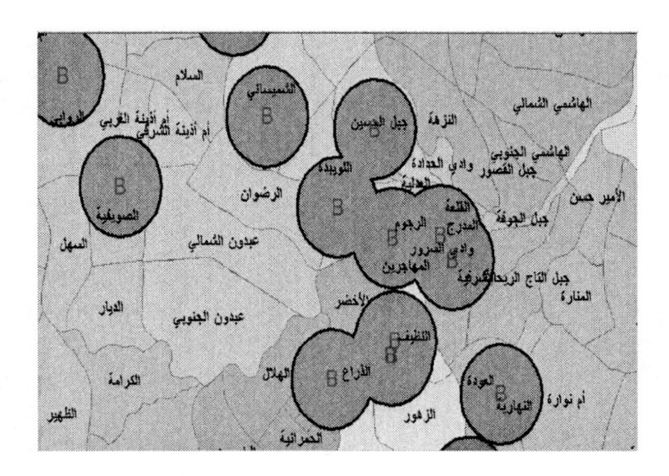

geospatial interactive data are now available for further analysis and planning of projects in all areas such as real estate and loan grants applications.

FUTURE TRENDS

GIS technology is an effective, accurate, and economical HCI tool that has received wide acceptance by numerous organizations and institutions and for all kinds of applications. Many financial institutions, banks, towns and cities worldwide have started or are in the process of using it in planning and engineering services. Today, there are many GIS software packages of various features and options that are available commercially. Using GIS software has become friendlier with the addition of new features in the software such as the animation, verification and validation features, and graphical user interfaces (GUIs). Verification helps to verify models in order to make sure that the model is a real representation of real systems under analysis. Validation is used to make sure that the assumptions, inputs, distribution, results, outputs, and conclusions are accurate. GUI is designed and used to provide an effective and user-friendly computer interactive environment. Animation helps provide a visual way to verify models and to make the GIS package more salable and enjoyable for potential users and buyers.

GIS technology is becoming a trendy technology for almost all applications that range from planning irrigation systems to planning ATM systems for banks and financial institutions. Recently, law enforcement agencies around the world have started using GIS to display, analyze, and battle crime. Computer-generated maps are replacing the traditional maps that used to cover the walls of law enforcement agencies. Such a tool has given police officers the power to classify and rearrange reams of data in an attempt to find patterns. Some police departments in developed countries employ GIS computer mapping to persuade residents to get

Figure 9. Proposed ATM locations in Amman City

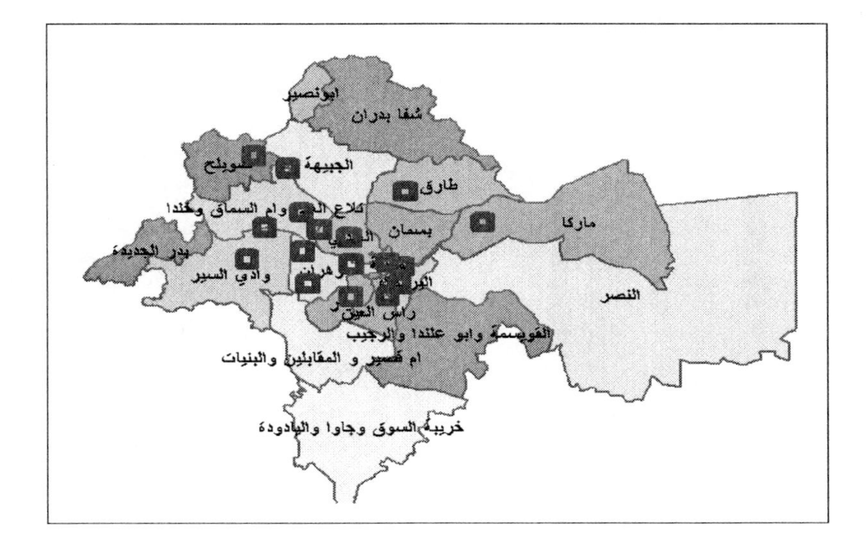

involved in community development. In India, GIS technology has been applied to analyze spatial information for the environmental change implications such as the work done for Delhi Ridge (Mohan, 2001; Joerin et al., 2001; Sadoun, 2003; Saleh & Sadoun, 2004; Golay et al., 2000; Jankowski, 1995).

GIS technology has been used to provide effective solutions for many of the emerging environmental problems at local, regional, national, and global levels. For example, GIS has been used to monitor vegetation cover over periods of time in order to evaluate environmental conditions. It can be used to predicate the increase in CO_2 emission and ozone depletion in heavily populated cities such as Tokyo, New York, Mexico City, Chicago, Los Angeles, and so forth. GIS technology can help provide all possible solutions and outcomes for different scenarios and settings.

Computer interactive tools such as GIS technology used to support decision-making activities can have different purposes: information management and retrieval, multi-criteria analysis, visualization, and simulation. The quality of human computer interface can be measured not only by its accuracy but also by its ease of use. Therefore, most state-of-the art GIS software tools seek to provide user-friendly interfaces using features such as the GUI and animation options.

Many other software packages have started to add GIS-like functionality; spreadsheets and their improved graphics capabilities in handling 2-D maps and 3-D visualizations are examples of such a trend. Software is being divided up on the desktop into basic modules that can be integrated in diverse ways, while other software is becoming increasingly generic in that manner. GIS is changing as more functions are embodied in hardware.

CONCLUSION, REMARKS, AND RECOMMENDATIONS

The tremendous advancement in information and computer technologies has changed the way of conducting all aspects of our daily life. Due to the amazing progress in these technologies, planning for developing a country has become more accurate, economical, effective, and quantitative. The availability of digital databases has helped in better decision making in urban planning. Such databases can help to convince the developed world to have business and trade with developing countries. Building a digital database or a GIS system can help in all domains. GIS has become a vital tool to solve administrative, security, health, commercial, and trade matters.

GIS technology offers an interactive and powerful tool to help in decision making and planning. It offers answers on maps that are easy to visualize and understand. Moreover, the user/planner/engineer can build his or her database once and then later can easily alter it as needed in almost no time. It can be used interactively and reduce the paper maps and allow customization of all specific encountered problems. Moreover, it allows great storage and fast access to data. Finally, the advancement of the World Wide Web has allowed public access to all needed information to help in better serving the world. The world is looking for standardization of database systems and centralization of the source to make it easy to find and use.

It is worth mentioning that development needs quantitative planning and informed decision making using modern technologies such as a digital database and a GIS, as such technologies provide quick, accurate, interactive, and convincing plans and recommendations.

In this work, we present recommendations based on the analysis of collected digital data using GIS techniques for building the needed maps for city planning. A case study is considered here, which deals with finding the proper locations of ATMs for the Bank of Jordan located in the capital city of Jordan, Amman. The registration and digitization processes, the GPS measurements integration, and the implementation of statistical data such as population and other related information are presented. Our work also presents the criteria used in the spatial analysis for modeling the process of the ATM sites' selection. Finally, the article provides a map of the new proposed ATM sites selected for the case study.

REFERENCES

Batty, M., et al. (1994a). Modeling inside GIS. Part 1: Model structures, exploratory spatial data analysis and aggregation. *International Journal of GIS, 8*, 291-307.

Batty, M., et al. (1994b). Modeling inside GIS. Part 2: Selecting and calibrating urban models using Arc/Info. *International Journal of GIS, 8*, 451-470.

Burrough, P., Evans, A., & Mc. Donned, A. (1980). *Principles of GIS.* Oxford: Oxford University Press.

Doucette, P., et al. (2000). Exploring the capability of some GIS surface interpolators for DEM gap fill. *PE & RS, 66*, 881-888.

Elmasri, R., & Navathe, S. (2004). *Fundamentals of database systems* (4th ed.). Boston: Addison-Wesley.

Golay, F., Gnerre, D., & Riedo, M. (2000). Towards flexible GIS user interfaces for creative engineering. *Proceedings of the International Workshop on Emerging Technologies for Geo-Based Applications*, Ascona.

Goodchild, M. F. (2003). The nature and value of geographic information. In M. Duckham, M. F. Goodchild, & M. F. Worboys (Eds.), *Foundations of geographic information science* (pp. 19-32). New York: Taylor and Francis.

Jankowski, P. (1995). Integrating geographical systems and multiple criteria decision-making methods. *International Journal of Geographical Information Systems, 9*, 251-273.

Longley, P. A., Goodchild, M., Maguire, D., & Rhind, D. (1999a). Introduction. In P. Longley et al. (Eds.), *Geographical information systems: Principles, techniques, applications and management* (pp. 1-20). New York: Wiley.

Longley, P. A., Goodchild, M., Maguire, D., & Rhind, D. (Eds.). (1999b). *Geographical information systems: Principles, techniques, applications and management.* New York: Wiley.

Mohan, M. (2001). Geographic information system application for monitoring environmental change and restoration of ecological sustainable development over Delhi, Ridge. *Proceedings of the International Conference on Spatial Information for Sustainable Development*, Nairobi, Kenya.

Sadoun, B. (2003). A simulation study for automated path planning for transportation systems. *Proceedings of the 2003 Summer Computer Simulation Conference*, Montreal, Canada.

Saleh, B., & Sadoun, B (2004). Design and implementation of a GIS system for planning. *International Journal of Digital Libraries*. Berlin, Germany: Springer-Verlag.

Shannon, R. (1976, December 6-8). Simulation modeling and methodology. *Proceedings of the 76th Bicentennial Conference on Winter Simulation,* Gaithersburg, MA (pp. 9-15).

KEY TERMS

Animation: A graphical representation of a simulation process. The major popularity of animation is its ability to communicate the essence of the model to managers and other key project personnel, greatly increasing the model's credibility. It is also used as a debugging and training tool.

ATM Systems: Automatic Teller Machines are installed by banks in different locations of the city or town in order to enable customers to access their bank accounts and draw cash from them.

City and Regional Planning/Engineering: The field that deals with the methods, designs, issues, and models used to have successful plans and designs for cities, towns, and regions.

Coordinate System: A reference system used to gauge horizontal and vertical distances on a planimetric map. It is usually defined by a map projection, a spheroid of reference, a datum, one or more standard parallels, a central meridian, and possible shifts in the x- and y-directions to locate x, y positions of point, line, and area features (e.g., in

ARC/INFO GIS system, a system with units and characteristics defined by a map projection). A common coordinate system is used to spatially register geographic data for the same area.

Data: A collection of attributes (numeric, alphanumeric, figures, pictures) about entities (things, events, activities). Spatial data represent tangible features (entities). Moreover, spatial data are usually an attribute (descriptor) of the spatial feature.

Database Management Systems (DBMS): Systems that store, organize, retrieve, and manipulate databases.

Digital Map: A digital map is a data set stored in a computer in digital form. It is not static, and the flexibility of digital maps is vastly greater than paper maps. Inherent in this concept is the point that data on which the map is based is available to examine or question. Digital maps can be manipulated easily in GIS package environments.

Digital Satellite Images: Digital images sent by satellite systems that are usually launched in special orbits such as the geostationary orbit. The latter type of satellite systems rotate at about 35,000 Km from the surface of the earth and is able to cover the same area of the earth 24 hours a day.

Digitation: The process of converting analog data to digital data where binary systems are usually used. Programmers find dealing with digital data is much easier than dealing with analog data.

GIS: A computer system that permits the user to examine and handle numerous layers of spatial data. The system is intended to solve problems and investigate relationships. The data symbolize real-world entities, including spatial and quantitative attributes of these entities.

GPS: Global Positioning System is a satellite-based navigation system that is formed from a constellation of 24 satellites and their ground stations. GPS uses these satellites as reference points to calculate positions accurate to a matter of meters. Actually, with advanced forms of GPS, you can make measurements to better than a centimeter! These days, GPS is finding its way into cars, boats, planes, construction equipment, movie-making gear, farm machinery, and even laptop computers.

Model: An abstraction of reality that is structured as a set of rules and procedures to derive new information that can be analyzed to aid in problem solving and planning. Analytical tools in a geographic information system (GIS) are used for building spatial models. Models can include a combination of logical expressions, mathematical procedures, and criteria, which are applied for the purpose of simulating a process, predicting an outcome, or characterizing a phenomenon. Shannon defined a model as "the process of designing a computerized model of a system (or a process) and conducting experiments with this model for the purpose either of understanding the behavior of the system or of evaluating various strategies for the operation of the system" (pp. 9-15).

Spatial Data: Spatial data represent tangible or located features such as a river, a 1,000 by 1,000 meter lot in a grid, a campus, a lake, a river, a road, and so forth.

Validation: Validation refers to ensuring that the assumptions used in developing the model are reasonable in that, if correctly implemented, the model would produce results close to that observed in real systems. Model validation consists of validating assumptions, input parameters and distributions, and output values and conclusions.

Verification: Verification is the process of finding out whether the model implements the assumptions considered. A verified computer program, in fact, can represent and invalid model.

Globalization, Culture, and Usability

Kerstin Röse
University of Kaiserslautern, Germany

INTRODUCTION

Globalization is a trend in the new industrial era. Global economy has seen a huge amount of product and technology exchanges all over the world. With the increase of export and resulting from that, with the increase of world-wide technical product exchange, a product will now be used by several international user groups. As a result, there is an increasing number of user groups with different cultural features and different cultural-based user philosophies. All these user groups and philosophies have to be taken into account by a product developer of human machine systems for a global market.

User requirements of product design have become much more valued than before because cultural background is an important influencing variable that represents abilities and qualities of a user (del Galdo & Nielsen, 1996). However, there is a gap in developers' knowledge when handling product design according to the culture-dependent user requirements of a foreign market (Röse & Zühlke, 2001), so the "*user-oriented*" product design has not always been fulfilled on the international market.

BACKGROUND

Usability is the key word to describe the design and engineering of usable products. The term describes also a systematic process of user-oriented design to engineer "easy-to-use" products (see ISO 13407, 1999). One key element for success in this field is to know the target groups and their requirements. Hence, in time of globalization, usability experts have to integrate intercultural aspects into their approaches (see Röse, 2002). Therefore, usability experts have to know their target group and requirements in this target culture.

For a foreign market, localized design (*local design* is for a specific culture and global design is for many cultures) is needed to address the target culture. "There is no denying that culture influences human-product interaction" (Hoft, 1996). This has caused a change in the design situation in a way that engineers nowadays have to face up to other user groups with different cultures, which they are not familiar with. It is now unrealistic for them to rely only on their intuition and personal experience gained from their own culture to cope with the localized design. Although, it is clear that cultural requirements should be well addressed in localized designs.

INTERCULTURAL HUMAN MACHINE SYSTEMS

Day (1996) pointed out that we have to recognize that "any technology should be assessed to determine its appropriateness for a given culture." This implies that, in time of globalization as far as user-oriented design is concerned, it must also be culture-oriented.

A good understanding of culture could provide designers with clues to answer these questions. A lot of cultural anthropologists and consultants have conducted many cultural studies and obtained plenty of cultural data, (e.g., Bourges-Waldegg & Scrivener, 1998; del Galdo, 1990; Honold, 2000; Marcus, 1996; Prabhu & Harel, 1999; Röse, 2002).

The Human Machine System [HMS] engineering process is influenced by the cultural context of the current developer and the former user. Developer will construct the future product for expected users. With the task and requirement analysis, he is be able to integrate his future user. The matching between developer and user model will influence the product and his construction. For the development of intercultural HMS, it means the following situation: developer from culture A has to construct/design a product for the user in culture B. Therefore, it is important to mention this fact and to analyze the culture-specific user-requirements. Honold (2000) describes intercultural influence on the user's interface engineering process in these main aspects: user-requirements, user interface design and user interface evaluation. In case of localization, it is necessary to know the specific user's needs of the cultural-oriented product/ system. *It is necessary to analyze the culture-specific user-requirements. Such an analysis is the basis for a culture-specific user interface design (see also ISO 13407, 1999). To get valid data from the evaluation of current systems or prototypes with a culture-specific user interface, an intercultural evaluation is necessary. Culture influences the whole user interface engineering process as well as the HMS engineering process. Through this influence, a management of intercultural usability engineering is necessary. This is a challenge for the future.*

Modern user-centered approaches include cultural diversity as one key aspect for user-friendly products (Röse, 2004). Liang (2003) has observed the multiple aspects of cultural diversity, the micro-view on the user and the macro-view on the engineering process.

Technology has changed the ways people doing their activities and accelerated the trend of globalization. The consequences are the increase of cultural diversity embedded in the interaction and communication and the pervasiveness of interactive systems applied in almost every human activities ... Therefore, when we look at cultural issues in interactive systems, we should consider not only human activities supported by the systems but also the activities or processes of the design, the implementation and the use. (Liang, 2003)

According to Röse (2002), the usage of intercultural user interfaces and intercultural human machine systems describes the internationalization and localization of products, and excludes global products. *Intercultural human machine systems* are defined as systems, where human and machine have the same target and the needed functions and information to reach the target are offered and displayed with ergonomic considerations based on ergonomic rules and guidelines. Beyond this, the intercultural human machine system takes into account the cultural diversity of users—according to culture-specific user requirements—and specific technical features as well as frame or context requirements based on cultural specifics. Hence, the intercultural human machine systems offering needed functions and information to realize a user-oriented human machine system, which is optimized for the target user and the used application in his/her culture determine usage context (Röse, 2004).

It has to be mentioned that there are cultural-based differences between user *and* developer. Therefore, the integration of cultural specifics is a natural tribute to the diversity of user and developer cultures in time of globalization. The mental model of a developer from Germany is mostly very different from the mental model of a developer in China. Differences between developers and users stem from differences of their implementation in a cultural context.

FUTURE TRENDS

New research or application fields offer new challenges. Intercultural human machine system design is a new field with huge challenges. Smith and Yetim (2004) state the following:

Effective strategies that address cultural issues in both the product and the process of information systems development now often are critical to system success. In relation to the product of development, cultural differences in signs, meanings, actions, conventions, norms or values, etc., raise new research issues ranging from technical usability to methodological and ethical issues of culture in information systems. In relation to the process of development, cultural differences affect the manner, in which users are able to participate in design and to act as subjects in evaluation studies. (Smith & Yetim, 2004)

The field of intercultural human machine systems, which is mainly mentioned in the research field of big global companies, is, in practice, a typical topic for global players. But the developers in different application areas have knowledge gaps in this field. An analysis of developers' requirements in production automation area has shown the following: more than half like to adapt an existing user system to foreign countries (internationalization) or to a specific user culture (localization). Forty percent think the globalization problem is solved, and more than 52% would be interested to support intercultural design, that is, by developing guidelines (Röse & Zühlke, 2001). Therefore, developing intercultural human machine systems is still a challenge for the future. Resolving such a challenge would inform developers and help them focus more on cultural aspects of design and engineering of products and systems (Evers, Röse, Honold, Coronado, & Day, 2003).

CONCLUSION

This article has shown the importance of globalization and culture to usability. An introduction to cultural diversity in the context of human machine systems and a definition of intercultural human machine systems were given. Usability experts need information about the target culture and the culture-specifics of the users. This information is a basis for culture- and user-oriented human machine system design.

Ergonomics human machine systems that accept and reflect cultural diversity of the targeted users are strongly needed.

Giving attention to users' and developers' needs in the context of globalization, usage, and engineering enable the creation of products and systems with high usability.

REFERENCES

Bourges-Waldegg, P., & Scrivener, S. A. R. (1998). Meaning, the central issue in cross-cultural HCI design. *Interacting with Computers: The Interdisciplinary Journal of Human-Computer-Interaction*, 9(February), 287-309.

Day, D. (1996). Cultural bases of interface acceptance: Foundations. In M. A. Sasse, R. J. Cunningham, & R. L. Winder (Eds.), *People and Computers XI: Proceeding of HCI '96* (pp. 35-47). London: Springer Verlag.

del Galdo, E. M. (1990). Internationalization and translation: Some guidelines for the design of human-computer interfaces. In J. Nielsen (Ed.), *Designing user interfaces for international use* (pp. 1-10). Amsterdam: Elsevier Science Publishers.

del Galdo, E. M., & Nielsen, J. (1996). *International user interface*. New York: John Wiley & Sons.

Evers, V., Röse, K., Honold, P., Coronado, J., & Day, D. (Eds.) (2003, July). Designing for global markets 5. *Workshop Proceedings of the Fifth International Workshop on Internationalisation of Products and Systems, IWIPS 2003*, Berlin, Germany (pp. 17-19).

Hoft, N. (1996). Developing a cultural model. In E. M. del Galdo & J. Nielsen (Eds), *International user interfaces* (pp. 41-73). New York: Wiley.

Honold, P. (2000, July 13-15). Intercultural usability engineering: Barriers and challenges from a German point of view. In D. Day, E. M. del Galdo, & G. V. Prabhu (Eds.), *Designing for global markets 2. Second International Workshop on Internationalisation of Products and Systems (IWIPS 2000)*, Baltimore (pp. 137-148). Backhouse Press.

ISO 13407 (1999). Benutzer-orientierte Gestaltung interaktiver Systeme (user-oriented design of interactive systems).

Liang, S. -F. M. (2003, August 24-29). Cross-cultural issues in interactive systems. In *Ergonomics in the Digital Age. Proceedings of the International Ergonomics Association and the 7th Join Conference of Ergonomics Society of Korea/ Japan Ergonomics Society*, Seoul, Korea.

Marcus, A. (1996). Icon and symbol design issues for graphical user interfaces. In E. M. del Galdo & J. Nielsen (Eds.), *International user interfaces* (pp. 257-270). New York: John Wiley & Sons.

Prabhu, G., & Harel, D. (1999, August 22-26). GUI design preference validation for Japan and China—A case for KANSEI engineering? In H. -J. Bullinger

& J. Ziegler (Eds.), *Human-Computer Interaction: Ergonomics and User Interfaces, Vol. 1, Proceedings 8th International Conference on Human-Computer Interaction (HCI International '99)*, Munich, Germany (pp. 521-525).

Röse, K. (2002). *Methodik zur Gestaltung Interkultureller Mensch-Maschine-Systeme in der Produktionstechnik* (Method for the design of intercultural human machine systems in the area of production automation). Dissertationsschrift zur Erlangung des Akademischen Grades, Doktor-Ingenieur' (Dr.-Ing.) im Fachbereich Maschinenbau und Verfahrenstechnik der Universität Kaiserslautern, Fortschritt-Bericht pak, Nr. 5, Universität Kaiserslautern.

Röse, K. (2004). The development of culture-oriented human-machine systems: Specification, analysis and integration of relevant intercultural variables. In M. Kaplan (Ed.), *Cultural ergonomics*. Published in the Elsevier Series, *Advances in Human Performance and Cognitive Engineering Research, Vol. 2* (pp. 61-103). Lawrence Erlbaum Associates, Oxford: Elsevier, UK.

Röse, K., & Zühlke, D. (2001, September 18-20). Culture-oriented design: Developers' knowledge gaps in this area. In G. Johannsen (Ed), *Preprints of 8th IFAC/IFIP/IFORS/IEA Symposium on Analysis, Design, and Evaluation of Human-Machine Systems*, Kassel, Germany (pp. 11-16).

Smith, A., & Yetim, F. (2004). Global human-computer systems: Cultural determinants of usability. *Interacting with Computers, The Interdisciplinary Journal of Human-Computer-Interaction, 16*(January), 1-5. Special Issue, *Global human-computer systems: Cultural determinants of usability*.

KEY TERMS

Culture: Common meaning and values of a group. Members of such a group share and use the accorded signs and roles as a basis for communication, behaviour, and technology usage. Mostly, a country is used as a compromise to refer or define rules and values, and is used often as a synonym for user's culture.

Culture-Oriented Design: Specific kind of user-oriented design, which focuses on the user as a central element of development, and also takes into account the cultural diversity of different target user groups.

Globalization: As one of three degrees of international products, it means a "look like" culture-less international standard for use in all markets (in accordance with Day, 1996).

Human Machine System (HMS): Based on the acceptance of an interaction between human and machine, it is a summary of all elements of the hard-, soft- and useware. The term includes the micro (UI) and macro (organization) aspects of a human machine system.

Intercultural Human Machine System: The intercultural human machine system takes into account the cultural diversity of human (different user requirements) and machine (variation of usage situations), in addition to a standard HMS.

Internationalization: As one of three degrees of international products, it means a base structure with the intent of later customizing and with structural and technical possibilities for it (in accordance with Day, 1996).

Localization: As one of three degrees of international products, it means a developing of culture specific packages for a particular (local) market (in accordance with Day, 1996).

User-Oriented Design: Development approach with a focus on users' requirements and users' needs as basis for a system or product development.

Grounding CSCW in Social Psychology

Umer Farooq
IBM T.J. Watson Research Center, USA

Peter G. Fairweather
IBM T.J. Watson Research Center, USA

Mark K. Singley
IBM T.J. Watson Research Center, USA

INTRODUCTION

Computer-Supported Cooperative Work (CSCW) is largely an applied discipline, technologically supporting multiple individuals, their group processes, their dynamics, and so on. CSCW is a research endeavor that studies the use of, designs, and evaluates computer technologies to support groups, organizations, communities, and societies. It is interdisciplinary, marshalling research from different disciplines such as anthropology, sociology, organizational psychology, cognitive psychology, social psychology, and information and computer sciences. Some examples of CSCW systems are group decision support systems (e.g., Nunamaker, Dennis, Valacich, Vogel, & George, 1991), group authoring systems (e.g., Guzdial, Rick, & Kerimbaev, 2000), and computer-mediated communication systems (e.g., Sproull & Kiesler, 1991).

Behavioral and social sciences provide a rich body of research and theory about principles of human behavior. However, researchers and developers have rarely taken advantage of this trove of empirical phenomena and theory (Kraut, 2003). Recently, at the 2004 Conference on CSCW, there was a panel discussion chaired by Sara Kiesler (Barley, Kiesler, Kraut, Dutton, Resnick, & Yates, 2004) on the topic of incorporating group and organization theory in CSCW. Broadly speaking, the panel discussed some theories applicable to CSCW and debated their usefulness.

In this article, we use the theory of small groups as complex systems from social psychology in a brief example to allude to how it can be used to inform CSCW methodologically and conceptually.

BACKGROUND

Preaching to the choir, Dan Shapiro at the 1994 Conference on CSCW made a strong call for a broader integration of the social sciences to better understand group- and organizational-level computer systems (Shapiro, 1994). Shapiro contrasted his proposal with the dominant use of ethnomethodology in CSCW research. As he noted, ethnomethodology implies a commitment to a worldview in which theories and other abstractions are rejected. Therefore, ethnographic accounts of behavior are driven not by explanation but "by the stringent discipline of observation and description" (p. 418). The result has been perhaps excellent designs, but typically, there is little sustained work to develop first principles that can be applied elsewhere (Barley et al., 2004).

Finholt and Teasley (1998) provided evidence of Shapiro's concern by analyzing citations in the ACM Proceedings of the Conference on CSCW. For example, examination of the 162 papers that appeared between 1990 and 1996 showed that each conference had a small number of papers with a psychological orientation. Overall, however, the proceedings indicated only modest attention to psychological questions, and this attention is diminishing. For instance, 77 out of 695 citations referenced the psychological literature in the 1990 Proceedings. By 1996, despite a 34% increase in the total number of citations, the number of references to the psychological literature decreased by 39% to 46 out of 933 citations. Thus, based on this study, the authors argue that the CSCW community should adopt a stronger orientation to social science disciplines.

Greater attention to psychological literature will offer well-validated principles about human behavior in group and organizational contexts, and convey data collection and analysis methods that identify salient and generalizable features of human behavior (Finholt & Teasley, 1998).

Kraut (2003, p. 354) warns of "disciplinary inbreeding", where researchers tend to cite work within their own community. For instance, contrasting with the earlier numbers on the decrease of citations to psychological literature, citations to the CSCW literature grew from 70 in 1990 to 233 in 1996, an increase of 330% (Finholt & Teasley, 1998). Kraut argues that unlike theories in cognitive psychology, social psychology as a theoretical base has been inadequately mined in the HCI and CSCW literatures (Barley et al., 2004). Part of the reason is the mismatch of goals and values of CSCW research with those of social psychology. CSCW is primarily an engineering discipline, whose goal is problem solving; in contrast, social psychology views itself as a behavioral science, whose mission is to uniquely determine the causes for social phenomena.

EXAMPLE

Social psychology has a rich body of theoretical literature that CSCW can build on (Beenen et al., 2004; Farooq, Singley, Fairweather, & Lam, 2004; Kraut, 2003). Let us take an example of a theory from social psychology that entrains implications for CSCW. Consider the theory of *small groups as complex systems* (for details of the theory, refer to Arrow, McGrath, & Berdahl, 2000). According to the theory, groups are intact social systems embedded within physical, temporal, socio-cultural, and organizational contexts. Effective study of groups requires attention to at least three system levels: individual members, the group as a system, and various layers of embedding contexts—both for the group as an entity and for its members. The following social psychological study illustrates how this theory can be leveraged in CSCW.

In the 1971 Stanford Prison Experiment (Bower, 2004), Zimbardo randomly assigned male college students to roles as either inmates or guards in a simulated prison. Within days, the young guards were stripping prisoners naked and denying them food. Zimbardo and his colleagues concluded that anyone given a guard's uniform and power over prisoners succumbs to that situation's siren call to abuse underlings. Currently, the validity and conclusions of these studies are being challenged on the grounds that the study used artificial settings and abuses by the guards stemmed from subtle cues given by experimenters (p. 106). In a recent and similar study to explore the dynamics of power in groups, Haslam and Reicher (2003) are indicating that tyranny does not arise simply from one group having power over another. Group members must share a definition of their social roles to identify with each other and promote group solidarity. In this study, volunteers assigned to be prison guards had trouble wielding power because they failed to develop common assumptions about their roles as guards. "It is the breakdown of groups and resulting sense of powerlessness that creates the conditions under which tyranny can triumph," (p. 108) Haslam holds.

In light of the above-mentioned study, the theory of groups as complex systems has at least two implications for CSCW. First, the theory warrants a research strategy that draws on both experimental and naturalistic traditions (Arrow et al., 2000). This will allow researchers to mitigate for difficulties of both laboratory experiments (e.g., lack of contextual realism) and field studies (e.g., lack of generalizability). Such a theory-driven research strategy can enrich current evaluation techniques in CSCW by increasing methodological robustness and validation (e.g., Convertino, Neale, Hobby, Carroll, & Rosson, 2004).

Second, the theory sheds light on the dynamics of power in a group. Arrow et al. (2000) assert that negotiations among members about power needs and goals typically involve both dyadic struggles to clarify relative power and collective norms about the status and influence structure (this was corroborated by Haslam and Reicher, 2003). This entails design implications for CSCW. Drawing on Arrow et al.'s (2000) theory, CSCW systems should then support, in general, design features that allow the fulfillment of group members' needs for attaining functional levels of agreement, explicit or implicit, regarding the following: (1) How membership status is established within the group (e.g., determined by a leader, based on the value of contributions to group projects, based on seniority); (2) the degree of power disparity between members allowed by the group; and (3) the acceptable uses of power to influence others in the group and how to sanction violations on these norms.

FUTURE TRENDS

CSCW is going to become increasingly inter-disciplinary. CSCW does and will continue to provide cultivating context for conflating multiple disciplines. To this end, marshalling theoretical literature from these disciplines will create a dynamic interplay in CSCW between theory, design, and practice. As CSCW co-evolves with its sister disciplines, we foresee a dialectical process of the former leveraging theoretical underpinnings of the latter, developing and refining these further, and in turn, also informing and enriching the theoretical base of its sister disciplines in context of collaborative technology and work.

CONCLUSION

To avoid producing unusable systems and badly mechanizing and distorting collaboration and other social activity, it is imperative to address the challenge of CSCW's *social-technical gap*: the divide between what we know we *must* support socially and what we *can* support technically (Ackerman, 2000). An understanding of this gap lies at the heart of CSCW's intellectual contribution that can be realized by fundamentally understanding the theoretical foundations of how people really work and live in groups, organizations, communities, and other forms of collective life (Ackerman, 2000). We certainly hold the view that science does and must continually strive to bring theory and fact into closer agreement (Kuhn, 1962). Perhaps the first step and challenge in this direction is to bridge the prescriptive discourse of CSCW with the descriptive disposition of the social sciences in order to arrive at scientifically satisfying and technically effective solutions.

REFERENCES

Ackerman, M. S. (2000). The intellectual challenge of CSCW: The gap between social requirements and technical feasibility. *Human-Computer Interaction, 15*(2-3), 179-204.

Arrow, H., McGrath, J. E., & Berdahl, J. L. (2000). *Small groups as complex systems: Formation, coordination, development, and adaptation.* Thousand Oaks, CA: Sage Publications.

Barley, S. R., Kiesler, S., Kraut, R. E., Dutton, W. H., Resnick, P., & Yates, J. (2004, November 6-10). Does CSCW need organization theory? *Proceedings of the ACM Conference on Computer Supported Cooperative Work*, Chicago (pp. 122-124). New York: ACM Press.

Beenen, G., Ling, K., Wang, X., Chang, K., Frankowski, D., Resnick, P., & Kraut, R.E. (2004, November 6-10). Using social psychology to motivate contributions to online communities. *Proceedings of ACM Conference on Computer Supported Cooperative Work*, Chicago (pp. 212-221). New York: ACM Press.

Bower, B. (2004). To err is human: Influential research on our social shortcomings attracts a scathing critique. *Science News, 166*(7), 106-108.

Convertino, G., Neale, D. C., Hobby, L., Carroll, J. M., & Rosson, M. B. (2004, October 23-27). A laboratory method for studying activity awareness. *Proceedings of ACM Nordic Conference on Human Computer Interaction*, Tampere, Finland (pp. 313-322). New York: ACM Press.

Farooq, U., Singley, M. K., Fairweather, P. G., & Lam, R. B. (2004, November 6-10). Toward a theory-driven taxonomy for groupware. *Proceedings of ACM Conference on Computer Supported Cooperative Work*, Conference Supplement, Chicago. New York: ACM Press. Retrieved October 16, 2005, from http://www.acm.org/cscw2004/prog_posters.html

Finholt, T. A., & Teasley, S. D. (1998). The need for psychology in research on computer-supported cooperative work. *Social Science Computer Review, 16*(1), 173-193.

Guzdial, M., Rick, J., & Kerimbaev, B. (2000, December 2-6). Recognizing and supporting roles in CSCW. *Proceedings of the ACM Conference on Computer Supported Cooperative Work*, Philadelphia (pp. 261-268). New York: ACM Press.

Haslam, S. A., & Reicher, S. (2003). A tale of two prison experiments: Beyond a role-based explanation of tyranny. *Psychology Review*, *9*, 2-6.

Kraut, R. E. (2003). Applying social psychological theory to the problems of group work. In J. M. Carroll (Ed.), *HCI models, theories, and frameworks: Toward a multidisciplinary science* (pp. 325-356). San Francisco: Morgan Kaufmann Publishers.

Kuhn, T. S. (1962). *The structure of scientific revolutions*. Chicago: University of Chicago Press.

Nunamaker, J. F., Dennis, A. R., Valacich, J. S., Vogel, D. R., & George, J. F. (1991). Electronic meeting systems to support group work. *Communications of the ACM*, *34*(7), 40-61.

Shapiro, D. (1994, October 22-26). The limits of ethnography: Combining social sciences for CSCW. *Proceedings of the Conference on Computer Supported Cooperative Work*, Chapel Hill, NC (pp. 417-428). New York: ACM Press.

Sproull, L., & Kiesler, S. (1991). *Connections: New ways of working in the networked organization*. Cambridge, MA: MIT Press.

KEY TERMS

Computer Supported Cooperative Work (CSCW): Research area that studies the design, evaluation, and deployment of computing technologies to support group and organizational activity.

Complex Systems: This concept is borrowed from Complexity Theory (see Arrow et al., 2000, for a detailed discussion). Complex systems are systems that are neither rigidly ordered nor highly disordered. System complexity is defined as the number and variety of identifiable regularities in the structure and behaviour of the group, given a description of that group at a fixed level of detail. Given this definition, Arrow et al. (2000) suggest that groups tend to increase in complexity over time, i.e. the number and variety of patterned regularities in the structure and behaviour of the group increase over time.

Disciplinary Inbreeding: A phrase coined by Kraut (2003, p. 354) to refer to the phenomenon of researchers citing academic work within their own community. He used it in specific context of the CSCW community.

Ethnomethodology: In the context of work, it is an approach to study how people actually order their working activities through mutual attentiveness to what has to be done. Ethnomethodology refuses any epistemological or ontological commitments, and limits its inquiry to what is directly observable and what can be plausibly inferred from observation.

Small Groups: According to the theory of small groups as complex systems (Arrow et al., 2000), a group is a complex, adaptive, dynamic, coordinated, and bounded set of patterned relations among members, task, and tools. Small groups generally have more than one dyadic link (although a dyad can comprise a group) but are bounded by larger collectives (e.g., an organization).

HCI in South Africa

Shawren Singh
University of South Africa, South Africa

INTRODUCTION

South Africa is a multi-lingual country with a population of about 40.5 million people. South Africa has more official languages at a national level than any other country in the world. Over and above English and Afrikaans, the eleven official languages include the indigenous languages: Southern Sotho, Northern Sotho, Tswana, Zulu, Xhosa, Swati, Ndebele, Tsonga, and Venda (Pretorius & Bosch, 2003). Figure 1 depicts the breakdown of the South African official languages as mother tongues for South African citizens.

Although English ranks fifth (9%) as a mother tongue, there is a tendency among national leaders, politicians, business people, and officials to use English more frequently than any of the other languages. In a national survey on language use and language interaction conducted by the Pan South African Language Board (Language Use and Board Interaction in South Africa, 2000), only 22% of the respondents indicated that they fully understand speeches and statements made in English, while 19% indicated that they seldom understand information conveyed in English.

The rate of electrification in South African is 66.1%. The total number of people with access to electricity is 28.3 million, and the total number of people without access to electricity is 14.5 million (International Energy Agency, 2002). Although the gap between the "haves" and "have-nots" is narrowing, a significant portion of the South African population is still without the basic amenities of life.

This unique environment sets the tone for a creative research agenda for HCI researchers and practitioners in South Africa.

BACKGROUND

E-Activities in South Africa

SA has been active in the e-revolution. The South African Green Paper on Electronic Commerce (EC) (Central Government, 2000) is divided into four categories. Each category contains key issues or areas of concern that need serious consideration in EC policy formulation:

- the need for confidence in the security and privacy of transactions performed electronically;
- the need to enhance the information infrastructure for electronic commerce;

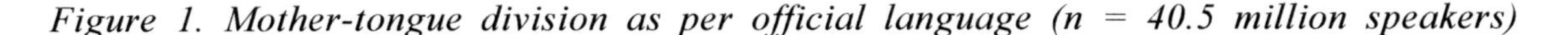

Figure 1. Mother-tongue division as per official language (n = 40.5 million speakers)

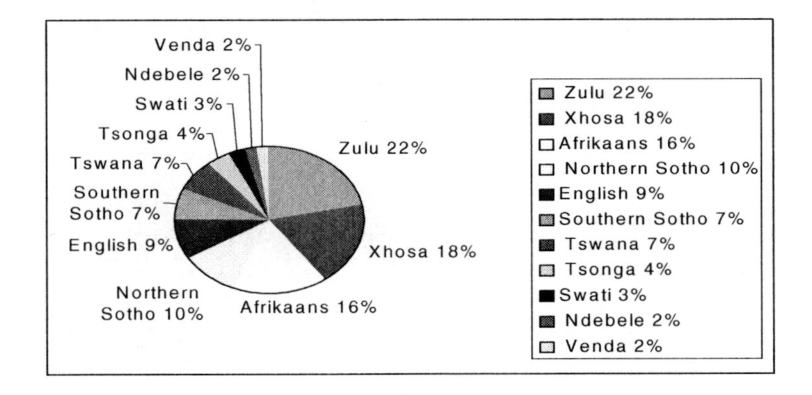

- the need to establish rules that will govern electronic commerce;
- the need to bring the opportunities of e-commerce to the entire population.

EC has not only affected government but has also actively moved into the mainstream South African's economy. Sectors of the economy that are using this technology are listed in Table 1, along with examples of companies using EC in each sector.

Electronic Communications and Transactions Bill

The Electronic Communications and Transactions Bill (2002) is an attempt by the Republic of South Africa to provide for the facilitation and regulation of electronic communications and transactions; to provide for the development of a national e-strategy for the Republic; to promote universal access to electronic communications and transactions and the use of electronic transactions by small, medium and micro enterprises (SMMEs); to provide for human resource development in electronic transactions; to prevent abuse of information systems; and to encourage the use of e-government services, and provide for matters connected therewith.

Some provisions of the bill are specifically directed at making policy and improving function in HCI-related areas. These are elucidated in the following bulleted items:

- To promote universal access primarily in under-serviced areas.
- To remove and prevent barriers to electronic communications and transactions in the Republic.
- To promote e-government services and electronic communications and transactions with public and private bodies, institutions, and citizens.
- To ensure that electronic transactions in the Republic conform to the highest international standards.
- To encourage investment and innovation in respect of electronic transactions in the Republic.
- To develop a safe, secure, and effective environment for the consumer, business and the government to conduct and use electronic transactions.
- To promote the development of electronic transaction services, which are responsive to the needs of users and consumers.
- To ensure that, in relation to the provision of electronic transactions services, the special needs of particular communities and areas, and the disabled are duly taken into account.
- To ensure compliance with accepted international technical standards in the provision and development of electronic communications and transactions.

Table 1. Sectors of the SA economy using EC, companies using EC within those sectors and their URLs

Sector	Company	URL
Banking-retail	ABSA	http://www.absa.co.za
Finance	SA Home Loans	http://www.sahomeloans.com/
Insurance	Liberty Life	MyLife.com
Media	Independent Newspapers Online	http://www.iol.co.za
Retail	Pick 'n Pay	http://www.pnp.co.za/
Travel	SAA	Kulula.com
Recruitment	Career Junction	http://www.careerjunction.co.za
Mining	Mincom	http://www.mincom.com
Automotive	Motoronline	http://www.motoronline.co.za
Data/telecomm	M-Web	http://www.mweb.co.za/
Health	Clickatell	http://www.clickatell.co.za

Though these objectives are Utopian, they are the first steps towards developing a manageable framework for the sustainable development of the e-community in South Africa. It is only by actions like these that we can get active role players involved in the development of a strategy for the e-community in South Africa.

THE STATE OF THE INTERNET IN SA

Internet access in South Africa continues to grow each year, but the rate of growth has slowed significantly. According to one study on Internet access in SA (Goldstuck, 2002), only 1 in 15 South Africans had access to the Internet at the end of last year. By the end of 2002, Internet access will have improved only marginally, to 1 in 14 South Africans. According to the report, the slow growth is largely the result of delays in licensing a second network operator, Telkom's own uncompromising attitude towards Internet service providers and market ignorance about the continued value of the Internet in the wake of the technology market crash of 2000 and 2001. South Africa will continue to lag behind the rest of the world in Internet use until the local telecommunication climate is more favorable (Worthington-Smith, 2002).

As Goldstuck (2002) points out, the educational environment in particular is poised for a boom in access, with numerous projects under way to connect schools to the Internet. That will not only be a positive intervention in the short term, but will provide a healthy underpinning for the long-term growth of Internet access in South Africa.

HCI IN SA

HCI is a discipline concerned with the design, evaluation, and implementation of interactive computer systems for human use. It also includes the study of phenomena surrounding interactive systems. Ultimately, HCI is about making products and systems easier to use and matching them more closely to users' needs and requirements. HCI is a highly active area for R&D and has applications in many countries. Concern has been expressed by some that SA is lagging seriously behind in this area. For instance, there is the worry that SA is not meeting the special challenges resulting from our multi-lingual society or the need to enable people with low levels of literacy to enjoy the benefits afforded by ICT. In short, there is a concern that the average South African will not gain value from rolling out current e-government initiatives. The opportunity for SA researchers and developers lies not just in meeting the specific needs of the country, but in positioning themselves as leaders in HCI in developing countries (Miller, 2003). According to Hugo (2003), the South African Usability community is small in size but large in quality. There are only a handful of practitioners, but their work is making a big impact on the local IT industry.

A number of educators are introducing HCI into the computer science curricula, and practitioners are working with application developers to integrate user-centered design (UCD) into the overall product development life cycle. In addition, the local HCI special interest group (CHI-SA, the South African ACM SIGCHI Chapter) is actively working with several stakeholders, including academics and non-government organizations to raise awareness of the impact of UCD on usability and overall organizational performance. In a few organizations, important progress has been made towards institutionalizing UCD.

Special attention is being given in South Africa to the relationship between multi-culturalism and technology dissemination. HCI practitioners are paying special attention to the processes of enculturation, acculturation, and cultural identities in the localization of software. They have also recognized the need to encourage developers to understand the role of the many cultural factors at work in the design of computing products, as well as the issues involved in intercultural and multi-cultural software design and internationalization, and how they affect the bottom line for organizations.

The usability community, which essentially consists of members of the CHI-SA, is currently a group of 116 people, 16 of whom are also members of ACM SIGCHI. After only one year as an official SIGCHI Chapter, CHI-SA distinguished itself in 2002 by presenting a workshop on multi-cultural HCI at the Development Consortium of CHI 2002 in Minneapolis.

Hugo (as reported by Wesson and Van Greunen, 2003) further characterized HCI in SA by highlighting the following:

- A shortage of qualified practitioners and educators. There are only a few people involved in HCI teaching and research with the majority being academics.
- A lack of awareness and implementation at industry level.
- Isolation and fragmentation between academia, industry, private research, development, and government.
- A lack of resources and inadequate training can result in inappropriate guidelines being adopted for the literature.
- A lack of knowledge of standards for usability and UCD such as ISO 9241 and ISO 13407 that exist in industry.

There is little evidence of SA commerce and industry taking up the principles and practice of HCI (Miller, 2003). Bekker (2003), a usability consultant with Test and Data Services, has noticed the following issues relating to HCI in the SA industry:

- Most of her clients use the Rational Unified Process and Microsoft Solutions Framework methodologies.
- Her clients are new to the concepts of HCI.
- SA companies are not designing usable software.
- People at the heart of the development process (the project managers and developers) fail to see the benefits of usability, and tend to ignore it ("I think it is because they hardly ever have contact with the users and have no idea how the users experience their system. They are highly computer literate people and cannot understand that anyone can battle using their system").
- Developers do not fully understand the nature of the users, which frustrate them.
- It is mostly the marketing and business departments of companies that request usability work from her.

FUTURE TRENDS

The Internet is becoming a significant tool in the SA economy but much more must be done to improve and broaden its use in SA. Research on HCI in SA is vital for both academics and practitioners. Several universities, notably, the University of South Africa, University of Cape Town, University of Kwa-Zulu Natal, University of Pretoria, University of Free State, and the Nelson Mandela Metropolitan University, are involved in teaching HCI. The Nelson Mandela Metropolitan University is the only university with a usability laboratory; the University of South Africa is in the process of building a usability laboratory and another usability laboratory.

CONCLUSION

Miller (2003) reports several barriers that are hindering the development of HCI in SA, such as:

- There is minimal awareness or a lack of appreciation within practitioners' community of what HCI is about and the advantages of better usability.
- There is a lack of HCI culture in the country, and therefore a lack of appreciation of its commercial benefits in industry.
- There is a lack of collaboration between the research and practitioner communities.
- There are concerns that the National Research Foundation (NRF) review panel lacks adequate HCI expertise.
- Only a small number of researchers are interested in HCI; ICT-related industry bodies are fragmented.
- There is a reliance on overseas provision of computer hardware such as high-resolution screens, and so forth.
- Government has offered little support other than the NRF Grant System and the Innovation Fund, which provide only low levels of funding.

REFERENCES

Bekker, A. (2003). Chi-SA tutorial. Personnel communication to S. Singh. Pretoria.

Central Government. (2000). *Green paper on electronic commerce*. Retrieved in July, 2004, from http://www.polity.org.za/govdocs/green_papers/greenpaper/index.html.

Electronic Communication and Transaction Bill, South Africa. (2002). Retrieved in July 2004, from http://www.polity.org.3a/html/gaudacs/bills/2002/b8-02.pdf

Goldstuck, A. (2002). *The Goldstuck report: Internet access in South Africa, 2002*. World Wide Worx. Retrieved from http://www.theworx.biz/accesso2.htm

Hugo, J. (2003, July). *Usability activities in South Africa*. UsabilityNet. Retrieved May 10, 2004, from www.usabilitynet.org

International Energy Agency. (2002). *World energy outlook: Energy and poverty*. Johannesburg, South Africa: International Energy Agency.

Language Use and Board Interaction in South Africa. (2000). *A sociolinguistic survey*. Pan South African Language Board Occasional Paper.

Miller, J. (2003). *Technology roadmap for the human-computer interface (HCI): Preliminary overview*. Miller Esselaar & Associates. Retrieved May 10, 2004, from www.milless.co.za

Pretorius, L., & Bosch, E. S. (2003). Enabling computer interaction in the indigenous languages of South Africa: The central role of computational morphology. *Interactions, 10*, 56-63.

Wesson, J., & van Greunen, D. (2003, July 1-5). *New horizons for HCI in South Africa*. Paper presented at the Human-Computer Interaction: Interact '03, Zurich, Switzerland.

Worthington-Smith, R. (Ed.). (2002). *The e-business handbook: The 2002 review of how South African companies are making the Internet work*. Cape Town: Trialogue.

KEY TERMS

E-Government: Is about re-engineering the current way of doing business, by using collaborative transactions and processes required by government departments to function effectively and economically, thus improving the quality of life for citizens and promoting competition and innovation. To put it simply, e-government is about empowering a country's citizens.

Electronic Commerce: Uses some form of transmission medium through which exchange of information takes place in order to conduct business.

ISO 13407: This standard provides guidance on human-centered design activities throughout the life cycle of interactive computer-based systems. It is a tool for those managing design processes and provides guidance on sources of information and standards relevant to the human-centered approach. It describes human-centered design as a multi-disciplinary activity, which incorporates human factors and ergonomics knowledge and techniques with the objective of enhancing effectiveness and efficiency, improving human working conditions, and counteracting possible adverse effects of use on human health, safety, and performance.

ISO 9241-11: A standard that describes ergonomic requirements for office work with visual display terminals. This standard defines how to specify and measure the usability of products, and defines the factors that have an effect on usability.

NRF: National Research Foundation, a government national agency that is responsible for promoting and supporting basic, applied research as well as innovation in South Africa.

Usability: The ISO 9241-11 standard definition for usability identifies three different aspects: (1) a specified set of users, (2) specified goals (asks) which have to be measurable in terms of effectiveness, efficiency, and satisfaction, and (3) the context in which the activity is carried out.

Hedonic, Emotional, and Experiential Perspectives on Product Quality

Marc Hassenzahl
Darmstadt University of Technology, Germany

QUALITY OF INTERACTIVE PRODUCTS

Human-computer interaction (HCI) can be defined as a discipline, which is concerned with the design, evaluation and implementation of interactive computing systems [products] for human use (Hewett et al, 1996). Evaluation and design require a definition of what constitutes a good or bad product and, thus, a definition of interactive product quality (IPQ). *Usability* is such a widely accepted definition. ISO 9241 Part 11 (ISO, 1998) defines it as the "extent to which a product can be used by specified users to achieve specified goals with effectiveness, efficiency and satisfaction in a specified context of use."

Although widely accepted, this definition's focus on tasks and goals, their efficient achievement and the involved cognitive information processes repeatedly caused criticism, as far back as Carroll and Thomas' (1988) emphatic plea not to forget the "fun" over simplicity and efficiency (see also Carroll, 2004).

Since then, several attempts have been made to broaden and enrich HCI's narrow, work-related view on IPQ (see, for example, Blythe, Overbeeke, Monk, & Wright, 2003; Green & Jordan, 2002; Helander & Tham, 2004). The objective of this article is to provide an overview of HCI current theoretical approaches to an enriched IPQ. Specifically, needs that go beyond the instrumental and the role of emotions, affect, and experiences are discussed.

BACKGROUND

Driven by the requirements of the consumer's product market, Logan (1994) was first to formulate a notion of *emotional usability*, which complements traditional, "behavioral" usability. He defined emotional usability as "the degree to which a product is desirable or serves a need beyond the [...] functional objective" (p. 61). It is to be understood as "an expanded definition of needs and requirements, such as fun, excitement and appeal" (Logan, Augaitis, & Renk, 1994, p. 369). Specifically, Logan and colleagues suggested a human need for novelty, change, and to express oneself through objects.

Other authors proposed alternative lists of needs to be addressed by an appealing and enjoyable interactive product. In an early attempt, Malone (1981, 1984) suggested a need for challenge, for curiosity, and for being emotionally bound by an appealing fantasy (metaphor). Jordan (2000) distinguished four groups of needs: physiological (e.g., touch, taste, smell), social (e.g., relationship with others, status), psychological (e.g., cognitive and emotional reactions), and Id-needs (e.g., aesthetics, embodied values). Gaver and Martin (2000) compiled a list of non-instrumental needs, such as novelty, surprise, diversion, mystery, influencing the environment, intimacy, and to understand and change one's self. Taken together, these approaches have at least two aspects in common: (a) they argue for a more holistic understanding of the human in HCI and (b) they seek to enrich HCI's narrow view on IPQ with non-instrumental needs to *complement* the traditional, task-oriented approach.

Although, the particular lists of needs differ from author to author, two broad categories—widely supported by psychological research and theory—can be identified, namely *competence/personal growth*, for example, the desire to perfect one's knowledge and skills, and *relatedness/self-expression*, for example, the desire to communicate a favorable identity to relevant others (see Hassenzahl, 2003).

A sense of *competence*, for example, to take on and master hard challenges, is one of the core needs in Ryan and Deci's (2000) *self-determination theory*, which formulates antecedents of personal well-being. Similarly, Csikszentmihalyi's (1997) *flow* theory, which became especially popular in the context of analyzing Internet use (see Chen, Wigand, & Nilan, 1999; Novak, Hoffman, & Yung, 2000), suggests that individuals will experience a positive psychological state (flow) as long as the challenge such an activity poses is met by the individuals' skills. Interactive products could tackle these challenges by opening up for novel and creative uses while, at the same time, providing appropriate means to master these challenges.

A second need identified by Ryan and Deci (2000) is *relatedness*—a sense of closeness with others. To experience relatedness requires social interaction and as Robinson (1993, cited in Leventhal, Teasley, Blumenthal, Instone, Stone, & Donskoy, 1996) noted, products are inevitably statements in the on-going interaction with relevant others. A product can be understood as an extension of an individual's self (Belk, 1988)—its possession and use serves self-expressive functions beyond the mere instrumental (e.g., Wicklund & Gollwitzer, 1982).

To summarize, an appealing interactive product may support needs beyond the mere instrumental. Needs that are likely to be important in the context of design and evaluation are *competence/personal growth*, which requires a balance between challenge and ability and *relatedness/self-expression*, which requires a product to communicate favorable messages to relevant others.

NEEDS BEYOND THE INSTRUMENTAL

In this article, the terms *instrumental* and *non-instrumental* are used to distinguish between HCI's traditional view on IPQ and newer additions. Repeatedly, authors refer to instrumental aspects of products as *utilitarian* (e.g., Batra & Ahtola, 1990), *functional* (e.g., Kempf, 1999) or *pragmatic* (e.g., Hassenzahl, 2003), and to non-instrumental as *hedonic*. However, hedonic can have two different meanings: some authors understand it as the affective quality (see section below) of a product, for example, pleasure, enjoyment, fun derived from possession or usage (e.g., Batra & Ahtola, 1990; Huang, 2003), while others see it as non-task related attributes, such as novelty or a product's ability to evoke memories (e.g., Hassenzahl, 2003). Beside these slight differences in meaning, instrumental and non-instrumental aspects are mostly viewed as separate but complementing constructs. Studies, for example, showed instrumental as well as non-instrumental aspects to be equally important predictors of product appeal (e.g., Hassenzahl, 2002a; Huang, 2003). A noteworthy exception to the general notion of ideally addressing instrumental and non-instrumental needs simultaneously is Gaver's et al. (2004b) concept of *ludic* products. According to them, a ludic product promotes curiosity, exploration and *de-emphasizes* the pursuit of external (instrumental) goals. Or as Gaver (personal communication) put it: Ludic products " ... aren't clearly useful, nor are they concerned with entertainment alone. Their usefulness is rather in prompting awareness and insight than in completing a given task." Gaver et al. (2004b) argue, then, for a new product category aimed at solely supporting *personal growth/competence* by providing a context for new, challenging and intrinsically interesting experiences and by deliberately turning the user's focus *away* from functionality.

A question closely related to instrumental and non-instrumental needs is their relative importance. Jordan (2000) argued for a hierarchical organization of needs (based on Maslow's [1954] hierarchical concept of human needs): The first level is product functionality, the second level is usability and the third level is "pleasure," which consists of his four non-instrumental aspects already presented earlier. Such a model assumes that the satisfaction of instrumental needs is a necessary precondition for valuing non-instrumental needs. A product must, thus, provide functionality, before, for example, being appreciated for its self-expressive quality.

This strict assumption can be questioned. Souvenirs, for example, are products, which satisfy a non-instrumental need (keeping a memory alive, see Hassenzahl, 2003) without providing functionality. However, for many products, functionality can be seen as a necessary precondition for acceptance. A mobile phone, for instance, which does not work will definitely fail on the market, regardless of its non-instrumental qualities.

One may, thus, understand a hierarchy as a *particular, context-dependent* prioritization of needs (Sheldon, Elliott, Kim, & Kasser, 2001). The relative importance of needs may vary with product categories (e.g., consumers' versus producers' goods), individuals (e.g., early versus late adopters) or specific usage situations. Hassenzahl, Kekez, and Burmester (2002), for example, found instrumental aspects of Web sites to be of value, only if participants were given explicit tasks to achieve. Instrumental aspects lost their importance for individuals with the instruction "to just have fun" with the Web site.

EMOTIONS, AFFECT, AND EXPERIENCE

Recently, the term *emotional design* (Norman, 2004) gained significant attention in the context of HCI. Many researchers and practitioners advocate the consideration of emotions in the design of interactive products—an interest probably triggered by science's general, newly aroused attention to emotions and their interplay with cognition (e.g., Damasio, 1994). In the context of HCI, Djajadiningrat, Overbeeke, and Wensveen (2000), for instance, argued for explicitly taking both into account, knowing *and* feeling. Desmet and Hekkert (2002) went a step further by presenting an explicit model of product emotions based on Ortony, Clore, and Collins' (1988) emotion theory.

In general, emotions in design are treated in two ways: some authors stress their importance as *consequences* of product use (e.g., Desmet & Hekkert, 2002, Hassenzahl, 2003; Kim & Moon, 1998; Tractinsky & Zmiri, in press), whereas others stress their importance as *antecedents* of product use and evaluative judgments (e.g., Singh & Dalal, 1999), *visceral level* in Norman (2004).

The "Emotions as consequences"—perspective views particular emotions as the result of a cognitive appraisal process (see Scherer, 2003). Initial affective reactions to objects, persons, or events are further elaborated by combining them with expectations or other cognitive content. Surprise, for example, may be felt, if an event deviates from expectations. In the case of a positive deviation, surprise may then give way to joy. An important aspect of emotions is their situatedness. They are the result of the complex interplay of an individual's psychological state (e.g., expectations, moods, saturation level) and the situation (product and particular context of use). Slight differences in one of the elements can lead to a different emotion. Another important aspect is that emotions are transient. They occur, are felt, and last only a relatively short period of time. Nevertheless, they are an important element of experience.

The ephemeral nature of emotions and the complexity of eliciting conditions may make it difficult to explicitly *design* them (Hassenzahl, 2004). Designers would need control over as many elements of an experience as possible. Good examples for environments with a high level of control from the designer's perspective are theme parks or movies. In product design, however, control is not as high and, thus, designers may have to be content with creating the possibility of an emotional reaction, for example, the context for an experience rather than the experience itself (Djajadiningrat et al., 2000; Wright, McCarthy, & Meekison, 2003).

In 1980, Zajonc (1980) questioned the view of emotions as consequences of a cognitive appraisal. He showed that emotional reactions could be instantaneous, automatic without cognitive processing. And indeed, neurophysiology discovered a neural shortcut that takes information from the senses directly to the part of the brain responsible for emotional reactions (amygdala) before higher order cognitive systems have had a chance to intervene (e.g., LeDoux, 1994). However, these instantaneous emotional reactions differ from complex emotions like hate, love, disappointment, or satisfaction. They are more diffuse, mainly representing a good/bad feeling of various intensities about an object, person, or event. To distinguish this type of emotional reaction from the more complex discussed earlier, they are often called *affective reactions* in contrast to *emotions*. Norman (2004) labeled the immediate reaction "visceral" (bodily) as opposed to the more "reflective."

Importantly, one's own immediate, unmediated affective reactions are often used as information (*feelings-as-information*, Schwarz & Clore, 1983), influencing and guiding future behavior. Damasio (1994) developed the notion of somatic markers attached to objects, persons, or events, which influence the way we make choices by signaling "good" or "bad". Research on persuasion, for example, has identified two ways of information processing: systematic (central) and heuristic (peripheral). Individuals not capable or motivated to process argument-related information, rely more strongly on peripheral cues, such as their own immediate affective reactions towards an argument (e.g., Petty & Cacioppo, 1986). These results emphasize the importance of a careful consideration of immediate, product-driven emotional reactions for HCI.

FUTURE TRENDS

To design a "hedonic" interactive product requires an understanding of the link between designable product features (e.g., functionality, presentational and interactional style, content), resulting product attributes (e.g., simple, sober, exciting, friendly) and the fulfillment of particular needs. In the same vein as a particular user interface layout may imply simplicity, which in turn promises fulfillment of the need to achieve behavioral goals, additional attributes able to signal and promote fulfillment of competency or self-expression needs (and ways to create these) have to be identified. We may, then, witness the emergence of principles for designing hedonic products comparable to existing principles for designing usable products.

As long as HCI strongly advocates a systematic, user-centered design process (*usability engineering*, e.g., Mayhew, 1999; Nielsen, 1993), tools and techniques will be developed to support the inclusion of non-instrumental needs and emotions. Some techniques have already emerged: measuring emotions in product development (e.g., Desmet, 2003), gathering holistic product perceptions (Hassenzahl, 2002b), assessing the fulfillment of non-instrumental needs (e.g., Hassenzahl, in press) or eliciting non-instrumental, "inspirational" data (e.g., Gaver, Boucher, Pennington, & Walker, 2004a). Others will surely follow.

CONCLUSION

Individuals have general needs, and products can play a role in their fulfillment. The actual fulfillment of needs (when attributed to the product) is perceived as quality. Certainly, individuals have instrumental goals and functional requirements that a product may fulfill; however, additional non-instrumental, hedonic needs are important, too. Two needs seem to be of particular relevance: *personal growth/competence* and *self-expression/relatedness*. Product attributes have to be identified, which signal and fulfill instrumental as well as non-instrumental needs. A beautiful product, for example, may be especially good for self-expression (Hassenzahl, in press; Tractinsky & Zmiri, in press); a product that balances simplicity/ease (usability) and novelty/stimulation may fulfill the need for personal growth.

Human needs are important, and individuals can certainly reach general conclusions about their relative importance (see Sheldon et al., 2001). However, quality is also rooted in the actual experience of a product. Experience consists of numerous elements (e.g., the product, the user's psychological states, their goals, other individuals, etc.) and their interplay (see Wright et al., 2003). The complexity of an experience makes it a unique event—hard to repeat and even harder to create deliberately. But experience nevertheless matters. Experiences are highly valued (Boven & Gilovich, 2003), and, consequentially, many products are now marketed as experiences rather than products (e.g., Schmitt, 1999). From an HCI perspective, it seems especially important to better understand experiences in the context of product use.

Definitions of quality have an enormous impact on the success of interactive products. Addressing human needs as a whole and providing rich experiences would enhance the role of interactive products in the future.

REFERENCES

Batra, R., & Ahtola, O. T. (1990). Measuring the hedonic and utilitarian sources of consumer choice. *Marketing Letters, 2,* 159-170.

Belk, R. W. (1988). Possessions and the extended self. *Journal of Consumer Research, 15,* 139-168.

Blythe, M., Overbeeke, C., Monk, A. F., & Wright, P. C. (2003). *Funology: From usability to enjoyment.* Dordrecht: Kluwer.

Boven, L. v., & Gilovich, T. (2003). To do or to have? That is the question. *Journal of Personality and Social Psychology, 85,* 1193-1202.

Carroll, J. M. (2004). Beyond fun. *Interactions, 11*(5), 38-40.

Carroll, J. M., & Thomas, J. C. (1988). Fun. *SIGCHI Bulletin, 19,* 21-24.

Chen, H., Wigand, R. T., & Nilan, M. S. (1999). Optimal experience of Web activities. *Computers in Human Behavior, 15,* 585-508.

Csikszentmihalyi, M. (1997). *Finding flow: The psychology of engagement with everyday life.* New York: Basic Books.

Damasio, A. R. (1994). *Descartes' error: Emotion, reason and the human brain.* New York: Grosset/Putnam.

Desmet, P. M. A. (2003). Measuring emotion: Development and application of an instrument to measure emotional responses to products. In M. Blythe, C. Overbeeke, A. F. Monk, & P. C. Wright (Eds.), *Funology: From usability to enjoyment* (pp. 111-124). Dordrecht: Kluwer.

Desmet, P. M. A., & Hekkert, P. (2002). The basis of product emotions. In W. Green, & P. Jordan (Eds.), *Pleasure with products: Beyond usability* (pp. 60-68). London: Taylor & Francis.

Djajadiningrat, J. P., Overbeeke, C. J., & Wensveen, S. A. G. (2000). Augmenting fun and beauty: A pamphlet. In *Proceedings of DARE 2000: Designing Augmented Reality Environments,* Helsingor, Denmark (pp. 131-134). New York: ACM Press. Retrieved October 14, 2005, from http://doi.acm.org/10.1145/354666.35680

Gaver, W. W., Boucher, A., Pennington, S., & Walker, B. (2004a). Cultural probes and the value of uncertainty. *Interactions, 11,* 53-56.

Gaver, W. W., Bowers, J., Boucher, A., Gellerson, H., Pennington, S., Schmidt, A., Steed, A., Villars, N., & Walker, B. (2004b, April 24-29). The drift table: Designing for ludic engagement. In *Proceedings of the CHI 2004 Conference on Human Factors in Computing Systems. Extended abstracts*, Vienna, Austria (pp. 885-900). New York: ACM Press.

Gaver, W. W., & Martin, H. (2000, April 1-6). Alternatives. Exploring information appliances through conceptual design proposals. In *Proceedings of the CHI 2000 Conference on Human Factors in Computing,* The Hague, The Netherlands (pp. 209-216). New York: ACM Press.

Green, W., & Jordan, P. (2002). *Pleasure with products: Beyond usability*. London: Taylor & Francis.

Hassenzahl, M. (2002a). The effect of perceived hedonic quality on product appealingness. *International Journal of Human-Computer Interaction, 13,* 479-497.

Hassenzahl, M. (2002b). Character grid: A simple repertory grid technique for Web site analysis and evaluation. In J. Ratner (Ed.), *Human factors and Web development* (2nd ed.) (pp. 183-206). Mahwah, NJ: Lawrence Erlbaum.

Hassenzahl, M. (2003). The thing and I: Understanding the relationship between user and product. In M. Blythe, C. Overbeeke, A. F. Monk, & P. C. Wright (Eds.), *Funology: From usability to enjoyment* (pp. 31-42). Dordrecht: Kluwer.

Hassenzahl, M. (2004). Emotions can be quite ephemeral. We cannot design them. *Interactions, 11,* 46-48.

Hassenzahl, M. (2004). The interplay of beauty, goodness and usability in interactive products. *Human Computer Interaction, 19*(4).

Hassenzahl, M., Kekez, R., & Burmester, M. (2002, May 22-25). The importance of a software's pragmatic quality depends on usage modes. In *Proceedings of the 6th international conference on Work With Display Units (WWDU 2002),* Berchtesgaden, Germany (pp. 275-276). Berlin: ERGONOMIC Institut für Arbeits- und Sozialforschung.

Helander, M. G., & Tham, M. P. (2004). Hedonomics—Affective human factors design [Special issue]. *Ergonomics, 46,* 13-14.

Hewett, T. T., Baecker, R., Card, S. K., Carey, T., Gasen, J., Mantei, M. M., Perlman, G., Strong, G., &

Verplank, W. (1996). *ACM SIGCHI Curricula for Human-Computer Interaction.* Retrieved October 20, 2004, from http://sigchi.org/cdg/cdg2.html.

Huang, M. -H. (2003). Designing Web site attributes to induce experiental encounters. *Computers in Human Behavior, 19,* 425-442.

ISO. (1998). *ISO 9241: Ergonomic requirements for office work with visual display terminals (VDTs)—Part 11: Guidance on usability.* Geneve: International Organization for Standardization.

Jordan, P. (2000). *Designing pleasurable products. An introduction to the new human factors.* London; New York: Taylor & Francis.

Kempf, D. S. (1999). Attitude formation from product trial: Distinct roles of cognition and affect for hedonic and functional products. *Psychology & Marketing, 16,* 35-50.

Kim, J. & Moon, J. Y. (1998). Designing towards emotional usability in customer interfaces—Trustworthiness of cyber-banking system interfaces. *Interacting with Computers, 10,* 1-29.

LeDoux, J. (1994, June). Emotion, memory and the brain [Special issue]. *Scientific American*, 50-57. Retrieved October 14, 2005, from http://www.sciamdigital.com/browse.cfm?sequencenameCHAR=item&methodnameCHAR=resource_getitembrowse&interfacenameCHAR=browse.cfm&ISSUEID_CHAR=A2EE5789-CDA4-4B4DC-A157AE26F64

Leventhal, L., Teasley, B., Blumenthal, B., Instone, K., Stone, D., & Donskoy, M. V. (1996). Assessing user interfaces for diverse user groups: Evaluation strategies and defining characteristics. *Behaviour & Information Technology, 15,* 127-137.

Logan, R. J. (1994). Behavioral and emotional usability: Thomson Consumer Electronics. In M. Wiklund (Ed.), *Usability in practice* (pp. 59-82). Cambridge, MA: Academic Press.

Logan, R. J., Augaitis, S., & Renk, T. (1994). Design of simplified television remote controls: a case for behavioral and emotional usability. In *Proceedings of the 38th Human Factors and Ergonomics Society Annual Meeting* (pp. 365-369). Santa Monica: Human Factors and Ergonomics Society.

Malone, T. W. (1981). Toward a theory of intrinsically motivating instruction. *Cognitive Science, 4,* 333-369.

Malone, T. W. (1984). Heuristics for designing enjoyable user interfaces: Lessons from computer games. In J. C. Thomas & M. L. Schneider (Eds.), *Human factors in computer systems* (pp. 1-12). Norwood, NJ: Ablex.

Maslow, A. H. (1954). *Motivation and personality.* New York: Harper.

Mayhew, D. L. (1999). *The usability engineering lifecycle. A practitioner's handbook for user interface design.* San Francisco: Morgan Kaufmann.

Nielsen, J. (1993). *Usability engineering.* Boston; San Diego, CA: Academic Press.

Norman, D. A. (2004). *Emotional design: Why we love (or hate) everyday thing.* New York: Basic Books.

Novak, T. P., Hoffman, D. L., & Yung, Y. F. (2000). Measuring the customer experience in online environments: a structural modeling approach. *Marketing Science, 19,* 22-44.

Ortony, A., Clore, G. L., & Collins, A. (1988). *The cognitive structure of emotions.* Cambridge, MA: Cambridge University Press.

Petty, R. E., & Cacioppo, J. T. (1986). *Communication and persuasion: Central and peripheral routes to attitude change.* New York: Springer.

Robinson, R. (1993). What to do with a human factor. *American Design Journal.*

Ryan, R. M., & Deci, E. L. (2000). Self-determination theory and the facilitation of intrinsic motivation, social development, and well-being. *American Psychologist, 55,* 68-78.

Scherer, K. R. (2003). Introduction: Cognitive components of emotion. In R. J. Davidson, K. R. Scherer, & H. H. Goldsmith (Eds.), *Handbook of affective science* (pp. 563-571). New York: Oxford University Press.

Schmitt, B. H. (1999). *Experiential marketing.* New York: Free Press.

Schwarz, N., & Clore, G. L. (1983). Mood, misattribution, and judgments of well-being: Informative and directive functions of affective states. *Journal of Personality and Social Psychology, 45,* 513-523.

Sheldon, K. M., Elliot, A. J., Kim, Y., & Kasser, T. (2001). What is satisfying about satisfying events? Testing 10 candidate psychological needs. *Journal of Personality and Social Psychology, 80,* 325-339.

Singh, S. N., & Dalal, N. P. (1999). Web home pages as advertisements. *Communications of the ACM, 42,* 91-98.

Tractinsky, N., & Zmiri, D. (in press). Exploring attributes of skins as potential antecedents of emotion in HCI. In P. Fishwick (Ed.), *Aesthetic computing.* MIT Press.

Wicklund, R. A., & Gollwitzer, P. M. (1982). *Symbolic self-completion.* Hillsdale, NJ: Lawrence Erlbaum.

Wright, P. C., McCarthy, J., & Meekison, L. (2003). Making sense of experience. In M. Blythe, C. Overbeeke, A. F. Monk, & P. C. Wright (Eds.), *Funology: From usability to enjoyment* (pp. 43-53). Dordrecht: Kluwer.

Zajonc, R. B. (1980). Feeling and thinking: Preferences need no inferences. *American Psychologist, 35,* 151-175.

KEY TERMS

Affect: An umbrella term used to refer to mood, emotion, and other processes, which address related phenomena. The present article more specifically uses the term "affective reaction" to distinguish an individual's initial, spontaneous, undifferentiated, and largely physiologically-driven response to an event, person, or object from the more cognitively differentiated "emotion."

Emotion: A transient psychological state, such as joy, sadness, anger. Most emotions are the consequence of a cognitive appraisal process, which links an initial affective reaction (see "Affect" term definition) to momentarily available "information", such as one's expectations, beliefs, situational cues, other individuals, and so forth.

Experience: A holistic account of a particular episode, which stretches over time, often with a definite beginning and ending. Examples of (positive) experiences are: visiting a theme park or consuming a bottle of wine. An experience consists of numerous elements (e.g., product, user's psychological states, beliefs, expectations, goals, other individuals, etc.) and their relation. It is assumed that humans constantly monitor their internal, psychological state. They are able to access their current state during an experience and to report it (i.e., experience sampling). Individuals are further able to form a summary, retrospective assessment of an experience. However, this retrospective assessment is not a one-to-one summary of everything that happened during the experience, but rather overemphasizes single outstanding moments and the end of the experience.

Instrumental Needs: Particular, momentarily relevant behavioral goals, such as making a telephone call, withdrawing money from one's bank account, or ordering a book in an online shop. Product attributes related to the achievement of behavioral goals are often referred to as "utilitarian," "pragmatic," or "functional."

Non-Instrumental Needs: Go beyond the mere achievement of behavioral goals, such as self-expression or personal growth. Product attributes related to the fulfillment of non-instrumental needs are often referred to as "hedonic." A more specific use of the term "hedonic" stresses the product's "affective" quality, for example, its ability to evoke positive affective reactions (mood, emotions, see "Affect" term definition).

Human Factors in the Development of Trend Detection and Tracking Techniques

Chaomei Chen
Drexel University, USA

Kaushal Toprani
Drexel University, USA

Natasha Lobo
Drexel University, USA

INTRODUCTION

Trend detection has been studied by researchers in many fields, such as statistics, economy, finance, information science, and computer science (Basseville & Nikiforov, 1993; Chen, 2004; Del Negro, 2001). Trend detection studies can be divided into two broad categories. At technical levels, the focus is on detecting and tracking emerging trends based on dedicated algorithms; at decision making and management levels, the focus is on the process in which algorithmically identified temporal patterns can be translated into elements of a decision making process.

Much of the work is concentrated in the first category, primarily focusing on the efficiency and effectiveness from an algorithmic perspective. In contrast, relatively fewer studies in the literature have addressed the role of human perceptual and cognitive system in interpreting and utilizing algorithmically detected trends and changes in their own working environments. In particular, human factors have not been adequately taken into account; trend detection and tracking, especially in text document processing and more recent emerging application areas, has not been studied as integral part of decision-making and related activities. However, rapidly growing technology, and research in the field of human-computer interaction has opened vast and, certainly, thought-provoking possibilities for incorporating usability and heuristic design into the areas of trend detection and tracking.

BACKGROUND

In this section, we briefly review trend detection and its dependence on time and context, topic detection and tracking, supported by instances of their impact in diverse fields, and the emerging trend detection especially for text data.

Trend Detection

A *trend* is typically defined as a continuous change of a variable over a period of time, for example, unemployment numbers increase as the economy enters a cycle of recession. Trend detection, in general, and topic detection techniques are groups of algorithmic tools designated to identify significant changes of quantitative metrics of underlying phenomena. The goal of detection is to enable users to identify the presence of such trends based on a spectrum of monitored variables. The response time of a detection technique can be measured by the time duration of the available input data and the identifiable trend; it is dependent on specific application domains. For example, anti-terrorism and national security may require highly responsive trend detection and change detection capabilities, whereas geological and astronomical applications require long-range detection tools. Other applications of this technology exist in the fields of business and medicine.

Much research has been done in the field of information retrieval, automatically grouping (clustering) documents, performing automated text summarization, and automatically labeling groups of documents.

Policymakers and investigators are, obviously, eager to know if there are ways that can reliably predict each turn in the economy. Economists have developed a wide variety of techniques to detect and monitor changes in economic activities. The concept of *business cycles* is defined as fluctuations in the aggregate economic activities of a nation. A business cycle includes a period of expansion, followed by recessions, contractions, and revivals. Three important characteristics are used when identifying a recession: duration, depth, and diffusion — the three Ds. A recession has to be long enough, from a year to 10 years; a recession has to be bad enough, involving a substantial decline in output; and a recession has to be broad enough, affecting several sectors of the economy.

Topic Detection and Tracking

Topic Detection and Tracking (TDT) is a sub-field primarily rooted in information retrieval. TDT aims to develop and evaluate technologies required to segment, detect, and track topical information in a stream consisting of news stories. TDT has five major task groups: (1) story segmentation, (2) topic detection, (3) topic tracking, (4) first story detection, and (5) story link detection. Topic detection focuses on discovering previously unseen topics, whereas topic tracking focuses on monitoring stories known to a TDT system. First story detection (FSD) aims to detect the first appearance of a new story in a time series of news associated with an event. Roy, Gevry, and Pottenger (2002) presented methodologies for trend detection. Kontostathis, Galitsky, Roy, Pottenger, and Phelps (2003) gave a comprehensive survey of emerging trend detection in textual data mining in terms of four distinct aspects: (1) input data and attributes, (2) learning algorithms, (3) visualization, and (4) evaluation.

TDT projects typically test their systems on TDT data sets, which contain news stories and event descriptors. The assessment of the performance of a TDT algorithm is based on *Relevance Judgment*, which indicates the relevancy between a story and an event. Take the event descriptor *Oklahoma City Bombing* as an example. If a matching story is about survivors' reaction after the bombing, the relevance judgment would be *Yes*. In contrast, the relevance judgment of the same story and a different event descriptor *U.S. Terrorism Response* would be *No*. Swan and Allan (1999) reported their work on extracting significant time varying features from text based on this type of data.

An interesting observation of news stories is that events are often reported in burst. Yang, Pierce, and Carbonell (1998) depicted a daily histogram of story counts over time. News stories about the same event tend to appear within a very narrow time frame. The gap between two bursts can be used to discriminate distinct events.

Kleinberg (2002) developed a burst detection algorithm and applied to the arrivals of e-mail and words used in titles of articles. Kleinberg was motivated by the need to filter his e-mail. He expected that whenever an important event occurs or is about to occur, there should be a sharp increase of certain words that characterize the event. He called such sharp increases *bursts*. Essentially, Kleinberg's burst detection algorithm analyzes the rate of increase of word frequencies and identifies the most rapidly growing words. He tested his algorithm on the full text of all the State of the Union addresses since 1790. The burst detection algorithm identified important events occurring at the time of some of the speeches. For example, *depression* and *recovery* were *bursty* words in 1930-1937, *fighting* and *Japanese* were bursty in 1942-1945, and *atomic* was the buzz word in 1947 and 1959.

EMERGING TREND DETECTION (ETD)

ETD for Text Data

Unlike financial and statistical data typically found in an economist's trend detection portfolio, ETD in computer science often refers to the detection of trends in textual data, such as a collection of text documents and a stream of news feed. ETD takes a large collection of textual data as input and identifies topic areas that are previously unseen or are growing in importance with in the corpus (Kontostathis et al., 2003). This type of data mining can be instrumental in supporting the discovery of emerging trends within an industry and improving the understanding of large volumes of information maintained by organizations (Aldana, 2000). In the past few years, many companies have been storing their data electronically. As the volume of data grows, it will hold information, which if analyzed in the form of trends and patterns, can be valuable to the company, provided it is appropriately and accurately extracted. By using ETD, companies can extract the meaningful data and use it to gain a competitive advantage (Aldana, 2000). ETD provides a viable way to analyze the evolution of a field. The problem switches from analyzing huge amounts of data to *how* to analyze huge amounts of data.

The Role of HCI in ETD

ETD systems are complicated to make and understand, thus there are many HCI issues that must be considered. First of all, the system should let the user define what an emerging trend is. In general, an emerging trend can be defined as a significant quantitative growth over time. However, what counts as significant should not be entirely determined by computer algorithms.

Many ETD algorithms have different threshold levels to define a topic as an emerging trend. Thus threshold levels should not be fixed and unchangeable for a system. Also, the user should be able to define what documents are in the data corpus. Additionally, the algorithm should be hidden from the user. Ideally, the system would take its inputs and produce the outputs. When the user is given information, pertaining to inputs and outputs, sufficient amounts of user guidance should be provided. The design of an ideal user interface of a computer-based information system should be intuitive and self-explanatory. Users should feel that they are in control and they can understand what is going on. Despite the technical complexity of an underlying algorithm, the user interface should clearly convey the functions to the user (Norman, 1998).

Once a new trend is found, the system should include some mechanisms to define the essence of the new trend. A text summarization algorithm is a possible solution to this problem. Text summarization is capturing the essence of a data set (a single paragraph, document, or cluster) after reviewing the entire data set and producing output that describes the data set.

Once the data corpus is scanned, the user should be provided with feedback about the corpus. The user should be provided with information like the number of documents found, number of topics (or trends) found, number of new trends found, and other related information. For example, the system studied by Masao and Kôiti (2000), produces an entity-relationship (ER) graph showing the relation of topics. This not only shows the user what new trends were found, but also shows how they are related. ETD systems should also support an adaptive search mechanism. Users should have the option of providing keywords to search for emerging trends in specific fields.

APPLICATIONS

Automatic trend detection has benefited a wide range of applications. An analyst will find emerging trend detection techniques useful in his area of work. The most generic application is to detect a new topic in a field and track its growth and use over time (Roy et al., 2002). Two examples are cited in the following sections.

European Monitoring Center for Drugs and Drug Addiction (EMCDDA)

The EMCDDA was disappointed when it realized that it failed to recognize the emerging trend in the use of the drug ecstasy. "...earlier identification of new drug consumption patterns would allow more time to assess the likely impact of such changes and, therefore, facilitate the earlier development of appropriate responses" (EMCDDA, 1999). With an effective trend detection system, agencies like the EMCDDA can prepare for and prevent the associated problems with a drug epidemic. However, with the number of documents in some databases reaching over 100,000 a manual review of the data is impossible.

XML

The emergence of XML in the 1990s is shown in Figure 1 in terms of the growing number of articles published each year on the second-generation language of the World Wide Web. Market and field

Figure 1. The growth of the number of articles published on the topic of XML (Kontostathis et al., 2003)

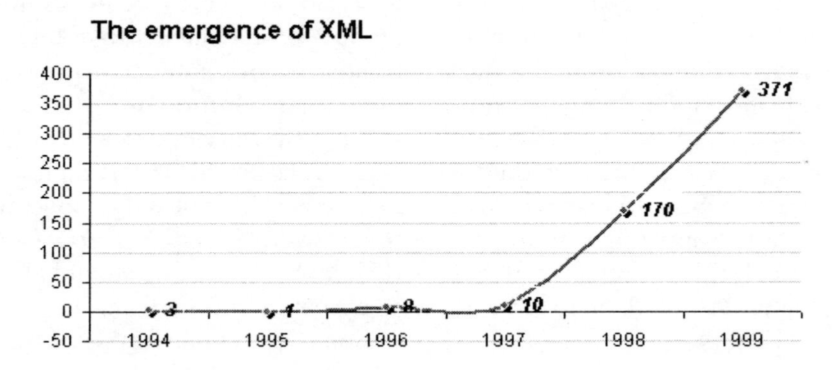

analysts will find such knowledge of an emerging trend particularly useful. For instance, a market-analyst watching a biotech firm will want to know about trends in the biotechnology field and how they affect companies in the field (Kontostathis et al., 2003).

Stock market analysts rely on patterns to observe market trends and make predictions. In general, the goal is to identify patterns from a corpus of data. In the past, analysts have relied on the human eye to discover these patterns. In the future, trend and topic tracking systems can take over this role, thus providing a more efficient and reliable method for stock market analysis.

FUTURE TRENDS

The future is promising for HCI concepts to be heavily embedded in the analysis and design of emerging trend detection (ETD) systems. Powerful data modeling techniques can make salient patterns clearer and in sharper contrast. Some of the major technical problems are how to make the changes over time easy to understand and how to preserve the overall context in which changes take place.

ThemeRiver is a visualization system that uses the metaphor of a river to depict thematic flows over time in a collection of documents (Havre, Hetzler, Whitney, & Nowell, 2002). The thematic changes

Table 1. Usability goals and how to apply them to ETD

Usability Goal	Definition	ETD Application
Learnability	"How easy the system is to learn" (Rozanski & Haake, 2003)	The system must be easy to learn for people from a wide variety of fields, including those with non-technical backgrounds.
Efficiency	"How quickly users can complete their tasks" (Rozanski & Haake, 2003)	The system should let the user focus on issues that are relevant to trend detection, without having to worry about issues with the system.
Memorability	"How easy the system is to remember" (Rozanski & Haake, 2003)	Users should not have to relearn the system each time they want to use it.
Control of Errors	Prevention and recovery from errors (Rozanski & Haake, 2003)	The system design should make errors less likely to happen, and when they do happen, the system should help the user out of the errors.
User Satisfaction	"How much users like the system" (Rozanski & Haake, 2003)	The users should be able to accomplish their goals without frustration.

Table 2. Nielson's heuristics and application to ETD

Heuristic	Application
Visibility of system status	While algorithms may take a while to process, there should be feedback so the user knows the progress of the system. (Nielson)
Match between real world and system	The system directions should be presented in language the user can understand; avoid complicated jargon and technical terms. (Nielson)
User control and freedom	The user should be able to set the various thresholds that go into defining topics and emerging trends. (Nielson)
Consistency and standards	Uniform color schemes and presentation of data are necessary. (Nielson)
Error prevention	Steps should be taken to prevent users from entering thresholds that do not work and starting processes without sufficient input. (Nielson)
Recognition rather than recall	Users should not have to remember long, complicated processes. The directions should be presented to them on the screen, or the setup should give the users clues on what to do next. (Nielson)
Flexibility and efficiency of use	The user should be able to easily change thresholds to compare results. There should be shortcuts available for more experiences users as well. (Nielson)
Aesthetic and minimalist design	The interface should be kept simple. (Nielson)
Help users recognize, diagnose, and recover from errors	Errors should be presented in a manner so that it does not look like regular data. (Nielson)
Help and documentation	Ample user manuals should be provided and should be presented in simple language. (Nielson)

are shown along a time line of corresponding external events. A thematic river consists of frequency streams of terms; the changing width of a stream over time indicates the changes of term occurrences. The occurrence of an external event may be followed by sudden changes of thematic strengths. On the one hand, searching for an abruptly widened thematic stream is a much more intuitive task to detect a new story than text-based TDT systems that can only report changes in terms of statistics.

There are many things to keep in mind while developing an HCI-friendly ETD system. The basic usability goals can be used as a guideline to producing a user-friendly ETD system. By striving to make a system learnable, efficient, memorable, keep errors under control, and give the user satisfaction from using the system, the foundation for an HCI friendly system is laid. Table 1 defines each of the usability goals and how they can be applied in an ETD system.

A set of usability heuristics, proposed by Jakob Nielson (n.d.), can also pose as a good rule of thumb (see Table 2).

Task analysis is *a detailed description of an operator's task, in terms of its components, to specify the detailed human activities involved, and their functional and temporal relationships* (HCI Glossary, 2004). By having users describe their process, step-by-step designers can learn much about the user's behavior. When conducting task analysis, have the users describe "the steps they would follow, the databases and tools they would use, and the decision points in the process" (Bartlett & Toms, 2003).

CONCLUSION

Emerging trend detection is a promising field that holds many applications. However, for ETD systems to reach their full potential, they must be effective, easy to learn, easy to understand, and easy to use. A poorly-designed system will shun users away from this technology. It is important to remember that ETD systems are interactive systems. An ETD system, that just takes a data corpus and scans it, is not an effective one. Users must be able to define and experiment with thresholds, view feedback about the data corpus, and be able to understand new trends.

REFERENCES

Aldana, W. A. (2000). *Data mining industry: Emerging trends and new opportunities*. Massachusetts Institute of Technology.

Bartlett, J. C., & Toms, E. G. (2003, July 28-August 1). Discovering and structuring information flow among bioinformatics resources. In the *Proceedings of the 26th annual international ACM SIGIR conference on Research and development in information retrieval*, Toronto, Canada (pp. 411-412). New York: ACM Press.

Basseville, M., & Nikiforov, I. V. (1993). *Detection of abrupt changes: Theory and application*. Englewood Cliffs, NJ: Prentice-Hall.

Chen, C. (2004). Searching for intellectual turning points. *PNAS, 101*(1), 5303-5310.

Del Negro, M. (2001). Turn, turn, turn: Predicting turning points in economic activity. *Economic Review, 86*(2), 1-12.

Dong, G., & Li, J. (1999, August 15-18). Efficient mining of emerging patterns: Discovering trends and differences. In the *Proceedings of the Fifth ACM SIGKDD International Conference on Knowledge Discovery and Data Mining*, San Diego, California (pp. 43-52). New York: ACM Press.

European Monitoring Centre for Drugs and Drug Addiction (EMCDDA). (1999). *Feasibility study on detecting, tracking & understanding emerging trends in drug use.*

Havre, S., Hetzler, E., Whitney, P., & Nowell, L. (2002). ThemeRiver: Visualizing thematic changes in large document collections. *IEEE Transactions on Visualization and Computer Graphics, 8*(1), 9-20. Retrieved June 2004, from http://www.emcdda.eu.int/index.cfm?fuseaction=public.AttachmentDownload&nNodeID=1829&slanguageISO=EN

HCI Glossary (2004). Retrieved on August 13, 2004, from http://www.stack.nl/~cva/index.php

Kleinberg, J. (2002, July 23-26). Bursty and hierarchical structure in streams. In the *Proceedings of the 8th ACM SIGKDD International Conference on Knowledge Discovery and Data Mining*, Edmonton, Alberta, Canada, July 23-26, 2002 (pp. 91-101). ACM Press.

Kontostathis, A., Galitsky, L., Roy, S., Pottenger, W., & Phelps, D. (2003). An overview of emerging trends detection (ETD). *Lecture Notes in Computer Science*. Springer-Verlag. Retrieved October 20, 2005 http://webpages.ursinus.edu/akontosathis/ETDSurvey/pdf

Masao, U., & Kôiti, H. (2000). Multi-topic multi-document summarization. *International Conference On Computational Linguisitics 2*, 892-898.

Nielson, J. (n.d.). Ten usability heuristics. Retrieved on August 20, 2004, from http://www.useit.com/papers/heuristic/heuristic_list.html

Norman, D. A. (1998). *The invisible computer: Why good products can fail, the personal computer is so complex, and information appliances are the solution*. Cambridge, MA: MIT Press.

Roy, S., Gevry, D., & Pottenger, W. (2002). Methodologies for trend detection in textual data mining. In the *Proceedings of the Textmine '02 Workshop, Second SIAM International Conference on Data Mining* (pp. 1-12).

Rozanski, E. P., & Haake, A. R. (2003, October 16-18) The many facets of HCI. In *Proceeding of the 4th Conference on Information Technology Curriculum,* Lafayette, Indiana (pp. 16-18, 2003). New York: ACM Press.

Swan, R., & Allan, J. (1999, November 2-6). Extracting significant time varying features from text. In the *Eighth International Conference on Information Knowledge Management (CIKM'99)*, Kansas City, Missouri (pp. 38-45). New York: ACM Press.

Yang, Y., Pierce, T., & Carbonell, J. (1998, August 24-28). A study on retrospective and online event detection. Paper presented at the *21st ACM International Conference on Research and Develop-*

ment in Information Retrieval (SIGIR'98), Melbourne, Australia (pp. 28-36). New York: ACM Press.

KEY TERMS

Burst Detection: The identification of sharp changes in a time series of values. Examples of bursts include the increasing use of certain words in association with given events.

Information Visualization: A field of study aims to utilize human's perceptual and cognitive abilities to enable and enhance our understanding of patterns and trends in complex and abstract information. Computer-generated 2- and 3-dimensional interactive graphical representations are among the most frequently used forms.

Intellectual Turning Points: Scientific work that has fundamentally changed the subsequence development in its field. Identifying intellectual turning points is one of the potentially beneficial areas of applications of trend detection techniques.

Paradigm Shift: A widely known model in philosophy of science proposed by Thomas Kuhn. Paradigm shift is regarded as the key mechanism that drives science. The core of science is the domination of a paradigm. Paradigm shift is necessary for a scientific revolution, which is how science advances.

Topic Detection and Tracking: A sub-field of information retrieval. The goal is to detect the first appearance of text that differs from a body of previously processed text, or to monitor the behaviour of some identified themes over time.

Trend: The continuous growth or decline of a variable over a period of time.

Trend Detection: Using quantitative methods to identify the presence of a trend. A number of domain-specific criteria may apply to determine what qualifies as a trend, for example, in terms of duration, diversity, and intensity. Primary quality measures of trend detection include sensitivity and accuracy.

Turning Point: A turning point marks the beginning or the end of a trend. For example, the point at which economy turns from recession to growth.

Human-Centered Conceptualization and Natural Language

Javier Andrade
University of A Coruña, Spain

Juan Ares
University of A Coruña, Spain

Rafael García
University of A Coruña, Spain

Santiago Rodríguez
University of A Coruña, Spain

Andrés Silva
Technical University of Madrid, Spain

INTRODUCTION

Conceptual modeling appears to be the heart of good software development (Jackson, 2000). The creation of a conceptual model helps to understand the problem raised and represents the human-centered/problem-oriented moment in the software process, as opposed to the computer-centered/software-oriented moment of the computational models (Blum, 1996). The main objective of human computer interaction (HCI) is also precisely to make human beings the focal point that technology should serve rather than the other way round.

The conceptual models are built with conceptual modeling languages (CMLs), whose specification involves constructors and rules on how to combine these constructors into meaningful statements about the problem.

Considering the criterion of the representation capability of the CMLs in software engineering, their main drawback is that they remain too close to the development aspects (Jackson, 1995). The constructors are too much oriented toward the computational solution of the problem, and therefore, the problem is modeled with implementation concepts (computer/software solution sensitivity) rather than concepts that are proper to human beings (human/problem sensitivity) (Andrade, Ares, García & Rodríguez, 2004). This stands in open opposition to what we have said about the moments in the software process and HCI. Moreover, this situation seriously complicates the essential validation of the achieved conceptual model, because it is drawn up in technical terms that are very difficult to understand by the person who faces the problem (Andrade et al., 2004).

The semantics of the constructors determines the representation capability (Wand, Monarchi, Parsons & Woo, 1995). Since the constructors are too close to implementation paradigms, the CMLs that currently are being used in software engineering are incapable of describing the problem accurately.

Suitable human/problem-related theoretical guidelines should determine which constructors must be included in a genuine CML. This article, subject to certain software-independent theoretical guidelines, proposes the conceptual elements that should be considered in the design of a real CML and, consequently, what constructors should be provided.

The Background section presents the software-independent guidelines that were taken into account to identify the above-mentioned conceptual elements. The Main Focus of the Article section discusses the study that identified those elements. Finally, the Future Trends section presents the most interesting future trends, and the final section concludes.

BACKGROUND

In generic conceptualization, concepts are logically the primary elements. Despite their importance, the nature of concepts remains one of the toughest philosophical questions. However, this does not stop us from establishing some hypotheses about concepts (Díez & Moulines, 1997):

- **HC1. Abstract Entities:** Concepts are identifiable abstract entities to which human beings have access, providing knowledge and guidance about the real world.
- **HC2. Contraposition of a System of Concepts with the Real World:** Real objects can be identified and recognized thanks to the available concepts. Several (real) objects are subsumed within one and the same (abstract) concept.
- **HC3. Connection Between a System of Concepts and a System of Language:** The relationship of expression establishes a connection between concepts and expressions, and these (physical entities) can be used to identify concepts (abstract entities).
- **HC4. Expression of Concepts by Non-Syncategorematic Terms:** Practically all non-syncategorematic terms introduced by an expert in a field express a concept.
- **HC5. Need for Set Theory:** For many purposes, the actual concepts should be substituted by the sets of subsumed objects to which set theory principles can be applied.

Likewise, from a general viewpoint, any conceptualization can be defined formally as a triplet of the form (*C*oncepts, *R*elationships, *F*unctions) (Genesereth & Nilsson, 1986), which includes, respectively, the concepts that are presumed or hypothesized to exist in the world, the relationships (in the formal sense) among concepts, and the functions (also in the formal sense) defined on the concepts.

This and the fact that natural language is the language *par excellence* for describing a problem (Chen, Thalheim & Wong, 1999) constitute the basis of our study.

MAIN FOCUS OF THE ARTICLE

It would certainly not be practical to structure a CML on the basis of the previous three formal elements, because (i) concepts are abstract entities (HC1); (ii) relationships and functions are defined on the concepts, which increases the complexity; and (iii) people naturally express themselves in natural language (HC3: connection between a system of concepts and a system of language).

Taking this and HC4 (expression of concepts by non-syncategorematic terms) into account, we propose defining the CMLs on the basis of the conceptual elements that result from the analysis of natural language. This procedure stems from the fact that there is a parallelism between natural language and the CML (Hoppenbrouwers, van der Vos & Hoppenbrouwers, 1997).

From the analysis detailed in this section, we find that the identified conceptual elements actually can be matched to some of the three elements of the previous formal triplet; that is, the generic and formal definition is not overlooked. However, ultimately, a functional information taxonomy can be established, which is much more natural and practical.

Analyzing Natural Language

Based on HC4, the conceptual elements were identified by analyzing the non-syncategorematic categories of nouns, adjectives, and verbs. Moreover, importance was also attached to adverbs, locutions, and other linguistic expressions, which, although many are syncategorematic terms, were considered relevant because of their conceptual load.

Nouns

Nouns can be divided into different groups according to different semantic traits. The most commonly used trait is the classification that determines whether the noun is common or proper.

Considering this latter trait, we notice a parallelism between nouns and elements that are handled in any conceptualization: common nouns can lead to concepts or properties, and proper nouns can lead to property values. The following subsections consider these elements.

Concepts

A concept can be defined as a mental structure, which, when applied to a problem, clarifies to the point of solving this problem. Here, this term is used in the sense of anything that is relevant in the problem domain and about which something is to be expressed by the involved individuals.

Interpreted in this manner, relationships, actions, and many other elements actually are concepts. However, here, we only consider the concepts that are proper to the problem domain; that is, concept means anything that is, strictly speaking, proper to the problem, which may refer to concrete or abstract, or elementary or compound things.

The concepts thus considered are included within C in the triplet (C, R, F), and bearing in mind HC2 (contraposition of a system of concepts with the real world), a concept subsumes a set of objects that are specific occurrences of it.

Properties

A property is a characteristic of a concept or a relationship, as a relationship could be considered as a compound concept (we will address this conceptual element later).

The set of all the properties of a concept/relationship describes the characteristics of the occurrences subsumed by this concept/relationship, each of which can be considered as functions or relationships of the triplet (C,R,F), depending on how the problem is to be conceptualized (Genesereth & Nilsson, 1986).

Property Values

The value(s) of a property is (are) selected from a range of values, which is the set of all the possible values of the property.

Considering the triplet (C,R,F), if a property is conceptualized as a function, the possible values of the property are within C. If it is conceptualized as a relationship, the actual property really disappears and unary relationships are considered for each possible property value instead (Genesereth & Nilsson, 1986).

Adjectives

The adjectives are used to determine the extent of the meaning of the noun (adjectival determiners) or to qualify the noun (adjectival modifiers).

The adjectival determiners always accompany a noun and do not have semantic traits that alter the semantics of the noun phrase. Consequently, these elements do not have to be considered in a conceptualization.

The adjectival modifiers can be descriptive or relational. The former refer to a property of the noun that they qualify. These adjectives are classed into different types according to their semantic trait (Miller, 1995): quality, size, type, and so forth. Only the type-related classification can lead to a new interpretation to be conceptualized—the relationship of generalization/specialization, which will be described next.

Finally, the relational adjectival modifiers allude to the scope of the noun, and therefore, the above interpretation is also possible.

Verbs

The linguistic theory that we have used to analyze verbs is the Case Grammar Theory (Cook, 1989), which describes a natural language sentence in terms of a predicate and a series of arguments called *cases* (agent, object, etc.).

This theory provides a semantic description of the verbs, which is the fundamental aspect to be considered here. It is precisely the semantic relationship between the verb and its cases that is interpreted and modeled in conceptual modeling. This theory interprets the relationship between concepts; the verb alludes to the relationship and the cases to the related concepts. This obviously equates with R in the (C, R, F) triplet.

Case Grammar theories establish a verb classification, depending on the cases that accompany the verb. The semantics and, therefore, the conceptualization of the relationship differ, depending on the type of verb that expresses the information. Nevertheless, these differences are always conceptual nuances of relationships.

Case Grammar theories do not establish just one verb classification that depends on the semantic nuances and the cases considered. We have used the classification established by Martínez (1998), because it places special emphasis on justifications based on natural language. Looking at this classification, we find that there are the following different types of relationships:

1. **Generalization/Specialization:** This represents the semantics "is a" in natural language, indicating the inclusion of a given concept in another more general one. This relationship takes the form of a hierarchy of concepts in which what is true for a set is also true for its subsets. In this respect, remember here HC5 (need for set theory) and the importance of determining whether the sets (concepts) are disjoint and total—disjoint or overlap and total or partial relationship, respectively.
2. **Aggregation:** This represents the natural language semantics "part of," making it possible to form a concept from other concepts of which the former is composed. Aggregation is a Relationship that Has the Property of Transitivity. However, it does not always appear to be transitive (e.g., the hands are part of a mechanic, and the mechanic is part of a company, but the hands are not part of the company). With the aim of resolving this paradox, several types of aggregations have been identified.
   There are two main types of aggregation, and the property of transitivity holds, provided that the aggregations are of the same type, although they do not always lead to intransitivity, even if they are different:
   a. **Member/Collection:** The parts or members are included in the collection because of their spatial proximity or social connection. The parts of which the whole is composed are of the same type, and the whole does not change if one is removed.
   b. **Compound/Component:** The components perform a function or have a relationship with respect to other components or the whole that they form. The parts are of different types, and the whole changes if a part is removed.
   c. **Defined by the Meaning of the Verb:** They are particular to each domain and generally differ from one domain to another. Therefore, a classification cannot be established as for the previous relationships, where the meaning remains unchanged in any domain.

Since a CML should be able to be used to represent reality as closely as possible and gather most of the semantics, it should provide different constructors for the different types of relationships. This distinction will make it possible to immediately assimilate all the underlying semantics for each type.

Other Grammatical Categories

There are linguistic structures that have not yet been analyzed and which actually introduce aspects to be conceptualized: adverbs, locutions, and other linguistic expressions that are very frequent in problem modeling. The conceptual elements that these structures introduce are as follows:

1. **Constraints:** No more than, like minimum, and so forth.
2. **Inferences and Calculations:** Calculate, if...then..., deduce, and so forth.
3. **Sequence of Actions:** First, second, after, and so forth.

Constraints

A constraint can be defined as a predicate whose values are true or false for a set of elements. Therefore, it can be viewed in the triplet (C,R,F) as a function that constrains the possible values of these elements.

Constraints are new elements to be conceptualized but which affect the elements already identified. Indeed, constraints can be classified according to the element they affect—constraints on properties, property values, or relationships. The first affect the actual properties, demanding compulsoriness or unicity. The second affect property values, restricting their possible values in the occurrences. The third restrict the occurrences that can participate in a relationship.

Inferences and Calculations

Considering the triplet (C,R,F), these elements can be placed within F. This is because they allude to the manipulation of known facts to output new facts.

Considering these elements involve providing the language with constructors to conceptualize inferences, which indicate what to infer and what to use for this purpose, and calculations, which indicate how to calculate something using a mathematical or algorithmic expression.

Sequence of Actions

This element indicates what steps the human beings take and when to solve the problem. This means that the modeling language should include the constructors necessary to represent the following:

- **Decomposition into Steps:** Human beings typically solve problems by breaking them down into steps. Logically, the non-decomposed steps should indicate exactly what function they have and how they are carried out (inferences and calculations).
- **Step Execution Order:** Clearly, the order in which the identified steps are taken is just as essential as the previous.

In the triplet (C,R,F), the actions or steps can be considered within C and their decomposition and order as relationships within R. For the latter type of relationships, constructors are needed to enable the bifurcations governed by the information known or derived from the domain.

Figure 1. Functional levels of information and their interrelationships

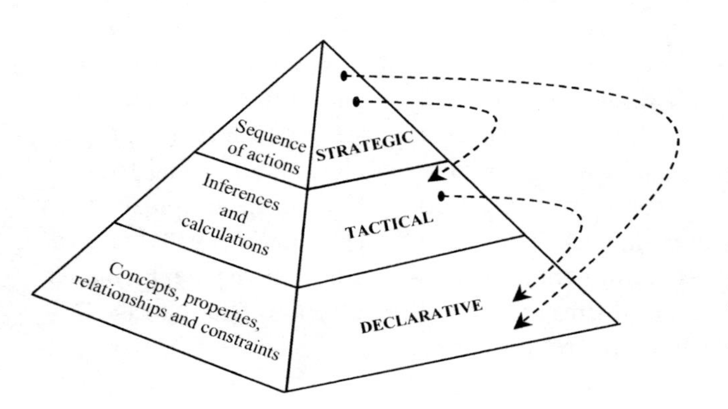

Establishing the Functional Information Taxonomy

Depending on the function that they fulfill in the problem, all of the previously identified conceptual elements can be grouped into the following information levels (see Figure 1):

- **Strategic:** It specifies what to do, when, and in what order.
- **Tactical:** It specifies how to obtain new declarative information.
- **Declarative:** It specifies the facts known about the problem.

Figure 1 also shows the previously mentioned interrelationships between the different levels. The declarative level is managed by the other two levels, as it specifies what is used to decide on the alternatives of execution in the step sequence and on what basis the inferences and calculations are made. Moreover, the strategic level manages the tactical level, as it has to specify what inferences and calculations have to be made for each non-decomposed step.

In a CML that accounts for the presented approach, constructors should be considered for the three submodels that jointly will conform a complete conceptual model: declarative, tactical and strategic.

FUTURE TRENDS

Based on the previous study, we have defined a graphic CML, detailed in Andrade (2002), with a view to getting optimum expressiveness and manageability. However, not all the information involved in a conceptualization can be detailed using a graphic notation. Any attempt to do so would complicate matters so much that it would relegate the benefits of the graphic capability to oblivion. For this reason, not all the model aspects are set out graphically. Thus, the previously mentioned CML was defined in the following way:

- **Declarative Submodel:** Constructs for concepts, relationships, and constraints on properties are graphic. Constructs for properties and constraints on relationships are half-graphic. Property values and constraints on property values are expressed through textual constructs.
- **Tactical Submodel:** Inferences are expressed through constructs that are half-graphic, whereas calculations are expressed through textual constructs.
- **Strategic Submodel:** Constructs for step decomposition are graphic, whereas constructs for step execution order are half-graphic.

To continue with the software process, research should focus now on identifying the criteria to (i) select the most suitable development paradigm(s) and (ii) map the conceptual constructions into the corresponding computational ones in that (those) selected paradigm(s). Moreover, a CASE tool to support the constructors defined and to facilitate the application of the previous criteria is really interesting. We are now working on both aspects.

CONCLUSION

The previously mentioned conceptual elements are considered in the conceptual modeling languages within software development today. However, (i) none of the languages considers them all; (ii) they are considered to represent certain technical development concepts; and (iii) as a result, the expressiveness—semantics—of their constructors is located in the software solution (computer) domain and not in the problem (human) domain, as should be the case.

The defined language has been applied to build the conceptual models in both software engineering (Andrade et al., 2004) and knowledge engineering (Ares & Pazos, 1998) conceptualization approaches, which demonstrates its generality and closeness to human beings and to the problem rather than to the software solution.

ACKNOWLEDGMENTS

The authors would like to thank Juan Pazos for his constructive suggestions, the University of A Coruña for providing financial support, and Valérie Bruynseraede for her help in translating the article into English.

REFERENCES

Andrade, J. (2002). *Un marco metodológico para el modelado conceptual* [doctoral thesis]. A Spain: University of A Coruña.

Andrade, J., Ares, J., García, R., & Rodríguez, S. (2004). Conceptual modelling in requirements engineering: Weaknesses and alternatives. In J.L. Maté & A. Silva (Eds.), *Requirements engineering for sociotechnical systems.* Hershey, PA: InfoSci.

Andrade, J., et al. (2004). A methodological framework for generic conceptualisation: Problem-sensitivity in software engineering. *Information and Software Technology, 46*(10), 635-649.

Ares, J., & Pazos, J. (1998). Conceptual modelling: An essential pillar for quality software development. *Knowledge-Based Systems, 11,* 87-104.

Blum, B.I. (1996). *Beyond programming. To a new era of design.* New York: Oxford University Press.

Chen, P.P., Thalheim, B., & Wong, L.Y. (1999). Future directions of conceptual modelling. In P.P. Chen, J. Akoka, H. Kangassalu, & B. Thalheim (Eds.), *Conceptual modelling* (pp. 287-301). Berlin: Springer-Verlag.

Cook, W.A. (1989). *Case grammar theory.* Washington, DC: Georgetown University Press.

Díez, J.A., & Moulines, C.U. (1997). *Fundamentos de filosofía de la ciencia.* Barcelona: Ariel.

Genesereth, M., & Nilsson, N. (1986). *Logical foundations of artificial intelligence.* CA: Morgan Kaufmann Publishers.

Hoppenbrouwers, J., van der Vos, B., & Hoppenbrouwers, S. (1997). NL structures and conceptual modelling: Grammalizing for KISS. *Data & Knowledge Engineering, 23*(1), 79-92.

Jackson, M. (1995). *Software requirements and specifications. A lexicon of practice, principles and prejudices.* New York: ACM Press.

Jackson, M. (2000). Problem analysis and structure. In T. Hoare, M. Broy, & R. Steinbruggen (Eds.), *Engineering theories of software construction-proceedings of NATO summer school* (pp. 3-20). Marktoberdorf: IOS Press.

Martínez, P. (1998). *Una propuesta de estructuración del conocimiento para el análisis de textos: Una aplicación a la adquisición de esquemas conceptuales de bases de datos* [doctoral thesis]. Madrid, Spain: Technical University of Madrid.

Miller, G.A. (1995). WordNet: A lexical database for English. *Communications of the ACM, 38*(11), 39-41.

Wand, Y., Monarchi, D., Parsons, J., & Woo, C. (1995). Theoretical foundations for conceptual modeling in information systems development. *Decision Support Systems, 15*(4), 285-304.

KEY TERMS

Concept: A mental structure derived from acquired information, which, when applied to a problem, clarifies to the point of solving this problem.

Conceptual Model: An abstraction of the problem as well as a possible model of a possible conceptual solution to the problem.

Conceptual Modeling: The use of concepts and their relationships to deal with and solve a problem.

Conceptual Modeling Language: A language used to represent conceptual models.

Human/Problem-Sensitivity: The proximity to the human-centered/problem-oriented concepts, as opposed to the computer-centered/software-oriented concepts (i.e., computer/software solution-sensitivity).

Natural Language: A language naturally spoken or written by human beings.

Non-Syncategorematic Terms: These linguistic terms (also known as categorematic terms) are capable of being employed by themselves as terms as opposed to syncategorematic terms.

Problem: A situation in which someone wants something and does not immediately know what actions to take to get it.

Syncategorematic Terms: These linguistic terms cannot stand as the subject or the predicate of a proposition. They must be used in conjunction with other terms, as they have meaning only in such combination.

Human–Computer Interaction and Security

Kai Richter
Computer Graphics Centre (ZGDV), Germany

Volker Roth
OGM Laboratory LLC, USA

INTRODUCTION

Historically, computer security has its roots in the military domain with its hierarchical structures and clear and normative rules that are expected to be obeyed (Adams & Sasse, 1999). The technical expertise necessary to administer most security tools stems back to the time where security was the matter of trained system administrators and expert users. A considerable amount of money and expertise is invested by companies and institutions to set up and maintain powerful security infrastructures. However, in many cases, it is the user's behavior that enables security breaches rather than shortcomings of the technology. This has led to the notion of the user as the weakest link in the chain (Schneier, 2000), implying that the user was to blame instead of technology. The engineer's attitude toward the fallible human and the ignorance of the fact that technology's primary goal was to serve human turned out to be hard to overcome (Sasse, Brostoff, & Weirich, 2001).

BACKGROUND

With the spreading of online work and networked collaboration, the economic damage caused by security-related problems has increased considerably (Sacha, Brostoff, & Sasse, 2000). Also, the increasing application of personal computers, personal networks, and mobile devices with their support of individual security configuration can be seen as one reason for the increasing problems with security (e.g., virus attacks from personal notebooks, leaks in the network due to personal wireless LANs, etc.) (Kent, 1997). During the past decade, the security research community has begun to acknowledge the importance of the human factor and has started to take research on human-computer interaction into consideration. The attitude has changed from blaming the user as a source of error toward a more user-centered approach trying to persuade and convince the user that security is worth the effort (Ackerman, Cranor, & Reagle, 1999; Adams & Sasse, 1999; Markotten, 2002; Smetters & Grinter, 2002; Whitten & Tygar, 1999; Yee, 2002).

In the following section, current research results concerning the implications of user attitude and compliance toward security systems are introduced and discussed. In the subsequent three sections, security-related issues from the main application areas, such as authentication, email security, and system security, are discussed. Before the concluding remarks, an outlook on future challenges in the security of distributed context-aware computing environments is given.

USER ATTITUDE

The security of a system cannot be determined only by its technical aspects but also by the attitude of the users of such a system. Dourish et al. (2003) distinguish between theoretical security (e.g., what is technologically possible) and effective security (e.g., what is practically achievable). Theoretical security to their terms can be considered as the upper bound of effective security. In order to improve effective security, the everyday usage of security has to be improved. In two field studies, Weirich and Sasse (2001) and Dourish et al. (2003) explored users' attitudes to security in working practice. The findings of both studies can be summarized under the following categories: perception of security, perception of threat, attitude toward security-related issues, and the social context of security.

Perception of security frequently is very inaccurate. Security mechanisms often are perceived as holistic tools that provide protection against threats, without any detailed knowledge about the actual scope. Therefore, specialized tools often are considered as insufficient, as they do not offer general protection. On the other hand, people might feel protected by a tool that does not address the relevant issue and thus remain unprotected (e.g., firewall protects against e-mail virus).

Perception of threats also reveals clear misconceptions. None of the users asked considered themselves as really endangered by attacks. As potential victims, other persons in their organization or other organizations were identified, such as leading personnel, people with important information, or high-profile institutions. Only a few of them realized the fact that they, even though not being the target, could be used as a stepping stone for an attack. The general attitude was that no one could do anything with the information on my computer or with my e-mails.

Potential attackers mainly were expected to be hackers or computer kids, with no explicit malevolent intentions but rather seeking fun. Notorious and disturbing but not really dangerous offenders, such as vandals, spammers, and marketers, were perceived as a frequent threat, while on the other hand, substantially dangerous attackers such as criminals were expected mainly in the context of online banking.

The attitude toward security technology was rather reserved. Generally, several studies reported three major types of attitudes toward security: privacy fundamentalists, privacy pragmatists, and privacy unconcerned (Ackerman et al., 1999). Users' experiences played a considerable role in their attitude, as experienced users more often considered security as a hindrance and tried to circumvent it in a pragmatic fashion in order to reach their work objectives. Weirich and Sasse (2001) report that none of the users absolutely obeyed the prescribed rules, but all were convinced that they would do the best they could for security.

Additionally, users' individual practices are often in disagreement with security technology. People use legal statements in e-mail footers or cryptic e-mails, not giving explicit information but using contextual cues instead. In conjunction with such subsidiary methods and the fact that people often seem to switch to the telephone when talking about important things (Grinter & Palen, 2002) indicates the poor perception users have of security technology.

The feeling of futility was reported with respect to the need for constantly upgrading security mechanisms in a rather evolutionary struggle (i.e., if somebody really wants to break in, he or she will). As a result, personal accountability was not too high, as users believed that in a situation where someone misused his or her account, personal credibility would weigh more than computer-generated evidence, in spite of the fact that the fallibility of passwords is generally agreed.

The social context has been reported to play an important role in day-by-day security, as users are not permanently vigilant and aware of possible threats but rather considered with getting their work done. Therefore, it is no wonder that users try to delegate responsibility to technical systems (encryption, firewalls, etc.), colleagues and friends (the friend as expert), an organization (they know what they do), or institutions (the bank cares for secure transfers). Most people have a strong belief in the security of their company's infrastructure. Delegation brings security out of the focus of the user and results in security unawareness, as security is not a part of the working procedure anymore.

Whenever no clear guidelines are available, people often base their practice on the judgments of others, making the system vulnerable to social engineering methods (Mitnick, Simon, & Wozniak, 2002). In some cases, collaboration appears to make it necessary or socially opportune to disclose one's password to others for practical reasons, technical reasons, or as a consequence of social behavior, since sharing a secret can be interpreted as a sign of trust. Such sharing is a significant problem, as it is used in social engineering in order to obtain passwords and to gain access to systems.

Dourish et al. (2003) came to the conclusion that "where security research has typically focused on theoretical and technical capabilities and opportunities, for end users carrying out their work on computer systems, the problems are more prosaic" (p. 12). The authors make the following recommendations for the improvement of security mechanisms in the system and in the organizational context:

- Users should be able to access security settings easily and as an integral part of the actions, not in the separated fashion as it is today; therefore, security issues should be integrated in the development of applications (Brostoff & Sasse, 2001; Gerd tom Markotten, 2002).
- It is necessary that people can monitor and understand the potential consequences of their actions (Irvine & Levin, 2000) and that they understand the security mechanisms employed by the organization.
- Security should be embedded into working practice and organizational arrangement, and visible and accessible in everyday physical and social environment (Ackerman & Cranor, 1999).
- Security should be part of the positive values in an organization. So-called social marketing could be used to establish a security culture in a company.
- The personal responsibility and the danger of personal embarrassment could increase the feeling of personal liability.
- The importance of security-aware acting should be made clear by emphasizing the relevance to the organization's reputation and financial dangers.

As has been shown, the design and implementation of security mechanisms are closely interlinked to the psychological and sociological aspects of the user's attitude and compliance toward the system. Any security system is in danger of becoming inefficient or even obsolete if it fails to provide adequate support and motivate users for its proper usage. The following sections discuss these findings in the context of the main application domains of computer security.

AUTHENTICATION

Information technology extends our ability to communicate, to store and retrieve information, and to process information. With this technology comes the need to control access to its applications for reasons of privacy and confidentiality, national security, or auditing and billing, to name a few. Access control in an IT system typically involves the identification of a subject, his or her subsequent authentication, and, upon success, his or her authorization to the IT system.

The crucial authentication step generally is carried out based on something the subject knows, has, or is. By far the most widespread means of authentication is based on what a subject has (e.g., a key). Keys unlock doors and provide access to cars, apartments, and contents of a chest in the attic. Keys are genuinely usable—four-year-olds can handle them. In the world of IT, something the subject knows (e.g., a password or a secret personal identification number [PIN]) is the prominent mechanism.

The exclusiveness of access to an IT system protected by a password rests on the security of the password against guessing, leaving aside other technical means by which it may or may not be broken. From an information theoretic standpoint, a uniformly and randomly chosen sequence of letters and other symbols principally provides the greatest security. However, such a random sequence of unrelated symbols also is hard to remember, a relation that is rooted in the limitation of humans' cognitive capabilities.

As a remedy, a variety of strategies were invented to construct passwords that humans can memorize more easily without substantially sacrificing the security of a password (e.g., passwords based on mnemonic phrases). For instance, Yan, et al. (2000) conducted a study with 400 students on the effect of three forms of advice on choosing passwords. They found, for example, that passwords based on pass phrases were remembered as easily as naively chosen passwords, while being as hard to crack as randomly chosen passwords. Insight into human cognition also has led to the investigation of alternatives such as cognitive passwords (Zviran & Haga, 1990), word associations (Smith, 1987), pass phrases (Spector & Ginzberg, 1994), images (Dhamija & Perrig, 2000), or pass faces (Brostoff & Sasse, 2000).

Authentication in public places, as is the case with automatic teller machines (ATM), has turned out to be vulnerable to attacks, where criminals obtain a user's PIN by using cameras or other methods of observation in so-called shoulder-surfing attacks (Colville, 2003). In order to obscure the numbers entered by the user and thus hamper the recording of the necessary PIN, several techniques have been proposed (Hopper & Blum, 2001; Wilfong, 1997). Recently, Roth, et al. (2004) suggested variants of cognitive trapdoor games to protect users against shoulder-surfing attacks. In this approach, the buttons on a PIN pad are colored either black or white, and the user has to decide whether the number to be entered is in the black or white group. As the colors are changing randomly, the user has to enter the same number three to four times to complete an input. By blurring the response set with so-called shadows, this method can be made resistant against camera attacks. Even though this approach is slightly more complicated than the classical approach, this technique has proven to be accepted by the user in an experimental setting.

E-MAIL SECURITY

Before the middle of the 1970s, cryptography was built entirely on symmetric ciphers. This meant that in order for enciphered communication to take place, a secret key needed to be exchanged beforehand over a secure out-of-band channel. One way of doing that was to send a trusted courier to the party with whom one intended to communicate securely. This procedure addressed two important issues: the secret key exchange and the implicit authentication of the exchanged keys. Once established, the keys could be used to secure communication against passive and active attacks until the key was expected to become or became compromised.

When asymmetric cryptography (Diffie & Hellman, 1976; Rivest, Shamir, & Adleman, 1978) was invented in the 1970s, it tremendously simplified that task of key exchange, and gave birth to the concept of digital signatures. Asymmetric cryptography did not solve the problem of authenticating keys per se. Although we now can exchange keys securely in the clear, how could one be certain that a key actually belonged to the alleged sender? Toward a solution to this problem, Loren Kohnfelder (1978) invented the public key certificate, which is a public key and an identity, signed together in a clever way with the private key of a key introducer whom the communicating parties need to trust. This idea gave rise to the notion of a public key infrastructure (PKI). Some existing models of public key infrastructures are the OpenPGP Web of Trust model (RFC 2440) and the increasingly complex ITU Recommendation X.509-based PKIX model (RFC 3280) (Davis, 1996; Ellison, 1996, 1997; Ellison & Schneier, 2000).

In applications such as electronic mail, building trust in certificates, exchanging keys, and managing keys account for the majority of the interactions and decisions that interfere with the goal-oriented tasks of a user and that the user has difficulty understanding (Davis, 1996; Gutmann, 2003; Whitten & Tygar, 1999). At the same time, the majority of users only has limited understanding of the underlying trust models and concepts (Davis, 1996) and a weak perception of threats (see previous discussion). Consequently, they avoid or improperly operate the security software (Whitten & Tygar, 1999).

In the safe staging approach, security functions may be grouped into stages of increasing complexity. A user may begin at a low stage and progress to a higher stage once the user understands and masters the security functions at the lower stages. The safe-staging approach was proposed by Whitten & Tygar (2003), who also pioneered research on the usability of mail security by analyzing users' performances when operating PGP (Whitten & Tygar, 1999).

SYSTEM SECURITY

Computer systems progressed from single user systems and multi-user batch processing systems to multi-user time-sharing systems, which brought the requirement to sharing the system resources and at the same time to tightly control the resource allocation as well as the information flow within the system. The principal approach to solving this is to establish a verified supervisor software also called the reference monitor (Anderson, 1972), which controls all security-relevant aspects in the system.

However, the Internet tremendously accelerated the production and distribution of software, some of which may be of dubious origin. Additionally, the increasing amounts of so-called malware that thrives on security flaws and programming errors lead to a situation where the granularity of access control in multi-user resource-sharing systems is no longer sufficient to cope with the imminent threats. Rather than separating user domains, applications themselves increasingly must be separated, even if they run on behalf of the same user. A flaw in a Web browser should not lead to a potential compromise of other applications and application data such as the user's e-mail client or word processor. Despite efforts to provide solutions to such problems (Goldberg et al., 1996) as well as the availability of off-the-shelf environments in different flavors of Unix, fine-grained application separation has not yet been included as a standard feature of a COTS operating system.

Even if such separation were available, malicious software may delude the user into believing, for example, that a graphical user interface (GUI) component of the malware belongs to a different trusted application. One means of achieving this is to mimic the visual appearance and responses of the genuine application. One typical example would be a fake login screen or window. Assurance that a certain GUI component actually belongs to a particular application or the operating system component requires a trusted path between the user and the system. For instance, a secure attention key that cannot be intercepted by the malware may switch to a secure login window. While this functionally is available in some COTS operating systems, current GUIs still provide ample opportunity for disguise, a problem that also is eminent on the Web (Felten, Balfanz, Dean, & Wallach, 1997). One approach to solving this problem for GUIs is to appropriately mark windows so that they can be associated with their parent application (Yee, 2002). One instance of a research prototype windowing system designed with such threats in mind is the EROS Trusted Window System (Shapiro, Vanderburgh, Northup, & Chizmadia, 2003).

FUTURE TRENDS

Mobile computing and the emergence of context-aware services progressively are integrating into new and powerful services that hold the promise of making life easier and safer. Contextual data will help the user to configure and select the services the user needs and even might elicit support proactively. Far beyond that, Ambient Intelligence (IST Advisory Group, 2003) is an emerging vision of dynamically communicating and cooperating appliances and devices in order to provide an intelligent surrounding for tomorrow's citizens. Radio frequency identification transmitters (RFID) already have been discussed with respect to their implications on privacy (Weis, 2004). Certainly, one person's contextual awareness is another person's lack of privacy (Hudson & Smith, 1996). In the future, the development of powerful and usable security concepts for applying personal information to the context and vice versa is one of the greatest challenges for today's security engineers and human-computer interaction research (Ackerman, Darell, & Weitzner, 2001). To accomplish this task seems crucial for future acceptance and for chances of such technologies without them becoming a "privacy Chernobyl" (Agre, 1999).

CONCLUSION

The view of the user as the weakest link and potential security danger finally has turned out to be an obsolescent model. Security engineers and perhaps, more importantly, those people who are responsible for IT security have noticed that working against the user will not do, and instead, they have decided to work with and for the user. During the past years, an increasing number of research has focused on the issue of making security usable, addressing the traditional fields of authentication, communication, and e-mail and system security. This article has given a brief overview of some of the work done so far. In order to make information technology more secure, the user is the central instance. The user must be able to properly use the security mechanisms provided. Therefore, understanding users' needs and identifying the reasons that technology fails to convince users to employ it is crucial. The first part of this article summarized some work done by Dourish, Weirich, and Sasse that provided important insights. But much work still has to be done.

Future technology will build even more on the integration and sharing of heterogeneous sources of information and services. The tendency toward distributed and location-based information infrastructures will lead to new security problems. Feeling safe is an important aspect of acceptance. The success of tomorrow's systems also will depend on the user's feeling safe while sharing information and using services, which has already been shown during the first stage of e-commerce. Therefore, making security usable is an important aspect of making security safe.

REFERENCES

Ackerman, M. S., & Cranor, L. (1999). Privacy critics: UI components to safeguard users' privacy. *Proceedings of the ACM SIGCHI Conference on Human Factors in Computing Systems*, Pittsburgh, Pennsylvania.

Ackerman, M. S., Cranor, L., & Reagle, J. (1999). Privacy in e-commerce: Examining user scenarios and privacy preferences. *Proceedings of the ACM Conference in Electronic Commerce*, New York.

Ackerman, M. S., Darell, T., & Weitzner, D. (2001). Privacy in context. *Human-Computer Interaction, 16*(2), 167-176.

Adams, A., & Sasse, M. A. (1999). Users are not the enemy: Why users compromise security mechanisms and how to take remedial measures. *Communications of the ACM, 42*(12), 40-46.

Agre, P. (1999). *Red rock eater digest—Notes and recommendations*. Retrieved from http://commons.somewhere.com/rre/1999/RRE.notes.and.recommenda14.html

Anderson, J. P. (1972). *Computer security technology planning study (No. ESD-TR-73-51)*. Bedford, MA: AFSC.

Brostoff, S., & Sasse, M. A. (2000). Are passfaces more usable than passwords? A field trial investigation. In S. McDonald, Y. Waern, & G. Cockton (Eds.), People and computers XIV—Usability or else! *Proceedings of HCI 2000*, Sunderland, UK.

Brostoff, S., & Sasse, M. A. (2001). Safe and sound: A safety-critical design approach to security. *Proceedings of the New Security Paradigms Workshop*, Cloudcroft, New Mexico.

Colville, J. (2003). ATM scam netted 620,000 Australian. *Risks Digest, 22*, 85.

Davis, D. (1996). Compliance defects in public key cryptography. *Proceedings of the 6th USENIX Security Symposium.*

Dhamija, R., & Perrig, A. (2000). Deja vu: A user study using images for authentication. *Proceedings of the 9th USENIX Security Symposium*, Denver, Colorado.

Diffie, W., & Hellman, M. E. (1976). New directions in cryptography. *IEEE Transactions on Information Theory, 22*(6), 644-654.

Dourish, P., Grinter, R. E., Dalal, B., Delgado de la Flor, J., & Dalal, M. (2003). *Security day-to-day: User strategies for managing security as an everyday, practical problem* (ISR Technical Report No. UCI-ISR-05-5). Irvine: University of California.

Ellison, C. (1996). Establishing identity without certification authorities. *Proceedings of the Sixth USENIX Security Symposium.*

Ellison, C., & Schneier, B. (2000). Ten risks of PKI: What you're not being told about public key infrastructure. *Computer Security Journal, 16*(1), 1-7.

Felten, E. W., Balfanz, D., Dean, D., & Wallach, D.S. (1997). *Web spoofing: An Internet con game* (technical report no. 540-96). Princeton, NJ: Princeton University.

Gerd tom Markotten, D. (2002). User-centered security engineering. *Proceedings of the 4th EurOpen/USENIX Conference*, Helsinki, Finland.

Goldberg, I., Wagner, D., Thomas, R., & Brewer, E.A. (1996). A secure environment for untrusted helper applications. *Proceedings of the 6th Usenix Security Symposium*, San Jose, California.

Grinter, R., & Palen, L. (2002). Instant messaging in teen life. *Proceedings of the ACM Conference on Computer-Supported Cooperative Work*, New Orleans, Louisiana.

Gutmann, P. (2003). Plug-and-play PKI: A PKI your mother can use. *Proceedings of the 12th USENIX Security Symposium*, Washington, DC.

Hopper, N. J., & Blum, M. (2001). Secure human identification protocols. *Proceedings of the ASIACRYPT.*

Hudson, S. E., & Smith, I. (1996). Techniques for addressing fundamental privacy and disruption tradeoffs in awareness support systems. *Proceedings of the ACM Conference on Computer Supported Cooperative Work*, Boston.

Irvine, C., & Levin, T. (2000). Towards quality of secure service in a resource management system benefit function. *Proceedings of the 9th Heterogeneous Computing Workshop*.

IST Advisory Group. (2003). *Ambient intelligence: From vision to reality* (draft). Brussels, Belgium: The European Commission.

Kent, S. (1997). *More than screen deep: Towards every citizen interfaces to the nation's information infrastructure*. Washington, DC: National Academy Press.

Kohnfelder, L. M. (1978). *Towards a practical public-key cryptosystem* [B.S. Thesis]. Cambridge, MA: MIT.

Mitnick, K. D., Simon, W. L., & Wozniak, S. (2002). *The art of deception: Controlling the human element of security*. New York: John Wiley & Sons.

Rivest, R.L., Shamir, A., & Adleman, L. (1978). A method for obtaining digital signatures and public-key cryptosystems. *Communications of the ACM, 21*(2), 120-126.

Roth, V., Richter, K., & Freidinger, R. (2004). A PIN-entry method resilient against shoulder surfing. *Proceedings of the CCS*, Washington, DC.

Sasse, M. A. (2003). Computer security: Anatomy of a usability disaster, and a plan for recovery. *Proceedings of the CHI2003 Workshop on Human-Computer Interaction and Security Systems*, Ft. Lauderdale, Florida.

Sasse, M. A., Brostoff, S., & Weirich, D. (2001). Transforming the "weakest link": A human-computer interaction approach to usable and effective security. *BT Technology Journal, 19*(3), 122-131.

Schneier, B. (2000). *Secrets and lies*. New York: Wiley & Sons.

Shapiro, J. S., Vanderburgh, J., Northup, E., & Chizmadia, D. (2003). *The EROS trusted window system* (technical report no. SRL2003-05). Baltimore: Johns Hopkins University.

Smetters, D. K., & Grinter, R. E. (2002). Moving from the design of usable security technologies to the design of useful secure applications. *Proceedings of the New Security Paradigms Workshop*, Virginia Beach.

Smith, S. L. (1987). Authenticating users by word association. *Computers & Security, 6*, 464-470.

Spector, Y., & Ginzberg, J. (1994). Pass-sentence—A new approach to computer code. *Computers & Security, 13*, 145-160.

Tom Markotten, D. G. (2002). User-centered security engineering. *Proceedings of the 4th NordU Conference*, Helsinki, Finland,

Weirich, D., & Sasse, M. A. (2001). Pretty good persuasion: A first step towards effective password security for the real world. *Proceedings of the New Security Paradigms Workshop 2001*, Cloudcroft, New Mexico.

Weis, S. A. (2004). RFID privacy workshop: Concerns, consensus, and questions. *IEEE Security and Privacy, 2*(2), 34-36.

Whitten, A., & Tygar, J. D. (1999). Why Johnny can't encrypt: A usability evaluation of PGP 5.0. *Proceedings of the 8th USENIX Security Symposium*, Washington, DC.

Whitten, A., & Tygar, J. D. (2003). Safe staging for computer security. *Proceedings of the CHI2003 Workshop on Human-Computer Interaction and Security Systems*, Ft. Lauderdale, Florida.

Wilfong, G. T. (1997). *Method and apparatus for secure PIN entry (patent #5,940,511)*. Lucent Technologies.

Yan, J., Blackwell, A., Anderson, R., & Grant, A. (2000). *The memorability and security of passwords—Some empirical results* (technical report). Cambridge, UK: Cambridge University.

Yee, K. -P. (2002). User interaction design for secure systems. *Proceedings of the 4th International Conference on Information and Communications Security*, Singapore.

Zviran, M., & Haga, W. J. (1990). Cognitive passwords: The key to easy access control. *Computers & Security, 9*, 723-736.

KEY TERMS

Asymmetric Cryptography: A data encryption system that uses two separate but related encryption keys. The private key is known only to its owner, while the public key is made available in a key repository or as part of a digital certificate. Asymmetric cryptography is the basis of digital signature systems.

Public Key Infrastructure (PKI): The public infrastructure that administers, distributes, and certifies electronic keys and certificates that are used to authenticate identity and encrypt information. Generally speaking, PKI is a system of digital certificates, certification authorities, and registration authorities that authenticate and verify the validity of the parties involved in electronic transactions.

Shoulder Surfing: The practice of observing persons while entering secret authentication information in order to obtain illegal access to money or services. This often occurs in the context of PIN numbers and banking transactions, where shoulder surfing occurs together with the stealthy duplication of credit or banking cards.

Social Engineering: The technique of exploiting the weakness of users rather than software by convincing users to disclose secrets or passwords by pretending to be authorized staff, network administrator, or the like.

Spoofing: The technique of obtaining or mimicking a fake identity in the network. This can be used for pretending to be a trustworthy Web site and for motivating users (e.g., entering banking information), pretending to be an authorized instance that requests the user's password, or making users accept information that is believed to come from a trusted instance.

Types of Authentication: Authentication generally can be based on three types of information: by some thing the user has (e.g., bank card, key, etc.), by something the user knows (e.g., password, number, etc.), or by something the user is (e.g., biometric methods like fingerprints, face recognition, etc.).

Iconic Interfaces for Assistive Communication[1]

Abhishek
Indian Institute of Technology, Kharagpur, India

Anupam Basu
Indian Institute of Technology, Kharapur, India

INTRODUCTION

A significant fraction of our society suffers from different types of physical and cognitive challenges. The seriousness of the problem can be gauged from the fact that approximately 54 million Americans are classified as disabled (Ross, 2001). In addition, approximately 7% of all school-age children experience moderate to severe difficulties in the comprehension and production of language due to cognitive, emotional, neurological, or social impairments (Evans, 2001). The problems faced by this community are diverse and might not be comprehended by their able-bodied counterparts. These people can become productive and independent, if aids and devices that facilitate mobility, communication, and activities of daily living can be designed.

Researchers in the human-computer interaction and rehabilitation engineering communities have made significant contributions in alleviating the problems of the physically challenged. The technology, that assists the physically challenged to lead a normal life is termed *assistive technology*. This article dwells on different aspects of assistive technology that have found application in real life.

One of the important approaches to assistive technology is the use of iconic environments that have proved their efficacy in dealing with some of the communication problems of the physically challenged. The second part of the article discusses the issues and methods of applying iconic interfaces to assist the communication needs of the physically challenged.

BACKGROUND

The problems faced by the disabled section of our society are huge and of a diverse nature. Disabilities can be classified into physical or cognitive disabilities. Physical disabilities like restricted mobility and loss of hearing, speaking, or visual acuity severely affect the normal life of some people. People suffering from such handicaps need the help of an assistant to help them to perform their routine activities and to use standard appliances.

The case of cognitively challenged people is even more serious. Their difficulties can range from deficits in vocabulary and word-finding to impairments in morphology, phonology, syntax, pragmatics, and memory (Evans, 2001). Persons suffering from autism show delay in language development; complete absence of spoken language; stereotyped, idiosyncratic, or repetitive use of language; or an inability to sustain a conversation with others. The problems faced by a dyslexic person can range from disabilities affecting spelling, number and letter recognition, punctuation problems, letter reversals, word recognition, and fixation problems (Gregor et al., 2000). Brain impairments can lead to learning, attention span, problem-solving, and language disorders (Rizzo et al., 2004).

Difficulties in using a standard computer stem from problems like finding command button prompts, operating a mouse, operating word processing, and finding prompts and information in complex displays. The complexity of a GUI display and the desktop metaphor creates severe problems (Poll et al., 1995). In the case of motor-impaired subjects, the rate of input is often noisy and extremely slow.

To solve these problems, which are inherently multi-disciplinary and non-trivial, researchers from different branches have come together and integrated their efforts. Assistive technology, therefore, is a multidisciplinary field and has integrated researchers from seemingly disparate interests like neuroscience, physiology, psychology, engineering, computer science, rehabilitation, and other technical and health-care disciplines. It aims at reaching an optimum mental, physical, and/or functional level (United Nations 1983). In the following, we look at the methodology adopted in this field in order to solve the aforementioned problems.

AN OVERVIEW OF ASSISTIVE DEVICES

Solving the whole gamut of problems faced by this community requires the construction of what are called *smart houses*. Smart houses are used by old people and people with disabilities. An extensive review of the issues concerning smart houses appears in Stefanov, et al. (2004). Broadly speaking, these houses contain a group of equipment that caters to the different needs of the inhabitants. The technology installed in these homes should be able to adapt to each person's needs and habits.

However, the construction of this technology is not a simple task. Assistive and Augmentative Communication (AAC) aims to use computers to simplify and quicken the means of interaction between the disabled community and their able-bodied counterparts. The tasks of AAC, therefore, can be seen as facilitating the interaction with the world and the use of computers. Interaction with the world is facilitated by one of the following methods:

- Design of robotic systems for assistance.
- Design of systems that help in indoor navigation, such as smart wheelchairs.
- Devices that convert normal speech to alphabetic or sign language (Waldron et al., 1995).
- Devices that convert sign language gestures into voice-synthesized speech, computer text, or electronic signals.
- Design of special software-like screen readers and text-to-speech systems for the blind population (Burger, 1994).

For physically disabled people, researchers have designed motorized wheelchairs that are capable of traversing uneven terrains and circumventing obstacles (Wellman et al., 1995). Robots have been used to assist users with their mundane tasks. Studies have shown that task priorities of users demand a mobile device capable of working in diverse and unfamiliar environments (Stanger et al., 1994). Smart wheelchairs solve some of these requirements. They are capable of avoiding obstacles and can operate in multiple modes, which can be identified as following a particular strategy of navigation (Levine et al., 1999).

The problem is quite different in the case of the visually disabled population. Visual data are rich and easily interpreted. Therefore, to encode visual data to any other form is not trivial. Haptic interface technology seeks to fill this gap by making digital information tangible. However, haptic interfaces are not as rich as visual interfaces in dissemination of information. To make these haptic environments richer and, hence, more useful, methods like speech output, friction, and texture have been added to highlight different variations in data, such as color (Fritz et al., 1999). Braille was devised in order for blind people to read and write words. Letters can be represented through tactile menus, auditory patterns, and speech in order to identify them.

As far as assistance for navigation in physical terrains is considered, visually impaired people can use dogs and canes to prevent obstacles. However, it is clear that these options are limited in many senses. For example, these might not help these people in avoiding higher obstacles like tree branches. In Voth (2004), the author explains the working of a low-cost, wearable vision aid that alerts its user of stationary objects. The technology is based on the observation that objects in the foreground reflect more light than those that are not. The luminance of different objects is tracked over several frames, and the relative luminance is compared in order to identify objects that are coming closer. The software informs the user when the object comes too close (i.e., at an arm's length), and a warning icon is displayed onto a mirror in front of the eyes of the user.

To help these people to use computers more effectively, three of the following types of problems must be handled:

- It should be noted that this population might be unable to provide input in the required form. This inability becomes critical when physical or cognitive challenges seriously inhibit the movement of limbs. This entails the design of special-purpose access mechanisms for such people. This leads to the design of hardware plug-ins, which can provide an interface between the subject and the computer. Scanning emulators have been used to help them to interact with the computer. Instruments like a head-operated joystick that uses infrared LEDs and photo detectors also have been designed (Evans et al., 2000).

- Cognitive impairments lead to slow input. To ensure that the communication runs at a practical speed, it is mandatory that we design strategies to quicken the rate of input. Two strategies are used to increase the rate of input. One of them is to work with partial input and infer its remaining constituents. This works because our language has many redundant and predictable constituents. Given a model of language, we can work with truncated and incomplete input. The second strategy, which increases the rate of the input, is adaptive display techniques. This, too, depends upon a model of the domain. This model is used to predict the next most likely input of the user, and the system adapts itself to make sure that the user makes less effort to choose this input. The same philosophy works in the case of toolkits like word prediction software.
- Language and cognitive disabilities also lead to noise in communication. Therefore, the methods used should be robust and fault-tolerant.

We now discuss some of the systems that implement the aforementioned ideas. Reactive keyboards (Darragh et al., 1990) aim to partially automate and accelerate communication by predicting the next word the user is going to type. This requires adaptive modeling of the task, based on the previously entered text or a language model. However, there is evidence that the cognitive load accrued in searching among the predicted list outweighs the keystroke savings (Koester et al., 1994). Prediction of texts also can be augmented with the help of a semantic network (Stocky et al., 2004). Signing, finger spelling and lip reading are used for communicating with the deaf. People also have attempted sign language communication through telephone lines (Manoranjan et al., 2000). Numerous applications, such as translation of word documents to Braille, have been done (Blenkhorn et al., 2001).

The most difficult problems facing the human-computer interaction community is to find appropriate solutions to problems faced by cognitively challenged people. For solutions to problems like autism and dyslexia, a detailed model of the brain deficiencies need to be known. Some efforts have been made to solve these problems. The case in which rapid speech resulted in problems of comprehension was dealt with in Nagarajan et al. (1998). Different speech modulation algorithms have been designed, which have proved to be effective for this population. Attempts have been made to design word processors for dyslexic people (Gregor et al., 2000).

The case of autistic people is similar. Generally, autistic conversation is considered as disconnected or unordered. Discourse strategies have been used to find out the patterns of problems in such people. These studies have concentrated on finding the typical features of conversation with an autistic patient. These include length and complexity of structure; categories of reference; ellipsis; and phonological, syntactic, and semantic interrelationships. Studies also have shown that autistic people have a tendency to turn to earlier topics within a conversation and turn to favored topics cross-conversationally. Solutions to the problems faced by these communities are important research areas. To diagnose their deficiencies and to use models of man-machine dialogue in order to comprehend and support the dialogue is a non-trivial and challenging task.

Another reason for difficulty in using computers stems from computer naivety, which can lead to problems in using and dealing with the desktop metaphor. This can lead to poor understanding of the knowledge of the keys, poor knowledge of the way in which applications work, and poor acquisition of the conceptual model for mouse-cursor position. Dealing with mouse pointers becomes even more cumbersome when the vision is imperfect. The mouse pointer may get lost in the complex background or be beyond the desktop boundaries without the user realizing it. The problem reaches epic proportions when small screens are used. Mouse pointer magnification and auditory signals can be used to keep the user aware of the precise location of the pointer. The software also can be configured to help the user ask for help in cases when the pointer cannot be found. To adapt to idiosyncrasies of the user, online learning algorithms have been designed that move the spatial position of the keys of the keyboard to adapt to these changes (Himberg et al., 2003). Studies also have analyzed how the trail of the mouse cursor can be used to analyze the nature of difficulties of the user (Keates et al., 2002).

The protocols and schemas of different applications might be cumbersome to learn for these people. For example, it might be difficult to understand and to learn to use an e-mail server (Sutcliffe et al., 2003). Models of behavior can be induced from data to finite state models in order to predict and assist users in routine tasks. However, the emphasis of researchers is toward removing the barriers of fixed protocols of behavior for these people and moving toward free, unordered input. However, this unordered input increases the search space and increases the order of complexity of the problem.

It is clear from the previous discussion that assistive devices must be reactive and attentive (i.e., they must view the subject over the user's shoulders, reason about the user's needs and motivations, and adapt the system to make sure that the user completes his or her work with minimum effort). This requires induction of a user model from the information available from the previous usage of the system. This user model then can be used to predict future user actions. In Neill et al. (2000), the authors describe how sequence profiling can be done to learn to predict the direction in which the user is likely to move on a wheelchair interface. Motor disabilities may lead to problems like pressing more than one key at a time. To prevent these, hardware solutions such as a keyguard can be used, or a software layer can be used, which can use a language model to detect and correct errors (Trewin, 1999, 2002).

Wisfids (Steriadis et al., 2003) can be seen as software agents that capture user events, drivers for accepting input signals that are addressed to them. They transform this input to useful information, which is further processed by software applications on which they are running. Row-column scanning can be made adaptive by making errors and reaction time as the parameters governing the scan delay (Simpson et al., 1999b).

ICONIC ENVIRONMENTS

Iconic interfaces are one of the popular methods that are used for communication by the cognitively challenged population. These interfaces have many advantages, vis-à-vis traditional text-based messaging systems. By being visual rather than textual, they are more intuitive and overcome the need to be literate in order to carry forward the communication. Due to the semantic uniformity of icons, they have been used for translation to Braille scripts (Burger, 1994) and for communication by deaf people (Petrie et al., 2004). The universal nature of icons avoids the idiosyncrasies of different languages. They also speed up the rate of input by removing inferable constituents of communication, such as prepositions. These advantages have made icons pervasive in modern computing systems and ubiquitous in communication and assistive aids.

However, this strength and richness comes at a cost. Use of icons for communication requires their interpretation, which is a non-trivial task. Use of simple icons makes the disambiguation easier. However, it increases the size of the vocabulary. Searching in a large icon set using a scroll bar is likely to be difficult for motor-impaired people. Unordered input and leaving out syntactical cues such as prepositions make the search space larger. Use of syntax-directed methods presupposes the knowledge of different case-roles. Therefore, if the iconic environments are to be useful and practical, they must be random and provide the facilities for overloading the meaning of icons. Semantically overloaded icons, being polysemous, reduce the size of the vocabulary. Small vocabulary implies less search overhead. This is possible only if these interfaces are supplemented by robust and rich inference mechanisms to disambiguate them.

Demasco et al. (1992) were probably the first to attempt the problem. The authors address the problem of expansion of compressed messages into complete intelligible natural language sentences. A semantic parser uses syntactic categories to make a conceptual representation of the sentence and passes it to a translation component.

Extension of the work by Demasco has been reported in Abhishek et al. (2004). They explained the working of a prototype system, which could generate the natural language sentences from a set of unordered semantically overloaded icons. In particular, they formulated the problem as a constraint satisfaction problem over the case roles of the verb and discussed different knowledge representation issues, which must be tackled in order to generate semantically correct iconic sentences. In another approach for disambiguation, Albacete, et al. (1998) used conceptual dependency to generate natural language sentences. The formalism has the strength that it can handle considerable ambiguity. New concepts are generated from different operations on different concepts. The strength of the formalism derives from the fact that the authors define primitive connectors, which encode rules, which define how differently the operators modify categories. The aforementioned systems can support the overloading of meanings to a large extent. While the former derives the overloading via use of complex icons, the latter uses operators to derive new concepts from existing concepts. These systems do not report empirical evaluation of their approaches. However, it can be seen as understanding the composition that operators might be difficult for people suffering from brain impairments and attention span problems.

To make a universal iconic system, it is important to consider cultural issues like symbols, colors, functionality, language, orthography, images, appearance, perception, cognition, and style of thinking. It has been verified experimentally that Minspeak icons need to be customized for users of different cultural contexts (Merwe et al., 2004). The design of a culturally universal icon set is an open issue.

FUTURE TRENDS

In the years to come, we will witness the trend to move toward affective and adaptive computing. If machines have to take over humans in critical, health-related domains, they must be able to assess the physiological and emotional state of the user in real time. Progress in this field has been slow and steady. See Picard (2003) for challenges facing affective computing.

As society ages, it is likely that people will suffer from many disabilities that come because of age. It also is known that most people suffer many problems in performing their day-to-day activities (Ross, 2001). Devices that are used for people with progressive debilitating diseases must have the flexibility to change with users' needs. In general, the need for these devices to be adaptive cannot be discounted. These devices should be able to accommodate a wide range of users' preferences and needs. Wayfinders and devices helping to navigate an uncertain environment are difficult to come by. For example, it might be difficult for a deaf and blind person to cross a street. Research for creating practical, real-time, embedded, intelligent systems that can operate in uncertain and dynamic environments is required.

Evaluation and scaling up of iconic environments have become pressing needs. Generation of semantically correct sentences is the core issue in this field. Knowledge-based or domain-specific systems do not help the cause of creating communication across languages. We foresee the use of richer reasoning and learning methods in order to move away from the bottleneck of encoding of huge and diverse world knowledge. This will require fundamental research in the semantics of language and our protocols of society.

CONCLUSION

In this article, we have discussed the issues concerning assistive devices. We then expounded the methods and techniques used by researchers to solve them. Iconic interfaces play a crucial role in one of the sectors of rehabilitation. However, their strength can be exploited only if rich inference methods can be designed to disambiguate them. We then compared some of the methods used for this task. We concluded that fundamental research in all areas of natural language and brain deficiencies must be done to solve these problems satisfactorily.

REFERENCES

Abhishek, & Basu, A. (2004). A framework for disambiguation in ambiguous iconic environments. *Proceedings of the 17th Australian Joint Conference on Artificial Intelligence,* Cairns, Australia.

Albacete, P.L., Chang, S.K., & Polese, G. (1998). Iconic language design for people with significant speech and multiple impairments. In V.O. Mittal, H.A. Yanco, J. Aronis, & R. Simpson (Eds.), *Assistive technology and artificial intelligence*. Springer-Verlag.

Blenkhorn, P., & Evans, G. (2001). Automated Braille production from word-processed documents. *IEEE Transactions on Neural Systems and Rehabilitation Engineering, 9*(1), 81-85.

Burger, D. (1994). Improved access to computers for the visually handicapped: New prospects and principles. *IEEE Transactions on Rehabilitation Engineering, 2*(3), 111-118.

Darragh, J. J., Witten, I. H., & James, M. L. (1990). The reactive keyboard: A predictive typing aid. *IEEE Computer, 23*(11), 41-49.

Demasco, P. W., & McCoy, K. (1992). Generating text from compressed input: An intelligent interface for people with severe motor impairments. *Communications of the ACM, 35*(5), 68-78.

Dobbinson, S., Perkins, M. R., & Boucher, J. (1998). Structural patterns in conversational with a woman who has autism. *Journal of Communication Disorders, 31*, 113-134.

Evans, D. G., Drew, R., & Blenkhorn, P. (2000). Controlling mouse pointer position using an infrared head-operated joystick. *IEEE Transactions on Rehabilitation Engineering, 8*(1), 107-117.

Evans, D. G., Diggle, T., Kurniawan, S. H., & Blenkhorn, P. (2003). An investigation into formatting and layout errors produced by blind word-processor users and an evaluation of prototype error prevention and correction techniques. *IEEE Transactions on Neural Systems and Rehabilitations Engineering, 11*(3), 257-268.

Evans, D. G., Pettitt, S., & Blenkhorn, P. (2002). A modified Perkins Brailler for text entry into Windows applications. *IEEE Transactions on Neural Systems and Rehabilitation Engineering, 10*(3), 204-206.

Evans, J. L. (2001). An emergent account of language impairments in children with SLI: Implications for assessment and intervention. *Journal of Communication Disorders, 34*, 39-54.

Fritz, J. P., & Barner, K. E. (1999). Design of a haptic data visualization system for people with visual impairments. *IEEE Transactions on Rehabilitations Engineering, 7*(3), 372-384.

Gregor, P., & Newell, A. F. (2000). An empirical investigation of ways in which some of the problems encountered by some dyslexics may be alleviated using computer techniques. *Proceedings of the ASSETS.*

Himberg, J., Hakkila, J., Manyyjarvi, J., & Kangas, P. (2003). On-line personalization of a touch screen based keyboard. *Proceedings of the IUI'03.*

Keates, S., Hwang, F., Robinson, P., Langdon, P., & Clarkson, P.J. (2002). Cursor measures for motion-impaired computer users. *Proceedings of the ASSETS-2002.*

Koester, H. H. (1994). Modeling the speed of text entry with a word prediction interface. *IEEE Transactions on Rehabilitation Engineering, 2*(3), 177-187.

Levine, S. P., et al. (1999). The NavChair assistive wheelchair navigation system. *IEEE Transactions on Rehabilitation Engineering, 7*(4), 443-451.

Manoranjan, M.D., & Robinson, J.A. (2000). Practical low-cost visual communication using binary images for deaf sign language. *IEEE Transactions on Rehabilitation Engineering, 8*(1), 81-88.

Mason, S. G., & Birch, G. E. (2003). A general framework for brain-computer interface design. *IEEE Transactions on Neural Systems and Rehabilitations Engineering, 11*(1), 70-85.

Merwe, E., & Alant, E. (2004). Associations with Minspeak Icons. *Journal of Communication Disorders, 37*, 255-274.

Nagarajan, S.S., et al. (1998). Speech modifications algorithms used for training language learning-impaired children. *IEEE Transactions on Rehabilitations Engineering, 6*(3), 257-268.

Neill, P. O., Roast, C., & Hawley, M. (2000). Evaluation of scanning user interfaces using real-time-data usage logs. *Proceedings of the ASSETS'00.*

Perelmouter, J., & Birbaumer, N. (2000). A binary spelling interface with random errors. *IEEE Transactions on Rehabilitation Engineering, 8*(2), 227-232.

Petrie, H., Fisher, W., Weimann, K., & Weber, G. (2004). Augmenting icons for deaf computer users. *Proceedings of the CHI-2004.*

Picard, R. W. (2003). Affective computing. *International Journal of Human Computer Studies, 59*, 55-64.

Poll, L. H. D., & Waterham, R.P. (1995). Graphical user interfaces and visually disabled users. *IEEE Transactions on Rehabilitation Engineering, 3*(1), 65-69.

Rizzo, A. A., et al. (2004). Diagnosing attention disorders in a virtual classroom. *IEEE Computer, 37*(6), 87-89.

Ross, D. A. (2001). Implementing assistive technology on wearable computers. *IEEE Intelligent Systems, 16*(3), 47-53.

Simpson, R. C., & Koester, H. H. (1999). Adaptive one-switch row-column scanning. *IEEE Transactions on Rehabilitation Engineering, 7*(4), 464-473.

Stanger, C. A., Anglin, C., Harwin, W. S., & Romilly, D. P. (1994). Devices for assisting manipulation: A summary of user task priorities. *IEEE Transactions on Rehabilitation Engineering, 2*(4), 256-265.

Stefanov, D. H., Bien, Z., & Bang, W. (2004). The smart house for older persons with physical disabilities: Structures, technology, arrangements and perspectives. *IEEE Transactions on Neural Systems and Rehabilitation Engineering, 12*(2), 228-250.

Steriadis, C. E., & Constantinou, P. (2003). Designing human-computer interfaces for quadriplegic people. *ACM Transactions on Computer-Human Interactions, 10*(2), 87-118.

Stocky, T., Faaborg, A., & Lieberman, H. (2004). A commonsense approach to predictive text entry. *Proceedings of the CHI 2004.*

Sutcliffe, A., Fickas, S., Sohlberg, M., & Ehlhardt, L. A. (2003). Investigating the usability of assistive user interfaces. *Interacting with Computers, 15*, 577-602.

Trewin, S. (2002). An invisible keyguard. *Proceedings of the ASSETS 2002.*

Trewin, S., & Pain, H. (1999). Keyboard and mouse errors due to motor disabilities. *International Journal of Human Computer Studies, 50*, 109-144.

United Nations. (1983). *World programme of action concerning disabled persons.* Available at http://www.un.org/documents/ga/res/38/a38r028.htm

Voth, D. (2004). Wearable aids for the visually impaired. *IEEE Pervasive Computing, 3*(3), 6-9.

Waldron, M. B., & Kim, S. (1995). Isolated ASL sign recognition system for deaf persons. *IEEE Transactions on Rehabilitation Engineering, 3*(3), 261-271.

Wellman, P., Krovi, V., Kumar, V., & Harwin, W. (1995). Design of a wheelchair with legs for people with motor disabilities. *IEEE Transactions on Rehabilitation Engineering, 3*(4), 343-353.

KEY TERMS

Adaptive User Interfaces: An interface that uses a user model to change its behavior or appearance to increase user satisfaction with time. These interfaces are used extensively in assistive devices.

Assistive and Augmentative Communication: A multi-disciplinary field that seeks to design devices and methods to alleviate the problems faced by physically challenged people running programs they don't know and/or trust.

Autism: A disease that leads to language disorders like delay in language development, repeated use of language, and inability to sustain conversation with others.

Disambiguation in Iconic Interfaces: The process of context-sensitive, on-the-fly semantic interpretation of a sequence of icons. The process is difficult because of the huge world knowledge required for comprehending and reasoning about natural language.

Smart Houses: Houses that are equipped with self-monitoring assistive devices of many types. Smart houses are popular with old people.

User Model: A model induced by machine-learning techniques from the available information and patterns of data from the user. This model is used by the system to predict future user actions.

Widgets: The way of using a physical input device to input a certain value. These are extensively used and are popular in the case of people with neuromotor disabilities.

ENDNOTE

1 This work is supported by grants from Media Lab Asia.

The Importance of Similarity in Empathic Interaction

Lynne Hall
University of Sunderland, UK

Sarah Woods
University of Hertfordshire, UK

INTRODUCTION

Empathy has been defined as, "An observer reacting emotionally because he perceives that another is experiencing or about to experience an emotion" (Stotland, Mathews, Sherman, Hannson, & Richardson, 1978). Synthetic characters (computer generated semi-autonomous agents corporeally embodied using multimedia and/or robotics, see Figure 1) are becoming increasingly widespread as a way to establish empathic interaction between users and computers. For example, Feelix, a simple humanoid LEGO robot, is able to display different emotions through facial expressions in response to physical contact. Similarly, Kismet was designed to be a sociable robot able to engage and interact with humans using different emotions and facial expressions. Carmen's Bright Ideas is an interactive multimedia computer program to teach a problem-solving methodology and uses the notion of empathic interactions. Research suggests that synthetic characters have particular relevance to domains with flexible and emergent tasks where empathy is crucial to the goals of the system (Marsella, Johnson, & LaBore, 2003).

Using empathic interaction maintains and builds user emotional involvement to create a coherent cognitive and emotional experience. This results in the development of empathic relations between the user and the synthetic character, meaning that the user perceives and models the emotion of the agent experiencing an appropriate emotion as a consequence.

Figure 1. Synthetic characters

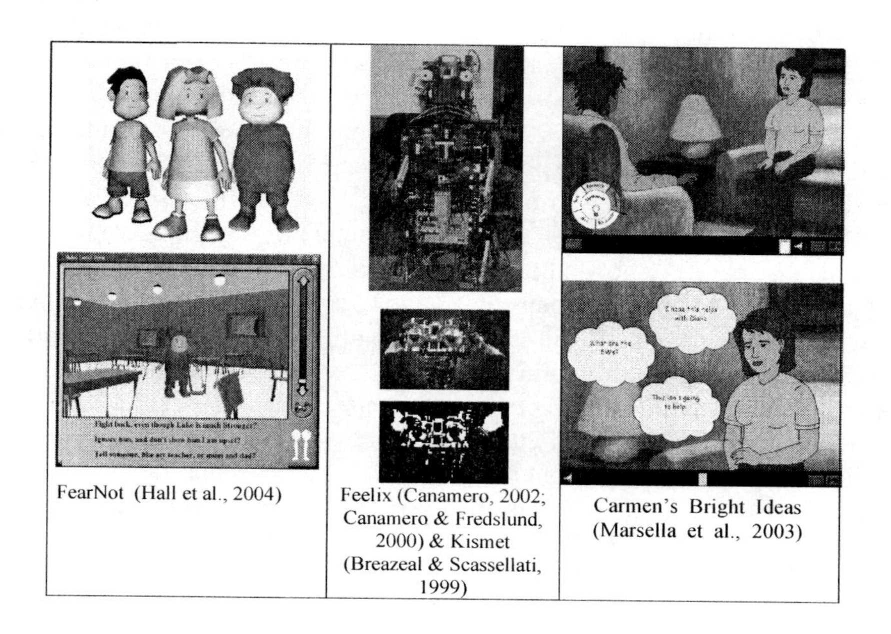

FearNot (Hall et al., 2004)

Feelix (Canamero, 2002; Canamero & Fredslund, 2000) & Kismet (Breazeal & Scassellati, 1999)

Carmen's Bright Ideas (Marsella et al., 2003)

BACKGROUND

A number of synthetic characters have been developed where empathy and the development of empathic relations have played a significant role, including theatre (Bates, 1994), storytelling (Machado, Paiva, & Prada, 2001) and personal, social, and health education (Silverman, Holmes, Kimmel, Ivins, & Weaver, 2002). Applications such as FearNot (Hall et al., 2004b) and Carmen's Bright Ideas (Marsella et al., 2003) highlight the potential of synthetic characters for exploring complex social and personal issues, through evoking empathic reactions in users.

In a similar vein, robotics research has started to explore both the physical and behavioural architecture necessary to create meaningful empathic interactions with humans. This has included examining robot personality traits and models necessary for empathic relations (Fong, Nourbakhsh, & Dautenhahn, 2003) and the design of robotic facial expressions eliciting basic emotions to create empathic interactions (e.g., Canamero, 2002). Empirical evaluations have shown that humans do express empathy towards robots and have the tendency to treat robots as living entities (e.g., Sparky, a social robot; Scheeff, Pinto, Rahardja, Snibbe, & Tow, 2002).

The results from research into empathic interaction with synthetic characters suggest that it is possible to evoke empathic reactions from users and that this can result in stimulating novel interactions. Further, research identifies that in empathising with characters a deeper exploration and understanding of sensitive social and personal issues is possible (Dautenhahn, Bond, Canamero, & Edmonds, 2002). This can lead to real-life impacts such as the development of constructive solutions, that is, Carmen's Bright Ideas (Marsella et al., 2003).

However, it remains unclear as to how empathy can be evoked by interaction and here, we focus on the impact of similarity on evoking empathy in child users. This article reports findings obtained in the VICTEC (Virtual ICT with Empathic Characters) project (Aylett, Paiva, Woods, Hall, & Zoll, 2005) that applied synthetic characters and emergent narrative to Personal and Health Social Education (PHSE) for children aged 8-12, in the UK, Portugal, and Germany, through using 3D self-animating characters to create improvised dramas. In this project, empathic interaction was supported using FearNot (Fun with Empathic Agents to Reach Novel Outcomes in Teaching). This prototype allowed children to explore physical and relational bullying issues, and coping strategies in a virtual school populated by synthetic characters. The main issue this article addresses is whether the level of similarity perceived by a child with a character has an impact on the degree of empathy that the child feels for the character.

WHY SIMILARITY MATTERS

Similarity is the core concept of identification (Lazowick, 1955) and a major factor in the development and maintenance of social relationships (Hogg & Abrams, 1988). The perception of similarity has significant implications for forming friendships, with studies identifying that where children perceive themselves as similar to another child, that they are more likely to choose them as friends (Aboud & Mendelson, 1998). The opposite has also been shown to be true, with children disliking those who are dissimilar to them in terms of social status and behavioural style (Nangle, Erdley, & Gold, 1996). This dislike of dissimilarity is especially evident for boys.

Perceived similarity as a basis for liking and empathising with someone is also seen in reactions to fictional characters, where the perception of a character as similar to oneself and identifying with them will typically result in liking that character, and empathising with their situation and actions. This can be frequently seen with characters portrayed in cinema and television (Hoffner & Cantor, 1991; Tannenbaum & Gaer, 1965). Further, people are more likely to feel sorry for someone (real or a character) if they perceive that person as similar to themselves (von Feilitzen & Linne, 1975).

To investigate the impact of similarity on children's empathic reactions to the synthetic characters in FearNot, we performed a large scale study, further discussed in Aylett et al. (2005). Liking someone is strongly influenced by perceived similarity and research suggests that if a child likes a character they are more likely to empathise with them. Thus, in considering the impact of similarity on the evocation of empathy we looked at perceived similarity of appearance and behaviour and their impact on the like/dislike of characters, as well as two empathic measures (feeling sorry for a character and feeling angry with a character).

THE FEARNOT STUDY

FearNot was trialed at the "Virtually Friends" event at the University of Hertfordshire, UK, in June 2004. Three hundred and forty-five children participated in the event: 172 male (49.9%) and 173 female (50.1%). The sample age range was 8 to 11, mean age of 9.95 (SD: 0.50). The sample comprised of children from a wide range of primary schools in the South of England.

Method

Two classes from different schools participated each day in the evaluation event. All children individually interacted with FearNot on standard PCs. FearNot began with a physical bullying scenario comprised of three episodes and children had the role of an advisor to help provide the victim character with coping strategies to try and stop the bullying behaviour. After the physical scenario, children had the opportunity to interact with the relational scenario showing the drama of bullying among four girls. After the interaction children completed the Agent Evaluation Questionnaire (AEQ). This was designed in order to evaluate children's perceptions and views of FearNot, see Table 1. This questionnaire is based on the Trailer Questionnaire (Woods, Hall, Sobral, Dautenhahn, & Wolke, 2003) that has been used extensively with a non-interactive FearNot prototype as is reported in Hall et al. (2004b). Questions relating to choosing characters were answered by selecting character names (posters of the characters were displayed with both a graphic

Table 1. Content of the agent evaluation questionnaire

Aspect	Nature of Questions
Character preference	• Character liked most • Character liked least • Prime character, who they would choose to be • Character with whom child would most like to be friends
Character attributes	• realism of movement (realistic to unrealistic) • smoothness of movement (smooth to jerky) • clothes appreciation (looked good to looked strange), liking (liked to did not like) and similar to own (similar to what you wear to different to what you wear) • character age
Character conversations	• conversation content (believable to unbelievable) • conversation interest (interesting to boring) • content similarity to own conversations (similar to different)
Interaction impact	• victims acceptance of advice (followed to paid no attention) • helping victim (helped a lot to not at all)
Bullying Storyline	• storyline believability (believable to unbelievable) • storyline length (right length to too long)
Similarity	• character that looks most and least like you • character that behaves most and least like you
Empathy towards characters	• Feeling sorry for characters and if yes which character • Feeling angry towards the characters and if yes which character • Ideomotoric empathy based on expected behaviour

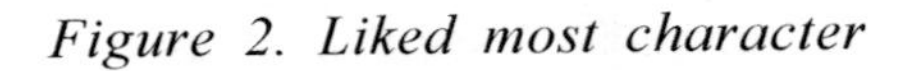

Figure 2. Liked most character

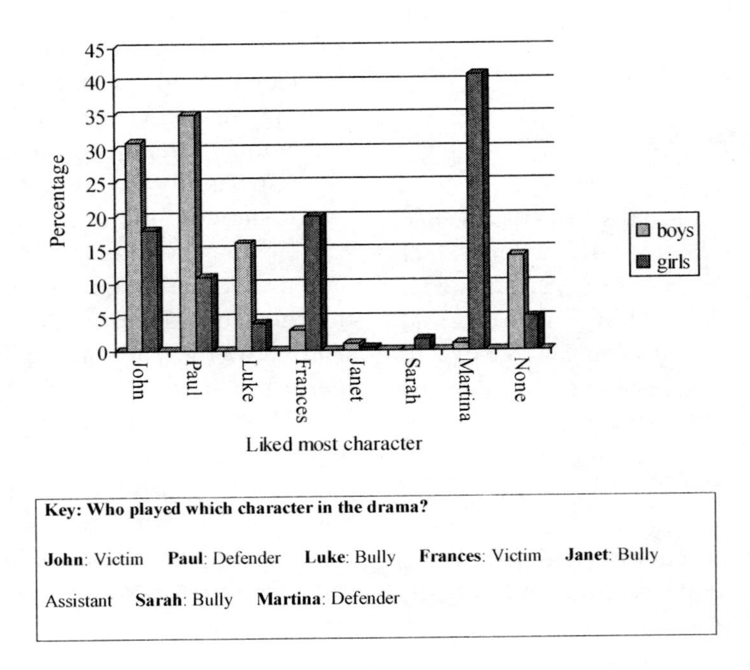

and the name as an aide memoire). Children's views were predominantly measured according to a 5 point Likert scale.

Results

Gender was a significant factor in the selection of which character was most similar in physical appearance to you, with almost all of the children choosing a same gender character or none. There was a significant association for those children who felt that a same gender character looked like them and also liked a same gender character: boys (X = 23.534, (8, 108), p = 0.001) girls (X = 24.4, (4, 89), p < 0.001), meaning that boys liked male characters that looked like them, and girls liked female characters that resembled them.

As can be seen from Figure 3, children liked those characters who looked the most similar to them, if the character played a defender, neutral or victim role. However, where the character was a bully, children were not as likely to like the character that they were similar to in appearance, particularly among the girls. Thirty-five percent of boys who looked like Luke liked him the most, although almost a third of the girls stated that they resembled a female bully character in appearance, only 4 (2.5%) liked them the most.

There were no significant differences related to whom you looked like and disliking characters, with the dislike clearly being based on alternative factors to appearance. Similar to the results of Courtney, Cohen, Deptula, and Kitzmann (2003), children disliked the bullies (aggressors) the most, followed by the victims and then the bystanders. Most children disliked Luke, the physical bullying protagonist followed by Sarah, the relational bully, then the victims. As in other results (Hall et al., 2004), children paid scant attention to the bully assistants, and only 5% of children disliked Janet the most.

A significant association was found between the character children felt looked the most like them and feeling sorry for characters in the drama. Looking like any of the female characters (e.g., being female) is more likely to result in feeling sorry for the victims, with over 80% of those who felt that they looked like any of the female characters feeling sorry for the victims. If children (mainly boys) felt that Luke (62%) looked the most like them, they expressed the least amount of empathy towards the characters in the dramas, however, only 67% of those who felt that they looked like John felt sorry for

Figure 3. Character child looked most similar to in appearance and liked the most

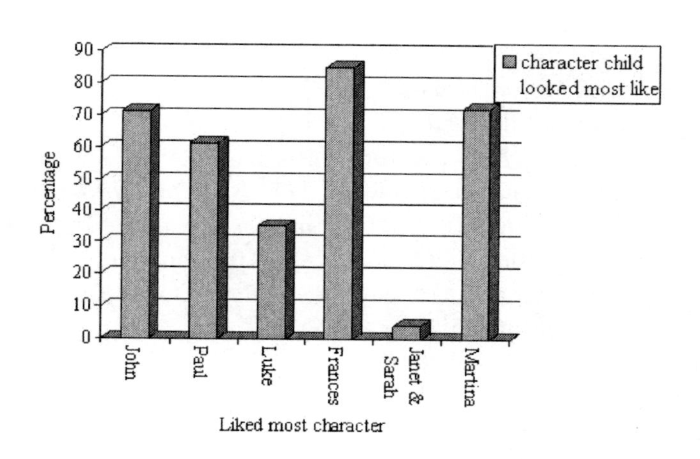

the victims, as compared to 87% of those (all female) who felt they looked like Frances.

A significant association was found between the character children felt looked most like them and feeling anger towards characters in the dramas. Again this result is related to gender, with significantly more girls than boys feeling anger towards the characters. However, the results still indicate that appearance similarity could have an impact on the evocation of anger. Boys who stated that Luke looked the most similar to them felt the least amount of anger towards characters (46%), followed by John (61%) and Paul (78%). For the girls, those who felt they looked most like Sarah the bully were most likely to be angry (95.5%) compared to 71% of those who looked most similar to Frances (the victim), suggesting that girls were more likely to be angry if the bully were similar to them, whereas boys were less likely to be angry if the bully were similar to them. For those children who stated that none of the characters looked like them, 66% identified that they felt angry, reflecting the higher number of boys than girls in this group.

DISCUSSION

The results indicate that greater levels of empathy are evoked in children if they perceive that they are similar to the characters. This would suggest that when developers seek to evoke empathic interaction that they should attempt to create synthetic characters that are similar to the intended users. Interestingly, our results also highlighted that while looking like a character may result in you being more inclined to like them, if they exhibit morals, ethics, and behaviours that are socially unacceptable, such as bullying, this can have a significant impact on your liking of that character. This reflects real-world behaviour, with all reported studies of children's reactions to aggressive behaviour/bullying supporting the view that children are more likely to dislike aggressors the most, followed by victims and then bystanders (Courtney et al., 2003). Our results supported this view.

Trusting and believing in synthetic characters and possible impact on real-life behaviour appears to be linked to perceived similarity. However, although perceived similarity may be a major factor in engagement with synthetic characters, there is also considerable evidence from the performing arts that engagement can readily occur with characters very dissimilar to oneself.

FUTURE TRENDS

This study has highlighted the potential for similarity and empathic interaction; however, further research is needed in this area. Future research directions include the impact of greater physical similarity on empathic interaction, with research in virtual humanoids considering more realistic and similar features and expressions (Fabri, Moore, & Hobbs, 2004). The importance of cultural similarity is also being investigated (Hayes-Roth, Maldonado, & Moraes, 2002) with results suggesting the need for high cultural homogeneity between characters and their users. While similarity may be of benefit, there remains the spectre of the "Uncanny Valley" (Woods, Dautenhahn, & Schulz, 2004), for example, a recent study examining children's perceptions of robot images revealed that "pure" human-like robots are viewed negatively compared to machine-human-like robots. Research is needed into determining what aspects of similarity need to be provided to enable higher levels of empathic interaction with synthetic characters, considering different modalities, senses, and interaction approaches.

CONCLUSION

This article has briefly considered empathic interaction with synthetic characters. The main focus of this article was on the impact of similarity on evoking empathic interaction with child users. Results suggest that if children perceive that they are similar to a synthetic character in appearance and/or behaviour, that they are more likely to like and empathise with the character. Future research is needed to gain greater understanding of the level and nature of similarity required to evoke an empathic interaction.

REFERENCES

Aboud, F. E., & Mendelson, M. J. (1998). Determinants of friendship selecton and quality: Developmental perspectives. In W. M. Bukowski, A. F. Newcomb, & W. W. Hartup (Eds.), *The company they keep: Friendship in childhood and adolescence* (pp. 87-112). New York: Cambridge.

Aylett, R. S., Paiva, A., Woods, S., Hall, L., & Zoll, C. (2005). Expressive characters in anti-bullying education. In L. Canamero & R. Aylett (Eds.), *Animating expressive characters for social interaction*. Amsterdamn, The Netherlands: John Benjamins.

Bates, J. (1994). The role of emotion in believable agents. *Communications of the ACM, 37*(7), 122-125.

Breazeal, C., & Scassellati, B. (1999, October 17-21). *How to build robots that make friends and influence people*. Paper presented at the IROS'99, Kyonjiu, Japan (pp. 363-390). Piscataway, NJ: IEEE Standards Office.

Canamero, L. (2002). Playing the emotion game with Feelix: What can a LEGO robot tell us about emotion? In B. Edmonds (Ed.), *Socially intelligent agents: Creating relationships with computers and robots* (pp. 69-76). MA: Kluwer Academic Publishers.

Canamero, L., & Fredslund, J. (2000, November 3-5). *How does it feel? Emotional interaction with a humanoid LEGO robot* (pp. 23-28). Paper presented at the Socially Intelligent Agents: The Human in the Loop: AAAI 2000 Fall Symposium, Menlo Park, CA (pp. 23-28). Dordrecht, The Netherlands: Kluwer Academic Publishers.

Courtney, M. L., Cohen, R., Deptula, D. P., & Kitzmann, K. M. (2003). An experimental analysis of children's dislike of aggressors and victims. *Social Development, 12*(1), 46-66.

Dautenhahn, K., Bond, A. H., Canamero, L., & Edmonds, B. (2002). *Socially intelligent agents: Creating relationships with computers and robots*. MA: Kluwer Academic Publishers.

Fabri, M., Moore, D. J., & Hobbs, D. J. (2004). Mediating the expression of emotion in educational collaborative virtual environments: An experimental study. *International Journal of Virtual Reality, 7*(2), 66-81.

Fong, T., Nourbakhsh, I., & Dautenhahn, K. (2003). A survey of socially interactive robots, *Robotics and Autonomous Systems,* 42, 143-166.

Hall, L., Woods, S., Dautenhahn, K., & Sobreperez, P. (2004, June 1-3). *Guiding virtual world design using storyboards*. Paper presented at the Interaction Design with Children (IDC), Maryland (pp. 125-126). New York: ACM Press.

Hall, L., Woods, S., Sobral, D., Paiva, A., Dautenhahn, K., Paiva, A., & Wolke, D. (2004b). *Designing empathic agents: Adults vs. kids.* Paper presented at the Intelligent Tutoring Systems 7th International Conference, ITS 2004, Maceio, Brazil, August 30-September 3 (pp. 125-126). Lecture notes in computer science 3220. New York: Springer.

Hayes-Roth, B., Maldonado, H., & Moraes, M. (2002, February 12-14). *Designing for diversity: Multi-cultural characters for a multi-cultural world.* Paper presented at the IMAGINA, Monte Carlo, Monaco (pp. 207-225). Available from http://www.stanford.edu/~kiky/Design4Diversity.pdf

Hoffner, C., & Cantor, J. (1991). Perceiving and responding to media characters. In J. Bryant, & D. Zillman (Eds.), *Responding to the screen: Reception and reaction processes* (pp. 63-101). Hillsdale, NJ: Erlbaum.

Hogg, M. A., & Abrams, D. (1988). *Social identifications.* New York: Routledge.

Lazowick, L. M. (1955). On the nature of identification. *Journal of Abnormal and Social Psychology, 51*, 175-183.

Machado, I., Paiva, A., & Prada, R. (2001, May 28-June 1). *Is the wolf angry or just hungry? Inspecting, modifying and sharing character's minds.* Paper presented at the 5th International Conference on Autonomous Agents (pp. 370-376). New York: ACM Press.

Marsella, S., Johnson, W. L., & LaBore, C. (2003, July 20-24). *Interactive pedagogical drama for health interventions.* Paper presented at the 11th International Conference on Artificial Intelligence in Education, Australia (pp. 169-176). Amsterdam, The Netherlands: ACM Press. Amsterdam, The Netherlands: IOS Press.

Nangle, D. W., Erdley, C. A., & Gold, J. A. (1996). A reflection on the popularity construct: The importance of who likes or dislikes a child. *Behaviour Therapy, 27*, 337-352.

Scheeff, M., Pinto, J. P., Rahardja, K., Snibbe, S., & Tow, R. (2002). Experiences with Sparky, a social robot. In B. Edmonds (Ed.), *Socially intelligent agents: Creating relationships with computers and robots* (pp. 173-180). MA: Kluwer Academic Publishers.

Silverman, B. G., Holmes, J., Kimmel, S., C., B., Ivins, D., & Weaver, R. (2002). The use of virtual worlds and animated personas to improve healthcare knowledge and self-care behavior: The case of the heart-sense game. In N. Ichalkaranje (Ed.), *Intelligent agents and their applications* (pp. 249-294). Heidelberg, Germany: Physica-Verlag GmbH.

Stotland, E., Mathews, K. E., Sherman, S. E., Hannson, R. O., & Richardson, B. Z. (1978). *Empathy, fantasy and helping.* Beverly Hills, CA: Sage.

Tannenbaum, P. H., & Gaer, E. P. (1965). Mood change as a function of stress of protagonist and degree of identification in a film viewing situation. *Journal of Personality and Social Psychology, 2*, 612-616.

von Feilitzen, C., & Linne, O. (1975). Identifying with television characters. *Journal of Communication, 25*(4), 51-55.

Woods, S., Dautenhahn, K., & Schulz, J. (2004, September 20-22). The design space of robots: Investigating children's views, *Ro-Man*, Kurashiki, Japan (pp. 47-52). New York: Springers Press.

Woods, S., Hall, L., Sobral, D., Dautenhahn, K., & Wolke, D. (2003, September 15-17). *A study into the believability of animated characters in the context of bullying intervention.* Paper presented at the IVA '03, Kloster Irsee, Germany (pp. 310-314). Lecture notes in computer science 2792. New York: Springer.

KEY TERMS

Autonomous Robot: A robot that is capable of existing independently from human control.

Emergent Narrative: Aims at solving and/or providing an answer to the narrative paradox observed in graphically represented virtual worlds. Involves participating users in a highly flexible real-time environment where authorial activities are minimised and the distinction between authoring-time and presentation-time is substantially removed.

Empathic Agent: A synthetic character that evokes an empathic reaction in the user.

Empathy: "An observer reacting emotionally because he perceived that another is experiencing or about to experience an emotion."

Synthetic Character: Computer generated semi-autonomous agent corporally embodied using multimedia and/or robotics.

Uncanny Valley: Feelings of unease, fear, or revulsion created by a robot or robotic device that appears to be, but is not quite, human-like.

Virtual Learning Environment: A set of teaching and learning tools designed to enhance a student's learning experience by including computers and the Internet in the learning process.

Improving Dynamic Decision Making through HCI Principles

Hassan Qudrat-Ullah
York University, Canada

INTRODUCTION

CSBILEs allow the compression of time and space and provide an opportunity for practicing managerial decision making in a non-threatening way (Issacs & Senge, 1994). In a computer simulation-based interactive learning environments (CSBILEs), decision makers can test their assumptions, practice exerting control over a business situation, and learn from the immediate feedback of their decisions. CSBILE's effectiveness is associated directly with decision-making effectiveness; that is, if one CSBILE improves decision-making effectiveness more than other CSBILEs, it is more effective than others. Despite an increasing interest in CSBILEs, empirical evidence to their effectiveness is inconclusive (Bakken, 1993; Diehl & Sterman, 1995; Moxnes, 1998). The aim of this article is to present a case for HCI design principles as a viable potential way to improve the design of CSBILEs and, hence, their effectiveness in improving decision makers' performance in dynamic tasks. This article is organized as follows: some background concepts are presented first; next, we present an assessment of the prior research on (i) DDM and CSBILE and (ii) HCI and dynamic decision making (DDM); the section on future trends presents some suggestion for future research. This article concludes with some conclusions.

BACKGROUND

Dynamic Decision Making

What is dynamic decision making (DDM)? Dynamic decision-making situations differ from those traditionally studied in static decision theory in at least three ways:

- A number of decisions are required rather than a single decision.
- Decisions are interdependent.
- The environment changes either as a result of decisions made or independently of them both (Edwards, 1962).

Recent research in system dynamics has characterized such decision tasks by multiple feedback processes, time delays, non-linearities in the relationships between decision task variables, and uncertainty (Bakken, 1993; Hsiao, 2000; Sengupta & Abdel-Hamid, 1993; Sterman, 1994).

We confront dynamic decision tasks quite routinely in our daily life. For example, driving a car, flying an airplane, managing a firm, and controlling money supply are all dynamic tasks (Diehl & Sterman, 1995). These dynamic tasks are different from static tasks such as gambling, locating a park on a city map, and counting money. In dynamic tasks, in contrast to static tasks, multiple and interactive decisions are made over several time periods whereby these decisions change the environment, giving rise to new information and leading to new decisions (Brehmer, 1990; Forrester, 1961; Sterman, 1989a, 1994).

CSBILEs

We use *CSBILE* as a term sufficiently general to include microworlds, management flight simulators, learning laboratories, and any other computer simulation-based environments. The domain of these terms is all forms of action whose general goal is the facilitation of decision making and learning in dynamic tasks. This conception of CSBILE embodies learning as the main purpose of a CSBILE (Davidsen, 2000; Lane, 1995; Moxnes, 1998; Sterman, 1994). Computer-simulation models, human intervention, and decision making are considered the essential components of a CSBILE (Bakken, 1993; Cox, 1992; Davidsen, 1996; Davidsen & Spector, 1997; Lane, 1995; Sterman, 1994).

Under this definition of CSBILE, learning goals are made explicit to decision makers. A computer-simulation model is built to represent adequately the domain or issue under study with which decision makers can induce and experience real worldlike responses (Lane, 1995). Human intervention refers to active keying in of the decisions by decision makers into the computer-simulation model via a decision-making environment or interface. Human intervention also arises when a decision maker interacts with a fellow decision maker during a group setting session of a CSBILE or when a facilitator intervenes either to interact with the simulated system or to facilitate the decision makers.

DDM AND CSBILEs

Business forces, such as intensifying competition, changing operating environments, and enormously advancing technology, have made organizational decision making a complex task (Diehl & Sterman, 1995; Moxnes, 1998; Sterman, 1989b), and all challenge traditional management practices and beliefs. The development of managerial skills to cope with dynamic decision tasks is ever in high demand. However, the acquisition of managerial decision-making capability in dynamic tasks has many barriers (Bakken, 1993). On the one hand, the complexity of corporate and economic systems does not lend itself well to real-world experimentation. On the other hand, most of the real-world decisions and their outcomes hardly are related in both time and space, which compounds the problem of decision making and learning in dynamic tasks.

However, computer technology, together with the advent of new simulation tools, provides a potential solution to this managerial need. For instance, CSBILEs are often used as decision support systems in order to improve decision making in dynamic tasks by facilitating user learning (Davidsen & Spector, 1997; Lane, 1995). CSBILEs allow the compression of time and space, providing an opportunity for managerial decision making in a non-threatening way (Issacs & Senge, 1994).

In the context of CSBILEs, how well do people perform in dynamic tasks? The literature on DDM (Funke, 1995; Hsiao, 2000; Kerstholt & Raaijmakers, 1997; Qudrat-Ullah, 2002, Sterman, 1989a, 1989b) and learning in CSBILEs (Bakken, 1993; Keys & Wolf, 1990; Lane, 1995; Langley & Morecroft) provides almost a categorical answer: very poorly. Very often, poor performance in dynamic tasks is attributed to subjects' misperceptions of feedback (Diehl & Sterman, 1995; Moxnes, 1998; Sterman, 1989b). The misperception of feedback (MOF) perspective concludes that subjects perform poorly because they ignore time delays and are insensitive to the feedback structure of the task system. The paramount question becomes the following: Are people inherently incapable of managing dynamic tasks? Contrary to Sterman's (1989a, 1989b) MOF hypothesis, an objective scan of real-world decisions would suggest that experts can deal efficiently with highly complex dynamic systems in real life; for example, maneuvering a ship through restricted waterways (Kerstholt & Raaijmakers, 1997). The expertise of river pilots seems to consist more of using specific knowledge (e.g., pile moorings, buoys, leading lines) that they have acquired over time than in being able to predict accurately a ship's movements (Schraagen, 1994). This example suggests that people are not inherently incapable of better performance in dynamic tasks but that decision makers need to acquire the requisite expertise. Thus, in the context of CSBILEs, equating learning as a progression toward a prototypic expertise (Sternberg, 1995) is a very appropriate measure. Then, the most fundamental research question for DDM research seems to be how to acquire prototypic expertise in dynamic tasks. A solution to this question effectively would provide a competing hypothesis to MOF hypothesis: people will perform better in dynamic tasks if they acquire the requisite expertise. We term this competing hypothesis as the acquisition-of-expertise (AOE) hypothesis. The following section explains how the human-computer interface (HCI) design may help to acquire prototypic expertise in dynamic tasks.

HCI AND DDM

The successes of the HCI design in disciplines such as management information system, information science, and psychology, all sharing the common goal of improving the organizational decision making, are considerable (Carey et al., 2004). However, despite the fact that all CSBILEs must have an HCI element, the role of HCI design in improving DDM has received little attention from DDM researchers. Only recently, Howie et al. (2000) and Qudrat-Ullah (2002) have directed DDM research to this dimension. Howie et al. (2000) has investigated empirically the impact of an HCI design based on human factor guidelines on DDM. The results revealed that the new interface design based on the following human-computer interaction principles led to improved performance in the dynamic task compared to the original interface:

- Taking advantage of people's natural tendencies (e.g., reading from left to right and from top to bottom);
- Taking advantage of people's prior knowledge (e.g., through the use of metaphors);
- Presenting information in a graphical manner to tap into people's pattern-recognition capabilities; and
- Making the relationship among data more salient so that people can develop a better mental model of the simulation.

In his empirical study, Qudrat-Ullah (2001) studied the effects of an HCI design based on learning principles (Gagné, 1995) on DDM. The results showed that the CSBILE with HCI design based on the following learning principles was effective on all four performance criteria:

- Gaining attention (e.g., the decision makers, at the very first screen of the CSBILE, are presented with a challenging task with the help of a text window and background pictures of relevant screens to grab their attention and arouse interest and curiosity);
- Informing of the objectives of the task (e.g., the objective is presented in clear terms: How does the tragedy of the commons occur?);
- Stimulating the recall of prior knowledge (e.g., the pre-play test helps stimulate recall of prior knowledge);
- Presenting the content systematically (e.g., text and objects are used in the CSBLE for material presentation);
- Providing learning guidance (e.g., the decision makers are led to an explanation interface as a guidance for learning);
- Eliciting performance (e.g., the navigational buttons of the CSBILE allow the decision maker to go back and forth from generic to specific explanation and vice versa, facilitating the performance elicitation);
- Providing feedback (e.g., the pop-up window messages provide feedback to the decision makers as such);
- Assessing performance (e.g., the post-test is designed to assess the performance of the decision maker);
- Enhancing retention and transfer (e.g., the debriefing session of the CSBILE augments the last instructional event, enhancing retention and transfer of knowledge);
- The new design improves task performance;
- Helps the user learn more about the decision domain;
- Develop heuristics; and
- Expends less cognitive effort, a support for AOE hypothesis.

FUTURE TRENDS

Although any generalization based on just two studies may not be that realistic, there appears a clear call to reassess the earlier studies on DDM supporting MOF hypothesis. By employing HCI design principles in CSBILEs, future studies should explore the following:

- **Cost Economics:** To what extent do the HCI design-based CSBILEs help dynamic decision makers to cope with limited information-processing capacities?
- **Reducing Misperception of Feedback:** Increasing task salience and task transparency in dynamic tasks results in improved perfor-

mance (Issacs & Senge, 1994). Are HCI design-based CSBILEs effective in reducing misperceptions of feedback?

- **Supporting Learning Strategies:** Successful decision makers in dynamic tasks develop perceptually oriented heuristics (Kirlik, 1995). To what extent do the HCI design-based CSBILEs help dynamic decision makers to develop perceptually oriented decision heuristics?

CONCLUSION

DDM research is highly relevant to the managerial practice (Diehl & Sterman, 1995; Kerstholt & Raaijmakers, 1997; Kleinmuntz, 1985). We need better tools and processes to help the managers cope with the ever-present dynamic tasks. This article makes a case for the inclusion of HCI design in any CSBILE model aimed at improving decision making in dynamic tasks. We believe that the lack of emphasis on HCI design in a CSBILE resulted, at least in part, in poor performance in dynamic tasks by people. Moreover, HCI design methods and techniques can be used to reduce the difficulties people have in dealing with dynamic tasks. At the same time, we have made the case to reassess the earlier studies on dynamic decision making supporting the MOF hypothesis. Perhaps by focusing more attention on improved interface design for CSBILEs, we can help people make better organizational decisions.

REFERENCES

Bakken, B. E. (1993). *Learning and transfer of understanding in dynamic decision environments* Doctoral thesis. Boston: MIT.

Brehmer, B. (1990). Strategies in real-time dynamic decision making. In R.M. Hogarth (Ed.), *Insights in decision making* (pp. 262-279). Chicago: University of Chicago Press.

Carey, J., et al. (2004) The role of human-computer interaction in management information systems curricula: A call to action. *Communications of the Association for Information Systems, 13,* 357-379.

Cox, R. J. (1992). Exploratory learning from computer-based systems. In S. Dijkstra, H.P.M. Krammer, & J. J. G. van Merrienboer (Eds.), *Instructional models in computer-based learning environments* (pp. 405-419). Berlin: Springer-Verlag.

Davidsen, P. I. (1996). Educational features of the system dynamics approach to modelling and simulation. *Journal of Structural Learning, 12*(4), 269-290.

Davidsen, P. I. (2000). Issues in the design and use of system-dynamics-based interactive learning environments. *Simulation & Gaming, 31*(2), 170-177.

Davidsen, P. I., & Spector, J. M. (1997). Cognitive complexity in system dynamics based learning environments. *Proceedings of the International System Dynamics Conference*, Istanbul, Turkey.

Diehl, E., & Sterman, J. D. (1995). Effects of feedback complexity on dynamic decision making. *Organizational Behavior and Human Decision Processes, 62*(2), 198-215.

Edwards, W. (1962). Dynamic decision theory and probabilistic information processing. *Human Factors, 4*, 59-73.

Forrester, J. W. (1961). *Industrial dynamics*. Cambridge, MA: Productivity Press.

Funke, J. (1995). Experimental research on complex problem solving. In P. Frensch, & J. Funke (Eds.), *Complex problem solving: The European perspective* (pp. 3-25). Hillsdale, NJ: Lawrence Erlbaum Associates.

Gagné, R. M. (1995). Learning processes and instruction. *Training Research Journal, 1*, 17-28.

Howie, E., Sy, S., Ford, L., & Vicente, K. J. (2000). Human-computer interface design can reduce misperceptions of feedback. *System Dynamics Review, 16*(3), 151-171.

Hsiao, N. (2000). *Exploration of outcome feedback for dynamic decision making* [doctoral thesis]. Albany: State University of New York at Albany.

Issacs, W., & Senge, P. (1994). Overcoming limits to learning in computer-based learning environments. In J. Morecroft, & J. Sterman (Eds.), *Modeling for learning organizations* (pp. 267-287). Portland, OR: Productivity Press.

Kerstholt, J. H., & Raaijmakers, J. G. W. (1997). Decision Making in Dynamic Task Environments. In R. Ranyard, R. W. Crozier, & O. Svenson (Eds.). *Decision Making: Cognitive Models and Explanations* (pp. 205-217). New York: Routledge.

Keys, J. B., & Wolfe, J. (1990). The role of management games and simulations in education and research. *Journal of Management, 16,* 307-336.

Kirlik, A. (1995). Requirements for psychological model to support design: Towards ecological task analysis. In J. M. Flach, P. A. Hancock, J. K. Caird, & K. J. Vicente (Eds.), *An ecological approach to human-machine systems 1: A global perspective* (pp. 68-120). Hillsdale, NJ: Erlbaum.

Kleinmuntz, D. N. (1985). Cognitive heuristics and feedback in a dynamic decision environment. *Management Science, 6*(31), 680-701.

Lane, D. C. (1995). On a resurgence of management simulations and games. *Journal of the Operational Research Society, 46,* 604-625.

Langley, P. A., & Morecroft, J. D. W. (1995). Learning from microworlds environments: A summary of the research issues. In G.P. Richardson, & J. D. Sterman (Eds.), *System dynamics '96,* (pp. 23-37). Cambridge, MA: System Dynamics Society.

Moxnes, E. (1998). Not only the tragedy of the commons: Misperceptions of bioeconomics. *Management Science, 44,* 1234-1248.

Qudrat-Ullah, H. (2001). Improving decision making and learning in dynamic tasks: An experimental investigation. *Proceedings of the Annual Faculty Research Conference*, Singapore.

Qudrat-Ullah, H. (2002). *Decision making and learning in complex dynamic environments* [doctoral thesis]. Singapore: National University of Singapore.

Schraagen, J. M. C. (1994). *What information do river pilots use?* [Report TNO TM-1994 C-10]. Soesterberg: Human Factors Research Institute.

Sengupta, K., & Abdel-Hamid, T. (1993). Alternative concepts of feedback in dynamic decision environments: An experimental investigation. *Management Science, 39*(4), 411-428.

Sterman, J. D. (1989a). Modeling managerial behavior: Misperceptions of feedback in a dynamic decision making experiment. *Management Science, 35*, 321-339.

Sterman, J. D. (1989b). Misperceptions of feedback in dynamic decision making. *Organizational Behavior and Human Decision Processes, 43*, 301-335.

Sterman, J. D. (1994). Learning in and around complex systems. *System Dynamics Review, 10*(2-3), 291-323.

Sternberg, R. J. (1995). Expertise in complex problem solving: A comparison of alternative conceptions. In P. Frensch, & J. Funke (Eds.), *Complex problem solving: The European perspective* (pp. 3-25). Hillsdale, NJ: Lawrence Erlbaum Associates.

KEY TERMS

Acquisition-of-Expertise Hypothesis: States that people will perform better in dynamic tasks, if they acquire the requisite expertise.

Feedback: It is a process whereby an input variable is fed back by the output variable. For example, an increased (or decreased) customer base leads to an increase (or decrease) in sales from word of mouth, which then is fed back to the customer base, increasingly or decreasingly.

Mental Model: A mental model is the collection of concepts and relationships about the image of the real-world things we carry in our heads. For example, one does not have a house, city, or gadget in his or her head, but a mental model about these items.

Non-Linearity: A non-linearity exists between a cause (decision) and effect (consequence), if the effect is not proportional to cause.

Prototypic Expertise: The concept of prototypic expertise views people neither as perfect experts nor as non-experts, but somewhere in between both extremes.

Requisite Expertise: Having an adequate understanding of the task that helps to manage the task successfully.

Simulated System: A simplified, computer simulation-based construction (model) of some real-world phenomenon (or the problem task).

Time Delays: Often, the decisions and their consequences are not closely related in time. For instance, the response of gasoline sales to the changes in price involves time delays. If prices go up, then after a while, sales may drop.

Including Users with Motor Impairments in Design

Simeon Keates
IBM T.J. Watson Research Center, USA

Shari Trewin
IBM T.J. Watson Research Center, USA

Jessica Paradise Elliott
Georgia Institute of Technology, USA

INTRODUCTION

For people with motor impairments, access to, and independent control of, a computer can be an important part of everyday life. However, in order to be of benefit, computer systems must be accessible.

Computer use often involves interaction with a graphical user interface (GUI), typically using a keyboard, mouse, and monitor. However, people with motor impairments often have difficulty with accurate control of standard input devices (Trewin & Pain, 1999). Conditions such as cerebral palsy, muscular dystrophy, and spinal injuries can give rise to symptoms such as tremor, spasm, restricted range of motion, and reduced strength. These symptoms may necessitate the use of specialized assistive technologies such as eye-gaze pointing or switch input (Alliance for Technology Access, 2000). At the same time, specialized technologies such as these can be expensive and many people simply prefer to use standard input devices (Edwards, 1995; Vanderheiden, 1985). Those who continue to use standard devices may expend considerable time and effort performing basic actions.

The key to developing truly effective means of computer access lies in a user-centered approach (Stephanidis, 2001). This article discusses methods appropriate for working with people with motor impairments to obtain information about their wants and needs, and making that data available to interface designers in usable formats.

BACKGROUND

In a recent research study commissioned by Microsoft, Forrester Research, Inc. (2003) found that 25% of all working-age adults in the United States had some form of dexterity difficulty or impairment and were likely to benefit from accessible technology. This equates to 43.7 million people in the United States, of whom 31.7 million have mild dexterity impairments and 12 million have moderate to severe impairments.

If retirees had been included in the data sample, the number of people who would benefit from accessible technology would be even higher as the prevalence of motor impairments, and thus the need for such assistance, increases noticeably with age (Keates & Clarkson, 2003). As the baby-boomer generation ages, the proportion of older adults is set to increase further.

The global aging population is growing inexorably (Laslett, 1996). By 2020, almost half the adult population in the United Kingdom will be over 50, with the over-80s being the most rapidly growing sector (Coleman, 1993). Governments are responding to this demographic change. Antidiscrimination legislation has been enacted in many countries such as the United States with the 1990 Americans with Disabilities Act, and the United Kingdom with the 1995 Disability Discrimination Act.

These pieces of legislation often allow users who are denied access to a service to litigate against the service provider. They are mechanisms for enforcing basic rights of access. A complementary "carrot" approach to this legislative "stick" is the change in governmental purchasing policy. In the United States, the Section 508 amendment to the 1998 Workforce Investment Act stipulates minimum levels of accessibility required for all computer systems purchased by the U.S. Federal Government, the world's largest purchaser of information-technology equipment. Many other national and regional governments are adopting similar purchasing policies.

Research Methods for Improving Accessibility

To provide truly accessible systems, it is necessary to examine the user experience as a whole and to adopt design best practices wherever possible. To this end, standards are being developed, such as the forthcoming British Standards Institute BS7000 Part 6, "Guide to Managing Inclusive Design," that focus on wider interpretations of accessibility throughout the complete lifetime of products.

In addition, heuristic evaluations of prototypes can reveal fundamental physical-access issues. Accessibility standards like the U.S. Section 508 guidelines (http://www.section508.gov/), or the W3C Web Accessibility Initiative Web Content Accessibility guidelines and checklists (Chisholm, Vanderheiden, & Jacobs, 1999) are readily available to assist with establishing the heuristics. Examples include testing whether keyboard-only access is possible and examining the size of targets the user is expected to click on. Addressing these issues in advance of user testing will allow the maximum benefit to be gained from the user sessions themselves.

Ideally, users with disabilities should be included in product design and usability testing early and often. Many user-interface designers are not adequately equipped to put this into practice (Dong, Cardoso, Cassim, Keates, & Clarkson, 2002). Most designers are unfamiliar with the needs of people with motor impairments and are unsure how to contact such users or include them in studies. The following sections outline some specific considerations and techniques for including this population in user studies.

SAMPLING USERS

For traditional user studies, the users would typically be customers or employees and would often be readily at hand. However, when considering users with a wide range of capabilities, it is often necessary to commit explicit effort and resource to seeking out potential participants.

Good sources of users include charitable organizations, social clubs, and support groups, which can be found in most towns and cities. However, even when sources of users have been identified, effort still needs to be expended in trying to identify candidate users who match the user-sampling profiles. Sample sizes are inevitably small since volunteers must be reasonably typical users of the product in addition to having a physical impairment.

Sampling Users by Condition

There are many possible approaches for identifying and sampling potential users. The most obvious is to identify users based on their medical condition. The advantage of this approach is that someone's medical condition is a convenient label for identifying potential users. Not only are most users aware of any serious condition, especially one that affects their motor capabilities, but it also makes locating users easier. For example, many charitable organizations are centered on specific medical conditions, such as cerebral palsy, muscular dystrophy, or Parkinson's disease.

The disadvantage of this approach is that many of these conditions are highly variable in terms of their impact on the user's functional capabilities, and so a degree of user-capability profiling is still required.

Sampling Users by Capability

The alternative approach to sampling users is not to focus on their medical condition, but to instead look at their capabilities. The advantage of this approach is that the accessibility of the resultant product should then be independent of the medical condition. The disadvantage of this approach is that more user-capability profiling is required at the outset to establish where each user sits in the capability continuum.

The most popular approaches to sampling issues are to either find users that represent a spread across the target population, or to find users that sit at the extremes of that population. The advantage of working with users that represent a spread across the population is that they ensure that the assessment takes the broadest range of needs into account. The disadvantage, though, is that there is not much depth of coverage of users who may experience difficulties in accessing the product.

The advantage of working with the extreme users is that the user-observation sessions will almost certainly discover difficulties and problems with the interaction. However, the disadvantage is that there is a real danger of discovering that particular users cannot use the product and little else beyond that. For example, giving a drawing program with an on-screen toolbox to a user who cannot use a mouse yields the obvious difficulty arising from the inability to choose a drawing tool. However, subsequent questions about the tools themselves are not possible because of the overriding difficulty of choosing them.

Of more use is to identify users who are more likely to be "edge" cases: those who are on the borderline of being able to use the product, and who would commonly be accepted as being able to use the interface (Cooper, 1999). Going back to the example of someone with a motor impairment attempting to use a drawing program, while someone unable to use a mouse at all would certainly not be able to use the drawing tools, someone with a moderate tremor may be able to do so. Even more interestingly, that person might be able to access some tools and not others, and thus it is possible to begin to infer a wide range of very useful data from such a user. On top of that, if the user cannot use the tools, then it may be inferred that any user with that level of motor impairment or worse will not be able to use them, automatically encompassing the users who cannot control a mouse in the assessment.

WORKING WITH USERS

As with all user studies, the participating users need to be treated with respect and courtesy at all times. When dealing with users with more severe impairments, it is especially important to be sensitive to their needs, and accommodations in study design may be necessary.

Location

Many usability tests are carried out in a laboratory. For people with physical impairments, this is not always ideal. Individuals may have made many modifications to their home or work environments to allow them to work comfortably and accurately, and this will often be difficult to reproduce in a lab session. The user may not be able or willing to bring assistive technologies they use at home. Furthermore, the user's impairment may make travel to sessions difficult and/or physically draining. When laboratory sessions are carried out, researchers should consider providing the following facilities, and plan to spend time at the start of each session making sure that the user is comfortable.

- Table whose height can be easily adjusted
- Moveable and adjustable chair
- Cordless keyboard and mouse that can be placed on a user's wheelchair tray
- Keyboard whose slope and orientation can be adjusted and then fixed in place on the table
- Alternative pointing devices such as a trackball
- Adjustments to the key-repeat delay, key-repeat rate, mouse gain, double-click speed, and any other software accessibility features to users' preferred settings

A compromise approach that can work well is to hold sessions at a center specializing in computer access for people with disabilities, where such equipment is already available. Users can also be encouraged to bring their own devices when practical (e.g., special keyboard and key guard).

If tests can be carried out at the user's own location, then a more realistic usability evaluation can be performed. For users who employ specialized assistive technologies such as head pointing or eye gaze, it may be useful to schedule time for the user to explain these technologies to the researchers as it may be difficult to understand what is happening if the operation of this device is unfamiliar.

Remote testing is an increasingly popular technique in which the user carries out a task from his or her own environment while the researcher is not physically present. This approach is sometimes necessary when local users cannot be found. Its efficacy depends on the kind of evaluation to be performed. Telephone, e-mail, or chat can be used as a means of communication, or the task can be entirely self-driven through a Web site or other software. Gathering information from users by e-mail has the advantage that users can take as long as they need to prepare responses, and can be useful for those whose speech is difficult to understand. The remote evaluation of products or product prototypes is also possible, but the quality of the information received will often be poorer. For example, if a user takes a long time to perform a task, the researcher does not know whether this is because they had trouble in deciding what to click on, trouble in clicking on the icon, or because they were interrupted by a family member. Detailed recordings of the user's input activities or the use of a camera can be helpful in this respect.

Methods for Gathering User Data

The following list represents a summary of typical methods used by researchers to elicit user wants and investigate product usability.

- **Questionnaires:** A series of preprepared questions asked either in writing or orally
- **Interviews:** Either prestructured or free-form
- **User Observation:** Watching the users perform a task, either using an existing product or a prototype
- **Focus Groups:** Discussion groups addressing a specified topic
- **Contextual Inquiry:** Interviewing and observing users in situ

All of the above methods are discussed in detail in many HCI- (human-computer interaction) design textbooks (e.g., Beyer & Holtzblatt, 1998; Nielsen & Mack, 1994). As when considering any technique or approach developed originally for the mainstream market, there are additional considerations that need to be borne in mind when adapting to designing for the whole population.

When including people with motor impairments, it is useful to plan for the following:

- Users may fatigue more quickly than the researcher expects. The user's fatigue level should be carefully monitored and the researcher should be prepared to give frequent breaks, end a session early, and split tasks over multiple sessions if necessary.
- Extra time should be allowed for computer-based activities. Allow 2 to 3 times as long for someone with a moderate impairment and be prepared for some individuals to spend longer.
- For those whose disability has caused a speech impairment in addition to motor impairment (e.g., cerebral palsy), additional time for communication will be necessary, and the researcher may need to ask the user to repeat statements multiple times. Users are generally happy to do this in order to be understood. Researchers should also repeat responses back to the user to check that they have understood. In some cases, the user may choose to type responses into a document open on the computer.
- Some users may have difficulty signing a consent form. Some may sign an *X* or use a stamp to sign, while others may wish to sign electronically. Be prepared for all of these.
- Users may prefer to respond to questionnaires verbally or electronically rather than use printed paper and pen.
- Some physical disabilities have highly variable symptoms or may cause additional health problems. Experimenters should expect higher-than-normal dropout rates, and be careful to confirm sessions near the time in case the participant is unable to attend.

PACKAGING THE USER DATA

Having discussed the issues that HCI researchers and practitioners have to consider when aiming to design for universal access, it is helpful to look at ways of packaging the user data in a succinct format.

Presenting User Profiles

There are a number of methods of packaging the user information for designers. For example, short videos of target users—perhaps depicting their lifestyles, or using or talking about particular products—provide designers with greater insights into the needs and aspirations of users. Such dynamic illustrations can be effective in inspiring designers to formulate inclusive solutions. They can also be very informative to designers who have never seen assistive technologies in use before.

Such accounts offer immediate means of assessing a variety of ways and situations in which a product or service will be used or accessed. It can be a powerful technique if care is taken when building up user profiles based on actual user data or amalgams of individual users constructed to represent the full range of target users and contexts of use.

The Application of Statistical Analyses

Statistical analyses are often useful ways to summarize and present quantitative data, but there are practical limitations to these techniques when including data from people with motor impairments. This is due to the variable availability of individual users, the small sample set, and considerable individual differences. It may be necessary to consider data gathered from individuals with motor impairments separately from other users. Because of the small number of users available, repeated measures designs should generally be employed. Obviously, these practical difficulties give rise to missing data problems resulting from incomplete conditions, caused by the loss of levels and factors from designs, and make the systematic varying of conditions in pilot studies difficult.

In addition, the increased range and skewed variability resulting from the range of motor impairments leads to increased noise and violation of the assumptions of statistical tests. Where statistical tests are possible without violation of standard assumptions, such as normality of distribution or homogeneity of variance, they should be carried out. However, even if the power of these experiments was unknown because of the reasons outlined and the small sample size, the effect sizes may still be large because of the sometimes radically different behaviours that are associated with different functional impairments. For this reason, some statistical results that do not appear significant should be analysed in terms of statistical power ($1 - \beta$, the probability of rejecting a false null hypothesis; Cohen, 1988) and estimates of the effect size given (Chin, 2000).

User Models

Another method for packaging quantitative user information is a user model. In this context, a user model is a quantitative description of a user's interaction behaviour that can be used to describe, predict, and/or simulate user performance on specific tasks. It has been used to model single-switch letter scanning and predict communication rates for scanning, and alternative and augmentative communication (AAC) devices (Horstmann, 1990; Horstmann & Levine, 1991). There are critics of the applicability of such models to motion-impaired users (Newell, Arnott, & Waller, 1992; Stephanidis, 1999) who object to the use of generalizations for a population with such great individual differences. However, models representing a specific individual, or a group of relatively similar individuals, can help designers to understand the effects of their design decisions and refine their designs for improved usability.

Claims Approach to Requirements

Where quantitative data is not available, another method of packaging the user information is that of claims (Sutcliffe & Carroll, 1999). For example, if an on-screen button is hard to press, then a claim could be made that increasing the size of the button would make it easier to operate. The claim also identifies the user and situation for which it applies, recognizing that there are often conflicting requirements that can lead to design compromises being sought.

FUTURE TRENDS

With antidiscrimination legislation being enacted by an increasing number of countries, designers are going to come under increasing pressure to ensure that all user interfaces, both hardware and software, are as accessible as possible. This means that in the future, designers will have to work more closely with users with all kinds of impairments, from vision and hearing to motor and cognitive.

One of the most time-consuming aspects of working with motor-impaired users is finding and recruiting them for the inevitable user trials required to ensure that the systems being developed meet their needs. One option that design teams may well begin to pursue is that of incorporating people with different impairment types into the design team itself. This approach offers the immediate advantage that detailed feedback on the effect of design choices on a system's accessibility can be determined rapidly. A potential further step is to train those people to actively drive the design process, taking their needs into consideration from the very outset. With many companies having to meet employee quota targets under disability legislation, this approach should become an increasingly attractive proposition to many design teams.

CONCLUSION

Including people with motor impairments in the design and evaluation of computer products is essential if those products are to be usable by this population. There are some special considerations and techniques for including this population in user studies. Researchers may need to expend some effort locating appropriate users. It is good practice to perform capability assessment to identify edge-case individuals who should, in principle, be able to use the product, but may be excluded by specific design features. Study materials and methodologies may need to be modified to meet the needs of users. Laboratories, user environments, and remote testing can all be used, although testing in the user's environment is preferred whenever possible. The statistical analysis of user data is possible in specific circumstances, and significant effects can be found even with small sample sizes, but care must be taken to use tests that do not rely on inappropriate assumptions. User data can be presented to designers quantitatively, as statistical summaries or user models, or qualitatively, as user profiles or claims.

REFERENCES

Alliance for Technology Access. (2000). *Computer and Web resources for people with disabilities: A guide to exploring today's assistive technologies* (3rd ed.). CA: Hunter House Inc.

Beyer, H., & Holtzblatt, K. (1998). *Contextual design: A customer-centered approach to systems designs.* San Francisco: Morgan Kaufmann.

Chin, D. N. (2000). *Empirical evaluation of user models and user-adapted systems.* Honolulu, HI: University of Hawaii, Department of Information and Computer Sciences.

Chisholm, W., Vanderheiden, G., & Jacobs, I. (1999). *Web content accessibility guidelines 1.0.* W3C. Retrieved from http://www.w3.org/TR/WAI-WEBCONTENT/

Cohen, J. (1988). *Statistical power analysis for the social sciences* (2nd ed.). Hillsdale: Erlbaum.

Coleman, R. (1993). A demographic overview of the ageing of first world populations. *Applied Ergonomics, 24*(1), 5-8.

Cooper, A. (1999). *The inmates are running the asylum.* Indianapolis, IN: SAMS Publishing.

Dong, H., Cardoso, C., Cassim, J., Keates, S., & Clarkson, P. J. (2002). *Inclusive design: Reflections on design practice* (Rep. No. CUED/C-EDC/TR118). Cambridge University Engineering Department.

Forrester Research, Inc. (2003). *The wide range of abilities and its impact on computer technology.* Cambridge, MA: Author.

Horstmann, H. M. (1990). Quantitative modeling in augmentative communication: A case study. *Proceedings of RESNA '90* (pp. 9-10).

Horstmann, H. M., & Levine, S. P. (1991). The effectiveness of word prediction. *Proceedings of RESNA '91* (pp. 100-102).

Keates, S., & Clarkson, P. J. (2003). *Countering design exclusion.* Springer-Verlag.

Keates, S., Clarkson, P. J., & Robinson, P. (2000). Investigating the applicability of user models for motion-impaired users. *Proceedings of ACM ASSETS 2000* (pp. 129-136).

Laslett, P. (1996). *A fresh map of life: The emergence of the third age* (2nd ed.). Weidenfeld and Nicolson.

Newell, A. F., Arnott, J. L., & Waller, A. (1992). On the validity of user modelling in AAC: Comments on Horstmann and Levine. *Augmentative and Alternative Communication, 8*, 89-92.

Nielsen, J., & Mack, R. L. (1994). *Usability inspection methods*. New York: John Wiley & Sons.

Stephanidis, C. (1999). Designing for all in the information society: Challenges towards universal access in the information age. *ERCIM ICST Research Report*, 21-24.

Stephanidis, C. (2001). User interfaces for all: New perspectives into human-computer interaction. In C. Stephanidis (Ed.), *User interfaces for all* (pp. 3-17). Lawrence Erlbaum.

Sutcliffe, A. G., & Carroll, J. M. (1999). Designing claims for reuse in interactive systems design. *International Journal of Human-Computer Studies, 50*(3), 213-241.

Trewin, S., & Pain, H. (1999). Keyboard and mouse errors due to motor disabilities. *International Journal of Human-Computer Studies, 50*, 109-144.

KEY TERMS

Accessibility: A characteristic of information technology that allows it to be used by people with different abilities. In more general terms, accessibility refers to the ability of people with disabilities to access public and private spaces.

Accessible Technology: Products, devices, or equipment that can be used, with or without assistive technology, by individuals with disabilities.

Assistive Technology: Products, devices, or equipment, whether acquired commercially, modified, or customized, that are used to maintain, increase, or improve the functional capabilities of individuals with disabilities.

Inclusive Design: The design of mainstream products and/or services that are accessible to, and usable by, as many people as reasonably possible on a global basis, in a wide variety of situations, and to the greatest extent possible without the need for special adaptation or specialized design.

Motor Impairment: A problem in body motor function or structure such as significant deviation or loss.

User-Centered Design: A method for designing ease of use into a product by involving end users at every stage of design and development.

User Model: A quantitative description of a user's interaction behaviour that can be used to describe, predict, and/or simulate user performance on specific tasks.

The Influence of Expressive Images for Computer Interaction

Zhe Xu
Bournemouth University, UK

David John
Bournemouth University, UK

Anthony C. Boucouvalas
Bournemouth University, UK

INTRODUCTION

The soul is divided into an immortal part, located in the head, and a mortal part, distributed over the body. Philosophical and intellectual loves of beauty are located in the immortal soul. Other "regular" emotions are located in the mortal soul. (Plato as cited in Koolhaas, 2001)

Emotion is one of the lovely gifts from nature. It is present not only in humans, but most species present sorts of emotions and expressions in daily behaviors. However, only human beings ask for explanations. Research into the mystery of emotion can be traced back to Heraclitus (500 BC), who claimed that "the emotional state is characterized by a mixture of body parameters such as temperature (hot/cold) and sweat amount (wet/dry)" (as cited in Koolhaas, 2001).

In the 21st century, technology has achieved a standard that Plato never dreamed about, but emotion is still an unsolved question. Although science needs more time to work out the mechanism, it does not keep emotion out of human communication.

With the commercial success of the Internet, more people spend their time with their box: the computer. Designing an attractive user interface is not only the objective of every software developer but also is crucial to the success of the product. Methods and guidelines (Newman & Lamming, 1995) have been published to design a "vivid" user interface. One of the most important methods is to add expressive images in the display (Marcus, 2003). For example, when a user finishes some operation, an emotional icon or *emoticon* (an industry term introduced in the 1980s by Meira Blattner) will pop up to communicate "well done" to the user.

Two widely accepted methods exist for displaying emotional feelings in software interfaces. One is the use of emotion-oriented icons; the other is using complex images, for example, a cartoon or a facial image (Boucouvalas, Xu, & John, 2003; Ekman, 1982).

Emotion icons cannot communicate complex feelings, and they are not usually customized. As the industry matures, perhaps emoticons will be replaced by expressive images as sophisticated as the computer-generated Golem of *The Lord of the Rings* movie fame.

Expressive images present emotional feelings to users. What internal factors (e.g., image intensity or people's mood) may influence the perceived emotional feelings? Will external factors (e.g., display duration) influence the perceived emotional feelings as well?

In this article, we are particularly interested in discussing the factors that may influence the perceived emotional feelings. Our conclusions are based on the findings from a series of experiments that demonstrate an empirical link between the level of expressive-image intensities and the perceived feelings. The detected factors include the following:

- Expression intensity
- Wear-down effect (display duration effect)

The test results demonstrate that increasing the expressive-image intensity can improve the perceived emotional feeling. However, when the intensity is increased to an extreme level, the perceived emotional feelings fall. The experiment results also indicate that the perceived emotional feelings are not affected by the length of time that users are exposed to the expressive images.

BACKGROUND

Emotion is not a concept that can be easily defined. Izard (1993) describes emotion as a set of motivational processes that influence cognition and action. Other researchers such as Zajonc (1980) argue that emotion is a particular feeling, a quality of conscious awareness, and a way of responding.

A widely accepted fact about emotion is that emotion can be classified into different categories and numerous intensities. One classification method divides emotions into elation, desire, hope, sadness, anger, frustration, and so forth (Koolhaas, 2001).

Emotional expressions not only present one's internal feelings, but also influence interpersonal feelings. Moffat and Frijda (1994) demonstrated that expressions are a means to influence others. Fridlund (1997) found that expressions occur most often during pivotal points in social interactions: during greetings, social crises, or times of appeasement. According to Azar (2000), "Thinking of facial expressions as tools for influencing social interactions provides an opportunity to begin predicting when certain facial expressions will occur and will allow more precise theories about social interactions" (p. 45).

The influences of emotions on the public domain have been examined for many years. Emotion is a powerful tool for reporters, editors, and politicians. The 9/11 New York attack may not have been experienced by all personally; however, most of us felt the same fear and pain when we saw the scenes. Strong links between the emotion of news and the importance individuals assign to issues have been suggested by a number of theories (Evatt, 1997).

Is emotion an important tool online as in daily life? Recent research argues that there is in fact a high degree of socioemotional content observed in computer-mediated communications (CMC; McCormick & McCormick, 1992; Rheingold, 1994), even in organizational and task-oriented settings (Lea & Spears, 1991). Even first-time users form impressions of other communicant's dispositions and personalities based on their communication style (Lea & Spears, 1991).

Multimodal presentations (e.g., animation, voice, and movie clips) for Internet communication are more popular than ever as the processing speed and bandwidth continues increasing. These new presentation styles make emotion expression easier to transmit than before.

Will users prefer emotional feelings to be presented pictorially on the computer interfaces? Will the expressive images influence the perceived feelings?

We have carried out a series of experiments to investigate these questions (Xu & Boucouvalas, 2002; Xu, John, & Boucouvalas, in press).

Xu and Boucouvalas (2002) demonstrated an effectiveness experiment. In that experiment, participants were asked to view three interfaces (an interface with an expressive image, voice, and text; an interface with an expressive image and text; and an interface with text only). The results show that most participants prefer the interface with the expressive image, voice, and text. A significant number of participants preferred the interface with the expressive image, voice, and text much more than the text-only interface. This means that with the expressive images, the effectiveness of the human-computer interface can be considerably improved.

Xu et al. (in press) presented a perceived-performance experiment, which demonstrated that emotion can affect the perceived performance of individuals. In that experiment, participants were asked to answer questions in an online quiz. A computer agent presented 10 questions (e.g., "What percentage of people wear contact lenses?" and choices A, 15%; B, 30%; C, 20%; D, 50%) to the participants. When the participants finished answering the questions, either the presenting agent himself (self-assessing) or a new agent checked the participants' answers (other-assessing). No matter what answers each participant provided, all were told that they answered the same 5 out of 10 questions correctly. For the other-assessing scenario, the assessing agent presented no emotional expressions positively related to participants' answers or emotional expressions negatively related to participants' answers. The results from the other-assessing scenario demonstrated that significant differences exist when comparing the positively-related-emotion situation with the negatively-related-emotion situation and the no-emotion situation. The participants in the positively-related-emotion situation believed that they achieved much better performances than the participants in the other situations.

FACTORS THAT MAY INFLUENCE EMOTION PRESENTATION

Are the influences found in the former experiments permanent or changeable in different situations? Will other factors affect the influence of emotion expression? Evatt (1997) discovered that the perceived salience of a public-policy issue will increase when news about the issue is presented in a highly emotion-evoking manner and decrease when the news about the same issue is presented in a less emotion-evoking manner. This demonstrates that when the intensity of the emotion-evoking manner increases, the salience the readers perceive will increase.

Hovland, Janis, and Kelley (1953) suggested that increasing the intensity of the emotion-evoking content might not always elevate salience. At a certain intensity level, the effect could start to drop off.

Hughes (1992) and Kinnick, Drugman, and Cameron (1996) demonstrated a wear-down phenomenon. Participants' favourable responses will be reduced after the emotion-evoking manner is repeated over a long period of time. As before, the measurement was based on pure text. However, Evatt (1997) demonstrated that the wear-down phenomenon is not always observed.

It can be seen that the influences of presenting textual information in an emotion-evoking manner will be affected by different factors. However, the above experiments are purely based on textual messages, which mean that the emotion-evoking manners used were pure text. Will the emotions presented by expressive images produce the same phenomena?

EMOTIONAL-INTERFACE DESIGN CONSIDERATIONS

In response to the previous experiments and background knowledge, it is necessary to identify the possible factors that may influence the perception of emotion over the Internet. In summary, three phenomena were observed.

- When the intensity of the expressive images increases, the perceived emotional feelings will increase. This phenomenon means that by increasing the intensity of the emotion-evoking manner (expressive images), the perceived feelings will increase even if the accompanied text remained the same.
- When the intensity of the expressive images rises beyond a realistic level, the perceived feelings will stop increasing and start to decrease. The levels of emotional feelings were predicted to fall as the participants were exposed to an extremely high intensity of the expressive images.
- After viewing three scenarios, the perceived feelings for the third scenario will be higher for people who view a scenario accompanied with medium-intensity expressive images following two scenarios accompanied without expressive images than for people who view the same scenarios each accompanied with medium-intensity expressive images (wear-down effect).

The above three phenomena have been observed by various researchers (see above discussion); however, some researchers doubt the existence of the phenomena, especially the wear-down effect. In this article, we developed two experiments to assess the applicability of the above phenomena.

THE INTENSITY EXPERIMENT

To assess the influences of expressive images with different intensities, a human-like agent was developed. The agent presented a story on the screen and offered facial expressions. To focus on the influences of expressive facial images, the story itself contained minimal emotional content.

Sixty students and staff from Bournemouth University participated in this online experiment. The experiment included two sessions. First, a cartoon human-faced agent presented a story to each participant. In the second section, participants answered a questionnaire about the emotional feelings they perceived.

A between-group experimental design was applied for this experiment. In all conditions, the agent presented the same stories to every participant. However, in the first condition (low-intensity condition), the agent presented facial images with low expressive intensity to the participants. In the second condition (medium-intensity condition), the agent presented facial images with medium expressive intensity. In the third condition, the agent presented extreme-expressive-intensity facial images to participants (extreme-intensity condition). Typical screens of all conditions are shown in Figure 1.

After viewing the story presentation session, all participants in the three groups were directed to the same questionnaire. The applied questionnaire was based on the Personal Involvement Inventory (PII) that was developed by Zaichkowsky (1986).

INTENSITY TEST RESULTS

The Shapiro-Wilks normality test (Norusis, 1998) was carried out and the result indicated that the observations of the emotion-intensity test were normally distributed, and therefore t-tests were carried out to determine whether the ratings of participants who viewed the different conditions were significantly different.

For the low-intensity condition, the mean value of the perceived emotional feeling was 3.22. In the medium-intensity condition, the mean perceived emotional feeling was 4.6. The t-test revealed a significant difference between the ratings of the low-intensity condition and the medium condition ($F=3.85, p=0.044$).

We were therefore able to accept the first phenomenon that states when the intensity of the emotionally expressive images increases, the perceived emotional feelings will increase as well.

For the extreme-intensity condition, the mean perceived emotional feeling was 3.7. The t-test showed a marginally significant difference between the ratings of the medium condition and the high condition ($F=4.25, p=0.08$). The result indicates that the second phenomenon is correct in asserting that when the intensity of the emotional-expression images rises beyond a realistic level, the perceived feelings will stop increasing and may fall.

THE WEAR-DOWN-FACTOR (THIRD PHENOMENON) EXPERIMENT

Will external factors influence the perceived emotional feelings? An experiment was carried out to test an external factor: wear-down. Wear-down (or wear-out in different literature) is described by Hughes (1992) as a reduction in the participant's favourable responses after repeated exposure to a message. For example, when an individual first meets an exciting stimulus, the excited feelings will be high. When the stimulus is repeated many times, the exciting feelings will not continue to rise; instead, the feelings will be stable or even fall if the stimulus is endless. The problem with assessing wear-down factors is that it is hard to predict the exact time that

Figure 1. Typical screens of the three conditions of the experiment

A. Low-intensity condition

B. Medium-intensity condition

C. Extreme-intensity condition

Figure 2. The wear-down factor

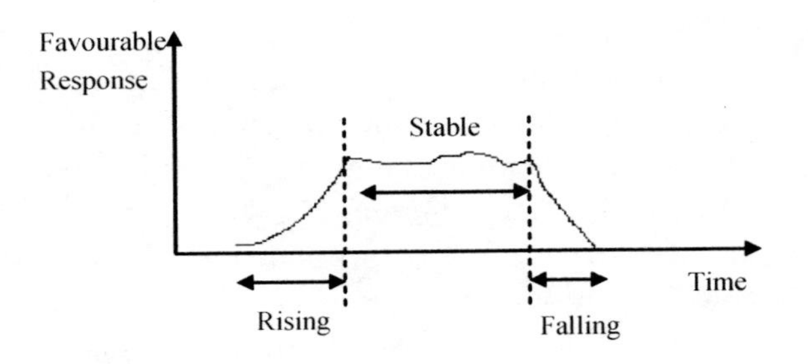

feelings will be stable or will fall. The wear-down factor is illustrated visually in Figure 2.

To assess the wear-down factor, we still relied on the agent-presenting environment. However, instead of presenting only one story, the agent presented three stories, all of which contained minimal emotional content in order to keep the focus on the expressive images.

Forty students and staff from Bournemouth University participated in this online experiment. The participants were divided into two groups. The cartoon human-faced agent presented three stories to each participant and then the participants answered a questionnaire about the perceived emotional feelings.

A between-group experimental design was applied. The stories were arranged in the same subject order and all the stories contained minimal emotional content themselves. The presentations to the two groups differed only in the intensity of the expressive facial images. In the first condition, the agent presented two stories without facial expressions followed by a story with medium-intensity facial expressions. In the second condition, the agent presented all three stories with medium-intensity expressions. The typical screens of Group 1 and Group 2 are shown in Figure 3.

After viewing the story presentation session, all participants in both groups were directed to the same questionnaire session. The applied questionnaire was also based on the Personal Involvement Inventory developed by Zaichkowsky (1986).

Although the story sets were the same, the third phenomenon predicted that the perceived emotional feelings would be higher when participants viewed medium expressive images after two sets of neutral expressive images. The third phenomenon is only concerned with the responses to the third story in each set. The design for each set of story presentations is shown in Table 1.

Figure 3. Typical screens of both conditions of the experiment

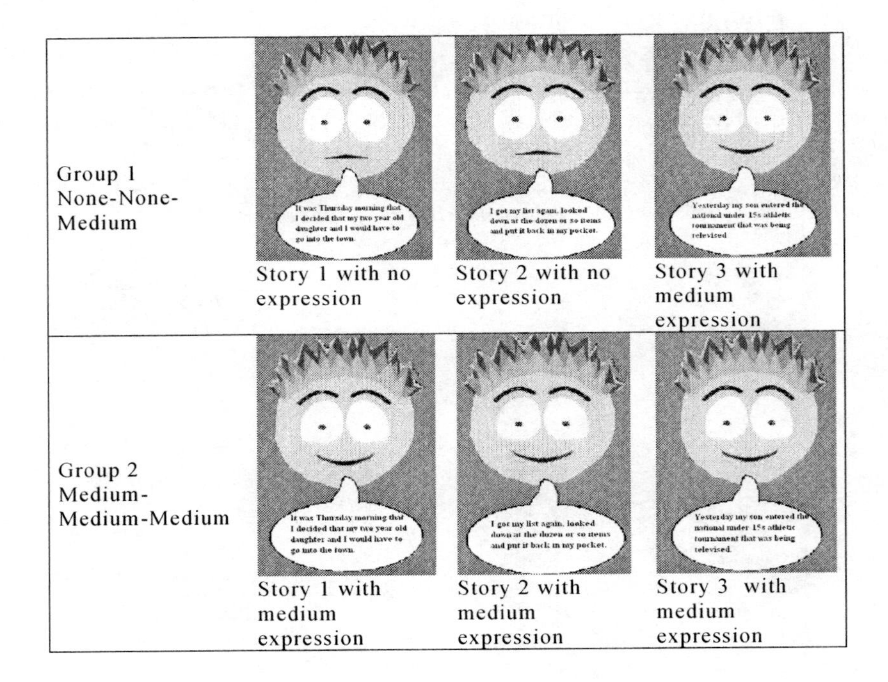

Table 1. Description of research protocol to test third phenomenon

Participants	Expressive-Image Level		
Group 1	None	None	Medium
Group 2	Medium	Medium	Medium

Table 2. Results of tests of phenomena

Phenomenon	Description of Test	Supported
1	Comparison of medium and low expressive images	Yes
2	Comparison of extreme and medium expressive images	Yes
3	Testing the wear-down effect	No

WEAR-DOWN-FACTOR EXPERIMENT RESULTS

First the Shapiro-Wilks normality test was carried out, which indicated that the observations of the wear-down-factor test were normally distributed. Therefore, t-tests were carried out to determine whether the results of the two groups of the participants were significantly different.

For the none-none-medium condition, the mean value of the perceived emotional feeling for Story 3 was 4.4. In the medium-medium-medium condition, the mean perceived emotional feeling was 4.7. The t-test revealed no significant difference between the ratings of the two groups. Thus, the third phenomenon was not supported by the test result.

DISCUSSION

A summary of the experiment results is presented in Table 2.

The experiment results support the first and second phenomena that predicted that the perceived emotional feelings from textual stories are strengthened when the story is accompanied with suitable expressive images.

It was first expected that when the agent presents a story with suitable expressions, the perceived emotional feelings will increase. As predicted, the participants who read the story with medium-expressive-intensity images perceive more emotional feelings than the participants who read the story with low-intensity images.

The next question is whether a ceiling exists. Will the gain achieved from increasing expression intensity be lost when the intensity reaches an unrealistic level? We predicted that the extremely high-intensity expressive images may decrease the perceived emotional feelings. The experiment result partially supported this phenomenon as a marginally significant difference between the two conditions was found.

Participants reported a significantly higher level of emotional feelings in the medium-intensity condition than in the low-intensity condition. When the expressive facial images are exaggerated to an unrealistic level, the perceived emotional feelings start to decrease.

The third phenomenon states that the influence of external factors, such as the wear-down effect, would affect the perceived emotional feelings. However, the results show that the perceived emotional feelings remain stable. This may suggest that the perceived emotional feelings are independent of external factors, in particular, the number of times expressive images are displayed. That is, the effect that an expressive image has on the viewer is not changed whether the viewer has seen the image many times or whether it is the first time it has been displayed. Another explanation is that the compared data are both within the stable phase, and we should keep the display longer to move to the wear-down phase.

EXPERIMENT IMPLICATIONS

The experiments indicate that emotion intensity has an influence on perceived feelings. It also indicates that external factors such as the wear-down effect do not influence perception. The influences of emotion intensity are consistent.

Some practical implications can be drawn for the design of affective computer interactions.

- To convey emotional feelings, expressive images should be provided during human-computer interaction.
- Medium-intensity expressive images can achieve the best performance. Both decreasing the intensity and increasing the intensity to an extreme level will show negative influences to the perceived feelings.
- The perceived emotional feelings may be independent of factors such as display duration, or the stable phase is long. This means that the expressive images can be shown as soon as appropriate, and there is no need to worry that the perceived emotional feelings may decrease with reasonable repeated use.

FUTURE TRENDS

Future trends in emotion systems are to create emotion-aware systems: systems that are aware of the social and emotional state of the users in determining the development of the interaction process. The development of systems interacting with and supporting the user in his or her tasks must consider important factors (e.g., expression intensity and timing) that may influence the perceived feelings.

This work presents the guidelines for displaying expressive images in a specific context. However, the result could be applicable to other contexts (e.g., emotional agents and human-human interface design). Further experiments can verify the results in other contexts and examine other factors (e.g., personality, application context, colours, etc.) that may influence expressive-image presentation. Then, a full and clear guideline of expressive-image presentation may be established.

CONCLUSION

A set of experiments that tested the factors that may influence perceived emotional feelings when users interact with emotional agents was conducted.

The experiment results demonstrated that internal factors such as the intensity of expressive images do significantly influence the perception of emotional feelings. The perceived emotional feelings do increase when the intensity of an expressive image increases. However, when the intensity of expressive images increases to an unrealistic level, the perceived emotional feelings will fall. The experiment examined external factors, such as the display time (wear-down effect), and found this does not produce a significant difference. It thus shows that the wear-down effect does not influence the perceived emotional feelings significantly.

The research indicates that expressive images do influence the perceived emotional feelings, and as long as the display is valid, the appropriate expressive images can be shown. There is either no decrease in perceived emotional feelings due to a wear-down effect, or the stable phase of the wear-down effect is very long.

REFERENCES

Azar, B. (2000). What's in a face? *Monitor on Psychology, 31*(1), 44-45.

Boucouvalas, A.C., Xu, Z., & John, D. (2003). Expressive image generator for an emotion extraction engine. *17th HCI 2003*, University of Bath, UK, Session 4, Track 2 Emotions and Computers (pp. 267-381).

Ekman, P. (1982). *Emotion in the human face.* New York: Cambridge University Press.

Evatt, D. L. (1997). *The influence of emotion-evoking content of news on issue salience.* Unpublished doctoral dissertation, University of Texas at Austin.

Fridlund, A. J. (1997). The new ethology of human facial expressions. In J. A. Russell & J. M. Fernandez-Dols (Eds.), *The psychology of facial expression* (pp. 103-129). New York: Cambridge University Press.

Hovland, C., Janis, I., & Kelley, H. (1953). *Communication and persuasion: Psychological studies of opinion change.* New Haven, CT: Yale University Press.

Hughes, G. D. (1992). Real-time response measures redefine advertising wearout. *Journal of Advertising Research, 32*(3), 61-77.

Izard, C. E. (1993). Four systems for emotion activation: Cognitive and noncognitive processes. *Psychological Review, 100*, 68-90.

Kinnick, K. N., Drugman, D. M., & Cameron, G. T. (1996). Compassion fatigue: Communication and burnout toward social problems. *Journalism and Mass Communication Quarterly, 73*(3), 687-707.

Koolhaas, J. M. (2001). *Theories of emotion.* Retrieved July 2, 2003, from http://www.bioledu.rug.nl/dierfysio1/coll-koolhaas/emotion/

Lea, M., & Spears, R. (1991). Computer-mediated communication, de-individuation and group decision-making. *International Journal of Man-Machine Studies, 34*, 283-301.

McCormick, N. B., & McCormick, J. W. (1992). Computer friends and foes: Content of undergraduates' electronic mail. *Computers in Human Behavior, 8*, 379-405.

Moffat, D., & Frijda, N. H. (1994). An agent architecture: Will. *Proceedings of the ECAI'94 Workshop on Agent Theories, Architectures and Languages* (pp. 216-225).

Newman, W., & Lamming, M. (1995). *Interactive system design.* Reading, MA: Addison-Wesley.

Norusis, M. J. (1998). *SPSS 8.0 guide to data analysis.* Englewood Cliff, NJ: Prentice Hall.

Prendinger, H., Saeyor, S., & Ishizuka, M. (2003). MPML and SCREAM: Scripting the bodies and minds of life-like characters. In H. Prendinger & M. Ishizuka (Eds.), *Life-like characters, tools, affective functions and applications* (pp. 231-241). London: Springer.

Rheingold, H. (1994). *The virtual community: Homesteading on the electronic frontier.* New York: Harper Perennial.

Xu, Z., & Boucouvalas, A. C. (2002). *Text-to-emotion engine for real time Internet communication.* International Symposium on Communication Systems, Networks and DSPs, (pp. 164-168).

Xu, Z., John, D., & Boucouvalas, A. C. (in press). Social norm, expression and agent interaction. *IEEE Journal.*

Zaichkowsky, J. K. (1986). The emotional side of product involvement. *Advances in Consumer Research, 14*, 32-35.

Zajonc, R. B. (1980). Feeling and thinking: Preferences need no inferences. *American Psychologist, 35*, 151-175.

KEY TERMS

Emotion: An excitement of the feelings caused by a specific exciting stimulus and manifested by some sensible effect on the body.

Emotion-Evoking Manner: The methods to make readers perceive emotions.

Emotion Icon: A combination of keyboard characters or small images meant to represent a facial expression.

Emotional Communication: The activity of communicating emotional feelings.

Personal Involvement Inventory: A measurement questionnaire developed by Zaichkowsky (1986).

Software Agent: A computer program that carries out tasks on behalf of another entity.

Wear-Down Effect: A reduction in the participant's favourable responses after repeated exposures to a message.

Information Interaction Beyond HCI

Philip Duchastel
Information Design Atelier, Canada

INTRODUCTION

HCI might well be poised to break out of its mould, as defined by its first half-century history, and to redefine itself in another mould that is at once more abstract and wider in scope. In the process, it would redefine its very name, HCI becoming a subset of the larger field of information interaction (II). This potential transformation is what is described here.

At this point in our technological era, we are in the process of symbolically modeling all aspects of reality such that our interactions with those aspects of the world around us that are most important are more digitally mediated. We are beginning to inhabit information environments and to interact ever more with artifacts, events, and processes that are pure information. This is the world of II, and what this means for HCI is what is examined here.

The presentation has a largely abstract character to it. Indeed, it seeks to reframe our discussion of the phenomenon of interaction under study in such a way as to go beyond the pitfalls of concrete problems usually associated with the field. By stepping back from the usual issues of concern and from the usual way of categorizing the elements of the field (Helander et al., 2000; Jacko & Sears, 2003), the goal is to contextualize HCI within a broader, necessarily philosophical plane of concern in order to look at it afresh and thereby see where it might be headed. The direction proposed is decidedly more englobing, more abstract, and, hence, more theoretical in its analysis.

BACKGROUND

HCI is a field that grew out of the expansion of computing beyond the early context of usage by technically inclined specialists, who were quite eager to access the potential of computing and did not mind the learning curve involved. The scope of HCI continues to expand, as computing becomes ever more pervasive and novice users expect to use computing artifacts without fuss, to put it bluntly. Thus, the goal of HCI is to ease usage while preserving the power of the artifact, effecting whatever compromises are possible in order to achieve a workable solution. That this goal is difficult not only to achieve but even to have accepted is well illustrated by Carroll's (1990, 1998) proposal for minimalism and by Norman's (1998) proposal for information appliances, building on the notion initially proposed by Raskin (see Norman).

So we continue to indulge in situations where complex system requirements are specified and HCI expertise is brought in to do what it may to perhaps ameliorate the situation somewhat. Attempts to break out of this design context (as through the various means presented in section II of the Handbook of HCI [Helander et al., 2000]) certainly point the way but may only succeed when computing itself is seen to disappear (in the spirit of Weiser and Brown's [1997] ubiquitous computing and Norman's [1998] "invisible" computer) into the larger context of human activity structures. Thus, how we view cognitive tasks is central to HCI past, present, and future, and needs to be considered in a high-level framework, as described next.

The most basic question of HCI is what the interaction is between. The three elements generally involved in the answer are the person (user), the system (computer and its interface), and the task (goal). An answer with more guts or more ambition would do away with the middle element and pursue analysis purely in terms of person and task. Doing away with the interface itself is, after all, the ultimate in the quest of transparency that drives all HCI design.

A computer system, represented to the person by its interface, is an artifact that mediates some specific process (i.e., supports the interfacing between person and task such that the person can realize the task). The person does not care about the interface (it is just a tool) but does care a great deal about the task. Transparency in HCI means forgetting about the interface.

Ubiquitous computing (Weiser & Brown, 1997) shares that same goal of transparency, although with a focus on having computers embedded everywhere within the environment. Here, the attention is not on computing itself (even if it is pervasive) but on accomplishing a task (i.e., interacting with the environment and more specifically with the information present in the environment).

A good example of transparency from a more familiar domain (Duchastel, 1996) is the steering wheel in a car. The steering wheel is the interface between oneself and the road (I never think about the steering wheel, but I observe the bends in the road). The steering wheel disappears, as an ideal interface should, and all that is left is the road and me (the task and the person).

A second aspect of the new HCI concerns interaction modalities and their concrete elements. Just as command modalities gave way to the WIMP paradigm of contemporary interfaces (Pew, 2003), the latter will give way to yet more natural interfaces involving speech and immersive technologies in the VR realm (see the following). The driver of this shift, beyond the developing feasibility of these technologies, is the HCI goal of adapting to humans through use of natural environmental settings (i.e., another facet of the transparency goal). The day when my interface will be an earpiece, lapel button, and ring (the button for sensory input of various kinds and for projection; the ring as a gestural device) may not be far off. Screens and wraparound glasses will be specialty devices, and keyboards and mice will be endangered species.

These evolutions (of process and gear) will make the person see computing as interfacing, with current gear long forgotten and the computer, while ubiquitous, nevertheless invisible. The disappearing computer will not leave great empty spaces, however. There will be agents to interact with (discussed later) and novel forms of interaction, discussed here.

The new landscapes include application areas such as communication, education, entertainment, and so forth (Shneiderman, 2003). They all involve interaction with information but also add to the mix the social aspect of interaction, thus creating a new and more complex cognitive context of action. The backdrop for HCI has changed suddenly, and the cognitive context has evolved to a sociocognitive one, as illustrated by the current interest in CSCW, itself only part of the new landscape.

The notion of interface can be reexamined (Carroll, 2003; Shneiderman, 2003). In a very broad definition (Duchastel, 1996), an interface can be considered as the locus of interaction between person and environment; more specifically, the information environment within which the person is inserted. In these general terms, interfaces can be viewed as abstract cognitive artifacts that constrain or direct the interaction between a person and that person's environment. In the end, the task itself is an interface, one that connects actor to goal through a structured process. Even the most archaic software is the concrete embodiment of a task structure. Thus, on the one hand, HCI deals with the person-information relation and is concerned with the design of information products; and on the other hand, it deals with the person-task relation and here is concerned with the guidance of process. It is the interplay between these two facets (product and process) that creates the richness of HCI as an applied field of the social sciences.

IMPLICATIONS FOR HCI

The constant novelty factor that we experience with technology generally and with computing in particular sets us up for fully using our intelligence to adapt. Not only do the tools (interfaces) change but so do the tasks and activities themselves, as witnessed, for instance, by the arrival of Web browsing and many other Web tasks. In this respect, then, HCI is faced with a losing battle with mounting diversity and complexity, and can only purport to alleviate some of the strain involved with these needs for humans to adapt. What has happened to HCI as the process of adapting computers to humans? HCI must find ways to assist human adaptation with general means, such as only gradually increasing the complexity of an artifact, forcing stability in contexts that may prove otherwise unmanageable, increasing monitoring of the user, and just-in-time learning support. All of these means are merely illustrative of a style of HCI design effort of which we likely will see more and more in response to computing complexity.

In reality, it is complexity of activity that has increased, not complexity of computing itself. Cars and telephones also have required adaptability for optimum usage recently. But as computing penetrates all areas more fully, and as the possibilities for more symbolic mediacy increase (e.g., the choices on the telephone now), the question to ask is how can HCI help? Are there general principles that can be applied? Perhaps not, for what we are witnessing here is the removal of the C from HCI. As computing becomes pervasive, it indeed disappears, as suggested earlier by Weiser and Brown (1997), and it is replaced by human-task interaction. Attention shifts up the scale of abstraction, and designers focus on task structure and context (Kyng & Mathiassen, 1997; Winograd, 1997) more than on operational task mediators, even though somewhere along the line, hard tool design is needed. A more human-focused HCI (away from the software, more toward the experience) evolves.

FUTURE TRENDS

Computer agents in the form of software that carries out specialized tasks for a user, such as handling one's telephoning, or in the form of softbots that seek out information and prepare transactions, are already well with us (Bradshaw, 1997). That their numbers and functions will grow seems quite natural, given their usefulness in an ever more digitized and networked world.

What will grow out of the agent phenomenon, however, has the potential to radically transform the context of our interactions, both digital and not, and, hence, the purview and nature of HCI. The natural evolution of the field of agent technology (Jennings & Wooldridge, 1998) leads to the creation, deployment, and adaptation of autonomous agents (AAs) (Luck et al., 2003; Sycara & Wooldridge, 1998). These agents are expected to operate (i.e., make reasoned decisions) on behalf of their owners in the absence of full or constant supervision. What is at play here is the autonomy of the agent, the degree of decision-making control invested in it by the owner, within the contextual limits imposed by the owner for the task at hand and within the natural limits of the software.

Seen from another perspective, the computer user removes himself or herself to an extent from the computer interactions that will unfold, knowing that the agent will take care of them appropriately and in the user's best interest. We witness here a limited removal of the human (the H) from HCI.

All this is relative, of course. Current stock management programs that activate a sale when given market conditions prevail already operate with a certain level of autonomy, as do process control programs that monitor and act upon industrial processes. Autonomy will largely increase, however, as we invest agents with abilities to learn (i.e., agents that learn a user's personal tastes from observation of choices made by the user) and to use knowledge appropriately within limited domains. As we also develop in agents the ability to evolve adaptation (from the research strand known as artificial life) (Adami & Wilke, 2004), we will be reaching out to an agent world where growing (albeit specialized) autonomy may be the rule. HCI will be complemented with AAI (Autonomous Agent Interaction), for these agents will become participants in the digital world just as we are, learning about one another through their autonomous interactions (Williams, 2004).

As we populate digital space with agents that are more autonomous, we create an environment that takes on a life of its own in the sense that we create uncertainty and open interaction up to adventure in a true social context. Not only will people have to learn how to react to the agents that they encounter, the latter also will have to react to people and to other autonomous agents (Glass & Grosz, 2003). The interfacing involved in this novel cognitive context is changing radically from its traditional meaning, with issues of understanding, trust, initiative, and influence coming to the fore. In discussing agents in the future of interfaces, Gentner and Nielsen (1996) talk of a shared world in which the user's environment no longer will be completely stable, and the user no longer will be totally in control; and they were talking of one's own assistive agents, not those of other people or of autonomous agents. The change occurring in HCI is merely reflecting the changing environment at large.

Perhaps an easy way to grasp what might be involved is to consider avatar interaction in VR worlds. Avatars are interfaces to other humans involved in a social interaction. Just as with authentic settings in which they mingle, humans in virtual settings must learn something about the others involved and learn to compose with them harmoniously in the accomplishment of their goals. The important consideration in this situation is that while the VR world may be artificial and may be experienced vicariously in physical terms, in psychological terms, the VR world can be just as genuine as the real world, as hinted by Turkle's (1995) interviews with digital world inhabitants (e.g., real life is just one more window). Interagent communication, just like its interpersonal counterpart, will be improvised and creative, with codes and norms emerging from the froth of the marketplace (Biocca & Levy, 1995). The potential for enhancing interaction certainly exists, particularly within VR worlds that not only reproduce but extend features of our regular world; new risks also appear, for instance, in the form of misrepresentation of agent intentions or outright deception (again, just as can occur in our normal interpersonal context) (Palmer, 1995).

The point is that the new cognitive context that is being created by both VR worlds and autonomous agents roaming cyberspace, all of which are but software artifacts, changes how we view interacting with computers. There still will exist the typical applications for assisting us in accomplishing specific creative tasks (and the associated HCI challenges), but the greater part of our interfacing with digital artifacts more generally will resemble our interfacing with others in our social world. In addition, interfacing specialists will be as concerned with the interface between AAs as with the interface between them and humans.

CONCLUSION

I foresee nothing short of a redefinition of the field, with classic HCI becoming a subset of a much wider-scoped field. This expansion shifts the focus of interfacing away from its traditional moorings in functionality and onto new landscapes that are much more sociocognitive in nature. The wider, more abstract notion of an interface being the locus of interaction between a person and his or her environment leads us to define the field in terms of information interaction (II). Indeed, the environment that a person inhabits is ever more symbolically and digitally mediated. While psychology broadly defines that interaction in general terms, II defines it in symbolic terms. Information constantly gleaned from the environment regulates our actions, which, in turn, are increasingly effected through information. We enter the age of interaction design (Preece et al., 2002; Winograd, 1997) and environment design (Pearce 1997).

This is particularly evident as we not only design interactions with information but also come to inhabit environments that are pure information (as VR worlds are). The added complexity resulting from the growth in autonomous agents potentially makes II all the more challenging, bringing, so to speak, a level of politics into what was hitherto a fairly individual and somewhat straightforward interaction. Agents can be both autonomous cognitive artifacts and assistive interfaces, depending on their design specifics.

Donald (1991) shows how cognitive inventions have led to cultural transitions in the evolution of the human mind and specifically how the invention of external memory devices, in expanding our natural biological memories, has fueled the modern age, leading us to digital realms. Autonomous agents lead us beyond out-of-the-skin memories to out-of-the-skin actions via the delegation with which we invest our assistive agents. The implications of this possibility are immense, even if only perceived hazily at this moment.

In sum, in the next few decades, HCI will transform itself into a much wider and more complex field based on information interaction. HCI will become a subset of the new field alongside AAI, dealing with interaction between autonomous agents. The new field will parallel the concerns of our own human-human interactions and thus involve social concerns alongside cognitive concerns.

REFERENCES

Adami, C., & Wilke, C. (2004). Experiments in digital evolution. *Artificial Life, 10*(2), 117-122.

Biocca, F., & Levy, M. (1995). Communication applications of virtual reality. In F. Biocca, & M. Levy (Eds.), *Communication in the age of virtual reality* (pp. 127-158). Hillsdale, NJ: Erlbaum.

Bradshaw, J. (Ed.). (1997). *Software agents*. Cambridge, MA: MIT Press.

Carroll, J. (1990). *The Nurnberg funnel*. Cambridge, MA: MIT Press.

Carroll, J. (1998). *Minimalism beyond the Nurnberg funnel*. Cambridge, MA: MIT Press.

Carroll, J. (2003). *HCI models, theories, and frameworks: Toward a multidisciplinary science*. San Francisco: Morgan Kaufmann.

Donald, M. (1991). *Origins of the modern mind*. Cambridge, MA: Harvard University Press.

Duchastel, P. (1996). Learning interfaces. In T. Liao (Ed.), *Advanced educational technology: Research issues and future potential*. New York: Springer Verlag.

Duchastel, P. (1998). Knowledge interfacing in cyberspace. *International Journal of Industrial Ergonomics, 22*, 267-274.

Duchastel, P. (2002). Information interaction. *Proceedings of the Third International Cyberspace Conference on Ergonomics*.

Gentner, D., & Nielsen, J. (1996). The anti-Mac interface. *Communications of the ACM*.

Glass, A., & Grosz, B. (2003). Socially conscious decision-Making. *Autonomous Agents and Multi-Agent Systems, 6*, 317-339.

Helander, M., Landauer, T., & Prabhu, P. (Eds.). (2000). *Handbook of human-computer interaction* (2nd ed.). Amsterdam: Elsevier.

Jacko, J., & Sears, A. (Eds.). (2003). *The human-computer interaction handbook: Fundamentals, evolving technologies and emerging applications*. Mahwah, NJ: Lawrence Erlbaum.

Jennings, N., & Wooldridge, M. (Eds.). (1998). *Agent technology*. Berlin: Springer.

Kyng, M., & Mathiassen, L. (Eds.). (1997). *Computers and design in context*. Cambridge, MA: MIT Press.

Luck, M., McBurney, P., & Preist, C. (2003). *Agent technology: Enabling next generation computing* [AgentLink report].

Newell, A., & Card, S. K. (1985). The prospects for psychological science in human-computer interaction. *Human Computer Interaction, 1*, 209-242.

Norman, D. (1990). *The invisible computer*. Cambridge, MA: MIT Press.

Olsen, G., & Olsen, J. (1991). User-centered design of collaboration technology. *Journal of Organizational Computing, 1*(1), 61-83.

Palmer, M. (1995). Interpersonal communication and virtual reality: Mediating interpersonal relationships. In F. Biocca, & M. Levy (Eds.), *Communication in the age of virtual reality* (pp. 277-302). Hillsdale, NJ: Erlbaum.

Pearce, C. (1997). *The interactive book*. Indianapolis, IN: Macmillan Technical Publishing.

Pew, R. (2003). The evolution of human-computer interaction: From Memex to Bluetooth and beyond. In J. Jacko, & A. Sears (Eds.), *The human-computer interaction handbook: Fundamentals, evolving technologies and emerging applications*, (pp. 1-17). Mahwah, NJ: Lawrence Erlbaum.

Preece, J., Rogers, Y., & Sharp, H. (2002). *Interaction design: Beyond human-computer interaction*. Hoboken, NJ: Wiley.

Shneiderman, B. (2003). *Leonardo's laptop: Human needs and the new computing technologies*. Cambridge, MA: MIT Press.

Sycara, K., & Wooldridge, M. (Eds.). (1998). *Proceedings of the Second International Conference on Autonomous Agents*, New York.

Turkle, S. (1995). *Life on the screen: Identity in the age of the Internet*. New York: Simon & Shuster.

Weiser, M., & Brown, J. S. (1997). The coming age of calm technology. In P. Denning, & R. Metcalfe (Eds.), *Beyond calculation* (pp. 75-86). New York: Springer-Verlag.

Williams, A. (2004). Learning to share meaning in a multi-agent system. *Autonomous Agents and Multi-Agent Systems, 8*, 165-193.

Winograd, T. (1997). Beyond interaction. In P. Denning, & R. Metcalfe (Eds.), *Beyond calculation* (pp. 149-162). New York: Springer-Verlag.

KEY TERMS

Agent and Autonomous Agent: Software that carries out specialized tasks for a user. Agents operate on behalf of their owners in the absence of full or constant supervision. Autonomous agents have a greater degree of decision-making control invested in them by their owners.

Artificial Life: The reproduction in digital models of certain aspects of organic life, particularly the ability of evolving adaptation through mutations that provide a better fit to the environment. In information sciences, artificial life is not concerned with the physico-chemical recreation of life.

Avatars: Computer-generated personas that are adopted by users to interface with other humans and agents involved in a social interaction, particularly in interacting in online virtual reality (VR) worlds.

Cognitive Artifacts: A class of designed objects that either can be considered in its concrete representations (interfaces, agents, software) or in its abstract mode as knowledge artifacts (contextualized functions, ideas, theories) (Duchastel, 2002).

Cognitive Task: An intellectual task, as opposed to a physical one. The range of such tasks has increased to the point that computers are involved more with communication than with computing; no longer do people only use computers to transact specific processes, but they also use them to stroll within new landscapes, as on the Web.

Information Interaction: The wider, more abstract notion of an interface, seen as the locus of interaction between a person and his or her environment. As that environment is ever more symbolically and digitally mediated, we are led to define more broadly the field in terms of information interaction (Duchastel, 2002).

Interface: A surface-level representation with which a user interacts in order to use a piece of equipment or a software application with a view to engage in some purposeful task. The purpose of an interface essentially is to facilitate access to the tool's functionality, whether we are dealing with physical tools or with mind tools. We can generalize this common notion of interface to define an interface as the locus of interaction between a person and his or her environment (Duchastel, 1996).

WIMP: A style of graphic user interface that involves windows, icons, menus, and pointers. It replaced the older textual command style interface, and the term is now of historical interest only.

Information Rich Systems and User's Goals and Information Needs

Michael J. Albers
The University of Memphis, USA

INTRODUCTION

Currently, most of the Web is designed from the viewpoint of helping people who know what they want but need help accomplishing it. User goals may range from buying a new computer to making vacation plans. Yet, these are simple tasks that can be accomplished with a linear sequence of events. With information-rich sites, the linear sequence breaks down, and a straightforward process to provide users with information in a useful format does not exist.

Users come to information-rich sites with complex problems they want to solve. Reaching a solution requires meeting goals and subgoals by finding the proper information. Complex problems are often ill-structured; realistically, the complete sequence can't even be defined because of users' tendencies to jump around within the data and to abandon the sequence at varying points (Klein, 1999). To reach the answer, people need the information properly positioned within the situation context (Albers, 2003; Mirel, 2003a). System support for such problems requires users to be given properly integrated information that will assist in problem solving and decision making.

Complex problems normally involve high-level reasoning and open-ended problem solving. Consequently, designer expectations of stable requirements and the ability to perform an exhaustive task analysis fall short of reality (Rouse & Valusek, 1993). While conventional task analysis works for well-defined domains, it fails for the ill-structured domains of information-rich sites (Albers, 2004). Instead of exhaustive task analysis, the designer must shift to an analysis focused on providing a clear understanding of the situation from the user's point of view and the user's goals and information needs.

BACKGROUND

In today's world, data almost invariably will come from a database. A major failing of many of these systems is that they never focus on the human-computer interaction. Instead, the internal structure of the software or database was reflected in both the interface operation and the output.

The problem is not lack of content. Information-rich sites normally have a high information content but inefficient design results in low information transmission. From the psychological standpoint, the information is disseminated ineffectively. The information is not designed for integration with other information but rather is optimized for its own presentation. As a result, users must look in multiple sources to find the information they need. While hypertext links serve to connect multiple sources, they often are not adequate. Johnson-Eilola and Selber (1996) argue that most hypertexts tend to maintain the traditional hierarchical organization of paper documents.

Mirel (1996, 2003b) examined the difficulties users have with current report design and found that sites often provide volumes of information but fail to effectively answer a user's questions. The information needed by professionals exists within the corporate database, but with complex problems, there are no ready-made answers that can be pulled out with simple information retrieval techniques. Thus, it cannot be expected that relevant information can be found by direct means, but it must be inferred. Interestingly (and complicating the design), inferring results is what experts do best. While all readers need information to be properly integrated, the amount of integration and coherence of the information required varies. McNamara and her colleagues (McNamara, 2001; McNamara & Kintsch, 1996) have found that users with a higher topic knowledge level perform better with less integrated information. Following the same idea, Woods and Roth (1998) define the critical question as "how knowledge is activated and utilized in the actual problem-solving environment" (p. 420).

Waern (1989) claims that one reason systems fail lies in the differences in perspective between the data generator and the information searcher. Much of the research on information structuring attempts to predefine user needs and, thus, the system breaks down when users try to go beyond the solution envisioned by the designers. Basden and Hibberd (1996) consider how current audience and task analysis methods tend to start with an assumption that all the information needed can be defined in advance and then collected into a database. In this view, the knowledge exists as external to the system and user. However, for systems that must support complex situations, the methods tend to break down. Spool (2003) found some designs drove people away by not answering their questions in the user's context.

DESIGN FOR INFORMATION-RICH SYSTEMS

Interface and content designers increasingly are being called upon to address information needs that go beyond step-by-step instruction and involve communicating information for open-ended questions and problems (Mirel, 1998, 2003b). Applying that approach to interface design can enhance user outcomes, as such systems can help to organize thinking rather than to suggest a course of action (Eden, 1988). The questions and problems that users bring to information-rich systems only can be addressed by providing information specific to a situation and presenting it in a way that supports various users' goals and information needs (Albers, 2003).

Addressing users' goals and information needs breaks with the fundamental philosophy of a design created to step a user through a sequence. Complex situations contain lots of ambiguity and subtle information nuances. That fact, if nothing more, forces the human into the process, since computers simply cannot handle ambiguity. From the computer's point of view, data are never ambiguous (if it has 256 shades of gray, then it can be assigned to one and only one of 256 little bins). The easiest design method, one that is much too prevalent, is to ignore the ambiguity. The system displays the information and leaves it up to the user to sort out the ambiguity. From the start, designers must accept that information, since a complex situation cannot be prestructured and must be designed to allow users to continuously adapt to it. Consequently, many of the standard considerations of stable requirements, exhaustive task analysis, and ignorance of cognitive interaction fail to apply and require reconsideration (Rouse & Valusek, 1993). This breakdown between the designer's and the user's thought processes explains why conventional task analysis works for well-defined domains but fails for the ill-structured domains of information-rich sites (Albers, 2004). Instead, the designer must have a clear understanding of the situation from the user's point of view, the user's goals, and the user's information needs.

Situation

The situation is the current world state that the user needs to understand. A situation always exists with the user embedded within it. To understand a situation, a user works within the situation by defining goals and searching for the information required to achieve the goals. An underlying assumption is that the user needs to interact with an information system in order to gain the necessary information and to understand the situation. In most cases, after understanding the situation, the user will interact with the situation, resulting in a change that must be reflected in an updated system.

Goal

User goals are the high-level view that allows the entire situation to be understood in context. To maximize understanding, the information should directly map onto the goal. Goals could be viewed from the user's viewpoint as plans and from the system's viewpoint as the road map detailing the possible routes to follow. Goals can consist of subgoals, which are solved in a recursive manner. Each goal gets broken into a group of subgoals, which may be broken down further, and each subgoal must be handled before the goal can be considered achieved. Goals should be considered from the user-situation viewpoint (what is happening and what does it mean to the user) rather than the system viewpoint (how can the system display a value for *x*). The interface provides a pathway for the user to obtain the information to achieve the goal. User goals provide the means of categorizing and arranging the information needs.

People set goals to guide them through a situation, but all people are not the same. Different people shape their goals differently and may set completely different goals. Rarely will an information-rich site be used by a homogeneous group of people sharing a common pool of goals. Instead, multiple user groups exist, with each group having a different pool of goals that must be addressed. These fundamental differences arise from the different goals of the user. In a highly structured environment, the user's basic goal is essentially one of efficiently completing the task, while in the unstructured information-rich environment, the user is goal-driven and focused on problem solving and decision making.

Information Needs

Information needs are the information required for the user to achieve a goal. A major aspect of good design is ensuring that the information is provided in an integrated format that matches the information needs to the user goals. Information needs focus on the content that users require to address their goals. Interestingly and perhaps unfortunately, the content often gets short-changed in many design discussions. The problem is that content normally is assumed to already exist, it can be used as is, and thus, it is outside the scope of the human-computer interaction. While the content is situation-specific, it never will just appear out of nowhere in a fully developed form. Also, as a person interacts with a situation, the information the person wants for any particular goal changes as he or she gets a better grasp of the goal and the situation (Albers, 2004).

The problem of addressing information needs extends well beyond having the information available and even having it well arranged. As users' information needs increase, they find it hard to figure out what information they need. One study found that approximately half of the participants failed to extract the proper information for ill-defined problems, even when the relevant graphs and illustrations were presented to them (Guthrie, Weber, & Kimmerly, 1993, as cited in van der Meij, Blijleve & Jensen, 2003). Consider how much more difficult this can be when a user either does not know or is not sure the information exists within the system. Yet, the designer's and technical writer's jobs are to ensure that the user knows that the information exists, extracts the proper information, and understands its relevance.

A good interface design must define the order in which the information must be presented, how it should be presented, and what makes it important to the situation and to the user's goal. It also must define what information is not needed or not relevant, even though at first glance it seems important. Since information-rich sites lend themselves to a high degree of freedom and a large amount of unpredictability, understanding how information relates to the goals is imperative to helping users address their situations.

Example: Marketing Analysis as a Complex Situation

Managers have access to a huge amount of data that they need to analyze in order to make informed decisions. Normally, rather than providing any help with interpreting the information, report designers take the view of just asking what information is desired and ensuring it is contained somewhere within the system.

For example, if a marketing analyst for a coffee manufacturer is inquiring into whether a new espresso product is likely to succeed in this specialized market, the analyst needs to view, process, and interact with a wide range of multi-scaled data. To figure out what it will take to break into and become competitive in the high-end espresso market, the analyst will examine as many markets, espresso products, and attributes of products as the analyst deems relevant to the company's goals, and as many as the technical tools and cognitive capacity enable the analyst to analyze. Looking at these products, the analyst will move back and forth in scale between the big picture and detailed views. The analyst will assess how espresso has fared over past and current quarters in different channels of distribution, regions, markets, and stores, and impose on the data his or her own knowledge of seasonal effects and unexpected market conditions. For different brands and products, including variations in product by attributes such as size, packaging, and flavor, the analyst might analyze 20 factors or more, including dollar sales, volume sales, market share, promotions, percent of households buying, customer demographics, and segmentation. The analyst will arrange and rearrange the data to find trends, correlations, and two-and three-way causal relationships; the analyst will filter data, bring back part of them, and compare different views. Each time, the analyst will get a different perspective on the lay of the land in the espresso world. Each path, tangent, and backtracking move will help the analyst to clarify the problem, the goal, and ultimately the strategic and tactical decisions (Mirel, 2003b).

By considering report analysis as a complex situation, the report interpretation methods do not have to be outside of the scope. The information the analyst needs exists. The problem is not a lack of data but a lack of clear methods and techniques to connect that data into an integrated presentation that fits the user's goals and information needs. Rather than simply supplying the analyst with a bunch of numbers, the report designers should have performed an analysis to gain a deeper understanding of how the numbers are used and should have provided support to enable the user to perform that analysis in an efficient manner.

FUTURE TRENDS

Information-rich Web sites will continue to increase as more people expect to gain information via the Internet. In general, the information-rich sites focus on complex situations that contain too many factors to be completely analyzed, so it is essentially impossible to provide a complete set of information or to fully define the paths through the situation.

In the near term, an artificial intelligence approach will not work; with current or near-term technology, the computer system cannot come close to understanding the situational context and resolving ambiguity. Rather, the system and interface design must provide proper support in order for users to gain a clear understanding of the solutions to their goals. Computers and people both excel at different tasks; effective design must balance the two and let each do what they do best.

Rather than being dominated by a tool mindset, we need to ensure that the technology does not override the communication aspects. Addressing designs specific to a user's goals means assuming a highly dynamic path with information being molded to fit each user group and each individual user. Rather than focusing on specific tasks that the system can perform, the analysis and design should focus on the user's situation and on the goals to be achieved. Understanding the user's goals, information needs, and information relationships provides a solid foundation for placing the entire situation in context and for solving the user's problem.

CONCLUSION

With a complex situation, the user's goal is one of problem solving and decision making, based on the user's goals and information needs. As such, the user has no single path to follow to accomplish a task (Albers, 1997). Unlike the clear stopping point of well-defined tasks, with complex tasks, the decision-making process continues until the user quits or feels confident enough to move forward.

Any complex situation contains an overabundance of data. As such, with complex situations, the user needs clearly structured information that helps to reveal solutions to the open-ended questions and provides connections across multiple-task procedures. Achieving an effective design requires knowing what information is required, how to manipulate the information to extract the required knowledge from it, and how to construct mental models of the situation that can be used to handle unanticipated problems (Brown, 1986).

Properly presented information with the proper content effectively addresses the user's goals. Users work within a complex situation with a set of open-ended goals that the system design must consider from the earliest stages (Belkin, 1980). The first step in meeting people's information needs requires initially defining their goals and needs. But more than just a list of goals and data, the analysis also reveals the social and cognitive aspects of information processing and the information relationships within the readers' mental models. Thus, the goal of an HCI designer is to develop a user-recognizable structure that maps onto both the user's mental model of the situation and the situation context and bridges between them. The collected goals and information needs create a vision of the users focused on what open-ended questions they want answered and why (Mirel, 1998). Everything the user sees contributes to the acceptance of the information and its ability to support the needs of understanding a complex situation.

As Mirel (2003b) states, "people's actual approaches to complex tasks and problems ... are contextually conditioned, emergent, opportunistic, and contingent. Therefore, complex work cannot be formalized into formulaic, rule-driven, context-free procedures" (p. 259). The analysis and design must consider the communication needs in complex situations and the highly dynamic situational context of information, with a focus on the user's goals and information needs as required to support the fundamental user wants and needs.

REFERENCES

Albers, M. (1997). Information engineering: Creating an integrated interface. *Proceedings of the 7th International Conference on Human-Computer Interaction.*

Albers, M. (2003). Complex problem solving and content analysis. In M. Albers & B. Mazur (Eds.), *Content and complexity: Information design in software development and documentation* (pp. 263-284). Mahwah, NJ: Lawrence Erlbaum Associates.

Albers, M. (2004). *Design for complex situations: Analysis and creation of dynamic Web information.* Mahwah, NJ: Lawrence Erlbaum Associates.

Basden, A., & Hibberd, P. (1996). User interface issues raised by knowledge refinement. *International Journal of Human-Computer Studies, 45,* 135-155.

Belkin, N. (1980). Anomalous states of knowledge as a basis for information retrieval. *The Canadian Journal of Information Science, 5,* 133-143.

Brown, J. (1986). From cognitive to social ergonomics and beyond. In D. Norman & S. Draper (Eds.), *User centered system design: New perspectives on human-computer interaction* (pp. 457-486). Mahwah, NJ: Lawrence Erlbaum Associates.

Eden, C. (1988). Cognitive mapping. *European Journal of Operational Research, 36,* 1-13.

Johnson-Eilola, J., & Selber, S. (1996). After automation: Hypertext and corporate structures. In P. Sullivan, & J. Dautermann. (Eds.), *Electronic literacies in the workplace* (pp. 115-141). Urbana, IL: NCTE.

Klein, G. (1999). *Sources of power: How people make decisions.* Cambridge, MA: MIT.

McNamara, D. (2001). Reading both high and low coherence texts: Effects of text sequence and prior knowledge. *Canadian Journal of Experimental Psychology, 55,* 51-62.

McNamara, D., & Kintsch, W. (1996). Learning from text: Effects of prior knowledge and text coherence. *Discourse Processes, 22,* 247-287.

Mirel, B. (1996). Writing and database technology: Extending the definition of writing in the workplace. In P. Sullivan & J. Dautermann. (Eds.), *Electronic literacies in the workplace* (pp. 91-114). Urbana, IL: NCTE.

Mirel, B. (1998). Applied constructivism for user documentation. *Journal of Business and Technical Communication, 12*(1), 7-49.

Mirel, B. (2003a). *Interaction design for complex problem solving: Getting the work right.* San Francisco: Morgan Kaufmann.

Mirel, B. (2003b). Design strategies for complex problem-solving software. In M. Albers & B. Mazur (Eds.), *Content and complexity: Information design in software development and documentation* (pp. 255-284). Hillsdale, NJ: Earlbaum.

Rouse, W., & Valusek, J. (1993). Evolutionary design of systems to support decision making. In G. Klein, J. Orasanu, R. Calderwood, & C. Zsambok (Eds.), *Decision making in action: Models and methods* (pp. 270-286). Norwood, NJ: Ablex.

Spool, J. (2003). *5 things to know about users.* Retrieved June 17, 2003, from http://www.uiconf.com/7west/five_things_to_know_article.htm

van der Meij, H., Blijleven, P., & Jansen, L. (2003). What makes up a procedure? In M. Albers & B. Mazur (Eds.), *Content and complexity: Information design in software development and documentation* (pp. 129-186). Mahwah, NJ: Lawrence Erlbaum Associates.

Waern, Y. (1989). *Cognitive aspects of computer supported tasks.* New York: Wiley.

Woods, D., & Roth. E. (1988). Cognitive engineering: Human problem solving with tools. *Human Factors, 30*(4), 415-430.

KEY TERMS

Complex Situation: The current world state that the user needs to understand. The understanding in a complex situation extends beyond procedural information and requires understanding the dynamic interrelationships of large amounts of information.

Information Needs: The information that contributes to solving a goal. This information should be properly integrated and focused on the goal.

Information-Rich Web Site: A Web site designed to provide the user with information about a topic, such as a medical site. In general, they contain more information than a user can be expected to read and understand.

Situational Context: The details that make the situation unique for the user.

User Goals: The specific objectives that a user wants to solve. In most complex situations, goals form a hierarchy with multiple tiers of subgoals that must be addressed as part of solving the primary goal.

Information Space

David Benyon
Napier University, UK

INTRODUCTION

Human-Computer Interaction (HCI) in the 21st century needs to look very different from its 20th-century origins. Computers are becoming ubiquitous; they are disappearing into everyday objects. They are becoming wearable. They are able to communicate with each other autonomously, and they are becoming self-adaptive. Even with something as ubiquitous as the mobile phone, we see a system that actively searches out a stronger signal and autonomously switches transmitters. Predictive techniques allow phones to adapt (e.g., anticipate long telephone numbers). These changes in technologies require us to change our view of what HCI is.

The typical view of how people interact with computers has been based primarily on a cognitive psychological analysis (Norman & Draper, 1986) of a single user using a single computer. This view sees the user as outside the computer. People have to translate their intentions into the language of the computer and interpret the computer's response in terms of how successful they were in achieving their aims. This view of HCI leads to the famous gulfs of execution (the difficulty of translating human intentions into computer speak) and evaluation (trying to interpret the computer's response).

With the ubiquity of information appliances (Norman, 1999) or information artifacts (Benyon et al. 1999), the single-person, single-computer view of HCI becomes inadequate. We need to design for people surrounded by information artefacts. People no longer are simply interacting with a computer; they are interacting with people using various combinations of computers and media. As computing devices become increasingly pervasive, adaptive, embedded in other systems, and able to communicate autonomously, the human moves from outside to inside an information space. In the near future, the standard graphical user interface will disappear for many applications, the desktop will disappear, and the keyboard and mouse will disappear. Information artefacts will be embedded both in the physical environment and carried or worn by people as they move through that environment.

This change in the nature of computing demands a change in the way we view HCI. We want to move people from outside a computer, looking in to the world of information, to seeing people as inside information space. When we think of having a meeting or having a meal, we do not see people as outside these activities. People are involved *in* the activity. They are engaged in the interactions. In an analogous fashion, we need to see people as inside the activities of information creation and exchange, as inside information space.

BACKGROUND

The notion that we can see people as existing in and navigating through an information space (or multiple information spaces) has been suggested as an alternative conceptualization of HCI (Benyon & Höök, 1997). Looking at HCI in this way means looking at HCI design as the creation of information spaces (Benyon, 1998). Information architects design information spaces. Navigation of information space is not a metaphor for HCI. It is a paradigm shift that changes the way that we look at HCI. The conception has influenced and been influenced by new approaches to systems design (McCall & Benyon, 2002), usability (Benyon, 2001), and information gathering (Macaulay et al., 2000).

The key concepts have developed over the years through experiences of developing databases and other information systems and through studying the difficulties and contradictions in traditional HCI. Within the literature, the closest ideas are those of writers on distributed cognition (Hutchins, 1995). A related set of ides can be found in notions of resources that aid action (Wright et al., 2000). In both of these, we see the recognition that cognition simply does not take place in a person's head. Cognition makes use of things in the world—cognitive artefacts, in Hutchins' terms. If you think about moving through an urban landscape, you may have a reasonable plan in mind. You have a reasonable representation of the environment in terms of a cognitive map (Tversky, 1993). But you constantly will be using cues and reacting to events. You may plan to cross the road at a particular place, but exactly where and when you cross the road depends on the traffic. Plans and mental models constantly are being reworked to take account of ongoing events. Navigation of information space seeks to make explicit the ways in which people move among sources of information and manage their activities in the world.

MAIN FOCUS OF THE ARTICLE

Navigation of information space is a new paradigm for thinking about HCI, just as direct manipulation was a new paradigm in the 1980s. Navigation of information space suggests that people are navigators and encourages us to look to approaches from physical geography, urban studies, gardening, and architecture in order to inspire designs. Navigation of information space requires us to explore the concept of an information space, which, in turn, requires us to look at something that is not an information space. We conceptualize the situation as follows. The activity space is the space of real-world activities. The activity space is the space of physical action and physical experiences. In order to undertake activities in the activity space, people need access to information. At one level of description, all our multifarious interactions with the experienced world are effected through the discovery, exchange, organization, and manipulation of information. Information spaces are not the province of computers. They are central to our everyday experiences and go from something as simple, for example, as a sign for a coffee machine, a public information kiosk, or a conversation with another person.

Information spaces often are created explicitly to provide certain data and certain functions to facilitate some activity—to help people plan, control, and monitor their undertakings. Information system designers create information artefacts by conceptualizing some aspect of an activity space and then selecting and structuring some signs in order to make the conceptualization available to other people. Users of the information artefact engage in activities by performing various processes on the signs. They might select items of interest, scan for some general patterns, search for a specific sign, calculate something, and so forth.

Both the conceptualization of the activity space and the presentation of the signs are crucial to the effectiveness of an information artefact to support some activity. Green and Benyon (1996) and Benyon, et al. (1999) provide many examples of both paper-based and computer-based information artefacts and the impact that the structuring and presentation have on the activities that can be supported with different conceptualizations of activity spaces and different presentations or interfaces on those conceptualizations. For example, they discuss the different activities that are supported by different reference styles used in academic publications, such as the Harvard style (the author's name and date of publication, as used as in this article) and the Numeric style (when a reference is presented in a numbered list). Another example is the difference between a paper train timetable and a talking timetable, or the activities that are supported by the dictionary facility in a word processor.

All information artefacts employ various signs structured in some fashion and provide functions to manipulate those signs (conceptually and physically). I can physically manipulate a paper timetable by marking it with a pen, which is something I cannot do with a talking timetable. I can conceptually manipulate it by scanning for arrival times, which is something I cannot do with a talking timetable. So, every information artefact constrains and defines an information space. This may be defined as the signs, structure, and functions that enable people to store, retrieve, and transform information. Information artefacts define information spaces, and information spaces include information artefacts. Information artefacts also are built on top of one another. Since an information artefact consists of a conceptualization of some aspect of the activity space and an interface that provides access to that conceptualization whenever a perceptual display (an interface) is created, it then becomes an object in the activity space. Consequently, it may have its own

Figure 1. Conceptualization of information space and activities

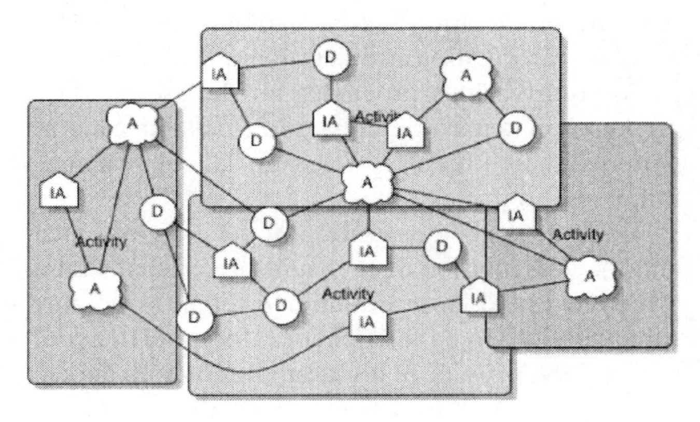

information artefact designed to reveal information about the display.

In addition to information artefacts, information spaces may include agents and devices. Agents are purposeful. Unlike information artefacts that wait to be accessed, agents actively pursue some goal. People are agents, and there are artificial agents such as spell checkers. Finally, there are devices. These are entities that do not deal with the semantics of any signals that they receive. They transfer or translate signs without dealing with meanings. Thus, we conceive of the situation as illustrated in Figure 1, where various activities are supported by information spaces. Note that a single activity rarely is supported by a single information artefact. Accordingly, people have to move across and between the agents, information artefacts, and devices in order to undertake their activities. They have to navigate the information space.

FUTURE TRENDS

The position that we are moving toward—and the reason that we need a new HCI—is people and other agents existing inside information spaces. Of course, there always will be a need to look at the interaction of people with a particular device. But in addition to these traditional HCI issues are those concerned with how people can know what a particular device can and cannot do and with devices knowing what other devices can process or display.

The conceptualization of HCI as the navigation of information spaces that are created from a network of interacting agents, devices, and information artefacts has some important repercussions for design. Rather than designing systems that support existing human tasks, we are entering an era in which we develop networks of interacting systems that support domain-oriented activities (Benyon, 1997). That is to say that we need to think about the big picture in HCI. We need to think about broad activities, such as going to work in the morning, cooking a meal for some friends, and how a collection of information artefacts both can support and make these activities enjoyable and rewarding.

In its turn, this different focus makes HCI shift attention from humans, computers, and tasks to communication, control, and the distribution of domain knowledge between the component agents and devices that establish the information space. We need to consider the transparency, visibility, and comprehensibility of agents and information artefacts, the distribution of trust, authority, and responsibility in the whole system, and issues of control, problem solving, and the pragmatics of communication. Users are empowered by having domain-oriented configurable agents and devices with which they communicate and share their knowledge.

CONCLUSION

Understanding people as living inside information spaces represents a new paradigm for thinking about HCI and, indeed, about cognition. There has been a failure of traditional cognitive science in its concept of mental representations both in terms of our ability to build intelligent machines and in our attempts to create really effective interactions between people and computers. This new conception draws upon our spatial skills and spatial knowledge as its source. We have learned much about how to design to help people move through the built environment that we can apply to the design of information spaces. We understand that views of cognition based exactly on a spatial conception (Lakoff & Johnson, 1999) are providing new insights. Navigation of information space can be seen as part of this development.

REFERENCES

Benyon, D.R. (1997). Communication and shared knowledge in HCI. *Proceedings of the 7th International Conference on Human-Computer Interaction*, Amsterdam, The Netherlands.

Benyon, D.R. (1998). Cognitive ergonomics as navigation in information space. *Ergonomics, 41*(2), 153-156.

Benyon, D.R. (2001). The new HCI? Navigation of information space. *Knowledge-Based Systems, 14*(8), 425-430.

Benyon, D.R., Green, T.R.G., & Bental, D. (1999). *Conceptual modelling for human computer interaction, using ERMIA*. London: Springer.

Benyon, D., & Höök, K. (1997). Navigation in information spaces: Supporting the individual. In S. Howard, J. Hammond, & G. Lindgaard (Eds.), *Human-computer interaction: INTERACT'97* (pp. 39-46). London: Chapman & Hall.

Green, T.R.G., & Benyon, D.R. (1996). The skull beneath the skin: Entity-relationship modelling of information artefacts. *International Journal of Human-Computer Studies, 44*(6), 801-828.

Hutchins, E. (1995). *Cognition in the wild*. Cambridge, MA: MIT Press.

Lakoff, G., & Johnson, M. (1999). *Philosophy in the flesh. The embodied mind and its challenge to Western thought*. New York: Basic Books.

Macaulay, C., Benyon, D.R., & Crerar, A. (2000). Ethnography, theory and systems design: From intuition to insight. *International Journal of Human Computer Studies, 53*(1), 35-60.

McCall, R., & Benyon, D.R. (2002). Navigation: Within and beyond the metaphor in interface design and evaluation. In K. Höök, D.R. Benyon, & A. Munro (Eds.), *Designing information spaces: The social navigation approach*. London: Springer.

Norman, D. (1999). *The Invisible Computer*. Cambridge, MA: MIT Press. Bradford Books.

Norman, D., & Draper, S. (1986). *User centred systems design*.

Tversky, B. (1993). Cognitive maps, cognitive collages and spatial mental models. *Proceedings of the European Conference COSIT '93*, Lecture notes on computer science, 716, (pp. 17-24). Berlin: Springer.

Wright, P. Fields. & Harrison. (2000). Analyzing human-computer interaction as distributed cognition: The resources model. *Human-Computer Interaction, 15*(1), 1-42.

KEY TERMS

Agent: An entity that possesses some function that can be described as goal-directed.

Device: An entity that does not deal with information storage, retrieval, or transmission but only deals with the exchange and transmission of data.

Information: Data that is associated with some system that enables meaning to be derived by some entity.

Information Artefact: Any artefact whose purpose is to allow information to be stored, retrieved, and possibly transformed.

Information Space: A collection of information artefacts and, optionally, agents and devices that enable information to be stored, retrieved, and possibly transformed.

Navigation of Information Space: (1) The movement through and between information artefacts, agents, and devices; (2) the activities designed to assist in the movement through and between information artefacts, agents, and devices.

Intelligent Multi-Agent Cooperative Learning System

Leen-Kiat Soh
University of Nebraska, USA

Hong Jiang
University of Nebraska, USA

INTRODUCTION

A computer-aided education environment not only extends education opportunities beyond the traditional classroom, but it also provides opportunities for intelligent interface based on agent-based technologies to better support teaching and learning within traditional classrooms. Advances in information technology, such as the Internet and multimedia technology, have dramatically enhanced the way that information and knowledge are represented and delivered to students. The application of agent-based technologies to education can be grouped into two primary categories, both of which are highly interactive interfaces: (1) intelligent tutoring systems (ITS) and (2) interactive learning environments (ILE) (McArthur, Lewis, & Bishay, 1993). Current research in this area has looked at the integration of agent technology into education systems. However, most agent-based education systems under utilize intelligent features of agents such as reactivity, pro-activeness, social ability (Wooldridge & Jennings, 1995) and machine learning capabilities. Moreover, most current agent-based education systems are simply a group of non-collaborative (i.e., non-interacting) individual agents. Finally, most of these systems do not peruse the multi-agent intelligence to enhance the quality of service in terms of content provided by the interfaces.

A *multi-agent system* is a group of agents where agents interact and cooperate to accomplish a task, thereby satisfying goals of the system design (Weiss, 1999). A group of agents that do not interact and do not peruse the information obtained from such interactions to help them make better decisions is simply a group of independent agents, not a multi-agent system. To illustrate this point, consider an ITS that has been interacting with a particular group of students and has been collecting data about these students. Next, consider another ITS which is invoked to deal with a similar group of students. If the second ITS could interact with the first ITS to obtain its data, then the second ITS would be able to handle its students more effectively, and together the two agents would comprise a multi-agent system.

Most ITS or ILE systems in the literature do not utilize the power of a multi-agent system. The Intelligent Multi-agent Infrastructure for Distributed Systems in Education (I-MINDS) is an exception. It is comprised of a multi-agent system (MAS) infrastructure that supports different high-performance distributed applications on heterogeneous systems to create a computer-aided, collaborative learning and teaching environment. In our current I-MINDS system, there are two types of agents: teacher agents and student agents. A *teacher agent* generally helps the instructor manage the real-time classroom. In I-MINDS, the teacher agent is unique in that it provides an automated ranking of questions from the students. This innovation presents ranked questions to the classroom instructor and keeps track of a profile of each class participant reflecting how they respond to the class lectures. A *student agent* supports a class participant's real-time classroom experience. In I-MINDS, student agents innovatively support the *buddy group* formation. A class participant's buddy group is his or her support group. The buddy group is a group of actual students that every student has access to during real-time classroom activities and with which they may discuss problems. Each of these agents has its interface which, on one hand, interacts with the user and, on the other hand, receives information from other agents and presents those to the user in a timely fashion.

In the following, we first present some background on the design choice of I-MINDS. Second, we describe the design and implementation of I-MINDS in greater detail, illustrating with concrete examples. We finalize with a discussion of future trends and some conclusions drawn from the current design.

BACKGROUND

In this section, we briefly describe some virtual classrooms, interactive learning environments (ILE), and intelligent tutoring systems (ITS)—listing some of the features available in these systems—and then compare these systems with I-MINDS. The objective of this section is to show that I-MINDS possesses most and, in some cases, more advanced functionalities and features than those found in other systems. What sets I-MINDS significantly apart from these systems is the multi-agent infrastructure where intelligent agents not only serve their users, but also interact among themselves to share data and information. Before moving further, we will provide some brief definitions of these systems. A *virtual classroom* is an environment where the students receive lectures from an instructor. An *ILE* is one where either the students interact among themselves, or with the instructor, or both to help them learn. An *ITS* is one where an individual student interacts with a computer system that acts as a tutor for that student. At its current design, I-MINDS is a full-fledged virtual classroom with an ILE, and has the infrastructure for further development into a system of intelligent tutors. I-MINDS currently has a complete suite of similar multimedia support features, important in virtual classrooms and interactive learning environments: live video and audio broadcasts, collaborative sessions, online forums, digital archival of lectures and discussions, text overlay on blackboard, and other media. The uniqueness of I-MINDS is that the features of its interactive learning environment and virtual classroom are supported by intelligent agents. These agents work individually to serve their users and collaboratively to support teaching and learning.

Most ITSs such as AutoTutor (Graesser, Wiemer-Hastings, Wiemer-Hastings, Kreuz, & the Tutoring Research Group, 1999) have not been considered in the context of a multi-agent system. For example, one ITS *A* may store useful information about the types of questions suitable for a certain type of student based on its own experience. Another ITS *B* encounters such a student but fails to provide questions that are suitable since it does not know yet how to handle this type of student. If the two ITSs can collaborate and share what they know, then *B* can learn from *A* to provide more suitable questions to the student. In systems such as AutoTutor, agents do not interact with other agents to exchange their experiences or knowledge bases. I-MINDS is different in this regard. First, an agent in I-MINDS is capable of machine learning. A teacher agent is able to learn how to rank questions better as it receives feedback from the environment. A student agent is able to learn to more effectively form a buddy group for its student. Further, these student agents interact with each other to exchange information and experience.

I-MINDS

The I-MINDS project has three primary areas of research: (a) distributed computing (i.e., the infrastructure and enabling technology), (b) intelligent agents, and (c) the specific domain application in education and instructional design. Our research on distributed computing examines consistency, scalability, and security in resource sharing among multiple processes. In our research on intelligent agents, we study interactions between teacher agent and student agents, and among student agents. For our application in education, we focus on automated question ranking by the teacher agent and buddy group formation by the student agents.

In this section, we will focus our discussions on the intelligent agents and the multi-agent system and briefly on the instructional design. Readers are referred to Liu, Zhang, Soh, Al-Jaroodi, and Jiang (2003) for a discussion on distributed computing in I-MINDS using a Java object-oriented approach, to Soh, Liu, Zhang, Al-Jaroodi, Jiang, and Vemuri (2003) for a discussion on a layered architecture and proxy-supported topology to maintain a flexible and scalable design at the system level, and to Zhang, Soh, Jiang, and Liu (2005) for a discussion on the multi-agent system infrastructure.

The most unique and innovative aspect of I-MINDS when applied to education is the usage of agents that work individually behind-the-scenes and that collaborate as a multi-agent system. There are two types of agents in I-MINDS: teacher agents and student agents. In general, a teacher agent serves an instructor, and a student agent serves a student.

Teacher Agent

In I-MINDS, a teacher agent interacts with the instructor and other student agents. The teacher agent ranks questions automatically for the instructor to answer, profiles the students through its interaction with their respective student agents, and improves its knowledge bases to better support the instructor.

In our current design, the teacher agent evaluates questions and profiles students. The teacher agent has a mechanism that scores a question based on the profile of the student who asks the question and the quality of the question itself. A student who has been asking good questions will be ranked higher than a student who has been asking poor questions. A good question is based on the number of weighted keywords that it contains and whether it is picked by the instructor to answer in real-time.

The teacher agent also has a self-learning component, which lends intelligence to its interface. In our current design, this component allows the agent to improve its own knowledge bases and its performance in evaluating and ranking questions. When a new question is asked, the teacher agent first evaluates the question and scores it. Then the teacher agent inserts the question into a ranked question list (based on the score of the question and the heuristic rules, to be described later) and displays the list to the instructor. The instructor may choose which questions to answer. Whenever the instructor answers a question, he or she effectively "teaches" the teacher agent that the question is indeed valuable. If the question had been scored and ranked high by the teacher agent, this selection reinforces the teacher agent's reasoning. This positive reinforcement leads to the increased weights for the heuristics and keywords that had contributed to the score and rank of the question, and vice versa.

Figure 1 shows a screen snapshot of our teacher agent's interface. The snapshot shows three components. First, the main window displays the lecture materials that could be a whiteboard (captured with a Mimios-based technology), a Web page, and any documents that appear on the computer screen.

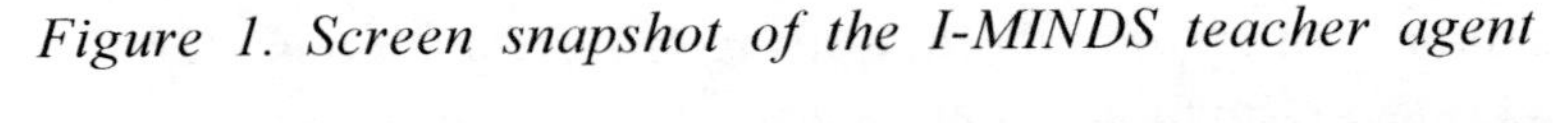
Figure 1. Screen snapshot of the I-MINDS teacher agent

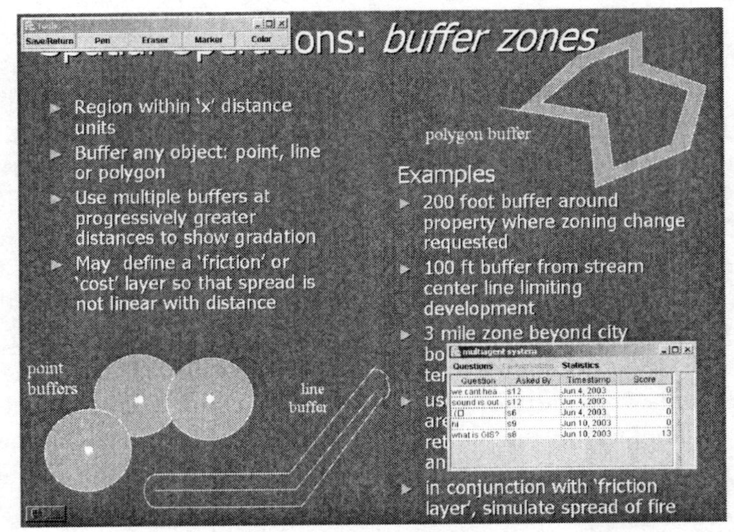

The Mimios-based technology transforms an ordinary whiteboard to a digital one. It comes with a sensor mounted on the ordinary whiteboard and a stylus that transmits a signal when pressed against the whiteboard when a person writes with the stylus. The sensor receives the signal and displays the movement of the handwriting on the whiteboard into the computer. In Figure 1, the lecture material happens to be a Microsoft PowerPoint slide on buffer zones, a topic in Geographic Information Systems (GIS). Second, the figure has a small toolbar, shown here at the top-left corner of the snapshot. Only an instructor can view and use this toolbar. This toolbar allows the instructor to save and/or transmit a learning material and change the annotation tools (pens, erasers, markers, and colors). Third, the snapshot shows a question display window at the bottom right corner. Once again, only the instructor can view and use this question display. The question display summarizes each question, ranked based on their scores. The display window also has several features. For example, an instructor may choose to answer or discard a question, may view the entire question, and may review the profile of the student who asked a particular question. Alternatively, the instructor may choose to hide the toolbar and the question's display window so as not to interfere with her/his lecture materials.

Student Agents

A student agent supports the student whom it serves, by interacting with the teacher agent and other student agents. It obtains student profiles from the teacher agent, forms the student's buddy group, tracks and records the student activities, and provides multimedia support for student collaboration. First, each student agent supports the formation of a "buddy group," which is a group of students with complementary characteristics (or profiles) who respond to each other and work together in online discussions. A student may choose to form his or her own buddy group if he or she knows about the other students and wants to include them in his or her buddy group. However, for students who do not have that knowledge,especially for remote students, the student agent will automatically form a buddy group for its student. I-MINDS also has two collaborative features that are used by the buddy groups: a forum and a whiteboard. The forum allows all buddies to ask and answer questions, with each message being color-coded. Also, the entire forum session is digitally archived, and the student may later review the session and annotate it through his or her student agent. The whiteboard allows all buddies to write, draw, and annotate on a community digital whiteboard. The actions on the whiteboard are also tracked and recorded by the student agent.

Note that the initial formation of a buddy group is based on the profile information queried from the teacher agent and preferences indicated by the student. Then, when a student performs a collaborative activity (initiating a forum discussion or a whiteboard discussion, or asking a question), the student agent informs other student agents identified as buddies within the student's buddy group of this activity. Thus, buddies may answer questions that the instructor does not have time to respond to in class. As the semester moves along, the student agent ranks the buddies based on their responsiveness and helpfulness. The student agent will drop buddies who have not been responsive from the buddy group. The student agent also uses heuristics to determine "when to invite/remove a buddy" and "which buddy to approach for help." The student agent adjusts its heuristic rules according to the current classroom environment.

Figure 2 shows a screen snapshot of the I-MINDS student agent, which is divided into four major quadrants. The top-left quadrant is the window that displays in real-time the lecture materials delivered from the teacher agent to each student agent. When the instructor changes a page, for example, the teacher agent will send the new page to the student agent. The student agent duly displays it. Further, when the instructor writes on a page, the teacher agent also transmits the changes to the student agent to display them for the student. The top-right quadrant is broken up into two sub-regions. On the top is a real-time video feed from the teacher agent. On the bottom is the digital archival repository of the lecture pages. A student may bring up and annotate each page. For example, he/she might paste a question onto a page and send it back to the instructor as a "question with a figure." On the bottom-left quadrant is the forum. Each message

Figure 2. Screen snapshot of the I-MINDS student agent

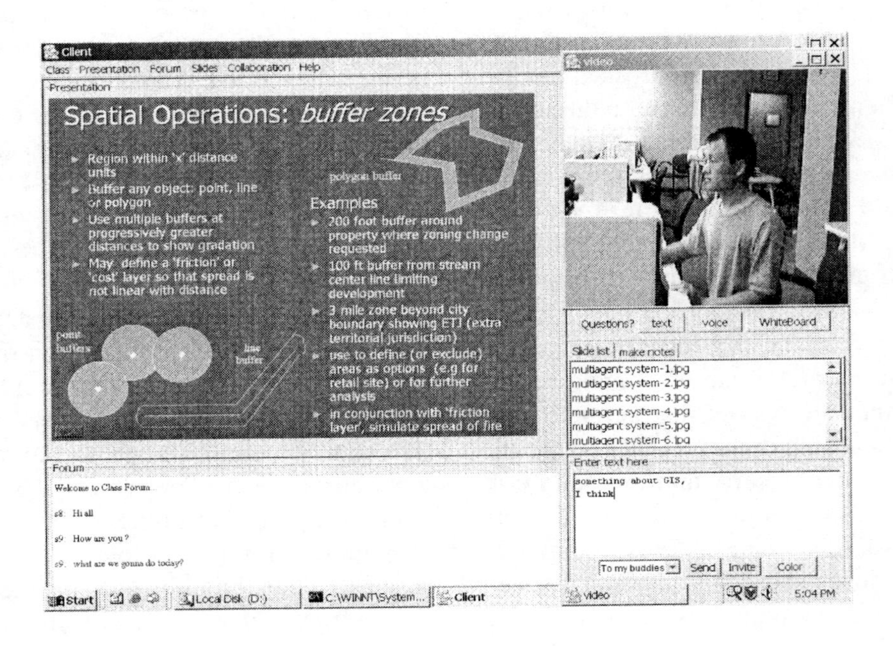

posted is colour-coded and labelled with the ID of the student who posted the message. On the bottom-right quadrant is the set of controls for asking questions. A student can type in his or her questions here, and then send the questions to the instructor, to the buddy group, to both the instructor and buddy group, or to a particular student in the buddy group. A student can also invite other students to join his or her buddy group through the "invite" function found in this quadrant.

The student agent interface has a menu bar on top, with menus in "Class," "Presentation," "Forum," "Slides," "Collaboration," and "Help." The "Class" menu has features pertinent to registration, login, and setup of a class lecture. The "Presentation" menu contains options on the lecture pages such as sound, colours, annotations, and so forth. The "Forum" menu allows a student to setup and manage his or her forums. The "Slides" menu allows a student to archive, search and retrieve, and in general, manage all the archived lecture pages. Finally, the "Collaboration" menu provides options on features that support collaborative activities—grabbing a token of the digital whiteboard, initiating a digital whiteboard discussion, turning off the automated buddy group formation, and so on.

FUTURE TRENDS

We see that, in the future, in the area of computer-aided education systems, multi-agent intelligence will play an important role in several aspects: (1) online cooperative or collaborative environment will become more active as personal student agents become more pro-active and social in exchanging information with other student agents to better serve the learners, (2) remote learners in the scenario of distance education will enjoy virtual classroom interfaces that can anticipate the needs and demands of the learners, and seamlessly situate the remote learners in a real classroom virtually, and (3) interfaces that can adapt their functions and arrangements based on the role of, the information gathered by, and the group activities participated by the agents operating behind the interfaces.

CONCLUSION

We have built a multi-agent infrastructure and interface, I-MINDS, aimed at helping instructors teach better and students learn better. The I-MINDS framework has many applications in education, due to its agent-based approach and real-time capabilities such as real-time in-class instructions with instant data gathering and information dissemination, unified agent and distributed computing architecture, group learning, and real-time student response monitoring. The individual interfaces are able to provide timely services and relevant information to their users with the support provided by the intelligent agents working behind-the-scenes. We have conducted a pilot study using two groups of students in actual lectures (Soh, Jiang, & Ansorge, 2004). One group is supported by I-MINDS in which the teacher delivered the lectures remotely and the students collaborated and interacted via the virtual classroom. The pilot study demonstrated some indicators of the effectiveness and feasibility of I-MINDS. Future work includes deploying I-MINDS in an actual classroom, incorporating cooperative learning and instructional design into the agents, and carrying out further studies on student learning.

REFERENCES

Graesser, A. C., Wiemer-Hastings, K., Wiemer-Hastings, P., Kreuz, R., & the Tutoring Research Group. (1999). AutoTutor: a simulation of a human tutor. *Journal of Cognitive Systems Research, 1*(1), 35-51.

Liu, X., Zhang, X., Soh, L. -K., Al-Jaroodi, J., & Jiang, H. (2003, December 2-5). A distributed, multiagent infrastructure for real-time, virtual classrooms. *Proceedings of the International Conference on Computers in Education (ICCE2003)*, Hong Kong, China (pp. 640-647).

McArthur, D., Lewis, M.W., & Bishay, M. (1993). The roles of artificial intelligence in education: current progress and future prospects. Technical Report, RAND, Santa Monica, CA, USA, DRU-472-NSF.

Soh, L.-K., Jiang, H., & Ansorge, C. (2004, March 3-7). Agent-based cooperative learning: a proof-of-concept experiment. *Proceedings of the 35th Technical Symposium on Computer Science Education (SIGCSE'2004)*, Norfolk, Virginia (pp. 368-372).

Soh, L.-K., Liu, X., Zhang, X., Al-Jaroodi, J., Jiang, H., & Vemuri, P. (2003, July 14-18). I-MINDS: An agent-oriented information system for applications in education. *Proceedings of the AAMAS2003 AOIS Workshop*, Melbourne, Australia (pp. 2-8).

Weiss, G. (Ed.) (1999). *Multiagent systems: A modern approach to distributed artificial intelligence.* MIT Press.

Wooldridge, M., & Jennings, N. R. (1995). Intelligent agents: Theory and practice. *The Knowledge Engineering Review, 10*(2), 115-152.

Zhang, X., Soh, L.-K., Jiang, H., & Liu, X. (2005). Using multiagent intelligence to support synchronous and asynchronous learning. To appear as a book chapter in L. Jain, & C. Ghaoui (Eds.), *Innovations in knowledge-based virtual education.* Springer-Verlag.

KEY TERMS

Agent: A module that is able to sense its environment, receive stimuli from the environment, make autonomous decisions and actuate the decisions, which in turn change the environment.

Computer-Supported Collaborative Learning: The process in which multiple learners work together on tasks using computer tools that leads to learning of a subject matter by the learners.

Intelligent Agent: An agent that is capable of flexible behaviour: responding to events timely, exhibiting goal-directed behaviour and social behaviour, and conducting machine learning to improve its own performance over time.

Intelligent Tutoring System (ITS): A software system that is capable of interacting with a student, providing guidance in the student's learning of a subject matter.

Interactive Learning Environment (ILE): A software system that interacts with a learner and may immerse the learner in an environment conducive to learning; it does not necessarily provide tutoring for the learner.

Machine Learning: The ability of a machine to improve its performance based on previous results.

Multi-Agent System: A group of agents where agents interact to accomplish tasks, thereby satisfying goals of the system design.

Virtual Classroom: An online learning space where students and instructors interact.

Interactive Speech Skimming via Time-Stretched Audio Replay

Wolfgang Hürst
Albert-Ludwigs-Universität Freiburg, Germany

Tobias Lauer
Albert-Ludwigs-Universität Freiburg, Germany

INTRODUCTION

Time stretching, sometimes also referred to as time scaling, is a term describing techniques for replaying speech signals faster (i.e., time compressed) or slower (i.e., time expanded) while preserving their characteristics, such as pitch and timbre. One example for such an approach is the SOLA (synchronous overlap and add) algorithm (Roucus & Wilgus, 1985), which is often used to avoid cartoon-character-like voices during faster replay. Many studies have been carried out in the past in order to evaluate the applicability and the usefulness of time stretching for different tasks in which users are dealing with recorded speech signals. One of the most obvious applications of time compression is speech skimming, which describes the actions involved in quickly going through a speech document in order to identify the overall topic or to locate some specific information. Since people can listen faster than they talk, time-compressed audio, within reasonable limits, can also make sense for normal listening, especially in view of He and Gupta (2001), who suggest that the future bottleneck for consuming multimedia contents will not be network bandwidth but people's limited time. In their study, they found that an upper bound for sustainable speedup during continuous listening is at about 1.6 to 1.7 times the normal speed. This is consistent with other studies such as Galbraith, Ausman, Liu, and Kirby (2003) or Harrigan (2000), indicating preferred speedup ratios between 1.3 and 1.8. Amir, Ponceleon, Blanchard, Petkovic, Srinivasan, and Cohen (2000) found that, depending on the text and speaker, the best speed for comprehension can also be slower than normal, especially for unknown or difficult contents.

BACKGROUND

While all the studies discussed in the previous section have shown the usefulness of time stretching, the question remains how this functionality is best presented to the user. Probably the most extensive and important study of time stretching in relation to user interfaces is the work done by Barry Arons in the early and mid 1990s. Based on detailed user studies, he introduced the SpeechSkimmer interface (Arons, 1994, 1997), which was designed in order to make speech skimming as easy as scanning printed text. To achieve this, the system incorporates time-stretching as well as content-compression techniques. Its interface allows the modification of speech replay in two dimensions. By moving a mark vertically, users can slow down replay (by moving the mark down) or make it faster (by moving the mark upward), thus enabling time-expanded or time-compressed replay. In the horizontal dimension, content-compression techniques are applied. With content compression, parts of the speech signal whose contents have been identified as less relevant or unimportant are removed in order to speed up replay. Importance is usually estimated based on automatic pause detection or the analysis of the emphasis used by the speaker. With SpeechSkimmer, users can choose between several discrete browsing levels, each of which removes more parts of the speech signal that have been identified as less relevant than the remaining ones. Both dimensions can be combined, thus enabling time as well as content compression during replay at the same time. In addition, SpeechSkimmer offers a modified type of backward playing in which small chunks of the signal are replayed in reverse order. It also offers some other features, such as bookmark-based navigation or jumps to some outstanding positions within the speech signal. The possibility to jump back a few seconds and switch back to normal replay has proven to be especially useful for search tasks. Parts of these techniques and interface design approaches have been successfully used in other systems (e.g., Schmandt, Kim, Lee, Vallejo, & Ackerman, 2002; Stifelman, Arons, & Schmandt, 2001).

Current media players have started integrating time stretching into their set of features as well. Here, faster and slower replay is usually provided in the interface by either offering some buttons that can be used to set replay speed to a fixed, discrete value, or by offering a slider-like widget to continuously modify replay speed in a specific range. It should be noted that if the content-compression part is removed from the SpeechSkimmer interface, the one-dimensional modification of replay speed by moving the corresponding mark vertically basically represents the same concept as the slider-like widget to continuously change replay speed in common media players (although a different orientation and visualization has been chosen).

Figure 1a illustrates an example of a slider-like interface, subsequently called a speed controller, which can be used to adapt speech replay to any value between 0.5 and 3.0 times the normal replay rate. Using such a slider to select a specific replay speed is very intuitive and useful if one wants to continuously listen to speech with a fixed time-compressed or time-expanded replay rate. However, this interface design might have limitations in more interactive scenarios such as information seeking, a task that is characterized by frequent speed changes together with other types of interaction such as skipping irrelevant parts or navigating back and forth. For example, one disadvantage of the usual speed controllers concerns the linear scale. The study by Amir et al. (2000) suggests that humans' perception of time-stretched audio is proportional to the logarithm of the speedup factor rather than linear in the factor itself. So, an increase from, say, 1.6 to 1.8 times the normal speed is perceived as more dramatic than changing the ratio from 1.2 to 1.4. Thus, the information provided by a linear slider scale may be irrelevant or even counterproductive. In any case, explicitly selecting a specific speedup factor does not seem to be the most intuitive procedure for information seeking.

INTERACTIVE SPEECH SKIMMING WITH THE ELASTIC AUDIO SLIDER

In addition to a speed controller, common media players generally include an audio-progress bar that indicates the current position during replay (see Figure 1b). By dragging the thumb on such a bar, users can directly access any random part within the file. However, audio replay is usually paused or continued normally while the bar's thumb is dragged. The reason why there is no immediate audio feedback is that the movements of the thumb performed by the users are usually too fast (if the thumb is moved quickly over larger distances), too slow (if the thumb is moved slowly or movement is paused), or too jerky (if the scrolling direction is changed quickly, if the user abruptly speeds up or jerks to a stop, etc.). Therefore, it is critical and sometimes impossible to achieve a comprehensible audio feedback, even if time-stretching techniques were applied to the signal or small snippets are replayed instead of single samples. On the other hand, such a slider-like inter-

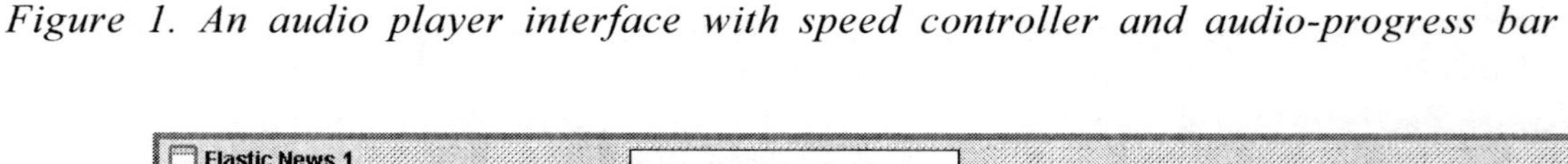

Figure 1. An audio player interface with speed controller and audio-progress bar

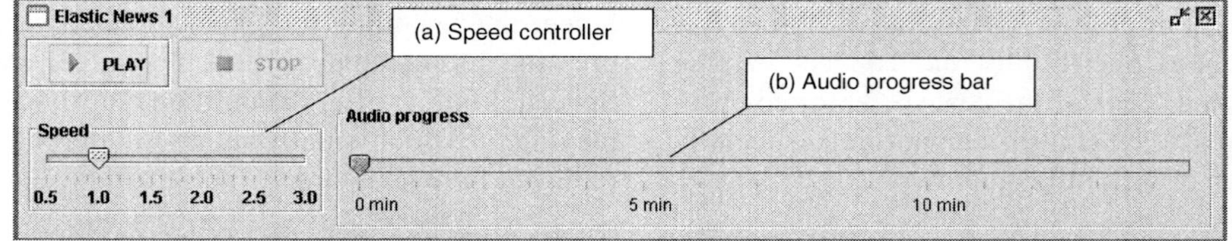

Figure 2. FineSlider (a): The scrolling speed depends on the distance between the slider thumb and mouse pointer, as is shown in the distance-to-speed mapping (b)

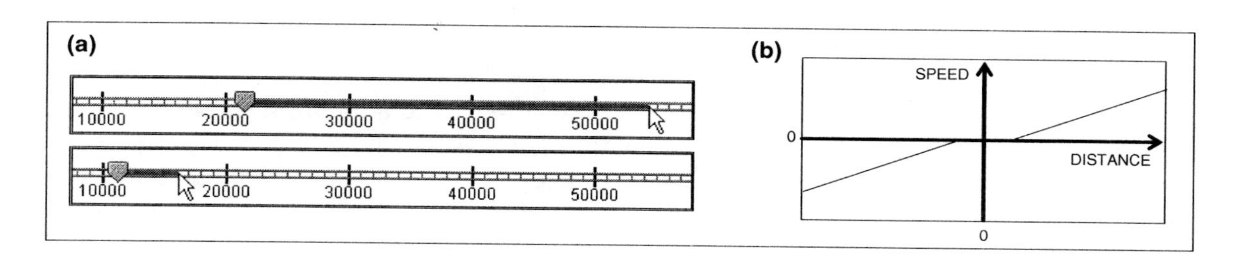

face—in which navigation is based on a modification of the current position within the file—could be very useful, especially in situations where navigation using a speed-controller interface is less intuitive and lacks flexibility.

Our approach for improved speech skimming is based on the idea of using time stretching in order to combine position-based navigation using the audio-progress bar with simultaneous audio feedback. It builds on the concept of elastic interfaces, which was originally introduced by Masui, Kashiwagi, and Borden (1995) for navigating and browsing discrete visual data. The basic idea of elastic interfaces is not to drag and move objects directly but instead to pull them along a straight line that connects the mouse pointer or cursor with the object (cf Hürst, in press). The speed with which the object follows the cursor's movements depends on the distance between the cursor and the object: If this distance is large, the object moves faster, and if it gets smaller, the object slows down. This behavior can be explained with the rubber-band metaphor, in which the direct connection between the object and the cursor is interpreted as a rubber band. Hence, if the cursor is moved away from the object, the tension on the rubber band gets stronger, thus pulling the object faster toward the cursor position. If the object and cursor get closer to each other, the rubber band loosens and the force pulling the object decreases; that is, its movement becomes slower.

One example for an elastic interface is the so-called FineSlider (Masui et al., 1995). Here, the distance between the mouse pointer and the slider thumb along a regular slider bar is mapped to movements of the slider thumb using a linear mapping function (see Figure 2). This distance can be interpreted as a rubber band, as described above. The main advantage of the FineSlider is that it allows a user to access any position within a document independent of its length, which is not necessarily true for regular sliders. Since the slider length represents the size of the document, moving the slider thumb one pixel might already result in a large jump in the file. This is because the number of pixels on the slider (and thus the number of positions it can access in the document) is limited by the corresponding window size and screen resolution, while on the other hand, the document can be arbitrarily long. By choosing an appropriate distance-to-speed mapping for the FineSlider, small distances between the slider thumb and the mouse pointer can be mapped to scrolling speeds that would otherwise only be possible with subpixel movements of the thumb. The FineSlider thus enables access to any random position of the file independent of its actual length.

If applied to a regular audio-progress bar, the concept of elastic interfaces offers two significant advantages. First, the thumb's movements are no longer mapped directly to the corresponding positions in the document, but are only considered indirectly via the distance-to-speed mapping. With regard to speech replay, this can be used to restrict scrolling speed and thus replay rates to a range in which audio feedback is still understandable and useful, such as 0.5 to 3.0 times the normal replay. Second, the motion of the thumb is much smoother because it is no longer directly controlled by the jerky movements of the mouse pointer (or the user's hand), thus resulting in a more reasonable, comprehensible audio feedback.

However, transferring the concept of elastic interfaces from visual to audio data is not straightforward. In the visual case (compare Figure 2b), a still image is considered the basic state while the mouse pointer is in the area around the slider thumb. Moving the pointer to the right continuously increases the forward replay speed, while moving it to the left enables backward playing with an increased replay rate. In contrast to this, there is no still or static state of a speech signal. In addition, strict backward playing of a speech signal makes no sense (unless it is done in a modified way such as realized in Arons, 1994). Therefore, we propose to adapt the distance-to-speed mapping for audio replay as illustrated in Figure 3a. Moving the pointer to the left enables time-expanded replay until the lower border of 0.5 times the normal replay speed is reached. Moving it to the right continuously increases replay speed up to a bound of 3.0 times the normal replay speed. The area around the slider thumb represents normal replay.

This functionality can be integrated into the audio-progress bar in order to create an elastic audio slider in the following way: If a user clicks anywhere on the bar's scale to the right of the current position of the thumb and subsequently moves the pointer to the right, replay speed is increased by the value resulting from the distance-to-speed mapping illustrated in Figure 3a. Moving the pointer to the left reduces replay speed accordingly. Placing or moving the pointer to the left of the actual position of the thumb results in an analogous but time-expanding replay behavior. The rubber-band metaphor still holds for this modification when the default state is not a pause mode but playback (i.e., slider movement) at normal speed, which can either be increased or decreased. The force of the rubber band pulls the slider thumb toward the right or left, thereby accelerating or braking it, respectively. Increasing the tension of the rubber band by dragging the mouse pointer further away from the thumb increases the force and thus the speed changes. If the band is loosened, the tension of the band and the accelerating or decelerating force on the slider is reduced.

The visualization of this functionality in the progress bar is illustrated in Figure 4. As soon as the user presses the mouse button anywhere on the slider scale, three areas of different color indicate the replay behavior. The purple area around the slider thumb represents normal speed. Green, to the right of the thumb, indicates accelerated playback (as in "Green light: Speed up!"), while the red color on the left stands for slower replay (as in "Red light: Slow down!"). A tool tip attached to the mouse pointer displays the current speedup factor. Further details on the design decisions and the technical implementation of the interface can be found in Hürst, Lauer, and Götz (2004b).

With such an elastic audio slider, users can quickly speed up replay or slow it down, depending on the current situation and demand. On the other hand, a traditional speed-controller design is better suited for situations in which the aim is not to interactively modify replay speed but instead to continuously listen to a speech recording at a different but fixed speed over a longer period. Both cases

Figure 3. Distance-to-speed mapping for the elastic audio slider without a speed controller (a) and coupled with the speed controller (b; with basic speed set to 1.5)

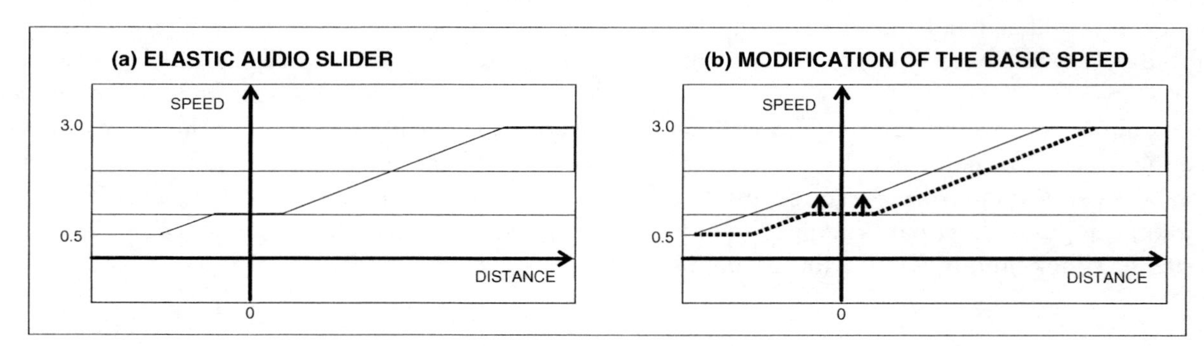

Figure 4. Visualization of the elastic audio slider (a), with views for normal (b), time-expanded (c), and time-compressed replay (d)

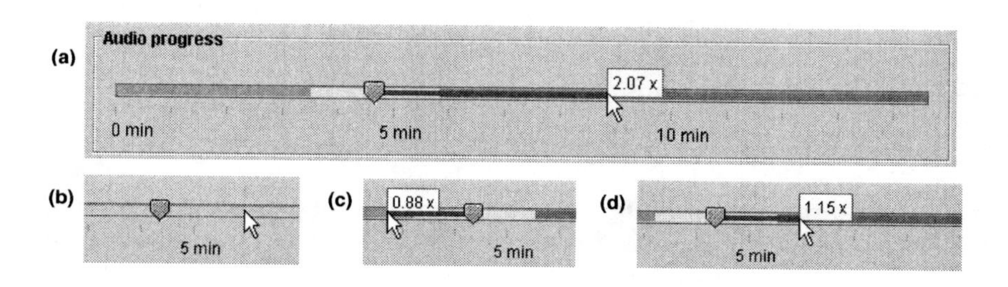

are very important applications of time stretching. Amir et al. (2000) present a study in which users' assessments of different replay speeds identified a natural speed for each file used in the study that the participants considered as the optimal speedup ratio. In addition, they identified unified speedup assessments once replay was normalized, that is, after replay was set to the natural speed of the corresponding file. Amir et al.'s study only gives initial evidence that must be confirmed through further user testing. Nonetheless, these findings argue for a combination of the speed controller with the elastic audio-slider functionality. The speed controller can be used to select a new basic speed. As a consequence, the area around the thumb in the audio-progress bar no longer represents the normal replay rate, but is adapted to the new basic speed. The area to the left and right of the thumb are adapted accordingly, as illustrated in the distance-to-speed mapping shown in Figure 3b. This modification enables users to select a basic replay speed that is optimal for the current document as well as for their personal preferences. Using the elastic audio slider, they can subsequently react directly to events in the speech signal and interactively speed up or slow down audio replay. Hence, interaction becomes more similar to navigation by using regular scroll bars or sliders and thus is more useful for interactive tasks such as information seeking.

FUTURE TRENDS

Our current work includes integrating additional time-compression techniques, such as pause reduction, into the system. Maybe the most challenging and exciting opportunity for future research will be the combination of the elastic audio-skimming functionality with the opportunity to skim visual data streams at the same time, thus providing real multimodal data browsing. While we already proved the usefulness of elastic browsing for video data (Hürst, Götz, & Lauer, 2004a), combined audiovisual browsing raises a whole range of new questions regarding issues such as how to handle the different upper and lower speedup bounds for audio and video, whether (and how) to provide audio feedback during visual skimming in the reverse direction, and how to maintain synchronized replay if pause reduction is used.

CONCLUSION

Common interface designs incorporating time-stretched speech replay in current media players are very useful to set the replay speed to a higher or lower rate. However, they lack the flexibility and intuitiveness needed for highly interactive tasks in which continuous changes of the replay speed, such as temporary speedups, are needed. For this reason, we introduced the elastic audio slider, a new interface design for the interactive manipulation of speech replay using time-stretching techniques. The proposed approach enhances the functionality of existing audio interfaces in a natural and intuitive way. Its integration into the audio-progress bar does not replace any of the typical features and functionalities offered by this widget, and it can be coupled in a gaining way with the commonly used tool for time stretching, that is, a speed-controller interface. Hence, it seamlessly integrates into common user-interface designs. The feasibility and usefulness of the proposed interface design was verified in a qualitative user study presented in Hürst, Lauer, and Götz (2004a). It proved that the elastic audio slider is a very useful tool for quickly speeding up audio replay (for example, to easily skip a part of minor interest) as well as to slow down replay temporarily (for example, to listen to a particular part more carefully). With the elastic audio slider, users are able to react quickly to events in the audio signal such as irrelevant passages or important parts and adapt the speed temporarily to the situation. This facilitates highly interactive tasks such as skimming and search.

REFERENCES

Amir, A., Ponceleon, D., Blanchard, B., Petkovic, D., Srinivasan, S., & Cohen, G. (2000). Using audio time scale modification for video browsing. *Proceedings of the 33rd Hawaii International Conference on System Sciences, HICCS 2000* (p. 3046).

Arons, B. (1994). *Interactively skimming recorded speech.* Unpublished doctoral dissertation, MIT, Cambridge, MA.

Arons, B. (1997). SpeechSkimmer: A system for interactively skimming recorded speech. *ACM Transactions on Computer Human Interaction, 4*(1), 3-38.

Galbraith, J., Ausman, B., Liu, Y., & Kirby, J. (2003). *The effects of time-compressed (accelerated) instructional video presentations on student recognition, recall and preference.* Anaheim, CA: Association for Educational Communications and Technology (AECT).

Harrigan, K. (2000). The SPECIAL system: Searching time-compressed digital video lectures. *Journal of Research on Technology in Education, 33*(1), 77-86.

He, L., & Gupta, A. (2001). Exploring benefits of non-linear time compression. *Proceedings of the 9th ACM International Conference on Multimedia, MM 01* (pp. 382-391).

Hürst, W. (in press). Elastic interfaces for visual data browsing. In C. Ghaoui (Ed.), *Encyclopedia of human computer interaction.* Hershey, PA: Information Science Publishing.

Hürst, W., Götz, G., & Lauer, T. (2004). New methods for visual information seeking through video browsing. *Proceedings of the 8th International Conference on Information Visualisation, IV 2004* (pp. 450-455).

Hürst, W., Lauer, T., & Götz, G. (2004a). An elastic audio slider for interactive speech skimming. *Proceedings of the Third Nordic Conference on Human-Computer Interaction, NordiCHI 2004,* Tampere, Finland.

Hürst, W., Lauer, T., & Götz, G. (2004b). Interactive manipulation of replay speed while listening to speech recordings. *Proceedings of the 12th ACM International Conference on Multimedia, MM 04,* New York.

Masui, T., Kashiwagi, K., & Borden, G. R., IV. (1995). Elastic graphical interfaces to precise data manipulation. *Conference Companion on Human Factors in Computing Systems, CHI 95* (pp. 143-144).

Roucus, S., & Wilgus, A. (1985). High quality time-scale modification for speech. *Proceedings of the IEEE International Conference on Acoustics, Speech, and Signal Processing, ICASSP 85* (Vol. 2, pp. 493-496).

Schmandt, C., Kim, J., Lee, K., Vallejo, G., & Ackerman, M. (2002). Mediated voice communication via mobile IP. *Proceedings of the 15th Annual ACM Symposiums on User Interface Software and Technology, UIST 2002* (pp. 141-150).

Stifelman, L., Arons, B., & Schmandt, C. (2001). The audio notebook: Paper and pen interaction with structured speech. *Proceedings of the SIGCHI Conference on Human Factors in Computing Systems, CHI 01* (pp. 182-189).

KEY TERMS

Content Compression: A term that describes approaches in which parts of a continuous media file are removed in order to speed up replay and data browsing or to automatically generate summaries or abstracts of the file. In relation to speech signals, content-compression techniques often shorten the signals by removing parts that have been identified as less relevant or unimportant based on pause detection and analysis of the emphasis used by the speakers.

Elastic Audio Slider: An interface design that enables the interactive manipulation of audio replay speed by incorporating the concept of elastic interfaces in common audio-progress bars.

Elastic Interfaces: Interfaces or widgets that manipulate an object, for example, a slider thumb, not by direct interaction but instead by moving it along a straight line that connects the object with the current position of the cursor. Movements of the object are a function of the length of this connection, thus following the rubber-band metaphor.

Rubber-Band Metaphor: A metaphor that is often used to describe the behavior of two objects that are connected by a straight line, the rubber band, in which one object is used to pull the other one toward a target position. The moving speed of the pulled objects depends on the length of the line between the two objects, that is, the tension on the rubber band. Longer distances result in faster movements, and shorter distances in slower movements.

SpeechSkimmer: A system developed by Barry Arons at the beginning of the '90s with the aim of making speech skimming as easy as scanning printed text. For this, its interface offers various options to modify replay speed, especially by applying time-stretching and content-compression techniques.

Speech Skimming: A term, sometimes also referred to as speech browsing or scanning, that describes the actions involved in skimming through a speech recording with the aim of classifying the overall topic of the content or localizing some particular information within it.

Time Compression: A term that describes the faster replay of continuous media files, such as audio or video signals. In the context of speech recordings, time compression usually assumes that special techniques are used to avoid pitch shifting, which otherwise results in unpleasant, very high voices.

Time Expansion: A term that describes the slower replay of continuous media files, such as audio or video signals. In the context of speech recordings, time expansion usually assumes that special techniques are used to avoid pitch shifting, which otherwise results in unpleasant, very low voices.

Time Stretching: Sometimes also referred to as time scaling, this term is often used to embrace techniques for the replay of continuous media files using time compression and time expansion, particularly in relation to speech signals in which faster or slower replay is achieved in a way that preserves the overall characteristics of the respective voices, such as pitch and timbre.

International Standards for HCI

Nigel Bevan
Serco Usability Services, UK

INTRODUCTION

The last 20 years have seen the development of a wide range of standards related to HCI (human-computer interaction). The initial work was by the ISO TC 159 ergonomics committee (see Stewart, 2000b), and most of these standards contain general principles from which appropriate interfaces and procedures can be derived. This makes the standards authoritative statements of good professional practice, but makes it difficult to know whether an interface conforms to the standard. Reed et al. (1999) discuss approaches to conformance in these standards.

ISO/IEC JTC1 has established SC35 for user interfaces, evolving out of work on keyboard layouts. This group has produced standards for icons, gestures, and cursor control, though these do not appear to have been widely adopted.

More recently, usability experts have worked with the ISO/IEC JTC1 SC7 software-engineering subcommittee to integrate usability into software engineering and software-quality standards. This has required some compromises: for example, reconciling different definitions of usability by adopting the new term *quality in use* to represent the ergonomic concept of usability (Bevan, 1999).

It is unfortunate that at a time of increasing expectations of easy access to information via the Internet, international standards are expensive and difficult to obtain. This is an inevitable consequence of the way standards bodies are financed. Information on how to obtain standards can be found in Table 4.

TYPES OF STANDARDS FOR HCI

Standards related to usability can be categorised as primarily concerned with the following.

1. The use of the product (effectiveness, efficiency, and satisfaction in a particular context of use)
2. The user interface and interaction
3. The process used to develop the product
4. The capability of an organisation to apply user-centred design

Figure 1 illustrates the logical relationships: The objective is for the product to be effective, efficient, and satisfying when used in the intended contexts. A prerequisite for this is an appropriate interface and interaction. This requires a user-centred design process, which to be achieved consistently, requires an

Figure 1. Categories of standards

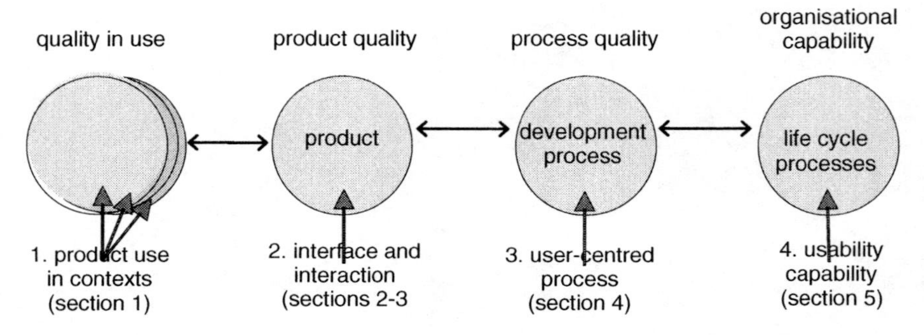

organisational capability to support user-centred design.

DEVELOPMENT OF ISO STANDARDS

International standards for HCI are developed under the auspices of the International Organisation for Standardisation (ISO) and the International Electrotechnical Commission (IEC). ISO and IEC comprise national standards bodies from member states. The technical work takes place in working groups of experts, nominated by national standards committees.

The standards are developed over a period of several years, and in the early stages, the published documents may change significantly from version to version until consensus is reached. As the standard becomes more mature, from the committee-draft stage onward, formal voting takes place by participating national member bodies.

The status of ISO and IEC documents is summarised in the title of the standard, as described in Table 1, and Table 2 shows the main stages of developing an international standard.

STANDARDS DESCRIBED IN THIS ARTICLE

Table 3 lists the international standards and technical reports related to HCI that were published or under development in 2004. The documents are divided into two categories: those containing general principles and recommendations, and those with detailed specifications. They are also grouped according to subject matter. All the standards are briefly described in Table 3.

APPROACHES TO HCI STANDARDS

HCI standards have been developed over the last 20 years. One function of standards is to impose consistency, and some attempt has been made to do this by ISO/IEC standards for interface components such as icons, PDA (personal digital assistant) scripts, and cursor control. However, in these areas, de facto industry standards have been more influential than ISO, and the ISO standards have not been widely adopted.

The ISO 9241 standards have had more impact (Stewart, 2000b; Stewart & Travis, 2002). Work on ergonomic requirements for VDT workstation hardware and the environment (ISO 9241, parts 3-9) began in 1983, and was soon followed by work on guidelines for the software interface and interaction (parts 10-17). The approach to software in ISO 9241 is based on detailed guidance and principles for design rather than precise interface specifications, thus permitting design flexibility.

More recently, standards and metrics for software quality have been defined by the software-engineering community.

The essential user-centred design activities needed to produce usable products are described in the ergonomic standard ISO 13407. These principles

Table 1. ISO and IEC document titles

Example	Explanation
ISO 1234 (2004)	ISO Standard 1234, published in 2004
ISO 1234-1 (2004)	Part 1 of ISO Standard 1234, published in 2004
ISO/IEC 1234 (2004)	Joint ISO/IEC Standard 1234, published in 2004
ISO TS 1234 (2004)	An ISO technical specification: A normative document that may later be revised and published as a standard
ISO PAS 1234 (2004)	An ISO publicly available specification: A normative document with less agreement than a TS that may later be revised and published as a standard
ISO TR 1234 (2004)	An ISO technical report: An informative document containing information of a different kind from that normally published in a normative standard
ISO *xx* 1234 (2004)	A draft standard of document type *xx* (see Table 2)

Table 2. Stages of development of draft ISO documents

Stage	Document type		Description
1	AWI	Approved work item	Prior to a working draft
2	WD	Working draft	Preliminary draft for discussion by working group
3	CD	Committee draft	Complete draft for vote and technical comment by national bodies
	CD TR or TS	Committee draft technical report/specification	
4	CDV	Committee draft for vote (IEC)	Final draft for vote and editorial comment by national bodies
	DIS	Draft international standard	
	FCD	Final committee draft (JTC1)	
	DTR or DTS	Draft technical report/specification	
5	FDIS	Final draft international standard	Intended text for publication for final approval

Table 3. Standards described in this article

Section	Principles and recommendations	Specifications
1. Context and test methods	ISO/IEC 9126-1: Software engineering - Product quality - Quality model	ISO DIS 20282-1: Ease of operation of everyday products - Context of use and user characteristics
	ISO/IEC TR 9126-4: Software engineering - Product quality - Quality-in-use metrics	ISO DTS 20282-2: Ease of operation of everyday products - Test method
	ISO 9241-11: Guidance on usability	ANSI/NCITS 354: Common Industry Format for usability test reports
	ISO/IEC PDTR 19764: Guidelines on methodology, and reference criteria for cultural and linguistic adaptability in information-technology products	Draft Common Industry Format for usability requirements
2. Software interface and interaction	ISO/IEC TR 9126-2: Software engineering - Product quality - External metrics	ISO/IEC 10741-1: Dialogue interaction - Cursor control for text editing
	ISO/IEC TR 9126-3: Software engineering - Product quality - Internal metrics	ISO/IEC 11581: Icon symbols and functions
	ISO 9241: Ergonomic requirements for office work with visual display terminals. Parts 10-17	ISO/IEC 18021: Information technology - User interface for mobile tools
	ISO 14915: Software ergonomics for multimedia user interfaces	ISO/IEC 18035: Icon symbols and functions for controlling multimedia software applications
	ISO TS 16071: Software accessibility	ISO/IEC 18036: Icon symbols and functions for World Wide Web browser toolbars
	ISO TR 19765: Survey of existing icons and symbols for elderly and disabled persons	ISO WD nnnn: Screen icons and symbols for personal, mobile, communications devices
	ISO TR 19766: Design requirements for icons and symbols for elderly and disabled persons	ISO WD nnnn: Icon symbols and functions for multimedia link attributes
	ISO CD 23974: Software ergonomics for World Wide Web user interfaces	ISO/IEC 25000 series: Software product–quality requirements and evaluation
	IEC TR 61997: Guidelines for the user interfaces in multimedia equipment for general-purpose use	
3. Hardware interface	ISO 11064: Ergonomic design of control centres	ISO 9241: Ergonomic requirements for office work with visual display terminals. Parts 3-9
	ISO/IEC CDTR 15440: Future keyboards and other associated input devices and related entry methods	ISO 13406: Ergonomic requirements for work with visual displays based on flat panels
		ISO/IEC 14754: Pen-based interfaces - Common gestures for text editing with pen-based systems
4. Development process	ISO 13407: Human-centred design processes for interactive systems	ISO/IEC 14598: Information technology - Evaluation of software products
	ISO TR 16982: Usability methods supporting human-centred design	
5. Usability capability	ISO TR 18529: Human-centred life-cycle process descriptions	
	ISO PAS 18152: A specification for the process assessment of human-system issues	
6. Other related standards	ISO 9241-1: General introduction	
	ISO 9241-2: Guidance on task requirements	
	ISO 10075-1: Ergonomic principles related to mental workload - General terms and definitions	

have been refined and extended in a model of usability maturity that can be used to assess the capability of an organisation to carry out user-centred design (ISO TR 18529). Burmester and Machate (2003) and Reed et al. (1999) discuss how different types of guidelines can be used to support the user-centred development process.

STANDARDS

Use in Context and Test Methods

1. **ISO 9241-11: Guidance on Usability (1998):** This standard (which is part of the ISO 9241 series) provides the definition of usability that is used in subsequent related ergonomic standards.
   - Usability: The extent to which a product can be used by specified users to achieve specified goals with effectiveness, efficiency, and satisfaction in a specified context of use
2. **ISO/IEC 9126-1: Software Engineering—Product Quality—Part 1: Quality Model (2001):** This standard describes six categories of software quality that are relevant during product development including quality in use (similar to the definition of usability in ISO 9241-11), with usability defined more narrowly as ease of use (Bevan, 1999).
3. **ISO/IEC CD TR 19764: Guidelines on Methodology, and Reference Criteria for Cultural and Linguistic Adaptability in Information-Technology Products (2003):** This defines a methodology and a guided checklist for the evaluation of cultural adaptability in software, hardware, and other IT products.
4. **ISO/IEC TR 9126-4: Software Engineering—Product Quality—Part 4: Quality-in-Use Metrics (2004):** Contains examples of metrics for effectiveness, productivity, safety, and satisfaction.
5. **ISO 20282: Ease of Operation of Everyday Products:** Ease of operation is concerned with the usability of the user interface of everyday products.
   - Part 1: Context of Use and User Characteristics (DIS: 2004): This part explains how to identify which aspects are relevant in the context of use and describes how to identify the characteristics that cause variance within the intended user population.
   - Part 2: Test Method (DTS: 2004): This specifies a test method for measuring the ease of operation of public walk-up-and-use products and of everyday consumer products.
6. **Common Industry Format**
   - ANSI/NCITS 354: Common Industry Format for Usability Test Reports (2001): This specifies a format for documenting summative usability test reports for use in contractual situations, and is expected to become an ISO standard (Bevan, Claridge, Maguire, & Athousaki, 2002).
   - Draft Common Industry Format for Usability Requirements (2004): Specifies a format for documenting summative usability requirements to aid communication early in development, and is expected to become an ISO standard.

Software Interface and Interaction

These standards can be used to support user-interface development in the following ways.

- To specify details of the appearance and behaviour of the user interface. ISO 14915 and IEC 61997 contain recommendations for multimedia interfaces. More specific guidance can be found for icons in ISO/IEC 11581, PDAs in ISO/IEC 18021, and cursor control in ISO/IEC 10741.
- To provide detailed guidance on the design of user interfaces (ISO 9241, parts 12-17).
- To provide criteria for the evaluation of user interfaces (ISO/IEC 9126, parts 2 and 3).

However, the attributes that a product requires for usability depend on the nature of the user, task, and environment. ISO 9241-11 can be used to help understand the context in which particular attributes may be required. Usable products can be designed by incorporating product features and attributes known to benefit users in particular contexts of use.

1. **ISO 9241: Ergonomic Requirements for Office Work with Visual Display Terminals:** ISO 9241 parts 10, and 12 to 17 provide requirements and recommendations relating to the attributes of the software.
   - Part 10: Dialogue Principles (1996): This contains general ergonomic principles that apply to the design of dialogues between humans and information systems: suitability for the task, suitability for learning, suitability for individualisation, conformity with user expectations, self-descriptiveness, controllability, and error tolerance.
   - Part 12: Presentation of Information (1998): This part includes guidance on ways of representing complex information using alphanumeric and graphical or symbolic codes, screen layout, and design, as well as the use of windows.
   - Part 13: User Guidance (1998): Part 13 provides recommendations for the design and evaluation of user-guidance attributes of software user interfaces including prompts, feedback, status, online help, and error management.
   - Part 14: Menu Dialogues (1997): It provides recommendations for the design of menus used in user-computer dialogues, including menu structure, navigation, option selection and execution, and menu presentation.
   - Part 15: Command Dialogues (1997): It provides recommendations for the design of command languages used in user-computer dialogues, including command-language structure and syntax, command representations, input and output considerations, and feedback and help.
   - Part 16: Direct Manipulation Dialogues (1999): This provides recommendations for the design of direct-manipulation dialogues, and includes the manipulation of objects and the design of metaphors, objects, and attributes.
   - Part 17: Form-Filling Dialogues (1998): It provides recommendations for the design of form-filling dialogues, including form structure and output considerations, input considerations, and form navigation.
2. **ISO/IEC 9126: Software Engineering—Product Quality:** ISO/IEC 9126-1 defines usability in terms of understandability, learnability, operability, and attractiveness. Parts 2 and 3 include examples of metrics for these characteristics.
   - Part 2: External Metrics (2003): Part 2 describes metrics that can be used to specify or evaluate the behaviour of the software when operated by the user.
   - Part 3: Internal Metrics (2003): Part 3 describes metrics that can be used to create requirements that describe static properties of the interface that can be evaluated by inspection without operating the software.
3. **Icon Symbols and Functions**
   - ISO/IEC 11581: Icon Symbols and Functions
     - Part 1: Icons—General (2000): This part contains a framework for the development and design of icons, including general requirements and recommendations applicable to all icons.
     - Part 2: Object Icons (2000)
     - Part 3: Pointer Icons (2000)
     - Part 4: Control Icons (CD: 1999)
     - Part 5: Tool Icons (2004)
     - Part 6: Action Icons (1999)
   - ISO/IEC 18035: Icon Symbols and Functions for Controlling Multimedia Software Applications (2003): This describes user interaction with and the appearance of multimedia control icons on the screen.
   - ISO/IEC 18036: Icon Symbols and Functions for World Wide Web Browser Toolbars (2003): This describes user interaction with and the appearance of World Wide Web toolbar icons on the screen.
   - ISO WD nnnn: Screen Icons and Symbols for Personal Mobile Communications Devices (2004): It defines a set of display-screen icons for personal mobile communication devices.
   - ISO WD nnnn: Icon Symbols and Functions for Multimedia Link Attributes (2004): It describes user interaction with and the appearance of link attribute icons on the screen.

- ISO CD TR 19765: Survey of Existing Icons and Symbols for Elderly and Disabled Persons (2003): It contains examples of icons for features and facilities used by people with disabilities.
- ISO TR 19766: Design Requirements for Icons and Symbols for Elderly and Disabled Persons

4. **ISO 14915: Software Ergonomics for Multimedia User Interfaces:**
   - Part 1: Design Principles and Framework (2002): This part provides an overall introduction to the standard.
   - Part 2: Multimedia Control and Navigation (2003): This part provides recommendations for navigation structures and aids, media controls, basic controls, media-control guidelines for dynamic media, and controls and navigation involving multiple media.
   - Part 3: Media Selection and Combination (2002): This part provides general guidelines for media selection and combination, media selection for information types, media combination and integration, and directing users' attention.
   - Part 4: Domain-Specific Multimedia Interfaces (AWI): This part is intended to cover computer-based training, computer-supported cooperative work, kiosk systems, online help, and testing and evaluation.
5. **IEC TR 61997: Guidelines for the User Interfaces in Multimedia Equipment for General-Purpose Use (2001):** This gives general principles and detailed design guidance for media selection, and for mechanical, graphical, and auditory user interfaces.
6. **ISO CD 23974: Software Ergonomics for World Wide Web User Interfaces (2004):** It provides recommendations and guidelines for the design of Web user interfaces.
7. **ISO/IEC 18021: Information Technology - User Interface for Mobile Tools for Management of Database Communications in a Client-Server Model (2002):** This standard contains user-interface specifications for PDAs with data-interchange capability with corresponding servers.
8. **ISO/IEC 10741-1: Dialogue Interaction—Cursor Control for Text Editing (1995):** This standard specifies how the cursor should move on the screen in response to the use of cursor control keys.

Hardware Interface

1. **ISO 9241: Ergonomic Requirements for Office Work with Visual Display Terminals:** Parts 3 to 9 contain hardware design requirements and guidance.
   - Part 3: Visual Display Requirements (1992): This specifies the ergonomics requirements for display screens that ensure that they can be read comfortably, safely, and efficiently to perform office tasks.
   - Part 4: Keyboard Requirements (1998): This specifies the ergonomics design characteristics of an alphanumeric keyboard that may be used comfortably, safely, and efficiently to perform office tasks. Keyboard layouts are dealt with separately in various parts of ISO/IEC 9995: Information Processing - Keyboard Layouts for Text and Office Systems (1994).
   - Part 5: Workstation Layout and Postural Requirements (1998): It specifies the ergonomics requirements for a workplace that will allow the user to adopt a comfortable and efficient posture.
   - Part 6: Guidance on the Work Environment (1999): This part provides guidance on the working environment (including lighting, noise, temperature, vibration, and electromagnetic fields) that will provide the user with comfortable, safe, and productive working conditions.
   - Part 7: Requirements for Display with Reflections (1998): It specifies methods of measuring glare and reflections from the surface of display screens to ensure that antireflection treatments do not detract from image quality.
   - Part 8: Requirements for Displayed Colours (1997): It specifies the requirements for multicolour displays.
   - Part 9: Requirements for Nonkeyboard Input Devices (2000): This specifies the

ergonomics requirements for nonkeyboard input devices that may be used in conjunction with a visual display terminal.

2. **ISO 13406: Ergonomic Requirements for Work with Visual Displays Based on Flat Panels**
   - Part 1: Introduction (1999)
   - Part 2: Ergonomic Requirements for Flat-Panel Displays (2001)
3. **ISO/IEC 14754: Pen-Based Interfaces — Common Gestures for Text Editing with Pen-Based Systems (1999)**
4. **ISO/IEC CD TR 15440: Future Keyboards and Other Associated Input Devices and Related Entry Methods (2003)**
5. **ISO 11064: Ergonomic Design of Control Centres:** This eight-part standard contains ergonomic principles, recommendations, and guidelines.
   - Part 1: Principles for the Design of Control Centres (2000)
   - Part 2: Principles of Control-Suite Arrangement (2000)
   - Part 3: Control-Room Layout (1999)
   - Part 4: Workstation Layout and Dimensions (2004)
   - Part 5: Human-System Interfaces (FCD: 2002)
   - Part 6: Environmental Requirements for Control Rooms (DIS: 2003)
   - Part 7: Principles for the Evaluation of Control Centres (DIS: 2004)
   - Part 8: Ergonomic Requirements for Specific Applications (WD: 2000)

The Development Process

ISO 13407 explains the activities required for user-centred design, and ISO 16982 outlines the types of methods that can be used. ISO/IEC 14598 gives a general framework for the evaluation of software products using the model in ISO/IEC 9126-1.

1. **ISO 13407: Human-Centred Design Processes for Interactive Systems (1999)**
2. **ISO TR 16982: Usability Methods Supporting Human-Centred Design (2002)**
3. **ISO/IEC 14598: Information Technology — Evaluation of Software Products (1998-2000)**

Usability Capability of the Organisation

The usability maturity model in ISO TR 18529 contains a structured set of processes derived from ISO 13407 and a survey of good practice. It can be used to assess the extent to which an organisation is capable of carrying out user-centred design (Earthy, Sherwood Jones, & Bevan, 2001). ISO PAS 18152 extends this to the assessment of the maturity of an organisation in performing the processes that make a system usable, healthy, and safe.

- ISO TR 18529: Ergonomics of Human-System Interaction—Human-Centred Life-Cycle Process Descriptions (2000)
- ISO PAS 18152: Ergonomics of Human-System Interaction—A Specification for the Process Assessment of Human-System Issues (2003)

Other Related Standards

1. **ISO 9241-2: Part 2: Guidance on Task Requirements (1992)**
2. **ISO 10075: Ergonomic Principles Related to Mental Workload**
   - Part 1: General Terms and Definitions (1994)
   - Part 2: Design Principles (1996)
   - Part 3: Principles and Requirements Concerning Methods for Measuring and Assessing Mental Workload (2004)
3. **ISO TS 16071: Guidance on Accessibility for Human-Computer Interfaces (2003):** This provides recommendations for the design of systems and software that will enable users with disabilities greater accessibility to computer systems (see Gulliksen & Harker, 2004).
6. **ISO AWI 9241-20: Accessibility Guideline for Information Communication Equipment and Services: General Guidelines (2004)**

WHERE TO OBTAIN INTERNATIONAL STANDARDS

ISO standards have to be purchased. They can be obtained directly from ISO, or from a national standards body (Table 4).

FUTURE OF HCI STANDARDS

Now that the fundamental principles have been defined, the ergonomics and software-quality standards groups are consolidating the wide range of standards into more organised collections. Some of the new series are already approved work items, CDs, or DISs.

ISO 9241: Ergonomics of Human-System Interaction

The parts of ISO 9241 are in the process of being revised into the new structure shown below.

- Part 1: Introduction
- Part 2: Job Design
- Part 11: Hardware & Software Usability
- Part 20: Accessibility and Human-System Interaction (AWI: 2004)

Software

- Part 100: Introduction to Software Ergonomics
- Part 110: Dialogue Principles (DIS: 2004, revision of ISO 9241-10)
- Part 112: Presentation Principles and Recommendations (part of ISO 9241-12)
- Part 113: User Guidance (ISO 9241-13, reference to ISO/IEC 18019)
- Part 114: Multimedia Principles (ISO 14915-1)
- Part 115: Dialogue Navigation (part of ISO 14915-2, reference to ISO/IEC 18035)
- Part 120: Software Accessibility (ISO/TS 16071)
- Part 130: GUI (graphical user interface) & Controls (does not yet exist, reference to ISO/IEC 11581)
- Part 131: Windowing Interfaces (part of ISO 9241-12)
- Part 132: Multimedia Controls (part of ISO 14915-2)
- Part 140: Selection and Combination of Dialogue Techniques (part of ISO 9241-1)
- Part 141: Menu Dialogues (ISO 9241-14)
- Part 142: Command Dialogues (ISO 9241-15)
- Part 143: Direct-Manipulation Dialogues (ISO 9241-16)
- Part 144: Form-Filling Dialogues (ISO 9241-17)
- Part 145: Natural-Language Dialogues
- Part 150: Media (ISO 14915-3)
- Part 160: Web Interfaces (ISO 23973, reference to ISO/IEC 18036)

Process

- Part 200: Human-System Interaction Processes
- Part 210: Human-Centred Design (ISO 13407)
- Part 211: HSL 16982, ISO PAS 18152

Ergonomic Requirements and Measurement Techniques for Electronic Visual Displays

- Part 301: Introduction (CD: 2004)
- Part 302: Terminology (CD: 2004)
- Part 303: Ergonomics Requirements (CD: 2004)
- Part 304: User-Performance Test Method (AWI: 2004)
- Part 305: Optical Laboratory Test Methods (CD: 2004)
- Part 306: Field Assessment Methods (CD: 2004)

Table 4. Sources of standards and further information

Information	URL (uniform resource locator)
Published ISO standards, and the status of standards under development	http://www.iso.org/iso/en/Standards_Search.StandardsQueryForm
ISO national member bodies	http://www.iso.ch/addresse/membodies.html
NSSN, a national resource for global standards	http://www.nssn.org

- Part 307: Analysis and Compliance Test Methods (CD: 2004)

Physical Input Devices

- Part 400: Ergonomic Principles (CD: 2004)
- Part 410: Design Criteria for Products (AWI: 2004)
- Part 411: Laboratory Test and Evaluation Methods
- Part 420: Ergonomic Selection Procedures (AWI: 2004)
- Part 421: Workplace Test and Evaluation Methods
- Part 500: Workplaces
- Part 600: Environment
- Part 700: Special Application Domains
- Part 710: Control Centre (in seven parts)

ISO/IEC 25000 Series: Software Product-Quality Requirements and Evaluation (SQuaRE)

The ISO/IEC 25000 series of standards will replace and extend ISO/IEC 9126, ISO/IEC 14598, and the Common Industry Format.

1. **ISO/IEC FCD 25000: Guide to SquaRE (2004)**
   - ISO/IEC AWI 25001: Planning and Management (ISO/IEC 14598-2)
   - ISO/IEC AWI 25010: Quality Model and Guide (ISO/IEC 9126-1)
2. **ISO/IEC CD 25020: Measurement Reference Model and Guide (2004)**
3. **ISO/IEC CD 25021: Measurement Primitives (2004)**
   - ISO/IEC AWI 25022: Measurement of Internal Quality (ISO/IEC 9126-3)
   - ISO/IEC AWI 25023: Measurement of External Quality (ISO/IEC 9126-2)
   - ISO/IEC AWI 25024: Measurement of Quality in Use (ISO/IEC 9126-3)
4. **ISO/IEC CD 25030: Quality Requirements and Guide (2004)**
   - ISO/IEC AWI 25040: Quality Evaluation-Process Overview & Guide (ISO/IEC 14598-1)
   - ISO/IEC AWI 25041: Evaluation Modules (ISO/IEC 14598-6)
   - ISO/IEC AWI 25042: Process for Developers (ISO/IEC 14598-3)
   - ISO/IEC AWI 25043: Process for Acquirers (ISO/IEC 14598-4)
   - ISO/IEC AWI 25044: Process for Evaluators (ISO/IEC 14598-5)
   - ISO/IEC 25051: Quality Requirements and Testing Instructions for Commercial Off-the-Shelf (COTS) Software (ISO/IEC 12119)
   - ISO/IEC 250nn: Common Industry Format (ANSI/NCITS 354)

CONCLUSION

The majority of effort in ergonomics standards has gone into developing conditional guidelines (Reed et al., 1999), following the pioneering work of Smith and Mosier (1986). Parts 12 to 17 of ISO 9241 contain a daunting 82 pages of guidelines. These documents provide an authoritative source of reference, but designers without usability experience have great difficulty applying these types of guidelines (de Souza & Bevan, 1990; Thovtrup & Nielsen, 1991). Several checklists have been prepared to help assess the conformance of software to the main principles in ISO 9241 (Gediga, Hamborg, & Düntsch, 1999; Oppermann & Reiterer, 1997; Prümper, 1999).

In the United States, there is continuing tension between producing national standards that meet the needs of the large U.S. market and contributing to the development of international standards. Having originally participated in the development of ISO 9241, the HFES decided to put subsequent effort into a national version: HFES-100 and HFES-200 (see Reed et al., 1999).

Standards are more widely accepted in Europe than in the United States, partly for cultural reasons, and partly to achieve harmonisation across European Union countries. Many international standards (including ISO 9241) have been adopted as European standards. The European Union (2004) Supplier's Directive requires that the technical specifications used for public procurement must be in the terms of any relevant European standards. Ergonomic standards such as ISO 9241 can also be used to support adherence to European regulations for the health and safety of display screens (Bevan, 1991; European Union, 1990; Stewart, 2000a).

Stewart and Travis (2002) differentiate between standards that are formal documents published by standards-making bodies and developed through a consensus and voting procedure, and those that are published guidelines that depend on the credibility of their authors. This gives standards authority, but it is not clear how many of the standards listed in this article are widely used. One weakness of most of the HCI standards is that they have been discussed around a table rather than being developed in a user-centred way, testing prototypes during development. The U.S. Common Industry Format is an exception, undergoing trials during its evolution outside ISO. There are ISO procedures to support this, and ISO 20282 is being issued initially as a technical specification so that trials can be organised before it is confirmed as a standard. This is an approach that should be encouraged in the future.

Another potential weakness of international standards is that the development process is slow, and the content depends on the voluntary effort of appropriate experts. Ad hoc groups can move more quickly, and when appropriately funded, can produce superior results, as with the U.S. National Cancer Institute Web design guidelines (Koyani, Bailey, & Nall, 2003), which as a consequence may remain more authoritative than the forthcoming ISO 23974.

Following the trends in software-engineering standards, the greatest benefits may be obtained from HCI standards that define the development process and the capability to apply that process. ISO 13407 provides an important foundation (Earthy et al., 2001), and the usability maturity of an organisation can be assessed using ISO TR 18529 or ISO PAS 18152, following the procedure in the software-process assessment standard ISO TR 15504-2 (Sherwood Jones & Earthy, 2003).

ACKNOWLEDGMENT

This article includes material adapted and updated from Bevan (2001).

REFERENCES

Bevan, N. (1991). Standards relevant to European directives for display terminals. *Proceedings of the 4th International Conference on Human Computer Interaction* (Vol. 1, pp. 533-537).

Bevan, N. (1999). Quality in use: Meeting user needs for quality. *Journal of Systems and Software, 49*(1), 89-96.

Bevan, N. (2001). International standards for HCI and usability. *International Journal of Human-Computer Studies, 55*(4), 533-552.

Bevan, N., Claridge, N., Maguire, M., & Athousaki, M. (2002). Specifying and evaluating usability requirements using the Common Industry Format. *Proceedings of IFIP 17th World Computer Congress* (pp. 33-148).

Burmester, M., & Machate, J. (2003). Creative design of interactive products and use of usability guidelines: A contradiction? *Proceedings of HCI International 2003: Vol. 1. Human-Computer Interaction: Theory and Practice (Part 1)* (pp. 434-438).

De Souza, F., & Bevan, N. (1990). The use of guidelines in menu interface design: Evaluation of a draft standard. *Human-Computer Interaction INTERACT'90*, 435-440.

Earthy, J., Sherwood Jones, B., & Bevan, N. (2001). The improvement of human-centred processes: Facing the challenge and reaping the benefit of ISO 13407. *International Journal of Human Computer Studies, 55*(4), 553-585.

European Union. (1990, June 21). Directive 90/270/EEC on the minimum safety and health requirements for work with display screen equipment. *Official Journal L 156*, 14-18.

European Union. (2004, April 30). Directive 2004/18/EC on the coordination of procedures for the award of public works contracts, public supply contracts and public service contracts. *Official Journal L 134*, 114-240.

Gediga, G., Hamborg, K., & Düntsch, I. (1999). The IsoMetrics usability inventory: An operationalisation of ISO 9241-10. *Behaviour and Information Technology, 18*(3), 151-164.

Gulliksen, J., & Harker, S. (2004). Software accessibility of human-computer interfaces: ISO Technical Specification 16071. *Universal Access in the Information Society, 3*(1), 6-16.

Koyani, S. J., Bailey, R. W., & Nall, J. R. (2003). *Research-based Web design and usability guidelines.* Retrieved from http://usability.gov/pdfs/guidelines.html

Oppermann, R., & Reiterer, R. (1997). Software evaluation using the 9241 evaluator. *Behaviour and Information Technology, 16*(4/5), 232-245.

Prümper, P. (1999, August 22-26). Human-computer interaction: Ergonomics and user interfaces. In *Proceedings of HCI International* '99, Munich, Germany (Vol. 1, pp. 1028-1032).

Reed, P., Holdaway, K., Isensee, S., Buie, E., Fox, J., Williams, J., et al. (1999). User interface guidelines and standards: Progress, issues, and prospects. *Interacting with Computers, 12*(2), 119-142.

Sherwood Jones, B., & Earthy, J. (2003). Human centred design: Opportunities and resources. *Human Systems IAC Gateway, 14*(3). Retrieved from http://iac.dtic.mil/hsiac/GW-docs/gw_xiv_3.pdf

Smith, S. L., & Mosier, J. N. (1986). *Guidelines for designing user interface software.* The MITRE Corporation. Retrieved from http://hcibib.org/sam/

Stewart, T. (2000a). *Display screen regulations: An employer's guide to getting the best out of regulations.* Retrieved from http://www.system-concepts.com/articles/updatesonstandards/

Stewart, T. (2000b). Ergonomics user interface standards: Are they more trouble than they are worth? *Ergonomics, 43*(7), 1030-1044.

Stewart, T., & Travis, D. (2002). Guidelines, standards and style guides. In J. A. Jacko & A. Sears (Eds.), *The human-computer interaction handbook* (pp. 991-1005). Mahwah, NJ: L.E.A.

Thovtrup, H., & Nielsen, J. (1991). Assessing the usability of a user interface standard. *Proceedings of CHI'91 Conference on Human Factors in Computing Systems* (pp. 335-341).

KEY TERMS

Context of Use: The users, tasks, equipment (hardware, software, and materials), and physical and social environments in which a product is used (ISO 9241-11).

Interaction: Bidirectional information exchange between users and equipment (IEC 61997).

Prototype: Representation of all or part of a product or system that, although limited in some way, can be used for evaluation (ISO 13407).

Task: The activities required to achieve a goal (ISO 9241-11).

Usability: The extent to which a product can be used by specified users to achieve specified goals with effectiveness, efficiency, and satisfaction in a specified context of use (ISO 9241-11).

User: Individual interacting with the system (ISO 9241-10).

User Interface: The control and information-giving elements of a product, and the sequence of interactions that enable the user to use it for its intended purpose (ISO DIS 20282-1).

Internet-Mediated Communication at the Cultural Interface

Leah P. Macfadyen
The University of British Columbia, Canada

INTRODUCTION

As individuals launch themselves into cyberspace via networked technologies, they must navigate more than just the human-computer interface. The rhetoric of the "global village"—a utopian vision of a harmonious multicultural virtual world—has tended to overlook the messier and potentially much more problematic social interfaces of cyberspace: the interface of the individual with cyberculture (Macfadyen, 2004), and the interface of culture with culture. To date, intercultural communications research has focused primarily on instances of physical (face-to-face) encounters between cultural groups, for example, in the classroom or in the workplace. However, virtual environments are increasingly common sites of encounter and communication for individuals and groups from multiple cultural backgrounds. This underscores the need for a better understanding of Internet-mediated intercultural communication.

BACKGROUND

Researchers from multiple disciplines (cultural studies, intercultural studies, linguistics, sociology, education, human-computer interaction, distance learning, learning technologies, philosophy, and others) have initiated studies to examine virtual intercultural communication. The interdisciplinarity of the field, however, offers distinct challenges: in addition to embracing different definitions of culture, investigators lack a common literature or vocabulary. Communicative encounters between groups and individuals from different cultures are variously described as cross-cultural, intercultural, multicultural, or even transcultural. Researchers use terms such as the Internet, the World Wide Web, cyberspace, and virtual (learning) environments (VLEs) to denote overlapping though slightly different perspectives on the world of networked digital communications. Others focus on CMC (computer-mediated communication), ICTs (Internet and communication technologies), HCI (human-computer interaction), CHI (computer-human interaction), or CSCW (computer-supported cooperative work) in explorations of technologies at the communicative interface.

This article offers an overview of existing theoretical and empirical approaches to examining what happens when culturally diverse individuals communicate with each other on the Internet: the publicly available, internationally interconnected system of computers (and the information and services they provide to their users) that uses the TCP/IP (transmission-control protocol/Internet protocol) suite of packet-switching communications protocols.

INVESTIGATING ONLINE INTERCULTURAL COMMUNICATION

Does Culture Influence Internet-Mediated Intercultural Communication?

What does current research tell us about the interplay between individuals, cultures, and communication online? A significant number of studies has begun to explore online intercultural communications between and within selected populations. Some have employed quantitative methods to investigate whether there are specific cultural differences in attitudes to technology and the use of technologies, in communication patterns and frequency, and in communication style or content (for detailed references to these quantitative studies, see Macfadyen, Roche, & Doff, 2004). Others (and especially those using qualitative approaches) focus less on the technology and instead seek evidence of cultural influences on interpersonal or intragroup processes, dynamics, and communications in cyberspace. For example, Chase, Macfadyen, Reeder, and Roche (2002) describe nine thematic clusters of apparent cultural mismatches that occurred in communications between culturally diverse individuals in a Web-based discussion forum: differences in the choices of participation format and frequency, differences in response to the forum culture, different levels of comfort with disembodied communication, differing levels of technoliteracy, differences in participant expectations, differing patterns of use of academic discourse vs. narrative, and differing attitudes to time and punctuality. To this list of discontinuities, Wilson (2001) adds "worldview, culturally specific vocabulary and concepts, linguistic characteristics...[and] cognition patterns, including reading behaviour" (p. 61). Kim and Bonk (2002) report cultural differences in online collaborative behaviours, and Rahmati (2000) and Thanasankit and Corbitt (2000) describe the different cultural values that selected cultural groups refer to in their approaches to decision making when working online.

Evidence is accumulating, then, that seems to suggest that cultural factors do impact communicative encounters in cyberspace. What is the most effective framework for exploring and explaining this phenomenon, and what role is played by the design of human-computer interfaces?

The Problem of Defining Culture

Perhaps not surprisingly, most intercultural communication researchers have begun by attempting to clarify and define what culture is to allow subsequent comparative analyses and examinations of cultural differences in communication practices. Given that culture "is one of the two or three most complicated words in the English language" (Williams, 1983, p. 87), this definitional quest is, unfortunately, beset with difficulty. The word itself is now used to represent distinct and important concepts in different intellectual disciplines and systems of thought, and decades of debate between scholars across the disciplines have not yielded a simple or uncontested understanding of the concept.

In reality, a majority of existing research and theory papers published to date that examine culture and communication in online environments implicitly define culture as ethnic or national culture, and examine online communication patterns among and between members of specific ethnic or linguistic groups; only a few attempt to broaden the concept of culture. Of these, Heaton (1998b) notes, "organizational and professional cultures are also vital elements in the mix" (pp. 262-263) and defines culture as "a dynamic mix of national/geographic, organizational and professional or disciplinary variables" (p. 263). Others highlight the importance of gender culture differences in online communications, or note the complicating influences of linguistic culture and linguistic ability, epistemological type, technical skill, literacy (Goodfellow, 2004), class, religion, and age (for detailed references, see Macfadyen et al., 2004).

The Problem of Essentialism

Even more problematic than the simplistic equating of culture with ethnicity is the persistent and uncritical application of essentialist theories of culture and cultural difference in intercultural communications research. These theories tend to characterize culture as an invariant and uncontested matrix of meanings and practices that are inherited by and shared within a group. They are commonly used either to develop testable hypotheses about the impact of culture on Internet-mediated intercultural communications, or to interpret data post hoc (or both). In particular, an increasing number of studies relies unquestioningly upon Hofstede's (1980, 1991) dimensions of (national) culture (Abdat & Pervan, 2000; Gunawardena, Nolla, Wilson, Lopez-Islas, Ramírez-Angel, & Megchun-Alpízar, 2001; Maitland, 1998; Marcus & Gould, 2000; Tully, 1998) even though serious questions have been raised about Hofstede's methodological assumptions that might make his subsequent conclusions less reliable (McSweeney, 2002). Also referenced frequently are Edward Hall's theory (1966) of high- and low-context communications (Buragga, 2002; Heaton, 1998a; Maitland) and the nationally delineated cultural models of Hampden-Turner and Trompenaars (2000).

Some researchers (Abdelnour-Nocera, 2002; Hewling, 2004; Reeder, Macfadyen, Roche, & Chase, 2004) are now offering critiques of the use of essentialist cultural theories in intercultural studies. Abdelnour-Nocera discusses, for example, the risks of using "ready made cultural models" such as

Hofstede's, arguing that one may miss "qualitative specific dimensions that don't fit certain pre-established parameters" (p. 516). The uncritical use of essentialist theories of culture carries with it additional and more fundamental problems. First, such theories tend to forget that cultures change, and instead imagine cultures as static, predictable, and unchanging. Second, the assumption of cultures as closed systems of meaning tends to ignore important questions of power and authority: how has one system of meaning, or discourse, come to dominate? Related to this, essentialist theories can sometimes be criticized as positioning individuals as simple enculturated players, lacking in agency, and do not allow for the possibility of choice, learning, and adaptation in new contexts. Street (1993) reminds us that "culture is not a thing" but that it is often "dressed up in social scientific discourse in order to be defined" (p. 25). Culture is, rather, an active process of meaning making: a verb rather than a noun.

Social Construction and Negotiation of Meaning

Asad (1980) argues that it is the production of essential meanings—in other words, the production of culture—in a given society that is the problem to be explained, not the definition of culture itself. In line with this, a number of recent studies has attempted to examine the negotiation of meaning and the processes of meaning making employed by different individuals or groups in cyberspace communications, and make use of less- (or non-) essentialist intercultural and/or communications theory in their research design and analysis. Reeder et al. (2004), for example, prefer a Vygotskyan social-constructivist stance in which the construction of identity is the starting point in their investigation of online intercultural communication. They interpret intercultural patterns of online communication in the light of cross-disciplinary theories from sociolinguistics, applied linguistics, genre and literacy theory, and aboriginal education. Belz (2003) brings a Hallidayan (1978, 1994) linguistic approach (appraisal theory) to her evaluation of intercultural e-mail communications without making a priori predications based on the communicators' nationalities. In his analysis of an intercultural e-mail exchange, O'Dowd (2003) builds upon Byram's (1997) notion of intercultural competence, another theoretical perspective that focuses on the negotiation of a communicative mode that is satisfactory to all interlocutors. Choi and Danowski (2002), meanwhile, base their research on theories of social networks; they discuss their findings with reference to core-periphery theories of network communication and to the online power-play negotiations of dominant and minority cultures. Alternatively, Gunawardena, Walsh, Reddinger, Gregory, Lake, and Davies (2002) explore the negotiation of the face online, building on face theory developed by theorists such as Ting-Toomey (1988). Yetim (2001) suggests that more attention must be paid to the importance of metacommunication as the site of clarification of meaning. Thorne (2003) offers another conceptual framework that draws on an assessment of "discursive orientation, communicative modality, communicative activity and emergent interpersonal dynamics" (p. 38) in the analysis of intercultural engagement online.

A few authors have recently developed new theoretical perspectives on online intercultural communication. Benson and Standing (2000) propose a systems theory of culture that emphasizes culture as a self-referential system of rules, conventions, and shared understandings rather than as a set of categories. They go on to explain perspectives on technology and communication as emergent properties of culture systems that express core attitudes relating to various social contexts. Postmodern theorists such as Poster (2001) argue that cyberspace requires a new and different social and cultural theory that takes into account the specific qualities of cyberspace. In cyberspace, he argues, individuals and groups are unable to position and identify their ethnicity via historically established relations to physical phenomena such as the body, land, physical environment, and social-political structures. Instead, cultural identities in cyberspace, constructed in new and fluid ways, are a temporary link to a rapidly evolving and creative process of cultural construction.

Understanding Cultural Influences by Examining the Interface

Not surprisingly, this intensive focus on defining the culture of online communicators has tended to distract attention from the original question: what is happening at the intercultural interface? Indeed, Thornton (1988) has argued, "Part of the problem that besets our current efforts to understand culture is the desire to define it" (p. 26). Abdelnour-Nocera (2002) has proposed a number of theoretical perspectives that he believes may be more effective for carefully examining the role the computer interface plays as a culturally meaningful tool in intercultural interaction. A situated-action perspective, already commonly referenced in the HCI literature and based on Suchman's work (1987), may, he suggests, place more useful emphasis on the unique context and circumstances of intercultural communication events online because it emphasizes the interrelationship between an action (here, communication) and its context of performance. Context here is not simply a vague backdrop to communication; rather, it is a complex constructed by users as they make selective sense of their environment of interaction and of each other based on their goals and resources (and skills).

Alternatively, what Abdelnour-Nocera (2002) calls the semiotic approach focuses on technological aspects of the online communicative interface that are subject to cultural interpretation (here, icons, headings, text, and pictures). The context of use and user culture are considered, but are understood to be the source of meaning-making, interpretive strategies that must be matched in efforts to construct meaningful technological interfaces for communication.

Also emphasizing the importance of situation and context, Bucher (2002) proposes that a more meaningful approach to understanding the relationship between Internet communication and culture must examine the role of an interactive audience, and especially their communicative and intercultural competence (although he does not define the latter). Bucher also explores the phenomenon of trust development in cyberspace communications as a key feature of the online context. The disembodied nature of communication in cyberspace, says Bucher, means a loss of control—of time, of space, of content, of communicators—and a sensation of informational risk that can only be overcome through trust. Trust is, however, at the disposal of the audience or listener, not the speaker.

Implications for Interface Design

Operationalizing our fragmentary understanding of computer-mediated intercultural communication processes is a challenge. Neat predictive formulas for cultural behaviours in online environments are attractive to managers and corporations because they seem to facilitate the easy modification of platforms, interfaces, and environments for different categories of users. And indeed, the technology-internationalization and -localization discourse continues to be dominated by the Hofstede model (Abdelnour-Nocera, 2002). Examples of this localization approach include Abdat and Pervan's (2000) recommendations for design elements that minimize the communicative challenges of high-power distance in Indonesian groups, Heaton's (1998b) commentary on technology and design preferences of Japanese and Scandinavian users, Onibere, Morgan, Busang, and Mpoeleng's (2001) unsuccessful attempt to identify localized interface design elements more attractive to Botswana users, Turk and Trees' (1998) methodology for designing culturally appropriate communication technologies for indigenous Australian populations, and Evers' (1998) portfolio of culturally appropriate metaphors for human-computer interface design.

Such design approaches, founded on essentialist classification theories of culture, though attractive, remain problematic because their foundational theories are problematic, as discussed. They do not offer any practical assistance to designers constructing online environments for culturally heterogeneous groups of cyberspace communicators, although such groups are rapidly becoming the cyberspace norm.

FUTURE TRENDS

Unfortunately, intercultural theories that highlight cultural fluidity, the role of context, and the intercultural negotiation of meaning are much more difficult to incorporate into approaches to interface design. Nevertheless, a few groups have initiated projects intended to make human-computer interfaces more culturally inclusive (rather than culturally specific). Foremost are Bourges-Waldegg and Scrivener (1998, 2000) who have developed an approach they call "meaning in mediated action" (MIMA), which builds on the semiotic perspective discussed by Abdelnour-Nocera (2002). Rather than attempting to design for individual cultures, this approach hopes to help designers understand context, representations, and meaning, and allow them to design interfaces that are more generally accessible.

One of the few proposed design strategies that makes use of situated-action perspectives (see above) and tries to accommodate the great variability in user context, changing user requirements, and fast-paced technology evolution is a scenario-based design approach proposed by Carroll (2000). Although this design approach does not address Internet-mediated intercultural communication issues explicitly, Carroll explains that it is a methodological tradition that "seeks to *exploit* the complexity and fluidity of design by trying to learn more about the structure and dynamics of the problem domain" rather than trying to control this complexity (p. 44).

As the design and evaluation of interfaces and environments for intercultural communication continue, it will also be important to explore whether different communicative technologies may actually constitute different kinds of cyberculture or mediate different kinds of intercultural exchange. In current literature, studies examining intercultural e-mail exchange predominate, although conclusions and implications are often extrapolated to all Internet and communication technologies. A smaller number of studies investigates intercultural communication in asynchronous forums and discussion boards, in group-conferencing platforms, in newsgroups, and via synchronous communications technologies. Even fewer discuss cultural implications for other human-Internet interfaces such as Web sites and graphics. As yet, little or no analysis exists of intercultural communication via current cutting-edge communication platforms such as Weblogs and wikis (for detailed references, see Macfadyen et al., 2004).

CONCLUSION

Ironically enough, for an endeavour dedicated to exploring the lived reality of cultural diversity and dynamic change, a focus on defining culture itself may actually be inhibiting our ability to examine and understand the real processes of intercultural exchange that occur in the virtual world of cyberspace. Instead, we may need to look beyond classification systems for individual communicators to the processes that occur at their interface; if we are lucky, more information about cultural identity may then become visible at the edge of our vision. Hewling (2004) makes use of the well-known optical-illusion image of two mirrored faces in profile—which can also be seen as a central goblet—to argue for another way of seeing intercultural encounters in cyberspace. She suggests that the use of Hofstedean ideas of culture can result in a focus on only the faces in the picture, while the more critical field of exploration is the mysterious space in between. Is it a goblet? Or, is it another kind of collectively shaped space? Bringing Street's (1993) ideas to the worlds of cyberspace, Raybourn, Kings, and Davies (2003) have suggested that intercultural interaction online involves the construction of a third culture: a process, not an entity in itself. While this culture (or multiplicity of situational cultures) may be influenced by cultures that communicators bring to each exchange, more insight may be gained by investigating the evolving processes and tools that individuals invoke and employ to negotiate and represent personal and group identity, and collective (communicative) construction of meaning. What role does language, literacy, and linguistic competence play? Does this creative process occur differently in text-only and media-rich online environments? Answers to such questions will surely be relevant to HCI practitioners because they will illuminate the need to stop viewing human-computer interfaces as tools for information exchange and will expose the persistent shortcomings of design approaches that attempt to match contextual features with supposedly static cultural preferences. These interfaces will more importantly be viewed as the supporting framework that may foster or hinder the creative communication processes in cyberspace that are the foundation of successful intercultural communication.

REFERENCES

Abdat, S., & Pervan, G. P. (2000). Reducing the negative effects of power distance during asynchronous pre-meeting without using anonymity in Indonesian culture. *Proceedings of Cultural Attitudes*

Towards Technology and Communication 2000, (pp. 209-215).

Abdelnour-Nocera, J. (2002). Context and culture in human computer interaction: "Usable" does not mean "senseful." *Proceedings of Cultural Attitudes Towards Technology and Communication 2002* (pp. 505-524).

Asad, T. (1980). Anthropology and the analysis of ideology. *Man, 14*, 607-627.

Belz, J. A. (2003). Linguistic perspectives on the development of intercultural competence in telecollaboration. *Language Learning and Technology, 7*(2), 68-117. Retrieved from http://llt.msu.edu/vol7num2/belz/

Benson, S., & Standing, C. (2000). A consideration of culture in national IT and e-commerce plans. *Proceedings of Cultural Attitudes Towards Technology and Communication 2000* (pp. 105-124).

Bourges-Waldegg, P., & Scrivener, S. A. R. (1998). Meaning: The central issue in cross-cultural HCI design. *Interacting with Computers, 9*(3), 287-309.

Bourges-Waldegg, P., & Scrivener, S. A. R. (2000). Applying and testing an approach to design for culturally diverse user groups. *Interacting with Computers, 13*(2), 111-126.

Bucher, H.-J. (2002). The power of the audience: Interculturality, interactivity and trust in Internet-communication. Theory, research design and empirical results. *Proceedings of Cultural Attitudes Towards Technology and Communication 2002* (pp. 3-14).

Buragga, K. A. (2002). An investigation of the relationship between cultural context and the use of computer-based information systems. *Proceedings of Cultural Attitudes Towards Technology and Communication 2002* (pp. 467-483).

Byram, M. (1997). *Teaching and assessing intercultural communicative competence*. Clevedon, UK: Multilingual Matters.

Carroll, J. M. (2000). Five reasons for scenario-based design. *Interacting with Computers, 13*(1), 43-60.

Chase, M., Macfadyen, L. P., Reeder, K., & Roche, J. (2002). Intercultural challenges in networked learning: Hard technologies meet soft skills. *First Monday, 7*(8). Retrieved from http://firstmonday.org/issues/issue7_8/chase/index.html

Choi, J. H., & Danowski, J. (2002). Making a global community on the Net: Global village or global metropolis? A network analysis of Usenet newsgroups. *Journal of Computer-Mediated Communication, 7*(3). Retrieved from http://www.ascusc.org/jcmc/vol7/issue3/choi.html

Evers, V. (1998). Designing interfaces for culturally diverse users. *Proceedings of Cultural Attitudes Towards Technology and Communication 1998* (pp. 261-262). Retrieved from http://www.it.murdoch.edu.au/catac/catac98/pdf/22_evers.pdf

Goodfellow, R. (2004). Online literacies and learning: Operational, cultural and critical dimensions. *Language & Education, 18*(5), 377-399.

Gunawardena, C. N., Nolla, A. C., Wilson, P. L., Lopez-Islas, J. R., Ramírez-Angel, N., & Megchun-Alpízar, R. M. (2001). A cross-cultural study of group process and development in online conferences. *Distance Education, 22*(1), 85-121.

Gunawardena, C. N., Walsh, S. L., Reddinger, L., Gregory, E., Lake, Y., & Davies, A. (2002). Negotiating "face" in a non-face-to-face learning environment. *Proceedings of Cultural Attitudes Towards Technology and Communication 2002* (pp. 89-106).

Hall, E. (1966). *The hidden dimension*. New York: Doubleday & Company.

Halliday, M. A. K. (1978). *Language as social semiotic: The social interpretation of language and meaning*. London: Edward Arnold.

Halliday, M. A. K. (1994). *An introduction to functional grammar* (2nd ed.). London: Edward Arnold.

Hampden-Turner, C., & Trompenaars, F. (2000). *Building cross-cultural competence: How to create wealth from conflicting values*. Chichester, UK: John Wiley & Sons.

Heaton, L. (1998a). Preserving communication context: Virtual workspace and interpersonal space in Japanese CSCW. *Proceedings of Cultural Attitudes Towards Technology and Communication 1998* (pp. 207-230). Retrieved from http://www.it.murdoch.edu.au/catac/catac98/pdf/18_heaton.pdf

Heaton, L. (1998b). Talking heads vs. virtual workspaces: A comparison of design across cultures. *Journal of Information Technology, 13*(4), 259-272.

Hewling, A. (2004). Foregrounding the goblet. *Proceedings of Cultural Attitudes Towards Technology and Communication 2004* (pp. 543-547).

Hofstede, G. (1980). *Culture's consequences: International differences in work-related values.* Beverly Hills, CA: Sage Publications.

Hofstede, G. (1991). *Cultures and organizations: Software of the mind.* London: McGraw-Hill.

Kim, K.-J., & Bonk, C. J. (2002). Cross-cultural comparisons of online collaboration. *Journal of Computer-Mediated Communication, 8*(1). Retrieved from http://www.ascusc.org/jcmc/vol8/issue1/kimandbonk.html

Lévy, P. (2001). *Cyberculture.* Minneapolis: University of Minnesota Press.

Macfadyen, L. P. (2004). The culture(s) of cyberspace. In C. Ghaoui (Ed.) *Encyclopedia of Human Computer Interaction.* Hershey, PA: Idea Group Reference.

Macfadyen, L. P., Roche, J., & Doff, S. (2006). *Communicating across cultures in cyberspace: A bibliographical review of online intercultural communication.* Hamburg, Germany: Lit-Verlag.

Maitland, C. (1998). Global diffusion of interactive networks: The impact of culture. *Proceedings of Cultural Attitudes Towards Technology and Communication 1998* (pp. 268-286). Retrieved from http://www.it.murdoch.edu.au/~sudweeks/catac98/pdf/24_maitland.pdf

Marcus, A., & Gould, E. W. (2000). Cultural dimensions and global Web user-interface design: What? So what? Now what? *Proceedings of the 6th Conference on Human Factors and the Web.* Retrieved from http://www.amanda.com/resources/hfweb2000/hfweb00.marcus.html

McSweeney, B. (2002). Hofstede's model of national cultural differences and their consequences: A triumph of faith, a failure of analysis. *Human Relations, 55*(1), 89-118.

O'Dowd, R. (2003). Understanding the "other side": Intercultural learning in a Spanish-English e-mail exchange. *Language Learning and Technology, 7*(2), 118-144. Retrieved from http://llt.msu.edu/vol7num2/odowd/default.html

Onibere, E. A., Morgan, S., Busang, E. M., & Mpoeleng, D. (2001). Human-computer interface design issues for a multi-cultural and multi-lingual English speaking country: Botswana. *Interacting with Computers, 13*(4), 497-512.

Poster, M. (2001). *What's the matter with the Internet?* Minneapolis: University of Minnesota Press.

Rahmati, N. (2000). The impact of cultural values on computer-mediated group work. *Proceedings of Cultural Attitudes Towards Technology and Communication 2000,* (pp. 257-274).

Raybourn, E. M., Kings, N., & Davies, J. (2003). Adding cultural signposts in adaptive community-based virtual environments. *Interacting with Computers, 15*(1), 91-107.

Reeder, K., Macfadyen, L. P., Roche, J., & Chase, M. (2004). Negotiating culture in cyberspace: Participation patterns and problematics. *Language Learning and Technology, 8*(2), 88-105. Retrieved from http://llt.msu.edu/vol8num2/reeder/default.html

Street, B. V. (1993). Culture is a verb: Anthropological aspects of language and cultural process. In D. Graddol, L. Thompson, & M. Byram (Eds.), *Language and culture* (pp. 23-43). Clevedon, UK: BAAL and Multilingual Matters.

Suchman, L. (1987). *Plans and situated actions.* Cambridge, UK: Cambridge University Press.

Thanasankit, T., & Corbitt, B. J. (2000). Thai culture and communication of decision making processes in requirements engineering. *Proceedings of Cultural Attitudes Towards Technology and Communication 2000* (pp. 217-242).

Thorne, S. (2003). Artifacts and cultures-of-use in intercultural communication. *Language Learning and Technology, 7*(2), 38-67. Retrieved from http://llt.msu.edu/vol7num2/thorne/default.html

Thornton, R. (1988). Culture: A contemporary definition. In E. Boonzaier & J. Sharp (Eds.), *South African keywords: Uses and abuses of political concepts* (pp. 17-28). Cape Town, South Africa: David Philip.

Ting-Toomey, S. (1988). Intercultural conflict style: A face-negotiation theory. In Y. Y. Kim & W. B. Gudykunst (Eds.), *Theories in intercultural communication* (pp. 213-235). Newsbury Park, CA: Sage.

Tully, P. (1998). Cross-cultural issues affecting information technology use in logistics. *Proceedings of Cultural Attitudes Towards Technology and Communication 1998* (pp. 317-320). Retrieved from http://www.it.murdoch.edu.au/catac/catac98/pdf/26_tully.pdf

Turk, A., & Trees, K. (1998). Culture and participation in development of CMC. *Proceedings of Cultural Attitudes Towards Communication and Technology 1998* (pp. 263-267). Retrieved from http://www.it.murdoch.edu.au/catac/catac98/pdf/23_turk.pdf

Williams, R. (1983). *Keywords: A vocabulary of culture and society.* Glasgow, UK: William Collins & Sons.

Wilson, M. S. (2001). Cultural considerations in online instruction and learning. *Distance Education, 22*(1), 52-64.

Yetim, F. (2001). A meta-communication model for structuring intercultural communication action patterns. *SIGGROUP Bulletin, 22*(2), 16-20.

KEY TERMS

Culture: Multiple definitions exist, including essentialist models that focus on shared patterns of learned values, beliefs, and behaviours, and social-constructivist views that emphasize culture as a shared system of problem solving or collective meaning making. The key to the understanding of online cultures—for which communication is as yet dominated by text—may be definitions of culture that emphasize the intimate and reciprocal relationship between culture and language.

Cyberculture: As a social space in which human beings interact and communicate, cyberspace can be assumed to possess an evolving culture or set of cultures (cybercultures) that may encompass beliefs, practices, attitudes, modes of thought, behaviours, and values.

Cyberspace: While the Internet refers more explicitly to the technological infrastructure of networked computers that make worldwide digital communications possible, cyberspace is understood as the virtual places in which human beings can communicate with each other, and that are made possible by Internet technologies. Lévy (2001) characterizes cyberspace as "not only the material infrastructure of digital communications but…the oceanic universe of information it holds, as well as the human beings who navigate and nourish that infrastructure."

Essentialism: The view that some properties are necessary properties of the object to which they belong. In the context of this article, essentialism implies a belief that an individual's cultural identity (nationality, ethnicity, race, class, etc.) determines and predicts that individual's values, communicative preferences, and behaviours.

Intercultural: In contrast to multicultural (which simply describes the heterogeneous cultural identities of a group), cross-cultural (which implies some kind of opposition), or transcultural (which has been used to suggest a cultural transition), intercultural is used to describe the creative interactive interface that is constructed and shared by communicating individuals from different cultural backgrounds.

Technoliteracy: A shorthand term referring to one's competence level, skill, and comfort with technology.

Knowledge Management as an E-Learning Tool

K

Javier Andrade
University of A Coruña, Spain

Juan Ares
University of A Coruña, Spain

Rafael García
University of A Coruña, Spain

Santiago Rodríguez
University of A Coruña, Spain

María Seoane
University of A Coruña, Spain

Sonia Suárez
University of A Coruña, Spain

INTRODUCTION

The goal of educational methods is to allow the pupil the acquisition of knowledge. Even so, the way in which this aim is pursued originates four different currents of methods sorted by two criteria: (1) who leads the educational process and (2) requirement of pupil physical attendance. Regarding the former criterion, the process may be conducted either by the teacher—Teaching-Oriented Process—or by the pupil—Learning-Oriented Process. Obviously, both processes have the same aim: the interiorization and comprehension of knowledge by the pupil. But the difference between them is based on the distinctive procedure followed in each case to achieve the common goal. Regarding the second criterion, the methods may or may not require pupil attendance.

Bearing in mind this classification, four different types of educational methods could be described:

1. **Teaching Method:** This includes the already known classic educational methods, the Conductivity Theory (Good & Brophy, 1990) being the foremost one. This method is characterized by the fact that the teacher has the heavier role during education—the transmission of knowledge.
2. **E-Teaching Method:** This second type comes from the expansion and popularity of communication networks, especially the Internet. This method brings the teacher to the physical location of the pupil; one of its most important representative elements is the videoconference.
3. **Learning Method:** This constitutes a new vision of the educational process, since the teacher acts as a guide and reinforcement for the pupil. The educational process has the heavier role in this method. In other words, the teacher creates a need for learning and afterwards provides the pupil with the necessary means in order to fill these created requests. Piaget Constructionist Theory is one of the most remarkable methods for this (Piaget, 1972, 1998).
4. **E-Learning Method:** This method is supported both by learning methods and by the expansion of communication networks in order to facilitate access to education with no physical or temporal dependence from pupil or teacher. As in learning methods, the pupil, not the teacher, is the one who sets the learning rhythm.

Table 1. Functional summary of main e-learning applications

	Moodle	Ilias	ATutor	WebCT	BlackBoard	QSTutor
Course Manager	√	√	√	√	√	√
Content Manager	√	√	√	√	√	√
Complementary Readings	√			√	√	
FAQs				√	√	√
Notebook	√	√	√		√	
Search		√	√			
Chat	√	√	√	√	√	√
Videoconference						√
Forum	√	√	√	√	√	√
E-mail		√	√			√

Each of these types of educational methods may be suitable for a given context, the e-learning systems being the preferred ones in the following circumstances:

1. When looking for a no-attendance-required educational method.
2. When the pupil, not the teacher, wants to set the educational rhythm. This choice might be based on several reasons, ranging from the need of adaptation to the availability of a pupil (i.e., to achieve temporal independence), to the consideration of learning as a more accurate approach than teaching, bearing in mind a particular application context (Pedreira, 2003).
3. When the knowledge to be transmitted is to be accessible to a high number of pupils. In teaching methods, the teacher is the one who transmits knowledge and supervises the pupils; therefore, the quality of the education is influenced by the number of pupils. Nevertheless, in e-learning the core of the educational process is the relationship between pupil and didactical material, with the teacher acting as a consultant. In this way, a teacher could pay attention to a higher number of pupils without causing any damage to the quality of the education.

This article is focused both on the study of e-learning systems and on the application procedure for this new discipline. The Background section is a brief discussion regarding currently used e-learning systems and their points of view. The Main Focus of the Article section suggests a new focus for this type of system in an attempt to solve some shortages detected in already existing systems. The Future Trends section introduces some guidelines that may conduct the evolution of this discipline in the future. Finally, the Conclusion section presents the conclusion obtained.

BACKGROUND

Nowadays, there are numerous applications that are self-named as e-learning tools or systems. Table 1 shows the results of the study regarding main identified applications such as Moodle (http://moodle.org), Ilias (http://www.ilias.uni-koeln.de/ios/index-e.html), ATutor (http://www.atutor.ca/atutor), WebCT (http://www.webct.com), BlackBoard (http://www.blackboard.net), and QSTutor (http://www.qsmedia.es). Each of these applications has been analyzed from the point of view of the functionality to which it gives support. As can be noticed in the table, these applications are based mainly on document management and provide a wide range of communication possibilities (especially forum and chat) and agendas.

Nevertheless, and despite the increasing appearance of e-learning applications, the point of view of this discipline currently is being discussed. This is due to the fact that, despite the important conceptual differences that e-learning has with classical teaching methods, the developers of that type of application usually operate with the same frame of mind as with classical methods; that is, an editorial mindset. In other words, it is common to find the situation in which an e-learning application merely is reduced to a simple digitalization and distribution of the same contents used in classical teaching (Martínez, 2003). In this scenario, pupils read content pages that have been structured in an analogous way to student books or traditional class notes, using multimedia applications with self-evaluating exercises in order to verify the assimilation of what previously has been read.

The systems developed in this way, and which should not be considered as e-learning but as e-reading (Martínez, 2002), are inappropriate, since technology must not be seen as a purpose in itself but as a means that eases the access to education. Then again, the docent material that has been elaborated as described needs attendance to an explanative lesson; therefore, it is not enough for pupils to auto-regulate their apprenticeship. All that has been said should induce a change in the existing orientation of this discipline, paying more attention instead to the elaboration and structuration of docent material.

MAIN FOCUS OF THE ARTICLE

In this situation, the present work intends to palliate the shortages previously identified by means of the definition of the basic structure; that is, any docent material should have to accurately achieve the goal of any e-learning application—the appropriate transmission of knowledge. In order to obtain this structure, the selected route has been the knowledge management (KM) discipline; one of its main purposes is the determination of knowledge representation intended for easier assimilation.

The following subsections detail not only the proposed structure for development of an e-learning application but also the defined ontology that should be used to structure one of its key aspects: the knowledge base.

Proposed Structure for E-Learning Systems

Three key elements may be identified at any educational process: pupil, teacher, and contents/docent material. E-learning is not an exception to this, therefore, any system that may give support to this discipline should be structured with the same triangular basis (Friss, 2003) by means of the definition of modules regarding the three described factors. All these modules should be supported by a communication module.

Pupil Information Module

This module provides pupils and teacher with information regarding the former. More specifically, this information should include for every pupil not only his or her personal information (especially contact data) together with an academic and/or professional profile, but also more dynamic aspects such as current courses and levels achieved, along with the evolution related to what was expected and a reflection of problems that might have been aroused.

Teacher Information Module

This module makes teacher information available for pupils. In this way, a given pupil should know not only how to contact a specific teacher but also the topics in which this teacher could help him or her.

Contents Module

The different resources available for pupils in order to acquire, consolidate, or increase their knowledge are contained in this module. The organization proposed for this module is based on the three basic pillars of KM discipline for the structure setting of the knowledge to be transmitted: knowledge base, lessons learned, and yellow pages (Andrade et al., 2003). These pillars, after subtle adaptations, are perfectly valid for the organization of this module:

1. The submodule named as knowledge base constitutes the central nucleus, not only of this module but also of e-learning, since it is the container for each course-specific content. Given its importance, this aspect will be approached later on.
2. The lessons learned (Van Heijst, Van der Spek & Kruizinga, 1997) submodule contains the experiences of both pupils and teacher regarding knowledge base. It is important to point out not only the positive experiences, like hints for a better solution, but also the negative ones, such as frequent mistakes during the application of knowledge.

Figure 1. Structure of e-learning systems

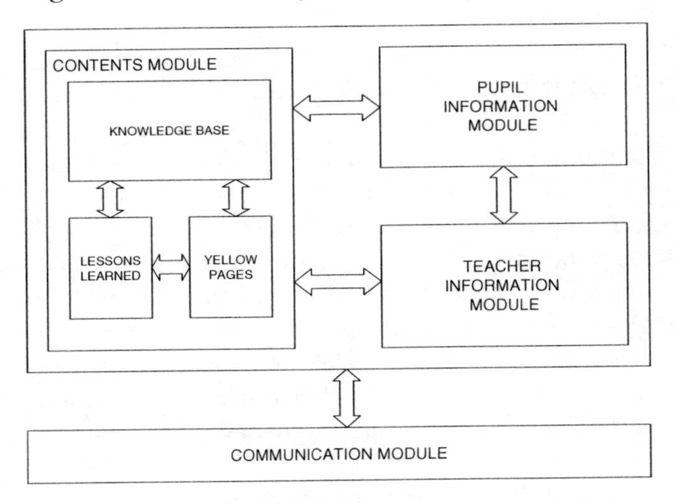

3. The yellow pages (Davenport & Prusak, 2000) help the pupil to identify the most suitable teacher in order to solve a particular question as well as to distinguish the appropriate resources (books, class notes, etc.) for digging out a specific topic.

Communication Module

This module, as shown in Figure 1, gives support to the previous ones. Its main task is giving users access to the e-learning system. By this means, teachers and pupils have access not only to the previous modules but also to communication and collaboration among the different system users. It is important to point out that this communication should not be limited to that between a pupil and a teacher, but in some domains, it also could be interesting to allow pupils and even teachers to intercommunicate.

Ontology for the Definition of Knowledge Base

As previously mentioned, the knowledge base acts like a storage space and a source of specific contents for the pupils to obtain the desired knowledge. The type of organization of these contents should allow significant learning in which pupils should be able to assimilate, conceptualize, and apply the acquired knowledge to new environments (Ausubel, David, Novak, & Hanesian, 1978; Michael, 2001).

In order to achieve this type of structure, the first step is the partition of the course into topics and subtopics for the identification of the specific lessons. This division will generate a subject tree that represents a useful tool for the pupil, so that he or she may understand the global structure of the course.

The following step should be the description of those lessons that have been identified. To achieve this, it is proposed that every lesson should be preceded by a brief introduction. Having done this, it is suggested to generate a genuine need of learning into the pupil, aiming for an increased receptiveness of the contents. Once the lesson has been displayed, the pupil should be guided with regard to the practical application of previously acquired theoretical knowledge. As a final stage, a verification of the evolution might be performed by means of an evaluation of the acquired knowledge.

Figure 2 shows the proposed ontology for definition and implementation of the knowledge base, supported by these components. As can be noticed, on some occasions, one or more components might not be applicable for a specific lesson. The necessary distinction of the relevant aspects for each case should be performed by the teacher. These components are detailed next.

Introduction of the Lesson

The aim of this component is to show the pupil a global vision of the lesson that is going to start. To achieve this, not only the purposes of its perfor-

Figure 2. Ontology for the development and implementation of a knowledge base

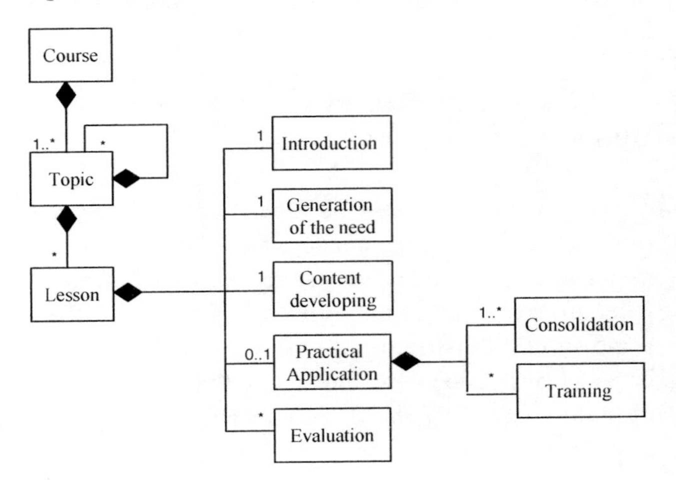

mance should be outlined clearly, but also the contents have to be described briefly. In addition, it should be made noticeable that lessons are codependent regarding themselves and/or the rest of the basic knowledge. It is also important to specify clearly what the basic knowledge requirements are for a successful approach.

Generation of the Need

The motivation of the pupil constitutes a key aspect of learning. A good strategy for achieving this is to stimulate the pupil's understanding of the usefulness of the knowledge that he or she is going to acquire (Wilkerson & Gijselaers, 1996). This purpose might be accomplished by generating need exercises, which consisting of a problem proposal whose method of solving is not known by the learner.

Content Developing

This is the component where the knowledge included in the lesson is maintained and transmitted. That knowledge may be dynamic or static, both constituting the so-called functional taxonomy of knowledge, which more specifically makes a distinction between them (Andrade et al., 2004a):

1. **Dynamic Knowledge:** Knowledge related to the behavior that exists in the domain; that is, functionality, action, processes, and control. This level can be divided into two sublevels:
   a. **Strategic:** Specifies what to do and where as well as when, in what order, and why to do it. This knowledge handles the functional decomposition of each operation in its constituent steps as well as the order in which they have to be undertaken.
   b. **Tactical:** Specifies how to do the tasks and under what circumstances they have to be done. This type of knowledge is associated with the execution process for each strategic step of the latest level.
2. **Static Knowledge:** Conforms the structural or declarative domain knowledge and specifies the elements—concepts, relations, and properties—that are handled when carrying out the tasks (i.e., handled by tactical knowledge) and the elements that are the basis for the decisions (i.e., implied in the decisions of strategic knowledge).

Therefore, a lesson should give support to the types of knowledge that have been identified. With this intention and given the different characteristics of each of them, they should be described on the basis of different parameters. In this way, Table 2 shows in schematic fashion those aspects that should be kept in mind when describing a knowledge asset, depending on the type of considered knowledge.

This taxonomy has been used by the authors not only for conceptualization and modeling of problems

Table 2. Aspects to be considered when describing a type of knowledge

Level	Characteristics to Consider	
Strategic	Functional decomposition of each operation in its constituent substeps.	
	Execution order for the identified steps for operation fulfilling.	
	Pre-conditions and post-conditions for the execution of every identified step.	
	Entries and exits of each identified step.	
	Responsible for each step.	
Tactical	Operation mode—algorithm, mathematical expression, or inference—of each identified step.	
	Limiting aspects in its application.	
	Elements—concepts, relations, and properties—that it handles and produces.	
Declarative	Concepts	Relevant properties.
		Relations in which it participates.
	Relations	Properties, concepts, and/or relations participating in the relation.
		Limiting aspects in its application.
	Properties	Type of value.
		Possible values that it can take.
		Source that provides its value(s).
		Limiting aspects in its application.

(Andrade et al., 2004b) but also for KM systems definition and implementation (Andrade et al., 2003). As a result, it has been concluded that the organization of knowledge, based on this taxonomy, facilitates its transmission, understanding, and assimilation. This statement is supported by the fact that functional taxonomy is consonant with human mindset, and therefore, it is sensitive to people (Andrade et al., 2004a).

Practical Application

Once the pupil has acquired the required knowledge assets, he or she should put them into practice for proper interiorization and assimilation. This task will be performed in two phases: consolidation and training.

The consolidation phase intends to reinforce theory by means of practice. With this aim, a group of examples will be made available for the pupil, who will be guided through his or her resolution by means of the rationalization of every decision made.

During the training phase, the aim is for the pupil to apply the resolution method straightforward by the use of a group of exercises increasingly complex.

It should be highlighted that when exercises are more similar to real circumstances, the results obtained will be enhanced; consequently, exercises and examples should be as realistic as possible.

Evaluation

The evaluation of the acquired knowledge will provide useful information for both pupils and teachers who would be able to verify a pupil's evolution regarding what was expected and whether pre-established levels have been accomplished or not. Likewise, the evolution would allow the detection of any learning problem in order to handle it appropriately.

FUTURE TRENDS

As previously mentioned in this article, e-learning discipline has been shown as an inappropriate approach; it is the mere digitalization and publication of the same docent material commonly used at attendance lessons, making technology not a tool but a goal in itself. The perception of this situation should induce a change when dealing with the development of e-learning systems in terms of fitting the definition and structuration of the docent material to the needs that human beings might present. In other words, as human computer interaction (HCI) emphasizes, human beings should not be servants of but served by technology.

Shown here is a first attempt toward the attainment of this objective. A lot of work remains to be done; therefore, it is predicted that future investigations will be focused on a more exhaustive definition regarding the way in which docent material should be elaborated in order to be more easily assimilated by pupils.

CONCLUSION

The present work has catalogued the different existing educational methods that attend both to who may lead the process and to whether physical attendance of the pupil is required or not. This classification has allowed, for every method and especially for e-learning, the identification of their inherent particulars.

Nonetheless, most of the so-called e-learning systems do not properly support the intrinsic characteristics of these types of systems, since they merely provide an electronic format for the docent material of classical teaching.

KM techniques have been used with the aim of providing an answer to this situation. This discipline tries to find the optimal strategies for the representation and transmission of knowledge so that its latter comprehension might be facilitated. Following this, a basic structure for e-learning systems was defined using modules and submodules. Similarly, after the identification of the knowledge base as one of the key aspects of the mentioned structure, a specific ontology was described for its definition and implementation.

Finally, it should be mentioned that the attainment of auto-content and auto-explanative docent material for an easier acquisition of knowledge could be achieved by the use of the defined ontology.

ACKNOWLEDGMENT

We would like to thank the Dirección Xeral de Investigación e Desenvolvemento da Xunta de Galicia (Autonomous Government of Galicia), for funding project PGIDIT03PXIA10501PR.

REFERENCES

Andrade, J., Ares, J., García, R., Pazos, J., Rodríguez, S., & Silva, A. (2004a). A methodological framework for generic conceptualisation: Problem-sensitivity in software engineering. *Information and Software Technology, 46*(10), 635-649.

Andrade, J., Ares, J., García, R., Pazos, J., Rodríguez, S., & Silva, A. (2004b). A methodological framework for viewpoint-oriented conceptual modeling. *IEEE Transactions on Software Engineering, 30*(5), 282-294.

Andrade, J., Ares, J., García, R., Rodríguez, S., Silva, A., & Suárez, S. (2003). Knowledge management systems development: A roadmap. In V. Palade, R.J. Howlett, & L. Jain (Eds.), *Knowledge-based intelligent information and engineering systems* (pp. 1008-1015). Berlin: Springer-Verlag.

Ausubel, D., David P., Novak, J.D., & Hanesian, H. (1978). *Educational psychology: A cognitive view* (2nd ed.). New York: Holt, Rinehart, and Winston.

Davenport, T.H., & Prusak, L. (2000). *Working knowledge: How organizations manage what they know.* Boston: Harvard Business School Press.

Friss de Kereki, I. (2003). Use of ontologies in a learning environment model. In V. Uskov (Ed.), *Computers and advanced technology in education* (pp. 550-555). Calgary: ACTA Press.

Good, T.L., & Brophy, J.E. (1990). *Educational psychology: A realistic approach* (4th ed.). New York: Longman.

Martínez, J. (2002). *E-learning: Nuevo medio, viejas costumbres.* Retrieved August 9, 2004, from http://www.gestiondelconocimiento.com/articulos.php

Martínez, J. (2003). *Aprendizaje efectivo-e(ffective) learning.* Retrieved August 9, 2004, from http://www.gestiondelconocimiento.com/articulos.php

Michael, J. (2001). In pursuit of meaningful learning. *American Journal in Physiology, 25,* 145-158.

Pedreira, N. (2003). *Un modelo de deuteroaprendizaje virtual* (doctoral thesis). Spain: University of A Coruña.

Piaget, J. (1972). *Psychology and epistemology: Towards a theory of knowledge.* New York: Viking Press.

Piaget, J. (1998). *Jean Piaget selected works.* New York: Routledge.

Van Heijst, G., Van der Spek, R., & Kruizinga, E. (1997). Corporate memories as a tool for knowledge management. *Expert Systems with Applications, 13,* 41-54.

Wilkerson, L., & Gijselaers, W.H. (1996). *The power of problem-based learning to higher education: Theory and practice.* San Francisco: Jossey-Bass.

KEY TERMS

Education: Formation or instruction process of an individual by means of the interiorization and assimilation of new assets of knowledge and capabilities.

E-Learning: Discipline that applies current information and communications technologies to the educational field. This discipline tries to facilitate the learning process, since its methods do not depend on physical location or timing circumstances of the pupil.

Knowledge: Pragmatic level of information that provides the capability of dealing with a problem or making a decision.

Knowledge Management: Discipline that intends to provide, at its most suitable level, the accurate information and knowledge for the right people whenever they may need it and at their best convenience.

Learning: Educational process for self education or instruction using the study or the experience.

Significative Learning: Type of learning in which contents are related in a substantial and not arbitrary fashion with what the pupil already knows.

Teaching: Educational process wherein a teacher, using the transmission of knowledge, educates or instructs someone.

Knowledgeable Navigation in Virtual Environments

Pedram Sadeghian
University of Louisville, USA

Mehmed Kantardzic
University of Louisville, USA

Sherif Rashad
University of Louisville, USA

INTRODUCTION

Virtual environments provide a computer-synthesized world in which users can interact with objects, perform various activities, and navigate the environment as if they were in the real world (Sherman & Craig, 2002). Research in a variety of fields (i.e., software engineering, artificial intelligence, computer graphics, human computer interactions, electrical engineering, psychology, perceptual science) has been critical to the advancement of the design and implementation of virtual environments. Applications for virtual environments are found in various domains, including medicine, engineering, oil exploration, and the military (Burdea & Coiffet, 2003).

Despite the advances, navigation in virtual environments remains problematic for users (Darken & Sibert, 1996). Users of virtual environments, without any navigational tools, often become disoriented and have extreme difficulty completing navigational tasks (Conroy, 2001; Darken & Sibert, 1996; Dijk et al., 2003; Modjeska & Waterworth, 2000). Even simple navigational tools are not enough to prevent users from becoming lost in virtual environments. Naturally, this leads to a sense of frustration on the part of users and decreases the quality of human-computer interactions. In order to enhance the experience of users of virtual environments and to overcome the problem of disorientation, new sophisticated tools are necessary to provide navigational assistance. We propose the design and use of navigational assistance systems that use models derived through data mining to provide assistance to users. Such systems formalize the experience of previous users and make them available to new users in order to improve the quality of new users' interactions with the virtual environment.

BACKGROUND

Before explaining any navigational tool design, it is important to understand some basic definitions about navigation. Wayfinding is the cognitive element of navigation. It is the strategic element that guides movement and deals with developing and using a cognitive map. Motion, or travel, is the motoric element of navigation. Navigation consists of both wayfinding and motion (Conroy, 2001; Darken & Peterson, 2002).

Wayfinding performance is improved by the accumulation of different types of spatial knowledge. Spatial knowledge is based on three levels of information: landmark knowledge, procedural knowledge, and survey knowledge (Darken & Sibert, 1996; Elvins et al., 2001). Before defining landmark knowledge, it is important to understand that a landmark refers to a distinctive and memorable object with a specific shape, size, color, and location. Landmark knowledge refers to information about the visual features of a landmark, such as shape, size, and texture. Procedural knowledge, also known as route knowledge, is encoded as the sequence of navigational actions required to follow a particular route to a destination. Landmarks play an important role in procedural knowledge. They mark decision points along a route and help a traveler recall the procedures required to get to a destination (Steck & Mallot, 2000; Vinson 1999).

A bird's eye view of a region is referred to as survey knowledge. This type of knowledge contains spatial information about the location, orientation, and size of regional features. However, object location and interobject distances are encoded in terms of a geocentric (i.e., global) frame of reference as opposed to an egocentric (i.e., first-person) frame of reference. Landmarks also play a role in survey knowledge. They provide regional anchors with which to calibrate distances and directions (Darken & Sibert, 1996; Elvins et al., 2001).

The quality of spatial knowledge that a user has about a virtual environment determines his or her performance on a wayfinding task. Any navigational assistance provided by a tool is aimed to assist the user to gain spatial knowledge about the environment. Therefore, a key element to the success of any navigational tool is how effective it is in representing and providing spatial knowledge that is easy to understand and useful from the perspective of the user.

In the past, different navigational tools and techniques to improve wayfinding have been included in the design of virtual environments. Maps and grids have been introduced to bring legibility to virtual environments and to improve wayfinding performance (Darken & Sibert, 1996). Personal agents have been used that can interact with the user and provide verbal navigational assistance (Dijk et al., 2003). Due to their popularity, there also has been tremendous focus on the use and design of landmarks to aid in wayfinding (Elvins et al., 2001; Steck & Mallot, 2000; Vinson, 1999). The achievement of previous researchers has been significant, but the area of navigation in virtual environments still remains an open research topic.

WAYFINDING: THE DATA-MINING APPROACH

Data mining is the process of discovering previously unknown patterns, rules, and relationships from data (Han & Kamber, 2001). A Knowledgeable Navigational Assistance System (KNAS) is a tool that employs models derived from mining the navigational records of previous users in order to aid other users in successfully completing navigational tasks. For example, the navigational records of previous users may be mined to form models about common navigational mistakes made by previous users. A KNAS could be designed to use these models to help users avoid backtracking and making loops. Another example would be to derive models of frequent routes taken by previous users. These frequent routes may defy traditional criteria for route selection but have hidden advantages. A KNAS could be designed to use these models and recommend these frequent routes to users (Kantardzic et al., 2004).

The process of designing a KNAS involves three distinct steps. The first step is recording the navigational data of users. Selection of the group of users that will have their navigational behavior recorded is dependent upon the application. Ideally, this group of users should be experienced with the system (Peterson & Darken, 2000; Peterson, et al., 2000). The data that are recorded can include both spatial and non-spatial attributes pertaining to the navigation of users. Examples of recorded data could include the landmarks and objects visited, routes traversed by users, as well as information on longitude, latitude, elevation, and time (Shekhar & Huang, 2001; Huang et al., 2003).

The second step is the actual mining of data. In most cases, data will need to be preprocessed before applying the mining algorithms (Kantardzic, 2003). The data-mining process will result in models that will be used by the KNAS.

The third and final step is actual implementation of the KNAS and the corresponding interface. The interface of the KNAS needs to allow the user to issue navigational queries, and the KNAS must use the derived data-mining models in order to formulate a reply. Figure 1 depicts the three steps necessary for construction of a KNAS: recording of the navigational data, the data mining process to form models, and the implementation of the KNAS.

To better demonstrate the concepts and ideas behind a KNAS, the construction of a KNAS is discussed, which is capable of recommending routes of travel from one landmark to another landmark. As previously discussed, landmarks are an important component in wayfinding and commonly are found in various virtual environments (Steck & Mallot, 2000; Vinson, 1999). Some may argue that relying on knowledge extracted from the records of previous users to derive routes is an overhead. After all, there are already tools that can produce route directions

Figure 1. A Knowledgeable Navigational Assistance System (KNAS) guiding inexperienced users based on the experience of previous users

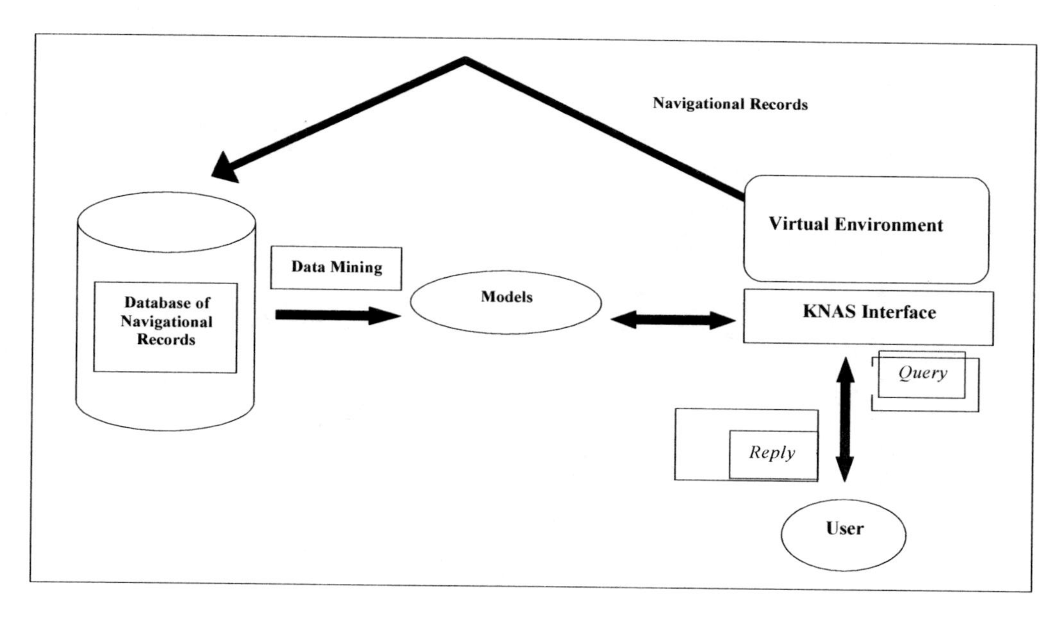

on how to get from one point to another without the data-mining process. An example would be using a tool on the Internet to get driving directions from one's home to a store. However, these tools usually take a traditional approach when producing route directions where the primary factor of consideration is distance. Therefore, the recommended route is the shortest possible route.

In a complex virtual environment, the shortest route is not necessarily the preferred route, because there are often other criteria included in defining an appropriate route. These other criteria include choosing a route based on the amount of scenery, the amount of obstacles encountered along the way, the educational value associated with the route, and so forth. If additional criteria besides distance are important for users of a virtual environment, then the model derived from the data-mining process will reflect this trend (Kantardzic, et al., 2004).

Figure 2 introduces a two-dimensional map of a simple virtual city that will be used to show the operations of the proposed KNAS. The virtual city has three landmarks (L1, L2, L3) and four roads (R1, R2, R3, R4) connecting the landmarks. The first step in designing the KNAS is to record the movements of several experienced users. For the sake of simplicity, each movement is recorded simply as a sequence of landmarks and roads traversed. A sequence of a user may be *L1 R1 L2*, which means that the user started at landmark L1 and used road R1 to get to landmark L2.

The next step is the discovery of the different routes that previous users have actually taken in getting from each landmark L_i to each landmark L_j, for $i \neq j$, and i and j from 1 to the total number of landmarks (i.e., three). Since the routes are stored as sequences of symbols, this can be accomplished by using a modified algorithm for sequence mining (Soliman, 2004). Figure 3a shows the result of mining routes from a hypothetical database for the city in Figure 2. The routes are followed by the corresponding count of occurrence and support. Figure 3b shows the final model of the most frequent routes. The KNAS will associate these most frequently used routes as the recommended routes of travel.

The final step is the design of the KNAS that can use this model. For this particular KNAS, the user interface should allow the user to issue a query such as, "What are the directions for traveling from Landmark 1 to Landmark 3?" The reply would be formulated by examining the final model, as seen in Figure 3b and translating the sequence into directions. For example, a simple reply would recommend, "Start at Landmark 1, travel on Road 1, reach Landmark 2, travel on Road 3, end at Landmark 3."

Figure 2. A 2-D map of a simple virtual city with three landmarks and four roads

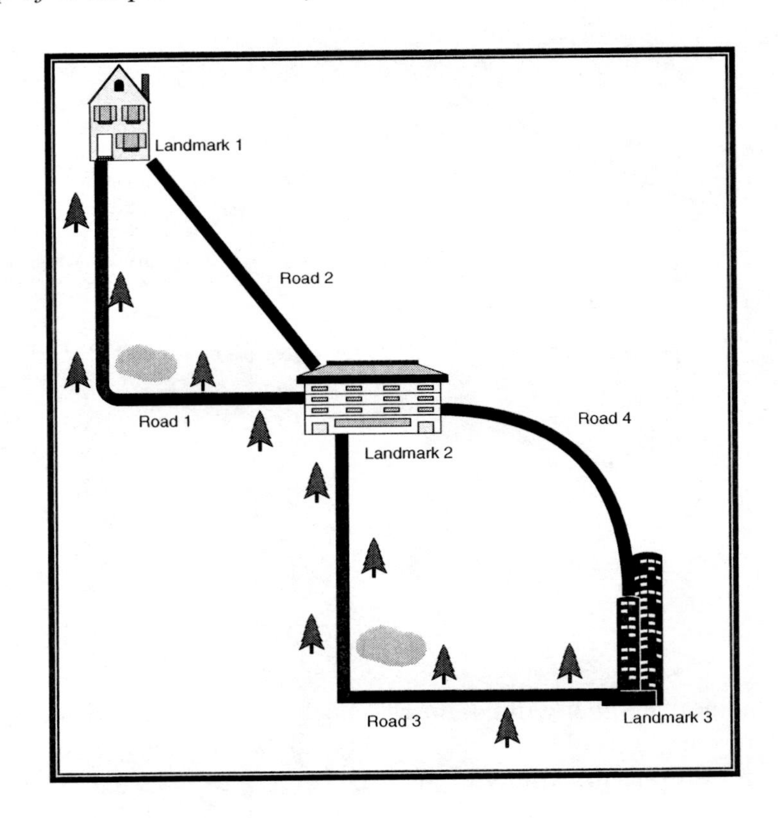

Notice that the recommended route is not the shortest possible route. Further analysis may reveal that previous users may have preferred this route, since it is more scenic.

This discussion has been kept simple for demonstrative proposes. In reality, the process of constructing a KNAS to recommend routes from one landmark to another requires much more work (Sadeghiam et al., 2005). For example, virtual environments usually do not record the movement of users as sequences of symbols but rather as three-dimensional coordinates. Pattern matching techniques must be used to translate these coordinates into sequences. More symbols would have to be introduced to account for such components as intersections and boundaries encountered during navigation. The sequences must be preprocessed in order to eliminate noisy data such as loops that correspond to backtracking and disorientation. The sequence mining algorithm must be efficient to deal with a large amount of data. When applying the sequence mining algorithm to find routes, all subsequences need to be considered. When making the final model of frequent sequences corresponding to recommended routes, more rigorous statistical methods such as confidence intervals should be used instead of simple percentages.

When the final model of frequent sequences is built, there is no guarantee that all possible combinations of landmarks will have a frequent sequence registered in the final model. This is true especially as the complexity of the environment increases. For example, a virtual environment with 500 landmarks would require the discovery of 124,750 recommended routes. This is often difficult if the amount of navigational data is limited and if rigorous statistical methods are used for determining recommended routes. Therefore, strategies must be implemented to combine several recommended routes, if the route needed is not mined explicitly. If this is not possible, the KNAS would have to resort to traditional criteria to come up with a recommendation.

Figure 3. Mining the navigational data: (a) The discovered routes; (b) Final model of frequent routes

K

(a)

Landmarks	Route	Count	Support
L1 <--> L2	L1 R1 L2	155	77.5%
	L1 R2 L2	45	22.5%
L1 <--> L3	L1 R1 L2 R3 L3	120	60%
	L1 R1 L2 R4 L3	35	17.5%
	L1 R2 L2 R3 L3	30	15%
	L1 R2 L2 R4 L3	15	7.5%
L2 <--> L3	L2 R3 L3	150	75%
	L2 R4 L3	50	25%

(b)

Landmarks	Recommended Route
L1 <--> L2	L1 R1 L2
L1 <--> L3	L1 R1 L2 R3 L3
L2 <--> L3	L2 R3 L3

FUTURE TRENDS

The technology of the KNAS is currently in its infancy. As virtual environments become increasingly more sophisticated, so does the database containing the data associated with the navigation and activities of the users. There is potentially a great deal of hidden knowledge contained within these data, and research is needed to discover ways to extract this knowledge. This is true especially for large complex virtual systems that support multiple users and distributed virtual environments. In addition, research is needed in designing a complex KNAS that takes advantage of stream mining techniques in order to update and modify the discovered models as more data is accumulated.

CONCLUSION

A common problem faced by users of virtual environments is a lack of spatial orientation. This is extremely problematic, since successful navigation is crucial to derive any benefit from most virtual environments. Numerous tools have been introduced in the past to aid in wayfinding, but much work remains in this field.

Knowledgeable navigational assistance systems offer an alternative to the traditional tools used in the past. Similar to traditional tools, a KNAS aids the user with navigational tasks, but the recommendation made by a KNAS is more likely to be viewed positively by the end user, since the recommendations have been formulated based on data of previous users. Therefore, a KNAS has the potential of enhancing the quality of human-computer interactions within virtual environments.

ACKNOWLEDGMENT

This research has been funded by the National Science Foundation (NSF) under grant #0318128.

REFERENCES

Burdea, G., & Coiffet, P. (2003). *Virtual reality technology*. NJ: John Wiley & Sons.

Conroy, R. (2001). *Spatial navigation in immersive virtual environments*. Doctoral thesis. London: University College London.

Darken, R., & Peterson, B. (2002). Spatial orientation, wayfinding, and representation. In K. Stanney (Ed.), *Handbook of virtual environments: Design, implementation, and applications*. Mahwah, NJ: Lawrence Erlbaum Associates.

Darken, R., & Sibert, J. (1996). Navigating large virtual spaces. *The International Journal of Human-Computer Interaction*, *8*(1), 49-72.

Dijk, B., Den, A., Rieks, N., & Zwiers, J. (2003). Navigation assistance in virtual worlds. *Informing Science Journal, 6*, 115-125.

Elvins, T., Nadeau, D., Schul, R., & Kirsh, D. (2001). Worldlets: 3D thumbnails for wayfinding in large virtual worlds. *Presence*, *10*(6), 565-582.

Han, J., & Kamber, M. (2001). *Data mining: Concepts and techniques*. San Francisco: Morgan Kaufmann Publishers.

Huang, Y., Xiong, H., Shekhar, S., & Pei, J. (2003). Mining confident co-location rules without a support threshold. *Proceedings of the 18th ACM Symposium on Applied Computing (SAC),* Melbourne.

Kantardzic, M. (2002). *Data mining: Concepts, models, methods, and algorithms*. Piscataway, NJ: IEEE Press.

Kantardzic, M., Rashad, S., & Sadeghian, P. (2004). Spatial navigation assistance system for large virtual environments: Data mining approach. *Proceedings of the Mathematical Methods for Learning*, Villa Geno, Como, Italy.

Modjeska, D., & Waterworth, J. (2000). Effects of desktop 3D world design on user navigation and search performance. *Proceedings of the IEEE Information Visualization,* Salt Lake City, Utah.

Peterson, B., & Darken, R. (2000). Knowledge representation as the core factor for developing computer generated skilled performers. *Proceedings of the I/ITSEC*, Orlando, Florida.

Peterson, B., Stine, J., & Darken, R. (2000). A process and representation for modeling expert navigators. *Proceedings of the 9th Conference on Computer Generated Forces and Behavioral Representation*, Orlando, Florida.

Shadeghian, P., Kantardzic, M., Lozitsky, O., & Sheta, W. (2005). Route recommendations: Navigation distance in complex virtual environments. *Proceedings of the 20th International Conference on Computers and Their Applications,* New Orleans, Louisiana.

Shekhar, S., & Huang Y. (2001). Discovering spatial co-location patterns: A summary of results. *Proceedings of the 7th International Symposium on Spatial and Temporal Databases (SSTD)*, Redondo Beach, California.

Sherman, W., & Craig, A. (2002). *Understanding virtual reality*. San Francisco: Morgan Kaufmann.

Soliman, M. (2004). *A model for mining distributed frequent sequences*. Doctoral thesis. Louisville, KY: University of Louisville.

Steck, S., & Mallot, H. (2000). The role of global and local landmarks in virtual environment navigation. *Presence, 9*(1), 69-83.

Vinson, N. G. (1999). Design guidelines for landmarks to support navigation in virtual environments. *Proceedings of the ACM Conference on Human Factors in Computing Systems*, Pittsburgh, Pennsylvania.

KEY TERMS

Data Mining: A process by which previously unknown patterns, rules, and relationships are discovered from data.

Knowledgeable Navigational Assistance System: A system that helps a user carry out navigational tasks in a virtual environment by using data mining models derived from analyzing the navigational data of previous users.

Landmark: A distinctive and memorable object.

Landmark Knowledge: A type of spatial knowledge dealing with information about visual features of landmarks.

Motion: The physical or motoric element of navigation.

Navigation: The aggregate of motion and wayfinding.

Procedural Knowledge: A type of spatial knowledge dealing with the navigational actions required in order to follow a particular route to a destination.

Survey Knowledge: A type of spatial knowledge dealing with information about location, orientation, and size of regional features.

Virtual Environment: A 3-D computer-synthesized world in which a user can navigate, interact with objects, and perform tasks.

Wayfinding: The cognitive element of navigation dealing with developing and using a cognitive map.

The Language of Cyberspace

Leah P. Macfadyen
The University of British Columbia, Canada

Sabine Doff
Ludwig Maximilians Universität, München, Germany

INTRODUCTION

Amid the many published pages of excited hyperbole regarding the potential of the Internet for human communications, one salient feature of current Internet communication technologies is frequently overlooked: the reality that Internet- and computer-mediated communications, to date, are communicative environments constructed through language (mostly text). In cyberspace, written language therefore mediates the human-computer interface as well as the human-human interface. What are the implications of the domination of Internet and computer-mediated communications by text?

Researchers from diverse disciplines—from distance educators to linguists to social scientists to postmodern philosophers—have begun to investigate this question. They ask: Who speaks online, and how? Is online language *really* text, or is it "speech"? How does culture affect the language of cyberspace? Approaching these questions from their own disciplinary perspectives, they variously position cyberlanguage as "text," as "semiotic system," as "socio-cultural discourse" or even as the medium of cultural hegemony (domination of one culture over another). These different perspectives necessarily shape their analytical and methodological approaches to investigating cyberlanguage, underlying decisions to examine, for example, the details of online text, the social contexts of cyberlanguage, and/or the social and cultural implications of English as Internet *lingua franca*. Not surprisingly, investigations of Internet communications cut across a number of pre-existing scholarly debates: on the nature and study of "discourse," on the relationships between language, technology and culture, on the meaning and significance of literacy, and on the literacy demands of new communication technologies.

BACKGROUND

The multiple meanings of the word "language"—both academic and colloquial—allow it to signify multiple phenomena in different analytical frameworks, and complicate any simple search for literature on the language of cyberspace. This article surveys the breadth of theoretical and empirical writing on the nature and significance of text, language, and literacy in Internet- and computer-mediated communications, and indicates the different theoretical approaches employed by current researchers. In particular, this article emphasizes research and theory relevant to conceptions of the Internet as a site of international and intercultural communications—the so-called "global village"—and offers some reflection on the importance of research on online language for the field of human-computer interaction.

PERSPECTIVES ON THE LANGUAGE OF CYBERSPACE

Cyberlanguage as Digital Text

Perhaps belying their perception of Internet communications as primarily *written* communication (a perspective contested by some (Collot & Belmore, 1996; Malone, 1995) a number of authors have focused on the features of digital text (and their impact on readers) as an approach to investigating cyberspace communications. A particular area of interest has been the development of hypertext, whose non-linear, non-sequential, non-hierarchical and multimodal (employing images, sound, and symbols as well as text) nature seemed to place it in stark contrast to traditional printed texts. Hypertext has been hailed as a postmodern textual reality (Burbules & Callister, 2000; Landow, 1997; Snyder, 1996), making fragmentation and complex-cross-referencing of text possible and easy. Researchers also argue that hypertext radically changes the nature of literacy, positioning the author as simply the "source," and favouring a new form of open-ended "associative" reading and thought (Burbules & Callister, 2000; Richards, 2000). One of the first researchers to focus on hypertext was Kaplan (1995), who described how it would "offer readers multiple trajectories through the textual domain" (¶ 1). Kaplan suggests that "each choice of direction a reader makes in her encounter with the emerging text, in effect, produces that text," and points out that while some hypertexts are printable, many new forms are native only to cyberspace, and have no printable equivalents. Douglas (2000), on the other hand, discusses ways in which hypertext may offer readers *less* autonomy than paper-based texts, a position supported by Harpold (2000) who argues that digital texts are "empirically fragile and ontologically inconsistent" (p. 129). Tuman (1995) offers a particularly strong critique of hypertext, which is, he argues, "ideally suited for the storing and accessing of diverse information, [but] not for sustained, critical analysis."

What are the implications of hypertext for human-computer interaction? Braga and Busnardo (2004) argue that while hypertext media encourage multimodal communications, some designers (especially novice designers) are not familiar enough with this type of communication because their own literate practices tend to be anchored in verbal language and print-based text. Construction of effective hypertext, they argue, calls for new and different approaches to organization of information and segmentation, a recognition of the challenges of screen-reading and navigation, and an understanding of the evolving conventions of different electronic contexts (Snyder, 1998) in which "electronically literate" and novice users may have different expectations.

Cyberlanguage as Semiotic System

A significant proportion of current studies of online language report on semiotics: the detailed and sometimes mechanistic features—signs and symbols—of the linguistic systems elaborated by users in a range of Internet and computer-mediated communication venues such as email, asynchronous discussion boards, computer conferencing, and synchronous "chat" platforms.

Many papers discuss evolving conventions of online communications: features, grammar, and lexicography. Most compare and contrast communications in different venues and/or with written or spoken language (almost always English). They generally conclude that online communication is an intermediate stage between oral and written modalities, and some studies (Collot & Belmore, 1996) differentiate further between features of synchronous (online) and asynchronous (offline) digital communications. A number of papers examine in particular the textual and graphical systems (such as emoticons) that users employ within online communications to add back some of the contextual features that are lost in electronic communications (for detailed references see Macfadyen, Roche, & Doff, 2004). Burbules (1997) meanwhile highlights the *hyperlink* as the key feature of digital texts, and explores some of the different roles links may play beyond their simple technical role as a shortcut: an interpretive symbol for readers, a bearer of the author's implicit ideational connections, an indicator of new juxtapositions of ideas.

Kress and Van Leeuwen (1996) propose that the shift to multimodality facilitated by digital texts necessitates a new theory of communication that is not simply based on language, but also takes into account semiotic domains and the multiplicity of semiotic resources made available by digital media. While such resources *are* significant features of online interfaces, we caution against simplistic extrapolation to interface design models that over-privilege signs and symbols as mediators of meaning in Internet- and computer-mediated interactions, and also against overly simple attempt to identify sets of "culturally-specific" signs and symbols for user groups based on essentialist notions of culture. (This phenomenon and its associated problems are discussed more extensively in Macfadyen, 2006).

New Literacies?

In a milieu where the interface is overwhelmingly dominated by text, it might seem self-evident that "literacy" be a key factor in determining the success of human-computer and human-human interaction. But what kind of "literacy" is required for Internet and computer-mediated communications? Traditionally, researchers concerned with literacy have defined it as the ability to read, write, and communicate, usually in a print-text-based environment. In the last decade, however, researchers in the field of New Literacy Studies (NLS) have challenged this perspective, and have initiated a focus on literacies as social practices, moving away from individual and cognitive-based models of literacy (Lea, 2004). A social practice model of literacy recognizes that language does not simply represent some kind of objective truth, but actually constitutes meaning in a given context. Writing and reading, it is argued, are key ways in which people negotiate meaning in particular contexts (Street, 1984).

Bringing NLS perspectives to the world of Internet and computer-mediated communication, some writers are now countering simplistic "operational" notions of electronic literacy that have tended to focus solely on "performance with the linguistic systems, procedures, tools and techniques involved in making or interpreting [digital] texts" (Goodfellow, 2004). Instead, they highlight discussions of the equal importance of "cultural" and "critical" aspects of literacy for online communications (Lankshear, Snyder, & Green, 2000), where the "cultural" dimension implies the ability to use operational skills in authentic social contexts and allow participation in social discourses, and the "critical" dimension refers to an even more sophisticated level of interaction with electronic discourses, including the ability to evaluate, critique, and redesign them.

Theoretical perspectives on "visual literacy," "digital literacy," "electronic literacy," and "computer literacy" have proliferated, with important contributions made by authors such as Warschauer (1999), Street (1984), Jones, Turner, and Street (1999), Snyder (1998) and Richards (2000). Hewling (2002) offers a detailed review of debates in this field. Next, a sampling of recent papers demonstrates that arguments tend to be polarized, positing electronic literacies either as continuous with existing human communication practices, or radically new and postmodern.

One group of research papers concentrates on the "new" skills required of users for communicative success in the online arenas of the Internet which, according to Thurstun (2000) comprise "entirely new skills and habits of thought" (p. 75). Gibbs (2000) extends this to suggest that new forms of communication are actually constructing "new forms of thinking, perceiving and recording" (p. 23). Kramarae (1999) discusses the new "visual literacy" (p. 51) required of Internet communicators, while Abdullah (1998) focuses on the differences in style and tone between electronic discourse and traditional academic prose.

On the other hand, writers such as Richards (2000) argue that dominant hypermedia models of electronic literacy are too limited, and rely too heavily on postmodern theories of representation and poststructuralist models which characterize writing and speaking as separate communication systems. Burbules (1997) similarly counters suggestions that the reading ("hyper-reading") practices required for hypertexts represent a postmodern break with old literacy traditions, reminding us that "there must be some continuity between this emergent practice and other, related practices with which we are familiar—it is reading, after all" (¶ 3). He continues by emphasizing the importance of the contexts and social relations in which reading takes place and agrees that "significant differences in those contexts and relations mean a *change* in [reading] practice" (¶ 2)—though he characterizes this change more as evolution than revolution.

Further highlighting the connections between language, literacy, and socio-cultural context, Warschauer (1999) points to the inutility of such simple binary models of "old" vs. "new" literacies, arguing that the roots of mainstream literacy lie in the "mastery of processes that are deemed valuable in particular societies, cultures and contexts" (p. 1). For this reason, he suggests, there will be no one electronic literacy, just as there is no one print literacy, and indeed, a number of studies point to the growing diversity of literacies, new and hybrid. For example, Cranny-Francis (2000) argues that users need a *complex* of literacies—old and new—to critically negotiate Internet communication. Dudfield (1999) agrees that students are increasingly engaging in what she calls "hybrid forms of literate behaviour" (¶ 1), while Schlickau (2003) specifically examines prerequisite literacies that learners need in order to make effective use of hypertext learning resources. In a more applied study, Williams and Meredith (1996) attempt to track development of electronic literacy in new Internet users.

Together, these studies highlight current thinking on which sets of skills users may need in order to communicate effectively online. A further level of debate centres, however, on theoretical claims that different kinds of technologies may demand different "kinds" of literacy—a position that Lea (2004) and others critique as excessively deterministic. Murray (2000) similarly explores claims of technologically induced socio-cultural paradigm shifts in greater detail, arguing that technology does not *impose* new literacy practices and communities, but merely facilitates social and cultural changes—including changes in literacy practices—that have already begun.

Cyberspace Contexts, Identities, and Discourses

Research from other disciplines have meanwhile begun to amass evidence of the great diversity of social and cultural contexts that influence and constitute Internet-mediated discourses, lending weight to theoretical perspectives that emphasize the importance of social and cultural contexts of literacy and communicative practices. Galvin (1995) samples and explores what he calls the "discourse of technoculture", and attempts to locate it in various social and political contexts. Gibbs (2000) considers a range of influences that have shaped Internet style and content, and the social implications of the phenomenon of cyberlanguage. Kinnaly's (1997) didactic essay on "netiquette" is included here as an example of the ways in which "the rules" of Internet culture (including language and behaviour) are normalized, maintained, and manifested via specific communicative practices. Wang and Hong (1996) examine the phenomenon of flaming in online communications and argue that this behaviour serves to reinforce cyberculture norms, as well as to encourage clear written communication. In the world of online education, Conrad (2002) reports on a code of etiquette that learners valued and constructed; these communally constructed "nice behaviours" contributed to group harmony and community, she argues. Jacobson (1996) investigates the structure of contexts and the dynamics of contextualizing communication and interaction in cyberspace, while Gibbs and Krause (2000) and Duncker (2002) investigate the range of metaphors in use in the virtual world, and their cultural roots. Collot and Belmore (1996) investigate elements such as informativity, narrativity, and elaboration; Condon and Cech (1996) examine decision-making schemata, and interactional functions such as metalanguage and repetition; and Crystal (2001) examines novel genres of Internet communications.

Internet language is also shaped by the relationships and identities of the communicators. For example, Voiskounsky's (1998) reports on ways in which culturally determined factors (status, position, rank) impact "holding the floor and turn-taking rules" (p. 100) in Internet communications, and Paolillo (1999) describes a highly structured relationship between participants' social positions and the linguistic variants they use in Internet Relay Chat. Other papers analyze crucial issues like the effects of emotion management, gender, and social factors on hostile types of communication within electronic chat room settings (Bellamy & Hanewicz, 1999) or compare the male-female schematic organization of electronic messages posted on academic mailing lists (Herring, 1996). De Oliveira (2003) assesses "politeness violations" in a Portuguese discussion list, concluding that in this context male communicators assert their traditional gender roles as "adjudicators of politeness" (¶ 1). Conversely, Panyametheekul and Herring (2003) conclude from a study of gender and turn allocation in a Web-based Thai chat-room that Thai women are relatively empowered in this context. Liu (2002) considers task-oriented and social-emotional-oriented aspects of computer-mediated communication, while Dery's 1994 anthology includes essays examining the communications of different cyberspace subcultures (hackers, technopagans) and the nature and function of cyberspace. A final group of contributions to the study of the language of cyberspace consider the implications of English-language domination of cyberspace and cyberculture (for extensive references, see Macfadyen et al., 2004). It is increasingly evident then, that the communicative contexts and socio-cultural configurations of online and networked communications are many and various, and deny simple generalization and classification.

FUTURE TRENDS

Lest we imagine that the current landscape of "read-only" Internet and computer-mediated communications—be it hypertext or e-mail—is static or established, it is important to recognize that technologies that are even more challenging to current conceptions of online literacy and language have already appeared on the horizon. To date, readers of Internet content have almost no opportunity to create or modify online text, while a limited number of authors or "producers" control all content selection and presentation. New forms of hypertext, such as those promised by wikis (a Web site or other hypertext document collection that allows any user to add content, and that also allows that content to be edited by any other user) will blur these clear distinctions between author and reader, producer and consumer of online text or "content" (Graddol, 2004). While it is already understood that reading involves the production of meaning, new open-access technologies permit multiple reader-authors to register different interpretations and analysis directly within text, and participate in a new and dynamic collaborative process of co-construction of meaning. As Braga and Busnardo (2004) suggest, these developments will offer an entire new challenge to communicators, greater than simply the navigation of non-linear texts. Readers will increasingly face multiply-authored texts that exist in a condition of constant change, a situation that radically challenges our existing notions of how knowledge is now produced, accessed, and disseminated.

CONCLUSION

Individuals employ a range of language and literacy practices in their interactions with textual—and increasingly multimodal—interfaces, as they construct and exchange meaning with others; these processes are further influenced by the social and cultural contexts and identities of the communicators. At the communicative interface offered by the Internet and other computer-mediated networks, language, literacy practices *and* technologies can all be seen as what Vygotsky (1962) calls the "cultural tools" that individuals can use to mediate (but not determine) meaning. This suggests that any meaningful theoretical approach to understanding language on the Internet must not privilege technology over literacy, or vice versa, but must integrate the two. With this in mind, Lea (2004) suggests that activity theory, actor network theory, and the concept of "communities of practice" may be particularly helpful perspectives from which to consider the literacy and how it impacts Internet- and computer-mediated communications. Exemplifying new approaches that are beginning to recognize the importance of communicator learning, agency and adaptation, Lea particularly highlights Russell's (2002) interpretation of activity theory—which focuses attention on "re-mediation" of meaning, or the ways in which individuals adopt *new* tools to mediate their communications with others—as a framework that may allow new explorations and understandings of the ways that humans adopt computers as new cultural tools in their interactions with each other.

REFERENCES

Abdullah, M. H. (1998). *Electronic discourse: Evolving conventions in online academic environments.* Bloomington, IN: ERIC Clearinghouse on Reading, English, and Communication.

Bellamy, A., & Hanewicz, C. (1999). Social psychological dimensions of electronic communication. *Electronic Journal of Sociology, 4*(1). Retrieved June 1, 2005, from http://www.sociology.org/content/vol004.001/bellamy.html

Braga, D. B., & Busnardo, J. (2004). Digital literacy for autonomous learning: Designer problems and learning choices. In I. Snyder, & C. Beavis (Eds.), *Doing literacy online: Teaching, learning and playing in an electronic world* (pp. 45-68). Cresskill, NJ: Hampton Press.

Burbules, N. C. (1997). Rhetorics of the Web: Hyperreading and critical literacy. In I. Snyder (Ed.), *From page to screen: Taking literacy into the electronic age*. New South Wales, Australia: Allen and Unwin. Retrieved June 1, 2005, from http://faculty.ed.uiuc.edu/burbules/papers/rhetorics.html

Burbules, N. C., & Callister, T. A., Jr. (2000). *Watch IT: The risks and promises of information technologies for education*. Oxford, UK: Westview Press.

Cicognani, A. (1998). On the linguistic nature of cyberspace and virtual communities. *Virtual Reality: Research, Development and Application, 3*(1), 16-24. Retrieved June 1, 2005, from http://www.arch.usyd.edu.au/~anna/papers/language.pdf

Collot, M., & Belmore, N. (1996). Electronic language: A new variety of English. In S. C. Herring, (Ed.), *Computer-mediated communication: Linguistic, social and cross-cultural perspectives* (pp. 13-28). Amsterdam: John Benjamin.

Condon, S. L., & Cech, C. G. (1996). Functional comparisons of face-to-face and computer-mediated decision making interactions. In S. C. Herring, (Ed.), *Computer-mediated communication: Linguistic, social and cross-cultural perspectives* (pp. 65-80). Amsterdam: John Benjamin.

Conrad, D. (2002). Inhibition, integrity and etiquette among online learners: The art of niceness. *Distance Education, 23*(2), 197-212.

Cranny-Francis, A. (2000). Connexions. In D. Gibbs, &. K.-L. Krause (Eds.), *Cyberlines. Languages and cultures of the Internet* (pp. 123-148). Albert Park, Australia: James Nicholas Publishers.

Crystal, D. (2001). *Language and the Internet*. Port Chester, NY: Cambridge University Press.

De Oliveira, S. M. (2003). Breaking conversational norms on a Portuguese users network: Men as adjudicators of politeness? *Journal of Computer-Mediated Communication, 9*(1). Retrieved June 1, 2005, from http://www.ascusc.org/jcmc/vol9/issue1/oliveira.html

Dery, M. (Ed.). (1994). *Flame wars: The discourse of cyberculture*. Durham, NC: Duke University Press.

Douglas, J. Y. (2000). "Nature" versus "nurture": The three paradoxes of hypertext. In S. B. Gibson, & O. Oviedo (Eds.), *The emerging cyberculture: Literacy, paradigm and paradox* (pp. 325-349). Cresskill, NJ: Hampton.

Dudfield, A. (1999). Literacy and cyberculture. *Reading Online*, July. Retrieved June 1, 2005, from http://www.readingonline.org/articles/art_index.asp?HREF=/articles/dudfield/index.html

Duncker, E. (2002, July 12-15). Cross-cultural usability of computing metaphors: Do we colonize the minds of indigenous Web users? In F. Sudweeks, & C. Ess (Eds.), *Proceedings, Cultural Attitudes Towards Technology and Communication, 2002*, Montreal, Canada (pp. 217-236). Murdoch, Australia: Murdoch University.

Foucault, M. (1972). *The archaeology of knowledge*. London: Tavistock Publications.

Galvin, M. (1995). Themes and variations in the discourse of technoculture. *Australian Journal of Communication, 22*(1), 62-76.

Gibbs, D. (2000). Cyberlanguage: What it is and what it does. In D. Gibbs, &. K.-L. Krause (Eds.), *Cyberlines. Languages and cultures of the Internet* (pp. 11-29). Albert Park, Australia: James Nicholas Publishers.

Gibbs, D., & Krause, K.-L. (2000). Metaphor and meaning: Values in a virtual world. In D. Gibbs, &. K.-L. Krause (Eds.), *Cyberlines. Languages and cultures of the Internet* (pp. 31-42). Albert Park, Australia: James Nicholas Publishers.

Goodfellow, R. (2004). Online literacies and learning: Operational, cultural and critical dimensions. *Language & Education, 18*(5), 379-399.

Graddol, D. (2004). The future of language. *Science, 303*, 1329-1331.

Harpold, T. (2000). The misfortunes of the digital text. In S. B. Gibson, & O. Oviedo (Eds.), *The emerging cyberculture: Literacy, paradigm and paradox* (pp. 129-149). Cresskill, NJ: Hampton.

Herring, S. C. (1996). Two variants of an electronic message schema. In S. C. Herring (Ed.), *Computer-mediated communication: Linguistic, social and cross-cultural perspectives* (pp. 81-108). Amsterdam: John Benjamin.

Hewling, A. (2002). *Elements of electronic literacy: A contextualised review of the literature*. Retrieved December 2003, from http://iet.open.ac.uk/pp/a.hewling/AnneHewling.htm

Jacobson, D. (1996). Contexts and cues in cyberspace: The pragmatics of naming in text-based virtual realities. *Journal of Anthropological Research, 52*(4), 461-479.

Jones, C., Turner, J., & Street, B. (1999). *Students writing in the university: Cultural and epistemological issues*. Amsterdam: John Benjamin.

Kaplan, N. (1995). *E-literacies: Politexts, hypertexts, and other cultural formations in the late age of print*. Retrieved December 2003, from http://iat.ubalt.edu/kaplan/lit/

Kinnaly, G. (1997) The rules of the road for effective Internet communication. *Library Mosaics, 8*(3), 10-16.

Kramarae, C. (1999). The language and nature of the Internet: The meaning of global. *New Media and Society, 1*(1), 47-53.

Kress, G., & van Leeuwen, T. (1996). *Reading images: The grammar of visual design*. London: Routledge.

Landow, G. P. (1997). *Hypertext 2.0: The convergence of contemporary critial theory and technology*. Baltimore: Johns Hopkins University Press.

Lankshear, C., Snyder, I., & Green, B. (2000). *Teachers and techno-literacy — Managing literacy, technology and learning in schools*. St. Leonards, Australia: Allen & Unwin.

Lea, M. R. (2004). The new literacy studies, ICTs and learning in higher education. In I. Snyder, & C. Beavis (Eds.), *Doing literacy online: Teaching, learning and playing in an electronic world* (pp. 3-24). Cresskill, NJ: Hampton Press.

Lévy, P. (2001). *Cyberculture*. Minneapolis: University of Minnesota Press.

Liu, Y. (2002). What does research say about the nature of computer-mediated communication: Task-oriented, social-emotion-oriented, or both? *Electronic Journal of Sociology, 6*(1). Retrieved from http://www.sociology.org/content/vol006.001/liu.html

Macfadyen, L. P. (2006). Internet-mediated communication at the cultural interface. *Encyclopedia of Human Computer Interaction*. Hershey, PA: Idea Group Reference.

Macfadyen, L. P., Roche, J., & Doff, S. (2004). *Communicating across cultures in cyberspace: A bibliographical review of online intercultural communication*. Hamburg, Germany: Lit-Verlag.

Malone, A. (1995). Orality and communication on the Internet. *Working Papers in Linguistics, 15*, 57-76.

Murray, D. E. (2000). Changing technologies, changing literacy communities? *Language Learning & Technology, 4*(2), 43-58. Retrieved June 1, 2005, from http://llt.msu.edu/vol4num2/murray/default.html

Panyametheekul, S., & Herring, S. C. (2003). Gender and turn allocation in a Thai chat room. *Journal of Computer-Mediated Communication, 9*(1). Retrieved from http://www.ascusc.org/jcmc/vol9/issue1/panya_herring.html

Paolillo, J. (1999). The virtual speech community: Social network and language variation on IRC. *Journal of Computer-Mediated Communication, 4*(4). Retrieved June 1, 2005, from http://www.ascusc.org/jcmc/vol4/issue4/paolillo.html

Richards, C. (2000). Hypermedia, Internet communication, and the challenge of redefining literacy in the electronic age. *Language, Learning & Technology, 4*(2), 59-77. Retrieved June 1, 2005, from http://llt.msu.edu/vol4num2/richards/default.html

Russell, D. R. (2002). Rethinking genre in school and society: An activity theory analysis. *Written Communication*, 14, 504-554.

Schlickau, S. (2003). *New media in teaching languages and cultures*. Professorial Dissertation, Ludwig Maximilians Universität, Munich.

Snyder, I. (1996). *Hypertext: The electronic labyrinth*. New York: University Press.

Snyder, I. (Ed.). (1998). *Page to screen*. London: Routledge.

Street, B. (1984). *Literacy in theory and practice*. Cambridge: Cambridge University Press.

Thurstun, J. (2000). Screenreading: Challenges of the new literacy. In D. Gibbs, &. K.-L. Krause (Eds.), *Cyberlines. Languages and cultures of the Internet* (pp. 61-77). Albert Park, Australia: James Nicholas Publishers.

Tuman, N. (1995). In response to M. Kaplan, E-literacies: Politexts, hypertexts, and other cultural formations in the late age of print. Retrieved June 2004, from http://iat.ubalt.edu/kaplan/lit/index.cfm?whichOne=Tuman_responds.cfm

Voiskounsky, A. (1998, August 1-3). Internet: Cultural diversity and unification. In C. Ess, & F. Sudweeks (Eds.), *Proceedings, Cultural Attitudes Towards Communication and Technology 1998,* London. Sydney, Australia: University of Sydney.

Vygotsky, L. (1962). *Thought and language*. Cambridge, MA: MIT Press.

Wang, H.-J., & Hong, Y. (1996). Flaming: More than a necessary evil for academic mailing lists. *Electronic Journal of Communication*, *6*(1). Retrieved June 1, 2005, from http://www.cios.org/getfile/wang_V6N196

Warschauer, M. (1999). *Electronic literacies: Language, culture and power in online education.* Mahwah, NJ: Lawrence Erlbaum Associates.

Williams, H. L., & Meredith, E. M. (1996). On-line communication patterns of novice Internet users. *Computers in the Schools, 12*(3), 21-31.

KEY TERMS

Cyberculture: As a social space in which human beings interact and communicate, cyberspace can be assumed to possess an evolving culture or set of cultures ("cybercultures") that may encompass beliefs, practices, attitudes, modes of thought, behaviours, and values.

Cyberlanguage: The collection of communicative practices employed by communicators in cyberspace, and guided by norms of cyberculture(s).

Cyberspace: While the "Internet" refers more explicitly to the technological infrastructure of networked computers that make worldwide digital communications possible, "cyberspace" is understood as the virtual "places" in which human beings can communicate with each other, and that are made possible by Internet technologies. Lévy (2001) characterizes cyberspace as "not only the material infrastructure of digital communications but…the oceanic universe of information it holds, as well as the human beings who navigate and nourish that infrastructure."

(Technological) Determinism: The belief that technology develops according to its own "internal" laws and must therefore be regarded as an autonomous system controlling, Permeating, and conditioning all areas of society.

Discourse: Characterized by linguists as units of language longer than a single sentence, such that *discourse analysis* is defined as the study of cohesion and other relationships between sentences in written or spoken discourse. Since the 1980s, however, anthropologists and others have treated discourse as "practices that systematically form the objects of which they speak" (Foucault, 1972, p. 49), and analysis has focused on discovering the power relations that shape these practices. Most significantly, the anthropological perspective on discourse has re-emphasized the importance of the *context* of communicative acts.

Literacy: Traditionally defined as "the ability to read, write and communicate," usually in a print-text-based environment. New literacy Studies researchers now view literacies as social practices, moving away from individual and cognitive-based models. This model of literacy recognizes that language does not simply represent some kind of objective truth, but actually constitutes meaning in a given context; literacy, therefore, represents an individual's ability to communicate effectively within a given socio-cultural context.

Postmodern: Theoretical approaches characterized as postmodern, conversely, have abandoned the belief that rational and universal social theories are desirable or exist. Postmodern theories also challenge foundational modernist assumptions such as "the idea of progress," or "freedom."

Semiotics: The study of signs and symbols, both visual and linguistic, and their function in communication.

Mobile Clinical Learning Tools Using Networked Personal Digital Assistants (PDAs)

Bernard Mark Garrett
University of British Columbia, Canada

INTRODUCTION

The School of Nursing at the University of British Columbia has more than 300 nursing students engaged in supervised clinical practice in hospital and community settings around Vancouver. Likewise, the Faculty of Medicine has more than 200 medical students undertaking supervised clinical experience locally and remotely in the Prince George and Vancouver Island regions. The management of these clinical experiences and the promotion of learning while in an active clinical setting is a complex process.

BACKGROUND

Supporting the students at a distance while undertaking their clinical experience is particularly resource-intensive. It requires the creation and maintenance of good communication links with the clinical and administrative staff, active management, clinical visits from faculty, and the provision and management of remotely based resources. However, there were few existing resources that helped to contextualize and embed clinical knowledge in the workplace in the practice setting (Landers, 2000). A technological solution was developed and implemented using several clinical applications designed for use on personal digital assistants (PDAs).

MOBILE CLINICAL LEARNING TOOLS

A suite of PDA-based tools were created for a pilot study with the involvement of nursing and medical students during the academic year of 2004-2005 to achieve the following objectives:

- To demonstrate the potential use of mobile networked technologies to support and improve clinical learning.
- To develop and evaluate a range of mobile PDA tools to promote reflective learning in practice and to engage students in the process of knowledge translation.
- To develop and evaluate a suite of pedagogic tools that help contextualize and embed clinical knowledge while in the workplace.
- To evaluate the value of networked PDA resources to help prevent the isolation of students while engaged in clinical practicum.

The tools developed provide a mobile clinical learning environment incorporating an e-portfolio interface for the Pocket PC/Windows Mobile (Microsoft, 2004) operating system. They were implemented on i-mate PDAs equipped with GSM/GPRS (Global System for Mobile Communications/General Packet Radio Service; GSM World, 2002). This platform offered considerable flexibility for the project. It supported the use of cellular telephone connectivity and Pocket Internet Explorer Web browser (which has a full Internet browser with support for HTML, XML/XSL, WML,cHTML, and SSL); the i-mate device had sufficient memory for the storage of text, audio, image, and video data, with a large screen and a user-friendly interface with an integrated digital camera.

The tools included a mobile e-portfolio (with a multimedia interface) designed to promote professional reflection (Chasin, 2001; Fischer et al., 2003; Hochschuler, 2001; Johns, 1995; Kolb, 1984). These mobile learning tools were designed to promote the skills of documentation of clinical learning, active reflection, and also to enable students to immediately access clinical expertise and resources remotely. Community clinical placements are being used for the testing domain, as there are currently no restrictions on using cellular network technology in these areas, whereas this is currently restricted in acute hospital settings in British Columbia and many other parts of the world.

THE PDA INTERFACE DESIGN

The main interface to the clinical tools was based on a clinical e-tools folder on the Pocket PC containing icon-based shortcuts to a number of specific applications (Figure 1).

The clinical e-portfolio tool represented the major focus for the project, allowing the student to access clinical placement information; log clinical hours; achieve clinical competencies; record portfolio entries in the form of text, pictures, or video clips; and record audio memos. This provides the user with a very adaptable interface, allowing them to choose how they input data. For example, a text-based entry describing a clinical procedure may be accompanied by a picture or audio memo.

The e-portfolio tool also incorporates a reflective practice wizard promoting the students to work through the stages of the Gibbs reflective cycle (Gibbs, 1988) when recording their experiences. This wizard also allows students to record their experiences with multimedia, including text, audio, digital images, or video input. Once the data have been recorded in the e-portfolio, they can be synchronized wirelessly (using the built-in GSM/GPRS or Bluetooth connectivity) with a Web-based portfolio. The data then can be reviewed and edited by the student or by clinical tutors.

The other icons represent the following applications:

- The synch portfolio icon initiates synchronization of the content of the student's e-portfolio on the PDA with that of a remote server.
- The University of British Columbia (UBC) library icon presents a shortcut to a Pocket Internet Explorer Web access to the UBC library bibliographic health care database search (CINAHL, Medline, etc.).
- The Pocket Explorer icon presents a shortcut to Pocket Internet Explorer for mobile Web access.
- The e-mail icon presents a shortcut to the Pocket PC mobile e-mail application.

The other icons on the screen (Diagnosaurus, ePocrates, etc.) represent third-party clinical software that was purchased and loaded onto the PDAs in order to support the students learning in the clinical area (e.g. a drug reference guide).

Figure 1. Screenshot of the clinical e-tools folder

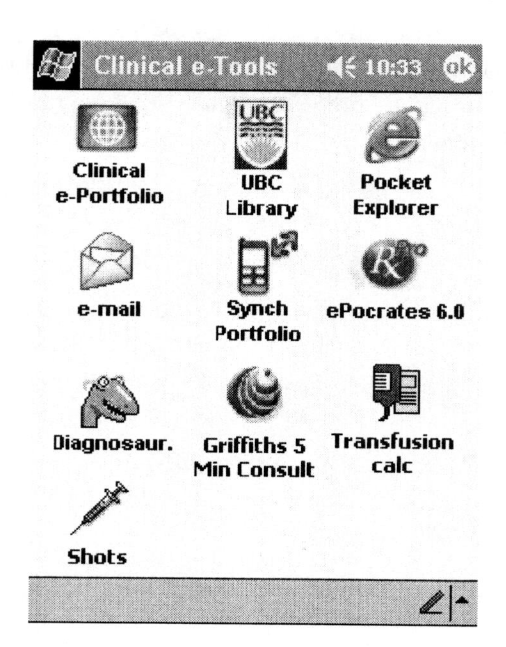

FUTURE TRENDS

In the future, the PDA will provide a one-stop resource to support clinical learning. Students also will be able to examine their learning objectives, record their achievements, and record notes/memos attached to specific clinical records for later review. Where students have particular concerns or questions that cannot be answered immediately in the clinical area, they will be able to contact their supervisors or faculty for support using e-mail, cell phone, or multimedia messaging service (MMS) communications.

The use of multimedia in PDA interfaces is likely to become much more widespread as the cost of these devices reduces and they become more accessible to a wider spectrum of the population. This already is occurring with the merging of cell phone and PDA technologies and the uptake of MMS and use of audio and video data entry on mobile devices (deHerra, 2003).

In the long term, multimedia mobile learning tools will encourage a more structured process of professional reflection among students in supervised clinical practice (Conway, 1994; Copa et al., 1999; Palmer et al., 1994; Reid, 1993; Sobral, 2000). When unexpected learning opportunities arise, students will be able to quickly review online materials in a variety of formats and prepare for their experience, record notes, record audio memos or images during their practice, and review materials following their experience.

An expansion in the use of such mobile clinical learning tools is envisaged, and there is considerable scope for the widespread application of such tools into areas where students are engaged in work-based learning. We are likely to see the integration of these technologies into mainstream educational practice in a wide variety of learning environments outside of the classroom.

CONCLUSION

The value of these new tools to students in clinical practice remains to be demonstrated, as the evaluation stage of the project has yet to be completed. The project also has highlighted the necessity of addressing some of the weaknesses of current PDA design, such as the small display screen and the need for more built-in data security. However, initial feedback appears promising, and the interface design appears to promote reflective learning in practice and engage students in the process of knowledge translation.

REFERENCES

Chasin, M.S. (2001). Computers: How a palm-top computer can help you at the point of care. *Family Practice Management, 8*(6), 50-51.

Conway, J. (1994). Profiling and nursing practice. *British Journal of Nursing, 3*(18), 941-946.

Copa, A., Lucinski, L., Olsen, E., & Wollenberg, K. (1999). Promoting professional and organizational development: A reflective practice model. *Zero to Three, 20*(1), 3-9.

deHerra, C. (2003). *What is in the future of Windows mobile pocket PCs*. Retrieved August 12, 2004, from http://www.cewindows.net/commentary/future-wm.htm

Fischer, S., Stewart, T.E., Mehta, S., Wax, R., & Lapinsky, S.E. (2003). Handheld computing in medicine. *Journal of the American Medical Informatics Association, 10*(2), 139-149.

Gibbs, G. (1988). *Learning by doing. A guide to teaching and learning methods*. Oxford: Oxford Polytechnic.

GSM World. (2002). GSM technology: GPRS platform? Retrieved June 24, 2004, from *http://www.gsmworld.com/technology/sms/intro.shtml*

Hochschuler, S.H. (2001). Handheld computers can give practitioners an edge. *Orthopedics Today, 21*(6), 56.

Johns, C. (1995). The value of reflective practice for nursing. *J. Clinical Nurs, 4*, 23-60.

Kolb, D.A. (1984). *Experiential learning*. Englewood Cliffs, NJ: Prentice Hall.

Landers, M.G. (2000). The theory-practice gap in nursing: The role of the nurse teacher. *Journal of Advanced Nursing, 32*(6), 1550-1556.

Microsoft. (2004). *Windows mobile based pocket PCs*. Retrieved August 10, 2004, from http://www.microsoft.com/windowsmobile/pocketpc/ppc/default.mspx

Palmer, A., Buns, S., & Bulman, C. (1994). *Reflective practice in nursing: The growth of the professional practitioner*. Oxford: Blackwell Science.

Reid, B. (1993). "But we're doing it already": Exploring a response to the concept of reflective practice in order to improve its facilitation. *Nurse Ed Today, 13*, 305-309.

Sobral, D.T. (2000). An appraisal of medical student reflection-in-learning. *Medical Education, 34*, 182-187.

KEY TERMS

Bluetooth: A short-range wireless radio standard aimed at enabling communications between digital devices. The technology supports data transfer at up to 2Mbps in the 2.45GHz band over a 10m range. It is used primarily for connecting PDAs, cell phones, PCs, and peripherals over short distances.

Digital Camera: A camera that stores images in a digital format rather than recording them on light-sensitive film. Pictures then may be downloaded to a computer system as digital files, where they can be stored, displayed, printed, or further manipulated.

e-Portfolio: An electronic (often Web-based) personal collection of selected evidence from coursework or work experience and reflective commentary related to those experiences. The e-portfolio is focused on personal (and often professional) learning and development and may include artefacts from curricular and extra-curricular activities.

General Packet Radio Service (GPRS): A standard for wireless communications that operates at speeds up to 115 kilobits per second. It is designed for efficiently sending and receiving small packets of data. Therefore, it is suited for wireless Internet connectivity and such applications as e-mail and Web browsing.

Global System for Mobile Communications (GSM): A digital cellular telephone system introduced in 1991 that is the major system in Europe and Asia and is increasing in its use in North America. GSM uses Time Division Multiple Access (TDMA) technology, which allows up to eight simultaneous calls on the same radio frequency.

i-Mate: A PDA device manufactured by Carrier Devices with an integrated GSM cellular phone and digital camera. The device also incorporates a built-in microphone and speaker, a Secure Digital (SD) expansion card slot, and Bluetooth wireless connectivity.

Personal Digital Assistant (PDA): A small handheld computing device with data input and display facilities and a range of software applications. Small keyboards and pen-based input systems are commonly used for user input.

Pocket PC: A Microsoft Windows-based operating system (OS) for PDAs and handheld digital devices. Versions have included Windows CE, Pocket PC, Pocket PC Phone Edition, and Windows Mobile. The system itself is not a cut-down version of the Windows PC OS but is a separately coded product designed to give a similar interface.

Wizard: A program within an application that helps the user perform a particular task within the application. For example, a setup wizard helps guide the user through the steps of installing software on his or her PC.

Moral Mediators in HCI

Lorenzo Magnani
University of Pavia, Italy

Emanuele Bardone
University of Pavia, Italy

Michele Bocchiola
LUISS University, Italy

INTRODUCTION

Our contention is that interactions between humans and computers have a moral dimension. That is to say, a computer cannot be taken as a neutral tool or a kind of neutral technology (Norman, 1993).[1] This conclusion seems a bit puzzling and surely paradoxical. How can a computer be moral?

All computational apparatuses can be generally considered as moral mediators, but for our considerations, computers are the best representative tools. First of all, they are the most widespread technological devices, they are relatively cheap in comparison to other technological utilities, and, very importantly, they can be easily interconnected all over the word through the Internet. This last feature allows people to keep in contact with each other and, consequently, to improve their relations. Computers require interactions with humans, but also allow interactions between humans. Since morality relates to how to treat other people within interactive behaviors, computers can help us to act morally in several ways. For instance, as the concept of moral mediators suggests, computers can help us to acquire new information useful to treat in a more satisfactory moral way other human beings.

BACKGROUND

In traditional ethics it is commonly claimed that the moral dimension primarily refers to human beings since they possess intentions, they can consciously choose, and they have beliefs. Also, artificial intelligence (AI) holds this view: Indeed, AI aims at creating a moral agent by smuggling and reproducing those features that make humans moral. On the contrary, our contention is that computer programs can also be considered moral agents even if their interfaces do not exhibit or try to explicitly reproduce any human moral feature.[2] As Magnani (2005) contends, computer programs can be defined as a particular kind of moral mediator.[3] More precisely, we claim that computers may have a moral impact because, for instance, they promote various kinds of relations among users, create new moral perspectives, and/or provide further support to old ones.[4]

MORAL MEDIATORS AND HCI

In order to shed light on this issue, the concept of moral mediator turns out to be a useful theoretical device. To clarify this point, consider, for instance, a cell phone: One of its common features is to ask for confirmation before sending text. This option affords the user to check his or her message not only for finding mistyping, but also for reflecting upon what he or she has written. In other words, it affords being patient and more thoughtful. For instance, after typing a nasty text message to a friend, receiving a confirmation message may affect a person's behavior to wait and discard the text message. The software not only affords a certain kind of reaction (being thoughtful), but it also mediates the user's response. The confirmation message functions as a mediator that uncovers reasons for avoiding the delivery of the text message. Just reading after a few seconds what one has furiously written may contribute to change one's mind. That is, a person might think that a friend does not deserve to receive the words just typed. Hence, new information is brought about. According to Magnani (2003), because of this behavior, we may call this kind of device a moral mediator.

Various kinds of moral mediators have been described that range from the role played by artifacts to the moral aspects that are delegated to natural objects and human collectives. In order to grasp the role of moral mediators, let us consider Magnani's example of endangered species.[5] When we consider animals as subjects requiring protection for their own existence, we are using them to depict new moral features of living objects previously unseen. In this case, endangered species can become a mediator that unearths and uncovers a new moral perspective expanding the notion of moral worth and dignity we can also attribute to human beings.[6]

AN EXAMPLE OF MORAL MEDIATOR: THE "PICOLA PROJECT"

This section will provide an exemplification of the moral mediation previously illustrated. In the following, we shall give a general description of a Web-based tool named PICOLA (Public Informed Citizen On-Line Assembly).[7] The PICOLA project, developed at Carnegie Mellon's Institute for the Study of Information and Technology (InSITeS) and at the Center for the Advancement of Applied Ethics (CAAE), aims at implementing an online environment for community consultation and problem solving using video, audio, and textual communication.

The appeal of deliberative democracy is mainly based on two ingredients: first, the idea of a free and equal discussion, and second, the consensus achieved by the force of the best argument (Fishkin & Laslett, 2003; Habermas, 1994, 1998). PICOLA can be considered a moral mediator because it implements those two ideas into a Web-based tool for enhancing deliberative democracy. Indeed, everyone has equal rights to speak and to be listened to, equal time for maintaining her or his position, equal weight in a poll, and so on. Besides this, it allows the formation of groups of discussion for assessing and deliberating about different issues. Within an actual framework, these two requirements are rarely matched. Even if everyone has the possibility to vote and be voted on, few persons can actually play a role in deliberative procedures. Web-based tools like PICOLA promote participation by allowing all interested citizens to be involved in a democratic process of discussion. It enables citizens to take part in democratic meetings wherever one may be. For instance, with PICOLA we do not need any actual location because it is possible to conceive virtual spaces where persons can discuss following the same rules.

Every Web site has to face the problem of trustworthiness. In a physical environment, people have access to a great deal of information about others: how people are, dress, or speak, and so on. This may provide additional information, which is difficult to obtain in a virtual environment. The need for trust is more urgent especially when people have to share common policies or deliberations.

PICOLA requires each user to create a profile for receiving additional information, as listed in Figure 1. As well as a user name and password, the user must insert the area where he or she lives, the issue he or she would like to discuss, and the role he or she wants to play: moderator, observer, or deliberator. This tool also allows users to add a picture to their profiles. Moreover, each user can employ so-called emoticons in order to display current feelings.

All these features work like moral mediators because they mediate the relation among users so they can get acquainted with each other. Being acquainted with each other is one of the most important conditions to enhance cooperation and to generate trust. Indeed, people are more inclined to reciprocate and to be engaged in cooperative behavior if they know to some extent the people they are interacting with. This diminishes prejudices and people are less afraid of the possible negative outcomes. In this sense, sharing user profiles could be viewed as a generator of social capital and trust (Putnam, 2000).

Each issue to be discussed during a session is introduced by an overview that provides general information (see Figure 2 on the issue of national security). Text is integrated with audio and video performances. The moral mediation mainly occurs for two reasons. First, people are sufficiently informed to facilitate sharing a starting point and giving up prejudices that may have arisen due to a lack of information. Second, the fact that different perspectives are presented helps users to weigh and consider the various opinions available. Moreover, the multimedia environment provides video and au-

Figure 1. User profile

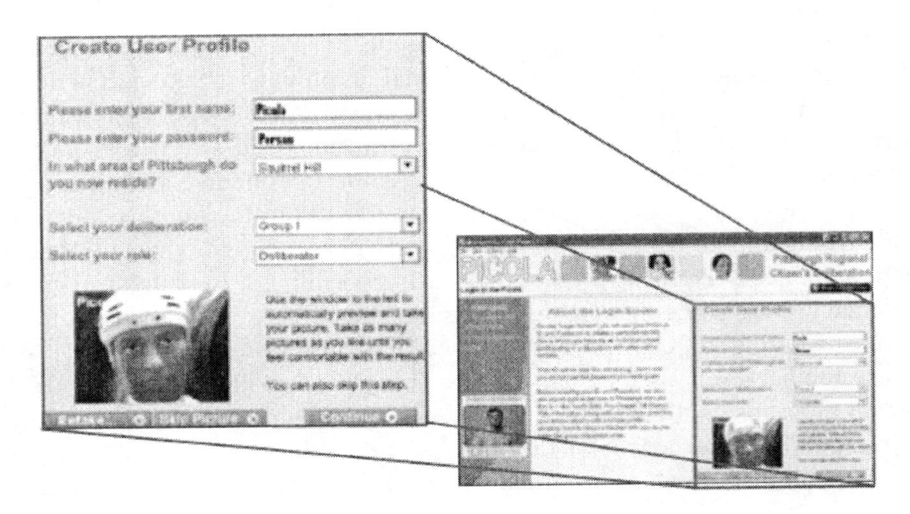

Figure 2. Discussion of the issues

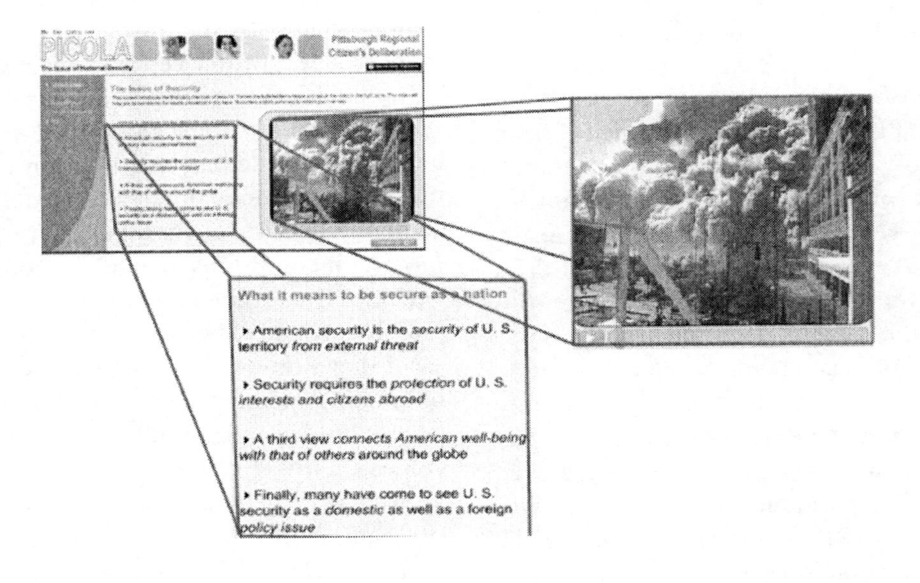

dio information so that the user may access additional resources. For instance, watching a video involves not only "cold" reasons, but also emotions that mediate the user's response to a particular issue.[8]

The discussion group is realized by a Web-based tool, which provides the possibility to visualize all the discussants sitting at a round table. The discussion is led by a moderator who has to establish the order of speakers. Each participant is represented with a profile showing his or her data, an emoticon, and a picture. Everyone has basically two minutes for maintaining a position and then has to vote at the end of the discussion (see Figure 3).

Computers can afford civic engagements (Davis, Elin, & Reeher, 2002). Creating a public online community is not just a way for allowing interaction

Figure 3. Group discussion

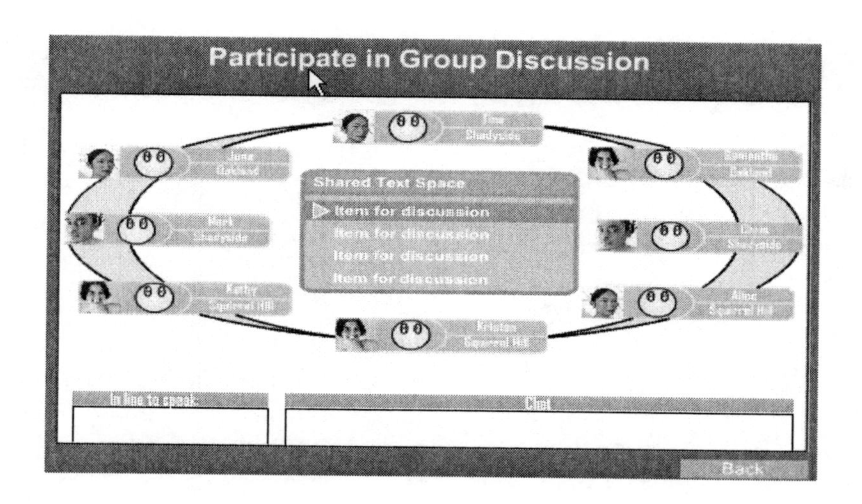

among people: It can improve the overall assessment of public interest policies, giving people more information, generating a public base for discussion, and implementing participation in the democratic life of a country. This is possible and welcome if we assume that the computer is, or should be, a very low-cost technology so that a larger amount of persons can exploit such a service.

FUTURE TRENDS

Moral mediators are widespread in our world and especially in our technological world. The moral-mediating role of human-computer interaction is just one of the possible interesting cases. However, other issues remain to be discussed and deepened.

First of all, the concept of moral mediator implicitly assumes that the effort of a moral or political deliberation is shared between users and computers. It could be useful to explore the details on how this distribution actually takes place to enhance the ethical mediation and/or to understand possible negative effects of the technological tools.

Second, an interesting line of research concerns the kind of moral affordance involved in such interaction. As we have argued, objects apparently are not relevant from the moral point of view, namely, they are inert. We are acquainted with the idea that a human being is morally important because of his or her intrinsic value and dignity. But how could a computer be intrinsically moral? Through the concept of moral mediator, we have suggested that nonhuman things can be morally useful in creating new moral devices about how to behave (the cellphone example or certain kinds of interaction with computer programs). Hence, their features afford a moral use. The concept of affordance, first introduced by Gibson (1979), might help to clarify this point about which features transform an object or artifact into a moral mediator.

CONCLUSION

In this article we have tried to show some aspects of the kind of moral mediation that is involved in human-computer interaction. We have argued that objects that seem inert from a moral point of view may contribute to enhance moral understanding and behavior. Tools and software may have moral impact because, for instance, they promote new kinds of relations among users, create new moral perspectives, and/or provide further support to old ones.

We have illustrated this approach showing how a Web-based tool such as PICOLA may provide useful support for enhancing deliberative democracy. As we have seen, a multimedia environment may be very useful in order to create consensus, to inform citizens about a given issue, and to develop considered beliefs.

REFERENCES

Davis, S., Elin, L., & Reeher, G. (2002). *Click on democracy: The Internet's power to change political apathy into civic action.* Cambridge, MA: Westview Press.

Dryzek, J. (2000). Discursive democracy in a reflexive modernity. In J. Dryzek (Ed.), *Deliberative democracy and beyond* (pp. 162-175). Oxford, UK: Oxford University Press.

Ermann, M. D., & Shauf, M. S. (2002). *Computers, ethics, and society.* Oxford, UK: Oxford University Press.

Estlund, D. (Ed.). (2002). *Democracy.* Oxford: Blackwell.

Fishkin, J. S., & Laslett, P. (Eds.). (2003). *Debating deliberative democracy.* Oxford, UK: Blackwell.

Gibson, J. (1979). *The ecological approach to visual perception.* New York: Lea.

Habermas, J. (1994). *Justification and application: Remarks on discourse ethics.* Cambridge, MA: The MIT Press.

Habermas, J. (1998). *The inclusion of the other: Studies in political theory.* Cambridge, MA: The MIT Press.

Johnson, D. G. (2000). *Computer ethics.* Englewood Cliffs, NJ: Prentice Hall.

Kirkman, R. (2002). Through the looking glass: Environmentalism and the problem of freedom. *The Journal of Value Inquiry, 36*(1), 27-41.

Magnani, L. (2001). *Abduction, reason, and science: Processes of discovery and explanation.* New York: Kluwer Academic/Plenum Publishers.

Magnani, L. (2003). Creative abduction as active shaping of knowledge: Epistemic and ethical mediators. *Proceedings of the Symposium on Artificial Intelligence and Creativity in Arts and Science, AISB'03 Convention on Cognition in Machines and Animals,* Wales, UK.

Magnani, L. (2005). *Knowledge as a duty.* Cambridge, MA: Cambridge University Press.

Nagle, J. C. (1998). Endangered species wannabees. *Seton Hall Law Review, 29,* 235-255.

Norman, D. (1993). *Things that make us smart.* Cambridge, MA: Perseus.

Norman, D. (2004). *Emotional design: Why we love or hate everyday things.* New York: Basic Books.

Picard, R. W., & Klein, J. (2002). Computers that recognise and respond to user emotion: Theoretical and practical implications. *Interacting with Computers, 14*(2), 141-169.

Putnam, R. (2000). *Bowling alone.* New York: Simon & Schuster.

Tractinsky, N., Katz, A. S., & Ikar, D. (2000). What is beautiful is usable. *Interacting with Computers, 13,* 127-145.

KEY TERMS

Civic Engagement: Describes the level of citizens' participation in all those activities that concern with fostering democratic values and public virtues such as trustworthiness, freedom of speech, and honesty.

Deliberative Democracy: Based on a decision-making consensus-oriented process, where parties can freely participate. The outcome of which is the result of reasoned and argumentative discussions. This model aims to achieve an impartial solution for political problems.

Morality: The complex system of principles given by cultural, traditional, and religious conceptions and beliefs, which human-beings employ for judging things as right or wrong.

Moral Agency: The capacity to express moral judgements and to abide by them.

Moral Mediators: External resources (artifacts, tools, etc.) can be defined as moral mediators, when they actively shape the moral task one's facing through uncovering valuable information that otherwise would remain unearthed and unattainable.

Online Community: A system of Internet users sharing interests and interacting frequently in the same areas, such as forums and chat Web sites.

PICOLA Project: The PICOLA Project (Public Informed Citizen On-Line Assembly) is an initiative spearheaded by Carnegie Mellon University to develop and implement through on-line tools, a virtual agora for public consultation regarding public policy issues.

Social Capital: Refers to connections among individuals—social networks and the norms of reciprocity and trustworthiness that arise from them (Putnam, 2002, p. 19).

ENDNOTES

1. Although Norman (2004) did not investigate the moral dimension of HCI (human-computer interaction), he has recently explored its emotional dimension.
2. For more information about the relation between computers and ethics, see Ermann and Shauf (2002) and Johnson (2000).
3. The idea of moral mediators is derived from the analysis of the so-called epistemic mediators Magnani (2001) introduced in a previous book.
4. Even if a tool or software can provide support or help, it may also contribute to create new ethical concerns. One of the most well-known problems related to the Web, and to the Internet in general, is the one concerning privacy. For further information on this issue, see Magnani (in press). For more information, visit the Electronic Privacy Information Center (EPIC) at http://www.epic.org/.
5. For further details about this issue, see Kirkman (2002) and Nagle (1998).
6. More details are in Magnani (2005, chs. 1 and 6).
7. Further information about the PICOLA project can be found at http://communityconnections.heinz.cmu.edu/picola/index.html.
8. On the role of emotions, see Norman (2004), Picard and Klein (2002), and Tractinsky, Katz, and Ikar (2000).

Motivation in Component-Based Software Development

G. Chroust
J. Kepler University Linz, Austria

INTRODUCTION

Information systems are designed for the people, by the people. The design of software systems with the help of software systems is another aspect of human-computer interfaces. New methods and their (non-)acceptance play an important role. Motivational factors of systems developers considerably influence the type and quality of the systems they develop (Arbaoui, Lonchamp & Montangero, 1999; Kumar & Bjoern-Andersen, 1990). To some extent, the quality of systems is a result of their developers' willingness to accept new and (supposedly) better technology (Jones, 1995). A typical example is component-based development methodology (Bachmann et al., 2000; Cheesman & Daniels, 2001). Despite considerable publication effort and public lip service, component-based software development (CBD) appears to be getting a slower start than anticipated and hoped for. One key reason stems from the psychological and motivational attitudes of software developers (Campell, 2001; Lynex & Layzell, 1997). We therefore analyze the attitudes that potentially hamper the adoption of the component-based software development approach. Maslow's Hierarchy of Need (Boeree, 1998; Maslow, 1943) is used for structuring the motives.

BACKGROUND

The Human Side of Software Engineering

Kunda and Brooks (1999) state that "software systems do not exist in isolation ... human, social and organizational considerations affect software processes and the introduction of software technology. The key to successful software development is still the individual software engineer" (Eason et al., 1974; Kraft, 1977; Weinberg, 1988). Different software engineers may account for a variance of productivity of up to 300% (Glass, 2001). On the other hand, any other single factor is not able to provide an improvement of more than 30%. The influence of an individual's motivation, ability, productivity, and creativity has the biggest influence by far on the quality of software development, irrespective of the level of technological or methodological support. Therefore, it is worthwhile investigating for what reasons many software engineers do not fullheartedly accept component-based methods (Lynex & Layzell, 1997).

Software development in general introduced a new type of engineers who show marked differences when compared to (classical) engineers (Badoo & Hall, 2001; Campell, 2001; Eason et al., 1974; Kraft, 1977; Kunda & Brooks, 1999; Lynex & Layzell, 1997). The phenomenon is not fully understood yet but seems to have to do with the peculiarities of software (Brooks, 1986), the type of processes and environments needed to develop software (Kraft, 1977), and especially to the proximity of software development to other mental processes (Balzert, 1996).

Maslow's Hierarchy of Needs

Maslow's theory (Boeree, 1998; Huitt, 2002; Maslow, 1943; McConnell, 2000) provides a practical classification of human needs by defining a five-level Hierarchy of Needs (Figure 1).

The five levels are as follows:

- **Basic Physiological Needs (Survival):** At this level, the individual is fighting for survival against an adverse environment, trying to avert hunger, thirst, cold, and inconvenient and detracting physical work environments.
- **Security (Physical, Economic ...):** On this level, the individual is concerned with the sta-

Figure 1. Maslow's hierachy of needs

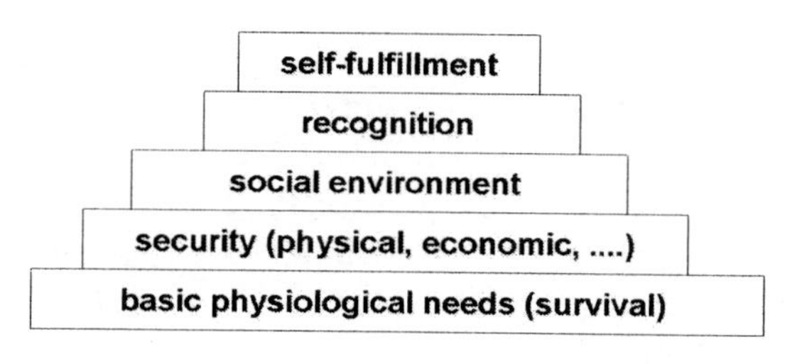

bility of his or her future and the safety of the environment. Worries include job security, loss of knowledge, loss of income, health, and so forth.

- **Social Environment:** This category includes the need to have friends, belong to a group, and to give and receive love.
- **Recognition:** Individuals strive to receive appropriate recognition and appreciation at work and to be recognized as having a valuable opinion.
- **Self-Fulfillment:** This level is considered the highest stage attainable in the development of a person, drawing satisfaction from the realization of one's own contribution to a goal and one's fullfillment of their full potential as a human being.

Reuse and Component-Based Software Development (CBD)

An old dream in software development is to avoid unnecessary duplication of work by consistently and systematically reusing existing artifacts. Reuse promises higher productivity, shorter time-to-market, and higher quality (Allen, 2001; Cheesman & Daniels, 2001). Initially, ready-made pieces of software were made available; these delivered a defined functionality in the form of a black box (i.e., without divulging the internal structure to the buyer/user). They were called COTS (commercials off the shelf) (Voas, 1998). Later, an improved and more restricted concept was employed: software components (Bachmann et al., 2000; Cheesman & Daniels, 2001; Woodman et al., 2001). Software components have to fulfill additional requirements, restrictions, and conventions beyond the properties of COTS. To a user of a software component, only its interfaces and functionality are known, together with the assurance that the component obeys a specific component model. This component model defines how the component can be integrated with other components, the conventions about the calling procedure, and so forth. The internal structure, code, procedures, and so forth are not divulged—it is a black box.

Systematic, institutionalized CBD needs a change in the attitude of software engineers, different work organization, and a different organization of the whole enterprise (Allen, 2001).

Component-Based Development and Software Engineers' Needs

The acceptance of a new technology often meets with strong opposition caused by psychological motives, which can be traced to Maslow's Hierarchy of Needs.

Basic Physiological Needs

This level does not have any strong relevance; software engineering is a desk-bound, safe, nonendangering activity. We have to recognize, however, that very often software engineers have to struggle with adverse infrastructure (floor space, noise, etc.) (deMarco, 1985).

Security

The desire for security is threatened by numerous factors. The fears can be categorized into four groups:

Losing the Job or Position

- **Job Redundancy:** CBD promises considerably higher productivity and less total effort as a result of removing the redundancy of reimplementing already existing functions. This carries the thread of making an individual redundant, especially since the development of components very often is outsourced to some distant organization (e.g., India).
- **Implementing vs. Composing:** deRemer (1976) stressed the difference between implementing a module/component (programming in the small) and building (composing) a system out of components (programming in the large).

He emphasized the need for a different view and for new approaches and tools. Programming in the large needs a systems view, making much of long-learned patterns of work obsolete, even counter-productive.

- **Changed Job Profile:** The necessity of integrating existing components requires a different mindset than one implementing some program from scratch (Vitharana, 2003). Does the software engineer have the ability or qualifications to fulfill the new profile of an integrator?
- **Loss of Knowledge and "Guru" Status:** In traditional development, considerable domain know-how and low-level development know-how rests in the heads of seasoned developers having developed software for many years. The use of components encapsulates and hides both implementation details and domain know-how. In addition, system development methods change. Much of the accumulated experience and know-how previously valuable to the employing institution becomes irrelevant.
- **De-Skilling:** In addition, certain de-skilling takes place at the lower level of software development (Kraft, 1977). The need for increased skills with respect to performing high-level composition activities often is not recognized and appreciated by the individuals.

Loss of Low-Level Flexibility

- **Pre-Conceived Expectations:** Components often do not provide exactly what the original requirements specified. A developer then is challenged to find a compromise between the user requirements and the available components—a job profile dramatically different from developing bespoke software (Vitharana, 2003). Engineers also have to live with good enough quality (Bach, 1997; ISO/IEC, 2004), as provided by the components and often are not able to achieve best quality. This is often difficult to accept emotionally.
- **Revision of Requirements:** Mismatches between stated requirements and available components make it necessary to revise and adapt requirements (Vitharana, 2003). Requirements are no longer set in stone, in contrast to the assumptions of classical development methods (e.g., the waterfall model).
- **Uncertainty About Functionality of Components:** By definition, the internal structure of a component is not revealed (Bachmann et al., 2000). Consequently, the developer has to rely on the description and claims provided by the component provider (Crnkovic & Larsson, 2002; Vitharana, 2003).

Lack of Confidence

- **Distrust in Component Quality:** Quality problems experienced in the past have created a climate of distrust with respect to other developers' software products. This feeling of distrust becomes stronger with respect to components, because their internals are not disclosed (Heineman. 2000; Vitharana, 2003). This situation becomes worse for so-called software of unknown provenance (SOUP) (Schoitsch, 2003). The current interest in open source programs is an indicator of a movement in the opposite direction.
- **Questions About Usability:** Besides the issue of quality, problems with portability and interoperability of components, as often experienced with COTS, also reduce the confidence in using components (Vecellio & Thomas, 2001).
- **Loss of Control of System:** Engineers usually like to understand fully the behavior of the designed system (the *why*) and exercise control over the system's behavior (the *how*). In CBD, due to the black-box character of the components, understanding and control can be achieved only to a limited extent, leaving a vague feeling of uncertainty.

Effort for Reuse vs. New Development

- **Uncertainty Concerning the Outcome of the Selection Process:** Successful CBD depends to a considerable extent on the probability and effectiveness of finding a component with (more or less) predefined properties (Vitharana, 2003). Occasionally, this search will not be successful, causing a delay in the development schedule and some lost effort spent in the search.

- **Effort Estimates:** In general, software engineers underestimate the effort needed to build a system from scratch and overestimate the cost and effort of adapting a system. The reasons seem to be the initially necessary effort to achieve a certain familiarity with the whole system before making even small adaptations, the learning curve (Boehm & Basili, 2000), the difficulty, and often also the unwillingness of becoming familiar with somebody else's thoughts and concepts (not-invented-here syndrome).

Social Environment

- **Reluctance to Utilize Outside Intellectual Property:** Our society extends the notion of ownership to immaterial products like ideas and intellectual achievements. They are protected by copyright, trademark, and patent legislation. Plagiarism is objected to and usually not sanctioned (Kock, 1999; Sonntag & Chroust, 2004). Reusing someone else's ideas is often deemed inappropriate.
- **Immorality of Copying:** In school, copying as a form of reuse and teamwork is usually discouraged. This might later cause some reluctance to actively share knowledge and to make use of someone else's achievements (Disterer, 2000).
- **Adopting a New Technology:** The adoption of a new technology seems to follow an exponential law (Jones, 1995). It starts with a few early adopters, and others follow primarily because of personal communication. The tendency of software developers to be introverts (Riemenschneider, Hardgrave, & Davis, 2002) might delay such a dissemination process.
- **Change of Work Organization:** CBD needs a different work organization (Allen, 2001; Chroust, 1996; Cusumano, 1991; Wasmund, 1995) resulting in rearranged areas of responsibility, power distribution, and status, potentially upsetting an established social climate and well-established conventions.

Recognition

A strong motivator for an individual is recognition by the relevant social or professional reference group, usually a peer group (Glass, 1983).

- **Gluing vs. Doing:** On the technical level, recognition usually is connected to a particular technical achievement. Gluing together existing components will achieve recognition only for spectacular new systems—and these are rare. Similarly, an original composer will gain recognition; whereas a person simply arranging music into potpourris usually goes unrecognized.
- **Shift of Influence and Power:** Successful CBD needs a change in organization (Allen, 2001; Kunda & Brooks, 1999), making persons gain or lose influence, power, and (job) prestige, a threat to the established pecking order.
- **The CBD Water Carrier:** Organizations heavily involved in CBD often separate the component development from component deployment (Cusumano, 1991). Component development to a large extent is based on making existing modules reusable "components as you go" and "components by opportunity" (Allen, 2001) and not on creating new ones from scratch "components in advance". Jobs in the reuse unit (similar to maintenance units) (Basili, 1990) might be considered to require less know-how and thus receive lower prestige, despite the fact that these jobs often require greater know-how and experience than designing components from scratch.
- **Contempt for the Work of Others:** The inherent individuality of software development, together with a multitude of different solutions to the same problem (i.e., there is always a better way), and the low quality of many software products have tempted many software developers into contempt for anyone else's methods and work (not-invented-here syndrome).
- **Rewarding Searching Over Writing:** As long as the amount of code produced (lines of code) is a major yard stick for both project size and programmer productivity searching, find-

ing and incorporating a component will be less attractive than writing it anew.

- **Accounting for Lost Search Effort:** There is no guarantee that even after an extensive (and time-consuming) search an appropriate component can be found (Vitharana, 2003). In this case, management must accept these occasional losses so as not to discourage searching for components (Fichman & Kemerer, 2001).

Self-Fulfillment

- **Not Invented Here:** The ability to design wonderful systems is a strong motivator for software engineers. This feeling goes beyond recognition of peers—one knows it oneself. This makes it difficult for developers to accept other people's work (Campell, 2001; Disterer, 2000) in the form of components.
- **No More Gold Plating:** The feeling of self-fulfillment often cannot live with the knowledge that a system still should or must be improved, leading to endless effort in gold plating a system before delivery (or even thereafter). Externally acquired components cannot be modified (i.e., gold plated) because of the inaccessibility of their code.
- **No Creative Challenge:** Gluing together components provided by somebody else does not fulfill many engineers' attempt for novelty and, thus, is not considered to be a creative challenge. The highly creative process of finding the best-fitting component, restructuring the system, and perhaps modifying the requirements for using existing components often is not appreciated.
- **No More Lone Artists:** Software engineers aspire to become a Beethoven or a Michelangelo and not the directors of a museum arranging a high-class exhibition. Someone remarked that many system features are not needed by the users but are just a monument of their designer's intellectual capability. Assembling components utilizes only someone else's achievement.
- **Lack of Freedom:** The limited choice of available components, the limitations of a component model, the need to obey predefined interfaces, and so forth restrict the freedom of development and often are seen as a limit to creativity.

FUTURE TRENDS

The fact that the software industry needs a large step forward with respect to productivity, quality, and time-to-market will increase the reuse of software artifacts and, as a consequence, will encourage the use of component-based development methods. Understanding the basic state of emotion of software developers will support efforts to overcome developers' reluctance to accept this methodology by emphasizing challenges and opportunities provided by the new methods, re-evaluating the importance and visibility of certain tasks, changing job profiles, and changing the reward and recognition structure.

The consequence might be that software designers wholeheartedly accept component-based methodologies not only as an economic necessity but also as a means of achieving the status of a great designer, as postulated by Brooks (1986). In turn, this could lead to a new level of professionalism in software development and would allow component-based development methods to be utilized fully in the field of software engineering.

CONCLUSION

Soft factors like motivation and psychological aspects often play a strong role even in a technical, seemingly rational field like software engineering. We have discussed and identified key soft factors that often account for the slow uptake of component-based software development methods and relate them to the framework of Maslow's Hierarchy of Needs. The users of software components were the focus of this discussion. There are some indications that for providers of components, a different emotional situation exists (Chroust & Hoyer, 2004).

Recognition of the different levels of resistance and their psychological background will, among other aspects, allow approaching the problems in a psychologically appropriate form. The need of the software industry to come to terms with its problems of quality, cost, and timeliness makes this a necessity.

REFERENCES

Abran, A., Moore, J., Bourque, P., Dupuis, R., & Tripp, L.L. (Eds.). (2004). *Guide to the software engineering body of knowledge: 2004 version tech. rep. DTR 19759.* International Organization for Standardization, Geneva, Switzerland.

Allen, P. (2001). *Realizing e-business with components.* Chicago: Addison-Wesley.

Arbaoui, S., Lonchamp, J., & Montangero, C. (1999). The human dimension of the software process. In J.-C. Derniame, D. Ali Kaba, & D.G. Wastell (Eds.), *Software process: Principles, methodology, and technology* (pp. 165-200). New York: Springer.

Bach, J. (1997). Good enough quality: Beyond the buzzword. *IEEE Computer, 30*(8), 96-98.

Bachmann, F., et al. (2000). *Volume II: Technical concepts of component-based software engineering* (technical report). CMU/SEI-2999-TR-008, ESC-TR-2000-007.

Badoo, N., & Hall, T. (2001). *Motivators of software process improvement: An analysis of practitioners' views* (technical report). Hertfordshire, UK: University of Hertfordshire.

Balzert, H. (1996). *Lehrbuch der Software-Technik, Software Entwicklung.* Heidelberg, Germany: Verlag.

Basili, V.R. (1990). Viewing maintenance as re-use oriented software development. *IEEE Software, 7*(1), 19-25.

Boehm, W., et al. (2000). *Software cost estimation with COCOMO II.* NJ: Prentice Hall.

Boeree, C.G. (1998). *Abraham Maslow, biography.* Retrieved November, 2001, from http://www.ship.edu/~cgboeree/maslow.html

Brooks, F. (1986). *No silver bullet—Essence and accidents of software engineering.* Information Processing 86, IFIP Congress, 1069-1076.

Campell, J. (2001). Course on reuse: Impediments to reuse. Retrieved from http://www.cs.qub.ac.uk/~J.Campbell/myweb/misd/ node8.html#section008 40000000000000000

Cheesman, J., & Daniels, J. (2001). *UML components.* Amsterdam, The Netherlands: Longman, Addison-Wesley.

Chroust, G. (1996). Software 2001—Ein weg in die Wiederverwendungswelt. In F. Lehner (Ed.), *Softwarewartung und reengineering—Erfahrungen und entwicklungen* (pp. 31-49). Germany: Deutscher Universitätsverlag.

Chroust, G., & Hoyer, C. (2004). Motivational issues in creating reusable software artifacts. In R. Trappl (Ed.), *Cybernetics and systems 2004* (pp. 417-422). Vienna: Austrian Soc. for Cybernetic Studies.

Crnkovic, I., & Larsson, S. (Eds.). (2002). *Building reliable component-based software systems.* Boston: Artech House Publshing.

Cusumano, M. (1991). Factory concepts and practices in software development. *IEEE Annals of the History of Computing, 13*(1), 3-31.

deRemer, F.K.H. (1976). Programming-in-the-large versus programming in-the-small. *IEEE Tr. on Software Eng.,* (292), 80-86.

Disterer, G. (2000). Individuelle und Soziale Barrieren Beim Aufbau von Wissenssammlungen. *Wirtschaftsinformatik, 42*(6), 539-546.

Eason, K., Domodaran, L., & Stewart, T. (1974). Interface problems in man-computer interaction. In E. Mumford, & H. Sackman (Eds.), *Human choice and computers* (pp. 91-105). North Holland Publshing Company.

Fichman, R.G., & Kemerer, C. (2001). Incentive compatibility and systematic software reuse. *Journal of Systems and Software, 57*(1), 54.

Glass, R. (1983). *Software runaways.* NJ: Prentice Hall.

Glass, R. (2001). Frequently forgotten fundamental facts about software engineering. *IEEE Software, 18*(3), 112-110.

Halaris, J., & Spiros, T. (1996). Reuse concepts and a reuse support repository. *Proceedings of the IEEE Symposium and Workshop on Engineering of Computer Based Systems*, Germany.

Heineman, G.T., et al. (2000). Component-based software engineering and the issue of trust. *Proceedings of the 22nd International Conference on Software Engineering*. Limerick, Holland.

ISO/IEC. (2004). *ISO 25000: Software and systems engineering: Software product quality requirements and evaluation (SQuaRE)—Guide to SQuaRE* [technical report]. International Organization for Standardization. Geneva, Switzerland.

Jones, C. (1995). Why is technology transfer so hard? *IEEE Computer, 28*(6), 86-87.

Kock, N. (1999). A case of academic plagiarism. *Comm. ACM, Vol. 42*(7), 94-104.

Kraft, P. (1977). *Programmers and managers*. Heidelberg, Germandy: Springer.

Kumar, K., & Bjoern-Andersen, N. (1990). A cross-cultural comparison of IS designer values. *Comm ACM, 33*(5), 528-538.

Kunda, D., & Brooks, L. (1999). *Human, social and organisational influences on component-based software engineering*. ICSE'99 Workshop on Component-Based Software Engineering, Los Angelos. Retrieved July, 2004, from http://www.sei.cmu.edu/cbs/icse99/papers/19/19.htm

Lynex, A., & Layzell, P. (1997). Understanding resistance to software reuse. *Proceedings of the 8th International Workshop on Software Technology and Engineering Practice*, London.

Maslow, A. (1943). A theory of human motivation. *Psychological Review, 50*, 370-396.

McConnell, S. (2000). Quantifying soft factors. *IEEE Software, 17*(6), 9-11.

Riemenschneider, C., Hardgrave, B.C., & Davis, F. (2002). Explaining software developer acceptance of methodologies: A comparison of five theoretical models. *IEEE Trans. on Software Engineering, 28*(12), 1135.

Schoitsch, E. (2003). Dependable embedded systems—Vision und roadmap. In G. Fiedler, & D. Donhoffer (Eds.), *Mikroelektronik 2003, Wien* (p. 33). Vienna, Austria: ÖVE Schriftenreihe.

Sonntag, M., & Chroust, G. (2004). Legal protection of component metadata and APIs. In R. Trappl (Ed.), *Cybernetics and systems 2004*. Vienna, Austria: Austrian Soc. for Cybernetic Studies.

Thomas, S., Hurley, S., & Barnes, D. (1996). Looking for the human factors in software quality management. *Proceedings of the 1996 International Conference on Software Engineering: Education and Practice (SE:EP '96)*. Otago, New Zealand.

Vecellio, G., & Thomas, W.M. (2001). Issues in the assurance of component based software. Retrieved July, 2001, from http://www.mitre.org/pubs/edge perspectives/ march 01/vecellio.html

Vitharana, P. (2003). Risks and challenges of component-based software development. *Comm. ACM, 46*(8), 67-72.

Voas, J. (1998). Certifying off-the-shelf software components. *IEEE Computer, 11*(6), 53-59.

Wasmund, M. (1995). The spin-off illusion: Reuse is not a by-product. *Proceedings of the 1995 Symposium on Software Reusability*, Seattle, Washington.

Weinberg, G. (1988). *Understanding the professional programmer*. New York: Dorset House.

Woodman, M., Benedictsson, O., Lefever, B., & Stallinger, F. (2001). Issues of CBD product quality and process quality. *Proceedings of the 4th ICSE Workshop: Component Certification and System Prediction, 23rd International Conference on Software Engineering (ICSE)*. Toronto, Canada.

KEY TERMS

Commercial Off the Shelf (COTS): Software products that an organization acquires from a third party with no access to the source code and for which there are multiple customers using identical copies of the component.

Component-Based Development (CBD): In contrast to classical development (waterfall-process and similar process models), CBD is concerned with the rapid assembly of systems from components (Bachmann et al., 2000) where:

- components and frameworks have certified properties; and

- these certified properties provide the basis for predicting the properties of systems built from components.

Component Model: A component model specifies the standards and conventions imposed on developers of components. This includes admissible ways of describing the functionality and other attributes of a component, admissible communication between components (protocols), and so forth.

Maslow's Hierarchy of Needs: Maslow's Hierarchy of Needs (Boeree, 1998; Maslow, 1943; McConnell, 2000) is used as a structuring means for the various factors. It defines five levels of need:

- self-fulfillment
- recognition
- social environment (community)
- basic physiological needs (survival)
- security (physical, economic ...)

In general, the needs of a lower level must be largely fulfilled before needs of a higher level arise.

Soft Factors: This concept comprises an ill-defined group of factors that are related to people, organizations, and environments like motivation, morale, organizational culture, power, politics, feelings, perceptions of environment, and so forth.

Software Component: A (software) component is (Bachmann et al., 2000):

- an opaque implementation of functionality
- subject to third-party composition
- in conformance with a component model

Software Engineering: (1) The application of a systematic, disciplined, quantifiable approach to development, operation, and maintenance of software; that is, the application of engineering to software and (2) the study of approaches as in (1) (Abran, Moore, Bourque, Dupuis & Tripp, 2004).

Obstacles for the Integration of HCI Practices into Software Engineering Development Processes

Xavier Ferre
Universidad Politécnica de Madrid, Spain

Natalia Juristo
Universidad Politécnica de Madrid, Spain

Ana M. Moreno
Universidad Politécnica de Madrid, Spain

INTRODUCTION

Usability has become a critical quality factor in software systems, and it has been receiving increasing attention over the last few years in the SE (software engineering) field. HCI techniques aim to increase the usability level of the final software product, but they are applied sparingly in mainstream software development, because there is very little knowledge about their existence and about how they can contribute to the activities already performed in the development process. There is a perception in the software development community that these usability-related techniques are to be applied only for the development of the visible part of the UI (user interface) after the most important part of the software system (the internals) has been designed and implemented.

Nevertheless, the different paths taken by HCI and SE regarding software development have recently started to converge. First, we have noted that HCI methods are being described more formally in the direction of SE software process descriptions. Second, usability is becoming an important issue on the SE agenda, since the software products user base is ever increasing and the degree of user computer literacy is decreasing, leading to a greater demand for usability improvements in the software market. However, the convergence of HCI and SE has uncovered the need for an integration of the practices of both disciplines. This integration is a must for the development of highly usable systems.

In the next two sections, we will look at how the SE field has viewed usability. Following upon this, we address the existing approaches to integration. We will then detail the pending issues that stand in the way of successful integration efforts, concluding with the presentation of an approach that might be successful in the integration endeavor.

Traditional View of Usability in Software Engineering

Even though usability was mentioned as a quality attribute in early software quality taxonomies (Boehm, 1978; McCall, Richards, & Walters, 1977), it has traditionally received less attention than other quality attributes like correctness, reliability, or efficiency. While the development team alone could deal with these attributes, a strong interaction with users is required to cater for usability. With SE's aim of making the development a systematic process, the human-induced unpredictability was to be avoided at all costs, thus reducing the interaction with users to a minimum.

The traditional relegation of usability in SE can be acknowledged by observing how interaction design is marginally present in the main software development process standards: ISO/IEC Std. 12207 (1995) and IEEE Std. 1074 (1997). The ISO 12207 standard does not mention usability and HCI activities directly. It says that possible user involvement should be planned, but this involvement is circumscribed to requirements setting exercises, prototype demonstrations, and evaluations. When users are mentioned, they play a passive role in the few activities in which they may participate. The IEEE standard 1074 only mentions usability in connection with UI requirements and risk management. Neither of the two standards addresses any of the activities needed to manage the usability of the software product.

Recent Changes Regarding Usability Awareness

There has been a noticeable shift in the attention paid to usability in the SE field in recent years, since important overlapping areas have been identified in the SWEBOK (Guide to the Software Engineering Body of Knowledge) (IEEE Software Engineering Coordinating Committee, 2001), for example, which is an effort to gather what is considered commonly accepted knowledge in the SE field. The SWEBOK requirements engineering knowledge area includes some techniques that are not identified by the authors as belonging to HCI, but they are indeed standard HCI techniques: interviews, scenarios, prototyping, and user observation. Additionally, good communication between system users and system developers is identified as one of the fundamental tenets of good SE. Communication with users is a traditional concern in HCI, so this is an overlapping area between HCI and SE. Usability is mentioned as part of the quality attributes and highlighted in the case of high dependability systems. It is also mentioned with regard to the software testing knowledge area. The work by Rubin (1994) is listed as part of the reference material for usability testing.

The approval of Amendment 1 to standard ISO/IEC 12207 (ISO/IEC, 2002), which includes a new process called Usability Process, by the ISO in 2002 represented a big change regarding the relevance of usability issues in the SE field. With the release of this amendment, the ISO recognized the importance of managing the usability of the software product throughout the life cycle. The main concepts in a human-centered process, as described in the ISO 13407 standard (1999), are addressed in the newly created usability process. The approach taken is to define in the usability process the activities to be carried out by the role of usability specialist. Some activities are the sole responsibility of the usability specialist, while others are to be applied in association with the role of developer. The first activity in the usability process is 6.9.1, Process implementation, which should specify how the human-centered activities fit into the whole system life cycle process, and should select usability methods and techniques. This amendment to the ISO/IEC 12207 standard highlights the importance of integrating usability techniques and activities into the software development process.

The fact that an international standard considers usability activities as part of the software development process is a clear indication that HCI and usability are coming onto the SE agenda, and that integrating HCI practices into the SE processes is a problem that the software development community needs to solve quite promptly.

BACKGROUND

This section details existing integration proposals. We will only consider works that are easily accessible for software practitioners, that is, mainly books, since average software practitioners do not usually consider conference proceedings and research journals as an information source.

Only a few of the numerous HCI methods give indications about how to integrate the usability activities with the other activities in the overall software development process. Of these works, some just offer some hints on the integration issue (Constantine & Lockwood, 1999; Costabile, 2001; Hix & Hartson, 1993), while others are more detailed (Lim & Long, 1994; Mayhew, 1999).

Hix and Hartson (1993) describe the communication paths that should be set up between usability activities (user interaction design) and software design. They strictly separate the development of the UI from the development of the rest of the software system, with two activities that connect the two parts: systems analysis and testing/evaluation. The systems analysis group feeds requirements to both the problem domain design group and the user interaction design group. It is a simplistic approach to HCI-SE integration, but the authors acknowledge that "research is needed to better understand and support the real communication needs of this complex process" (p. 112).

Constantine and Lockwood (1999) offer some advice for integrating usability and UI design into the product development cycle, acknowledging that there is no one single way of approaching this introduction. Therefore, they leave the issue of integration to be solved on a case-by-case basis. They state that "good strategies for integrating usability into the life cycle fit new practices and old practices together, modifying present practices to incorporate usability into analysis and design processes, while also tailoring usage-centered design to the organization and its practices" (p. 529).

Costabile (2001) proposes a way of modifying the software life cycle to include usability. The basis chosen for such modifications is the waterfall life cycle. The choice of the waterfall life cycle as a "standard" software life cycle is an important drawback of Costabile's proposal, as it goes against the user-centered aim of evaluating usability from the very beginning and iterating to a satisfactory solution. In the waterfall life cycle, paths that go back are defined for error correction, not for completely changing the approach if it proves to be wrong, since it is based on frozen requirements (Larman, 2001). Glass (2003) acknowledges that "requirements frequently changed as product development goes under way [...]. The experts knew that waterfall was an unachievable ideal" (p. 66).

MUSE (Lim & Long, 1994) is a method for designing the UI, and the work by Lim and Long includes its detailed integration with the JSD (Jackson System Development) method. The authors state that MUSE, as a structured method, emphasizes a design analysis and documentation phase prior to the specification of a "first-best-guess" solution. Therefore, MUSE follows a waterfall approach, not a truly iterative approach. Regarding its integration with other processes, JSD is presented in this work as a method that is mainly used for the development of real-time systems. Real-time systems account for a very small part of interactive systems, so the integration of MUSE with JSD is not very useful from a generic point of view. Additionally, structured design techniques like structured diagrams or semantic nets make it difficult to adapt to processes based on other approaches, in particular to object-oriented development.

Mayhew (1999) proposes the Usability Engineering Life cycle for the development of usable UIs. This approach to the process follows a waterfall life cycle mindset: an initial Analysis phase, followed by a Design/Test/Development phase, and finally an Installation phase. The Analysis stage is only returned to if not all functionality is addressed, and this is, therefore, not a truly iterative approach to software development. Nevertheless, it is one of the more complete HCI processes from the SE point of view. Although Mayhew claims that the method is aimed at the development of the UI only, the activities included in this life cycle embrace an important part of requirements-related activities (like, for example, contextual task analysis). Links with the OOSE (object-oriented software engineering) method (Jacobson, Christerson, Jonsson, & Övergaard, 1993) and with rapid prototyping methods are identified, but the author acknowledges that the integration of usability engineering with SE must be tailored and that the overlap between usability and SE activities is not completely clear. Accordingly, Mayhew presents UI development as an activity that is quite independent from the development of the rest of the system.

PENDING ISSUES FOR INTEGRATION

Having studied the existing integration proposals and considering the most widespread perception of usability issues among developers, we can identify four main obstacles that need to be overcome in order to successfully integrate HCI practices into the overall software development process: UI design vs. interaction design, integration with requirements engineering, iterative development, and user-centered focus throughout the development process.

UI Design vs. Interaction Design

One of the biggest obstacles for HCI-SE integration is the existing terminology breach and the disparity in the concepts handled. These differences are especially noticeable in the denomination of what can be considered the main HCI area of expertise: UI design. As it is understood in the HCI field, UI design represents a wider concept than in SE terminology.

SE refers by UI design to just the design of the concrete visual elements that will form the UI and its response behavior (in visual terms). It does not include any activity related to requirements engineering. On top of this, there is a widely-accepted principle in SE stating that the part of the system that manages the visual elements of the UI should be separated from the business logic (the internal part of the system). The strict application of this principle results in a UI design that is not directly related to the design of the internal system processes. On the graphical side, UI design is produced by graphic designers, whose work is governed by aesthetic principles. It is this conception of UI design that makes SE regard it as part of a related discipline, not as one of the core activities that matter most for any software development project.

On the other hand, HCI literature uses the term to represent a broader set of activities. Most HCI methods label themselves as methods for the design of the UI (Hix & Hartson, 1993; Lim & Long, 1994; Mayhew, 1999), while including activities that are outside the scope of UI design in SE terms (like user and task analysis).

With the aim of a successful integration, we suggest the use of a different term for what HCI considers UI design in order to raise SE receptiveness to the integration efforts. Specifically, we propose the term *interaction design*, meaning the coordination of information exchange between the user and the system (Ferre, Juristo, Windl, & Constantine, 2001). Software engineers may then understand that usability is not just related to the visible part of the UI, since activities that study the best suited system conception, user needs and expectations and the way tasks should be performed need to be undertaken to perform interaction design. All these additional issues belong to the requirements engineering subfield of SE, as detailed in the next subsection.

Integration with Requirements Engineering

Some integration proposals considered in the previous section are based on two development processes carried out in parallel: the interaction design process (following an HCI method) and the process that develops the rest of the system (a SE process). The underlying hypothesis is that the issues with which each process deals are not directly related, that is, some coordination between the two processes is needed, but they may be basically carried out separately.

Nevertheless, looking at the key tasks for final product usability, like user and task analysis, user observation in their usual environment, needs analysis and the development of a product concept that can better support such needs, we find that they are all activities that, to a lesser or greater extent, have been traditionally carried out within the framework of requirements engineering, a SE subdiscipline. HCI can provide a user-centered perspective to assure that these activities are performed in the software process with positive results regarding the usability of the final software product, emphasizing this perspective throughout the development process. In short, there is a big enough overlap between the two disciplines to call for a tight integration of activities and techniques from each discipline.

Some HCI authors, like Mayhew (1999), defend that requirements-related activities should be performed by HCI experts instead of software engineers and that the rest of the development process should build upon the HCI experts' work. This approach may be valid in organizations where usability is the main (or only) quality attribute to be aimed for and where a usability department is one of the leading departments in the organization. For other organizations that are not so committed to usability, it is not worth their while to completely abandon their way of performing requirements engineering (a recognized cornerstone of any software development project in SE) when the only gain is an improvement in just one quality attribute (usability) of the resulting product. We take the view that HCI experts may work together with software engineers, and *the user-centered flavor of HCI techniques may greatly enrich requirements engineering techniques*, but the complete substitution of SE for HCI practices in this area is not acceptable from a SE point of view. Additionally, the big overlap between HCI and SE regarding the requirements-related issues makes the approach of undertaking two separate processes (HCI and SE) communicating through specific channels ineffective, because performing SE activities without a user-centered focus could invalidate the results from a usability point of view.

Iterative Development

An iterative approach to software development is one of the basic principles of user-centered development according to HCI literature (Constantine & Lockwood, 1999; Hix & Hartson, 1993; ISO, 1999; Nielsen, 1993; Preece, Rogers, Sharp, Benyon, Holland, & Carey, 1994; Shneiderman, 1998). The complexity of the human side in human-computer interaction makes it almost impossible to create a correct design at the first go.

On one hand, SE literature has gradually come to accept that an iterative as opposed to a waterfall life cycle approach is the best for medium to high complexity problems when the development team does not have in-depth domain knowledge. Nevertheless, a waterfall mindset is still deeply rooted in day-to-day practice among software developers. The reason for this is that the waterfall is a very attractive software development model from a structural viewpoint, because it gives the illusion of order and simplicity within such a complex activity (software systems development). Therefore, although SE acknowledges the virtues of the iterative approach, which would appear to facilitate the integration of a user-centered perspective, this approach is not usually applied in practice, which is a major deterrent. Additionally, as mentioned earlier, a common mistake in the efforts for integrating HCI practices into software development has been to use the waterfall life cycle as a starting point. We defend that *a truly iterative approach should be highlighted as one of greatest possible contributions of HCI practice to overall software development*, as it has been part of its core practices for a long time.

A User-Centered Perspective throughout Development

When usability activities are performed independently from the rest of development activities, there is a risk of losing the user-centered perspective somewhere along the way. This perspective underlies the entire development process in HCI, since it is necessary for producing a usable software system. Therefore, *a user-centered perspective needs to be conveyed to the developers that are to undertake all the activities that are not strictly usability-related*. This will ensure that usability is considered throughout the development process, as other quality attributes (like, for example, reliability) are.

When because of the specific circumstances of an existing software development organization, it is impossible or undesirable to hire a lot of HCI experts to apply the HCI techniques, developers will need to apply some of the techniques themselves. Indeed, we think that some common HCI techniques, like card sorting, user modeling or navigation design, could be undertaken by the software engineering team, provided that they receive adequate usability training. Some other HCI techniques that require a lot of usability expertise would still need to be applied by HCI experts.

FUTURE TRENDS

As a discipline, SE is pervasive in software development organizations all over the world. Its concepts are the ones with which the majority of developers are familiar, and this is especially true of senior management at software development organizations. HCI, on the other hand, has been traditionally considered as a specialist field, but there is an increasing demand within the SE field for effective integration of its activities and techniques into the overall software process. Therefore, the trend is towards an effective integration of HCI techniques and activities into SE development practices. Teams will include usability specialists, and software engineers will acquire the basic usability concepts in order to improve team communication. Some software engineers may even be able to apply some HCI techniques.

For multidisciplinary teams with a SE leadership to be workable, the terminology breach needs to be surmounted. For this purpose, we suggest that a usability roadmap aimed at software developers be drawn up. This roadmap would serve as a toolbox for software engineers who want to include HCI practices in the development process currently applied at their software development organizations. It should then be expressed according to SE terminology and concepts, and it should include information for each HCI technique about what kind of activity it is applied for and about when in an iterative process its application most contributes to the usability of the final software product. Software developers may then manage usability activities and techniques along with SE ones. The only requirement for the existing development process would be that it should be truly iterative, since a waterfall approach would make any introduction of usability techniques almost irrelevant.

CONCLUSION

HCI and SE take different but complementary views of software development. Both have been applied separately in most projects to the date, but overlapping areas between both disciplines have been identified and the software development field is claiming a tighter integration of HCI aspects into SE development processes.

Existing integration proposals suffer important shortcomings, such as not being truly iterative or advocating a separate HCI process. There is a terminology breach between SE and HCI, apparent in the denomination of HCI's main concern, UI design, which could be expressed as interaction design to assure better communication with software engineers. Additionally, it is crucial for usability to be present throughout the whole development process in order to maintain a proper user-centered focus.

A usability roadmap expressed using SE terminology and concepts may help software developers to overcome these obstacles and to perform a successful integration of HCI aspects into the software development process.

REFERENCES

Boehm, B. (1978). *Characteristics of software quality*. New York: North Holland Publishing Co.

Constantine, L. L., & Lockwood, L. A. D. (1999). *Software for use: A practical guide to the models and methods of usage-centered design*. Reading, MA: Addison-Wesley.

Costabile, M. F. (2001). Usability in the software life cycle. In S. K. Chang (Ed.), *Handbook of software engineering and knowledge engineering* (pp. 179-192). Singapore: World Scientific Publishing.

Ferre, X., Juristo, N., Windl, H., & Constantine, L. (2001, January/February). Usability basics for software developers. *IEEE Software*, *18*(1), 22-29.

Glass, R. L. (2003). *Facts and fallacies of software engineering*. Boston: Addison-Wesley.

Hix, D., & Hartson, H. R. (1993). *Developing user interfaces: Ensuring usability through product and process*. New York: John Wiley & Sons.

IEEE. (1997). *IEEE Std 1074-1997. IEEE Standard 1074 for Developing Software Life Cycle Processes*. New York: IEEE.

IEEE Software Engineering Coordinating Committee. (2001, May). *Guide to the software engineering body of knowledge—Trial Version 1.00*. Los Alamitos, CA: IEEE Computer Society.

ISO. (1999). *ISO 13407. Human-centred design processes for interactive systems*. Geneva, Switzerland: ISO.

ISO/IEC. (1995). *International Standard: Information Technology. Software Life Cycle Processes, ISO/IEC Standard 12207:1995*. Geneva, Switzerland: ISO/IEC.

ISO/IEC. (2002). *ISO/IEC International Standard: Information Technology. Software Life Cycle Processes. Amendment 1. ISO/IEC 12207:1995/Amd. 1:2002*. Geneva, Switzerland: ISO/IEC.

Jacobson, I., Christerson, M., Jonsson, P., & Övergaard, G. (1993). *Object-oriented software engineering. A use-case driven approach* (Revised Printing). Harlow, UK: ACM Press—Addison-Wesley.

Larman, C. (2001). *UML and patterns: An introduction to object-oriented analysis and design and the unified process* (2nd ed.). Upper Saddle River, NJ: Prentice Hall PTR.

Lim, K. Y., & Long, J. (1994). *The MUSE method for usability engineering*. Glasgow, UK: Cambridge University Press.

Mayhew, D. J. (1999). *The usability engineering lifecycle*. San Francisco: Morgan Kaufmann.

McCall, J. A., Richards, P. K., & Walters, G. F. (1977). *Factors in software quality, vol. 1, 2, and 3. AD/A-049-014/015/055*. National Tech. Information Service.

Nielsen, J. (1993). *Usability engineering*. Boston: AP Professional.

Preece, J., Rogers, Y., Sharp, H., Benyon, D., Holland, S., & Carey, T. (1994). *Human-computer interaction*. Harlow, UK: Addison-Wesley.

Rubin, J. (1994). *Handbook of usability testing: How to plan, design, and conduct effective tests*. New York: John Wiley & Sons.

Shneiderman, B. (1998). *Designing the user interface: Strategies for effective human-computer interaction*. Reading, MA: Addison-Wesley.

KEY TERMS

Interaction Design: The coordination of information exchange between the user and the system.

Iterative Development: An approach to software development where the overall life cycle is composed of several iterations in sequence.

Requirements Engineering: The systematic handling of requirements.

Software Engineering: The application of a systematic, disciplined, quantifiable approach to the development, operation, and maintenance of software.

Software Process: The development roadmap followed by an organization to produce software systems, that is, the series of activities undertaken to develop and maintain software systems.

Software Requirements: An expression of the needs and constraints that are placed upon a software product that contribute to the satisfaction of some real world application.

User-Centered Development: An approach to software development that advocates maintaining a continuous user focus during development, with the aim of producing a software system with a good usability level.

On Not Designing Tools

Sarah Kettley
Napier University, UK

INTRODUCTION

The reader is no doubt well aware of HCI's emphasis on the analysis of systems in which the computer plays the role of tool. The field encompasses positivist and pragmatic approaches in analyzing the products and the trajectories of use of technology (Coyne, 1995; Ihde, 2002; Preece et al., 1994), and many useful guidelines for the design of task-oriented tools have been produced as a result. However, use value and efficiency increasingly are leaving consumers cold; society has always needed things other than tools, and expectations of personal digital products are changing. Once utilitarian, they are now approached as experience, and Pat Jordan, for example, has successfully plotted the progression from functionality to usability to pleasure (Jordan, 2000). A precedent set by the Doors of Perception community (van Hinte, 1997) has seen slow social movements becoming more prevalent, design symposia dedicated to emotion, and traditional market research challenged by the suggestion that the new consumer values something other than speed and work ethics. This search for authenticity appears to be resistive to demographic methodologies (Boyle, 2003; Brand, 2000; Lewis & Bridger, 2000) yet underpins important new approaches to sustainable consumption (Brand, 2000; Bunnell, 2002; Csikzsentmihalyi & Rochberg-Halton, 1981; Fuad-Luke, 2002; van Hinte, 1997). The next section introduces pragmatic and critical approaches to HCI before examining the importance of the artwork as authentic experience.

BACKGROUND

Pragmatism

HCI's activity revolves around tools. Its philosophical framework traditionally has been one of usefulness, demonstrated in terms of the workplace; it can show "tangible benefits that can be talked of in cash terms ... providing clear cut examples of case studies where ... costs have been reduced, work levels improved, and absenteeism reduced" (Preece et al., 1994, p. 19). Winograd and Flores (1986) defined the scope of their investigation as being primarily "what people do in their work" and saw the issues arising from this study to be pertinent to "home-life as well" (p. 143). Interaction design is focused similarly on optimizing the efficiency of the tool: "Users want a site that is easy to use, that has a minimum of download time, and that allows them to complete their tasks in a minimal amount of time with a minimal amount of frustration" (Lazar, 2001, p. 3). Both disciplines are increasingly taking into account the social situation of communities of users, and the constitutive nature of technology itself; that is, it is understood that the introduction of a technology into society often is merely the beginning rather than the culmination of the cycle of appropriation. It is this socially constitutive aspect of technology that requires HCI to embrace not only pragmatism but also critical design practices.

A Critical View

A critical stance questions the role of technology with respect to social and political structures and inquires into the future of humankind in light of its appropriation. Design carries with it the ethical implications of its impact on communities, no matter that trajectories cannot be predetermined: "Design ... imposes the interests of a few on the many" and is "a political activity" (Coyne, 1995, pp. 11-12). It raises questions about human activity in meaning making in contrast with passivity. McCarthy & Wright (2003), for example, evoke the Apple Mac as an "object to be with" but go on to ask whether we "passively consume this message or complete the experience ourselves" (p. 88). The situation at the moment is such that any experience of computers that throws into relief the nature of the computer itself is critical in nature. In challenging pragmatism, the critical position raises questions about the need for socially grounded performative meaning making and about how truth is often seen to be embodied and presented by the technological reasoning of the machine. In practical terms, pragmatism in interaction design is characterized by an emphasis on the transparent interface, as championed by Winograd & Flores (1986) in the early visions for ubiquitous computing (Weiser, 1991) and by cognitive psychologist Donald Norman (1999); the critical nature of the artwork for HCI lies in its re-physicalization of technology. This physicality, or obstinacy, is dependent on the user's awareness of the interface in interaction, which platonic design seeks to minimize if not erase.

Phenomenology: Disappearance and Obstinacy

The notion of the tool is challenged by awareness; tools by definition disappear (Baudrillard, 1968; Heidegger, 1962). The phenomenologically invisible interface was described first by Winograd and Flores (1986) in their seminal book, *Understanding Computers and Cognition*, further elucidated by Steve Weiser (1991) in his visions for the paradigm of ubiquitous computing, and finally popularized by Donald Norman's (1999) *The Disappearing Computer*. These texts take as a starting point a phenomenological view of action in the world; that is, as soon as tools become conspicuous, to use a Heideggerian term, they are no longer available for the specific task in mind (or ready-to-hand), instead becoming obstinate and obtrusive (present-at-hand). As long as we approach a tool with the goal of using it for a specific task, such obstinacy will remain negative, but there does exist a different class of artifact where it becomes interesting, positive, and even necessary for the existence of the artifact in the first place. *Objection-able* might be an alternative to Heidegger's terminology, embodying the idea of a thing regaining its materiality, that existence that is dependent on performative human perception. Baudrillard (19698) talks about objects as being non-tools, about their being ready for appreciation, part of a value system created through appreciation. They are the noticed artifacts in our lives, and as such are positioned to accrue the personal meaning that underlies truly authentic experience. This approach takes the concept of phenomenological disappearance and shifts the focus from the transparent interface to that of the visible object. The difference lies in the location of breakdown and in its recasting as an essentially positive part of experience. Winograd and Flores (1986) point out that meaning arises out of "how we talk about the world," emerging in "recurrent patterns of breakdown and the potential for discourse about grounding" (p. 68); in the design of transparent, seamless experiences, breakdown is something to be prepared against rather than encouraged. The authors apply Heidegger's readiness-to-hand to the design of systems that support problem solving in the face of inevitable and undesirable breakdown situations. This article presents a case for an alternative application of an understanding of the same phenomenological concepts towards the production of visible, objection-able artifacts. The following section introduces art as process and product, defined by objection-ability, and examines the human need for art in light of this quality.

ART

Philosophical Importance of Art

Art objects are those that are created expressly to spark cognition through a combination of order and disorder, through continuity and discontinuity (Pepperell, 2002). New languages are formed in expressing aspects of being in new ways. The artifact acts as a medium for expression (even if the intent of the artist is to erase authorship); but it is in the gap for open subjective reading, in the active articulation of pre-linguistic apprehension, that meaning is co-created (Eldridge, 2003). Thus, to conceive of a meaningful digital product is to intentionally invert the paradigm of the invisible computer. If we are designing products to become meaningful objects, then in order to trigger that articulation, we must introduce discontinuity, even Heideggerian breakdown, opening up the space for the inter-subjective co-production of meaning (Baudrillard, 1968; Eco, 1989; Greenhalgh, 2002; Heidegger, 1962;

Ihde, 2002; Pepperell, 2002). In the use of denotative symbolism, this space is increasingly closed to the reader. Even in more contextual design practice where connotative meaning is taken into account, the end goal remains one of seamlessness in context. It should be useful instead to approach the design process as an attempt to build coherent new vocabularies with computational materials in the manner of artists. This challenges the notion of design patterns in particular, which limit subjective reading; artists, in contrast, embark on an "obsessive, intense search" (Greenhalgh, 2002, p. 7), "working through the subject matter" of their emotions in order to objectify subjective impulses" (Eldridge, 2003, p. 70). The goal of the artwork is not transparency but reflection, and Bolter and Gromala (2003) show us how digital arts practice illustrates this. In their reflection, digital artworks become material in interaction, questioning the very technologies they depend upon for their existence. It is proposed that we now seek to complement the current paradigms of computing through the conscious use of computational materials to create new expressions, new objects, that are not efficient feature-rich tools but that instead may play a different, rich social role for human beings.

MeAoW (MEDIA ART OR WHATEVER)

The Center for Advanced Technology at New York University initiated a lecture series named *The CAT's MeAoW* to "facilitate artists' engagement with technologies and technologists" (Mitchell et al., 2003, p. 156); the title was chosen intentionally to reflect the lack of consensus on terminology pervading the field. Artists always have been involved and often instrumental in the development of technology, using it toward their own expressive ends and asking different sorts of questions through it to those of the scientists. The myriad uses of the computer in art reflect its plasticity, creating various interconnected fields of artistic endeavor, including graphics, video, music, interactive art, and practice making use of immersive technology, embedded and wearable systems, and tangible computing (see, for example, the work of Thecla Schiphorst and Susan Kozel on the Whisper project, and Hiroshi Ishii and the work of the Tangible computing group at MIT's Media Lab). Issues of temporality, perception, authorship, and surveillance continue to engage artists using these media as well as hardware, coding, and output as expressive materials in their own right. Steve Mann's work with wearable systems stems from his days as a photographer and interests in issues of surveillance—his wearable computing devices give power back to the user in their ability to survey for themselves. Major communities of practice at the intersections of art and technology can be found centered on organizations such as SIGGRAPH (since the mid-1960s), festivals such as Ars Electronica (since 1979), and MIT's journal, *Leonardo* (since 1968). The human-computer interaction community, in contrast, has been served by its special interest group, SIGCHI, since a relatively recent 1982. It is only recently that a few HCI researchers and practitioners have begun to approach arts practices and outcomes as rigorously as those methodologies adopted from the social sciences. Anthony Dunne's (1999) concept of the post-optimal object led him to explore methods for the design and dissemination of genotypes, artifacts intended not as prototypical models for production but as props for critical debate; Dunne's work with Fiona Raby went on to place more or less functioning artifacts in volunteers' households (Dunne & Raby, 2002), while his work with William Gaver and Elena Pacenti resulted in the influential *Cultural Probes*, a collection of arts-based methods designed to inform the designers in the project in a far less directional way than user-centered design methodologies have been used to doing (Gaver et al., 1999).

FUTURE TRENDS

Arts-based methods of evaluation and production are increasing in importance for human-computer interaction, which, in turn, indicates a rethinking of the end goals of interaction and computational product design to take account of the critical. In order to deal with theses changes, HCI is having to add new transdisciplinary methodologies to complement its more comfortable user-centered approaches. Noting the temptation for practitioners in HCI to bend the cultural probes to more quantitative ends, Gaver (2004) has renamed the approach *probology* in an effort to restate its initial aims and to reiterate the importance of asking the right type of questions through it. John Haworth's (2003) Arts and Humanities Research Board-funded project, Creativity and Embodied Mind in Digital Fine Art (2002-2003), produced critical products for the public realm and was based on an "innovative interlocking" of methods, including "creative practice and reflection, literature and gallery research, interviews with artists, seminar-workshops, and an interactive website," emphasizing "the importance of both pre-reflexive and reflexive thought in guiding action" (pp. 1-3) (Candy et al., 2002). Mitchell, et al. (2003) extrapolate the broadening range of qualitative methodologies that HCI is encompassing to suggest a future inclusion of non-utilitarian evaluation techniques more typically employed by artists. The editors say these "differ radically from those of computer scientists," making the important point that artists "seek to provoke as well as to understand the user" (Mitchell et al., 2003, p. 111). They do not underestimate the fundamental rethinking this will require of user tests and correctly assert that these less formal methods offer more reward in terms of understanding "social impact, cultural meaning, and the potential political implications of a technology" (Mitchell et al., 2003, pp. 111-112). While these evaluative methods facilitate a different kind of understanding of the technology in context, the corresponding arts-based design process, through provocation, delivers a different kind of value to the user in the first place. Most elegantly, Bolter and Gromala (2003) have elucidated the apparent paradox of the visible tool in their concept of a rhythm between the transparency made possible by a mastery of techniques and the reflectivity of the framing that gives meaning to its content. They call for a greater understanding of the nature of this rhythm in use, and this author adds to this the need for its connection to the experience of the design process itself.

CONCLUSION

There are compelling reasons for the presentation of an alternative to the digital product as information appliance, as recent concern over the status of the authentic points to a need for performative meaning making rather than passive acceptance of spectacle. Art is presented as a model for this process and its product, requiring, in turn, an inversion of the primacy of disappearance over materiality. Drawing attention to the object itself means introducing disorder and breakdown necessary for dialogue and the socially based co-creation of meaning. It is suggested that an answer may lie in other design disciplines beyond product design and within the exploratory processes of art.

REFERENCES

Baudrillard, J. (1968). *The system of objects* (Benedict, J., trans.). London: Verso.

Blythe, M. A., Overbeeke, K., Monk, A. F., & Wright, P.C. (Eds.) (n.d.). *Funology, from usability to enjoyment*. London: Kluwer Academic Publishers.

Bolter, J. D., & Gromala, D. (2003). *Windows and mirrors—Interaction design, digital art, and the myth of transparency*. Cambridge, MA: MIT Press.

Boyle, D. (2003). *Authenticity; Brands, fakes, spin and the lust for real life*. London: Flamingo.

Brand, S. (2000). *The clock of the long now*. Phoenix, AZ: Basic Books.

Bunnell, K. (2002). *Craft technology and sustainability. Proceedings of the Crafts in the 21st Century Conference*, Edinburgh, Scotland.

Candy, L., et al. (2002). Panel: Research into art and technology. *Proceedings of the 6th International Conference on Creativity and Cognition.*

Coyne, R. (1995). *Designing information technology in the postmodern age*. Cambridge, MA: MIT Press.

Csikzsentmihalyi, M., & Rochberg-Halton, E. (1981). *The meaning of things*. Cambridge: Cambridge University Press.

Dunne, A. (1999). *Hertzian tales*. London: RCA CRD.

Dunne, A., & Raby, F. (2002). The placebo project. *Proceedings of the DIS2002.*

Eco, U. (1989). *The open work*. Cambridge, MA: Harvard University Press.

Eldridge, R. (2003). *An introduction to the philosophy of art*. Cambridge: Cambridge University Press.

Fuad-Luke, A. (2002). Slow design—A paradigm shift in design philosophy? *Proceedings of the Development by Design Conference*.

Gaver, W., Boucher, A., Pennington, S., & Walker, B. (2004). Cultural probes and the value of uncertainty. *Interactions, 6*(5), 53-56.

Gaver, W., Dunne, A., & Pacenti, E. (1999). Cultural probes. *Interactions, 4*(1), 21-29.

Greenhalgh, P. (2002). Craft in a changing world. In P. Greenhalgh (Ed.), *The persistence of craft*. London: A&C Black.

Haworth, J. (2003). *Creativity and embodied mind in digital fine art*. Retrieved January 9, 2004, from http://www.haworthjt.com/cemdfa/contents.html

Heidegger, M. (1962). *Being and time*. Oxford: Blackwell Publishing.

Ihde, D. (2002). *Bodies in technology*. Minneapolis: University of Minnesota Press.

Jordan, P. (2000). *Designing pleasurable products*. London: Taylor & Francis.

Lazar, J. (2001). User-centered Web development. Sudbury, MA: Jones and Bartlett.

Lewis, D., & Bridger, D. (2000). *The soul of the new consumer*. London: Nicholas Brealey Publishing.

McCarthy, J., & Wright, P. (2003). The enchantments of technology. In M. A. Blythe, K. Overbeeke, A. F. Monk, & P. C. Wright (Eds.), *Funology, from usability to enjoyment* (pp. 81-91). London: Kluwer Academic Publishers.

Mitchell, W., Inouye, A. S., & Blumenthal, M. S. (Eds.). (2003). *Beyond creativity—Information technology, innovation, and creativity*. Washington, DC: The National Academies Press.

Norman, D. (1999). *The invisible computer*. Cambridge, MA: MIT Press.

Pepperell, R. (2002). *The post-human condition: Consciousness beyond the brain*. Bristol: Intellect.

Preece, J., et al. (Eds.). (1994). *Human-computer interaction*. Harlow, UK: Addison-Wesley.

van Hinte, E. (1997). *Eternally yours—Visions on product endurance*. Rotterdam, The Netherlands: 010 Publishers.

Weiser, S. (1991). The computer for the 21st century. *Scientific American, 94*(110), 94-104.

Winograd, T., & Flores, F. (1986). *Understanding computers and cognition*. Norwood, NJ: Ablex Corporation.

KEY TERMS

Art: A coherent system or articulate form of human communication using elements of expression, and the search for new expressions articulating the human condition. This can include all forms of expression; for example, the visual and plastic arts, drama, music, poetry, and literature, and covers both process and product. Art may, in its own right, be conservative, pragmatic, critical, or radical.

Authenticity: The agentive participation in meaning making, as opposed to passive reception. This is the only way in which an individual can relate incoming information to the context of his or her own lifeworld, without which meaning does not exist for that person. We often sense the lack of authenticity in interaction without necessarily understanding our own misgivings.

Breakdown: A term used by German philosopher Martin Heidegger, originally with negative connotations to describe any cognitive interruption to a smooth interaction, or coping, in a situation. It is in breakdown that opportunities for human communication arise.

Critical Stance: Any approach to an accepted system that intentionally highlights issues of power structures supported by it, often emancipatory in nature and always political.

Expression: The utterance through any language system of prelinguistic emotion or understanding toward the creation of consensual meaning between people.

Invisible Computer: A computer or computer interface that disappears cognitively either through user expertise or by direct mapping of the relationship between interface elements, and the actions afforded by them. Other current terms are *transparency* and *seamlessness* and their antonyms, *reflection* and *seamfulness*.

Meaning Making: The constant goal of humans is to understand the world we find ourselves in. Meaning is arrived at continuously through social interactions with other individuals.

Pragmatism: The thoroughly practical view of praxis in which theory is not separate from action but a component of useful action in its application to a certain situation. In HCI, this takes into account the hermeneutic nature of product or system development and appropriation.

Sustainable Consumption: A recent movement in product design and consumer research on the need for a change in our patterns of consumption. The work cited here focuses particularly on the meaningfulness of the products and services we consume as integral to this shift in attitude.

Tool: An artifact used to achieve specific, predetermined goals. Defined by HCI and certain branches of philosophy by its disappearance in use.

Online Learning

John F. Clayton
Waikato Institute of Technology, New Zealand

INTRODUCTION

By looking closely at the term *online learning*, we could arrive at a simple definition, which could be the use by students of connected (online) computers to participate in educational activities (learning). While this definition is technically correct, it fails to explain the full range and use of connected computers in the classroom. Historically, the term appears to have evolved as new information and communication tools have been developed and deployed. For example, in the early stages of development, Radford, (1997) used the term *online learning* to denote material that was accessible via a computer using networks or telecommunications rather than material accessed on paper or other non-networked media. Chang and Fisher (1999) described a Web-based learning environment as consisting of digitally formatted content resources and communication devices to allow interaction. Zhu and McKnight (2001) described online instruction as any formal educational process in which the instruction occurs when the learner and the instructor are not in the same place and Internet technology is used to provide a communication link among the instructor and students. Chin and Ng Kon (2003) identified eight dimensions that constructed an e-learning framework. The range of definitions of online learning is not only a reflection of technological advancement but also a reflection of the variety of ways educationalists at all levels use connected computers in learning.

BACKGROUND

Examples of Online Learning Activities

In one learning scenario, a group of 10-year-old students following a pre-prepared unit in a supervised computer laboratory may use the information storage capacity of the World Wide Web (WWW) to gather additional resources to prepare a presentation on weather patterns. In a second scenario, a group of 14-year-olds studying the same topic in a classroom with a dedicated computer work station situated by the teacher's desk could use the communicative functions of the Internet to establish mail lists with metrological staff to follow studies being undertaken on weather patterns in a region. In a third scenario, a group of 18-year-olds consisting of small pockets of learners in isolated locations using home-based connected workstations may use an educational courseware package, incorporating information storage and communicative functions to participate in a complete distance unit, studying impacts and implications of climate change. In each of the scenarios described, students and teachers have used connected computers in distinct ways to achieve varied objectives. The technical competencies required, the learning support provided, and the physical location of the students in each scenario is different and distinct. In each scenario, a definable learning environment can be identified for each group of learners.

LEVELS OF ONLINE LEARNING

Educational institutions, from elementary schools to universities, are using the WWW and the Internet in a variety of ways. For example, institutions may establish simple Web sites that provide potential students with information on staff roles and responsibilities; physical resources and layout of the institution; past, present, and upcoming events; and a range of policy documents. Other institutions may use a range of Web-based applications such as e-mail, file storage, and exams to make available separate course units or entire programs to a global market (Bonk, 2001; Bonk et al., 1999). To classify levels of Web integration that are educational in nature, we should look closely at the uses of the Web for learning. Online educationalists have identified a number of different forms of online instruction, including sharing information on a Web site, communicating one-to-one or one-to-many via e-mail, delivering library resources via the Internet (e.g., electronic databases), or submitting assignments electronically (e.g., e-mail attachments, message board postings) (Dalziel, 2003; Ho & Tabata, 2001; Rata Skudder et al., 2003; Zhu & McKnight, 2001). However, the range of possibilities highlighted by these educationalists does not fully identify, explain, or describe the interactions, the teaching, or the learning that occurs within these environments. For best practice guidelines to be created for e-environments, the common features and activities of the Internet or computer-connected courses affecting all students, regardless of Web tools used or how information is structured and stored, need to be identified and described.

LEARNING ENVIRONMENTS

In researching and evaluating the success or failure of time spent in educational settings, researchers could use a number of quantitative measures, such as grades allocated or total number of credits earned, participation rate in activities, graduation rate, standardized test scores, proficiency in subjects, and other valued learning outcomes (Dean, 1998; Fraser & Fisher, 1994). However, these measures are somewhat limited and cannot provide a full picture of the education process (Fraser, 1998, 2001). There are other measures that can be used that are just as effective; for example, student and teacher impressions of the environment in which they operate are vital. The investigation in and of learning environments has its roots nourished by the Lewinian formula, B=*f*(P,E). This formula identifies that behavior (B) is considered to be a function of (*f*), the person (P), and the environment (E). It recognizes that both the environment and its interaction with personal characteristics of the individual are potent determinants of human behavior (Fraser, 1998).

PERCEPTUAL MEASURES

In the past, it has been common to use pencil and paper forms with the administrator supervising data entry in learning environment research (Fisher & Fraser, 1990; Fraser et al., 1992; Fraser & Walberg, 1995). Instruments are carefully designed and ask students to select an appropriate response from a range of options. For example, the Science Laboratory Environment Inventory (SLEI) begins by providing students with directions on how to complete the questionnaire. They are informed that the form is designed to gauge opinion and that there is no right or wrong answers. Students are asked to think about a statement and draw a circle around a numbered response. The range of responses is from 1 to 5, and the meaning of each response is explained carefully; for example, 1 is that the practice takes place almost never, while 5 indicates the practice occurs very often (Fraser & Fisher, 1994; Fraser & Tobin, 1998). Data are analyzed by obtaining a total score for a specific scale. This scoring is often completed manually. Advancements in computer technologies have made it possible to explore the disposal of paper-and-pencil instruments and manual data entry. Increasingly, traditional instruments are being replaced by electronic versions delivered through the Internet (Maor, 2000; Joiner et al., 2002; Walker, 2002).

FUTURE TRENDS

Setting the Scene

Three connected computer- or WWW-based educational activities on the weather were described in section one. The first scenario illustrated how the information storage and retrieval functions of the WWW could be used to expand available student resources. In this scenario, students could be supervised directly and assisted in their tasks by a teacher responsible for a dedicated computer suite established at the school. The second scenario demonstrated how the communication features of connected computers could be used to provide authentic examples to enrich student understanding. In this scenario, students could work independently of the teacher, who was present, however, to offer guidance and support. The third scenario described how Web-based educational management platforms could be used to provide educational opportunities for isolated pockets of students. In this scenario, students are completely independent, and they rely on the information and communication technologies provided by their tutor for guidance and support.

On the surface, it would appear that the online learning environments created in each of the three scenarios are distinct and that no common interactions or relationships can be identified for investigation. For example, tutor-student interactions appear to be different. In the first scenario, students are guided by the continual physical presence of a tutor mentoring progress. In the second scenario, the tutor, on occasion, is physically present, offering guidance and support. In the third scenario, there is no physical relationship established, and the tutor's interactions with students are virtual. Also, the physical environment created in each scenario appears to be distinct. For example, in the first scenario, all students are located physically in a dedicated laboratory. In the second scenario, the computer is located in an existing teaching space, possibly in a strategic place close to the teacher's desk. The environment in the third scenario is dependent on the physical layout of the individual student's home.

It could be argued, given these differences, that it would not be possible to investigate each environment created using a single instrument. However, is this really the case? In each of the scenarios described, it is assumed that students have a functional knowledge of computer operations. For example, there is the assumption that students will be able to:

- know if the computer is turned on or turned off
- use a keyboard and computer mouse
- view information presented on a visual display unit and;
- select and/or use appropriate software applications.

A more complex example focuses on our understanding of the process of learning. As mentioned in each of the examples, the students engage with the computer, and the tutor facilitates this engagement. It can be argued that there is in online environments a tutor-student relationship. We then can ask these questions: How do these relationships function? Are the students satisfied or frustrated by the relationships created? Does the tutor feel the relationships created are beneficial?

These two examples—tutor-student and student-computer relationships—demonstrate how it may be possible to identify and describe common features of connected computer and online activities. It then can be argued that if it is possible to identify and describe these relationships, it is also possible to investigate and explore them. It logically follows that if we can investigate and explore relationships, it is also possible to create best practice guidelines for institutions and individuals to follow, thereby raising the standard of educational activities for all participants.

Investigation of Relationships in Online Learning

As noted, when reviewing educational activities in the online environment, we can immediately raise various questions about the nature of teacher-student and student-computer interactions. These two features have been expanded by Morihara (2001) to include student-student interaction, student-media interaction (an expansion to include other components rather than simply text) and the outcomes of the learning that take place in the environment created. Haynes (2002) has refined these relationships and identified four relationships within online environments that are outlined as follows:

1. student interface relationship
2. student-tutor relationships
3. student-student relationships
4. student-content relationships

These four broad areas appear to identify the crucial relationships and interactions that occur within online environments. However, they do not help in clarifying how the student as an individual reacts to and reflects on his or her experiences in this environment.

The importance of creating time for and encouraging self-reflection of the learning process is well-documented by constructivists (Gilbert, 1993; Gunstone, 1994; Hewson, 1996; Posner et al., 1982), and it would appear to be crucial to investigate if, when, and how this reflection occurs. Therefore, there appear to be five broad areas of online learning interaction outlined as follows:

1. **Student-Media Interaction:** How are students engaged with digitally stored information, and how do they relate to the information presented?
2. **Student-Student Relationships:** How, why, and when dp students communicate with each other, and what is the nature of this communication?
3. **Student-Tutor Relationships:** How, why, and when do students communicate with their tutor, and what is the nature of this communication?
4. **Student-Interface Interaction:** What are the features of the interface created that enhance/ inhibit student learning and navigation?
5. **Student Reflection Activities:** How are students encouraged to reflect on their learning, are they satisfied with the environment, and how do they relate to the environment created?

These relationships and interactions should form the development framework for the identification of scales and items to construct an online learning survey. Data generated from this instrument should guide online learning activities and help to shape online interactions. The best-practice guidelines generated will serve to raise the standard of online educational activities for all participants.

CONCLUSION

The growth of connected computing technologies, the creation of the Internet, and the introduction of the World Wide Web have led to a number of educationalists and educational institutions becoming involved in the development and delivery of courses using these technologies. While the range, depth, and breadth of potential uses of these technologies is vast and forever growing, and while it may appear that this divergent use of technologies creates a range of different, describable online learning environments with little or no commonality, it can be argued that there are indeed common relationships and interactions. Five relationships and interactions have been identified and described in this article: Student-Media Interaction, Student-Student Relationships, Student-Tutor Relationships, Student-Interface Interaction, and Student Reflection Activities. This article also argued that if relationships and interactions can be identified and described, it is logical to assume that they can be explored and investigated. These investigations ultimately should lead to the creation of best-practice guidelines for online learning. These guidelines then could be used by educational institutions and individuals to raise the standard of online educational activities.

REFERENCES

Bonk, C. (2001). *Online teaching in an online world*. Retrieved September 12, 2005, from http://PublicationShare.com

Bonk, C., Cummings, J., Hara, N., Fischler, R., & Lee, S. (1999). *A ten level Web integration continuum for higher education: New resources, partners, courses, and markets*. Retrieved May 1, 2002, from http://php.indiana.edu/~cjbonk/paper/edmdia99.html

Chang, V., & Fisher, D. (1999). Students' perceptions of the efficacy of Web-based learning environment: The emergence of a new learning instrument. *Proceedings of the Herdsa Annual International Conference*, Melbourne, Australia.

Chin, K.L., & Ng Kon, P. (2003). Key factors for a fully online e-learning mode: A Delphi study. *Proceedings of the 20th Annual Conference of the Australasian Society for Computers in Learning in Tertiary Education*, Adelaide, Australia.

Dalziel, J. (2003). Implementing learning design: The learning activity management system (LAMS). *Proceedings of the 20th Annual Conference of the Australasian Society for Computers in Learning in Tertiary Education*, Adelaide, Australia.

Dean, A.M. (1998). *Defining and achieving university student success: Faculty and student perceptions* [master's thesis]. Virginia Polytechnic Institute and State University, Blacksburg.

Fisher, D., & Fraser, B. (1990). *School climate* (SET research information for teachers No. 2). Melbourne, Australian Council for Educational Research.

Fraser, B. (1998). Classroom environment instruments: Development, validity and applications. *Learning Environments Research: An International Journal, 1*(1), 68-93.

Fraser, B. (2001). Twenty thousand hours: Editors introduction. *Learning Environments Research: An International Journal, 4*(1), 1-5.

Fraser, B., & Fisher, D. (1994). Assessing and researching the classroom environment. In D. Fisher (Ed.), *The study of learning environments* (pp. 23-39). Perth: Curtin University of Technology.

Fraser, B., Giddings, G.J., & McRobbie, C.J. (1992). *Assessing the climate of science laboratory classes* (what research says, no. 8). Perth: Curtin University of Technology.

Fraser, B., & Tobin, K. (Eds.). (1998). *International handbook of science education.* Dordrecht, The Netherlands: Kluwer Academic Publishers.

Fraser, B., & Walberg, H. (Eds.). (1995). *Improving science education.* The University of Chicago Press.

Gilbert, J. (1993). Teacher development: A literature review. In B. Bell (Ed.), *I know about LISP but how do I put it into practice?* (pp. 15-39). Hamilton: CSMER University of Waikato.

Gunstone, R. (1994). The importance of specific science content in the enhancement of metacognition. In P. Fensham, R. Gunstone, & R. White (Eds.), *The content of science: A constructivist approach to its teaching and learning* (pp. 131-147). London: The Falmer Press.

Haynes, D. (2002). The social dimensions of online learning: Perceptions, theories and practical responses. *Proceedings of the the Distance Education Association of New Zealand*, Wellington, New Zealand.

Hewson, P. (1996). Teaching for conceptual change. In D. Treagust, D. Duit, & B. Fraser (Eds.), *Improving teaching and learning in science and mathematics* (pp. 131-141). New York: Teachers College Press.

Ho, C.P., & Tabata, L.N. (2001). *Strategies for designing online courses to engage student learning.* Retrieved March 15, 2003, from http://leahi.kcc.hawaii.edu/org/tcon01/index.html

Joiner, K., Malone, J., & Haimes, D. (2002). Assessment of classroom environments in reformed calculus education. *Learning Environments Research: An International Journal, 5*(1), 51-76.

Maor, D. (2000). *Constructivist virtual learning environment survey.* Retrieved March 15, 2003, from http://www.curtin.edu.au/learn/unit/05474/forms/CVLES_form.html

Morihara, B. (2001). *Practice and pedagogy in university Web teaching.* Retrieved March 15, 2003, from http://leahi.kcc.hawaii.edu/org/tcon01/index.html

Posner, G., Strike, K., Hewson, P., & Gertzog, W. (1982). Accommodation of scientific conception: Toward a theory of conceptual change. *Science Education, 66*(2), 211-227.

Radford, A.J. (1997). *The future of multimedia in education.* Retrieved June 24, 2002, from http://www.firstmonday.dk/issues/index.html

Rata Skudder, N., Angeth, D., & Clayton, J. (2003). All aboard the online express: Issues and implications for Pasefica e-learners. *Proceedings of the 20th Annual Conference of the Australasian Society for Computers in Learning in Tertiary Education*, Adelaide, Australia.

Walker, S. (2002). Measuring the distance education psychosocial environment. *Proceedings of the TCC 2002: Hybrid Dreams: The Next Leap for Internet-Mediated Learning Virtual Conference.*

Zhu, E., & McKnight, R. (2001). *Principles of online design.* Retrieved March 15, 2003, from http://www.fgcu.edu/onlinedesign/

KEY TERMS

Internet: An internet (note the small i) is any set of networks interconnected with routers forwarding data. The Internet (with a capital I) is the largest internet in the world.

Intranet: A computer network that provides services within an organization.

Learning Environment: A term used to describe the interactions that occur among individuals and groups and the setting within which they operate.

Learning Management System: A broad term used to describe a wide range of systems that organize and provide access to e-learning environments for students, tutors, and administrators.

Online Learning: The use by students of connected (online) computers to participate in educational activities (learning).

Perceptual Measure: An instrument used to investigate identified relationships in learning environments.

World Wide Web: A virtual space of electronic information storage.

O

An Overview of an Evaluation Framework for E-Learning

Maria Alexandra Rentroia-Bonito
Technical University of Lisbon, Portugal

Joaquim A. Jorge
Technical University of Lisbon, Portugal

Claude Ghaoui
Liverpool John Moores University, UK

INTRODUCTION

Technology-based education is taken as an effective tool to support structured learning content dissemination within pre-defined learning environments. However, effectiveness and efficacy of this paradigm relate to how well designers and developers address the specificities of users' learning needs, preferences, goals, and priorities taking into account their immediate work, social, and personal context. This is required in order to focus development efforts on the design of e-learning experiences that would satisfy identified needs. Thus, studying and assessing the human computer interaction side of such projects is a critical factor to designing holistic and productive e-learning experiences.

Literature does not show consistent and integrated findings to support the effectiveness of e-learning as a strategic tool to develop knowledge and skill acquisition (Rosenberg, 2001; Shih & Gamon, 2001). The objective of this article is to develop on one hand, main identified issues of an integrated evaluation framework, focusing on key variables from people and technology standpoint within context of use, and, on the other hand, to summarize the relevant tasks involved in designing e-learning experiences. Main identified issues of an integrated evaluation framework include: (i) some relevant context-specific factors, and (ii) other issues that are identified when people interact with technology. Context-specifics factors such as culture, organization of work, management practices, technology, and working processes may influence the quality of interaction (Laudon & Laudon, 2002) and may also help define the organizational readiness to sustain the acceptance and evolution of e-learning within organizational dynamics. Thus we propose an e-learning evaluation framework to be used as a diagnostic and managerial tool that can be based on: (a) an observed individual vari able, as a visible sign of implicit intentions, to support development effort during instructional design and initial users' engagement, and/or (b) usability and accessibility as key identified technology variables addressing acceptance and usage.

The Background section presents our proposed theoretical evaluation framework to guide our analysis based upon the reviewed li es arising from the proposed framework. Last, we elaborate on some future work and general conclusion.

BACKGROUND

Natural, effective, and also affective interactions between humans and computers are still open research issues due to the complexity and interdependency of the dynamic nature of people, technology, and their interactions overtime (Baudisch, DeCarlo, Duchowski, & Gesiler, 2003; Cohen, Dalrymple, Moran, Pereira, & Sullivan, 1989; Gentner & Nielsen, 1996; Horvtiz & Apacible, 2003; Preece, Rogers, & Sharp, 2002). Despite last-decade advancements in principles associated to usability design, there is still an ever-present need to better understand people-technology relationship in their context of use in order to design more natural, effective, satisfying and enjoyable users' experiences. Multimodal interactions, smart, ambient, and collaborative technologies are some current issues that are driving new interaction paradigms (Dix, Finlay, Abowd, & Beale, 1998; Oviatt, 1999). New skills and methods to perform work-related tasks at operational and strategic levels within organizational dynamics, plus societal attitudes, individual lifestyles, priorities, preferences, physical and cognitive capabilities and locations require more innovative approaches to designing user experiences. In addition, technical and users' feedback coming from different evaluation sources require workable methods and tools to capture and analyse quantitative and qualitative data in a systematic, consistent, integrated, and useful way. This situation makes e-learning evaluation process a complex one (Garrett, 2004; Janvier & Ghaoui, 2004; Preece et al., 2002; Rosson & Carroll, 2002). Moreover, interpretation of an evaluation outcome requires an additional set of skills. Figure 1 shows three main aspects to consider when evaluation e-learning experiences: (1) people-related issues (learning preferences), (2) instruction-related issues (instruction design), and (3) system-related issues (usability and accessibility).

Organizational context and individual learning preferences aim at improving people-task fit. This means that people's skills and related learning objectives are defined by: (a) their preferred ways of learning, and (b) the tasks individuals have to perform within the scope of their organizational roles and specifics contexts. Principles and practices of instructional design and multimodal feasible choices are taken into account to structure, organize, and present learning content and related tasks (Clark & Mayer, 2003). This way, contextual and work-relatedness of content is ensured.

Usability and accessibility, as quality attributes of system performance, address the acceptance and usage of a system by the intended users. Learning outcomes, namely performance and satisfaction after being analyzed, would drive initiatives for improvement or new developments at operational and strategic levels. These issues are further described in the next sections.

Evaluating People-Related Issues

From a people standpoint, learning styles are identified by researchers, among the multiple individual traits that influence learning process, as a key component to design and evaluate effective and satisfactory instructional methodologies and education-oriented technologies. Reviewed literature on learning styles and individual differences (Atkins, Moore, Sharpe, & Hobbs, 2000; Bajraktarevic, Hall, & Fullick, 2003; Bernardes & O'Donoghue, 2003; Leuthold, 1999; McLaughlin, 1999; Sadler-Smith & Riding, 2000; Storey, Phillips, Maczewski, & Wang, 2002; Shih & Gamon, 2001) show that most research findings are not conclusive and often contradictory regarding the impact of learning styles on outcomes of e-learning (McLaughlin, 1999; Shih & Gamon,

Figure 1. Designing e-learning experiences: People and technology aspects

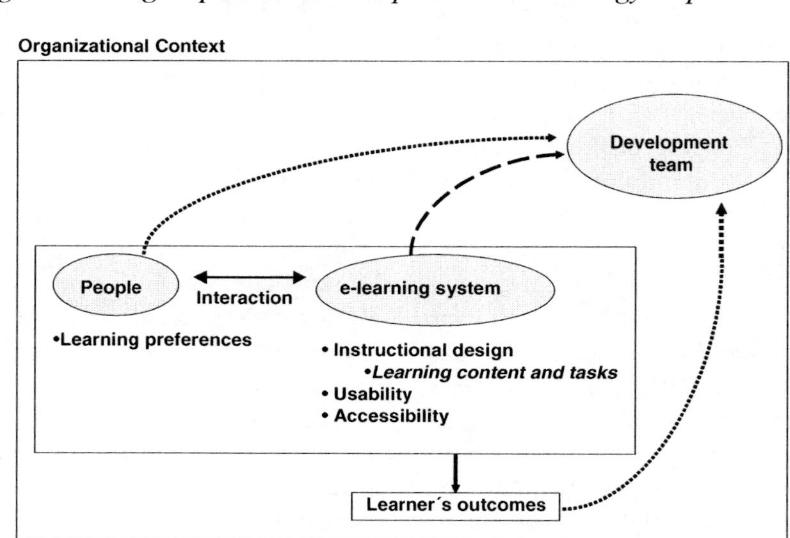

2001). Still, many researchers agree that learning styles: (a) are a relevant factor to the learning experience, and (b) influence learning behaviors likely affecting the degree to which individuals will engage in particular types of learning (Sadler-Smith & Riding, 2000). However, the measurement of learning styles is complex and time-consuming, because they are assessed by using questionnaire or psychometric test. Consequently, its usage raises individual's concerns about data privacy and protection. To motivate a workable approach, we focus our theoretical framework on learning preferences, which is defined as an observable individual trait that shows what tasks or objects people favor over others of the same kind (McLaughlin, 1999). Hence, learning preferences, in our approach, would support the designing of adequate learning content.

Learning preferences are revealed through choices or actions, and can be validated by using ethnographic techniques, self-reporting, or log analysis. This kind of systematic observation helps also perceive learning and cognitive strategies in context. This way, they can input a gradually evolving intelligent e-learning system. This knowledge on users' actions and patterns, based on observations and inquiries of users, would help development teams to understand the patterns of their favoring when completing learning tasks and to interact with related material across different modalities, media, type of learning, context, and other actors involved, namely peers, instructors, and support staff.

Upon this perspective, quality of the interactivity within an e-learning environment is defined by: (a) the adequacy of learning content to individual learning preferences and the quality of service of technical conditions, and (b) the quality of relationships with humans involved in the learning experience. The former depend on the available choices for specific user groups' based upon their preferences and technical conditions. The latter depends on three specific roles: Instructors, Social Science practitioners, and Administrative/Helpdesk staff. What each of these roles should involve to ensure quality of interactivity? Table 1 summarizes our view on the main responsibilities of these three roles.

Evaluating Instruction-Related Aspects

A second relevant aspect in this theoretical framework is instruction and design (methodology, content, and related tasks). This aspect raises key issues related to allocation of organizational resources to learning content creation and updating in terms of: (a) matching pedagogically learning contents and related tasks to suit diverse learning needs, preferences, objectives, increasingly evolv-

Table 1. Relevant roles for ensuring interactivity within e-learning environments

Role	Main responsibilities
Instructor	(a) Defining, and monitoring, the degree of adequacy between users' learning preferences, method, modalities, and media across space and time taking into account local and remote specificities of physical and technical contexts of learning, (b) Reinforcing social aspects within learning communities, contributing to habit formation, expected performance, and conformance to social norms and practices (Preece et al., 2002), and (c) Being aware, and actively exercise, his or her prominent role as members of development teams supporting systems by constantly and systematically matching them to user groups' involvement, participation, learning, and capabilities.
Social Science practitioners	(a) Understanding the dynamic socio-cultural nature of the People-System interaction, and (b) Defining actions to develop human potential within socio-technological contexts across business and academic sectors.
Administrative/Helpdesk staff	Ensuring quality levels in operational technical support to smooth transition phase of a changing management process.

Table 2. Instruction-related aspects

What to evaluate....
• To what extent, do learning outcomes relate to business strategies? • How well does content structure facilitate internal and external navigation? • To what extent, are content organization and graphic layout are effective to achieve learning objectives? • How well do frequency patterns and learning outcomes justify investment? • To what extent, is this way of learning accommodating the identified needs of a diverse and disperse population? • What are the most cost-effective media and modalities for distributing specific content to users in their context of use? • What are the most frequent and effective learning tasks across media and modalities? • How well do learning preferences and learning tasks correlate? • How effective is it to use organizational experts as coaches?

ing education-oriented technology, and business strategies; and (b) generating structuring, organizing, and presenting subject matter or task knowledge within task execution's scope. Interactivity with learning content depends on these two issues. Table 2 shows some instruction-related items to evaluate.

Evaluating System-Related Aspects

We assume the role of an e-learning system as an intermediary agent between instructors and learners. As an intermediary agent, an e-learning system should be designed not only to be effective, efficient, but also affective and social. Effectiveness is concerned with learning objectives, methods, and usability goals. Efficiency is concerned with measuring the usage of resources to achieve defined objectives. Affectivity measures users' feelings and satisfaction during e-learning experiences (Dix et al., 1998; Rosson & Carroll, 2002). Sociality is perceived as part of a working group (Preece et al., 2002; Reeves & Nass, 1996). If any of these attributes are missing, e-learners would not be able to engage in the experience and profit from its outcomes. Also, quality of the interaction is affected by the quality of service supplied by the system and related technological and physical infrastructure.

Regarding quality of service, our e-learning evaluation framework addresses the technological and physical specificities of the experience, such as system performance, downloading times, traffic flows, access profiling, backups, learning facilities and equipments, among others. Table 3 shows some items of what to evaluate.

To holistically evaluate, we do not only evaluate usability, but also the social implication of interaction on the organizational context and its level of accessibility. Usability is defined as the extent to which a system can be used to achieve defined goals by intended users in an effective, efficient, and satisfactory manner in a specified context of use (Dix et al., 1998). Usability evaluation has been mainly based on prototyping, heuristic evaluations, observing users, and user testing by using different types of methods and tools with a strong quantitative orientation. Web-based applications have brought the need to cost-effectively evaluate usability among distributed applications and by geographically disperse and diverse users. Thus, automated usability evaluation tools are a promise to achieve cost-effectively usability goals (Ivory & Hearst, 2002).

Regarding social aspect of People-System interaction, Ågerfalk and Cronholm (2001) stated that actability: (a) is "*...an information system's ability to perform actions, and to permit, promote and facilitate the performance of actions by users, both through the system and based on information from the system, in some business context...*"; (b) is a quality metric; (c) focuses on the social context of interaction within business context; (d) is of a more qualitative nature; and (e) its definition reinforces the role of information systems as communication actors and the need of pre-existing users' knowledge and skills in IT and business tasks. The potential implication of this definition is that actable information systems are usable systems that make explicit the business actions that they can support within a specific organizational context. Potential complementarities between usability and actability are still to be shown. This is still being researched (Ågerfalk & Cronholm, 2001), and to our knowledge, there is no reliable and valid measuring

Table 3. System-related aspects: Quality of service

What to evaluate....
• How well does face-to-face (e.g., live streaming video) or remote component (e.g., pre-recorded lectures, courseware, etc.) run in each learning session? • What, from where, and by whom, are the most frequent learning material downloaded? • How does the system support interactivity between and within groups across time and place? • How does the system learn useful and relevant information to achieve learning objectives? • To what extent, would information system's outputs change current business practices?

instrument yet. Nevertheless, this is a promising area to increase the success rate of information system implementation (Xia & Lee, 2004). However, its importance as a complement to the quantitative orientation of usability testing is clear if assuming that interactions take place within specific social contexts (communities, groups of people, families, or organizations). Within any group's context, conformance to social rules is required from each of the group's members and strongly affects individual and surrounding social dynamics.

Regarding accessibility, it means that any potential users can access contents regardless of their cognitive and physical capabilities (Chilsolm, Vanderheiden, & Jacobs, 2001). Feasible goals in making accessibility a reality is a trade-off between flexibility in design, personalization to specific needs, usage of assistive technology (Arion & Tutuianu, 2003; Sloan, Gibson, Milne, & Gregor, 2003), and organizational readiness to create and sustain accessibility as a strategic issue.

Assuming that an extended version of the definition of usability includes actability, thus this integrated evaluation would cover efficiency, efficacy, satisfaction, and accessibility, in addition to conformance to social norms, legislation, ethical, and current business practices. Integrated feedback on these key five issues would make users and development teams, (working in a participatory-based methodology) fully aware about, and responsible for, the impact of pre-defined rules on organizational climate and dynamic. In addition, this kind of feedback would indicate areas for improving flexibility in design but in a controlled way, namely, giving more options closely focused on users' capabilities and their task needs. Table 4 shows some items of what to evaluate regarding usability and accessibility.

MAIN FOCUS OF THE EVALUATION FRAMEWORK FOR E-LEARNING

Main issues associated with three aspects are: (a) e-learning personalization mainly in terms of individual learning preferences taking other relevant background variables as control variables (e.g., goals,

Table 4. System-related aspects: Usability and accessibility

What to evaluate....
• How well is the system easy to learn and used across user groups in their respective context? • To what extent, do users perceive the system to contribute easily to their interaction with: (a) Content, (b) Peers, (c) Instructors, (d) Support staff? • To what extent, is organizational dynamics affected by the use of the system? In what user group is the influence most significant? • Is the system accessible to potential users regardless of its cognitive and physical capabilities? • How well do learning results justify investment in achieving usability and accessibility goals in terms of learning outcomes across users' groups? • Can the identified communication patterns reinforce dynamically organizational values and expected performance level and behaviors?

learning priorities, type of learning need, background, previous IT experience, etc.) , (b) Coordinating, Monitoring, and Controlling the learning process and e-learning strategy, (c) Degree of participation of users in designing experience, and (d) Integrating quantitative and qualitative feedback on effectiveness, efficiency, accessibility, satisfaction, and conformance to social context's rules to improve human computer interaction and its outcomes. Table 5 shows these issues summarizing investment areas and some examples of organizational programs to help implementing and ensuring e-learning effectiveness. This considers two basic scenarios created by geographical locations of users participating in an e-learning experience.

It is worth noting two points. First, that IT experience or basic skills can be acquired not necessarily within organizational settings. Exposure at this level should be an orchestrated effort in society and, at a political level, to ensure competitiveness. When business context does not sustain that, specific organizational interventions should be in place to ensure engagement and habit formation, such as orientation and coaching programs. Second, coordination and monitoring of efforts is key to ensure consistent methods across different participating geographical locations.

Given the increasing users' diversity in knowledge, skill levels, needs, and contexts, we believe that applying automated or semi-automated evaluation tools and ethnographic techniques within development cycle could be cost-effective in improving gradually the "intelligence" of e-learning systems. Namely, this would help e-learning systems to adapt to: (a) the dynamic development of users' competence and formation of habits, expectations, involvement; and (b) observed choices and actions. This could foster a gradual alignment between, on one hand, learning outcomes and technology with individual expectations, learning preferences and, on the other hand, optimizing allocated organizational resources within e-learning environments.

To do so, the development team should have additional set of skills and tasks. Based on reviewed literature (Ågerfalk & Cronholm, 2001; Bernardes & O'Donoghue, 2003; Clark & Mayer, 2003; Preece et al., 2002; Rosenberg, 2001; Rosson & Carroll, 2002; Sloan et al., 2003) and insights from multidisciplinary practitioners, Table 6 summarizes some main tasks and suggested techniques or tools required from team's member.

FUTURE TRENDS

Flexible design (rooted in universal principles and dedicated design) appears to be continuing orientation for designing interfaces during coming years. In this sense, development teams should be prepared to approach flexibility in design based upon tested,

Table 5. Relevant context-specific aspects required for designing an e-learning experience

Context-specific aspects	
1. Organization	Orientation programs
	Investment in Connectivity, Communication, Vocational Counseling and Content
2. Management practices	(a) Definition of level of investment on skill and system development articulated with business strategic objectives; (b) Identifying key competencies and strategic options to develop them, (c) Managing change and involved partners. (d) IT skill development
3. Business processes	Coordination, Controlling, and Monitoring quality of: (a) instructional design, (b) content production, (c) service and (d) learning outcomes
4. Technology	(a) Usability and Accessibility (b) Monitoring Connectivity, system performance and functionalities

Table 6. Development-team members' main tasks and suggested evaluation techniques or tools

Development team main roles	Main tasks	Suggested technique(s) or tool(s)
1. System Developer	(a) Contextual task and user modelling (b) Linking business words and actions within the e-learning system (c) Defining interaction metaphor and scenarios for the different users' groups considering their geographical location and time convenience (c) Effective, satisfactory and enjoyable user experience to speed up learning curves (d) Matching learning content with people's learning preferences, learning tasks, modalities, media and assistive technology, if needed (e) Ensuring proper flow of information among users' groups (f) Monitoring acceptance and usage levels (g) Administering user profiles	• Observing users during People-System interaction • Field studies • Focus groups • Surveys and questionnaires • Structured interviews • Prototyping • Heuristic evaluation • Usability testing • Pluralistic evaluation • Content analysis • Statistical analysis • Log analysis
2. Information architect	(a) Structuring work-related knowledge structures in terms of business language, (b) Matching people's learning preferences and presentation of information (c) Matching modality to structured content	• Technical reports • Statistical techniques • Content analysis • Log analysis • Prototyping • Heuristics evaluation • Usability testing
3. Content manager	(a) Generating and distributing content cost-effective	• Statistical techniques • Log analysis • Social Network Analysis • Structured interviews • Brainstorming
4. Training-process manager	(a) Identification of key skills' gaps and needs, and (b) Learning cost-effectiveness (c) Supporting expected business behaviours and performance levels (d) Efficacy of intervention programs (e) Improving procedural justice in distributing content to proper target (f) Monitoring learning efficacy and productivity levels, and development of required IT skills	• Model-based evaluation • Descriptive statistics • Surveys • Focus groups • Structured interviews • Questionnaires
5. Instructor	(a) Definition of learning objectives regarding identified skill gaps (b) Matching instructional design and teaching methodology with defined learning objectives and users' learning preferences, physical or cognitive capabilities, background and previous experience (c) Structure, organize and present learning content regarding users' needs and learning preferences (d) Matching communication patterns up to students needs	• Review-based evaluation • Descriptive statistics • Surveys • Focus groups • Structured interviews • Questionnaires • Log analysis

Table 6. Development-team members' main tasks and suggested evaluation techniques or tools (cont.)

6. Social-sciences staff	(a) Assessment of ergonomic, social and cultural impact of technology usage in order to minimize health-related problems and costs (b) Assessing needs regarding cognitive and physical capabilities and learning preferences to efficiently, accessibly and flexibly design for users (c) Defining organizational interventions or programs such as Counselling and Coaching (d) Assuring users' confidence during People-System interaction	• Observing users in context • Descriptive statistics • Surveys • Focus groups • Storytelling • Structured interviews • Questionnaires • Social Network Analysis • Wizard of Oz
7. Administrativ e & Helpdesk staff	Administrative and technical diagnosis to provide high-quality assistance	• Opinion polls • Questionnaires
8. Representativ e users	(a) Efficiency and effectiveness in performing work-related or learning tasks easily (b) Higher people-system fit (c) Less waste of personal and organizational resources (d) Eventually, less work-related conflict and stress	• Cognitive walkthrough • Role Playing • Think aloud protocol

simplified, and valid evaluation models. Further developments may include the following:

First, flexibility demands the definition of a more affective- and socially-oriented heuristics, which would require smart tools and techniques to improve the quality of the interaction across learning preferences.

Second, flexibility may benefit from having user representatives as part of a development team. Research work should identity what conditions and stages of such involvement could be both more cost-effective and ensure better success for e-learning. These results could guide human computer interaction curricula changes to perhaps disseminate practices among other related professions. Third, the increased use of portable equipments are facilitating virtual classroom environment, where people, in any place at any time, can participate. Research should explore the effectiveness and convenience of different modalities during the learning process's cycle and stages. For instance, it may be efficient to consult any information of interest when researching a new topic of interest by using a portable device assistant (PDA) from anywhere. However, developing that content requires specific physical conditions that cannot exist any where any time. Thus current and emergent habits across generations of people should be explored in terms of convenience and learning effectiveness.

CONCLUSION

We discussed a holistic framework to evaluate e-learning experiences taking into account people, technology, and instructional aspects. Learning preferences, usability, including social and affective aspects involved into the human computer interaction (Bernardes & O'Donoghue, 2003; Laudon & Laudon, 2002; McLaughlin, 1999; Picard, 1997; Preece et al., 2002; Rentroia-Bonito & Jorge, 2003; Rosson & Carroll, 2002) and accessibility (Chilsolm, et al., 2001) were identified as a set of key issues to evaluate e-learning experiences. The expected results of our integrated evaluation approach would allow development-team members to achieve a better understanding of how the identified issues affect learning outcomes to identify future improvements and developments.

REFERENCES

Ågerfalk, P. J., & Cronholm, S. (2001). *Usability versus actability: A conceptual comparative analysis*. Retrieved July 2004, from http://citeseer.ist.psu.edu/452356.html

Arion, M., & Tutuianu, M. (2003). Online learning for the visually impaired. In C. Ghaoui (Ed.), *Usability evaluation of online learning programs* (pp. 387-408). Hershey, PA: Idea Group Publishing.

Atkins, H., Moore, D., Sharpe, S., & Hobbs, D. (2000). Learning style theory and computer mediated communication. *World Conference on Educational Multimedia, Hypermedia and Telecommunications, 2001*(1), 71-75.

Bajraktarevic, N., Hall, W., & Fullick, P. (2003). *Incorporating learning styles in hypermedia environment: Empirical evaluation*. Retrieved June 2004, from http://wwwis.win.tue.nl:8080/ah2003/proceedings/www-4/

Baudisch, P., DeCarlo, D., Duchowski, A., & Gesiler, W. (2003). Focusing on the essential: Considering attention in display design. *Communications of the ACM, 46*(3), 60-66.

Bernardes, J., & O'Donoghue, J. (2003). Implementing online delivery and learning support systems: Issues, evaluation and lessons. In C. Ghaoui (Ed.), *Usability evaluation of online learning programs* (pp. 19-39). Hershey, PA: Idea Group Publishing.

Chilsolm, W., Vanderheiden, & Jacobs, I. (2001). Web content accessibility guidelines. *Interactions, 8*(4), 35-54.

Clark, R., & Mayer, R. E. (2003). *E-learning and the science of instruction*. CA: Pfeiffer.

Cohen, P., Dalrymple, M., Moran, D. B., Pereira, F. C., & Sullivan, J. W. (1989). Synergistic use of direct manipulation and natural language. *Conference on Human Factors in Computing Systems: Wings for the Mind. Proceedings of the SIGCHI,* Austin, Texas, April 30-May 4, 1989 (pp. 227-233). Retrieved April 2004, from http://portal.acm.org/results.cfm?coll=GUIDE&CFID=58548886&CFTOKEN=52867714

Dix, A., Finlay, J., Abowd, G., & Beale, R. (1998). *Human-computer interaction*. Essex, UK: Prentice Hall.

Garrett, B. M. (2004). Employing an experimental approach to evaluate the impact of an intelligent agent. *Interactive Technology & Smart Education Journal, 1*(1), 41-53.

Gentner, D., & Nielsen, J. (1996). The Anti-Mac interface. *Communications of the ACM, 39*(8), 70-82.

Horvtiz, E., & Apacible, J. (2003). Learning and reasoning about interruption. *Proceedings of the 5th International Conference on Multimodal Interfaces,* November 5-7, 2003 (pp. 20-27). New York: ACM Press. Retrieved April 2004, from http://portal.acm.org/results.cfm?coll=GUIDE&CFID=58548295&CFTOKEN=42279624

Ivory, M., & Hearst, M. A. (2002). Improving Web site design. *IEEE Internet Computing* (pp. 56-63). Retrieved July 2004, from http://webtango.berkeley.edu/papers/

Janvier, W. A., & Ghaoui, C. (2004). Case study: An evaluation of the learner model in WISDEM. *Interactive Technology & Smart Education Journal, 1*(1), 55-65.

Laudon, K., & Laudon J. (2002). *Managing information systems, managing the digital firm* (7th ed.). New Jersey: Prentice-Hall International.

Leuthold, J. (1999, September 3-5). Is computer-based learning right for everyone? *Proceedings of the Thirty-Second Annual Hawaii International Conference on System Sciences,* 1015 (vol, 1, pp. 208-214). Retrieved April 2004, from www.alnresearch.org/Data_Files/articles/full_text/leuthold.pdf

McLaughlin, C. (1999). The implications of the research literature on learning styles for the design of instructional material. *Australian Journal of Educational Technology, 15*(3), 222-241.

Oviatt, S. (1999). Ten myths of multimodal interaction. *Communications of the ACM, 42*(11), 74-81.

Picard, R. (1997). *Affective computing*. Massachusetts: The MIT Press.

Preece, J., Rogers, Y., & Sharp, H. (2002). *Interaction design: Beyond human-computer interaction*. New York, NY: John Wiley & Sons.

Reeves, B., & Nass, C. (1996). *The media equation: How people treat computers, television, and new media like real people and places*. New York: Cambridge University Press.

Rentroia-Bonito, M. A., & Jorge, J. (2003). An integrated courseware usability evaluation method. *Proceedings of 7th Knowledge-Based Intelligent Information and Engineering Systems International Conference, Part II,* Oxford, UK (pp. 208-214). Berlin, Germany: Springer-Verlag.

Rosenberg, M. (2001). *e-Learning strategies for delivering knowledge in the digital age*. Ohio: McGraw-Hill.

Rosson, M. B., & Carroll, J. M. (2002). *Usability engineering. Scenario-based development of human-computer interaction.* California: Morgan Kaufmann Publishers.

Sadler-Smith, E., & Riding, R. (2000). The implications of cognitive style for management education & development: Some evidence from the UK. Retrieved April 2004, from http://www.elsinnet.org.uk/abstracts/aom/sad-aom.htm

Shih, C., & Gamon, J. (2001). Web-based learning: Relationships among student motivation, attitude, learning styles and achievement. *Journal of Agricultural Education, 42*(4), 12-20.

Sloan, D., Gibson, L., Milne, S., & Gregor, P. (2003). Ensuring optimal accessibility of online learning resources. In C. Ghaoui (Ed.), *Usability evaluation of online learning programs* (pp. 371-386). Hershey, PA: Idea Group Publishing.

Storey, M. A., Phillips B., Maczewski, M., & Wang, M. (2002). Evaluating the usability of Web-based learning tools. *Educational Technology and Society, 5*(3), 91-100. Retrieved April 2004, from http://ifets.ieee.org/periodical/Vol_3_2002/ storey.html

Xia, W., & Lee, G. (2004). Grasping the complexity of IS development projects. *Communication of the ACM*, 5, 69-74.

KEY TERMS

Context-Specifics Aspects: Cover the most important factors that shaped and become characteristics of organizational dynamics such as culture, business strategies, organization of work, management practices, current technology, workforce competency level, working processes, among others.

E-Learning Development Team: The set of multi-disciplinary professionals required to develop and evaluate an integrated e-learning evaluation. Each team should include designers, developers, instructors, process managers, social-science staff professionals (e.g., psychology, sociology, human resources practitioners, and managers, among others), and Helpdesk staff and eventually user representatives of target population.

E-Learning Evaluation Framework: Comprises an integrated feedback based on people, system, and context-specifics aspects.

Individual Styles (Learning and Cognitive Styles): Relate to implicit main individual modes of acquiring information, organizing, and processing information in memory. They are assessed by using questionnaire or psychometric test.

Learning Preferences: Individual favoring of one teaching method over another, which can be consistently observed through individual choices or actions.

People Aspects: In this evaluation framework, basically covers background, individual learning preferences, goals, and priorities.

System Aspects: Cover the technological and physical specificities of the e-learning experience at server and data layers and the usability and accessibility issues of the presentation layer of the e-learning system.

O

An Overview of Multimodal Interaction Techniques and Applications

Marie-Luce Bourguet
Queen Mary University of London, UK

INTRODUCTION

Desktop multimedia (multimedia personal computers) dates from the early 1970s. At that time, the enabling force behind multimedia was the emergence of the new *digital technologies* in the form of digital text, sound, animation, photography, and, more recently, video. Nowadays, multimedia systems mostly are concerned with the compression and transmission of data over networks, large capacity and miniaturized storage devices, and quality of services; however, what fundamentally characterizes a multimedia application is that it does not understand the data (sound, graphics, video, etc.) that it manipulates. In contrast, intelligent multimedia systems at the crossing of the artificial intelligence and multimedia disciplines gradually have gained the ability to understand, interpret, and generate data with respect to content.

Multimodal interfaces are a class of intelligent multimedia systems that make use of multiple and natural means of communication (modalities), such as speech, handwriting, gestures, and gaze, to support human-machine interaction. More specifically, the term *modality* describes human perception on one of the three following perception channels: visual, auditive, and tactile. Multimodality qualifies interactions that comprise more than one modality on either the input (from the human to the machine) or the output (from the machine to the human) and the use of more than one device on either side (e.g., microphone, camera, display, keyboard, mouse, pen, track ball, data glove). Some of the technologies used for implementing multimodal interaction come from speech processing and computer vision; for example, speech recognition, gaze tracking, recognition of facial expressions and gestures, perception of sounds for localization purposes, lip movement analysis (to improve speech recognition), and integration of speech and gesture information.

In 1980, the put-that-there system (Bolt, 1980) was developed at the Massachusetts Institute of Technology and was one of the first multimodal systems. In this system, users simultaneously could speak and point at a large-screen graphics display surface in order to manipulate simple shapes. In the 1990s, multimodal interfaces started to depart from the rather simple speech-and-point paradigm to integrate more powerful modalities such as pen gestures and handwriting input (Vo, 1996) or haptic output. Currently, multimodal interfaces have started to understand 3D hand gestures, body postures, and facial expressions (Ko, 2003), thanks to recent progress in computer vision techniques.

BACKGROUND

In this section, we briefly review the different types of modality combinations, the user benefits brought by multimodality, and multimodal software architectures.

Combinations of Modalities

Multimodality does not consist in the mere juxtaposition of several modalities in the user interface; it enables the synergistic use of different combinations of modalities. Modality combinations can take several forms (e.g., redundancy and complementarity) and fulfill several roles (e.g., disambiguation, support, and modulation).

Two modalities are said to be redundant when they convey the same information. Redundancy is well illustrated by speech and lip movements. The redundancy of signals can be used to increase the accuracy of signal recognition and the overall robustness of the interaction (Duchnowski, 1994).

Two modalities are said to be complementary when each of them conveys only part of a message but their integration results in a complete message. Complementarity allows for increased flexibility and efficiency, because a user can select the modality of communication that is the most appropriate for a given type of information.

Mutual disambiguation occurs when the integration of ambiguous messages results in the resolution of the ambiguity. Let us imagine a user pointing at two overlapped figures on a screen, a circle and a square, while saying "the square." The gesture is ambiguous because of the overlap of the figures, and the speech also may be ambiguous if there is more than one square visible on the screen. However, the integration of these two signals yields a perfectly unambiguous message.

Support describes the role taken by one modality to enhance another modality that is said to be dominant; for example, speech often is accompanied by hand gestures that simply support the speech production and help to smooth the communication process.

Finally, modulation occurs when a message that is conveyed by one modality alters the content of a message conveyed by another modality. A person's facial expression, for example, can greatly alter the meaning of the words he or she pronounces.

User Benefits

It is widely recognized that multimodal interfaces, when carefully designed and implemented, have the potential to greatly improve human-computer interaction, because they can be more intuitive, natural, efficient, and robust.

Flexibility is obtained when users can use the modality of their choice, which presupposes that the different modalities are equivalent (i.e., they can convey the same information). Increased robustness can result from the integration of redundant, complementary, or disambiguating inputs. A good example is that of visual speech recognition, where audio signals and visual signals are combined to increase the accuracy of speech recognition. Naturalness results from the fact that the types of modalities implemented are close to the ones used in human-human communication (i.e., speech, gestures, facial expressions, etc.).

Software Architectures

In order to enable modality combinations in the user interface, adapted software architectures are needed. There are two fundamental types of multimodal software architectures, depending on the types of modalities. In feature level architectures, the integration of modalities is performed during the recognition process, whereas in semantic level architectures, each modality is processed or recognized independently of the others (Figure 1).

Feature-level architectures generally are considered appropriate for tightly related and synchronized modalities, such as speech and lip movements (Duchnowski et al., 1994). In this type of architecture, connectionist models can be used for processing modalities because of their good performance as pattern classifiers and because they easily can integrate heterogeneous features. However, a truly multimodal connectionist approach is dependent on the availability of multimodal training data, and such data currently is not available.

When the interdependency between modalities implies complementarity or disambiguation (e.g., speech and gesture inputs), information typically is integrated into semantic-level architectures (Nigay et al., 1995). In this type of architecture, the main approach for modality integration is based on the use

Figure 1. Multimodal software architectures

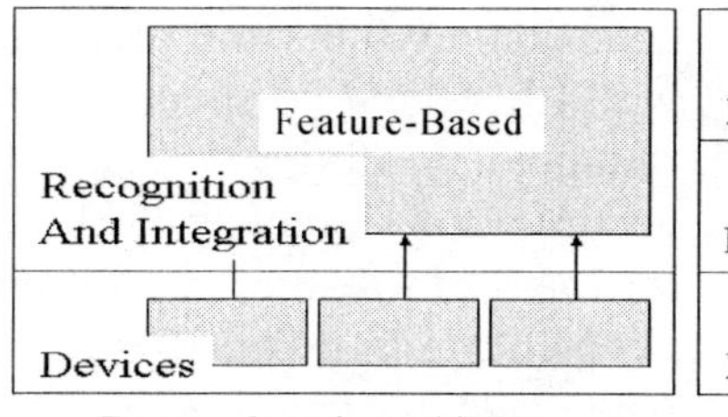

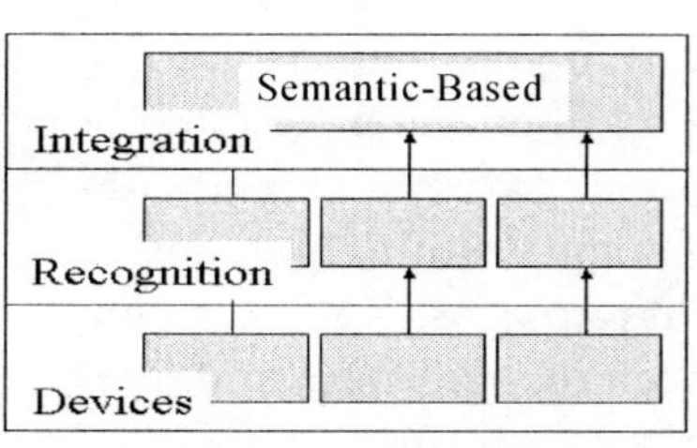

of data structures called *frames*. Frames are used to represent meaning and knowledge and to merge information that results from different modality streams.

MAIN ISSUES IN MULTIMODAL INTERACTION

Designing Multimodal Interaction

Recent developments in recognition-based interaction technologies (e.g., speech and gesture recognition) have opened a myriad of new possibilities for the design and implementation of multimodal interfaces. However, designing systems that take advantage of these new interaction techniques are difficult. Our lack of understanding of how different modes of interaction can be combined best into the user interface often leads to interface designs with poor usability. Most studies to understand natural integration of communication modes are found in the experimental psychology research literature, but they tend to qualitatively describe human-to-human communication modes. Very few attempts have been made so far to qualitatively or quantitatively describe multimodal human-computer interaction (Bourguet, 1998; Nigay, 1995; Oviatt, 1997). Much more work is still needed in this area.

Implementing Multimodality

Developers still face major technical challenges for the implementation of multimodality, as indeed, the multimodal dimension of a user interface raises numerous challenges that are not present in more traditional interfaces (Bourguet, 2004). These challenges include the need to process inputs from different and heterogeneous streams; the coordination and integration of several communication channels (input modalities) that operate in parallel (modality fusion); the partition of information sets across several output modalities for the generation of efficient multimodal presentations (modality fission); dealing with uncertainty and recognition errors; and implementing distributed interfaces over networks (e.g., when speech and gesture recognition are performed on different processors). There is a general lack of appropriate tools to guide the design and implementation of multimodal interfaces.

Bourguet (2003a, 2003b) has proposed a simple framework, based on the finite state machine formalism, for describing multimodal interaction designs and for combining sets of user inputs of different modalities. The proposed framework can help designers in reasoning about synchronization patterns problems and testing interaction robustness.

Uncertainty in Multimodal Interfaces

Natural modalities of interaction, such as speech and gestures, typically rely on recognition-based technologies that are inherently error prone. Speech recognition systems, for example, are sensitive to vocabulary size, quality of audio signal, and variability of voice parameters (Halverson, 1999). Signal and noise separation also remains a major challenge in speech recognition technology, as current systems are extremely sensitive to background noise and to the presence of more than one speaker. In addition, slight changes in voice quality (due, for example, to the speaker having a cold) can significantly affect the performance of a recognizer, even after the user has trained it.

Several possible user strategies to prevent or correct recognition errors have been uncovered. Oviatt (2000) shows that in order to avoid recognition errors, users tend to spontaneously select the input mode they recognize as being the most robust for a certain type of content (modality selection strategy). When recognition errors occurr, Suhm (2001) suggests that users be willing to repeat their input at least once, after which they will tend to switch to another modality (modality switching strategy). Finally, Oviat (2000) reports cases of linguistic adaptation, where users choose to reformulate their speech in the belief that it can influence error resolution—a word may be substituted for another, or a simpler syntactic structure may be chosen. Overall, much more research is still needed to increase the robustness of recognition-based modalities.

APPLICATIONS

Two applications of multimodal interaction are described.

Augmented Reality

Augmented reality is a new form of multimodal interface in which the user interacts with real-world objects and, at the same time, is given supplementary visual information about these objects (e.g., via a head mounted display). This supplementary information is context-dependent (i.e., it is drawn from the real objects and fitted to them). The virtual world is intended to complement the real world on which it is overlaid. Augmented reality makes use of the latest computer vision techniques and sensor technologies, cameras, and head-mounted displays. It has been demonstrated, for example, in a prototype to enhance medical surgery (Dubois, 1999).

Tangible Interfaces

People are good at sensing and manipulating physical objects, but these skills seldom are used in human-computer interaction. Tangible interfaces are multimodal interfaces that exploit the tactile modalities by giving physical form to digital information (Ishii, 1997). They implement physical objects, surfaces, and textures as tangible embodiments of digital information. The tangible query interface, for example, proposes a new means for querying relational databases through the manipulation of physical tokens on a series of sliding racks.

FUTURE TRENDS

Ubiquitous Computing

Ubiquitous computing describes a world from which the personal computer has disappeared and has been replaced by a multitude of wireless, small computing devices embodied in everyday objects (e.g., watches, clothes, or refrigerators). The emergence of these new devices has brought new challenges for human-computer interaction. A fundamentally new class of modalities has emerged—the so-called passive modalities—that corresponds to information that is automatically captured by the multimodal interface without any voluntary action from the user. Passive modalities complement the active modalities such as voice command or pen gestures.

Compared with desktop computers, the screens of ubiquitous computing devices are small or non-existent; small keyboards and touch panels are hard to use when on the move, and processing powers are limited. In response to this interaction challenge, new modalities of interaction (e.g., non-speech sounds) (Brewster, 1998) have been proposed, and the multimodal interaction research community has started to adapt traditional multimodal interaction techniques to the constraints of ubiquitous computing devices (Branco, 2001; Schaefer, 2003; Schneider, 2001).

CONCLUSION

Multimodal interfaces are a class of intelligent multimedia systems that extends the sensory-motor capabilities of computer systems to better match the natural communication means of human beings. As recognition-based technologies such as speech recognition and computer vision techniques continue to improve, multimodal interaction should become widespread and eventually may replace traditional styles of human-computer interaction (e.g., keyboard and mice). However, much research still is needed to better understand users' multimodal behaviors in order to help designers and developers to build natural and robust multimodal interfaces. In particular, ubiquitous computing is a new important trend in computing that will necessitate the design of innovative and robust multimodal interfaces that will allow users to interact naturally with a multitude of embedded and invisible computing devices.

REFERENCES

Bolt, R. A. (1980). Put-that-there: Voice and gesture at the graphics interface. *Proceedings of the 7th Annual Conference on Computer Graphics and Interactive Techniques*. Seattle, Washington.

Bourguet, M. L. (2003a). Designing and prototyping multimodal commands. *Proceedings of the IFIP TC13 International Conference on Human-Computer Interaction, INTERACT'03*, Zurich, Switzerland.

Bourguet, M. L. (2003b). How finite state machines can be used to build error free multimodal interaction systems. *Proceedings of the 17th British Group Annual Conference*, Bath, UK.

Bourguet, M. L. (2004). Software design and development of multimodal interaction. *Proceedings of the IFIP 18th World Computer Congress Topical Days*.

Bourguet, M. L., & Ando, A. (1998). Synchronization of speech and hand gestures during multimodal human-computer interaction. *Proceedings of the Conference on Human Factors in Computing Systems,* Los Angeles, California.

Branco, P. (2001). Challenges for multimodal interfaces towards anyone anywhere accessibility: A position paper. *Proceedings of the 2001 Workshop on Universal Accessibility of Ubiquitous Computing: Providing for the Elderly*. Alcácer do Sal, Portugal.

Brewster, S., Leplatres, G., & Crease, M. (1998). Using non-speech sounds in mobile computing devices. *Proceedings of the 1st Workshop on Human Computer Interaction with Mobile Devices*, Glasgow, Scotland.

Dubois, E., Nigay, L., Troccaz, J., Chavanon, O., & Carrat, L. (1999). Classification space for augmented surgery: An augmented reality case study. *Proceedings of Seventh IFIP TC13 International Conference on Human-Computer Interaction*, Edinburgh, UK.

Duchnowski, P., Meier, U., & Waibel, A. (1994). See me, hear me: Integrating automatic speech recognition and lipreading. *Proceeding of the International Conference on Spoken Language Processing*, Yokohama, Japan.

Halverson, C., Horn, D., Karat, C., & Karat, J. (1999). The beauty of errors: Patterns of error correction in desktop speech systems. *Proceedings of the Seventh IFIP TC13 International Conference on Human-Computer Interaction*, Edinburgh, Scotland.

Ishii, H., & Ullmer, B. (1997). Tangible bits: Towards seamless interfaces between people, bits and atoms. *Proceedings of the ACM Conference on Human Factors in Computing Systems,* Atlanta, Georgia.

Ko, T., Demirdjian, D., & Darrell, T. (2003). Untethered gesture acquisition and recognition for a multimodal conversational system. *Proceedings of the 5th International Conference on Multimodal Interfaces*, Vancouver, Canada.

Nigay, L., & Coutaz, J. (1995). A generic platform for addressing the multimodal challenge. *Proceedings of the Conference on Human Factors in Computing Systems.*

Oviatt, S. (2000). Taming recognition errors with a multimodal interface. *Communications of the ACM, 43*(9), 45-51.

Oviatt, S., De Angeli, A., & Kuhn, K. (1997). Integration and synchronisation of input modes during multimodal human-computer interaction. *Proceedings of the ACM Conference on Human Factors in Computing Systems,* Atlanta, Georgia.

Schaefer, R., & Mueller, W. (2003). Multimodal interactive user interfaces for mobile multi-device environments. *Proceedings of the Ubicomp 2003 Workshop Multi-Device Interfaces for Ubiquitous Peripheral Interaction*, Seattle, Washington.

Schneider, G., Djennane, S., Pham, T. L., & Goose, S. (2001). Multimodal multi device UIs for ubiquitous access to multimedia gateways. *Proceedings of the 17th International Joint Conference on Artificial Intelligence and Workshop Artificial Intelligence In Mobile Systems,* Seattle, Washington.

Suhm, B., Myers, B., & Waibel, A. (2001). Multimodal error correction for speech user interfaces. *ACM Transactions on Computer-Human Interaction, 8*(1), 60-98.

Vo, M. T., & Wood, C. (1996). Building an application framework for speech and pen input integration in multimodal learning interfaces. *Proceedings of the IEEE International Conference on Acoustics, Speech, and Signal Processing.* Atlanta, Georgia.

KEY TERMS

Active Modality: Modality voluntarily and consciously used by users to issue a command to the computer; for example, a voice command or a pen gesture.

Feature-Level Architecture: In this type of architecture, modality fusion operates at a low level of modality processing. The recognition process in one modality can influence the recognition process in another modality. Feature-level architectures generally are considered appropriate for tightly related and synchronized modalities, such as speech and lip movements.

Haptic Output: Devices that produce a tactile or force output. Nearly all devices with tactile output have been developed for graphical or robotic applications.

Modality Fission: The partition of information sets across several modality outputs for the generation of efficient multimodal presentations.

Modality Fusion: Integration of several modality inputs in the multimodal architecture to reconstruct a user's command.

Mutual Disambiguation: The phenomenon in which an input signal in one modality allows recovery from recognition error or ambiguity in a second signal in a different modality is called mutual disambiguation of input modes.

Passive Modality: Information that is captured automatically by the multimodal interface; for example, to track a user's location via a microphone, a camera, or data sensors.

Semantic-Level Architecture: In semantic level architectures, modalities are integrated at higher levels of processing. Speech and gestures, for example, are recognized in parallel and independently. The results are stored in meaning representations that then are fused by the multimodal integration component.

Visual Speech Recognition: Computer vision techniques are used to extract information about the lips' shape. This information is compared with information extracted from the speech acoustic signal to determine the most probable speech recognition output.

P

PDA Usability for Telemedicine Support*

Shirley Ann Becker
Florida Institute of Technology, USA

INTRODUCTION

Telemedicine is broadly defined as the use of information and communications technology to provide medical information and services (Perednia & Allen, 1995). Telemedicine offers an unprecedented means of bringing healthcare to anyone regardless of geographic remoteness. It promotes the use of ICT for healthcare when physical distance separates the provider from the patient (Institute of Medicine, 1996). In addition, it provides for real-time feedback, thus eliminating the waiting time associated with a traditional healthcare visit.

Telemedicine has been pursued for over three decades as researchers, healthcare providers, and clinicians search for a way to reach patients living in remote and isolated areas (Norris, 2001). Early implementation of telemedicine made use of the telephone in order for healthcare providers and patients to interact. Over time, fax machines were introduced along with interactive multimedia, thus supporting teleconferencing among participants. Unfortunately, many of the early telemedicine projects did not survive because of high costs and insurmountable barriers associated with the use of technology.

Telemedicine has been resurrected during the last decade as a means to help rural healthcare facilities. Advances in information and communications technology have initiated partnerships between rural healthcare facilities and larger ones. The Internet in particular has changed the way in which medical consultations can be provided (Coiera, 1997). Personal computers (PCs) and supporting peripherals, acting as clients, can be linked to medical databases residing virtually in any geographic space. Multimedia data types, video, audio, text, imaging, and graphics promote the rapid diagnosis and treatment of casualties and diseases.

Innovations in ICT offer unprecedented healthcare opportunities in remote regions throughout the world. Mobile devices using wireless connectivity are growing in popularity as thin clients that can be linked to centralized or distributed medical-data sources. These devices provide for local data storage of medical data, which can be retrieved and sent back to a centralized source when Internet access becomes available. Those working in nomadic environments are connected to data sources that in the past were inaccessible due to a lack of telephone and cable lines. For the military, paramedics, social workers, and other healthcare providers in the field, ICT advances have removed technology barriers that made mobility difficult if not impossible.

Personal digital assistants (PDAs)[1] are mobile devices that continue to grow in popularity. PDAs are typically considered more usable for multimedia data than smaller wireless devices (e.g., cell phones) because of larger screens, fully functional keyboards, and operating systems that support many desktop features. Over the past several years, PDAs have become far less costly than personal-computing technology. They are portable, lightweight, and mobile when compared to desktop computers. Yet, they offer similar functionality scaled back to accommodate the differences in user-interface designs, data transmission speed, memory, processing power, data storage capacity, and battery life.

BACKGROUND

Computing experts predicted that PDAs would supplant the personal computer as ubiquitous technology (Chen, 1999; Weiser as cited in Kim & Albers, 2001). Though this has not yet happened, PDA usage continues to grow with advances in operating systems, database technology, and add-on features such as digital cameras. They are being used in sales, field engineering, education, healthcare, and other areas that require mobility. In the medical field, for example, they are being used to record and track patient data (Du Bois & McCright, 2000). This mobility is made possible by enterprise servers push-

Table 1. User-interface design constraints for PDA devices (Paelke, Reimann, & Rosenbach, 2003)

Limited resolution	Typical resolution of a PDA is low (240*320 pixels). This impacts the visibility of content, objects and images.
Small display size	The small screen size of a PDA limits the number of objects and the amount of text on a screen page. This limitation impacts design layout in terms of font size, white space, links, text, images, and graphics, among others.
Navigational structure	Navigation is impacted by the increased number of screen pages required to accommodate text and objects that on a desktop or laptop would fit on one screen page. Design choices include a long page with a flat navigation hierarchy versus the design of multiple short pages with a deeper navigational hierarchy.
Limited use of color	A PDA uses a gray scale or a color palette limited to several thousand color choices (compared to millions of color choices for desktop applications). Readability and comprehension may be impacted when color is used to relay information or color combinations are insufficient in contrast.
Limited processing power	Limited processing power impacts the quality of graphical displays and imaging. It also restricts the use of interactive real-time animation.
Mouse is replaced with stylus pen	A PDA does not use a mouse, which has become a standard peripheral in a desktop environment. As a result, there is a learning curve associated with the use of a stylus pen, which replaces mouse functionality.
Small keyboard size	The PDA keyboard size and layout impacts data entry. As a result, it is more difficult for users to entered lengthy and complex medical data in a real-time environment.

ing data onto these devices without user intervention. Enterprise servers are also capable of pulling data from a localized (PDA) database such that centralized data sources are readily updated.

A PDA synchronizes with laptops and desktop computers, making data sharing transparent. This is made possible by a user interface and functionality that are compatible in terms of computing capabilities and input and output devices (Myers, 2001). Compatibility is a major issue in telemedicine given that medical and patient data gathered or stored on a PDA is typically sent to a centralized data source. Nomadic use of PDAs mandates this type of data integration whether it is real-time or batched data when wireless connectivity is temporarily inaccessible (Huston & Huston, 2000). In addition, telemedicine data sharing is typically asymmetric in that the enterprise server transmits a larger volume of medical data to the PDA. In turn, the PDA transmits only a small volume of patient data to the server (Murthy & Krishnamurthy, 2004).

Though PDAs hold great promise in promoting healthcare in remote regions, the usability of these devices continues to be an issue. There are physical constraints that typically do not apply to a laptop or desktop computer (Table 1 describes these constraints). The user interface of a PDA is modeled after a desktop environment with little consideration for physical and environmental differences (Sacher & Loudon, 2002). Yet, these differences are significant in terms of usability given the small screen and keyboard sizes and limited screen resources in terms of memory and power reduction (Brewster, 2002).

There has been important research on PDA usability, primarily in the effective use of its limited screen area. Early research focused primarily on the display of contextual information in order to minimize waste of the screen space while maximizing content (Kamba, Elson, Harpold, Stamper, & Sukariya as cited in Buchanan, Farrant, Jones, Thimbleby, Marsden, & Pazzani, 2001). More recent efforts are taking into account not only screen size, but navigation, download time, scrolling, and input mechanisms (Kaikkonen & Roto, 2003).

PDA USABILITY AND TELEMEDICINE

An important finding of usability research associated with mobile technology is the need for usability testing beyond a simulated environment. Waterson, Landay, and Matthews (2002), in their study of the usability of a PDA, found that usability testing should include both content and device design. Chittaro and Dal Cin (2002) studied the navigational structures of mobile user interfaces. Their research also identified the need for actual devices to be used in usability testing. Real-world constraints would take into account screen-size and page-design issues, date entry using a built-in keypad, wireless accessibility, data transmission speeds, visual glare, background noise, and battery power, among others.

Our initial findings also reflected the need for usability testing in the telemedical environment in which technology is used. We initiated research on the use of PDAs for monitoring diabetic patients living in remote regions of the United States (Becker, Sugumaran, & Pannu, 2004). Figure 1 illustrates one of our user-interface designs for the Viewsonic® PocketPC. This screen shows part of a foot form completed by a healthcare provider during a home visit. The data entered by the user is stored in a local database that can be transmitted wirelessly to an enterprise server.

The PocketPC is used in this research because of its low cost and its support of relational database technology. It has a built-in digital camera, which is important because of the physical distance between a patient and a healthcare facility. Images of foot sores are taken during a home visit and stored in the local database residing on the PDA. These images become part of the patient's history when transmitted to the enterprise server. Later, the images can be viewed by a clinician for the timely diagnosis and treatment of the sores.

Our research has shown the technical feasibility of using PDA technology to gather data in the field during a home visit. However, more research is needed to address usability issues uncovered during the use of the PDA during simulated home visits. A significant finding in the use of PDA technology is that usability is tightly integrated with the technological challenges associated with it. One such challenge is the heavy reliance on battery power when PDAs are deployed in the field. When the battery no longer holds a charge, critically stored relational data may be irretrievable due to pull technology used to transmit data from a local source to a central one.

As part of this research, the usability of multimedia data formats is being studied to improve information access in a nomadic environment. For rapid diagnosis and treatment of casualties, multimedia data formats may prove critical. In our work, images are being used to replace textual descriptions that would consume valuable screen space. Figure 1 illustrates this concept of using color codes to represent physical areas of the foot. As such, foot problems can be reported for each area by clicking on the list appearing on the right side of the screen. Audio capabilities are also being explored in providing helpful information that otherwise would be text based. Both of these multimedia capabilities are in the design phase and will be tested in future field studies.

Figure 1. PDA used to gather data about a diabetic patient's foot health

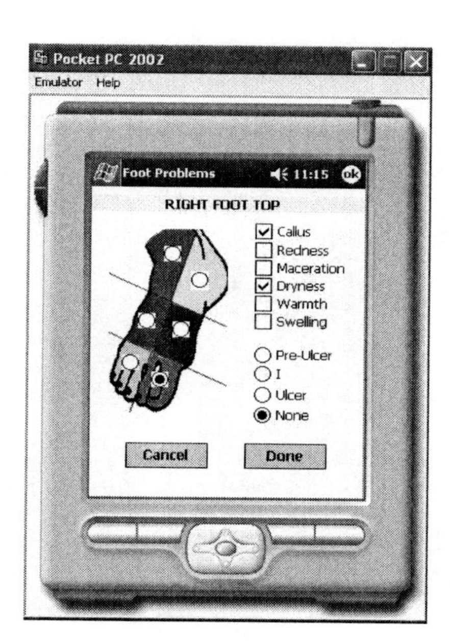

FUTURE TRENDS

Table 2 identifies research opportunities associated with the use of PDAs in telemedicine. Much of what has been done in this area has focused on tracking patient histories. However, there are significant opportunities for real-time data retrieval and transmission using PDA technology. Clinicians could use a PDA, for example, to send prescriptions to pharmacies, receive lab reports, and review medical data for the diagnosis and treatment of

Table 2. Telemedicine research opportunities using PDA technology (Wachter, 2003)

Diagnosis and Treatment	Mobile decision support software would allow for data entry of patient symptoms with output providing a diagnosis and treatment plan.
Patient Tracking	Synchronizing a PDA with a hospital's centralized data source would allow vital signs and other information to be gathered in real-time at the point of care. A clinician would have the capability of obtaining lab reports and test results once they have been entered into the system.
Prescriptions	A PDA would be used by a clinician to send a patient prescription to a pharmacy. This would minimize human error associated with interpreting handwritten prescriptions. It would also provide a centralized tracking system in order to identify drug interactions when multiple prescriptions for a patient are filled.
Medical Information	Clinicians would have access to information on medical research, drug treatments, treatment protocols, and other supporting materials. According to Wachter (2003), a leading clinical PDA technology vendor has converted more than 260 medical texts into PDA formats thus supporting this effort.
Dictation	PDAs support multimedia data including audio, images, and text. As such, clinicians would have an opportunity to record multimedia patient data directly linked to patient history data in a centralized source.
Charge Capture	Data entry into a PDA that is transmitted to a centralized source would provide the means for efficient billing of medical charges to a patient.

patients. These devices could also be used to minimize human error associated with more traditional mechanisms of recording patient data.

There are infrastructure challenges associated with the use of telemedicine in terms of technology acceptance and utilization. Chau and Hu (2004) point out that although telemedicine is experiencing rapid growth, there are organizational issues pertaining to technology and management. It is critical that organizational support is available throughout the implementation stages of telemedicine. Past experience in the use of ICT with no infrastructural support resulted in failure. The effective management of telemedicine systems and supporting technologies is needed to address barriers to ICT acceptance by healthcare personnel and patients. As such, there are research opportunities in the organizational acceptance and use of PDAs in a telemedical environment.

Security, safety, and social concerns have also been identified by Tarasewich (2003) as research challenges in the use of mobile technology. Though encryption and other security technologies can readily be used during the transmission of data, there remains the issue of security associated with lost or stolen PDAs. Given the memory, data storage, and other technological constraints of a PDA, research is needed on developing security mechanisms for localized data. Research is also needed on ensuring localized data remains private and is accessible only by authorized personnel.

CONCLUSION

The exponential growth of wireless and PDA technologies has brought unprecedented opportunities in providing managed healthcare. For the military and others working in nomadic environments, PDA technology offers the capability for rapid diagnosis and treatment of casualties. Regardless of location, healthcare personnel could be provided with real-time access to reference materials, patient lab reports, and patient history data.

Though there is great promise in the use of PDAs for providing telemedical services, there is research needed in the usability of these devices. Multimedia data formats offer alternative interfaces to accessing data, and research is needed to assess their impact on ease of use and understandability. In addition, technological constraints need to be studied in terms of their impact on device usability. Memory, data storage, transmission speeds, and battery life need to be considered as part of usability testing to assess the impact on rapid medical diagnosis and treatment.

There is a major challenge of moving from traditional medical services and resources to an environment that promotes PDA technology and telemedicine. The potential benefits are great in terms of ubiquitous helth care with no time or space constraints. However, widespread acceptance of PDA technology in a telemedical environment will only become achievable through the design of usable interfaces.

REFERENCES

Becker, S. A., Sugumaran, R., & Pannu, K. (2004). *The use of mobile technology for proactive healthcare in tribal communities.* 2004 National Conference on Digital Government Research (pp. 297-298).

Brewster, S. (2002). Overcoming the lack of screen space on mobile computers. *Personal and Ubiquitous Computing, 6*, 188-205.

Buchanan, G., Farrant, S., Jones, M., Thimbleby, H., Marsden, G., & Pazzani, M. (2001). Improving mobile Internet usability. *Proceedings of the Tenth International Conference on World Wide Web* (pp. 673-680).

Chau, P. Y. K., & Hu, J. H. P. (2004). Technology implementation for telemedicine programs. *Communications of the ACM, 47*(2), 87-92.

Chen, A. (1999, September 27). Handhelds on deck. *InfoWeek*, 67-72.

Chittaro, L., & Dal Cin, P. (2002). Evaluating interface design choices on WAP phones: Navigation and selection. *Personal and Ubiquitous Computing, 6*(4), 237-244.

Coiera, E. (1997). *Guide to medical informatics, the Internet and telemedicine.* London: Chapman and Hall.

Du Bois, G., & McCright, J. (2000, September 4). Doctors are on the move. *eWeek*, 31.

Huston, T. L., & Huston, J. L. (2000). Is telemedicine a practical reality? *Communications of the ACM, 43*(6), 91-95.

Institute of Electrical and Electronics Engineers (IEEE). (1990). *IEEE standard computer dictionary: A compilation of IEEE standard computer glossaries.* New York: Institute of Electrical and Electronics Engineers.

Institute of Medicine. (1996). *Telemedicine: A guide to assessing telecommunications in health care.* Washington, DC: National Academy Press.

Kaikkonen, A., & Roto, V. (2003). Navigating in a mobile XHTML application. *Proceedings of the Conference on Human Factors in Computing Systems* (pp. 329-336).

Kamba, T., Elson, S. A., Harpold, T., Stamper, T., & Sukariya, P. (1996). Using small screen space more efficiently. *Proceedings of the SIGCHI Conference on Human Factors in Computing Systems* (pp. 383-390).

Kim, L., & Albers, M. J. (2001). Web design issues when searching for information in a small screen display. *Proceedings of the 19th Annual International Conference on Computer Documentation* (pp. 193-200).

Murthy, V. K., & Krishnamurthy, E. (2004). *Multimedia computing environment for telemedical applications.* Poster presentation at the IADIS International Conference on Web Based Communities, Lisbon, Portugal.

Myers, B. A. (2001). Using handhelds and PCs together. *Communications of the ACM, 44*(11), 34-41.

Norris, A. C. (2001). *Essentials of telemedicine and telecare.* New York: John Wiley & Sons.

Paelke, V., Reimann, C., & Rosenbach, W. (2003). A visualization design repository for mobile devices. *Proceedings of the Second International Conference on Computer Graphics, Virtual Reality, Visualization and Interaction in Africa* (pp. 57-62).

Perednia, D. A., & Allen, A. (1995). Telemedicine technology and clinical applications. *Journal of the American Medical Association, 273*(6), 483-487.

Sacher, H., & Loudon, G. (2002). Uncovering the new wireless interaction paradigm. *Interactions, 9*(1), 17-23.

Tarasewich, P. (2003). Designing mobile commerce applications. *Communications of the ACM, 46*(12), 57-60.

Wachter, G. W. (2003). Making rounds with handheld technology. *Telehealth Practices Report, 8*(5), 7-9.

Waterson, S., Landay, J. A., & Matthews, T. (2002). In the lab and out in the wild: Remote Web usability testing for mobile devices. *Conference on Human Factors in Computing Systems* (pp. 796-797).

Weiser, M. (1998). The future of ubiquitous computing on campus. *Communications of the ACM, 41*(1), 41-42.

KEY TERMS

Compatibility: The ability to transmit data from one source to another without losses or modifications to the data or additional programming requirements.

Interoperability: The ability of two or more systems or components to exchange information and to use the information that has been exchanged (Institute of Electrical and Electronics Engineers [IEEE], 1990).

Peripheral Devices: Hardware devices, separate from the computer's central processing unit (CPU), which add communication or other capabilities to the computer.

Personal Digital Assistant (PDA): A personal digital assistant is a handheld device that integrates computing, telephone, Internet, and networking technologies.

Telecare: The use of information and communications technology to provide medical services and resources directly to a patient in his or her home.

Telehealth: The use of information and communications technologies to provide a broader set of healthcare services including medical, clinical, administrative, and educational ones.

Telemedicine: The use of information and communications technologies to provide medical services and resources.

Wireless Application Protocol (WAP): The wireless application protocol promotes the interoperability of wireless networks, supporting devices, and applications by using a common set of applications and protocols (http://www.wapforum.org).

ENDNOTES

* This article is based on work supported by the National Science Foundation under Grant No. 0443599. Any opinions, findings, and conclusions or recommendations expressed in this content are those of the author(s) and do not necessarily reflect the views of the National Science Foundation.

[1] PocketPCs, Palm Pilots, and other handheld devices are referred to as PDAs in this article.

Pen-Based Digital Screen Interaction

Khaireel A. Mohamed
Albert-Ludwigs-Universität Freiburg, Germany

Thomas Ottman
Albert-Ludwigs-Universität Freiburg, Germany

INTRODUCTION

Through a transducer device and the movements effected from a digital pen, we have a pen-based interface that captures digital ink. This information can be relayed on to domain-specific application software that interpret the pen input as appropriate computer actions or archive them as ink documents, notes, or messages for later retrieval and exchanges through telecommunications means.

Pen-based interfaces have rapidly advanced since the commercial popularity of personal digital assistants (PDAs) not only because they are conveniently portable, but more so for their easy-to-use freehand input modal that appeals to a wide range of users. Research efforts aimed at the latter reason led to modern products such as the personal tablet PCs (personal computers; Microsoft Corporation, 2003), corporate wall-sized interactive boards (SMART Technologies, 2003), and the communal tabletop displays (Shen, Everitt, & Ryall, 2003).

Classical interaction methodologies adopted for the desktop, which essentially utilize the conventional pull-down menu systems by means of a keyboard and a mouse, may no longer seem appropriate; screens are getting bigger, the interactivity dimension is increasing, and users tend to insist on a one-to-one relation with the hardware whenever the pen is used (Anderson, Anderson, Simon, Wolfman, VanDeGrift, & Yasuhara, 2004; Chong & Sakauchi, 2000). So, instead of combining the keyboard, mouse, and pen inputs to conform to the classical interaction methodologies for these modern products, our ultimate goal is then to do away with the conventional GUIs (graphical user interfaces) and concentrate on perceptual starting points in the design space for pen-based user interfaces (Turk & Robertson, 2000).

BACKGROUND

If we attempt to recognize the digital pen as the only sole input modal for digital screens, for both interfacing and archival modes purported within the same writing domain, we then require the conceptualization of a true perceptual user interface (PUI) model. Turk and Robertson (2000) discuss the main idea of having an alternative (graphical user) interface through the PUI paradigm as a nonhassled and natural way of communicating with the background operating system. It is subjective, and it concerns finding out and (to a certain extent) anticipating what users expect from their application environment. There are several reasons to utilize the PUI as an interactive model for the digital screen. Amongst some of the more prominent ones are the following:

- To reintroduce the natural concept of communication between users and their devices
- To present an intelligent interface that is able to react accordingly (as dictated by the objective of the application program) to any input ink strokes
- To redesign the GUI exclusively for perceptual conceptions

Modern and networked interactive digital screens utilize the electronic pen's digital ink as a convenient way of interfacing with specially developed application programs, and go on to offer the visual communication of opinions for multiple users. This is as a result of taking advantage of the pen-based environment. For example, we want to reproduce the simple, customary blackboard and still be able to include all other functionalities that an e-board can offer. But by minimizing the number of static menus and buttons (to accommodate new perceptual designs in accordance to the PUI standards), the resultant "clean slate" becomes the only perceptual input available to users to relate to the background systems. Here, we see two distinct domains merged into one: the domain to receive handwritings (or drawings) as the symbolic representation of information (termed technically as traces), and the domain to react to user commands issued through pull-down menus and command buttons.

Based purely on the input ink traces, we must be able to decipher users' intentions in order to correctly classify which of the two domains it is likely to be in: either as primitive symbolic traces, or some sort of system command. Often, these two domains overlap and pose the problem of ambiguousness, a gray area that cannot be simply classified by means of straightforward algorithms. For instance, the background system may interpret a circle drawn in a clockwise direction over some preexisting ink traces as a select command when in fact the user had simply intended to leave the circle as a primitive ink trace to emphasize the importance of his or her previously written points. Fortunately, this problem can be solved if the program can anticipate the intentions of its users (Wooldridge, 2002); however, this method necessitates the constant tracking of the perceptual environment and would require a more stringent and somewhat parallel structural construct in order to run efficiently (Mohamed, 2004b; Mohamed, Belenkaia, & Ottman, 2004).

There are currently many works by authors that describe vividly the interpretations of these traces exclusively in either domain as well as in combination of the two. In the trace-only domain, Aref, Barbara, and Lopresti (1996) and Lopresti, Tomkins, and Zhou's (1996) collective research in dealing with a concentrated area of deciphering digital inks as hand-drawn sketches and handwritings, and then performing pictorial queries on them, is the result of their effective categorization of ink as a "first-class" data type in multimedia databases. Others like Bargeron and Moscovich (2003) and Götze, Schlechtweg, and Strothotte (2002) analyze users' rough annotations and open-ended ink markings on formal documents and then provide methods for resetting these traces in a more orderly, cross-referenced manner. On the opposite perspective, we see pilot works on pen gestures, which began even before the introduction of styluses for digital screens. They are purported on ideas of generating system commands from an input sequence of predetermined mouse moves (Rubine, 1991). Moyle and Cockburn (2003) built simple gestures for the conventional mouse to browse Web pages quickly, as users would with the digital pen. As gesturing with the pen gained increasing popularity over the years, Long, Landay, Rowe, and Michiels (2000) described an exhaustive computational model for predicting the similarity of perceived gestures in order to create better and more comfortable user-based gesture designs.

For reasons of practicality and application suitability, but not necessarily for the simplicity of implementation, well-developed tool kits such as SATIN (Hong & Landay, 2000) and TEDDY (Igarashi, Matsuoka, & Tanaka, 1999) combine the pen input modality for two modes: sketching and gesturing. The automatic classification of ink inputs directed for either mode do not usually include too many gestures, and these tools normally place heavier cognition loads on the sketching mode. We agree that incorporating a pen-based command gesture recognition engine, as a further evaluation of the input traces and as an alternative to issuing system commands for addressing this scenario, is indeed one of the most practical ways to solve the new paradigm problem.

ISSUES ON REPRESENTING DIGITAL INK TRACES

A trace refers to a trail of digital ink data made between a successive pair of pen-down and pen-up events representing a sequence of contiguous ink points: the X and Y coordinates of the pen's position. Sometimes, we may find it advantageous to also include time stamps for each pair of the sampled coordinates if the sampling property of the transducer device is not constant. A sequence of traces accumulates to meaningful graphics, forming what we (humans) perceive as characters, words, drawings, or commands.

In its simplest form, we define a trace as a set of (x_i, y_i, t_i) tuples, deducing them directly from each complete pair of pen-down and pen-up events. Each trace must be considered unique and should be identifiable by its trace ID (identification). Figure 1 depicts the object-oriented relations a trace has with its predecessors, which can fundamentally be de-

Figure 1. Hierarchical object-oriented instances that define a trace

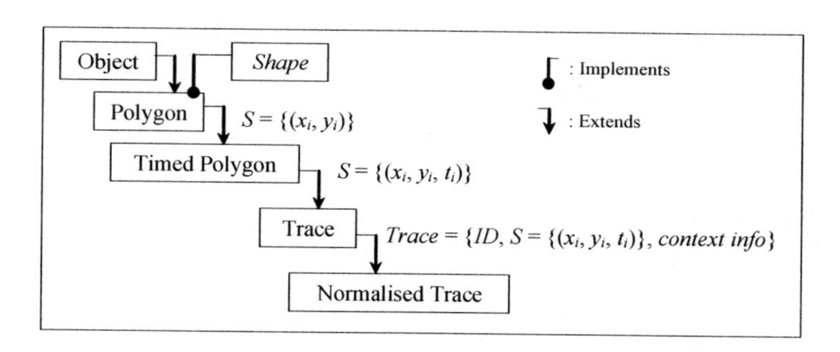

scribed as that of a shape interface (following the Java OOP conventions).

A trace also consists of rendering information such as pen color, brush style, the bounding box, center of gravity, and so forth for matters of visual interfacings. These are represented inside the context information of a trace. Traces with similar context information can later be assembled (or classified) together as trace groups. A normalised trace, on the other hand, is a filtered trace with removed noise and rationalized contents. It is used entirely in comparing techniques during the process of identifying and classifying pen gestures.On the temporal front, the timing associated when writing in free hand can be categorized as follows:

- the duration of the trace,
- its lead time, and
- its lag time.

Lead time refers to the time taken before an ink trace is scribed, and lag time refers to the time taken after an ink trace is scribed. This is illustrated in Figure 2. For a set of contiguous ink components $S_j = \{c_0, c_1, c_2, \ldots, c_n\}$ in a freehand sentence made up of n traces, we note that the lag time for the ith component is exactly the same as the lead time of the $(i+1)^{th}$ component; that is, $\text{lag}(c_i) = \text{lead}(c_{i+1})$. Consequently, the timings that separate one set of ink components apart from another are the first lead time $\text{lead}(c_0)$ and the last lag time $\text{lag}(c_n)$ in S_j. These times are significantly longer than their in-between neighbors c_1 to c_{n-1}.

Most people write rather fast, such that the time intervals between intermediate ink components in one word are very short. If we observe a complete freehand sentence made up of a group of freehand words, we can categorize each ink component within those words into one of the following four groups.

- **Beginnings:** Ink components found at the start of a freehand word

Figure 2. Splitting up freehand writing into ink components on the time line

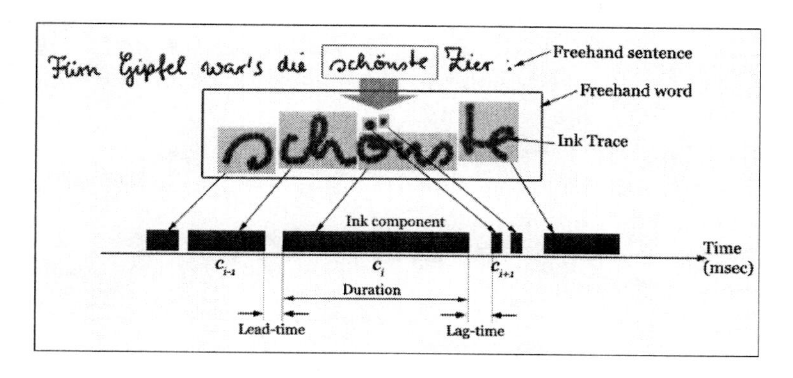

- **Endings:** Ink components found at the end of a freehand word
- **In-Betweens:** Ink components found in the middle of a freehand word
- **Stand-Alones:** Disjointed ink components

The groups differ in the demarcations of their lead and lag times, and as such, provide for a way in which a perceptual system can identify them. Other forms of freehand writings include mathematical equations, alphabets or characters of various languages, and signatures.

W3C's (World Wide Web Consortium's) current InkML specification defines a set of primitive elements sufficient for all basic ink applications (Russell et al., 2004). Few semantics are attached to these elements. All content of an InkML document is contained within a single <ink> element, and the fundamental data element in an InkML file is the <trace> element.

ISSUES ON REPRESENTING PEN GESTURES

Pen gestures are the direct consequence of interpreting primitive ink traces as system commands or as appropriate computer actions. A pen gesture is not, however, an instance of a trace.

While it is entirely up to the interpreter program to extract the meaning from the inputs and application contexts it received, our guidelines to the above claim are based on the fundamentals of the recognition algorithm of classifying gestures. Rubine's (1991) linear classifier algorithm is a straightforward dot product of the coefficient weights of a set of trained feature values with the same set of extracted features from a raw input trace (see Figure 3).

Essentially, this means that gestures do not require the storing of (x_i, y_i, t_i) tuples, but rather they should store the trained coefficient weights $\{c_0, c_1, \ldots, c_n\}$, which were negotiated and agreed upon by all parties attempting to synchronize the generality of the interpretation mechanism. That is, we need to ensure that the numbers, types, and techniques of features agreed upon for extraction are standardized across the board before we can be sure of issues of portability between applications.

Figure 3. Relationship between a gesture and a trace with n features

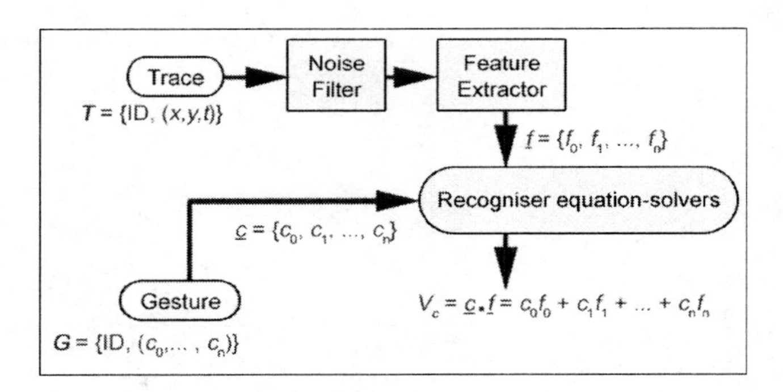

The temporal relation that singles out stand-alone components and freehand gestures from their continuous writing and drawing counterparts (traces) is the longer-than-average lead and lag times of a single-stroke ink trace, as shown in Figure 4. In this case, there is a pause period between samplings of ink components that results in significantly longer lead and lag times.

Based on the facts so far, we can now realize that it is possible to tackle the recognition process of the overlapping domains by focusing on the temporal sphere of influence. It dictates the very beings of the digital inks as either traces or gestures without the need to segregate the common input writing canvas.

ISSUES OF MOUNTING A PUI FOR THE INTERACTIVE DIGITAL-PEN ENVIRONMENT

We require a robust architecture that can provide the necessary natural feedback loop between the

Figure 4. Stand-alone ink components that can be interpreted as possible gestures

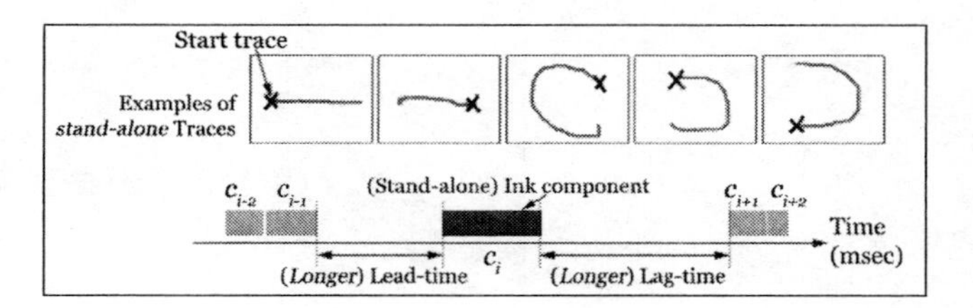

Figure 5. PUI model serving a sketching environment made up of a transducer and the digital pen

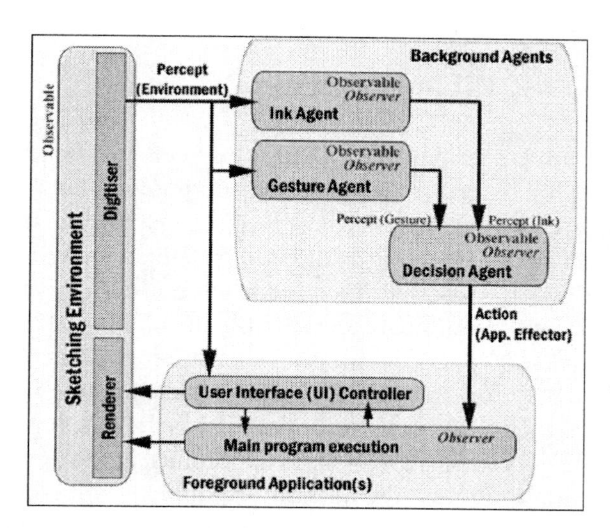

interfacing and interpreting mechanisms, the users, and the foreground application in order to affect the process of anticipation in a PUI environment. The problem of ambiguousness of the overlapping between two distinct ink domains (described previously) still stands and needs to be solved. We point out here again that if a program is able to intelligently anticipate the intention of its users through the constant tracking of the perceptual input environment, then that problem can be overcome.

This brings about a typical agent-oriented approach similar to that of a system utilizing interface agents. A notion that emphasizes adaptability, cooperation, proactiveness, and autonomy in both design and run times engages agents for abstruse software development (Mohamed, 2003; Mohamed & Ottmann, 2003; Wooldridge, 2002). In our case, we tasked two semiautonomous agents to process input digital inks in parallel, with one serving in the trace-based domain (Ink agent) and the other in the gesture-based domain (Gesture agent). Both are expected to simultaneously track the input digital ink in the temporal sphere of influence.

Figure 5 demonstrates the PUI model that we mounted to successfully work for the interactive digital-pen environment for the digital screen. It incorporates all of our previous discussions to ensure that the continuous tracking of all input digital inks is efficiently executed. A Decision agent is added between the two domain-specific agents and the foreground application for an added strength of decision making when drawing up percepts from the frontline agents.

It is not very often that we see people gesturing to a control system in the middle of writing a sentence or drawing a diagram. So we can anticipate, rather convincingly based on the lead and lag times obtained, that the latest ink component might be an instance of a trace rather than a gesture.

Our analyses (Mohamed, 2004a) on handling the lead and lag times of the beginnings, endings, in-betweens, and stand-alones of the ink components led to the following findings. Based on key statistical concepts in the context of determining the best solution for our problem definition, we establish the alternative hypothesis H_1 and its nullifying opposite H_0 as stated below.

H_0: An ink component is not a symbolic trace.
H_1: An ink component is a symbolic trace.

The darkened regions in Figure 6, depicting the lead and lag time relationship and the probability of

Figure 6. Lead and lag time relationships between traces and gestures

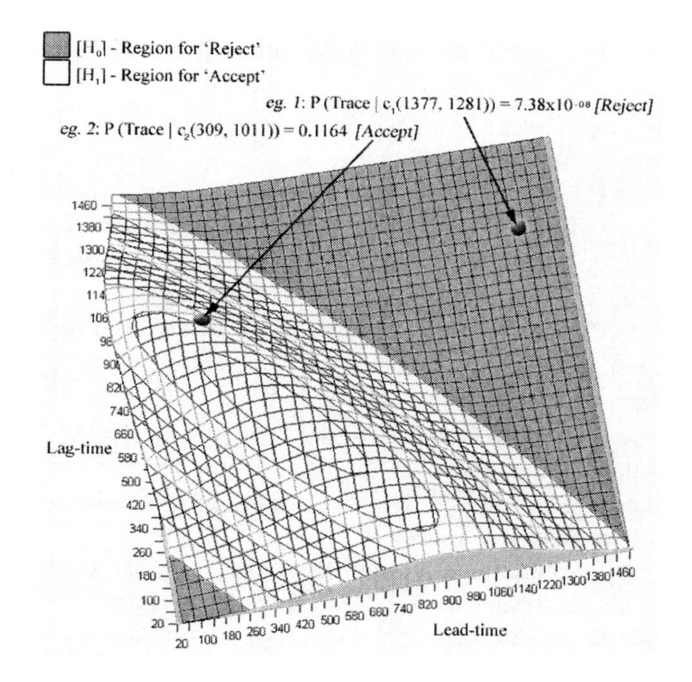

an ink component being a trace, are the areas for the rejection of the alternative hypothesis H_1, while the lighter region is the area for acceptance of H_1. Figure 6 is made to correspond directly as a lookup table; one can use the input parameter c_g (t_{lead}, t_{lag}) and retrieve from the table an output probability of whether any ink components should be considered as symbolic traces given its lead and lag times (i.e., P (Trace | c_g (t_{lead}, t_{lag}))) with an attached H_1 (strong acceptance) or H_0 (strong rejection).

Two examples are given in Figure 6. The first, with an input parameter of c_1 (1377, 1281), gets a probability value of $P = 7.38x10^{-08}$ and is recommended for rejection. This means that the likelihood of the ink component c_1 being a symbolic trace is very slim, and further tests should be made to check if we could indeed upgrade it to a command gesture. The second, with an input parameter of c_2 (309, 1011), receives $P = 0.1164$ and is recommended for acceptance. This is a clear-cut case that the ink component c_2 should definitely be considered as a symbolic trace.

FUTURE TRENDS

The lead and lag times resulting from freehand writings on digital boards are part of the ongoing process of managing, analysing, and reacting to all primitive ink data perceived from a writing environment. We currently have in place a background process model (Figure 5) designed to actively assist running foreground applications tailored for the PUI paradigm. We believe that the temporal methods highlighted are statistically strong for influencing future decisions down the communication chains within the PUI model. As we expect to branch out from currently working with single-stroke gestures to incorporating multistroke gestures in our (interactive) PUI model, it is essential that we observe the constituents of any affected ink components through their geometric properties. Most of the literature reviewed so far point toward the further segmentation of ink components into their primitive forms of lines and arcs for symbol recognition on a 2-D (two-dimensional) surface.

By being able to understand what is represented on-screen from the sketches made out by the primitive ink traces, and being able to issue multistroke gesture commands, we can expect a better perceptual environment for more interactive pen-based digital-screen interfaces.

CONCLUSION

The advent of pen-based input devices has clearly revealed a need for new interaction models that are different from the classical desktop paradigm. Instead of the keyboard and mouse, the electronic pen's digital ink is utilized for a commodious way to visually communicate ideas within the vicinity of the digital screen. By developing a true perceptual user interface for interaction with these screens, by means of only having the digital pen as the sole input modal, we achieve a robust architecture made up of semiautonomous agents that are able to correctly anticipate users' intentions on an invisible graphical user interface, treating the inputs convincingly as either written traces or gesture commands.

REFERENCES

Anderson, R., Anderson, R., Simon, B., Wolfman, S. A., VanDeGrift, T., & Yasuhara, K. (2004). Experiences with a tablet PC based lecture presentation system in computer science courses. *Proceedings of the 35th SIGCSE Technical Symposium on Computer Science Education* (pp. 56-60).

Aref, W. G., Barbarà, D., & Lopresti, D. (1996). Ink as a first-class datatype in multimedia databases. *Multimedia Database Systems: Issues and Research Directions* (pp. 113-141). New York: Springer-Verlag.

Bargeron, D., & Moscovich, T. (2003). Reflowing digital ink annotations. *Proceedings of the Conference on Human Factors in Computing Systems* (pp. 385-393).

Chong, N. S. T., & Sakauchi, M. (2000). Back to the basics: A first class chalkboard and more. *Proceedings of the 2000 ACM Symposium on Applied Computing* (pp. 131-136).

Götze, M., Schlechtweg, S., & Strothotte, T. (2002). The intelligent pen: Toward a uniform treatment of electronic documents. *Proceedings of the 2nd In-*

ternational Symposium on Smart Graphics (pp. 129-135).

Hong, J. I., & Landay, J. A. (2000). SATIN: A toolkit for informal ink-based applications. *Proceedings of the 13th Annual ACM Symposium on User Interface Software and Technology* (pp. 63-72).

Igarashi, T., Matsuoka, S., & Tanaka, H. (1999). TEDDY: A sketching interface for 3D freeform design. *Proceedings of the 26th Annual Conference on Computer Graphics and Interactive Techniques* (pp. 409-416).

Long, J. A. C., Landay, J. A., Rowe, L. A., & Michiels, J. (2000). Visual similarity of pen gestures. *Proceedings of the CHI 2000 Conference on Human Factors in Computing Systems* (pp. 360-367).

Lopresti, D., Tomkins, A., & Zhou, J. (1996). Algorithms for matching hand-drawn sketches. *Proceedings of the Fifth International Workshop on Frontiers in Handwriting Recognition* (pp. 233-238).

Microsoft Corporation. (2003). *Windows XP tablet PC edition.* Retrieved from http://www.microsoft.com/windowsxp/tabletpc

Mohamed, K. A. (2003). *Ingrid, Geraldine and Amidala: Competent agents for pen gestures interactivity* (Tech. Rep. No. 209). Germany: Albert-Lugwigs-Universität Freiburg, Institut für Informatik.

Mohamed, K. A. (2004a, February). *Restructuring data constructs in overlapping digital ink domains for agent-oriented approaches* (Tech. Rep. No. 211). Germany: Albert-Lugwigs-Universität Freiburg, Institut für Informatik.

Mohamed, K. A. (2004b, November). Increasing the accuracy of anticipation with lead-lag timing analysis of digital freehand writings for the perceptual environment. *Proceedings of the International Conference of Computational Methods in Sciences and Engineering 2004* (pp. 387-390).

Mohamed, K. A., Belenkaia, L., & Ottmann, T. (2004 June). Post-processing InkML for random-access navigation of voluminous handwritten ink documents. *Proceedings of the Thirteenth International World Wide Web Conference* (pp. 266-267).

Mohamed, K. A., & Ottmann, T. (2003). Fast interpretation of pen gestures with competent agents. *Proceedings of the IEEE Second International Conference on Computational Intelligence, Robotics and Autonomous Systems*, PS09-4.

Moyle, M., & Cockburn, A. (2003). The design and evaluation of a flick gesture for "back" and "forward" in Web browsers. *Proceedings of the Fourth Australasian User Interface Conference (AUIC2003)* (pp. 39-46).

Rubine, D. (1991). Specifying gestures by example. *Proceedings of the 18th Annual Conference on Computer Graphics and Interactive Techniques* (pp. 329-337).

Russell, G., Chee, Y.-M., Seni, G., Yaeger, L., Tremblay, C., Franke, K., et al. (2004). Ink markup language. *W3C Working Draft.* Retrieved from http://www.w3.org/tr/inkml

Shen, C., Everitt, K., & Ryall, K. (2003). UbiTable: Impromptu face-to-face collaboration on horizontal interactive surfaces. *Proceedings of the UbiComp 2003: Ubiquitous Computing* (pp. 281-288).

SMART Technologies. (2003). *SMART Boards™ interactive whiteboard.* Retrieved from http://www.smarttech.com/products/smartboard

Turk, M., & Robertson, G. (2000). Perceptual user interfaces (Introduction). *Communications of the ACM, 43*(3), 32-34.

Wooldridge, M. J. (2002). *An introduction to multiagent systems.* West Sussex, UK: John Wiley & Sons.

KEY TERMS

Features: Information that can be gathered to describe a raw trace such as angles between sampled points, lengths, and the speed of the sketched trace.

Gestures: Refers in our case to digital-pen gestures, which are movements of the hands while writing onto digital screens that are interpreted as system commands.

GUI (Graphical User Interface): Specifically involves pull-down menus for keyboard and mouse inputs.

InkML: An XML (extensible markup language) data format for representing digital-ink data.

Interactive Session: A period of communication for the exchange of ideas by an assembly of people for a common purpose.

Interactivity Dimension: The number of users that a single transducer, display system, or application software can support (by means of complete hardware or software simulations) during one particular interactive session.

Interface Agents: Semiautonomous agents that assist users with, or partially automate, their tasks.

PUI (Perceptual User Interface): An invisible graphical user interface that engages perceptual starting points.

Temporal Sphere of Influence: Tracking actions while within a specified time domain.

Trace: The resultant digital-ink representation made by movements of the hand using a digital pen on a digital screen.

W3C (World Wide Web Consortium): An international consortium of companies involved with the Internet and the Web.

The Prospects for Identity and Community in Cyberspace

Leah P. Macfadyen
The University of British Columbia, Canada

INTRODUCTION

Before you read on make sure you have a photo...I will not answer to anyone I cannot imagine physically. Thanks. (Message posted to an online discussion forum)

Individuals are increasingly employing Internet and communication technologies (ICTs) to mediate their communications with individuals and groups, both locally and internationally. Elsewhere, I have discussed current perspectives on the origins and impact of cyberculture(s) (Macfadyen, 2006a), theoretical arguments regarding the challenges of intercultural communication in online environments (Macfadyen, 2006b), and recent approaches to studying the language of cyberspace (Macfadyen, 2006c)—the very medium of interpersonal and intragroup communication in what is, as yet, the largely text-based environment of cyberspace. Virtual environments might in some sense be viewed as a communicative "bottleneck"—a milieu in which visual and oral cues or well-developed relationships may be lacking, and in which culturally diverse individuals may hold widely different expectations of how to establish credibility, exchange information, motivate others, give and receive feedback, or critique or evaluate information (Reeder, Macfadyen, Roche, & Chase, 2004).

Anecdotal evidence, and a growing body of research data, indicate that the greatest challenge that online communicators (and especially novice online communicators) experience is that of constructing what they consider to be a satisfactory or "authentic" identity in cyberspace, and in interpreting those online identities created by others. Rutter and Smith (1998) note, for example, that in their study of a regionally-based social newsgroup in the UK, communicators showed a real desire to paint "physical pictures" of themselves in the process of identity construction, and frequently included details of physical attributes, age, and marital status. Moreover, authentic identity construction and presentation also appears to contribute to communicator's perceptions of the possibility for construction of authentic "community" online.

BACKGROUND

As with the literature on many other aspects of ICTs (Macfadyen, Roche, & Doff, 2004), current literature on the possibilities for "authentic" identity and community in cyberspace tends to offer either simple pessimistic condemnation (e.g., Blanco, 1999; Miah, 2000) or optimistic enthusiasm (e.g., Lévy, 2001, 2001a; Michaelson, 1996; Rheingold, 2000; Sy, 2000). Perhaps it is not surprising that feelings run so high, however, when we consider that human questions of identity are central to the phenomenon of cyberspace. Lévy reminds us that cyberspace is not merely the "material infrastructure of digital communications," but also encompasses "the human beings who navigate and nourish that infrastructure" (2001, p. XVI). Who are these humans? How can we be sure? And how is the capacity for global communications impacting interpersonal and group culture, communications and relationships? This article surveys recent theoretical and empirical approaches to thinking about identity and community in cyberspace, and the implications for future work in the field of human-computer interaction.

VIRTUAL IDENTITY AND COMMUNITY: CRITICAL THEMES

Virtual Identity, Virtual Ethnicity, and Disembodiment

Does the reality of "disembodied being" in cyberspace present a challenge to construction of identity? Key theoretical arguments regarding identity in cyberspace revolve around questions of human *agency*: the degree to which individuals shape, or are shaped by the structures and constraints of the virtual world. Holmes (1998) argues, "human agency has radically changed in spatial, temporal and technological existence" (p. 7); the emergence of cybercultures and virtual environments means, he suggests, that previous perspectives on individuality as constituted by cognitive and social psychology may be less meaningful, especially as they do not consider aspects of space and time in the consideration of community and behaviour. Building on Holmes rethinking of social relations, other contributors to his edited collection *Virtual Politics: Identity and Community in Cyberspace* suggest that alterations in the nature of identity and agency, the relation of self to other, and the structure of community and political representation by new technologies have resulted in a loss of *political* identity and agency for the individual. Jones (1997) similarly questions whether public unity and rational discourse can occur in a space (cyberspace) that is populated by multiple identities and random juxtapositions of distant communicators. Fernanda Zambrano (1998) characterizes individuals in virtual society as "technological terminals" for whom state and nation are irrelevant but actually sees disembodiment and deterritorialization of the individual as a strength, offering the possibility for "productive insertion in the world" beyond traditional geographically-bound notions of citizenship. Offering decidedly more postmodern perspectives, Turkle (1995) suggests that a model of fragmented (decentred) selves may be more useful for understanding virtual identity, using theoretical perspectives on identity from psychology, sociology, psychoanalysis, philosophy, aesthetics, artificial intelligence, and virtuality, and Poster (2001) proposed a new vision of fluid online identity that functions simply as a temporary and ever-changing link to the evolving cultures and communities of cyberspace. Others (see, for example, Miah, 2000; Orvell, 1998) are, however, less willing to accept virtual identity as a postmodern break with traditional notions of identity, and instead argue that virtual reality is simply a further "sophistication of virtualness that has always reflected the human, embodied experience" (Miah, 2000, p. 211).

This latter author, and others, point out that regardless of theoretical standpoint, virtuality poses a real and practical challenge to identity construction, and a number of recent studies have attempted to examine tools and strategies that individuals employ as they select or construct identity or *personae* online (Burbules, 2000; Jones, 1997; Smith & Kollock, 1998). Rutter and Smith (1998) offer a case study of identity creation in an online setting, examining elements such as addressivity (who talks to whom) and self-disclosure, and how these elements contribute to sociability and community. Jordan (1999) examines elements of progressive identity construction: online names, online bios and self-descriptions.

Interestingly, a number of authors focus explicitly on the notion of "virtual ethnicity:" how individuals represent cultural identity or membership in cyberspace. Foremost among these is Poster (1998, 2001) who theorizes about "the fate of ethnicity in an age of virtual presence" (p. 151). He asks whether ethnicity requires bodies—inscribed as they are with rituals, customs, traditions, and hierarchies—for true representation. Wong (2000) meanwhile reports on ways that disembodied individuals use language in the process of cultural identity formation on the Internet, and similarly, Reeder et al. (2004) attempt to analyze and record cultural differences in self-presentation in an online setting. In a related discussion, contributors to the collection edited by Smith and Kollock (1998) offer counter-arguments to the suggestion that as a site of disembodied identity, cyberspace may eliminate consideration of racial identity; instead, they suggest that cyberindividuals may simply develop new nonvisual criteria for people to judge (or misjudge) the races of others.

Online identities may therefore be multiple, fluid, manipulated and/or may have little to do with the "real selves" of the persons behind them (Fernanda Zambrano, 1998; Jones, 1997; Jordan, 1999; Rheingold, 2000; Wong, 2000). Is "identity deception" a special problem on the Internet? Some theorists believe so. Jones (1997) examines in detail the way that assumed identities can lead to "virtual crime", while Jordan suggests that identity fluidity can lead to harassment and deception in cyberspace. Lévy (2001), on the other hand, argues that deception is no more likely in cyberspace than via any other medium, and even suggests that the cultures of virtual communities actively *discourage* the irresponsibility of anonymity.

Virtual Community, Virtual Culture, and Deterritorialization

Are *all* communities—online and offline—virtual to some degree? In his classic text *Imagined Communities*, Anderson (1991) argues that most national and ethnic communities are *imagined* because members "will never know most of their fellow-members, meet them, or even hear them, yet in the minds of each lives the image of their communion" (p. 5). Ribeiro (1995) extends Anderson's model to argue that cyberculture, computer English and "electronic capitalism" are necessary internal characteristics of a developing *virtual* transnational community. Burbules (2000) nevertheless cautions us to remember that "imagined" communities are also "real", and undertakes a careful analysis of the notion of community that ultimately situates virtual communities as "actual."

In the same way that virtual identities are disembodied, virtual communities are (usually) deterritorialized—a feature highlighted by a number of writers (Blanco, 1999; Sudweeks, 1998). Interestingly, Poster (2001) draws parallels between online virtual communities and other ethnicities that have survived in the absence of "a grounded space"—such as Jewishness.

What, then, are the defining features of virtual communities? A number of theorists posit that virtual communities can best be described as a constantly evolving "collective intelligence" or "collective consciousness" that has been actualized by Internet technologies (Abdelnour-Nocera, 2002b; Guedon, 1997; Lévy, 2001a; Poster, 2001; Sudweeks, 1998).

More common, however, are theoretical discussions of the construction of a group *culture*—and of shared identity and meaning—as a feature of virtual community (Abdelnour-Nocera, 2002a; Baym, 1998; Blanco, 1999; Lévy, 2001; Porter, 1997; Walz, 2000). Essays contained in the collection edited by Shields (1996) examine the socio-cultural complexities of virtual reality and questions of identity, belonging and consciousness in virtual worlds. Abdelnour-Nocera (1998) suggests that Geertz's (1973) idea of culture as a "web of meaning that he (man) himself has spun" (p. 194) is most useful when considering the construction of shared meaning in a community where language is the main expressive and interpretative resource.

Virtual communities share a number of other common internal features, including: use and development of specific language (Abdelnour-Nocera, 2002b); style, group purpose, and participant characteristics (Baym, 1998); privacy, property, protection, and privilege (Jones, 1998); forms of communication (Jordan, 1999); customary laws (e.g., reciprocity), social morality, freedom of speech, opposition to censorship, frequent conflicts, flaming as "punishment" for rule-breaking, formation of strong affinities and friendships (Lévy, 2001); unique forms of immediacy, asynchronicity, and anonymity (Michaelson, 1996); and internal structure and dynamics (Smith & Kollock, 1998).

Strikingly absent from most discussions of the creation and nature of online communities is much mention of the role of the language of communication, and most contributions apparently assume that English is *the* language of cyberspace. If, as Adam, Van Zyl Slabbert, and Moodley (1998) argue, "language policy goes beyond issues of communication to questions of collective identity" (p. 107), we might expect to see more careful examination of the choice of language that different users and communities make, and how this contributes to the sense of community online. Only very recently in a special issue of the *Journal of Computer-Mediated Communication* have research reports on the "multilingual internet" and discussions of online language and community begun to appear (see Danet & Herring [2003] and references therein).

The Promises of Cybertechnology for Identity and Community: Hopes and Fears

Papers assembled in the recent edited collection of Stald and Tufte (2002) present a diverse selection of developing engagements of cultural groups in what is characterized as the "global metropolis" of the Internet. Contributing authors report on media use by young Danes, by rural black African males in a South African university, by South Asian families in London, by women in Indian communities, by

Iranian immigrants in London, and by young immigrant Danes. In particular, these contributions focus on the ways that these minority groups and communities understand themselves vis-à-vis a majority culture, and the different ways that these groups utilize Internet technologies in the construction of complex identities. In Canada, Hampton and Wellman (2003) document an increase in social contact, and increased discussion of and mobilization around local issues in a networked suburban community.

Does technology *limit* virtual identity and virtual community? Reeder et al. (2004) point to ways that (culturally biased) technological design of the spaces of virtual encounters implicitly shapes the nature of the communications that occur there. Poster (1998) examines the technological barriers to portrayal of ethnicity in different online settings, and Rheingold (2000) discusses how technology affects our social constructs. Blanco (1999) similarly argues that virtual communities are "rooted in the sociotechnological configuration that the Internet provides" (p. 193) but suggests that sociocultural innovations *are* in fact allowing a reciprocal alteration of technology design.

In addition to the worry that Internet technologies may change the nature of social interactions, a number of contributors raise other fears. Blanco (1999) worries that "communication is becoming an end in itself instead of a tool for political, social and cultural action" (p. 193). Jones (1998) also concludes that efforts to recapture lost community online are only partly successful, and that cybercommunities bring with them new and distinctive difficulties, and Miah (2000) similarly argues that the claimed emancipatory functions of cyberspace are over-stated and counter-balanced by the challenges to identity construction. Michaelson (1997) worries that "participation in online groups has the potential to diminish commitment to local communities" (p. 57), and also that the technological and cultural resources required for such new forms of community may contribute to new forms of stratification. Poster (2001) reports on angst about cyberspace as a destructive mass market that can potentially remove ownership of culture from ethnic groups. (Interestingly, LaFargue (2002) pursues this same question, asking "does the commodification of a cultural product, such as an exotic handicraft, safeguard social conventions within the communities of their producers?" (p. 317). This author does not, however, explicitly view technologically-driven cultural change as a cause for concern, but rather develops a theoretical argument relating microeconomic engagement of handicraft producers with mass markets to ongoing negotiation of individual and cultural identity.)

On the other hand, optimists look to the potential of online community as a uniting force. Sy (2000) describes how new Filipino virtual communities represent a form of cultural resistance to Western hegemonic encroachment. Bickel (2003) reports how the Internet has allowed the voices of otherwise silenced Afghan women to be heard, and the new leadership identities for women that this has brought about. Michaelson (1996) agrees that for some, virtual communities offer opportunities for greater participation in public life. Lévy (2001a) is enthusiastic about the nonhierarchical and free nature of deterritorialized human relationships, and Rheingold (2000) offers a number of examples of positive social actions and developments that have emerged from the establishment of virtual communities.

FUTURE TRENDS

Perhaps more significant for future research directions are perspectives that highlight the real live complexities of technological implications for identity and community. A number of authors, for example, caution against the notion of "simple substitution" of virtual relationships for physical relationships, and undertake a comparison of "real" and "virtual" communities and relationships (Burbules, 2000; Davis, 1997; Lévy, 2001; Miah, 2000; Porter, 1997; Smith & Kollock, 1998). Importantly, Hampton and colleagues (Hampton, 2002; Hampton, & Wellman, 2003) have very recently undertaken critical and ambitious empirical studies of networked communities, in an attempt to test some of the optimistic and pessimistic predictions offered by theorists. Future work by these and other investigators should illuminate in finer detail the differential uses, impacts and benefits of ICTs for diverse communities and populations over time, as ICTs cease to be "new" and become increasingly woven into the fabric of human societies.

CONCLUSION

If, as many theorists argue, Internet and communication technologies represent a genuine paradigm shift in human communications, or a transition from the modern to the postmodern, strong positive and negative reactions might be expected as individuals and communities grapple with the social implications. Indeed, Guedon (1997) takes care to show how this polarized pattern of responses has been repeated, historically, with the successive appearances of new communications technologies (such as the telephone). Other writers also (Davis, 1997; Michaelson, 1996; Poster, 2001; Rutter & Smith, 1998) explicitly compare and contrast optimistic and pessimistic perspectives on virtual identity and community.

As Holmes (1998) argues, social activity can now no longer be reduced to simple relations in space and time — a realization that offers a new challenge to the more positivistic social science approaches, since the object of study (social activity) is now "eclipsed by the surfaces of electronically mediated identities" (p.8). While once can still study interaction with computers in situ, such studies will fail to examine the reality that individuals are now able to participate in multiple worlds whose borders and norms radically exceed those previously available.

Perhaps most relevant for the field of HCI is Jones' (1998) contention that many (most?) current perspectives on ICT-mediated communications are rooted in a "transportation model" of communication in which control over movement of information is central. Mitigating against such models of ICT-mediated communication is the reality that "people *like* people", and actively seek to maximize human interaction. Developers of ICTs succeed best, he argues, when they recognize this reality and "put technology in service of conversation rather than communication, in service of connection between people rather than connection between machines" (p. 32).

REFERENCES

Abdelnour-Nocera, J. (1998). Virtual environments as spaces of symbolic construction and cultural identity: Latin American virtual communities. In C. Ess, & F. Sudweeks (Eds.), *Proceedings, Cultural Attitudes Towards Communication and Technology 1998* (pp. 193-195). Sydney, Australia: University of Sydney. Retrieved June 1, 2005, from http://www.it.murdoch.edu.au/catac/catac98/pdf/15_nocera.pdf

Abdelnour-Nocera, J. (2002a, July 12-15). Context and culture in human computer interaction: "Usable" does not mean "senseful". In F. Sudweeks, & C. Ess (Eds.), *Proceedings, Cultural Attitudes Towards Technology and Communication 2002,* Montreal, Canada (pp. 505-524). Murdoch, Australia: Murdoch University.

Abdelnour-Nocera, J. (2002b). Ethnography and hermeneutics in cybercultural research. accessing IRC virtual communities. *Journal of Computer-Mediated Communication*, *7*(2). Retrieved June 1, 2005, from http://www.ascusc.org/jcmc/vol7/issue2/nocera.html

Adam, H., Van Zyl Slabbert, F., & Moodley, K. (1998). *Comrades in business*. Utrecht, The Netherlands: International Books.

Anderson, B. (1991). *Imagined communities* (2nd ed.). New York: Verso.

Baym, N. (1998). The emergence of on-line community. In S. Jones (Ed.), *CyberSociety 2.0: Revisiting computer-mediated communication and community* (pp. 35-68). Newbury Park, CA: Sage.

Bickel, B. (2003). Weapons of magic: Afghan women asserting voice via the Net. *Journal of Computer-Mediated Communication*, *8*(2). Retrieved June 1, 2005, from http://www.ascusc.org/jcmc/vol8/issue2/bickel.html

Blanco, J. R. (1999). About virtual subjects and digital worlds: The case of virtual communities. *Politica y Sociedad, 30,* 193-211.

Burbules, N. C. (2000). Does the Internet constitute a global educational community? In N. C. Burbules, & C. A. Torres (Eds.), *Globalization and education: Critical perspectives* (pp. 323-356). New York: Routledge.

Danet, B., & Herring, S. C. (2003). Introduction: The multilingual Internet. *Journal of Computer-*

Mediated Communication, *9*(1). Retrieved June 1, 2005, from http://www.ascusc.org/jcmc/vol9/issue1/intro.html

Davis, M. (1997). Fragmented by technologies: A community in cyberspace. *Interpersonal Computing and Technology*, *5*(1-2), 7-18. Retrieved from June 1, 2005, http://www.helsinki.fi/science/optek/1997/n1/davis.txt

Fernanda Zambrano, L. (1998). Identities? When history has an influence. *Tierra Firme*, *16*(64), 755-766.

Geertz, C. (1973). *The interpretation of cultures*. New York: Basic Books

Guedon, J.-C. (1997, July). The cyberworld or how to surmount the wall of the individual. *Français dans le Monde*, Special Issue, 14-25.

Hampton, K. N. & Wellman, B. (2003). Neighboring in Netville: How the Internet supports community and social capital in a wired suburb. *City and Community, 2*(4), 277-311.

Hampton, K. N. (2002). Place-based and IT mediated "community." *Planning Theory & Practice, 3*(2), 228-231.

Holmes, D. (1998). Introduction: Virtual politics – Identity and community in cyberspace. In D. Holmes (Ed.), *Virtual politics: Identity and community in cyberspace* (pp. 1-25). New York: Sage Publications.

Jones, S. (1997). *Virtual culture: Identity and communication in cybersociety*. New York: Sage Publications.

Jones, S. (1998, August 1-3). Understanding micropolis and compunity. In C. Ess, & F. Sudweeks (Eds.), *Proceedings, Cultural Attitudes Towards Communication and Technology 1998,* London, August 1-3 (pp. 21-23). Sydney, Australia: University of Sydney. Retrieved June 1, 2005, from http://www.it.murdoch.edu.au/catac/catac98/pdf/02_jones.pdf

Jordan, T. (1999). *Cyberculture: The culture and politics of cyberspace and the Internet*. London: Routledge.

LaFargue, C. (2002, July 12-15). Electronic commerce and identity construction in an era of globalization: Suggestions for handicraft producers in developing regions. In F. Sudweeks, & C. Ess (Eds.), *Proceedings, Cultural Attitudes Towards Technology and Communication 2002,* Montreal, Canada, July 12-15 (pp. 317-335). Murdoch, Australia: Murdoch University.

Lévy, P. (2001). *Cyberculture*. Minneapolis: University of Minnesota Press.

Lévy, P. (2001a). *The impact of technology in cyberculture*. Minneapolis: University of Minnesota Press.

Macfadyen, L. P. (2006a). The culture(s) of cyberspace. In C. Ghaoui (Ed.), *Encyclopedia of human computer interaction.* Hershey, PA: Idea Group Reference.

Macfadyen, L. P. (2006b). Internet-mediated communication at the cultural interface. In C. Ghaoui (Ed.), *Encyclopedia of human computer interaction.* Hershey, PA: Idea Group Reference.

Macfadyen, L. P. (2006c). The language of cyberspace. In C. Ghaoui (Ed.), *Encyclopedia of human computer interaction.* Hershey, PA: Idea Group Reference.

Macfadyen, L. P., Roche, J., & Doff, S. (2004). *Communicating across cultures in cyberspace: A bibliographical review of online intercultural communication*. Hamburg, Germany: Lit-Verlag.

Miah, A. (2000). Virtually nothing: Re-evaluating the significance of cyberspace. *Leisure Studies, 19*(3), 211-225.

Michaelson, K. L. (1996). Information, community and access. *Social Science Computer Review, 14*(1), 57-59.

Orvell, M. (1998). Virtual culture and the logic of American technology. *Revue française d'études americaines*, *76*, 12-27.

Porter, D. (Ed.). (1997). *Internet culture.* New York: Routledge.

Poster, M. (1998). Virtual ethnicity: Tribal identity in an age of global communications. In S. Jones (Ed.), *CyberSociety 2.0: Revisiting computer-mediated communication and community* (pp. 184-211). Newbury Park, CA: Sage.

Poster, M. (2001). *What's the matter with the Internet?* Minneapolis: University of Minnesota Press.

Reeder, K., Macfadyen, L. P., Roche, J. & Chase, M. (2004). *Language Learning and Technology, 8*(2), 88-105. Retrieved June 1, 2005, from http://llt.msu.edu/vol8num2/reeder/default.html

Rheingold, H. (2000). *The virtual community: Homesteading on the electronic frontier.* Boston: MIT Press. Retrieved Retrieved June 1, 2005, from http://www.well.com/user/hlr/vcbook/

Ribeiro, G. L. (1995). On the Internet and the emergence of an imagined transnational community. *Sociedad e Estado, 10*(1), 181-191.

Rutter, J., & Smith, G. (1998, August 1-3). Addressivity and sociability in 'Celtic Men'. In C. Ess, & F. Sudweeks (Eds.), *Proceedings, Cultural Attitudes Towards Communication and Technology 1998,* London, August 1-3 (pp. 196-201). Sydney, Australia: University of Sydney. Retrieved from http://www.it.murdoch.edu.au/catac/catac98/pdf/16_rutter.pdf

Shields, R. (Ed.). (1996). *Cultures of Internet: Virtual spaces, real histories, and living bodies.* London: Sage Publications.

Smith, M. A., & Kollock, P. (Eds.). (1998). *Communities in cyberspace.* New York: Routledge.

Stald, G., & Tufte, T. (Eds.). (2002). *Global encounters: Media and cultural transformation.* Luton, UK: University of Luton Press.

Sudweeks, F. (1998, August 1-3). Cybersocialism: Group consciousness in culturally diverse collaborations. In C. Ess, & F. Sudweeks (Eds.), *Proceedings, Cultural Attitudes Towards Communication and Technology, 1998* (pp. 127-128). Sydney, Australia: University of Sydney. Retrieved from http://www.it.murdoch.edu.au/catac/catac98/pdf/09_sudweeks.pdf

Sy, P. A. (2000, July 12-15) Barangays of IT. In F. Sudweeks, & C. Ess (Eds.). *Proceedings, Cultural Attitudes Towards Technology and Communication 2000,* Perth, Australia (pp. 47-56). Murdoch, Australia: Murdoch University.

Turkle, S. (1995). *Life on the screen: Identity in the age of the Internet.* New York: Simon and Schuster.

Walz, S. P. (2000). Symbiotic interface contingency. In F. Sudweeks, & C. Ess (Eds.), *Proceedings, Cultural Attitudes Towards Technology and Communication 2000* (pp. 125-143). Murdoch, Australia: Murdoch University.

Wong, A. (2000). Cyberself: Identity, language and stylization on the Internet. In D. Gibbs, & K.-L. Krause (Eds.), *Cyberlines. Languages and cultures of the Internet* (pp.175-206). Albert Park, Australia: James Nicholas Publishers.

KEY TERMS

Culture: Multiple definitions exist, including essentialist models that focus on shared patterns of learned values, beliefs, and behaviours, and social constructivist views that emphasize culture as a shared system of problem-solving or of making collective meaning. Key to the understanding of online cultures—where communication is as yet dominated by text—may be definitions of culture that emphasize the intimate and reciprocal relationship between culture and language.

Cyberculture: As a social space in which human beings interact and communicate, cyberspace can be assumed to possess an evolving culture or set of cultures ("cybercultures") that may encompass beliefs, practices, attitudes, modes of thought, behaviours and values.

Deterritorialized: Separated from or existing without physical land or territory.

Disembodied: Separated from or existing without the body.

Modern: In the social sciences, "modern" refers to the political, cultural, and economic forms (and their philosophical and social underpinnings) that characterize contemporary Western and, arguably, industrialized society. In particular, modernist cultural theories have sought to develop rational and universal theories that can describe and explain human societies.

Postmodern: Theoretical approaches characterized as postmodern, conversely, have abandoned the belief that rational and universal social theories are desirable or exist. Postmodern theories also challenge foundational modernist assumptions such as "the idea of progress," or "freedom".

Question Answering from Procedural Semantics to Model Discovery

John Kontos
National and Kapodistrian University of Athens, Greece

Ioanna Malagardi
National and Kapodistrian University of Athens, Greece

INTRODUCTION

Question Answering (QA) is one of the branches of Artificial Intelligence (AI) that involves the processing of human language by computer. QA systems accept questions in natural language and generate answers often in natural language. The answers are derived from databases, text collections, and knowledge bases. The main aim of QA systems is to generate a short answer to a question rather than a list of possibly relevant documents. As it becomes more and more difficult to find answers on the World Wide Web (WWW) using standard search engines, the technology of QA systems will become increasingly important. A series of systems that can answer questions from various data or knowledge sources are briefly described. These systems provide a friendly interface to the user of information systems that is particularly important for users who are not computer experts. The line of development of ideas starts with procedural semantics and leads to interfaces that support researchers for the discovery of parameter values of causal models of systems under scientific study. QA systems historically developed roughly during the 1960-1970 decade (Simmons, 1970). A few of the QA systems that were implemented during this decade are:

- The BASEBALL system (Green et al., 1961)
- The FACT RETRIEVAL System (Cooper, 1964)
- The DELFI systems (Kontos & Kossidas, 1971; Kontos & Papakontantinou, 1970)

The BASEBALL System

This system was implemented in the Lincoln Laboratory and was the first QA system reported in the literature according to the references cited in the first book with a collection of AI papers (Feigenbaum & Feldman, 1963). The inputs were questions in English about games played by baseball teams. The system transformed the sentences to a form that permitted search of a systematically organized memory store for the answers. Both the data and the dictionary were list structures, and questions were limited to a single clause.

The FACT RETRIEVAL System

The system was implemented using the COMIT compiler-interpreter system as programming language. A translation algorithm was incorporated into the input routines. This algorithm generated the translation of all information sentences and all question sentences into their logical equivalents.

The DELFI System

The DELFI system answers natural language questions about the space relations between a set of objects. These are questions with unlimited nesting of relative clauses that were automatically translated into retrieval procedures consisting of general-purpose procedural components that retrieved information from the database that contained data about the properties of the objects and their space relations. The system was a QA system based on procedural semantics. The following is an example of a question put to the DELFI system:

Figure 1. The objects of the DELFI I example application database

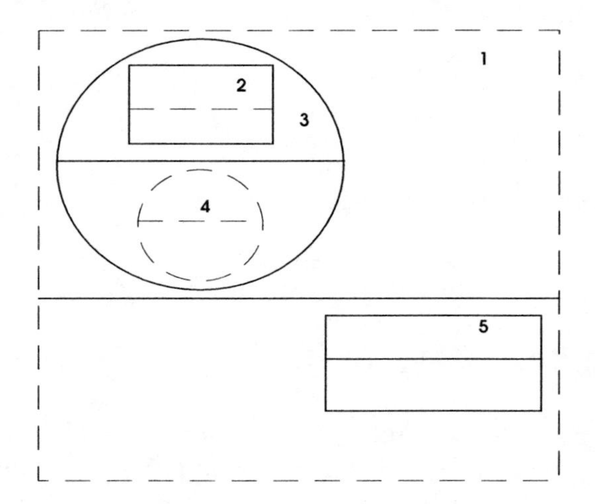

"Is an object that has dotted contour below the object number 2?"

The answer is, "Yes, object no. 4," given the objects with numbers 1, 2, 3, 4, and 5 as shown in Figure 1 (Kontos, 2004; Kontos & Papakonstantinou, 1970).

The DELFI II System

The DELFI II system (Kontos & Kossidas, 1971) was an implementation of the second edition of the system DELFI augmented by deductive capabilities. In this system, the procedural semantics of the questions are expressed using macro-instructions that are submitted to a macro-processor that expands them with a set of macro-definitions into full programs. Every macro-instruction corresponded to a procedural semantic component. In this way, a program was generated that corresponded to the question and could be compiled and executed in order to generate the answer. DELFI II was used in two new applications. These applications concerned the processing of the database of the personnel of an organization and the answering of questions by deduction from a database with airline flight schedules using the following rules:

- If flight F1 flies to city C1, and flight F2 departs from city C1, then F2 follows F1.
- If flight F1 follows flight F2, and the time of departure of F1 is at least two hours later than the time of arrival of F2, then F1 connects with F2.
- If flight F1 connects with flight F2, and F2 departs from city C1, and F1 flies to city C2, then C2 is reachable from C1.

Given a database that contains the following data:

- F1 departs from Athens at 9 and arrives at Rome at 11
- F2 departs from Rome at 14 and arrives at Paris at 15
- F3 departs from Rome at 10 and arrives at London at 12

If the question "Is Paris reachable from Athens?" is submitted to the system, then the answer it gives is *yes*, because F2 follows F1, and the time of departure of F2 is three hours later than the time of arrival of F1. It should be noted also that F1 departs from Athens, and F2 flies to Paris.

If the question "Is London reachable from Athens?" is submitted to the system, then the answer it gives is *no*, because F3 follows F1, but the time of departure of F3 is one hour earlier than the time of arrival of F1. It should be noted here that F1 departs from Athens, and F3 flies to London.

In Figure 2, the relations between the flights and the cities are shown diagrammatically.

Figure 2. The relations between the flights and cities of the DELFI II example application (Kontos, 2003)

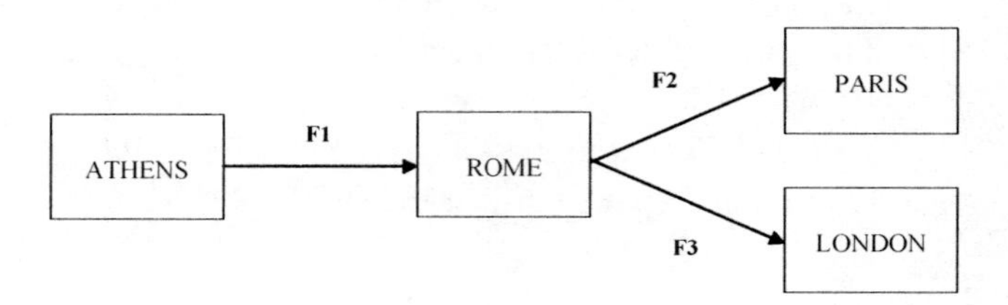

BACKGROUND

The Structured Query Language (SQL) QA Systems

In order to facilitate the commercial application of the results of research work like the one described so far, it was necessary to adapt the methods used to the industrial database environment. One important adaptation was the implementation of the procedural semantics interpretation of natural language questions using a commercially available database retrieval language. The SQL QA systems implemented by different groups, including the author's, followed this direction by using SQL so that the questions can be answered from any commercial database system.

The domain of an illustrative application of our SQL QA system involves information about different countries. The representation of the knowledge of the domain of application connected a verb like *exports* or *has capital* to the corresponding table of the database to which the verb is related. This connection between the verbs and the tables provided the facility of the system to locate the table a question refers to using the verbs of the question. During the analysis of questions by the system, an ontology related to the domain of application may be used for the correct translation of ambiguous questions to appropriate SQL queries. Some theoretical analysis of SQL QA systems has appeared recently (Popescu et al., 2003), and a recent system with a relational database is described in Samsonova, et al. (2003).

QA From Texts Systems

Some QA systems use collections of texts instead of databases for extracting answers. Most such systems are able to answer simple factoid questions only. Factoid questions seek an entity involved in a single fact. Some recent publications on QA from texts are Diekema (2003), Doan-Nguyen and Kosseim (2004), Harabagiu et al. (2003), Kosseim et al. (2003), Nyberg et al. (2002), Plamondon and Kosseim (2002), Ramakrishnan (2004), Roussinof and Robles-Flores (2004), and Waldinger et al. (2003). Some future directions of QA from texts are proposed in Maybury (2003). An international competition between question answering systems from texts has been organized by NIST (National Institute of Standards and Technology) (Voorhees, 2001).

What follows describes how the information extracted from scientific and technical texts may be used by future systems for the answering of complex questions concerning the behavior of causal models using appropriate linguistic and deduction mechanisms. An important function of such systems is the automatic generation of a justification or explanation of the answer provided.

The ARISTA System

The implementation of the ARISTA system is a QA system that answers questions by knowledge acquisition from natural language texts and was first presented in Kontos (1992). The ARISTA system was based on the representation independent method also called ARISTA for finding the appropriate causal sentences from a text and chaining them by the operation of the system for the discovery of causal chains.

This method achieves causal knowledge extraction through deductive reasoning performed in response to a user's question. This method is an alternative to the traditional method of translating texts into a formal representation before using their content for deductive question answering from texts. The main advantage of the ARISTA method is that since texts are not translated into any representation, formalism retranslation is avoided whenever new linguistic or extra linguistic prerequisite knowledge has to be used for improving the text processing required for question answering.

An example text that is an extract from a medical physiology book in the domain of pneumonology and, in particular, of lung mechanics enhanced by a few general knowledge sentences was used as a first illustrative example of primitive knowledge discovery from texts (Kontos, 1992). The ARISTA system was able to answer questions from text that required the chaining of causal knowledge acquired from the text and produced answers that were not explicitly stated in the input texts.

The Use of Information Extraction

A system using information extraction from texts for QA was presented in Kontos and Malagardi

(1999). The system described had as its ultimate aim the creation of flexible information extraction tools capable of accepting natural language questions and generating answers that contained information either directly extracted from the text or extracted after applying deductive inference. The domains examined were oceanography, medical physiology, and ancient Greek law (Kontos & Malagardi, 1999). The system consisted of two main subsystems. The first subsystem achieved the extraction of knowledge from individual sentences, which was similar to traditional information extraction from texts (Cowie & Lehnert, 1996; Grishman, 1997), while the second subsystem was based on a reasoning process that combines knowledge extracted by the first subsystem for answering questions without the use of a template representation.

QUESTION ANSWERING FOR MODEL DISCOVERY

The AROMA System

A modern development in the area of QA that points to the future is our implementation of the AROMA (**AR**ISTA **O**riented **M**odel **A**daptation) system. This system is a model-based QA system that may support researchers for the discovery of parameter values of procedural models of systems by answering *what if* questions (Kontos et al., 2002). The concept of *what if* questions are considered here to involve the computation data of describing the behavior of a simulated model of a system.

The knowledge discovery process relies on the search for causal chains, which in turn relies on the search for sentences containing appropriate natural language phrases. In order to speed up the whole knowledge acquisition process, the search algorithm described in Kontos and Malagardi (2001) was used for finding the appropriate sentences for chaining. The increase in speed results because the repeated sentence search is made a function of the number of words in the connecting phrases. This number is usually smaller than the number of sentences of the text that may be arbitrarily large.

The general architecture of the AROMA system is shown in Figure 3 and consists of three subsystems; namely, the Knowledge Extraction Subsystem, the Causal Reasoning Subsystem, and the Simulation Subsystem. All of these subsystems have been implemented by our group and tested with a few biomedical examples. The subsystems of the AROMA system are briefly described next.

Figure 3. The AROMA system architecture (Kontos et al., 2003)

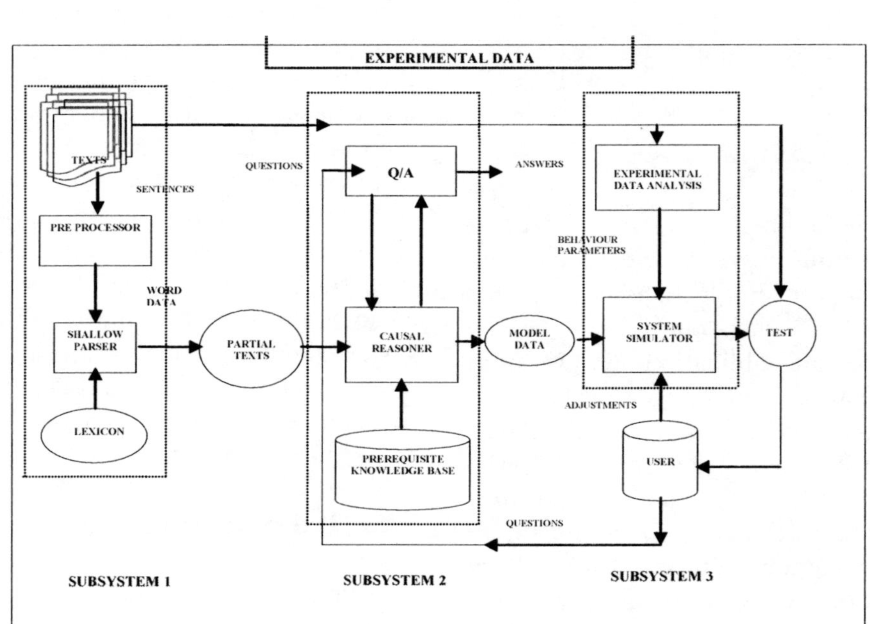

The Knowledge Extraction Subsystem

This subsystem integrates partial causal knowledge extracted from a number of different texts. This knowledge is expressed in natural language using causal verbs such as *regulate*, *enhance*, and *inhibit*. These verbs usually take as arguments entities such as entity names and process names that occur in the texts that we use for the applications. In this way, causal relations are expressed between the entities, processes, or entity-process pairs.

The input texts are submitted first to a preprocessing module of the subsystem, which automatically converts each sentence into a form that shows word data with numerical information concerning the identification of the sentence that contains the word and its position in that sentence. This conversion has nothing to do with logical representation of the content of the sentences. It should be emphasized that we do not deviate from our ARISTA method with this conversion. We simply annotate each word with information concerning its position within the text. This form of sentences is then parsed, and partial texts with causal knowledge are generated.

The Causal Reasoning Subsystem

The output of the first subsystem is used as input to the second subsystem, which combines causal knowledge in natural language form to produce answers and model data by deduction not mentioned explicitly in the input text. The operation of this subsystem is based on the ARISTA method. The sentence fragments containing causal knowledge are parsed, and the entity-process pairs are recognized. The user questions are processed, and reasoning goals are extracted from them. The answers to the user questions that are generated automatically by the reasoning process contain explanations in natural language form. All this is accomplished by the chaining of causal statements using prerequisite knowledge such as ontology to support the reasoning process.

THE SIMULATION SUBSYSTEM

The third subsystem is used for modeling the dynamics of a system specified on the basis of the texts processed by the first and second subsystems. The data of the model, such as structure and parameter values, are extracted from the input texts combined with prerequisite knowledge, such as ontology and default process and entity knowledge. The solution of the equations describing the system is accomplished with a program that provides an interface with which the user may test the simulation outputs and manipulate the structure and parameters of the model.

FUTURE TRENDS

The architecture of the AROMA system is pointing to future trends in the field of QA by serving, among other things, the processing of *what if* questions. These are questions about what will happen to a system under certain conditions. Implementing systems for answering *what if* questions will be an important research goal in the future (Maybury, 2003).

Another future trend is the development of systems that may conduct an explanatory dialog with their human user by answering *why* questions using the simulated behavior of system models. A *why* question seeks the reason for the occurrence of certain system behaviors.

The work on model discovery QA systems paves the way toward important developments and justifies effort leading to the development of tools and resources, aiming at the solution of the problems of model discovery based on larger and more complex texts. These texts may report experimental data that may be used to support the discovery and adaptation of models with computer systems.

CONCLUSION

A series of systems that can answer questions from various data or knowledge sources was briefly described. These systems provide a friendly interface to the user of information systems that is particularly important for users that are not computer experts. The line of development of systems starts with procedural semantics systems and leads to interfaces that support researchers for the discovery of model parameter values of simulated systems. If these efforts for more sophisticated human-computer interfaces succeed, then a revolution may take place in the way research and development are conducted in many scientific fields. This revolution will make computer systems even more useful for research and development.

REFERENCES

Cooper, W.S. (1964). Fact retrieval and deductive question—Answering information retrieval systems. *Journal of the ACM, 11*(2), 117-137.

Cowie, J., & Lehnert, W. (1996). Information extraction. *Communications of the ACM, 39*(1), 80-91.

Diekema, A.R. (2003). What do you mean? Finding answers to complex questions. *Proceedings of the 2003 AAAI Spring Symposium*. Palo Alto, California.

Doan-Nguyen, H., & Kosseim, L. (2004). Improving the precision of a closed-domain question answering system with semantic information. *Proceedings of Researche d' Information Assistee Ordinateur (RIAO-2004)*, Avignon, France.

Feigenbaum, E.A., & Feldman, J. (1963). *Computers and thought*. New York: McGraw Hill.

Green, B.F., et al. (1961). BASEBALL: An automatic question answerer. *Proceedings of the Western Joint Computer Conference 19*, (pp. 219-224).

Grishman R. (1997). Information extraction: Techniques and challenges. In M.T. Pazienza (Ed.), *Information extraction* (pp. 10-27). Berlin Heidelberg, Germany: Springer-Verlag.

Harabagiu, S.M., Maiorano, S.J., & Pasca, M.A. (2003). Open-domain textual question answering techniques. *Natural Language Engineering, 9*(3), 231-267.

Kontos, J. (1992). ARISTA: Knowledge engineering with scientific texts. *Information and Software Technology, 34*(9), 611-616.

Kontos, J. (2004). *Cognitive science: The new science of the mind* [in Greek]. Athens, Greece: Gutenbeng.

Kontos, J., Elmaoglou, A., & Malagardi, I. (2002). ARISTA causal knowledge discovery from texts. *Proceedings of the 5th International Conference on Discovery Science DS 2002*, Luebeck, Germany.

Kontos, J., & Kossidas, A. (1971). On the question-answering system DELFI and its application. *Proceedings of the Symposium on Artificial Intelligence*, Rome, Italy.

Kontos, J., & Malagardi, I. (1999). Information extraction and knowledge acquisition from texts using bilingual question-answering. *Journal of Intelligent and Robotic Systems, 26*(2), 103-122.

Kontos, J., & Malagardi, I. (2001). A search algorithm for knowledge acquisition from texts. *Proceedings of the 5th Hellenic European Research on Computer Mathematics & Its Applications Conference*, Athens, Greece.

Kontos, J., Malagardi, I., & Peros, J. (2003). The AROMA system for intelligent text mining. *HERMIS International Journal of Computers Mathematics and Its Applications, 4*, 163-173.

Kontos, J., & Papakonstantinou, G. (1970). A question-answering system using program generation. *Proceedings of the ACM International Computing Symposium*, Bonn, Germany.

Kosseim, L., Plamondon, L., & Guillemette, L.J. (2003). Answer formulation for question-answering. *Proceedings of the Sixteenth Conference of the Canadian Society for Computational Studies of Intelligence*, Halifax, Canada.

Maybury, M.T. (2003). Toward a question answering roadmap. *Proceedings of the AAAI Spring Symposium*. Palo Alto, California.

Nyberg, E., et al. (2002). The JAVELIN question-answering system at TREC 2002. *Proceedings of the 11th Text Retrieval Conference*. Gaithersburg, Maryland.

Plamondon, L., & Kosseim, L. (2002). QUANTUM: A function-based question answering system. *Proceedings of the Fifteenth Canadian Conference on Artificial Intelligence*, Calgary, Canada.

Popescu, A., Etzioni, O., & Kautz, H. (2003). Towards a theory of natural language interfaces to databases. *Proceedings of the IUI'03*, Miami, Florida.

Ramakrishnan, G., et al. (2004). Is question answering an acquired skill? *Proceedings of the WWW2004*, New York.

Roussinof, D., Robles-Flores, J.A. (2004). Web question answering: Technology and business applications. *Proceedings of the Tenth Americas Conference on Information Systems*, New York.

Samsonova, M., Pisarev, A., & Blagov, M. (2003). Processing of natural language queries to a relational database. *Bioinformatics, 19*(1), i241-i249.

Simmons, R.F. (1970). Natural language question-answering systems: 1969. *Computational Linguistics, 13*(1), 15-30.

Voorhees, E.M. (2001). The trec question answering track. *Natural Language Engineering, 7*(4), 361-378.

Waldinger, R., et al. (2003). Deductive question answering from multiple resources. In M. Maybury AAAI 2004, *Proceedings of New Directions in Question Answering, AAAI 2003.*

KEY TERMS

Causal Chain: A sequence of instances of causal relations such that the effect of each instance except the last one is the cause of the next one in sequence.

Causal Relation: A relation between the members of an entity-process pair, where the first member is the cause of the second member, which is the effect of the first member.

Explanation: A sequence of statements of the reasons for the behavior of the model of a system.

Model: A set of causal relations that specify the dynamic behavior of a system.

Model Discovery: The discovery of a set of causal relations that predict the behavior of a system.

Ontology: A structure that represents taxonomic or meronomic relations between entities.

Procedural Semantics: A method for the translation of a question by a computer program into a sequence of actions that retrieve or combine parts of information necessary for answering the question.

Question Answering System: A computer system that can answer a question posed to it by a human being using prestored information from a database, a text collection, or a knowledge base.

***What If* Question:** A question about what will happen to a system under given conditions or inputs.

***Why* Question:** A question about the reason for the occurrence of a certain system behavior.

Recommender Systems in E-Commerce

A. B. Gil
University of Salamanca, Spain

F. J. García
University of Salamanca, Spain

INTRODUCTION

Electronic commerce (EC) is, at first sight, an electronic means to exchange large amounts of product information between users and sites. This information must be clearly written since any users who accesses the site must understand it. Given the large amounts of information available at the site, interaction with an e-market site becomes an effort. It is also time-consuming, and the user feels disoriented as products and clients are always on the increase. One solution to make online shopping easier is to endow the EC site with a recommender system. Recommender systems are implanted in EC sites to suggest services and provide consumers with the information they need in order to decide about possible purchases. These tools act as a specialized salesperson for the customer, and they are usually enhanced with customization capabilities; thus they adapt themselves to the users, basing themselves on the analysis of their preferences and interests. Recommenders rely mainly on user interfaces, marketing techniques, and large amounts of information about other customers and products; all this is done, of course, in an effort to propose the right item to the right customer. Besides, recommenders are fundamental elements in sustaining usability and site confidence (Egger, 2001); that's the reason why e-market sites give them an important role in their design (Spiekermann & Paraschiv, 2002).

If a recommender system is to be perceived as useful by its users, it must address several problems, such as the lack of user knowledge in a specific domain, information overload, and a minimization of the cost of interaction.

EC recommenders are gradually becoming powerful tools for EC business (Gil & García, 2003) making use of complex mechanisms mainly in order to support the user's decision process by allowing the analogical reasoning by the human being, and avoiding the disorientation process that occurs when one has large amounts of information to analyse and compare. This article describes some fundamental aspects in building real recommenders for EC.

We will first set up the scenario by exposing the importance of recommender systems in EC, as well as the stages involved in a recommender-assisted purchase. Next, we will describe the main issues along three main axes: first, how recommender systems require a careful elicitation of user requirements; after that, the development and tuning of the recommendation algorithms; and, finally, the design and usability testing of the user interfaces. Lastly, we will show some future trends in recommenders and a conclusion.

BACKGROUND

E-commerce sites try to mimic the buying and selling protocols of the real world. At these virtual shops, we find metaphors of real trade, such as catalogues of products, shopping carts, shop windows, and even "salespersons" that help us along the process (see http://www.vervots.com).

There exist quite a number of models proposed to describe the real world customer-buying process applied to electronic trade; among these, we might propose the Bettman model (Bettman, 1979), the Howard-Sheth model (Howard & Sheth, 1994) or the AIDCA (Attention Interest Desire Conviction Action) model (Shimazu, 2002). The theory of purchase decision involves many complex aspects, among which one might include the psychological ones, those of marketing, social environment, and so forth. The behaviour of the buyer (Schiffman & Kanuk, 1997) includes besides a wide spectrum of experiences associated with the use and consumption of products and services: attitudes, lifestyles, sense of ownership, satisfaction, pleasure inside groups, entertainment, and so forth.

Therefore, the fundamental goal today in EC is that of providing the virtual shops with all of the capabilities of physical trade, thus becoming a natural extension of the traditional processes of buying and selling. One must provide these applications with dynamism, social and adaptive capacities in order to emulate traditional trade. The recommender system can supply the user with information related to the particular kind of shopping technique he or she is using. The most important phases that support the user's decision can be resumed as follows:

- **Requirement of Identification:** It permits the entry of every user into the system in an individual way, thus making it possible for the recommender to make use of a customized behaviour.
- **Product Brokering:** The user, thus properly identified, interacts with the site in search of certain products and/or services; the searching process is facilitated by recommender systems, which relieves the user from information overload and helps each concrete user to locate the desired product.
- **Merchant Brokering:** This type of buying mechanism comes into play when users want to acquire a certain product already known to them; at this moment, they look for the best offer for this precise item. The recommender systems make use of a comparison process, carry out the extraction of necessary information from different virtual shops, and work towards the goals established by the buyer (best price, best condition, etc.).
- **Negotiation:** This aspect reflects the customized interaction between the buyer and the site in the process of pre-acquisition, as well as the maintenance of these relations in post-sale process. This process is performed in the transaction between the user and the site, and it depends on the purchasing needs of the user and on the sales policy of the site. The user must perceive negotiation as a transparent process. In order to benefit and activate the relationship with the site, one must facilitate fidelity policies. Also, one should avoid pervasive recommendation or cross-sell, which the user could see as obtrusive and abusive methods. This will consolidate the success of the virtual shop.
- **Confidence and Evaluation:** Recommender Systems work with relevant information about the user. A significant part of this phase of approximation to the user is related to the safety in the transactions and the privacy of the information that the user hands over to the company. Besides, post-sale service is critical in virtual shops, both in the more straightforward sense (an item is not acceptable) and in the sense of confirming the results of a given recommendation. This reinforces the confidence of the user in the site and integrates them in a natural way in the sales protocol. Confidence on a recommendation system relies on three fundamental premises (Hayes, Massa, Avesani, & Cunningham, 2002). Confidence in the recommender, assuming that it has sufficient information on our tastes and needs, also accepts that the recommender has knowledge on other possible alternatives.

There exists a large number of recommenders over the Internet. These systems have succeeded in domains as diverse as movies, news articles, Web pages, or wines; especially well-known examples are the ones that we find in Amazon.com or BarnesAndNoble.com.

MAIN ISSUES IN RECOMMENDATION

The user's interaction with the recommender system can be seen as two different but related processes. There is a first stage in which the system builds a knowledge base about the user (Input Stage we could say) and then a second stage in which a recommendation is made, at the same time taking notice of the user's preferences (Output Stage). Normally, systems offer a set of possible products that try, perhaps without previous knowledge about the particular user, to extract some sort of first-approximation ratings. When the user makes further visits, both the data extracted from a first contact and the information garnered from any purchases made will be of use when a recommendation is produced. By analyzing this information by means of different techniques (which we explain later), the systems are able to create profiles that are later to be used for recommendations.

The second phase is the output stage, in which the user gets a number of recommendations. This is a complex phase, and one must take into account the fact that the user gets information about a given product, the ease with which the customer gets new recommendations, the actual set of new recommendations produces, and so forth.

Let us now describe the real-world data that are extracted from the user and their importance from the point of view of usability.

User Data

The information about both user and domain defines the context in recommendation; it establishes how the various concepts can be extracted from a list of attributes that are produced by the interaction of the user with the system. These attributes must carefully be chosen for the sake of brevity, and also because they are the essential and effective information employed as a user model in the specialized EC domain.

User actions at the user interface, such as requiring some kind of information, scrolling or drilling down group hierarchies, and so forth, are translated into user preferences for different parts of the result, and fed back into the system to prioritize further processing.

The following table contains the kinds of data that one can hope to extract from the user in the different phases of interaction with the system.

A recommender system may take input for users implicitly or explicitly, or as a combination of both. Table 1 summarizes many of the attributes used for building a customized recommendation. This information, attending to the complex elaboration of the data extracted about users in the domain, can be divided into three categories:

- **Explicit:** Expressed by the user directly (e.g., registration data as name, job, address and any other question and direct answer attributes)
- **Implicit:** Captured by user interactivity with the system (e.g., purchase history, navigational history).
- **Synthetic:** Added by contextual techniques mixing explicit and implicit ones. The principle behind the elaboration of these data is that

Table 1. Summary of information in EC for context building in recommendation

Kind of data	Information extracted from	Yields...
Explicit	Personal data	Name Gender Age Job Income Address
	Specific queries	User's level of expertise User's areas of interest Related interests
Implicit	Navigational aspects	Number of visits Time spent in each visit/page Sequence of URLs visited Search process
	Purchases	Number of articles Amount payed Date of purchase Preferences for particular services Gender
Synthetics	Contextual techniques combining explicit and implicit attributes	Trust between users Similar purchases related to content based
		Price sensitivity
		Level of expertise in the service domain
		Purchasing possibilities, ...

consumers attach values to all of the attributes of a service. The total value of a service is a function of the values of its components (e.g., parameters adding time-delay aspects in the model, semantic content traits in the items, etc.). The choice of the user and the recommendation is a function of the absence or incorporation of specific characteristics of the domain. These characteristics are taken into account using information regarding the content of items, mainly of semantic content, in order to infer the reasons behind a user's preferences. Recommendations suitable to a user will be context-dependent. The context of a user's search often has a significant bearing on what should be recommended.

Techniques Used for Recommendation

Several different approaches have been considered for automated recommendation systems (Konstant, 2004; Sarwar, Karypis, Konstan, & Riedl, 2000). These can be classified into three major categories: those based on user-to-user matching and referred to as collaborative filtering, those based on item content information, and hybrid methods referred to as knowledge-based systems.

- **Collaborative-social-filtering** systems, which build the recommendation by aggregation of consumers' preferences: These kinds of systems try to find a match to other users, basing themselves on similarity in either behavioral or social patterns. The statistical analysis of data or data mining and knowledge discovery in databases (KDD) techniques (monitoring the behavior of the user over the system, ratings over the products, purchase history, etc.) build the recommendation by analogies with many other users (Breese, Heckerman, & Kadie, 1998). Similarity between users is computed mainly using the so-called user-to-user correlation. This technique finds a set of "nearest neighbors" for each user in order to identify similar likes and dislikes. Some collaborative filtering systems are Ringo (Shardanand & Maes, 1995) or GroupLens (Konstant, Miller, Maltz, Herlocker, Gordon, & Riedl, 1997). This technique suffers mainly from a problem of sparsity due to the need for a large volume of users in relation to the volume of items offered (critical mass) for providing appropriate suggestions.
- **Content-based-filtering systems**, which extract the information for suggestions basing themselves on items the user has purchased in the past: These kinds of systems use supervised machine learning to induce a classifier to discriminate between interesting or uninteresting products for the user due to her purchase history. Classifiers may be implemented using many different techniques from artificial intelligence, such as neural networks, Bayesian networks, inducted rules, decision trees, etc. The user model is represented by the classifier that allows the system to ponder the like or dislike for the item. This information identifies the more weighted items that will be recommended to the user. Some content-based systems also use item-to-item correlation in order to identify association rules between items, implementing the co-purchase item or cross-sell. As an example, one could mention (Mooney & Roy, 2000) Syskill & Webert (Pazzani, Muramatsu, & Billsus, 1996), where a decision tree is used for classifying Web documents attending some content domain on a binary scale (hot and cold) or the well-known recommendation mechanism for the second or third item in Amazon. This technique suffers mainly from the problem of over-specialization; the consumer is driven to the same kind of items he has already purchased. Another important problem comes also for recommending new articles in the store, as no consumers have bought this item before; hence, the system can't identify this new item in any purchase history, and it cannot be recommended till at least one user buys this new article.
- **Knowledge-based systems** could be understood as a hybrid extended between collaborative-filtering and content-based systems. It builds the knowledge about users linked also with the products knowledge. This information is used to find out which product meets the user's requirements. The cross-relationships

between products and clients produce inferences that build the knowledge in the EC engine. Several papers (Balabanovic & Shoham, 1997; Paulson & Tzanavari, 2003; Shafer, Konstan, & Riedl, 2001) show the benefits of these systems.

Usability in Recommender Systems

User satisfaction with a recommender system is only partly determined by the accuracy of the algorithm behind it. The design of a recommender system is that of a user-centered tool, where personalization appears as one of the aspects of major weight in the capacities that the user perceives when interacting with the virtual shop. The development of a recommender system is the sum of several complex tasks, and usability questions arise. For this purpose, some standard questionnaires were created. IBM (Lewis, 1995) has suggested some of these: the Post-Study Systems Usability Questionnaire (PSSUQ), the Computer System Usability Questionnaire (CSUQ), or the After-Scenario Questionnaire (ASQ). Other examples were developed as the Questionnaire for User Interface Satisfaction (QUIS) (Chin, Diehl, & Norman, 1988), the System Usability Scales (SUS[1]), developed in 1996 and whose questions all address different aspects of the user's reaction to the Web site as a whole or the Web site analysis. Finally, one could also mention Measurement Inventory (WAMMI[2]), developed in 1999.

Nowadays, there is a growing number of studies that examine the interaction design for recommender systems in order to develop general design guidelines (Hayes et al., 2002; Swearingen & Sinha, 2001) and to test usability. Their aim is to find the factors which mostly influence the usage of the recommender systems. They consider such aspects as design and layout, functionality, or ease of use (Zins, Bauernfeind, Del Missier, Venturini, & Rumetshofer, 2004a, 2004b).

The evaluation procedures that recommender systems must satisfy to obtain a level of usability are complex to carry out due to the various aspects to be measured. Some of the evaluation procedures apply techniques that comprise several other steps well known (Nielsen & Mack, 1994), such as concepts tests, cognitive walkthrough, heuristic evaluations, or experimental evaluations by system users. This could only be achieved by a cooperation of usability experts, real users, and technology providers.

Usability in recommender systems can be measured by objective and subjective variables. Objective measures include the task completion time, the number of queries, or the error rate. Subjective measures include all other measures, such as user's feedback or level of confidence in recommendations, and the transparency level which the user perceives in recommendations (Sinha & Swearingen, 2002).

FUTURE TRENDS

Significant research effort is being invested in building support tools to ensure that the right information is delivered to the right people at the right time. A positive understanding of the needs and expectations of customers is the core of any development. There is no universally best method for all users in all situations, and so flexibility and customization will continue to always be the engine of the development in recommender systems. The future in recommender systems in EC will be governed mainly by three fundamental aspects: the need for supporting more complex and heterogeneous dialog models, the need for attending internationalization and the standardization, and also the support in the nascent of a ubiquitous electronic commerce environment.

Different players in the EC place operate with these contents. The problem increases with dialog models, all the more so between tools and automation in the purchase or attending the analogical reasoning of human beings. Recommenders have to be sufficiently flexible to accomplish any ways of purchase. One of the key components for developing an electronic commerce environment will be interoperability that could support an expertise degree in anything that e-market offers and a heterogeneous environment; this will also require the use of international standards based on open systems.

The EC market needs an international descriptive classification, as an open and interoperable standard that can be used for all the players in the interchange. The goal is to produce an appropriate mechanism to increase the efficiency of the management in the EC site as well as the personalization and specialization customer services offered in the recommendation.

For the immediate future, the recommenders in EC need to incorporate new trends in the description of items and their connection with users (Hofmann, 2004). The integration of the forceful expansion Web services infrastructure with the richer semantics of the semantic Web, in particular through the use of more expressive languages for service marks the beginning of a new era in EC, as it endows recommender tools with powerful content knowledge capabilities.

These points for the future will work together also into the forceful ubiquitous electronic commerce environment. These will require deploying a network capable of providing connectivity to a large user and service provider community through new devices. Applications beyond those currently envisioned will evolve, and sophisticated user interfaces that conveniently provide users with information about the services offered by this new environment will emerge. The ability to obtain user data in an unobtrusive way will determine the success of recommendations in environments as daily as a tourist visit or just going to shop in the supermarket.

CONCLUSION

EC sites are making a big effort to supply the customer with tools that ease and enhance shopping on the Net. The effort to facilitate the user's tasks in EC necessitates an understanding of consumer's behavior in order to facilitate a personalized access to the large amount of information one needs to search and assimilate before making any purchases. The user-centered design in recommender systems improves the study of realistic models on discovering and maintaining the user decision processes narrated by inputs that help build user models. The need to build recommender systems whose aim is to improve effectiveness and perceived user satisfaction has produced a surge in usability studies; these are carried out by means of different procedures and through the application of various techniques.

We point out the future importance of recommenders in EC sites, all the more so due to the inclusion of semantic Web models in EC and of new interface paradigms.

REFERENCES

Balabanovic, M., & Shoham, Y. (1997). Fab: Content-based, collaborative recommendation. *Communication of the ACM, 40*(3), 66-72.

Bettman, J. (1979). *An information processing theory to consumer choice*. Addison-Wesley.

Breese, J., Heckerman, D., & Kadie, C. (1998, July 24-26). Empirical analysis of predictive algorithms for collaborative filtering. *Proceedings of the Fourteenth Conference on Uncertainly in Artificial Intelligence*, Madison, WI (pp. 43-52). University of Wisconsin Business School, Madison: Morgan Kaufmann Publisher.

Chin, J. P., Diehl, V.A., & Norman, K. (1988, May 15-19). Development of an instrument measuring user satisfaction of the human-computer interface. *Proceedings of CHI'88 Conference of Human Factors in Computing Systems*, Washington, D.C, (pp. 213-218). New York: ACM Press.

Egger F. N. (2001, June 27-29). Affective design of e-commerce user interfaces: How to maximize percived trustworthiness. In M. Helander, H.M. Khalid, & Tham (Eds.) *Proceedings of the International Conference on Affective Human Factors Design*, Singapore, June 27-29 (pp. 317-324). London: ASEAN Academic Press.

Gil, A., & García, F. (2003). E-commerce recommenders: Powerful tools for e-business. *Crossroads—The ACM Student Magazine, 10*(2), 24-28.

Hayes, C., Massa, P., Avesani, P., & Cunningham, P. (2002, May). An on-line evaluation framework for recommender systems. In F. Ricci & B. Smyth (Eds.) *Workshop on Recommendation and Personalization Systems*, Malaga, Spain (pp. 50-59). Spring Verlag.

Hofmann, T. (2004, January). Latent semantic models for collaborative filtering. *ACM Transactions on Information Systems (TOIS), 22*(1), 89-115.

Howard, J., & Sheth, J.N. (1994). *The theory of buyer behavior*. New York: Wiley.

Konstant, J. (2004, January). Introduction to recommender systems: Algorithms and evaluation.

ACM Transactions on Information Systems (TOIS), 22(1), 1-4.

Konstant, J., Miller, B., Maltz, D., Herlocker, J., Gordon, L., & Riedl, J. (1997). Grouplens: Applying collaborative filtering to Usenet news. *Communications of the ACM, 40*(3), 77-87.

Lewis, J. R. (1995). IBM Computer usability satisfaction questionnaires: psychometric evaluation and instructions for use. *International Journal of Human Computer Interaction, 7*(1), 57-58.

Mooney, R. J., & Roy, L. (2000, June 2-7). Content-based book recommending using learning for text categorization. *Proceedings of the V ACM Conference on Digital Libraries*, San Antonio, Texas (pp. 195-204). New York: ACM Press.

Nielsen, J., & Mack, R. L. (Eds.). (1994). *Usability inspection methods*. New York: John Wiley & Sons.

Paulson, P., & Tzanavari, A. (2003). Combining collaborative and content-based filtering using conceptual graphs. In J. Lawry, J. Shanahan, & A. Ralescu (Eds.), *Modeling with words. Lecture Notes in Artificial Intelligence Series* (pp. 168-185). Springer-Verlag.

Pazzani, M., Muramatsu, J., & Billsus, D. (1996, August 4-8). Syskill & Webert: Identifying interesting Web sites. In *Proceedings of the Thirteenth National Conference on Artificial Intelligence*, Portland, Oregon (pp. 54-46). Menlo Park, CA: AAAI Press.

Sarwar, B., Karypis, G., Konstan, J., & Riedl, J. (2000, October 17-20). Analysis of recommendation algorithms for e-commerce. *ACM Conference on Electronic Commerce*, Minneapolis, Minnesota (pp.158-167). ACM Press.

Schiffman, L., & Kanuk, L. (1997). *Consumer behavior* (6th ed.). Upper Saddle River, NJ: Prentice Hall International.

Shafer, J., Konstan, J. A., & Riedl, J. (2001). E-commerce recommendation applications. *Data Mining and Knowledge Discovery, 5*(1/2), 115-153.

Shardanand, U., & Maes, P. (1995, May 7-11). Social information filtering: Algorithm for automating "word of mouth". *Proceedings of ACM CHI'95 Conference on Human Factors in Computing Systems*, Denver, Colorado (pp. 210-217). ACM Press.

Shimazu, H. (2002). ExpertClerk: A conversational case-based reasoning tool for developing salesclerk agents in e-commerce Webshops. *Artificial Intelligence Review, 18*(3-4), 223-244.

Sinha, R., & Swearingen, K. (2002, April 20-25).The role of transparency in recommender systems. *Proceedings of Conference on Human Factors in Computing Systems, CHI'02*, Minneapolis, Minnesota (pp. 830-831). New York: ACM Press.

Spiekermann, S., & Paraschiv, C. (2002). Motivating human–agent interaction: Transferring insights from behavioral marketing to interface design. *Electronic Commerce Research,* 2, 255-285.

Swearingen, K., & Sinha, R. (2001, September 13). Beyond algorithms: An HCI perspective on recommender systems. In *Proceedings of the ACM SIGIR 2001 Workshop on Recommender Systems*, New Orleans, Louisiana.

Zins, A., Bauernfeind, U., Del Missier, F., Venturini, A., & Rumetshofer, H. (2004a, January). An experimental usability test for different destination recommender systems. *Proceedings of ENTER 2004*, Cairo, Egypt.

Zins, A., Bauernfeind, U., Del Missier, F., Venturini, A., & Rumetshofer, H. (2004b, January 26-28). Prototype testing for a destination recommender system: Steps, procedures and implications. In A.J. Frew (Ed.) *Proceedings of ENTER 2004*, Cairo, Egypt (pp. 249-258). New York: Springer Verlag.

KEY TERMS

Collaborative-Social-Filtering Recommender Systems: Technique based on the correlation between users' interest. This technique creates interest groups between users, based on the selection of the same.

Content-Based-Filtering Recommender Systems: Technique based on the correlation between item contents by statistical studies about different characteristics. These techniques compute user-purchase histories in order to identify association rules between items.

Direct Recommendation: This kind of recommendation is based on a simple user request mechanism in datasets. The user interacts directly with the system that helps him in the search of the item through a list with the n-articles that are closest to his or her request in relation to a previously-known profile.

Information Overload: Undesirable or irrelevant information that disturbs the user and distracts him or her from the main objective. This kind of problem usually occurs in contexts that offer excessive amounts of information, badly handled due to low usability systems.

Knowledge-Based Systems: A hybrid extended technique between collaborative-filtering and content-based systems. It builds knowledge about users by linking their information with knowledge about products.

Pervasive Recommendation: Unsolicited information about products or services related with the one requested. They are usually shown as advertising or secondary recommendations, acting as fillers for the page or as new elements in the interface; they could be perceived as disturbing elements. The system of inner marketing establishes a policy of publicity for each product destined to given segments of consumers. This provides a method to perform cross-sell marketing.

Recommender Systems in E-Commerce: Tools implanted in EC sites for suggesting services and in order to provide consumers with the needed information to decide about services to acquire. They are usually domain-specialized tools.

ENDNOTES

[1] http://www.usability.serco.com/trump/documents/Suschapt.doc

[2] http://www.ucc.ie/hfrg/questionnaires/wammi/

Replicating Human Interaction to Support E-Learning

William A. Janvier
Liverpool John Moores University, UK

Claude Ghaoui
Liverpool John Moores University, UK

INTRODUCTION

HCI-related subjects need to be considered to make e-learning more effective; examples of such subjects are: psychology, sociology, cognitive science, ergonomics, computer science, software engineering, users, design, usability evaluation, learning styles, teaching styles, communication preference, personality types, and neuro-linguistic programming language patterns. This article discusses the way some components of HI can be introduced to increase the effectiveness of e-learning by using an intuitive interactive e-learning tool that incorporates communication preference (CP), specific learning styles (LS), neurolinguistic programming (NLP) language patterns, and subliminal text messaging. The article starts by looking at the current state of distance learning tools (DLTs), intelligent tutoring systems (ITS) and "the way we learn". It then discusses HI and shows how this was implemented to enhance the learning experience.

BACKGROUND

In this section, we briefly review the current states in DLT and ITS.

The general accepted standard, with current DLTs, is that the learner must be able to experience self-directed learning, asynchronous and synchronous communication (Janvier & Ghaoui, 2002a, 2003a).

Bouras and Philopulos (2000) in their article consider that "distributed virtual learning environment," using a combination of HTML, Java, and the VRML (Virtual Reality Modelling Language), makes acquiring knowledge easier by providing such facilities as virtual chat rooms for student-student-teacher interaction, lectures using the virtual environment, announcement boards, slide presentations, and links to Web pages.

People's experience (including ours) of a number of DLTs was that, while they achieved an objective of containing and presenting knowledge extremely well, the experience of using them fell far short of normal HI, was flat, and gave no rewarding motivation. The user had to accept a standard presentation that did not vary from user to user; briefly there was no real system that approached HI, and, thus, the learning experience lacked the quality that was required to make it as effective as it should be.

Similarly with ITS, they are normally built for a specific purpose with student modelling being developed from the interaction between the student and the system.

Murray (1997) postulates that while ITS, also called knowledge-based tutors, are becoming more common and proving to be increasingly effective, each one must still be built from scratch at a significant cost. Domain independent tools for authoring all aspects of ITS (the domain model, the teaching strategies, the learner model, and the learning environment) have been developed. They go beyond traditional computer-based instruction in trying to build models of subject matter expertise, instructional expertise, and/or diagnostic expertise. They can be powerful and effective learning environments; however, they are very expensive in time and cost, and difficult to build.

Nkambou and Kabanza (2001) report that most recent ITS architectures have focused on the tutor or curriculum components but with little attention being paid to planning and intelligent collaboration between the different components. They suggest that the ideal architecture contains a curriculum model, a tutor (pedagogical) model, and a learner model: This last is central to an ITS.

To move forward, e-learning requires a combination of both; however, Murray (1999), in common with many other researchers, believes that ITS are too complex for the untrained user and that:

we should expect users to have a reasonable degree of training in how to use them, on the order of database programming, CAD-CAM authoring, 3-D modelling, or spreadsheet macro scripting.

In e-learning, the development has taken two routes: that of the DLT and that of the ITS. With both, there is no effort to pre-determine the student's psyche before the system is used, and thus the basic problem of HI replication in HCI has not been instigated at the inception of an e-learning session.

MAIN ISSUES IN HUMAN INTERACTION

In this section, we discuss communication preference, personality types, neurolinguistic programming, NLP language patterns, and subliminal text messaging.

Communication Preference (CP)

Each person has a preference in the way he or she communicates with others; they also have preferences in the way to learn or pass on information to someone else: This is called communication preference. Learning is introduced by one of the five senses (touch, sight, taste, hearing, and smell) and initially passes into the subconscious sensual memory from their sensual memory to short-term memory and then, usually via rehearsal to long-term memory. All input into short-term memory is filtered, interpreted, and assessed against previously input, beliefs, and concepts using perceptual constancy, perceptual organization, perceptual selectivity, and perceptual readiness. Cue recognition allows for memory to pick out the key points that link to further memory recall and practice a skill using cognitive, psychomotor and perceptual skills (Cotton, 1995).

Stored instances (single items of memory) do not necessarily represent actuality due to the fact that they have already been distorted by the subject's own interpretation of the facts as perceived by their "inner voice, eye, ear, nose, and taste." Initially, instances are stored in short-term memory where the first and last inputs of a stream of instances are easier to recall: These can then be transferred to long-term memory by rehearsal. Different individuals use their preferred inner sense to aid perception. For learning to be effective, new instances are associated with existing instances. The use of the working memory constantly improves and refines long-term memory; indeed, practical "day-dreaming" is a form of forward planning that can improve retention (Cotton, 1995).

Iconic sensory input is the most important for remembering and learning. Cotton (1995) showed that using an A4 sheet divided into sections with questions and answers aided storing—linking this with sound, further increased retention in long-term memory. Research shows a link between good recall and good recognition, and that memory is seldom completely lost: It only requires specific cues to connect linked instances and bring them to the

Figure 1. Memory transfers

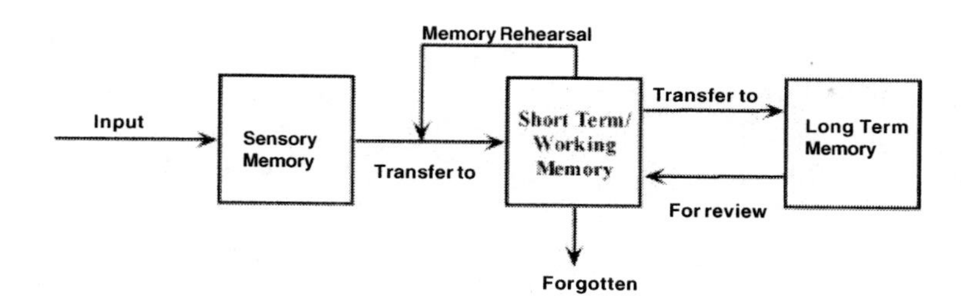

level of conscious thought. Here associations, self-testing, sectional learning, rhyme rules, mnemonics, spelling associations, memory systems, networks of memory patterns (both linear and lateral) are used to develop cue recognition (Catania, 1992; Cotton, 1995).

Borchert, Jensen, and Yates (1999) report that the visually (**V**) oriented students prefer to receive information, via their eyes, in charts, graphs, flow charts, and symbolic representation, the aural (**A**) orientated indicates a preference for hearing information, and the kinaesthetic (**K**) orientated student prefers "learning by doing" either learning by simulated or real-world experience and practice.

Personality Types

Different personality types require different communication treatment. Extroverts (E) and introverts (I) can be viewed as one couplet, slow (S) and quick (Q) decision makers as another couplet. All personality types can be plotted somewhere on the resultant EI/SQ scale.

- **ES types** prefer the company of others, are slow to make decisions, they take their time, and sometimes will not make a final commitment. *They can be categorized as "arty."*
- **IS types** prefer their own company, are slow to make decisions, they take their time, and sometimes will not make a final commitment; they are very precise. *They are often categorized as "analytical."*
- **EQ types** prefer the company of others, are fast to make decisions, and often make a commitment before thinking it through. *They are often categorized as "salesmen."*
- **IQ types** prefer their own company, are fast to make decisions, and often make a commitment before thinking it through. *They are often categorized as "businessmen."* (Fuller, Norby, Pearce, & Strand, 2000; Janvier & Ghaoui, 2003b; Myers & Myers, 1995)

Research has shown that when a student joins higher education, his or her primary personality type and thus learning style has been established (Wilson, Dugan, & Buckle, 2002). By this time, the introvert has usually learned to use their auxiliary or, maybe, even their tertiary Jungian Function and, thus, tend to hide their true primary personality type: The unwary tutor can use inappropriate communication techniques with resulting frustration (Janvier & Ghaoui, 2003b; Wilson et al., 2002).

Neurolinguistic Programming

The name **neurolinguistic programming (NLP)** comes from the disciplines that influenced the early development of its field, beginning as an exploration of the relationship between **neurology, linguistics**, and observable patterns (**programs**) of behaviour. Combining these disciplines, NLP can be defined as:

The reception, via our nervous system, of instances received and processed by the five senses (iconic, echoic, haptic, gustatory, and olfactory), the resultant use of language and nonverbal communication system through which neural representation are coded, ordered, and given meaning using our ability to organize our communication and neurological systems to achieve specific desired goals and results,

Or more succinctly, "The Study of the Structure of Subjective Experience and what can be calculated from it" (Janvier & Ghaoui, 2002b; Pasztor, 1998b; Sadowski & Stanney, 1999; Slater, Usoh, & Steed, 1994).

John Grinder, a professor at UC Santa Cruz and *Richard Bandler*, a graduate student, developed NLP in the mid-70s. They were interested in how people influence one another, in the possibility of being able to duplicate the behavior, and thus the way people could be influenced. They carried out their early research in the *University of California at Santa Cruz* where they incorporated technology from linguistics and information science, knowledge from behavioural psychology and general systems theory developing their theories on effective communication. As most people use the term today, NLP is a set of models of how *communication* impacts and is impacted by *subjective experience*. It's more a collection of tools than any overarching theory. Much of early NLP was based on the work of *Virginia Satir*, a family therapist; *Fritz Perls*, founder of Gestalt therapy; *Gregory Bateson*, anthropologist; and *Milton Erickson*, hypnotist. - Stever Robbins, NLP Trainer (Bandler & Grinder, 1981).

NLP Language Patterns

Craft (2001) explores relationships between NLP and established learning theory and draws a distinction between models, strategies, and theories. Craft argues that, while NLP has begun to make an impact in education, it still remains a set of strategies rather than a theory or model. NLP research has shown that this set of strategies results in increased memory retention and recall, for example:

Pasztor (1998a) quotes the example of a student with a visual NLP style whose tutorial learning strategy was based on "listen, self-talk" and sport-learning strategy was "listen, picture, self-talk." The former did not achieve memory store/recall while the latter did. She also reports that rapport with a partner is the key to effective communication and that incorporating NLP in intelligent agents will allow customization of the personal assistant to the particular habits and interests of the user thus making the user more comfortable with the system.

Introducing the correct sub-modality (visual, auditory, kinaesthetic) will enable the subject to more easily store and recall instances in/from memory. It is argued that inviting a subject to *"see" invokes iconic*, to *"hear" invokes auditory* and to *"feel" invokes kinaesthetic* recall (Pasztor, 1997).

Subliminal Text Messaging

A subliminal text message is one that is below the threshold of conscious perception and relates to iconic memory (the persisting effects of visual stimuli). After-effects of visual stimuli are called icons. Iconic memory deals with their time courses after the event. Subliminal images and text (instance input that the conscious mind does not observe but the subconscious does) can have a powerful effect on memory and cognitive memory. "*Unconscious words are pouring into awareness where conscious thought is experienced, which could from then on be spoken [the lips] and/or written down*" (Gustavsson, 1994). The time course of iconic memory is measured over fractions of seconds, but, in this time, the subject retains images that no longer are there (e.g., the imposition of fast changing still images on the retina create the effect of motion).

Johnson and Jones (1999) affirm, *"participants in text based chats rooms enjoy their anonymity and their ability to* ***control the interaction precisely by careful use of text****"* and that the nature of such interactions may well change dramatically if the participants could observe each other and their mind state (Pasztor, 1997, 1998b; Sadowski & Stanney, 1999).

AN INTUITIVE INTERACTIVE MODEL TO SUPPORT E-LEARNING

In this section, we discuss a model called WISDeM and its evaluation, the results, and the statistical comparison of these results.

WISDeM

WISDeM (Web intuitive/interactive student e-learning model) has further been developed into an intuitive tool to support e-learning; it has combined these HI factors: CP, personality types, learning styles (LS), NPL language patterns, motivational factors, novice|expert factor, and subliminal text messaging.

LS research has shown that there are more facets to learning than a simple communication preference; Keefe (1987) defines LS as,

> *...the set of cognitive, emotional, characteristic and physiological factors that serve as relatively stable indicators of how a learner perceives, interacts with, and responds to the learning environment...*

Knowing your LS and matching it with the correct teaching strategies can result in more effective learning and greater academic achievement (Hoover & Connor, 2001).

Fuller et al. (2000) in their research posed the question, *"Does personality type and preferred teaching style influence the comfort level for providing online instruction?";* their answer was "Yes". They outlined the teaching styles preferences for the Myers-Briggs Type Indicators®—

Extroversion|**I**ntroversion, **S**ensing|i**N**tuition, **T**hinking|**F**eeling, **J**udgement|**P**erception, and provided some suggestions for faculty development for seven of the sixteen MBTI[0] types (ESTJ, ESTP, ESFJ, ESFP, ENTJ, ENTP, ENFJ, ISTJ, ISTP, ISFJ, ISFP, INTJ, INTP, INFJ, INTJ).

Montgomery and Groat (1998) point out that *"matching teaching styles to LS is not a panacea that solves all classroom conflicts,"* that other factors such as the student's motivation, pre-conceptions, multicultural issues, and so forth, also impinge on the student's quality of learning; but, that, nonetheless, understanding and reacting to LS in teaching enhances the quality of learning and rewards teaching.

Initially, WISDeM uses two psychometric questionnaires based on the concepts, principles researched covering VARK (Fleming, 2001), Jungian Functions (Murphy, Newman, Jolosky, & Swank, 2002; Wilson et al., 2002) and MBTI® (Larkin-Hein & Budny, 2000; Murphy et al., 2002; Myers & Myers, 1995).

It creates the initial student profile "before" the student accesses module learning material to enable effective HCI interaction. After the initial login, the student completes the CP questionnaire from which the system establishes if he or she is **v**isual, **a**uditory or **k**inaesthetic. A relevant report is output and the student opens the LS questionnaire. The questions in this are couched using text that matches his or her CP. Upon completion of this questionnaire, an LS report is produced, and, provided the student agrees, the initial student profile is saved in the CPLS database.

As the student uses the system, his or her CPLS, together with a novice|expert (NE) is used to create and update a unique student model (SM). The system's algorithms use this SM and the updated student's knowledge state to retrieve and display relevant messages and information in the interface. The NE factor is dynamically moderated as a student moves through a topic and reverts to the default value when a new topic is started.

The tool built allows a student to select any topic for revision. In topic revision, the student can select either "LEARN" or "TEST" knowledge for anyone or a series of lectures as the module develops.

The system's use of repetitive header messages invokes subliminal text messaging: The student skips over previously noticed information, his or her "I've seen this before," or "Have I seen something like it before?" filter kicks in leading to conscious or subconscious rejection (Catania, 1992; Gustavsson, 1994). The unconscious observation of the NLP language patterns matching his CP is effective: his or her eyes scan the page, take in the displayed information at the subliminal level, while he or she consciously notices what he or she wants to see.

Evaluation

The evaluation was a systematic and objective examination of the completed project. It aimed to answer specific questions and to judge the overall value of the system, to provide information to test the hypothesis, *"Matching neurolinguistic language patterns in an online learning tool, with the learner's communication preference and learning styles, will provide an intuitive tutoring system that will enhance Human-Computer Interaction and communication and, thus, enhance the storing and recall of instances to and from the learner's memory; thus enhancing learning,"* supporting the hypothesis (H_a) or not, the null hypothesis (H_o).

Statistically, the null hypothesis ($H_o: \mu = 0$) states that there is no effect or change in the population and that the sample results where obtained by chance, whereas the alternative hypothesis ($H_a : \mu > 0$) affirms that the hypothesis is true (H = hypothesis, μ = population). *P-values* were used to test the hypothesis, where a null hypothesis [H_o] is accepted or failed. The p-value represents the probability of making a Type 1 error, which is rejecting H_o when it is true. The smaller the p-value, the smaller is the probability that a mistake is being made by rejecting H_o. A cut-off value of 0.05 is used: values >= 0.05 means that the H_o should be accepted, values < 0.05 suggest that the alternative hypothesis [H_a] needs consideration and that further investigation is warranted, values <= 0.01 are a strong indication that H_a is valid.

To ensure the maximum integrity in sampling, *simple random sampling* (sampling without replacement) was used (*"Simple random sampling is the sampling design in which n distinct units are selected from N units in the population in such a way that every possible combination of the n*

Figure 2. Interactive group evaluation flow chart

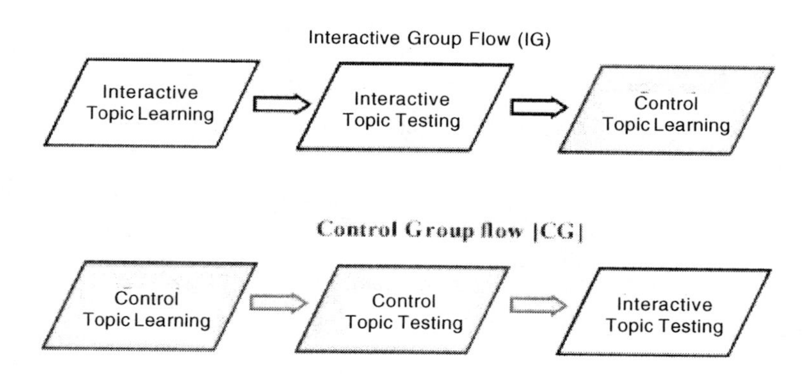

units is equally likely to be the sample selected" (Thompson, 1992)). This provided the probability that any student can participate without any preference, and, therefore, the sample was more likely to reflect the whole rather than if any other sampling method had been used (Clarke & Cooke, 1978; Thompson, 1992; Yates, Moore, & Starnes, 2002).

The evaluation required two tests, *control* and *interactive*: Thus, the statistical analysis was required to *compare two means*. The importance of the sample selection is paramount: Stronger results are obtained where each group is matched. Thus, the subjects were selected from students in one particular module; this ensured that there was a match in year, course, age group (20's), but not sex (the random selection of students reflected the class spread of 84.13% male and 15.87% female). Another factor considered was the fact that the sample sizes would be unequal: The test was run in two sections requiring circa two hours in total. Risk factor analysis suggested that some students would complete only one part of the evaluation: Thus, Two-Sample T-test was used.

CPLS Evaluation Results

The CPLS evaluation had 97 responders (86 male, 11 female). Their % *communication preferences* were:

Visual 67.01 **Auditory** 27.84 **Kinaesthetic** 5.15

Communication Preference is reported generally as:

V = 60% **A** = 30% **K** = 10%.

The evaluation group's averages were:

V = 67.01% **A** = 27.84% **K** = 5.15%.

These compare quite well with previously reported research (Brown, 2001; Catania, 1992; Cotton, 1995; Janvier & Ghaoui, 2003b). As the group shows a stronger tendency to visualization, this should reflect in memory retention data being stronger for this group than for the average due to the fact that the lecture content is based mainly on a visual presentation and auditory delivery styles. Hence, comparative results in the future could well be skewed where the group balance was more kinaesthetic.

Completion time varied from 10 minutes to 30 minutes with the majority being very close to the group average of 15 minutes. The figures squared well with the fact that decision style affects the speed of completion of a task: Judgemental types tend to complete tasks faster than Perceptual: Perceptive types take longer with a task being *more curious than decisive* and have the tendency to lose interest and not complete the task (see p. 76 in Myers & Myers, 1995).

The totals of each type containing the J-type [ESFJ-ESTJ-ENSJ-ENTJ-ISFJ-ISTJ-INSJ-INTJ] as compared with the P-type [ESFP-ESTJ-ENSP-ENTJ-ISFP-ISTJ-INSP-INTP] was:

Judgement 71.13% **P**erceptual 28.87%

Personality Types provided a split between extroverts (59.79%) and introverts (40.21%).

Each type was rated from 0 to 5. The average rating for the dominant type, from a possible rating of 3 to 5 presented as an average rating number, were:

Interpersonal Communication	**E** 3.55	**I** 3.49
Information Processing	**S** 3.97	**N** 3.23
Information Evaluation	**T** 3.49	**F** 3.62
Decision Style	**J** 4.38	**P** 3.48

Evaluation Results

The interactive group (IG) completed the interactive topic learning, then the interactive topic testing, and then the control topic testing sections; whereas, the control group (CG) completed the control topic learning, then the control topic testing, and then the interactive topic testing sections. This ensured that a set of comparative results were available. Due to the fact that the students were completing topic testing twice, it was anticipated that there would be an improvement in marks: There was for both with the IGs gaining more than the CGs.

The intuitive section had 50 students log into the system of which 33 answered questions:

- 33 completed the Interactive Group [IG] Multi-choice Q&A,
- 27 completed the Control Group [CG] Multi-choice Q&A,
- 27 completed both types.

The average time taken for the evaluation/exercise was 94 minutes: varying from 50 min. to 140 min.

Comparing the Marks for Both Sets of Students

The *Two-Sample T-test* for *Interactive and Control Student Marks* used a confidence level of 95% with a pooled StDev of 19.8. It produced a *P-Value of 0.036*. This *P-Value* indicates that H_a is valid; however, due to the fact that the P-Value is not below 0.01, the degree of probability requires more sampling to harden: More research results need to be gathered and assessed to enable this section of the results analysis to be viewed as proof of the hypothesis, at this time, the results, provided a strong indication that the hypothesis is true.

Comparing the Gains Made by Students

The *Two-Sample T-test Interactive and Control students Gain* used a confidence level of 95% and a pooled StDev of 7.50. The results gave a *P-Value of 0.005*. This *P-Value* indicates that H_a is valid, in particular well below 0.01. This indicates that the degree of probability is very strong demonstrating probable improvement in memory retention and recall: The results of the analysis provide a strong indication that the hypothesis is true.

CONCLUSION

The evaluation results indicated that the model implemented:

- Is likely to make a significant improvement to student learning and remembering.
- Produced more rehearsal from students than the control system and improved their marks.
- Supported a general belief in the system, that it did indeed assist knowledge retention. This in itself is an important factor for the students' psyche.
- As compared with the control system, the interactive system held interest longer and was more capable of interacting at the student's own level than the control system.

The evaluation indicates that it does in fact aid memory retention and recall and, thus, remembering and learning, that the use of NLP language patterns can affect the way students recall instances. It also indicates that CP and LS used in HCI and established "before" the student starts using an e-learning system is an important message to take forward.

FUTURE TRENDS

While communication preference, assessing personality types, and the conscious observance of body language and reacting with these using neurolinguistic programming language patterns have been used since the mid 1970s very effectively and enhanced human communication and learning (remembering) in the sales industry and management training (Bandler & Grinder, 1981), there has not been the reciprocal development in human-computer interaction development. Gustavsson (1994) and Johnson and Jones (1999) have indicated that subliminal messaging are affectively used in the advertising industry; once again, there has been no such reciprocal development in human-computer interaction. At this time, much research is looking at improving HCI interaction with the use of avatars (e.g., ADELE (Ganeshan, Johnson, Shaw, & Wood, 2000; Shaw, Ganeshon, Johnson, & Millar, 2000)). The future development in HCI is likely to slowly encompass HI; however, the inclusion of the basic tenants of HI as demonstrated in these needs to be introduced earlier rather than later.

REFERENCES

Bandler, R., & Grinder, J. (1981). *Frogs into princes: Neuro linguistic programming*. Real People Pr.

Borchert, R., Jensen, D., & Yates, D. (1999, June 20-23). *Hands-on and visualization modules for enhancement of learning in mechanics: Development and assessment in the context of Myers Briggs Types and VARK learning styles*. Paper presented at the ASEE Annual Conference, Charlotte, NC, USA.

Bouras, C., & Philopulos, A. (2000, September 5-7). *Distributed virtual learning environment: A Web-based approach*. Paper presented at the 26th EUROMICRO Conference (EUROMICRO'00), Maastricht, The Netherlands.

Brown, B. L. (2001). *Memory*. Professor of Psychology, Department of Psychology, Georgia Perimeter College, USA.

Catania, A. C. (1992). *Learning—Remembering* (3rd ed.). Upper Saddle River, NJ: Prentice-Hall International Editions.

Clarke, G. M., & Cooke, D. (1978). *A basic course in statistics* (4th ed., vol. 1). Nicki Dennis. UK: Hodder Arnold.

Cotton, J. (1995). *The theory of learning—An introduction*. London: Kogan Page.

Craft, A. (2001). Neuro-linguistic programming and learning theory. *Curriculum Journal, 12*(1), 125-136.

Fleming, N. (2001). *Teaching and learning styles: VARK strategies*. New Zealand: Neil Fleming.

Fuller, D., Norby, R. F., Pearce, K., & Strand, S. (2000). Internet teaching by style: Profiling the online professor. *Educational Technology & Society, 3*(2), 71-85.

Ganeshan, R., Johnson, W. L., Shaw, E., & Wood, B. P. (2000, June 19-23). *Tutoring diagnostic problem solving*. Paper presented at the 5th International Conference, ITS 2000, Montreal, Canada.

Gustavsson, B. (1994, March 21-22). *Technologizing of consciousness—Problems in textualizing organizations*. Paper presented at the Workshop on Writing, Rationality and Organization, Brussels, Belgium.

Hoover, T., & Connor, N. J. (2001). Preferred learning styles of Florida Association for Family and Community Education Volunteers: Implications for professional development. *Extension Journal, 39*(3, June). Electronic Journal. Retrieved from http://www.joe.org/joe/2001june/a3.html

Janvier, W. A., & Ghaoui, C. (2002a, September 26-27). *WISDeM: Communication preference and learning styles in HCI*. Paper presented at the HCT2002 Workshop—Tools for Thought: Communication and Learning Through Digital Technology, Brighton, UK.

Janvier, W. A., & Ghaoui, C. (2002b, November 1-4). *WISDeM—Student profiling using communication preference and learning styles mapping to teaching styles*. Paper presented at the APCHI 2002—5th Asia Pacific Conference on Computer Human Interaction, Beijing, China.

Janvier, W. A., & Ghaoui, C. (2003a, May 18-21). *WISDeM and e-learning system interaction issues.* Paper presented at the 2003 IRMA International Conference, Philadelphia, USA.

Janvier, W. A., & Ghaoui, C. (2003b, September 3-5). *Using communication preference and mapping learning styles to teaching styles in the distance learning intelligent tutoring system—WISDeM.* Paper presented at the Knowledge-Based Intelligent Information and Engineering Systems—7th International Conference—KES 2003, Oxford, UK.

Johnson, C. G., & Jones, G. J. F. (1999). *Effecting affective communication in virtual environments* (No. 389). Department of Computer Science, University of Exeter, UK.

Keefe, J. W. (1987). *Learning style theory and practice.* Natl Assn of Secondary School, Reston, VA.

Larkin-Hein, T., & Budny, D. D. (2000, June 18-21). *Why bother learning about learning styles and psychological types?* Paper presented at the ASEE Annual Conference, St. Louis, MO.

Montgomery, S. M., & Groat, L. N. (1998). *Student learning styles and their implications for teaching.* Retrieved June 2002, from http://www.crlt.umich.edu/occ10.html

Murphy, E., Newman, J., Jolosky, T., & Swank, P. (2002). *What is the Myers-Briggs Type Indicator (MBTI)®.* Retrieved October 2002, from http://www.aptcentral.org/

Murray, T. (1997). Expanding the knowledge acquisition bottleneck for intelligent tutoring systems. *International Journal of Artificial Intelligence in Education, 8*, 222-232.

Murray, T. (1999). Authoring intelligent tutoring systems: An analysis of the state of the art. *International Journal of Artificial Intelligence in Education, 10*, 98-129.

Myers, I. B., & Myers, P. B. (1995). *Gifts differing: Understanding personality type.* Palo Alto, CA: Financial Times Prentice Hall.

Nkambou, R., & Kabanza, F. (2001). *Designing intelligent tutoring systems: A multiagent planning approach, 27,* 46-60. Sigue Outlook.

Pasztor, A. (1997, August 7-10). *Intelligent agents with subjective experience.* Paper presented at the 19th Annual Conference of the Cognitive Science Society, Stanford University, Stanford, CA, USA.

Pasztor, A. (1998a, August 7-10). *Intelligent agents with subjective experience.* Paper presented at the 19th Annual Conference of the Cognitive Science Society, Stanford University, CA.

Pasztor, A. (1998b). Subjective experience divided and conquered, communication and cognition. In E. Myin (Ed.), *Approaching consciousness, Part II* (pp. 73-102). Retrieved May 2002, from http://citeseer.nj.nec.com/pasztor98subjective.html

Sadowski, W., & Stanney, K. (1999). *Measuring and managing presence in virtual environments.* Retrieved January 2002, from http://vehand.engr.ucf.edu/handbook/Chapters/Chapter45.html

Shaw, E., Ganeshan, R., Johnson, W. L., & Millar, D. (2000, July 31-August 6). *Building a case for agent-assisted learning as a catalyst for curriculum reform in medical education.* Paper presented at the SSGRR 2000 Computer & e-Business Conference, International Conference on Advances in Infrastructure for Electronic Business, Science, and Education on the Internet, L'Aquila, Rome, Italy.

Slater, M., Usoh, M., & Steed, A. (1994). Depth of presence in virtual environments - Body centred interaction in immersive virtual environments. *Presence: Teleoperators and Virtual Environments, 3.2*, 130-144.

Thompson, S. K. (1992). *Sampling* (vol. 1). New York: John Wiley & Sons.

Wilson, K., Dugan, S., & Buckle, P. (2002). *Understanding personality functioning without forced choice: Expanding the possibilities for management education based on empirical evidence.* Canada: Haskayne School of Business, University of Calgary.

Yates, D. S., Moore, D. S., & Starnes, D. S. (2002). *The practice of statistics* (2nd ed.). New York: W H Freeman and Company.

KEY TERMS

Communication Preference: The selection of your own way in the art and technique of using words effectively to impart information or ideas.

E-Learning (Online Learning): Learning using electronic media.

Human-Computer Interaction (HCI): The study of how humans interact with computers, and how to design computer systems that are usable, easy, quick, and productive for humans to use.

Learning Styles: The sixteen styles made up out of from four couplets types: Extrovert|Introvert, Sensing|iNtuition, Thinking|Feeling, and Perception|Judgement.

Modality:

- **Auditory:** Use of auditory imagery: hearing, tonality, pitch, melody, volume, and tempo.
- **Kinaesthetic:** Use of emotional, feeling, movement imagery: intensity, temperature.
- **Visual:** Use of visual imagery: sight, colour, brightness, contrast, focus, size, location, and movement.

Neurolinguistic Programming (NLP): The study of the structure of subjective experience and what can be calculated from it.

Neurolinguistic Programming Language Patterns: The use of the words, or similar constructs, "See" for iconic, "Hear" for auditory, and "Feel" for kinaesthetic subjects both in language and text at the relevant times.

A Semantic Learning Objects Authoring Tool

F. J. García
University of Salamanca, Spain

A. J. Berlanga
University of Salamanca, Spain

J. García
University of Salamanca, Spain

INTRODUCTION

The introduction of computers is recreating a new criterion of differentiation between those who become integrated as a matter of course in the technocratic trend deriving from the daily use of these machines and those who become isolated by not using them. This difference increases when computer science and communications merge to introduce virtual education areas, where the conjunction of teacher and student in the space-time dimension is no longer an essential requirement and where the written text becomes replaced (or rather complemented) by the digital text (García & García, 2005).

In order to rescue those educators who have much to offer in an educational system, whether virtual or presential, as authors of teaching resources, suitable authoring tools should be designed, thinking more in the pedagogical process than in the technical aspects.

Hypertext Composer, or simply HyCo, is one of these authoring tools, which presents a pedagogical interaction model that makes easier the creation of educational resources for every teacher/author, independently of his or her computer expertise level. At the same time, HyCo is an authoring tool and a retrieval tool, in that it encapsulates all the complexity in handling current tools within the facilities that the author needs and offers, as a result, a hypermedia teaching product that can be distributed in different formats for the user's access.

HyCo has an important semantic basis that nears this tool to the Semantic Web concept (Berners-Lee, Hendler & Lassila, 2001) and allows creating Semantic Learning Objects (SLO) that could be imported for more specialized Learning Management Systems (LMS). In order to achieve the semantic definition of the created educational resources, HyCo uses Learning Technology Standards or Specifications (LTS), looking for obtaining contents that are able to work in other systems (interoperability), follow-up information about learners and contents (manageability), usability in other contexts (reusability), and avoiding obsolescence (durability).

This article is devoted to introducing HyCo as an authoring/retrieval tool of SLOs, which presents an interaction model that hides all the technical complexity to the authors but, at same time, offers all the power of semantic definitions in order to publish or use the contents in advanced e-learning environments. The rest of the article is organized as follows: the Background section establishes the background of the presented topic, making a comparison with related works; the HyCo Authoring Tool section presents the HyCo authoring tool; finally, the sections Future Trends and Conclusion provide the future trends and the remarks of the article, respectively.

BACKGROUND

There are many different hypermedia authoring tools that could be used in order to produce hypermedia systems for the education domain. Some of them are commercial ones, whereas many others have been developed for educational and research goals. HyCo has no commercial ambitions for now, and we decided to develop our own solution in order to achieve our research goals, which include semantic, adaptive, and collaborative issues; some of them are presented in the actual version, some are in working prototypes, and others will appear in future versions.

First, it is important to say that HyCo inherits properties from the two main trends in hypermedia systems: closed and open hypermedia systems. The former ones store both content and hypermedia structures internally (monolithic systems) or in a database. External application or information cannot participate easily or be included in the hypertext system. These systems produce self-contained hypermedia systems, but they do not support heterogeneity and particularly do not support hypertext distributed over multiple heterogeneous managers, while the open hypermedia systems have the ability to integrate distributed information and the property to store their content outside the hyperbase, especially keeping linking information separate from documents and allowing for more powerful link structures.

HyCo presents a reader mode, in which the hypertext can be navigated within the tool in a self-contained way like in classic authoring systems such as IRIS Intermedia (Yankelovich, Haan, Meyrowitz, & Drucker, 1988) or Storyspace (Bernstein, 1991, 2002). These two systems are significant representatives of the so-called closed hypermedia systems, which store both content and hypermedia structures internally (monolithic systems) or in a database. In addition, HyCo has voice synthesis capabilities in order to make more accessible the developed hypertext system. The differentiation of the author and reader roles in the same authoring tool differs from other systems, which only present authoring capabilities as MS FrontPage (http://www.microsoft.com/frontpage).

About the use of external vs. internals links, HyCo follows a compromise between these approaches by storing links internally but representing them externally. Links are stored inside the educational resource; in this way, users do not have separate link files that could cause wrong opening operations. But HyCo links are represented separately and compactly rather than being spread implicitly throughout the system. This idea is based on the link system of Storyspace v2 (Bernstein, 2002) and Chimera (Anderson, Taylor, & Whitehead, 2000) instead of the embedded link model of HTML.

Related to the semantic characteristics, a similar proposal can be found in HYLOS (Hypermedia Learning Object System) (Feustel & Schmidt, 2001). This system is devoted to creating ELearning Objects (ELOs) instead of HTML pages. In this case, they complete the contents with its metadata to compound an ELO. The used metadata are a subset of the LOM (Learning Object Metadata) (IEEE, 2003) instead of the IMS Metadata (IMS, 2003c) used in HyCo.

HyCo AUTHORING TOOL

HyCo is a powerful authoring tool for educational purposes, which means that an author can create hypermedia educational resources with it. But the same tool also could be used to access created contents in a read-only mode by a student or reader.

HyCo is a multiplatform tool—it does not force the use of one concrete platform. The idea is that if we want teachers to use it, they should work in the context in which they feel good. The actual version of HyCo works in the wider range of operating systems; for this reason, Java 2 Standard Edition technology (Sun, 2004) was chosen as a development base.

The main goal of the authoring tool is the creation of educational contents while trying to achieve an independence of the final publication format. There exists a clear separation between the contents and its presentation. In this way, the educator writes the contents once and reuses them every time he or she needs them. In order to achieve this goal, HyCo tool uses an internal XML-based format (Bray et al., 2004) to store the educational contents of the produced electronic books. Precisely, the HyCo XML-based format allows the introduction of LTSs in this authoring tool; specifically, HyCo supports IMS specifications (IMS, 2003a, 2003b, 2003c) and EML (Educational Modeling Language) (Koper, 2001).

Separating the content and the presentation forces offers authors a way to generate an independent result of the authoring tool. In this way, HyCo has an output gallery that supports HTML, PDF, TXT, RTF, SVG, and PS output formats.

HyCo's user interface has two main facilities that improve its usability. First, this authoring tool has an internationalized interface that actually supports two languages—Spanish and English. Its interface also allows voice synthesis, which permits the generation of an artificial voice that reads the contents to the user. This capability is very interesting in order to make presentations and also to facilitate access to the educational contents to handicapped people (e.g., blind people).

However, more precisely, the HyCo authoring tool comprises three main components: an editor for linear educational resources, which provides an indexed tree structure; an editor for composite semantic learning objects by inserting values for the appropriate metadata elements; and retrieval and management facilities for multimedia information of the learning resources.

Editor for Educational Resources

This editor is based on the content index metaphor that reproduces a hierarchical structure that guides us in our creative process. The following step is to associate contents with each index entry, an index that may vary as the contents take shape, by inserting, eliminating, or changing entries. Each index entry gives rise to a thematic unit, or lexia, that can contain text, multimedia material, and links with other units or documents.

This indexed or tree structure facilitates the authoring of the hypertext, but having only an index as a navigation tool is not acceptable in order to create real pedagogical hypermedia resources where the student may construct his or her own knowledge. The hyperdocuments should be designed in such a way to encourage the readers to see the same text in as many useful contexts as possible. This means placing texts within the contexts of other texts, including different views of the same text (Jones & Spiro, 1992).

For this reason, HyCo allows associating links to the multimedia elements that compose an index entry (i.e., a hypertext node). In this way, the hypertext can be followed by its index-structure, but when a node is selected, the reader may choose navigating by an existing link. Thus, HyCo documents combine both content index and Web-like structures.

The content index metaphor is supported directly in the user interface, which is frame-structured. The left part of the screen shows links to every part of the hypertext structure, and the main frame is the writing/displaying area. The Web metaphor is supported by two buttons that allow creating or modifying the links. The main interface is completed with a toolbox area, which allows inserting, erasing, or renaming the entries of the structure, and with an information area at the lower-right corner, where the characteristics of the selected link (e.g., type of link, name, description) are displayed, as shown in Figure 1.

Editor for Composite Semantic Learning Objects

An SLO is a learning resource that is wrapped with a set of metadata and can be used in the instructional design process. In HyCo, every SLO should be compliant with IMS Metadata (IMS, 2003c). Then, every section of every educational resource or e-book created in HyCo can be converted to an SLO.

To do this, HyCo executes a two-step process, where the first step is an automatic process and the second step is a manual process. In the automatic

Figure 1. HyCo main interface

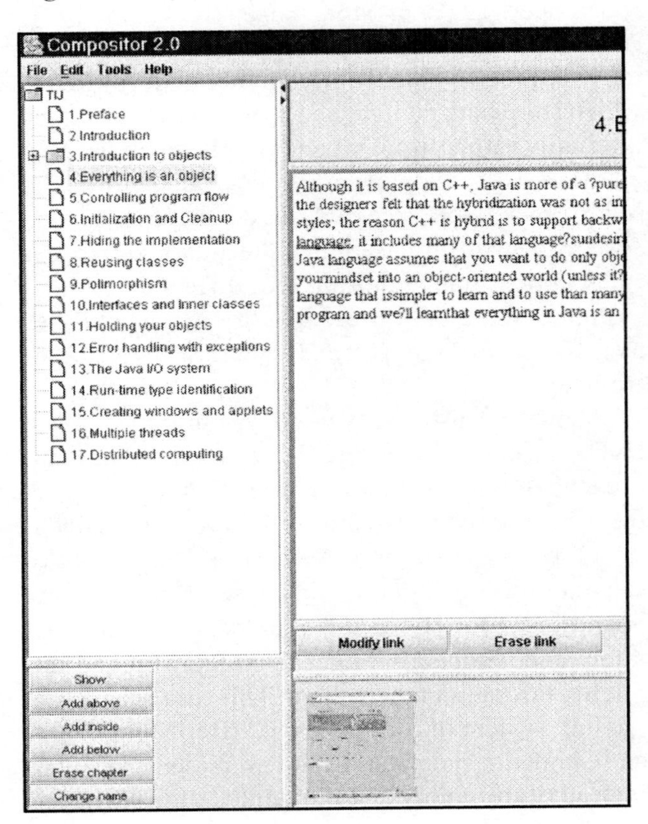

process, HyCo sets all the IMS metadata elements that can be inferred from other data or that are liable to have default values.

Once this process is over, HyCo executes the manual process, where it presents to the user the elements that cannot be generated automatically and/or that require reexamination, modification, or addition (see Figure 2).

When the two-step process is finished, an XML file is generated for each new SLO (each one of them corresponds to each educational resource, section, or subsection) and stored in an IMS metadata SLO repository.

Retrieval and Management Facilities for Multimedia Information

All the elements that the user has to manage and link to the text are organized in information repositories that are so-called galleries. These galleries present very intuitive interfaces to manage the concrete elements, images, or videos, for example. These galleries have thumbnails and descriptions of the elements but also offer simple search engines that allow the user to find the right element in the collection. The gallery metaphor was initially thought to manage the multimedia elements, but the success of this metaphor for users of HyCo has prompted the extension of this concept in order to manage other properties of the document, specifically the styles and the output formats. As an example of the gallery metaphor, Figure 3 shows the sound gallery interface.

FUTURE TRENDS

The educational hypermedia systems have evolved clearly and unstoppably toward the Web. The needed authoring tools for creating them are now more mature; the notions of reusability and interoperability of semantic learning objects are presented in the most important e-learning management systems, but there are many weak points, too. The definition of educational standards, as IMS, is an important advancement for the real reusability and interoperability of learning components, but these standards should be more present in the authoring process. Another

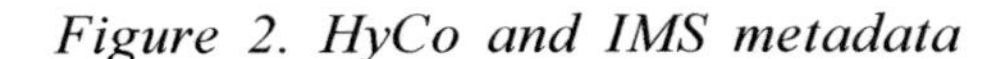

Figure 2. HyCo and IMS metadata

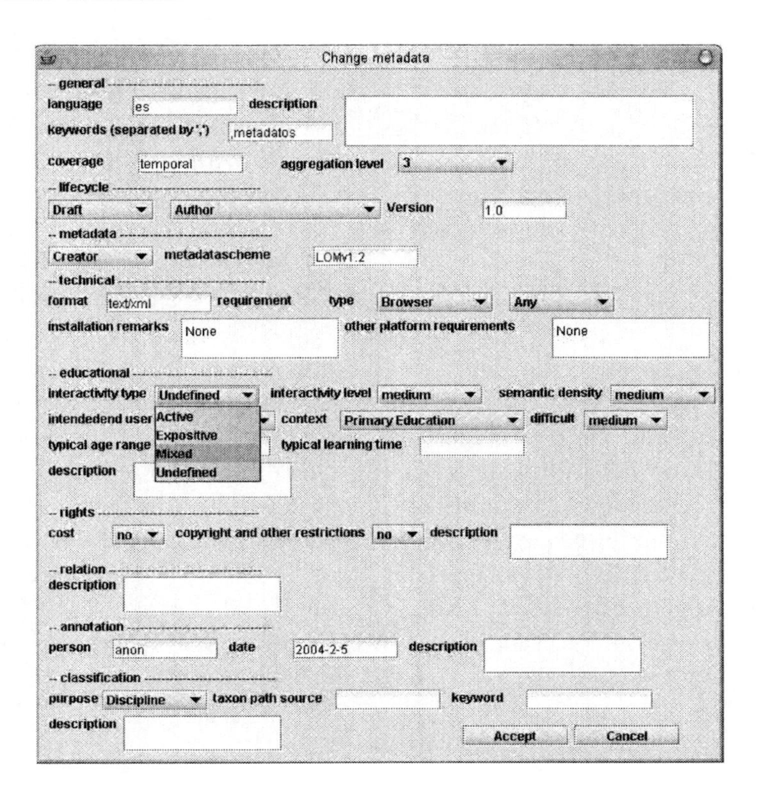

Figure 3. Sound gallery

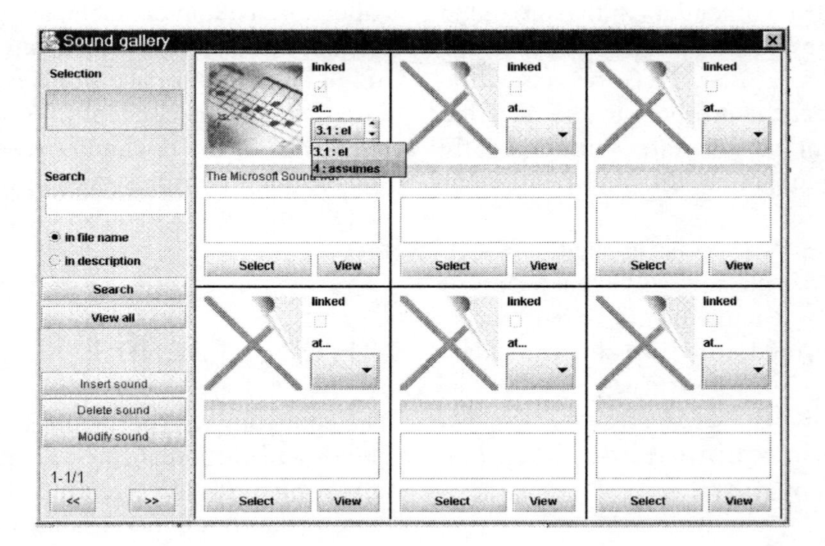

important improvement area is the pedagogical model that authoring tools should support; it is compulsory that the learning components creation process will be guided by correct pedagogical guides. Adaptivity is other important key in success of the educational hypermedia systems; in order to achieve a personalized learning process, the author must have the adequate resources to define the rules that will guide the individualized learning. Finally, collaborative and cooperative authoring capabilities are interesting in all creational processes, and the case of the educational hypermedia systems is not an exception.

The HyCo authoring tool presents characteristics related to the semantic and pedagogical topics introduced previously. The actual research and development efforts are directed to the adaptive and collaborative facilities.

CONCLUSION

In this article, we have introduced HyCo as an authoring tool that allows the definition of both learning resources and learning components, or SLOs (i.e., semantic educational resources based on XML specifications), which could be delivered in diverse LMSs. Specifically, HyCo supports two LTSs, EML, and IMS. The first one was our first attempt in this field, but it was exceeded by IMS.

The success of the HyCo authoring process has been proved with three educational Web-based systems. Two of them are drafts devoted to testing the authoring tool—one about computer history and the other about software engineering. But the third one is a complete electronic book in hypermedia format about cardiovascular surgery that consists of 14 chapters, more than 500 sections, and over 1,000 images. This book is successfully used in lectures of this subject at the University of Salamanca. Recently, we have made an SUS (System Usability Scale) (Brooke, 1996) usability test with the HyCo authoring tool. The obtained results were 83.495 over 100 with a population of 14 users that are not related to the HyCo development team.

ACKNOWLEDGMENTS

We would like to thank the AWEG (Adaptive Web Engineering Group) of the University of Salamanca for its ideas and support to the elaboration of this article.

This study was partly financed by the Regional Government of Castile and Leon through research project SA017/02. Also, it is supported by the European Union Project ODISEAME (Open distance Interuniversity Synergies between Europe, Africa and Middle East), ref. EUMEDIS B7-4100/2000/2165-79 P546.

REFERENCES

Anderson, K. M., Taylor, R. N., & Whitehead, J. (2000). Chimera: Hypermedia for heterogeneous software development environments. *ACM Transactions on Information Systems, 18*(3), 211-245.

Berners-Lee, T., Hendler, J., & Lassila, O. (2001). The semantic Web. *Scientific American.*

Bernstein, M. (1991). Storyspace: Hypertext and the Process of Writing. In E. Berk, & J. Devlin (Eds.), *Hypertext/hypermedia handbook* (pp. 529-533). New York: McGraw-Hill.

Bernstein, M. (2002). Storyspace 1. *Proceedings of the Thirteenth ACM Conference on Hypertext and Hypermedia.*

Bray, T., Paoli, J., Sperberg-MacQueen, C.M., Maler, E., & Yergeau, F. (Eds.). (2004). *Extensible markup language (XML) 1.0* (3rd edition). Retrieved from http://www.w3.org/TR/2004/REC-xml-20040204

Brooke, J. (1996). SUS: A quick and dirty usability scale. In P. W. Jordan, B. Thomas, B. A. Weerdmeester, & I. L. McClelland (Eds.), *Usability evaluation in industry* (pp. 189-194). London: Taylor & Francis.

Feustel, B., & Schmidt, T.C. (2001). Media objects in time—A multimedia streaming system. *Computer Networks, 37*(6), 729-737.

García, F. J., & García, J. (2004). Educational hypermedia resources facilitator. *Computers & Education, 44*(3), 301-325. In press.

IEEE. (2003). *LOM final draft v6.4.* Retrieved from http://ltsc.ieee.org/wg12/index.html

IMS. (2003a). *IMS content packaging.* Retrieved from http://www.imsglobal.org/content/packaging/index.cfm

IMS. (2003b). *IMS learning design specification.* Retrieved from http://www.imsglobal.org/learningdesign/index.cfm

IMS. (2003c). *IMS learning resource metadata specification.* Retrieved from http://www.imsglobal.org/metadata/index.cfm

Jones, R.A., & Spiro, R. (1992). Imagined conversations: The relevance of hypertext, pragmatism, and cognitive flexibility theory to the interpretation of "classic texts" in intellectual history. *Proceedings of the 4th ACM Conference on Hypertext—ECHT'92.*

Koper, R. (2001). *Modelling units of study from a pedagogical perspective: The pedagogical metamodel behind EML.* Retrieved from http://eml.ou.nl/introduction/docs/ped-metamodel.pdf

Sun Microsystems. (2004). *The Java tutorial: A practical guide for programmers.* Retrieved from http://java.sun.com/docs/books/tutorial/index.html

Yankelovich, N., Haan, B. J., Meyrowitz, N. K., & Drucker, S. M. (1988). Intermedia: The concept and the construction of a seamless information environment. *IEEE Computer, 21*(1), 81-96.

KEY TERMS

Educational Modeling Language (EML): Developed by the Open University of the Netherlands, since 1998, EML is a notational method for e-learning environments based on a pedagogical metamodel that considers that didactic design plays a main role.

Hypermedia: The style of building systems for information representation and management around the network of multimedia nodes connected together by typed links.

Hypermedia Authoring Tools: Authoring tools for hypermedia systems are meant to provide environments where authors may create their own hypermedia systems in varying domains.

Hypertext: A body of written or pictorial material interconnected in such a complex way that it could not conveniently be presented or represented on paper (Theodor H. Nelson).

IMS Specifications: IMS was born in 1997 as a project of the National Learning Infrastructure Initiative at Educause. In 2000, it became a non-profit organization. Its mission is to promote distributed learning environments.

Learning Technology Standards or Specifications (LTS): Agreements about the characteristics that a learning element should have in order to be compatible, interchangeable, and interoperable into other learning systems. The use of standards ensures instructional technologies' interoperability and their learning objects for universities and corporations around the globe. Examples of LTS are IMS, EML (Educational Modeling Language), and LTSC IEEE LOM (Learning Technology Standard Committee of the IEEE—Learning Object Metadata).

Metadata: Information about data or other information.

Metaphor: An understandable mental image of real objects. The knowledge and the relationships among elements in a known domain are translated to a non-familiar domain.

Semantic Learning Object: A learning resource that is wrapped with a set of standardized metadata and can be used in the instructional design process.

Semantic Web: A Web that includes documents or portions of documents describing explicit relationships among things and containing semantic information intended for automatic processing by our machines.

Voice Synthesis: The process that allows the transformation of the text to sound.

Sense of Presence

Corina Sas
Lancaster University, UK

INTRODUCTION

Sense of presence is one of the most interesting phenomena that enriches users' experiences of interacting with any type of system. It allows users to be there (Schloerb & Sheridan, 1995) and to perceive the virtual world as another world in which they really exist.

The interest in presence phenomenon is not novel (Gerrig, 1993), but it has grown lately due to the advent of virtual reality (VR) technology. The specific characteristics of virtual environments (VEs) transform them into suitable experimental testbeds for studies in various research areas. This also resuscitated the interest in presence, and much work has focused on the development of a theoretical body of knowledge and on a whole set of experimental studies aimed at understanding, explaining, measuring, or predicting presence. All of these efforts have been made to increase the understanding of how presence can be manipulated within the VEs, particularly within the application areas where presence potential has been acknowledged.

Probably one of the most important reasons motivating presence research is the relationship it holds with task performance. This debatable relationship together with the more obvious one between presence and user satisfaction suggest that presence may play an important role in the perceived system usability.

Since presence may act as a catalyst for the learning potential of VEs, it can be harnessed for the training and transfer of skills (Mantovani & Castelnuovo, 1998; Schank, 1997). The potential of presence to increase the pervasive power of the delivered content motivates research on presence impact on e-marketing and advertising (Grigorovici, 2003). Another promising application area for presence research is within the realm of cognitive therapy of phobias (Strickland et al., 1997).

The highly subjective nature of presence continues to challenge researchers to find appropriate methodologies and instruments for measuring it. This is reflected in the ongoing theoretical work of conceptualizing a sense of presence. The difficulties related to investigating presence led to a large set of definitions and measuring tools.

The purpose of this article is to introduce the concept of presence. The first section offers some conceptual delimitations related to presence construct. The second section describes its main determinants along two dimensions (i.e., technological factors and human factors). The third section addresses the challenges of measuring presence, offering also an overview of the main methods, tools, and instruments developed for assessing it. The fourth section presents the complex relationship between presence and task performance.

BACKGROUND

Attempts to define presence have been numerous, and the lack of a unanimously accepted definition suggests the multi-dimensional nature of this construct and its not yet mature understanding.

Presence has been described as a sense of being physically present at the remote site (Schloerb & Sheridan, 1995; Sheridan, 1992), a basic state of consciousness consisting of the attribution of sensation to some distal stimuli (Loomis, 1992), a suspension of disbelief experienced by users while being in a remote world and not the physical one (Slater & Usoh, 1993), or the perceptual illusion of non-mediation (Lombard & Ditton, 1997). After analyzing various presence definitions, we proposed the following one (Sas & O'Hare, 2001, 2003):

Presence is a psychological phenomenon, through which one's cognitive processes are oriented toward another world, either technologically mediated or imaginary, to such an extent that he or she experiences mentally the state of being (there), similar to one in the physical reality, together with an imperceptible shifting of focus of consciousness to the proximal stimulus located in that other world.

Any attempt to conceptualize a construct also should consider its discriminant validity by contrasting it with other close concepts in the field. Furthermore, three other constructs—telepresence, immersion, and flow—are introduced, and their relationships with presence are outlined briefly.

The term of telepresence was coined by Marvin Minsky (1980), emphasizes the meaning of mediation, and denotes a sense of being physically present at a remote world. Draper, et al. (1998) defined it as the perception of presence within a remote environment. This concept precedes and is closely related to the presence construct. Despite often being taken as synonyms, there is, however, a subtle difference between presence and telepresence, rooted in the proximity to the site where one perceives, acts, and ultimately experiences presence.

Another distinction often mentioned in presence literature is that between presence and immersion. Immersion is usually associated with technological factors referring to the extent to which computer generated worlds are extensive (able to accommodate a large set of sensory systems), surrounding (able to provide information from any virtual direction), inclusive (able to shut out all information from the physical world), vivid (able to provide rich information content, resolution, and display quality), and matching (able to accurately reproduce the body movements previously tracked) (Slater et al., 1995, 1996). In contrast, presence relates more to user characteristics, whose impact is unfortunately less explored.

The last useful distinction is the one between presence and flow, defined as a state of optimal experience that occurs when people attempt tasks that challenge their skills (Csikszentmihalyi, 1990). Flow assumes a match between the task difficulty and one's abilities, highly focused attention that leads to enjoyment, feeling of control, and an altered perception of time. From this, several distinctions emerge with respect to both the experience itself and its results. The experience in the case of flow, as opposed to presence, always requires intense concentration and focus of attention, a sense of control, and usually an intense and active participation in the task, usually perceived more narrowly through only some of its characteristics (Fontaine, 1992). With respect to the results, since presence is not an optimal experience, it does not necessarily lead to pleasant and fulfilling experience. However, it is possible that during flow, someone will experience a strong sense of presence, but the latter also can occur outside the flow (Heeter, 2003).

Despite the diversity characterizing the definitions proposed for capturing the presence construct, there seems to be a common ground shared by researchers in the presence field, which refers to presence determinants.

PRESENCE DETERMINANTS

Several presence theories have been developed in the attempt to extend the understanding of presence. Draper (1998) identified a first group consisting of psychological models of presence and a second one consisting of technological models of presence. The first class of theories includes telepresence as flow experience developed by Csikszentmihalyi (1990), behavioral cybernetics theory (Smith & Smith, 1985), and a structured attentional resource model for teleoperation (Schloerb & Sheridan, 1995). The second class of theories groups different models, such as those elaborated by Sheridan (1992), Steuer (1992), Schloerb (1995), Zeltzer (1992), Witmer and Singer (1998), and Slater and Usoh (1993).

The factors affecting presence can be grouped into technological factors that consider the system and its characteristics, and human factors referring to users' cognitive and personality aspects (Lombard & Ditton, 1997; Lessiter et al., 2000).

Technological Factors

A large amount of work has been carried out in the area of technological factors affecting presence. Lombard and Ditton (1997) provided a detailed account of this. Some of these factors are visual display characteristics such as image quality; image size; viewing distance; visual angle; motion; color; dimensionality; camera techniques; and aural presentation characteristics such as frequency range, dynamic range, signal to noise ratio, and high quality audio. As stimuli for other senses, Lombard and

Ditton (1997) referred to olfactory output, body movement, tactile stimuli, and force feedback.

Media and user characteristics often were mentioned as having a particular impact on the level of sense of presence experienced by the users. However, there is little empirical research supporting this (Lessiter et al., 2000).

Human Factors

Psotka and Davison (1993) considered two categories of factors determinant of immersion, such as susceptibility to immersion and quality of immersion. The first set refers to user characteristics with an emphasis on cognitive aspects such as imagination, vivid imagery, concentration, attention, and self-control, while the second set is concerned primarily with technological factors like affordances of VR, distractions from the real world, and physiological effects.

Kaber, Draper, and Usher (2002) summarized user characteristics that seem to impact presence experienced within VEs. Broadly categorized in immersive tendencies and attention, these factors are suggestibility of immersion, tendency to daydream, becoming lost in novels, concentration, and robustness to distracting events.

Other personality factors impacting on presence are empathy, absorption, creative imagination, personality, cognitive style, and willingness to be transported in the VE (Heeter, 1992; Lombard & Ditton, 1997; Sas & O'Hare, 2001, 2003; Sas et al., 2003).

Conceptualizing presence is the initial stage of understanding this construct. It has been followed by the attempts of measuring presence. Different methods and measurement instruments have been proposed for offering quantitative indicators of the degree of presence that one can experience.

MEASURING PRESENCE

Despite its significance, measuring presence raises significant challenges, primarily related to the nature of presence. Presence is a psychological phenomenon, subjectively experienced inside the inner world of one's consciousness. Therefore, capturing and analyzing it requires a certain degree of introspection, together with one's understanding of what presence means. In addition, presence is a state or a transient psychological condition that is context-dependent and that, accordingly, could vary within the same individual during an experiment.

Therefore, participants could encounter difficulties in assessing their level of presence after the task has been completed and the experiment has ended. Even more difficult is measuring presence during the experiment. This involves asking somebody to be permanently aware of each change occurring in his or her level of presence. Such a requirement adds itself to those involved in the execution of the task, therefore inducing cognitive overload. This either could prevent the subjects from experiencing presence or could affect the task performance. Either case impacts on the measurement validity.

Another difficulty in measuring presence is related to the complexity and multi-dimensionality of this construct (Lombard, 2003), which is reflected in the different definitions and theories trying to explain presence. In addition, presence research seems to be an interdisciplinary field that benefits from inputs from various disciplines, such as psychology, philosophy, computer science, media studies or drama studies to enumerate the most important ones. These multiple perspectives provide valuable insights into understanding presence, but at the same time they come at a cost. A fully articulated and accepted theory of presence requires a common understanding of presence.

Lombard (2003) identified two general approaches to measuring presence: subjective measurements and objective measurements. Subjective measurements usually consist of self-rating questionnaires that require participants to evaluate the experienced level of presence. The main advantage of the subjective measurements consists of their accessibility. They also come at a low cost, and very important, appear to be valid and reliable measures (Lombard, 2003; Prothero et al., 1995). Such questionnaires have been developed by Lessiter et al. (2000), Lombard (2000), Schubert (1999), Witmer (1998), and Slater et al. (2000).

Limitations of this approach are related mainly to the inner and versatile nature of presence and to the level of introspection that participants are assumed to be able to achieve. Such types of information could be elicited post-experiment or during the experiment.

In order to overcome some of the limitations related to subjective measures of presence, another approach started to emerge. At the core of objective measurements lies the hypothesis that, while users experience presence, a series of physiological and behavioral modifications occurred in their bodies. The particular physiological modifications that were considered to reflect presence were skin conductance, blood pressure, heart rate, muscle tension, respiration, eye movement, posture, and so forth (Lombard, 2003).

These measures involve the recording of such modifications in real time and present the considerable advantage of being unobtrusive. They also can be carried out without requiring subjects' involvements in these measurements. The objective measurements have their own limitations, such as high cost and difficulties in administrating them. However, their main drawback concerns the limited evidences of the fact that physiological modifications correlate with presence (Prothero et al., 1995).

Another aspect of major interest regarding presence is its relationship to task performance. The significance of this relationship justifies the efforts invested in defining and measuring presence. At the same time, this issue has generated serious theoretical treatments and empirical investigations.

PRESENCE AND TASK PERFORMANCE

The existence of a relationship between presence and task performance is arguable and has given rise to a long-standing debate in the presence research area. More empirical studies are required in order to refute or support this dependency. Theoretical work and empirical studies have highlighted two possible research positions. The first position states that presence is merely an epiphenomenon (Ellis, 1996; Welch et al., 1996), and consequently, its impact upon task performance is limited. According to this position, the role of presence consists only of affectively coloring the user's experience. The second position argues that presence impacts on the performance of tasks carried out within the VEs. There are two perspectives on this position.

The first one views it as a mediated relationship. In other words, presence and task performance could be related, in fact, to a third extraneous variable or set of variables (Slater et al., 1996; Stanney et al., 1998) that impacts both presence and task performance. These extraneous variables were considered to be related to the technological aspects of VEs, such as improved VEs (Stanney et al., 1998) or immersion (Slater et al., 1996).

The second and probably most important explanation of this dependency between presence and task performance argues for a causal relationship (Sadowski & Stanney, 2002). This perspective has fueled most of the research in the field. However, the issue of causal relationship presents a twofold problem. First, it is a challenge to design an experiment for highlighting the causal relationship, and this relationship, if it exists, would seem to be highly task-dependent (Slater et al., 1996; Stanney, 1995).

The significance of the content being delivered through any mediated experience has been related to the nature of activity or tasks in which the user participates, which, in turn, seems to impact presence (Lombard & Ditton, 1997). Heeter (1992) distinguished between two potential groups of tasks that could impact on presence differently and are related to two fundamental types of activity: learning and playing. Particularly in the case of tasks involving a ludic component, the sense of presence is likely correlated with enjoyment, which, in turn, is likely correlated with task performance (Barfield et al., 1995). Tasks or activities that involve ambiguous verbal and nonverbal social cues and sensitive personal information better exploits the medium's potential to offer presence than do simple nonpersonal tasks (Lombard & Ditton, 1997). Correlations between performance improvement and presence appear to be positive. However, they are usually weak, since less than 10% of variance in the performance seems to account for perceived presence (Snow, 1996).

Despite this limitation, the causal relationship of presence and task performance has increased face validity based on the perceptual and cognitive psychology of skills transfer (Stanney et al., 1998). In this light, an additional benefit of understanding this relationship consists of the transfer of skills from the VE to the real world. Slater et al. (1996) considered presence merely as a facilitator whose main contribution consists of enabling the user to perform naturally in a way similar to the real world or, in other words, inducing one's natural reactions.

FUTURE TRENDS

The article highlights the uneven interest manifested in this research area; that particularly favors technological factors. For this, it advocates a shift of interest that would motivate studies focusing primarily on user characteristics (e.g., personality or cognitive factors rather than bodily-related aspects). Indeed, it appears that almost half of the variance in sense of presence is covered by personality factors (Sas & O'Hare, 2003).

The efforts invested for bridging this gap could be efficiently exploited for the development of hybrid theories. Such theories can provide a comprehensive explanation of presence by focusing simultaneously on technological and human factors and on the relationship between them.

CONCLUSION

This article introduces the presence construct, offering at the same time a review of presence determinants. Presence determinants are organized along two fundamental groups such as technological and human factors. However, these two groups of factors impactin on presence and, taken as a basis for grouping presence theories, should be seen on a continuum rather than as a dichotomy. Both human and technological factors should be seen as part of a wider equation whose addressing increases the potential of understanding and possibly manipulating presence.

The inner nature of this phenomenon poses a series of serious problems for investigating it at both theoretical and empirical levels. The article outlines the methods and instruments developed for assessing presence, with an emphasis on the challenges and difficulties of measuring.

Apart from the application areas that harnessed its potential, the interest in presence also is supported by the frequency of this phenomenon, the relationship it holds with task performance, and its likelihood to effectively color a user's experience, which, in turn, can contribute to increased satisfaction. The latter two aspects (performance and satisfaction) suggest the impact that presence may have on the perceived usability of a system.

REFERENCES

Barfield, W., Zeltzer, D., Sheridan, T., & Slater, M. (1995). Presence and performance with virtual environments. In W. Barfield, & T. Furness (Eds.), *Virtual environments and advanced interface design* (pp. 473-513). New York: Oxford University Press.

Csikszentmihalyi, M. (1990). *The psychology of optimal experience.* New York: Harper and Row.

Draper, J., Kaber, D., & Usher, J. (1998). Telepresence. *Human Factors, 40*(3), 354-375.

Ellis, S. (1996). Presence of mind: A reaction to Thomas Sheridan's "Further musings on the psychophysics of presence." *Presence: Teleoperators and Virtual Environments, 5*(2), 247-259.

Fontaine, G. (1992). The experience of a sense of presence in intercultural and international encounters. *Presence: Teleoperators and Virtual Environments, 1*(4), 482-490.

Gerrig, R. (1993). *Experiencing narrative worlds: On the psychological activities of reading.* New Haven, CT: Yale University Press.

Grigorovici, D. (2003). Pervasive effects of presence in immersive virtual environments. In G. Riva, F. Davide, & W. IJsselsteijn (Eds.), *Being there: Concepts, effects and measurement of user presence in synthetic environments* (pp. 191-206). Amsterdam, The Netherlands: Ios Press.

Heeter, C. (1992). Being there: The subjective experience of presence. *Presence: Teleoperators and Virtual Environments, 1*(2), 262-271.

Heeter, C. (2003). Reflections in real presence by a virtual person. *Presence: Teleoperators and Virtual Environments, 12*(4), 335-345.

Kaber, D., Draper, J., & Usher, J. (2002). Influence of individual differences on virtual reality application design for individual and collaborative immersive virtual environments. In K. Stanney (Ed.), *The handbook of virtual environments: Design, implementation and applications* (pp. 379-402). Mahwah, NJ: Lawrence Erlbaum & Associates.

Lessiter, J., Keogh, J., & Davidoff, J. (2000). A cross-media presence questionnaire: The ITC sense of presence inventory. *Presence: Teleoperators and Virtual Environments*, *10*(3), 282-297.

Lombard, M. (2003). *Resources for the study of presence: How do we measure presence?* Retrieved September 28, 2003, from www.temple.edu/ispr/measure.htm

Lombard, M., & Ditton, T. (1997). At the heart of it all: The concept of presence. *Journal of Computer—Mediated Communication*, *3*(2). Deft, The Netherlands.

Lombard, M., et al. (2000). Measuring presence: A literature-based approach to the development of a standardized paper-and-pencil instrument. *Proceedings of the Third International Workshop on Presence*. Retrieved from http://astro.temple.edu/~lombard/research/p2_P2000.html

Loomis, J. (1992). Distal attribution and presence. *Presence: Teleoperators and Virtual Environments*, *1*(1), 113-118.

Mantovani, F., & Castelnuovo, G. (1998). Sense of presence in virtual training: Enhance skills acquisition and transfer of knowledge through learning experience in virtual environments. In G. Riva, B. Wiederhold, & E. Molinari (Eds.), *Virtual environments in clinical psychology and neuroscience: Methods and techniques in advanced patient-therapist interaction* (pp. 167-181). Amsterdam, The Netherlands: IOS Press.

Minsky, M. (1980). Telepresence. *Omni*, *2*, 44-52.

Prothero, J., Parker, D., Furness, T., & Wells, M. (1995). Towards a robust, quantitative measure for presence. *Proceedings of the Conference on Experimental Analysis and Measurement of Situation Awareness*. Daytona Beach, Florida, (pp. 359-366).

Psotka, J., & Davison, S. (1993). Cognitive factors associated with immersion in virtual environments. *Proceedings of the Conference on Intelligent Computer-Aided Training and Virtual Environment Technology*.

Sadowski, W., & Stanney, K. (2002). Measuring and managing presence in virtual environments. In K. Stanney (Ed.), *Handbook of virtual environments: Design, implementation, and applications*. Lawrence Erlbaum Associates.

Sas, C., & O'Hare, G. (2001). The presence equation: An investigation into cognitive factors underlying presence within non-immersive virtual environments. *Proceedings of the Fourth International Workshop on Presence,* Philadelphia.

Sas, C., & O'Hare, G. (2003). Presence equation: An investigation into cognitive factor underlying presence. *Presence: Teleoperators and Virtual Environments*, *12*(5), 523-537.

Sas, C., O'Hare, G., & Reilly, R. (2004). Presence and task performance: An approach in the light of cognitive style. *International Journal of Cognition Technology and Work, 6*(1), 53-56.

Schank, R. (1997). *Virtual learning: A revolutionary approach to building a highly skilled workforce*. New York: McGraw-Hill.

Schloerb, D., & Sheridan, T. (1995). Experimental investigation of the relationship between subjective telepresence and performance in hand-eye tasks. In *Telemanipulator and telepresence technologies* (pp. 62-73). Bellingham, WA: Society of Photo-Optical Instrumentation Engineers.

Schubert, T., Friedmann, F., & Regenbrecht, H. (1999). Decomposing the sense of presence: Factor analytic insights. *Proceedings of the Second International Workshop on Presence*.

Sheridan, T. (1992). Musings on telepresence and virtual presence. *Presence: Teleoperators and Virtual Environments*, *1*(1), 120-126.

Slater, M., Linakis, V., Usoh, M., & Kooper, R. (1996). Immersion, presence and performance in virtual environments: An experiment with tri-dimensional chess. In M. Green (Ed.), *Virtual reality software and technology* (pp. 163-172). Hong Kong: ACM Press.

Slater, M., & Steed, A. (2000). A virtual presence counter. *Presence: Teleoperators and Virtual Environments*, *9*(5), 413-434.

Slater, M., & Usoh, M. (1993). Representations systems, perceptual position, and presence in immersive virtual environments. *Presence: Teleoperators and Virtual Environments*, *2*(3), 221-233.

Slater, M., Usoh, M., & Steed, A. (1995). Taking steps: The influence of a walking metaphor on presence in virtual reality. *ACM Transactions on Computer-Human Interaction (TOCHI)*, *2*(3), 201-219.

Smith, T., & Smith, K. (1985). Cybernetic factors in motor performance and development. In D. Goodman, R. Wilberg, & I. Franks (Eds.), *Differing perspectives in motor learning, memory, and control* (pp. 239-283). Amsterdam: North-Holland.

Snow, M. (1996). *Charting presence in virtual environments and its effects on performance* [doctoral thesis]. Virginia Polytechnic Institute and State University, Blacksburg, VA.

Stanney, K. (1995). Realizing the full potential of virtual reality: Human factors issues that could stand in the way. *Proceedings of IEEE Virtual Reality Annual International Symposium*.

Stanney, K., et al. (1998). Aftereffects and sense of presence in virtual environments: Formulation of a research and development agenda. *International Journal of Human-Computer Interaction*, *10*(2), 135-187.

Steuer, J. (1992). Defining virtual reality: Dimensions determining telepresence. *Journal of Communication*, *4*(2), 73-93.

Strickland, D., Hodges, L., North, M., & Weghorst, S. (1997). Overcoming phobias by virtual exposure. *Communications of the ACM*, *40*(8), 34-39.

Welch, R., Blackmon, T., Liu, A., Mellers, B., & Stark, L. (1996). The effects of pictorial realism, delay of visual feedback, and observer interactivity on the subjective sense of presence. *Presence: Teleoperators and Virtual Environments*, *5*(3), 263-273.

Witmer, B., & Singer, M. (1998). Measuring presence in virtual environments: A presence questionnaire. *Presence: Teleoperators and Virtual Environments*, *7*(3), 225-240.

Zeltzer, D. (1992). Autonomy, interaction and presence. *Presence: Teleoperators and Virtual Environments*, *1*(1), 127-132.

KEY TERMS

Flow: A psychological state experienced when there is a match between task requirements and user's skills, a state that involves high attention and leads to feelings of control and enjoyment.

Human Factors: User characteristics in terms of personality and cognitive factors that impact the task performance and the quality of interaction with any artifact.

Immersion: A quality of a system, usually computer-generated world consisting of a set of technological factors that enable users to experience the virtual world vividly and exclusively.

Presence: A psychological phenomenon enabling the mental state of being there in either technologically mediated or imaginary spaces.

Task Performance: The proficiency of accomplishing a task that allows discriminating the users (i.e., experts, novices).

Technological Factors: Aspects characterizing a technical system (i.e., computer-generated world) and its components that impact the quality of interaction and task performance.

Telepresence: A psychological phenomenon of being mentally present at a technologically mediated remote world.

Site Maps for Hypertext

Amy M. Shapiro
University of Massachusetts Dartmouth, USA

INTRODUCTION

Hypertexts are electronic presentations of information comprised of any number of documents connected by electronic links that allow users to move between them with a mouse click. In addition to text, the documents also may contain pictures, videos, demonstrations, or sound resources. With the addition of such media, hypertext often is referred to as hypermedia. A hypertext can present information contained in a college course or the products offered by a cleaning supply company. A hypertext can contain as little as two documents or as much as the holdings of an entire library. Because hypertexts can be quite large, site maps often are used to provide users with an overview of a site's content and structure. While they may appear as simple tables of content, they also can provide a graphical representation of the site's documents and even the network of links connecting them. Regardless of the form a site map takes, it may appear as a simple overview or, more commonly, as an interactive tool in which each entry serves as a link to the page it represents. Site maps may appear on a hypertext homepage or on a separate page, often as a help menu option.

BACKGROUND

Indexes and tables of contents, the precursors to site maps, have been in use for hundreds of years. Well before anyone envisioned a technology as remarkable as hypertext, readers were using tables of contents and indexes to glean summaries of printed texts and to find specific pieces of information. Modern psychologists and educational theorists became interested in such devices as educational tools and have published a number of formal studies on that topic. The overriding conclusion drawn by that body of research is that outlines, indexes, and tables of content, called advance organizers in the psychology literature, can augment what is learned from traditional text (Glover & Krug, 1988; Kraiger, Salas, & Cannon-Bowers, 1995; Snapp & Glover, 1990; Townsend & Clarihew, 1989). One reason advance organizers work is that they cue the reader to access existing memories that may help to organize or anchor new information from the text (Mayer, 1979).

It also has been shown, though, that advance organizers augment learning on the part of domain novices, who have little stored knowledge from which to draw. In this situation, advance organizers appear to work, because they provide a structure in which to organize new information (Townsend & Clarihew, 1989). For example, Mannes, and Kintsch (1987) gave novice learners a text accompanied by an advance organizer that either mirrored the structure of the text or provided a different organization. When tested for both recall and recognition of the text content, those in the compatible condition outperformed those in the mismatched condition. While these results indicate that the compatible advance organizer provided a structure in which the text content was more easily stored, it should be noted that those in the inconsistent condition performed better on a problem-solving posttest. The authors argue that the additional work required to create an organized understanding of the text in the face of an inconsistent organizer may have resulted in deeper understanding, which is reflected in the problem-solving task.

With such evidence pointing to the educational benefits of advance organizers, psychologists naturally became interested in the potential of their electronic cousins—site maps—to promote hypertext-based learning. Much of the work on advance organizers has transferred well to learning with site maps. Additionally, site maps have been used to remedy the problem of getting lost in the information space, a problem not generally encountered with traditional text. The following sections summarize what is known about the use of site maps for staying oriented and for augmenting learning.

Site Maps for Staying Oriented and Finding Information

When working in a large hypertext, it is not uncommon for users to find themselves lost or disoriented. When this happens, users are pulled away from their primary task, whether that be searching for a specific piece of information or learning the global content. This experience commonly is referred to as being lost in hyperspace. The danger posed by that cognitive state is that users become so focused on finding their way through the system that they are unable to achieve their intended goals. Site maps keep users oriented by providing them with a view of the system contents as a whole. The effect is similar to the familiar you-are-here maps often available at museums or large shopping malls. It is well accepted that site maps are effective for remedying the experience of getting lost and allowing users to find more quickly and more easily their way and return to their primary goals (Hammond & Allinson, 1989; Monesson, 2002). Indeed, in a review of studies exploring the effectiveness of educational hypertext, Chen and Rada (1996) conclude that site maps "appear to be necessary for users dealing with large and complex information structures and to be useful to resolve the problems of disorientation and high cognitive overhead" (p. 149). The more accurately the site map represents the hypertext structure and content, the more useful it will be in orienting lost explorers.

Site maps also serve a similar purpose to a traditional table of contents in that they inform users about the topics represented on the site. Another purpose of site maps results as a consequence of those already described here. Specifically, once users are aware of where they are and what information is contained in the hypertext, a site map can help users to find their way to a desired location on the site. That is, they can enable users in planning a best route to a given page. Indeed, site maps have been shown to alter learners' search performances and browsing behaviors (Chou, Lin, & Sun, 2000; McDonald & Stevenson, 1999; Monesson, 2002; Puntambekar, Stylianou, & Hübscher, 2003).

Site Maps for Learning

While the ability of site maps to orient users is fairly clear, their effectiveness as learning tools is less certain. Specifically, the research on learning outcomes using site maps paints a picture of a tool that, at first glance, appears unpredictable. Some studies conclude that there is no educational benefit of using site maps. For example, Wenger, and Payne (1994) found that learners using site maps increased the amount of a hypertext they visited but observed no accompanying increase in learning outcomes. Others, such as Neiderhauser, Reynolds, Salmen, and Skolmoski (2000) initially found some effect of site maps on learning, only to determine through regression analyses that the impact was minimal. Likewise, Nilsson, and Mayer (2002) have shown that the effect of a map can be contingent on user characteristics such as spatial ability.

Other studies, however, have found more significant educational benefit from site map use but that benefit only has been observed for domain novices (Potelle & Rouet, 2003; Puntambekar, Stylianou, & Hübscher, 2003). For example, Potelle, and Rouet (2003) provided novice and more advanced learners with a hypertext accompanied by one of three site maps, either a hierarchy, a graphical network representing the system's nodes and links, or an alphabetical index. No differences in the advanced learners' performances were detected among site map conditions. The novices, however, performed best on learning posttests when they used the hierarchical map. The authors concluded that the novices benefited from the clear structure and transparent organization of the site map. It may be the case that the site map's structure allowed the novices to create a similarly organized mental model for the material that enhanced their understanding. Since domain experts are understood to possess well organized knowledge structures (Chase & Simon, 1973; Chi & Koeske, 1983; West & Pines, 1985), it is understandable that the more advanced learners did not benefit from the hierarchical site map. These learners were less likely to require the outside organizer.

Shapiro (1998) also found that prior knowledge is a mediating factor in determining whether users benefit from site maps. She presented learners with a hypertext about animals and their ecosystems. While the hypertext was identical between conditions, subjects were assigned to hierarchical site map conditions that represented the system as structured either by ecosystems or by animal families. All subjects were pretested for their knowledge of the topic to ensure that they had moderate to high knowledge of animal families but poor knowledge of how ecosystems function. All subjects performed equivalently on the animal families posttest, regardless of which site map they used. Those who used the ecosystems site map, however, outperformed their counterparts on the ecosystems posttest questions. As in Potelle and Rouet (2003), then, these results indicate that site maps benefit learners primarily when they lack sufficient prior knowledge about a topic.

Prior knowledge is not the only variable that determines the educational effectiveness of site maps. Their compatibility with learners' goals is also important. This point is demonstrated in a series of studies by Dee-Lucas and Larkin (1995). When subjects in their first study were given no specific learning goal, all learned equivalently regardless of whether they read a hypertext equipped with a hierarchical site map or an alphabetical index (a third control group read the same information as traditional text but performed less well than both hypertext groups). When subjects in a second experiment were given the explicit goal of summarizing the hypertext's information, those exposed to the hierarchical site map outperformed the other subject groups. The act of summarizing a hypertext requires that one understand the relationships between ideas and the content of each page. Because hierarchies define relationships between topics, it is understandable that the ability to summarize would be enhanced by exposure to a hierarchical site map. In sum, when the site map's structure matched the learners' goal, the learning goal was better achieved.

FUTURE TRENDS

At the present time, there is good reason to use site maps for the purpose of staying oriented and, in the case of beginning learners, for promoting good learning outcomes. As the World Wide Web becomes more commonplace in the everyday functioning of the classroom, site maps should prove to be important components of educational sites.

A great deal of research on site maps is required before their utility will be fully understood and can be capitalized on. Once the relationship among site maps' structures, features of the hypertext they represent, and user characteristics is understood, designers will be ready to equip hypertexts with adaptive site map modules. With adaptive modules, site maps can be generated automatically and configured to best meet the needs of any given user. To take advantage of adaptive site maps, educators or users themselves may be able to enter information about users' knowledge states, their goals, and other relevant information. The system then will generate a site map tailored for that specific context. It even may be possible to enter in a learning goal and have the system generate a site map that illustrates a path through the system tailored to that goal. Much as drivers are able to call up a recommended route using services such as www.MapQuest.com, learners may one day be able to call up recommended routes through large databases (i.e., the U.S. Library of Congress) to gain an understanding of a topic or domain.

CONCLUSION

Site maps are useful for keeping users oriented in a hypertext, informing them about the nature of the site's content and helping them to determine how to find their way to a desired page or topic. Site maps have not been found to enhance learning outcomes for users already knowledgeable in a domain, although that question is somewhat understudied. Site maps have been found to enhance learning for domain novices. Their effectiveness for this group, however, also is contingent on how well the structure of the site map coheres with a user's learning goals. Much more research is needed, however, to understand fully how to capitalize best on site maps for educational purposes. Specifically, much remains to be learned about the factors mediating their ability to enhance learning.

REFERENCES

Chase, W. G., & Simon, H. A. (1973). Perception in chess. *Cognitive Psychology, 4*, 55-81.

Chen, C., & Rada, R. (1996). Interacting with hypertext: A meta-analysis of experimental studies. *Human-Computer Interaction, 11*, 125-156.

Chi, M. T. H., & Koeske, R. D. (1983). Network representation of a child's dinosaur knowledge. *Developmental Psychology, 19*, 29-39.

Chou, C., Lin, H., & Sun, C. -T. (2000). Navigation maps in hierarchical-structured hypertext courseware. *International Journal of Instructional Media, 27*, 165-182.

Dee-Lucas, D., & Larkin, J. H. (1995). Learning from electronic texts: Effects of interactive overviews for information access. *Cognition and Instruction, 13*(3), 431-468.

Glover, J., & Krug, D. (1988). Detecting false statements in text: The role of outlines and inserted headings. *British Journal of Educational Psychology, 58*(3), 310-306.

Hammond, N., & Allinson, L. (1989). Extending hypertext for learning: An investigation of access and guidance tools. *Proceedings of the CHI'89.*

Kraiger, K., Salas, E., & Cannon-Bowers, J. (1995). Measuring knowledge organization as a method for assessing learning during training. *Human Factors, 37*(4), 804-816.

Mannes, B., & Kintsch, W. (1987). Knowledge organization and text organization. *Cognition and Instruction, 4*, 91-115.

Mayer, R. E. (1979). Twenty years of research on advance organizers: Assimilation theory is still the best predictor of results. *Instructional Science, 8*, 133-167.

McDonald, S., & Stevenson, R. J. (1999). Spatial versus conceptual maps as learning tools in hypertext. *Journal of Educational Multimedia and Hypermedia, 8*, 43-64.

Monesson, J. (2002). Topictracker: An investigation of a graphical map for use in revisiting previously viewed Web pages. *Humanities & Social Sciences, 62*, 3977.

Niederhauser, D. S., Reynolds, R. E., Salmen, D. J., & Skolmoski, P. (2000). The influence of cognitive load on learning from hypertext. *Journal of Educational Computing Research, 23*(3), 237-255.

Nilsson, R., & Mayer, R. (2002). The effects of graphic organizers giving cues to the structure of a hypertext document on users' navigation strategies and performance. *International Journal of Human-Computer Studies, 57*, 1-26.

Potelle, H., & Rouet, J. -F. (2003). Effects of content representation and readers' prior knowledge on the comprehension of hypertext. *International Journal of Human-Computer Studies, 58*, 327-345.

Puntambekar, S., Stylianou, A., & Hübscher, R. (2003). Improving navigation and learning in hypertext environments with navigable concept maps. *Human-Computer Interaction, 18*, 395-429.

Shapiro, A. M. (1998). The relationship between prior knowledge and interactive organizers during hypermedia-aided learning. *Journal of Educational Computing Research, 20*(2), 143-163.

Snapp, J., & Glover, J. (1990). Advance organizers and study questions. *Journal of Educational Research, 83*(5), 266-271.

Townsend, M., & Clarihew, A. (1989). Facilitating children's comprehension through the use of advance organizers. *Journal of Reading Behavior, 21*(1), 15-35.

Wenger, M. J., & Payne, D. G. (1994). Effects of a graphical browser on readers' efficiency in reading hypertext. *Journal of the Society for Technical Communication, 41*, 224-233.

West, L. H. T., & Pines, A. L. (1985). *Cognitive structure and conceptual change*. Orlando, FL: Academic Press.

KEY TERMS

Advance Organizer: Any presentation of information that displays and represents the content and structure of a hypertext or text.

Hypertext: A collection of electronic texts connected through electronic links. In addition to text, the documents also may contain pictures, videos, demonstrations, or sound resources. With the addition of such media, hypertext often is referred to as hypermedia.

Mental Model: The content and structure of an individual's knowledge.

Prior Knowledge: An individual's collected store of knowledge prior to exposure to a hypertext or other body of information.

Site Map: An electronic representation of the documents in a hypertext and sometimes the links connecting them. Site maps may appear as simple overviews or as interactive tools in which each entry serves as a link to the page it represents.

Social Factors and Interface Design Guidelines

S

Zhe Xu
Bournemouth University, UK

David John
Bournemouth University, UK

Anthony C. Boucouvalas
Bournemouth University, UK

INTRODUCTION

Designing an attractive user interface for Internet communication is the objective of every software developer. However, it is not an easy task as the interface will be accessed by an uncertain number of users with various purposes. To interact with users, text, sounds, images, and animations can be provided according to different situations. Originally, text was the only medium available for a user to communicate over the Internet. With technology development, multimedia channels (e.g., video and audio) emerged into the online context.

Individuals' sociability may influence human behaviour. Some people prefer a quiet environment and others enjoy more liveliness. On the other hand, the activity purpose influences the environment preference as well. Following usability principles and task analysis (Badre, 2002; Cato, 2001; Dix, Finlay, Abowd, & Beale, 1998; McCraken & Wolfe, 2004; Neilsen, 2000; Nielsen & Tahir, 2002; Preece, Rogers, & Sharp, 2002), we can predict that business-oriented systems and informal systems will require different types of interfaces: Business systems are concerned with the efficiency of performing tasks, while the effectiveness of informal systems depend more on the user's satisfaction with the experience of interacting with the system.

Suppose you are an Internet application designer; should you provide a vivid and multichannel interface or a concise and clear appearance? When individuals' sociability and the activity purpose contradict, should the interface design follow the sociability requirement, the purpose of the activity, or even neither of them?

To answer these questions, the characteristics of communication interfaces should be examined. For face-to-face communications, sounds, voices, various facial expressions, and physical movements are the most important contributing factors. These features are named physical and social presence (Loomis, Golledge, & Klatzky, 1998).

In the virtual world, real physical presence does not exist anymore; however, emotional feelings, group feelings, and other social feelings are existent but vary in quantity. The essential differences of interfaces are the quantity of the presented social feelings. For example, a three-dimensional (3-D) interface may provide more geographical and social feelings than a two-dimensional (2-D) chat room may present.

To assess the different feelings that may emerge from different interfaces, a two-dimensional chat room and a three-dimensional chatting environment were developed. The identification of social feelings present in the different interface styles is presented first. Then an experiment that was carried out to measure the influence the activity styles and the individuals' sociability have on the interface preferences is discussed.

The questions raised in this article are "What are the social feelings that may differ between the two interfaces (2-D vs. 3-D)?" and "Will users prefer different interfaces for different types of activities?"

Table 1. Different activities

Business Oriented	Social Oriented
Do math homework	Take a break from work
Schedule technical meetings	Fill up free time
Seek technical advice	Gossip and chat

BACKGROUND

Graphically, Internet communication interfaces can be classified into two categories: two dimensional and three dimensional. A 2-D interface is an acceptable choice for our flat monitor. 3-D interfaces apply various graphical algorithms to simulate the sense of depth in 2-D interfaces; hence, most 3-D interfaces can be defined as 2.5-D. In this article, the 3-D interfaces mentioned below can actually be classified into 2.5-D.

Social Presence

Communication channels are vivid in face-to-face communication. Physical movement, facial expressions, and variations of sound create the diversity. Computers and the Internet cannot provide the physical presence of users. Instead, people feel that they are chatting directly with other users. This is called social presence.

Social presence is defined as the "degree of salience of the other person in the interaction and the consequent salience (and perceived intimacy and immediacy) of the interpersonal relationships" (Short, Williams, & Christie, 1976, p. 65).

Communication researchers (Bailenson, Blascovich, Beall, & Loomis, 2001; Short et al., 1976) argue that even in a text-dominated environment, social presence still exists and provides important functions.

Interfaces with rich or poor communication channels may lead to different amounts of perceived social feelings. Witmer and Singer (1998) discussed some factors influencing social presence. These factors include the degree of control, environmental richness, multimodal presentation, scene realism, immediacy of control, anticipation, mode of control, physical modifiability, sensory modality, degree of movement perception, active search, isolation, selective attention, interface awareness, and meaningfulness of the experience.

With social-presence theory, different interfaces can be classified and assessed by the amount of social feelings presented.

Human Sociability Style

Sociability is defined as the quality or state of being sociable. The Merriam-Webster online dictionary (1996) defines sociable as the inclination by nature to companionship with others of the same species.

Personality is an important factor that differentiates humans (Nye & Brower, 1996). The same events may trigger significantly different feelings and actions according to different sociabilities.

An individual's sociability may influence his or her actions and scene preferences. Some people may enjoy going out and socializing with friends while others prefer reading a book alone. Their different social preferences may further influence their choice of Internet communication interface and their preference of the quantity of social-presence feelings.

Activity Style

The purpose of communication can be classified into two general categories: business oriented and social oriented. For business-oriented communication, people intend to grasp the information they need as soon as possible. On other hand, people use social-oriented communication to make friends, set up relationships, and create social networks. Table 1 lists some typical business-oriented activities and social-oriented activities.

Business-oriented activities may require an easy-to-use and concise environment, for example, an office, a conference room, or a classroom. In this kind of environment, people know who is in charge, know the problems they are trying to discuss, and intend to work out solutions as soon as possible.

Social-oriented activities demand a relaxing, free, and highly sociable context, for example, a restaurant, a bar, or a private garden. In this kind of environment, people can relax and enjoy their time.

Figure 1. The 2-D interface

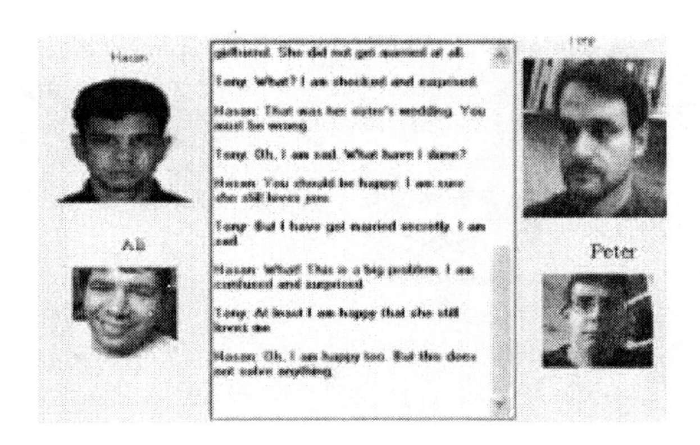

2-D vs. 3-D INTERFACES

One 2-D and one 3-D prototype interface were developed. The 2-D and 3-D interfaces only apply commonly available techniques to satisfy the universality requirements for interface and experiment assessment.

The 2-D interface is designed as a relatively simple environment. Similar to chatting environments over the Internet, the 2-D interface is based on textual message transmission. However, the user's image can be displayed as well. After logging onto the 2-D interface, a user can easily find others logged onto the server and get an overview of the whole environment. A typical screen of the 2-D interface is shown in Figure 1.

The design idea of the 3-D interface emerged from real-time 3-D games, and the interface is converted from a 3-D maze. The interface is divided into different spaces by walls and users can move around with the aid of cursors. The 3-D interface displays at a glance within a position panel the current relative positions of all users. To engage in conversation with others, a user needs to walk close enough to another user. When a conversation starts, textual messages will be shown both within a message box outside of the 3-D space and in a dialog bubble within the environment. To represent each user, facial images are displayed. However, users have a choice to use cartoon images to represent themselves. Automatic facial-expression display is achieved in the 3-D interface by integrating a text-to-emotion engine. A user's text input is sent to the emotion-extraction engine to examine emotion information that is embedded. The 3-D interface receives the output from the emotion-extraction engine and displays corresponding expression images. Further discussion about emotion-extraction engines can be found in Boucouvalas, Xu, and John (2003), Xu and Boucouvalas (2002), and Xu, John, and Boucouvalas (2003, in press).

A typical screen of the 3-D interface is shown in Figure 2.

Figure 2. The 3-D interface

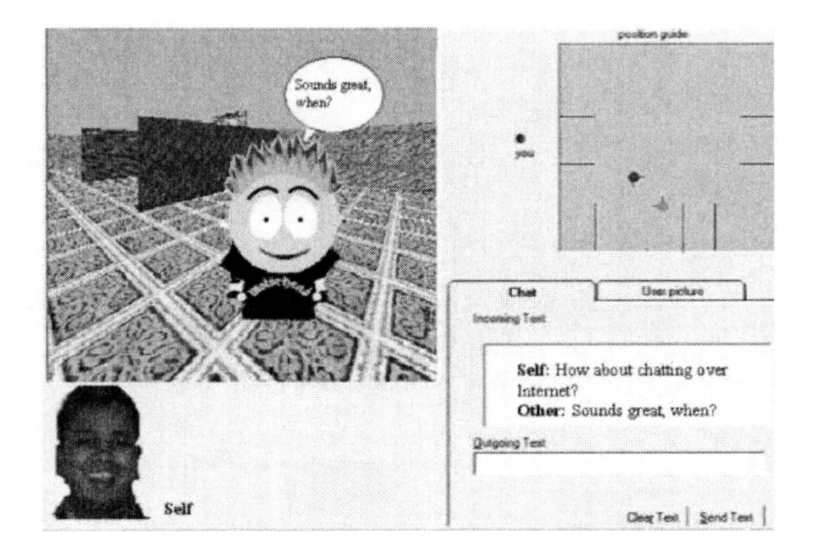

Figure 3. User-viewing component showing the movement of a user

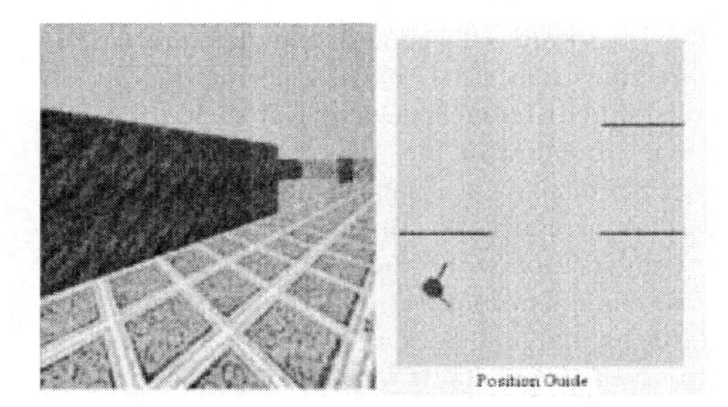

SOCIAL FEELINGS DISCUSSIONS

In this section, we describe the possible different social feelings perceived by users. A total of eight different senses are discussed here.

- **Movement Senses:** Similar to most chatting environments, every user in the 2-D chat prototype is in a fixed position. In contrast, the 3-D interface provides some aspects of movement. Users not only move around the space, but also have a specific field of view and can look for the spatial guidelines from the position-guide panel.
- **Geographic Senses:** Unlike 2-D chatting interfaces, a 3-D interface allows for various complex geographic entities such as a city or something as simple as a tree. Users may perceive geographic-movement phenomena in the virtual movement. Figure 3 shows the field of view and the position-guide component of the 3-D interface.
- **Sense of 3-D Depth of Space:** Space depth is a widely presented feature in both 3-D games and in real life. However, most 2-D interfaces cannot provide the same feelings.An example of the sense of depth can be found in Figure 3, which shows the use of perspective.
- **Exploration of Space:** For 2-D interfaces, the whole interface is presented to users. Users know at the beginning who is in the environment, whom they are talking to, and what functions the interface provides. For 3-D interfaces, users need to explore the space to meet others or to access the assistant functions provided by the system (e.g., buy something from a virtual shop). A position panel can only provide some aspects of the overall location and limited user information.
- **Eye Contact:** Eye contact is also very important. It is impolite to turn our backs to people talking to us. For the 2-D text-based interface, users cannot move their positions, and the images representing them are fixed. Thus, no virtual eye contact can be established. The 3-D interface provides the possibility of making virtual eye contact as users move around the 3-D interface. Figure 4 demonstrates the viewing component of the 3-D interface, which shows a direct glance and side-glance.
- **Communication Efficiency:** In the 2-D text-based system, a user's input sentences can be viewed by everyone. However, when a large number of users are exchanging messages quickly, it is difficult to follow the messages of a particular user. For 2-D chatting interfaces,

Figure 4. Different eye-contact angle

(a)
Direct glance

(b)
Side glance

Figure 5. Group discussion in different interfaces

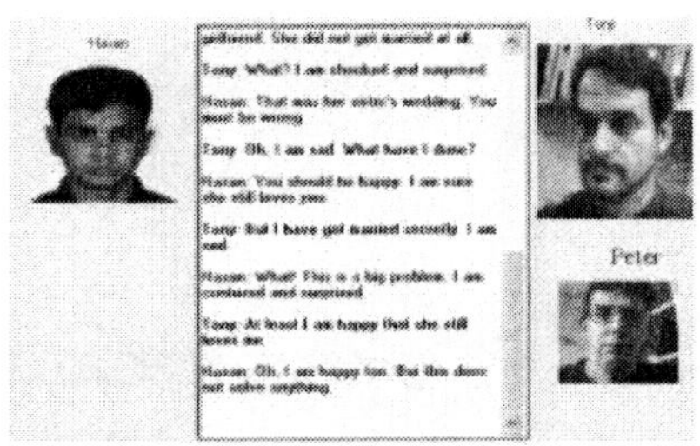

(a) The 2-D interface

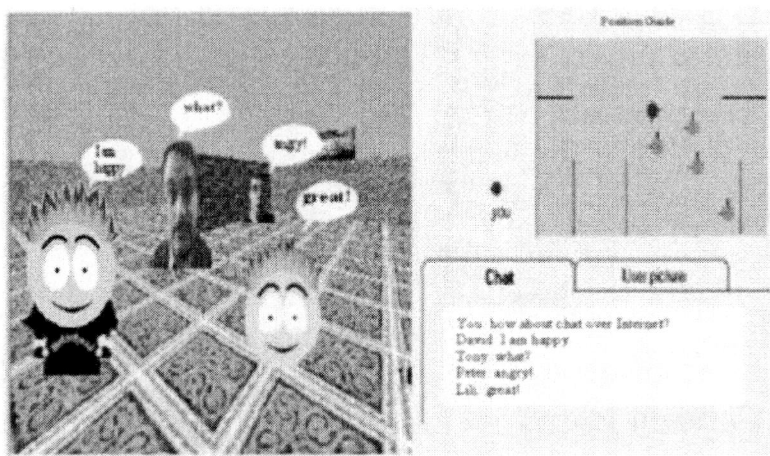

(b) The 3-D interface

communication will not be efficient for a large number of people gathering in the same room. With the speech bubbles, users in a 3-D system can identify others' chatting messages relatively easily. Users can concentrate on one user's speech by moving and changing the eye angles. Figure 5 demonstrates a busy chat environment with a 2-D interface and a chat environment with a 3-D interface.

- **Social-Attraction Feeling:** When some people gather together to discuss something, we may assume that interesting events or urgent situations occurred. The social-attraction feeling still applies to computer communication. For 2-D interfaces, users can judge the number of users discussing a topic only by scrolling the text. For 3-D interfaces, users can find this information visually by glancing for a cluster of gathered users from the position panel and the viewing component. Figure 5b demonstrated this feeling in a 3-D interface.
- **Movement Plus Talk:** In the 2-D text-based system, a user's position is fixed. For the 3-D system, movement is a fundamental element. It is quite possible that some users may chat while moving.

DESIGN CONSIDERATIONS

Social feelings are important factors for Internet communication interfaces, and they may vary in different environments. First, our article focuses on answering the question, What kinds of social feelings may be perceived in different environments? A 2-D interface and a 3-D interface will be compared in order to answer this question.

Second, both activity styles and individuals' sociability styles influence interface preferences and have different requirements on social-presence feelings. Which one is more important: the activity style or sociability—in other words, the activity or the human? We expect that when an individual has a particular aim, for example, solving a crucial problem or finding friends to talk to, then the activity style will dominate the preference of interface instead of the individual's sociability style.

The following phenomena are observed in Internet communications.

1. The activity style will strongly influence the preference of social presence.
2. Human sociability will strongly influence the overall preference of social presence.
3. An individual's sociability style will not strongly influence the preference of individual interfaces when the activity is chosen, hence the activity style is the dominant power for the interface preference.

The following experiments will present a detailed discussion about the phenomena.

SOCIABILITY AND ACTIVITY STYLE EXPERIMENT

There are two main aims of this experiment.

1. To examine the preference of participants for different interface styles when performing different types of activities.
2. To assess the effect that an individual's sociability undertakes on the satisfaction rating of each style of interface.

Two styles of interfaces are presented to participants.

1. A 2-D interface that is a less sociable environment (Users are split into different rooms. Each room lists the current users online. Users formally request connections before joining conversations.)
2. A 3-D interface that is a more sociable environment (All users explore the same 3-D space. All users are free to explore, approach other users, and engage them with conversation.)

There are two types of activities considered.

1. Business oriented (e.g., solving a technical problem, etc.)
2. Social oriented (e.g., having a tea-break chat, etc.)

THE EXPERIMENT PROCEDURE

A total of 50 students and staff from Bournemouth University participated in the experiments. The gender of each participant is recorded and a questionnaire assessing the sociability of each participant was shown. The questionnaire was developed by Bellamy and Hanewicz (1999) and contains seven items with five scale points ranging from *agree* to *disagree* in order to measure the sociability of the participants.

Participants then viewed the two interfaces (2-D and 3-D). After viewing them, participants were shown a list of 12 activities that can be performed in both environments. The participants were instructed to select the style of interface that was best suited for specific activities.

The 12 activities can be divided into two styles: business oriented and social oriented. Six of them belong to the business-oriented group and the other six are classified into the social-oriented group. The activities are shown in Table 2.

EXPERIMENT RESULTS

According to the activity style, we classify the results into two categories. The results of the business-oriented activities are shown in Figure 6, and Figure 7 presents the results of the social-oriented activities.

The charts show that more participants chose the 2-D interfaces for the business-oriented activities, and the 3-D interfaces were chosen for most social-oriented activities.

Table 3 lists the summary of dependent variables (2×3).

Table 2. The name and style of the activities

Activities	Style
Conduct a technical meeting over the Internet (Business 1)	Business oriented
Seek technical advice about your computer (Business 2)	Business oriented
Monitor your employee's progress (Business 3)	Business oriented
Study online (Business 4)	Business oriented
Chat about the latest celebrity gossip (Social 1)	Social oriented
Seek new friends (Social 2)	Social oriented
Watch an animation (Social 3)	Social oriented
Play a multiplayer game (e.g., football) (Social 4)	Social oriented
Privately chat with your good friends (Social 5)	Social oriented
Do your math homework (e.g., 3x + 5y = 70) (Business 5)	Business oriented
Discuss stock-market news (Business 6)	Business oriented
Display an exhibition of your paintings (Social 6)	Social oriented

Figure 6. The business-oriented activities

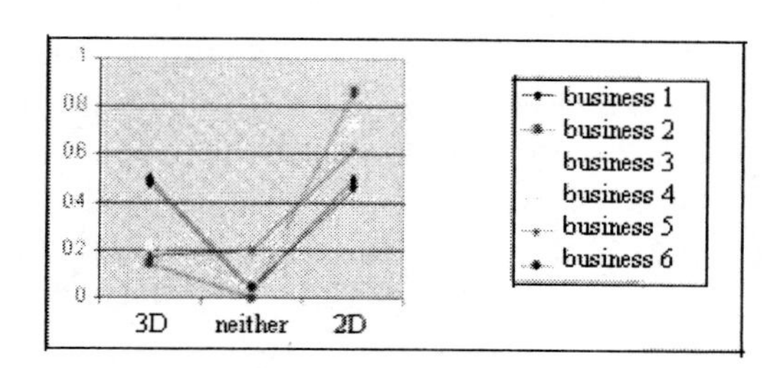

Figure 7. The social-oriented activities

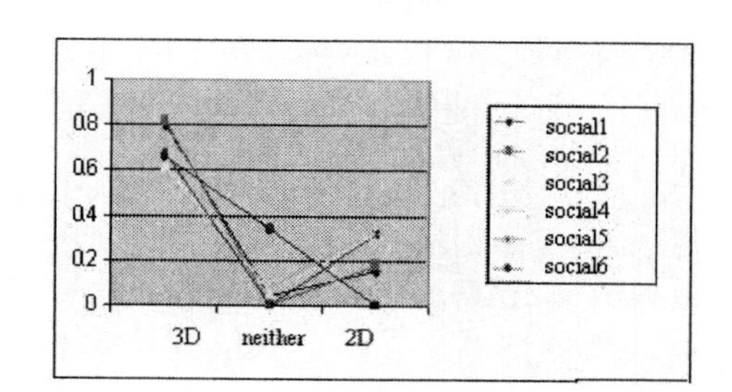

Table 3. Value explanation

Activity Style	Interface Choice
0: Business oriented 1: Social oriented	2: 2-D interface 3: 3-D interface 0: Neither

- **Activity Style vs. the Choice of Interfaces:** Correlation tests were carried out to compare the correspondence between the choice of interface and the activity style. The test result is 0.597, which is significant at p = 0.01. The results demonstrate that there is a significant relationship between activity style and the preference of interface. When users need to carry out a business-oriented activity, for example, finding an emergency telephone number, most people will prefer a simple interface that presents a low level of social feelings. When users want to spend their spare time, for example, playing an online game, the preferred interface is a relative complex and vivid multichannel environment. The practical hint for interface design is if the purpose of the online communication is to provide technical help or technical discussions for the users, a simple, straightforward, and uncluttered environment will be preferred by users. If the purpose of the online communication is to relax and to enjoy the online lifestyle, a vivid video- or audio-assisted environment will suit most online surfers.
- **Sociability vs. Choice of Interface:** The correlation test does not find any significant link between the individuals' sociability and interface preference. The results of the t-test show that the distribution of the preference ratings of highly sociable participants were not significantly different from the less sociable participants for the 12 activities. However, a marginally significant difference was found in a t-test between the two groups of the participants for Business 1 and Social 2, in which p = 0.08 and p = 0.09, correspondingly. To further explore the function of human sociability, we calculated the means of the 12 choices and repeated the t-tests. The results show that there is a marginally significant difference between the ratings of the sociable and less sociable participants (p = 0.06) for the means of the 12 choices. The results show that of the overall level, sociability has a significant influence on the preference of interface style. On average, low-sociability persons prefer simple and clean interfaces, and high-sociability persons prefer complex and realistic interfaces. However, for specific activities, human sociability has very limited influence on interface-style preference. The influence of an individual's sociability is much weaker than the influence of the activity style. This analysis provides another design criterion for online communication. Designers should pay more consideration to the activity that may be carried out on the Web. However, if a series of online communication interfaces will be presented to a specific user, human sociability should be considered and designers should adopt sociability into the design consideration.
- **Gender vs. the Choice of Interface:** Are there any differences in interface preference between genders? Will a female prefer a vivid online environment more than a male will? To answer this question, a t-test and correlation test were carried out. The result shows that there is one marginally significant difference (p = 0.075) between the ratings of males and females that is found for Business 2, and one significant difference (p = 0.04) that is found for Social 3. However, there is no significant correlation that supports these effects. This analysis indicates that gender has a very limited amount of influence on the preference of interface.

- **Revisiting the Phenomena:** It is now time to reexamine the phenomena. It can be seen that the activity style has a strong influence on the preference of the interface and the corresponding social-presence feelings. Overall, the sociability style of individuals strongly influences the preferences of social feelings, but not at the individual activity level. The three phenomena were observed in the experiment.

FUTURE TRENDS

Adaptivity is an extremely important interface design criteria. To create adaptive systems, human factors (e.g., emotion, cognition, and personality) need to be considered carefully. To attract the targeted audience, the system should analyse the potential users' customs and hobbies. As the purpose of a software system varies, the purpose may influence the preference of the interface. This article shows some general design guidelines and presents experiments to demonstrate the guidelines' accuracy. Trends in human-computer interaction (HCI) design are the adoption of more human factors into design consideration and the development of new guidelines (e.g., clear guidelines for sociability, gender, and age).

CONCLUSION

Social presence exists everywhere in the virtual world. The more social-presence feelings presented, the more realistic and more sensible the environment is. The argument of our article is the necessity to increase social presence everywhere in Internet communication.

A 2-D interface and a 3-D interface were developed. The feature comparisons between the two interfaces illustrate that a 2-D text-based interface is straightforward, and a 3-D interface provides some aspects of virtual-reality feelings, which are more complex.

The experiment results show that significant differences exist in the preference of social presence for different activities. The experiment results strongly show that social presence should be considered for interface design.

The individual's sociability may also influence his or her preference of social presence. However, a significantly different preference is only revealed at the overall level, not for most individual activities. This indicates that an individual's sociability does impact his or her preference of social presence. However, when dealing with specific activities, the influence of the activity style is much stronger than the individual's sociability.

Gender does influence the social preference in some specific activities. However, no significant preference difference can be found for the majority of activities and the overall level. This means the impact of gender needs further investigation.

As high social presence may be linked with a vivid or multichannel (e.g., video, audio, or animation) communication interface and a simple or text-dominated interface may present low social presence, the experiment results also provide guidelines for HCI design.

REFERENCES

Badre, A. N. (2002). *Shaping Web usability: Interaction design in context.* Boston, MA: Addison-Wesley.

Bailenson, J. N, Blascovich, J., Beall, A. C., & Loomis, J. M. (2001). Equilibrium theory revisited: Mutual gaze and personal space in virtual environments. *Presence: Teleoperators and Virtual Environments, 10*(6), 583-598.

Bellamy, A., & Hanewicz, C. (1999). Social psychological dimensions of electronic communication. *Electronic Journal of Sociology, 4*(1). Retrieved from http://www.sociology.org/content/vol004.001/bellamy.html

Boucouvalas, A. C., Xu, Z., & John, D. (2003). Expressive image generator for an emotion extraction engine. *Proceedings of the 17th Annual Human-Computer Interaction Conference,* Bath University, UK.

Cato, J. (2001). *User-centred Web design.* London: Addison-Wesley.

Dix, A., Finlay, J., Abowd, G., & Beale, R. (1998). *Human computer interaction* (2nd ed.). London: Prentice Hall.

Loomis, J. M., Golledge, R. G., & Klatzky, R. L. (1998). Navigation system for the blind: Auditory display modes and guidance. *Presence: Teleoperators and Virtual Environments, 7*(2), 192-203.

McCraken, D. D., & Wolfe, R. J. (2004). *User-centred Website development: A human-computer interaction approach.* Upper Saddle River, NJ: Pearson.

Merriam-Webster online. (1996). Retrieved July 7, 2003, from http://www.m-w.com/home.htm

Neilsen, J. (2000). *Designing Web usability.* Indianapolis, IN: New Riders.

Nielsen, J., & Tahir, M. (2002). *Homepage usability: 50 Websites deconstructed.* Indianapolis, IN: New Riders.

Nye, J. L., & Brower, A. M. (1996). *What's social about social cognition?* London: Sage Publications.

Preece, J., Rogers, Y., & Sharp, H. (2002). *Interaction design: Beyond human-computer interaction.* New York: John Wiley & Sons.

Short, J., Williams, E., & Christie, B. (1976). *The social psychology of telecommunications.* London: John Wiley & Sons.

Witmer, B. G., & Singer, M. J. (1998). Measuring presence in virtual environments: A presence questionnaire. *Presence: Teleoperators and virtual environments, 7*(3), 225-240.

Xu, Z., & Boucouvalas, A. C. (2002). Text-to-emotion engine for real time Internet communication. *International Symposium on Communication Systems, Networks and DSPs*, 164-168.

Xu, Z., John, D., & Boucouvalas, A. C. (2003). Emotion extraction engine: Expressive image generator. *Proceedings of EUROMEDIA,* Plymouth, UK.

Xu, Z., John, D., & Boucouvalas, A. C. (in press). Emotion analyzer. *Emotion Journal.*

KEY TERMS

2-D Interface: An interfaces in which text and lines appear to be on the same flat level.

2.5-D Interface: An interface that applies various graphical algorithms to simulate the sense of depth on a 2-D interface.

3-D Interface: An interface in which text and images are not all on the same flat level.

Emotion-Extraction Engine: A software system that can extract emotions embedded in textual messages.

Sociability: The relative tendency or disposition to be sociable or associate with one's fellows.

Social Presence: The extent to which a person is perceived as a real person in computer-mediated communication.

Task Analysis: A method of providing an extraction of the tasks users undertake when interacting with a system.

Social-Technical Systems

Brian Whitworth
New Jersey Institute of Technology, USA

INTRODUCTION

System Levels

Computer systems have long been seen as more than just mechanical systems (Boulding, 1956). They seem to be systems in a general sense (Churchman, 1979), with system elements, like a boundary, common to other systems (Whitworth & Zaic, 2003). A computer system of chips and circuits is also a software system of information exchanges. Today, the system is also the human-computer combination (Alter, 1999); for example, a plane is mechanical, its computer controls are informational, but the plane plus pilot is also a system: a human-computer system. Human-computer interaction (HCI) sees computers as more than just technology (hardware and software). Computing has reinvented itself each decade or so, from hardware in the 1950s and 1960s, to commercial information processors in the 1970s, to personal computers in the 1980s, to computers as communication tools in the 1990s. At each stage, system performance increased. This decade seems to be that of social computing, in which software serves not just people but society, and systems like e-mail, chat rooms, and bulletin boards have a social level. Human-factors research has expanded from computer usability (individual), to computer-mediated communication (largely dyads), to virtual communities (social groups). The infrastructure is technology, but the overall system is personal and social, with all that implies. Do social systems mediated by technology differ from those mediated by the natural world? The means of interaction, a computer network, is virtual, but the people involved are real. One can be as upset by an e-mail as by a letter. Online and physical communities have a different architectural base, but the social level is still people communicating with people. This suggests computer-mediated communities operate by the same principles as physical communities; that is, virtual society is still a society, and friendships cross seamlessly from face-to-face to e-mail interaction.

Table 1 suggests four computer system levels, matching the idea of an information system as hardware, software, people, and business processes (Alter, 2001). Social-technical systems arise when cognitive and social interaction is mediated by information technology rather than the natural world.

BACKGROUND

The Social-Technical Gap

The levels of Table 1 are not different systems, but overlapping views of the same system. Higher levels depend on lower levels, so lower level failure implies failure at all levels above it; for example, if the hardware fails, the software does too as does the user interface. Higher levels are more efficient ways of operating the system as well as observing it. For example, social systems can generate enormous productivity. For this to occur, system design must recognize higher system-level needs. For example, usability drops when software design contradicts users' cognitive needs.

Table 1. Information system levels

Level	Examples	Discipline
Social	Norms, culture, laws, zeitgeist, sanctions, roles	Sociology
Cognitive	Semantics, attitudes, beliefs, opinions, ideas, morals	Psychology
Information	Software programs, data, bandwidth, memory, processing	Computing
Mechanical	Hardware, computer, telephone, fax, physical space	Engineering

In physical society, architecture normally fits social norms; for example, you may not legally enter my house, and I can physically lock you out. In cyberspace, the architecture of interaction is the computer code that "makes cyberspace as it is" (Lessig, 2000). If this architecture ignores social requirements, there is a social-technical gap between what computers do and what society wants (Figure 1). This seems a major problem facing social software today (Ackerman, 2000). Value-centered computing counters this gap by making software more social (Preece, 2000).

Antisocial Interaction

Social evolution involves specialization and cooperation on a larger and larger scale (Diamond, 1998). Villages became towns, then cities and metropolitan centers. The roving bands of 40,000 years ago formed tribes, chiefdoms, nation states, and megastates like Europe and the United States. Driving this evolution are the larger synergies that larger societies allow. The Internet offers the largest society of all—global humanity—and potentially enormous synergies. To realize this social potential, software designers may need to recognize how societies generate *nonzero-sum gains* (Wright, 2001). While nonzero sum is an unpleasant term, Wright's argument that increasing the shared social pie is the key to social prosperity is strong. The logic that society can benefit everyone seems simple, yet communities have taken thousands of years to stabilize nonzero-sum benefits. Obviously, there is some resistance to social synergy.

If social interactions are classified by the expected outcome for the self and others (Table 2), situations where individuals gain at others' expense are antisocial. Most illegal acts, like stealing, fall into this category. The equilibrium of antisocial interaction is that all parties defect when nonzero-sum gains are lost. Antisocial acts destabilize the nonzero-sum gains of society, so to prosper, society must reduce antisocial acts. This applies equally to online society. Users see an Internet filled with pop-up ads, spam, pornography, viruses, phishing, spoofs, spyware, browser hijacks, scams, and identity theft. These can be forgiven by seeing the Internet as an uncivilized place, a stone-age culture built on space-age technology, inhabited by the "hunter-gatherers of the information age" (Meyrowitz, 1985, p. 315). This is the "dark side" of the Internet, a worldwide "tangled web" for the unwary (Power, 2000), a superhighway of misinformation, a social dystopia beyond laws where antisocial acts reign.

Figure 1. Social-technical gap

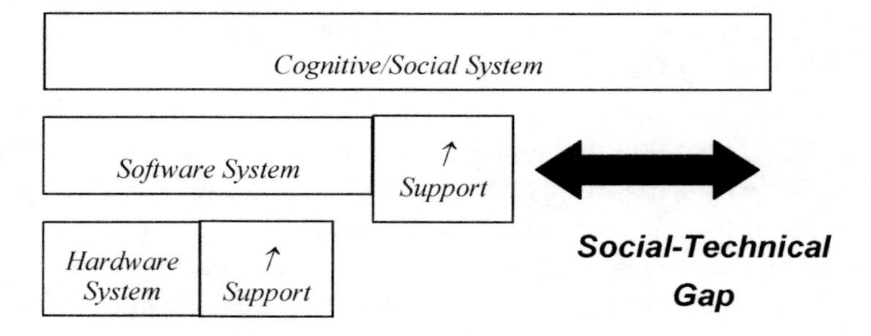

Table 2. Expected interaction outcomes

Other(s)→ Self ↓	Gain	Minor Effect	Loss
Gain	*Synergy*	*Opportunity*	*Antisocial*
Minor Effect	*Service*	*Null*	*Malice*
Loss	*Sacrifice*	*Suicide*	*Conflict*

Users are naturally wary of such a society; that is, they do not trust it. Trust has been defined as expecting that another's action will be beneficial rather than detrimental (Creed & Miles, 1996). Antisocial acts, by definition, do not create trust. Lack of trust reduces interaction, especially if there is a less risky alternative. For example, while electronic commerce is a billion-dollar industry, it has consistently performed below expectations, though in online trade sellers reduce costs and buyers gain choice at a lower price. E-commerce benefits both customers and companies, so why is it not the majority of trade? Every day millions of customers who want to buy things browse thousands of Web sites for products and services, yet the majority purchase from brick-and-mortar, not online, sources (Salam, Rao, & Pegels, 2003). If online society does not prevent antisocial acts, users will not trust it, and if they do not trust it, they will use it less.

In the tragedy of the commons, acts that benefit individuals harm the social group, whose loss affects the individuals in it (Poundstone, 1992). If farmers graze a common grass area, a valuable common resource is destroyed (from overgrazing), yet if one farmer does not graze, another will. The tragedy occurs if individual economics drives the group to destroy a useful common resource. Most animal species are barely able to cross this individual-gain barrier to social synergy. Only insect colonies compare to humans in size, but each community is one genetic family, allowing selection for cooperative behavior (Ridley, 1996). Humanity has created social benefits without genetic selection. How did we cross the zero-sum barrier? The answer seems to be our ability to develop social systems.

If the commons farmers form a village, it makes no sense for the village to destroy its own resource. If the village social system, of norms, rules, and sanctions, can stop individuals from overgrazing, the village keeps its commons and the benefits thereof. If only the village chief grazes the commons, there is an inherent instability between individual and community gain. However, if the commons is shared, say by a grazing roster, both village and members benefit. As society has evolved, bigger communities have produced more but also shared more. Social systems that spread social benefits fairly seem to stabilize nonzero-sum benefits better than those in which society's benefits accrue only to a few. The social concept of fairness seems to reconcile the conflict between private benefit and public good.

LEGITIMATE INTERACTION: A SOCIAL REQUIREMENT

The fact that social systems of law and justice are primarily about reducing unfairness in society (Rawls, 2001) is necessary because in society, one person's failure can cause another's loss, and one person's contribution can be another's gain, for example, in software piracy. One way to reduce antisocial acts is to make people accountable for the effects of their acts not just on themselves but also on others. Without such accountability, perceptions of unfairness arise, for example, when people take benefits others earned, or pay no price for harming others. Unfairness is not just the unequal distribution of outcomes, but the failure to distribute outcomes according to action contributions. Studies suggest people react strongly to unfairness, tend to avoid unfair situations (Adams, 1965), and even prefer fairness to personal benefit (Lind & Tyler, 1988). This natural justice perception seems to underlie our ability to form positive societies. Progress in legitimate rights seems to correlate with social wealth, as does social corruption with community poverty (Eigen, 2003). Perhaps people in fair societies contribute more work, ideas, and research because others do not steal it, or self-regulate more, which reduces security costs. Either way, accountability (or justice) seems a requirement for social prosperity.

The social goal has been defined as legitimate interaction that is fair to individuals and beneficial to the social group (Whitworth & deMoor, 2003). Legitimacy is a complex social concept. Fairness alone does not define it as conflict can also be fair. A duel is a fair fight, but duels are still outlawed as being against society. Legitimate interaction includes public-good benefits as well as individual fairness. In sociology, the term legitimate applies to governments that are justified to their people, not coerced (Barker, 1990). It can mean having the sanction of law, but legitimacy is more than legality. Mill (1859/1995, p. 1) talks of the "limits of power that can be legitimately exercised by society over the individual." Jefferson wrote, "... the mass of mankind has not been born with saddles on their backs, nor a favored few booted and spurred, ready to ride them legitimately..." (Somerville & Santoni, 1963, p. 246). Fukuyama (1992) argues that legitimate communities prosper, while those that ignore it do so at their peril. These statements have no meaning if legitimacy and legality are the same, as then no law-setting government could act illegitimately.

The social requirement of legitimacy complements that of security. Security ensures a system is used as intended, while legitimacy defines that intent. Whether a user is who he or she says (authentication) is a security issue. What rights he or she should have (authority) is a legitimacy issue. In generating trust and business, no amount of security can compensate for a lack of legitimacy. Dictatorships have powerful security forces, but their citizens distrust them, reducing social synergy. In prosperous modern societies, security is directed by legitimacy, and legitimacy depends on security.

Online Legitimacy

Physical society uses various means to prevent antisocial acts from destabilizing social benefits, including the following.

1. **Ethics:** Supports right acts by religion or custom
2. **Barriers:** Fences, doors, or locks to prevent unfair acts
3. **Revenge:** Individuals "pay back" those that cheat
4. **Norms:** Community laws, sanctions, and police

All have also been tried in cyberspace, with varying degrees of success.

Arguably the best means to legitimate interaction is to have moral, ethical people, who choose not to cheat. But while most agree altruism is good and selfishness bad, we often do not practice what we preach (Ridley, 1996). Will online society make people more ethical than physical society?

Barriers, like a locked door, can prevent unfairness, but any barrier raised can be overcome. Online security is a continual battle between those who create and those who cross barriers. Also, barriers can reduce as well as increase fairness. Do we really want a cyber society built on the model of medieval fortresses?

A third way to legitimate interaction is through revenge: to repay actions in kind, or cheat the cheaters (Boyd, 1992). In Axelrod's (1984) prisoner's dilemma tournament, the most successful program was TIT-FOR-TAT, which began cooperating, then copied whatever the other did. If people who are cheated today will take revenge tomorrow, cheating may not be worth it, but do we want cyber society run under a vigilante justice system?

A fourth way for society to support legitimate interaction is by norms and laws. If laws oppose antisocial acts, why not apply laws online? This approach is popular, but old means may fail in new system environments (Whitworth & deMoor, 2003). Laws assume a physical-world architecture so may not easily transfer to virtual worlds that work differently from the physical world (Burk, 2001). Legal processes may suffice for physical change, but while laws can take years to pass, the Internet can change in a month. New cases, like cookies, can arise faster than laws can be formed, like weeds growing faster than they are culled. Also, the programmers who define cyberspace can bypass any law. The Internet, once thought innately ungovernable, could easily become a system of perfect regulation and control (Lessig, 1999) as once software is written, issues of law may have already been decided. Finally, laws are limited by jurisdiction, as attempts to legislate telemarketers illustrate. U.S. law applies to U.S. soil, but cyberspace does not exist inside America. The many laws of many nations do not apply to a global Internet. For these reasons, the long arm of the law struggles to reach into cyberspace. The case is still out, but many are pessimistic. Traditional law seems too physical, too slow, too impotent, and too restricted for the challenge of a global information society.

FUTURE TRENDS

That the social needs of online society are not yet met suggests two things. First, Internet growth may be just beginning, and second, meeting social needs is the way to achieve that growth. Perhaps we are only seeing the start of a major human social evolution. We may be no more able to envisage a global information society than people in the middle ages could conceive today's global trade system. The differences are not just technical, like ships and airplanes, but also social, differences in how we interact. Traders today send millions of dollars to foreigners they have never seen for goods they have not touched to arrive at unknown times. Past traders would have seen that as mere folly, but today's market economy has social as well as technical support:

To participate in a market economy, to be willing to ship goods to distant destinations and to invest in projects that will come to fruition or pay dividends only in the future, requires confidence, the confidence that ownership is secure and payment dependable…knowing that if the other reneges, the state will step in… (Mandelbaum, 2002, p. 272)

Social benefits require the influence of social entities, like the state. Individual parties in an interaction are biased to their own benefit. Only a community can embody legitimate rules above individuals, yet these must be manifested as well as conceived. The concept of the state assumes physical boundaries that do not exist in cyberspace. For online society to flourish, the gap between social right and software might must be closed, but stretching physical law into cyberspace is problematic (Samuelson, 2003). Physical laws operate after the fact for practical reasons: To punish unfairness, it must first occur. Yet in cyberspace, we write the code that defines all interaction. It is as if we could write the laws of physics in the physical world. Hence, a new possibility arises. Why not focus on the solution (legitimacy) rather than the problem (unfairness)? Why let antisocial acts like spam develop, then try ineffectually to punish them when we can design for social fairness in the first place? When societies move from punishing unfairness to encouraging legitimacy, it is a major advance, from the laws of Moses or Hammurabai to visionary statements of social opportunity like the French Declaration of Human Rights or the United States constitution. Cyberspace is a chance to apply several thousand years of social learning to the global electronic village; designing social software in a social vacuum may condemn us to relearn the social lessons of physical history in cyberspace.

In physical society, it was the push for distributed ownership that created social rights; the original pursuers of rights were British elite seeking property rights from their King: "It was the protection of property that gave birth, historically, to political rights" (Mandelbaum, 2002, p. 271). Over time, the right to own was extended to all citizens, as giving today's freedoms proved profitable. Ownership as a concept can be applied online. Twenty years ago, issues of "Who owns the material entered in a group communication space?" (Hiltz & Turoff, 1993, p. 505) were raised. If information objects can be owned, a social property-rights framework can be applied to information systems (Rose, 2001). Analysing who owns what can translate social statements into IS specifications and vice versa (Whitworth & deMoor, 2003; Figure 2).

Future social-software designers may face questions of what should be done, not what can be done. There seems no reason why software should not support what society believes. If society believes people should be free, our Hotmail avatars should belong to us. If society gives a right not to communicate (Warren & Brandeis, 1890), we should be able to refuse spam (Whitworth & Whitworth, 2004). If society supports privacy, we should be able to remove personal data from online lists. If society gives creators rights to the fruits of their labors (Locke, 1963), we should be able to sign and own electronic items. If society believes in democracy, online bulletin boards should be able to elect their leaders. Such suggestions do not mean the mechanization of online interaction: Social rights do not work that way. Society grants people privacy, but does not force them to be private. Likewise, owning a bulletin-board item means you may delete it, not that you must delete it. Software support for social rights would allocate rights to act, not automate right acts, giving choice to people to not to program code.

Figure 2. Social-requirements analysis

CONCLUSION

The core Internet architecture was designed over 30 years ago to engineering requirements existing when a global electronic society was not even envisaged. It seems due for an overhaul to meet the social needs of virtual society. Architecture, whether physical or electronic, affects everything, and social systems require precisely such general changes. The marriage of society and technology needs respect on both sides. To close the social-technical gap, technologists cannot stand on the sidelines: They must help. System designers must recognize accepted social concepts, like freedom, privacy, and democracy, that is, specify social requirements as they do technical ones. Translating social requirements into technical specifications is a daunting task, but the alternative is an antisocial cyber society that is not a nice place to be. If human society is to expand into cyberspace, with all the benefits that implies, technology must support social requirements. The new user of social-technical software is society, and the user requirement of society is legitimate interaction.

REFERENCES

Ackerman, M. S. (2000). The intellectual challenge of CSCW: The gap between social requirements and technical feasibility. *Human Computer Interaction, 15*, 179-203.

Adams, J. S. (1965). Inequity in social exchange. In L. Berkowitz (Ed.), *Advances in experimental social psychology* (Vol. 2, pp. 267-299). New York: Academic Press.

Alberts, B., Bray, D., Lewis, J., Raff, M., Roberts, K., & Watson, J.D. (1994). *Molecular biology of the cell.* New York: Garland Publishing, Inc.

Alter, S. (1999). A general, yet useful theory of information systems. *Communications of the AIS, 1,* 13-60.

Alter, S. (2001). Which life cycle: Work system, information system, or software? *Communications of the AIS, 7*(17), 1-52.

Axelrod, R. (1984). *The evolution of cooperation.* New York: Basic Books.

Barker, R. (1990). *Political legitimacy and the state.* Oxford, UK: Oxford University Press.

Boulding, K. E. (1956). General systems theory: The skeleton of a science. *Management Science, 2*(3), 197-208.

Boyd, R. (1992). The evolution of reciprocity when conditions vary. In A. H. Harcourt & F. B. M. de Waal (Eds.), *Coalitions and alliances in humans and other animals* (473-492). Oxford, UK: Oxford University Press.

Burk, D. L. (2001). Copyrightable functions and patentable speech. *Communications of the ACM, 44*(2), 69-75.

Chung, L., Nixon, B. A., Yu, E., & Mylopoulos, J. (1999). *Non-functional requirements in software engineering.* Boston: Kluwer Academic.

Churchman, C. W. (1979). *The systems approach.* New York: Dell Publishing.

Creed, W. E., & Miles, R. E. (1996). Trust in organizations: A conceptual framework linking organizational forms, managerial philosophies, and the opportunity costs of control. In R. M. Kramer, M. Roderick, & T. Tyler (Eds.), *Trust in organizations: Frontiers of theory and research* (pp. 16-38). London: Sage.

Diamond, J. (1998). *Guns, germs and steel.* London: Vintage.

Eigen, P. (2003). *Transparency international corruption perceptions index 2003* [Speech]. London: Foreign Press Association. Retrieved from http://www.transparency.org/cpi/2003/cpi2003.pe_statement_en.html

Fukuyama, F. (1992). *The end of history and the last man.* New York: Avon Books Inc.

Goodwin, N. C. (1987, March). Functionality and usability. *Communications of the ACM,* 229-233.

Hiltz, S. R., & Turoff, M. (1993). *The network nation: Human communication via computer* (Rev. ed.). Cambridge, MA: MIT Press.

Lessig, L. (1999). *Code and other laws of cyberspace.* New York: Basic Books.

Lessig, L. (2000). Cyberspace's constitution. *Lecture given at the American Academy, Berlin, Germany.* Retrieved from http://cyber.law.harvard.edu/works/lessig/AmAcd1.pdf

Lind, E. A., & Tyler, T. R. (1988). *The social psychology of procedural justice.* New York: Plenum Press.

Locke, J. (1963). An essay concerning the true original extent and end of civil government. In J. Somerville & R. E. Santoni (Eds.), *Social and political philosophy: Readings from Plato to Ghandi* (chap. 5, section 27, pp. 169-204). New York: Anchor Books.

Mandelbaum, M. (2002). *The ideas that conquered the world.* New York: Public Affairs.

Meyrowitz, J. (1985). *No sense of place: The impact of electronic media on social behavior.* New York: Oxford University Press.

Mill, J. S. (1955). *On liberty.* Chicago: The Great Books Foundation. (Reprinted from 1859)

Poundstone, W. (1992). *Prisoner's dilemma.* New York: Doubleday, Anchor.

Power, R. (2000). *Tangled web: Tales of digital crime from the shadows of cyberspace.* Indianapolis, IN: QUE Corporation.

Preece, J. (2000). *Online communities: Designing usability, supporting sociability.* Chichester, UK: John Wiley & Sons.

Rawls, J. (2001). *Justice as fairness.* Cambridge, MA: Harvard University Press.

Ridley, M. (1996). *The origins of virtue: Human instincts and the evolution of cooperation.* New York: Penguin.

Rose, E. (2001, March). Balancing Internet marketing needs with consumer concerns: A property rights framework. *Computers and Society,* 17-21.

Salam, A. F., Rao, H. R., & Pegels, C. C. (2003). Consumer-perceived risk in e-commerce transactions. *CACM, 46*(12), 325-331.

Samuelson, P. (2003). Unsolicited communications as trespass. *Communications of the ACM, 46*(10), 15-20.

Sanders, M. S., & McCormick, E. J. (1993). *Human factors in engineering and design.* New York: McGraw-Hill.

Somerville, J., & Santoni, R. E. (1963). *Social and political philosophy: Readings from Plato to Ghandi.* New York: Anchor Books.

Warren, S. D., & Brandeis, L. D. (1890). The right to privacy. *Harvard Law Review, 4*(5), 193-220.

Whitworth, B., & deMoor, A. (2003). Legitimate by design: Towards trusted virtual community environments. *Behaviour & Information Technology, 22*(1), 31-51.

Whitworth, B., & Whitworth, E. (2004, October). Reducing spam by closing the social-technical gap. *IEEE Computer*, 38-45.

Whitworth, B., & Zaic, M. (2003). The WOSP model: Balanced information system design and evaluation. *Communications of the Association for Information Systems, 12*, 258-282.

Wright, R. (2001). *Nonzero: The logic of human destiny.* New York: Vintage Books.

KEY TERMS

Avatar: An information object that represents a person in cyberspace, whether a Hotmail text ID or a graphical multimedia image in an online multiplayer game.

Information System: A general system that may include hardware, software, people, and business or community structures and processes (Alter, 1999, 2001), vs. a social-technical system, which must include all four levels.

Nonzero Sum: In zero-sum interaction, one party gains at another's expense so the parties compete. Negative acts that harm others but benefit the actor give an "equilibrium" point at which everyone defects and everyone loses (Poundstone, 1992). In contrast, in nonzero-sum interaction, parties cooperate to increase the shared resource pie, so they gain more than they could have working alone: It is a win-win situation. The synergistic benefits of society seem based on nonzero-sum gains (Wright, 2001).

Social System: Physical society is not just mechanics nor is it just information, as without people information has no meaning. Yet it is also more than people. Countries with people of similar nature and abilities, like North and South Korea, or East and West Germany, performed differently as societies. As people come and go, we say the society continues. Jewish individuals of 2,000 years ago have died just as the Romans of that time, yet we say the Jews survived while the Romans did not. What survived was not buildings, information, or people, but a manner of interaction: their social system. A social system is a general form of human interaction that persists despite changes in individuals, communications, or architecture (Whitworth & deMoor, 2003) based on persistent common cognitions regarding ethics, social structures, roles, and norms.

System: A system must exist within a world and cannot exist if its world is undefined: No world means no system. Existence is a property a system derives from the world around it. The nature of a system is the nature of the world that contains it; for example, a physical world, a world of ideas, and a social world may contain physical systems, idea systems, and social systems, respectively. A system that exists still needs an identity to define what is a system and what is not a system. A system indistinguishable from its world is not a system; for example, a crystal of sugar that dissolves in water still has existence as sugar, but is no longer a separate macroscopic system. The point separating system from nonsystem is the system boundary. Existence and identity seem two basic requirements of any system.

System Elements: An advanced system has a boundary, an internal structure, environment effectors, and receptors (Whitworth & Zaic, 2003). Simple biological systems (cells) formed a cell-wall boundary and organelles for internal cell functions (Alberts et al., 1994). Simple cells like Giardia developed flagella to effect movement, and protozoa developed light-sensitive receptors. We ourselves, though more complex, still have a boundary (skin), an internal structure of organs, muscle effectors, and sense receptors. Computer systems have the same elements: a physical-case boundary, an internal architecture, printer and screen effectors, and keyboard and mouse receptors. These elements apply at different levels; for example, software systems have memory boundaries, internal program structures, specialized input analysers, and specialized output driver units.

System Environment: In a changing world, changes outside a system may cause changes within it, and changes within may cause changes without. A system's environment is that part of a world that can change the system or be affected by it. What succeeds in the system-environment interaction depends on the environment. In Darwinian evolution, the environment defines system performance. Three things seem relevant: opportunities, threats, and the rates by which these change. In an opportunistic environment, right action can give great benefit. In a risky environment, wrong action can give great loss. In a dynamic environment, risk and opportunity change quickly, giving turbulence (sudden risk) or luck (sudden opportunity). An environment can be of any combination, for example, opportunistic, risky, and dynamic.

System Levels: Is the physical world the only real world? Are physical systems the only possible systems? The term information system suggests otherwise. Philosophers propose idea systems in logical worlds. Sociologists propose social systems. Psychologists propose cognitive mental models. Software designers propose data entity relationship models quite apart from hardware. Software cannot exist without a hardware system of chips and circuits, but the software world of data records and files is not equivalent to the hardware world. It is a different system level. Initially, computer problems were mainly hardware problems, like overheating. Solving these led to software problems, like infinite loops. Informational requirements began to drive chip development, for example, network and database protocol needs. HCI added cognitive requirements to the mix. Usability demands are now part of engineering-requirements analysis (Sanders & McCormick, 1993) because Web sites fail if people reject them (Goodwin, 1987). Finally, a computer-mediated community can also be seen as a social system. An information system can be conceived on four levels: mechanical, informational, cognitive, and social. Each emerges from the previous, not in some mystical way, but as a different framing of the same thing. For example, information derives from mechanics, human cognitions from information, and society from a sum of human cognitions (Whitworth & Zaic, 2003). If all levels derive from hardware, why not just use that perspective? Describing modern computers by chip and line events is possible but inefficient, like describing World War II in terms of atoms and electrons. As higher levels come into play, systems become more complex but also offer higher performance efficiencies.

System Performance: A traditional information system's performance is its functionality, but functions people cannot use do not add performance. If system performance is how successfully a system interacts with its environment, usability can join nonfunctional IS requirements, like security and reliability, as part of system performance. The four advanced system elements (boundary, internal structure, effectors, and receptors) can maximize opportunity or minimize risk in a system environment. A multidimensional approach to system performance, as suggested by Chung, Nixon, Yu, and Mylopoulos (1999), suggests eight general system goals applicable to modern software: functionality, usability, reliability, flexibility, security, extendibility, connectivity, and confidentiality (Whitworth & Zaic, 2003).

Socio-Cognitive Engineering

Mike Sharples
University of Nottingham, Jubilee Campus, UK

INTRODUCTION

Socio-cognitive engineering is a framework for the systematic design of socio-technical systems (people and their interaction with technology), based on the study and analysis of how people think, learn, perceive, work, and interact. The framework has been applied to the design of a broad range of human-centered technologies, including a Writer's Assistant (Sharples, Goodlet, & Pemberton, 1992), a training system for neuroradiologists (Sharples et al., 2000), and a mobile learning device for children (Sharples, Corlett, & Westmancott, 2002). It has been adopted by the European MOBIlearn project (www.mobilearn.org) to develop mobile technology for learning. It also has been taught to undergraduate and postgraduate students to guide their interactive systems projects. An overview of the framework can be found in Sharples et al. (2002).

BACKGROUND

The approach of socio-cognitive engineering is similar to user-centered design (Norman & Draper, 1986) in that it builds on studies of potential users of the technology and involves them in the design process. But users are not always reliable informants. They may idealize their methods, describing ways in which they would like to or have been told to work, rather than their actual practices. Although users may be able to describe their own styles and strategies of working, they may not be aware of how other people can perform a task differently and possibly more effectively. Surveys of user preferences can result in new technology that is simply an accumulation of features rather than an integrated system.

Thus, socio-cognitive engineering is critical for the reliability for user reports. It extends beyond individual users to form a composite picture of the human knowledge and activity, including cognitive processes and social interactions, styles and strategies of working, and language and patterns of communication. The term *actor* is used rather than *user* to indicate that the design may involve people who are stakeholders in the new technology but are not direct users of it.

The framework extends previous work in soft systems (Checkland & Scholes, 1990), socio-technical and cooperative design (Greenbaum & Kyng, 1991; Mumford, 1995; Sachs, 1995), and the application of ethnography to system design (see Rogers & Bellotti [1997] for a review). It incorporates existing methods of knowledge engineering, task analysis, and object-oriented design, but integrates them into a coherent methodology that places equal emphasis on software, task, knowledge, and organizational engineering.

The framework also clearly distinguishes studying everyday activities using existing technology from studying how the activity changes with proposed technology. It emphasizes the dialectic between people and artefacts; using artefacts changes people's activities, which, in turn, leads to new needs and opportunities for design.

FRAMEWORK

Figure 1 gives a picture of the flow and main products of the design process. It is in two main parts: a phase of *activity analysis* to interpret how people work and interact with their current tools and technologies, and a phase of *systems design* to build and implement new interactive technology. The bridge between the two is the relationship between the Task Model and the Design Concept. Each phase comprises *stages* of analysis and design that are implemented through specific *methods*. The framework does not prescribe which methods to use; the choice depends on the type and scale of the project.

Figure 1. Overview of the flow and main products of the design process

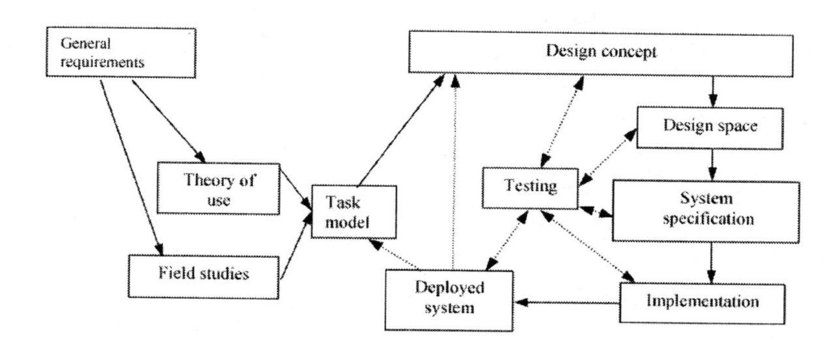

It is important to note that the process is not a simple sequence but involves a dialogue between the stages. Earlier decisions and outcomes may need to be revised in order to take account of later findings. When the system is deployed, it will enable and support new activities, requiring another cycle of analysis, revision of the Task Model, and further opportunities for design.

The elements of socio-cognitive engineering are as follows:

- **Project:** The diagram shows the process of design, implementation, and deployment for a single project.
- **Actors:** Different types of people may be involved in or affected by the design and deployment, including (depending on the scale of the project) design, marketing and technical support teams, direct users of the system, and other people affected by it (e.g., administrative staff).
- **Roles:** The actors take on roles (e.g., team leader), which may change during the project.
- **Stage:** Each box represents one stage of the project.
- **Methods:** Each stage can be carried out by one or more methods of analysis and design, which need to be specified before starting the stage.
- **Tools:** Each method has associated tools (for activity analysis, software specification, systems design, and evaluation) in order to carry out the method.
- **Outcomes:** Each stage has outcomes that must be documented, and these are used to inform and validate the system design.
- **Measures:** Each design decision must be validated by reference to outcomes from one of the stages.

The general sequence for socio-cognitive engineering is as follows:

1. Form a project team.
2. Produce General Requirements for the project.
3. Decide which methods and tools will be used for each stage of the project.
4. Decide how the process and outcomes will be documented.
5. Decide how the project will be evaluated.
6. Carry out each stage of the project, ensuring that the requirements match the design.
7. Carry out a continuous process of documentation and evaluation.

The process starts by specifying the *General Requirements* for the system to be designed. These provide broad yet precise initial requirements and constraints for the proposed system in language that designers and customers can understand. They are used to guide the design and to provide a reference for validation of the system. The requirements normally should indicate:

- The scope of the project;
- The main actors involved in designing, deploying, using, and maintaining the system;

- The market need and business case; and
- General attributes and constraints of the proposed system (i.e., whether it aims to support individual or collaborative working).

The requirements will be extended and made more precise as the project progresses.

This leads to two parallel studies: a theory-based study of the underlying cognitive processes and social activities, and an investigation into how everyday activities are performed in their normal contexts. The *Theory of Use* involves an analysis of relevant literature from cognitive psychology, social sciences, and business management to form a rich picture of the human knowledge and activity. It is essential that this should offer a clear guide to system design. Thus, it must be relevant to the intended use of the system and extend the requirements in a form that can be interpreted by software designers and engineers.

The aim of carrying out *Field Studies* is to uncover how people interact with current technology in their normal contexts. The role of the fieldworker is both to interpret activity and to assist technology design and organizational change. This addresses the widely recognized problem of ethnographic approaches that, while they can provide an understanding of current work practices, are not intended to explore the consequences of socio-technical change.

Table 1 shows a multi-level structure for field studies, with level 1 consisting of a survey of the existing organizational structures and schedules, levels 2 and 3 providing an analysis of situated practices and interactions of those for whom the technology is intended, and level 4 offering a synthesis of the findings in terms of designs for new socio-technical systems. The four levels give an overview of activity, leading to a more detailed investigation of particular problem areas, with each level illuminating the situated practices and also providing a set of issues to be addressed for the next level. These piece together into a composite picture of how people interact with technology in their everyday lives, the limitations of existing practices, and ways in which they could be improved by new technology.

The outcomes of these two studies are synthesized into a *Task Model*. This is a synthesis of theory and practice related to how people perform relevant activities with their existing technologies. It is the least intuitive aspect of socio-cognitive engineering; it is tempting to reduce it to a set of bullet-point issues, yet it provides a foundation for the systems design. It could indicate:

- The main actors and their activity systems;
- How the actors employ tools and resources to mediate their interaction and to externalize cognition;
- How the actors represent knowledge to themselves and to others;

Table 1. Multi-level structure for field studies

Level 1	*Activity structures and schedules*
Activity:	Study work plans, organizational structures, syllabuses, resources.
Purpose:	To discover how the activities are supposed to be conducted.
Outcome:	Description of the existing organizational and workplace structures; identification of significant events.
Level 2	*Significant events*
Activity:	Observe representative formal and informal meetings and forms of communication.
Purpose:	To discover how activities, communication, and social interaction are conducted in practice.
Outcome:	A description and analysis of events that might be important to system design; identification of mismatches between how activity has been scheduled and how it is has been observed to happen.
Level 3	*Conceptions and conflicts*
Activity:	Conduct interviews with participants to discuss areas of activity needing support, breakdowns, issues, differences in conception.
Purpose:	To determine people's differing conceptions of their activity; uncover issues of concern in relation to new technology; explore mismatches between what is perceived to happen and what has been observed.
Outcome:	Issues in everyday life and interactions with existing technology that could be addressed by new technology and working practices.
Level 4	*Determining designs*
Activity:	Elicitation of requirements; design space mapping; formative evaluation of prototypes.
Purpose:	To develop new system designs.
Outcome:	Prototype technologies and recommendations for deployment.

- The methods and techniques that the actors employ, including differences in approach and strategy;
- The contexts in which the activities occur;
- The implicit conventions and constraints that influence the activity; and
- The actors' conceptions of their work, including sources of difficulty and breakdown in activity and their attitudes toward the introduction of new technology.

The *Design Concept* needs to be developed in relation to the Task Model. It should indicate how the activities identified by the Task Model could be transformed or enhanced with the new technology. It should:

- Indicate how limitations from the Task Model will be addressed by new technology;
- Outline a system image (Norman, 1986) for the new technology;
- Show the look and feel of the proposed technology;
- Indicate the contexts of use of the enhanced activity and technology; and
- Propose any further requirements that have been produced as a result of constructing the Design Concept.

The Design Concept should result in a set of detailed design requirements and options that can be explored through the design space.

The relationship between the Task Model and Design Concept provides the bridge to a cycle of iterative design that includes:

- Generating a space of possible system designs, systematically exploring design option and justifying design decisions;
- Specifying the functional and non-functional aspects of the system;
- Implementing the system; and
- Deploying and maintaining the system.

Although these stages are based on a conventional process of interactive systems design (see Preece, Rogers, & Sharp [2002] for an overview), they give equal emphasis to cognitive and organizational factors as well as to task and software specifications. The stages shown in Figure 1 are an aid to project planning but are not sufficiently detailed to show all the design activities. Nor does the figure make clear that to construct a successful integrated system requires the designers to integrate software engineering with design for human cognition, social interaction, and organizational management. The 'building-block diagram in Table 2 gives a more detailed picture of the system's design process.

The four "pillars" indicate the main processes of software, task, knowledge, and organizational engineering. Each "brick" in the diagram shows one outcome of a design stage, but it is not necessary to build systematically from the bottom up. A design team may work on one pillar (e.g., knowledge engi-

Table 2. A building-block framework for socio-cognitive system design

	Software Engineering	*Task Engineering*	*Knowledge Engineering*	*Organizational Engineering*
Maintain	Installed system	New task structure	Augmented knowledge	New organizational structure
Evaluate	Debugging	Usability	Conceptual change, skill development	Organizational change
Integrate	**Prototype System**			
Implement	Prototypes, Documentation	Interfaces, Cognitive Tools	Knowledge Representation	Communications, Network Resources
Design	Algorithms and Heuristics	Human-Computer Interaction	Domain Map, User Model	Socio-Technical System
Interpret	**Task Model**			
Analyze	Requirements	Tasks: Goals, Objects, Methods	Knowledge: Concepts, Skills	Workplace: Practices, Interactions
Survey	Existing Systems	Conventional Task Structures and Processes	Domain Knowledge	Organizational Structures and Schedules
Propose	**General Requirements**			

neering) up to the stage of system requirements, or ot may develop an early prototype based on a detailed task analysis but without a systematic approach to software engineering. How each activity is carried out depends on the particular application domain, actors, and contexts of use.

The design activities are modular, allowing the designer to select one or more methods of conducting the activity according to the problem and domain. For example, the usability evaluation could include an appropriate selection of general methods for assessing usability, or it could include an evaluation designed for the particular domain.

It should be emphasized that the blocks are not fixed entities. As each level of the system is developed and deployed, it will affect the levels that follow (e.g., building a prototype system may lead to revising the documentation or re-evaluating the human-computer interaction; deploying the system will create new activities). These changes need to be analyzed and supported through a combination of new technology and new work practices. Thus, the building blocks must be revisited both individually to analyze and update the technology in use, and through a larger process of iterative redesign.

Although Table 1 shows system evaluation as a distinct phase, there also will be a continual process of testing to verify and validate the design, as shown in Figure 1. Testing is an integral part of the entire design process, and it is important to see it as a lifecycle process (Meek & Sharples, 2001) with the results of testing early designs and prototypes being passed forward to provide an understanding of how to deploy and implement the system, and the outcomes of user trials being fed back to assist in fixing bugs and improving the design choices.

The result of the socio-cognitive engineering process is a new socio-technical system consisting of new technology, its associated documentation, and proposed methods of use. When this is deployed in the workplace, home, or other location, it not only should produce bugs and limitations that need to be addressed but also engender new patterns of work, social, and organizational structures that become contexts for further analysis and design.

FUTURE TRENDS

The computer and communications industries are starting to recognize the importance of adopting a human-centered approach to the design of new socio-technical systems. They are merging their existing engineering, business, industrial design, and marketing methods into an integrated process, underpinned by rigorous techniques to capture requirements, define goals, predict costs, plan activities, specify designs, and evaluate outcomes. IBM, for example, has developed the method of User Engineering to design for the total user experience (IBM, 2004). As Web-based technology becomes embedded into everyday life, it increasingly will be important to understand and design distributed systems for which there are no clear boundaries between people and technology.

CONCLUSION

Socio-cognitive engineering forms part of an historic progression from user-centered design and soft systems analysis toward a comprehensive and rigorous process of socio-technical systems design and evaluation. It has been applied through a broad range of projects for innovative human technology and is still being developed, most recently as part of the European MOBIlearn project.

REFERENCES

Checkland, P., & Scholes, J. (1990). *Soft systems methodology in action*. Chichester, UK: John Wiley & Sons.

Greenbaum, J., & Kyng, M. (Eds.). (1991). *Design at work: Cooperative design of computer systems*. Hillsdale, NJ: Lawrence Erlbaum Associates.

IBM. (2004). *User engineering*. Retrieved August 12, 2004, from http://www-306.ibm.com/ibm/easy/eou_ext.nsf/publish/1996

Meek, J., & Sharples, M. (2001). A lifecycle approach to the evaluation of learning technology.

Proceedings of the CAL 2001 Conference, Warwick, UK.

Mumford, E. (1995). *Effective systems design and requirements analysis: The ETHICS approach*. Basingstoke, Hampshire, UK: Macmillan.

Norman, D.A. (1986). Cognitive engineering. In D.A. Norman, & S.W. Draper (Eds.), *User centred system design*. Hillsdale, NJ: Lawrence Erlbaum.

Norman, D.A., & Draper, S. (1986). *User centered system design: New perspectives on human-computer interaction*. Hillsdale, NJ: Lawrence Erlbaum Associates.

Preece, J., Rogers, Y., & Sharp, H. (2002). *Interaction design: Beyond human-computer interaction*. New York: John Wiley & Sons.

Rogers, Y., & Bellotti, V. (1997). Grounding blue-sky research: How can ethnography help? *Interactions, 4*(3), 58-63.

Sachs, P. (1995). Transforming work: Collaboration, learning and design. *Communications of the ACM, 38*(9), 36-44.

Sharples, M., Corlett, D., & Westmancott, O. (2002). The design and implementation of a mobile learning resource. *Personal and Ubiquitous Computing, 6*, 220-234.

Sharples, M., Goodlet, J., & Pemberton, L. (1992). Developing a writer's assistant. In J. Hartley (Ed.), *Technology and writing: Readings in the psychology of written communication* (pp. 209-220). London: Jessica Kingsley.

Sharples, M., et al. (2000). Structured computer-based training and decision support in the interpretation of neuroradiological images. *International Journal of Medical Informatics, 60*(30), 263-228.

Sharples, M., et al. (2002). Socio-cognitive engineering: A methodology for the design of human-centred technology. *European Journal of Operational Research, 132*(2), 310-323.

KEY TERMS

Activity System: The assembly and interaction of people and artefacts considered as a holistic system that performs purposeful activities. See http://www.edu.helsinki.fi/activity/pages/chatanddwr/activitysystem/

Human-Centred Design: The process of designing socio-technical systems (people in interaction with technology) based on an analysis of how people think, learn, perceive, work, and interact.

Socio-Technical System: A system comprising people and their interactions with technology (e.g., the World Wide Web).

Soft Systems Methodology: An approach developed by Peter Checkland to analyze complex problem situations containing social, organizational, and political activities.

System Image: A term coined by Don Norman (1986) to describe the guiding metaphor or model of the system that a designer presents to users (e.g., the desktop metaphor or the telephone as a "speaking tube"). The designer should aim to create a system image that is consistent and familiar (where possible) and enables the user to make productive analogies.

Task Analysis: An analysis of the actions and/or knowledge and thinking that a user performs to achieve a task. See http://www.usabilitynet.org/tools/taskanalysis.htm

User-Centered Design: A well-established process of designing technology that meets users' expectations or that involves potential users in the design process.

User Engineering: A phrase used by IBM to describe an integrated process of developing products that satisfy and delight users.

Software Engineering and HCI

Shawren Singh
University of South Africa, South Africa

Alan Dix
Lancaster University, UK

INTRODUCTION

Technology Affecting CBISs

As computer technology continues to leapfrog forward, CBISs are changing rapidly. These changes are having an enormous impact on the capabilities of organizational systems (Turban, Rainer, & Potter, 2001). The major ICT developments affecting CBISs can be categorized in three groupings: hardware-related, software-related, and hybrid cooperative environments.

Hardware-Related

Hardware consists of everything in the "physical layer" of the CBISs. For example, hardware can include servers, workstations, networks, telecommunication equipment, fiber-optic cables, handheld computers, scanners, digital capture devices, and other technology-based infrastructure (Shelly, Cashman, & Rosenblatt, 2003). Hardware-related developments relate to the ongoing advances in the hardware aspects of CBISs.

Software-Related

Software refers to the programs that control the hardware and produce the desired information or results (Shelly et al., 2003). Software-related developments in CBIS are related to the ongoing advances in the software aspects of computing technology.

Hybrid Cooperative Environments

Hybrid cooperative environments developments are related to the ongoing advance in the hardware and software aspects of computing technology. These technologies create new opportunities on the Web (e.g., multimedia and virtual reality) while others fulfill specific needs on the Web (e.g., electronic commerce (EC) and integrated home computing).

These ICT developments are important components to be considered in the development of CBIS's. As new types of technology are developed, new standards are set for future development. The advent of hand-held computer devices is one such example.

BACKGROUND

A Software Engineering View

In an effort to increase the success rate of information systems implementation, the field of software engineering (SE) has developed many techniques. Despite many software success stories, a considerable amount of software is still being delivered late, over budget, and with residual faults (Schach, 2002).

The field of SE is concerned with the development of software systems using sound engineering principles for both technical and non-technical aspects. Over and above the use of specification, and design and implementation techniques, human factors and software management should also be addressed. Well-engineered software provides the service required by its users. Such software should be produced in a cost-effective way and should be appropriately functional, maintainable, reliable, efficient, and provide a relevant user interface (Pressman, 2000a; Shneiderman, 1992; Whitten, Bentley, & Dittman, 2001).

There are two major development methodologies that are used to develop IS applications: the *traditional systems* development methodology and the *object-oriented* (OO) development approach.

The *traditional systems* approaches have the following phases:

- **Planning:** this involves identifying business value, analysing feasibility, developing a work plan, staffing the project, and controlling and directing the project.
- **Analysis:** this involves information gathering (*requirements gathering*), process modeling and data modeling.
- **Design:** this step is comprised of physical design, architecture design, interface design, database and file design, and program design.
- **Implementation:** this step requires both construction and installation.

There are various *OO* methodologies. Although diverse in approach, most *OO* development methodologies follow a defined system development life cycle. The various phases are intrinsically equivalent for all of the approaches, typically proceeding as follows:

- **OO Analysis Phase** (determining what the product is going to do) and extracting the objects (**requirements gathering**), **OO design phase, OO programming phase** (implemented in appropriate OO programming language), **integration phase, maintenance phase and retirement** (Schach, 2002).

One phase of the SE life cycle that is common to both the traditional development approach and the OO approach is *requirements gathering*. Requirements' gathering is the process of eliciting the overall requirements of the product from the customer (user). These requirements encompass information and control need, product function and behavior, overall product performance, design and interface constraints, and other special needs. The requirements-gathering phase has the following process: requirements elicitation; requirements analysis and negotiation; requirements specification; system modeling; requirements validation; and requirements management (Pressman, 2000a).

Despite the concerted efforts to develop a successful process for developing software, Schach (2002) identifies the following pitfalls:

- Traditional engineering techniques cannot be successfully applied to software development, causing the software depression (software crisis). Mullet (1999) summarizes the software crisis by noting that software development is seen as a craft rather than an engineering discipline. The approach to education taken by most higher education institutions encourages that "craft" mentality; lack of professionalism within the SE world (e.g., the failure of treating an operating system's crash as seriously as a civil engineer would treat the collapse of a bridge); the high acceptance of fault tolerance by software engineers (e.g., if the operating system crashes; reboot hopefully with minimal damage); the mismatch between hardware and software developments. Hardware and software developments are both taking place at a rapid pace but independently of each other. Both hardware and software developments have a maturation time to be compatible with each other, but by that time everything has changed. The final problem for software engineers is the constant shifting of the goalposts. Customers initially think they want one thing but frequently change their requirements.

Notwithstanding these pitfalls, Pressman (2000b) argues that SE principles *always* work. It is never inappropriate to stress the principles of solid problem solving, good design, thorough testing, control of change, and emphasis on quality.

The Web is an intricate and complex combination of technologies (both hardware and software) that are at different levels of maturity. Engineering Web-based EC software, therefore, has its own unique challenges. In essence, the network becomes a massive computer that provides an almost unlimited software resource that can be accessed by anyone with a modem (Pressman, 2000a). We illustrate these intricacies in Figure 1, which is a representation of a home computer that is attached to the Internet. It depicts the underlying operating system (the base platform), the method of connection to the Internet (dial up, the technology that supports Web activities), browser, an example of a Web communication language (HTML), and additional technology that may be required to be Web active.

Figure 1. EC Web application platform (adapted from Hurst & Gellady, 2000)

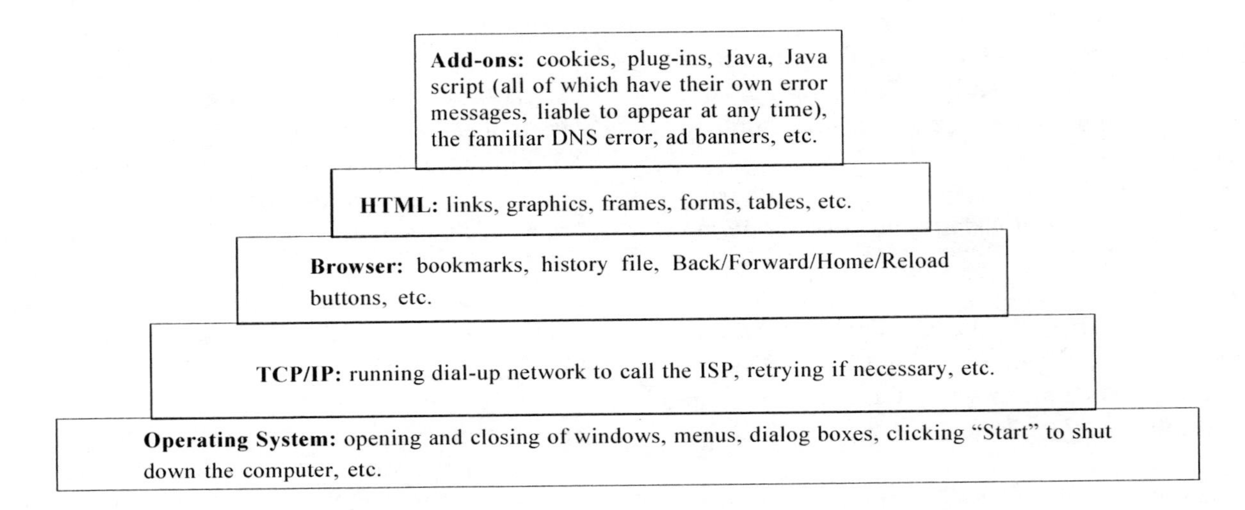

All the aspects of Figure 1 will support EC software in some way or another. An SE defect in any of these five layers would create a problem. For example, if the operating system is poorly engineered, the technology that sits on this platform will give piecemeal functionality at best. The problem is further complicated by piecemeal "patch" solutions. These piecemeal solutions can severely affect the usability of the Web, for example by giving cryptic error messages, installing add-ons that affect some unknown setting that the users do not understand, or installing add-ons that require a particular bit of hardware or software to be present.

The View of HCI Advocates

Human-computer interaction is concerned with the way in which computers can be used to support human beings engaged in particular activities. HCI thus involves the specification, design, implementation, and evaluation of interactive software systems in the context of the user's task and work (Preece, Rogers, & Sharp, 2002; Preece, Rogers, Sharp, Benyon, Holland, & Carey, 1994; Shneiderman, 1998).

An aspect related to HCI is *interaction design*. Interaction design is the process of designing interactive products to support people in their everyday and work lives. In particular, it is about creating user experiences that enhance and extend the way people work, communicate, and interact (Preece et al., 2002).

As stated earlier, it is the users' experience that affects their activities on the Web. The advocates of HCI are intent on discovering the key to successful user experiences and so the concept of usability is intensively investigated in HCI. The ISO 9241-11 standard (1999) defines usability as the following: the extent to which a product can be used by a specified set of users, to achieve specified goals (tasks) with effectiveness, efficiency, and satisfaction in a specified context of use.

INTEGRATED USABILITY

Several researchers have produced sets of generic usability principles, which can be used in improving software (e.g., Mayhew, 1999; Preece et al., 1994; Shneiderman, 1998, 2000). Some of these usability principles are: learnability, visibility, consistency and standards, flexibility, robustness, responsiveness, feedback, constraints, mappings, affordances, stability, simplicity, help, and documentation. Unfortunately, the definitions of such design and usability principles are mostly too broad or general, and, in some cases, very vague. Some of these principles have been adapted for EC (see for example Badre, 2002). It has been shown repeatedly that general usability advice is not effective on its own when designing systems for a context-specific environment. Therefore, it is generally difficult for a non-usability expert or a novice to apply these principles in a particular domain and situation, taking into account the unique factors that give rise to problems in that domain.

We argue that usability advice should be linked to a context-specific environment. For example, if a designer is interested in enticing surfers to stop browsing and engage in transactions, the designer would be well advised to make different design choices for an Internet banking site than for an online library. So, the design of a site for Pick 'n Pay (supermarket chain), ABSA (commercial bank), and the University of South Africa's library should therefore be approached differently.

The HCI proponents also propose certain life cycle models. Williges, Williges, and Elkerton (1987), for example, have produced a model of development to rectify some of the problems in the "classic" life cycle model of SE. In this approach, HCI principles and interface design drive the whole process. Other such life cycle models include the Star model of Hartson and Hix (1989), the Usability Engineering life cycle of Mayhew (1999), and the Interaction Design model of Preece et al. (2002). These methods also introduce various strategies for the development of effective user interfaces. The argument for putting forward these alternative development models is that by spotting user requirements early on in the development cycle, there will be less of a demand for code generation and modification in the later stages of systems development.

FUTURE TRENDS

Standards can serve as good anchor points to focus the dialogues and collaborative activities. However, the existing standards are rather inconsistent and thus confusing. More efforts should be invested to render these tools more usable and useful. Specifically, it is worthy to develop implementation strategies for situating or localizing the standards so that they can be applied effectively in particular contexts (Law, 2003).

CONCLUSION

Both the SE proponents and the HCI proponents have a point with regard to their approach. SE proponents try to produce a workable solution and HCI proponents try to develop a usable solution. The two approaches are not mutually exclusive. A workable solution may not be a usable solution, and a usable solution may not be a workable solution. The problem is that the HCI advocates are isolated from their SE colleagues, who in turn ignore the HCI advocates. The HCI advocates use a "blinder approach" in their attempt to develop software by only focusing on the HCI aspects of the design of software, while the SE developers are concerned with a satisfactory solution. The aspects of Figure 1 will in effect influence the HCI advocates' approaches as well as the SE advocates' approaches for designing software for the Web. The uncertainty aspect has to be factored into the design process.

REFERENCES

Badre, N. A. (2002). *Shaping Web usability: Interaction design in context*. Boston: Addison-Wesley.

Hartson, H. R., & Hix, D. (1989). Human-computer interface development: Concepts and systems for its management. *ACM Computing Surveys, 21,* 5-92.

Hurst, M., & Gellady, E. (2000). *White paper one: Building a great customer experience to develop brand, increase loyalty and grow revenues*. Creativegood. Retrieved September 14, 2003, from http://www.creativegood.com

ISO 9241-11 (1999). *Ergonomic requirements for office work with visual display terminals*. Part 11 Guidance on usability. Retrieved from http://www.iso.org./iso/en/catalogueDetailPage.catalogueDetail?csnumber=16883&icsi=13&lcsz=180&lc53=

Law, E. L. C. (2003). *Bridging the HCI-SE gap: Historical and epistemological perspectives.* Paper presented at the INTERACT Workshop, Zürich, Switzerland.

Mayhew, D. J. (1999). *The usability engineering lifecycle: A practitioner's handbook for user interface design*. San Francisco: Morgan Kaufmann.

Mullet, D. (1999). *The software crisis*. University of North Texas. Retrieved October 20, 2003, from http://www.unt.edu/benchmarks/archives/1999/july99/crisis.htm

Preece, J., Rogers, Y., & Sharp, H. (2002). *Interaction design: Beyond human-computer interaction*. New York: John Wiley & Sons.

Preece, J., Rogers, Y., Sharp, H., Benyon, D., Holland, S., & Carey, T. (1994). *Human-computer interaction*. Harlow, UK: Addison-Wesley.

Pressman, R. S. (2000a). *Software engineering: A practitioner's approach* (5th ed.). London: McGraw Hill.

Pressman, R. S. (2000b). What a tangled Web we weave. *IEEE Software,* (January/February), 18-21.

Schach, S. R. (2002). *Object-oriented and classical software engineering* (5th ed.). Boston: McGraw Hill.

Shelly, B. G., Cashman, J. T., & Rosenblatt, J. H. (2003). *Systems analysis and design* (5th ed.). Australia: Thomson: Course Technology.

Shneiderman, B. (1992). *Designing the user interface: Strategies for effective human-computer interaction*. Reading, MA: Addison Wesley.

Shneiderman, B. (1998). *Design the user interface: Strategies for effective human-computer interaction* (3rd ed.). Reading, MA: Addison-Wesley.

Shneiderman, B. (2000). Universal usability. *Communication of the ACM, 43*(5), 84-91.

Stair, R. M. (1992). *Principles of information systems: A managerial approach*. Boston: Boyd & Fraser.

Turban, E., Rainer, R. K., & Potter, E. R. (2001). *Introduction to information technology*. New York: John Wiley & Sons.

Vanderdonckt, J., & Harning, M. B. (2003, September 1-2). *Closing the gaps: Software engineering and human-computer interaction.* Interact 2003 Workshop, Zurich, Switzerland. Retrieved in July 2004, from http://www.interact2003.org/workshops/ws9-description.html

Whitten, L. J., Bentley, D. L., & Dittman, C. K. (2001). *Systems analysis and design methods* (5th ed.). Boston: McGraw-Hill Irwin.

Williges, R. C., Williges, B. H., & Elkerton, J. (1987). Software interface design. In G. Salvendy (Ed.), *Handbook of human factors* (pp. 1414-1449). New York: John Wiley & Sons.

KEY TERMS

Information Systems: First known as business data processing (BDP) and later as management information systems (MIS). The operative word is "system" because it combines technology, people, processes, and organizational mechanisms for the purpose of improving organizational performance.

Interaction Design: The process of designing interactive products to support people in their everyday and work lives.

ISO 9241-11: This part of ISO 9241 introduces the concept of usability but does not make specific recommendations in terms of product attributes. Instead, it defines usability as the "extent to which a product can be used by specified users to achieve specified goals with effectiveness, efficiency and satisfaction in a specified context of use."

Requirements' Gathering: The process of eliciting the overall requirements of a product from the customer.

Software Engineering: Concerned with the development of software systems using sound engineering principles for both technical and non-technical aspects. Over and above the use of specification, design and implementation techniques, human factors and software management should also be addressed.

Usability: The ISO 9241-11 standard definition of usability identifies three different aspects: (1) a specified set of users, (2) specified goals (asks) which have to be measurable in terms of effectiveness, efficiency, and satisfaction, and (3) the context in which the activity is carried out.

Spam

Sara de Freitas
Birbeck College, University of London, UK

Mark Levene
Birbeck College, University of London, UK

INTRODUCTION

With the advent of the electronic mail system in the 1970s, a new opportunity for direct marketing using unsolicited electronic mail became apparent. In 1978, Gary Thuerk compiled a list of those on the Arpanet and then sent out a huge mailing publicising Digital Equipment Corporation (DEC—now Compaq) systems. The reaction from the Defense Communications Agency (DCA), who ran Arpanet, was very negative, and it was this negative reaction that ensured that it was a long time before unsolicited e-mail was used again (Templeton, 2003). As long as the U.S. government controlled a major part of the backbone, most forms of commercial activity were forbidden (Hayes, 2003). However, in 1993, the Internet Network Information Center was privatized, and with no central government controls, spam, as it is now called, came into wider use.

The term *spam* was taken from the Monty Python Flying Circus (a UK comedy group) and their comedy skit that featured the ironic spam song sung in praise of spam (luncheon meat)—"spam, spam, spam, lovely spam"—and it came to mean mail that was unsolicited. Conversely, the term *ham* came to mean e-mail that was wanted. Brad Templeton, a UseNet pioneer and chair of the Electronic Frontier Foundation, has traced the first usage of the term *spam* back to MUDs (Multi User Dungeons), or real-time multi-person shared environment, and the MUD community. These groups introduced the term *spam* to the early chat rooms (Internet Relay Chats).

The first major UseNet (the world's largest online conferencing system) spam sent in January 1994 and was a religious posting: "Global alert for all: Jesus is coming soon." The term *spam* was more broadly popularised in April 1994, when two lawyers, Canter and Siegel from Arizona, posted a message that advertized their information and legal services for immigrants applying for the U.S. Green Card scheme. The message was posted to every newsgroup on UseNet, and after this incident, the term *spam* became synonymous with junk or unsolicited e-mail. Spam spread quickly among the UseNet groups who were easy targets for spammers simply because the e-mail addresses of members were widely available (Templeton, 2003).

BACKGROUND

At present, the practice of spamming is pervasive; however, due to the relative recent nature of the problem and due to its fast changing nature, the discussion about the topic has been limited to academic literature. While in computer science literature there has been a concentration of work on the technical features and solutions designed to prevent or ameliorate the practice (Androutsopoulos et al., 2000; Gburzynski & Maitan, 2004; Goodman & Rounthwaite, 2004), the more general scientific discussion has been provided by a few scientific commentators (Gleick, 2003; Hayes, 2003), and the few books written on the subject (Schwartz & Garfinkel, 1998) have become outdated in a relatively short span of time. In other academic areas, there is some literature available concerning the legal implications of spam (Crichard, 2003) and the marketing dimension of spamming (Nettleton, 2003; Sipior et al., 2004); however, these, too, have suffered from the fast changing and global scope of the problem. Furthermore, aspects such as the social and political implications of spamming have been restricted to journalistic commentary in newspaper articles (BBC News, 2003, 2004; Gleick, 2003; Krim, 2004). In order to provide a broader focus in this article, therefore, the authors have supplemented this literature with interviews conducted with specialists in the field in order to provide the most up-to-date information, including interviews with Enrique Salem, CEO of Brightmail; Mikko Hyponnen of F-Secure; and Steve Linford of the Spamhaus Project.

However, while the broader issues of spamming have been discussed in the general literature reviewed, in the area of human-computer interaction, there has been a paucity of discussion, although this may change with the wider take-up of mobile devices with their context awareness. Notable articles that have touched on related issues in the human-computer interaction field have included those that have considered issues of privacy (Ackerman et al., 2001) and usability in particular difficulties with using computer technology (Kiesler et al., 2000). However, this is not to say that spamming does not play a role in reversing the convenience that many experience when using e-mail on their desktop, laptop, or mobile device, and it is often the most vulnerable that are affected adversely by spamming practice.

The mass appeal and use of electronic mail over the Internet has brought with it the practice of spamming or sending unsolicited bulk e-mail advertising services. This has become an established aspect of direct marketing, whereby marketers can reach many millions of people around the world with the touch of a button. However, this form of direct marketing or spamming, as it has come to be called, has become an increasing problem for many, wasting people's time as they delete unwanted e-mail and slowing down the movement of electronic traffic over local and wide area networks (Salem interview, 2004; Goodman & Rounthwaite, 2004).

Graph 1. The escalation of spam worldwide, 2001 to July 2004 (Source: Brightmail)

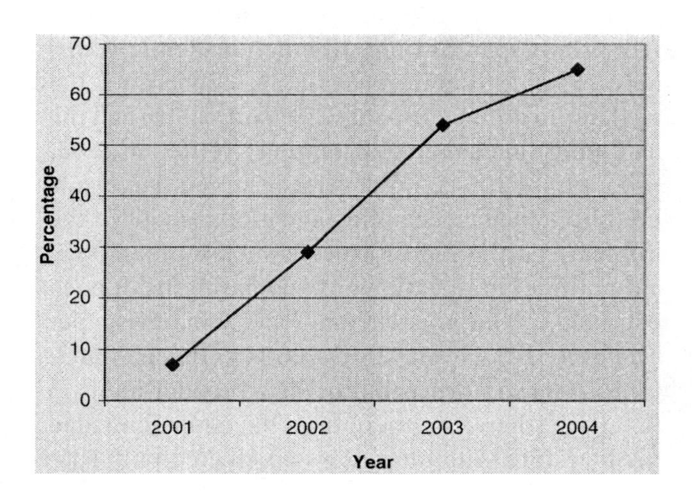

The scale of the problem has become particularly concerning in recent months; unsolicited e-mail—or spam—currently accounts for 65% of all e-mail received in July 2004 (Brightmail, 2004; Enrique Salem, CEO of Brightmail, interview 2004). Of the 70 million e-mails that Brightmail filtered in September 2003 alone, 54% was unsolicited, and that percentage is increasing year after year (see Graph 1). But although there are a number of different ways to filter unwanted e-mail, which may lead to a significant reduction of spam in the short term, many experts in the field are concerned that spam will never be completely eradicated (Hypönnen, F-Secure interview, 2004; Linford, Spamhaus interview, 2004).

CRITICAL ISSUES OF SPAM

So who are the spammers? The spammers can be identified in three main groups: (1) legitimate commercial direct marketers, who want to make commercial gain from sending bulk e-mails about products and services; (2) criminal groups, including fraudsters, who are using spam to "legitimise"' their activities and to defraud others (Gleick, 2003; Levy, 2004; Linford interview, 2004); and (3) disaffected individuals—crackers—who want to disrupt Internet services and who, in many cases, may have inside information about how the systems are structured. The criminal group is potentially the most dangerous, and while spam is not an illegal activity, this practice is set to spread to the criminal fraternity in China, Russia, and South America. This trend is becoming more widespread with the ease of obtaining spam kits over the Internet, which allows the potential spammer to set up quickly (Thomson, 2003).

Increasingly, illegitimate spammers, fraudsters, and crackers are joining forces to introduce fraud schemes such as the 419 scam and phishing (sending e-mails as if they came from trusted organisations) to convince unsuspecting victims to reveal sensitive personal information; in particular, to gain information about users' credit card information or to gain access details of online transaction services (Levy, 2004).

WAYS OF COMBATING SPAM

In the light of this increasing problem, a series of attempts, both technological and non-technological, have been made to try to combat the annoyance of full mailboxes in order to counter the heavyweight of unwanted e-mail traffic and to deter criminal activity (Goodman & Rounthwaite, 2004). Hand in hand with the push for tighter legislation to tackle the problem, several technical solutions have been deployed, and new ones are being proposed.

Before an e-mail arrives in your mailbox it passes through a mail server, which is either hosted within your organization or through an Internet Service Provider (ISP). Filtering out spam at this early stage (pre-receipt) before the message arrives at your machine is obviously desirable, and many IT departments and ISPs already have installed anti-spam software on their servers. Tools also exist that are user-based and filter out e-mail that already has arrived at your mailbox (post-receipt). Due to the flood of spam that is relentlessly sent to us, for now, it is probably best to have filtering tools both at the server and the user ends.

Two problems that need to be addressed by any spam filtering system are the rates of false positives and false negatives. A false positive is a mail message that the filter tags as spam but is actually ham, while a false negative is a mail message that the filter tags as ham but is actually spam. Having no filter at all is the case of 0% false positives and 100% false negatives, and a filter that blocks everything is one with 100% false positives and 0% false negatives. Ideally, we want 0% false positive (i.e., all ham gets through the filter) and 0% false negatives (i.e. all spam is blocked).

The methods for combating spam include the following, which are summarized in tabular form (see Table 1).

- Blocklisting
- Protocol change
- Economic solutions
- Computational solutions
- E-mail aliasing
- Sender warranted e-mail
- Collaborative filtering
- Rule-based solutions
- Statistical solutions
- Legislative solutions

All these methods for combating spam impede the usability of e-mail and necessitate extra technical and administrative support; however, the safety and security for individuals using Internet and e-mail-based services is reliant upon controlling the misuse of the systems; therefore, these methods are a trade-off between free and open access and secure and safe systems. Of course, there are social and political implications for employing these preventative methods; however, there is clearly a need to address these failings using more than one listed method.

There is clearly a need to consider the problem of spam in the human-computer interaction field, particularly relating to issues of increasing usability for more vulnerable user groups, such as those with particular disabilities, frailties, and illnesses, who may be particularly susceptible to particular scams and fraudulent deceits.

FUTURE TRENDS

Future areas of development for spamming may center upon relatively unprotected mobile phones and devices (Sipior et al., 2004; Syntegra, 2003). To date, the practice known as *wardriving*, where individuals drive around until they detect wireless connectivity and then bombard the unprotected network with spam, provides a real indication about the potential dangers of spamming for the future. Another concerning trend has been the use of spam to send out viruses (Stewart, 2003), the SoBig virus attack, for example, that used this method.

In addition, the cheap and easy availability of spam kits that provide mailing lists and the spamming software on the Internet have spread the practice to new territories, in particular to China, Russia, and South America, making the practice more widespread and leading to an escalation in the rate of spamming.

Other adaptations of the spamming practice recently have included the use of malicious code, using worms and trojans spam relays are created; the MyDoom worm operated in this way, installing proxies that spammers could then exploit.

Table 1. Methods for combating spam

Solution	Method	Benefits	Limitations
Block listing	Use of lists of IP addresses of known sources of spam (e.g., SBL and RBL)	Blocks a significant volume of spam	Cannot block all spam and needs to be updated on a regular basis
Protocol change	To provide a method of tracking the source of an e-mail	Will help to identify spammers and add spam addresses to block lists	Will not prevent spam as such
Economic solutions	Impose a fee for sending e-mail	Will deter spammers from sending large volumes of junk e-mail	Will be difficult and costly to implement a worldwide standard for collecting the fee
Computational solutions	Impose an indirect payment in the form of a machine computation prior to sending e-mail	It is a viable alternative to the economic solution, without needing the infrastructure to collect a fee	A protocol involving cryptographic techniques will need to be put in place and software developed to implement the method
E-mail aliasing	Set up e-mail aliases for different groups of people with different acceptance criteria	Will reduce spam through an authentication process	This method involves an extension to current e-mail servers and the management of e-mail aliases
Sender warranted e-mail	Use of a special header to certify the e-mail as valid	No need for additional software or e-mail protocol	Will probably not deter spammers if widely adopted, and wide licensing of the technology will be problematic
Collaborative filtering	Communities collaborate to fight spam using a collaborative tool that is an add-on to e-mail software	Possible eradication of large volumes of spam through collaborative reporting of spam	Still vulnerable to random changes in spam e-mail, and there are problems with scalability of this method
Rule-based solutions	These filters maintain a collection of patterns to be matched against incoming spam, as in SpamAssassin	It is easy to install and effective in blocking a large percentage of spam, and in the case of SpamAssassin, it is free	It needs a lot of tuning and should be combined with other methods to filter out a larger volume of spam
Statistical solutions	Often deployed as a post-receipt spam filter using Bayesian text classification to tag e-mail as spam or ham	It is very effective and also adaptive, so it is hard to fool	Most effective when used with other pre-receipt filter systems
Legislative solutions	National and global legislation to enforce anti-spam laws	Prosecution of individual spammers	Problems of enforcement, not least due to crossing of different jurisdictional boundaries

The increase in the technical sophistication of spammers also is evidenced by the use of so-called reputation attacks, where spammers use a worm to launch a denial of service attack against anti-spamming organisations. One such example was the Mimail attacks (Levy, 2004) that specifically targeted anti-spam organisations seeking to block out spam. Clearly, spamming is becoming more refined and will evolve to adapt to any perceived weaknesses in network security.

CONCLUSION

This article has highlighted the scale and depth of the spamming problem, and while many are committed to the eradication of all Internet-based fraud and illegitimate activity, it seems unlikely that spam will completely disappear. It is more likely that the practice will continue to evolve and transmute to adapt to new vulnerabilities in the systems and to exploit users who are not fully aware of how they can be exploited through impersonations of familiar Web sites and services. With the current force behind the anti-spam movement gaining momentum, we can expect to see less spam in the future, but only with preventative measures such as those described in this article being put in place. In the near future however, the cat-and-mouse game among spammers and anti-spammers is set to continue. In particular, the new routes for spammers clearly lie in reaching users through mobile devices, which need to become better protected by virus and spam software, if these cyber crimes are to be controlled and ameliorated.

REFERENCES

Ackerman, M., Darrell, T., & Weitzner, D.J. (2001). Privacy in context. *Human-Computer Interaction, 16*, 167-176.

Androutsopoulos, I., Koutsias, J., Chandrinos, K.V., & Spyopoulos, C.D. (2000). An experimental comparison of naïve Bayesian and keyword-based anti-spam filtering with personal e-meal messages. *Proceedings of the 23rd Annual International ACM SIGIR Conference on Research and Development in Information Retrieval*, Athens, Greece.

BBC News. (2003). *Top UK sites fail privacy test.* Retrieved August 1, 2004, from http://news.bbc.co.uk/2/hi/technology/3307705.stm

BBC News. (2004). *US anti-spam law fails to bite.* Retrieved August 1, 2004, from http://news.bbc.co.uk/2/low/technology/3465307.stm

Brightmail. (2004). Retrieved August 1, 2004, from http://www.brightmail.com/spamstats.html

Crichard, M. (2003). Privacy and electronic communications. *Computer Law and Security Report, 19*(4), 299-303.

Gburzynski, P., & Maitan, J. (2004). Fighting the spam wars: A re-mailer approach with restrictive aliasing. *ACM Transactions on Internet Technology, 4*(1), 1-30.

Gleick, J. (2003). Get out of my box. *Guardian Review,* 1-2.

Goodman, J., & Rounthwaite, R. (2004). Stopping outgoing spam. *Proceedings of the 5th ACM Conference on Electronic Commerce*, New York.

Hayes, B. (2003). Spam, spam, spam, lovely spam. *American Scientist, 91*(3), 200-204.

Hypönnen, M. (personal communication, February 27, 2004)

Kiesler, S., Zdaniuk, B., Lundmark, V., & Kraut, R. (2000). Troubles with the Internet: The dynamics of help at home. *Human-Computer Interaction, 15*, 323-351.

Krim, J. (2004, May 21). Senate hears mixed reviews of anti-spam law. *Washington Post.* Retrieved August 1, 2004, from http://www.washingtonpost.com/wp-dyn/articles/A43622-2004May20.html

Levy, E. (2004). Criminals become tech savvy. *IEEE Security and Privacy, 2*(2), 65-68.

Linford, S. (personal communication, January 9, 2004)

Nettleton, E. (2003). Electronic marketing and the new anti-spam regulations. *The Journal of Database Marketing and Customer Strategy Management, 11*(3), 235-240.

Salem, E. (personal communication, February 23, 2004)

Schwartz, A., & Garfinkel, S. (1998). *Stopping spam.* Beijing: O'Reilly and Associates.

Sipior, J.C., Ward, B.T., & Bonner, P.G. (2004). Should spam be on the menu? *Communications of the ACM, 47*(6), 59-63.

Stewart, J. (2003). Spam and SoBig: Arm in arm. *Network Security,* 12-16.

Syntegra. (2003). *Can spam kill the mobile messaging market?* [white paper]. Retrieved August 1, 2004, from http://www.us.syntegra.com/acrobat208950.pdf

Templeton, B. (2003). Brad Templeton Home Page. Retrieved October 29, 2003, from www.templetons.com/brad/spam react.html

Thomson, I. (2003). *Mafia muscles in on spam and viruses.* Vnunet.com. Retrieved Aptil 26, 2004, from http://www.vnunet.com/News/1151421

KEY TERMS

Blocklisting: Set up as an approach for blocking unsolicited or junk e-mail. Blocklists provide lists of URLs or Web addresses from which spammers operate. The blocklists therefore provide a way of ameliorating or preventing spam from reaching the intended destination.

E-Mail Aliasing: Where an individual has more than one e-mail address, the practice allows the user to use different addresses for different tasks; for example, one address for Internet communications and another for business.

False Negatives: A false negative is a mail message that the filter tags as ham but is actually spam.

False Positives: A false positive is a mail message that the filter tags as spam but is actually ham.

Phishing: Short for password harvest fishing, it is the process of impersonating another trusted person or organization in order to obtain sensitive personal information, such as credit card details, passwords, or access information.

Sender Warranted E-Mail: This method allows the sender to use a special header to certify that the e-mail is genuine. The process could help to prevent spam scams.

Spam: Otherwise termed unsolicited e-mail, unsolicited commercial e-mail, junk mail, or unwanted mail, it has been used in opposition to the term *ham*, which is wanted e-mail. The term was developed from a Monty Python comedy sketch depicting spam as useless and ham as lovely, albeit in ironic terms.

Wardriving: Also termed WiLDing—Wireless Lan Driving, it is an activity whereby individuals drive around an area detecting Wi-Fi wireless networks, which they then can access with a laptop.

S

Spam as a Symptom of Electronic Communication Technologies that Ignore Social Requirements

Brian Whitworth
New Jersey Institute of Technology, USA

INTRODUCTION

Spam, undesired and usually unsolicited e-mail, has been a growing problem for some time. A 2003 Sunbelt Software poll found spam (or junk mail) has surpassed viruses as the number-one unwanted network intrusion (Townsend & Taphouse, 2003). *Time* magazine reports that for major e-mail providers, 40 to 70% of all incoming mail is deleted at the server (Taylor, 2003), and AOL reports that 80% of its inbound e-mail, 1.5 to 1.9 billion messages a day, is spam the company blocks. Spam is the e-mail consumer's number-one complaint (Davidson, 2003). Despite Internet service provider (ISP) filtering, up to 30% of in-box messages are spam. While each of us may only take seconds (or minutes) to deal with such mail, over billions of cases the losses are significant. A Ferris Research report estimates spam 2003 costs for U.S. companies at $10 billion (Bekker, 2003).

While improved filters send more spam to trash cans, ever more spam is sent, consuming an increasing proportion of network resources. Users shielded behind spam filters may notice little change, but the Internet transmitted-spam percentage has been steadily growing. It was 8% in 2001, grew from 20% to 40% in 6 months over 2002 to 2003, and continues to grow (Weiss, 2003). In May 2003, the amount of spam e-mail exceeded nonspam for the first time, that is, over 50% of transmitted e-mail is now spam (Vaughan-Nichols, 2003). Informal estimates for 2004 are over 60%, with some as high as 80%. In practical terms, an ISP needing one server for customers must buy another just for spam almost no one reads. This cost passes on to users in increased connection fees.

Pretransmission filtering could reduce this waste, but creates another problem: spam false positives, that is, valid e-mail filtered as spam. If you accidentally use spam words, like *enlarge*, your e-mail may be filtered. Currently, receivers can recover false rejects from their spam filter's quarantine area, but filtering before transmission means the message never arrives at all, so neither sender nor receiver knows there is an error. Imagine if the postal mail system shredded unwanted mail and lost mail in the process. People could lose confidence that the mail will get through. If a communication environment cannot be trusted, confidence in it can collapse.

Electronic communication systems sit on the horns of a dilemma. Reducing spam increases delivery failure rate, while guaranteeing delivery increases spam rates. Either way, by social failure of confidence or technical failure of capability, spam threatens the transmission system itself (Weinstein, 2003). As the percentage of transmitted spam increases, both problems increase. If spam were 99% of sent mail, a small false-positive percentage becomes a much higher percentage of valid e-mail that failed. The growing spam problem is recognized ambivalently by IT writers who espouse new Bayesian spam filters but note, "The problem with spam is that it is almost impossible to define" (Vaughan-Nichols, 2003, p. 142), or who advocate legal solutions but say none have worked so far. The technical community seems to be in a state of denial regarding spam. Despite some successes, transmitted spam is increasing. Moral outrage, spam blockers, spamming the spammers, black and white lists, and legal responses have slowed but not stopped it. Spam blockers, by hiding the problem from users, may be making it worse, as a Band-Aid covers but does not cure a systemic sore. Asking for a technical tool to stop spam may be asking the wrong question. If spam is a social problem, it may require a social solution, which in cyberspace means technical support for social requirements (Whitworth & Whitworth, 2004).

BACKGROUND

Why Spam Works

Spam arises from the online social situation technology creates. First, it costs no more to send a million e-mails than to send one. Second, "hits" are a percentage of transmissions, so the more spam sent means more sender profit. Hence, it pays individuals to spam. The logical goal of spam generators is to reach all users to maximize hits at no extra cost. Yet the system cannot sustain this. With 23 million businesses in America alone, if each sent just one unsolicited message a year to all users, that is over 63,000 e-mails per person per day. Spam seems the electronic equivalent of the "tragedy of the commons" (Hardin, 1968), where some farmers, each with some cows and land, live near a common grass area. The tragedy is that if the farmers calculate their benefits, they all graze the commons, which is destroyed from overuse. In this situation, individual temptation can undermine a public-good commons.

For spam, the public good is free online communication, and the commons is the wires, storage, and processors of the Internet. The individual temptation is to use the commons for personal gain. E-mail creates value by exchanging meaning between people. As spam increases, e-mail gives less meaning for more effort, that is, less value. Losses include wasted processing, storage, and lines; "ignore time" (time to reject spam); antispam software costs; time to resolve spam false positives; time to confirm spam challenges; important messages lost by spam; and unknown lost opportunity costs from messages not sent because spam raises the user cost to send a message (Reid, Malinek, Stott, & T., 1996). E-mail lowered this communication threshold, but spam makes communication harder by degrading the e-mail commons. If half of Internet traffic is spam, the Internet is half wasted, and for practical purposes, half destroyed. Spam seems to be an electronic tragedy of the commons.

SOME SPAM RESPONSES

If spam is a traditional social problem in electronic clothes, why not use traditional social responses?

Ignore It

One answer to spam is to ignore it: After all, if no one bought, spam would stop. However, a "handful of positive responses is enough to make a mailing pay off, and there will always be a handful of suckers out there" (Ivey, 1998, p. 15). There are always spam responders; a new one is born on the Internet every minute.

Ethics

Online society seems unlikely to make people more ethical than they are in physical society, so it seems unlikely spammers will "see the light" any time soon.

Barriers

Currently the most popular response to spam is spam filters, but spammers need only 100 takers per 10 million requests to earn a profit (Weiss, 2003), much less than a 0.01% hit rate. So even with 99.99% successful spam blockers, spam transmission will increase.

Revenge

One way users handled companies faxing annoying unsolicited messages was by "bombing" them with return faxes, shutting down their fax machines. For e-mail, ISPs, not senders, are registered. If we isolate ISPs that allow spam, this penalizes valid users as well as spammers. Lessig (1999) argued before the U.S. Supreme Court for a bounty on spammers, "like bounty hunters in the Old West" (Bazeley, 2003). However, the cyberspace "Wild West" is not inside America, nor under U.S. courts. And do we really want an online vigilante society?

Third-Party Guarantees

Another approach is for a trusted third party to validate all e-mail. The Tripoli method requires all e-mails to contain an encrypted guarantee from a third party that it is not spam (Weinstein, 2003). However, custodian methods require significant coordination and raise Juvenal's question, "Quis custodiet ipsos custodies [Who will watch our watchers]?" Will stakeholders like the Direct Marketing Association or Microsoft guarantee against spam? If spam is in the eye of the beholder, such companies may consider their spam not spam at all.

Legal Responses

Why not just pass a law against spam? This approach may not work for several reasons (Whitworth & deMoor, 2003). First, virtual worlds work differently from the physical world. Applying laws online creates problems; for example, financial and health-care organizations by law must archive all communications so must not only receive spam, but also store it (Paulson, 2003). It is difficult to stretch physical law into cyberspace (Samuelson, 2003). Legal prosecutions require physical evidence, an accused, and a plaintiff, yet spam evidence is in a malleable cyberspace, e-mail sources are easily "spoofed" to hide the accused, and the plaintiff is everyone with an e-mail address. What penalties apply when each individual loses so little? Second, virtual worlds change faster than physical worlds. Spam can mutate in form, for example, Internet messaging spam or "spim." Any spam variant would require new laws, yet while society takes years to pass laws, the Internet can change monthly. Third, in cyberspace, code is law (Mitchell, 1995). Software can make spammers anonymous or generate new addresses so quickly that bans have no effect. Finally, laws are limited by jurisdiction; for example, state laws against telemarketers were ineffective against out-of-state calls, and the U.S. nationwide do-not-call list is ineffective against overseas calls. U.S. law applies to U.S. soil, but spam can come from any country. Traditional law seems too physically constrained, too slow, and too impotent to deal with the spam challenge. As Ken Korman (2003, p. 3) concedes, "Though legislative efforts to control spam continue, it is unlikely that new laws will have any real effect on the problem." *PC World* adds, "By all accounts, CAN-SPAM has failed to stop the e-mail inundation" (Spring, 2004, p. 24).

Challenge Systems

Challenge systems, like MailBlocks (2003), ask e-mail senders, "Are you really a person? If so, type the number shown in this graphic." Since most spam is computer generated, and most spammers will not accept replies (lest they be spammed in return), such methods work well, but users communicate twice to receive once.

An E-Mail Charge

One way to change the communication environment is to charge for e-mail. This would hit spammer's pockets, but also reduce general usage by increasing the communication threshold (Kraut, Shyam, Morris, Telang, Filer, & Cronin, 2002). What would be the purpose of a charge, however small? An Internet toll would add no new service as e-mail already works without such charges. Its sole purpose would be to punish spammers by slowing the flow for everyone. A variant is that all senders compute a time-costly function (Dwork & Naor, 1993), but the effect is still to increase the transmission cost. Increasing across-the-board e-mail costs seems like burning down your house to prevent break-ins. If e-mail were metered, we would all pay for something already paid for. Who would receive each payment? If senders paid receivers, each e-mail would be a money transfer. The cost of administering such a system could outweigh its benefit, and who would set the charge rate? If e-mail providers took the charge, it would be an e-mail tax, but what global entity can legitimately claim it? Making the Internet a field of profit could open it to corruption. Spam works because e-mail costs so little, but that is also why the Internet works. Fast, easy, and free communication has benefited us all. To raise the communication threshold by charging for what we already have seems retrogressive. A solution that reduces spam but leaves the Internet advantage intact is to design for fair communication in the first place.

LEGITIMATE COMMUNICATION

Spam is an opportunity as well as a threat. The challenge is to close the social-technical gap (Ackerman, 2000) between society and technology. Traditional social methods, like the law, are struggling to do this. An alternative is for technology to support society rather than being impartial to social needs. The Internet was once thought to be innately ungovernable, but it could just as easily become a system of perfect regulation and control (Lessig, 1999). If in cyberspace code, not law, makes the rules, it makes sense to design social software to support legitimate interaction, that is, social exchanges that are both fair to individuals and beneficial to the social group (Whitworth & Whitworth, 2004). This raises the question of whether spam is legitimate communication.

Is Spam Legitimate Communication?

Spam is unfair because senders have all the transmission choices, just like telemarketers who have your home phone number but invariably refuse to give you theirs. They call you at home, but you cannot call them at home. Spammers waste others' time, but this is irrelevant to them because it is not their loss. Yet the loss is still real, and it is unfair that those who cause it do not bear it, that those who suffer spam are not its creators.

Spam is unprofitable to society if its total losses exceed its total profit. If 90% of people spammed do not buy, do their losses balance the gains of the 10% who buy? What if 99.9% do not buy? There is a saturation point when spam's losses outweigh its benefits. We seem well past that point already. By one estimate, it costs about $250 to send a million e-mails, which cause about $2,800 in lost wages to society in general (Emery, 2003). Spammers steal time, which in today's world equates to money. Some see it as a mild crime, like littering on the Internet, but when litter blocks the streets, there is concern. Over millions of people the productivity loss is significant, as a cyber thief taking a few cents from millions of bank accounts can steal a sizable sum.

If spam is unfair to individuals and harmful to online society, it is illegitimate communication on two counts.

Communication Rights

The method of legitimacy analysis (Whitworth & deMoor, 2003) asks, Who owns the elements of e-mail communication: the messages, channels, and addresses?

Who Owns E-Mail Messages?

From a social-rights perspective, e-mail is a request, not a requirement, to receive a message. Receivers should be able to refuse ownership after reading it, perhaps via an e-mail toolbar rejection button. The receiver does not own a rejected message (by definition), and the transmission system does not own it, so it belongs to the sender who created it and, as with postal mail, should be returned to the sender. This does not happen because e-mail was designed as a forward-and-forget system, so replies to spammers may go nowhere (Cranor & LaMacchia, 1998), one reason the spam-the-spammer approach does not work (Held, 1998).

The social logic that communication is a two-way process implies that receiving back rejected e-mail should be a necessary condition of transmission. Rejected spam would then return down the sender's communication lines to their computer, creating spammer disk and channel costs. It seems inefficient to return rejected messages that can be deleted at delivery, but supporting social accountability in the long term both reduces waste and tells senders an e-mail was rejected. Currently, spammers do not know who reads their messages and who does not. If rejected e-mail were returned, it would pay spammers to reduce their lists and give them the information needed to do so. The right to reject e-mail is a social requirement. Implementing it is an engineering problem. The e-mail transmission system controls both the pieces of the communication game and the board itself. It should be able to enforce a rule that to send into the system, one must also receive from it.

Who Owns Communication Channels?

Current systems give any sender the automatic right to open a channel to another. Yet society gives no such right to communicate, but rather the right to be left alone (Warren & Brandeis, 1890). The social concept is that one is not forced to communicate. To pursue undesired interaction is to harass or stalk. If someone knocks on our door, we need not answer. If they telephone, we need not pick up. But we get e-mail in our inbox, like it or not.

E-mail systems could present new messages in two parts: an initial "Can I talk to you?" channel request, then the messages and content. Channel requests could give channel properties like the sender, title, and reciprocity (if replies are accepted; Rice, 1994), but not message content. Microsoft's plan to offer caller ID for e-mail seems a step in the right direction as it gives some channel information to receivers, but why not give all channel information? Receivers could then only receive messages from those who also receive. Current challenge-spam defenses offer this service but transmit content multiple times, and if the challenge bounces, they multiply spam.

Channel requests would send no content, only channel properties. The receiver can choose to open the channel or not. No third party need guarantee anything. No tedious challenges to sender humanity are needed. Sending messages is as before, except one could get a "channel unavailable" response. This is not a message rejection, but an unwillingness to talk at all. To receivers, messaging would also look the same, except unknown messages (like spam) would appear in a separate "Request to Converse" in-box, where users must double-click them to get content. Since most people do not click on spam, transmission volumes would reduce. Such handshaking occurs in data networks and could occur for e-mail. Giving known senders a permanent channel would create a self-generated list of known communicants (Hall, 1998).

Who Owns E-Mail Addresses?

The social concept of privacy suggests that people own their personal data. Good companies already include in their messages phrases such as, "To stop further e-mail, reply to this message." Yet these voluntary acts are not enough. Spammers can feign them, or worse, use your reply to confirm an active e-mail and sell your address to others. Requesting removal could put you on even more lists, becoming what *PC World* magazine calls "spam bait" (Spring, 2003): "By now, most computer users know that replying to most spam only generates more spam" (Woellert, 2003, p. 56). Yet if users managed their own online data records, they could save companies data-maintenance costs.

FUTURE TRENDS

Currently, spam is tolerated by technology as the bandwidth can handle it. However, this may not continue. Some hope technology will continue to expand bandwidth and processing beyond the spam challenge, but simple arithmetic suggests otherwise. The spam potential increases as the square of the number of users, which grows each day. In a future with billions of people online, the potential interactions are beyond any technology we can presently conceive. The predictions are gloomy. Given current trends, it seems there is nothing to stop spam from becoming over 95% of Internet transmissions in a decade. Meanwhile, society's laws still struggle with telephone spam (telemarketing), let alone computer spam. The question seems to be not if e-mail will fail, but when.

Some experts suggest e-mail is already "broken," but will be replaced by new, and better, forms of communication (Boutin, 2004). Time will tell if this is true. If spam is a general social disease, it may cross application boundaries. Already, spim, a spam version of Internet instant messaging (Hamilton, 2004), is growing faster than spam ever did. Technology may not insulate us from antisocial acts in computer-mediated communication (CMC).

Spam seems to be a watershed moment, a critical point at which traditional social values and technology power confront. The stakes are high. If human society loses its way in cyberspace, the vision of an electronic global society may fade. A brighter scenario is that the legitimate-communication requirement will be recognized and technology redesigned accordingly; that is, the social-technical gap will close. Currently, the unity of global society is not political or legal, but technical. Society lets people return postal mail, but e-mail does not let people return messages. Society recognizes the right not to communicate, but e-mail gives a right to communicate. Society would let people remove themselves from marketing lists, but one cannot remove oneself from e-mail lists. Technology has the social requirements backward. Spammers force messages upon us that we should be able to reject. They access in-boxes we should own. They control e-mail addresses that should be ours. Technology gives spammers every reason to do what they are doing, and no reason to stop.

If the social-technical gap were reduced, spam would also reduce. If e-mail could be returned to the sender and really arrive there, spam would reduce. If spammers had to "knock" before entering an inbox, spam would reduce. If e-mail users could remove themselves from e-mail lists, spam would reduce.

Such legitimacy-based changes have a unique property: They do not selectively discriminate spam or spammers. They would apply to all of us equally. Everyone's personal data would be their personal property. Anyone could converse or not. Any e-mail could be rejected, not just spam. The goal is legitimate interaction, not punishment or revenge, to reduce unwanted mail from all of us, not just spammers.

CONCLUSION

These conclusions can be summarized as follows.

1. Technology advances alone, like filters, will not in the long run reduce spam.
2. Traditional social solutions alone, like the law, will work poorly in cyberspace.
3. Spam is a social problem that requires a social solution.
4. The technical architecture of social-technical systems must support social requirements for social solutions to work.

The growing flood of spam from spam-generating to spam-filtering machines—information without meaning sent from no one to no one—seems a good place to start facing the social-technical challenge.

REFERENCES

Ackerman, M. S. (2000). The intellectual challenge of CSCW: The gap between social requirements and technical feasibility. *Human Computer Interaction, 15*, 179-203.

Albert, R., Jeong, H., & Barabassi, A. (1999). The diameter of the World Wide Web. *Nature, 401*, 130.

Bazeley, M. (2003, April 26). New weapon for spam: Bounty. *Mercury News*. Retrieved from http://www.siliconvalley.com/mld/siliconvalley/5725404.htm

Bekker, S. (2003, October 14). Spam to cost U.S. companies $10 billion in 2003. *ENTNews*. Retrieved from http://www.entmag.com/news/article.asp?EditorialsID=5651

Boutin, P. (2004, April 19). Can e-mail be saved? *Infoworld*, pp. 41-53.

Cranor, L. F., & LaMacchia, B. A. (1998). Spam! *Communications of the ACM, 41*(8), 74-83.

Davidson, P. (2003, April 17). Facing dip in subscribers, America Online steps up efforts to block spam. *USA Today*, p. 3B.

Dennis, A. R., & Valacich, J. S. (1999). Rethinking media richness: Towards a theory of media synchronicity. *Proceedings of the 32nd Hawaii International Conference on System Sciences.*

Dodge, M., & Kitchin, R. (2001). *Mapping cyberspace.* London: Routledge.

Dwork, C., & Naor, M. (Eds.). (1993). Pricing via processing or combating junk mail. In *Lecture notes in computer science: Vol. 74. Advances in cryptology: Crypto '92* (pp. 139-147). New York: SpringerVerlag.

Emery, T. (2003, January 27). Meeting takes aim at spam. *The Beacon Journal*. Retrived from http://www.ohio.com/mld/beaconjournal/business/5028845.htm

Gibson, W. (1984). *Neuromancer.* London: HarperCollins.

Hall, R. J. (1998). How to avoid unwanted e-mail. *Communications of the ACM, 3*(41), 88-95.

Hamilton, A. (2004, February 23). You've got spam! Spam not annoying enough? Now junk instant messages are on the rise. *Time*, pp. 1.

Hardin, G. (1968). The tragedy of the commons. *Science, 162*, 1243-1248.

Hauben, M. (1995). *The Net and netizens: The impact the Net has on people's lives* (Preface).

Retrieved from http://www.cs.columbia.edu/~hauben/netbook/

Held, G. (1998). Spam the spammer. *International Journal of Network Management, 8*, 69.

Hiltz, S. R., & Turoff, M. (1985). Structuring computer-mediated communication systems to avoid information overload. *Communications of the ACM, 28*(7), 680-689.

Ivey, K. C. (1998). Spam: The plague of junk e-mail. *IEEE Computer Applications in Power, 11*(2), 15-16.

Korman, K. (2003). Canning spam. *netWorker, 7*(2), 3.

Kraut, R. E., Shyam, S., Morris, J., Telang, R., Filer, D., & Cronin, M. (2002). Markets for attention: Will postage for email help? *Proceedings of CSCW 02* (pp. 206-215).

Lessig, L. (1999). *Code and other laws of cyberspace.* New York: Basic Books.

MailBlocks. (2003). *MailBlocks is the ultimate spam-blocking email service.* Retrieved from http://about.mailblocks.com/

Mitchell, W. J. (1995). *City of bits space, place and the infobahn.* Cambridge, MA: MIT Press.

Murrell, L. (2003). *Spam is now number one source of unwanted network intrusions.* Sunbelt Software. Retrieved from http://www.itsecurity.com/tecsnews/jul2003/jul141.htm

Paulson, L. D. (2003). Group considers drastic methods to stop spam. *Computer, 36*(7), 21-22.

Reid, F. J. M., Malinek, V., Stott, C. J. T. E., & Evans, J. S. B. T. (1996). The messaging threshold in computer-mediated communication. *Ergonomics, 39*(8), 1017-1037.

Rice, R. (1994). Network analysis and computer-mediated communication systems. In S. Wasserman & J. Galaskiewicz (Eds.), *Advances in social network analysis* (pp. 167-2003). Newbury Park, CA: Sage.

Samuelson, P. (2003). Unsolicited communications as trespass. *Communications of the ACM, 46*(10), 15-20.

Spring, T. (2003, November 11). Spam slayer: Laws won't solve everything. *PC World.* Retreived from http://www.pcworld.com/news/article/0,aid,113329,tk,dnWknd,00.asp

Spring, T. (2004, April). Spam wars rage. *PC World*, pp. 24-26.

Taylor, C. (2003, June 16). Spam's big bang. *Time*, pp. 50-53.

Vaughan-Nichols, S. J. (2003). Saving private e-mail. *IEEE Spectrum, 40*(8), 40-44.

Warren, S. D., & Brandeis, L. D. (1890). The right to privacy. *Harvard Law Review, 4*(5), 193-220.

Weinstein, L. (2003). Inside risks: Spam wars. *Communications of the ACM, 46*(8), 136.

Weiss, A. (2003). Ending spam's free ride. *netWorker, 7*(2), 18-24.

Whitworth, B., & deMoor, A. (2003). Legitimate by design: Towards trusted virtual community environments. *Behaviour & Information Technology, 22*(1), 31-51.

Whitworth, B., Gallupe, R. B., & McQueen, R. (2001). Generating agreement in computer-mediated groups. *Small Group Research, 32*(5), 625-665.

Whitworth, B., & Whitworth, E. (2004). Reducing spam by closing the social-technical gap. *IEEE Computer*, 38-45.

Woellert, L. (2003, August 11). Out, out damned spam. *Business Week*, pp. 54-56.

KEY TERMS

Asynchronous Communication: E-mail is normally considered asynchronous communication. Synchrony has been defined as "the extent to which individuals work together on the same activity at the same time" (Dennis & Valacich, 1999), but is e-mail synchronous if e-mail communicants are online at the same time? Another view is that synchrony requires instant transmission, but if e-mail became instantaneous, would it then be synchronous? Conversely, consider a telephone (synchronous) conversation during which one party boards a rocket to

Mars; as the rocket leaves, there is a transmission delay of several minutes. Is the telephone now asynchronous communication? That the same medium is both synchronous and asynchronous is undesirable. Media properties should only change when the medium changes; that is, they should be defined in media terms, not sender-receiver or transmission terms. The asynchronous-synchronous difference is whether the medium stores the message or not. In this, e-mail remains asynchronous no matter how fast it is, and telephone synchronous no matter how slow it is. The asynchrony is between receiver and medium, not receiver and sender. The opposite is ephemerality, in which signals must be processed on arrival.

Communication Environment: In one sense, technology operates in a physical environment, but for computer-mediated communication, technology is the environment, that is, that through which communication occurs. Telephone, CMC, and face to face (FTF) are all equally communication environments. FTF is mediated by the physical world just as CMC is mediated by technology. One cannot compare environments as one does objects in an environment. To judge one environment by another is like saying the problem with America is that it is not England. Describing e-mail as distributed rather than colocated is like this. If distributed e-mail correspondents magically colocate in the same room, what changes? In their environment, nothing changes at all. E-mail is not distributed or colocated because physical space does not exist in cyberspace. Nor do environments perform as objects do. Imagine a new environment called "underwater." Users find walking underwater painfully slow, then find a new way of moving (swimming) that fits the environment better, inventing flippers to support it. Now the new world seems better. Asking which environment is better at walking is inappropriate. Cross-media studies (CMC vs. FTF) make this mistake of analysing electronic communication in face-to-face terms (Hiltz & Turoff, 1985). A better approach is within-environment research designs (Whitworth, Gallupe, & McQueen, 2001).

Communication Threshold: The acceptable user cost to send a message (Reid et al., 1996). If the cost to send a message is less than the individual's messaging threshold, it is sent. Otherwise, it is not. E-mail lowered the messaging threshold so more messages were sent than otherwise would be.

Computer-Mediated Communication: CMC, like e-mail, is one-to-one, asynchronous communication mediated by electronic means. List e-mail seems to be many-to-many communication, but the transmission system simply duplicates one-to-one transmissions. In true one-to-many transmissions, like a bulletin board, one communication operation is transmitted to many people (e.g., posting a message).

Computer-Mediated Interaction: Computer-mediated interaction (CMI) is interaction mediated by electronic means, whether between people or computer agents.

Cyberspace: Space is central to our lives, whether virtual or physical (Dodge & Kitchin, 2001). Gibson (1984) coined the term cyberspace from the Greek kyber (to navigate), describing a nonphysical space (the "matrix") that substituted for reality. Today, it means the electronic environment that enables computer-mediated interaction. Cyberspace removes the physical space constraints of human interaction (Hauben, 1995) but is still a space, albeit of a different kind. Physical space locates us to a three-number coordinate position. Cyberspace also locates us to a unique URL (uniform resource locator) position. While physical locations have differing distances between them, points in cyberspace seem equally distant. If one moves through cyberspace by mouse clicks, cyberspace points could have distances between them. In theory, every cyberspace point is one click from every other, but in practice, this is not so. Research on the diameter of the World Wide Web suggests an average of 19 links between random points (Albert, Jeong, & Barabassi, 1999).

False Positive: A filtering system can make two types of errors: false acceptance and false rejection. The latter is a false positive. A spam filter can wrongly let spam through, or wrongly filter real e-mail as spam. In false acceptance, it is not doing its job, while in false positives, it is doing it too well. Decreasing one type of error tends to increase the other, as with Type I and Type II errors in experimental design. As the spam-filter catch rate rises above 99.99%, the number of false positives also rises.

Supporting Culture in Computer-Supported Cooperative Work

Lu Xiao
The Pennsylvania State University, USA

Gregorio Convertino
The Pennsylvania State University, USA

Eileen Trauth
The Pennsylvania State University, USA

John M. Carroll
The Pennsylvania State University, USA

Mary Beth Rosson
The Pennsylvania State University, USA

INTRODUCTION

Computer Supported Collaborative Work (CSCW)

Information Technology (IT) has a significant impact on our lives beyond mere information access and distribution. IT shapes access to services, technology, and people. The design and use of IT can change people's communication styles and the way they work, either individually or in a group. The recent introduction of groupware and Computer Supported Collaborative Work (CSCW) systems enables people to collaborate with fewer time and space constraints and affects people's lives and their cultures in the long term.

CSCW is a new and fast developing research field. The terms *groupware* and *CSCW* were coined in the mid-1980s. The study of CSCW and groupware could be defined as a middle field of research between the study of single user applications (e.g., human-computer interaction [HCI] research) and applications for organizations (e.g., information systems [IS] or management information system [MIS] research) (Grudin, 1994). CSCW studies the way people work in groups as well as technological solutions that pertain to computer networking with associated hardware, software, services, and techniques (Wilson, 1991). There are several alternative labels used to denominate CSCW applications: groupware, group support systems (GSS), collaborative computing, workgroup computing, and multiuse applications.

Some of the key issues studied in CSCW include commuter-mediated communication, awareness and coordination, and multi-user interfaces. However, there has been very limited research to account for culture in CSCW. In this article, we discuss the role of culture in the design and implementation of CSCW systems that support work in cross-cultural contexts. We first present two different perspectives on culture in the literature. We then review prior research in both HCI and IS fields and follow with a summary of preliminary research work in CSCW about cross-cultural group work. We conclude by discussing alternative approaches to design and by suggesting a theoretical tool that may inform future research on the cultural factors in CSCW.

CULTURE

Culture is "an integrated system of learned behavior patterns that are characteristic of the members of any given society. Culture refers to the total way of life of particular groups of people. It includes everything that a group of people thinks, says, does and makes—its systems of attitudes and feelings. Culture is learned and transmitted from generation to generation" (Kohls, 1996, p. 23). Two distinct perspectives on culture are represented in the literature: culture is relatively constant vs. culture is variable and situated. The major advocate of the first perspective (i.e., culture is a constant entity based on shared assumptions) is Hofstede (1980), who defines culture as "the collective programming of the mind which distinguishes the members of one group or category of people from another" (p. 25). Researchers who hold the first perspective on culture also define culture as beliefs, values, and assumptions that are reflected in artifacts, symbols, and behaviors (Kroeber & Kluckhohn, 1963). Schein (1992) defined organizational culture as a set of implicit assumptions shared within the group that determines its perspective of and reaction to various environments.

The other perspective on culture characterizes it as variable, historically situated, and evolving with the context. Rather than being a holistic and relatively stable entity, culture is seen as fragmented, variable, contentious, and in-the-making (Brightman, 1985; Prus, 1997). The values and attitudes of the working group affect the behavior of the group, whose collective patterns of behavior contributes to the group culture. The group culture, in return, has significant impact on the values and attitudes of the group. This cyclic relationship is true for not only working groups or organizations but also for nations (Davison & Jordan, 1996).

BACKGROUND

Culture: A Research Issue in Multiple Disciplines

In this section, we review studies from different research fields that have investigated the role of culture in computer technology. We first describe prior research in HCI and IS (or MIS) literature. Then, we focus on studies that have accounted for cultural factors in CSCW and groupware.

Current Research in HCI and Information System

HCI researchers have investigated how cultural factors may affect design and evaluation of single-user applications (Barber & Badre, 1998; Marcus, 2000; Marcus & Gould, 2000; Sheppard & Scholtz, 1999). The research in this domain has focused on research issues such as cultural usability (Barber & Badre, 1998) and the design of intercultural user interfaces (UI) (Marcus, 2000). An instance of the impact of culture on UI design pertains to the meaning of colors. The color red, for example, in some cultures is associated with danger, anger, and so forth (Dix & Mynatt, 2004). In other cultures, such as in China, it is more commonly associated with happiness and good luck. Designing UI for multicultural audiences may require interfaces that adapt the standards to the cultural context of the specific audiences.

Several IS (MIS) studies have investigated the influence of cultural factors on the use of information systems. Table 1, reproduced from Ward and Ward (2002), summarizes a number of studies on GSS and culture. Setting future agendas for IS research at the group level of analysis, Walsham (2000) observed, "There are clear agendas here for IS researchers to investigate in more detail the role of groupware in multi-cultural contexts" (p. 204).

Culture Issue in CSCW and Groupware

Located between HCI and IS research, CSCW has given increasing attention to cultural factors in CSCW and groupware. CSCW researchers have acknowledged the relevance of culture to appropriately design groupware and to successfully support cooperative work. For example, Olson and Olson (2001) have observed that remote teams misunderstand each other because of cultural differences. Dix and his colleagues have observed that lack of consideration for different cultural perceptions and habits about personal space (proxemics) may have unpleasant effects in cross-cultural meetings (Dix & Mynatt, 2004). The following section discusses two distinctive examples of system design that support cross-cultural communication.

Table 1. Research on GSS and culture (reproduced from Ward and Ward, 2002)

Author	Activity	Results	Groups Researched
Tan et al. (1993)	Influence of minority source	Status influence altered	
Aiken et al. (1993)	Effective use of technology	Effective regardless of culture of language	Malaysia and American groups
Watson et al. (1994)	Adoption of technology	Culture will shape adoption of GSS features Meeting designers need to match tools and communication to meeting goals and cultural norms	Singapore and US groups
Niederman (1997)	New technology New meeting norms	Reaction similar Some differences	
Aitkinson and Pervan (1998)	Anonymity	Higher productivity	Four national groups
Abdat and Pervan (1999)			Indonesian groups
Anderson (2000)	Cognitive conflict task	No difference for pre-meeting consensus, influence equality, and post-meeting consensus No difference for consensus change Higher levels of perceived process gains, perceived decision satisfaction, perceived decision process satisfaction, and perceived quality of discussion	Multicultural and US groups

DESIGN APPROACHES OF SUPPORTING CULTURE IN COLLABORATION

Okamoto, Isbister, Nakanishi, and Ishida (2002) have designed and implemented large screen systems that support cross-cultural communication that happens synchronously with communicators either at the same location or in remote locations. In their large screen systems, communicators' real images can be seen from the large screen, thus enabling their communication through nonverbal cues. Communicators' cultural backgrounds and shared information based on their profiles are presented on the large screen, including language knowledge, culture literacy and experience (e.g., how long the person has been immersed in the culture), and culture affinity and ties (e.g., how many friends the person has from certain countries). The idea of the system is to provide support for culture awareness to improve communication.

Grill, Kronsteiner, and Kotsis (2003) suggested creating a culture translation agent to support cross-cultural communication information. Using Hofstede's (1980) definition of culture as collective programming of the mind, Grill et al. (2003) assume that different programming of the minds leads to alternative code bases (i.e., alternative common ground) in communication (Clark & Brennan, 1991). The authors propose the idea of implementing a cultural translation system that helps overcome the misunderstanding in communication due to different code bases. In such a culture translation system, a culture translation agent (CTA) is created as a modular agent. Such an agent functions as a communication support tool that monitors whether messages sent between communicators might cause misunderstandings due to culture difference and notifies communicators about it. The CTA uses a matching algorithm to compare phrases and terms in the message with a code base constructed on code bases of the relevant cultures.

Although the idea of implementing a CTA to support cross-cultural communication seems to be promising, overall, we consider the design approach of Okamoto et al. (2002) more favorable. Privileging the perspective that culture is dynamic and context-dependent, we argue that a static code base cannot reflect the dynamic features of culture (specifically referring to the cultural factors that significantly affect group collaboration). For example, things that normally would cause miscommunication because of cultural differences between the communicators rather may be understood well because one has been exposed to the other's culture for an extended period of time. In this case, a culture translator may not be useful for communication. Instead, the existence of a translator may be an obstacle for communicators to learn each other's culture, which could be a positive outcome of cross-cultural communication. Compared to CTA, the approach of Okamoto et al. (2002) takes into account the dynamic features of culture (e.g., an individual's culture literacy and experience are provided on the large screen, and communicators are able to see each other and communicate directly). Thus, the system supports cross-cultural communication by providing individual cultural background information while simultaneously enabling face-to-face communication.

We believe that appropriately supporting cross-cultural coordination represents a new challenge for CSCW design. In fact, people from different cultures may have different value systems and attitudes toward the same activity (e.g., expectations and assumptions on labor division and deadlines), different understanding of rules of the group, and so forth. Such differences generally affect both work relationships and group performance.

FUTURE TRENDS

Activity Theory: A Useful Theoretical Tool

Activity theory is a useful tool to understand cultural mediation in human activities. In agreement with ecological approaches to HCI and in contrast to individual-centric theories, activity theory emphasizes the connection rather than the separation between human cognition and human action (Bødker, 2003). Culture is viewed as a primary mediator in human activities.

The cultural-historical approach put forth by Russian cultural-historical scholars Leontiev, Luria, and Vygotsky draws on Marx's historical materialism and focuses on the function of culture in human development by considering the contributions of cultural artifacts, historical development, and practical activity (Cole, 1998). Activity theory was born from this perspective, where the primary unit of analysis is the activity (i.e., the fundamental type of context) (Bødker, 1991; Korpela, Mursu & Soriyan, 2001). Building on this basis, Engeström (1987) has depicted the intertwined relationships among subject, object, and community of the activity through a triangular model (Figure 1). The central subject-object-community triangle then is extended to include sociocultural forms of mediation: instruments, rules, and division of labor (Engeström, 1987).

Supporting collaborators' awareness has been a central concern for CSCW researchers. Globally, three major forms of awareness have been studied in CSCW research: social awareness (who is present), action awareness (what are they doing), and the more general awareness of the entire activity (Carroll, Neale, Isenhour, Rosson & McCrickard, 2003). With the aim of accounting for cultural factors in group cooperation, we suggest the inclusion of cultural mediation as part of the activity awareness concept. Specifically, drawing on Engeström's (1987) activity model, we propose a comprehensive concept of awareness in CSCW, which accounts for collaborators' awareness of cultural mediators, such as group norms and rules, division of labor, and collaborative tools.

However, Engeström's (1987) model is based on the assumption of a single, shared cultural context. This model needs to be extended in order to describe and explain collaborative phenomena among people of different cultures. In fact, different cultures generally imply different artifacts, rules, and ways of dividing labor.

Cross-cultural collaboration requires the additional task of negotiating meanings at a cultural level. Future research issues about awareness of cultural mediation in CSCW include the study of awareness breakdowns due to lack of visibility or misunderstandings about cultural differences; the study of the process of building common ground (Clark & Brennan, 1991) in cross-cultural settings; and the study of the influence of cultural background information on group performance.

Figure 1. Engeström's model of human activity (1987)

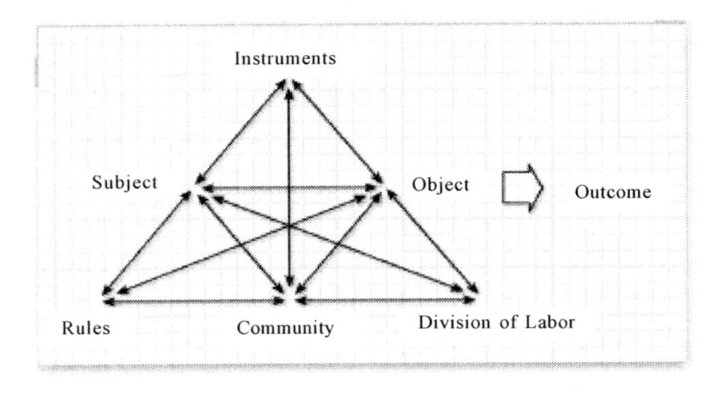

CONCLUSION

In this article, we have reviewed the current understanding of culture as a factor in CSCW. Using two examples to illustrate different approaches to design CSCW systems that support cross-cultural communication (culture translation system vs. support for cross-cultural communication and awareness), we gave suggestions for system design that takes into account the culture factor. We have also suggested directions of future research on the culture factor in CSCW and groupware. We suggested the introduction of culture mediation awareness to the concept of activity awareness. The best solution is CSCW systems that support culture mediation awareness by providing information to users about group culture.

REFERENCES

Abdat, S., & Pervan, G.P. (1999). Developing group support systems for Indonesian culture at the pre-meeting stage of strategy formulation: A proposal and some initial requirements. *Proceedings of the 5th International Conference of the International Society for Decision Support Systems*, Melbourne, Australia.

Aiken, M., Hwang, M.R., & Martin, J. (1993). A comparison of Malaysian and American groups using a group decision support system. *Journal of Information Science Principles and Practice, 19*(6), 489-491.

Barber, W., & Badre, A. (1998). Culturability: The merging of culture and usability. *Proceedings of the 4th Conference on the Human Factors and the Web*, Basking Ridge, New Jersey.

Bødker, S. (1991). Activity theory as a challenge to systems design. In H.E. Nissen, H. Klein, & R. Hirschheim (Eds.), *Information systems research: Contemporary approaches and emergent traditions* (pp. 551-564). North Holland.

Brightman, R. (1985). Forget culture: Replacement, transcendence and relexification. *Cultural Anthropology, 10*(4), 509-546.

Carroll, J.M., Neale, D.C., Isenhour, P.L., Rosson, M.B., & McCrickard, D.S. (2003). Notification and awareness: Synchronizing task-oriented collaborative activity. *International Journal of Human-Computer Studies, 58*(5), 605-632.

Clark, H.H., & Brennan, S.E. (1991). Grounding in communication. In L.B. Resnick, J. Levine, & S.D. Teasley (Eds.), *Perspectives on socially shared cognition* (pp. 127-149). Hyattsville, MD: American Psychological Association.

Cole, M. (1996). Cultural psychology: A once and future discipline. Cambridge, MA: Harvard University Publisher.

Davison, R., & Jordan, E. (1996). Cultural factors in the adoption and use of GSS. *Proceedings of the Information Federation for Information Processing Working Group 8.4 2nd International Office of the Future Conference*, Tucson, Arizona.

Dix, A., & Mynatt, B. (2004). *Human-computer interaction* (3rd ed.). Prentice Hall.

Dourish, P., & Bellotti, V. (1992). Awareness and coordination in shared workspaces. *Proceedings of the ACM CSCW '92 Conference on Computer Supported Cooperative Work.* New York.

Ellis, C.A., Gibbs, S.J., & Rein, G.L. (1991). Groupware. *Communications of the ACM, 1,* 1991, 38-58.

Engeström, Y. (1987). *Learning by expanding: An activity-theoretical approach to developmental research*. Helsinki: Orienta-Konsultit.

Giddens, A. (1984). *The constitution of society: Outline of the theory of structuration*. Berkeley: University of California.

Grill, T., Kronsteiner, R., & Kotsis, G. (2003). Sharing culture: Enabling technologies for communication support. *Proceedings of the 2003 International Conference on Cyberworlds*, Singapore.

Grudin, J. (1994). Groupware and social dynamics: Eight challenges for developers. *Communications of the ACM, 37*(1), 92-105.

Hofstede, G. (1980). *Culture's consequences: International differences in work-related values*. Beverley Hills, CA: Sage Publications.

Kohls, L.R. (1996). *Survival kit for overseas living*. Yarmouth, ME: Intercultural Press.

Korpela, M., Mursu, A., & Soriyan, H.A. (2001). Two times four integrative levels of analysis: A framework. *Proceedings of the IFIP TC8/WG8.2 Working Conference on Realigning Research and Practice in Information Systems Development: The Social and Organizational Perspective*, Boise, Idaho.

Kroeber, A.L., & Kluckhohn, C. (1963). *Culture: A critical review of concepts and definitions*. New York: Vintage Books.

Kuutti, K. (1991). The concept of activity as a basic unit for CSCW research. *Proceedings of the 2nd ECSCW*, Amsterdam, The Netherlands.

Marcus, A. (2000). International and intercultural user interfaces. In C. Stephandis (Ed.), *User interfaces for all* (pp. 47-63). New York: Lawrence Erlbaum.

Marcus, A. (2001). Cross-cultural user-interface design for work, home, play, and on the way. *Proceedings of the 19th Annual International Conference on Computer Documentation,* Sante Fe, New Mexico.

Marcus, A., & Gould, E.W. (2000). Crosscurrents: Cultural dimensions and global Web user-interface design, interactions. *ACM Publisher, 7*(4), 32-46.

Niederman, F. (1997). Facilitating computer-supported meetings: An exploratory comparison of U.S. and Mexican facilitators. *Journal of Global Information Management, 5*(1), 17-26.

Okamoto, M., Isbister, K., Nakanishi, H., & Ishida, T. (2002). Supporting cross-cultural communication with a large-screen system. *New Generation Computing, 20*(2), 165-185.

Olson, G.M., & Olson, J.S. (2002). Distance matters. In J.M. Carroll (Ed.), *Interaction in the new millennium* (pp. 397-417). Addison Wesley Professional.

Prus, R.C. (1997). *Subcultural mosaics and intersubjective realities: An ethnographic research agenda for pragmatizing the social sciences*. Albany: State University of New York Press.

Schein, E. (1992). *Organizational culture and leadership* (2nd ed.). San Francisco: Jossey-Bass.

Sheppard, C., & Scholtz, J. (1999). The effects of cultural markers on Website use. *Proceedings of the 5th Conference on Human Factors and the Web*, Gaithersburg, Maryland.

Walsham, G. (2000). Globalization and IT: Agenda for research. *Proceedings of the International Conference on Home Oriented Informatics and Telematics*.

Ward, T., & Ward, S.A. (2002). Using the GSS for cross cultural business networking. *Proceedings of the Industrial Marketing and Purchasing Conference*, Perth, Australia.

Watson, R.T., Ho, T.H., & Raman, K.S. (1994). A fourth dimension of group support systems. *Communications of the ACM, 37*(10), 45-55.

Wilson, P. (1991). *Computer supported cooperative work: An introduction*. Oxford: Intellect Books.

KEY TERMS

Activity Theory: Construes activity as a collective phenomenon. Activity is pursued by individuals or groups within a community working toward shared objectives or motives and recruiting and transforming the material environment, including shared tools, data, social and cultural structures, and work practices (Kuutti, 1991).

Awareness: "An understanding of the activities of others, which provides a context for your own activity" (Dourish & Bellotti, 1992, p. 107).

Computer-Supported Cooperative Work (CSCW): A field located between HCI and IS research fields, CSCW studies the way people work in groups as well as technological solutions that pertain to computer networking with associated hardware, software, services, and techniques (Wilson, 1991).

Context: The structure or environment where special interactions occur (Giddens, 1984).

Culture: "An integrated system of learned behavior patterns that are characteristic of the members of any given society. Culture refers to the total way of life of particular groups of people. It includes everything that a group of people thinks, says, does and makes—its systems of attitudes and feelings. Culture is learned and transmitted from generation to generation (Kohls, 1996, p. 23).

Groupware: "Computer-based systems that support groups of people engaged in a common task (or goal) and that provide an interface to a shared environment" (Ellis et al., 1991, p. 40).

Hofstede's Cultural Dimensions: Hofstede identified five dimensions of national culture: power distance, uncertainty avoidance, individualism, masculinity, and long-term time vs. short-term orientation.

Supporting Navigation and Learning in Educational Hypermedia

Patricia M. Boechler
University of Alberta, Canada

INTRODUCTION

Computers have become commonplace tools in educational environments and are used to provide both basic and supplemental instruction to students on a variety of topics. Searching for information in hypermedia documents, whether on the Web or through individual educational sites, is a common task in learning activities. Previous research has identified a number of variables that impact how students use electronic documents. Individual differences such as learning style or cognitive style (Andris, 1996; Fitzgerald & Semrau, 1998), prior topic knowledge (Ford & Chen, 2000), level of interest (Lawless & Kulikowich, 1998), and gender (Beasley & Vila, 1992) all influence performance. Additionally, characteristics of the document such as the inherent structure of the material, the linking structure (Korthauer & Koubek, 1994), and the types of navigation tools that accompany the document can affect student performance and behaviour (Boechler & Dawson, 2002; McDonald & Stevenson, 1998, 1999). In short, the effective use of hypermedia documents in educational settings depends on complex interactions between individual skills (e.g., spatial and reading skills) and the features of the document itself.

BACKGROUND

Previous research has suggested that one way of addressing ability differences in hypermedia users is to follow a compensatory strategy in which users are provided with mediators, modalities, or organizing structures that make up for a deficit in a particular ability (Messick, 1976). One kind of organizing structure that can help users make sense of material is a spatial structure that illustrates how different parts of the material are related. A spatial map, spatial overview, or graphic organizer is a visual representation of the structure of the document. These are usually in a diagrammatic form such as block diagrams, diagrams organized around a central term (spider map), or hierarchically ordered tree diagrams. For example, see Figure 1.

Figure 1. An example of a hierarchically ordered tree diagram

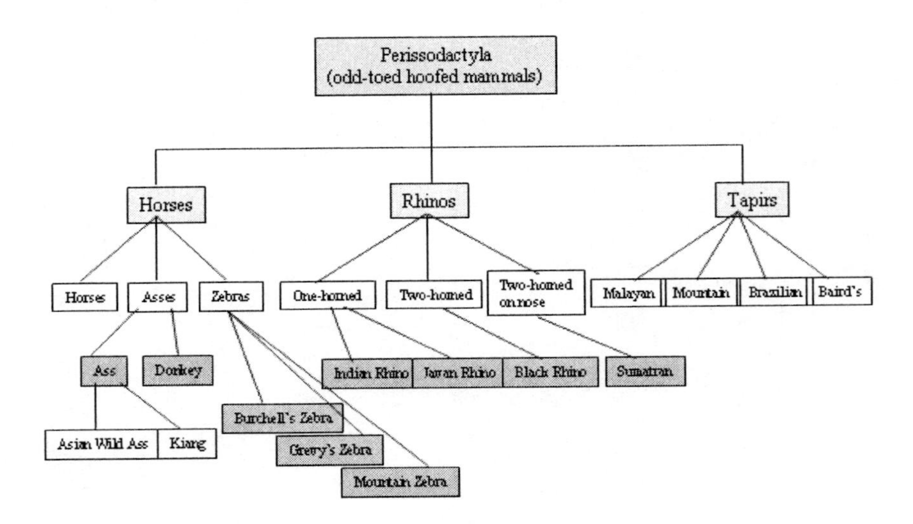

Learning depends on the construction of stable and usable mental representations of knowledge. From an educational perspective, how do we induce such representations? When presenting factual (e.g., some apples are red) or demonstrable (e.g., gravity makes things fall downward) information, creating an appropriate mental representation is a matter of relying on these physical aspects of the world to stand as mental representations to be stored, manipulated, or retrieved. The creation of mental representations for abstract and complex ideas is not as straightforward. In the cognitive-psychology literature, it is suggested that people often use spatial structures as metaphors to reason out the relational attributes of a set of abstract elements, attributes that are not observable (Gattis, 2001). Research across several bodies of literature (educational psychology, information science, instructional technology) suggests that the arrangement of visual information in particular in a hypermedia interface can impact both navigation (Allen, 2000; Boechler & Dawson, 2002; Chen, 2000; Westerman & Cribbin, 2000) and learning (Boechler & Shaddock, 2004; Mayer & Sims, 1994; Moreno & Mayer, 1999). In both cases, a successful spatial or visual arrangement should make salient the relations between semantic elements in the document. Concerning navigation, in the information-science literature, Dillon (2000) proposes a spatial and semantic model to explain hypermedia navigation processes. The spatial and semantic model assumes all information spaces convey structural cues that are both spatial and semantic in nature, and that different user characteristics and contexts determine which type of cues will be relied on in relation to one another. Similarly, in the educational-psychology literature, Mayer and Sims propose a dual-coding theory to explain learning in hypermedia. In the dual-coding theory of multimedia learning, learners construct referential connections between the mental representations of the verbal and visual information presented within a hypermedia document. Hence, the underlying assumption is that, for both navigation and learning, the impact of the visual arrangement lies in the degree to which it preserves the meaning relations between different parts of the document material. This mapping of verbal and visual elements can be accomplished using multitudes of diverse visual cues (spatial separation, clustering, bordering, connecting lines, etc.).

THE EFFECTIVENESS OF GRAPHIC ORGANIZERS

Although not all studies support the positive effects of graphic organizers (e.g., Farris, Jones, & Elgin, 2002; Stanton, Taylor, & Tweedie, 1992), in the hypermedia literature, there are many examples of the usefulness of graphic organizers. For instance, Stanny and Salvendy (1995) found that the performance of low-spatial-ability users could be improved to the level of high-spatial-ability users by providing a 2-D (two-dimensional) hierarchical structure as a guide for users. Allen (2000) found that low-spatial-ability users performed better when provided with a word map: a configuration that showed the relationships between words in a bibliographic collection.

McDonald and Stevenson (1999) reported on two studies examining the effects of navigational aids on navigation and learning. The first study indicated that providing a spatial map improved navigation performance over using a content list or no navigation tool. In this case, the map consisted of labels with connecting lines indicating the links between nodes. Navigation performance was measured by task time and the number of extraneous pages accessed. However, this type of spatial map did not improve recall for the document material. The second study showed that providing a spatial map that also included link descriptions that showed the conceptual relations between the pages improved learning.

Boechler and Shaddock (2004) found that the presence of visual links between page labels in a navigation tool predicted incidental learning of material during an information-search task. Whether the navigation tool was two dimensional or three dimensional did not predict these learning outcomes.

Nilsson and Mayer (2002) reported two studies using graphic organizers. They concluded that there are benefits to graphic organizers, but that such benefits come at the expense of other aspects of performance. Specifically, a graphic organizer can assist users in navigation, but if the organizers make the task of navigating too easy, it is less likely users will integrate the information they have viewed. Clearly, not all features exhibited in graphic organizers are as effective at enhancing learning but, in general, graphic organizers do seem to assist in the navigation process and in some instances assist learning as well.

Why do Graphic Organizers Help?

Researchers have suggested two reasons that graphic organizers assist learners (Nilsson & Mayer, 2002). First, they reduce the cognitive overhead that students must expend by providing a framework for people to take in new information. When learners use hypermedia documents, they must remember which material was shown and determine how the material is related. In other words, they must form a meaningful representation in memory of how the material is organized. Providing this organization up front through a graphic organizer lessens the effort that learners need to expend to understand the meaning of the material. Second, graphic organizers help learners to not become disoriented as they move through the document. Disorientation occurs when the user does not know where to go next, the user knows where to go but not how to get there, or the user does not know where he or she is in relation to the overall structure of the document. The less cognitive resources a learner needs to use to navigate the document, the more resources are available to actually learn the material.

FUTURE TRENDS

Many studies, such as those reviewed above, that evaluate the usefulness of graphic organizers or spatial maps report positive effects. Other studies report gains in some areas of performance accompanied by losses in different areas. What is true for all such studies is that these studies have many diverse characteristics in the features of the graphic organizer, in the task that users are required to perform, and in the different cognitive abilities of the users themselves. Future research must seek to reveal and synthesize how these different variables interact before a complete understanding of the role of graphic organizers can be achieved.

CONCLUSION

People may need different kinds of interface support to learn effectively in hypermedia environments. Understanding the relationships between individual skills and types of support will help educational designers provide the optimum interface for students of different needs. Graphic or spatial organizers may be one useful tool for providing such support. Cognitive and learning theories can provide guidance for exploring the interactions that occur between the interface characteristics and individual differences for both navigation and learning outcomes.

REFERENCES

Allen, B. (2000). Individual differences and the conundrums of student-centred design: Two experiments. *Journal of the American Society for Information Science, 51*(6), 508-520.

Andris, J. F. (1996). The relationship of indices of student navigational patterns in a hypermedia geology lab simulation to two measures of learning style. *Journal of Educational Multimedia and Hypermedia, 5*(3/4), 303-315.

Beasley, R. E., & Vila, J. A. (1992). The identification of navigation patterns in a multimedia environment: A case study. *Journal of Educational Multimedia and Hypermedia, 1*, 209-222.

Boechler, P. M., & Dawson, M. R. W. (2002). The effects of navigation tool information on hypermedia navigation behavior: A configural analysis of page-transition data. *Journal of Educational Multimedia and Hypermedia, 11*(2), 95-115.

Boechler, P. M., & Shaddock, C. (2004). The effects of visual cues and spatial skill on incidental learning in a hypermedia information search task. *Proceedings of Ed-Media 2004* (pp. 4317-4321).

Chen, C. (2000). Individual differences in a spatial-semantic virtual environment. *Journal of the American Society for Information Science, 51*(6), 529-542.

Dillon, A. (2000). Spatial-semantics: How students derive shape from information space. *Journal of the American Society for Information Science, 51*(6), 521-528.

Farris, J. S., Jones, K. S., & Elgin, P. D. (2002). *Users' schemata of hypermedia: What is so "spatial" about a Website?* Interacting with *Computers, 14*, 487-502.

Fitzgerald, G., & Semrau, L. (1998). The effects of learner differences on usage patterns and learning outcomes with hypermedia case studies. *Journal of Educational Multimedia and Hypermedia, 7*, 309-332.

Ford, N., & Chen, S. Y. (2000). Individual differences, hypermedia navigation and learning: An empirical study. *Journal of Multimedia and Hypermedia, 9*(4), 281-311.

Gattis, M. (2001). Space as a basis for abstract thought. In M. Gattis (Ed.), *Spatial schemas and abstract thought* (pp. 1-12). Cambridge, MA: MIT Press.

Korthauer, R. D., & Koubek, R. J. (1994). An empirical evaluation of knowledge, cognitive style, and structure upon the performance of a hypermedia task. *International Journal of Human-Computer Interaction, 6*(4), 373-390.

Lawless, K. A., & Kulikowich, J. M. (1998). Domain knowledge, interest and hypermedia navigation: A study of individual differences. *Journal of Educational Multimedia and Hypermedia, 7*(1), 51-69.

Mayer, R.E., & Sims, V. (1994). For whom is a picture worth a thousand words? Extensions of a dual coding theory of multimedia learning. *Journal of Educational Psychology, 86*(3), 389-401.

McDonald, S., & Stevenson, R. (1998). Spatial versus conceptual maps as learning tools in hypermedia. *Journal of Educational Hypermedia and Multimedia, 8*(1), 43-64.

McDonald, S., & Stevenson, R. (1999). Navigation in hyperspace: An evaluation of the effects of navigational tools and subject matter expertise on browsing and information retrieval in hypermedia. *Interacting with Computers, 10*, 129-142.

Messick, S. (1976). *Individuality in learning.* San Francisco: Jossey-Bass.

Moreno, R., & Mayer, R. E. (1999). Cognitive principles of multimedia learning: The role of modality and contiguity. *Journal of Educational Psychology, 91*(2), 358-368.

Nilsson, R. M, & Mayer, R. (2002). The effects of graphic organizers giving cues to the structure of a hypertext document on users' navigation strategies and performance. *International Journal of Human-Computer Studies, 57*, 1-26.

Stanny, K. M., & Salvendy, G. (1995). Information visualization: Assisting low spatial individuals with information access tasks through the use of visual mediators. *Ergonomics, 38*(6), 1184-1198.

Stanton, N. A., Taylor, R. G., & Tweedie, L. A. (1992). Maps as navigational aids in hypertext environments: An empirical evaluation. *Journal of Educational Multimedia and Hypermedia, 1*(4), 431-444.

Tennant, M. (1988). *Psychology and adult learning.* London: Routledge.

Westerman, S. J., & Cribbin, T. (2000). Mapping semantic information in diagrams. *Contemporary Educational Psychology, 18*, 162-185.

KEY TERMS

Cognitive Overhead: The amount of mental resources that need to be expended to complete a given task.

Cognitive Style: Cognitive style has been defined as "an individual's characteristic and consistent approach to organizing and processing information" (Tennant, 1988, p. 89).

Compensatory Strategy: An educational approach that focuses on providing structures that support and enhance learners' weaknesses rather than exploiting their strengths.

Disorientation: The sensation of feeling lost in a hypermedia document, characterized by three categories of the user's experience: (a) The user does not know where to go next, (b) the user knows where to go but not how to get there, or (c) the user does not know where he or she is in relation to the overall structure of the document.

Dual-Coding Theory of Multimedia Learning: A theory of learning in hypermedia (Mayer & Sims, 1994) that is a process account of how learners build mental connections between the verbal material, the visual material (e.g., images or diagrams), and the meaning that links the two together.

Spatial Ability: Spatial ability refers to a person's ability to perceive, retain, and mentally manipulate different kinds of spatial information. There are numerous types of spatial ability (e.g., scanning ability, visualization) that can be measured by standardized tests to detect ability differences between learners.

Spatial and Semantic Model: A model of hypermedia navigation proposed by Dillon (2000) that is based on the notion of an information space for users: "The concept of shape assumes that an information space of any size has both spatial and semantic characteristics. That is, as well as identifying placement and layout, users directly recognize and respond to content and meaning" (p. 523).

Task Analysis at the Heart of Human-Computer Interaction

Dan Diaper
Middlesex University, UK

INTRODUCTION

The history of task analysis is nearly a century old, with its roots in the work of Gilbreth (1911) and Taylor (1912). Taylor's scientific management provided the theoretical basis for production-line manufacturing. The ancient manufacturing approach using craft skill involved an individual, or a small group, undertaking, from start to finish, many different operations so as to produce a single or small number of manufactured objects. Indeed, the craftsperson often made his or her own tools with which to make end products. Of course, with the growth of civilisation came specialisation, so that the carpenter did not fell the trees or the potter actually dig the clay, but still each craft involved many different operations by each person. Scientific management's novelty was the degree of specialisation it engendered: each person doing the same small number of things repeatedly.

Taylorism thus involved some large operation, subsequently called a task, that could be broken down into smaller operations, called subtasks. Task analysis came into being as the method that, according to Anderson, Carroll, Grudin, McGrew, and Scapin (1990), "refers to schemes for hierarchical decomposition of what people do." The definition of a task remains a "classic and under-addressed problem" (Diaper, 1989b). Tasks have been differently defined with respect to their scope: from the very large and complex, such as document production (Wilson, Barnard, & MacLean, 1986), to the very small, for example, tasks that "may involve only one or two activities which take less than a second to complete, for example, moving a cursor" (Johnson & Johnson, 1987). Rather than trying to define what is a task by size, Diaper's (1989b) alternative is borrowed from conversation analysis (Levinson, 1983). Diaper suggests that tasks always have well-defined starts and finishes, and clearly related activities in between. The advantage of such a definition is that it allows tasks to be interrupted or to be carried out in parallel.

Task analysis was always involved with the concept of work, and successful work is usually defined as achieving some goal. While initially applied to observable, physical work, as the field of ergonomics developed from World War II, the task concept was applied more widely to cover all types of work that "refocused attention on the information processing aspect of tasks and the role of the human operator as a controller, planner, diagnostician and problem solver in complex systems" (Annett & Stanton, 1998). With some notable exceptions discussed below, tasks are still generally defined with people as the agents that perform work. For example, Annett and Stanton defined task analysis as "[m]ethods of collecting, classifying and interpreting data on human performance."

BACKGROUND

Stanton (2004) suggests that "[s]implistically, most task analysis involves (1) identifying tasks, (2) collecting task data, (3) analyzing this data so that the tasks are understood, and then (4) producing a documented representation of the analyzed tasks (5) suitable for some engineering purpose." While there are many similar such simplistic descriptions, Stanton's five-item list provides an adequate description of the stages involved in task analysis, although the third and fourth are, in practice, usually combined. The following four subsections deal with them in more detail, but with two provisos. First, one should always start with Stanton's final item of establishing the purpose of undertaking a task analysis. Second, an iterative approach is always desirable because how tasks are performed is complicated.

The Purpose of a Task Analysis

Task analysis has many applications that have nothing to do with computer systems. Even when used in HCI (human-computer interaction), however, task analysis can contribute to all the stages of the software-development life cycle. In addition, task analysis can make major contributions to other elements associated with software development, in particular the preparation of user-support systems such as manuals and help systems, and for training, which was the original application of hierarchical task analysis (HTA; Annett & Duncan, 1967; Annett, Duncan, Stammers, & Gray, 1971). HTA was the first method that attempted to model some of the psychology of people performing tasks.

Although infrequently documented, identifying the purposes for using task analysis in a software project must be the first step (Diaper, 1989a) because this will determine the task selection, the method to be used, the nature of the outputs, and the level of analysis detail necessary. The latter is vital because too much detailed data that does not subsequently contribute to a project will have been expensive to collect, and too high a level will require further iterations to allow more detailed analysis (Diaper, 1989b, 2004). Decomposition-orientated methods such as HTA partially overcome the level-of-detail problem, but at the expense of collecting more task data during analysis. Collecting task data is often an expensive business, and access to the relevant people is not always easy (Coronado & Casey, 2004; Degen & Pedell, 2004; Greenberg, 2004). Within a software-development life cycle, Diaper (2004) has suggested that one identify all the stages to which a task analysis will contribute and then make selections on the basis of where its contribution will be greatest.

Identifying Tasks

In the context of task scenarios, which Diaper (2002a, 2002b) describes as "low fidelity task simulations," Carroll (2000) rightly points out that "there is an infinity of possible usage scenarios." Thus, only a sample of tasks can be analysed. The tasks chosen will depend on the task analysis' purpose. For new systems, one usually starts with typical tasks. For existing systems and well-developed prototypes, one is more likely to be concerned with complex and difficult tasks, and important and critical ones, and, when a system is in use, tasks during which failures or problems have occurred. Wong (2004) describes his critical decision method as one way of dealing with the latter types of tasks.

Unless there are overriding constraints within a software project, then task analysts should expect, and accept, the need to be iterative and repeatedly select more tasks for analysis. Since the coverage of all possible tasks can rarely be complete, there is a need for a systematic task selection approach. There are two issues of coverage: first, the range of tasks selected, and second, the range of different ways that tasks may be carried out, both successfully and unsuccessfully.

One criticism of task analysis is that it requires extant tasks. On the other hand, all tasks subjected to task analysis are only simulations as, even when observed in situ, a Hiesenberg effect (Diaper, 1989b) can occur whereby the act of observation changes the task. Often, it is desirable to simulate tasks so that unusual, exceptional, and/or important task instances can be studied and, of course, when a new system or prototype is not available.

Collecting Task Data

There are many myths about task analysis (Diaper et al., 2003), and one of the most persistent involves the detailed observation of people performing tasks. Sometimes, task-analysis data do involve such observation, but they need not, and often it is inappropriate even with an existing system and experienced users.

Johnson, Diaper, and Long (1984; see also Diaper, 1989b, 2001) claim that one of the major strengths traditionally associated with task analysis is its capability to integrate different data types collected using different methods. The critical concept is that of fidelity. According to Diaper (2002a, 2002b), "fidelity, a close synonym is validity, is the degree of mapping that exists between the real world and the world modelled by the (task) simulation," although as he says parenthetically, "N.B. slightly more accurately perhaps, from a solipsistic position, it is the mapping between one model of the assumed real world and another."

At one end of the task-fidelity spectrum there is careful, detailed task observation, and at the other, when using scenarios of novel future systems, task data may exist only in task analysts' imagination. Between, there is virtually every possible way of collecting data: by interviews, questionnaires, classification methods such as card sorting, ethnography, participative design, and so forth. Cordingley (1989) provides a reasonable summary of many such methods. The primary constraint on such methods is one of perspective, maintaining a focus on task performance. For example, Diaper (1990) describes the use of task-focused interviews as an appropriate source of data for a requirements analysis of a new generation of specialised computer systems that were some years away from development.

Task Analysis and Task Representation

The main representation used by virtually all task analysis methods is the activity list, although it goes by many other names such as a task protocol or interaction script. An activity list is a prose description of one or more tasks presented as a list that usually has a single action performed by an agent on each line. Each action on an activity-list line may involve one or more objects, either as the target of the action or as support for the action, that is, as a tool. An important component of an activity list should be the identification of triggers (Dix, Ramduny-Ellis, & Wilkinson, 2004). While most tasks do possess some sequences of activity list lines in which the successful completion of an action performed on one line triggers the next, there are many cases when some event, either physical or psychological, causes one of two or more possible alternatives to occur.

Diaper (2004) suggests that an activity list is sometimes sufficient to meet a task analysis' purposes. He suggests that one of the main reasons for the plethora of task analysis methods is the volume of data represented in the activity list format, often tens, if not hundreds, of pages. As Benyon and Macaulay (2002) discuss, the role of task analysis methods applied to activity lists is not only to reduce the sheer amount of the data, but to allow the data to be abstracted to create a conceptual model for designers.

The two oldest and most widely cited task-analysis methods are HTA (Annett, 2003, 2004; Shepherd, 2001), and goals, operators, methods, and selection rules (GOMS; Card, Moran, & Newell, 1983; John & Kieras, 1996; Kieras, 2004). HTA is often misunderstood in that it produces, by top-down decomposition, a hierarchy of goals, and these are often confused with physical or other cognitive activities. HTA uses rules to allow the goal hierarchy to be traversed. Analyses such as HTA provide a basic analysis (Kieras) that can then be used by methods such as GOMS. While often perceived as too complicated, it is claimed that GOMS provides good predictive adequacy of both task times and errors.

There are between 20 and 200 task analysis methods depending on how one counts them. This presents a problem as different methods have different properties and are suitable for different purposes. An agreed taxonomy of methods for method selection is still unavailable. In Diaper and Stanton (2004a), there are half a dozen different taxonomies. Diaper (2004), rather depressingly, suggests, "in practice, people either choose a task analysis method with which they are familiar or they use something that looks like HTA."

Limbourg and Vanderdonkt (2004) produced a taxonomy of nine task analysis methods, abridged in Table 1. The methods have been reorganized so that they increase in both complexity and expressiveness down the table. References and further descriptions can be found in Diaper and Stanton (2004a).

As can be seen from Table 1, there is no accepted terminology across task analysis methods. An exception, noted by Diaper and Stanton (2004b), is that of goals and their decomposition and generalisation.

A number of recent attempts have been made to classify tasks into a small number of subtasks (Carroll, 2000; Sutcliffe, 2003; Ormerod & Shepherd, 2004). The latter's subgoal template (SGT) method, for example, classifies all information handling tasks into just four types: act, exchange, navigate, and monitor. Underneath this level, they have then identified 11 task elements. The general idea is to simplify analysis by allowing the easy identification of subtasks, which can sometimes be reused from previous analyses.

TASK ANALYSIS AT THE HEART OF HUMAN-COMPUTER INTERACTION

Diaper and Stanton (2004b) claim that "[t]oday, task analysis is a mess." Introducing Diaper (2002c),

Table 1. An abridged classification of some task analysis methods (based on Limbourg & Vanderdonkt, 2004)

Method	Origin	Planning	Operational-isation	Hierarchy Leaves	Operational Level
HTA	Cognitive analysis	Plans			Tasks
GOMS	Cognitive analysis	Operators	Methods & selection rules	Unit tasks	Operators
MAD*	Psychology	Constructors	Pre- & postconditions		Tasks
GTA	Computer-supported cooperative work	Constructors		Basic tasks	Actions & system operations
MUSE	Software engineering & human factors	Goals & constructors		Actions	Tasks
TKS	Cognitive analysis & software engineering	Plans & constructors	Procedures	Actions	
CTT	Software engineering	Operators	Scenarios	Basic tasks	Actions
Dianne+	Software engineering & process control	Goals	Procedures		Operations
TOOD	Process control	Input/output transitions			Task

Kilgour suggests that Diaper should consider the "rise, fall and renaissance of task analysis." While Diaper argues that really there has been no such fall, Kilgour is right that there was a cultural shift within HCI in the 1990s away from explicitly referring to task analysis. Diaper's oft-repeated argument has been that whatever it is called, analysing tasks has remained essential and at the heart of virtually all HCI work. Diaper (2002a, 2002b) comments, "It may well be that Carroll is correct if he believes that many in the software industry are disenchanted with task analysis…It may well be that the semantic legacy of the term "task analysis" is such that alternatives are now preferable."

TASK ANALYSIS TODAY

Central to Diaper's current definition of task analysis, and the primary reason why task analysis is at the heart of virtually all HCI work, is the concept of performance. His definition (Diaper, 2004; Diaper et al., 2003) is as follows:

Work is achieved by the work system making changes to the application domain. The application domain is that part of the assumed real world that is relevant to the functioning of the work system. A work system in HCI consists of one or more human and computer components and usually many other sorts of thing as well. Tasks are the means by which the work system changes the application domain. Goals are desired future states of the application domain that the work system should achieve by the tasks it carries out. The work system's performance is deemed satisfactory as long as it continues to achieve its goals in the application domain. Task analysis is the study of how work is achieved by tasks.

Most models and representations used in software engineering and HCI are declarative; that is, they describe things and some of the relationships between things, but not the processes that transform things over time. For example, data-flow diagrams are atemporal and acausal and specify only that data may flow, but not when and under what circumstances. In contrast, it is essential, for all successful task-analytic approaches, that performance is modeled because tasks are about achieving work.

Based on Dowell and Long's (1989) and Long's (1997) general HCI design problem, Diaper's (2004) systemic task analysis (STA) approach emphasizes the performance of systems. While STA is offered as a method, it is more of an approach in that it deals with the basics of undertaking the early stages of a task analysis and then allows other analysis methods and their representations to be generated from its activity list output. The advantage of STA over most other task analysis methods is that it models systems and, particularly, the performance of the work system. STA allows the boundary definition of a work system to change during a task so that different constituent subtasks involve differently defined work systems, and these may also be differently defined for the same events, thus allowing alternative perspectives.

The novelty of STA's view of work systems is threefold. First, as the agent of change that performs work, a work system in HCI applications is not usually anthropocentric, but a collection of things, only some of them human, that operate together to change the application domain. Second, it is the work system that possesses goals concerning the desired changes to the application domain rather than the goals being exclusively possessed by people. Third, STA is not monoteleological, insisting that work is never achieved to satisfy a single goal, but rather it states that there are always multiple goals that combine, trade off, and interact in subtle, complex ways.

STA's modeling of complex work systems has recently been supported by Hollnagel (2003b) in cognitive task design (CTD); he claims that "cognition is not defined as a psychological process unique to humans, but as a characteristic of systems performance, namely the ability to maintain control. The focus of CTD is therefore the joint cognitive system, rather than the individual user." Hollnagel's formulation of CTD is more conservative than STA's; for example, CTD is sometimes monoteleological when he refers to a single goal, and he restricts nonhuman goals to a limited number of things, albeit "a growing number of technological artefacts" capable of cognitive tasks. In STA, it is not some limited number of technological artefacts that possess goals and other cognitive properties, but the work system, which usually has both human and nonhuman components.

FUTURE TRENDS

While recognising the difficulty, perhaps impossibility, of reliably predicting the future, Diaper and Stanton (2004b) suggest that one can reasonably predict possible futures, plural. They propose that "[f]our clusters of simulated future scenarios for task analysis organized post hoc by whether an agreed theory, vocabulary, etc., for task analysis emerges and whether task analysis methods become more integrated in the future." While not predicting which future or combination will occur, or when, they are however confident that "[p]eople will always be interested in task analysis, for task analysis is about the *performance* of work," even though they admit that "[l]ess certain is whether it will be called task analysis in the future."

Probably because of its long history, there is an undoubted need for the theoretical basics that underpin the task concept and task analysis to be revisited, as Diaper (2004) attempts to do for the development of STA. Diaper and Stanton (2004b) also suggest that some metamethod of task analysis needs to be developed and that more attention needs to be placed on a wide range of types of validation, theory, methods, and content, and also on methods' predictive capability to support design and for other engineering purposes (Annett, 2002; Stanton, 2002; Stanton & Young, 1999). At least two other areas need to be addressed in the future: first, how work is defined, and second, the currently ubiquitous concept of goals.

Task analysis has always been concerned with the achievement of work. The work concept, however, has previously been primarily concerned with employment of some sort. What is needed, as Karat, Karat, and Vergo (2004) argue, is a broader definition of work. Their proposals are consistent with STA's definition of work being about the work system changing the application domain. They persuasively argue for nonemployment application domains, for example, domestic ones. Thus, a home entertainment system, television, or video game, for example, could be components of a work system, and the goals to be achieved would be to induce pleasure, fun, or similar feelings in their users. That such application domains are psychological and internal to such work systems' users, rather than the more traditional changes to things separate and external to some work system's components, is also consistent with STA's conceptualisation of task analysis.

Finally, Diaper and Stanton (2004b) broach, indeed they attempt to capsize, the concept of goals. They question whether the goals concept is necessary, either as what causes behavior or as an explanation for behavior, which they suggest, based on several decades of social psychological research, is actually usually post hoc; that is, people explain why they have behaved in some manner after the event with reference to one or more goals that they erroneously claim to have possessed prior to the behavior. Not only in all task analysis work, but in virtually every area of human endeavour, the concept of goals is used. Abandoning the concept as unnecessary and unhelpful is one that will continue to meet with fierce resistance since it seems to be a cornerstone of people's understanding of their own psychology and, hence, their understanding of the world. On the other hand, academic researchers have a moral duty to question what may be widely held shibboleths. Currently, goal abandonment is undoubtedly a bridge too far for nearly everyone, which is why STA still uses the goals concept, but greater success, if not happiness, may result in some distant future if the concept is abandoned. At the least, it is time to question the truth and usefulness of the goals concept.

CONCLUSION

Two handbooks (although at about 700 pages each, neither is particularly handy) on task analysis have recently become available: Diaper and Stanton (2004a) and Hollnagel (2003a). Both are highly recommended and, while naturally the author prefers the former because of his personal involvement, he also prefers the Diaper and Stanton tome because it provides more introductory material, is better indexed and the chapters more thoroughly cross-referenced, comes with a CD-ROM of the entire book, and, in paperback, is substantially cheaper than Hollnagel's book. No apology is made for citing the Diaper and Stanton book frequently in this article, or for the number of references below, although they are a fraction of the vast literature explicitly about task analysis. Moreover, as task analysis is at the heart of virtually all HCI because it is fundamentally about the performance of systems, then whether called task analysis or not, nearly all the published HCI literature is concerned in some way with the concept of tasks and their analysis.

REFERENCES

Anderson, R., Carroll, J., Grudin, J., McGrew, J., & Scapin, D. (1990). Task analysis: The oft missed step in the development of computer-human interfaces. Its desirable nature, value and role. *Human-Computer Interaction: Interact'90*, 1051-1054.

Annett, J. (2002). A note on the validity and reliability of ergonomic methods. *Theoretical Issues in Ergonomics Science, 3*, 228-232.

Annett, J. (2003). Hierarchical task analysis. In E. Hollnagel (Ed.), *Handbook of cognitive task design* (pp. 17-36). Mahwah, NJ: Lawrence Erlbaum Associates.

Annett, J. (2004). Hierarchical task analysis. In D. Diaper & N. A. Stanton (Eds.), *The handbook of task analysis for human-computer interaction* (pp. 67-82). Mahwah, NJ: Lawrence Erlbaum Associates.

Annett, J., & Duncan, K. D. (1967). Task analysis and training design. *Occupational Psychology, 41*, 211-221.

Annett, J., Duncan, K. D., Stammers, R. B., & Gray, M. J. (1971). *Task analysis* (Training Information Paper No. 6). London: HMSO.

Annett, J., & Stanton, N. A. (1998). Special issue: Task analysis [Editorial]. *Ergonomics, 41*(11), 1529-1536.

Benyon, D., & Macaulay, C. (2002). Scenarios and the HCI-SE design problem. *Interacting with Computers, 14*(4), 397-405.

Card, S., Moran, T., & Newell, A. (1983). *The psychology of human-computer interaction.* Mahwah, NJ: Lawrence Erlbaum Associates.

Carroll, J. M. (2000). *Making use: Scenario-based design for human-computer interactions.* Cambridge: MA: MIT Press.

Cordingley, E. (1989). Knowledge elicitation techniques for knowledge-based systems. In D. Diaper (Ed.), *Knowledge elicitation: Principles, techniques and applications* (pp. 87-176). West Sussex, UK: Ellis Horwood.

Coronado, J., & Casey, B. (2004). A multicultural approach to task analysis: Capturing user requirements for a global software application. In D. Diaper & N. A. Stanton (Eds.), *The handbook of task analysis for human-computer interaction* (pp. 179-192). Mahwah, NJ: Lawrence Erlbaum Associates.

Degen, H., & Pedell, S. (2004). The JIET design for e-business applications. In D. Diaper & N. A. Stanton (Eds.), *The handbook of task analysis for human-computer interaction* (pp. 193-220). Mahwah, NJ: Lawrence Erlbaum Associates.

Diaper, D. (1989a). Task analysis for knowledge descriptions (TAKD): The method and an example. In D. Diaper (Ed.), *Task analysis for human-computer interaction* (pp. 108-159). West Sussex, UK: Ellis Horwood.

Diaper, D. (1989b). Task observation for human-computer interaction. In D. Diaper (Ed.), *Task analysis for human-computer interaction* (pp. 210-237). West Sussex, UK: Ellis Horwood.

Diaper, D. (1990). Analysing focused interview data with task analysis for knowledge descriptions (TAKD). *Human-Computer Interaction: Interact'90*, 277-282.

Diaper, D. (2001). Task analysis for knowledge descriptions (TAKD): A requiem for a method. *Behaviour and Information Technology, 20*(3), 199-212.

Diaper, D. (2002a). Scenarios and task analysis. *Interacting with Computers, 14*(4), 379-395.

Diaper, D. (2002b). Task scenarios and thought. *Interacting with Computers, 14*(5), 629-638.

Diaper, D. (2002c). Waves of task analysis. *Interfaces, 50*, 8-10.

Diaper, D. (2004). Understanding task analysis for human-computer interaction. In D. Diaper & N. A. Stanton (Eds.), *The handbook of task analysis for human-computer interaction* (pp. 5-47). Mahwah, NJ: Lawrence Erlbaum Associates.

Diaper, D., May, J., Cockton, G., Dray, S., Benyon, D., Bevan, N., et al. (2003). Exposing, exploring, exploding task analysis myths. *HCI2003 Proceedings* (Vol. 2, pp. 225-226).

Diaper, D., & Stanton, N. A. (2004a). *The handbook of task analysis for human-computer interaction.* Mahwah, NJ: Lawrence Erlbaum Associates.

Diaper, D., & Stanton, N. A. (2004b). Wishing on a star: The future of task analysis for human-computer interaction. In D. Diaper & N. A. Stanton (Eds.), *The handbook of task analysis for human-computer interaction* (pp. 603-619). Mahwah, NJ: Lawrence Erlbaum Associates.

Dix, A., Ramduny-Ellis, D., & Wilkinson, J. (2004). Trigger analysis: Understanding broken tasks. In D. Diaper & N. A. Stanton (Eds.), *The handbook of task analysis for human-computer interaction* (pp. 381-400). Mahwah, NJ: Lawrence Erlbaum Associates.

Dowell, J., & Long, J. (1989). Towards a conception for an engineering discipline of human factors. *Ergonomics, 32*(11), 1513-1535.

Gilbreth, F. B. (1911). *Motion study.* Princeton, NJ: Van Nostrand.

Greenberg, S. (2004). Working through task-centred system design. In D. Diaper & N. A. Stanton (Eds.), *The handbook of task analysis for human-computer interaction* (pp. 49-68). Mahwah, NJ: Lawrence Erlbaum Associates.

Hollnagel, E. (2003a). *Handbook of cognitive task design.* Mahwah, NJ: Lawrence Erlbaum Associates.

Hollnagel, E. (2003b). Prolegomenon to cognitive task design. In E. Hollnagel (Ed.), *Handbook of cognitive task design* (pp. 3-15). Mahwah, NJ: Lawrence Erlbaum Associates.

John, B. E., & Kieras, D. E. (1996). Using GOMS for user interface design and evaluation: Which technique? *ACM Transactions on Computer-Human Interaction, 3*, 320-351.

Johnson, H., & Johnson, P. (1987). *The development of task analysis as a design tool: A method for carrying out task analysis* (ICL Report). Unpublished manuscript.

Johnson, P., Diaper, D., & Long, J. (1984). Tasks, skills and knowledge: Task analysis for knowledge based descriptions. *Interact'84: First IFIP Conference on Human-Computer Interaction, 1*, 23-27.

Karat, J., Karat, C.-M., & Vergo, J. (2004). Experiences people value: The new frontier for task analysis. In D. Diaper & N. A. Stanton (Eds.), *The handbook of task analysis for human-computer interaction* (pp. 585-602). Mahwah, NJ: Lawrence Erlbaum Associates.

Kieras, D. (2004). GOMS models for task analysis. In D. Diaper & N. A. Stanton (Eds.), *The handbook of task analysis for human-computer interaction* (pp. 83-116). Lawrence Erlbaum Associates.

Levinson, S. C. (1983). *Pragmatics.* Cambridge, MA: Cambridge University Press.

Limbourg, Q., & Vanderdonkt, J. (2004). Comparing task models for user interface design. In D. Diaper & N. A. Stanton (Eds.), *The handbook of task analysis for human-computer interaction* (pp. 135-154). Mahwah, NJ: Lawrence Erlbaum Associates.

Long, J. (1997). Research and the design of human-computer interactions or "What happened to validation?" In H. Thimbleby, B. O'Conaill, & P. Thomas (Eds.), *People and computers XII* (pp. 223-243). New York: Springer.

Ormerod, T. C., & Shepherd, A. (2004). Using task analysis for information requirements specification: The sub-goal template (SGT) method. In D. Diaper & N. A. Stanton (Eds.), *The handbook of task analysis for human-computer interaction* (pp. 347-366). Lawrence Erlbaum Associates.

Shepherd, A. (2001). *Hierarchical task analysis.* London: Taylor and Francis.

Stanton, N. A. (2002). Developing and validating theory in ergonomics. *Theoretical Issues in Ergonomics Science, 3*, 111-114.

Stanton, N. A. (2004). The psychology of task analysis today. In D. Diaper & N. A. Stanton (Eds.), *The handbook of task analysis for human-computer interaction* (pp. 569-584). Mahwah, NJ: Lawrence Erlbaum Associates.

Stanton, N. A., & Young, M. S. (1999). What price ergonomics? *Nature, 399*, 197-198.

Sutcliffe, A. (2003). Symbiosis and synergy? Scenarios, task analysis and reuse of HCI knowledge. *Interacting with Computers, 15*(2), 245-264.

Taylor, F. W. (1912). *Principles of scientific management.* New York: Harper and Row.

Wilson, M., Barnard, P., & MacLean, A. (1986). *Task analysis in human-computer interaction* (Rep. No. HF122). IBM Hursely Human Factors.

Wong, B .L. W. (2004). Critical decision method data analysis. In D. Diaper & N. A. Stanton (Eds.), *The handbook of task analysis for human-computer interaction* (pp. 569-584). Mahwah, NJ: Lawrence Erlbaum Associates.

KEY TERMS

Activity List: A prose description of a task or subtask divided into lines to represent separate task behaviors and that usually has only one main agent and one action per line.

Application Domain: That part of the assumed real world that is changed by a work system to achieve the work system's goals.

Goal: A specification of the desired changes a work system attempts to achieve in an application domain.

Performance: The quality, with respect to both errors and time, of work.

Subtask: A discrete part of a task.

Task: The mechanism by which an application domain is changed by a work system to achieve the work system's goals.

Work: The change to an application domain by a work system to achieve the work system's goals.

Work System: That part of the assumed real world that attempts to change an application domain to achieve the work system's goals.

Task Ontology-Based Human-Computer Interaction

Kazuhisa Seta
Osaka Prefecture University, Japan

INTRODUCTION

In ontological engineering research field, the concept of "task ontology" is well-known as a useful technology to systemize and accumulate the knowledge to perform problem-solving tasks (e.g., diagnosis, design, scheduling, and so on). A task ontology refers to a system of a vocabulary/concepts used as building blocks to perform a problem-solving task in a machine readable manner, so that the system and humans can collaboratively solve a problem based on it.

The concept of task ontology was proposed by Mizoguchi (Mizoguchi, Tijerino, & Ikeda, 1992, 1995) and its validity is substantiated by development of many practical knowledge-based systems (Hori & Yoshida, 1998; Ikeda, Seta, & Mizoguchi, 1997; Izumi &Yamaguchi, 2002; Schreiber et al., 2000; Seta, Ikeda, Kakusho, & Mizoguchi, 1997). He stated:

> *...task ontology characterizes the computational architecture of a knowledge-based system which performs a task. The idea of task ontology which serves as a system of the vocabulary/concepts used as building blocks for knowledge-based systems might provide an effective methodology and vocabulary for both analyzing and synthesizing knowledge-based systems. It is useful for describing inherent problem-solving structure of the existing tasks domain-independently. It is obtained by analyzing task structures of real world problem. ... The ultimate goal of task ontology research is to provide a theory of all the vocabulary/concepts necessary for building a model of human problem solving processes.* (Mizoguchi, 2003)

We can also recognize task ontology as a static user model (Seta et al., 1997), which captures the meaning of problem-solving processes, that is, the input/output relation of each activity in a problem-solving task and its effects on the real world as well as on the humans' mind.

BACKGROUND

Necessity of Building Task Ontologies as a Basis of HCI

It is extremely difficult to develop an automatic problem-solving system that can cope with a variety of problems. The main reason is that the knowledge for solving a problem varies considerably depending on the nature of the problems. This engenders a fact that is sometimes ignored: Users have more knowledge than computers. From this point of view, the importance of a user-centric system (DeBells, 1995) is now widely recognized by many researchers. Such framework follows a collaborative, problem-solving-based approach between human and computer by establishing harmonious interaction between human and computer.

Many researchers implement such a framework with a human-friendly interface using multimedia network technologies. Needless to say, it is important not only to apply the design principles of the human interface but also principle knowledge for exchanging meaningful information between humans and computers.

Systems have been developed to employ research results of the cognitive science field in order to design usable interfaces that are acceptable to humans. However, regarding the content-oriented view, it is required that the system can understand the meaning of human's cognitive activities in order to capture a human's mind.

We, therefore, need to define a cognitive model, that is, to define the cognitive activities humans perform in a problem-solving/decision-making process and the information they infer, and then systemize them as task ontologies in a machine understandable manner in order to develop an effective human-computer interaction.

Problem-Solving Oriented Learning

A task with complicated decision making is referred to as "Problem-Solving Oriented Learning (PSOL) task" (Seta, Tachibana, Umano, & Ikeda, 2003; Seta & Umano, 2002). Specifically, this refers to a task that does not only require learning to build up sufficient understanding for planning and performing problem-solving processes but also to gain the ability/skill of making efficient problem-solving decisions based on sophisticated strategies.

Consider for example, a learner who is not very familiar with Java and XML programming and tries to develop an XML-based document retrieval system. A novice learner in a problem-solving domain tries to gather information from Web resources, investigates and builds up his/her own understanding of the target area, and makes plans to solve the problem at hand and then perform problem-solving and learning processes. Needless to say, a complete plan cannot be made at once, but is detailed gradually by iterating, spirally, those processes while applying a "trial and error" approach. Thus, it is important for a learner to control his/her own cognitive activities.

Facilitating Learners' Meta Cognition through HCI

In general, most learners in PSOL tend to work in an *ad hoc* manner without explicit awareness of meaning, goals and roles of their activities. Therefore, it is important to prompt construction of a rational spiral towards making and performing efficient problem-solving processes by giving significant direction using HCI.

Many researchers in the cognitive science field proposed a concept whereby metacognition plays an important role to acquire and transfer expertise (Brown, Bransford, Ferrara, & Campione, 1983; Flavell, 1976; Okamoto, 1999). Furthermore, repeated interaction loops between metacognition activities and cognition activities play an important role in forming an efficient plan for problem-solving and learning processes.

Figure 1 shows the plan being gradually detailed and refined along the time axis. Figure 1(a) is a planning process when a learner has explicit awareness of interactions and iterate metacognition activities and cognition activities spirally, while Figure 1(b) is a planning process with implicit awareness of them. In PSOL, monitor and control of problem-solving/learning processes are typical activities of metacognition while their performances are ones of cognition. It is natural that the former case allows efficient plans for problem-solving workflow more

Figure 1. The interaction helps the effective planning process

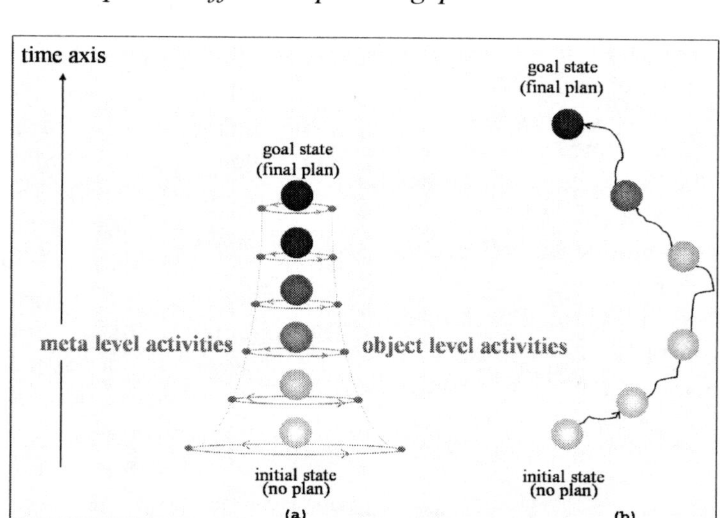

Figure 2. Rasmussen's (1986) cognitive model

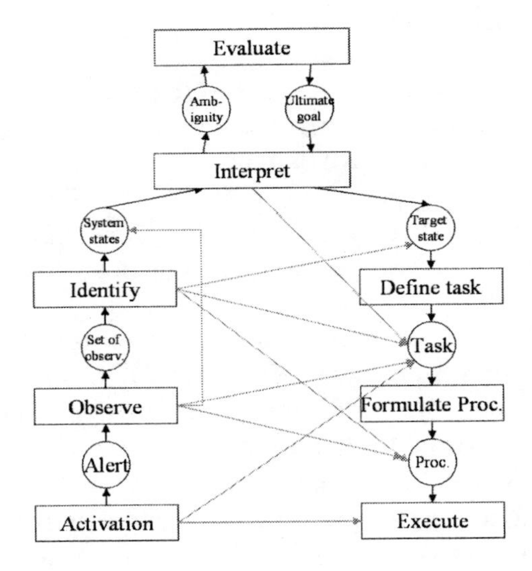

rapidly than the latter. Without explicit awareness of interaction loops, a learner tends to get confused and lose his/her way because nested structures of his/her work and new information of the target world impose heavy loads.

Therefore, it is important to implement an HCI framework that enables effective PSOL by positioning a learner at the center of the system as a subject of problem solving or learning, and providing appropriate information to prompt the learner's metacognition effectively.

MAIN ISSUES IN TASK ONTOLOGY-BASED HCI

In this section, we introduce our approach to supporting PSOL to understand task ontology based HCI framework.

Rasmussen's (1986) cognitive model is adopted as a reference model in the construction of the task ontology for supporting PSOL. It simulates the process of human cognition in problem-solving based on cognitive psychology. Cognitive activity in PSOL is related to this model based on which PSOL task ontology is constructed. This provides a learner with useful information for effective performance of cognitive activity at each state, according to the theoretical framework that was revealed in the cognitive psychology.

Rasmussen's (1986) Cognitive Model

Figure 2 represents an outline of Rasmussen's cognitive model known as the ladder model.

Activities in PSOL broadly comprise activities in connection with a problem-solving act and activities in connection with a learning act (see Figure 3 in the next section).

An *Activation activity* in Rasmussen's cognitive model corresponds to the situation in which a problem is given in problem-solving activities, or one in which a learner detects change in the real world. An *Observe activity* corresponds to observing the details of the change or a gap from the problem-solving goal. An *Identify activity* corresponds to identifying its possible cause. An *Interpret activity* corresponds to interpreting the influence of the change on problem solving and deciding the problem-solving goal. A *Define Task activity* corresponds to determine a problem-solving task for implementing it based on the problem-solving goal. A *Formulate Procedure activity* corresponds to setting up a problem-solving plan to solve the problem-solving task.

Although basically the same correspondence applies in learning activities as the case of problem-solving activities, the object of learning activities, mainly focuses on the state of one's own knowledge or understanding, that is, metacognition activities. Namely, the Activation activity in Rasmussen's cognitive model corresponds to detecting the change of one's own knowledge state. The Observe activity corresponds to observing details or a gap from its own understanding state (goal state) decided as a goal of learning. The Identify activity corresponds to identifying its possible cause. The Interpret activity corresponds to interpreting the influence of its own understanding state, especially the influence on problem solving in PSOL, and deciding the goal of learning. The Define Task activity corresponds to setting up a learning task for implementing it based on the problem-solving goal. The Formulate Procedure activity corresponds to setting up a learning plan to solve the problem-solving task.

Clarifying a correspondence relationship between the cognitive activity by a learner in PSOL and the cognitive activity in Rasmussen's cognitive model permits construction of a problem-solving-oriented learning task ontology as a basis of human-

Figure 3. Learner's work in problem-solving-oriented learning (PSOL)

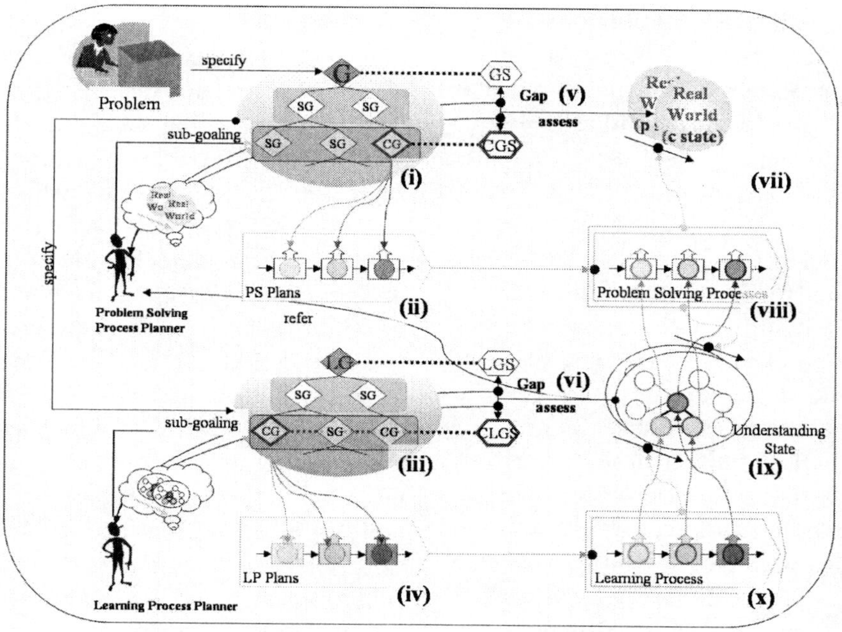

computer interaction comprehending the properties of PSOL appropriately. Implementing an interaction between a system and a learner based on this allows the system to show effective information to encourage the learner's appropriate decision-making.

Cognitive Model in Problem-Solving Oriented Learning

Figure 3 shows a cognitive model that captures detailed working processes of a learner. This model is PSOL task specific while Rasmussen's model is a task independent one. By making the correspondence between these models, we can define an HCI framework based on Rasmussen's theory.

Figures 3(i) and 3(iii) represent the planning process of the problem-solving plan and learning plan, respectively, and 3(viii) and 3(x) represent problem-solving and learning processes in Figure 1, respectively. Figures 3(v) and 3(vi) represent the monitoring process.

We have presented a problem, say, "developing an XML based document retrieval system" in the upper left corner. Two virtual persons, a problem-solving planner and learning process planner in the learner, play roles of planning, monitoring, and controlling problem-solving, and learning processes, respectively.

With PSOL, a learner first defines a problem-solving goal and refines it to sub-goals which contribute to achieving goal G (Figure 3(i)). They are refined to feasible problem-solving plans (Figure 3(ii)); thereafter, the learner performs them to solve the problem (Figure 3(viii)).

If the learner recognizes a lack of knowledge in the sub goals and performs problem-solving plans, we can generate an adequate learning goal (LG) to get knowledge (Figure 3(iii)) and refine it to learning process plans (Figure 3(iv)). In learning processes (Figure 3(x)), s/he constructs knowledge (Figure 3(iv)) to be required to plan and perform the problem-solving process. Based on constructed knowledge, she or he specifies and performs the problem-solving processes (Figure 3(viii)), to change the real world (Figure 3(vii)). The learner assesses gaps among goal states (GS), current goal states (CGS) of problem-solving process plans, and current state (c-state) of the real-world (Figure 3(v)) and ones among learning goal states (LGS), current learning goal states (CLGS) of learning process plans and understanding state (Figure 3(vi)). She or he continuously iterates these processes until the c-state of the real world satisfies the GS of problem solving.

It is notable that learners in PSOL have to make and perform not only problem-solving plans, but also learning plans in the process of problem solving. Furthermore, it is important for the learner to monitor real-world changes by performing problem-solving processes and to monitor his/her own understanding states by performing learning processes and checking and analyzing whether states of the real world and understanding states satisfy defined goal states (Figures 3(v) and 3(vi)). The gap between current states and goal states causes the definition of new goals to be dissolved.

Consequently, PSOL impels a learner to perform complicated tasks with heavy cognitive loads. A learner needs to manage and allocate the attentional capacity adequately because of limited human attentional capacity. This explains why a novice learner tends to get confused and lose his/her way.

Task Ontology for Problem-Solving-Oriented Learning

Figure 4 presents an outline of the PSOL Task Ontology (Problem-Solving-Oriented Learning Task Ontology). Ovals in the figure express a cognitive activity performed by a learner in which a link represents an "is-a" relationship.

The PSOL task ontology defines eight cognitive processes modeled in Rasmussen's cognitive model as lower concepts (portions in rectangular box (a) in the figure). They are refined through an is-a hierarchy to cognitive activities on the meta-level (meta activity), and cognitive activities on the object level (base activity). Moreover, they are further refined in detail as their lower concepts: a cognitive activity in connection with learning activities and a cognitive activity in connection with problem-solving activities. Thereby, a conceptual system is constructed that reflects the task structure of PSOL. For example, typical metacognition activities that a learner

Figure 4. A hierarchy of problem-solving- oriented learning task ontology

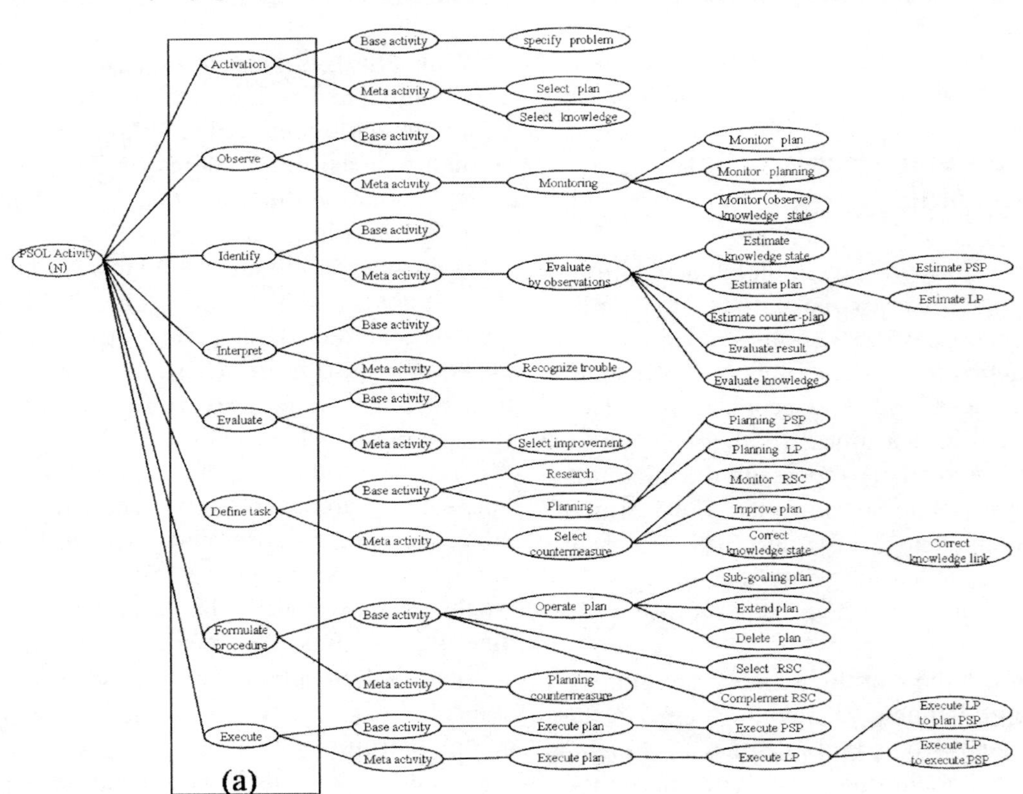

Figure 5. A definition of "identify c-state of NT in executable plan"

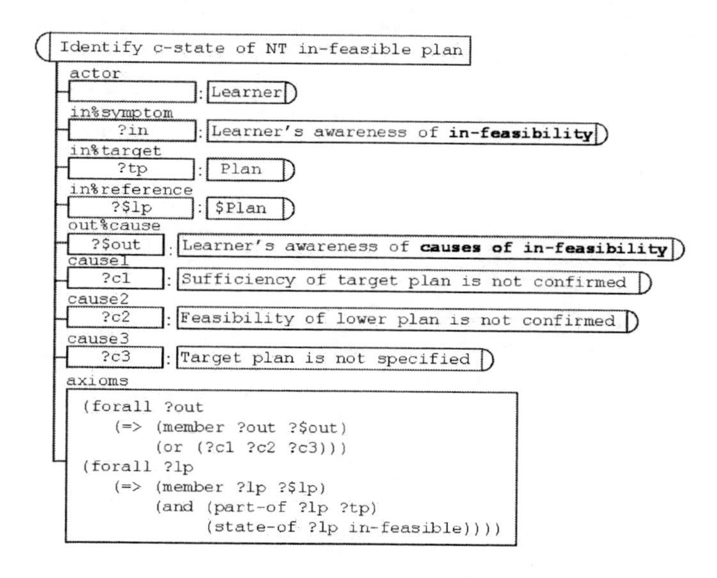

performs in PSOL, such as "Monitor knowledge state" and "Monitor learning plan," are systematized as lower concepts of metacognition activities in the Observe activity.

Figure 5 shows a conceptual definition of an act that identifies a possible cause of why a plan is infeasible. All the concepts in Figure 5 have a conceptual definition in a machine readable manner like this, thus, the system can understand what the learner tries to do and what information he/she needs.

Cause identification activities defined include: the actor of the activity is a learner; a learner's awareness of infeasibility becomes an input (in%symptom in Figure 5); the lower plan of an target plan that the learner tries to make it feasible now is made into a reference information (in%reference in Figure 5). Moreover, this cognitive activity stipulates that a learner's awareness of causes of infeasiblity is output (out%cause in Figure 5). The definition also specifies that the causes of the infeasibility include (axioms in Figure 5): that the sufficiency of that target plan is not confirmed (cause1 in Figure 5); that the feasibility of a lower plan, small grained plan that contributes to realize the target plan, is not confirmed (cause2 in Figure 5); and that the target plan is not specified (cause3 in Figure 5). Based on this machine understandable definition, the system can suggest the candidate causes of infeasibility of the object plan, and the information the learner should focus on.

Making this PSOL task ontology into the basis of a system offers useful information in the situation that encourages appropriate decision-making. This is one of the strong advantages using PSOL task ontology.

An Application: Planning Navigation as an Example

The screen image of Kassist, a system based on the PSOL Task Ontology, is shown in Figure 6. Kassist is an interactive open learner-modeling environment. The system consists of six panels. A learner describes a problem-solving plan, own knowledge state about the object domain, and a learning process in each panels of (a), (b), and (c), respectively. Furthermore, a learner can describe the correspondence relationship between the problem-solving process placed at (a) and the concept of (b), that is, the correspondence relationship with the knowledge of the object domain required for carrying out the process of (a); and the correspondence relationship between the learning process placed at (c), and the concept of (b), that is, the correspondence relationship with the learning process of (c) which con-

Figure 6. Interactive navigation based on problem solving oriented learning task ontology

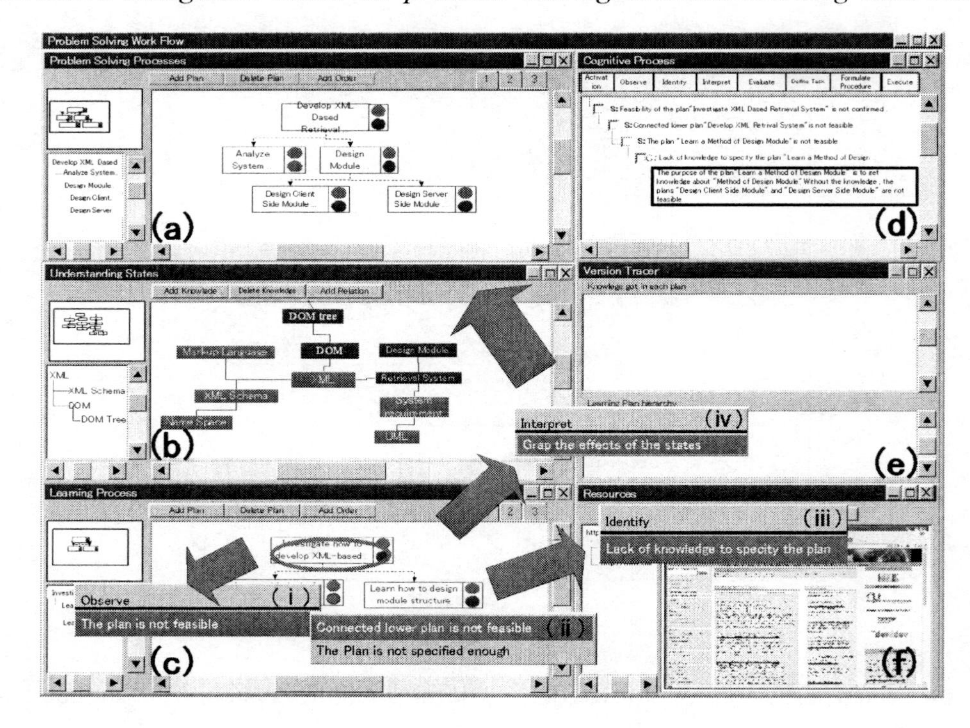

structs an understanding on the concept of (b). Each shaded node in (b) represents either "knowing" or "not knowing" the concept. A learner can describe the correspondence of the concepts and processes in the object world placed on (a), (b), and (c) with resource (f), used as those reference information, so that appropriate information can be referred to when required.

This provides a learner with an environment in which she or he can externalize and express her own knowledge; it then encourages his or her spontaneous metacognition activities such as the Activation activity and Observation activity in Rasmussen's model. Moreover, we can implement a more positive navigation function that encourages a learner's metacognition activity in the subsequent cognitive process by making ontology the basis of a system.

Consider, for example, this task "Investigate how to develop XML-based document retrieval system". Assume a situation where a learner does not know how to tackle this task:

i. In this situation, a learner clicks the "Investigate how to develop XML-based document retrieval system" node on (c); among the lower learning plans connected by the "part-of" links, a plan "Learn how to design module structure", whose feasibility is not secured is highlighted based on the ontology; and a learner is shown the causes of infeasibility with a message "connected lower plan is not feasible".
ii. Then, it shows the cause as a learner has lack of knowledge to specify the plan.
iii. Moreover, plans influenced by the infeasibility of this learning plan are displayed in the interpretation process.
iv. Here, a problem-solving plan "Design module structure" is highlighted. Such navigation allows a learner to comprehend knowledge required to carry out problem-solving and to understand at what stage in a problem-solving process such knowledge is needed and their influence.

Thus, a learner can conduct appropriate decision-making by acquiring detailed knowledge based on this modular design method.

A series of cognitive activities are typical metacognition activities in PSOL. They include: a learner's awareness of feasibility of a learning process as a start; monitoring one's own knowledge state; comprehending its influence on a problem-solving plan; and building a learning plan for mastering knowledge required for problem solving. Reference to the appropriate information offered by a system to a learner encourages his or her appropriate metacognition activities, which help implement effective PSOL.

FUTURE TRENDS

Ontology-Aware System

The systems which support users to perform intelligent tasks based on the understanding of ontologies are called "ontology aware systems" (Hayashi, Tsumoto, Ikeda, & Mizoguchi, 2003). Systemizing ontologies contributes to providing theories and models, which are human-orientated to enhance systems' abilities of explanation and reasoning. Furthermore, from the viewpoint of system development, building systems with explicit ontologies would enhance their maintainability and extendability. Therefore, future work in this field should continue developing systems that integrate ontology and HCI more effectively.

CONCLUSION

This article introduced a task ontology based human computer interaction framework and discussed various related issues. However, it is still difficult and time consuming to build high quality sharable ontologies that are based on the analysis of users' task activities. Thus, it is important to continue building new methodologies for analyzing users' tasks. This issue should be carefully addressed in the future, and we hope more progress can be achieved through collaboration between researchers in the fields of ontology engineering and human computer interaction.

REFERENCES

Brown, A. L., Bransford, J. D., Ferrara, R. A., & Campione, J. C. (1983). Learning, remembering, and understanding. In E. M. Markman, & J. H. Flavell (Eds.), *Handbook of child psychology* (4th ed.), *Cognitive development, 30* (pp. 515-629). New York: John Wiley & Sons.

DeBells, M. (1995). User-centric software engineering, *IEEE Expert*, 34-41.

Flavell, J. H. (1976). Metacognitive aspects of problem solving. In L. Resnick (Ed.), *The nature of intelligence* (pp. 231-235). Hillsdale, NJ: Lawrence Erlbaum Associates.

Gruber, T. R. (1993). A translation approach to portable ontologies. *Knowledge Acquisition*, *5*(2), 199-220.

Hayashi, Y., Tsumoto, H., Ikeda, M., & Mizoguchi, R. (2003). Kfarm: An ontology-aware support environment for learning-oriented knowledge management. *The Journal of Information and Systems in Education*, *1*(1), 80-89.

Hori, M., & Yoshida, T. (1998). Domain-oriented library of scheduling methods: Design principle and real-life application. *International Journal of Human-Computer Studies*, *49*(4), 601-626.

Ikeda, M., Seta, K., & Mizoguchi, R. (1997). Task ontology makes it easier to use authoring tools. In the *Proceedings of the Fifteenth International Joint Conference on Artificial Intelligence '97 (IJCAI-97)* (pp. 342-347).

Izumi, N., & Yamaguchi, T. (2002). Integration of heterogeneous repositories based on ontologies for EC applications development. *International Journal of Electronic Commerce Research and Applications*, *1*(1), 77-91.

Mizoguchi, R. (2003). Tutorial on ontological engineering - Part 1: Introduction to ontological engineering. *New Generation Computing*, *21*(4), 365-384.

Mizoguchi, R., Tijerino, Y., & Ikeda, M. (1992). Task ontology and its use in a task analysis interview system—Two-level mediating representation in MULTIS. In the *Proceedings of the 2nd Japanese Knowledge Acquisition Workshop (JKAW)* (pp. 185-198).

Mizoguchi, R., Tijerino, Y., & Ikeda, M. (1995). Task analysis interview based on task ontology. *Expert Systems with Applications*, *9*(1), 15-25.

Okamoto, M. (1999). The study of metacognition in arithmetic word problem solving. Tokyo: Kazama-shobo Publishing (in Japanese).

Rasmussen, J. (1986). A framework for cognitive task analysis. In *Information processing and human-machine interaction: An approach to cognitive engineering* (pp. 5-8). New York: North-Holland.

Schreiber, G., Akkermans, H., Anjewierden, A., De Hong, R., Shadbolt, N., Van de Veide, W., & Wielinga, B. (2000). *Knowledge engineering and management—The common KADS methodology*. Cambridge, MA: The MIT Press.

Seta, K., Ikeda, M., Kakusho, O., & Mizoguchi, R. (1997). Capturing a conceptual model for end-user programming—Task ontology as a static user model. In the *Proceedings of the Sixth International Conference on User Modeling, UM '97* (pp. 203-214).

Seta, K., Tachibana, K., Umano, M., & Ikeda, M. (2003). Basic consideration on reflection support for problem-solving oriented learning. In the *Proceedings of the International Conference on Computers in Education (ICCE 2003)*, Hong Kong (pp. 160-168).

Seta, K., & Umano, M. (2002). A support system for planning problem solving workflow. In the *Proceedings of the International Conference on Computers in Education (ICCE-02), Workshop on Concepts and Ontologies for Web-based Educational Systems*, Auckland, New Zealand (pp. 27-33).

KEY TERMS

Attentional Capacity: Cognitive capacity divided and allocated to perform cognitive task.

Metacognition: Cognition about cognition. It includes monitoring the progress of learning, checking the status of self-knowledge, correcting self-errors, analyzing the effectiveness of the learning strategies, controlling and changing self-learning strategies, and so on.

Ontology: A specification of a conceptualization (Gruber, 1993).

Problem-Solving Oriented Learning (PSOL): Learning not only to build up sufficient understanding for planning and performing problem-solving processes but also to gain the capacity of making efficient problem-solving processes according to a sophisticated strategy.

Rasmussen's Ladder Model: A cognitive model that models human's decision-making processes. This model is often used for human error analysis.

Task Ontology: A system of vocabulary/concepts used as building blocks for knowledge-based systems.

The Think Aloud Method and User Interface Design

M.W.M. Jaspers
University of Amsterdam, The Netherlands

INTRODUCTION

Daily use of computer systems often has been hampered by poorly designed user interfaces. Since the functionality of a computer system is made available through its user interface, its design has a huge influence on the usability of these systems (Carroll, 2002; Preece, 2002). From the user's perspective, the user interface is the only visible and, hence, most important part of the computer system; thus, it receives high priority in designing computer systems.

A plea for human-oriented design in which the potentials of computer systems are tuned to the intended user in the context of their utilization has been made (Rossen & Carroll, 2002).

An analysis of the strategies that humans use in performing tasks that are to be computer-supported is a key issue in human-oriented design of user interfaces. Good interface design thus requires a deep understanding of how humans perform a task that finally will be computer-supported. These insights then may be used to design a user interface that directly refers to their information processing activities. A variety of methodologies and techniques can be applied to analyze end users' information processing activities in the context of a specific task environment among user-centered design methodologies. More specifically, cognitive engineering techniques are promoted to improve computer systems' usability (Gerhardt-Powels, 1996; Stary & Peschl, 1998).

Cognitive engineering as a field aims at understanding the fundamental principles behind human activities that are relevant in the context of designing a system that supports these activities (Stary & Peschl, 1998). The ultimate goal is to develop end versions of computer systems that support users of these systems to the maximum in performing tasks in such a way that the intended tasks can be accomplished with minimal cognitive effort. Empirical research has indeed shown that cognitively engineered interfaces are considered superior by users in terms of supporting task performance, workload, and satisfaction, compared to non-cognitively engineered interfaces (Gerhardt-Powels, 1996). Methods such as the think aloud method, verbal protocol analysis, or cognitive task analysis are used to analyze in detail the way in which humans perform tasks, mostly in interaction with a prototype computer system.

BACKGROUND

In this section, we describe how the think aloud method can be used to analyze a user's task behavior in daily life situations or in interaction with a computer system and how these insights may be used to improve the design of computer systems. Thereafter, we will go into the pros and cons of the think aloud method.

The Think Aloud Method

Thinking aloud is a method that requires subjects to talk aloud while solving a problem or performing a task (Ericsson & Simon, 1993). This method traditionally had applications in psychological and educational research on cognitive processes. It is based on the idea that one can observe human thought processes that take place in consciousness. Thinking aloud, therefore, may be used to know more about these cognitive processes and to build computer systems on the basis of these insights. Overall, the method consists of (1) collecting think aloud reports in a systematic way and (2) analyzing these reports to gain a deeper understanding of the cognitive processes that take place in tackling a problem. These reports are collected by instructing subjects to solve a problem while thinking aloud; that is, stating directly what they think. The data so gathered are very direct; there is no delay. These verbal utterances are transcribed, resulting in verbal protocols, which require substantial analysis and interpretation to gain deep insight into the way subjects perform tasks (Deffner, 1990).

The Use of the Think Aloud Method in Computer System Design

In designing computer systems, the think aloud method can be used in two ways: (1) to analyze users' task behaviors in (simulated) working practices, after which a computer system is actually built that will support the user in executing similar tasks in future; or (2) to reveal usability problems that a user encounters in interaction with a (prototype) computer system that already supports the user in performing certain tasks.

In both situations, the identification and selection of a representative sample of (potential) end users is crucial. The subject sample should consist of persons who are representative of those end users who will actually use the system in the future. This requires a clearly defined user profile, which describes the range of relevant skills of system users. Computer expertise, roles of subjects in the workplace, and a person's expertise in the domain of work that the computer system will support are useful dimensions in this respect (Kushnirek & Patel, 2004). A questionnaire may be given either before or after the session to obtain this information. As the think aloud method provides a rich source of data, a small sample of subjects (eight to 10) suffices to gain a thorough understanding of task behavior (Ericsson & Simon, 1993) or to identify the main usability problems with a computer system (Boren & Ramey, 2000). A representative sample of the tasks to be used in the think aloud study is likewise essential. Tasks should be selected that end users are expected to perform while using the (future) computer system. This requirement asks for a careful design of tasks to be used in the study to assure that tasks are realistic and representative of daily life situations. It is recommended that task cases be developed from real-life task examples (Kushnirek & Patel, 2004).

Instructions to the subjects about the task at hand should be given routinely. The instruction on thinking aloud is straightforward. The essence is that the subject performs the task at hand, possibly supported by a computer, and says out loud what comes to mind.

A typical instruction would be, "I will give you a task. Please keep talking out loud while performing the task." Although most people do not have much difficulty rendering their thoughts, they should be given an opportunity to practice talking aloud while performing an example task. Example tasks should not be too different from the target task. As soon as the subject is working on the task, the role of the instructor is a restrained one. Interference should occur only when the subject stops talking. Then, the instructor should prompt the subject by the following instruction: "Keep on talking" (Ericsson & Simon, 1993).

Full audiotaping and/or videorecording of the subject's concurrent utterances during task performance and, if relevant, videorecording of the computer screens are required to capture all the verbal data and user/computer interactions in detail. After the session has been recorded, it has to be transcribed. Typing out complete verbal protocols is inevitable to be able to analyze the data in detail (Dix et al., 1998). Videorecordings may be viewed informally, or they may be analyzed formally to understand fully the way the subject performed the task or to detect the type and number of user-computer interaction problems.

The use of computer-supported tools that are able to link the verbal transcriptions to the corresponding video sequences may be considered to facilitate the analysis of the video data (Preece, 2002).

Prior to analyzing the audio and/or video data, it is usually necessary to develop a coding scheme to identify step-by-step how the subject tackled the task and/or to identify specific user/computer interaction problems in detail. Coding schemes may be developed bottom-up or top-down. In a bottom-up procedure, one would use part of the protocols to generate codes by taking every new occurrence of a cognitive subprocess code. For example, one could assign the code *guessing* to the following verbal statements: "Could it be X?" or "Let's try X." The remaining protocols then would be analyzed by using

Figure 1. Excerpt from a coded verbal protocol for analyzing human task behavior

Code	Verbal Protocol Segment	Explanation
NPSCR04	How can I exit this screen?	Navigation problem screen04
MBT012	What does this button mean?	Meaning of button012
RTACT002	It has been busy a very long time	Response time after action002
VSSACT006	What is it doing now?	Visibility of system status after action006
MSSACT009	What does *fatal error098* mean?	Meaning of system feedback after action009

this coding scheme. An excerpt from a coded verbal protocol is given in Figure 1. Note that the verbal protocol is marked up with annotations from the coding scheme. Otherwise, categories in the coding scheme may be developed top-down, for example, from examination of categories of interactions from the human/computer interaction literature (Kushnirek & Patel, 2004). Before it is applied, a coding scheme must be evaluated on its intercoder reliability.

To prevent experimenter bias, it is best to leave the actual coding of the protocols to a minimum of two independent coders. Correspondence among codes assigned by different coders to the same verbal statements must be found, for which the Kappa mostly is used (Altman, 1991).

The coded protocols and/or videos can be compiled and summarized in various ways, depending on the goal of the study. If the goal is to gain a deep insight into the way humans perform a certain task in order to use these insights for developing a computer system to support task performance, then the protocol and video analyses can be used as input for a cognitive task model. Based on this model, a first version of a computer system then may be designed. If the aim is to evaluate the usability of a (prototype) computer system, the results may summarize any type and number of usability problems revealed. If the computer system under study is still under development, these insights then may be used to better the system.

PROS AND CONS OF THE THINK ALOUD METHOD

The think aloud method, preferably used in combination with audio- and/or videorecording, is one of the most useful methods to gain a deep understanding of the way humans perform tasks and of the specific user problems that occur in interaction with a computer system. As opposed to other inquiry techniques, the think aloud method requires little expertise, while it provides detailed insights regarding human task behavior and/or user problems with a computer system (Preece, 2002). On the other hand, the information provided by the subjects is subjective and may be selective. Therefore, a careful selection of the subjects who will participate and the tasks that will be used in the study is crucial. In addition, the usefulness of the think aloud method is highly dependent on the effectiveness of the recording method. For instance, with audiotaping only, it may be difficult to record information that is relevant to identify step-by-step what the subjects were doing while performing a task, whether computer-supported or not (Preece, 2002).

Another factor distinguishing the think aloud method from other inquiry techniques is the promptness of the response it provides. The think aloud method records the subject's task behavior at the time of performing the task. Other inquiry techniques, such as interviews and questionnaires, rely on the subject's recollection of events afterwards. Subjects may not be aware of what they actually are doing while performing a task or interacting with a computer, which limits the usefulness of evaluation measures that rely on retrospective self-reports (Boren & Ramey, 2000; Preece, 2002). The advantage of thinking aloud, whether audio- or videotaped, as a data eliciting method includes the fact that the resulting reports provide a detailed account of the whole process of a subject executing a task.

Although using the think aloud method is rather straightforward and requires little expertise, analyzing the verbal protocols can be very time-consuming and requires that studies are well planned in order to avoid wasting time (Dix et al., 1998).

The think aloud method has been criticized, particularly with respect to the validity and completeness of the reports it generates (Boren & Ramey, 2000; Goguen & Linde, 1993).

An argument made against the use of the think aloud method as a tool for system design is that humans do not have access to their own mental processes and, therefore, cannot be asked to report on these. With this notion, verbalizing thoughts is viewed as a cognitive process on its own. Since humans are poor at dividing attention between two different tasks (i.e., performing the task under consideration and verbalizing their thoughts), it is argued that thinking aloud may lead to incomplete reports (Nisbett & Wilson, 1997).

However, this critique seems to bear on some types of tasks that subjects are asked to perform in certain think aloud studies. As Ericsson and Simon (1993) point out, in general, talking out loud does not interfere with task performance and, therefore, does not lead to much disturbance of the thought processes. If reasoning takes place in verbal form, then verbalizing thoughts is easy and uses no extra human memory capacity. However, if the information is nonverbal and complicated, verbalization will not only cost time but also extra human memory capacity. Verbalization of thoughts then becomes a cognitive process by itself. This will cause the report of the original task processing to be incomplete, and sometimes, it even may disrupt this process (Ericsson & Simon, 1993). Therefore, the think aloud method only may be used on a restricted set of tasks. Tasks for which the information can be reproduced verbally and for which no information is asked that is not directly used by the subject in performing the task under attention are suitable for introspection by the think aloud method (Boren & Ramey, 2000).

The fact that the experimenter may interrupt the subject during task behavior is considered another source of error, leading to distorted reports (Goguen & Linde, 1993). It has been shown, however, that as long as the experimenter minimizes interventions in the process of verbalizing and merely reminds the subject to keep talking when a subject stops verbalizing his or her thoughts, the ongoing cognitive processes are no more disturbed than by other inspection techniques (Ericsson & Simon, 1993).

The think aloud method, if applied under prescribed conditions and preferably in combination with audio- and/or videorecording, is a valuable information source of human task/behavior and, as such, a useful technique in designing and evaluating computer systems.

FUTURE TRENDS

The think aloud method is propagated and far more often used as a method for system usability testing than as a user requirements eliciting method. In evaluating (prototype) computer systems, thinking aloud is used to gain insight into end users' usability problems in interaction with a system to better the design of these systems. The use of think aloud and video analyses, however, may be helpful not merely in evaluating the usability of (prototype) computer systems but also in analyzing in detail how end users tackle tasks in daily life that in the end will be computer supported. The outcomes of these kinds of analyses may be used to develop a first version of a computer system that directly and fully supports users in performing these kinds of tasks. Such an approach may reduce the time spent in iterative design of the system, as the manner in which potential end users process tasks is taken into account in building the system.

Although a deep understanding of users' task behaviors in daily settings is indispensable in designing intuitive systems, we should keep in mind that the implementation of computer applications in real-life settings may change and may have unforeseen consequences for work practices. So, besides involving potential user groups in an early phase of system design and in usability testing, it is crucial to gain insight into how these systems may change these work practices to evaluate whether and how these systems are being used. This adds to our understanding of why systems may or may not be adopted into routine practice.

Today, a plea for qualitative studies for studying a variety of human and contextual factors that likewise may influence system appraisal is made in literature (Aarts et al., 2004; Ammenwerth et al.,

2003; Berg et al., 1998; Orlikowski, 2000; Patton, 2002). In this context, sociotechnical system design approaches are promoted (Aarts et al., 2004; Berg et al., 1998; Orlikowski, 2000). Sociotechnical system design approaches are concerned not only with human/computer interaction aspects of system design but also take psychological, social, technical, and organizational aspects of system design into consideration. These approaches take an even broader view of system design and implementation than cognitive engineering approaches—the organization is viewed as a system with people and technology as components within this system. With sociotechnical system design approaches, it can be determined which changes are necessary and beneficial to the system as a whole, and these insights then may be used to decide on the actions to effect these changes (Aarts et al., 2004; Berg et al., 1998). This process of change never stops; even when the implementation of a computer system is formally finished, users still will ask for system improvements to fit their particular requirements or interests (Orlikowski, 2000).

CONCLUSION

The use of the think aloud method may aid in designing intuitive computer interfaces, because using thinking aloud provides us with a more thorough understanding of work practices than do conventional techniques such as interviews and questionnaires.

Until now, thinking aloud was used mostly to evaluate prototype computer systems. Thinking aloud, however, likewise may be used in an earlier phase of system design, even before a first version of the system is available. It then can be used to elicit every step taken by potential end users to process a task in daily work settings. These insights then may be used as input to the design of a computer system's first version.

Development, fine-tuning, testing, and final implementation of computer systems take a lot of time and resources. User involvement in the whole life cycle of information systems is crucial, because only when we really try to understand end users' needs and the way they work, think, and communicate with each other in daily practice can we hope to improve computer systems.

REFERENCES

Aarts, J., Doorewaard, H., & Berg, M. (2004). Understanding implementation: The case of a computerized physician order entry system in a large Dutch university medical center. *Journal of the American Medical Informatics Association, 11*(3), 207-216.

Altman, D.G. (1991). *Practical statistics for medical research*. London: Chapman and Hall.

Ammenwerth, E., Gräber, S., Herrmann, G., Bürkle, T., & König, J. (2003). Evaluation of health information systems—Problems and challenges. *International Journal of Medical Informatics, 71*(2-3), 125-135.

Berg, M., Langenberg, C., Van Der Berg, I., & Kwakkernaat, J. (1998). Considerations for sociotechnical design: Experiences with an electronic patient record in a clinical context. *International Journal of Medical Informatics, 52*(3), 243-251.

Boren, M.T., & Ramey, J. (2000). Thinking aloud: Reconciling theory and practice. *IEEE Transactions on Professional Communication, 43*(3), 261-278.

Carroll, J.M. (2002). *Human computer interaction in the new millennium*. New York: Addison Wesley.

Deffner, G. (1990). Verbal protocols as a research tool in human factors. *Proceedings of the Human Factors Society, 34th Annual Meeting*.

Dix, A., Finlay, J., Abowd, G., & Beale, R. (1998). *Human computer interaction*. London: Prentice Hall.

Ericsson, K.A., & Simon, H.A. (1993). *Protocol analysis: Verbal reports as data*. Cambridge, MA: MIT Press.

Gerhardt-Powals, J. (1996). Cognitive engineering principles for enhancing human computer performance. *International Journal of Human Computer Interaction, 8*(2), 189-211.

Goguen, J.A., & Linde, C.L. (1993). Techniques for requirements elicitation. *Proceedings of the IEEE International Conference on Requirements Engineering*.

Kaplan, B. (2001). Evaluating informatics applications—Some alternative approaches: Theory, social interactionism, and call for methodological pluralism. *International Journal of Medical Informatics, 64*(1), 39-56.

Kushnirek, A.W., & Patel V.L. (2004). Cognitive and usability engineering methods for the evaluation of clinical information systems. *Journal of Biomedical Informatics, 37*, 56-76.

Nielsen, J. (1993). *Usability engineering*. Toronto: Academic Press.

Nisbett, R.E., & Wilson, T.D. (1997). Telling more than we know: Verbal reports on mental processes. *Psychological Review, 3*, 231-241.

Orlikowski, W.J. (2000). Using technology and constituting structures: A practice lens for studying technology in organizations. *Organizational Science, 11*, 404-428.

Patton, M.Q. (2002). *Qualitative research and evaluation methods*. Thousand Oaks, CA: Sage Publications.

Preece, J. Rogers, Y., & Sharp, H. (2002). *Interaction design: Beyond human computer interaction*. New York: Wiley.

Rossen, M.B., & Carroll, J.M. (2002). *Usability engineering*. New York: Academic Press.

Stary, C., & Peschl, M.F. (1998). Representation still matters: Cognitive engineering and user interface design. *Behavior and Information Technology, 17*(6), 338-360.

KEY TERMS

Cognitive Engineering: A field aiming at understanding the fundamental principles behind human activities that are relevant in context of designing a system that supports these activities.

Cognitive Task Analysis: The study of the way people perform tasks cognitively.

Cognitive Task Model: A model representing the cognitive behavior of people performing a certain task.

Sociotechnical System Design Approach: System design approach that focuses on a sociological understanding of the complex practices in which a computer system is to function.

Think Aloud Method: A method that requires subjects to talk aloud while solving a problem or performing a task.

User Profile: A description of the range of relevant skills of potential end users of a system.

Verbal Protocol: Transcription of the verbal utterances of a test person performing a certain task.

Verbal Protocol Analysis: Systematic analysis of the transcribed verbal utterances to develop a model of the subject's task behavior that then may be used as input to system design specifications.

Video Analysis: Analysis of videorecordings of the user/computer interactions with the aim to detect usability problems of the computer system.

T

Tool Support for Interactive Prototyping of Safety-Critical Interactive Applications

Rémi Bastide
Université Paul Sabatier, France

David Navarre
Université Paul Sabatier, France

Philippe Palanque
Université Paul Sabatier, France

INTRODUCTION

The complete specification of interactive applications is now increasingly considered a requirement in the field of software for safety-critical systems due to their use as the main control interface for such systems. The reason for putting effort in the use and the deployment of formal description techniques lies in the fact that they are the only means for both modeling in a precise and unambiguous way all the components of an interactive application (presentation, dialogue, and functional core; Pfaff, 1985) and proposing techniques for reasoning about (and also verifying) the models (Palanque & Bastide, 1995).

Formal description techniques are usually applied to early phases in the development process (requirements analysis and elicitation) and clearly show their limits when it comes to evaluation (testing).

When the emphasis is on validation, iterative design processes (Hix & Hartson, 1993) are generally put forward with the support of prototyping as a critical tool (Rettig, 1994). However, if used in a nonstructured way and without links to the classical phases of the development process, results produced using such iterative processes are usually weak in terms of reliability. They can also be unacceptable when interfaces for safety-critical applications are concerned.

If we consider interfaces such as the ones developed in the field of air traffic control (ATC), a new characteristic appears, which is the dynamics of interaction objects in terms of existence, reactivity, and interrelations (Jacob, 1999). In opposition to WIMP (windows, icons, menus, and pointing) interfaces, in which the interaction space is predetermined, these interfaces may include new interactors (for instance, graphical representations of planes) at any time during the use of the application (Beaudouin-Lafon, 2000). Even though this kind of problem is easily mastered by programming languages, it is hard to tackle in terms of modeling. This is why classical description techniques must be improved in order to be able to describe in a complete way highly interactive applications.

BACKGROUND

Several approaches propose solutions for the reconciliation of the specification and the validation phases in the field of interactive applications, but these solutions are often incomplete according to three different viewpoints.

- **Interaction Style Viewpoint:** Post-WIMP user interfaces are not yet widely developed. For this reason, most of the approaches (see, for instance, Hussey & Carrington, 1999) only deal with WIMP interfaces, that is, static interfaces for which the set and the number of interactors is known beforehand. The behaviour and the role of these interactors are standardised (typically windows and buttons belong to this category).
- **Development Phase Viewpoint:** We often find disparate solutions that do not integrate the various phases in a consistent manner (Märtin,

Figure 1. Iterative development process with PetShop

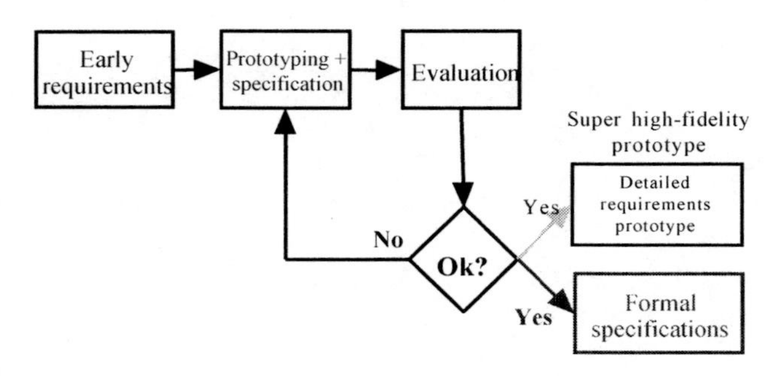

1999). So, most often, several gaps remain to be bridged manually by the teams involved in the development process.

- **Reliability of Results Viewpoint:** Several integrated approaches have been proposed for WIMP-interactive applications. Among them, we find TRIDENT (Bodart, Hennebert, Leheureux, & Vanderdonckt, 1993), which is the more successful one as it handles both data and dialogue description and as it also incorporates ergonomic evaluation by means of embedded ergonomic rules. However, specification techniques used in the project have not been provided with analysis techniques for verifying models and the consistency between models.

PROTOTYPING CAN BE FORMAL, TOO

The PetShop (Petri Nets Workshop) CASE (computer-aided software engineering) tool promotes an iterative development process articulated around the use of a formal description technique of the dialogue of the interactive application.

This formal description technique (based on the petri nets) was developed at LIIHS in the early '90s (Bastide & Palanque, 1990) and has been refined since then (Bastide & Palanque, 1999). The use of this kind of modeling technique provides extended benefits with respect to those less formal. Indeed, analysis tools, exploiting the mathematical background of formalism, allow the validation of the application before its implementation.

Figure 2. A menu opened on the radar for a selected plane

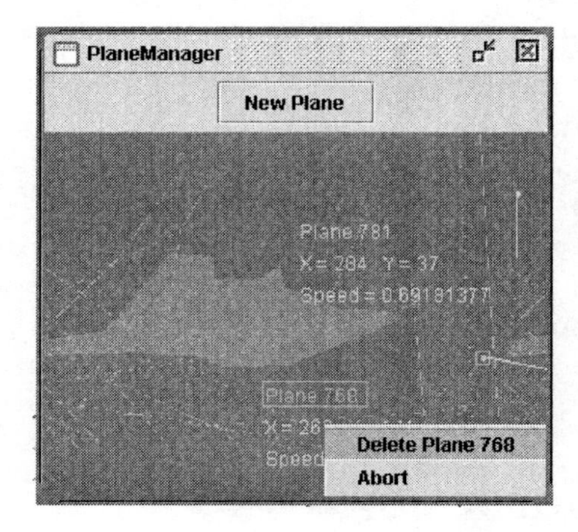

A SAFETY-CRITICAL CASE STUDY

The example presented in this article is extracted from a complex application studied in the context of the European project Mefisto (http://giove.cnuce.cnr.it/mefisto.html).

This project is dedicated to formal description techniques and focuses on the field of air traffic control. This example comes from an en route air traffic control application focusing on the impact of data-link technologies in the ATC field. Using such applications, air traffic controllers can direct pilots in a sector (a decomposition of the airspace).

The radar image is shown in Figure 2. On the radar image, each plane is represented by a graphical element providing air traffic controllers with useful information for handling air traffic in a sector.

Figure 3 presents the general architecture of PetShop. The rectangles represent the functional modules of PetShop. The document-like shapes represent the models produced and used by the modules.

PetShop features an object petri-net editor that allows for the editing and executing of the ObCSs (object control structures) of the classes. At run time, the designer can both interact with the specification and the actual application. These are presented in two different windows overlapping in Figure 4. The window PlaneManager corresponds to the execution of the window with the object petri net underneath.

Figure 3. Architecture of the PetShop environment

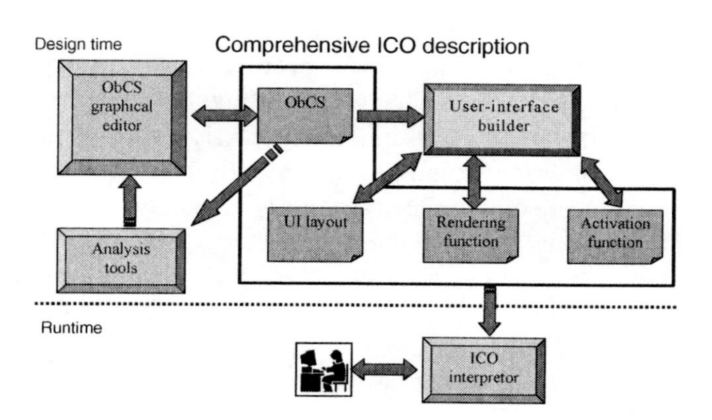

A well-known advantage of petri nets is their executability. This is highly beneficial to our approach since as soon as a behavioural specification is provided in terms of ObCSs, this specification can be executed to provide additional insights on the possible evolutions of the system.

Figure 4 shows the execution of the specification of the ATC application in PetShop. The ICO specification is embedded at run time according to the interpreted execution of the ICO (see Bastide & Palanque, 1995, 1996, for more details about both data structures and execution algorithms).

At run time, the user can look at both the specification and the actual application. They are in two different windows overlapping in Figure 4. The window PlaneManager corresponds to the execution of the window with the object petri net underneath.

Figure 4. Interactive prototyping with PetShop

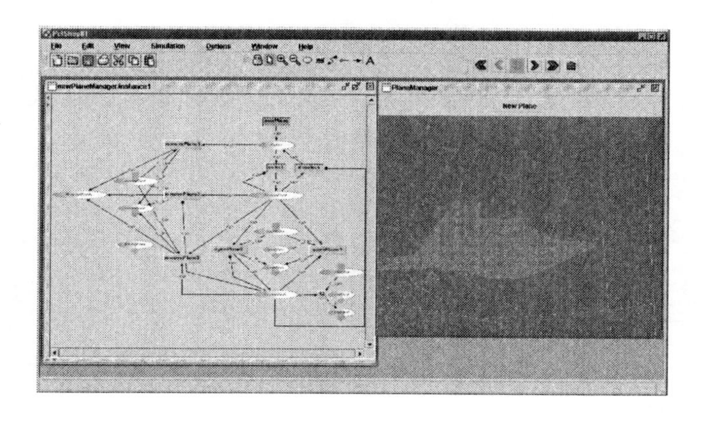

In this window, we can see the set of transitions that are currently enabled (represented in dark grey and the other ones in light grey). This is automatically calculated from the current marking of the object petri net.

Each time the user acts on the PlaneManager, the event is passed onto the interpreter. If the corresponding transition is enabled, then the interpreter fires it, performs its action (if any), changes the marking of the input and output places, and performs the rendering associated (if any).

INTERACTIVE PROTOTYPING

Within PetShop, prototyping from a specification is performed in an interactive way. At any time during the design process, it is possible to introduce modifications either to make the specification more precise or to change it. The advantage of model-based prototyping is that it allows designers to immediately evaluate the impact of a modification.

We have identified two different kinds of modifications that can be performed using PetShop, namely lexical and syntactic.

Lexical Modifications

The lexical part of the user interface gathers elementary elements of the presentation (for instance, the drawing of a button) and all the elementary actions offered to the user (such as clicking on a button). Lexical modifications are concerned with the addition, removal, or modification of these kinds of elements.

- Changing the rendering of a plane. When selected, the colour of a plane changes to green. As a lexical modification, we propose to change it to red.
  - At the specification level, nothing changes in the specification. Only the content of the method *showSelected* must be modified, and this must be done using the JBuilder environment.
- Changing the event triggering the selection of a plane. The currently used event is *Left*

Button Shift Click. We propose to use the event *Left Button Click* instead. Therefore, we need to perform the following modifications.

- At the specification level, the corresponding event must be changed in the activation function.
- At the code level, the Java code must be modified to change the adapter (representing the activation function) of the widget plane.

Syntactic Modifications

The syntactic part of the user interface describes the links and relationships between the lexical elements (for instance, pressing shift and clicking on a plane, then right clicking on the plane to open the menu and delete the plane).

- Modifying the selection mechanism. Currently, only one plane can be selected at a time. In order to allow multiple selections, the following modifications must be performed.
    - At the specification level, the inhibitor arc (the arc terminated by a black circle) linking the transition *select* to the place SelectedPlane (see Figure 4) must be removed.
    - At the code level, there is no modification.
- Defining an upper limit for the number of planes in the sector. In the initial informal specification, there is no limit on the number of planes. Adding a maximum limit of 20 planes (number of planes normally controlled by a controller) requires the following modifications.
    - At the specification level, a new place must be added in the PlaneManager ObCS (Figure 4). Initially, this place will hold 20 tokens. This place has to be connected by an arc to the transition NewPlane of the same ObCS. When a plane leaves a sector (or is deleted using the menu), the corresponding transition must add a new token to this place.
    - At the code level, there is no modification.

FUTURE TRENDS

This article has presented the use of PetShop as a CASE tool for the interactive prototyping of safety-critical software. We have shown how PetShop can deal with a specific kind of interactive system in which interactors can be dynamically instantiated, and thus the dialogue part of the system may consist in a potentially infinite number of states. This kind of application takes full advantage of the expressive power of the ICO's formalism, which is based on high-level petri nets and thus is able to deal both with an infinite number of states and concurrent behaviours.

Sophisticated interaction techniques such as the multimodal ones bring new challenges such as temporal constraints and the description of fusion mechanisms. ICOs have been extended to deal with multimodal interactive systems (Palanque & Schyn 2003), and these extensions are currently under integration within the PetShop environment.

CONCLUSION

Prototyping is now recognized as a cornerstone of the successful construction of interactive systems as it allows making users the centre of the development process. However, prototyping tends to produce low-quality software as no specification or global design is undertaken. We have shown in this article how formal specification techniques can contribute to the development process of interactive systems through prototyping activities.

While the ICO formal specification technique has reached a maturity level allowing coping with real size dynamic, interactive applications, the PetShop environment is currently made available on the Web (http://liihs.irit.fr/petshop/start/pubs.html). This CASE tool allows designers to build formal descriptions in a modeless and interactive way, thus allowing them to continuously assess their design with respect to the actual execution of their specification and the actual verification results from integrated analysis tools. A real size application has been completely specified in the field of the European project Mefisto (http://giove.cnuce.cnr.it/mefisto.html).

However, the work done on this air traffic control application has also shown the amount of work that is still required before the environment can be used by other people than the ones that took part in its development.

This article has presented a CASE tool called PetShop, dedicated to the formal description of interactive systems.

REFERENCES

Bastide, R., & Palanque, P. (1990). *Petri net objects for the design, validation and prototyping of user-driver interfaces*. Third IFIP Conference on Human-Computer Interaction: Interact'90, (pp. 625-631).

Bastide, R., & Palanque, P. (1995). A petri net based environment for the design of event-driven interfaces. *Proceedings of the 16th International Conference on Application and Theory of Petri Nets (ATPN'95)* (pp. 153-167).

Bastide, R., & Palanque, P. (1996). Implementation techniques for petri net based specifications of human-computer dialogues. *Proceedings of the 2nd Conference on Computer Aided Design of User Interfaces (CADUI'96)* (pp. 153-167).

Bastide, R., & Palanque, P. (1999). A visual and formal glue between application and interaction. *Journal of Visual Language and Computing, 10*(3), 129-153.

Bastide, R., Sy, O., Palanque, P., & Navarre, D. (2000). *Formal specification of CORBA services: Experience and lessons learned*. ACM Conference on Object-Oriented Programming, Systems, Languages, and Applications (OOPSLA), (pp. 153-163).

Beaudouin-Lafon, M. (2000). *Instrumental interaction: An interaction model for designing post-WIMP user interfaces*. ACM CHI'2000 Conference on Human Factors in Computing Systems, (pp. 153-161).

Bodart, F., Hennebert, A. M., Leheureux, J. M., & Vanderdonckt, J. (1993). Encapsulating knowledge for intelligent automatic interaction objects selection. *Human Factors in Computing Systems INTERCHI'93*, 424-429.

Hix, D., & Hartson, R. (1993). *Developing user interfaces: Ensuring usability through product and process*. New York: John Wiley & Sons.

Hussey, A., & Carrington, D. (1999). Model-based design of user interfaces using object-Z. *Proceedings of the 3rd International Conference on Computer-Aided Design of User Interfaces* (pp. 153-179).

Jacob, R. (1999). A software model and specification language for non-WIMP user interfaces. *ACM Transactions on Computer-Human Interaction, 6*(1), 1-46.

Märtin, C. (1999). *A method engineering framework for modelling and generating interactive applications*. Third International Conference on Computer-Aided Design of User Interfaces, (pp. 50-63).

Palanque, P., & Bastide, R. (1995). Verification of an interactive software by analysis of its formal specification. *Proceedings of Interact'95* (pp. 191-196).

Palanque, P., & Schyn, A. (2003). A model-based approach for engineering multimodal interactive systems. *Proceedings of INTERACT 2003, IFIP TC 13th Conference on Human Computer Interaction* (pp. 543-550).

Pfaff, G. (Ed.). (1985). *User interface management systems: Eurographics seminar*. Seeheim, Germany: Springer Verlag.

Rettig, M. (1994). Prototyping for tiny fingers. *Communications of the ACM, 37*(4), 21-27.

KEY TERMS

ATC (Air Traffic Control): This acronym refers to both the activities and the systems involved in the management of flights by air-traffic controllers. The air traffic controllers' main task is to ensure flight safety with an efficient, secure, and ordered air traffic flow. ATC systems are dedicated to the support of these tasks.

CASE (Computer-Aided Software Engineering): This acronym refers to a set of tools dedicated to support various phases in the development process of software systems. Usually, they support modeling activities and the refinement of models toward implementation.

Interactors: Elementary interactive components such as push buttons, text fields, list boxes, and so forth.

ObCS (Object Control Structure): A behavioural description of objects and classes.

PetShop (Petri Nets Workshop): A CASE tool dedicated to the formal design and specification of interactive safety-critical software.

WIMP (Windows, Icons, Menus, and Pointing): This is a classical interaction technique found in most window managers like Microsoft Windows.

Toward an HCI Theory of Cultural Cognition

Anthony Faiola
Indiana University, USA

INTRODUCTION

With the increasing demand for global communication between countries, it is imperative that we understand the importance of national culture in human communication on the World Wide Web (WWW). As we consider the vast array of differences in the way we think, behave, assign value, and interact with others, culture becomes a focal point in research of online communication. More than ever, culture has become an important human-computer interaction (HCI) issue, because it impacts both the substance and the vehicle of communication via communication technologies. Global economics and information delivery is leading to even greater diversification among individuals and groups of users who employ the WWW as a key resource for accessing information and purchasing products. Companies will depend more on the Internet as an integral component of their communication infrastructure. With a shift toward online services for information, business professionals have identified international Web usability as an increasingly relevant area of HCI research. What must be addressed are the cultural factors surrounding Web site design. Specifically argued is that culture is a discernible variable in international Web site design, and as such, should better accommodate global users who seek to access online information or products. There are still many unresolved questions regarding cross-cultural HCI and communication and the delivery of information via the Web. To date, there has been no significant connection made between culture context and cognition, cross-cultural Web design, and related issues of HCI. This correlation is relevant for identifying new knowledge in cross-cultural Web design theory and practice.

BACKGROUND

In order to maximize the easy access of online information and products, the building of Web sites should accommodate for more than multilingual communication (Sheridan, 2001). Much rather, Web site developers responsible for design and testing, should have an equal concern for the complexity inherent in cultural diversity, that is, with factors such as social and psychology development. Past and recent cross-cultural studies and theoretical models have made direct links between culture, context, and related preferences (Chau, Cole, Massey, Montoya-Weiss, & O'Keefe, 2002; Hall, 1959, 1966; Hofstede, 1997; Trompenaars, 1997), but with a high emphasis placed especially on behavior.

Nisbett and Norenzayan (2002) argued that most psychologists in the 20th century continue to hold erroneous assumptions about the relationship between culture and cognition. This, they say, is fostered from theoretical positions in learning theory as seen by the work of Miller and Johnson-Laird (1976), as well as other cognitive scientists who embrace Piaget's position of extreme formalism and content independence of inferential rules found in culture. The problem with this view is that formalist theory assumes that cognitive processes are universal and all normal humans are equipped with the same set of attentional, memorial, learning, and inferential procedures, regardless of the content they operate on. However, the landmark work of cultural psychologist Nisbett (Nisbett, 2003; Nisbett & Norenzayan, 2002; Nisbett, Peng, Choi, & Norenzayan, 2001) provides a significant rebuttal to these assumptions about the independent relationship between culture and cognition. He includes a theoretical model of significant depth on which to build support for a new theory for international Web design that addresses the complexity of cognition in cultural context. Nisbett and Norenzayan (2002) state that the idea that culture profoundly influences the contents of thought through shared knowledge structures has been a central theme in modern cognitive psychology.

Nisbett's perspective on culture and cognition is derived from a range of studies and knowledge claims based on the earliest work of Vygotsky

(1979, 1989) and carried on by Luria (1971, 1976) in the 1960s-1970s. The central theory of the Russian School argues that cognitive processes emerge from practical activity that is culturally constrained and historically developing. Nisbett makes note of the significance of this early Vygotskian research (Luria, 1971; Vygotsky, 1979) to promote the idea that culture fundamentally shapes thought. This latter claim has provided a theoretical model for cultural cognition theory (CCT), which goes to the root of human information processing and other complex cognitive systems that are affected by cultural context. Nisbett and Norenzayan (2002) present evidence concerning assumptions about universality and content independence, concluding that multiple studies support their view of the relationship between culture and cognition, casting substantial doubt on the standard assumptions held by many psychologists. At the same time, the vast majority of empirical studies in cross-cultural psychology and cultural anthropology support the position that cognition is dependent upon cultural context, especially where formal education is present.

MAIN FOCUS

Over the last ten years, usability theory and testing have dominated the discussion among HCI and information technologists in academia and industry, setting the stage for culture to become the next frontier of Web design research (Dalal, Quible, & Wyatt, 2000; Eveland & Dunwoody, 2000; Fernandes, 1995; Kim & Allen, 2002; Marcus & Gould, 2000; Sears, Jacko, & Dubach, 2000; Wheeler, 1998; Zahedi, Van Pelt, & Song, 2001; Zassoursky, 1991). At present, many technologists neglect the impact of culture on communication, content delivery, and information structure. For them, technology is often used in place of creative and research-based solutions for overcoming limitations to human communication. However, as the strategic planning of Web design has fallen into the hands of HCI designers, social scientists, and communication experts, technology has not been seen as the panacea to online communication issues between cultures. Rather, in the process of investigating the most appropriate ways to maximize online information delivery to international users, these specialists are exploring ways to confront an array of cultural contexts that are both vast and complex.

In regard to HCI, there are multiple usability studies that address cross-cultural Web site design from a socio-behavioral perspective (Barber & Badre, 1998; Dalal et al., 2000; Eveland & Dunwoody, 2000; Fernandes, 1995; Honold, 2000; Kim & Allen, 2002; Marcus, 2000, 2003; Marcus & Gould, 2000; Sears et al., 2000; Wheeler, 1998; Zahedi et al., 2001). However, limited focus on the relationship between culture and cognition as a theoretical model has been adequately explored, especially when we consider user preferences that are culturally determined by social-cognitive development. Hence, this article presents the view of cultural cognition theory (CCT) from the earliest work of Vygotsky (1979, 1989) and Luria (1971, 1976). Their work was further developed through the contemporary research of Richard E. Nisbett (Nisbett et al., 2001, 2002; Nisbett, Fong, Lehman, & Cheng, 1987; Nisbett & Ross, 1980) and colleagues. A relationship is drawn between CCT and cross-cultural Web development as a means to identify cognitive differences among designers that ultimately influence site design and user-Web interaction. The collective works outlined earlier identify propositions that argue how culture shapes cognitive phenomena, influencing the content of thought through shared knowledge, and subsequently learning, cognitive development, and processes that may impact Web design.

One proposition purported by Nisbett and Norenzayan (2002) is that "cultures differ markedly in the sort of inferential procedures (cognitive processes) they typically use for a given problem" (p. 2). To support this claim, they spend considerable time outlining a range of studies dealing with linguistics and mathematics that show cultural differences in basic knowledge structures and inferential procedures (Lucy, 1992; Luria, 1971; Miller, Smith, Zhu, & Zhang, 1995; Miller & Stigler, 1987; Wynn, 1990). Specifically, these studies show infinitely variable differences in knowledge domains, analytical processes, learning skills, and inferential procedures (such as deductive rules and schemes for induction and causal analysis), among diverse cultures. This is because these processes operate on different inputs for different people in different situations and cultures. A summary of these studies showed repeated evidence to support the linguistic differences that affect thought. Nisbett and Norenzayan (2002) recommend that more research is needed to examine the pervasiveness of the influence of language on thought.

Hence, HCI research that focuses on social and cognitive science and cross-cultural communication methodologies may inform international Web site designers. A range of Web design components needs to be considered during the development stage of cross-cultural sites. Besides the explicit cultural differences of text, numbers, dates, symbol-sets, and time, more critical are the implicit and less formal dimensions of page format, imagery, colour, information architecture, and system functionality. Russo and Boor (1993) discuss the intuitive behaviour that influences the use of these elements. Studies focused on information technology also show considerable cultural differences in attitudes toward computers (Choong & Salvendy, 1998, 1999; Igbaria & Zviran, 1996; Omar, 1992). However, Van Peurssen (1991) suggests that culture is a concept that is far too complex for mere description. In fact, variations in the implicit aspects of Web design may be too subtle to discern their cultural origin, and therefore demand a more rigorous investigation of their relationship to culture and cognition.

Based on this theoretical foundation, future HCI research must identify the knowledge structures, logic, and analytical approaches that constitute a specific Web design orientation based on the cognitive structures influenced by cultural context. By observing the way Web designers organize online information, findings should show cultural orientations in design styles, preferences, and strategic planning.

FUTURE TRENDS

New research opportunities in communication technology are emerging worldwide that are working towards universal access for all international Web users. However, the rapid transition to online delivery of information products has forced corporations to confront the polarizing effect of culture and communication as they attempt to build Web sites that cross the barriers of language and a broad range of more subtle cultural issues. To work through these challenges, technological trends and technologies will continue to address complex issues surrounding semantics and language delivery through the Web. Language technologies are showing promising capabilities in addressing the challenges of unrestricted cross-cultural communication.

Research in the United States and Europe (Gast, 2003; Sierra, Wooldridge, Sadeh, Conte, Klusch, & Treur, 2000; Wagner, Yezril, & Hassel, 2000) is being funded to support cooperatively built technologies to better facilitate cross-cultural communication. Cross-cultural sites may attempt to appeal to international users by using human-like anthropomorphic agents that simulate their cultural profiles. This could include cultural-centric related attributes such as body language, gestures and facial expressions that mimic human characteristics. The intention would be to adapt to the users' emotional world in order to enhance cognitive capabilities and interactions while using a site. Web site developers will increasingly take into account the full complexity of the human emotional apparatus, including the human-like responses that users often seek while engaging interactive systems.

Recent research (Burnett & Buerkle, 2004; Dou, Nielsen, & Tan, 2002; Faiola, 2002; Faiola & Matei, 2005; Hillier, 2003; Yetim & Raybourn, 2003; Zahedi et al., 2001) continues to examine the influence of culture on Web design by comparing subjects from a range of diverse cultures. These studies provide computational models that show trends and comparisons of the data that can help to draw conclusions regarding the influence of cultural cognition on local developers and subjects who interacted with Web sites. Future studies must also address cultural differences in: (1) task times, navigational paths, interface design, and information architecture, and (2) user preferences for sites created by designers from their own cultures, that is, will users demonstrate bias toward color, design, and information structure by designers from their own culture?

CONCLUSION

From a cultural cognition perspective, we have addressed how cognitive processes are susceptible to cultural variation, adding that cultural orientation has a direct impact on designers of Web sites. As a result, Web developers need to design online sites from an understanding of cultural cognitive theory and knowledge as part of their basic strategy. Future trends in CCT research may prove the assumption that cultural differences need to drive variations in Web site design and development.

With increasing dependence on the World Wide Web (WWW) for international communication, the need for effective delivery of content will force Web developers to: (1) move away from homogeneous design models and routine time-on-task usability testing, (2) devise, design, and test models that can account for the complexity of online communication and information exchange between a diversity of national cultures, and (3) consider the significant influence of cultural context on cognition and cognitive development of Web site designers from a diversity of cultural orientations.

REFERENCES

Barber, W., & Badre, A. (1998). Culturability: The merging of culture and usability. In the *Proceedings of the 4th Conference on Human Factors and the Web*. Retrieved on January 10, 2003, from http://research.microsoft.com/users/marycz/hfweb98/barber/index.htm

Burnett, G., & Buerkle, H. (2004). Information exchange in virtual communities: A comparative study. *Journal of Computer-Mediated Communication, 9*(2).

Chau, P. Y. K., Cole, M., Massey, A. P., Montoya-Weiss, M., & O'Keefe, R. M. (2002). Cultural differences in the online behavior of consumers. *Communications of the ACM, 45*(10), 138-143.

Choong, Y. Y., & Salvendy, G. (1998). Design of icons for use by Chinese in mainland China. *Interacting with Computers, 9*, 417-430.

Choong, Y. Y., & Salvendy, G. (1999). Implications for design of computer interfaces for Chinese users in mainland China. *International Journal of Human–Computer Interaction, 11*(1), 29-46.

Dalal, N. P., Quible, Z., & Wyatt, K. (2000). Cognitive design of home pages: An experimental study of comprehension on the World Wide Web. *Information Processing and Management, 36*(4), 607-621.

Dou, W., Nielsen, U., & Tan, C. (2002). Using corporate websites for export marketing. *Journal of Advertising Research, 42*(5), 105-116.

Eveland W. P., & Dunwoody, S. (2000). Examining information processing on the World Wide Web using think aloud protocols. *Media Psychology, 2*(3), 219-244.

Faiola, A. (2002). A visualization pilot study for hypermedia: Developing cross-cultural user profiles for new media interfaces. *The Journal of Educational Multimedia and Hypermedia, 11*(1), 51-70.

Faiola, A., & Matei, S. (2005). Cultural cognitive sytle and Web design: Beyond a behavioral inquiry of computer-mediated communication. *Journal of Computer-Mediated Communication, 11*(1).

Fernandes, T. (1995). *Global interface design*. Boston: AP Professional.

Gast, A. P. (2003). *The impact of restricting information access on science and technology*. Cambridge, MA: MIT Press.

Hall, E. T. (1959). *The silent language*. New York: Doubleday.

Hall, E. T. (1966). *The hidden dimension*. New York: Doubleday.

Hillier, M. (2003). The role of cultural context in multilingual website usability. *Electronic Commerce Research and Applications, 2*(1), 2-14.

Hofstede, G. (1997). *Cultures and organizations: Software of the mind*. London: McGraw-Hill.

Honold, P. (2000). Cultural and context: An empirical study for the development of a framework for the elicitation of cultural influence in product usage. *The International Journal of Human-Computer Interaction, 12*, 327-345.

Igbaria, M. & Zviran, M. (1996). Comparison of end-user computing characteristics in the U.S., Israel, and Taiwan. *Information and Management, 30*, 1-13.

Kim, K. S., & Allen, B. (2002). Cognitive and task influences on Web searching behavior. *Journal of the American Society for Information Science and Technology, 2*, 109-119.

Lucy, J. A. (1992). *Grammatical categories and cognition: A case study of the linguistic relativity hypothesis*. Cambridge, New York: Cambridge University Press.

Luria, A. R. (1971). Towards the problem of the historical nature of psychological processes. *International Journal of Psychology, 6*, 259-272.

Luria, A. R. (1976). *Cognitive development: Its cultural and social foundations.* Cambridge, MA: Harvard University Press.

Marcus, A. (2000). International and intercultural user-interface design. In C. Stephanidis (Ed.), *User interfaces for all,* (pp. 47-63). New York: Erlbaum.

Marcus, A. (2003). Fast forward: User-interface design and China: A great leap forward. *Interactions, 10*(1), 21-25.

Marcus, A., & Gould, E. W. (2000). Crosscurrents cultural dimensions and global web user-interface design. *Interactions, 7*(4), 32-46.

Miller, G. A., & Johnson-Laird, P. N. (1976). *Language and perception.* Cambridge, MA: Belknap Press.

Miller, K. F., Smith, C. M., Zhu, J., & Zhang, H. (1995). Preschool origins of cross national differences in mathematical competence: The role of number naming systems. *Psychological Science, 6*, 56-60.

Miller, K. F., & Stigler, J. W. (1987). Counting in Chinese: Cultural variation in basic cognitive skill. *Cognitive Development, 2*, 279-305.

Nisbett, R. E. (2003). *The geography of thought: How Asians and Westerners think differently...and why*. New York: The Free Press.

Nisbett, R. E., Fong, G. T., Lehman, D. R., & Cheng, P. W. (1987). Teaching reasoning. *Science, 238,* 625-631.

Nisbett, R. E., & Norenzayan, A. (2002). Culture and cognition. In H. Pashler & D. L. Medin (Eds.), *Stevens' handbook of experimental psychology: Cognition* (3rd ed., Vol. 2, pp. 561-597). New York: John Wiley & Sons.

Nisbett, R. E., Peng, K., Choi, I., & Norenzayan, A. (2001). Culture and systems of thought: Holistic vs. analytic cognition. *Psychological Review, 8*, 291-310.

Nisbett, R. E., & Ross, L. (1980). *Human inference: Strategies and shortcomings of social judgement.* Englewood Cliffs: Prentice-Hall.

Omar, M. (1992). Attitudes of college students towards computers: A comparative study in the Middle East. *Computers in Human Behavior, 8*, 249-257.

Russo, P., & Boor, S. (1993, April 24-29). How fluent is your interface? Designing for interational users. In S. Ashlund, K. Mullet, A. Hunderson, E. Hollnagel, & T. White (Eds.), *Proceedings of ACM CHI '93: Human Factors in Computing Systems Conference*, Amsterdam, The Netherlands (pp. 342-347). New York: ACM Press.

Samovar, L., & Porter, R. E. (2003). Understanding intercultural communication: An introduction and overview. In L. Samovar & R. E. Porter (Eds.), *Intercultural communication: A reader* (10th ed., pp. 6-17). Belmont, CA: Wadsworth.

Sears, A., Jacko, J., & Dubach, E. M. (2000). International aspects of World-Wide-Web Usability and the role of high-end graphical enhancements. *International Journal of Human-Computer Interaction, 12*(2), 241-261.

Sheridan, E. F. (2001). Cross-cultural Web-site design. *Multilingual Computing and Technology, 12*(7), 53-56.

Sierra, C., Wooldridge, M., Sadeh, N., Conte, R., Klusch, M., & Treur, J. (2000). Agent research and development in Europe. *Information Services and Use, 24*(4), 189-202.

Trompenaars, F. (1997). *Riding the waves of culture: Understanding cultural diversity in business.* London: Nicholas Brealey.

Van Peurssen, C. (1991). *The new management: Developments within a changing culture*. Kampen, Netherlands: Kok Agora.

Vygotsky, L. S. (1979). *Mind in society: The development of higher psychological processes.* Cambridge, MA: Harvard University Press.

Vygotsky, L. S. (1989). *Thought and language.* Cambridge, MA: MIT Press.

Wagner, C., Yezril, A., & Hassel, S., (2000). *International cooperation in research and development: An update to an inventory of U.S. government spending.* Washington, DC: Rand Science & Technology Policy Institute.

Wheeler, D. L. (1998). Global culture or culture clash: New information technologies in the Islamic world - A view from Kuwait. *Communication Research, 25,* 359-376.

Wynn, K. (1990). Children's understanding of counting. *Cognition, 36,* 155-193.

Yetim, F., & Rayborun, E. M. (2003, April 6). Supporting intercultural-computer-medaited discourse: Methods, models, and architectures. *ACM CHI Extended Abstracts 2003,* Fort Lauderdale, Florida (pp. 1054-1055).

Zahedi, F., Van Pelt, W. V., & Song, J. (2001). A conceptual framework for international Web design. *IEEE Transactions on Professional Communication, 44*(2), 83-103.

Zassoursky, Y. N. (1991). Mass culture as market culture. *Journal of Communication, 41*(2), 12-18.

KEY TERMS

Anthropomorphic: An attribution of human characteristic or behavior to natural phenomena or inanimate objects, that is, the embodiment of a graphic user interface agent can be anthropomorphic since the representation may take on human attributes or qualities. Anthropomorphism is an attempt to design technologies to be user-friendly.

Cognition: The mental processes of an individual, including internal thoughts, perceptions, understanding, and reasoning. It includes the way we organize, store, and process information, as well as make sense of the environment. It can also include processes that involve knowledge and the act of knowing, and may be interpreted in a social or cultural sense to describe the development of knowledge.

Cultural Cognition Theory: A theory that frames the concept that culture profoundly influences the contents of thought through shared knowledge structures and ultimately impact the design and development of interactive systems, whether software or Web sites.

Cultural Psychology: An interdisciplinary field within the social sciences that brings together general and cognitive psychology, cross-cultural communication, anthropology, as well as linguistics and philosophy. Cultural psychologists study how cultural context, meaning, practice, and established institutions might impact individual human psychology.

Culture: A deposit of knowledge, experience, beliefs, values, attitudes, meanings, social hierarchies, religion, notions of time, roles, spatial relationships, concepts of the universe, and material objects and possessions acquired by a group of people in the course of generations through individual and group development (Samovar & Porter, 2003).

Usability: The *effectiveness*, *efficiency*, and *satisfaction* with which specified users achieve specified goals in particular environments. (From the International Organisation for Standardisation (ISO) code: ISO/IS 9294-11, http://www.iso.org/iso/en/ISOOnline.openerpage.)

Vygotskian: A general theory of cognitive development, developed by Vygotsky (1979, 1989) in the 1920s and 1930s in Russia, suggests that: (1) social interaction plays a fundamental role in the development of cognition, and (2) consciousness is the end product of socialization, for example, that cognitive development depends upon the zone (contextual) of proximal development.

T

Tracking Attention through Browser Mouse Tracking

Robert S. Owen
Texas A&M University-Texarkana, USA

INTRODUCTION

There are different ways in which we have used the concept of attention with regard to human information processing and behavior (cf Kahneman, 1973). Attention could be taken to mean whatever one is thinking about, as when a student is lost in the thoughts of daydreaming rather than paying attention to the teacher's lesson. Attention can also be associated with where we are looking or that for which we are looking (cf Moray, 1969), as when a flashing Web advertisement takes your attention or when one is mentally focused on searching through a Web page to find information.

This attention switching or attention movement perspective on attention (cf Broadbent, 1957) is of most interest in this article. A flashing Web banner advertisement could, by design, take our attention from where we had intended to focus, or a Web page could be designed such that it draws our interest and leads us to seek further information. If a person is looking in the wrong place to find what he or she wants, then it would be good for us to know about this. This article will review some theories of attention that are relevant to understanding how human attention processing mechanisms work with regard to these issues, and will review the basics of a method that can be used to track attention movement by tracking mouse movements in a browser. This method has grounding in well-established theory, and it can be used in a laboratory or can be used remotely with data saved to a server for replay.

BACKGROUND

Serial, Parallel, and Hardwired Systems of Attention

A little over a century ago, interest in the idea of attention emerged as researchers began studying various mechanisms that might affect human mental processing limitations (e.g., Bryan & Harter, 1899; Jastrow, 1892; Solomons & Stein, 1896; Welch, 1898). Psychologists lost interest in this line of research to the study of behaviorism for several decades, but renewed interest emerged again in the 1950s (e.g., Adiseshiah, 1957; Bahrick, Noble, & Fitts, 1954; Broadbent, 1957; Garvey & Knowles, 1954). Throughout the period of the 1950s through the 1970s, researchers were in part attempting to understand why and how processing limitations occur.

Single-Channel Hypothesis

One early view in this rebirth was the *single-channel hypothesis*, which viewed the processing system as something like a single-channel, serial transmission line (Welford, 1967). In an attempt to locate the bottleneck in this communication channel, Broadbent (e.g., 1957) proposed that there is a many-to-one selection switch in the channel. It is difficult, for example, to comprehend multiple conversations at a time even though we can understand one conversation out of many and can switch our attention to another. The single-channel hypothesis, however, was not able to explain the observation that people can in other kinds of situations apparently process multiple tasks concurrently. We can, for example, comprehend only one conversation out of many, yet can concurrently drive an automobile while listening.

Undifferentiated-Capacity Hypothesis

Moray (1967) proposed that some of the problems with the single-channel hypothesis could be explained by a flexible central processor of limited capacity. Popularized by Kahneman (1973) and labeled the *undifferentiated-capacity hypothesis* by Kerr (1973), this model viewed the processing system as possessing a very general pool of resources that can be allocated to the performance of various concurrent tasks. This model attempts to explain how limitations to process a particular task will change depending on what other processing tasks might also compete for resources from the central processor. For example, some of us can talk while typing, but our typing speed and accuracy often suffers when doing so. Neither of these two models was viewed by Kahneman as adequate alone; Kahneman viewed the single-channel idea as associated with processes that have structural limitations. Our visual system, for example, can only point at and process one single view at a time.

Multiple-Resource Theory

The undifferentiated-capacity hypothesis is also not completely adequate. Researchers found, for example, that it is easier to attend to auditory and visual messages concurrently than to two concurrent audio messages (Rollins & Hendricks, 1980; Triesman & Davies, 1973). This could be due in part to the existence of more than one flexible processor operating in parallel, for example, one limited-capacity processor for visual messages and one limited-capacity processor for auditory ones, both operating in parallel and feeding into a flexible limited-capacity central processor. Friedman, Polson, and Dafoe (1988) found that there are differences in processing degradation between tasks processed in each cerebral hemisphere and a common second (concurrently performed) task, further suggesting evidence of multiple capacity- or resource-limited processors.

Automatism and Skilled Processing

A problem with the capacity explanations is that processing can sometimes appear to be resource free, or to consume from a processor that has no apparent bottlenecks or resource limitations. Early researchers such as Bryan and Harter (1899) were finding that practice could lead to the automatization of task performance, or skill acquisition. The early dual-task studies were finding that when two tasks are performed concurrently, they tend to interfere with each other less and less with continued practice. It appears that with practice, some processes become hardwired outside of the control of the flexible processing systems, and so the person can effortlessly do these automatic processes in parallel with the controlled or effortful processes that require the use of the flexible general-purpose processor (cf Shiffrin & Schneider, 1977).

The discussion above suggests that there are at least three general mechanisms involved in how people process information.

1. System components composed of a flexible, general-purpose central processor and other more specialized, but flexible, processors. These resources can process different tasks concurrently or in parallel.
2. Serial system components and structurally limited components that must be switched from one task to another. Eyes can only be pointed in one direction at a time and must be physically moved if we want to pay attention to something else. Ears can receive many conversations at once, but the preprocessor associated with them can only process a single conversation at a time.
3. Hardwired system components that do not consume the resources of these flexible parallel and serial processing components. Processes become hardwired through practice. Learning to ride a bicycle, for example, requires all of a child's attention at first, and the slightest distraction can cause the child to fall. With practice, however, the child will be able to ride effortlessly, concurrently carrying on a conversation or thinking about something else.

Voluntary Attention

The notion that we have a flexible central processor or a set of processors and can choose where to focus our thinking is associated with what is called *voluntary attention* (e.g., Hunt & Kingstone, 2003; James, 1899). A student may choose to daydream rather than listen to the teacher: Both tasks can be performed concurrently, but the student consciously and deliberately allocates most attentional resources toward thinking about something while allocating some resources to listen just enough to pick out anything important that should be written in the notebook. An online shopper consciously and deliberately chooses to use attentional resources toward seeking out the best deal on a new computer rather than to attend to e-mail or an online forum discussion with friends.

Selective Attention

Related to voluntary attention is *selective attention*, whereby we can choose what to process when presented with choices (e.g., Camels, Berthoumieux, & d'Arripe-Lontueville, 2004; Moray, 1969). An automobile driver may choose to look at scenery outside of the car window rather than to focus on traffic ahead. An online shopper can choose to physically shift the eyes to one brand's ad rather than the ads for other brands on a single Web page. Some of this might be due to allocating resources to different tasks within a capacity-limited processor, and some of this might be due to switching our attention between serial or structurally limited system resources.

Involuntary Attention

In some cases, our attention is taken without desire on our part. *Involuntary attention* can occur when a stimulus is, say, novel and unusual, surprising, or highly contrasted with the background (e.g., Berlyne, 1960; Berti & Schroger, 2003). Web advertisers attempt to do this with flashing banner ads and pop-up windows. It is possible to become habituated to a stimulus, however, and to no longer notice it; flashing banner ads, for example, might be less effective as a user experiences them more and more. It is possible that with practice, some tasks become automated in helping us to ignore some stimuli, as when a person automatically clicks off a pop-up banner ad without much conscious thought.

METHODS OF TRACKING ATTENTION WITH A BROWSER MOUSE

Many methods and devices have been devised for the measurement and tracking of attention, but the focus of the rest of this article is on how we might be able to track attention by using a mouse and a Web browser. With a mouse, JavaScript, and a Web browser, we can detect the following:

- "Click" events (mouse-button presses), as when a person does a click-through on a banner ad. We know where a person was looking when the mouse button was pressed.
- X- and y-coordinates of the mouse pointer in the browser window. We can collect the position of the mouse at any point in time, and so by periodically sampling position and time information, we can track mouse movements within a screen. In this way, we can know where a person was considering a click, or we can instruct the person to move the pointer continuously to show us where he or she was looking.
- "Mouse-over" events, as when the mouse pointer is moved over or through an item on a menu list, over a button in a button list, or over an image in a Web page. If we know the location of such items in the page, we can collect the mouse-over information along with the time associated with the mouse-over event, thereby allowing us to track mouse movements within a screen.

Collecting Data to Track Attention

In all the cases above, we are ultimately collecting a measure of the screen position of the mouse pointer, and in all cases we could additionally take a measure of clock time. With an indication of time, we then have real-time data that can be used to remotely replay where a user's mouse pointer was positioned during a visit to a Web site or within a single page. An assumption is that if the user clicks on something, he or she presumably has an interest in that thing and is therefore paying attention to it. By watching where a mouse pointer moves from and to, and where clicks are made we could track a person's attention.

JavaScript, a programming language that is included with the mainstream Web browsers, can be asked to perform a programmed routine whenever some particular user action occurs. For example, clicking the mouse button when the pointer is over a particular hyperlink can send the user to another page, open a pop-up window, or (of interest here) save data associated with the circumstances of that user action. If we are attempting to track a person's attention in real time, then we want to save mouse position and mouse clicks along with time information (Owen, 2002, discusses some issues with time-measurement resolution and with saving data).

X- and Y-Coordinates

Using JavaScript, it is possible to capture the values of the pointer's x- and y-coordinates when the mouse is moving (an author's example is at http://mousEye.SyKronix.net). If the mouse moves, then the x- and y-coordinates of the pointer are captured and saved along with the clock time. This makes it possible to track movements of the mouse in real time, for which differences between the clock time on each capture suggest the speed of movement. If the user is instructed to move the mouse pointer to indicate where he or she is looking, then we could capture data that tells us where the person was looking in real time. There are two concerns with this method, however. One is that a very high amount of data would be collected when using this method. Another is that the user might not actually move the mouse pointer in tune with where he or she is looking. An alternative method that solves these problems is presented next.

Mouse-Over Events

A simpler and more powerful alternative to detecting where the user is focusing attention is to use JavaScript to detect whenever the mouse pointer moves over some particular position on the screen. JavaScript can be used to detect whenever the user moves the mouse pointer over a hyperlinked portion of text or graphic image. This will trigger an event in the program that causes the program to save which link or image the mouse pointer moved over. This is an improvement over the method of saving the x- and y-coordinates because the program function is only called whenever the user rolls over the target area of interest, not simply whenever the mouse is moved. We might only be interested, for example, in a user's movement between four different quadrants of the screen. Detecting when the mouse pointer moves out of one quadrant into another leaves us with a much simpler task of data collection and substantially less data to send to the server, to store, and to later analyze.

The most important advantage of this method, however, is that the program function that is triggered by the mouse-over event allows us to change the image. This is a crucially important advantage in a study that wants to track where the user is looking. If the image changes, then we can mimic the way that people really do look at a Web page, with the center of vision, the area of concentration, being more clear, and the peripheral areas being less clear. The author has used this method in several ways:

- **No change in the image:** The movement of the mouse pointer over various graphic-image fields merely sends back information regarding where the pointer was positioned in real time. The user is instructed to move the mouse pointer wherever he or she is looking. For example, if the user is participating in a Web usability study, we might ask the user to find something within the Web site (an example is at http://mousEye.SyKronix.net/demos/web.html). The user might move the pointer over a button on the left column, but then decide to move the pointer over an index item along the bottom and click. In this way, we did not simply collect information about what was ultimately clicked, but collected information about the hesitancy in selecting another choice. If time information is collected with the trigger of these events, then we would also have information regarding the amount of time that was associated with the hesitancy in considering one choice before moving to make another. Such data can be collected remotely, sent to a server, and played back in real time later as if the researcher is actually standing over the shoulder of the user.
- **Change in focus:** The entire page of, say, a one-page advertisement is presented somewhat out of focus. Moving the pointer over the place where the user is looking makes that part more clear. In this way, the user is forced to reveal where he or she is looking (an example

is at http://mousEye.SyKronix.net/demos/drive.html). Again, this data can be sent to a remote server for storage, and can be played back later (in the style of a movie) in real time.

- **Change in contrast:** The entire page of, say, a one-page advertisement is washed of color and low in contrast. Moving the pointer over the place where the user is looking makes that part brighten in color and contrast (an example is at http://mousEye.SyKronix.net/demos/phren.html). Again, this forces the user to reveal where he or she is looking, and we can play this back remotely.

FUTURE TRENDS

Eye tracking (see Porta, 2002, for a review of methods) can be used in such studies as reading (e.g., Stewart, Pickering, & Sturt, 2004) and Web usability (e.g., Karn, Ellis, & Juliano, 2000; Schiessl, Duda, Thölke, & Fischer, 2003). The adoption of eye tracking equipment for practical applications, however, does not appear to be widely accepted. This is possibly in part due to the cost and complexity of the equipment, the intrusiveness of the equipment in market or usability testing, and the lack of acceptance of such methods in real-world marketing industries that are more comfortable with focus-group methods and such.

The use of mouse tracking as a substitute for eye-tracking equipment could be an answer to some of these issues. The author developed these methods in 1997 and adapted them for a marketing research agency in 2000, but salespeople believed that clients were too comfortable with focus group methods. The mouse tracking method obtained reliable results in testing, but development was abandoned for lack of sellability to clients (nonproprietary examples were posted at http://mousEye.SyKronix.net). A few others have since reported experiments with mouse-tracking methods, suggesting that there is hope to see greater use in the future. Mueller and Lockerd (2001) describe a use that appears to use the x- and y-coordinates method. Tarasewich and Fillion (2004) and Ullrich, Wallach, and Melis (2003) describe methods that change the focus of the area under the mouse pointer.

CONCLUSION

Mouse tracking, which has reasonable grounding in attention theory, is advocated as a means to track user attention remotely as well as in a lab. Unlike eye tracking equipment, it is low in cost, relatively unobtrusive, and can be done remotely in a Web user's own natural environment as well as in a laboratory setting. Using the mouse-over method that allows us to change the focus and contrast of objects in the periphery of vision, it is possible to encourage the user to move a mouse pointer to indicate the location of his or her attention. The notion of automatism suggests that with a little practice, this would not interfere with the user's task of browsing through a Web page or site. In this way, we could study voluntary attention, selective attention, and involuntary attention in studies of Web usability and Web advertising.

REFERENCES

Adiseshiah, W. T. V. (1957). Speed in decision making under single channel display conditions. *Indian Journal of Psychology, 32*, 105-108.

Bahrick, H. P., Noble, M., & Fitts, P. M. (1954). Extra-task performance as a measure of learning a primary task. *Journal of Experimental Psychology, 48*(4), 298-302.

Berlyne, D. E. (1960). *Conflict, arousal and curiosity.* New York: McGraw Hill.

Berti, S., & Schroger, E. (2003). Working memory controls involuntary attention switching: Evidence from an auditory distraction paradigm. *European Journal of Neuroscience, 17*, 1119-1122.

Broadbent, D. E. (1957). A mechanical model for human attention and immediate memory. *Psychological Review, 64*, 205-215.

Bryan, W. L., & Harter, N. (1899). Studies in the telegraphic language: The acquisition of a hierarchy of habits. *Psychological Review, 6*, 345-375.

Camels, C., Berthoumieux, C., & d'Arripe-Lontueville, F. (2004). Effects of an imagery training program on selective attention of national softball players. *The Sport Psychologist, 18*, 272-296.

Friedman, A., Polson, M. C., & Dafoe, C. G. (1988). Dividing attention between the hands and head: Performance trade-offs between rapid finger tapping and verbal memory. *Journal of Experimental Psychology: Human Perception and Performance, 14*, 60-68.

Garvey, W. D., & Knowles, W. B. (1954). Response time pattern associated with various display-control relationships. *Journal of Experimental Psychology, 47*, 315-322.

Hunt, A. R., & Kingstone, A. (2003). Covert and overt voluntary attention: Linked or independent? *Cognitive Brain Research, 18*, 102-105.

James, W. (1899). *Talks with teachers.* Retrieved October 10, 2004, from http://www.emory.edu/EDUCATION/mfp/james.html#talks

Jastrow, J. (1892). Studies from the laboratory of experimental psychology of the University of Wisconsin: The interference of mental processes. A preliminary survey. *American Journal of Psychology, 4*, 198-223.

Kahneman, D. (1973). *Attention and effort.* Englewood Cliffs: Prentice Hall, Inc.

Karn, K. S., Ellis, S., & Juliano, C. (2000). *The hunt for usability: Tracking eye movements.* Retrieved September 1, 2004, from http://www.acm.org/sigchi/bulletin/2000.5/eye.html

Kerr, B. (1973). Processing demands during mental operations. *Memory and Cognition, 1*, 401-412.

Moray, N. (1967). Where is capacity limited? A survey and a model. In A. F. Sanders (Ed.), *Attention and performance* (pp. 84-92). Amsterdam: North-Holland Publishing Co.

Moray, N. (1969). *Attention: Selective processes in vision and hearing.* London: Hutchinson Educational.

Mueller, F., & Lockerd, A. (2001). Cheese: Tracking mouse movement activity on Web sites. A tool for user modeling. In *CHI '01 extended abstracts on human factors in computing systems* (pp. 279-280). New York: ACM Press.

Owen, R. S. (2002). Detecting attention with a Web browser. *Proceedings of the 5th Asia Pacific Conference on Computer Human Interaction* (Vol. 1, pp. 328-338).

Porta, M. (2002). Vision-based user interfaces: Methods and applications. *International Journal of Human-Computer Studies, 57,* 27-73.

Rollins, H. A., & Hendricks, R. (1980). Processing of words presented simultaneously to eye and ear. *Journal of Experimental Psychology: Human Perception and Performance, 6,* 99-109.

Schiessl, M., Duda, S., Thölke, A., & Fischer, R. (2003). Eye tracking and its application in usability and media research. *NMI-Interakti, 6*, 41-50.

Shiffrin, R. M., & Schneider, W. (1977). Controlled and automatic human information processing: II. Perceptual learning, automatic attending, and a general theory. *Psychological Review, 84*, 127-190.

Solomons, L., & Stein, G. (1896). Normal motor automatism. *Psychological Review, 3*, 492-512.

Stewart, A. J., Pickering, M. J., & Sturt, P. (2004). Using eye movements during reading as an implicit measure of the acceptability of brand extensions. *Applied Cognitive Psychology, 18*, 697-709.

Tarasewich, P., & Fillion, S. (2004). Discount eye tracking: The enhanced restricted focus viewer. *Proceedings of the Tenth Americas Conference on Information Systems* (pp. 1-9).

Triesman, A. M., & Davies, A. A. (1973). Divided attention to ear and eye. *Attention and Performance, 4*, 101-125.

Ullrich, C., Wallach, D., & Melis, E. (2003). What is poor man's eye tracking good for? *Designing for Society: Proceedings of the 17th British HCI Group Annual Human-Computer Interaction Conference* (Vol. 2, pp. 61-64).

Welch, J. C. (1898). On the measurement of mental activity through muscular activity and the determination of a constant of attention. *American Journal of Physiology, 1*(3), 283-306.

Welford, A. T. (1967). Single-channel operation in the brain. In A. F. Sanders (Ed.), *Attention and performance* (pp. 5-52). Amsterdam: North-Holland Publishing Co.

KEY TERMS

Attention: Mental processing that consumes our conscious thinking. Through a variety of mechanisms, there is a limit to the amount of processing that can take place in our consciousness.

Automatism: An attention mechanism established through practice whereby the performance of a task apparently no longer interferes with the concurrent performance of other tasks.

Habituation: The suppression of a response to or attention to a stimulus after repeated exposures.

Involuntary Attention: The idea that something can take a person's attention by being novel, contrasting, startling, and so forth.

Multiple-Resource Theory: A model of attention that assumes many specialized preprocessors (e.g., visual system, auditory system) that can function in parallel.

Selective Attention: The idea that a person can actively choose to attend to one of multiple stimuli that are present while ignoring others.

Single-Channel Hypothesis: A model of attention that assumes that some mechanisms can process only one task at a time in a serial fashion. Some mechanisms have structural limitations (e.g., eyes can only point to one place). Multiple tasks are processed within some time frame by attention switching between tasks.

Undifferentiated-Capacity Hypothesis: A model of attention that assumes a flexible, multipurpose central processor that can process multiple tasks concurrently. This processor has a limited amount of resources, however, that can be allocated across all tasks.

Voluntary Attention: The idea that a person can actively seek out information or things to think about.

Traditional vs. Pull-Down Menus

Mary Henley
Flow Interactive Ltd., London, UK

Jan Noyes
University of Bristol, UK

INTRODUCTION

Human interactions with computers are often via menus, and "in order to make information retrieval more efficient, it is necessary that indexes, menus and links be carefully designed" (Zaphris, Shneiderman, & Norman, 2002, p. 201). There are a number of alternatives to menus, such as icons, question-and-answer formats, and dynamic lists, but most graphical user interfaces are almost entirely menu-driven (Hall & Bescos, 1995). Menu systems have many advantages. For example, Norman (1991) identified low memory load, ease of learning and use, and reduced error rates as advantages of menu-driven interfaces. They frequently form the main part of a WYSIWYG (What You See Is What You Get) interface, providing most of the functionality in the more common operating systems such as Microsoft Windows. Consequently, familiarity also can be added to the list of advantages of using menus when accessing computer systems. These aspects are particularly important when considering public-access technologies, where individuals from across the population exhibiting a range of ages, skills, and experience levels will attempt to use the systems. Further, training or the opportunities for training will be minimal and, most likely, non-existent.

BACKGROUND

Two main types of menu designs are commonly found: traditional and pull-down. Traditional menus occupy the whole screen. Secondary and further menu levels also appear and, again, take up the whole screen. Once the final option choice has been taken, the screen is cleared for work. This type of menu is common in public access technologies such as cash points and multimedia information kiosks. Traditional menus are thought to be easier for first-time/novice users, because they are explicit in terms of operation. This is in contrast to pull-down menus, where operation is usually via a mouse or the enter and cursor keys. Pull-down menus have an initial main menu in the form of a bar at the top of the screen from which further lists of options may be seen and selected, thus leaving the majority of the remaining screen area for other purposes. This comprises their primary advantage: the user is able to stay in the same workspace/screen. However, this form of cascading menu hides information until the user activates the menu item, which can be viewed as a disadvantage (Walker, 2000). Pull-down menus form the main method for option selection in the most commonly available packages from Microsoft and Macintosh. There are a number of variations of pull-down menus. For example, horizontal and vertical menus (Dong & Salvendy, 1999) and split and folded menus (Straub, 2004). Split menus present frequently accessed items at the top of the menu, while folded menus give the high frequency items first and on their own. After a short delay, the complete menu appears.

The comparison of traditional and pull-down menu types has been a somewhat neglected area, with much work focusing on the comparison of menus with other styles of interface, such as command languages (Mahach, Boehm-Davis & Holt, 1995). As a further example, Benbasat and Todd (1993) compared icons with text and direct manipulation with menus. Direct manipulation was defined in this context as the "physical manipulation of a system of interrelated objects which are analogous to objects found in the real world" (Benbasat & Todd, 1993, p. 375). These objects are usually represented as icons. Benbasat and Todd (1993) found no difference between the use of icons and text, and a speed advantage of direct manipulation over menus. This advantage, however, was diminished when the task was repeated for a third time, indicating that there may be a learning effect occurring in menu interactions. However, studies such as this do not serve to indicate the basic type of menu layout that is most beneficial to the user.

Given the importance of navigation in computer-based tasks, many studies have been carried out on menu design. For example, Yu and Roh (2002) investigated the effects of searching using a simple menu with a hierarchal structure, a global and local navigation menu, and a pull-down menu. They found search speeds differed significantly, with the pull-down menu being faster than the other two.

Carey, Mizzi, and Lindstrom (1996) compared traditional and pull-down menu formats and found that experienced users completed menu search tasks faster than novice users, regardless of the menu style used, although there was no significant difference between the two user groups in the number and type of errors made. The traditional menus elicited fewer errors than the pull-down menus for both experienced and novice users, but there was no time difference for task completion between the two menu types. Carey et al. (1996) also found that users preferred the traditional style menu, with this preference being stronger for novices than for experienced users. They suggested the fact that using a cash point application may have skewed the results in favor of the traditional menus due to a familiarity effect. A further bias in favor of the traditional menu condition lies in the fact that it required fewer key presses per transaction than the pull-down menus. This is an intrinsic feature of the two menu designs—the pull-down system by definition will require an additional action at the start to open the menu from the top of the screen, while the traditional menu would already be occupying the majority of the screen.

Bernard and Hamblin (2003) compared cascading menus in horizontal and vertical forms with a categorical indexed menu design. Although the terminology is different, the categorical indexed menu appears to be equivalent to the traditional menu, and the cascading menus seem to be pull-down menus. They found that searching was faster using the indexed menu than the cascading, pull-down menus. Their results indicated that using a categorical menu would be 4.27 minutes quicker when accessing 40 pages on the Internet. (This figure was derived from Nielsen [2003], who suggested that a user accesses 40 pages of information in a typical surf of the Internet.) Bernard and Hamblin (2003) also found that the indexed menu was preferred by participants who chose this design more as a first choice over the two cascading menu designs.

In a study we conducted comparing traditional and pull-down menus with older and younger adults, time differences between the menu types were found for both age groups, with traditional menus eliciting shorter times than pull-down menus. Carey et al. (1996) found that traditional menus elicited fewer errors than pull-down menus and found no evidence for their hypothesis that experienced users would commit fewer errors than novices. The difference in error rates between the menu types was replicated in our work, although the effect was only present for the older age group.

In terms of participant opinions about the two menu types, younger respondents expressed a preference for pull-down menus; older adults preferred traditional menus. Both menu styles were shown to be easy to use by both age groups. There was one significant difference—young participants found the traditional menus hard to search by trial and error compared to their ratings for pull-down menus and to the older adults' ratings of both menu types. This may have been because younger participants are more familiar with pull-down menus. However, this finding is not supported by Bernard and Hamblin (2003). Their participants were relatively young with a mean age of 32.6 years (SD = 8.2) but indicated a preference in use for the indexed menu.

FUTURE TRENDS

These experimental studies have demonstrated a number of points. First, older adults were more disadvantaged in their use of pull-down menus compared to traditional menus, relative to younger adults. This was true of the time taken to complete the task, the number of errors, and the steps required. Secondly, the type of searching used by participants in searching the two different menu structures was the same—most searching was carried out using semantic knowledge. This was possible because of the strong semantic consistency within the menus in this experiment. Finally, older users expressed a preference for traditional menus, and younger users preferred pull-down menus. This may be due to a familiarity effect of pull-down menus amongst younger, experienced computer users. As stated in the introduction, pull-down menus are used more commonly in PC and Macintosh software, meaning that people who use computers regularly will be more accustomed to them than traditional menus. These findings, therefore, have implications for the future design of systems, as more people become familiar with pull-down menus, and thus, the age effect associated with traditional menus will start to diminish.

Looking to the future, an adaptive menu system that responds to the needs of a particular user may prove useful. Lee and Yoon (2004) pointed out that as menus become longer, it is more difficult for users to locate specific items. However, some items will be selected more frequently than others, and systems could be designed to recognize this. Public access technologies could utilize this feature and adjust the order of presentation of menu items so that more frequently accessed items were presented near the top of menus.

CONCLUSION

When deciding between traditional and pull-down menu styles in an application, there are other factors to take into account. For some applications, only one menu style may be suitable for practical reasons. Although there does seem to be an advantage to traditional menus in terms of speed of use and reduced error rates, in particular for older and less experienced computer users, this type of menu takes up much more screen space than pull-down styles. In many interfaces, this will be impractical, particularly when a lot of information must be available on the screen. However, if everything else is equivalent, it is suggested that traditional menus be used in interface design, especially when the user group comprises older adults and/or novice computer users.

ACKNOWLEDGMENT

This research was funded by Research into Aging and AgeNET, and their support is gratefully acknowledged.

REFERENCES

Benbasat, I., & Todd, P. (1993). An experimental investigation of interface design alternatives: Icon vs. text and direct manipulation vs. menus. *International Journal of Man-Machine Studies, 38*, 369-402.

Bernard, M., & Hamblin, C. (2003). Cascading versus indexed menu design. *Usability News*, *5*(1), 8.

Carey, J. M., Mizzi, P. J., & Lindstrom, L. C. (1996). Pull-down versus traditional menu types: An empirical comparison. *Behaviour and Information Technology, 15*(2), 84-95.

Dong, J., & Salvendy, G. (1999). Designing menus for the Chinese population: Horizontal or vertical? *Behaviour and Information Technology, 18*(6), 467-471.

Hall, L.E., & Bescos, X. (1995). Menu—What menu? *Interacting with Computers, 7*(4), 383-394.

Houghton, R. C. (1986). Designing user interfaces: A key to system success. *Journal of Information Systems Management*, *3*, 56-62.

Lee, D. -S., & Yoon, W. C. (2004). Quantitative results assessing design issues of selection-supportive menus. *International Journal of Industrial Ergonomics*, *33*, 41-52.

Mahach, K. R., Boehm-Davis, D., & Holt, R. (1995). The effects of mice and pull-down menus versus command-driven interfaces on writing. *International Journal of Human-Computer Interaction, 7*(3), 213-234.

Nielsen, J. (2003). *September 2002 global Internet index average usage.* Retrieved from http://www.Nielsen-netratings.com/hot_off_the_net.jsp

Norman, K. L. (1991). *The psychology of menu selection: Designing cognitive control at the human computer interface*. Norwood, NJ: Ablex.

Straub, K. (2004, July). Adaptive menu design. *UI Design Update Newsletter*, 6.

Walker, D. (2000). *A flying menu attack can wound your navigation.* Retrieved from http://www.shorewalker.com/pages/flying_menu-1.html

Yu, B. -M., & Roh, S. -Z. (2002). The effects of design on information-seeking performance and user's attitude on the World Wide Web. *Journal of the American Society for Information Science, 53*(11), 923-933.

Zaphris, P., Shneiderman, B., & Norman, K. L. (2002). Expandable indexes vs. sequential menus for searching hierarchies on the World Wide Web. *Behaviour and Information Technology, 21*(3), 201-207.

KEY TERMS

Graphical User Interface: Commonly abbreviated GUI, this type of interface is a standardized way of presenting information and opportunities for interaction on the computer screen using the graphical capabilities of modern computers. GUIs use windows, icons, standard menu formats, and so forth, and most are used with a mouse as well as a keyboard.

Icons: Icons are a principal feature of GUIs and are small pictures representing objects, files, programs, users, and so forth. Clicking on the icon will open whatever is being represented.

Menu: A list of commands or options that appear on the computer screen and from which the user makes a selection. Most software applications now have a menu-driven component in contrast to a command-driven system; that is, where explicit commands must be entered as opposed to selecting items from a menu.

Mouse: In computer terms, this is a hand-operated electronic device that moves the cursor on a computer screen. A mouse is essentially an upside-down trackball, although the former needs more room during operation as it moves around a horizontal surface.

Public Access Technologies: These are computer-based technologies that are designed to be used by the general public. This has implications for their design, since they will be used by a diverse and unknown user group drawn from the human population.

Pull-Down Menu: When the user points at a word with either a keystroke or a mouse, a full menu appears (i.e., is pulled down, usually from the top of the display screen), and the user then can select the required option. A cascading menu (i.e., a submenu) may open when you select a choice from the pull-down menu.

Traditional Menu: This type of menu is essentially a series of display screens that appear sequentially as the user responds to the requests detailed on each screen.

WIMP: This acronym stands for Windows, Icons, Mouse, Pointing device, and is a form of GUI.

WYSIWYG: This acronym stands for What You See Is What You Get; it is pronounced Wiz-zee-wig. A WYSIWYG application is one where you see on the screen exactly what will appear on the printed document (i.e., text, graphics, and colors will show a one-to-one correspondence). It is particularly popular for desktop publishing.

Turning the Usability Fraternity into a Thriving Industry

Pradeep Henry
Cognizant Technology Solutions, India

INTRODUCTION

Many business and IT executives today think that usability is an important aspect of *software applications* that are used in enterprises (Orenstein, 1999). However, the term *usability* represents different things to different people. And, to most people, usability does not sound like an aspect that could really impact enterprise performance and bottom-line.

Literature suggests that the usability fraternity has failed to make an impact so far. For example, Bias and Mayhew (1994) ask "... given that the Human Factors Society (now the Human Factors and Ergonomics Society) is a quarter of a century old, why is it taking so long for usability engineering to achieve its place alongside the other accepted disciplines?"

Later, this article looks at some reasons why, and what to do about it.

BACKGROUND

There are thousands of advertising agencies in the world and many of them have a large staff and huge revenues. Advertising is a recognized industry. Is usability a recognized industry? How many *usability firms* are there? How many usability firms have over 100 people or 10 million dollar revenues? How many are listed in the stock market?

One U.S.-based organization says that though their usability engineering group strength of 18 specialists is small, this number is still larger than what many independent usability groups have. That gives us an idea of the average size of usability firms.

What are some of the problems that are stopping this field from growing big? Here are some: Practitioners are not picking up the right skills. Practitioners are not doing the right "usability" things. Practitioners are not impacting the business world. And practitioners are not promoting the right things. Of course, there are exceptions, but they are few. The following sections look at each problem in detail.

DEVELOP GOOD CREDENTIALS

Many usability practitioners are believed to not have the right kind of training. Shneiderman, Tremaine, Card, Norman, and Waldrop (2002) say that CHI (computer-human interaction) fails because its practitioners are badly trained. And Mauro (n.d.) says: "This important new science (*usability engineering*) has in many instances been dramatically misrepresented by pseudo-practitioners, who claim to have such expertise but often do not. As a result, many corporations and government agencies that retained such experts often found the experience unsatisfying and the promises of creating significantly more usable products and services illusive."

What is the *education* or *skill-set* that usability practitioners bring to their profession? Well, some bring expertise limited to the *human* side of users. Some others bring visual design or graphic tools expertise. Sure, those skills are required, but they are not enough. Practitioners need to be well-trained in technology and business. These are often the missing skills.

Being technology-literate is important for practitioners. Technology-literate would mean having a degree in computer science or software engineering. Technology-literate practitioners will know if their design can be implemented using the chosen application development software. They will know the technical impact of the design solutions they come up with (say, on system performance). When they speak the language of developers, they will also be trusted by those professionals, who will implement the design solutions.

Being business-literate is important for practitioners. Business-literate would mean knowing how enterprises in various industries (banking, insurance, retail, etc.) perform their business functions. Also, since business-literate practitioners understand the business reasons why an enterprise invests in an application, they will know the impact of *user-performance* (or usability) on the enterprise rather than on the users alone.

FOCUS ON DESIGN, NOT TESTING

Literature—from the earliest to the most recent—has always recommended conducting usability tests, unfortunately with little or no emphasis on design. Authors have included well-known usability gurus. No wonder, most usability practitioners appear to be focused on testing. Testing, of course, is a useful technique to discover certain types of design problems. However, the point is that a test-fix-test-fix kind of approach is not going to result in a *highly usable user interface*. This argument could be best appreciated by imagining the approach of *usability-testing* a building that was not designed correctly in the first place. Shneiderman et al. (2002) call this orientation to evaluation "The first human factors limitation."

Shneiderman et al. (2002) say, "... we do not contribute anything of substance: we are critics, able to say what is wrong, unable to move the product line forward." He goes on to say that usability practitioners must become *designers*. Yes, practitioners should apply strong design skills using a strong design-driven process (Henry, 2003) that preferably has testing as one of the evaluation methods.

MAKE HIGH-IMPACT CONTRIBUTIONS

Here are a few reasons why *low-impact user interfaces* are rampant.

- Most usability practitioners often fight only for screen-level improvements to user interfaces. Such improvements do not make a significant impact on the performance or bottom-line of the enterprise using the application. On the other hand, an improvement in the structure of the application's navigation is likely to make a significant improvement in the user's performance thereby impacting things like *enterprise workforce productivity* (Henry, 2003).
- One candidate the author evaluated for recruitment into his usability group had an MS degree in Human Factors and three years of *HCI* experience. As always, the candidate was given a test to evaluate his skills. The candidate's design recommendation sheet was filled with terms such as memory load, mental load, conceptual load, syntactic learning load, and cognitive load. Such a narrow focus on the *human* side of users too does not help make a significant impact on the enterprise.
- There is another advice (and therefore practice) that leads to low-impact contribution. Usability practitioners have been inspired into believing that even a small usability improvement is better than no improvement at all. That might sound like good advice. But, following this advice only results in mediocre practitioners, mediocre applications, and therefore a poor image for the whole usability fraternity.

If usability practitioners only deliver low-impact contributions, how will enterprises take them seriously? Practitioners need to rethink the current thinking and practices in usability.

PROMOTE HIGH-IMPACT CONTRIBUTIONS

Most of the time, the usability fraternity just talks about the small improvements that it creates. These "small improvements" are things that do not significantly impact the enterprise. These are things that are not perceived as significant by the enterprise.

Instead, practitioners should start talking about big things (of course, assuming they have achieved big things). For example, if they redesigned an application user interface to significantly reduce expenses on user-training, they should talk about it and preferably in dollar terms.

Usability practitioners know that users get *confused*, *frustrated*, *dissatisfied*, and so forth while interacting with poorly designed user interfaces.

However, those are not the terms that typically get the attention of business folks. So, practitioners should understand and articulate the business impact of such user reactions.

Practitioners, however, should not use a pretentious word to describe what they do. One company the author worked with had a user interface "Architect," who pulled out her screen visual layout template — when in fact the author was talking about user interaction architecture!

Also, if a practicing team has the capability to design *high-impact user interfaces*, it should not try to cost-justify its efforts. Instead, it should talk about *Return on Investment (ROI)*. The phrase cost-justification might not sound different from ROI, but it is. Sure, methods like cost-benefit analysis, *cost-justification*, and ROI projection all attempt to predict the financial and other business consequences of an action. However, when terms like cost-justification are mentioned, attention is directed to cost. On the other hand, when a term like ROI is mentioned, attention is directed to returns or gains. Merholz and Hirsch (2003) say, "The key strategy is to get enterprises to recognize that user experience is not simply a cost of doing business, but an investment — that with appropriate expenditure, you can expect a financial return."

FUTURE TRENDS

Usability practitioners will benefit from reports that are based on *research projects* regularly conducted on how well and how frequently they touch enterprise bottom-line. That would help the practitioners know how well the field is doing from the enterprise perspective so they could focus on areas that matter most to the enterprise.

CONCLUSION

It is sad that usability, which has the potential to bring a big business value, has so far not made an impact on (and therefore is not taken seriously by) enterprises that invest in IT. Therefore it has remained a small and struggling field for many decades. The blame is on the usability practitioners. The good news, however, is that with the right skills and the right approach, usability practitioners will not only keep their jobs at difficult economic times, but actually grow the field into a thriving industry.

REFERENCES

Bias, R. G., & Mayhew, D. J. (1994). Summary: A place at the table. In R. G. Bias, & D. J. Mayhew (Ed.), *Cost-justifying usability* (p. 324). San Diego, CA: Academic Press.

Henry, P. (2003, March & April). Advancing UCD while facing challenges working from offshore. *Interactions,* 38-47.

Mauro, C. L. (n.d.). *More science than art.* Retrieved September 4, 2004, from http://www.taskz.com/ usability_indepth.php

Merholz, P., & Hirsch, S. (2003). *Report review: Nielsen/Norman group's usability return on investment.* Retrieved eptember 4, 2004, from http://www.boxesandarrows.com/archives/ report_review_nielsennorman_groups_usability_return_on_investment.php

Orenstein, D. (1999, August). Is software too hard to use? *CNN.com.* Retrieved October 4, 2004, from http://www.cnn.com/TECH/computing/9908/25/easyuse.ent.idg/index.html

Shneiderman, B., Tremaine, M., Card, S., Norman, D. A., & Waldrop, M. M. (2002, April 20-25). CHI@20: Fighting our way from marginality to power. *Extended Abstracts: Conference on Human Factors in Computing Systems,* Minneapolis, Minnesota (pp. 688-691). New York: ACM Press.

KEY TERMS

Application: Computer software meant for a specific use such as payroll processing.

Business and IT Executives: Senior people at a business organization, such as the chief information officer (CIO), chief technology officer (CTO), and chief executive officer (CEO).

Business Process: A series of related activities performed by staff in a business organization to achieve a specific output (for example: loan processing).

Enterprise: A business organization.

Enterprise Performance and Bottom-Line: Data critical for business success, such as employee productivity, customer satisfaction, and revenues.

Usability Practitioner: A person who designs and evaluates software user interfaces.

Usability Fraternity: A group of people that designs and evaluates software user interfaces.

Ubiquitous Computing and the Concept of Context

Antti Oulasvirta
Helsinki Institute for Information Technology, Finland

Antti Salovaara
Helsinki Institute for Information Technology, Finland

INTRODUCTION

Mark Weiser (1991) envisioned in the beginning of the 1990s that ubiquitous computing, intelligent small-scale technology embedded in the physical environment, would provide useful services in the everyday context of people without disturbing the natural flow of their activities.

From the technological point of view, this vision is based on recent advances in hardware and software technologies. Processors, memories, wireless networking, sensors, actuators, power, packing and integration, optoelectronics, and biomaterials have seen rapid increases in efficiency with simultaneous decreases in size. Moore's law on capacity of microchips doubling every 18 months and growing an order of magnitude every five years has been more or less accurate for the last three decades. Similarly, fixed network transfer capacity grows an order of magnitude every three years, wireless network transfer capacity every 5 to 10 years, and mass storage every 3 years. Significant progress in power consumption is less likely, however. Innovations and breakthroughs in distributed operating environments, ad hoc networking, middleware, and platform technologies recently have begun to add to the ubiquitous computing vision on the software side.

Altogether, these technological advances have a potential to make technology fade into the background, into the woodwork and fabric of everyday life, and incorporate what Weiser (1991) called natural user interfaces. Awareness of situational factors (henceforth, the context) consequently was deemed necessary for this enterprise. This article looks at the history of the concept of context in ubiquitous computing and relates the conceptual advances to advances in envisioning human-computer interaction with ubiquitous computing.

BACKGROUND

Ubiquitous Computing Transforms Human-Computer Interaction

Human-computer interaction currently is shifting its focus from desktop-based interaction to interaction with ubiquitous computing beyond the desktop. Context-aware services and user interface adaptation are the two main application classes for context awareness. Many recent prototypes have demonstrated how context-aware devices could be used in homes, lecture halls, gardens, schools, city streets, cars, buses, trams, shops, malls, and so forth.

With the emergence of so many different ways of making use of situational data, the question of what context is and how it should be acted upon has received a lot of attention from researchers in HCI and computer science. The answer to this question, as will be argued later, has wide ramifications for the design of interaction and innovation of use purposes for ubiquitous computing.

HISTORY

Context as Location

In Weiser's (1991) proposal, ubiquitous computing was realized through small computers distributed throughout the office. Tabs, pads, and boards helped office workers to access virtual information associated to physical places as well as to collaborate over disconnected locations and to share information using interfaces that take locational constraints sensitively into account. Although Weiser (1991) never intended to confine context to mean merely location, the following five years of research mostly focused on location-based adaptation. Want et al. (1992) described the ActiveBadge, a wearable badge for office workers that could be used to find and notify people in an office. Weiser (1993) continued by exploring systems for sharing drawings between disconnected places (the Tivoli system). Schilit et al. (1994) defined context to encompass more than location—to include people and resources as well—but their application examples were still mostly related to location sensing (i.e., proximate selection, location-triggered reconfiguration, location-triggered information, and location-triggered actions). Want, et al. (1995) added physical parameters like time and temperature to the definition. Perhaps the best-known mobile application developed during this location paradigm era was the CyberGuide (Long et al., 1996), an intelligent mobile guide that could be used to search for nearby services in a city. This paradigm was also influential in the research on Smart Spaces, such as intelligent meeting rooms.

The Representational Approach to Context

Although the idea that location equals context was eventually dismissed, many researchers coming from computer science still believed that contexts were something that should be recognized, labeled, and acted upon (Schmidt et al., 1998). Here, context was supposed to be recognized from sensor data, labeled, and given to applications that would use it as a basis for adaptation. Dey et al.'s (1999) five Ws of context—Who, Where, When, What, and Why—extended this approach and demonstrated convincing examples of how a labeled context could be used for presenting, executing, and tagging information. Tennenhouse's (2000) proactive computing paradigm endorsed a similar way of thinking about context, emphasizing the role of computers in doing real-time decisions on behalf of (or pro) the user. A somewhat similar approach that also attempts to delegate decision-making responsibility to intelligent systems is taken by the Ambient Intelligence (AmI) technology program of the European Union (ISTAG). One part of the AmI vision entails intelligent agents that assume some of the control responsibility from the users.

The latest widely referred to definition was given by Dey et al. (2001), who defined context as "any information that characterizes a situation related to the interaction between users, applications, and the surrounding environment" (p. 106). Satyanarayanan's (2001) formulation of pervasive computing also belongs to this line of thinking, but the author has chosen to avoid defining context and merely admits that it is rich and varies.

In his review of context definitions over the years, Dourish (2004) calls this the *representational approach* to context. Recent work within this branch has come close to finding the limits to recognizing and labeling contexts. For example, simple physical activities of a person in a home environment can be recognized with about 80-85% accuracy (Intille et al., 2004), as can be the interruptability of a person working in an office (Fogarty et al., 2004). Some critics have drawn parallels from this enterprise to problems encountered in strong AI (Erickson, 2002).

FUTURE TRENDS

New Directions Inspired by Human and Social Sciences

By the year 1996, other approaches to context were beginning to emerge. Wearable computing (Mann, 1996) looked at personal wearable computers able to help us remember and capture our everyday experiences through video and sound recording of context. Tangible bits (Ishii & Ullmer, 1997), although inspired by ubiquitous computing, looked at context not as something that had to be reacted upon but as surroundings of the user that could be augmented with tangible (i.e., graspable) computers and ambient media that display digital information using distraction-free output channels.

More recently, researchers have started to emphasize the social context and issues in people's practices and everyday conduct. These approaches give special consideration to activities that people engage in and highlight their dynamic nature, different from the labeling-oriented representational approach. Activity-centered approaches emphasize both turn taking in communication between the user and the applications (Fischer, 2001) and acknowledge the situated and time-varying nature of the needs that a user has in his or her life (Greenberg, 2001). This line of research highlights the difficulties that exist in making correct inferences about a user's tasks through sensor information. Considerations of social issues in ubiquitous computing design include questions of how to fit computation intelligence into people's routines in an unremarkable manner (Tolmie et al., 2002) and how people's patterns of interaction with humans and computers change when computationally augmented artifacts are adopted into use. Yet another emerging idea from HCI addresses specifically the aim to be free from distraction; that is, when it is appropriate to interrupt the user at his or her present task. Some call systems with such inferring capabilities *attentive user interfaces* (Vertegaal, 2003).

CONCLUSION

HCI in ubiquitous computing has been both inspired and constrained by conceptual developments regarding the concept of context. Weiser's (1991) initial work caused researchers to conceive context narrowly as encompassing mainly location and other static, easily measurable features of a user's context. After about five years of research, the restrictiveness of this definition was realized, and broader definitions were formulated. Still, context mainly was pursued by computer scientists and seen as something that must be labeled and reacted upon to adapt user interfaces. More recent work by human and social scientists has emphasized the role of user studies and theoretical reasoning in understanding what context entails in a particular application.

REFERENCES

Dey, A. K., & Abowd, G. D. (1999). *Towards a better understanding of context and context–awareness [technical report].* Atlanta: Georgia Institute of Technology.

Dey, A. K., Salber, D., & Abowd, G. A. (2001). A conceptual framework and a toolkit for supporting the rapid prototyping of context-aware applications. *Human-Computer Interaction, 16*(2-4), 97-166.

Dourish, P. (2004). What we talk about when we talk about context. *Personal and Ubiquitous Computing, 8*(1), 19-30.

Erickson, T. (2002). Some problems with the notion of context-aware computing. *Communications of the ACM, 25*(2), 102-104.

Fischer, G. (2001). Articulating the task at hand and making information relevant to it. *Human-Computer Interaction, 16*(2-4), 243-256.

Fogarty, J., Hudson, S. E., & Lai, J. (2004). Examining the robustness of sensor-based statistical models of human interruptibility. *Proceedings of the SIGCHI Conference on Human Factors in Computing Systems (CHI'04).*

Greenberg, S. (2001). Context as a dynamic construct. *Human-Computer Interaction, 16*(2-4), 257-268.

Intille, S., Bao, L., Munguia Tapia, E., & Rondoni, J. (2004). Acquiring in situ training data for context-aware ubiquitous computing applications. *Proceedings of the SIGCHI Conference on Human Factors in Computing Systems (CHI'04).*

Ishii, H., & Ullmer, B. (1997). Tangible bits: Towards seamless interfaces between people, bits and atoms. *Proceedings of the SIGCHI Conference on Human Factors in Computing Systems (CHI'97).*

Long, S., Aust, D., Abowd, G., & Atkeson, C. (1996). Cyberguide: Prototyping context-aware mobile applications. *Proceedings of the Conference on Companion on Human Factors in Computing Systems (CHI'96).*

Mann, S. (1996). "Smart clothing": Wearable multimedia computing and "personal imaging" to restore the technological balance between people and their environments. *Proceedings of the Fourth ACM International Conference on Multimedia.*

Satyanarayanan, M. (2001). Pervasive computing: Vision and challenges. *IEEE Personal Communications, 8*(4), 10-17.

Schilit, B., Adams, N., & Want, R. (1994). Context-aware computing applications. *Proceedings of the IEEE Workshop on Mobile Computing Systems and Applications*.

Schmidt, A., Beigl, M., & Gellersen, H. -W. (1998). There is more to context than location. *Computers & Graphics, 23*(6), 893-901.

Tennenhouse, D. (2000). Proactive computing. *Communications of the ACM, 43*(5), 43-50.

Tolmie, P., Pycock, J., Diggins, T., MacLean, A., & Karsenty, A. (2002). Unremarkable computing. *Proceedings of the SIGCHI Conference on Human Factors in Computing Systems (CHI'02).*

Vertegaal, R. (2003). Introduction to a special issue on attentive user interfaces. *Communications of the ACM, 46*(3), 31-33.

Want, R., Hopper, A., Falcão, V., & Gibbons, J. (1992). Active badge location system. *ACM Transactions on Information Systems, 10*(1), 91-102.

Want, R., et al. (1995). An overview of the ParcTab ubiquitous computing experiment. *IEEE Personal Communications, 2*(6), 28-43.

Weiser, M. (1991). The computer for the 21st century. *Scientific American, 265*(3), 66-75.

Weiser, M. (1993). Some computer science issues in ubiquitous computing. *Communications of the ACM, 36*(7), 75-84.

KEY TERMS

Attentive User Interfaces: AUIs are based on the idea that modeling the deployment of user attention and task preferences is the key for minimizing the disruptive effects of interruptions. By monitoring the user's physical proximity, body orientation, eye fixations, and the like, AUIs can determine what device, person, or task the user is attending to. Knowing the focus of attention makes it possible in some situations to avoid interrupting the users in tasks that are more important or time-critical than the interrupting one.

Context: That which surrounds and gives meaning to something else. (Source: The Free On-line Dictionary of Computing, http://foldoc.doc.ic.ac.uk/foldoc/)

Peripheral Computing: The interface attempts to provide attentionally peripheral awareness of people and events. Ambient channels provide a steady flow of auditory cues (i.e., a sound like rain) or gradually changing lighting conditions.

Pervasive Computing: Technology that provides easy access to information and other people anytime and anywhere through a mobile and scalable information access infrastructure.

Proactive Computing: A research agenda of developing interconnected devices and agents, equipped with faster-than-human-speed computing capabilities and means to affect real-world phenomena that a user can monitor and steer without a need to actively intervene in all decision-making situations. By raising the user above the traditional human-computer interaction loop, efficiency and freedom from distraction are expected to be enhanced.

Tangible Bits: According to Hiroshi Ishii of MIT, "the smooth transition of users' focus of attention between background and foreground using ambient media and graspable objects is a key challenge of Tangible Bits" (p. 235).

Tangible User Interfaces: Systems that give a physical form to digital information through augmenting tools and graspable objects with computing capabilities, thus allowing for smooth transitions between the background and foreground of the user's focus of attention.

Unremarkable Computing: An approach that focuses on designing domestic devices that are unremarkable to users. Here, unremarkable is understood as the use of a device being a part of a routine, because, it is believed, routines are invisible in use for those who are involved in them.

Wearable Computing: Technology that moves with a user and is able to track the user's motions both in time and space, providing real-time information that can extend the user's knowledge and perception of the environment.

Ubiquitous Internet Environments

Anxo Cereijo Roibás
University of Brighton, UK

INTRODUCTION

Let's remember the first films that started to show the broad public futuristic communication scenarios, where users were able to exchange almost any kind of information to communicate with anyone at any place and at any time, like Marc Daniels' "Star Trek" in the 1960s and James Cameron's "Terminator" in the 1970s, for example. The consequence of this was that impersonalized spaces (e.g., airports) (Auge, 1992) could easily become a personalized environment for working or leisure, according to the specific needs of each user.

These kinds of scenarios recently have been defined as ubiquitous communication environments. These environments are characterized by a system of interfaces that can be or fixed in allocated positions or portable (and/or wearable) devices. According to our experience with 2G technologies, we can foresee that the incoming 3G communication technologies will make sure, however, that the second typology of interfaces will become more and more protagonist in our daily lives. The reason is that portable and wearable devices represent a sort of prosthesis, and therefore, they reflect more than ever the definition of interface as an extension of the human body. When in 1973 Martin Cooper from Motorola patented an interface called Radio Telephone System (which can be defined as the first mobile phone), he probably didn't suspect the substantial repercussion of his invention in the human microenvironment and in its social sphere. The mobile phone, enabling an interpersonal communication that is time- and place-independent, has changed humans' habits and their way of making relationships (Rheingold, 1993). This system made possible a permanent and ubiquitous connection among users. At the same time, it has made users free to decide whether to be available or not in any moment and in any place they might be (Hunter, 2002).

This article is based on empirical work in the field with network operators (Vodafone) and handset manufacturers (Nokia) and research at the Politecnico di Milano University, the University of Lapland, and the University of Brighton. The intention is to give a practical approach to the design of interfaces in ubiquitous communication scenarios.

BACKGROUND

The methodologies and guidelines for the HCI design for handhelds initially were imported from the general theories of HCI for Web (Nielsen, 2000). Only after 1999 did this issue start to gain relevance as a research area. This can be reflected in the proliferation of focused conferences such as Mobile HCI started as a workshop in 1999 and has been explicitly treated in more holistic HCI conferences (e.g., CHI, HCI, Interact , Ubicomp, etc.). Unfortunately, the literature in this area is still scarce (Beaulieu, 2002; Bergman, 2000; Burkhardt, 2002; Hunter, 2002; Stanton, 2001; Weiss, 2002).

INTERNET MOBILE AND MOBILE COMMUNICATION

The Internet is related to a virtual space in which it is possible to interact with information. Mobile Internet, however, has represented an evolution of the concept of utopical (no real space) interaction to the concept of topical interaction, in which interaction (still with a virtual information space) happens in real places (Benedikt, 1991). This simultaneous presence of utopical and topical interaction makes necessary a direct relationship between both ambits (e.g., thanks to the GPS, what happens in real space must have an effect on the virtual one and vice versa). The communication now becomes space-sensitive or, better, context-sensitive.

Mobile communication is a broader concept than mobile Internet, as it embraces not only the connection to the net (intranet or extranet) but also voice and messaging (SMS, EMS, MMS) (Cereijo, 2001).

USAGE OF MOBILE COMMUNICATION

In the 1980s, the first generation (1G) of mobile communication systems revolutionized the TLC world, as users could carry a phone in their pocket. The 2G communication system and its new protocols to access the Internet, beyond just voice calls, provided users in mobility with a whole range of interactive services based on wireless data transmission.

Today, the market is characterized by different technologies—in America and Japan, the IS95 network based on CDMA (Code Division Multiple Access); in Europe, Asia, and Africa, the GSM (Global System for Mobile Communication) using TDMA (Time Division Multiple Access). Analysts foresee a relevant growth of mobile Internet users in the upcoming years—by 2005, more mobile phones will be connected to the Internet than PCs (Ovum, 3G Mobile, 2000).

Now we see the advent of 3G and 4G systems offering unprecedented bandwidth and speed connection up to 2 Mbps for data transmission with audio and video streaming capabilities directly on the phone. The variety and difference of the services offered are a challenge for today's service and application developers, and the battlefield is usability and effectiveness (Cereijo, 2002).

MULTI-ACCESS AND MULTI-CHANNEL CONVERGENCE

3G will be able to merge (at least) four media (Internet, SMS, TV, Smart-home). It is obvious that it will be crucial to offer an integrated system of new services with a perceived added value for the user in mobility. This integration also is called convergence and implies that all the information exchanged in the system (independently from the device of access) somehow must be centralized. The concept of convergence is related to that of the interoperability of the components of the same platform (e.g., the agenda, e-mail, block notes must share the same information) and between different multilingual devices available. That means that a user's transaction with a certain interface (e.g., flight booking from an iTV setup box) also must appear in real time, if the user accesses the related site afterwards with a different interface (e.g., PC or Pocket PC). The problem of the convergence implies some other ones, such as the information must be optimized according to the physical and technical features of each interface (Cereijo, 2003).

One of the main consequences of 3G will be an enhanced interaction with information (companies and institutions), people (personal and group communication), the smart-house and the automated office. This context of ubiquitous communication (across mobile phones, iTV, palms, pocket PCs, PDAs, etc.) will have applications in domotics, videoconferencing, commerce, iTV, entertainment, learning, finance, medicine, and so forth (Burkhardt, 2002).

MULTI-CHANNEL IDENTITY

One of the challenges of 3G will be the design of the multi-channel identity. Each type of device has different technical and physical features that condition the design decisions (regarding architecture, navigation, contents, and graphics).

This requires a coordinated graphic and interaction design that takes these issues into account. At the same time, the peculiarities of each interface of the system (Figure 1) might make the achievement of desired design homogeneity difficult (from both the functional and visual point of view) (Bergman, 2000).

Figure 1.

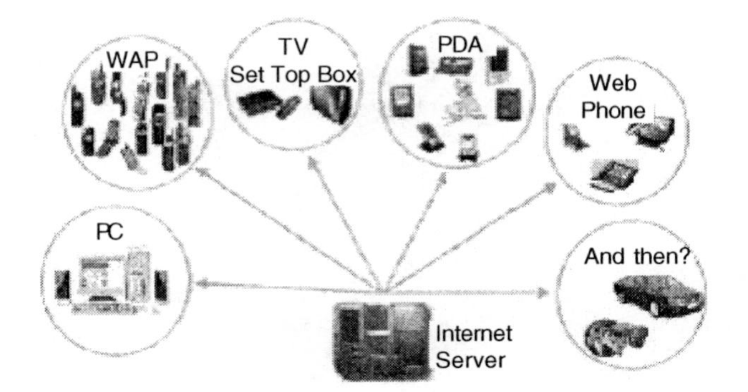

USABILITY AND SELF-USABILITY

The usability of a system can be defined as the effectiveness, efficiency, and satisfaction with which specified users achieve specified goals in particular environments (ISO 9241, ergonomic requirements for office work with visual display, Part 11). This means that the measurement of users' performance in mobility is not only important, but also their satisfaction, which is related to their perceived relevance of the mobile services provided in the particular contexts of use. Now more than ever, a participative design process (synergetic cooperation between the designer and the user during the whole cycle of the creation of the product) can be appropriated to reduce design and development costs as well as to provide successful services (Jordan, 2000).

There is another aspect that concerns usability that deserves special regard. It can be defined as self-usability and is related to the emotional effects of this extended way of multi-access services communication (e.g., iTV users by interacting with distributed interfaces). In order to make technology friendlier, mobile users have developed the so-called new m-language of the mobile interactive community. It is made of slang words, neologisms, acronyms, numbers, and icons of cyber communication. This new way of expressing was born as a consequence of interacting with SMS, e-mail, chat, forums, and newsgroups, and evolved with WAP and GPRS communication (Macleod, 1994).

MICRODESIGN

The interface is not a physical object but the space in which the interaction among the human body, devices, and aim of the action happens. Therefore, in an optic of human-centered microdesign, the interest of the design director is not the single button, display, or screen, but the global project of the human-machine interface (Nielsen, 1992). Interacting with a wired device is related to a specific user experience; usually, the contents are fat, and audio-visual factors are relevant. The navigation here has several possible schemes due to the large size and rich color of the screen. Moreover, animated images and videos can enrich and facilitate interaction. Accessing the Internet through mobile devices instead does not replicate the PC-based access. In fact, compared with the PC, in this kind of interface, we encounter important limitations. We can mention the small size of the interactive area, the low resolution, the high cost of the device and that of the connection, the slowness of data entry, the reduced memory storage, the limited processing, and the short battery autonomy (Weiss, 2000).

Another aspect that contributes to increase the aforementioned complexity in microdesign is related to the shape of the device. In fact, two main trends can be denoted—in one case, navigation tools have been gradually transferred to the graphic interface and voice commands (i.e., PDAs, Palms, and table PCs, where keyboards have been almost eliminated in favor of the screen [Figure 2a]); and in the other case, the device tends to configure itself as a typical mobile phone (Figure 2b).

In any case, it is crucial to be able to develop services that offer positive experiences to end users independently from these technology limitations. Information retrieval will no more be a frustrating experience, but users will be satisfied in less time throughout a logical sequence with minimum effort (Picard, 2000).

Figure 2(a).

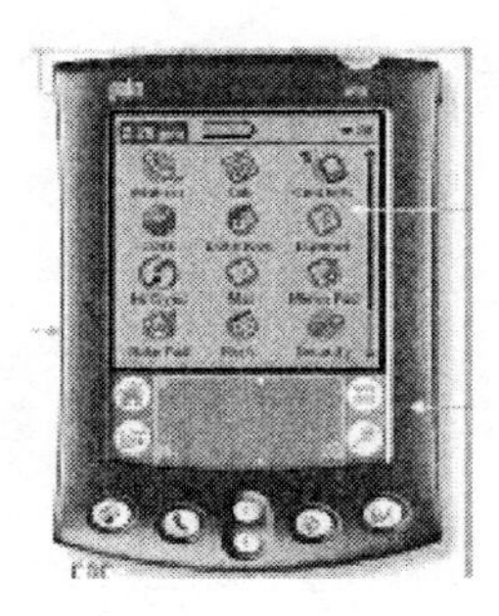

Figure 2(b).

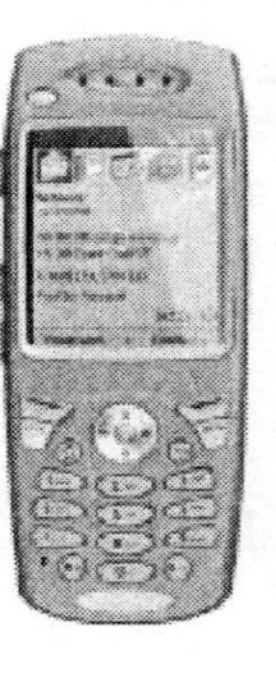

From GSM to 3G systems, wireless devices are driving the mobile Internet evolution, opening up innovative ways of communicating and interacting with content. The knowledge of this interaction context is crucial in order to be able to design and develop multi-access services. The uncertainty factor regarding any technology's most appropriate future use in terms of usage or shape can be reduced with a good knowledge of the utilization scenario.

Wireless communication introduces an innovative way to get the best out of the information world. The key aspect is mobility—anytime-anywhere access means quick and successful end-user experience in getting the right information at the right time and in the right place (Leed, 1991). On the other hand, mobile devices are personal tools, they contain personal information (agenda, phone book), and they can be personalized. Moreover, they remain on most of the day and night and are carried anywhere as an absolutely needed connecting device. They not only enhance communication, they also change our social relations (Ravy, 2000).

The first step will be the definition of scenarios of ubiquitous interactive communication (UMTS, MMS, iTV, etc.) to prospect multi-access communication contexts where users will be able to interact with people, information, TV, other devices and machines, and so forth—independently of place and time—thanks to a system of distributed interfaces (PC, TV, PDA, pocket PC, mobile phone, etc.) (Cereijo, 2003). Once the 3G scenario and the context of use are well defined, the designer should:

- Know the expectances, limitations, and behavior of users in mobility.
- Set a user-centered design.
- Design right interaction models and patterns.
- Foresee transcultural adaptability.
- Ensure a good communication of the services provided in order to guarantee a satisfactory added value to the user.
- Keep a coherent and relevant multi-channel identity.
- Recognize the level of usability of an interface according to:
  - Self-learning and length of the training phase.
  - Satisfaction of users' expectations and users' perceived value of the services provided.
  - Speed in task completion.
  - Number of irreversible mistakes.
  - Degree of interaction enabled.

The designer must know who is the user in mobility in order to know the user's limitations (Wharton, 1992), needs, and expectations; to hypothesize the user's behavior; to predict a model of multi-access interaction; to measure (Sears, 1993) the user's performance and the emotive level provided; to optimize human interface (Shneiderman, 1987); and to find a balance between automation and creative-affective interaction (deals with the pleasure of the user's "savoir faire").

According to the European experience with mobile interactive communities that use WAP and SMS, users appreciate enhanced interaction, personal, always-on and immediate communication (information delivered with sensibility to time and context), new emotional experiences, new ways of socializing and sharing experiences, and new expressions of entertainment (e.g., multiplayer games, group iTV interaction). An adequate targetization of users in terms of lifestyle, needs, behavior, and role in a social group will lead to a successful personalization of the devices.

In order to provide users with a high perceived value of mobile interfaces, it is crucial to develop a human-centered microdesign approach that leads to innovative interaction schemes (accessible via multidevices: PC, smart phones, PDA) and site architecture as well as intuitive navigation. A scant attention to human factors and behavioral science principles (learning ergonomics, social psychology, biomechanics, and HCI), which definitely is detrimental in Web design, is fatal in the mobile Internet. For example, users' frustrations in WAP is undoubtedly more dangerous than on the Web, as mobile devices interfaces are much smaller, and delivery of service is slower and more expensive (Cooper, 1995).

Even though there still are no standards for usability in mobile Internet, it is possible to indicate some basic guidelines for a user-centered microdesign.

INFODESIGN

Due to space economy, naming is a basic issue in microdesign (the use of a concise but relevant, auto-evident and consistent language for contents and the name of the navigation elements). It also requires a strong effort of organization and hierarchy of content (providing the user with progressive levels of deepening). Especially for mobile phones, where graphics have a discrete presence, most of the attention of design is focused on contents. This brief way of communicating must consider, however, that users have a limited memory, and not all information in the world can be laid down. All sections must be complete with all the information that users would expect to find (e.g., companys address, etc.). The Infodesign also must guarantee a maximum level of interaction (e.g., if, after carrying out a business search for a hotel in London, both the e-mail address and the telephone number displayed must be perceived intuitively by the user as interactive links in order to indicate that the e-mail can be sent directly and/or the number can be called just by clicking it). Figure 3 shows a case of auto-evident content; a number in brackets expresses how many sections are contained in each link.

MULTI-PLATFORM ARCHITECTURE

The structure of the site is a complex framework that reflects the multi-device access (e.g., WEB and WAP). The mobile Internet site's architecture must be in harmony with the user's workflow, and its layout must make the content easily accessible. Internet designers build the site's structure once the information has been hierarchized and organized into different levels. This process is stressed in mobile Internet. In fact, in microdesign, it is crucial to reach a good balance between horizontal and vertical organization (according to our research, for mobile phones and PDAs, not more than three mouse clicks—stages—are advisable in order to reach any service). As a general rule, we can say that vertical scrolling is easy on most devices; horizontal is not (Brewster S., 1999).

Figure 3.

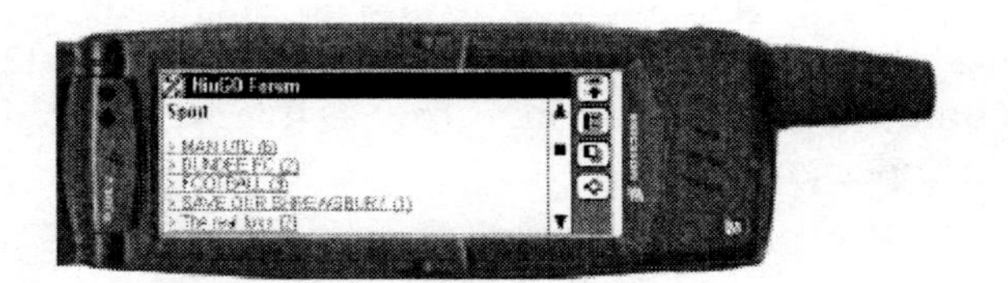

GRAPHICS

Ergonomically, such a small screen implies that more than in any other interface, graphics must aid navigation (the relationship between a user's intentions and the outcome must be intuitive) and comprehension of contents. Graphics that reduce readability must be avoided. Using multi-device graphics that are independent of the device to keep communication coherence compels us to use some crucial tricks (e.g., using contrasted monochrome icons, avoiding scrolling, etc.) in order to not lose in efficiency and efficacy when using the same graphic elements with different interfaces. It is common to find sites for PDAs with unreadable text (i.e., type too small, color of the type, color of the background, etc.). Moreover, the use of some graphic elements can be useful to differentiate different kinds of contents (e.g., HiuGO's WAP site uses the string to entitle a related section). Finally, graphic elements should contribute to service personalization (Reyes, 2001).

NAVIGATION

First, it can be denoted an evolution from the first models of interfaces for mobile Internet that allowed only one way (sequential menu) scrolling to the 3G devices that permit bidirectional movement (matrix menu) by means of a rocket, central key, joystick, or keypad. In any case, basic principles of mobile Internet navigation as evidence, fluidity, and quickness won't suffer many changes. In fact, navigation must allow users to satisfy easily their expectations and needs (relevance) in order to reach easily and quickly any section of the site. All pages must have the basic navigation elements (back, top, home, etc.); they must be evident and it must avoid confusion between buttons and non-linkable icons (this problem is more visible in the case of devices with touch screens, where linkable icons undoubtedly can help navigation). Mobile users hate to lose their way within the site, and they don't appreciate dead links, which can affect a site's reliability).

Some recent researches on users' behaviors put into evidence a reluctance of utilization of the physical buttons of mobile devices as function keys (only some expert users are prone to make current use of them). Therefore, only a minimum number of navigation functions or commands should be placed in the options menu (the most frequent ones); these commands also should be placed (with intentioned redundancy) on the screen, together with the rest of the functions and links or in physical buttons.

FUTURE TRENDS

According to the outputs in conferences such as Mobile HCI and Ubicomp, there are three main trends for the future of mobile HCI. The first one still remains the improvement usability and accessibility of mobile interfaces that are more and more "rich brains in poor bodies." The second one regards the solution of new dilemmas as the dichotomies personalization/privacy, ubiquity/security, context-awareness/confidentiality, info-accessibility/info-overload, effective-communication/affective-communication, and so forth. Finally, multi-platform systems imply that synchronization of information among multiple devices presents a challenge. This makes more realistic the dream of ubiquitous communication, but the consequent complexity from the interactive point of view needs to be hidden.

CONCLUSION

The incoming 3G-communication scenario will place users more and more in the center of a holistic communication network, which implies being physically surrounded by devices that will interact with users and machines. This environment of ubiquitous communication is becoming a mass-consuming phenomenon, where users might interact with different interfaces at the same time, which implies the need to create shared universal interaction codes, a coherent language between all holistic systems of interfaces. Innovative design patterns, human factor studies, behavioral theories, and evaluation techniques in ubiquitous communication scenarios have been investigated in order for these technologies to enjoy widespread popularity and usage.

REFERENCES

Augé, M. (1992). *Non-lieux*. Paris: Seuil.

Beaulieu, M. (2002). *Wireless Internet: Applications and architecture*. Boston: Addison Wesley.

Benedikt, M. (1991). *Cyberspace. First step*. Cambridge, MA: MIT Press.

Bergman, E. (2000). *Information appliances and beyond: Interaction design for consumer product*. San Francisco: Morgan Kaufmann.

Brewster, S., & Cryer, P. (1999). Maximising screen-space on mobile computing devices. *Proceedings of the CHI'99*.

Burkhardt, J., et al. (2002). *Pervasive computing: Technology and architecture of mobile Internet applications*. Reading, MA: Addison Wesley.

Cereijo Roibas, A. (2001a). Examining the UI design challenges of supporting multimedia applications on next generation terminals and identifying ways to overcome these. *Proceedings of the User Interface Design for Mobile Handsets Conference*, London.

Cereijo Roibas, A. (2001b). Physical and cognitive skills/limitations of mobile users. *Proceedings of the Mobile User Interface Conference*, London.

Cereijo Roibas, A. (2001c). Usability for microdesign: The HiuGO experience. *Proceedings of the Usability for Mobile Devices and Services Conference*, London.

Cereijo Roibas, A., et al. (2003). How will mobile devices contribute to an accessible ubiquitous iTV scenario. *Proceedings of the 2nd International Conference on Universal Access in Human-Computer Interaction (ICUAHCI)*, Crete.

Hunter, R. (2002). *World without secrets: Business, crime and privacy in the age of ubiquitous computing.* New York: John Wiley.

Jordan, P. (2000). *Designing pleasurable products—An introduction to the new human factors.* London: Taylor & Francis.

Leed, E.J. (1991). *The mind of the traveler.* Basic Book.

Lévy, P. (1994). The mind of the traveller. *From gilgamesh to global tourism.* New York: Basic Books.

Macleod, M. (1994). Usability in context: Improving quality of use. *Proceedings of the International Ergonomics Association 4th International Symposium on Human Factors in Organizational Design and Management*, Stockholm, Sweden.

Nielsen, J. (1992). Finding usability problems through heuristic evaluation. *Proceedings of the ACM CHI '92*, Monterey, California.

Nielsen, J. (2000). *Designing Web usability: The practice of simplicity.* Indianapolis, IN: New Riders Publishing.

Picard, R. (1997). *Affective computing.* Cambridge: MA: MIT Press.

Raby, F. (2000). *Project #26765—Flirt: Flexible information and recreation for mobile users.* London: RCA CRD Research Publications.

Reyes, A., et al. (2001). *Flash design for mobile devices: A style guide for the wireless revolution.* New York: John Wiley.

Sears, A. (1993). Layout appropriateness: A metric for evaluating user interface widget layouts. *IEEE Transactions on Software Engineering, 19*(1), 707-719.

Shneiderman, B. (1987). *Designing the user interface: Strategies for effective human-computer interaction.* Reading, MA: Addison-Wesley.

Stanton, N. (Ed.). (2001). Ubiquitous computing: Anytime, anyplace, anywhere? *International Journal of Human-Computer Interaction, 13*(2), 107-111.

Weiss, S. (2002). *Handheld usability.* New York: John Wiley.

Wharton, C., Bradford, J., Jefferies, R., & Franzke, M. (1992). Applying cognitive walkthroughs to more complex user interfaces: Experiences, issues, and recommendations. *Proceedings of the CHI 92.*

KEY TERMS

Infodesign: A broad term for the design tasks of deciding how to structure, select, and present information.

Microdesign: The process of designing how a user will be able to interact with a small artifact.

Mistake: An error of reasoning or inappropriate subgoals, such as making a bad choice or failing to think through the full implications of an action.

Multi-Access: Interacting with a computer using more than one input or output channel at a time, usually suggesting drastically different input channels being used simultaneously (e.g., voice, typing, scanning, photo, etc.).

Multi-Channel: Different interfaces that can be available to the user for data entry in a multi-platform system (iTV, PC, mobile phone, smartphone, pocket-PC, etc.).

Multi-Channel Identity: A perceived communicational coherence for each service provided through the whole system of interfaces.

Reversible Actions: Any action that can be undone. Reversibility is a design principle that says people should be able to recover from their inevitable mistakes.

Self-Usability: Sort of mechanisms set by users (e.g., use of acronyms in a SMS) in order to make more usable the interaction with complex human artifacts.

Understanding and Improving Usability Inspection Methods

Alan Woolrych
University of Sunderland, UK

Mark Hindmarch
University of Sunderland, UK

INTRODUCTION

Usability inspection method (UIM) is the term used for a variety of analytical methods designed to "find" usability problems in an interface design. The basic principle involves analysts inspecting the interface against a set of pre-determined rules, standards or requirements. Analysts inspect the interface and predict potential usability problems based on breaches of these rules. None of the UIMs currently in use are capable of detecting all of the problems associated with an interface. After describing some of the UIMs in use, this article will look at the authors' work on improving these methods by focusing on the resources analysts bring to an inspection.

BACKGROUND

In order to better explain the work we have done on improving UIMs, three of the more commonly used UIMs will be described. These are by no means the only usability inspection methods; other examples being claims analysis (Carroll & Rosson, 1991) and pluralistic walkthroughs (Bias, 1994).

Heuristic Evaluation

This method was developed by Nielsen (1992). The basis of the method is the comparison of an interface with a set of usability guidelines, known as the heuristics. Originally nine, there are currently 10 guidelines dealing with areas such as visibility of system status, user control and freedom and error prevention.

A heuristic evaluation is carried out by a number of different evaluators; five is recommended as the optimal number (Nielsen & Landauer, 1993), and the problems identified by the individual evaluators are then merged into a master problem set.

This technique has been used at numerous stages in the development process from paper prototypes to full software packages (Nielsen, 1990). The advantages of the technique, and the reason the method is so popular, are that it can be used by novices as well as experts, although novices find fewer problems than experts (Nielsen, 1992) and the technique is comparatively quick and inexpensive to employ. The disadvantage is that it tends to only uncover more superficial problems with an interface; problems that require complex interaction on the part of the user are more likely to be missed by heuristic evaluation.

Cognitive Walkthrough

This technique is based on the CE+ [This is a combination of Cognitive Complexity Theory (**C**CT) (Kieras & Polson, 1985) and Explanation-based Learning (**E**XPL) (Polson & Lewis, 1990)] theory of exploratory learning. This theory states that users exploring a new interface are guided by general task goals, and they search for interface elements that promise to move them closer to these goals. Cognitive Walkthrough is a practical technique for applying CE+ in an evaluation and was fully outlined in Wharton, Rieman, Lewis, and Polson (1994). In contrast to Heuristic Evaluation, Cognitive Walkthrough can only be performed by experts.

The technique focuses on how well an interface can support a novice user without formal training. A Cognitive Walkthrough is usually performed by the interface designer with a small group of colleagues. It requires that certain information be available to the evaluators to be successful, including a description of the users and their knowledge resources, a description of the tasks to be performed and the correct sequence of actions necessary to carry out the tasks. In performing the walkthrough, for each step in a task, the evaluation team asks a series of questions including:

- Is the correct action obvious to the user?
- Will the user match the system's response with the chosen action?

Cognitive Walkthrough has been criticized for being too time consuming and requiring large amounts of paperwork to be completed, although attempts have been made to streamline the method (Rowley & Rhoades, 1992; Spencer, 2000).

Heuristic Walkthrough

Sears (1997) proposed a method which combines aspects of cognitive walkthroughs and heuristic evaluation to address the weaknesses of both.

Heuristic walkthrough is a two phase technique. The first phase has similarities with cognitive walkthrough, in that evaluators have a set of questions to guide their exploration of the interface as well as a set of common user tasks; this is designed to expose the evaluators to the core functionality of the interface. During the second phase, the evaluators use usability heuristics to assess problems with the interface. However, unlike a straightforward heuristic evaluation which is relatively unstructured, the use of the heuristics in a heuristic walkthrough is focused on those areas of the interface identified as important in the first phase of the evaluation.

It is claimed that the major advantage of heuristic walkthrough is its ability to identify severe usability problems compared to heuristic evaluation while avoiding the narrow focus commonly associated with cognitive walkthroughs.

Despite variations in the strengths and weaknesses of the various UIMs, the unreliability of the assessment of all such methods is well documented (e.g., Gray & Salzman, 1998). In practice, these methods fail to predict *all* of the usability problems in a design; not all of the analysts' predictions are true predictions. Such false predictions are commonly known as false positives. For a variety of reasons, analysts will make predictions about usability problems that in reality cause no problems to the users (Cockton & Woolrych, 2001).

For example, UIMs such as Heuristic Evaluation (Nielsen, 1992) are simply not good enough in their current state. The negative outcome of the use of such inspection methods is two-fold. First of all, because such methods are not thorough (they fail to find all of the usability problems), designs subjected to them can result in poor usability, especially if the nature of their flaws is not fully understood. For example, is there a type of usability problem that the method is typically good at finding? Or more importantly, is there a type of problem the method is particularly bad at finding, and are these problems likely to be severe ones? Second, if the false positives are addressed as real usability problems, time and money is wasted in the redesign of usable features.

Although the assessment of UIMs has been very poor (Gray & Saltzman, 1998), this has improved recently (e.g., Cockton, Woolrych, Hall, & Hindmarch, 2003). Research must, therefore, focus on the reliable assessment of inspection methods before work on inspection method improvement can begin.

Despite these problems with inspection methods, there is still a place for reliable UIMs given the original rationale for their development—saving valuable resources such as time and costs. The challenge is to improve UIM quality without increasing costs.

IMPROVING USABILITY INSPECTION METHODS

Thorough assessment of UIMs is reliant on accurate coding of analyst (non)-predictions. UIMs are commonly assessed by their validity, thoroughness, and effectiveness (Sears, 1997), even though percentages fail to comprehensively assess UIMs (Woolrych & Cockton, 2000).

Validity drops as the number of problems found with a UIM exceeds the real problems found. Analysts make false predictions (false positives) as well as successful ones. Fewer false positives mean a more valid UIM.

$$\text{Validity} = \frac{\text{Count of real problems found using UIM}}{\text{Count of problems predicted by UIM}}$$

The thoroughness of a UIM improves as more of the real problems that exist are actually found.

$$\text{Thoroughness} = \frac{\text{Count of real problems found using UIM}}{\text{Count of known usability problems}}$$

The effectiveness of a UIM is the weighted product of its thoroughness and validity (Hartson, Andre, & Williges, 2003). To calculate this accurately, we must correctly code all analysts' predictions. This is not just to get the right percentages. To understand how and why false positives and genuine misses arise, we must first be able to properly code analysts' predictions.

The main focus of analysts' (non)-predictions in previous assessments of UIMs has concentrated on just three outcomes: accurate predictions (hits), problems missed by the analysts (misses) and false positives (false alarms).

Our research (Cockton et al., 2003) has led to the development of the DARe (**D**iscovery **A**nalyst **Re**sources) model of analyst performance in usability evaluation. The model is simple. At its heart is the rigid distinction between the knowledge resources and methods used for problem discovery and problem analysis, where possible problems are confirmed or eliminated. It is called the DARe Model, after those knowledge resources that are critical to success with UIMs. Developments in the research pointed toward an analyst-centered approach. Several issues thus far were not being addressed in previous research.

For instance, when the method and environment remained the same, why did some analysts discover some problems while other analysts did not? Did some analysts correctly eliminate some false positives while others "kept" them? Did some analysts discover real usability problems and incorrectly eliminate them (false negative, previously coded as a genuine miss).

In order to answer these questions, we have to consider the *five* possible outcomes from inspection and accurately code them (Cockton, Woolrych, & Hindmarch, 2004), while analyzing the problem discovery, strategy, and analysis resources adopted by individual analysts for potential problem analysis. The five possible outcomes from inspection are as follows:

1. Predicted (true positive) real problem
2. Predicted, unconfirmed (false positive)
3. Not Predicted (discovered and correctly eliminated) not a problem (true negative)
4. Not Predicted (discovered and incorrectly eliminated) real problem (false negative)
5. Genuine Miss (undiscovered real problem)

One and five are easy to explain. A true positive is simply an accurate prediction of a usability problem by the analyst that in reality is a real usability problem to the user. A genuine miss is a real usability problem that has not been considered at all by the analyst; it has simply not been discovered. The third outcome (#2) is a false positive, where analysts believe an element or feature could cause user difficulties, but in real usage these difficulties do not occur.

The final two outcomes (#3 and #4) are true and false negatives respectively. A true negative is where an analyst discovers a potential problem, analyses it, and correctly eliminates it (effectively eliminating a false positive before it is reported). A false negative is a potential problem that is discovered and is eliminated in analysis, but in real usage is a real problem.

It is paramount to the accurate and thorough assessment of usability inspection methods that these outcomes are not miscoded. To address the issue of miscoding, two tools were developed. These were Extended Structured Problem Report Formats (ESPRFs) and falsification testing (see next section). Without these tools there was a high risk of miscoding analyst (non)-predictions. For example, without ESPRFs, there was a risk of coding a *genuine miss* as a *false negative* and vice-versa. The ESPRF requires analysts to record *all* discovered problems, even if they are subsequently eliminated during analysis. Moreover, without ESPRFs, false negatives could not be coded at all, as analysts would not normally report improbable problems eliminated during inspection.

Falsification testing (in detail next) is necessary to accurately code all of the possible outcomes. In simple terms, task sets are developed based on analyst (non)-predictions extracted from the ESPRFs. Each (non)-prediction is rigorously "stressed" in user testing in order to ensure accurate coding; in other words, we can be confident that if a problem exists it will be found!

Extended Structured Problem Report Formats (ESPRFs)

ESPRFs were developed as research support tools. The original version was the Structured Problem Report Format (SPRF) (Cockton & Woolrych, 2001). This version was developed to aid problem matching and extraction and intended to be used in both research and in practical use. The SPRF was derived from an analysis of usability problems, following from difficulties faced in the assessment of UIMs (Lavery & Cockton, 1997; Cockton, & Atkinson, 1997, Cockton & lavery 1999). General principles from SUPEX such as transcription, segmentation, difficulty isolation, and generalization were all applied.

The purpose of the SPRF was to address the issues associated with constructing a master problem set from multiple analyst inspections. A simple description of reported problems was not enough. An individual's description of the same predicted usability problem can vary greatly even to the point of appearing to be describing completely different events. The SPRF allowed for multiple points of reference to achieve accurate problem merging for the creation of a master problem set by requiring the analyst to report further detail about the reported problem. The SPRF required the analyst to provide a problem description but also to record the likely/actual difficulties associated with the problem and also any specific contexts in which the problem occurred (if applicable) and the assumed causes of the problem.

Experience has shown that, in many cases, problem merging is possible using the problem description alone. However, there were many instances where the problem description alone was insufficient to confirm a confident match, but reference to other descriptions did enable problem matching.

The current ESPRF (Cockton & Woolrych, 2001) was developed purely as a research tool. It is an extension of the SPRF, which is designed to extract analyst search strategy, knowledge resources used in problem analysis, and aid in the accurate coding of analyst (non)-predictions. There are four parts to the ESPRF, in which Part 1 consists of the original SPRF, for problem extraction and matching.

Part 2 of the ESPRF addresses discovery resources and methods. Analysts were required to explain their discovery strategy, such as system, or user-centered, unstructured or structured. Essentially there are four strategies for problem discovery: system scanning, system searching, goal playing, and method following. The first two are system-centered, the first and third are unstructured. Different knowledge resources are required: little if any for system scanning; product knowledge for system searching; user/domain knowledge for goal playing; and task knowledge for method following. Part 2 also addresses confirmation rationales for problems that analysts decide to retain as probable problems.

Part 3 deals specifically with heuristic application to individual problems. Analysts were required to provide evidence of conformance rather than just name a heuristic (as was the case in the SPRF).

Part 4 requires analysts to justify any problem elimination, with specific reference to user impact and behavior. The initial extensions focused on information that was clearly missing from the SPRF, that is, how analysts approach discovery and whether/why elimination occurs.

As well as preventing the miscoding of analyst (non)-predictions, the ESPRF also allows for the identification of analyst discovery strategy and relevant resources used in discovered problem analysis. ESPRFs are essential tools in usability inspection method assessment; however, ESPRFs need the support of falsification testing for thorough analysis of usability inspection methods.

FALSIFICATION TESTING

The method for falsification testing (Woolrych, Cockton, & Hindmarch, 2004) involves the rigorous testing of analyst predictions via user testing. Analyst predictions are analyzed and merged into a master problem set. Thorough analysis of the predicted problem determines the individual difficulties with each problem.

Within the context of the test application, task sets are systematically derived to expose these likely difficulties, that is, if the prediction is valid. Put simply, the individual's predicted problems are "stressed" via user testing to ensure a high level of confidence in final coding.

Falsification testing is a fixed task user testing method. Users are restricted in their choice of task approach and execution. The goal of falsification testing is an accurate assessment of validity, that is, to accurately determine if a prediction is a "hit" or "false positive", and, with the assistance of ESPRFs, to accurately code true/false negatives.

The principle is simple, if a prediction is accurate, then it will be confirmed by user testing. If a prediction does not materialize as a problem, we can have confidence that it does not exist, and that the particular prediction can be confidently coded as a false positive. Falsification testing ensures that false positive coding of predictions is not a consequence of incomplete coverage in user testing.

FUTURE TRENDS

Previous attempts to "streamline" and "fix" usability inspection methods have been unsuccessful primarily due to the fact that assessment of UIMs was inadequate. The DARe model for analyst behavior and associated research has provided a more rigorous assessment of UIMs. Much more is known about the limitations of UIMs in general although work in this area is not yet complete. The weight of evidence in current research points towards an analyst-centered approach to usability inspection method improvement.

Future work will concentrate on identifying the ideal search strategy for problem discovery during inspection. Also, further work is needed to identify and understand the knowledge resources that help analysts find and correctly analyze discovered usability problems.

CONCLUSION

As long as practitioners continue to use usability inspection methods in their current state, a likely outcome is products with unusable features, which is a discredit to the HCI community. One way forward is to better understand analyst behavior during inspection. Understanding such behavior starts with confidence in the accurate coding of *all* analysts' (non)-predictions. Recorded rationales for problem confirmation/elimination extracted from ESPRFs can be used to identify appropriate knowledge resources for potential problem analysis. Analysis of such knowledge resources can identify those that aid analysts in appropriate problem confirmation/elimination analysis.

This is essentially the framework for the DARe model for analyst performance in inspection. Further work is necessary, however the DARe model can provide the focus of future research that will lead to improved assessment of usability inspection. In turn, positive inroads can be made in better analyst training with emphasis on knowledge resources for both problem discovery and analysis. Then we can consider improving individual usability inspection methods.

ACKNOWLEDGMENT

The authors would like to thank John Rieman for his explanation of the origins of CE+.

REFERENCES

Bias, R. G. (1994). The pluralistic usability walkthrough: Coordinated empathies. In J. Nielsen, & R. L. Mack (Eds.), *Usability inspection methods* (pp. 65-78). New York: John Wiley & Sons.

Carroll, J. M., & Rosson, M. B. (1991). Deliberated evolution: Stalking the view matcher in design space. *Human-Computer Interaction*, *6*(3-4), 281-318.

Cockton, G., & Lavery, D. (1999). *A framework for usability problem extraction.* Paper presented at INTERACT 99 (pp. 347-355).

Cockton, G., & Woolrych, A. (2001). *Understanding inspection methods: Lessons from an assessment of heuristic evaluation.* Paper presented at HCI 2001 (pp. 171-192).

Cockton, G., Woolrych, A., Hall, L., & Hindmarch, M. (2003). *Changing analysts' tunes: The surprising impact of a new instrument for usability inspection method assessment.* Paper presented at HCI 2003 (pp. 145-161),

Cockton, G., Woolrych, A., & Hindmarch, M. (2004). Reconditioned merchandise: *Extended structured problem report formats in usability inspection.* Paper presented at CHI 2004 (pp. 1433-1436).

Gray, W. D., & Salzman, M. (1998). Damaged merchandise? A review of experiments that compare usability evaluation methods. *Human-Computer Interaction, 13*(3), 203-261.

Hartson, R. H., Andre, T. S., & Williges, R. C. (2003). Criteria for evaluating usability evaluation methods. *International Journal of HCI, 15*(1), 145-181.

Kieras, D., & Polson, P. G. (1985). An approach to the formal analysis of user complexity. *International Journal of Man Machine Studies*, *22*, 365-394.

Lavery, D., & Cockton, G. (1997). *Representing predicted and actual usability problems.* Paper presented at Workshop on Representations in Interactive Software Development (pp. 97-108).

Lavery, D., Cockton, G., & Atkinson, M. P. (1997). Comparison of evaluation methods using structured usability problem reports. *Behaviour and Information Technology, 16*(4), 246-266.

Nielsen, J. (1990). *Paper versus computer implementations as mockup scenarios for heuristic evaluation.* Paper presented at INTERACT '90 (pp. 315-320).

Nielsen, J. (1992). *Finding usability problems through heuristic evaluation.* Paper presented at CHI '92 (pp. 373-380).

Nielsen, J., & Landauer, T.K. (1993). *A mathematical model of the finding of usability problems.* Paper presented at INTERCHI '93 (pp. 206-213).

Polson, P., & Lewis, C. H. (1990) Theory-based design of easily-learned interfaces. *Human-Computer Interaction, 5,* 191-220.

Rowley, D. E., & Rhoades, D.G. (1992). T*he cognitive jogthrough: A fast-paced user interface evaluation procedure.* Paper presented at CHI '92 (pp. 389-395).

Sears, A. (1997). Heuristic walkthroughs: Finding the problems without the noise. *International Journal of Human-Computer Interaction, 9*(3), 213-223.

Spencer, R. (2000). *The streamlined cognitive walkthrough method, working around social constraints encountered in a software development company.* Paper presented at CHI 2000 (pp. 353-359).

Wharton, C., Rieman, J., Lewis, C., & Polson, P. (1994). The cognitive walkthrough method: A practitioner's guide. In J. Nielsen, & R. L. Mack (Eds.), *Usability inspection methods* (pp. 105-140). New York: John Wiley & Sons.

Woolrych, A., & Cockton, G. (2000). *Assessing heuristic evaluation: Mind the quality, not just percentages.* Paper presented at HCI 2000 (pp. 35-36).

Woolrych, A., Cockton, G., & Hindmarch, M. (2004). *Falsification testing for usability inspection method assessment.* Paper presented at HCI 2004 (pp. 137-140).

KEY TERMS

False Negative: A potential usability problem discovered in a usability inspection that upon analysis is incorrectly eliminated by the analyst as an improbable problem. The discovered problem is confirmed in real use as causing difficulties to users.

False Positive: A prediction of a usability problem reported in a usability inspection that in reality is not a problem to the real users.

Falsification Testing: A method for testing the accuracy of predictions made during usability inspections.

Genuine Miss: A usability problem that causes user difficulties that remains undiscovered in usability inspection.

Thoroughness: A measure for assessing usability inspection methods. Determined by dividing the number real problems found by the UIM by the number of known problems.

True Negative: A potential usability problem discovered in a usability inspection that upon analy-

sis is correctly eliminated by the analyst as an improbable problem. The discovered problem is confirmed in real use as causing the user no difficulties.

True Positive: A prediction of a usability problem reported in a usability inspection that is proven to be a real problem in actual use with real users.

Usability Inspection Methods (UIM): The term given to a variety of analytical methods for predicting usability problems in designs.

Validity: A measure for assessing usability inspection methods. Determined by dividing the number of *real* problems found by the UIM by the number of problems predicted by the UIM.

Understanding Cognitive Processes in Educational Hypermedia

Patricia M. Boechler
University of Alberta, Canada

INTRODUCTION

Cognitive load theory (CLT) is currently the most prominent cognitive theory pertaining to instructional design and is referred to in numerous empirical articles in the educational literature (for example, Brünken, Plass, & Leutner, 2003; Chandler & Sweller, 1991; Paas, Tuovinen, Tabbers, & Van Gerven, 2003; Sweller, van Merri¸nboer, & Paas, 1998). CLT was developed to assist educators in designing optimal presentations of information to encourage learning. CLT has also been extended and applied to the design of educational hypermedia and multimedia (Mayer & Moreno, 2003). The theory is built around the idea that the human cognitive architecture has inherent limitations related to capacity, in particular, the limitations of human working memory. As Sweller et al. (pp. 252-253) state:

The implications of working memory limitations on instructional design cannot be overstated. All conscious cognitive activity learners engage in occurs in a structure whose limitations seem to preclude all but the most basic processes. Anything beyond the simplest cognitive activities appear to overwhelm working memory. Prima facie, any instructional design that flouts or merely ignores working memory limitations inevitably is deficient. It is this factor that provides a central claim to cognitive load theory.

In order to understand the full implications of cognitive load theory, an overview of the human memory system is necessary.

BACKGROUND

The Human Memory System: The Modal Model of Memory

It has long been accepted that the human memory system is made up of two storage units: long-term memory and working memory. There is an abundance of behavioral (for example, Deese & Kaufman, 1957; Postmand & Phillips, 1965) and neurological evidence (Milner, Corkin, & Tueber, 1968; Warrington & Shallice, 1969) to support this theory. Long-term memory is a repository for information and knowledge that we have been exposed to repetitively or that has sufficient meaning to us. Long-term memory is a memory store that has an indefinable duration but is not conscious; that is, any information in long-term memory must first be retrieved into working memory for us to be aware of it. Hence, any conscious manipulation of information or intentional thinking can only occur when this information is available to working memory. The depth and duration of processing in working memory determines whether information is passed on to long-term memory. Once knowledge is stored in long-term memory, we can say that enduring learning has occurred.

Working Memory Limitations

Unfortunately, working memory has some very definite limitations. First, there is a limit of volume. Baddeley, Thomson, and Buchanan (1975) reported that the size of working memory is equal to the amount of information that can be verbally rehearsed in approximately 2 seconds. A second limitation of working memory concerns time. When information is attended to and enters working memory, if it is not consciously processed, it will decay in approximately 20 seconds.

CLT AND EDUCATIONAL HYPERMEDIA

The modal model of human memory, specifically these limitations of working memory, is the basis for CLT. A version of CLT, Mayer and Moreno's (2003) selecting-organizing-integrating theory of active learning, is specifically targeted to learning in hypermedia environments. The theory is built upon three core assumptions from the modal model of memory: the dual channel assumption, the limited capacity assumption, and the active processing assumption. The dual channel assumption is based on the notion that working memory has two sensory channels, each responsible for processing different types of input. The auditory or verbal channel processes written and spoken language. The visual channel processes images. The limited capacity assumption applies to these two channels; that is, each of these channels has a limit as to the amount of information that can be processed at one time. The active processing assumption is derived from Wittrock's (1989) generative learning theory and asserts that substantial intentional processing is required for meaningful learning. With these assumptions as a foundation, Mayer and Moreno have focused on three key mental activities that can place demands on available cognitive resources: attention, mental organization, and integration.

Improving Working Memory Capacity Directly

How does CLT advocate improving working memory limitations? To date, the solution for reducing cognitive load has focused on directly reducing the demands on working memory. Mayer and Moreno (2003) outline a number of methods for reducing cognitive load in hypermedia: (a) Resting on the dual channel assumption, cognitive load on one channel can be relieved by spreading information across both modalities, that is, by providing information in both a visual and auditory format, (b) presenting material in segments and providing pretraining on some material can reduce overload, (c) the redundancy of information can be eliminated, and (d) visual and auditory information can be synchronized.

Mayer and Moreno (2003) also refer to "incidental processing" as "cognitive processes that are not required for making sense of the presented material but are primed by the design of the learning task" (p. 45). Incidental processing is considered undesirable as it relates to the cognitive resources that are needed to process extraneous, irrelevant material that may be included on the presentation. Mayer and Moreno advocate weeding out this extraneous material to reduce cognitive load.

Measuring Cognitive Load

If the premise of cognitive load theory is correct, then certainly a primary activity in designing instructional materials must be the meaningful measurement of cognitive load. This is not a simple task as the method of measurement is dependent on the constructs that different researchers use to describe cognitive load. For example, Paas et al. (2003) propose that three constructs define cognitive load: mental load, which reflects the interaction between task and subject characteristics; mental effort, which reflects the actual cognitive reserves that are expended on the task; and performance, which can be defined as the learner's achievements. Previous research in cognitive load measurement has relied on three types of measures to assess the cognitive load of the user: (a) physiological measures such as heart rate and pupillary responses, (b) performance data on primary and secondary tasks, and (c) self-reported ratings (Paas et al.). These tasks have been used in various configurations to measure overall cognitive load (Brünken et al., 2003; Chandler & Sweller, 1996; Gimino, 2002; Paas, 1992). To date, most efforts to measure cognitive load have focused on self-reported ratings (see Paas et al.).

FUTURE TRENDS

Our ability to reduce cognitive load in educational hypermedia rests on our thorough definition of the underlying constructs of cognitive load as well as the design of test mechanisms that allow us to measure cognitive load and detect situations where cognitive resources are overtaxed. Future research directed at these two issues will contribute to the explanatory power of the theory and allow us to apply these theoretical principles to educational settings that make use of hypermedia materials.

CONCLUSION

The cognitive load theory for educational hypermedia has emerged as a prominent theory for guiding instructional designers in the creation of educational hypermedia. It is based on the modal model of human memory, which posits that there are limits to the working memory store that impact the amount of cognitive effort that can be expended on a given task. When available cognitive resources are surpassed, performance on memory and learning tasks is degraded, a condition referred to as cognitive overload. CLT for educational hypermedia advocates that educational materials must be designed that take into account these limitations. In order to do this, two obstacles to using CLT to its full advantage must be resolved: (a) the diversity of the descriptions of its underlying constructs and (b) the lack of valid and reliable methods for the measurement of cognitive load.

REFERENCES

Baddeley, A. (1992). Working memory. *Science, 255*, 556-559.

Baddeley, A., Thomson, N., & Buchanan, M. (1975). Word length and the structure of short-term memory. *Journal of Verbal Learning & Verbal Behavior, 14*, 575-589.

Brünken, R., Plass, J. L., & Leutner, D. (2003). Direct measurement of cognitive load in multidimensional learning. *Educational Psychologist, 38*(1), 53-61.

Chandler, P., & Sweller, J. (1991). Cognitive load theory and the format of instruction. *Cognition and Instruction, 8*, 293-332.

Chandler, P., & Sweller, J. (1996). Cognitive load while learning to use a computer program. *Applied Cognitive Psychology, 10*, 151-170.

Deese, J., & Kaufman, R. A. (1957). Serial effects in recall of unorganized and sequentially organized verbal material. *Journal of Experimental Psychology, 54*, 180-187.

Frensch, P. A., & Miner, C. S. (1994). Effects of presentation rate and individual differences in short-term memory capacity on an indirect measure of serial learning. *Memory and Cognition, 22*, 95-110.

Gimino, A. (2002). *Students' investment of mental effort.* Paper presented at the annual meeting of the American Educational Research Association, New Orleans, LA.

Mayer, R., & Moreno, R. (2003). Nine ways to reduce cognitive load in multimedia learning. *Educational Psychologist, 38*(1), 43-52.

Milner, B. S., Corkin, S., & Tueber, H. L. (1968). Further analysis of the hippocampal amnesic syndrome: 14 year follow-up study of H. M. *Neuropsychologica, 6*, 215-234.

Paas, F. G. (1992). Training strategies for attaining transfer of problem-solving skill in statistics: A cognitive approach. *Journal of Educational Psychology, 84*, 429-434.

Paas, F. G., Tuovinen, J. E., Tabbers, H., & Van Gerven, P. W. M. (2003). Cognitive load measurement as a means to advance cognitive theory. *Educational Psychologist, 38*(1), 63-71.

Postmand, L., & Phillips, L. W. (1965). Short-term temporal changes in free recall. *Quarterly Journal of Experimental Psychology, 17*, 132-138.

Reber, A. S. (1993). *Implicit learning and tacit knowledge: An essay on the cognitive unconscious* (Oxford Psychology Series No. 19). New York: Oxford University Press.

Shanks, D. R., & St. John, M. F. (1994). Characteristics of dissociable learning systems. *Behavioral & Brain Sciences, 17*, 367-395.

Squire, L. R. (1992). Memory and the hippocampus: A synthesis from findings with rats, monkeys and humans. *Psychological Review, 99*, 195-231.

Sweller, J., van Merri¸nboer, J. J., & Paas, F. G. (1998). Cognitive architecture and instructional design. *Educational Psychology Review, 10*(3), 251-296.

Tulving, E. (2000). Concepts of memory. In E. Tulving & F. I. M. Craik (Eds.), *The Oxford handbook of memory* (pp. 33-43). New York: Oxford University Press.

Tulving, E., & Schacter, D. L. (1990). Primary and human memory systems. *Science, 247*, 301-306.

Warrington, E. K., & Shallice, T. (1969). The elective impairments of auditory verbal short-term memory. *Brain, 92*, 885-896.

Wittrock, M.C. (1989). Generative processes of comprehension. *Educational Psychologist, 24*, 345-376.

KEY TERMS

Active Processing Assumption: The active processing assumption asserts that intentional and significant mental processing of information must occur for enduring and meaningful learning to take place.

Cognitive Load Theory: Cognitive Load Theory asserts that the capacities and limitations of the human memory system must be taken into account during the process of instructional design in order to produce optimal learning materials and environments.

Dual Channel Assumption: The dual channel assumption is based on the notion that working memory has two sensory channels, each responsible for processing different types of input. The auditory or verbal channel processes written and spoken language. The visual channel processes images.

Limited Capacity Assumption: The limited capacity assumption applies to the dual channels of verbal and auditory processing. The assumption is that each of these channels has a limit as to the amount of information that can be processed at one time.

Long-Term Memory: Long-term memory is a repository for information and knowledge that we have been exposed to repetitively or that has sufficient meaning to us. Long-term memory is a memory store that has an indefinable duration but is not conscious; that is, any information in long-term memory must first be retrieved into working memory for us to be aware of it.

Mental Effort: A second construct related to measuring cognitive load, "mental effort is the aspect of cognitive capacity that is actually allocated to accommodate the demands imposed by the task" (Paas et al., 2003, pp. 64).

Mental Load: One of three constructs devised by Paas et al. (2003) to assist in the measurement of cognitive load. Mental load reflects the interaction between task and subject characteristics. According to Paas et al. (2003), " it provides an indication of the expected cognitive capacity demands and can be considered an a priori estimate of cognitive load"(pp. 64).

Performance: Performance is the third construct in Paas et al.'s (2003) definition of cognitive load and is reflected in the learner's measured achievement. Aspects of performance are speed of completing a task, number of correct answers and number of errors.

Short-Term or Working Memory: Short-Term or Working Memory refers to a type of memory store where conscious mental processing occurs, that is, thinking. Short-term memory has a limited capacity and can be overwhelmed by too much information.

Usability Barriers

David R. Danielson
Stanford University, USA

INTRODUCTION

Numerous technical, cognitive, social, and organizational constraints and biases can reduce the quality of usability data, preventing optimal responses to a system's usability deficiencies. Detecting and appropriately responding to a system's usability deficiencies requires powerful collection methods and tools, skilled analysts, and successful interaction amongst usability specialists, developers, and other stakeholders in applying available resources to producing an improved system design. The detection of usability deficiencies is largely a matter of analyzing a system's characteristics and observing its performance in use. Appropriate response involves the translation of collected data into usability problem descriptions, the production of potential design solutions, and the prioritization of these solutions to account for pressures orthogonal to usability improvements. These activities are constrained by the effectiveness and availability of methods, tools, and organizational support for user-centered design processes. The quality of data used to inform system design can, for example, be limited by a collection tool's ability to record user and system performance, an end user's ability to accurately recall past interactions with a system, an analyst's ability to persuade developers to implement changes, and an organization's commitment to devoting resources to user-centered design processes.

The remainder of this article (a) briefly reviews basic usability concepts, (b) discusses common barriers to successfully collecting, analyzing, and reacting to usability data, and (c) suggests future trends in usability research.

BACKGROUND

Usability barriers hinder data collection processes, reduce the quality of usability data, and therefore hinder the detection of and response to a system's deficiencies. Barriers to system usability are necessarily barriers to one or more dimensions of usability. Usability dimensions are commonly taken to include at least user efficiency, effectiveness, and subjective satisfaction with a system in performing a specified task in a specified context (ISO 9241-11, 1998), and frequently also include system memorability and learnability (Nielsen, 1993).

Usability data are defined by Hilbert and Redmiles (2000) as any information used to measure or identify factors affecting the usability of a system being evaluated. Such data are collected via *usability evaluation methods* (UEMs), methods or techniques that can assign values to usability dimensions (J. Karat, 1997) and/or indicate usability deficiencies in a system (Hartson, Andre, & Williges, 2003). Usability evaluation may be analytic (based on interface design attributes, independent of actual usage) or empirical (based on observations of system performance in actual use; Hix & Hartson, 1993), and may be formative (employed during system development) or summative (employed after system deployment; Scriven, 1967).

Usability data quality refers to the extent to which the data efficiently and effectively predicts system usability in actual usage, can be efficiently and effectively analyzed, and can be efficiently and effectively reacted to. High-quality usability data indicate real system deficiencies (validity) that will be repeatedly encountered by individual users (reliability) and by a wide range of users (representativeness); represent deficiencies in their entirety (completeness); can be easily translated by usability analysts into problem descriptions that accurately represent the underlying deficiencies (communicative effectiveness and efficiency); indicate problems that seriously influence the quality of users' experiences with the system (severity); and persuade developers and other stakeholders to implement design changes (downstream utility) that verifiably improve system usability (impact) at low cost (cost effectiveness). (For a discussion of each of these dimensions, see the article titled "Usability Data Quality" in this encyclopedia.)

BARRIERS TO USABILITY-DATA QUALITY

The successful collection, analysis, and reaction to usability data are hindered in practice by numerous constraints and biases. Far more empirical work identifying barriers to data quality has focused on collection than analysis and reaction for the obvious reasons: Collection processes are more amenable to experimental control and more accessible to researchers (i.e., easier to simulate or observe in entirety). Nonetheless, in recent years, barriers throughout the development process have been identified, as discussed in this section.

Resource Constraints

If representative customers and end users are distributed (especially internationally), costs become the primary barrier to (empirical) collection, which will tend to drive the selection of methods (Englefield, 2003; Stanton & Baber, 1996; Vasalou, Ng, Wiemer-Hastings, & Oshlyansky, 2004) and affect data quality. As a result, informal data-collection methods are more frequently employed in practice than formal methods (Vredenberg, Mao, Smith, & Carey, 2002).

Perhaps the most common constraint arises from the timing of data collection in the development cycle. Not surprisingly, the general finding is that the later usability data are collected, the less likely they are to result in design changes (Bias & Mayhew, 1994). This problem can be exacerbated when a short development cycle is demanded by concerns orthogonal to usability.

When data collection is performed at low cost (for example, by using nonintrusive remote collection methods), the resource burden is often not avoided but rather shifted to analysis since such methods can result in more data than are possible to translate into problem descriptions within the development cycle.

User Ability and Motivation

One of the most widely employed collection methods, think-aloud usability testing, requires users to engage in a highly unnatural activity, namely, verbally unloading a stream of consciousness while interacting with a system (Nielsen, 1993). Lin, Choong, and Salvendy (1997) point out that many users have difficulty in keeping cognitive processes verbalized while performing tasks, and that expert users in particular find it difficult to verbalize their (often automatic) processes. When activities are routine or would not normally require attention, concurrent verbalization is not only difficult, but can affect cognitive processes (Birns, Joffre, Leclerc, & Paulsen, 2002; Ericsson & Simon, 1980) and therefore hinder the validity of behavioral observations made during testing.

Remote methods in which the setting of data collection is more realistic do not avoid these barriers. Fundamentally, data collection is limited by the ease of use of the collection instrument (Hartson & Castillo, 1998) and users' ability to notice usability problems as they occur (Galdes & Halgren, 2001), ability to evaluate incomplete prototypes with missing functionality, ability to remember and articulate the context of a previously encountered problem (J. Karat, 1997), and willingness to accept the cost of providing feedback.

Selective Feedback and Feedback Bias

Under many circumstances, usability data that could drive system improvements are simply never collected. Even when mechanisms are in place for reporting critical incidents during actual use, users will choose which problems to report, often neglecting those they deem unimportant (Costabile, 2001). Neglecting low-severity problems can in some cases be a benefit to data quality, but only to the extent that users are able to recognize which problems recur and to tune their feedback activities effectively. Users conversely will often neglect reporting high-severity problems, naturally in favor of focusing their attention on correcting such problems and getting their work done.

Neglecting feedback altogether may in some cases be the lesser of two feedback evils, the other being speculation. Hilbert and Redmiles (1999) have noted speculative feedback from both novices and experts that can affect data quality. Novices will in some cases speculate that a real usability problem would not interfere with expert usage, and consequently neglect to report serious system learnability deficiencies, while experts will incorrectly speculate on the usability of interface features for novice users rather than focus on real problems they encounter. User assumptions about the usefulness of their own feedback is a particularly difficult problem to tackle in part because such assumptions may not be articulated in the context of data collection; when customers filter feedback to indicate only those problems they foresee as feasibly (and quickly) being addressed by a software vendor, for example, such filtering is unlikely to be made explicit without prompting.

Indirect Data

A key effect of attempting to maximize the cost effectiveness of data collection is an increased reliance on indirect sources of usability data during formative evaluation stages. Despite its shortcomings, face-to-face usability testing has the advantage of allowing analysts to view problems as they occur and to clarify end-user comments and reasoning while interacting with a prototype or live system. When cost constraints require remote collection, the resulting data frequently lack context (Hilbert & Redmiles, 1998). Moreover, users will frequently delay providing usability feedback until well after an important incident has occurred (Hartson & Castillo, 1998). Birns et al. (2002) describe instances of users encountering usability problems only to later blame themselves for the problem after subsequent interactions with the system. Thus, the goal of collecting data uninfluenced by such factors can be compromised. Not only are indirect data often influenced by subsequent interactions, but they tend to more heavily focus on users' subjective preferences rather than objective descriptions of problems. Were user preferences consistent predictors of performance deficiencies, data quality would be unaffected, but frequently they are not (Frøkjær, Hertzum, & Hornbæk, 2000; Nielsen & Levy, 1994).

Preference feedback and descriptions of problems from memory may not directly represent encountered usability deficiencies, but they are still at least "from the horse's mouth." Another key effect of cost constraints is a reliance on filtered usability feedback from sales professionals or from customers making the buying decisions (who may have varying levels of engagement with end users) rather than directly from the end users themselves. The quality of such filtered data remains largely uninvestigated.

A similar filtering that has been investigated is data collection via analytic methods. Such methods often explicitly require usability specialists to take on the role of the end user. Hertzum and Jacobsen (2001) distinguish between two types of barriers to data quality in these cases: (a) *anchoring*, in which a system is evaluated with respect to users too similar to the evaluator to be representative of the user population, and (b) *stereotyping*, in which the system is evaluated with respect to a homogenous catchall user not accounting for a wide-enough range of user characteristics. Not surprisingly, such biases and differences in technique lead to differences in the results of analytic methods applied by different evaluators (Andre, Hartson, Belz, & McCreary, 2001; Cockton, Woolrych, Hall, & Hindmarch, 2003); more problems tend to be discovered by a team of evaluators specializing their focus (i.e., looking for only certain types of problems in the interface; Zhang, Basili, & Schneiderman, 1999). As a result, multiple evaluators are commonly recommended in employing analytic methods.

Method Scope

A shift toward preference data for particular collection methods is one example of limited method scope; the types of problems typically indicated by different collection methods can vary (John & Kieras, 1996), one of the primary reasons they traditionally supplement one another in user-centered design processes. Englefield (2003) distinguishes between the breadth of a usability method's data collection capabilities (similar to method thoroughness, or the extent to which it is capable of detecting all usability deficiencies) and its sensitivity to particular types of problems, claiming for example that empirical methods tend to be sensitive to sociotechnical design problems that expert inspections have difficulty identifying.

An important consequence is that methods vary in their appropriateness at different stages in the development cycle (Lewis & Wharton, 1997; Rubin, 1994) and for different types of prototypes or systems; the fidelity of the prototype used can have subtle effects on the sensitivity of data collection (Virzi, Sokolov, & Karis, 1996). Coordinating the application of multiple collection methods is a matter of achieving optimal scope so as not to focus too heavily on particular types of problems at the expense of others, for example, by devoting too many resources to analytic methods such as cognitive walk-throughs to address system learnability, but doing so at the expense of other usability dimensions.

Stimulus and Simulation Effects

Ecological validity is of primary concern for usability data collection given the goal of discovering deficiencies that occur for real users in their actual working environments. Remote collection allows end users to evaluate systems under more realistic social and technical constraints (Krauss, 2003), but in many cases, the user is nonetheless evaluating an incomplete prototype. Particularly in the application of empirical methods, usability specialists have long been aware of subtle and unintended effects of prototypes on the quality of the data they collect. Usability engineering work frequently involves the assessment of systems that range in fidelity from fully functional products to paper prototypes and sketches. Matching fidelity attributes to data collection goals is often a difficult balancing act. Have too high fidelity, and an end user may focus on color schemes or branding logos that were never intended to represent the definite final product. Have too low fidelity, and normally attention-consuming aspects of the interface may fail to have their realistic impact. Have the product be too vertical (deep functionality for only a few features or tasks), and the user may lose focus when attempting to explore nonfunctioning areas of the prototype. Have it be too horizontal (shallow functionality across many features and tasks), and collected data may be of little use in driving design decisions.

A more subtle difficulty occurs in producing mock data for prototypes and simulated usability tests (Kantner, Sova, & Rosenbaum, 2003). Both the realism and the credibility of the test itself can hinge on this activity, particularly if the tasks of interest to the evaluator are focused on information gathering and usage. The realism of such mock data can be difficult to achieve in part because prototype designers are typically not subject-matter experts in the system's domain and are often still learning about the domain during development.

Finally, evaluators often must accept the practical limits of task simulation. Tasks that span days or even weeks, for example, require the application of less direct methods to gather data (Galdes & Halgren, 2001). Systems for which tasks frequently involve safety risks similarly present simulation difficulties.

Evaluation Effects

Some of the advantages of having an expert evaluator present to interact with end users during empirical data collection have been previously mentioned, such as the ability to clarify user comments and reasoning. However, numerous aspects of evaluator intervention affect usability data, including the amount of such intervention (Held & Biers, 1992), the type of observation and type of evaluator prompts (such as leading questions and task guidance; Galdes & Halgren, 2001; Kjeldskov & Skov, 2003), and the presence of recording devices (Nielsen, 1993). Boren and Ramey (2000) investigated the usage of verbal protocols in usability testing, finding widespread inconsistencies in the method's application. These effects are perhaps impossible to avoid altogether; as C. Karat (1994) puts it, collection methods act as "filters" on user-system interactions.

The effects of evaluator differences have largely been investigated by noting a substantial lack of overlap between usability problem sets produced by multiple evaluators (Hertzum, Jacobsen, & Molich, 2002; Jacobsen, Hertzum, & John, 1998; Molich, Ede, Kaasgaard, & Karyukin, 2004). These effects cannot be avoided when the testing work must be divided for practical reasons, for example, when a system has an international user base, and cultural and language barriers require the expertise of multiple team members (Vasalou et al., 2004). The clear implication is that usability tests conducted by different evaluators lack reliability and consistency. However, because such consistency is not the only goal of usability data collection, these effects can turn out to be a blessing. The secondary effect of inconsistent think-aloud techniques is an increase in the scope and sensitivity of the method; similar to analytic methods, optimal empirical data collection for usability purposes is likely a team effort.

Analyst Ability and Analytical Bias

Placing usability specialists with varying backgrounds and perspectives in the picture affects not only the data resulting from evaluation methods, but the interpretation of that data and resulting work products in the development cycle. Analyst skills come into play in two important areas: (a) the organization and prioritization of usability deficiencies and potential solutions, and (b) the reporting of these conclusions to development.

Determining problem severity is partially a judgment call. Jacobsen et al. (1998) point out that the criteria employed by different analysts in estimating severity can vary, and that such estimates may be biased when analysts judge problems that they themselves observed. On the other hand, severity judgments by analysts who did not observe the problem are limited by a lack of information about the details and context of the deficiency. Here again, an individual analyst is probably nonoptimal. Similarly, analyst expertise is critical in the identification of underlying causes of usability problems and their translation into design solutions (John & Marks, 1997). Such expertise is also important in the recognition of similar problems previously encountered and applicable solution patterns. Expertise is also important in the successful prioritization of problems, which is typically the result of considering estimated problem severity and representativeness (Gediga, Hamborg, & Düntsch, 1999) as well as likely implementation costs.

Once problems and potential solutions have been identified, analysts are responsible for effectively reporting a plan of action for improving system usability. This activity is of course set against a background of a project commitment and incentives (or lack thereof) for achieving usability goals, and a previous relationship between usability and development teams (Bias & Mayhew, 1994; Mirel, 2000), but additionally depends on analyst ability and standardized reporting (Bevan, 1998). Andre et al. (2001) argue, however, that even armed with standard reports, descriptions of design solutions are often vague or incomplete, and there is inevitable loss of information as developers interpret these documents. Analysts must pick the right set of problems to present as too large a set may hinder persuasiveness (Dumas & Redish, 1993). They must also effectively leverage face-to-face meetings with development (Galdes & Halgren, 2001).

Development Conflicts

Solutions to usability deficiencies can be at odds with one another, forcing a reliance on prioritization, but even the highest priority and most persuasive data can be thwarted by concerns orthogonal to usability improvements. Some of these concerns are simply outside development control, such as imposed corporate standards (Hertzum, 1999). Others require trade-off decisions, such as in considering legacy concerns, whether potential solutions to usability problems may conflict with other aspects of software quality, and the architectural changes needed to fix high-severity problems (Folmer & Bosch, 2004). The timing of data collection is again critical since addressing usability concerns late in the cycle is likely to increase development costs (Folmer & Bosch, 2003).

Process Bias

Conceptually, there are two types of potential process biases than can hinder usability data impact: (a) if a type of data is cost effective and tends to achieve high impact when reacted to, but nonetheless tends not to be persuasive in the eyes of developers and other stakeholders ("untapped potential"), and (b) if a type of data tends not to have impact but has high persuasiveness ("false prophet"). For example, suppose a development team highly respects one type of data (say, empirical usability test data) and tends to focus on implementing changes suggested by it at the expense of other types of data, such as heuristic reviews. If it turns out that heuristic reviews produce cost-effective, high-impact data, or if the usability testing data tends to be of low quality, impact on system usability suffers from a process bias. The first type (high impact and cost effectiveness, but low persuasiveness) only indicates a potential bias

since it is at least possible that downstream stakeholders will tune responses to data from a particular method based on an effective understanding of when it will likely have impact.

FUTURE TRENDS

The range of factors influencing usability data's likelihood of impacting system design, in conjunction with concerns orthogonal to system usability, leave many challenges open for usability research. This section identifies two general emerging research areas important to usability theory and practice.

Balancing Data Collection Needs and Effects

The attractiveness of data with high communicative effectiveness leads many evaluators to value users' design ideas and rationalizations about their behaviors, but the validity of such data is a serious concern. In practice, this problem can be exacerbated by the persuasiveness and appeal of design sketches straight from end users. The appeal is certainly understandable. When limited resources are spent visiting a small set of customers or end users, there is an organizational stake in ensuring that data at a premium is put to good use; from a customer relationship perspective, there is, perhaps more importantly, value in (directly) demonstrating that customer feedback and ideas are listened to. A valuable direction for usability science may be to map out the problems and potentials of balancing usability data quality concerns with needs that are created by the simple act of engaging customers and end users during development.

Balancing the Present and Future

New companies often have small sets of early adopters who are the primary sources for evaluating systems in actual use. Responding to their needs is critical, and not surprisingly, they have significant impact on the early design of the product. However, effective response to high-quality data from these customers now may create an unavoidable hole later; who representative users are in the initial stages of a system can change dramatically in a short time. A similar problem arises with any relatively new product, for which power users simply do not yet exist. How does one most effectively design for nonexistent experts? How are usability concerns most effectively balanced to meet current needs but not create usability legacy problems?

CONCLUSION

User-centered design processes are subject to a number of basic constraints limiting the quality of usability data collected within these processes. The inherent limitations of widely employed methods and tools hinder the successful collection of high-quality usability data. End users, analysts, developers, and other stakeholders frequently introduce biases and other unintended effects into data collection, analysis, and development processes. Such effects often inevitably result in less-than-optimal responses to a system's deficiencies, but can in some cases increase the scope and sensitivity of data collection. Barriers to usability data quality can additionally arise from concerns orthogonal to system usability, or from factors not easily predicted during system development.

REFERENCES

Andre, T. S., Hartson, H. R., Belz, S. M., & McCreary, F. A. (2001). The user action framework: A reliable foundation for usability engineering support tools. *International Journal of Human-Computer Studies, 54*, 107-136.

Bevan, N. (1998). Common industry format usability tests. Paper presented at the *Seventh Annual Meeting of the Usability Professionals' Association.*

Bias, R. G., & Mayhew, D. J. (Eds.). (1994). *Cost-justifying usability.* Boston: Academic Press.

Birns, J. H., Joffre, K. A., Leclerc, J. F., & Paulsen, C. A. (2002). *Getting the whole picture: Collecting usability data using two methods—concurrent think aloud and retrospective probing*. Paper presented at the Eleventh Annual Meeting of the Usability Professionals' Association.

Boren, M. T., & Ramey, J. (2000). Thinking aloud: Reconciling theory and practice. *IEEE Transactions on Professional Communication, 43*(3), 261-278.

Cockton, G., Woolrych, A., Hall, L., & Hindmarch, M. (2003). Changing analysts' tunes: The surprising impact of a new instrument for usability inspection method assessment. *People and Computers XVII: Designing for Society (Proceedings of HCI'03)* (pp. 145-161).

Costabile, M. F. (2001). Usability in the software life cycle. In S. K. Chang (Ed.), *Handbook of software engineering and knowledge engineering* (pp. 179-192). Singapore: World Science.

Dumas, J. S., & Redish, J. C. (1993). *A practical guide to usability testing.* Norwood, NJ: Ablex.

Englefield, P. (2003). *A pragmatic framework for selecting empirical or inspection methods to evaluate usability.* IBM. Retrieved from http://www.ibm.com/easy/

Ericsson, K. A., & Simon, H. A. (1980). Verbal reports as data. *Psychological Review, 87*(3), 215-251.

Folmer, E., & Bosch, J. (2003). Usability patterns in software architecture. *Proceedings of HCII2003, International Conference on Human-Computer Interaction.* Paper presented at the 10th International Conference on Human-Computer Interaction, HCII2003, Crete, Greece.

Folmer, E., & Bosch, J. (2004). *Cost effective development of usable systems: Gaps between HCI and SE. Proceedings of ICSE Workshop "Bridging the Gaps Between Software Engineering and Human-Computer Interaction, II."*

Frøkjær, E., Hertzum, M., & Hornbæk, K. (2000). Measuring usability: Are effectiveness, efficiency, and satisfaction really correlated? *Proceedings of CHI'00, Human Factors in Computing Systems* (pp. 345-352).

Galdes, D., & Halgren, S. L. (2001). *Collecting usability data through customer journals: A successful technique for feature-rich applications.* Paper presented at the Tenth Annual Meeting of the Usability Professionals' Association.

Gediga, G., Hamborg, K. -C., & Düntsch, I. (1999). The IsoMetrics usability inventory: An operationalization of ISO 9241/10 supporting summative and formative evaluation of software systems. *Behaviour and Information Technology, 18*, 151-164.

Hartson, H. R., Andre, T. S., & Williges, R. C. (2003). Criteria for evaluating usability evaluation methods. *International Journal of Human-Computer Interaction, 15*(1), 145-181.

Hartson, H. R., & Castillo, J. C. (1998). Remote evaluation for post-deployment usability improvement. *Proceedings of AVI'98, Advanced Visual Interfaces* (pp. 22-29).

Held, J. E., & Biers, D. W. (1992). Software usability testing: Do evaluator intervention and task structure make any difference? *Proceedings of the Human Factors and Ergonomics Society 36th Annual Meeting* (pp. 1215-1219).

Hertzum, M. (1999). User testing in industry: A case study of laboratory, workshop, and field tests. *User Interfaces for All: Proceedings of the 5th ERCIM Workshop* (pp. 59-72).

Hertzum, M., & Jacobsen, N. E. (2001). The evaluator effect: A chilling fact about usability evaluation methods. *International Journal of Human-Computer Interaction, 13*(4), 421-443.

Hertzum, M., Jacobsen, N. E., & Molich, R. (2002). Usability inspections by groups of specialists: Perceived agreement in spite of disparate observations. *Proceedings of CHI'02, Human Factors in Computer Systems Extended Abstracts* (pp. 662-663).

Hilbert, D. M., & Redmiles, D. F. (1998). Agents for collecting application usage data over the Internet. *Proceedings of Agents'98, International Conference on Autonomous Agents* (pp. 149-156).

Hilbert, D. M., & Redmiles, D. F. (1999). Separating the wheat from the chaff in Internet-mediated user feedback expectation-driven event monitoring. *ACM SIGGROUP Bulletin, 20*(1), 35-40.

Hilbert, D. M., & Redmiles, D. F. (2000). Extracting usability information from user interface events. *ACM Computing Surveys, 32*(4), 384-421.

Hix, D., & Hartson, H. R. (1993). *Developing user interfaces.* New York: John Wiley & Sons.

ISO 9241-11. (1998). *Ergonomic requirements for office work with visual display terminals (VDTs): Pt. 11. Guidance on usability.* Geneva, Switzerland: International Organization fo Standardization.

Jacobsen, N. E., Hertzum, M., & John, B. E. (1998). The evaluator effect in usability studies: Problem detection and severity judgments. *Proceedings of the Human Factors and Ergonomics Society 42nd Annual Meeting* (pp. 1336-1340).

John, B. E., & Kieras, D. E. (1996). The GOMS family of user interface analysis techniques: Comparison and contrast. *ACM Transactions on Computer-Human Interaction, 3*, 320-351.

John, B. E., & Marks, S. J. (1997). Tracking the effectiveness of usability evaluation methods. *Behavior and Information Technology, 16*, 188-202.

Kantner, L., Sova, D. H., & Rosenbaum, S. (2003). Alternative methods for field usability research. *Proceedings of SIGDOC'03* (pp. 68-72).

Karat, C. (1994). A comparison of user interface evaluation methods. In J. Nielsen & R. L. Mack (Eds.), *Usability inspection methods* (pp. 203-234). New York: John Wiley & Sons.

Karat, J. (1997). User-centered software evaluation methodologies. In M. Helander, T. K. Landauer, & P. Prabhu (Eds.), *Handbook of human-computer interaction* (2nd ed., pp. 689-704). Amsterdam, The Netherlands: Elsevier Science.

Kjeldskov, J., & Skov, M. B. (2003). Creating realistic laboratory settings: Comparative studies of three think-aloud usability evaluations of a mobile system. *Proceedings of INTERACT'03* (pp. 663-670).

Krauss, F. S. H. (2003). Methodology for remote usability activities: A case study. *IBM Systems Journal, 42*(4), 582-593.

Lewis, C., & Wharton, C. (1997). Cognitive walkthroughs. In M. Helander, T. K. Landauer, & P. Prabhu (Eds.), *Handbook of human-computer interaction* (2nd ed., pp. 717-732). Amsterdam, The Netherlands: Elsevier Science.

Lin, H. X., Choong, Y. -Y., & Salvendy, G. (1997). A proposed index of usability: A method for comparing the relative usability of different software systems. *Behaviour and Information Technology, 16*, 267-278.

Mirel, B. (2000). Product, process, and profit: The politics of usability in a software venture. *ACM Journal of Computer Documentation, 24*(4), 185-203.

Molich, R., Ede, M. R., Kaasgaard, K., & Karyukin, B. (2004). Comparative usability evaluation. *Behaviour and Information Technology, 23*(1), 65-74.

Nielsen, J. (1993). *Usability engineering.* Boston: Academic Press.

Nielsen, J., & Levy, J. (1994). Measuring usability: Preference vs. performance. *Communications of the ACM, 37*(4), 66-75.

Rubin, J. (1994). *Handbook of usability testing: How to plan, design, and conduct effective tests.* New York: John Wiley & Sons.

Scriven, M. (1967). The methodology of evaluation. In R. W. Tyler, R. M. Gagne, & M. Scriven (Eds.), *Perspectives in curriculum evaluation* (pp. 39-83). Skokie, IL: Rand McNally.

Stanton, N. A., & Baber, C. (1996). Factors affecting the selection of methods and techniques prior to conducting a usability evaluation. In P. W. Jordan, B. Thomas, B. A. Weerdmeester, & I. McClelland (Eds.), *Usability evaluation in industry* (pp. 39-48). London: Taylor & Francis.

Vasalou, A., Ng, B., Wiemer-Hastings, P., & Oshlyansky, L. (2004). Human-moderated remote user testing: Protocols and applications. *Eighth ERCIM Workshop, Adjunct Proceedings.* Retrieved from http://ui4all.ics.forth.gr/workshop2004/

Virzi, R. A., Sokolov, J. L., & Karis, D. (1996). Usability problem identification using both low- and high-fidelity prototypes. *Proceedings of CHI'96, Human Factors in Computing Systems* (pp. 236-243).

Vredenberg, K., Mao, J.-Y., Smith, P., & Carey, T. (2002). A survey of user-centered design practice.

Proceedings of CHI'02, Human Factors in Computing Systems (pp. 471-478).

Zhang, Z., Basili, V., & Schneiderman, B. (1999). Perspective-based usability inspection: An empirical validation of efficacy. *Empirical Software Engineering, 4*(1), 43-69.

KEY TERMS

Anchoring: An evaluator bias in analytic methods in which a system is evaluated with respect to users too similar to the evaluator to be representative of the user population.

Method Breadth: Extent to which a usability-evaluation method is capable of detecting all of a system's usability deficiencies.

Method Sensitivity: Extent to which a usability-evaluation method is capable of detecting a particular type of usability deficiency.

Stereotyping: An evaluator bias in analytic methods in which a system is evaluated with respect to a homogenous catchall user not accounting for a wide-enough range of user characteristics.

Usability Barrier: Technical, cognitive, social, or organizational constraint or bias that decreases usability-data quality, consequently hindering the optimal detection of, and response to, a system's usability deficiencies.

Usability Data: Any information used to measure or identify factors affecting the usability of a system being evaluated.

Usability Data Quality: Extent to which usability data efficiently and effectively predict system usability in actual usage, can be efficiently and effectively analyzed, and can be efficiently and effectively reacted to.

Usability Evaluation Method (UEM): Method or technique that can assign values to usability dimensions and/or indicate usability deficiencies in a system.

Usability Data Quality

David R. Danielson
Stanford University, USA

INTRODUCTION

A substantial portion of usability work involves the coordinated collection of data by a team of specialists with varied backgrounds, employing multiple collection methods, and observing users with a wide range of skills, work contexts, goals, and responsibilities. The desired result is an improved system design, and the means to that end are the successful detection of, and reaction to, real deficiencies in system usability that severely impact the quality of experience for a range of users.

In the context of user-centered design processes, valid and reliable data from a representative user sample is simply not enough. High-quality usability data is not just representative of reality. It is useful. It is persuasive in the eyes of the right stakeholders. It results in verifiable improvements to the system for which it is intended to represent a deficiency. The data must be efficiently and effectively translated into development action items with appropriate priority levels, and it must result in effective work products downstream, leading to cost-effective design changes.

The remainder of this article (a) briefly reviews basic usability data collection concepts, (b) examines the dimensions that make up high-quality usability data, and (c) suggests future trends in usability data quality research.

BACKGROUND

Usability data are critical to the successful design of systems intended for human use, and are defined by Hilbert and Redmiles (2000) as any information used to measure or identify factors affecting the usability of a system being evaluated. Such data are collected via *usability evaluation methods* (UEMs), methods or techniques that can assign values to usability dimensions (J. Karat, 1997) and/or indicate usability deficiencies in a system (Hartson, Andre, & Williges, 2003). Usability dimensions are commonly taken to include at least user efficiency, effectiveness, and subjective satisfaction with a system in performing a specified task in a specified context (ISO 9241-11, 1998), and frequently also include system memorability and learnability (Nielsen, 1993a).

Usability data are collected using either *analytic methods*, in which the system is evaluated based on its interface design attributes (typically by a usability expert), or *empirical methods*, in which the system is evaluated based on observed performance in actual use (Hix & Hartson, 1993). In *formative evaluation*, data are collected during the development of a system in order to guide iterative design. In *summative evaluation*, data are collected to evaluate a completed system in use (Scriven, 1967). Usability data have been classified in numerous other models and frameworks frequently focusing on the procedure for producing the data (including the resources expended and the level of the formality of the method), the (relative) physical location of the people and artifacts involved, the nature and fidelity of the artifact being evaluated, and the goal of the collection process.

DIMENSIONS OF USABILITY DATA QUALITY

Usability-data quality refers to the extent to which usability data efficiently and effectively (a) predict system usability in actual usage (validity, reliability, representativeness, and completeness), (b) can be analyzed (communicative effectiveness and efficiency, and analyst estimates of severity), and (c) can be reacted to (downstream utility, impact, and cost effectiveness). This section discusses the dimensions of usability data quality and their assessment.

Validity

High-quality usability data are predictive of a real deficiency in one or more usability attributes for a given system. End-user behavior and comments may be perfectly unbiased or unaffected by the collection process, yet still lack validity from the perspective of usability science. Strict performance measures (such as time on task) may be viewed as lacking validity primarily because they often fail to, on their own, demonstrate an underlying problem (Gediga, Hamborg, & Düntsch, 2002). Qualitative data more often do point directly to a deficiency, but if a user comments on a system feature that will never be used, for example, the comment may truly reflect the user's attitudes but nonetheless lack validity.

Verifying usability data validity lies in comparing the data's predicted problems to the actual system performance in use (John & Marks, 1997; Nielsen & Phillips, 1993). In practice, assessing validity is nontrivial for three fundamental reasons. First, there is not widespread agreement on how to operationalize ultimate usability criteria into actual criteria (Gray & Salzman, 1998; Hartson et al., 2003); that is, agreeing on standard measures (and measurement procedures) for the underlying dimensions of usability itself is a long-standing difficulty. Second, observing the system in use and recording deficiencies is itself a usability data collection process, and thus at best the actual criterion is subject to possible validity concerns of its own. While these first two problems are by no means unique to usability research, they illustrate the difficulty in assessing usability data quality without a widely agreed upon method for identifying what will be accepted as the system's real deficiencies. Finally, individual pieces of usability data are often difficult to translate into underlying problems, and this step is necessary if validity is to be assessed.

To make the problem slightly more tractable, researchers have by and large elected to evaluate validity using usability testing as a benchmark for comparison, as it is assumed to most closely reflect system performance in use (Cuomo & Bowen, 1994; Desurvire, 1994; Jacobsen, Hertzum, & John, 1998). There are of course potential problems with this approach as usability testing has at least ecological validity concerns (Thomas & Kellogg, 1989). Indeed, this problem generally makes the literature comparing UEM effectiveness difficult to interpret (Gediga et al., 2002; Gray & Salzman, 1998). Ideally, a standard method is applied to assessing live system performance, producing a usability problem set. Validity is then assessed by comparing the problem set produced by a UEM to the standard set (Sears, 1997).

Reliability and Representativeness

High-quality usability data not only indicate real problems, but indicate problems that will be repeatedly encountered by individual users (reliable) and by a wide range of users (representative). As with many disciplines, data collected for usability purposes vary in the extent to which the repeated exposure to a problem is a good predictor of validity. While subjective satisfaction ratings that vary one day to the next put validity in question, encountering only occasional difficulty in executing a system action or completing a task, for example, does not since user errors indicating real interface problems commonly vary in frequency of occurrence. Unlike research in many other disciplines, representativeness across participants is not simply a question to be investigated, but a contributor to problem importance and therefore a dimension of data quality.

Measuring reliability and representativeness is a matter of identifying the recurrence of specific problems (Jeffries, Miller, Wharton, & Uyeda, 1991). Such measurement is nontrivial because problem reports may differ in verbiage but still indicate the same underlying problem, or conversely may be similar in their qualitative descriptions but indicate different deficiencies (Andre, Hartson, Belz, & McCreary, 2001; Hartson et al., 2003).

Completeness

High-quality usability data represent usability problems in their entirety. One of the critical difficulties in analyzing pure behavioral data is their lack of contextual information about the user's current task, attention level, and cognitive processes while a problem takes place (Hilbert & Redmiles, 1999); another problem is their flood of extraneous data that are not useful in evaluating the deficiency (Hartson & Castillo, 1998). Ideal usability data predict a

system deficiency without requiring analysts to fill in the blanks and without noise. Like validity, completeness cannot be properly assessed without standard problem descriptions for comparison.

Researchers have instead shown interest in assessing analogous concepts at the UEM level, namely, the thoroughness of a method (the extent to which it uncovers all known usability problems, often with usability testing as the benchmark; Sears, 1997) and false positives, noting predicted problems that lack validity (Gediga et al., 2002).

Communicative Effectiveness and Efficiency

High-quality usability data can be easily translated by usability analysts into problem descriptions that faithfully represent the underlying deficiencies indicated by the data. The practical importance of completeness (and lack of noise) is clear to analysts who must decipher large amounts of data within tight development cycles. Usability data often suggest multiple possible design responses and vary in the amount of analysis time required to produce problem descriptions (Preece et al., 1994), the amount of detail and surrounding context made available by the collection method or tool (Hartson & Castillo, 1998), and the extent to which they refer to causes or effects of a deficiency (Boren & Ramey, 2000; van Welie, van der Veer, & Eliëns, 1999). Each of these variables contributes to a usability analyst's ability to quickly and accurately translate the data into problem descriptions and ultimately potential solutions. Empirical usability-data collection is largely an attempt to invoke feedback more specific than "looks good to me," and this can be particularly difficult with remote methods in which the opportunity for follow-up may not exist.

Severity

High-quality usability data indicate deficiencies that are not simply annoyances, but seriously impact the quality of users' experiences with the system and ability to carry out their work. In many cases, data collection processes are expected to ignore the noisy, less serious (even if real) problems to maximize cost effectiveness. Severity metrics are in some sense predeployment predictions of the data's likely effects on usability as well as attempts to motivate the prioritization of development resources in implementing design changes. Following Nielsen (1994), such metrics typically incorporate reliability and representativeness with the predicted impact of the problem (i.e., whether a work-around exists and can be easily discovered, or if the problem will be a "showstopper" and prevent task completion or further use of the system). Thus, they often combine objective measures with an analyst's subjective assessment.

Downstream Utility

High-quality usability data persuades developers and other stakeholders in the product development cycle to implement design changes. Downstream utility (or persuasiveness) is conceptualized in terms of the likelihood of usability data contributing to a change in a system's interface (John & Marks, 1997; Sawyer, Flanders, & Wixon, 1996). Because the tracking of usability data in real development cycles and the assessment of that data's influence on design activities are difficult tasks, little research has investigated persuasiveness at the granularity of individual user comments or problem reports (Ebling & John, 2000). However, the focus on UEMs for this particular dimension of data quality is also reflective of two things. First, it recognizes that within user-centered design processes, data come in packages that often center around individual usability tests or heuristic reviews; that is, work products often simply report the results of individual tests that apply a single method. Second, it recognizes that stakeholders downstream have different levels of confidence in usability work based on the method. Severe usability problems sandwiched between low-quality data or appearing in reports for methods that have not gained stakeholder respect may be less persuasive by association.

Impact

High-quality usability data result in design changes that improve system usability. Data that is of high quality along the dimensions already discussed, prior to deployment, are typically assumed to ensure impact. While this assumption is imperfect (what made a problem severe or who a system's representative users were during the development cycle, for example, may change by the time the system is in live use), it is in practice an effective way of ensuring usability improvements. John and Marks (1997) first formalized the "design-change effectiveness" of usability data by tracking the data's influence on design changes and subsequently observing the effects of these implemented changes in a development cycle, using a live system usability test as the benchmark for comparison. Their work remains a rare attempt to observe data impact.

Usability-data impact is of course not restricted to individual process cycles and products, and is not always geared toward correcting specific problems. Performance measures, for example, are often explicitly meant to determine general pain points in an interface to guide subsequent data collection and generally drive usability resource allocation (Dillon & Morris, 1999; Gray & Salzman, 1998). Such data can also guide selection amongst design alternatives (Nielsen & Phillips, 1993). Usability data can result in long-term guidelines, standards, and design patterns applied to subsequent products (Henninger, 2001), validate system improvements leading to increased usability funding and visibility, increase organizational acceptance of usability processes (C. Karat, 1994), indicate needed change in these processes, and open the eyes of developers and other stakeholders (particularly empirical data such as video from usability tests) to end-user needs and behaviors (Englefield, 2003). Finally, as user-centered design processes are iterative and involve attempts to improve prototypes as much as possible in each iteration, quantitative data can be instrumental in indicating when to stop iterating to maximize cost effectiveness.

Cost (Effectiveness)

High-quality usability data achieve impact at relatively low cost. Costs include the resources necessary to collect, analyze, and react to the data by implementing design changes. In more constrained experiments, time is often used as a simple proxy for cost, comparing the quality of the collected data to the time taken in collection (Englefield, 2003).

To be fair, while comparing usability improvements to cost is useful, it is in some sense meaningless outside the perspective of usability. For this reason, researchers have attempted to analyze the return on investment (ROI) of usability work, looking at broader impacts of usability data. Usability engineering methods have been argued to reduce development time and cost, reduce call center and support costs due to decreased usability deficiencies, reduce system training costs (Nielsen, 1993b), increase the customer base, retain customers due to satisfaction with the system, and ultimately increase product sales (Bias & Mayhew, 1994).

These effects appropriately do not refer directly to the usability of the system; the target of ROI analysis is the entity incurring the cost. Benefits to end users are relevant only insofar as they result in benefits to those investing in usability-data collection. As a result, the connection between system usability and ROI depends in part on the type of system being evaluated. It is commonly noted that in e-commerce, the effect of usability and buying behavior is relatively straightforward since a showstopper in usability is necessarily a showstopper in completing a transaction. Similarly, the usability of internal systems such as intranets leads to increased employee productivity and satisfaction for the company footing the usability bill. But in contexts where buying decisions are not made by end users and instead by customers who have varying levels of engagement with end users, the connection, and the ROI of usability-data collection, is less clear-cut.

Even in instances in which the end user makes the buying decision, the typical context of those decisions likely impacts the relationship between usability and ROI. Lesk (1998) gives the example of end users making buying decisions at trade shows (or computer stores), in which case actual system usability may take a backseat to the perceived usability achieved by a quick interaction with a demo or display unit. Generally, the connection between usability and system acceptance is not entirely clear. Dillon and Morris (1999) review system acceptance models indicating perceived usefulness to be a more powerful predictor, but as one might intuitively predict, usability may over time influence continued use; that is, actual usability begins to impact the user perceptions that under many circumstances powerfully influence system acceptance.

FUTURE TRENDS

The range of factors influencing usability data's likelihood of influencing system design, in conjunction with concerns orthogonal to system usability, leave many challenges open for usability research. This section identifies a few important research areas for usability-data-quality assessment.

Customization Benefits and Unforeseen Pitfalls

Systems that are customized (whether at the organizational or individual level) frequently introduce usability benefits to end users. Organizations may configure systems to most efficiently support their business processes, and end users may find increased satisfaction with a system personalized to fit their work styles and tastes. However, customization can clearly open a can of worms for usability processes since ill-guided customization introduces usability deficiencies. In some sense, highly customizable systems have the potential to take system usability out of the hands of user-centered design. Assessing these systems requires accounting for the extent to which they are likely to prevent the introduction of usability deficiencies through customization. The more flexible the customization, the less trivial this assessment becomes as it requires performing formative evaluations of a system to which customers and end users will employ frequent and unforeseen adjustments.

From Methods to Individual Data

The focus on usability evaluation methods as the unit of analysis for several data-quality dimensions leaves open a number of interesting questions. While a piece of data's collection method is likely a critical attribute for data quality (as mentioned for downstream utility in particular), more fine-grained analyses investigating how user comments and problem reports make their way through development cycles and influence system design is an open area for future usability research.

Usability Process Assessments

In contrast to investigating the influence of individual pieces of usability data, arguably the most critical area for usability data quality research lies in identifying optimal usability processes. While many studies have attempted to address the relative merits of individual collection methods and the features of these methods (such as the optimal number of participants for empirical testing), little empirical work has attempted to piece together the most effective aggregations of these methods for collecting usability data of maximal scope and the optimal coordination of these methods. While a good deal of conventional wisdom exists regarding such coordination, usability research can offer to validate and refine this wisdom.

CONCLUSION

The quality of usability data depends on how well they predict real (and severe) system deficiencies experienced by a wide range of users, how easily and successfully they can be analyzed by usability experts, and how easily they can be reacted to in producing an improved system design. Usability research attempts to accurately assess the quality of usability data by observing their effects throughout product development cycles. Difficulties in assessing usability data quality arise from numerous sources, including often unforeseeable mismatches between the assessment and actual system-usage environments.

REFERENCES

Andre, T. S., Hartson, H. R., Belz, S. M., & McCreary, F. A. (2001). The user action framework: A reliable foundation for usability engineering support tools. *International Journal of Human-Computer Studies, 54*, 107-136.

Bias, R. G., & Mayhew, D. J. (Eds.). (1994). *Cost justifying usability.* Boston: Academic Press.

Boren, M. T., & Ramey, J. (2000). Thinking aloud: Reconciling theory and practice. *IEEE Transactions on Professional Communication, 43*(3), 261-278.

Cuomo, D. L., & Bowen, C. D. (1994). Understanding usability issues addressed by three user-system interface evaluation techniques. *Interacting with Computers, 6*, 86-108.

Desurvire, H. W. (1994). Faster, cheaper!! Are usability inspection methods as effective as empirical testing? In J. Nielsen & R. L. Mack (Eds.), *Usability inspection methods* (pp. 173-202). New York: John Wiley & Sons.

Dillon, A., & Morris, M. (1999). Power, perception, and performance: From usability engineering to technology acceptance with the P3 model of user response. *Proceedings of the Human Factors and Ergonomics Society 42nd Annual Meeting* (pp. 1017-1021).

Ebling, M. R., & John, B. E. (2000). On the contributions of different empirical data in usability testing. *Proceedings of DIS'00, Design of Interactive Systems* (pp. 289-296).

Englefield, P. (2003). *A pragmatic framework for selecting empirical or inspection methods to evaluate usability*. IBM. Retrieved from http://www.ibm.com/easy/

Gediga, G., Hamborg, K.-C., & Düntsch, I. (2002). Evaluation of software systems. In A. Kent & J. G. Williams (Eds.), *Encyclopedia of library and information science* (Vol. 72, pp. 166-192). New York: Marcel Dekker.

Gray, W. D., & Salzman, M. C. (1998). Damaged merchandise? A review of experiments that compare usability evaluation methods. *Human-Computer Interaction, 13*, 203-261.

Hartson, H. R., Andre, T. S., & Williges, R. C. (2003). Criteria for evaluating usability evaluation methods. *International Journal of Human-Computer Interaction, 15*(1), 145-181.

Hartson, H. R., & Castillo, J. C. (1998). Remote evaluation for post-deployment usability improvement. *Proceedings of AVI'98, Advanced Visual Interfaces* (pp. 22-29).

Henninger, S. (2001). An organizational learning method for applying usability guidelines and patterns. *Proceedings of EHCI'01, Engineering for Human-Computer Interaction* (pp. 249-262).

Hilbert, D. M., & Redmiles, D. F. (1999). Separating the wheat from the chaff in Internet-mediated user feedback expectation-driven event monitoring. *ACM SIGGROUP Bulletin* (vol. 20, no. 1, pp. 35-40).

Hilbert, D. M., & Redmiles, D. F. (2000). Extracting usability information from user interface events. *ACM Computing Surveys, 32*(4), 384-421.

Hix, D., & Hartson, H. R. (1993). *Developing user interfaces*. New York: John Wiley & Sons.

ISO 9241-11. (1998). Ergonomic requirements for office work with visual display terminals (VDTs): Pt. 11. Guidance on usability. *International Organization for Standardization,* Geneva, Switzerland.

Jacobsen, N. E., Hertzum, M., & John, B. E. (1998). The evaluator effect in usability studies: Problem detection and severity judgments. *Proceedings of the Human Factors and Ergonomics Society 42nd Annual Meeting* (pp. 1336-1340).

Jeffries, R., Miller, J. R., Wharton, C., & Uyeda, K. M. (1991). User interface evaluation in the real world: A comparison of four techniques. *Proceedings of CHI'91, Human Factors in Computing Systems* (pp. 119-124).

John, B. E., & Marks, S. J. (1997). Tracking the effectiveness of usability evaluation methods. *Behavior and Information Technology, 16*, 188-202.

Karat, C. (1994). A comparison of user interface evaluation methods. In J. Nielsen & R. L. Mack (Eds.), *Usability inspection methods* (pp. 203-234). New York: John Wiley & Sons.

Karat, J. (1997). User-centered software evaluation methodologies. In M. Helander, T. K. Landauer, & P. Prabhu (Eds.), *Handbook of human-computer interaction* (2nd ed., pp. 689-704). Amsterdam, The Netherlands: Elsevier Science.

Lesk, M. (1998). *Practical digital libraries: Books, bytes, and bucks*. San Francisco: Morgan Kaufmann.

Nielsen, J. (1993a). *Usability engineering.* Boston: Academic Press.

Nielsen, J. (1993b). Is usability engineering really worth it? *IEEE Software, 10*(6), 90-92.

Nielsen, J. (1994). Heuristic evaluation. In J. Nielsen & R. L. Mack (Eds.), *Usability inspection methods* (pp. 59-83). New York: John Wiley & Sons.

Nielsen, J., & Phillips, V. L. (1993). Estimating the relative usability of two interfaces: Heuristic, formal, and empirical methods compared. *Proceedings of INTERCHI'93*, 214-221.

Preece, J., Rogers, Y., Sharp, H., Benyon, D., Holland, S., & Carey, T. (1994). *Human-computer interaction*. Wokingham, UK: Addison-Wesley.

Sawyer, P., Flanders, A., & Wixon, D. (1996). Making a difference: The impact of inspections. *Proceedings of CHI'96, Human Factors in Computing Systems* (pp. 376-382).

Scriven, M. (1967). The methodology of evaluation. In R. W. Tyler, R. M. Gagne, & M. Scriven (Eds.), *Perspectives in curriculum evaluation* (pp. 39-83). Skokie, IL: Rand McNally.

Sears, A. (1997). Heuristic walkthroughs: Finding the problems without the noise. *International Journal of Human-Computer Interaction, 9*(3), 213-224.

Thomas, J. C., & Kellogg, W. A. (1989). Minimizing ecological gaps in interface design. *IEEE Software, 6*(1), 78-86.

Van Welie, M., van der Veer, G. C., & Eliëns, A. (1999). Breaking down usability. *Proceedings of INTERACT'99* (pp. 613-620).

KEY TERMS

Analytic Method: Method in which a system is evaluated based on its interface design attributes (typically by a usability expert).

Empirical Method: Method in which a system is evaluated based on observed performance in actual use.

Formative Evaluation: The collection of usability data during the development of a system in order to guide iterative design.

Summative Evaluation: The collection of usability data to evaluate a completed system in use.

Usability Data: Any information used to measure or identify factors affecting the usability of a system being evaluated.

Usability Data Quality: Extent to which usability data efficiently and effectively predict system usability in actual usage, can be efficiently and effectively analyzed, and can be efficiently and effectively reacted to.

Usability Evaluation Method (UEM): Method or technique that can assign values to usability dimensions and/or indicate usability deficiencies in a system.

The Use and Evolution of Affordance in HCI

Georgios S. Christou
Cyprus College, Cyprus

INTRODUCTION

The term *affordance* was coined by Gibson (1977, 1979) to define properties of objects that allow an actor to act upon them. Norman (1988) expanded on this concept and presented the concepts of real and perceptual affordances in his book *The Psychology of Everyday Things*. Norman was essentially the first to present the concept of affordance to the field of human-computer interaction (HCI).

Since then, affordance as a term has been used by many designers and researchers. But as Norman (1999) explained, many of the uses of the term are vague or unclear, which prompted the writing of his 1999 article in the *Interactions* periodical. In fact, there have been many publications that try to elucidate the term (see Hartson, 2003; McGrenere & Ho, 2000).

This article will try to provide a brief overview of the term and its many subclasses. It will try to give the reader a clear idea about what affordance is and how the concept can be used to allow designers and researchers to create better user interfaces and better interaction devices. The article however, does not try to clear up any ambiguities in the usage of the term in the literature or present a new way of viewing affordance. Rather, it tries to provide a short overview of the literature around affordance and guide the reader to a correct understanding of how to use affordance in HCI.

BACKGROUND

This section presents the evolution of the concept of affordance. It presents the creation of the term by Gibson (1977, 1979), and the way that affordance was incorporated into HCI.

Gibson's Affordance

As mentioned in the introduction, Gibson (1977, 1979) was the one who coined the term affordance to refer to the actionable properties between the world and an actor (whatever that actor may be; Gibson as cited in Norman, 1999). Gibson did not create the term to refer to any property that may be observable by the actor. Rather, he referred to all the properties that allow the actor to manipulate the world, be they perceivable or not. Thus, in Gibson's view, an affordance is just a characteristic of the environment that happens to allow an actor to act upon the environment. In this view, saying that a designer has added an affordance to a device or an interface does not immediately mean that the device or the interface has becosme more usable, or that the user would be able to sense the affordance in any way that would help him or her understand the usage of that device or interface. In fact, in Gibson's definition, an affordance is not there to be perceived. The affordance just exists and it is up to the actor to discover the functionality that is offered by the affordance. It is just a feature of the environment.

Norman's Affordance

Norman (1988) took the term affordance from Gibson (1977, 1979), and in his book *The Psychology of Everyday Things*, he elaborated upon it, creating something quite different from the original definition. Norman did not change the original term. Rather, he introduced the concept of perceived affordance, which defines the clues that a device or user interface gives to the user as to the functionality of an object. He also distinguished it from Gibson's affordance, which he named real affordance. We will mention probably the most used example of affordance in HCI to clarify the difference between a real affordance and a perceived affordance. Consider a door that opens when pushed having a flat plate that takes the place of the door handle (Figure 1b). The design of the door handle gives out the clue that the door is not supposed to be pulled since there is no handle that the actor can grab in order to pull the door. Conversely, a door handle that can be grabbed (Figure 1a) gives out the clue that the door opens when pulled. However, as Norman (1988) points out, this convention is not always followed, resulting in people thinking that they cannot figure out how to open a door whereas the problem lies in bad design and bad use of a perceived affordance. The difference between the real affordance, or the affordance as defined by Gibson, and the perceived affordance in Norman's definition is that the door affords to be opened in some way but the perceived affordance that the flat panel gives out is that the door can be opened by pushing on the panel.

Norman (1988) concludes that well-designed artifacts should have perceived affordances that give out the correct clues as to the artifacts' usage and functionality.

Gaver's Affordance

Gaver (1991) wrote an article in which he also creates a definition of affordance, but he breaks affordance down into four different categories. Gaver defines perceptible affordance, false affordance, correct rejections, and hidden affordance (Figure 2). Perceptible affordance is the affordance for which there is perceptual information for the actor to perceive. This type of affordance would fall under Norman's (1988) perceived-affordance definition. Conversely, if there is information that suggests that an affordance is there when there is none, then that is a false affordance. A hidden affordance is an affordance for which no perceptual information exists. Finally, a correct rejection is the case when there is no perceptual information and no affordance.

In Gaver's (1991) terms, affordance is the existence of a special configuration of properties so that:

> *physical attributes of the thing to be acted upon are compatible with those of an actor, that information about those attributes is available in a form compatible with a perceptual system, and (implicitly) that these attributes and the action they make possible are relevant to a culture and a perceiver.* (Gaver, 1991, p. 81)

In fact, Gaver (1991) united the two concepts of real and perceived affordance, and named the system of the property of an object and the ability of that property to be perceived as affordance.

Hartson's Affordance

Hartson (2003) used the concept of affordance to create the User Action Framework (UAF). He used the concept by basing it on Norman's (1988) definition, but also redefining it to make the distinction

Figure 1. Two door handles, one (a) very confusing as to its usage, and one (b) which gives clues as to its usage

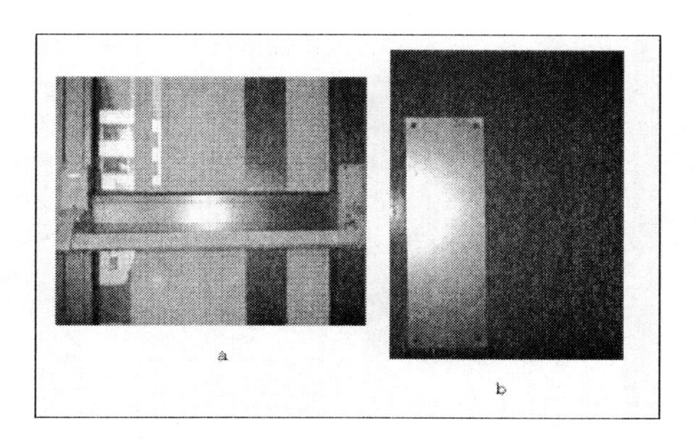

Figure 2. Separating types of affordance from information available about them (Gaver, 1991)

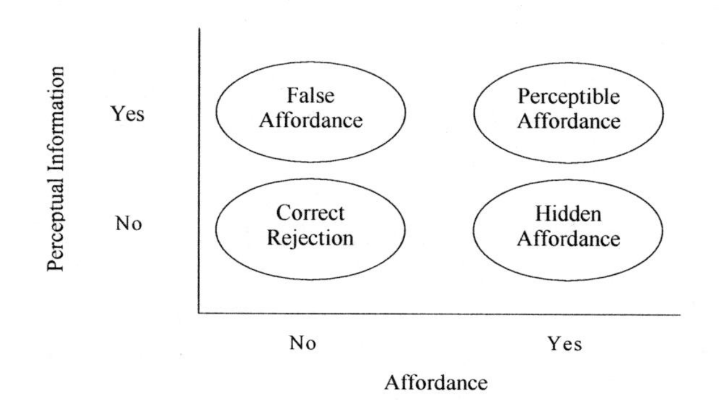

between each type of affordance that can be encountered clearer. He refers to four different types of affordance: physical, cognitive, sensory, and functional. He defines physical affordance as a feature of the artifact that allows the actor to do something with it. A cognitive affordance is the information that allows the actor to realize that the physical affordance is there. A sensory affordance is the information that the actor gets before it gets processed at a cognitive level, and a functional affordance is the usefulness that the physical affordance gives to the actor. The metal plate on the door from the previous example can be used to elucidate the differences between each affordance type that Hartson proposes. The physical affordance of the plate is the feature that allows the placement of the hand of the user on the door so that the user can open the door. The cognitive affordance is the combination of information from the user's knowledge and the appearance of the plate that allows the user to realize whether the door is opened by pulling or pushing. If one assumes that the metal plate has *Push* engraved on it, then the clarity of the lettering and the size and shape of the letters that allow the user to clearly make them out is the sensory affordance. Finally, the functional affordance is the placement of the plate at the correct position on the door as to allow for the easiest opening of the door if the user pushes on the metal plate.

Hartson (2003) goes on to propose the UAF, which is a framework for designing systems and artifacts. The framework is based on the four types of affordance that he proposes. For more information on the UAF, the reader is referred to Andre, Belz, McCreary, and Hartson (2000), Andre, Hartson, Belz, and McCreary (2001), and Hartson, Andre, Williges, and Van Rens (1999).

DISCUSSION

Affordance is perhaps one of the most exciting concepts in HCI. The introduction of the concept by Norman (1988) created a big stir and, consequently, it created a lot of discussion because of its inconsistent usage. Many, along with Norman, claim that affordance as was presented in the POET book was not understood correctly by the HCI community, something that triggered a lot of discussion around the concept and a lot of literature trying to elucidate the meaning of this term.

People like Gaver (1991), Hartson (2003), and McGrenere and Ho (2000) have also provided frameworks or theories that are based on affordance that can help designers design better user interfaces and interaction devices. For example, a designer could create a user interface and make sure that the buttons are clearly labeled so that the user can easily read the labels (sensory affordance) to understand what the button does (cognitive affordance) in order to use it correctly (functional affordance; Hartson). This example uses the types of affordance that Hartson proposes and shows how one could think about the different types of affordance in Hartson's definition to create a better interface.

The concept of affordance is indeed useful in HCI because, at the very least, it forces the designer to think about the information that he or she is giving to the user by the very design of the user interface or interaction device.

Another short example that uses Norman's (1988) definition of affordance may elucidate the usage of the concept in interaction-device design. Suppose the design of a keyboard for a palm device (much like the one on the Handspring Treo devices). The keys on this keyboard are restricted by the size of the device, which means the physical affordance of a user pressing the buttons is hindered. However, the designer who thinks about this in the design stage will have already taken this affordance into account and may allow for slightly bigger buttons by designing the device with slightly bigger dimensions at the bottom where the keyboard lies. Other examples in interaction-device design that take account of affordance can be found in the tangible-user-interfaces (Ishii & Ullmer, 1997) literature.

When thinking about affordance, sometimes it does not matter which affordance definition one might use as long as the definition allows for the design to become more usable. In both cases above, one can see that by taking into account the concept of affordance, the design of a device becomes a little more usable and maybe even a little more foolproof.

FUTURE TRENDS

Affordance is an always-evolving concept. Researchers try to elucidate the term and create frameworks that clarify its usage to practitioners. There are also attempts to create formal mathematical notations for describing affordance (Steedman, 2002). More and more practitioners are trying to apply the affordance design concepts to non-WIMP (windows, icons, menus, pointer) interaction styles, like tangible user interfaces, virtual-reality interfaces, and so forth, and there have been some great results, like the Bottles interface (Ishii, Mazalek, & Lee, 2001) and Sound Canvas (Cheung, 2002). There are many more avenues for research in the concept of affordance. As the literature shows, affordance research may lead to the understanding of other concepts that should also be considered in the design of interfaces (see Overbeeke & Wensveen, 2003).

Another future direction for research in affordance is the production of case studies that show how affordance can be used in improving interaction design (Hartson, 2003).

CONCLUSION

This article presented a brief overview of the usage of the concept of affordance in HCI. A brief history of the creation of the term was provided, along with some of the major contributions to the evolution and clarification of the meaning of the concept. It briefly discussed how the concept could be used by designers in order to create more usable user interfaces and interaction devices. It also mentioned some of the frameworks that have been created to facilitate the design process based on this concept.

By reviewing the most prevalent definitions, a foundation was provided upon which one could build a solid understanding of what affordance is. When used correctly, affordance can provide the user of a user interface or interaction device clues as to how to use it correctly even if the user has little training with the interface or device.

Finally, some examples were provided to demonstrate how a designer could incorporate thinking about affordance at the design stage of an interface or interaction device so that the artifact created would be made more usable in some way.

REFERENCES

Andre, T. S., Belz, S. M., McCreary, F. A., & Hartson, H. R. (2000). Testing a framework for reliable classification of usability problems. *Proceedings of the Human Factors and Ergonomics Society 44th Annual Meeting* (pp. 573-577).

Andre, T., Hartson, H. R., Belz, S., & McCreary, F. (2001). The user action framework: A reliable foundation for usability engineering support tools. *International Journal of Human-Computer Studies, 54*(1), 107-136.

Cheung, P. (2002). Designing sound canvas: The role of expectation and discrimination. *CHI '02 Extended Abstracts on Human Factors in Computing Systems* (pp. 848-849).

Gaver, W. W. (1991). Technology affordances. *Proceedings of the SIGCHI Conference on Human Factors in Computing Systems* (pp. 79-84).

Gibson, J. J. (1977). The theory of affordances. In R. E. Shaw & J. Bransford (Eds.), *Perceiving, acting, and knowing,* (pp. 127-143). Hillsdale, NJ: Lawrence Erlbaum and Associates.

Gibson, J. J. (1979). *The ecological approach to visual perception.* Boston: Houghton Mifflin Co.

Hartson, R. (2003). Cognitive, physical, sensory and functional affordances in interaction design. *Behaviour & Information Technology, 22*(5), 315-338.

Hartson, H. R., Andre, T. S., Williges, R. C., & Van Rens, L. (1999). The user action framework: A theory-based foundation for inspection and classification of usability problems. *Human Computer Interaction: Ergonomics and User Interfaces (Proceedings of the 8th International Conference on Human Computer Interaction, HCI International '99)* (pp. 1058-1062).

Ishii, H., Mazalek, A., & Lee, J. (2001). Bottles as a minimal interface to access digital information. *CHI '01 Extended Abstracts on Human Factors in Computing Systems* (pp. 187-188).

Ishii, H., & Ullmer, B. (1997). Tangible bits: Towards seamless interfaces between people, bits and atoms. *Proceedings of the SIGCHI Conference on Human Factors and Computing Systems* (pp. 234-241).

McGrenere, J., & Ho, W. (2000). Affordances: Clarifying and evolving a concept. *Proceedings of Graphics Interface 2000*, Montreal, Canada (pp. 179-186).

Norman, D. A. (1988). *The psychology of everyday things.* New York: Basic Books.

Norman, D. A. (1990). *The design of everyday things.* New York: Doubleday.

Norman, D. A. (1999). Affordance, conventions and design. *Interactions, 6*(3), 38-42.

Overbeeke, K., & Wensveen, S. (2003). From perception to experience, from affordances to irresistibles. *Proceedings of the 2003 International Conference on Designing Pleasurable Products and Interfaces* (pp. 92-97).

Steedman, M. (2002). Formalizing affordance. *Proceedings of the 24th Annual Meeting of the Cognitive Science Society* (pp. 834-839).

KEY TERMS

Cognitive Affordance: According to Hartson (2003), this type of affordance is the combination of the information that allows the user to understand what the purpose of the artifact that has this affordance is.

False Affordance: Gaver (1991) used this term to refer to information that makes the actor think that there is an affordance when in fact there is none.

Functional Affordance: The last of Hartson's (2003) definitions, this type of affordance is the feature of the artifact that allows the actor to actually accomplish the work that the artifact is supposed to perform (the usefulness of the artifact).

Hidden Affordance: Gaver (1991) used this term to represent the affordance of an artifact that the user cannot perceive. Thus, while the affordance is there, there is no perceptible information for the actor to realize that the affordance is there.

Perceived or Perceptible Affordance: The term perceived affordance was created by Norman (1988), whereas perceptible affordance was coined by Gaver (1991). They both refer to a property of an artifact that provides observable cognitive clues as to its usage and function by an actor.

Real or Physical Affordance: The term affordance was first proposed by Gibson (1977). The term real affordance was proposed by Norman (1988), and the term physical affordance was proposed by Hartson (2003). They all refer to the same definition proposed by Gibson, which is that affordance is an actionable property between the world and an actor. An affordance does not have to be perceptible by the actor.

Sensory Affordance: Again, one of Hartson's (2003) definitions, this affordance is a feature of an artifact that helps the user sense something about the artifact.

U

Use of the Secondary Task Technique for Tracking User Attention

Robert S. Owen
Texas A&M University-Texarkana, USA

INTRODUCTION

The notion that the human information processing system has a limit in resource capacity has been used for over 100 years as the basis for the investigation of a variety of constructs and processes, such as mental workload, mental effort, attention, elaboration, information overload, and such. The *dual task* or *secondary task technique* presumes that the consumption of processing capacity by one task will leave less capacity available for the processing of a second concurrent task. When both tasks attempt to consume more capacity than is available, the performance of one or both tasks must suffer, and this will presumably result in the observation of degraded task performance.

Consider, for example, the amount of mental effort devoted to solving a difficult arithmetic problem. If a person is asked to tap a pattern with a finger while solving the problem, we might be able to discover the more difficult parts of the problem solving process by observing changes in the performance of the secondary task of finger tapping. While a participant is reading a chapter of text in a book or on a Web browser, we might be able to use this same technique to find the more interesting, involving, or confusing passages of the text. Many implementations of the secondary task technique have been used for more than a century, such as the maintenance of hand pressure (Lechner, Bradbury, & Bradley, 1998; Welch, 1898), the maintenance of finger tapping patterns (Friedman, Polson, & Dafoe, 1988; Jastrow, 1892; Kantowitz & Knight, 1976), the performance of mental arithmetic (Bahrick, Noble, & Fitts, 1954; Wogalter & Usher, 1999), and the speed of reaction time to an occasional flash of light, a beep, or a clicking sound (e.g., Bourdin, Teasdale, & Nourgier, 1998; Owen, Lord, & Cooper, 1995; Posener & Bois, 1971).

In using the secondary task technique, the participant is asked to perform a secondary task, such as tapping a finger in a pattern, while performing the primary task of interest. By tracking changes in secondary task performance (e.g., observing erratic finger tapping), we can track changes in processing resources being consumed by the primary task. This technique has been used in a wide variety of disciplines and situations. It has been used in advertising to study the effects of more or less suspenseful parts of a TV program on commercials (Owen et al., 1995) and in studying the effects of time-compressed audio commercials (Moore, Hausknecht, & Thamodaran, 1986). It has been used in sports to detect attention demands during horseshoe pitching (Prezuhy & Etnier, 2001) and rock climbing (Bourdin et al., 1998), while others have used it to study attention associated with posture control in patients who are older or suffering from brain disease (e.g., Maylor & Wing, 1996; Muller, Redfern, Furman, & Jennings, 2004). Murray, Holland, and Beason (1998) used a dual task study to detect the attention demands of speaking in people who suffer from aphasia after a stroke. Others have used the secondary task technique to study the attention demands of automobile driving (e.g., Baron & Kalsher, 1998), including the effects of distractions such as mobile telephones (Patten, Kircher, Ostlund, & Nilsson, 2004) and the potential of a fragrance to improve alertness (Schieber, Werner, & Larsen, 2000). Koukounas and McCabe (2001) and Koukounas and Over (1999) have used it to study the allocation of attention resources during sexual arousal.

The notion of decreased secondary task performance due to a limited-capacity processing system is not simply a laboratory curiosity. Consider, for example, the crash of a Jetstream 3101 airplane as it was approaching for landing, killing all on board. The airplane had deviated slightly from its course, and shortly after, the flight crew declared an emergency to the approach controller, attributing engine failure as the cause. The U.S. National Transportation Safety Board (NTSB, 2000), however, concluded that the airplane simply ran out of fuel and that the crew had not considered this possibility. The airplane's performance capabilities and simulator tests suggested that the flight crew still should have been able to land the airplane with the first engine out, with the second engine erratic, or with both engines out. The NTSB report surmised that the failure of the first engine could have caused the pilots to "fixate on instruments such as the altitude indicator and airspeed indicator and to allow the course heading to wander" (NTSB).

In the same way, we can observe erratic or degraded performance on tasks that are performed concurrently with other ordinary, everyday tasks, such as watching TV, reading from a book or computer screen, or driving a car. If we can observe erratic or degraded performance on a secondary task, then we can presume that the primary task of watching TV, reading, or browsing a Web site is consuming quite a lot of the person's mental processing capacity. There are three conclusions that we can draw from such observations.

1. We need to consider this limited capacity of the human processing system and the potential for dysfunctional performance when designing human-machine systems such as aircraft, automobiles, ordinary and everyday office computer applications, Web sites, and so forth.
2. We can use this observation of degraded performance on concurrent tasks as a way to identify human overload or failure points in a human-machine system.
3. We can use this observation of degraded performance as a measure of a variety of human mental processes, such as attention, mental effort, information overload, and such.

The first issue is the motivation behind this article. The remainder of this article, however, will focus on the latter two issues. First will be a brief theoretical discussion on how interference or dysfunctional mental processing performance occurs from a black-box perspective of the system. This will be followed by a discussion of how this interference can be observed with the so-called dual task or secondary task technique, used in the measure of mental overload, mental attention, mental effort, and such.

BACKGROUND

The concept of information overload is based on the assumption that the human information processing system has a limit in its capacity to process information. Most of us could effortlessly add two 2-digit numbers, but would experience extreme difficulty in attempting to add three 10-digit numbers without some additional scratch-pad memory in the form of a pencil and paper. The operationalization of evidence for information overload relies on the probability of errors in task performance.

Studies in the 1950s and 1960s attempted to locate a bottleneck in the processing system as if it was a single-channel serial transmission line (cf Welford, 1967). Broadbent (1954, 1957) proposed that there was a many-to-one selection switch in the channel, with throughput limited by how fast this switch could operate in selecting parallel input signals. Moray (1967), however, proposed that the system behaved instead like a flexible central processor of limited capacity.

The idea of a limited-capacity central processor was furthered by Kahneman (1973), Kerr (1973), and others. The idea was that the processing system is very flexible in the kinds of tasks that it can process concurrently at any given instant, but that it is very limited in its overall size. Kahneman viewed the earlier models of processing as explanations of structural limitations in processing. We cannot, for example, focus our eyes on two objects simultaneously. The limited-capacity processor model was proposed by Kahneman and contemporaries as an explanation of how some mental processing tasks can be performed concurrently.

There is currently no single correct view of the mechanisms that cause the human processing system to be limited in its ability to process information. Importantly, we know that the human processing system is not just a single-resource processor, but that there are multiple resources that can be limited (cf Friedman et al., 1988; Rollins & Hendricks, 1980; Triesman & Davies, 1973). From a practical perspective, however, we can still observe degraded performance when the processing system is asked to perform too much work. Whether the actual cause is due to a bottleneck in a serial system, division or sharing in a single-processor system, or division or sharing in a multiple-resource system, we can assume that the system is being swamped somehow, somewhere if we can observe concurrent task interference.

SECONDARY TASK METHODS THAT CAN BE USED TO MEASURE ATTENTION

RT Probe

In using the RT (reaction time) probe, the participant's reaction time in responding to a secondary stimulus is of interest. As the demands for processing the primary task increase, we begin to see interference with the performance of the secondary task, manifested by (often in this order) increased reaction times, greater variance in reaction times, and misses (failure to react to the stimulus) and false alarms (reacting in absence of a stimulus). When using this procedure, the quality of responses to the secondary stimulus serves as a probe into, or sensor of, the processing demands of the primary task. Measures of decreased performance on the secondary task are taken to indicate increased consumption of processing resources by the primary task.

Secondary stimuli are typically implemented as randomly spaced beep sounds (e.g., Owen et al., 1995) or brief flashes of light at random intervals (e.g., Moore et al., 1986; Stapleford, 1973). Variations on these methods could also be used. In the Baron and Kalsher (1998) study, while participants performed a simulated automobile driving task on a computer, they were to push a button as rapidly as possible after the presentation of a stop sign that appeared on the screen at random intervals. Participants are typically asked to press a handheld button in response to secondary stimuli, but vocal responses are also often used. The Bourdin et al. (1998) rock climbing study took vocal reaction times through a helmet microphone in response to auditory beeps.

Tapping Task

Jastrow (1892) describes the use of finger tapping tasks (secondary tasks) performed concurrently with such processes as mental math and reading under various conditions (primary tasks). Jastrow listed the following as secondary tasks.

- Tapping a finger at a regular rate of whatever the participant chooses
- Tapping a finger at a regular rate but as quickly as possible
- Tapping along with (paced by) a metronome
- Tapping in groups of twos, threes, fours, and so forth
- Tapping in alternate groups of threes and twos, of sixes, fours, and twos, and so forth

By using secondary tasks that were more or less difficult, Jastrow was able to take attention measures that worked under different conditions.

Kantowitz and Knight (1976) used finger tapping paced visually with a computer-timed light blink. Friedman et al. (1988) used rapid finger tapping. Note that some such secondary tasks could be difficult enough that they themselves interfere with primary task performance. In the study of Friedman et al., interest was not so much in the performance of the finger tapping task, but in the way that the finger tapping task interfered with the ability of participants to recall nonsense words that had been displayed during the finger tapping. In some dual task studies such as this, we might expect both tasks to interfere with each other. Often, when dual task studies refer to the use of the secondary task technique, however, the design attempts to keep the secondary task from interfering with the performance of the primary task such that degradations in the performance of the secondary task (and not vice versa) serve as a probe into processes associated with the primary task.

Grip Maintenance

Welch (1898) describes a device that she was using to take quantitative measures of attention in the 1800s. The participant was asked to hold a constant grip on a spring-loaded handle. The handle was attached to a lever; at the end of the lever was a pen that left a mark on a revolving drum. As the participant loosened or tightened his or her grip on the handle, the pen moved one way or the other on the revolving drum. With this apparatus, Welch could trace physical changes in grip over time. Welch observed that error in maintaining a constant grip corresponded with an increase in effort and with an increase in the number of simultaneous tasks that the participant was asked to perform.

Lechner et al. (1998) describe grip maintenance as a method that has seen recent use in studies of sincerity of effort in physical therapy. It otherwise does not seem to be in common use as a secondary task probe. However, it seems that it would not be especially difficult to implement this method through a mouse on a computer. A spring could be attached to the mouse, or a weight could be tied to the mouse cord hanging over the back edge of the table. The secondary task would be to maintain the mouse in a constant position against the force of the weight or spring.

Other Secondary Task Measures

Mental arithmetic is sometimes used as a secondary task (e.g., Bahrick et al., 1954). Wogalter and Usher (1999) asked participants to say answers to math problems aloud while attempting to install a computer hard-disk drive according to the instruction manual. Brown, McDonald, Brown, and Carr (1988) paired handwriting with listening. In a simulated automobile driving task, Young and Stanton (2002) asked participants to judge whether a pair of geometric shapes in the lower left corner of the screen was the same or different by pressing buttons attached to the steering stalk.

Note that we can pair almost any set of tasks in which performance for one of the tasks runs across a range from baseline performance to degraded performance, with degraded performance taken as a measure of increased resource consumption by the other task. Schieber et al. (2000), for example, paired the tasks of tuning a radio while driving a car. Although our interest might be associated with issues of automobile driving, we are actually interested in radio tuning as the primary task while observing degradations in driving performance as a secondary task measure of the amount of processing resources being consumed by radio tuning.

FUTURE TRENDS

This article has described some uses of the secondary task technique in the measure of attention, mental effort, and such over the past century. Although the technique can be relatively simple to implement and relatively low tech in many (but certainly not all) cases, it nonetheless remains useful. In recent years, there have been attempts to combine the dual task technique with more sophisticated methods such as magnetic resonance imaging (use of magnets and radio waves to construct brain pictures; see discussion in Corbetta & Shulman, 2002), but it is unlikely that more complicated methods will replace the secondary task technique in the foreseeable future. The secondary task technique is appealing in that it is very portable and relatively unobtrusive to the environment of the task under study.

Greater use of the secondary task technique is therefore advocated for use in the study of ordinary and everyday human-computer systems. For example, it could easily and unobtrusively be incorporated into the usability testing of Web sites. While the participant is involved in an assigned task during a Web site usability test, he or she could be asked to simply tap a finger at a regular pace. The casual observation of changes in the performance of this secondary task could be taken as an objective observation that a particular step in the primary task is consuming a substantial amount of processing resources. Nothing in the usability study needs to be altered in order to implement the secondary task technique in this way.

CONCLUSION

The secondary task technique is portable, relatively uncomplicated, and relatively unobtrusive as a probe into a variety of mental processes associated with attention. It has been in use for over a century and continues to be used in a variety of disciplines. Although it has some limitations (see Owen, 1991), these are not likely to be of issue in practical applications such as usability testing. This article has not detailed the how-to aspects in the use of this technique, but the implementation of some of the simpler uses, such as the observation of degraded performance in finger tapping, should be reasonably obvious. A more detailed account of how to implement the RT-probe technique, useful in settings that might require more rigorous tests, can be found in Owen et al. (1995). Simpler implementations such as finger tapping tasks, however, should be adequate in many applied situations such as Web usability testing.

REFERENCES

Bahrick, H. P., Noble, M., & Fitts, P. M. (1954). Extra-task performance as a measure of learning a primary task. *Journal of Experimental Psychology, 48*(4), 298-302.

Baron, R. A., & Kalsner, M. J. (1998). Effects of a pleasant ambient fragrance on simulated driving performance: The sweet smell of...safety? *Environment and Behavior, 30*(4), 535-548.

Bourdin, C., Teasdale, N., & Nourgier, V. (1998). Attentional demands and the organization of reaching movements in rock climbing. *Research Quarterly for Exercise and Sport, 69*(4), 406-410.

Broadbent, D. E. (1954). The role of localization in attention and memory span. *Journal of Experimental Psychology, 47*, 191-196.

Broadbent, D. E. (1957). A mechanical model for human attention and immediate memory. *Psychological Review, 64*, 205-215.

Brown, J. S., McDonald, J. L., Brown, T. L., & Carr, T. H. (1988). Adapting to processing demands in discourse production: The case of handwriting. *Journal of Experimental Psychology, 14*(1), 45-59.

Corbetta, M., & Shulman, G. L. (2002). Control of goal-directed and stimulus-driven attention in the brain. *Nature Reviews Neuroscience, 3*(3), 201-215.

Friedman, A., Polson, M. C., & Dafoe, C. G. (1988). Dividing attention between the hands and head: Performance trade-offs between rapid finger tapping and verbal memory. *Journal of Experimental Psychology: Human Perception and Performance, 14*, 60-68.

Jastrow, J. (1892). Studies from the laboratory of experimental psychology of the University of Wisconsin: The interference of mental processes. A preliminary survey. *American Journal of Psychology, 4*, 198-223.

Kahneman, D. (1973). *Attention and effort.* Englewood Cliffs: Prentice Hall.

Kantowitz, B. H., & Knight, J. L., Jr. (1976). Testing tapping timesharing, II: Auditory secondary task. *Acta Psychologica, 40*, 343-362.

Kerr, B. (1973). Processing demands during mental operations. *Memory and Cognition, 1*, 401-412.

Koukounas, E., & McCabe, M. P. (2001). Sexual emotional variables influencing sexual response to erotica: A psychophysiological investigation. *Archives of Sexual Behavior, 30*(4), 393-408.

Koukounas, E., & Over, R. (1999). Allocation of attentional resources during habituation and dishabituation of male sexual arousal. *Archives of Sexual Behavior, 28*(6), 539-552.

Lechner, D. E., Bradbury, S. F., & Bradley, L. A. (1998). Detecting sincerity of effort: A summary of methods and approaches. *Physical Therapy, 78*(8), 867-888.

Maylor, E. A., & Wing, A. M. (1996). Age differences in postural stability are increased by additional cognitive demands. *The Journals of Gerontology, 51B*(3), 143-154.

Moore, D. L., Hausknecht, D., & Thamodaran, K. (1986). Time compression response opportunity in persuasion. *Journal of Consumer Research, 13*, 85-99.

Moray, N. (1967). Where is capacity limited? A survey and a model. In A. F. Sanders (Ed.), *Attention and performance* (pp. 84-92). Amsterdam: North-Holland Publishing Co.

Muller, M. L. T. M., Redfern, M. S., Furman, J. M., & Jennings, J. R. (2004). Effect of preparation on dual-task performance in postural control. *Journal of Motor Behavior, 36*(2), 137-146.

Murray, L. L., Holland, A. L., & Beason, P. M. (1998). Spoken language of individuals with mild fluent aphasia under focused and divided-attention conditions. *Journal of Speech, Language & Hearing Research, 41*(1), 1092-4388.

National Transportation Safety Board (NTSB). (2000). *Aircraft accident brief, accident no. DCA00ma052.* Author.

Owen, R. S. (1991). Clarifying the simple assumption of the secondary task technique. *Advances in Consumer Research, 18*, 552-557.

Owen, R. S., Lord, K. R., & Cooper, M. C. (1995). Using computerized response time measurement for detecting secondary distractions in advertising. *Advances in Consumer Research, 22*, 84-88.

Patten, C. J. D., Kircher, A., Ostlund, J., & Nilsson, L. (2004). Using mobile telephones: Cognitive workload and attention resource allocation. *Accident Analysis and Prevention, 36*, 341-350.

Posner, M. I., & Boies, S. J. (1971). The components of attention. *Psychological Review, 78*(5), 391-408.

Prezuhy, A. M., & Etnier, J. L. (2001). Attentional patterns of horseshoe pitchers at two levels of task difficulty. *Research Quarterly for Exercise and Sport, 72*(3), 293-298.

Rollins, H. A., & Hendricks, R. (1980). Processing of words presented simultaneously to eye and ear. *Journal of Experimental Psychology: Human Perception and Performance, 6*, 99-109.

Schieber, F., Werner, K. J., & Larsen, J. M. (2000). Real-time assessment of secondary task intrusiveness upon real-world driver steering performance using a fiber-optic gyroscope. *Proceedings of the Human Factors and Ergonomics Society Annual Meeting 2000* (Vol. 3, pp. 3-306–3-307).

Stapleford, R. L. (1973). Comparisons of population subgroups performance on a keyboard psychomotor task. *Proceedings of the 9th Annual Conference on Manual Control* (pp. 245-255).

Triesman, A. M., & Davies, A. A. (1973). Divided attention to ear and eye. *Attention and Performance, 4*, 101-125.

Welch, J. C. (1898). On the measurement of mental activity through muscular activity and the determination of a constant of attention. *American Journal of Physiology, 1*(3), 283-306.

Welford, A. T. (1967). Single-channel operation in the brain. In A. F. Sanders (Ed.), *Attention and performance* (pp. 5-22). Amsterdam: North-Holland Publishing Co.

Wolgalter, M. S., & Usher, M. O. (1999). Effects of concurrent cognitive task loading on warning compliance behavior. *Proceedings of the Human Factors and Ergonomics Society 1999 Annual Meeting* (pp. 525-529).

Young, M. S., & Stanton, N. A. (2002). Malleable attentional resources theory: A new explanation for the effects of mental underload on performance. *Human Factors, 44*(3), 365-375.

KEY TERMS

Attention: Mental processing that consumes our conscious thinking. This is associated with a variety of more specific constructs such as mental effort, mental focus, mental elaboration, and such. Processes associated with the attention-related constructs are what we presume to be detecting in dual task studies.

Dual Task Study: A study in which two tasks are performed concurrently to observe changes in task interference. Usually, the participant is expected to or asked to focus on the primary task so that interference is observed only in the secondary task, but this is not necessarily always the objective. Observations of task interference are taken to suggest that the limits of the processing system are being reached.

Limited-Resource Model: The idea that the human information processing system has a limited pool of resources available for the concurrent performance of any number of tasks. The observation of degraded performance in one or more processing tasks is taken to suggest that the capacity of the system is being approached.

Primary Task: The resource consumption task that is of interest in a secondary task study. Often

(but not necessarily), the participant is asked to focus on the performance of this task while concurrently performing the secondary task. For example, the participant could be asked to focus on reading a passage of text on successive screens of a computer display (primary task) while concurrently pressing a handheld button switch whenever a random beep sound is heard (secondary task).

RT Probe (Reaction Time Probe): A commonly used secondary task in which changes in reaction time performance of the secondary task are of interest.

Secondary Task: The task that is used as a probe in a secondary task study. Changes in the performance of the secondary task are taken to suggest changes in the processing of the primary task or the detection of processing system overload.

Secondary Task Technique: A dual task study in which one task is designated as the primary task of interest while a secondary task is concurrently performed as a probe to test the consumption of processing resources by the primary task. Changes in secondary task performance are taken to indicate changes in resource consumption by the primary task.

Using Mobile Communication Technology in Student Mentoring

Jonna Häkkilä
University of Oulu, Finland

Jenine Beekhuyzen
Griffith University, Australia

INTRODUCTION

Information technology (IT), computer science, and other related disciplines have become significant both in society and within the field of education. Resulting from the last decades' considerable developments towards a global information society, the demand for a qualified IT workforce has increased. The integration of information technology into the different sectors of every day life is increasing the need for large numbers of IT professionals. Additionally, the need for nearly all workers to have general computing skills suggests possibilities for an individual to face inequality or suffer from displacement in modern society if they lack these skills, further contributing to the digital divide. Thus, the importance of IT education has a greater importance than ever for the whole of society.

Despite the advances and mass adoption of new technologies, IT and computing education continually suffers from low participant numbers, and high dropout and transfer rates. This problem has been somewhat addressed by introducing mentoring programs (von Hellens, Nielsen, Doyle, & Greenhill, 1999) where a student is given a support person, a mentor, who has a similar education background but has graduated and is employed in industry. Although the majority of these programs have been considered successful, it is important to note that it is difficult to easily measure success in this context.

In this article, we introduce a novel approach to mentoring which was adopted as part of an ongoing, traditional-type mentoring program in a large Australian university. The approach involved introducing modern communications technology, specifically mobile phones having an integrated camera and the capability to make use of multimedia messaging services (MMS). As mobile phones have become an integrated part of our everyday life (with high adoption rates) and are an especially common media of communication among young people, it was expected that the use of the phones could be easily employed to the mentoring program (phones were provided for the participants). Short message service (SMS), for example text messaging, has become a frequently used communication channel (Grinter & Eldridge 2003). In addition to text, photo sharing has also quickly taken off with MMS capable mobile phones becoming more widespread. The ability to exchange photos increases the feeling of presence (Counts & Fellheimer, 2004), and the possibility to send multimedia messages with mobile phones has created a new form of interactive storytelling (Kurvinen, 2003). Cole and Stanton (2003) found the pictorial information exchange as a potential tool for children's collaboration during their activities in story telling, adventure gaming and for field trip tasks.

Encouraged by these experiences, we introduced mobile mentoring as part of a traditional mentoring program, and present the experiences. It is hoped that these experiences can affirm the legitimacy of phone mentoring as a credible approach to mentoring. The positive and negative experiences presented in this article can help to shape the development of future phone mentoring programs.

BACKGROUND

Current education programs relating to information technology continue to suffer from low applicant numbers in relation to the available enrollment positions. In the USA alone, the number of computer science graduates dropped from a high of 50,000 in 1986 to 36,000 in 1994, reported by the Office of Technology Policy in 1998 (von Hellens et al., 1999). Many general IT degrees also have high dropout rates, particularly in the transition from the first to second year of undergraduate studies. Student statistics also show that university IT degree programs are not attracting the high achieving students, some possible reasons include the low entrance level scores needed to enter the program, the attraction to high-entrance level degree programs such as medicine, law, and psychology and the confusion and uncertainty relating to what a career in IT will entail (ASTEC, 1995).

Misconceptions associated with understanding IT as a field specialized for those with masculine attributes exist and are reinforced by the teachings at secondary school level (Beekhuyzen & Clayton, 2004; Greenhill, von Hellens, Nielsen, & Pringle, 1997), thus often having a negative effect on students, particularly on females. Consistent results have been obtained in studies concerning high school physics, which faces similar difficulties and biased ideas as IT (Häkkilä, Kärkäs, Aksela, Sunnari, & Kylli, 1998). A remarkable number of university students choose their area of study without any preliminary experience in the particular field. With information technology, the students also often have unclear or distorted perceptions of what to expect later in their studies or after graduation, including what kind of employment their area of study can offer (Nielsen, von Hellens, Pringle, & Greenhill, 1999).

Within the IT context, university student mentoring has been introduced to offer students insight into the industry and to employment possibilities enabling them to have them a closer look at the everyday life of working in the field. The aim is to dispel some of the misconceptions associated with what IT work is all about. When entering into this mentoring program, the student is matched with a personal mentor who has a similar educational background and is currently employed in the IT industry. Conventionally, mentoring is carried out with face-to-face meetings, e-mail and telephone conversations between mentor and mentee. In line with many published studies, early results from our studies suggest that mentoring can provide valuable information on career possibilities, thus increasing the motivation of study and working in the area. It also clarifies and enhances student perceptions concerning the realities of the field. Note: all participation in the program is of a voluntary basis, and no financial benefits are obtained.

When commencing the traditional part of the mentoring program, mentors and mentees participate in an initial short training session. In this session, the mentoring partners are introduced, and the role and expectations of mentors and mentees is discussed. Mentor and mentee generally meet thereafter on a regular basis during one semester period (usually 13-15 weeks) which is arranged as suits best for both parties. Face-to-face communication is also usually complimented by e-mail conversations. A mid-program event is organized by the Alumni Association, usually with a presentation by an industry representative on a pertinent topic such as networking (in terms of meeting people, making contacts, etc.—a skill particularly useful within the IT industry). A final session is held to close the program and gather together all program participants to discuss their experiences.

ENHANCING COMMUNICATION WITH MOBILE TECHNOLOGY

In addition to the traditional mentoring methods being employed by the mentoring program in the university, we have introduced the use of mobile communication technology into the mentoring program. The primary aim in introducing the novel approach was to augment communication during the mentoring process. There was no aim to replace the conventional communication mediums but to add value with features offered by the mobile communication device. A pilot study was conducted in 2003. Due to positive feedback, the approach has continued to be integrated in the traditional program in 2004.

The equipment used in the experiment consists of two Nokia 7650 Mobile Phones, of which one was given to the mentee and one to the mentor for the duration of the program. The mentor was advised to communicate with the student about all which (s)he felt was a relevant part of their work and leisure, and

Figure 1. Two multimedia messages describing the work tasks of a mentor

I have a conference paper deadline tomorrow, i just finished working with it. Here's a pile of articles and other stuff on my desk. I'm very messy, heh..! Anyway, i just heard i got a scholarship and i can go to that conference in switzerland, yes!! Well, good night now...

Mentor, 1:04 am

I do paper prototyping and interviewing for my research project. One thing i like in this field is that it offers so different kinds of tasks. I also get to talk and meet people quite a lot – funny that people think IT is not social!

Mentor, 9:48 pm

especially to use picture messaging as an illustrative supplement. Using this type of technology to communicate brings about many issues relating to human-computer interaction. For example, the size of the screen, the structure of the information being viewed/ sent (Chae & Kim, 2003), and the increasing complexity of functionality can lead to ineffective use of the mobile device. However, benefits include the ability to access information from anywhere without the need to physically sit at a computer workstation (Chae & Kim, 2003).

The student mentee was given a certain monetary amount (AUD $15—Australian dollars) of pre-paid credit on the mobile phone, which they were allowed to use during the study. For example, the price for sending an SMS message and MMS message were 0.20 AUD (20 cents) and 0.75 AUD (75 cents), respectively. The phone mentoring period lasted for one week for one mentor-mentee pair. In the beginning of the experiment period, the functions of the phone were explored together to ensure seamless communication. At the completion of the one-week period, the participants gave their feedback about the experience via a questionnaire.

The media used in communications between the mentor and mentee were short messages (SMS) and multimedia messages (MMS), the latter to be more common. Conversations consisted mainly of one message or a message and a reply, where the reply included feedback or comment to the previous messages. The typical number of sent messages was two per day from mentor to mentee, and one from mentee to the mentor, although more messages were exchanged if a message gave rise to a longer, more detailed conversation. The time for messaging was found to be varied from morning hours to past midnight and also sometimes during the weekend, as shown in Figures 1 and 2.

The majority of messages sent contained a short description of the work task the mentor was currently involved in, accompanied by a picture. In addition to the actual task, the messages often described the atmosphere at that particular moment and also included short opinions (see Figure 1). Some of the messages were not primarily related to the work tasks, but described more the mentor's personal interests—free time, hobbies, and personal preferences.

The initiative for conversations containing professional information was taken by the mentor, and was not motivated by, for example, a question from a mentee. However, mentees took initiative in reporting about their duties related to studying. For instance, assignments and projects they were working on. Conversations relating to free time were initiated equally by both parties. Examples of messages relating to free time are illustrated in Figure 2.

Figure 2. Two free time orientated multimedia messages from a mentor (company name of the employer replaced with asterisk)

I was rock climbing tonight, I do it every Tuesdays. I really like it! Feels just a bit hard to type now :)
Mentor, 11:46 am
Aah, it was nice to sleep in! By the way, we have pretty flexible working hours in ********. If there's no meetings or anything in the morning, u can just come whenever and stay later in the end of the day. Suits me!
Mentor, 10:27 am

FEEDBACK

The results obtained from both sets of participating parties were positive and encouraging. The most positive aspects reported by the mentees were on increasing the frequency of the communications and thus developing a closer relationship and gaining a deeper insight for the mentor's work. Comments collected from two students at the end of the mobile mentoring period are presented in the following:

- **Mentee #1:** "I believe that mobile communication is quick and easy. It gives you an opportunity to learn more about your mentor and what they do and vice versa. It is especially good when both parties are unable to meet on a regular basis due to time constraints, commitments, etc."
- **Mentee #2:** "The best thing about the phone mentoring was that I was able to see how another person, in the field I want to work in, interacts with their life as well as being able to share aspects of my life with my mentor. It helped break the ice, enabling my mentor and myself to get to know each other."

Positive feedback obtained from mentors particularly concerned the flexibility in regard to the place and time of the communication, ease of use, and the extra personal touch it gave to the conversations. The amount of credit was reported to be sufficient for a one-week experimental period. Overall, the integration of mobile technology was suggested to offer a valuable tool for the mentoring program.

Other reported positive aspects of this novel approach include:

- Easy way to communicate
- Minimal effort required
- You can do it at any time and you don't miss the person
- (The phone is) very popular with people, so it is an advantage to use it with mentoring,
- Quicker, more efficient, because people have their phone on them more than they check their emails
- Teaches responsibility, taking care of the phone
- More informative, a picture can say a thousands words.

As a weakness, mentees referred to the short length of the experimental period, and were suggesting it to be elongated from one to, for example, two weeks. This was argued by explaining that it would offer a longer period of time to get used to both using the technology (the MMS phone) and the mode of communication. A wish for a longer lasting experimental period was also mentioned by mentors, as one week was not considered to be a long enough time to cover the different aspects related to the diverse work in IT. The suggestion of a longer experimental period is also supported when examining the messages, as the communication become relaxed towards the end of the week. This feedback from participants in 2003 was integrated into the program run in 2004, as longer experimental periods (1 week/10 working days) are used.

The technical barriers noted were battery time/length and lack of network coverage in some areas, which were described to limit the communication in some instances. These, in addition to small screen size, low bandwidths, limited storage and cumbersome input facilities are common barriers that have been presented in the literature (Chae & Kim, 2003; Tarasewich, Nickerson, & Warke, 2001). However, the participants' perception was that the technical barriers did not have significant impact to the overall experiment. A criticism from mentors was that if the use of mobile communications would be the only medium of interaction in mentoring, the conversations between mentor and a mentee would remain too light and no deep knowledge or "big picture" would be obtained from the short communications. For instance, the following comment was obtained from a mentor when asked the weaknesses of mobile mentoring:

- **Mentor:** "The nature of conversations differs a great deal in comparison to ones had in face-to-face meetings and e-mail exchange, where the discussion is held for longer. However, I highly recommend this system as an additional part of communication, as it offers a possibility for more intense and frequent interaction and highlights the aspects which otherwise hardly were considered, e.g., the work environment, task descriptions and the time schedule of the day."

FUTURE TRENDS

When the mobile mentoring program first began in early 2003, the number of multimedia messaging capable phones was minimal. It is expected that when they become more common, mobile mentoring can be adopted on a larger scale, and it may come a natural part of the interaction process. However, as a starting point, it was important to lend out the MMS capable phones and get people actively involved in the process.

In 2004, yet another novel approach to mentoring was added to the ongoing Alumni Association mentoring program in the form of international mentoring. In addition to a local mentor to communicate with (either traditionally and/or phone), volunteering student mentees were given contact persons working abroad as mentors. The mentoring contacts were obtained through the university department's connections to the international IT industry. The communication employs mainly e-mail, but also additional mobile messaging techniques. Due to difficulties in connectivity between mobile phone operators, the MMS between two phones was found inoperative, thus picture messages were exchanged by sending a MMS from a phone to e-mail. Initial results of this additional experiment will be reported.

CONCLUSION

This article introduces how mobile communication technology has been embedded into a university student mentoring program which was held among first-year information technology students within an Australian university. The study was implemented by giving mentor-mentee pairs mobile phones with MMS functionality. Participants communicated with each other over a one-week period. Participants were advised to incorporate visual information into the communication by a form of multimedia messaging.

Although effective in many situations, mentoring can be rather unproductive and thus unsuccessful for many reasons. One common reason for failure is a lack of structure. Many communications between mentor and mentee are adhoc and generally unplanned which can and often does result in long periods between communications. Lack of structure can also distort perceptions of outcomes and results, with no clear aim being achieved.

The results show that integrating mobile communication into the mentoring process has provided added value to the traditional program. Participants suggest that it enhances the mentoring experience and that it can be regarded as a valuable tool in communications between the mentor and mentee. Positive aspects of the program were identified as increased frequency and flexibility in communication, which are highly valued because of the time constraints of both mentor and mentee. Mentors emphasised also the easy access and speed of use, as sending a message with mobile phone was regarded as more easy and flexible than e-mail, which took more time and was limited to the work situations and a computer. Both parties reported on the development of a deeper personal relationship and relaxed communication between mentor and mentee over time. Including visual information to the communication in a form of MMS, new aspects of both mentor's work and her/his lifestyle were highlighted.

Generally, mobile communication technology was found to offer a valuable tool for a mentoring program as a supporting tool of communication, even though some weaknesses were identified. The results have encouraged the authors to continue the integration of mobile communication into mentoring to enhance the information exchanged between mentor and mentee. Continuing and future research in this area includes the continuation of the study using both MMS and conventional styles, with improvements according to feedback received from this pilot phase concerning issues such as the time period devoted to the experiment. The aim is to increase the amount of participants from a relatively small sample to larger numbers of mentee-mentor pairs and to attempt to better measure the benefits of the program.

REFERENCES

ASTEC. (1995, September). *The science and engineering base for development of Australia's information technology and communication sec-*

tor. A discussion paper by Australian Science and Technology Council (ASTEC).

Beekhuyzen, J., & Clayton, K. (2004, July 28-30). ICT career perceptions: Not so geeky?!?. In the *Proceedings of the 1st International Conference on Research on Women in IT*, Kuala Lumpur.

Chae, M., & Kim, J. (2003, December). What's so different about the mobile Internet? *Communication of the ACM, 46*(12), 240-247.

Cole, H., & Stanton, D. (2003). Designing mobile technologies to support co-present collaboration. *Personal and Ubiquitous Computing, 7,* 365-371.

Counts, S., & Fellheimer, E. (2004). Supporting social presence through lightweight photo sharing on and off the desktop. In the *Proceedings of the Conference on Computers and Human Interaction (CHI)* (pp. 599-606).

Greenhill, A., von Hellens, L., Nielsen, S., & Pringle, R. (1997, May). Australian women in IT education: Multiple meanings and multiculturalism. In the *Proceedings of the 6th IFIP Conference on Women, Work and Computerization*, Bonn, Germany (pp. 387-397).

Grinter, R. E., & Eldridge, M. (2003). Wan2tlk? Everyday text messaging. In the *Proceedings of the Conference on Computers and Human Interaction (CHI)* (pp. 441-448).

Häkkilä, J., Kärkäs, M., Aksela, H., Sunnari, V., & Kylli, T. (1998). *Tytot, pojat ja fysiikka. Lukiolaisten käsityksiä fysiikasta oppiaineena*. University of Oulu, Finland. Abstract in English: Girls, Boys and Physics.

Kurvinen, E. (2003). Only when Miss Universe snatches me: Teasing in MMS messaging. In the *Proceedings of the Conference on Designing Pleasurable Products and Interfaces* (pp. 98-102). ACM Press.

Nielsen, S. H., von Hellens, L., Pringle, R., & Greenhill, A. (1999). Students' perceptions of information technology careers. Conceptualising the influence of cultural and gender factors for IT education. *GATES (Greater Access to Technology Education and Science) Journal, 5*(1), 30-38.

Tarasewich, P., Nickerson, R. C., & Warke, M. (2001, August). Wireless/mobile e-commerce: Technologies, applications, and issues. In the *Proceedings of the Seventh Americas Conference on Information Systems (AMCIS)*, Boston.

von Hellens, L., Nielsen, S., Doyle, R., & Greenhill, A. (1999, December). Bridging the IT skills gap. A strategy to improve the recruitment and success of IT students. In the *Proceedings of the 10th Australasian Conference on Information Systems,* Wellington, New Zealand (pp. 1129-1143).

KEY TERMS

Mentee: Participant of the mentoring program; student or equivalent; being "advised."

Mentor: Participant of mentoring program; the "advisor."

Mentoring Program: A process where a mentee is given a personal guide, a mentor, who has professional or otherwise advanced experience and can advise the mentee on the specifics about the particular field of study and work in the industry.

Mobile Communication Technology: A medium to communicate via mobile devices.

Mobile Mentoring: Mentoring which uses mobile communication technology as an integrated part of the communication between mentor and mentee.

Multimedia Messaging Service (MMS): A form of mobile communication, where each message can contain picture, audio, video, and text material with certain data size limitations. A multimedia message is typically sent from one camera phone to another.

Short Message Service (SMS): A form of mobile communication, where mobile phone user is able to send and receive text messages typically limited to 160 characters. A short message is typically sent from one mobile phone to another.

Various Views on Digital Interactivity

Julie Thomas
American University of Paris, France

Claudia Roda
American University of Paris, France

INTRODUCTION

As Kress and Van Leeuwen (2001) state, there is no communication without interaction. Broadly, levels of "interactivity" can be recognized as depending on quality of feedback and control and exchange of discourse according to the mode or modes ("multimodal discourse") involved. Important constraints that operate to modify interactivity of any kind can be identified as the amount of "common ground" (Clark, 1996), constraints of space and time, relative embodiment, and choice of or control over the means, manner, and/or medium of feedback.

Ha and James (1998) emphasize the element of response as characterized by playfulness, choice, connectedness, information collection, and reciprocal communication.

BACKGROUND: SELECTED ELEMENTS OF DIGITAL INTERACTIVITY

Feedback

Any evaluation of feedback, as defined by Kiousis (2002), should take into account various factors. For example, feedback should not be just two-way, but should encompass several different avenues and facets of expression; it can be linear and/or nonlinear. Hyperlinks should offer the element of choice, and the ability to modify the mediated environment must exist. Individual perception of interactivity depends on the quality of media (form, content, structure, relation to user) but also on "social presence" (Short, Williams, & Christie, 1976) or "telepresence" (awareness of mediated environment), perceived speed, timing, and flexibility. Kiousis adds to these factors the concepts of "proximity"—how "near" the user feels—and "sensory activation"—the involvement of the user's senses.

Immersion and Engagement

The qualities of "immersion" and "engagement," referred to by Douglas and Hargadon (2000) as "The Pleasure Principle" and equated by Laurel (1993) with the "willing suspension of disbelief," appear to be crucial in creating the illusion of interaction.

The role of immersion and engagement is obvious with reference to simulations, the use of links, and user perception of control and decision-making.

Simulation

Simulation (particularly as in Game format) privileges a sensation of control, a sense of presence, and entry into mediated environments as "active" rather than "passive" through manipulating time (speed involved in decision making), agency, the spatial orientation of the user, and what Darley (2000) describes as "vicarious kinaesthesia:" the feeling of "direct physical involvement"(p. 157). Perhaps we might add to this list the element of "surprise," the "unexpected," the *apparently* random, necessitating a response and therefore creating an impression of responsive dialogue and mutual discourse, a perception of feedback and engagement.

Play

In all questions of interactivity, the target audience must be considered (McMillan, 2002), and the nature of links must be examined. Manovich (2001) com-

plains that by following "pre-programmed, objectively existing associations," users of interactive media are being asked to mistake the structure of somebody else's mind for their own (p. 61).

One of the characteristics of interactivity is the nature of "play" involved. The importance of play in performing identity and social structure has long been recognized (Huizinga, 1955), and, as Zimmerman (2004) has more recently noted, play both expresses and simultaneously resists the structure of the system within which it exists. Within any interactive system, this element of play could perhaps be seen as a crucial factor in removing the impression of a predictable structure, which stifles user individuality and involvement. Although choices, or links, are indeed programmed, there can be no play without constraints; games always have "rules" that cannot be changed without creating a different "game" (unless, of course, this is a device of the game creator to produce engagement and thus reinforce the nature and structure of the game!)

This consistency of "world" or "play" further contributes to the "willing suspension of disbelief". As Douglas (2000) remarks, ambiguity is always embedded in the interactive, but this ambiguity can be harnessed in service to the sense of play, which of itself both provides and subverts the structural framework.

Hypertext: Interactivity as Narrative and/or Drama

No consideration of digital interactivity is possible without a discussion of interactive hypertext, often characterized as "multidimensional." It is necessary to remember that multidimensional does not mean "random explorations," but what Douglas (2000) calls "polysequential" rather than Nelson's "nonsequential" writing (Nelson, 1992), or even Bush's 1945 "encyclopedia of associative trails" for Memex (Bush, 1992), for in such an "encyclopedia," although the associations of the reader will be used to construct individual unique meaning or personal narrative, the "encyclopedia" has not *necessarily* been structured for this purpose by the author; this is the difference between constructed narrative and information retrieval.

Multidimensional hypertext at its best takes advantage of and exploits the human tendency to construct narratives to make sense of the world, relying on individual human selection of appropriate stimuli and human ability not simply to choose links but to create connections, rather than simply following pre-ordained paths. Joyce (1995) remarks that the user/reader's task is to make meaning by perceiving order in space, so that the meaning is orderly but there is a continual replacement of meaningful structures throughout the text: the narrative is constantly evolving in time and space.

Murray (1997) identifies three qualities (which she calls "pleasures") that characterize the interactive audience: immersion, agency, and transformation. Immersion, meaning engagement of the imagination and the senses, has already been discussed as a property of interactivity. Murray emphasizes the active audience and differentiates between the role of the interactive user/reader and the role of the author by describing the user/reader as agent. Her emphasis on various points of view as one technique for incorporating multi-sequencing in hypertext is typical of a narrative approach.

An alternative approach is that of Laurel (1993), who suggests drama as a model for interactivity, and emphasizes three features:

1. Enactment (to act out) rather than to read. Narrative is description; drama is action.
2. Intensification. incidents are selected, arranged, and represented to intensify emotion and condense time.
3. Unity of action versus episodic structure. In the narrative, incidents tend to be connected by theme rather than by cause to the whole; in drama, there is a strong central action with separate incidents causally linked to that action. Drama is thus more intense and economical.

When Laurel advocates strategies for designing interactive media, she emphasizes that the conceptual structure should encourage the potential for action. Laurel outlines several key points for designing interactive media, and emphasizes that tight linkage between visual, kinesthetic, and auditory modalities is the key to immersion.

ENHANCING INTERACTION: CREATIVE LINKING AND INTERACTIVE SPACE

Link Authoring

Every interface asks the audience to participate in its construction, and creative link authoring is one of the most important factors determining whether the audience will perceive this interface as interactive.

Early on, Nelson (1992) proposed different simple "styles" of guiding the sequencing of hypertext: planned variations, which focus on the transmission of a message, representing interconnections, representing the structure of the subject for the reader to explore. Golovchinsky and Marshall (2000) point out that the quality and quantity of the reader's choices are confined by the fixity of the links and that the "trick" of creating interactive hypertexts is to subvert this "fixity." Choices as to the use of fixed links, variation of links, query-mediated links, provide a "hidden" structure, which conditions the audience's choices and reactions to the text as well as the level of perceived interaction. Further, linking "reconfigures" the text and is crucial to creating the placement in space, which gives the text its multidimensional aspect and "aligns" and "realigns" meaning, both visual and verbal. As Garrand (1997) remarks, there must be a balance between the viewer's freedom and narrative coherence (the constraints of the game further the sense of play!), and subtle and appropriate linking creates that balance. Garrand, writing with reference to interactive multimedia, emphasizes that linking for interaction must be "vertical" as well as "horizontal," that interactive writing is 3-D writing.

Links both emphasize the visual element of the text itself—the text as a visual feature—and help to create an "enactment" of three dimensional space in the spatial relations of "navigation" (up/down, left/right, etc.) and in the impression of "layering." In hypermedia link authoring, where hypertext is linked with images, videos, sounds, animations, and so forth, linking makes clear that verbal text is only one kind of content, and that a link does not just "match" verbal text, sound, image, and so forth, but reveals content from different perspectives. Although links used in the course of interactive exploration can give the impression of what Douglas (2000) refers to as an "unlimited database," too much detail and too many links detract from immersion.

Interactive Space: Visual and Verbal

One aspect of digital interactivity is about creating the impression of the enactment of an infinite possibility of sequencing through creative linking; structure and content are formed by and equated with space "traveled." The physical action of "clicking" to select links is combined with the mental action of "connecting" links; both serve to structure and layer digital space, and to produce the sensation of movement through space. As noted before, users do not visualize themselves traveling up and down a line, or even back and forth on branching lines to and from a center of meaning, but navigating through 3D space. Further, identification of what might be considered as being "inside" or "outside" the text loses meaning and importance. This "virtual" space is self-contained but through linking and association can contain more than the "sum of its parts."

As Wertheim (1999) has remarked, the frescoes of Giotto in the Arena Chapel of Padua (1305) provide a visual parallel and enactment of this kind of Memory Palace, and also a precedent for the layering of meaning in space, which has come to be seen as characteristic.

Livingstone (1999) also points out that the physical movement of the human agent (in clicking, choosing paths, etc.) manipulates objects, which exist only in digital space, as if they existed within physical space. He compares this to Lakoff and Johnson's "embodied interaction" (1980), and, as we "drag" objects onto and around the screen, the conceptual relationships we make between the *real* and the *digital* form the foundations of a completely new interactive space with its own specific characteristics, and its own formulae for conveying meaning.

"Paths" of reading are also important for the creation of interactive space. As Kress (2003) makes clear, reading paths are culturally dictated (left to right/ right to left, etc.). "Multimodal" texts open the question of reading paths—in terms of "directionality" (which direction?) and in terms of which elements the reader chooses as "points" along the reading path. What are the elements to be read together? (Just as children learning to read do not make the assumptions about "ordered" reading space that trained adults do.) Is the reader looking at a text to be "read" as a conventional text, a text to be "read" as an image, an image to be "read" as part of a text? Thus the "reading" of an interactive verbal/visual text "screen" implies that the reader establish the order through his/her own preferences as to relevance, thereby constructing a personalized meaningful space.

The creation of interactive multimodal discourse thus demands that authors and designers consider carefully the interplay between visual and verbal units of meaning and their placement, not simply in terms of the space of the screen, but in terms of the relative value of that space, and how juxtaposition in that space affects the relative values of text and image. Not only do text and image provide different possibilities for the creation of meaning and "engagement," but verbal text on-screen becomes another aspect of the visual (fonts, graphics, visual sculpting of blocks of text, layout, etc.)—and this should be taken into account by creators to capitalize on capacity for interactivity.

De Certeau (1988) suggests that, through the "spatial practice" of walking, the pedestrian learns to create and inhabit his own city by the paths he chooses. A similar creation of personal space in virtual space is important for immersion and engagement, which is why Johnson-Sheehan and Baehr (2001) place such importance on the use of "design metaphors"—architectural, physical spaces such as cafés, museums, and so forth, to involve the user physically—and why the use of visual features (frames, icons, images) to create possibilities for the navigator, rather than simply as "dead" links, is also relevant to user perception of screen space as interactive space. (Laine, 2002)

Darley (2000) has proposed that the interactive element of visual digital culture is best thought of as related to earlier forms of entertainment—like the amusement park, or music hall for example, which demand active participation from the audience—rather than more contemporary media, like television or cinema. This comparison highlights an aspect of visual digital interactivity which often is not considered adequately because it is so obvious—the screen is not a television, not only in the aspect of viewer control or "interactivity", but also in the way that images are presented, sequenced, used, and "valued."

FUTURE TRENDS

An increased implementation of techniques to enhance the impression of interactivity is important for every aspect of digital media. Some interesting future applications include "Peer-to-Peer Communications/Visualizing Community" (Burnett, 2004), design practice in humanities-based applications (Strain & VanHoosier-Carey, 2003), and the field of interaction design as a whole. As Lowgren (2002) remarks: "Interaction Design is a fairly recent concept...It clearly owes part of its heritage to HCI, even though the turns within established design fields—such as graphic design, product design and architecture—towards the digital material are every bit as important." Further, as McCullough (2004) notes, "the goal of natural interaction drives the movement toward pervasive computing and embedded systems" (p. 70).

Techniques of narrative characteristic of interactive hypertext are being exploited to increase user involvement in a variety of commercial and web applications (Broden, Gallagher, & Woytek, 2004).

The digital design identity of corporations and brands offers another area for future application. McCullough (2004) has underlined the prospective value of interactive media for developing new relationships between the brand and the market, and particularly emphasized the expected future diversification of interactive systems by digital brands and services as a way of manifesting and performing brand identity.

CONCLUSION

As Aarseth (2003) suggests, "attempts to clarify what interactivity means should start by acknowledging that the term's meaning is constantly shifting and probably without descriptive power and then try to argue why we need it, in spite of this" (p. 426). We need interactivity and all the various points of view that coexist within the shifting meaning of this term because successful interaction transforms the passive receiver of information into the active participant in communication.

REFERENCES

Aarseth, E. (2003). We all want to change the world: The ideology of innovation in digital media. In Liestol, g., Morrison, A., & Rasmussen, T. (Eds.), *Digital media revisited* (pp. 415-439). Cambridge, MA; London: MIT Press.

Arata, L. (1999). Reflections about interactivity. MIT Communications Forum. Retrieved on November 26, 2004, from http://web.mit.edu/comm-forum/papers/arata.html

Broden, N., Gallagher, M., & Woytek, J. (2004). *Use of narrative in interactive design. Boxes and Arrows.* Retrieved on November 23, 2004, from http://www.boxesandarrows.com/archives/use_of_narrative_in_interactive_design.php

Burnett, R. (2004). *How images think*. Cambridge, MA; London: MIT Press.

Bush, V. (1992). As we may think. In T. Nelson (Ed.), *Literary machines 93.1* (pp. 39-54). Sausalito, CA: Mindful Press.

Clark, H. H. (1996). *Using language*. Cambridge: Cambridge University Press.

Darley, A. (2000). *Visual digital culture*. London; New York: Routledge.

de Certeau, M. (1988) tr. Rendall. *The practice of everyday life*. Berkeley; London: University of California Press.

Douglas, J. Y. (2000). *The end of books - Or books without end? Reading interactive narratives*. Ann Arbor: University of Michigan Press.

Douglas, J. Y., & Hargadon, A. (2000, May 30-June 4). The pleasure principle: immersion, engagement, flow. In the *Proceedings, Eleventh ACM on Hypertext and Hypermedia 2000*, San Antonio, Texas (pp. 153-160). New York: ACM Press.

Garrand, T. (1997). Scripting narratives for interactive multimedia. *Journal of Film and Video 49*(1/2), 66-79.

Golovchinsky, G., & Marshall, C. C. (2000, May 30-June 4). Hypertext interaction revisited. In the *Proceedings, Eleventh ACM on Hypertext and Hypermedia 2000*, San Antonio, Texas (pp. 171-179). New York: ACM Press.

Ha, L., & James, E.L. (1998). Interactivity reexamined: A baseline analysis of early business Web sites. *Journal of Broadcasting & Electronic Media, 42*(4), 457-474.

Huizinga, J. (1955). *Homo Ludens*. Boston: Beacon Press.

Johnson-Sheehan, R., & Baehr, C. (2001). Visual-spatial thinking in hypertexts. *Technical Communication, 48*(1), 22-30.

Joyce, M. (1995). *Of two minds*. Ann Arbor: University of Michigan Press.

Kiousis, S. (2002). Interactivity: A concept explication. *New Media & Society, 4*(3), 355-383.

Kress, G. (2003). *Literacy in the new media age*. London; New York: Routledge.

Kress, G., & Van Leeuwen, T. (2001). *Multimodal discourse: The modes and media of contemporary communication*. London: Arnold.

Laine, P. (2002, June 11-15). How do interactive texts reflect interactive functions? In the *Proceedings, Thirteenth Conference on Hypertext and Hypermedia, 2002*, College Park, Maryland (pp. 67-68). New York: ACM Press.

Lakoff & Johnson (1980). *Metaphors we live by*. Chicago; London: University of Chicago Press.

Laurel, B. (1993). *Computers as theatre*. Reading, MA: Addison-Wesley.

Livingstone, D. (1999). The space between the assumed real and the digital virtual. In Ascott, R. (Ed.), *Reframing consciousness* (pp. 138-143). Exeter; Portland, OR: Intellect Press.

Lowgren, J. (2002). *Just how far beyond HCI is interaction design? Boxes and Arrows.* Retrieved

on November 24, 2004, from http/ www.boxesandarrows.com/archives/just_how_far_beyond_hci_is_interaction_design.php

Manovich, L. (2001). *The language of new media.* Cambridge, MA: MIT Press.

McCullough, M. (2004). *Digital ground.* Cambridge, MA; London: MIT Press.

McMillan, S. J. (2002). A four-part model of cyber-interactivity. *New Media and Society, 4*(2), 271-291.

Murray, J. (1997). *Hamlet on the holodeck: The future of narrative in cyberspace.* Cambridge, MA: MIT Press.

Nelson, T. (1992). *Literary machines 93.1.* Sausalito, CA: Mindful Press.

Short, J. A., Williams, E., & Christie, B. (1976). *The social psychology of telecommunications.* New York: John Wiley & Sons.

Strain, & VanHoosier-Carey (2003). Eloquent interfaces: Humanities-based analysis in the age of hypermedia. In Hocks, M. & Kendrick, M. (Eds.), *Eloquent images: Word and image in the age of new media* (pp. 257-281). Cambridge, MA; London: MIT Press.

Wertheim, M. (1999). *The pearly gates of cyberspace: A history of space from Dante to the Internet.* London: Virago Press.

Zimmerman, E. (2004). Narrative, interactivity, play and games: Four naughty concepts in need of discipline. In Wardrip-Fruin, N. & Harrigan, P. (Eds.), *First person: New media as story, performance, and game* (pp. 154-164). Cambridge, MA; London: MIT Press.

KEY TERMS

Common Ground: Shared knowledge and experience common to both sender and receiver. This "common ground" enables the references and context of the message to be deciphered successfully and meaning to be communicated.

Digital Interactivity: Despite the fact that interactivity as a blanket concept cannot be precisely defined, the quality of interactivity defined by the user generally depends on the amount of "common ground", the user's *perceived* ability to control and influence form and content of the mediated environment, to be "engaged" in mediated space (in terms of belief and/or in terms of sensory stimulation or displaced physical enactment or embodiment), and to participate in multidimensional feedback which offers choice in real time.

Hypertext: Text (and we use the term here in the broad sense to include "text" that may be verbal and/or visual) which is constructed as "polysequential" (Douglas, 2000) and multidimensional through a network of associational links.

Interaction Design: "There is no commonly agreed definition of interaction design; most people in the field, however, would probably subscribe to a general orientation towards shaping software, Web sites, video games and other digital artefacts, with particular attention to the qualities of the experiences they provide to users" (Lowgren, 2002).

Multimodal Discourse: Discourses are "socially situated forms of knowledge about (aspects of) reality. This includes knowledge of the events constituting that reality...as well as a set of related evaluations, purposes, interpretations and legitimations." Modes are "semiotic resources which allow the simultaneous realization of discourses and types of (inter)action... Modes can be realized in more than one production medium. Narrative is a mode because it allows discourses to be formulated in particular ways...because it constitutes a particular kind of interaction, and because it can be realized in a range of different media" (Kress & Van Leeuwen, 2001, pp. 20-22).

Telepresence: Telepresence has been successfully achieved when the mediated environment is perceived by the user as having similar "presence" and importance as the physical environment (Kiousis, 2002).

Vicarious Kinaesthesia: The dimension of direct physical involvement which gives the user in a mediated environment the impression of agency, of controlling events that are taking place in the present (Darley, 2000, p. 157).

Visual and Physical Interfaces for Computer and Video Games

Barry Ip
University of Wales Swansea, UK

Gabriel Jacobs
University of Wales Swansea, UK

INTRODUCTION

Over the last three decades and, above all, during the last few years, advances in areas that have been crucial for the success of the now multi-billion-dollar computer and video game industry (in particular, those of graphics and gameplay complexity) have been nothing short of breathtaking. Present-day console games run on machines offering quite remarkable possibilities to game developers. Their stylish presentation and compelling interactivity continue to set exceedingly high standards to which many serious applications running on desktop computers can only aspire. In spite of their adolescent image, games (particularly, console games) have continually raised general computer-user expectations.

BACKGROUND

In August 2004, 128-bit consoles (Playstation2, Xbox, Gamecube) were approaching the end of their product lifecycles and were due to be replaced by 256-bit systems. It is inevitable that games for the new machines will offer even greater sophistication in their user interfaces, especially with respect to graphics. It is not surprising, then, that interest in this area is intensifying, not only within the games development community (as evidenced in dedicated Web-based resources for game design, such as those at Gamasutra—www.gamasutra.com) but academia with the increase in the number of universities delivering game-design courses paralleling the growing quantity of research devoted to the topic. It is also in the field of on-screen visual interface, as opposed to physical interface (hardware such as the now common joypad games controllers), that most progress has been made and on which most research currently is centered.

Visual Interface

Screen displays have improved beyond recognition since the dawn of commercially available computer games in the 1970s. *Spacewar* (Figure 1), released in 1962 for the PDP-1 mainframe computer, often is referred to as the first graphics-based computer game, but it was not until the advent of Atari's *Pong* in 1975 (Figure 2) that computer games entered the

Figure 1. Spacewar (1962 – PDP-1)

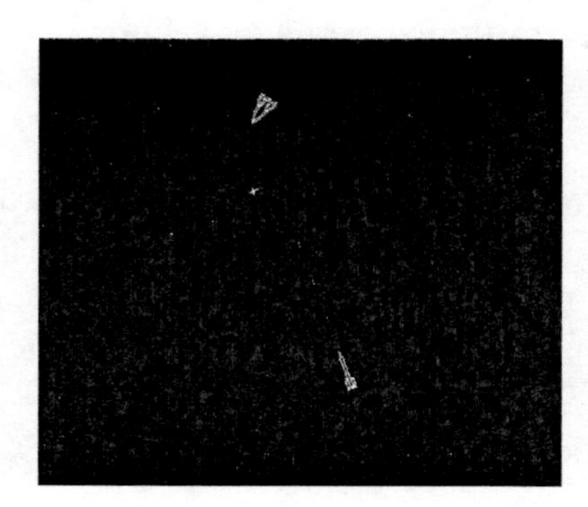

Figure 2. Pong (1975 – Atari)

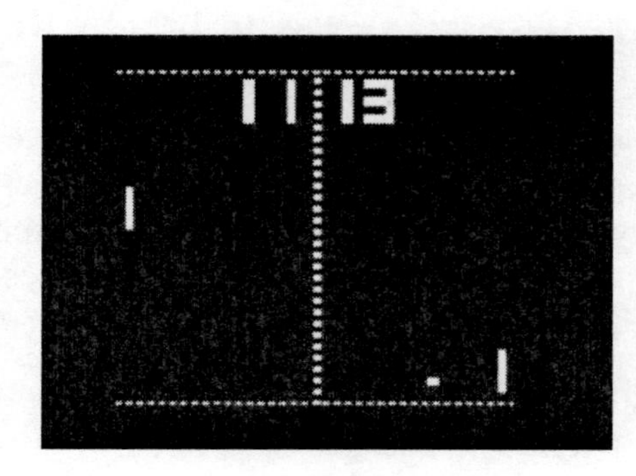

home, and the real computer-games industry began. The visual interface of *Spacewar* nonetheless typified that of the 1960s and 1970s in both its graphical simplicity and the undemanding nature of the user control it offered; gamers had only four options—rotate left, rotate right, thrust, and shoot. Still less advanced, even given the 13-year age gap, was the interface of *Pong*—players merely moved a block of pixels up and down; the block was supposed to represent a table-tennis bat that sent a square "ball" to the other side of the screen at an angle determined by the position of the bat and the previous stroke. Jump forward in time to 1989, and there was something of a transformation in the norm for the games interface. *Super Mario World 3* (Figure 3) characterized games of its period with its basic two-dimensional platform-style visuals, but the interface was, in fact, far more sophisticated than the games of the 1970s and early 1980s; in addition to colour, it offered dynamic on-screen textual information, including options between stages and more complex controls. The visual interfaces of the current generation of games have taken on even greater complexity, as exemplified by *Mario Sunshine* (Figure 4), where 3D rendering, a wide array of controls, and changes in visual perspectives (e.g., from first- to third-person and 360-degree camera angles) are the order of the day. Thus, whereas a quarter of a century ago, even inexperienced gamers were able to play a game to the maximum of what it had to offer with barely any learning involved, this is no longer the case with most of today's games, given the degree of familiarity required for understanding and making full use of a typical game's interface. This is particularly true of the strategy, simulation, and role-play genres, where the emphasis on information management necessitates an intricate visual interface (see Figure 5 for one example).

Figure 3. Super Mario World 3 (1989 – Nintendo Entertainment System)

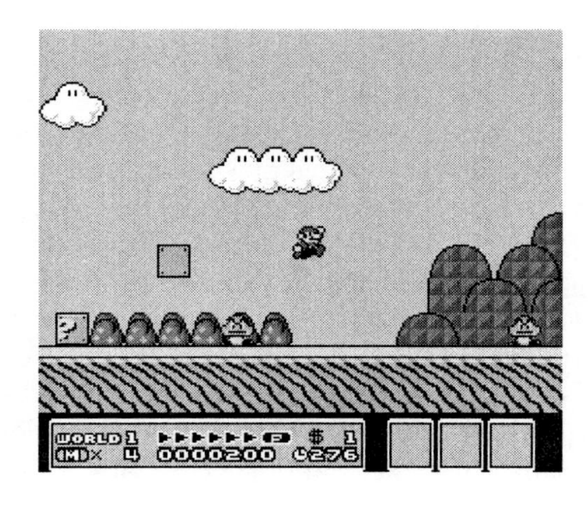

Yet, despite the extent of the revolution in visual interfaces, game designers still need to adhere to certain conventions (Poole, 2000). While the task of designing effective interfaces in most cases is linked inevitably to the type of game being developed, research has shown that the design of any game generally considered to be good tends to conform to a fixed pattern (Cousins, 2003; Fabricatore et al, 2002; Ip & Jacobs, 2004). Accordingly, even with all the opportunities offered by today's hardware, there is not as much freedom in design as might be

Figure 4. Mario Sunshine (2002 – Nintendo Gamecube)

Figure 5. Warcraft 3: Frozen Throne (2003 – PC)

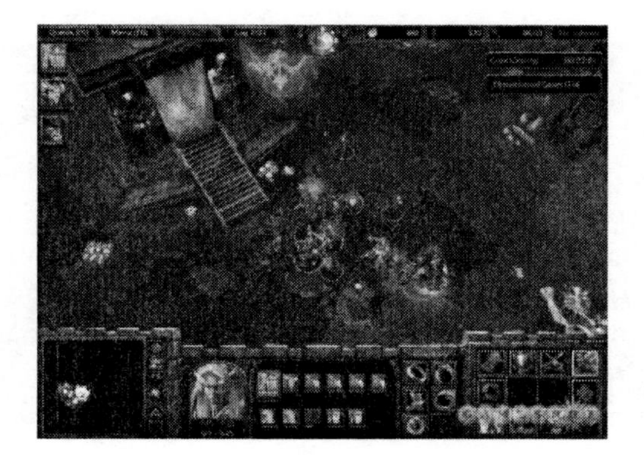

imagined. Studies by academic researchers such as Taylor (2002) and Warren (2003) and those by hands-on games designers such as Dalmau (1999) and Caminos and Stellmach (2004) have shown convincingly that good visual interface design is now a prerequisite for those attempting to develop a truly immersive gaming experience, but that using the full power of the hardware does not in itself lead to effective design. Johnson and Wiles (2001) have proposed that the most successful games are those whose user interfaces invoke a deep sense of concentration, enjoyment, and absorption, whether or not they are innovative. Caminos and Stellmach (2004) discuss the issues surrounding the development of an intuitive user interface for games, noting that, despite what may appear to be simple design problems, getting the basic interface right is extraordinarily difficult. However, the full capabilities of the hardware cannot be ignored, of course. In addition to what one might call the conventional aspects of visual-interface design (basic screen layout, menu design, etc.), modern game designers must also take account of more complex possibilities, such as the point-of-view (camera angle) delivered to the gamer as smoothly as possible, or realistic graphical effects (Adams, 1999; Poole, 2000; Schell & Shochet, 2001). Taylor (2002) has examined the delicate relationship between first- and third-person perspectives and how these influence gameplay, while Federoff (2002) has proposed various methods for the evaluation of a game's playability based on its visual appearance.

Physical Interface

Research on game user interfaces so far has converged above all on visuals. This is somewhat surprising, since while for the majority of non-game computer applications the hardware interface consists of nothing more than a keyboard and a mouse, the present generation of computer and video games can benefit significantly from a broader range of peripherals. In spite of this, hardware games interfaces have changed far less over the years than their on-screen counterpart.

The most common devices for domestic gaming, as defined by the standard mode of interaction with the game, are joypads, joysticks, and keyboards, which typically are bundled with the initial purchase of the hardware. Many other optional devices have also been used over the past two decades, and items such as steering wheels and light guns have been fairly popular for use with appropriate games. Recently, dance mats and pressure pads, which enable users to interact with their entire bodies, have also been gaining some ground. Peripherals designed for the arcade market include innovative hardware interfaces, such as skiing platforms and driving or flying simulators, but the high cost of producing these as well as the bulkiness of the cabinets they use have more or less confined them to their arcades; there has been little or no transference to the domestic market.

Nevertheless, there have been numerous attempts to introduce elaborate peripherals into the home market, some of which have been spectacular failures. The promise of virtual-reality (VR) hardware, head-mounted displays above all, which in the 1990s showed all the signs of offering the ultimate gaming experience, has remained largely unfulfilled. Head-mounted displays fell short of expectations in a number of ways, including comparatively poor graphic resolution and their tendency to cause so-called VR sickness, the equivalent of motion sickness (Adams, 1998; Edge, 1999; Vaughan, 1999), not to mention the fact that players generally do not want to wear cumbersome equipment (Hecker, 1999). Thus, VR has not yet been able to establish a significant foothold in the game industry, whereas over a relatively long period of time, there have been numerous successful applications of it in other fields (Kalawsky, 1993). Two other examples of advanced physical interfaces that have not achieved widespread use are Mattel's Powerglove for the Nintendo Entertainment System (NES) and Nintendo's Virtual Boy. The Powerglove, introduced in 1989, was a potentially winning VR data glove that could track hand motion in three dimensions; it was soon withdrawn because of serious technical problems (Gardner, 1989). Virtual Boy was a stereoscopic device that claimed on its release in 1995 to be ushering in a new era of video games. It failed badly as a result of a rather clumsy design and relatively poor interaction possibilities (Herman, 1997).

Conventional joypads, keyboards, and mice, then, persist as the dominant physical interfaces. As can be seen in Table 1 (a comparison among the most popular platforms and their hardware interfaces since the 1970s), in spite of the availability of a wide selection of peripherals, the principal physical interface for console games has remained the joypad and, for computer games, the keyboard and mouse. Indeed, such is the emphasis on these interfaces, that even titles that would appear to be ideally suited to genre-specific devices (i.e., driving and shooting games) are designed primarily with joypads and keyboards in mind, and comparatively common additional interfaces such as steering wheels and light guns are still seen largely as novelty items. Evident in Table 1 also is that the number of buttons on joypads has increased steadily with each new generation since the 1970s (except between generations 4 and 5, and in the case of the Nintendo Gamecube, the number of buttons actually decreased from the previous N64 machine). Yet the increase in the number of buttons aside, it is clear that the predominant physical interface (the joypad for console games, the keyboard and mouse for the desktop or laptop computer) has actually altered little since the 1970s.

FUTURE TRENDS

The fact that over the years there has been little change in standard physical game interfaces is now being taken seriously by manufacturers, because it may be only a matter of time before joypads, keyboards, and mice simply will no longer fit the bill. Hence, we are seeing developments such as Sony's EyeToy (in which a camera attached to Playstation

Table 1. Most popular physical interfaces of game platforms

Generation /decade	Platform	Standard interface	Common optional interfaces
1/1970s	Magnavox Odyssey	Analogue dial controller	/
1/1970s	Atari VCS/2600	Single-button joystick	/
2/1980s	Nintendo NES	Two-button joypad	Joystick, infra-red gun
2/1980s	Sega Master System	Two-button joypad	Joystick, infra-red gun
2/1980s	NEC PC-Engine	Two-button joypad	/
3/1980s	Sega Megadrive	Three-button joypad	Joystick, six-button joypad, infra-red gun
3/1990s	Nintendo Super NES	Six-button joypad	Joystick, infra-red gun
3/1990s	SNK Neo Geo	Four-button joystick	Joypad
4/1990s	Playstation1	8-button joypad	Infra-red gun, steering wheel, dance mat
4/1990s	Sega Saturn	8-button joypad	Infra-red gun, steering wheel
4/1990s	Nintendo N64	9-button joypad	/
5/late 1990s to present	Playstation2	8-button joypad with additional analogue sticks and built-in rumble	Infra-red gun, steering wheel, dance mat, EyeToy
5/late 1990s to present	Xbox	8-button joypad with additional analogue sticks and built-in rumble	Infra-red gun, steering wheel, dance mat
5/late 1990s to present	Gamecube	7-button joypad with additional analogue sticks and built-in rumble	Steering wheel
1970s to present	PC	Keyboard, mouse	Joypad, joystick, steering wheels/flight simulation controllers, speech recognition

2 allows the player to be immersed in a game by becoming a character on the screen) and hands-free gaming devices based on relatively simple Web cams that can be used as a substitute for the mouse (Gorodnichy & Roth, 2004). Further, despite the obstacles presented by VR, research is on the increase into the possibilities for domestic use offered by immersive VR-type full-body interaction (Warren, 2003). One of the most intriguing among the emerging ideas is an affective interface that has been described as "computing that relates to, arises from, or deliberately influences emotions" (Picard, 1997, p. 3). A study conducted by Scheirer et al (2002) has demonstrated how user emotions can be measured and taken into account in order to facilitate the design of the user interface and, even more importantly, to enable certain factors (e.g., screen layout or number of button presses before the next option is presented) to respond by making changes in real time that depend on the behavior of the user (e.g., when frustration is detected). It may be that such developments will be necessary in order for the videogame industry to continue to flourish, since, even with the advent of 256-bit machines capable of photo-realistic 3D graphics and fully controllable, interactive, seamless motion video, improvements in the visual interface alone may prove insufficient to ward off the player dissatisfaction, which, since the early 1990s and in the face of all the success, has been a central factor in holding back what might have been an even greater market volatility (Edge, 1993, 2002, 2003, 2004). Indeed, it may be that the very existence of ever-more advanced visual interfaces, if unaccompanied by parallel developments in physical interfaces, may cause such a disparity between the two that design creativity will suffer, while the market witnesses even greater consumer resistance than has been the case so far.

CONCLUSION

We have seen that the visual interface for games has come a long way in terms of both capabilities and design complexity, while the physical interface has not kept pace. Of course, it may be that even if they lack realism, the joypad controllers of Gamecube, Playstation, Xbox are the most natural command mechanisms for games requiring users to move objects or characters around a virtual space and to make them perform actions. In driving games, a steering wheel and pedals; in first-person shooters, a gun; in golf games, a golf club; and so forth, would seem to be obvious standard replacements for current controllers, but in practice, joypads have proven so far to retain a solidly entrenched position as the dominant vehicle of interaction between player and game. This is partly because many other peripherals have not reached a stage at which they can be used with the ease and accuracy of a joypad (drivers of real cars find, for example, that even in the most expensive games, steering wheels do not behave like real steering wheels, even when properly calibrated). In any event, while the joypad can be used across many different types of game, specialized peripherals are self-evidently unsuitable for all types. In fact, the joypad has become so much a part of the videogame culture that it well may persist in more or less its present form, regardless of future visual-interface developments, while the keyboard and the mouse, however clumsy they may be when used for PC games, so far have shown considerable resistance even against the joystick, notwithstanding the low cost of the latter and specific joystick ports on many desktop computers.

On the other hand, the recent spate of releases of compact and relatively inexpensive interactive control devices (e.g., the motion-sensitive EyeToy, headphones, and microphones offering voice commands and audio feedback; pressure pads for skateboarding and snowboarding games; increasingly accurate light guns with long reach; and infrared pads for football games, which recognize shooting, tackles, and so on) still may see substantial growth as they break free of the confines of arcades. Needless to say, the obstacle to mass sales, which is the specificity of such devices, always will remain, but as prices fall and as visual interfaces make greater demands on their corresponding controllers, so the all-purpose joypad eventually may become more of a secondary device than it is at present.

If this does not prove to be the case, one wonders how joypads, keyboards, and mice will be able to cope with the coming advances in visual interfaces and to what extent we shall see consumer resistance as a result. That said, who could have predicted a few years ago that Nintendo's Gameboy, with its diminutive screen and limited control functions (let alone mobile phones with their even smaller screens and more primitive control functions) would have been so well received by so many consumers as hand-held games devices?

REFERENCES

Adams, E. (1998). *The VR gorilla-rhino test.* Retrieved August 2004, from www.gamasutra.com/features/designers_notebook/19980814.htm

Adams, E. (1999). *Designing and developing sports games.* Retrieved August 2004, from www.gamasutra.com/features/designers_notebook/19990924.htm

Caminos R., & Stellmach, T. (2004). *Cross-platform user-interface development.* Retrieved August 2004, from www.gamasutra.com/gdc2004/features/20040326/caminos_02.shtml

Cousins, B. (2002). Mind your language. *Develop, 20*, 14-16.

Dalmau, S. (1999). *Learn faster to play better: How to shorten the learning cycle.* Retrieved August 2004, from www.gamasutra.com/features/19991108/dalmau_02.htm

East is Eden. (2002). *Edge, 108*, 52-63.

Fabricatore, C., Nussbaum, M., & Rosas, R. (2002). Playability in action videogames: A qualitative design model. *Human-Computer Interaction, 17*(4), 311-368.

Federoff, M. (2002). *Heuristics and usability guidelines for the creation and evaluation of fun in video games*. Master's thesis. University of Indiana, Bloomington.

Game over. (1993). *Edge 3*, 62-67.

Gardner, D.L. (1989). The power glove. *Design News, 4*, 63-68.

Gorodnichy, D., & Roth, G. (2004). Nouse, use your nose as a mouse: Perceptual vision technology for hands-free games and interfaces. *Image and Vision Computing, 22*(12), 931-942.

Hecker, C. (1999). Whatever happened to virtual reality? *Edge, 71*, 62.

Herman, L. (1997). *Phoenix: The fall and rise of videogames*. Union, NJ: Rolenta Press.

Ip, B., & Jacobs, G. (2004). Quantifying games design. *Design Studies,* 25, 607-624.

Johnson, D., & Wiles, J. (2001). Effective affective user-interface design in games. *Proceedings of the International Conference on Affective Human Factors Design*, Singapore.

Kalawsky, R. (1993). *The science of virtual reality and virtual environments*. Wokingham, UK: Addison-Wesley.

Picard, R. (1997). *Affective computing*. Cambridge, MA: MIT Press.

Poole, S. (2000). *Trigger happy*. New York: Arcade.

Scheirer, J., Fernandez, R., Klein, J., & Picard, R. (2002). Frustrating the user on purpose: A step toward building an affective computer. *Interacting with Computers, 14*, 93-118.

Schell, J., & Shochet, J. (2001). *Designing interactive theme park rides: Lessons from Disney's battle for the buccaneer gold.* Retrieved August 2004, from www.gamasutra.com/features/20010706/schell_01.htm

Taylor, L. (2002). *Video games: Perspective, point-of-view, and immersion*. Master's thesis. University of Florida.

Vaughan, B. (1999). Whatever happened to virtual reality? *Edge, 71*, 64.

Warren, J. (2003). *Unencumbered full body interaction in video game*. Master's thesis. New York: Parsons School of Design New York.

Whatever happened to virtual reality? (1999). *Edge, 71*, 60-67.

KEY TERMS

Computer Game: An interactive game played on a computer.

First-Person Perspective: The visualization of the gaming environment through the eyes of the character.

Game(s) User Interface: Elements and devices through which the user interacts with the game.

Joypad: A palm-sized device designed for use with both hands to interact with the game. Its layout is typified by directional keys on the left and buttons on the right and top sections of the pad. Modern pads incorporate additional analogue sticks on the left or on both the left and right sides.

Joystick: A 360-degree stick mounted on a sturdy platform of buttons used for interacting with the game; used predominantly in stand-alone arcade machines and early home consoles.

Light Gun: A device used for shooting games, which allows the user to target objects on screen; used predominantly in stand-alone arcade machines and some home consoles.

Physical Interface: Tangible devices for interaction with the game.

Third-Person Perspective: The visualization of the gaming environment through an external body of the character.

Videogame: An interactive game played on a stand-alone arcade machine or home console.

Visual Interface: Visual on-screen elements that can be altered or that provide information to the user during interaction with the game.

WAP Applications in Ubiquitous Scenarios of Work

Anxo Cereijo Roibás
University of Brighton, UK

INTRODUCTION

European users have eagerly adopted novel forms of digital media and related information and communications technologies (Stanton, 2001), making them a part of their increasingly varied and segmented cultures (Brown, Green, & Harper, 2001). For example, the young are active consumers of music, videos, movies, and games; businessmen on the other hand need more and more working tools and applications that enable connectivity when they are on the move. A not very dissimilar scenario is envisaged on troops in action where work on tactical and strategic information and mission management, command, and control, including real-time mission replanning, are essential. All these users rely on the Internet, i-TV, and mobile phones, and they have adapted all of these into the fabric of their lifestyles, or in short, their mobile life. But, functionality cannot be the main driver for design as mobile life is also deeply founded upon shared values and worldviews of the users, pleasure, enjoyment, culture, safety, trust, desire, and so forth (Rheingold, 1993).

For example, WAP (wireless application protocol) technologies seemed to provide a powerful tool to the mobile worker. However, it is well known the fraud of WAP mainly due to the scarce usability, high usage cost, and inadequate range of the services provided together with intrinsic limitations of the device itself (insufficient memory storage, low battery autonomy, poor screen resolution, etc. [Cereijo Roibás, 2001]). However, some WAP applications have been widely used by Italian users. The success of this system of applications is due to its efficiency, effectiveness, and relevance for some specific work purposes. Each of the services will be analysed, describing the expected use of each service and the actual use of it by Italian users.

BACKGROUND

Many efforts have been devoted to design valuable tools for the mobile worker, but so far only a few of them have been successful. Surprisingly, most of the mobile applications originally designed as work tools (chat, message board, etc.) have found a fertile market in entertainment. There seem to be two main causes of this failure: the lack of usability of the applications provided and, above all, the failure to create realistic usage scenarios.

The European Commission (EC) and European Space Agency (ESA) jointly set up an expert group on collaborative working environments that met for the first time in Brussels on May 4, 2004. The expert group discussed the vision of next-generation collaborative working environments (NGCWEs). The vision drawn by the expert group was that NGCWEs will deliver a high quality of experience to coworkers, and will be based on flexible service components and customized to different communities. Mobility, interaction among peers (systems and persons), utility-like computing capacity and connectivity, contextualization and content, security, privacy, and trust were among the RTD challenges in nine areas identified by the experts.

FROM THE MOBILE ENTERTAINMENT COMMUNITY TO THE NOMADIC WORK TEAM

If there is always an uncertainty about the most suitable use of a new service, the case of wireless applications is not an exception (Flynn, 2002). As will be explained for each case, the following applications have been created to work in different platforms: WAP, WEB, and SMS (short-message service). They were supposed to find wide use within the increasing Italian mobile community for entertainment purposes. However, they have been used more and more as work tools. Obviously, some services such as the multimode chat have had and still have a strong use for entertainment purposes ("Now There is the 'Wappario,' 2000). Users that need to communicate with their colleagues in real time when they are out of the office are largely using mobile chat as a working tool. There is no experience of the above-mentioned phenomenon in the classical Web-only chat services.

THE USE OF WAP AND SMS APPLICATIONS AS WORKING TOOLS

Who, Where, When, Why, and How

Supposed target users of multimode applications (Burkhardt et al., 2002) were thought to be teenagers, but, as recent technology history has shown, consumers' behaviour and use of technology have contradicted predictions. Mobile technologies, ranging from WAP to SMS, from GPRS (general packet radio service) to MMS (multimedia message service), were born to meet the desire of teens, the same people who had made text messaging their preferred medium. However, as had happened for short messages, multimode applications have been used for a purpose that contradicts its unique selling proposition, confirming once more the inner limits of today's marketing of new technology. Outside cubicles, mobility is at the heart of multimode applications, allowing users to make a real personal use of technology. People stopped being slaves of given and prepackaged software: They want technology when it shows to be relevant in real life—not a utopian Internet where they are living nowhere but in wires because without wires, people are free. Relevance is the reason for multimode applications: People want technology relevant to work, dating, participating in TV voting, and chatting. What counts is that technology is at their hands when they need it, and they are the ones giving meaning to it. It is not a chat software waiting for them at a URL (uniform resource locator), but it is a person at home or in the office with a need and a device that can help satisfy it. How these technologies are used is a matter of context: There is no optimized path to be imposed on end users. Multimode applications impose new challenges because end users get more and more demanding. End users do not want technology, but services. Technology is going back behind the scenes.

Mobile Applications for Work and Sharing Knowledge

The services that will be discussed below were launched by the Italian mobile-services provider HiuGO in early 2001 as part of a Blu-branded offering. All applications fully exploited the potentiality of the mobile medium by combining messaging with WAP and later on GPRS browsing. The applications have been tested and corrected according to the company's usability standards both before their launch and after it as data from consumers were collected. Continuous interaction with mobile users included the following activities: monitoring usage and traffic patterns, controlling services requirements, polling end users' expectations, analysing end users' interactions (Schneiderman, 1987), and checking users' satisfaction.

The main methodology consisted of the following:

- Users' perceived value of the services provided (questionnaires)
- Self-training with a quick learning phase (usability test)
- Time for task completion (usability test)
- Number of users' irreversible mistakes (usability test)
- Satisfaction of users' expectations (questionnaires)
- Level of users' interaction (usability test)
- Flexibility toward users' personalization (usability test and questionnaires)

Message Board

A message board is a thematic forum with file-sharing options. Users can access the forum via SMS or WAP/GPRS. All functionalities are available and optimized for mobile devices. Users sub-

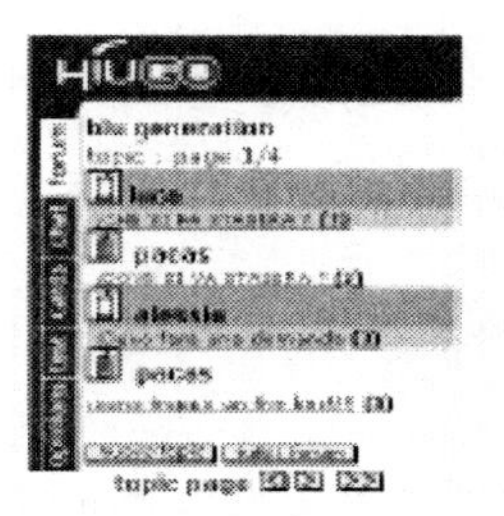

Figure 1. Message board on an IPAQ display

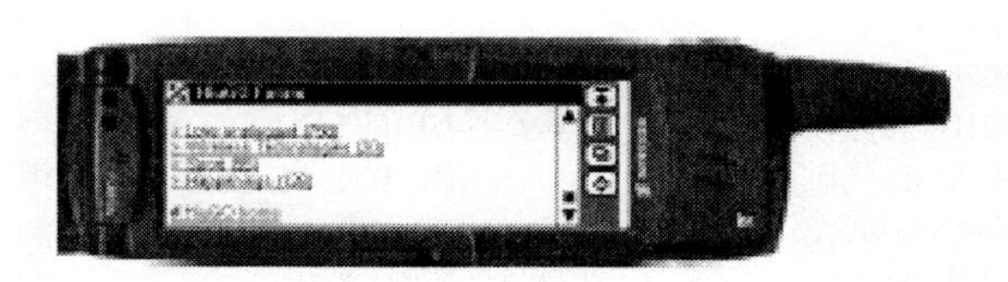

Figure 2. Multimode chat on a SmartPhone display

scribe to one or more forums choosing a nickname and password. Once successfully registered, they can consult content browsing through topics, publish messages, and receive alerts when a topic they are interested in is updated simply by setting a keyword alert. For example, by sending a text message with the following keyword and search parameters, <update food supplies>, the user will get an alert every time a message is sent. The mobile device turns out to be essential in getting critical updates. This tool has revealed to be ideal also for knowledge sharing among working groups (Figure 1). In distance learning courses and for nomadic groups of workers and journalists, message boards turn out to be a vital resource. Journalists who need to move from one place to another can keep in touch with each other and with the editorial team thanks to closed message boards where they can consult last-minute updates and information as well as rumours, or send short previews of their articles in real time as events are happening.

Multimode Chat

Multimode chat is a real-time communication service that works via SMS (asynchronous), the Internet, and WAP/GPRS (synchronous): The same interface is accessible from a PC (personal computer) and from a mobile phone (each chatter has an icon or avatar that evidences the device he or she is using in that moment). The chat service is organised in thematic rooms and permits one-to-many and one-to-one messages. Users can also create their buddy lists and be alerted via SMS when one of them is available (Figure 2). Multimode chat has showed itself to be essential in working communities that need to keep in touch on the fly, for example, among study groups and in distance learning courses. When public chats are held with a teacher, students part of a buddy group get reminders of appointments. Those who cannot access the Net via PCs can participate in the discussion via mobile devices. Again, an example of usage is given by journalists who log in the chat via mobile phones and can send information to their teams and colleagues in real time while interviewing someone or during press conferences. Mobile chats are also useful for security guards who can keep in touch with each other through SMS or WAP/GPRS: They cannot be heard because they do not need to talk, allowing them to share secure information in a simple way.

PageMaker

PageMaker is a publishing tool that makes the creation of personal WAP pages easy and immediate. Users can publish their own content (text and images, and now also colour images with new devices) and protect it with a password. They can also use advanced interactive features such as personal chats and message boards accessible via the Internet or mobile devices. PageMaker has been adopted by all kinds of professionals willing to promote their activities. Lawyers promote their studios as do shoppers. Users searching for info can get localised results thanks to various positioning systems that have been implemented. Information sent to users can be enriched by interactive maps to help reach places. Potentialities of the medium have been exploited by marketing managers: Coupons and special offers can now reach the target in an unprecedented way: reaching them on a personal device with relevant information—a special offer responding to a specific need.

Multimode Mail

Multimode mail uses personal mailboxes accessible via PC and mobile devices. SMS is used to get alerts on new mails. All mail functionalities are available from the WAP interface. This service proves to be very useful for working communities of every kind.

Event Enhancer

Event enhancer is a complete suite of multimode software facilitating attendees and exhibitors during events. Users who subscribe to the service can receive information on locations and alerts on special events of their interest. A dedicated matchmaking engine also allows them to save time and effort in finding the right person at the right time: By inserting your profile and needs, you will be put in contact with the person or company you need to meet. Users who have been matched can also chat via SMS or WAP before meeting, exchange business cards, and download commercial information on Bluetooth-enabled handsets. Users can also book interesting events via SMS or WAP. The application has been adopted by schools: Courses have their own schedules available via mobile devices, teachers who give their contact information can be reached at any moment, and students can enroll in classes, seminars, or special courses at the last minute (in the Italian school system). But the event enhancer has also turned out to be a very useful and successful application for companies. It has been adopted as a marketing tool to optimize ROI on fairs and events: Procter & Gamble first adopted it at the international beauty fair in 2001, setting an example for others. Event Enhancer in fact helped in driving traffic to the stand and offered personalized service: Attendees were given the chance to set an appointment and get an SMS reminder, receive personalized advice in their mobile mail or via SMS, get a mobile coupon, and participate in an instant-win competition. Many other applications may be mentioned. Sometimes a simple SMS can improve productivity or facilitate work. An example is given by referees. When the match ends, each referee has to send the official score to the national federation. Once referees had to use faxes, but now a simple text message to a dedicated service number is all they have to give.

Transactions and Error Rates

In the following, we provide some interesting statistics (April 2001).

- **Message Board:** There are 150,000 active users, and 78% of them access the boards from mobile devices via WAP or SMS. The average number of transactions via SMS are 18 per month. The error rate is 7%. There are 20,000 new monthly users, and the churn rate is 7%.
- **Multimode Chat:** There are 100,000 active registered users, with 90% accessing from mobile devices via WAP or SMS. The average number of monthly transactions via SMS is 20. There is an error rate of 5.5%. There are 20,000 new monthly users, and there is a churn rate of 15%.
- **Multimode Mail:** There are 250,000 active users of multimode mail. The average number of transactions via SMS is 24 per month. An error rate of 3% exists. The number of new monthly users is 22,000. The churn rate is 3%.
- **PageMaker:** PageMaker has 70,000 active users, with 60% accessing from mobile devices via WAP or SMS. The average number of monthly transactions via SMS is 10. The error rate is 7%. The number of new monthly users is 6,000, and the churn rate is 7%.
- **Event Enhancer:** There are 120,000 active users, and 78% of them access the service from mobile devices via WAP or SMS. The average number of transactions via SMS is 10 per month. There is an error rate of 4%. The number of new monthly users is 4,000. The churn rate is 11%.

FUTURE TRENDS

Small personal interfaces, such as those on mobile phones, interconnected with other surrounding platforms (e.g., interactive TV, PCs, PDAs [personal digital assistants], in-car navigators, smart-house appliances, etc.) and particularly suitable for context-awareness applications (Schilit, Adams, & Want, 1994) will strongly stimulate the development and diffusion of the prospected ubiquitous communication scenarios. These new scenarios will imply the need to rethink new kinds of services and applications and of course new forms of content. A fertile research area in this sense regards the design of applications and services for the mobile worker (Winslow & Bramer, 1994).

Designing complex ubiquitous communication scenarios for work involving cross-platform customer technologies (ranging from I-TV, radio, music, and mobile phones to portable or wearable information devices) for different users and contexts requires an original way of conceiving the interactive user experience. This need to design novel ubiquitous and mobile services and products that will address the new demands, requirements, and potentials of mobile workers in critical situations implies a new approach to design that goes beyond the existing conventions. This design for innovation will lead to the identification of novel experience models and their social, cultural, and regulatory implications, allowing us to explore new and relevant interactive forms and paradigms

Potential challenges will be the creation of enhanced network-enabled capability in distributed intelligent systems through superior context awareness (Tamminen, Oulasvirta, & Toiskallio, 2004), collaborative planning and replanning and coherency in asynchronous joint and collaborative work (Luff & Heath, 1998), and the improvement of human communication-systems effectiveness in general. These integrated systems should contribute to improving information operator uptake under stress, augmenting cognition and decision making, achieving information and knowledge advantage, increasing decision agility and decision efficacy, and maintaining good strategic and supervisory control of mixed distributed autonomous and manned assets (Suchman, 1995).

In this sense, we can anticipate the use of intelligent agents (acting as information brokers) embedded in ubiquitous systems (Maes, 1991) aimed to improve the effectiveness and accessibility of human interaction with context-awareness technologies. These agents need to have the following characteristics: the capabilities of learning, organising, carrying out routine tasks, and taking autonomous decisions for support in case of unexpected events.

CONCLUSION

As the Italian case shows, it was not necessary to wait until the arrival of the more promising 3G (third-generation) technologies to have successful mobile interactive services that enhance communications in work environments (Cereijo Roibás et al., 2002). This case demonstrates that it is possible to design useful and usable services for a starting poor technology despite its HCI (human-computer interaction) limitations (Cereijo Roibás et al.). This experience shows how extensive attention to the mobile user is essential in order to envision realistic and relevant scenarios of use (Kleinrock, 1996).

REFERENCES

Brown, B., Green, N., & Harper, R. (Eds.). (2001). *Wireless world: Social, cultural and interactional issues in mobile communications and computing.* London: Springer Verlag.

Burkhardt, J., Henn, H., Hepper, S., Rindtorff, K., & Schack, T. (2002). *Pervasive computing: Technology and architecture of mobile Internet applications.* Addison Wesley.

Cereijo Roibás, A. (2001). Examining the UI design challenges of supporting multimedia applications on next generation terminals and identifying ways to overcome these. *Proceedings of the 2002 User Interface Design for Mobile Devices Conference,* London.

Cereijo Roibás, A., & Stellacci, A. (2002). Factors of success of UCommerce. *Proceedings of the Fifth International Conference on Electronic Commerce Research,* Montreal, Canada.

Cereijo Roibás, A., Stellacci, A., & Aloe, A. (2001). Physical and cognitive skills/limitations of mobile users. *Proceedings of the 2001 Mobile User Interface Conference,* London.

Kleinrock, L. (1996). Nomadicity: Anytime, anywhere in a disconnected world. *Mobile Networks and Applications, 1*, 351-357.

Luff, P., & Heath, C. (1998). Mobility in collaboration. *Proceedings of the Conference on Computer Supported Cooperative Work* (pp. 305-314).

Maes, P. (Ed.). (1991). Situated agents can have goals. In *Designing autonomous agents.* Cambridge, MA: MIT-Bradford Press.

Now there is the "Wappario." (2000). *Dictionary of the Cybernauts,* p. 28. Rome: La Republica.

Rheingold, H. (1993). *The virtual community: Homesteading on the electronic frontier.* Reading, MA: Addison-Wesley.

Rosenberg, N. (1996). Uncertainty and technological change. In R. Landau, T. Taylor, & G. Wright (Eds.), The *mosaic of economic growth.* Stanford University Press.

Schilit, W. N., Adams, N. I., & Want, R. (1994). Context-aware computing applications. *Proceedings of the Workshop on Mobile Computing Systems and Applications* (pp. 85-90).

Schneiderman, B. (1987). *Designing the user interface: Strategies for effective human-computer interaction.* Reading, MA: Addison-Wesley.

Stanton, N. (Ed.). (2001). Ubiquitous computing: Anytime, anyplace, anywhere [Special issue]? *International Journal of Human-Computer Interaction, 13*(2), 107-111.

Suchman, L. (1995). Making work visible. *Communications of the ACM, 39*(9), 56.

Tamminen, S., Oulasvirta, A., Toiskallio, K., & Kankainen, A. (2004). Understanding mobile contexts. *Personal and Ubiquitous Computing, 8*(2), 135-143.

Winslow, C. D., & Bramer, W. L. (1994). *Futurework: Putting knowledge to work in the knowledge economy.* Anderson Consulting. New York: The Free Press.

KEY TERMS

Decision-Support Systems: Software designed to facilitate decision making, particularly group decision making.

Ethnography: An approach to research that involves in-depth study through observation, interviews, and artefact analysis in an attempt to gain a thorough understanding from many perspectives.

GPRS (General Packet Radio Service): A standard for wireless communications that runs at speeds up to 115 kilobits per second, compared to current GSM (global system for mobile communications) systems' 9.6 kilobits. GPRS supports a wide range of bandwidths, is an efficient use of limited bandwidth, and is particularly suited for sending and receiving small bursts of data, such as e-mail and Web browsing as well as large volumes of data.

MMS (Multimedia Message Service): A store-and-forward method of transmitting graphics, video clips, sound files, and short text messages over wireless networks using the WAP protocol. Carriers deploy special servers, dubbed MMS centers (MMSCs), to implement the offerings on their systems. MMS also supports e-mail addressing so the device can send e-mails directly to an e-mail address. The most common use of MMS is for communication between mobile phones. MMS, however, is not the same as e-mail. MMS is based on the concept of multimedia messaging. The presentation of the message is coded into the presentation file so that the images, sounds, and text are displayed in a predetermined order as one singular message. MMS does not support attachments as e-mail does.

Multimode: Service that can be accessed and used with different interfaces in a multiplatform system (e.g., a chat that is available across handhelds and PCs).

Shared-Window System: System that allows a single-user application to be shared among multiple users without modifying the original application. Such a system shows identical views of the application to the users and combines the input from the users or allows only one user to input at a time.

SMS (Short-Message Service): A text-message service offered by the GSM digital cellular-telephone system. Using SMS, a short alphanumeric message can be sent to a mobile phone to be displayed there, much like in an alphanumeric pager system. The message is buffered by the GSM network until the phone becomes active. Messages must be no longer than 160 alphanumeric characters and contain no images or graphics.

Usability Lab: A lab designed for user testing, typically a quiet room with computer equipment and a space for an observer to sit, along with a special observation area.

User Studies: Any of the wide variety of methods for understanding the usability of a system based on examining actual users or other people who are representative of the target user population.

User Testing: A family of methods for evaluating a user interface by collecting data from people actually using the system.

WAP (Wireless Application Protocol): A protocol used with small handheld devices and small file sizes.

Wearable and Mobile Devices

Sara de Freitas
Birkbeck College, University of London, UK

Mark Levene
Birkbeck College, University of London, UK

INTRODUCTION

Information and Communication Technologies, known as ICT, have undergone dramatic changes in the last 25 years. The 1980s was the decade of the Personal Computer (PC), which brought computing into the home and, in an educational setting, into the classroom. The 1990s gave us the World Wide Web (the Web), building on the infrastructure of the Internet, which has revolutionized the availability and delivery of information. In the midst of this information revolution, we are now confronted with a third wave of novel technologies (i.e., mobile and wearable computing), where computing devices already are becoming small enough so that we can carry them around at all times, and, in addition, they have the ability to interact with devices embedded in the environment.

The development of wearable technology is perhaps a logical product of the convergence between the miniaturization of microchips (nanotechnology) and an increasing interest in pervasive computing, where mobility is the main objective. The miniaturization of computers is largely due to the decreasing size of semiconductors and switches; molecular manufacturing will allow for "not only molecular-scale switches but also nanoscale motors, pumps, pipes, machinery that could mimic skin" (Page, 2003, p. 2). This shift in the size of computers has obvious implications for the human-computer interaction introducing the next generation of interfaces. Neil Gershenfeld, the director of the Media Lab's Physics and Media Group, argues, "The world is becoming the interface. Computers as distinguishable devices will disappear as the objects themselves become the means we use to interact with both the physical and the virtual worlds" (Page, 2003, p. 3). Ultimately, this will lead to a move away from desktop user interfaces and toward mobile interfaces and pervasive computing.

BACKGROUND

Mobile computing supports the paradigm of anytime-anywhere access (Perry et al., 2001), meaning that users have continuous access to computing and Web resources at all times and where ever they may be. Used in a wide range of contexts, mobile computing allows:

1. The extension of mobile communications and data access beyond a desktop and static location.
2. Access to electronic resources in situations when a desktop/laptop is not available.
3. Communication with a community of users beyond the spatio/temporal boundaries of the work or home location.
4. The ability to do field work; for example, data collection, experience recording, and notetaking.
5. Location sensing facilities and access to administrative information.

Mobile devices have several limitations due to their small size (form factor) that need to be considered when developing applications:

1. **Small Screen Size:** This can be very limited, for example, on mobile phones. Solutions to this problem necessitate innovative human-computer interaction design.
2. **Limited Performance:** In terms of processor capability, available memory, storage space, and battery life. Such performance issues are

continuously being improved, but to counter this, users' expectations also are growing.

3. **Slow Connectivity:** Relatively slow at the moment for anywhere Internet connectivity; 3G technologies promise to improve the situation. Wireless LAN connectivity, such as 802.11, provides simple and reliable performance for localized communication.

Mobile devices generally support multimodal interfaces, which ease usability within the anytime-anywhere paradigm of computing. Such support should include:

- Pen input and handwriting recognition software.
- Voice input and speech recognition software.
- Touch screen, supporting color, graphics, and audio where necessary.

In order to take advantage of the promise of mobile computing devices, they need to have operating systems support such as:

- A version of Microsoft Windows for mobile devices.
- Linux for mobile devices.
- Palm for PDAs.
- Symbian for mobile phones.

In addition, mobile devices need to support applications-development technologies such as:

- Wireless Application Protocol (WAP), where in the current version content is developed in XHTML, which extends HTML and enforces strict adherence to XML (eXtensible Markup Language).
- J2ME (Sun Java 2 Micro Edition), which is a general platform for programming embedded devices.
- .NET framework, which includes Microsoft's C# language as an alternative to Java.
- NTT DoCoMo's i-mode, which currently covers almost all of Japan with well over 30 million subscribers. Phones that support i-mode have access to several services such as e-mail, banking, news, train schedules, and maps.

Standard software tools also should be available on mobile devices to support, among other applications:

- E-mail.
- Web browsing and other Web services.
- Document and data handling, including compression software.
- Synchronization of data with other devices.
- Security and authentication.
- Personalization and collaboration agents.
- eLearning content management and delivery, which is normally delivered on mobile devices via its Web services capability.

Apart from the last two, these tools are widely available, although the different platforms are not always compatible. This is not a major problem, since communication occurs through standard Web and e-mail protocols. Current personalization and collaboration tools are based mainly on static profiling, while what is needed is a more dynamic and adaptive approach. There are still outstanding issues regarding content management and delivery of eLearning materials, since these technologies, which we assume will be XML-centric, are still evolving.

HCI AND MOBILE AND WEARABLE DEVICES

This article will highlight some of the central HCI issues regarding the design, development, and use of mobile and wearable devices. Our review pertains to devices such as mobile phones, personal digital assistants (PDAs), and wearable devices, and less to mobile devices such as laptops and tablet PCs that generally are larger in size.

Several main issues regarding the HCI issues of using mobile and wearable devices have been posited in the literature, including contextual concerns (Lumsden & Brewster, 2003; Sun, 2003), limitations of the interface (Brewster, 2002), and their convergence with other technologies and systems. These devices reflect the range of different contexts that mobile and wearable technology can be used for interfacing with data sets, interactive content, and enhanced visual display that augment activities and exploration within physical environments.

Table 1. Summary of a selection of mobile and wearable interaction tools and interfaces

Interaction Tool/Interface Type	Example	Description	Reference
Gestural interfaces	Georgia Tech Gesture toolkit	The Georgia Tech toolkit allows for those developing gesture-based recognition components of larger systems. The toolkit is based upon Cambridge University's voice recognition toolkit and uses hidden Markov models.	Westeyn et al. (2003)
Voice input devices	Wearable Microphone Array (WMA)	The Wearable Microphone Array provides an interface between context aware speech and the wearable computer. The system is specially adapted for mobile use and is worn on a tie or shirt.	Xu et al. (2004)
Wearable orientation interfaces	Wearable orientation system	The wearable orientation system tested three different interfaces: a virtual sonic beacon, speech output, and a shoulder-tapping system. The latter two interfaces were found to be helpful for those with sight impairments.	Ross and Blasch (2002)
Wearable orientation interfaces	CyberJacket and Tourist Guide	The CyberJacket incorporates a tourist guide for allowing visitors to the area to orientate more rapidly. The system incorporates an accelerometer device, a GPS location sensor, a sound card, and a processor with Web browser.	Randell and Muller (2002)
Mobile augmented reality	Outdoor Virtual Reality	Outdoor Virtual Reality combines an HMD, Tinmith-evo5 software architecture, and a tracking device to allow virtual and real objects to be interacted with on the move and outside. The authors have developed two applications from their system: a 3D visualization tool and an outdoor game (ARQuake).	Thomas et al. (2002a).
Audio interfaces	Ensemble	Ensemble uses garments fitted with light sensors, accelerometers, and pressure sensors as an interface for children learning about music. MIDI controllers and electronic musical instruments also are integrated. The system allows the children to explore the relation between actions and sounds.	Andersen (2004)
Smart clothing	WearARM	The WearARM provides computation power with a design that blends into existent clothing, strapping around the arm underneath your clothing. Intended mainly as a research platform, it will be integrated into the MIThrill (see the following).	Anliker et al. (2002)

According to some commentators, "the design of interaction techniques for use with mobile and wearable systems has to address complex contextual concerns" (Lumsden & Brewster, 2003, p. 197). While the physical environment that the mobile user inhabits is constantly changing, there is a host of environmental issues to contend with, including privacy, noise levels, and general interruptions to the flow of communications and data access. While many current wearable systems are built on mobile technology components such as PDAs, these do not always provide the best interfaces for maximizing wearability, relying as they do upon graphical and visual interfaces.

Developers have met this challenge by designing a whole range of new and adapted interfaces in order to provide eyes- and hands- free interaction. A review of some of the recent mobile and wear-

Table 1. Summary of a selection of mobile and wearable interaction tools and interfaces, cont.

Smart clothing	Smart clothing prototype for the Arctic environment	The smart clothing prototype for the Arctic includes a suit with communication, global positioning and navigation, user and environment monitoring, and heating.	Rantanen et al. (2002)
Touch pad interface	Touchpad mouse Wearable computers	The touchpad mouse can be used as a component with other wearable computers (i.e., with wearable computer and HMD). The touchpad can be worn in a number of different positions on the body; however, testing has shown that the preferred place is on the thigh.	Thomas et al. (2002b)
Peephole displays that combine pen input with spatially aware displays	PDAs	Peephole displays that combine pen input with spatially aware displays, enabling navigation through objects that are larger than the screen.	Yee (2003)
Body area computing system	Wearable Unit with Reconfigurable Modules (WURM)	Plessl et al. argue that future wearable computing systems should be regarded as embedded systems and suggest the development of a body area computing system composed of distributed nodes around a central communications network. Sensors are distributed around the body using field-programmable arrays (FPGAs).	Plessl et al. (2003)

able technology interfaces has found the following interaction tools and interfaces (see Table 1).

As these divergent interfaces indicate, there is as yet no preferred interface for wearable technology, and the scope for HCI input into design issues clearly is needed to inform future integrated systems. In addition to providing more mobile and embedded interfaces, other design parameters have attempted to address individual user difficulties inherent in traversing the physical environment while communicating, and some have been aimed specifically at user groups, including those with hearing or sight impairments (Ross & Blasch, 2002).

EXAMPLES OF WEARABLE AND MOBILE DEVICES

Wearable devices are distinctive from other mobile devices by allowing hands-free interaction or by at least minimizing the use of a keyboard or pen input when using the device. This is achieved by devices that are worn on the body, such as a headset that allows voice interaction and a head mounted display that replaces a computer screen. The area of wearable devices is currently a hot research topic with potential applications in many fields (e.g. aiding people with disabilities). In addition to the interfaces that we already have mentioned, we have reviewed three examples of mobile and wearable devices.

The IBM Linux Watch

(www.research.ibm.com/WearableComputing/factsheet.html)

IBM recently has developed a wristwatch computer that they collaboratively are commercializing with Citizen under the name of *WatchPad*. Apart from telling the time, WatchPad supports calendar scheduling, address book functionality, to-do-lists, the ability to send and receive short e-mail messages, Bluetooth wireless connectivity, and wireless access to Web services. WatchPad runs a version of the Linux operating system allowing a very flexible software applications development platform. It is possible to design WatchPad for specific users (e.g., a student's watch could hold various schedules and provide location sensing and messaging capabilities). A recent commercial product with overlapping functionally, called Wrist Net Watch (www.fossil.com/tech), has been developed by Fossil. Current information such as news headlines and weather is delivered in real time to the watch through the MSN Direct service.

Xybernaut Mobile Assistant

(www.xybernaut.com/Solutions/product/mav_product.htm)

This commercial product is the most widely available multi-purpose wearable device currently on the market. It is a lightweight wearable computer with desktop/laptop capabilities, including wireless Web connectivity and e-mail, location sensing, hands-free voice recognition and activation, access to data in various forms, and other PC-compatible software. It has a processor module that can be worn in different ways, a head-mounted display unit, a flat-panel display that is touch-screen activated and allows pen input, and a wrist-strapped mini-keyboard. Xybernaut is currently trialling the use of the mobile assistant in an educational context, concentrating on students with special needs. It allows the student full computing access beyond the classroom, including the ability to do standard computing functions such as calculations, word processing, and multi-media display and, in addition, has continuous Internet connectivity and voice synthesis capabilities. It also supports leisure activities, such as listening to music and playing games.

iButtons

(www.ibutton.com/ibuttons/index.html)

iButtons developed by Dallas Semiconductor Corporation/Maxim currently are being piloted in a range of educational institutions. An iButton is a computer chip enclosed in a durable stainless steel can. Each can of an iButton has a data contact (called the lid) and a ground contact (called the base) that are connected to the chip inside the can. By touching each of the two contacts, it is possible to communicate with an iButton, and iButtons are distinguished from each other by each having a unique identification address. By adding different functionality to the basic iButton (i.e. memory, a real-time clock, security, and temperature sensing), several different products are being offered. There are many applications for this technology, including authentication and access control, eCash, and a range of other services. In educational contexts, these smart buttons allow registration of students as well as access to classrooms, Web pages, and computers.

MIThril: A Platform for Context-Aware Wearable Computing

(www.media.mit.edu/wearables/mithril/)

MIThril is a wearable research platform developed at the MIT Media Lab (DeVaul et al., 2001). Although not a commercial product, MIThril is indicative of the functionality that we can expect in next-generation wearable devices. Apart from the hardware requirements, it includes a wide range of sensors with sufficient computing and communication resources and the support for different kinds of interfaces for user interaction, including a vest. There are also ergonomic requirements that include wearability (i.e. the device should blend with the user's ordinary clothing) and flexibility (i.e., the device should be suitable for a wide range of user behaviors and situations).

As an application of this architecture, a reminder delivery system called Memory Glasses was developed, which acts on user-specified reminders (e.g. "During my next lecture, remind me to give additional examples of the applications of wearable computers") and requires a minimum of the wearer's attention. Memory Glasses uses a proactive reminder system model that takes into account time, location, and the user's current activities based on daily events that can be detected (i.e. entering or leaving an office).

FUTURE TRENDS

Wearable and mobile devices currently are being used in a range of contexts, but they also are being used in conjunction with a range of other technologies that may have implications for the evolution of human-computer interfaces. Possible uses might include the use of wearable and mobile devices for outdoor activities; for example, Cheok, et al. (2004) consider the use of wearable devices in conjunction with game play that links virtual and real spaces (Xu et al., 2003). Wearables also might allow users to explore access to a range of personalized information services integrating access through portal systems, although this might have implications for security and privacy issues (Di Pietro & Mancini, 2003). Continued development in terms of commercial applications currently are being researched, which may lead to more personalized methods of retail ordering and customer tracking. Others have noted the use of wearable technology in conjunction with augmented reality (Piekarski & Thomas, 2004; Thomas et al., 2002a; Xu et al., 2003).

CONCLUSION

At this moment in time, the innovations seem to be progressing at such a rapid pace that often suppliers of these devices are trying to create a new demand for products at a relatively early stage of their development. It is not hard to predict that the technological issues we have touched upon will continue to be addressed and improved. Regarding standards, we expect current ones to evolve in parallel with new developments, but due to the experimental nature of some of these devices, there will be periods where non-standard appliances will be piloted.

Personalization of user interaction is also an important issue, where adaptation to the user behavior is critical, easing the customization of the interface to suit users' specific needs within the context of the device being used (Weld et al., 2003). Advances in machine learning and artificial intelligence on the one hand and information overload on the other have led to a new challenge of building enduring personalized cognitive assistants that adapt to their users by sensing the user's interaction with the environment; it can respond intelligently to a range of scenarios that may not have been encountered previously and also can anticipate what is the next action to be taken (Brachman, 2002).

Finally, it is also important to investigate the social potential and impact of wearable and mobile devices (Kortuem & Segall, 2003) so that collaborative systems can be developed to facilitate and encourage interaction among members of the community. One possible educational application of such a collaborative system may be an interactive learning environment that supports a range of mobile and wearable devices in addition to integrating a range of learning services.

REFERENCES

Andersen, K. (2004). ensemble: Playing with sensors and sound. *Proceedings of the Conference on Human Factors and Computing Systems,* Vienna, Austria.

Anliker, U., Lukowicz, P., Troester, G., Schwartz, S. & DeVaul, R.W. (2002). The WearARM: Modular, high performance, low power clothing platform designed for integration into everyday clothing. *Proceedings of the 5th International Symposium on Wearable Computers.*

Brachman, R. (2002). Systems that know what they are doing. *IEEE Intelligent Systems, 17*(6), 67-71.

Brewster, S. (2002). Overcoming the lack of screen space on mobile computers. *Pervasive and Ubiquitous Computing, 6*(3), 188-205.

Cheok, A.D., et al. (2004). *Human personal & ubiquitous computing, 8*(2), 71-81.

DeVaul, R.W., Schwartz, S., & Pentland, A. (2001). MIThrill: Context-aware computing for daily life. Retrieved August 1, 2004, from http://www.media.mit.edu/wearables/mithril/MIThrill.pdf

Di Pietro, R., & Mancini, L.V. (2003). Security and privacy issues of handheld and wearable wireless devices. *Communications of the ACM, 43*(9), 75-79.

Kortuem, G., & Segall, Z. (2003). Wearable communities: Augmenting social networks with wearable computers. *IEEE Pervasive Computing, 2*(1), 71-78.

Lumsden, J., & Brewster, S. (2003). A paradigm shift: Alternative interaction techniques for use with mobile and wearable devices. *Proceedings of the 2003 Conference of the Centre for Advanced Studies on Collaborative Research*, Toronto, Canada.

Page, D. (2003). Computer Ready to Wear. *High Technology Careers.* Retrieved May 6, 2003, from www.hightechcareers.com/doc799/readytowear799.html

Perry, M., O'Hara, K., Sellen, A., Brown, B. & Harper, R. (2001). Dealing with mobility: Understanding access anytime, anywhere. *ACM Transactions on Computer-Human Interaction, 8*(4), 323-347.

Piekarski, W., & Thomas, B.H. (2004). Interactive augmented reality techniques for construction at a distance of 3D geometry. *Proceedings of the Workshop on Virtual Environments 2004,* Zurich, Switzerland.

Plessl, C., et al. (2003). The case for reconfigurable hardware in wearable computing. *Personal and Ubiquitous Computing, 7*, 299-308.

Randell, C., & Muller, H.L. (2002). The well-mannered wearable computer. *Personal and Ubiquitous Computing, 6*, 31-36.

Rantanen, J., et al. (2002). Smart clothing prototype for the Arctic environment. *IEEE Pervasive and Ubiquitous Computing, 6*, 3-16.

Ross, D.A., & Blasch, B.B. (2002). Development of a wearable computer orientation system. *Personal and Ubiquitous Computing, 6*, 49-63.

Sun, J. (2003). Information requirement elicitation in mobile commerce. *Communications of the ACM, 46*(12), 45-47.

Thomas, B., et al. (2002a). First person indoor/outdoor augmented reality application: ARQuake. *Personal and Ubiquitous Computing, 6*, 75-86.

Thomas, B., Grimmer, K., Zucco, J., & Milanese, S. (2002b). Where does the mouse go? An investigation into the placement of a body-attached touch-pad mouse for wearable computers. *Personal and Ubiquitous Computing, 6*, 97-112.

Weld, D., et al. (2003). Automatically personalizing user interfaces. Retrieved May 12, 2003, from www.cs.washington.edu/homes/weld/papers/weld-ijcai03.pdf

Westeyn, T., Brashear, H., Atrash, A., & Starner, T. (2003). Multimodal architectures and frameworks: Georgia Tech gesture toolkit: Supporting experiments in gesture recognition. *Proceedings of the 5th International Conference on Multimodal Interfaces*, Vancouver, Canada (pp. 85-92).

Xu, K., Prince, J.D., Cheok, A.D., Qiu, Y., & Kumar, K. G. (2003). Visual registration for unprepared augmented reality environments. *Personal and Ubiquitous Computing, 7*(5), 287-298.

Xu, Y., Yang, M., Yan, Y., & Chen, J. (2004). Wearable array as user interface. *Proceedings of the 5th Australasian User Interface Conference (AUIC2004).*

Yee, K-P. (2003). Peephole displays: Pen interaction on spatially aware handheld computers. *Proceedings of the ACM Conference on Human Factors in Computing Systems 2003*, Ft. Lauderdale, Florida.

KEY TERMS

Hands-Free Operation: Allows the user to interact with data and information without the use of hands.

Head-Mounted Displays (HMDs): Visual display units that are worn on the head as in the use of VR systems.

Head-Up Displays (HUPs): Displays of data and information that are superimposed upon the user's field of view.

Mobile Devices: Can include a range of portable devices, including mobile phones and PDAs, but also can include wearable devices, such as HMDs and smart clothing, that incorporate sensors and location tracking devices.

Multimodal Interaction: Uses more than one mode of interaction and often uses visual, auditory, and tactile perceptual channels of interaction.

Pervasive and Context-Aware Computing: Allows mobile devices to affect everyday life in a pervasive and context-specific way.

Wearable Devices: May include microprocessors worn as a wristwatch or as part of clothing.

Wearable Sensors: Can be worn and detected by local computing systems.

Web Credibility

David R. Danielson
Stanford University, USA

INTRODUCTION

Credibility evaluation processes on the World Wide Web are subject to a number of unique selective pressures. The Web's potential for supplying timely, accurate, and comprehensive information contrasts with its lack of centralized quality control mechanisms, resulting in its simultaneous potential for doing more harm than good to information seekers. Web users must balance the problems and potentials of accepting Web content and do so in an environment for which traditional, familiar ways of evaluating credibility do not always apply. Web credibility research aims to better understand this delicate balance and the resulting evaluation processes employed by Web users.

This article reviews credibility conceptualizations utilized in the field, unique characteristics of the Web relevant to credibility, theoretical perspectives on Web credibility evaluation processes, factors influencing Web credibility assessments, and future trends.

BACKGROUND

Credibility is one of several dimensions that influence message persuasiveness (Petty & Cacioppo, 1986), attitudes toward an information source (Sundar, 1999), and behaviors relevant to message content (Petty & Cacioppo, 1981). While credibility is largely viewed as a source characteristic, attitudinal assessments relevant to credibility, including those made on the Web, are directed at messages (content), sources (information providers), and media (the Web itself).

Conceptualizations of source credibility have traditionally focused on two primary source attributes, expertise and trustworthiness (Hovland & Weiss, 1951), and these conceptualizations have been influential in Web credibility research (Fogg & Tseng, 1999; Wathen & Burkell, 2002). Expertise refers to a source's perceived ability to provide information that is accurate and valid (based on attributes such as perceived knowledge and skill), while trustworthiness refers to a source's perceived willingness to provide accurate information given the ability (based on attributes such as perceived honesty and lack of bias; Hovland, Jannis, & Kelley, 1953). Thus, the underlying dimensions in conceptualizations of credibility predominantly refer to perceived qualities. Particularly with respect to interactive systems, including the Web, existing research has focused primarily on factors influencing the perception of credibility as opposed to factors predicting objective measures of accuracy.

Numerous related constructs have been investigated in the Web credibility literature, including believability (Flanagin & Metzger, 2000), which is arguably a synonymous construct of credibility (Tseng & Fogg, 1999); information completeness (Dutta-Bergman, 2004), referring to the extent to which necessary elements for confirming message accuracy are present; cognitive authority (Rieh, 2002), referring to the extent to which users believe they can trust the information; and reputation (Toms & Taves, 2004), referring to future expectations of information quality and credibility.

Attitudes toward messages that are relevant to credibility and its related constructs are determined at least by the characteristics of (and interactions amongst) the source, message, and receiver (Self, 1996; Slater & Rouner, 1996). Such assessments are often extensions of source credibility: Credible sources are viewed as likely to produce credible messages. Particularly when constraints such as limited time, lack of ability, or low motivation force the user to focus on surface or peripheral features of the message, source, or medium in processing Web content, one may expect source credibility to heavily influence perceptions of message accuracy and information quality (see Petty & Cacioppo, 1986).

In recognizing the frequent need for computer users to balance a range of information-seeking goals with the need for efficiency and productivity,

Fogg and Tseng (1999) have proposed four types of credibility in assessing interactive systems: presumed, reputed, surface, and experienced. *Presumed credibility* assessments are based upon general underlying assumptions about the system, for example, in assuming that Web sites in the dot-org domain are more credible than those in the dot-com domain. *Reputed credibility* assessments are based upon third-party reports or endorsements, for example, in finding pages linked to by a credible site as likely to provide accurate information. *Surface credibility* assessments are based upon features observable via simple inspection, for example, in using visual design or interface usability as an indicator of credibility. Finally, *experienced credibility* is based upon first-hand experience with the system, for example, in returning to a Web site that has previously provided information verified by the user to be accurate.

Conceptualizations and taxonomies of credibility recognize the construct as not only referring to source characteristics, but as referring to attributes relevant to the perceived likelihood of message accuracy and validity. In so doing, they distinguish credibility from another related construct: trust. Trust relates more properly to the perceived likelihood of behavioral intentions, reliability, and dependability rather than message accuracy, and as Fogg and Tseng (1999) point out, the word is often used in phrases referring to credibility, such as "trust the information" and "trust the advice."

Given a grounding in the credibility concept, Web-credibility researchers have set out to operationalize the construct in a number of ways. As Wathen and Burkell (2002) point out, credibility may be operationalized by either direct or indirect assessment methods, both of which have been applied to Web credibility research. Researchers employ direct assessment methods by asking users to rate the extent to which the source, message, or medium is described by the underlying dimensions of credibility. Indirect methods in the field include measuring attitude and behavior changes as a result of stimulus Web content. Moreover, the field is by no means limited in its range of methodological approaches. Experimental, quasi-experimental, and traditional and Web survey methods are all commonly employed. Qualitative analyses, including interviews, case studies, and thinking-aloud protocols, are also employed to investigate user reasoning about credibility.

UNIQUENESS OF THE WEB

The types of needs that trigger usage of the Web may be relatively similar to other media (Rieh & Belkin, 1998), and Sundar (1999) has found the underlying dimensions of Web and traditional media credibility assessments to be similar. Rieh (2002), on the other hand, has since found that the range of evidence Web users consider in making these assessments is much wider than for other media, and even in cases where the factors considered are similar, they may be weighed differentially across media (Payne, Dozier, & Nomai, 2001). Moreover, the Web may be less credible than print newspapers (Flanagin & Metzger, 2000), but in some cases, more credible than traditional media counterparts such as television, radio, and magazines (Flanagin & Metzger; Johnson & Kaye, 1998). Finally, Klein (2001) has found users to be generally aware of credibility differences between the Web and other media.

Given these differences, one may ask, What is special about the Web with respect to credibility? Researchers have theorized or empirically identified a number of ways in which features of the Web may give rise to differences between online credibility assessments and those made with traditional media. These explanations tend to focus on four general characteristics of the Web: (a) the relative lack of filtering and gatekeeping mechanisms, (b) the form of the medium, including interaction techniques and interface attributes either inherent to the Web and other hypertext systems or emergent from common design practices, (c) a preponderance of source ambiguity and relative lack of source attributions, and (d) the newness of the Web as a medium in conjunction with its lack of evaluation standards.

Filtering Mechanisms

Perhaps the most critical feature of the Web with respect to user credibility evaluations is its relative lack of centralized information filtering or quality

control mechanisms (Abdulla, Garrison, Salwen, Driscoll, & Casey, 2002; Andie, 1997; Flanagin & Metzger, 2000; Johnson & Kaye, 1998). In contrast to traditional media, Web users are free to upload information irrespective of scrutiny (Johnson & Kaye), and content is frequently made available without the benefit of editorials, reviews, and other gatekeeping procedures (Flanagin & Metzger). This lack of quality control can affect perceptions of credibility for the Web as a medium (Johnson & Kaye) and users' evaluative processes, shifting their attribute focus (Rieh, 2002).

The Web may have a property analogous to gatekeeping procedures, primarily in the form of ranking systems evaluating link structures as these structures provide predictive power over credibility assessments (Toms & Taves, 2004). Simultaneously, however, the Web fosters the incidental arrival at sites, and this may increase the likelihood of encountering inaccurate information (Andie, 1997).

Form

The Web, much like the television, offers new form factors and interactive characteristics previously unavailable in information-seeking environments. Just as the television's multimodal properties altered evaluative processes and credibility perceptions (Newhagen & Nass, 1989), the Web's unique interactive features, in conjunction with emerging design practices that further distinguish it from traditional media, may result in fundamentally different credibility-evaluation processes.

The relative ease of data manipulation, duplication, and dissemination is one critical characteristic of digital information systems. Web content is susceptible to frequent alteration (Metzger, Flanagin, & Zwarun, 2003), can easily be tailored to individual recipients (Campbell et al., 1999), and is easily duplicated and widely replicated. This last attribute potentially has significant implications for within-medium verification procedures since inaccurate content may be replicated by its recipients with extraordinary ease. Unique evaluation processes may also result from the diverse and relatively unstructured organization of content (Rieh, 2002), the lack of organizational conventions (Burbules, 2001), and a relative difficulty in distinguishing content from advertising (Flanagin & Metzger, 2000) due in part to both browser display and design conventions.

Source Ambiguity

Web content often lacks a clear source. In many cases, the source is not present at all (Burbules, 2001; Eastin, 2001), and in others it is present but not easy to ascertain (Toms & Taves, 2004). This problem is accompanied by users' generally high reliance on source identity as a criterion for assessing information quality and credibility (Rieh & Belkin, 1998). Such ambiguity also coincides with Rieh's finding that a group of scholars showed a greater reliance on source identity at the institutional level (such as URL [uniform resource locator] domain type) than at the individual level (such as author credentials), which contrasts with findings regarding traditional media. Additionally, Web users often lack information about source reputation (Toms & Taves), a potentially unavoidable problem due to the number and diversity of online sources.

Infancy as a Medium

Finally, one indicator of the Web's uniqueness with respect to credibility is the general concern over fostering Internet literacy (Greer, 2003) in addition to the acknowledged need for evaluation guidelines and assessment standards (Tate & Alexander, 1996; Wathen & Burkell, 2002). As Greer points out, evaluating credibility on the Web is not an easy task, and this difficulty is due in part to the need to learn new evaluative skills, such as checking URL domains. Not only must users master new evaluative techniques, but they must do so in an environment in which familiar ways of assessing credibility are less applicable (Burbules, 2001) and in contexts that may require them to rethink previous evaluative strategies. As Graefe (2003) points out, information objects on the Web (such as product descriptions) are removed from sensory information typically used to verify claims in real-world evaluative contexts. While this is of course equally true of some non-Web contexts, e-commerce ups the ante: Decisions about credibility and relevant behaviors must frequently be made wholly independent of normally available sensory information.

EVALUATIVE PROCESSES

While relatively few theories of Web credibility evaluation have been proposed thus far, existing frameworks and perspectives provide useful ways of conceptualizing online credibility assessment processes. Fogg (2002, 2003a, 2003b) views credibility assessment as an iterative process resulting in the coordination of several component assessments of noticeable elements. *Prominence interpretation theory* posits two aspects of credibility assessments: (a) the likelihood of an element related to the source or message under evaluation being noticed (prominence), and (b) the value assigned to the noticed element based on the user's judgment about how the element affects the likelihood of information accuracy (interpretation). Fogg identifies five factors affecting prominence: user involvement, information topic, the task, experience level, and other individual differences such as the need for cognition. Three factors affecting interpretation are identified: user assumptions, skills and knowledge, and contextual factors such as the environment in which the assessment is made. The process of noticing prominent interface and message elements and assigning evaluative judgments to each occurs iteratively until the user reaches satisfaction with an overall credibility assessment or reaches a constraint, such as lack of time. Fogg points out that seemingly discrepant findings in the Web-credibility literature on the effects of a particular factor (for example, whether privacy policies impact credibility assessments) may be explained parsimoniously if one of the studies is found to have focused on element prominence and the other on interpretation (Fogg, 2003a).

Wathen and Burkell (2002) have conceptualized evaluative processes of Web credibility in terms of a stage model (with the caveat that their proposed stages may represent simultaneous evaluations). They distinguish between evaluations of surface credibility, message credibility, and content. In surface credibility assessments, users focus on presentational and organizational characteristics of a Web site, deciding whether the site is likely to provide the desired content. In message credibility assessments, users more thoroughly review indicators of source and message credibility, deciding whether the provided information is likely to be believable. Finally, in content assessments, users integrate source evaluations with self-knowledge about their own expertise, domain knowledge, and information needs, deciding if and how to act on the information. If failure occurs at either the surface or message credibility assessment stages, the user is likely to leave the site. Influenced by the elaboration likelihood model (Petty & Cacioppo, 1986), Wathen and Burkell further suggest that the probability of leaving interacts with individual differences, such as the need for cognition, need for the information, and motivation; if the user is highly motivated, surface features may be overlooked.

Noticeably, verification procedures are not explicitly included in Fogg's (2002, 2003a, 2003b) or Wathen and Burkell's (2002) theories, and this absence is supported by empirical research finding the verification of Web content to be infrequent (Metzger et al., 2003; Nozato, 2002). Their theories are complemented by a few theoretical perspectives in the Web-credibility literature. Rieh (2002), based on judgment and decision-making research, suggests Web users make at least two types of assessments: (a) *predictive judgments* prior to encountering an information object based on existing knowledge and assumptions, and (b) *evaluative judgments* based on characteristics of the information object. Predictive assessments may also be based on characteristics of an information object's surrogate, such as hyperlinked text.

In the context of consumer assessments of product quality, Graefe (2003), following Nelson (1974), provides a conceptualization more explicitly accounting for verification procedures, distinguishing between *search qualities* (discovered during inspection of the information object), *experience qualities* (discovered only after use of the information object), and *credence qualities* (such as an information provider's intentions) that cannot be verified and that introduce inherent risk into the assessment process. Graefe further points out that on the Web, search qualities often must take the place of experience qualities due to verification difficulties.

FACTORS AFFECTING WEB CREDIBILITY

The influences of a number of Web content and individual site characteristics on credibility have been investigated in the research literature, with a few important general trends focusing on the impact of interface attractiveness, site-operator identity, advertising, individual differences, and the topic of the site's content.

Visual Design

Credibility judgments often have a striking dependence on the surface assessments of visual appearance and interface characteristics, with professional-looking design that is appropriate to site content significantly increasing credibility (Eysenbach & Köhler, 2002; Fogg, Soohoo, Danielson, Marable, Stanford, & Tauber, 2003; Kim & Moon, 1998). The likability of Web sites is significantly affected by interface attractiveness (Roberts, Rankin, Moore, Plunkett, Washburn, & Wilch-Ringen, 2003), and likability in turn impacts credibility assessments (Cialdini, 2001). As Fogg, Soohoo, et al. (2003) point out, this is consistent with social psychological research indicating that attractiveness increases the credibility of human communicators.

Identity

A number of site characteristics impacting credibility center around demonstrating the identity, contact information, and credentials of real individuals associated with the site (Fogg, Marshall, Laraki, et al., 2001; Rieh, 2002). While personal photos can have either a negative or positive impact on trustworthiness depending on contextual factors (Fogg, Marshall, Kameda, et al., 2001; Riegelsberger, Sasse, & McCarthy, 2003), indicators of real human beings behind the site tend to increase credibility. The importance of identity in Web-credibility assessments is consistent with a strong reliance on source authority (Rieh & Belkin, 1998).

Advertising

The impact of advertising on site credibility appears to arise from two competing pressures. On the one hand, users have motivation to ignore advertising in favor of information-seeking goals (Greer, 2003), and on the other, they have motivation to examine advertising content as an indicator of source credibility (Fogg, Soohoo, et al., 2003). Advertising can negatively impact credibility assessments, particularly when it is not clearly distinguished from site content (Fogg, Marshall, Laraki, et al., 2001). While relevance between advertising and site content can positively influence attitudes toward the advertisement (Cho, 1999; Shamdasani, Stanaland, & Tan, 2001), it is less clear if the effect occurs in the opposite direction (Choi & Rifon, 2002).

Individual Differences

A number of user characteristics impact credibility assessments of the Web as a medium and of individual sites. Web credibility research suggests females generally find the Web more credible than males (Robinson & Kaye, 2000), older users find online news less credible than younger users (Johnson & Kaye, 1998), and college students find the Web more credible than the general adult population (Metzger et al., 2003). Younger users additionally tend to be more critical of typographical errors and broken links than older users (Fogg, Marshall, Laraki, et al., 2001). Both an experiment by Greer (2003) and a survey study by Nozato (2002) found a significant positive relationship between usage and perceptions of online news credibility.

In addition to age, gender, and Web usage, a critical factor in Web credibility assessments is content-domain expertise. In a study comparing the assessments of health and finance experts to those of general Web users, Stanford, Tauber, Fogg, and Marable (2002) found health experts to focus on name reputation, source attributions, and company motive more than general Web users, and finance experts to focus more on the quantity of available information, company motive, and potential biases than general Web users.

Content Domain

Stanford et al. (2002) point out that there are inherent differences in the types of information provided within varying domains, including how established the information commonly tends to be and the typical goals of information providers within that domain. These may lead to different expectations when users evaluate content. Rieh (2002) found evaluations of computer-related and medical information to rely more heavily on assessments of trustworthiness than for research and travel. Moreover, Rieh's work indicated less focus on the source in making credibility assessments of travel sites, consistent with Fogg, Soohoo, et al. (2003), who also found the effect for search-engine sites. This last finding may reflect the unique nature of sites acting primarily as gatekeepers to other brand-name sites; the recognizable airlines and high-ranking sites pointed to may cause users to overlook the reputation of the gatekeeper itself, leading to the effect.

FUTURE TRENDS

Web credibility research has identified factors influencing credibility and evaluative strategies employed by users, but the area remains ripe for further empirical work. This section suggests three emerging and important areas in the field: the relationship between network structures on the Web and credibility, the effects of user motivation, and further theory development.

Networks of Credibility

Hypertext structures like the Web offer the opportunity to understand the connection between network structures (both actual structure and the structure perceived by users) and end-user credibility assessments. Toms and Taves (2004) provide an important step in this direction, showing that link structures are powerful indicators of not only relevance, but of credibility. The extent to which users explicitly or implicitly recognize these structures and employ them in credibility assessments is unknown.

User Motivation

Although influenced by Petty and Cacioppo's (1986) elaboration likelihood model and their notions of central and peripheral routes to persuasion, little work has focused on the impact of motivation on Web credibility assessments. As Dutta-Bergman (2004) points out, the potential analogy between the notions of directed search and browsing and the notions of central and peripheral processing point the way to useful research.

Theory Development

Finally, it is worth noting that because Web credibility research is a relatively new field of inquiry, theoretical frameworks that can drive systematic programs of empirical work are only recently beginning to appear. These early frameworks are critical, and there remains a need for the further development and empirical testing of Web-specific theories of credibility assessment.

CONCLUSION

Evaluative processes and credibility assessments on the Web arise out of complex interactions between characteristics of the user, the site under evaluation, and the Web as a medium. Fundamental characteristics of the Web act as pressures on information seekers, shaping which interface elements will be noticed, how they will be interpreted, and the evaluative processes users will employ in making credibility assessments.

REFERENCES

Abdulla, R. A., Garrison, B., Salwen, M., Driscoll, P., & Casey, D. (2002). *The credibility of newspapers, television news, and online news.* Paper presented at the Association for Education in Journalism and Mass Communication Annual Convention, Miami Beach, FL.

Andie, T. (1997). Why Web warriors might worry. *Columbia Journalism Review, 36*, 35-39.

Burbules, N. C. (2001). Paradoxes of the Web: The ethical dimensions of credibility. *Library Trends, 49*, 441-453.

Campbell, M. K., Bernhardt, J. M., Waldmiller, M., Jackson, B., Potenziani, D., Weathers, B., et al.

(1999). Varying the message source in computer-tailored nutrition education. *Patient Education & Counseling, 36*, 157-169.

Cho, C. (1999). How advertising works on the WWW: Modified elaboration likelihood model. *Journal of Current Issues and Research in Advertising, 21*, 34-50.

Choi, S. M., & Rifon, N. J. (2002). Antecedents and consequences of Web advertising credibility: A study of consumer response to banner ads. *Journal of Interactive Advertising, 3*(1). Available at http://www.jiad.org/vol3/no1/chio/

Cialdini, R. (2001). *Influence: Science and practice.* Boston: Allyn & Bacon.

Dutta-Bergman, M. J. (2004). The impact of completeness and Web use motivation on the credibility of e-health information. *Journal of Communication, 54*, 253-269.

Eastin, M. S. (2001). Credibility assessments of online health information: The effects of source expertise and knowledge of content. *Journal of Computer-Mediated Communication, 6*(4). Available at http://jcmc.indiana.edu/vol6/issue4/eastin.html

Eysenbach, G., & Köhler, C. (2002). How do consumers search for and appraise health information on the World Wide Web? Qualitative study using focus groups, usability tests, and in-depth interviews. *BMJ, 324*, 573-577.

Flanagin, A. J., & Metzger, M. J. (2000). Perceptions of Internet information credibility. *Journalism and Mass Communication Quarterly, 77*, 515-540.

Fogg, B. J. (2002). *Prominence-interpretation theory: Explaining how people assess credibility* [Research report]. Stanford University, Stanford Persuasive Technology Lab, CA. Retrieved from http://credibility.stanford.edu/

Fogg, B. J. (2003a). *Persuasive technology: Using computers to change what we think and do.* San Francisco: Morgan Kaufmann.

Fogg, B. J. (2003b). Prominence-interpretation theory: Explaining how people assess credibility online. *Proceedings of CHI'03, Extended Abstracts on Human Factors in Computing Systems* (pp. 722-723).

Fogg, B. J., Marshall, J., Kameda, T., Solomon, J., Rangnekar, A., Boyd, J., et al. (2001). Web credibility research: A method for online experiments and early study results. *Proceedings of CHI'01, Extended Abstracts on Human Factors in Computing Systems* (pp. 295-296).

Fogg, B. J., Marshall, J., Laraki, O., Osipovich, A., Varma, C., Fang, N., et al. (2001). What makes Web sites credible? A report on a large quantitative study. *Proceedings of CHI'01, Human Factors in Computing Systems* (pp. 61-68).

Fogg, B. J., Soohoo, C., Danielson, D. R., Marable, L., Stanford, J., & Tauber, E. R. (2003). How do users evaluate the credibility of Web sites? A study with over 2,500 participants [Extended *Consumer WebWatch* report]. *Proceedings of DUX2003, Designing for User Experiences Conference.* Retrieved from http://www.consumerwebwatch.org/news/report3_credibilityresearch/stanfordPTL_abstract.htm

Fogg, B. J., & Tseng, H. (1999). The elements of computer credibility. *Proceedings of CHI'01, Human Factors in Computing Systems* (pp. 80-87).

Graefe, G. (2003). Incredible information on the Internet: Biased information provision and lack of credibility as a cause of insufficient information quality. *Proceedings of the 8th International Conference on Information Quality* (pp. 133-146).

Greer, J. D. (2003). Evaluating the credibility of online information: A test of source and advertising influence. *Mass Communication & Society, 6*(1), 11-28.

Hovland, C. I., Jannis, I. L., & Kelley, H. H. (1953). *Communication and persuasion.* New Haven, CT: Yale University Press.

Hovland, C. I., & Weiss, W. (1951). The influence of source credibility on communication effectiveness. *Public Opinion Quarterly, 15*, 635-650.

Johnson, T. J., & Kaye, B. K. (1998). Cruising is believing? Comparing Internet and traditional sources on media credibility measures. *Journalism and Mass Communication Quarterly, 75*, 325-340.

Kim, J., & Moon, J. (1998). Designing towards emotional usability in customer interfaces: Trustworthiness of cyber-banking system interfaces. *Interacting with Computers, 10*, 1-29.

Klein, K. B. (2001). User perceptions of data quality: Internet and traditional text sources. *Journal of Information Systems, 41*(4), 5-15.

Metzger, M. J., Flanagin, A. J., & Zwarun, L. (2003). College student Web use, perceptions of information credibility, and verification behavior. *Computers & Education, 41*, 271-290.

Nelson, P. (1974). Advertising as information. *Journal of Political Economy, 82*, 729-754.

Newhagen, J., & Nass, C. I. (1989). Differential criteria for evaluating credibility of newspapers and TV news. *Journalism Quarterly, 66*, 277-284.

Nozato, Y. (2002). *Credibility of online newspapers.* Paper presented at the Association for Education in Journalism and Mass Communication Annual Convention, Washington, DC.

Payne, G. A., Dozier, D. M., & Nomai, A. J. (2001). *Newspapers and the Internet: A comparative assessment of news credibility.* Paper presented at the Association for Education in Journalism and Mass Communication Annual Convention, Miami Beach, FL.

Petty, R. E., & Cacioppo, J. T. (1981). *Attitudes and persuasion: Classic and contemporary approaches.* Dubuque, IL: W. C. Brown Co. Publishers.

Petty, R. E., & Cacioppo, J. T. (1986). Elaboration likelihood model. In L. Berkowitz (Ed.), *Advances in experimental social psychology* (Vol. 19, pp. 123-205).

Riegelsberger, J., Sasse, M. A., & McCarthy, J. D. (2003). Shiny, happy people building trust? Photos on e-commerce Websites and consumer trust. *Proceedings of CHI'03, Human Factors in Computing Systems* (pp. 121-128).

Rieh, S. Y. (2002). Judgment of information quality and cognitive authority in the Web. *Journal of the American Society for Information Science and Technology, 53*, 145-161.

Rieh, S. Y., & Belkin, N. J. (1998). Understanding judgment of information quality and cognitive authority in the WWW. *Proceedings of the 61st Annual Meeting for the American Society for Information Science,* 35.

Roberts, L., Rankin, L., Moore, D., Plunkett, S., Washburn, D., & Wilch-Ringen, B. (2003). Looks good to me. *Proceedings of CHI'03, Extended Abstracts on Human Factors in Computing Systems* (pp. 818-819).

Robinson, T. J., & Kaye, B. K. (2000). Using is believing: The influence of reliance on the credibility of online political information among politically interested Internet users. *Journalism and Mass Communication Quarterly, 77*, 865-879.

Self, C. S. (1996). Credibility. In M. Salwen & D. Stacks (Eds.), *An integrated approach to communication theory and research* (pp. 421-441). Mahwah, NJ: Lawrence Erlbaum Associates.

Shamdasani, P. N., Stanaland, A., & Tan, J. (2001). Location, location, location: Insights for advertising placement on the Web. *Journal of Advertising Research, 41*, 7-21.

Slater, M. D., & Rouner, D. (1996). How message evaluation and source attributes may influence credibility assessment and belief change. *Journalism and Mass Communication Quarterly, 73*, 974-991.

Stanford, J., Tauber, E. R., Fogg, B. J., & Marable, L. (2002). Experts vs. online consumers: A comparative credibility study of health and finance Web sites. *Consumer WebWatch.* Retrieved from http://www.consumerwebwatch.org/news/report3_credibilityresearch/slicedbread_abstract.htm

Sundar, S. (1999). Exploring receivers' criteria for perception of print and online news. *Journalism and Mass Communication Quarterly, 76*, 373-386.

Tate, M., & Alexander, J. (1996). Teaching critical evaluation skills for World Wide Web resources. *Computers in Libraries, 16*, 49-55.

Toms, E. G., & Taves, A. R. (2004). Measuring user perceptions of Web site reputation. *Information Processing and Management, 40*, 291-317.

Tseng, H., & Fogg, B. J. (1999). Credibility and computing technology. *Communications of the ACM, 42*(5), 39-44.

Wathen, C. N., & Burkell, J. (2002). Believe it or not: Factors influencing credibility on the Web. *Journal of the American Society for Information Science and Technology, 53*, 134-144.

KEY TERMS

Credibility: A characteristic of information sources that influences message persuasiveness, attitudes toward the information source, and behaviors relevant to message content, consisting of two primary attributes: expertise and trustworthiness.

Evaluative Judgment: An assessment based on characteristics of an information object independent of assessments based on information prior to encountering the object (predictive judgments).

Experienced Credibility: A credibility assessment based upon first-hand experience with a system.

Expertise: A source's perceived ability to provide information that is accurate and valid (based on attributes such as perceived knowledge and skill); with trustworthiness, it is one of two primary attributes of credibility.

Predictive Judgment: An assessment (prior to encountering an information object) based on existing knowledge, assumptions, or the information object's surrogate independent of assessments based on characteristics of the object (evaluative judgments).

Presumed Credibility: A credibility assessment based upon general underlying assumptions about a system.

Reputed Credibility: A credibility assessment based upon third-party reports or endorsements.

Surface Credibility: A credibility assessment based upon features observable via simple inspection.

Trustworthiness: A source's perceived willingness to provide accurate information given the ability (based on attributes such as perceived honesty and lack of bias); with expertise, it is one of two primary attributes of credibility.

Web-Based Human Machine Interaction in Manufacturing

Thorsten Blecker
Hamburg University of Technology, Germany

Günter Graf
University of Klagenfurt, Austria

INTRODUCTION

The quality of HMI in automation is an important issue in manufacturing. This special form of interaction occurs when the combination of human abilities and machine features are necessary in order to perform the tasks in manufacturing. Balint (1995) has identified three categories of such human-machine systems:

1. Machines might do the job without human involvement, but the feasibility is questionable. For example, weld seams in car assembly are made mostly autonomously by robots, but in many cases, humans have to guide the robot to the weld point, because the robot is not able to locate the point correctly, which is a relatively easy task for a human.
2. Humans might do the job without machines, but the efficiency/reliability is questionable. This is the case in almost all cases of automation (e.g., the varnishing of cars).
3. HMI is necessary (no purely machine- or human-based execution is possible), although robots today are widely in use; in many cases, they cannot substitute humans completely, because the possible conflicts that can occur are so diverse that a robot alone cannot manage them.

The term HMI is used widely for the interaction of a human and a somewhat artificial, automated facility, which is true in many situations, including HCI. In this article, we speak of HMI in industrial settings. We term the machine especially for industrial facilities for producing a certain (physical) output; in this case, the term man-machine interaction also is used synonymously for HMI. We define HMI as the relation between a human operator and one or more machines via an interface for embracing the functions of machine handling, programming, simulation, maintenance, diagnosis, and initialization.

BACKGROUND

The interface between humans and machines generally influences the quality of HMI, especially in the third category of the previously presented human-machine systems. The design of the interface between humans and the machines has evolved dramatically in recent decades (Nagamachi, 1992). The first step was mechanically controlled machines. With the rise of numerical control, the interaction between human and machines changed. In the second step, the operator no longer has an exact knowledge about how the machine is programmed and cannot influence the processes in the machine. The third step is computerized machines, where the operator can influence and program a wide array of parameters in the machine. In this step, computerized HMI becomes a central aspect in manufacturing on the shop floor.

The advances of computerized techniques for enriching the interface allow a human-centered modification of HMIs. This enables an effective use of the skills and abilities of the operators of machines and the features of the machines themselves. Such a human-centered design of manufacturing technologies should obey the following steps (Stahre, 1995):

1. Consider existing skills of the user.
2. Facilitate the maximizing of operator choice and control.
3. Integrate the planning, execution, and monitoring components.
4. Design to maximize the operator's knowledge.
5. Encourage social communications and interaction.

RISE OF WEB-BASED HMI

The usage of interoperable, adaptive, and standardized information technologies on the shop floor is essential to solve the problems in human-centered manufacturing, in which the previously mentioned fulfill the criteria. Due to restrictions in the capability of computers and their associated technologies in the 1980s and 1990s, the computer interfaces were built upon those technological limits and were not oriented to an optimized effectiveness of the human machine interaction on the shop floor. In addition, HMI has been machine-specific up until now and bounded on the implementation by the facility vendor. The diffusion of Internet technologies within automation and new trends in automation technologies provide the necessary infrastructure (Blecker, 2003). The following trends are essential:

1. **Mobilization of Computers:** For example, Web pads enable the mobilization of all interactions between humans and machines as well as between humans on the shop floor.
2. **Embedded Computing:** Every machine may have an integrated full-featured computer that stores data, which provide a front end; it autonomously can sense and respond to the environment (by blinking, e-mail messages, software calls, etc.) and offers services for machine maintenance and control. Embedded computers in machines and facilities on the shop floor induce the development of intelligent systems in every machine. Here, intelligence means that the system can set a wide array of autonomous (clearly predefined) actions on the occurrence of certain events.
3. **Standardization of Networks:** (Industrial) Ethernet replaces common field busses and proprietary networking. It is also compatible with wireless networks, which enable wireless communication on the shop floor.

Consequently, Internet technologies have become ubiquitously available on the shop floor. The data and computation services will be portably accessible from many, if not most, locations on the shop floor. Internet technologies also trigger a standardization of the screen design and content distribution. This leads to a major change in the traditional HMI, especially for blue-collar workers. In fact, the interaction between workers and machines approximates the common screen handling of the office world. Therefore, we state that the human machine interaction is converging into a Web-Based Human Machine Interaction.

Web-based HMI is an advanced and extended form of computerized HMI characterized by the logical separation of the computer unit from the machine itself. Internet technologies integrate the human as well as the machine within a corporate network. They make the entirely Web-based information infrastructure and all of the interaction partners connected to it available for the employees as well as the information systems on the shop floor. By using Web-based interfaces for user input, screens can be implemented or modified rapidly. Cost savings are realized, since any device (mobile or fixed) that can support a browser becomes a personal computer. The enhancements due to the use of Web-based HMI in manufacturing can be summarized in the following groups:

1. An ergonomic visualization in many variants (colored, high resolution screens and standardized visualization technologies enable an appealing and effective representation of data from the shop floor and data, for example, from the ERP-System).
2. Hardware and software advancements enable more efficient input- and data-manipulation processes.
3. The contents and screen designs are easily updatable und changeable.
4. The visualization is not bounded to the computer in the machine but connects via the Internet, which enables the delocalization of the interaction in various scenarios.

Web-based HMI changes and enhances several workflows, especially in manufacturing information processing (AWK, 1999). This triggers several consequences.

CONSEQUENCES OF WEB-BASED HMI

Through the standardized technologies used in Web-based HMI, all other forms of applications that build upon Internet technologies are distributable on the shop floor to every worker. The consequences affect the following fields of activity with extended HMI processes:

1. Collaboration on the shop floor
2. Data collection
3. Communication and coordination with the management
4. HMI processes themselves

The availability of a full networked computer enables collaboration applications (e.g., workflow management systems, instant messaging, and voice-over Internet Protocol (IP) on the shop floor. Internet technologies enable the intuitive integration of those technologies and interfaces that support the worker in his or her special environment. For example, workers in the assembly may interact directly with engineers via the IP-based speech and video connections to solve special problems or to learn specific work processes cooperatively. The aerospace industry uses such methods for the assembly of complicated parts of planes. Those intraorganizational virtual relationships help to reduce costs through Web-switched communication by reducing or, in the case of wireless techniques, by replacing the necessary cable lanes and a markedly eased setup of infrastructure through the use of open standards.

The Web-based interfaces enable an accompanying data collection through the integration of data-entry screens into the normal workflow screens of the interface. The mobilization of computers allows the worker to have a personalized pad, which enables mobile data collection in various applications (e.g., logistics or quality assurance). Those pads or devices have sufficient computation power and offer connectivity to use specialized equipment, such as Bluetooth-headsets for speech entry. The networked local computer may use a Web service on a remote server for speech recognition.

The high resolution of screens and the integration into an intranet on the shop floor push the setup of *Web-based training* on the job in manufacturing. Especially relearned, low-educated workers show good results, if they are trained in short lessons during their work hours (Schmidt/Stark, 1996). Furthermore, an effect of the extended use of computers and the training on abilities to handle computers has positive effects on the diffusion of new information systems and the resulting processes (Rozell & Gardner, 1999).

The extended Web-based interaction abilities also virtualize the communication and coordination with the management of the organization. Operators have access to upper level information via Web browsers, and the top-down communication becomes more intuitive, which directly simplifies the coordination structures (Eberts, 1997).

Although the improved interfaces have widespread effects on information handling on the shop floor, the most important aspect remains the HMI itself. HMI has several similarities to human computer interaction in the office world, although there are important differences. These include the extensive application of touch-screen interaction and the feedback via the activities of the controlled machine. The other interaction scenarios are comparable with common HCI scenarios. Indeed, there are differences in the work environment (industrial settings), the design of the computers (use of touch screens, no keyboards or mice) and the abilities of the workers (Fakun& Greenough, 2002). These differences require an analysis of Web-based HMI on the shop floor that differentiates from the results of common HCI. The Web-enabled facilities induce two contrary consequences:

1. Web-based information distribution leads to a more intuitive and efficient HMI, which decreases the interaction complexity for the human. This induces reduced qualification requirements, because the handling of machines requires less specialized knowledge. This leads to the hypothesis that there are lesser skills necessary for workers interacting with machines.

2. The diffusion of Internet technologies enables a networking of machines and information systems, which demands the usage of those optimization possibilities for competitive improvements. This results in an increasing HMI complexity, because much more information is to be handled on the shop floor, which requires additional skills of the workers. Furthermore, the span of control of a single worker over different machines may increase. This leads to the hypothesis that additional skills of workers are necessary.

Workers on the shop floor may not have the necessary skills for the extended screen-oriented information handling, although they are often specialists and well-trained (Mikkelsen et al., 2002). Therefore, cooperation between the human resources and planning departments in manufacturing has to clarify whether the workers should receive extended training or whether the screen and information design has to be adapted according to the user's abilities. To evaluate the consequences in practical cases, we have to consider the resulting behavior that is necessary for fulfilling the tasks within manufacturing. Those behaviors can be categorized as follows (Strahe, 1995):

1. **Skill-Based Behavior:** Well-learned, sensory-motor behavior analogous to nearly instinctive hand and foot actions while driving a car.
2. **Rule-Based Behavior:** Actions triggered by a certain pattern of stimuli. A computer using an if-then algorithm to initiate an appropriate response could execute these actions.
3. **Knowledge-Based Behavior:** Responding to new situations. High-level situation assessment and evaluation, consideration of alternative actions in light of various goals (making decisions and multifactor scheduling of actions).

HMI has shifted dramatically the possible behaviors in operating machines. Skill-based behavior has dominated the pre-computerized HMI, where machines only were usable based on the skills of workers. The rise of numeric control pushed the rule-based behavior into the forefront. Workers have only assisted the machines by inserting punch cards, which have been prepared by engineers. The diffusion of computerized, programmable control architectures enabled the direct influence of skilled workers again (Strahe, 1995) and is promoted through the upcoming Web-based HMI knowledge-based behavior on the shop floor.

Compromising the technological advances in HMI has changed the machine control itself as well as the interaction with all actors on and off the shop floor. To benefit from those changes, a coordinated implementation of the technology and organizational processes is required (Wu, 2002).

FUTURE TRENDS

The realization of the potential of Web-based HMI requires an adequate implementation of technical and organizational structures. First, management has to assure whether a ubiquitous Web-based HMI infrastructure is desirable. The additional benefits of Web-based HMI are reasonable only if there is a demand for it (Stolovitch, 1999). Therefore, an implementation of the described technologies and organizational changes should be accomplished if:

1. Extended knowledge-based behaviors are required; and
2. Complex manufacturing tasks with extended information processing requirements on the shop floor are necessary.

If those tasks are not necessary, an isolated application of Web-based HMI will bring forth some benefits on the existing work processes. However, the gain of the full potentials of Web-based HMI requires an integration of the various information systems on the shop floor, the implementation of adequate organizational structures, managerial processes, as well as education strategies for online training on the job. Those action fields induce bundled measures in many aspects of the factory and at the same time form the main future trends in Web-based HMI. We concentrate here on the issues concerning the interactions of humans and (Web-enabled) machines.

Adaptation Information Systems and Machining Infrastructure

Technological barriers are critical. Web-based HMI scenarios require adequate machinery that has embedded computation power. Moreover, there has to be a networking infrastructure. Barriers result from the existing infrastructure. Technology management has to ensure the implementation of Internet technologies within the production system. Facilities as well as information systems have to be strategically equipped with Internet technologies (Blecker, 2006). This also means that existing information systems may be extended to meet the new requirements.

Development of an Education Strategy

Indirect communication over different Internet-based communication technologies requires employees to have sufficient knowledge in handling information technologies. They also should have a basic understanding of how the omnipresent network operates; otherwise, they are likely to see it as a black box. This would lead to a passive use of the information network, where an active use really is required. Therefore, human resource management has to train employees to meet the requirements. The training also should reduce the resistance of employees. The suggested mechanisms make the work environment more transparent. Indeed, this transparency has to be dealt with carefully, because it also allows the detailed reconstruction of the usage and the spying of the interaction behavior of the employees.

Ergonomics and Motivation

Yi and Hwang (2003) have shown that application-specific self-efficacy, enjoyment, and learning-goal orientation all determine the actual usage of a Web-based information system. Those aspects have to be considered during the setup of a Web-based human-machine infrastructure. Especially in the exposed areas on the shop floor, the design of the devices and the interaction possibilities beyond traditional HCI are important. Therefore, the distributed content has to be adopted for use on the shop floor, although the representation also has to satisfy the requirements of normal screen design, as is shown in Ozok and Salvendy (2004). The adoption should boil down the information to the most important messages. This can be assured using semantic technologies (Geroimenko & Chen, 2003) and the use of short abstracts and keywords. The input workflows should be implemented in wizard style, for example, so that scrolling and additional mouse-like movements on the screen can be omitted.

Production Portals for Visual Representation

To design enterprise-wide screen guidelines based on the information system integration, it is necessary to set up a strategy for visual integration of information systems as well as the machine control for the workers on the shop floor. Production portals are a solution to those challenges. A production portal is a digital enterprise portal that is used by a manufacturing organization or plan as a means to assist its decision-making activities (Huang & Mak, 2003). These portals are able to deliver adapted interfaces, for example, for experts or beginners with the help of dynamically generated pages based on Web technologies. Through the dynamic linking capabilities of Web technologies (e.g., the use of Web services for the delivery of information from enterprise resource planning systems), the integration of all information sources into one screen design can be realized. Due to the characteristics of work on the shop floor, multitasking also is not a desired feature. An explorer-like tree (Botsch & Kunz, 2001) organizes all of the personalized features that are relevant for the worker. In this case, workers do not have to work with different application windows but can navigate in one browser window between information sources and data entry forms through relatively simple links.

CONCLUSION

The evolution in human machine systems will be driven in the future by new information technologies. Management has to react to those changes by the application of the latest technological advancements in interface design. Special attention is to be further placed on input technologies such as augmented reality or, for example, data gloves, which will be integrated into the human-machine system through Internet technologies (Roco & Bainbridge, 2003). Furthermore, human-centered aspects, such as cognitive models of workers, that are psychologically tested also have to be integrated into screen and/or input interface design. Web-based infrastructures enable the necessary flexibility and adaptability of interfaces.

REFERENCES

AWK, Aachener Werkzeugmaschinen-Kolloquium. (1999). Internet-Technologien für die Produktion—Neue Arbeitswelt in Werkstatt und Betrieb. In AWK (Hrsg.), *Wettbewerbsfaktor Produktionstechnik, Aachener Perspektiven* (pp. 357-398). Aachen.

Balint, L. (1995). Adaptive human-computer interfaces for man-machine interaction in computer-integrated systems. *Computer Integrated Manufacturing Systems, 8*(2), 133-142.

Blecker, Th. (2003). Towards a production concept based on Internet technologies. *Proceedings of the 6th International Conference on Industrial Engineering and Production Management - IEPM'03,* Port, Portugal.

Blecker, Th. (2006). *Web based manufacturing.* Berlin: Erich Schmidt Verlag.University of Klagenfurt.

Botsch, V., & Kunz, C. (2001). *Visualization and navigation of networked information spaces, the Matrix BrowserI.* Retrieved February 6, 2004, from http://www.hci.iao.fraunhofer.de/ fileadmin/user_upload/BotschKunz2002_Matrix Browser.pdf

Chittaro, L. (2003). *Human-computer interaction with mobile devices and services.* Berlin: Springer.

Eberts, R.E. (1997). Computer based instruction. In M. Helander, T.K. Landauer, & P. Prabhu (Eds.), *Handbook of human-computer interaction* (pp. 825-841). Amsterdam: Elsevier.

Fakun, D., & Greenough, R.M. (2002). User-interface design heuristics for developing usable industrial hypermedia applications. *Human Factors and Ergonomics in Manufacturing, 12*(2), 127-149.

Geroimenko, V., & Chen, C. (2003). *Visualizing the semantic Web—XML-based Internet and information visualizationI.* Berlin: Springer.

Huang, G.Q., & Mak, K.L. (2003). *Internet applications in product design and manufacturing.* Berlin: Springer.

Mikkelseen, et al. (2002). Job characteristics and computer anxiety in the production industry. *Computers in Human Behaviour, 18,* 223-239.

Nagamachi, M. (Ed.). (1992). *Design for manufacturability: A systems approach to concurrent engineering and ergonomics.* Washington, DC: CRC Press.

Ozok, A.A., & Salvendy, G. (2004). Twenty guidelines for the design of Web-based interfaces with consistent language. *Computers in Human Behavior, 20,* 149-161.

Roco, M., & Bainbridge, W.S. (Eds.). (2003). *Converging technologies for improving human performance.* Dordrecht.

Rozell, E.J., & Gardner III, W.L. (1999). Computer-related success and failure: A longitudinal field study of the factors influencing computer-related performance. *Computers in Human Behavior, 15,* 1-10.

Schmidt, H., & Stark, G. (1996). *Computer-based training in der betrieblichen lernkultur.* Bielefeld.

Stahre, J. (1995). Evaluating human/machine interaction problems in advanced manufacturing. *Computer Integrated Manufacturing Systems, 8*(2), 143-150.

Stolovitch, D. (Ed.). 1999). *Handbook of human performance technology: Improving individual and organizational performance worldwide.* San Francisco: Pfeiffer.

Wu, B. (Ed.). (2002). *Handbook of manufacturing and supply systems design: From strategy formulations to system operation.* Washington, DC: CRC Press.

Yi, M.Y., & Hwang, Y. (2003). Predicting the use of Web-based information systems: Self-efficacy, enjoyment, learning goal orientation, and the technology acceptance model. *International Journal of Human-Computer Studies 59,* 431-449.

KEY TERMS

Embedded Devices: Full-featured computers that are integrated into machines.

HMI in Manufacturing: Relation between a human operator and one or more machines via an interface for embracing the functions of machine handling, programming, simulation, maintenance, diagnosis, and initialization.

Industrial Ethernet: Ethernet technology that is adjusted to specific environmental conditions (e.g., regarding electromagnetic compatibility, shaking, moisture, and chemical resistance in manufacturing).

Production Portal: The linking of all available information systems into one standardized screen. Production portals aggregate heterogeneous systems in manufacturing and provide secure, structured, and personalized information for individual users (e.g., based on job functions).

Ubiquitous Computing: Trend to integrate information and communication technologies into all devices.

Voice-Over IP: Standard for making telephone calls via an Internet connection. It enables the flexible use of different input devices, including video telephone applications.

Web-Based HMI: An advanced and extended form of computerized HMI characterized by the logical separation of the computer unit from the machine itself.

Web Pad (or Handheld PC): Devices that are connected via wireless technologies to an intranet (WLAN, Bluetooth, GPRS/UMTS) and offer a full-featured operating system with a Web browser.

Web Service: The term *Web services* describes a standardized way of integrating Web-based applications using the XML, SOAP, WSDL, and UDDI open standards over an Internet protocol backbone. XML is used to tag the data, SOAP is used to transfer the data, WSDL is used to describe the services available, and UDDI is used to list what services are available. Used primarily as a means for businesses to communicate with each other and with clients, Web services allows organizations to communicate data without intimate knowledge of each other's IT systems behind the firewall.

Web-Based Instructional Systems

George D. Magoulas
University of London, UK

INTRODUCTION

Information and communication technologies have played a fundamental role in teaching and learning for many years. Technologies, such as radio and TV, were used during the 50s and 60s for delivering instructional material in audio and/or video format. More recently, the spread of computer-based educational systems has transformed the processes of teaching and learning (Squires, Conole, & Jacobs, 2000). Potential benefits to learners include richer and more effective learning resources using multimedia and a more flexible pace of learning. In the last few years, the emergence of the Internet and the World Wide Web (WWW) have offered users a new instructional delivery system that connects learners with educational resources and has led to a tremendous growth in Web-based instruction.

Web-based instruction (WBI) can be defined as using the WWW as the medium to deliver course material, manage a course (registrations, supervision, etc.), and communicate with learners. A more elaborate definition is due to Khan (1997), who defines a Web-based instructional system (WIS) as "...a hypermedia-based instructional program which utilises the attributes and resources of the World Wide Web to create a meaningful learning environment where learning is fostered and supported." Relan and Gillani (1997) have also provided an alternative definition that incorporates pedagogical elements by considering WBI as "...the application of a repertoire of cognitively oriented instructional strategies within a constructivist and collaborative learning environment, utilising the attributes and resources of the World Wide Web."

Nowadays, WISs can take various forms depending on the aim they serve:

- Distance-learning (DL) systems' goal is providing remote access to learning resources at a reduced cost. The concept of DL (Rowntree, 1993) is based on: (i) learning alone, or in small groups, at the learner's pace and in their own time and place, and (ii) providing active learning rather than passive with less frequent help from a teacher.
- Web-based systems, such as intelligent tutoring systems (Wenger, 1987), educational hypermedia, games and simulators (Granlund, Berglund, & Eriksson, 2000), aim at improving the learning experience by offering a high level of interactivity and exploratory activities, but require a significant amount of time for development. The inherent interactivity of this approach leads learners to analyse material at a deeper conceptual level than would normally follow from just studying the theory and generates frequently cognitive conflicts that help learners to discover their possible misunderstandings and reconstruct their own cognitive models of the task under consideration.
- Electronic books provide a convenient way to structure learning materials and reach a large market (Eklund & Brusilovsky, 1999).
- Providers of training aim to offer innovative educational services to organisations for workplace training and learning, such as to supplement and support training in advance of live training, update employee skills, develop new skills.

The main difference between WBI and the traditional computer-based instructional programs lies in the way information is presented to the user. The WISs' approach to e-learning does not only provide "active learning," which according to Bates (1991) is the most effective way to learn, but also interactivity, which is a well-known facilitator of the learning experience (Mason & Kaye, 1989). Thus, we have, on the one hand, traditional instructional programs which present educational content in a linear fashion using a static structure, and on the other hand, WISs that exploit the hypermedia capabilities, for example, offering flexibility in the delivery of instruction through the use of hyperlinks (Federico, 1999). As a consequence, WBI has led to a new model for teaching and learning that focuses on the learner not as passive recipient of knowledge but as an active, self-directed participant in the learning process. Nevertheless, this approach to instruction has also created a series of challenges that users of educational technology, such as teachers, learners, providers of educational content, educational institutions and so forth, have to meet: (i) ensure the improvement of learning experience, as usual technology-driven innovations consume prodigious amounts of time and money to little educational effect; (ii) bring a real and substantial change in education by improving their understanding of learning and teaching with the use of this new technology.

This article presents the main features of Web-based instructional systems, including their advantages and disadvantages. It discusses critical factors that influence the success and effectiveness of WISs. It stresses the importance of pedagogy on WBI and explores the pedagogical dimensions of the interface tools and functionalities of WISs. Lastly, it summarises future trends in Web-based instruction.

BACKGROUND

The appeal of WISs lies in their ability to actively engage learners in the acquisition and use of information, support multiple different instructional uses (tutoring, exploration, collaboration, etc.), support different learning styles and promote the acquisition of different representations that underlie expert-level reasoning in complex, ill-structured domains (Selker, 1994). Learners select the knowledge they perceive as being most suited to their needs. But, although the act of browsing is a pleasing experience, browsing in an unknown domain is not likely to lead to satisfactory knowledge acquisition at all (Jonassen, Mayes, & McAleese, 1993). Thus, navigational aids, such as a pre-defined hierarchical structure of the subject matter, are necessary especially in large domains. The pre-defined structure of the domain knowledge provides learners (especially novices) with guidance during their study, offering them a sense of safety and a reliable navigation path. In this way, learners are supported in constructing their own individualised model of the knowledge space and are able to follow paths through the subject content produced by designers, or to develop their own routes according to individually-prescribed requirements (Large, 1996).

Another attractive element is the flexibility to access course contents through intranets and the Internet at any time and from different places, which is considered as the main reason many educators have tried to develop distance learning programs on the WWW. This flexibility creates many opportunities for exploration, discovery, exchange/sharing of information and learning according to learners' individual needs. Flexibility, however, comes at a price:

- The complexity of the system may increase (Ellis & Kurniawan, 2000). Users may need more time to search for the information (Ng & Gunstone, 2002), and the dynamism and richness of the content may negatively affect learners' level of comprehension (Power & Roth, 1999).
- Despite the plethora of communication tools, learners sometimes find feedback insufficient, feel isolated or not supported enough, and drop out of the course (Quintana, 1996).
- It is unlikely all learners are equally able to performing their own sequencing, pacing, and navigation. Moreover, the learner is not always going to choose the content to study next in a way that will lead to effective learning (Hammond, 1992; Leuthold, 1999).
- Previous knowledge of the domain content varies for different learners, and indeed knowledge may grow differently through the interaction with the system (Winkels, 1992).
- Learners tend to get lost, especially when the educational content is large and/or when they are novices. This can lead to *disorientation* experienced when users do not know where they are within hypertext documents and how to move towards the desired location, commonly known as *"lost in hyperspace"* (Brusilovsky, 2001).
- Learners may fail to get an overview of how all the information fits together when browsing. In the absence of information that might help them formulate knowledge goals and find relevant information, learners may stumble through the content in a disorganised and instructionally

inefficient manner (Hammond, 1992). Furthermore, if learners are too accustomed to memorising and are faced with multiple explanations of the same knowledge, they may attempt to memorise them all. This is one of the aspects of a problem known as "*information overload*" which is usually experienced by users of WISs (McCormack & Jones, 1998).

CRITICAL ISSUES FOR DESIGNING AND DELIVERING WEB-BASED INSTRUCTION

The Role of the Users

In WBI, the roles of teachers and learners are different from their classic definitions. Thus, teachers design educational content that is attractive to learners in order to motivate them, interact with the learners, and act as *facilitators* of the learning process. Learners are mainly responsible for their own learning, assessment of knowledge goals and objectives. As a consequence, learners need to be able to form their own ideas about the content and understand the educational material in their own way. That change of roles requires course broadening of skills and competencies for teachers and learners. Table 1 highlights the differences in users' roles and the impact of WBI.

The Pedagogy of Web-Based Instruction

The need for changing instructional methods has come partly in response to demands of the workplace and partly because of re-assessment of instructional methodologies. Individuals are now expected to be adaptable to modern ways of communication, such as e-mail systems, the Internet, intranets, the WWW, conferencing systems. They are also expected to apply high cognitive skills, such as analysing, summarising, and synthesising information as well as engaging in creative and critical thinking (Vogel & Klassen, 2001). In principle, WISs can serve this purpose but the greatest benefits of their use can occur via a pedagogic approach that most effectively uses the characteristics of this technology to increase the quality of the learning experience as already explained earlier.

As a result, a number of educational trends emerged in recent years have played a particularly important role in Web-base instruction; three of these are presented in the following:

- **Individualised Learning:** This approach provides learners the capability to select the mode of delivery and timing of module material. For example, learners can choose a blended way for learning which consists of lectures, participation in traditional face-to-face communication in a classroom, and collaborative work in a remote environment on the WWW.
- **Constructivist Theory:** The constructivist perspective describes learning as change in meaning constructed from experience (Newby, Stepich, Lehman, & Russell, 1996). Constructivism covers a wide diversity of perspectives that consider learning as an active process of constructing rather than acquiring knowledge and instruction as a process of supporting that construction rather

Table 1. Users, roles, and Web-based instruction

Teachers' role		Learners' role	WBI
Instructor	Facilitator		
Identification of learning outcomes; structuring and sequencing of domain knowledge; designing educational activities and assessments	Response to questions; providing consistent and timely feedback; encouraging discussion among learners; motivating learners and reinforcing effective study habits	Study the educational content; undertake responsibility for their learning; adopt new forms of communication and new ways of learning	Interactive tools; information sharing and communication mechanisms; individualised assessment; distributed educational resources

than communicating knowledge (Duffy & Cunningham, 1996).

- **Experiential Learning:** According to Kolb (1984), experiential learning involves the following steps: concrete experience; observation and reflection; formulation of abstract concepts and generalisations; testing of the implications of the concepts in new situations. Experiential learning can take different forms: *learning by doing* (Graf & Kellogg, 1990); *experience-based learning*, *trial and error* and *applied experiential learning* (Gentry, 1990); *reflection in action* (Senge, 1995); *action learning* (Pedler, 1997). But experience must be accompanied by reflection, as experience alone does not automatically lead to learning. This is important for both teachers and students.

The WWW and especially hypermedia provide an eminently suitable environment for the development of educational systems that adopt these instructional models; that is, educational hypermedia are considered as excellent representations of constructivist approaches in theory (Jonassen et al., 1993). To set up a WIS to facilitate these forms of learning, one should ensure it contains the set of elements such as attraction of attention, recall of prior knowledge, consistent presentation style and structure, group work or individual tasks, self-assessment questions, practice/exercises, feedback, review, learning guidance, post knowledge. The use of the WWW adds extra dimensions to teaching and learning. But for learning to take place, the learner has to be not only active but also engaged in the learning process. Table 2 provides a, example, making a link between learner's involvement and acquired skills (following Bloom's (1956) taxonomy of intellectual behaviour) with the types of educational content in a WIS.

Planning, designing, and implementing WBI includes several dimensions, which of course contribute to the effectiveness of this approach. Among a number of factors, the *user interface* of the educational system, the *communication facilities* offered and the *educational content* are of particular importance. Table 3 gives an overview of pedagogical considerations for designing a WIS.

The considerations for the components in Table 3 show that WBI strives to create environments that favour a constructivist model of learning that allows: learners actively construct, transform and extend

Table 2. Pedagogical aspects of Web-based instruction

Skills/abilities	Learner's involvement	Type of content
Knowledge: recall studied content	Memorisation of knowledge (from specific facts to complete theories).	Hypertext and images
Comprehension: grasp the meaning of the content	Interpreting, explaining or summarising the material; estimating future trends (predicting consequences or effects). Taking up tests about knowledge of facts, theories, procedures, etc.	Hypertext and images, self-assessment questions
Application: apply learned material to new and concrete situations	Applying principles, concepts, laws, and theories.	Examples, self-assessment questions
Analysis: break down material into components and understand the organisational structure of the content	Identification of components, analysis of their interrelationships, and recognition of the organisational principles involved. An understanding of the content and the structural form of the material is required.	Examples, self-assessment questions
Synthesis: put parts of material together to form a new whole	Production of a unique communication, a work plan or set of abstract relations between concepts. Develop creative behaviours with major emphasis on the formulation of new patterns or structure.	Interactive tools, simulations, case studies, self-assessment questions
Evaluation: judge the value of the content for a given purpose/task	Making conscious judgements based on clearly defined criteria or goals.	Interactive tools, simulations, case studies, self-assessment questions

Table 3. Pedagogical dimension of system's components in Web-based instruction

Component	Pedagogical Role
User interface	– *Reduce learner's anxiety*: consistent and easy-to-use. – *Support learners and teachers in tasks completion*: provide tools based on users' profile.
Communication facilities	– *Enhance cognitive skills*: help formulate ideas, elaborate on the subject matter. – *Support collaboration and interaction*: among learners themselves and/or between learners and educators.
Educational content	– *Main source of information*: use of a user-friendly language, accessible, easily understandable. – *Support different learning styles*: include types of content, various levels of difficulty. – *Emphasise exploration*: adopt a hypermedia form of presentation, provide different types of resources, simulations, learning by discovery. – *Enhance social skills*: include group work, projects. – *Evaluate knowledge*: self-assessment questions, projects, various types of assessment.

their knowledge; active engagement in the interpretation of the content and reflection on their interpretations; linking educational content with real-world situations (Jonassen, 1994). Thus, through exploration of educational material, which addresses different knowledge levels, learning objectives, and learning styles, learners take the responsibility of their learning.

Individual Differences

Learners differ in traits such as skills, aptitudes and preferences for processing information, constructing meaning from information, and applying it to real-world situations. Recent approaches to WBI try to take into account various dimensions of individual differences, such as the level of knowledge or literacy, gender, culture, spatial abilities, cognitive styles, learning styles, accessibility issues for the disabled and elderly. To this end, learner-centered approaches, which have been motivated by socio-cultural and constructivist theories of learning (Soloway et al., 1996), have been proposed. Learner-centered design acknowledges that understanding of learners needs is of primary importance to provide effective WBI to heterogeneous student populations (Soloway, Guzdial, & Hay, 1994; Quintana, Krajcik, & Soloway, 2000).

The impacts of individual differences on WBI have been investigated along different dimensions:

- **Cognitive and learning styles** that refer to a user's information processing habits have an impact on user's skills and abilities, such as preferred modes of perceiving, thinking, remembering, and problem solving (Ford & Chen, 2000, 2001; Shih & Gamon, 2002).
- **Gender differences** affect WBI in the sense that males and females have different requirements with respect to navigation support and interface features. The preferences of males and females also differentiate remarkably in terms of information seeking strategies, media preferences, and learning performance (Campbell, 2000; Large, Beheshti, et al., 2002; Leong & Hawamdeh, 1999; Liu, 2003).
- **Prior knowledge and system experience** affect learners' interactions with the WIS and their level of knowledge of the educational content. The impact of this individual differences' dimension depends on learners' previous understanding of the educational content, that is, because of relevant studies, and their familiarity with the WIS's features and functionalities, that is, familiarity with distance-learning systems (Reed & Oughton, 1997; Lawless & Kulikowich, 1998; Last, O'Donnell, & Kelly, 2001).

The empirical evaluation of the effects of individual differences on the degree of success or failure experienced by learners needs to be explored in more detail to fully understand their impact on the quality of learning attained within WISs (Magoulas, Papanikolaou, & Grigoriadou, 2003). Actually, the main problems in exploiting such information in a WIS is to determine which characteristics should be used (are worth modelling) and how (what can be done differently for learners with different preferences or styles) (Brusilovsky, 2001). In the next section, this problem is addressed in the context of personalisation technologies, which are considered a promising approach to accommodate individual differences.

FUTURE TRENDS

Personalised learning environments (PLEs) have instantiated a relatively recent area of research that aims at alleviating the information overload and lost in hyperspace problems by integrating two distinct technologies in WBI: intelligent tutoring systems (ITS) and educational hypermedia systems. This is in effect a combination of two approaches to WISs: the more directive *tutor-centred* style of traditional tutoring systems and the flexible *learner-centred* browsing approach of educational hypermedia systems (Brusilovsky, 2001).

PLEs adapt the content, structure, and/or presentation to each individual user's characteristics, usage behaviour, and/or usage environment. Personalisation usually takes place at three different levels: content level, presentation level, and navigation level. For example, in a system with personalisation at the content level, the educational content is generated or assembled from various pieces depending on the user. Thus, advanced learners may receive more detailed and deep information, while novices will be provided with additional explanation. At the presentation level, adaptive text and adaptive layout are two widely used techniques. Adaptive text implies that the same Web page is assembled from different texts following learner's current need, such as removing some information from a piece of text or inserting extra information to suit the current user. Adaptive layout aims to differentiate levels of the subject content by changing the layout of the page, instead of the text, such as font type and size, and background colour. At the navigation level, the most popular techniques include direct guidance, adaptive ordering, link hiding, and link annotation.

Table 4. Web-based instructional systems that employ personalisation features (adapted from Magoulas et al., 2003)

System and Subject domain	Individual Differences Dimension	Level of Personalisation	Pedagogical Approach
CS383 (Carver et al., 1996) *Computer Systems*	Learning style	Presentation	Media selection based on learners' learning style
AST (Specht et al., 1997) *Introductory Statistics*	Knowledge level; Learning style; User preferences	Content; Navigation	Multiple teaching strategies
ELM-ART II (Weber & Specht, 1997) *Programming in Lisp*	Knowledge level; User preferences	Content; Navigation	Example-based programming
DCG (Vassileva, 1997, 1998) *Domain Independent*	Knowledge level; Learning goal; User preferences	Content	Generic Task Model Theory
INTERBOOK (Brusilovsky et al., 1998) *Domain Independent*	Knowledge level	Content; Navigation	N/A
KBS-HYPERBOOK (Henze et al., 1999) *Introduction to Programming using Java*	Knowledge level; Learning goals	Navigation	Project-based learning
ARTHUR (Gilbert & Han, 1999) *Computer Science Programming*	Learning style	Content	Multiple instructional styles: visual-interactive, auditory-text, auditory-lecture, text style
INSPIRE (Papanikolaou et al., 2003) *Computer Architecture*	Knowledge level; Learning style	Presentation; Content; Navigation	Component Display Theory; Elaboration Theory
AES-CS (Triantafillou, Pomportsis, & Demetriadis, 2003) *Multimedia Technology Systems*	Knowledge level, Cognitive style	Presentation; Content; Navigation	Multiple instructional strategies

Table 4 (adapted from Magoulas et al., 2003) presents the features of several PLEs with respect to: the individual student characteristics used to guide the personalisation (see "Individual Differences" column), the type of personalisation provided (see "Level of Personalisation" column), and the teaching/learning approach or theory (see "Pedagogical Approach" column).

Several approaches to evaluate the performance of PLEs have been proposed in the literature, and the empirical results look really promising (Weibelzahl, Lippitsch, & Weber, 2002). However, many questions are still open in this context. Among the most critical ones are questions related to the level of tutor and learner control over the PLE, the development of appropriate methods of assessing information about the behaviour of the learner in the course of learner-system interaction, and the systematic evaluation of the effectiveness of personalisation.

CONCLUSION

Advances in technology are increasingly impacting the way in which the curriculum is delivered and assessed. The ever-increasing learner needs make particularly important for Web services to provide learning tools. The attraction of WISs lies in their capability to actively engage the learner in the learning process, support multiple instructional uses (tutoring, exploration, research, etc.) and different learning styles, provide feedback mechanisms and promote the acquisition of various skills. There are of course some critical factors that influence the use of WISs in an educational setting. This article covered issues related to teachers' and learners' new roles, learner-centered design and pedagogical considerations, which in our opinion are the most important ones to fully exploit the benefits of WBI in education.

REFERENCES

Bates, A. W. (1991). Interactivity as a criterion for media selection in distance education. *Never Too Far*, *16*, 5-9.

Bloom, B. S. (1956). *Taxonomy of educational objectives: Cognitive domain* (Handbook I). New York: McKay.

Brusilovsky, P. (2001). Adaptive hypermedia. *User Modeling and User-adapted Interaction*, *11*, 87-110.

Campbell, K. (2000). Gender and educational technologies: Relational frameworks for learning design. *Journal of Educational Multimedia and Hypermedia*, *9*, 131-149.

Duffy, T., & Cunningham, D. (1996). Constructivism: implications for the design an delivery of instruction. In D. Johansson (Ed.), *Handbook of research for educational communications and technology* (pp. 170-198). New York: Macmillan.

Eklund, J., & Brusilovsky, P. (1999, March). InterBook: An adaptive tutoring system. *UniServe Science News*, *12*, 8-13.

Ellis, R. D., & Kurniawan, S. H. (2000). Increasing the usability of online information for older users. A case study in participatory design. *International Journal of Human-Computer Interaction, 12*(2), 263-276.

Federico, P.-A. (1999). Hypermedia environments and adaptive instruction. *Computers in Human Behavior*, *15*, 653-692.

Ford, N., & Chen, S. (2000). Individual differences, hypermedia navigation, and learning: An empirical study. *Journal of Educational Multimedia and Hypermedia*, *9*, 281-311.

Ford, N., & Chen, S. (2001). Matching/mismatching revisited: An empirical study of learning and teaching styles. *British Journal of Educational Technology, 32*(1), 5-22.

Gentry, J. (1990). What is experiential learning? In J. Gentry (Ed.), *The ABSEL guide to business gaming and experiential learning* (pp. 9-20). New York: Nichols.

Graf, L., & Kellogg, C. (1990). Evolution of experiential learning approaches and future developments. In J. Gentry (Ed.), *The ABSEL guide to business gaming and experiential learning* (pp. 231-250). New York: Nichols.

Granlund, R., Berglund, E., & Eriksson, H. (2000). Designing Web-based simulation for learning. *Future Generation Computer Systems, 17*(2), 171-185.

Hammond, N. V. (1992). Tailoring hypertext for the learner. In P. Kommers, D. Jonassen, & J. T. Mayes (Eds.), *Cognitive tools for learning*. Heidelberg, FRG: Springer Verlag.

Jonassen, D., Mayes, T., & McAleese, R. (1993). A manifesto for a constructivist approach to uses of technology in higher education in designing environments for constructive learning. In T. Duffy, J. Lowyck, & D. Jonassen (Eds.), *NATO/ASI Series F, vol. 105*. Berlin: Springer-Verlag.

Khan, B. H. (1997). *Web-based instruction*. Englewood Cliffs, NJ: Educational Technology Publications.

Kolb, D. (1984). *Experiential learning: Experiences as a source of learning and development*. Englewood Cliffs: Prentice-Hall.

Large, A. (1996). Hypertext instructional programs and learner control: A research review. *Education for Information*, 4, 95-106.

Large, A., Beheshti, J., et al. (2002). Design criteria for children's Web portals: The users speak out. *Journal of The American Society For Information Science And Technology, 53*(2), 79-94.

Last, D. A., O'Donnell, A. M., & Kelly, A. E. (2001). The effects of prior knowledge and goal strength on the use of hypermedia. *Journal of Educational Multimedia and Hypermedia, 10*(1), 3-25.

Lawless, K. A., & Kulikowich, J. M. (1998). Domain knowledge, interest and hypertext navigation: A study of individual differences. *Journal of Educational Multimedia and Hypermedia, 7*(1), 51-69.

Leong, S., & Hawamdeh, S. (1999). Gender and learning attitudes in using Web-based science lessons. *Information Research, 5*(1). Retrieved October 5, 2005 from http://www.shef.ac.uk/is/publications/infres/paper66.html

Leuthold, J. H. (1999). Is computer-based learning right for everyone? In the *Proceedings of the 32nd Hawaii International Conference on System Sciences* (pp. 1-8). Piscataway, NJ: IEEE Pages.

Liu, M. (2004). Examining the performance and attitudes of sixth graders during their use of a problem-based hypermedia learning environment. *Computers in Human Behavior, 20*(3), 357-379.

Magoulas, G. D., Papanikolaou, K. A., & Grigoriadou, M. (2003). Adaptive Web-based learning: Accommodating individual differences through system's adaptation. *British Journal of Educational Technology, 34*(4), 511-527.

Mason, R., & Kaye, A. (1989). *Mindweave, communication, computers and distance education*. Pergamon Press.

McCormack, C., & Jones, D. (1998). *Building a Web-based education system*. John Wiley & Sons.

Newby, T., Stepich, D., Lehman, J., & Russell, J. (1996). *Instructional technology for teaching and learning: Designing instruction, integrating computers and using media*. NJ: Prentice-Hall.

Ng, W., & Gunstone, R. (2002). Students' perceptions of the effectiveness of the World Wide Web as a research and teaching tool in science learning. *Research in Science Education, 32*(4), 489-510.

Power, D. J., & Roth, R. M. (1999) Issues in designing and using Web-based teaching cases. *Proceedings of the Fifth Americas Conference on Information System (AMCIS 1999)* (pp. 936-938).

Quintana, C., Krajcik, J., & Soloway, E. (2000). Exploring a structured definition for learner-centered design. *Fourth International Conference of the Learning Sciences* (pp. 256-263). Mahwah, NJ: Erlbaum.

Quintana, Y. (1996). *Evaluating the value and effectiveness of internet-based learning*. Paper presented at the Sixth Annual Conference of the Internet Society, Montreal, Canada. Retrieved on September 15, 2004, from http://www.isoc.org/inet96/proceedings/c1/c1_4.htm

Reed, W. M., & Oughton, J. M. (1997). Computer experience and interval-based hypermedia navigation. *Journal of Research on Computing in Education, 30*, 38-52.

Relan, A., & Gillani, B. B. (1997). Web-based information and the traditional classroom: Similarities and differences. In B. H. Khan (Ed.), *Web-based instruction*. Englewood Cliffs, NJ: Educational Technology Publications.

Rowntree, D. (1992). *Exploring open and distance learning*. London: Kogan Page.

Selker, T. (1994). Coach: A teaching agent that learns. *Communications of the ACM, 37*(7).

Senge, P. (1995). *The fifth discipline*. New York: Doubleday Dell Publishers group.

Shih, C., & Gamon, J. (2002). Relationships among learning strategies, patterns, styles, and achievement in Web-based courses. *Journal of Agricultural Education, 43*(4), 1-11.

Soloway, E., Guzdial, M., & Hay, K. E. (1994). Learner-centered design: The challenge for HCI in the 21st century. In R. Bilger, S. Guest, & M. J. Tauber (Eds.) *Interactions, 1*(2), 36-48. Vancouver, British Columbia: Association for Computing Machinery, Inc.

Soloway, E., Jackson, S. L., Klein, J., Quintana, C., Reed, J., Spitulnik, J., Stratford, S. J., Studer, S., Jul, S., Eng, J., & Scala, N. (1996). Learning theory in practice: Case studies of learner-centered design. *Proceedings of CHI 96* (pp. 189-196).

Squires, D., Conole, G., & Jacobs, G. (2000). *The changing face of learning technologies*. Cardiff: University of Wales Press.

Triantafillou, E., Pomportsis, A., & Demetriadis, S. (2003). The design and formative evaluation of an adaptive educational system based on cognitive styles. *Computers & Education, 41*(1), 87-103.

Vogel, D., & Klassen, J. (2001). Technology-supported learning: Status, issues and trends. *Journal of Computer Assisted Learning, 17*, 104-114.

Weibelzahl, S., Lippitsch, S., & Weber, G. (2002). Advantages, opportunities, and limits of empirical evaluations: Evaluating adaptive systems. *Künstliche Intelligenz, 3/02*, 17-20.

Wenger, E. (1987). *Artificial intelligence and tutoring systems: Computational and cognitive approaches to the communication of knowledge*. Morgan Kaufmann.

Winkels, R. (1992). *Explorations in intelligent tutoring and help*. Amsterdam: IOS Press.

KEY TERMS

Computer-Assisted Instruction: The use of computers in educational settings, that is, tutorials, simulations, exercises. It usually refers either to stand-alone computer learning activities or to activities which reinforce educational material introduced and taught by teachers.

Constructivism: Teaching model that considers learning as the active process of constructing knowledge, and instruction as the process of supporting that construction.

Educational Hypermedia: Web-based learning environments that offer learners browsing through the educational content supported by flexible user interfaces and communication abilities.

Educational Technology: The use of technology to enhance individual learning and to achieve widespread education.

Individual Differences: In the context of Web-based instruction, this term is usually used to denote a number of important human factors, such as gender differences, learning styles, attitudes, abilities, personality factors, cultural backgrounds, prior knowledge, knowledge level, aptitudes and preferences for processing information, constructing meaning from information, and applying it to real-world situations.

Information Overload: Learners face the information overload problem when acquiring increasing amounts of information from a hypermedia system. It causes learners frustration with the technology and anxiety that inhibits the creative aspects of the learning experience.

Instructional/Pedagogical Design/Approach: In the context of Web-based instruction, this usually relates to pedagogical decision-making, which concerns two different aspects of the system design:

planning the educational content (what concepts should be the focus of the course) and planning the delivery of instruction (how to present these concepts).

Lost in Hyperspace: This is a feeling experienced by learners when losing any sense of location and direction in the hyperspace. It is also called *disorientation* and is caused by badly-designed systems that do not provide users with navigation tools, signposting, or any information about their structure.

Web-Based Instruction: Can be defined as using the Web as the medium to deliver course material, administer a course (registrations, supervision, etc.), and communicate with learners.

Index

Symbols

A

B

C

D

E

F

G

J

K

L

M

N

O

P

Q

R

S

T

X

Index of Key Terms

C

M

N

O

P

T

U

V

W